Lecture Notes in Computer Science 5264

Commenced Publication in 1973
Founding and Former Series Editors:
Gerhard Goos, Juris Hartmanis, and Jan van Leeuwen

Fuchun Sun Jianwei Zhang Ying Tan
Jinde Cao Wen Yu (Eds.)

Advances in Neural Networks – ISNN 2008

5th International Symposium
on Neural Networks, ISNN 2008
Beijing, China, September 24-28, 2008
Proceedings, Part II

Springer

Volume Editors

Fuchun Sun
Tsinghua University, Dept. of Computer Science and Technology
Beijing 100084, China
E-mail: fcsun@mail.tsinghua.edu.cn

Jianwei Zhang
University of Hamburg, Institute TAMS
22527 Hamburg, Germany
E-mail: zhang@informatik.uni-hamburg.de

Ying Tan
Peking University, Department of Machine Intelligence
Beijing 100871, China
E-mail: ytan@pku.edu.cn

Jinde Cao
Southeast University, Department of Mathematics
Nanjing 210096, China
E-mail: jdcao@seu.edu.cn

Wen Yu
Departamento de Control Automático, CINVESTAV-IPN
México D.F., 07360, México
E-mail: yuw@ctrl.cinvestav.mx

Library of Congress Control Number: 2008934862

CR Subject Classification (1998): F.1.1, I.2.6, I.5.1, H.2.8, G.1.6

LNCS Sublibrary: SL 1 – Theoretical Computer Science and General Issues

ISSN 0302-9743
ISBN-10 3-540-87733-9 Springer Berlin Heidelberg New York
ISBN-13 978-3-540-87733-2 Springer Berlin Heidelberg New York

Springer is a part of Springer Science+Business Media

springer.com

Printed in Germany

Typesetting: Camera-ready by author, data conversion by Scientific Publishing Services, Chennai, India
Printed on acid-free paper SPIN: 12529940 06/3180 5 4 3 2 1 0

Preface

This book and its companion volume, LNCS vols. 5263 and 5264, constitute the proceedings of the 5th International Symposium on Neural Networks (ISNN 2008) held in Beijing, the capital of China, during September 24–28, 2008. ISNN is a prestigious annual symposium on neural networks with past events held in Dalian (2004), Chongqing (2005), Chengdu (2006), and Nanjing (2007). Over the past few years, ISNN has matured into a well-established series of international symposiums on neural networks and related fields. Following the tradition, ISNN 2008 provided an academic forum for the participants to disseminate their new research findings and discuss emerging areas of research. It also created a stimulating environment for participants to interact with each other and exchange information on future challenges and opportunities of neural network research.

ISNN 2008 received 522 submissions from about 1,306 authors in 34 countries and regions (Australia, Bangladesh, Belgium, Brazil, Canada, China, Czech Republic, Egypt, Finland, France, Germany, Hong Kong, India, Iran, Italy, Japan, South Korea, Malaysia, Mexico, The Netherlands, New Zealand, Poland, Qatar, Romania, Russia, Singapore, South Africa, Spain, Switzerland, Taiwan, Turkey, UK, USA, Virgin Islands (UK)) across six continents (Asia, Europe, North America, South America, Africa, and Oceania). Based on rigorous reviews by the Program Committee members and reviewers, 192 high-quality papers were selected for publication in the proceedings with an acceptance rate of 36.7%. These papers were organized in 18 cohesive sections covering all major topics of neural network research and development. In addition to the contributed papers, the ISNN 2008 technical program included four plenary speeches by Dimitri P. Bertsekas (Massachusetts Institute of Technology, USA), Helge Ritter (Bayreuth University, Germany), Jennie Si (Arizona State University, USA), and Hang Li (Microsoft Research Asia, China). Besides the regular sessions and panels, ISNN 2008 also featured four special sessions focusing on some emerging topics.

As organizers of ISNN 2008, we would like to express our sincere thanks to Tsinghua University, Peking University, The Chinese University of Hong Kong, and Institute of Automation at the Chinese Academy of Sciences for their sponsorship, to the IEEE Computational Intelligence Society, International Neural Network Society, European Neural Network Society, Asia Pacific Neural Network Assembly, the China Neural Networks Council, and the National Natural Science Foundation of China for their technical co-sponsorship. We thank the National Natural Science Foundation of China and Microsoft Research Asia for their financial and logistic support.

We would also like to thank the members of the Advisory Committee for their guidance, the members of the International Program Committee and additional reviewers for reviewing the papers, and members of the Publications Committee for checking the accepted papers in a short period of time. In particular, we would like to thank Springer for publishing the proceedings in the prestigious series of *Lecture Notes in Computer Science*. Meanwhile, we wish to express our heartfelt appreciation to the plenary and panel speakers, special session organizers, session Chairs, and student helpers. In addition, there are still many more colleagues, associates, friends, and supporters who helped us in immeasurable ways; we express our sincere gratitude to them all. Last but not the least, we would like to thank all the speakers, authors, and participants for their great contributions that made ISNN 2008 successful and all the hard work worthwhile.

September 2008

Fuchun Sun
Jianwei Zhang
Ying Tan
Jinde Cao
Wen Yu

Organization

General Chair

Bo Zhang, China

General Co-chair

Jianwei Zhang, Germany

Advisory Committee Chairs

Xingui He, China
Yanda Li, China
Shoujue Wang, China

Advisory Committee Members

Hojjat Adeli, USA
Shun-ichi Amari, Japan
Zheng Bao, China
Tianyou Chai, China
Guoliang Chen, China
Ruwei Dai, China
Wlodzislaw Duch, Poland
Chunbo Feng, China
Walter J. Freeman, USA
Kunihiko Fukushima, Japan
Aike Guo, China
Zhenya He, China
Frank L. Lewis, USA
Ruqian Lu, China
Robert J. Marks II, USA
Erkki Oja, Finland
Nikhil R. Pal, India
Marios M. Polycarpou, USA
Leszek Rutkowski, Poland
DeLiang Wang, USA
Paul J. Werbos, USA
Youshou Wu, China
Donald C. Wunsch II, USA
Youlun Xiong, China

Lei Xu, Hong Kong
Shuzi Yang, China
Xin Yao, UK
Gary G. Yen, USA
Bo Zhang, China
Nanning Zheng, China
Jacek M. Zurada, USA

Program Committee Chairs

Ying Tan, China
Jinde Cao, China
Wen Yu, Mexico

Steering Committee Chairs

Zengqi Sun, China
Jun Wang, China

Organizing Committee Chairs

Fuchun Sun, China
Zengguang Hou, China

Plenary Sessions Chair

Derong Liu, USA

Special Sessions Chairs

Xiaoou Li, Mexico
Changyin Sun, China
Cong Wang, China

Publications Chairs

Zhigang Zeng, China
Yunong Zhang, China

Publicity Chairs

Andrzej Cichocki, Japan
Alois Knoll, Germany
Yi Shen, China

Finance Chair

Yujie Ding, China
Huaping Liu, China

Registration Chair

Fengge Wu, China

Local Arrangements Chairs

Lei Guo, China
Minsheng Zhao, China

Electronic Review Chair

Xiaofeng Liao, China

Steering Committee Members

Shumin Fei, China
Chengan Guo, China
Min Han, China
Xiaofeng Liao, China
Baoliang Lu, China
Zongben Xu, China
Zhang Yi, China
Hujun Yin, UK
Huaguang Zhang, China
Ling Zhang, China
Chunguang Zhou, China

Program Committee Members

Ah-Hwee Tan, Singapore
Alan Liew, Australia
Amir Hussain, UK
Andreas Stafylopatis, Greece
Andries Engelbrecht, South Africa
Andrzej Cichocki, Japan
Bruno Apolloni, Italy
Cheng Xiang, Singapore
Chengan Guo, China
Christos Tjortjis, UK

Chuandong Li, China
Dacheng Tao, Hong Kong
Daming Shi, Singapore
Danchi Jiang, Australia
Dewen Hu, China
Dianhui Wang, Australia
Erol Gelenbe, UK
Fengli Ren, China
Fuchun Sun, China
Gerald Schaefer, UK
Guangbin Huang, Singapore
Haibo He, USA
Haijun Jiang, China
He Huang, Hong Kong
Hon Keung Kwan, Canada
Hongtao Lu, China
Hongyong Zhao, China
Hualou Liang, USA
Huosheng Hu, UK
James Lam, Hong Kong
Jianquan Lu, China
Jie Zhang, UK
Jinde Cao, China
Jinglu Hu, Japan
Jinling Liang, China
Jinwen Ma, China
John Qiang Gan, UK
Jonathan H. Chan, Thailand
José Alfredo F. Costa, Brazil
Ju Liu, China
K. Vijayan Asari, USA
Kang Li, UK
Khurshid Ahmad, UK
Kun Yuan, China
Liqing Zhang, China
Luonan Chen, Japan
Malik Ismail, USA
Marco Gilli, Italy
Martin Middendorf, Germany
Matthew Casey, UK
Meiqin Liu, China
Michael Li, Australia
Michel Verleysen, Belgium
Mingcong Deng, Japan
Nian Zhang, USA

Nikola Kasabov, New Zealand
Norikazu Takahashi, Japan
Okyay Kaynak, Turkey
Paul S. Pang, New Zealand
Péter Érdi, USA
Peter Tino, UK
Ping Guo, China
Ping Li, Hong Kong
Qiankun Song, China
Qing Ma, Japan
Qing Tao, China
Qinglong Han, Australia
Qingshan Liu, China
Quanmin Zhu, UK
Rhee Man Kil, Korea
Rubin Wang, China
Sabri Arik, Turkey
Seiichi Ozawa, Japan
Sheng Chen, UK
Shunshoku Kanae, Japan
Shuxue Ding, Japan
Stanislaw Osowski, Poland
Stefan Wermter, UK
Sungshin Kim, Korea
Tingwen Huang, Qatar
Wai Keung Fung, Canada
Wei Wu, China
Wen Yu, Mexico
Wenjia Wang, UK
Wenlian Lu, China
Wenwu Yu, Hong Kong
Xiaochun Cheng, UK
Xiaoli Li, UK
Xiaoqin Zeng, China
Yan Liu, USA
Yanchun Liang, China
Yangmin Li, Macao
Yangquan Chen, USA
Yanqing Zhang, USA
Yi Shen, China
Ying Tan, China
Yingjie Yang, UK
Zheru Chi, Hong Kong

Reviewers

Dario Aloise
Ricardo de A. Araujo
Swarna Arniker
Mohammadreza Asghari Oskoei
Haibo Bao
simone Bassis
Shuhui Bi
Rongfang Bie
Liu Bo
Ni Bu
Heloisa Camargo
Liting Cao
Jinde Cao
Lin Chai
Fangyue Chen
Yangquan Chen
Xiaofeng Chen
Benhui Chen
Sheng Chen
Xinyu Chen
Songcan Chen
Long Cheng
Xiaochun Cheng
Zunshui Cheng
Jungik Cho
Chuandong Li
Antonio J. Conejo
Yaping Dai
Jayanta Kumar Debnath
Jianguo Du
Mark Elshaw
Christos Emmanouilidis
Tolga Ensari
Yulei Fan
Mauricio Figueiredo
Carlos H. Q. Foster
Sabrina Gaito
Xinbo Gao
Zaiwu Gong
Adilson Gonzaga
Shenshen Gu
Dongbing Gu
Suicheng Gu
Qianjin Guo
Jun Guo
Chengan Guo
Hong He
Fengqing Han
Wangli He
Xiangnan He
Yunzhang Hou
Wei Hu
Jin Hu
Jun Hu
Jinglu Hu
Yichung Hu
Xi Huang
Chuangxia Huang
Chi Huang
Gan Huang
He Huang
Chihli Hung
Amir Hussain
Lei Jia
Qiang Jia
Danchi Jiang
Minghui Jiang
Lihua Jiang
Changan Jinag
Chi-Hyuck Jun
Shunshoku Kanae
Deok-Hwan Kim
Tomoaki Kobayashi
Darong Lai
James Lam
Bing Li
Liping Li
Chuandong Li
Yueheng Li
Xiaolin Li
Kelin Li
Dayou Li
Jianwu Li
Ping Li
Wei Li
Xiaoli Li
Yongmin Li
Yan Li

Rong Li
Guanjun Li
Jiguo Li
Lulu Li
Xuechen Li
Jinling Liang
Clodoaldo Aparecido de Moraes Lima
Yurong Liu
Li Liu
Maoxing Liu
Nan Liu
Chao Liu
Honghai Liu
Xiangyang Liu
Fei Liu
Lixiong Liu
Xiwei Liu
Xiaoyang Liu
Yang Liu
Gabriele Lombardo
Xuyang Lou
Jianquan Lu
Wenlian Lu
Xiaojun Lu
Wei Lu
Ying Luo
Lili Ma
Shingo Mabu
Xiangyu Meng
Zhaohui Meng
Cristian Mesiano
Xiaobing Nie
Yoshihiro Okada
Zeynep Orman
Stanislaw Osowski
Tsuyoshi Otake
Seiichi Ozawa
Neyir Ozcan
Zhifang Pan
Yunpeng Pan
Zhifang Pang
Federico Pedersini
Gang Peng
Ling Ping
Chenkun Qi
Jianlong Qiu
Jianbin Qiu
Dummy Reviewer
Zhihai Rong
Guangchen Ruan
Hossein Sahoolizadeh
Ruya Samli
Sibel Senan
Zhan Shu
Qiankun Song
Wei Su
Yonghui Sun
Junfeng Sun
Yuan Tan
Lorenzo Valerio
Li Wan
Lili Wang
Xiaofeng Wang
Jinlian Wang
Min Wang
Lan Wang
Qiuping Wang
Guanjun Wang
Duan Wang
Weiwei Wang
Bin Wang
Zhengxia Wang
Haikun Wei
Shengjun Wen
Stefan Wermter
Xiangjun Wu
Wei Wu
Mianhong Wu
Weiguo Xia
Yonghui Xia
Tao Xiang
Min Xiao
Huaitie Xiao
Dan Xiao
Wenjun Xiong
Junlin Xiong
Weijun Xu
Yan Xu
Rui Xu
Jianhua Xu

Gang Yan
Zijiang Yang
Taicheng Yang
Zaiyue Yang
Yongqing Yang
Bo Yang
Kun Yang
Qian Yin
Xiuxia Yang
Xu Yiqiong
Simin Yu
Wenwu Yu
Kun Yuan
Zhiyong Yuan
Eylem Yucel
Yong Yue
Jianfang Zeng
Junyong Zhai
Yunong Zhang
Ping Zhang
Libao Zhang
Baoyong Zhang
Houxiang Zhang
Jun Zhang
Qingfu Zhang
Daoqiang Zhang
Jiacai Zhang
Yuanbin Zhang
Kanjian Zhang
Leina Zhao
Yan Zhao
Cong Zheng
Chunhou Zheng
Shuiming Zhong
Jin Zhou
Bin Zhou
Qingbao Zhu
Wei Zhu
Antonio Zippo
Yanli Zou
Yang Zou
Yuanyuan Zou
Zhenjiang Zhao

Table of Contents – Part II

Machine Learning and Data Mining

Intelligent Control and Robotics

Pattern Recognition

Audio, Image Processing and Computer Vision

Fault Diagnosis

Other Applications and Implementations

Applications of Neural Networks in Electronic Engineering

Cellular Neural Networks and Advanced Control with Neural Networks

Nature Inspired Methods of High-dimensional Discrete Data Analysis

Pattern Recognition and Information Processing Using Neural Networks

Table of Contents – Part I

Computational Neuroscience

Cognitive Science

Mathematical Modeling of Neural Systems

Stability and Nonlinear Analysis

Probabilistic Methods

Supervised Learning

Unsupervised Learning

Support Vector Machine and Kernel Methods

Hybrid Optimisation Algorithms

Rough Set Combine BP Neural Network in Next Day Load Curve Forcasting

Chun-Xiang Li[1], Dong-Xiao Niu[2], and Li-Min Meng[1]

[1] Information and Network Management Center, North China Electric Power University, 071003 Baoding, Hebei, China
licx@ncepubd.edu.cn

[2] Department of Economics Management, North China Electric Power University,071003 Baoding, Hebei, China
ndx@ncepu.edu.cn

Abstract. Artificial neural network (ANN) is used in load forecasting widely. However, there are still some difficulties in choosing the input variables and selecting one appropriate architecture of the neural networks. According to the characteristics of electric short-term load forecasting, presents on a BPANN basing on rough set. Rough set theory is first used to perform input attributes selection. The initial decision table involves factors of weather and date which are able to affect load curve. Then K-Nearest Neighbor method is taken into selecting of most similar data to the target day as the training set of BPANN. Reduced input data of BPANN can avoid over-training and improved performance of BPANN and decreases times of training. The forecasting practice in Baoding Electric Power Company shows that the proposed model is feasible and has a good forecasting precision.

Keywords: Load forecasting, Rough set, Artificial neural network, BP ANN.

1 Introduction

Short-term load forecasting plays an important role in power system planning and operation. It is very important in enhancing the operating efficiency of distribution network, improving the quality power of supply and so on. Precise forecasting benefits enhancing the security and stability of the power system, and reduce the cost of the electricity generation. Therefore, many traditional forecasting models have been proposed and implemented in this field [1],[2], such as multiple linear regression, general exponential smoothing, stochastic process, auto-regressive moving-average model and so on. Yet, the complexity and indeterminacy of the load makes the traditional models which are based on analytic former and numerical arithmetic hard to get the precise forecasting [3],[4],[5]. Recently some new method and theory had been taken in short-term load forecasting, for instance artificial neural network (ANN), support vector machine (SVM), fuzzy set, rough set, etc[6],[7]. Back-Propagation ANN (BPANN) has

F. Sun et al. (Eds.): ISNN 2008, Part II, LNCS 5264, pp. 1–10, 2008.

ability of learn complex and nonlinear relationships which is difficult to model with conventional techniques [8],[9],[10],[11],[12],[13]. When we do forecasting with BPANN, how many factors would be taken in is a difficult decision. Too more factors will lead the capability of network to be low. Rough sets are an implement of math to deal with the information of indefiniteness, impreciseness and incompleteness effectively. We adopt rough set theory to reduce relevant factor sets of power load, and get the reduction expression of knowledge and reveal the dependency and relevancy of conditional attributions. Choosing the original attribute and inputting the reduction fruit to the neural network and it will get a good solution result in load forecasting [14],[15],[17].

2 Rough Set Theory

Rough Sets theory was proposed by Pawlak [16] in 1982, as a new mathematics tool. It is widely applied to dispose incomplete and uncertain information, whose main aim is that under the precondition without any change in keep classification capabilities, the classification rules of the concept can be acquired through RS reduction. Trough over 10 years development this theory has triumphantly used in decision-support system, process control, machine study and so on.

2.1 Decision Table and Reduction

In rough sets theory, knowledge denotation system may be described by formula 1

$$S =< U, A, V, F > \tag{1}$$

where U is universe and expresses a set with finite objects, A is attribute set composed of condition attribute C and decision attribute $D, A = C \cup D, a \in A, C \cap D = \emptyset, V \in V_a, V_a$ is range of a, $f : U \times A \to V$ is a information function, it specifies attribute values of every object in U.

The decision table can be considered as a group of defined equivalence relation, this is knowledge base. Information systems based on rough sets definition can be denoted by the use of table format, where columns express attributes and rows represent objects, and every row describes information of an object. The attributes can be divided into condition attributes and decision attributes.

Not all the condition attributes in decision are necessary. Some are redundant, and when these are eliminated, the expression effect is not changed.

In rough sets, binary indivisible relationship ind(P) determined by $P \in A$ can be expressed by formula 2

$$\text{ind}(P) = \{(x, y) \in U \times U \,||\, \forall a \in P, f(x, a) = f(y, x)\} \tag{2}$$

It is very clearly that if $(x, y) \in$ind(P), then x and y can not be differentiated according to existing information, ind(P) are an equivalent relation in U for $\forall P \subseteq A$.

Set $S =< U, C \cup D >$, if $C_1 \in C$, $C_1 \neq \emptyset$, and the following two conditions hold:

$$\text{A1): } \mathrm{ind}_{C_1}(D) = \mathrm{ind}_C(D)$$

$$\text{A2): } \forall C_2 \subseteq C_1, \mathrm{ind}_{C_2}(D) \neq \mathrm{ind}_{C_1}(D)$$

Base on A1)-A2) we can say C_1 is a reduction of C with regard to D, the intersection of all reductions is called core, defined as $\mathrm{core}_{C_1} = \cap \mathrm{red}_D(C)$.

2.2 Rough Subjection Degree Function

Another denotation means of rough sets is rough subjection degree function $\mu_x^R(a)$, and is expressed by formula 3

$$\mu_x^R(a) = \frac{\mathrm{card}|X \cap [\ a\ ]_R|}{\mathrm{card}|[\ a\ ]_R|} \tag{3}$$

where $\mu_x^R(a)$ expresses degree that the element x belongs to X based on indivisible relationship R. Obviously, rough subjection degree function $\mu_x^R(a)$, satisfies 0¡$\mu_x^R(a)$¡1. In the fault diagnosis a and X correspond fault symptom and fault category, while rough subjection degree function $\mu_x^R(a)$ is the accuracy of decision.

3 The BP Artificial Neural Network

BPANN is a kind of feed forward neural network which is the most popular networks, as show in Fig.1. It consists of three layers: the input layer, the hidden layer and the output layer. The nodes in the input layer are relevant factors about load, and the node in the output layer provides the forecasting result. The number of nodes in the hidden layer is chosen by trial and error, selecting a few alternative numbers and then running simulations to find out one that gave the best predictive performance. The activation function for the hidden nodes is Sigmoid function, and for the output node is linear function as 4.

$$O = \sum_{i=1}^{n} w_i y_i \tag{4}$$

w_i is the conjunction weights between hidden layer cell and output layer cell, y_i is the output of the hidden layer cell, n is the number of hidden layer cells. We adopt joining momentum coefficient and forgetting coefficient in 5 to enhance the network convergence speed.

$$\Delta w_{ij}^l(t) = \eta \delta_{ik}^l o_{jk}^{l-1} + \phi \Delta w_{ij}^l(t-1) + \varphi \Delta w_{ij}^l(t-2) \tag{5}$$

The energy function to evaluate the performance of the network is defined as 6.

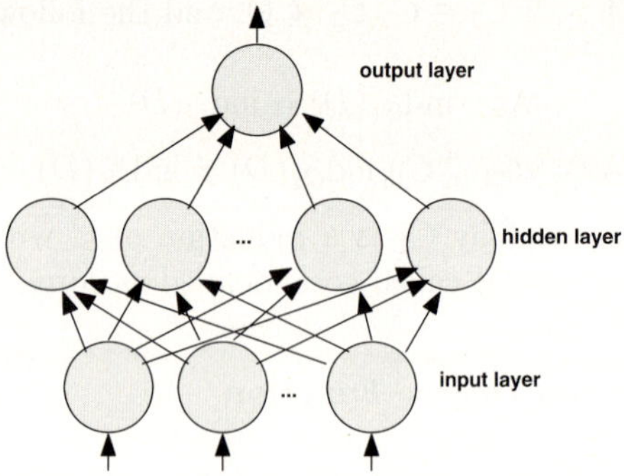

Fig. 1. Structure of three-layer BP network

$$E(t) = \frac{1}{N}\sum_{i=1}^{N}\frac{|O_i^* - O_i^t|}{O_i^*} \times 100\% \tag{6}$$

O_i^* is the actual load and O_i^t is the forecast load, N is the number of data.

4 The Forecasting Model

The training input data of BPANN are obviously important. The main factors of effecting load are weather and date, where weather factor involves temperature, humidity, wind, rainfalletc, while date factor involves month, week, day. In those factors, temperature can remarkably affect power load. The effect includes two aspects: one is the variety of temperature in one day should change the day load curve [18]; the other one is the temperature of someday before today will affect the load too. To the former, we pick up six temperature points in one day to indicate the characteristic of the day's temperature curve. And to the latter, we use the max-temperature, min-temperature and average-temperature of seven days before today. The maximum load and maximum temperature of every day in one year, as show in Fig.2, is very different. April, AugustNovember and December often has the maximum load in a year. We separate date to month, week and holiday. The initial decision table show in Table1.

Before reduction, the attributes must be discretization. The attributes relate to temperature variable is discretized as 7.Other attributes' discretization show in Table2.

$$u_i = \begin{cases} 0 & T \le -15\,^\circ\mathrm{C} \\ 1 & -15\,^\circ\mathrm{C} < T \le 0\,^\circ\mathrm{C} \\ 2 & 0\,^\circ\mathrm{C} < T \le 15\,^\circ\mathrm{C} \\ 3 & 15\,^\circ\mathrm{C} < T \le 25\,^\circ\mathrm{C} \\ 4 & 25\,^\circ\mathrm{C} < T \end{cases} \qquad (i = 1, 2, \ldots, 27) \tag{7}$$

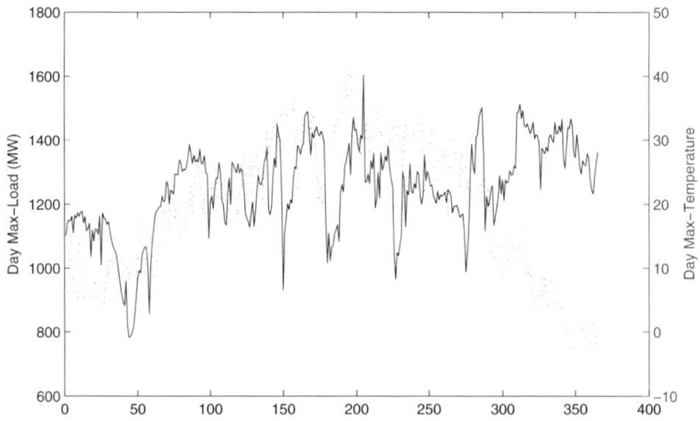

Fig. 2. Day max-load and max-temperature contrast of Baoding in 2002

Table 1. Initial decision table

Attribute name	Attribute meaning
$u_1,\ldots,u_6$	T_d^l,(l=0,4,8,12,16,20), express d day's six temperature point on l clock
$u_7,\ldots,u_{13}$	$T_{d-i}^{\max}$(i=1,...,7), express $d-i$ day's max-temperature
$u_{14},\ldots,u_{20}$	$T_{d-i}^{\min}$(i=1,...,7), express $d-i$ day's min-temperature
$u_{21},\ldots,u_{27}$	T_{d-i}^{avg}(i=1,...,7), express $d-i$ day's average temperature
u_{28}	Hum_d, express d day's humidity
u_{29}	R_d, express d day's rainfall
u_{30}	Wd_d, express d day's wind
u_{31}	M_d, express d day's month
u_{32}	Wk_d, express d day's week
u_{33}	H_d, express whether d day is holiday

Table 2. Attribute discretization

Attribute name	Discretization value
u_{28}	0:(-∞,30]; 1:(30,60]; 2:(60,+∞)
u_{29}	0:[0,7]; 1:(7,15]; 2:(15,25]; 3:(25,+∞)
u_{30}	0:[0,3]; 1:(3,6]; 2:(6,+∞)
u_{31}	The number of variable's month
u_{32}	The number of variable's week
u_{33}	0:is holiday; 1:is not holiday

(Remark: 0:(-,30] means u=0 where the variable in (-,30]).

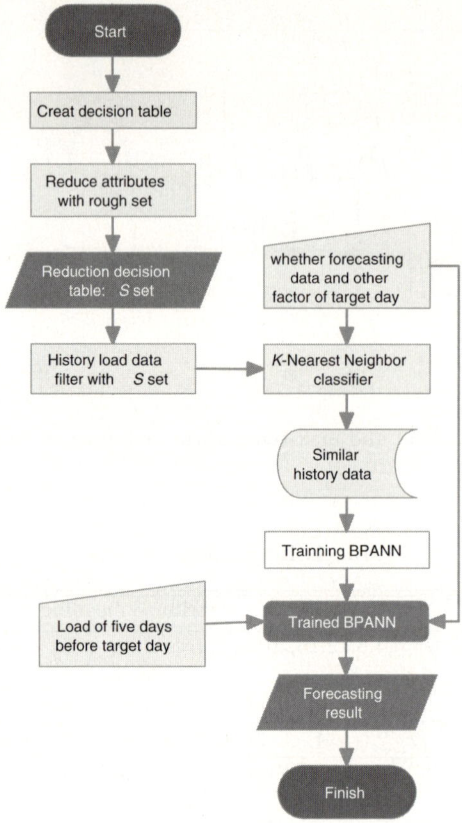

Fig. 3. Flow chart with rough set BPANN forecasting model

With load data of two years ago, the day max-load is separated into three grades: high, middle and low as showing in Tab3 where

$$A_1 = L_{max} - \tfrac{1}{3}(L_{max} - L_{min})$$
$$A_2 = L_{max} - \tfrac{2}{3}(L_{max} - L_{min}) \quad (8)$$

(L_{max} is the maximum load and L_{min} is the minimum load in the year)

Using rough set theory to reduce attributes in Tab1 with the discretization values in Tab2. Some attributes may interrelate to other attributes. For example, the rain is heavier, the temperature is lower. So the sequence of adding attributes in the process of reduce may be confirmed with the weightiness of attributes from high to low. We use the order below:$u_1, \ldots, u_6, u_{31}, u_{32}, u_{33}, u_7, \ldots, u_{27}, u_{29}, u_{28}, u_{30}$. The reduction result is S=$\{u_1, \ldots, u_6, u_{31}, u_{32}, u_7, u_8, u_{14}, u_{21}, u_{22}, u_{28}, u_{29}\}$.

We introduce K-Nearest Neighbor[19] to classify history data set into three classes, just like Tab3 showing, with attributes in S set and can get three centroids of the whole set. Computing the distance between S_d set, the

Table 3. Classifying day max-load

Grade	Max-load value of one day
High	$L_d \geq A_1$
Middle	$A_2 \leq L_d < A_1$
Low	$L_d < A_2$

forecasting target day's attributes set, to three centroids respectively and finding the nearest centroid to S_d. Then select some data in the set affiliated with this centroid as the training input data of BPANN. The input data vector is $I = S \cup L_{d-i}(i = 1, \ldots, 5)$, where L_d express the load of d day. The flow shows in Fig.3.

5 Test Result

We practice load forecasting for Baoding electric power company with three methods which are regression model, ANN model and RS-BPANN model. Regression model uses the last ten days' load data to build the model and forecasting. ANN model adopts the last month load data and interrelated attributes

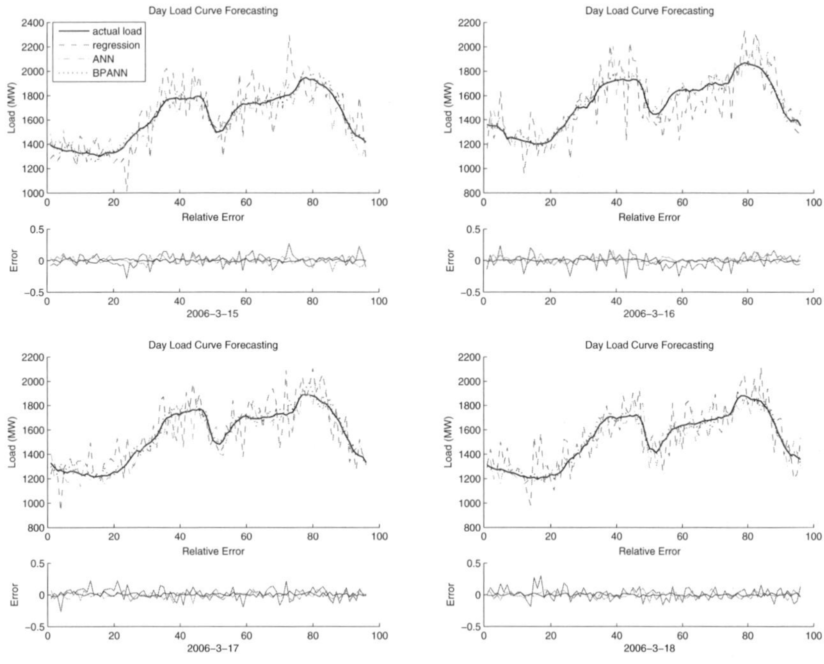

Fig. 4. Load forecasting curve for Baoding of March 15 to 18 in 2006 and error contrast

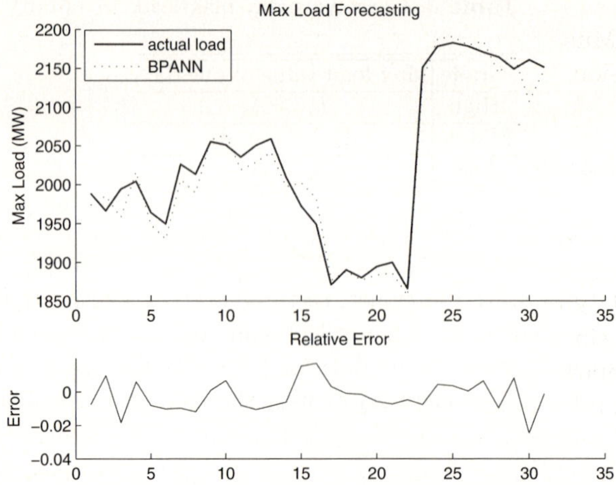

Fig. 5. Day max-load forecasting for March 2006

Table 4. Contrast of forecasting relative error from March 12 to 18 in 2006

	Regression model			ANN			BPANN		
Date	Max-err	Min-err	Avg-err	Max-err	Min-err	Avg-err	Max-err	Min-err	Avg-err
3-11	0.2514	0.0581	0.0939	0.1409	0.0474	0.0790	0.0563	0.0100	0.0224
3-13	0.2088	0.1098	0.1024	0.1231	0.0412	0.0573	0.0941	0.0125	0.0497
3-14	0.2453	0.0014	0.0972	0.1401	0.0381	0.0634	0.0301	0.0098	0.0213
3-15	0.2777	0.1011	0.1063	0.1010	0.0183	0.0321	0.0572	0.0076	0.0099
3-16	0.2812	0.0921	0.1025	0.1156	0.0341	0.0109	0.0609	0.0192	0.0142
3-17	0.2692	0.1031	0.1323	0.1198	0.0540	0.0209	0.1041	0.0083	0.0106
3-18	0.3012	0.0899	0.1503	0.1364	0.0253	0.0928	0.0709	0.0079	0.0164

involved in S set as the training data. The difference of ANN to RS-BPANN is that the training data are nearest days' data but not most similar days'. Fig.4 is the forecasting results of 96 point day load curve of March 15 to 18 in 2006. Regression model has serious wave and forecasting relative errors are very big. ANN model has smaller forecasting relative errors. Due to picking up training data using the forenamed method, RS-BPANN forecasting result can fit good to real load curve.

Using these three models to forecast 96 point day load curve from March 12 to 18 in 2006, and the Tab4 shows the contrast of relative error. Regression model's average errors are more than 10 percent commonly, which is too big for short-term load forecasting popularly. ANN's forecasting average errors have little swing and less than 8 percent mainly. RS-BPANN has lest forecasting errors and average errors, which are less than 3 percent usually. It is obviously that RS-BPANN can gain a higher forecasting precision.

Fig.5 shows the max-load forecasting for March 2006 with BPANN and the relative error. It can be seen that the forecasting result curve is similar to the actual load curve. The forecasting relative errors less than 3% and average relative is 1.2%. Multi-forecasting practice shows that the BPANN can gain a satisfying precision.

6 Conclusion

Rough set theory combining BPANN can solve the problem of input data selection. Though analyze history load data and correlative factors, create initial decision table and reduce it with rough set. Reduction factor set should be helpful to pick up most similar history load to train the BPANN. Temperature and data are the most important factors to affect load curve, so whether the forecasting weather information is exact will influence the forecasting load result. In view of it, the model is applicable only in short-term forecasting. Contrast to regression model and traditional ANN model, the model here present on is more exact.

References

1. Shyh, J.H., Kuang, R.S.: Short-term Load Forecasting Via ARMA Model Identification Including Non-Gaussian Process Considerations. IEEE Transactions on Power Systems 18, 673–679 (2003)
2. Yong, H.L., Pin, C.L.: Novel High-precision Grey Forecasting Model. Automation in Construction 16, 771–777 (2007)
3. Chorng, S.O., Jih, J.H., Gwo, H.T.: Model Identification of ARIMA Family Using Genetic Algorithms. Applied Mathematics and Computation 164, 885–912 (2005)
4. Senjyu, T., Andal, P., Uezato, K., Funabashi, T.: Next Day Load Curve Forecasting Using Recurrent Neural Network Structure. IEEE Proceedings Generation, Transmission and Distribution 151, 388–394 (2004)
5. Baczynski, D., Parol, M.: Influence of Artificial Neural Network Structure on Quality of Short-term Electric Energy Consumption Forecast. IEEE Proceedings Generation, Transmission and Distribution 151, 241–245 (2004)
6. Wang, N., Zhang, W.X.: A Restricted Least Squares Estimation for Fuzzy Linear Regression Models. Fuzzy Systems and Mathematics 20, 17–124 (2006)
7. Song, K.B., Baek, Y.S., Hong, D.H., Jan, G.: Short-Term Load Forecasting for the Holidays Using Fuzzy Linear Regression Method. IEEE transactions on power systems 20, 96–101 (2005)
8. Saksornchai, T., Lee, W.J., Methaprayoon, K.: Improve the Unit Commitment Scheduling by Using the Neural-Network-Based Short-Term Load Forecasting. IEEE Transactions on Industry Applications 41, 169–179 (2005)
9. Abdel, A.R.E.: Short-term Hourly Load Forecasting Using Abductive Networks. IEEE Transactions on Power Systems 19, 164–173 (2004)
10. Ming, M., Lu, J.C., Sun, W.: Short-Term Load Forecasting Based on Ant Colony Clustering and Improved BP Neural Networks. In: 2006 International Conference on Machine Learning and Cybernetics, vol. 2, pp. 3012–3015 (2006)
11. Naresh, R., Dubey, J., Sharma, J.: Two-phase Neural Network Based Modelling Framework of Constrained economic load dispatch. IEEE Proceedings Generation, Transmission and Distribution 151, 373–378 (2004)

12. Yu, S.W., Zhu, K.J., Diao, F.Q.: A Dynamic All Parameters Adaptive BP Neural Networks Model and Its Application on Oil Reservoir Prediction. Applied Mathematics and Computation 195, 66–75 (2008)
13. Ivan, N.D.S., Rogerio, A.F.: An Approach Based on Neural Networks for Estimation and Generalization of Crossflow Filtration Processes. Applied Soft Computing 8, 590–598 (2008)
14. Al-Hamadi, H.M., Soliman, S.A.: Fuzzy Short-term Electric Load Forecasting Using Kalman Filter. IEEE Proc. Gener. Transm. Distrib. 153, 217–227 (2006)
15. Niu, D.X., Chen, Z.Y., Xing, M., Xie, H.: Combined Optimum Gray Neural Network Model of The Seasonal Power Load Forecasting With the Double Trends. Proceeding of the CSEE 22, 29–32 (2002)
16. Paw, L.Z.: Rough sets. International Journal of Computer InformationScience 5, 341–356 (1982)
17. Stephen, A.B., Wei, H.L., Michael, A.B.: Generalized Multiscale Radial Basis Function Networks. Neural Networks 20, 1081–1094 (2007)
18. Chen, H.J., Du, Y.J., Jiang, J.N.: Weather Sensitive Short-Term Load Forecasting Using Knowledge-Based ARX Models. IEEE Power Engineering Society General Meeting 1, 1190–1196 (2005)
19. Kuan, J., Lewis, P.: Fast k nearest neighbour search for R-tree family. In: International Conference on Information Communications and Signal Processing ICICS 1997, vol. 2, pp. 924–928 (1997)

Improved Fuzzy Clustering Method Based on Entropy Coefficient and Its Application

Li Liu, Jianzhong Zhou*, Xueli An, Yinghai Li, and Qiang Liu

College of Hydroelectric and Digitalization Engineering, Huazhong University of Science and Technology, Wuhan Hubei 430074, China
liuli0324@163.com, jz.zhou@hust.edu.cn

Abstract. Based on the principle of fuzzy clustering analysis and the theory of entropy, an improved fuzzy clustering method is given by improving the method of establishing the membership function, combining the clustering weight with the entropy coefficient, and replacing the Zadeh operator $M(\vee,\wedge)$ with the weight average operator $M(+,\bullet)$. With the improved method, the zero-weight problem is addressed effectively, the weights of each factor are modified properly and the phenomenon of Major Factor Dominating is also alleviated appropriately. Finally, an illustrative example is given to clarify the method, which shows that the improved fuzzy clustering method is reasonable, feasible, simple and practical.

Keywords: Fuzzy clustering, Entropy coefficient, Membership function, Weight.

1 Introduction

Cluster analysis is a multivariate analysis method in mathematical statistics which studies that things always cluster according to their categories. Since some feathers of objective things have no strict and apparent bounds, it is absolutely suitable to introduce fuzzy mathematics into the study of cluster analysis.

Fuzzy clustering method (FCM), which based on the fuzzy set theory [1], is one of the most important methods of unsupervised learning and has significant advantages over traditional clustering. FCM is used to deal with ill-defined boundaries between clusters, and the membership of data points are interpreted as degrees of sharing. However, the membership do not always correspond to the intuitive concept of degree of belonging or compatibility because the linear membership function is used in classical method, in which the zero-weight problem is existent and the warped results might be made. Otherwise, the clustering weight has a deep effect on the evaluation results, so how to determine it appropriately is the key problem of the FCM. But the determination of the weight of every indicator in classical method is by calculating the super scale which is the ratio of the value of every indicator at each monitoring point over corresponding standard [2], it only contains the information of the individual indicator, which has nothing to do with the relationship among evaluating objects,

* Corresponding author.

F. Sun et al. (Eds.): ISNN 2008, Part II, LNCS 5264, pp. 11–20, 2008.

the result may deviate significantly from actual values. In addition, the phenomenon of Major Factor Dominating may occurs if the Zadeh operator M(∨,∧) [3] is still used in the classical method, which is likely to lead the result with deflection.

By analyzing the classical method, an improved FCM based on entropy coefficient has been put forward. Finally in this paper, the improved method has been used to assess the status of water quality as example and the results are satisfactory.

2 Fuzzy Clustering Method

As a kind of cluster analysis, FCM can be classified as the fuzzy comprehensive evaluation method, the fuzzy probability method, the fuzzy complex index method, and so on. The most models of fuzzy clustering contain three steps: the first is data standardization; the second is to establish fuzzy similar matrix and the last is clustering.

Usually, i=1, 2 … n is taken to stand for the clustering object set, j=1, 2 … m for the clustering indicator set, k=1, 2 … K for the membership degree set.

2.1 Data Standardization

In order to minimize the effect of each factor due to the difference of dimension and unit, the values of different clustering indicators of the clustering objects should be standardized and made into the interval [0, 1] according to the normalization method before the fuzzy similarity relation is constructed:

$$d_{i,j} = \left(x_{i,j} - \min_{1\le i\le n}\{x_{i,j}\}\right) \Big/ \left(\max_{1\le i\le n}\{x_{i,j}\} - \min_{1\le i\le n}\{x_{i,j}\}\right). \tag{1}$$

There, $x_{i,j}$ is the measured value from i-th object for j-th indicator, $d_{i,j}$ is the standardized value of $x_{i,j}$, $\max\{x_{i,j}\}$ and $\min\{x_{i,j}\}$ is the maximum and the minimum of all $x_{i,j}$ about the same indicator.

2.2 Establishing Membership Function

When calculating the weight, the influence between the non-neighboring grades has always been neglected in the classical FCM, it means that the zero-weight problem is existent and the warped results might be made. So, the exponential membership function is put forward in the method as follow:

$$g_{j,1}(d_{i,j}) = \begin{cases} 1 & x_{i,j} \in (0, \lambda_{j,1}] \\ e^{\frac{\lambda_{j,1} - x_{i,j}}{\lambda_{j,1}}} & x_{i,j} \in (\lambda_{j,1}, +\infty) \end{cases}, \tag{2}$$

$$g_{j,k}(d_{i,j}) = \begin{cases} e^{\frac{x_{i,j} - \lambda_{j,k}}{x_{i,j}}} & x_{i,j} \in (0, \lambda_{j,k}] \\ e^{\frac{\lambda_{j,k} - x_{i,j}}{\lambda_{j,k}}} & x_{i,j} \in (\lambda_{j,k}, +\infty) \end{cases}, \tag{3}$$

$$g_{j,K}\left(d_{i,j}\right)=\begin{cases} e^{\frac{x_{i,j}-\lambda_{j,K}}{x_{i,j}}} & x_{i,j}\in\left(0,\lambda_{j,K}\right] \\ 1 & x_{i,j}\in\left(\lambda_{j,K},+\infty\right) \end{cases}. \tag{4}$$

There, $g_{j,k}$ is the exponential membership function from j-th indicator for k-th degree, $\lambda_{j,k}$ is the threshold of $g_{j,k}$.

Each value of certain factor is unique corresponding with each nonzero weight of various grades, so that the zero-weight problem can be addressed.

2.3 Determining Clustering Weight

The clustering weight is a kind of relative weight which reflects the dangerous degree of each indicator. In the classical FCM, the clustering weight of evaluating indicators is determined by the monitoring data compared to standard:

$$r_{i,j}=x_{i,j}\Big/\frac{1}{K}\sum_{k=1}^{K}S_{j,k}, \tag{5}$$

$$\omega_{i,j}=r_{i,j}\Big/\sum_{j=1}^{m}r_{i,j}. \tag{6}$$

There, $S_{j,k}$ is the certified value from j-th indicator for k-th degree, $\omega_{i,j}$ is the clustering weight from i-th object for j-th indicator.

In order to make the clustering more practical, much regard should be paid to the indicator of a system if its value fluctuates greatly, so the concept of entropy is introduced.

Entropy, which was introduced into information by Shannon in 1948 [16], is a concept of thermodynamics and has been successfully applied to measure the complexity of system. It can be calculated by the following formula:

$$f_{i,j}=d_{i,j}\Big/\sum_{i=1}^{n}d_{i,j}, \tag{7}$$

$$h_j=-c\cdot\sum_{i=1}^{n}\left(f_{i,j}\cdot\ln f_{i,j}\right). \tag{8}$$

There, c=1/ln$n$, h_j is the entropy of j-th indicator, and suppose $f_{i,j}\bullet\ln f_{i,j}=0$ if $f_{i,j}=0$.

Then, the entropy coefficient can be defined as:

$$\theta_j=\left(1-h_j\right)\Big/\left(m-\sum_{j=1}^{m}h_j\right). \tag{9}$$

There, θ_j is the entropy coefficient of j-th indicator, $\theta_j\in[0,1]$ and $\sum\theta_j=1$.

The new clustering weight is modified by combined with subjective weight and entropy coefficient as fellow:

$$\omega'_{i,j}=\theta_j\cdot\omega_{i,j}\Big/\sum_{j=1}^{m}\theta_j\cdot\omega_{i,j}. \tag{10}$$

There, $\omega'_{i,j}$ is the correction value of $\omega_{i,j}$.

With this method, when the quality indicators fluctuate greatly, in another word, when $f_{i,j}$ differ greatly, there will be a smaller h_j and a bigger θ_j. On the contrary, there will be smaller weights when the quality indicators have a smaller fluctuation. So it is rational to determine the weights of the quality indicators combining with the entropy coefficient.

2.4 Calculating Clustering Coefficient

The Zadeh operator M(∨,∧), which use "∧" and "∨" to denote intersection and union operators respectively, is a synthetic evaluation mathematic model of Major Factor Dominating and always used in the classical FCM. This model might neglect lots of useful information, especially the information of those non-main factors, when qualitative factors are many and each weight is small. So, the weight average operator M(+,•), which use "+" and "•" to replace "∨" and "∧" respectively, has been introduced:

$$\varepsilon_{i,k} = \sum_{j=1}^{m} \omega'_{i,j} \cdot g_{j,k}\left(d_{i,j}\right). \tag{11}$$

There, $\varepsilon_{i,k}$ is the clustering coefficient from i-th object for k-th degree.

According to the value of weight, all factors have been maintained and considered in the clustering coefficient, it is suitable for synthetic evaluation with multiple factors.

2.5 Judgment Object League

According to the Maximum Membership Principle, the object degree is the membership degree which the maximum clustering coefficient corresponds to.

3 Application and Analysis

As an important aspect of the water resources assessment, the assessment of water quality is a fuzzy concept with multiple factors and levels. The FCM about it has been studied and put into practice extensively in recent year.

As the criterion of the water quality data in China, 《GB3838-2002》 is adopted and presented in Table 1:

Table 1. Assessment standard of water quality data (mg/L)

Test Indicator	Water Type				
	I	II	III	IV	V
DO≥	7.5	6	5	3	2
COD_{Mn}≤	2	4	6	10	15
COD≤	15	15	20	30	40
BOD≤	3	3	4	6	10
TP≤	0.02	0.1	0.2	0.3	0.4
NH_3-N≤	0.15	0.5	1	1.5	2

Applied the improved FCM to assess the status of water quality and the indicators of samples are shown in Table 2:

Table 2. Water quality data of control section (mg/L)

Sample	Test Indicator					
	DO	COD_{Mn}	COD	BOD	TP	NH_3-N
1	6.74	3.04	12.84	1.63	0.015	0.15
2	8.14	3.97	22.4	2.57	0.154	0.59
3	8.3	3.46	27.21	2.16	0.086	0.19
4	8.39	2.89	15.99	1.88	0.021	0.08
5	7.98	3.24	17.8	2.01	0.015	0.26
6	5.92	1.71	5	1.1	0.012	0.036
7	5.72	2.62	5	2.04	0.025	0.093
8	6.44	4.07	5	1.82	0.193	0.41
9	6.45	4.7	9.01	1.25	0.135	0.248
10	5.72	2.21	8.45	1.7	0.104	0.052
11	6.05	3.06	6.49	1.58	0.094	0.168

In order to make fuzzy assessment, it is necessary to normalize the data of Table 2 first.

According to the formula (1) (with the water type ascends, the value of DO decreases, so this indicator should be picked the reciprocal first), the sample matrix is given and shown in Table 3:

Table 3. Normalization results of water quality data

Sample	Test Indicator					
	DO	COD_{Mn}	COD	BOD	TP	NH_3-N
1	0.524	0.445	0.353	0.361	0.017	0.206
2	0.066	0.756	0.783	1	0.785	1
3	0.023	0.585	1	0.721	0.409	0.278
4	0	0.395	0.495	0.531	0.05	0.079
5	0.11	0.512	0.576	0.619	0.017	0.404
6	0.894	0	0	0	0	0
7	1	0.304	0	0.639	0.072	0.103
8	0.649	0.789	0	0.49	1	0.675
9	0.644	1	0.181	0.102	0.68	0.383
10	1	0.167	0.155	0.408	0.508	0.029
11	0.829	0.452	0.067	0.327	0.453	0.238

Meanwhile, based on the values of Table 1, the thresholds $\lambda_{j,k}$ of each membership functions are given. Then, according to the formula (2), (3) and (4), the membership functions matrix g can be obtained. Considering the length of paper, the membership functions matrix of each samples are not shown here.

The design of weight is one of the important parts in the FCM. With formula (5) and (6), the clustering weight matrix ω is work out:

$$\omega = \begin{bmatrix} 0.273 & 0.202 & 0.263 & 0.154 & 0.036 & 0.072 \\ 0.123 & 0.143 & 0.249 & 0.132 & 0.201 & 0.153 \\ 0.147 & 0.152 & 0.369 & 0.135 & 0.137 & 0.06 \\ 0.218 & 0.191 & 0.326 & 0.177 & 0.05 & 0.038 \\ 0.199 & 0.185 & 0.314 & 0.164 & 0.031 & 0.107 \\ 0.46 & 0.168 & 0.151 & 0.153 & 0.043 & 0.025 \\ 0.36 & 0.194 & 0.114 & 0.215 & 0.067 & 0.05 \\ 0.192 & 0.181 & 0.069 & 0.115 & 0.312 & 0.131 \\ 0.213 & 0.232 & 0.137 & 0.088 & 0.242 & 0.088 \\ 0.299 & 0.136 & 0.161 & 0.149 & 0.232 & 0.023 \\ 0.278 & 0.185 & 0.121 & 0.136 & 0.206 & 0.073 \end{bmatrix}$$

And the entropy sequence h is calculated according to the formula (7) and (8)

$$h = \begin{bmatrix} 0.85 & 0.92 & 0.767 & 0.915 & 0.786 & 0.824 \end{bmatrix}$$

Then the entropy weight coefficient θ is attained by formula (9):

$$\theta = \begin{bmatrix} 0.16 & 0.085 & 0.248 & 0.091 & 0.228 & 0.187 \end{bmatrix}$$

So, with the formula (10), the new clustering weight matrix ω' is work out:

$$\omega' = \begin{bmatrix} 0.27 & 0.106 & 0.403 & 0.086 & 0.051 & 0.083 \\ 0.109 & 0.068 & 0.343 & 0.066 & 0.255 & 0.159 \\ 0.129 & 0.071 & 0.501 & 0.067 & 0.171 & 0.061 \\ 0.21 & 0.098 & 0.485 & 0.096 & 0.069 & 0.043 \\ 0.19 & 0.094 & 0.465 & 0.089 & 0.042 & 0.119 \\ 0.478 & 0.093 & 0.244 & 0.09 & 0.063 & 0.031 \\ 0.392 & 0.113 & 0.194 & 0.133 & 0.105 & 0.063 \\ 0.181 & 0.091 & 0.101 & 0.062 & 0.42 & 0.145 \\ 0.203 & 0.118 & 0.203 & 0.048 & 0.33 & 0.098 \\ 0.281 & 0.068 & 0.234 & 0.079 & 0.312 & 0.025 \\ 0.272 & 0.097 & 0.184 & 0.075 & 0.288 & 0.084 \end{bmatrix}$$

Finally, the clustering coefficient matrix ε is worked out by the formula (11):

The clustering coefficient matrix reflects the intimacy degree of the clustering object with the membership degree. According to the Maximum Membership Principle, the results of evaluation are shown in Table 4:

$$\varepsilon = \begin{bmatrix} 0.928 & 0.703 & 0.482 & 0.2 & 0.076 \\ 0.419 & 0.69 & 0.71 & 0.429 & 0.235 \\ 0.505 & 0.573 & 0.525 & 0.51 & 0.327 \\ 0.931 & 0.716 & 0.548 & 0.256 & 0.119 \\ 0.815 & 0.699 & 0.596 & 0.295 & 0.148 \\ 0.888 & 0.545 & 0.424 & 0.184 & 0.068 \\ 0.842 & 0.556 & 0.431 & 0.186 & 0.065 \\ 0.374 & 0.586 & 0.656 & 0.337 & 0.173 \\ 0.506 & 0.672 & 0.515 & 0.221 & 0.088 \\ 0.611 & 0.742 & 0.46 & 0.187 & 0.068 \\ 0.612 & 0.703 & 0.39 & 0.15 & 0.05 \end{bmatrix}$$

Table 4. The results of assessment

Simple	Improved FCM	BP Network Method	Gray Clustering Method	Classical FCM
1	I	I	I	II
2	III	III	II	III
3	II	II	II	III
4	I	I	I	I
5	I	I	I	II
6	I	I	I	II
7	I	I	I	II
8	III	III	II	III
9	II	III	II	II
10	II	II	II	II
11	II	II	II	II

There also have results of evaluation by BP network method, gray clustering method and classical FCM in Table 4. Compared with the other methods, the improved FCM have the same results of evaluation with sample 4, 10, 11 and the most similar results of evaluation with sample other.

The results support the notion that the phenomenon of Major Factor Dominating has been greatly alleviated in the improved FCM. Sample as sample 3:

- the content of COD exceed the criterion of water typeIII and belong to the water typeIV
- the content of TP, COD_{Mn} and NH_3-N belong to the water type II
- the content of DO and BOD belong to water type I

This sample has been classified to water typeIII by the classical FCM, but been classified to water type II by the improved FCM and other methods. That is because the classical FCM has its defect that the comprehensive evaluation result might on the high side when the certain indicator exceed standard significantly. This phenomenon has been alleviated well in the improved FCM by using the weight average operator, which not only maintains all the information of single factor evaluating, but also considers the effects of all factors. It is clear that the bad effect from some abnormal values are weakens. Therefore, the evaluation result from the improved method is more accurate and reasonable.

It is also found that the influence of the factor of which the value fluctuates greatly has been emphasized in improved FCM by adopting the new value of weight which combining with the entropy coefficient. Sample as sample 8:

- the content of TP belong to water typeIII and almost achieve the criterion of water typeIV
- the content of COD_{Mn} belong to water typeIII
- the content of DO and NH_3-N belong to water type II and almost achieve the criterion of water typeIII
- the content of COD and BOD belong to water type I

This sample has been classified to the water type II by the gray clustering method, and been classified to the water typeIII by the improved FCM and other methods. The improved FCM for determination of weight considers adequately the information of values all the monitoring sections provided to balance the relationship among numerous evaluating object, it is more reasonable in water quality evaluation model because the special influence of the factor of which the value fluctuates greatly will be considered enough. Thus, the result of evaluation is more close to the true.

4 Conclusions

After researching and analyzing, 3 suggestions have been given when FCM is applied in comprehensive evaluation model:

1. when establishing membership functions, the influence between the non-neighboring grades should be considered, so exponential membership function should be adopted to address the zero-weight problem
2. when determining clustering weight, the influence of the factors of which the value fluctuates greatly also should be emphasized, it can be solved by combining the clustering weight with the entropy coefficient
3. the phenomena of Major Factor Dominating should be alleviated by replacing the Zadeh operator $M(\vee,\wedge)$ with the weight average operator $M(+,\bullet)$

FCM has lot of advantages like simple principle, high precision and utilization of information. The improved FCM has large advantage when it is applied in the comprehensive evaluation model, and it is worth of popularization.

Acknowledgements

This work is supported by the National Basic Research Program of China (973 Program) (No.2007CB714107), the Special Research Foundation for the Public Welfare Industry of the Ministry of Science and Technology and the Ministry of Water Resources (No.200701008) and the National Natural Science Foundation of China (No.50579022).

References

1. Zadeh, L.A.: Fuzzy Sets. Information and Control 8, 338–353 (1965)
2. Fan, B.D.: Fuzzy Comprehensive Evaluation Model for Groundwater Quality. China Rural Water and Hydropower, 29–32 (1998)
3. Guo, J.S., Long, T.R., Huo, G.Y., Wang, H.: A Comparison of Four Methods of Water Quality Assessment. Journal of Chongqing Jianzhu University 22, 6–12 (2000)
4. Zou, Z.H., Yun, Y., Sun, J.N.: Entropy Method for Determination of Weight of Evaluating Indicators in Fuzzy Synthetic Evaluation for Water Quality Assessment. Journal of Environmental Sciences 18, 1020–1023 (2006)
5. State Environmental Protection Administration of China, General Administration of Quality Supervision, Inspection and Quarantine of China. GB3838-2002 Environmental Quality Standards for Surface Water. Environmental Science Press, Beijing (2002)
6. Dahiya, S., Singh, B., Gaur, S., Garg, V.K., Kushwaha, H.S.: Analysis of Groundwater Quality Using Fuzzy Synthetic Evaluation. Journal of Hazardous Materials 147, 938–946 (2007)
7. Zadeh, L.A.: Fuzzy Logic Computing with Words. IEEE Transactions-Fuzzy Systems 4, 103–111 (1996)
8. Klir, G.J., Yuan, B.: Fuzzy Sets and Fuzzy Logic-Theory and Applications. Prentice-Hall, Englewood Cliffs (1995)
9. Liu, L., Zhou, J.Z., An, X.L., Yang, L., Liu, S.Q.: Improvement of the Grey Clustering Method and Its Application in Water Quality Assessment. In: Proceedings of the 2007 International Conference on Wavelet Analysis and Pattern Recognition, Beijing, China, pp. 907–911 (2007)
10. Chen, L., Wang, Y.Z.: Research on TOPSIS integrated evaluation and decision method based on entropy coefficient. Control and Decision 18, 456–459 (2003)
11. Chang, N.B., Chen, H.W., Ning, S.K.: Identification of River Water Quality Using the Fuzzy Synthetic Evaluation approach. Journal of Environmental Management 63, 293–305 (2001)
12. Lu, R.S., Lo, S.L., Hu, J.Y.: Analysis of Reservoir Water Quality Using Fuzzy Synthetic Evaluation. Stochastic Environmental Research and Risk Assessment 13, 327–336 (1999)
13. Dojlido, J., Raniszewski, J., Woyciechowska, J.: Water Quality Index-application for Rivers in Vistula River Basin in Poland. Water Science and Technology 30, 57–64 (1994)
14. Heinonen, P., Herve, S.: The Development of a New Water Quality Classification System for Finland. Water Science and Technology 30, 21–24 (1994)
15. Delgado, M., Gomez-Skarmeta, A.F., Martin, F.: A Methodology to Model Fuzzy Systems Using Fuzzy Clustering in a Rapid-prototyping Approach. Fuzzy Sets and Systems 97, 287–301 (1998)

16. Shannon, C.E.: A Mathematical Theory of Communications. Bell System Technical Journal 27, 379–423 (1948)
17. Karmakar, S., Mujumdar, P.P.: Grey Fuzzy Optimization Model for Water Quality Management of a River System. Advances in Water Resources 29, 1088–1105 (2006)
18. Icaga, Y.: Fuzzy Evaluation of Water Quality Classification. Ecological Indicators 7, 710–718 (2007)
19. Tang, R.L., Guo, C.Z., Dong, X.J.: An Optimization Model with Entropic Coefficients for Management in Irrigation Water Resources. Journal of Hohai University 28, 18–21 (2000)
20. Tian, Q.H., Du, Y.X.: Study of Performance Evaluation for Mechanical Products Based on Entropy Fuzzy Comprehensive Review. China Manufacturing Information 33, 97–99 (2004)

An Algorithm of Constrained Spatial Association Rules Based on Binary

Gang Fang[1], Zukuan Wei[1], and Qian Yin[2]

[1] School of Computer Science and Engineering, University of Electronic Science and Technology of China, 610054 Chengdu, China
[2] College of Information Science and Technology, Beijing Normal University, 100875 Beijing, China
fg_7@163.com, anlexwee@uestc.edu.cn, yinqian@bnu.edu.cn

Abstract. An algorithm of constrained association rules mining was presented in order to search for some items expected by people. Since some presented algorithms of association rules mining based on binary are complicated to generate frequent candidate itemsets, they may pay out heavy cost when these algorithms are used to extract constrained spatial association rules. And so this paper proposes an algorithm of constrained spatial association rules based on binary, the algorithm is suitable for mining constrained association among some different spatial objects under the same spatial pattern, which uses the way of ascending value to generates frequent candidate itemsets and digital character to reduce the number of scanned transaction in order to improve the efficiency. The experiment indicates that the algorithm is faster and more efficient than theses presented algorithms based on binary when mining constrained spatial association rules from spatial database.

Keywords: Spatial association rules, Constrained items, Ascending value, Digital character, Binary.

1 Introduction

It is the key of spatial data mining for user to extract amusing spatial pattern and character and universal association between space data and non-space data, and potential character of data from spatial database [1]. Mining spatial association rules is one of main tasks in spatial data mining. Now the research mainly focuses on solving two types of spatial association, including lengthways and transverse association. Lengthways association is among these attributes of congener objects under the same pattern of association [2]; transverse association includes two aspects, one is among some different objects under the same pattern, the other is among some different objects under these diversiform patterns. As we all know, it is efficient and fast to use traditional association rules mining algorithms [3, 4] and the method of spatial analysis [1] to extract lengthways spatial association rules. However, when these presented algorithms are used to extract transverse association with constrained items, the efficiency will be badly affected since the first method don't easily generate frequent candidate itemsets and fast calculate support of itemsets, such as Separate[4], which is

F. Sun et al. (Eds.): ISNN 2008, Part II, LNCS 5264, pp. 21–29, 2008.

an algorithm of mining constrained association rules based on Apriori. In addition, the efficiency of the latter will be also badly affected if spatial data are very plentiful and spatial association became comparatively complicated. Subsequently, the algorithm of association rules mining based on binary is presented to easily generate frequent candidate itemsets and fast calculate support of itemsets, such as B_Apriori [5] and Armab [6], but these algorithms aren't fast and efficient when they are used to extract constrained spatial association rules. Hence, this paper proposes an algorithm of constrained spatial association rules mining based on binary, denoted by ACSARMB, which is suitable for extracting constrained association rules among some different spatial objects under the same spatial pattern. The experiment indicates that the algorithm is fast and efficient by comparing with Separate and Armab.

2 Forming Spatial Mining Database

Aiming to every objective, each of spatial objects under the same spatial pattern has two attributes, namely, the association between objective and object is expressed by the spatial pattern, or the spatial pattern doesn't express the association between them at all. Now for an example where the value of predicate about spatial pattern is expressed as **close_to(*x*, *y*)** by spatial predicate, the process of forming spatial database adapted to mining algorithm is expressed as follows:

Step1: Ascertaining the objective and corresponding spatial objects in spatial original database. The objective is denoted by O_i and all objects are denoted by A, B, C, D, E…and so on.

Step2: Extracting the value of spatial predicate. Aiming to each objective O_i, the way of buffer-analysis is used to ascertain the spatial association between objective and object. If some or other objects A exists in the buffer whose centre is O_i, the value of spatial predicate will be expressed as **close_to** (O_i, A).

Step3: Forming transaction database. Database contains *TID* and *ID*, all different objectives are made of *TID*, all the value of spatial predicate are made of *ID*. Transaction database is expressed as table 1.

Table 1. Transaction database

TID	List of item(*ID*)
O_1	close_to (O_1, A), close_to (O_1, B), close_to (O_1, D), close_to (O_1, E)……
O_2	close_to (O_2, A), close_to (O_2, C), close_to (O_2, D), close_to (O_1, F)……
…	…
O_n	close_to (O_n, A), close_to (O_n, C), close_to (O_n, E), close_to (O_1, F)……

Step4: Changing transaction database into normal mining database. Normal mining database contains *TID* and *ID*, all different objectives are made of *TID*, two digitals including "1" and "0" are made of *ID*, aiming to O_i, if *ID* in table 1 contain the value of spatial predicate, such as close_to (O_1, X), make "1" locate for the location corresponding close_to (O_1, X), or let "0" locate for the location, and so transaction database denoted by table 1 is turned into normal mining database denoted by table 2.

Table 2. Normal mining database

TID	List of item(*ID*) A,B,C,D,E,F......	Binary
T_1	1, 1, 0, 1, 1, 0......	110110...
T_2	1, 0, 1, 1, 0, 1......	101101...
...	...	...
T_n	1, 0, 1, 0, 1, 1......	101011...

3 The Algorithm of Constrained Spatial Association Rules Based on Binary

Let I= $\{i_1, i_2...i_m\}$ be a set of items, if i_k ($i_k \in I$), let $T= \{i_1 \wedge i_2 \wedge ... \wedge i_m\}$ ($T_k \subseteq I$) be a subset of items, named a Transaction. For example, let $T_k= \{i_1, i_2, i_3\}$ be a subset of items, called a transaction. And then let $D=\{T_1, T_2...T_n\}$, let $T_k \subseteq I$, (k=1...n) be a set of transactions, called Transaction Database (*TD*).

3.1 Definitions and Theorems

Definition 1. Binary Transaction (*BT*), a transaction is expressed as binary, binary transaction of transaction T is expressed as $BT=(b_1\ b_2...b_m)$, $b_k \in [0,1], k=1...m$, if $i_k \in T_i$, and then b_k=1, otherwise b_k=0.

Example. Let I={1,2,3,4,5} be a set of items, if a transaction is expressed as T_i={2,3,5}, and then BT_i=(01101).

Definition 2. Digital Transaction (*DT*), which is the integer, the value of which would be obtained by turning binary of transaction into algorism.

Example. If *BT*=01101, and then *DT*=13.

Definition 3. Constrained Digital Transaction (*CDT*), which is a digital transaction including only some items expected by people.

Definition 4. Relation of digital transaction accord with relation of transaction set.

Example. Suppose digital transaction of a transaction T_1 is denoted by DT_1, digital transaction of a transaction T_2 is denoted by DT_2. If $T_1 \subseteq T_2$, and then $DT_1 \subseteq DT_2$, DT_1 is regarded as the subset of DT_2, which is regarded as the superset of DT_1.

Definition 5. Frequent Constrained Digital Transaction (*FCDT*), which is a digital transaction including constrained digital transaction, its support surpasses minimal support given by users.

Definition 6. Candidate Digital Transaction Section (*CDTS*), which is an integral section from *CDT* to max, each power of 2 does not belong to *CDTS*.

Max=$BT_{i1} \vee BT_{i2} \vee ... \vee BT_{jk}$, each *BT* only expresses a kind of item, their support surpass minimal support given. Min=*CDT* is regarded as initialization.

Example. If support of BT_j (j=1..4) surpasses minimal support, BT_1=(01000), BT_2=(00100), BT_3=(00010), BT_4=(00001), max=$BT_1 \vee BT_2 \vee BT_3 \vee BT_4$ =15, CDT=4, and selective digital transaction section is expressed as $CDTS$ = (4,15].

Theorem 1. The given binary transaction uniquely corresponds to a digital transaction, vice versa.

Theorem 2. Let p and q be binary transactions with m bits, let T_p be transaction about p, let T_q be transaction about q, then $T_p \subseteq T_q \Leftrightarrow p$ and $q=p$.

Proof. is expressed as follows:

Suppose digit 1 locates each bit of binary p from i_1 to i_k ($k \leq m$), digit 0 locates other ones. If p and $q=p$, then digit 1 locates each bit of binary q from i_1 to i_k ($k \leq m$) (otherwise these bits must occur digit 0 with logic "and" operation), other ones will be either 0 or 1, so $T_p \subseteq T_q$ according to definition 1 and 4.

And via the hypothesis about p as before, since $T_p \subseteq T_q$, then digit 1 must locate each bits of binary q from i_1 to i_k ($k \leq m$) (otherwise, $\exists\, i_k$, so $i_k \in T_p$, $i_k \notin T_q$, the result is contrary to premises as $T_p \subseteq T_q$), other ones will be either 0 or 1, so p and $q=p$.

Via theorems as before, two conclusions deduced are expressed as follows:

Conclusion 1. Let p and q be binary transactions with m bits, let DT_p be digital transaction about p, let DT_q be digital transaction about q. If p and $q=p$, and then $DT_p \leq DT_q$, namely, digital character.

Conclusion 2. Let p and q be binary transactions with m bits, let T_p be transaction about p, let DT_p be digital transaction about p, let T_q be transaction about q, let DT_q be digital transaction about q. If $DT_p < DT_q$, then $T_q \not\subseteq T_p$, namely, digital character.

Theorem 3. Let p and q be binary transactions with m bits, let T_p be transaction about p, let T_q be transaction about q, and $p \wedge q=p$, let F be frequent itemsets.

①If $T_q \subseteq F$, then $T_p \subseteq F$. ②If $T_p \not\subseteq F$, then $T_q \not\subseteq F$

Proof is expressed as follows:

Since $p \wedge q=p$, via Theorem 2, then $T_p \subseteq T_q$, and $T_q \subseteq F$, based on set theory, then $T_p \subseteq F$. If $T_p \not\subseteq F$, then $T_q \not\subseteq F$ as the same theory.

3.2 The Process of Constrained Association Rules Mining

Firstly, we define some signs as follows:

DB: *DB* is used to save digital transaction, there are *N* digital transactions.

D: data-domain of *D* contain "value" and "count", "value" is used to save digital transaction and "count" is used to save the number of digital transaction.

FCDT: data-domain of *FCDT* contains "value" and "support", "value" is used to save digital transaction and "support" is used to save support of digital transaction, which surpasses minimal support given by users.

NFDT: data-domain of *NFDT* only contains "value" to save digital transaction, the support of which is under minimal support given.

CDT: saving constrained digital transaction which only contains items expected by people.

Step 1: Data transformations. Transaction would be transformed into digital transaction from normal mining database to *DB* via definition 1 and 2, and then digital transactions in *DB* would be saved in *D* on descending by digital transaction.

Step 2: Creating candidate digital transaction section (*CDTS*). Frequent digital transactions that gained by scanning *D*, are used to create *CDTS* via definition 6, which only expresses a kind of item, their support surpass minimal support given.

Step 3: Forming sets of frequent constrained digital transaction (*FCDT*). All frequent constrained digital transactions are searched from *CDTS* to save in *FCDT*.

Step 4: Creating constrained digital association rules. Constrained digital association rules are created from *FCDT* when confidence surpass given minimum.

3.3 The Algorithm of Constrained Association Rules Mining

Let (*CDT*, max] be a *CDTS*, and there are *n* digital transactions saved in *D*, where data aren't repeated. The algorithm is expressed as follows:

```
While (DT∈(CDT, max]) {
    If (all NFDT ⊄DT) &&(CDT⊆DT) ){
     While ((D_i.value≥ DT)&&(i≤n)) {
       If (DT⊆D_i.value)  s_count+= D_i.count;
         i++;
      }//computing support of DT
     If (s_count/N≥min-support) {
         Delete all FCDT_k (FCDT_k⊂DT) from FDT;
         Write DT and s_count to FCDT;
     }//Saving DT and deleting its subset in FCDT
     Else
        Write DT to NFDT;
    }
    DT++;
 }//searching all FCDT satisfied constrained items
 For (all FCDT_j∈ FCDT) {
     DT =FCDT_j.value;
     s_count= FCDT_j.support;
     Create_Rules (DT, s_count);
 }//Mining constrained digital association rules
Create_Rules (DT, s_count);
 While ((D_i.value≥ DT) && (i≤n)) {
   Vary= DT& (~CDT);
   If (Vary⊆D_i.value) c_count+= D_i.count;
     i++;
  }//computing support of sub
 If (s_count/c_count≥confidence)
      Display Vary⇒ CDT;
```

3.4 The Process of Generating Constrained Spatial Association Rules

Step 1: Constrained digital association rules are transformed into binary, if digital "1" exists in some binary bits, transaction-item (i_j) related to each bit should exist according to definition 1, and then comprehensible constrained association rules are expressed as $\{i_1,i_2\}\rightarrow\{i_4\}$, here is $\{i_4\}$ denoted by constrained items.

Step 2: Item (i_j) of comprehensible constrained association rules would be renewed into spatial predicate **close_to** (T, O_j), and then the constrained spatial association rules are expressed as follows: close_to (T, A) $\wedge$close_to (T, B) $\rightarrow$close_to (T, D).

Step 3: The normal constrained spatial association rules are expressed as follows:

is_a (X, T) $\wedge$close_to(X, A) $\wedge$close_to(X, B) $\rightarrow$close_to (X, D) [40%, 60%]

Let X is an objective, so above rule is explained as follows:

When percent 60 of hotel are close to A and B, simultaneity they are also close to D, there are percent 40 of data accord with the rule in transaction database.

4 Analyzing Capability of Algorithms

4.1 Analyzing Capability of Three Algorithms from Two Aspects

We compare **ACSARMB** with **B_Separate** and **Armab**, according to two aspects including the way of generating frequent candidate items and the number of scanned transaction when counting support of itemsets. **B_Separate** is an algorithm of mining constrained association rules based on binary, which is similar to B_Apriori. **Armab** is an algorithm of mining association rules based on binary.

Firstly, **ACSARMB** is even easier than other to generate frequent candidate itemsets. The ways of these algorithms are expressed as follows:

B_Separate: Separate is changed into B_Separate by interrelated theories in B_Apriori. The key theory of generating frequent candidate itemsets is that frequent candidate itemsets including (k+1) items is formed by the way of generating superset of frequent itemsets including k items, the other way is that joining frequent items including k items with frequent 1 item, and then computing the support of frequent candidate itemsets to search frequent itemsets.

For example, frequent itemsets including two items are expressed as {ab, ac, bd}, then frequent candidate itemsets including three items are expressed as {abc, abd}, then algorithm computes support.

Armab: the theory of generating frequent candidate itemsets is that frequent candidate itemsets is formed by generating subset of binary transaction aiming to every transaction, and then computing the support of frequent candidate itemsets. But the algorithm repeatedly computes support when two transactions have intersection.

For example, I= {a, b, c, d, e}, aiming to a digital transaction 22 denoted by {acd}, frequent candidate itemsets are pure subset about 22 which is expressed as {2, 4, 6, 16, 18, 20} denoted by {d, c, dc, a, ad, ac}, then the algorithm computes support.

ACSARMB: the theory of generating frequent candidate itemsets is that frequent candidate itemsets is formed by increasing value of digital transaction from *CDTS*, and then algorithm computes the support of frequent candidate itemsets.

For example, $CDTS=(CDT, 31]$, $CDT=2$, frequent candidate itemsets is denoted by CFI_i, and then process of generating CFI_i is expressed as follows:

$CFI_1=3$ (computing support, CFI_1 is *FCDT*),
$CFI=4$ (pruning 4 by *CDTS*),
$CFI=5$ (pruning 5 by *CDT*),
$CFI_2=6$(computing support, CFI_2 isn't *FCDT*),
$CFI=7$ (pruning 7 by CFI_2),
$CFI=8$ (pruning 8 by *CDT*),
$CFI=9$ (pruning 9 by *CDT*),
$CFI_3=10$ (computing support, CFI_3 is *FCDT*),
$CFI_4=11$ (computing support, CFI_4 is *FCDT*, and then deleting its subset {3, 10}),
……

Secondly, the number of scanned transaction of **ACSARMB** is smaller than other when counting support of itemsets.

Suppose $I= \{i_1, i_2 \ldots i_6\}$, there are ten transactions which aren't repeated, these transactions are expressed as {63, 62, 61, 59, 58, 51, 31, 30, 29, 15}, frequent candidate itemsets is expressed as $\{i_1, i_5, i_6\}$, its *DT* is expressed as 35.

B_Separate: the number of scanned transaction is equal to 10 when computing support of candidate frequent itemsets.

Armab: the number of scanned transaction is the same as **B_Separate**.

ACSARMB: the number of scanned transaction is equal to 6 when computing support of frequent candidate itemsets according to **Conclusion 2**, namely digital character, means it is impossible that $DT_1 \subseteq DT_2$, if $DT_1 > DT_2$.

4.2 Comparing Capability of Algorithms by Experiment

Now we use result of experiment to testify above analyses. Three mining algorithms are used to generate frequent itemsets from these digital transactions, which are expresses as digital from 3 to 4095, these digital transaction don't include any single items, and so $m=12$, $N=4083$, $CDT=4$. Our experimental environment are expressed as follow: Intel(R) Celeron(R) M CPU 420 @ 1.60 GHz, 512MB, language of the procedure is Visual C# 2005.NET, OS is Windows XP Professional. The experimental result of three algorithms is expressed as figure 1, where support is absolute and

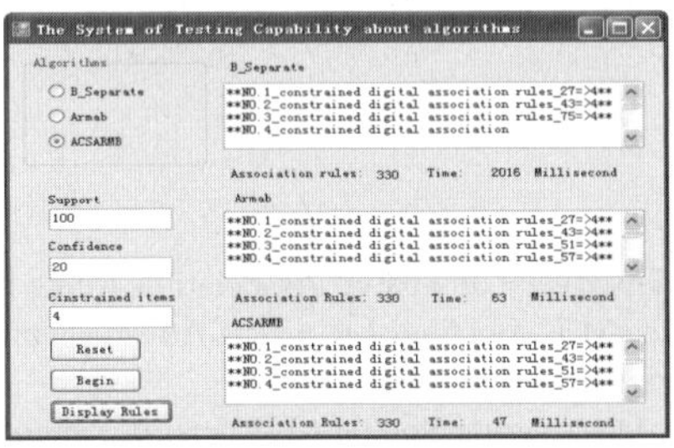

Fig. 1. The experimental result of three algorithms

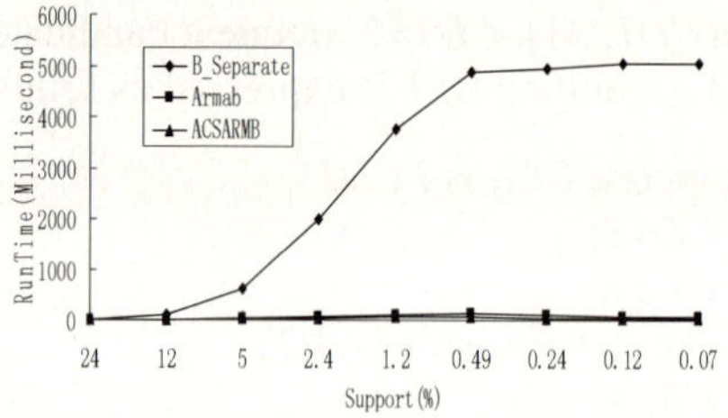

Fig. 2. The executing time of three algorithms as support of itemsets change

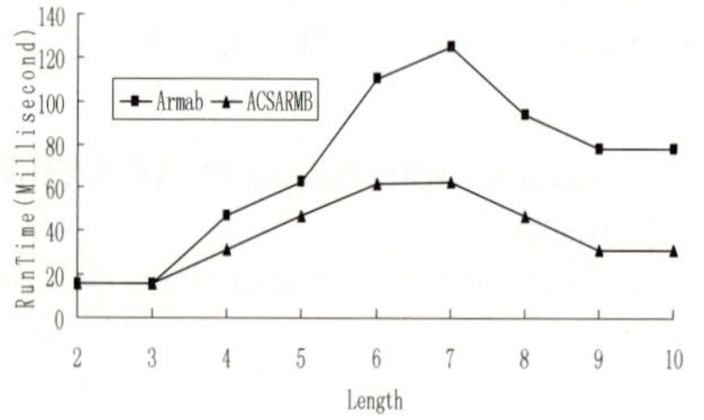

Fig. 3. The executing time of two algorithms as length of itemsets change

confidence is relative. The executing time of three algorithms is expressed as figure 2 as support of itemsets change. The executing time of Armab and ACSARMB is expressed as figure 3 as length of itemsets change.

5 Conclusion

The presented ACSARMB is suitable for extracting transverse constrained spatial association from spatial database which is among these different spatial objects under the same spatial pattern. The result of experiment indicates that the algorithm is fast and efficient by comparing with B_Separate and Armab.

Acknowledgments

This work was fully supported by a grant from the S&T Foundation of Chengdu Sci.&Tech. Bureau. (Project No. 06GGYB801GX-032).

References

1. Koperski, K., Han, J.: Discovery of Spatial Association Rules in Geographic Information Databases. In: Egenhofer, M.J., Herring, J.R. (eds.) Advances in Spatial Databases. LNCS, vol. 951, pp. 47–66. Springer, Berlin (1995)
2. Shekhar, S., Huang, Y.: Discovering Spatial Co-Location Patterns: A Summary of Results. In: Jensen, C.S., Schneider, M., Seeger, B., Tsotras, V.J. (eds.) SSTD 2001. LNCS, vol. 2121, pp. 1–19. Springer, Heidelberg (2001)

3. Agrawal, R., Srikant, R.: Fast Algorithms for Mining Association Rules in Large Database. In: Proc of the 20th NTL conf on Very Large Databases, pp. 487–499. Morgan Kaufmann, San Francisco (1994)
4. Shao, F.J., Yu, Z.Q.C.: Principle and Algorithm of Data Mining, pp. 117–120. Water and Electricity Publication of China press, Beijing (2003)
5. Chen, G., Zhu, Y.Q., Yang, H.B.: Study of Some Key Techniques in Mining Association Rule. Journal of Computer Research and Development 42, 1785–1789 (2005)
6. Fan, P., Liang, J.R., Li, T.Z., Gong, J.M.: Association Rules Mining Algorithm Based on Binary. Journal of Application Research of Computers 24, 79–81 (2007)
7. Han, J.W., Pei, J., Yin, Y.W.: Mining Frequent Patterns without Candidate Generation. In: ACM Proceedings of the 2000 ACM SIGMOD international conference on Management of data Dallas, pp. 1–12. ACM press, Netherlands (2000)

Sequential Proximity-Based Clustering for Telecommunication Network Alarm Correlation

Yan Liu[1], Jing Zhang[1], Xin Meng[2], and John Strassner[1]

[1] Motorola Labs, Schaumburg, IL 60193, USA
{yanliu,j.zhang,john.strassner}@motorola.com
[2] Motorola Inc., Beijing, China
xin.meng@motorola.com

Abstract. Alarm correlation for fault management in large telecommunication networks demands scalable and reliable algorithms. In this paper, we propose a clustering based alarm correlation approach using sequential proximity between alarms. We define two novel distance metrics appropriate for measuring similarity between alarm sequences obtained from interval-based division: 1) the two-digit binary metric that values the occurrences of two alarms in neighboring intervals to tolerate the false separation of alarms due to interval-based alarm sequence division, and 2) the sequential ordering-based distance metric that considers the time of arrival for different alarms within the same interval. We validate both metrics by applying them with hierarchical clustering using real-world cellular network alarm data. The efficacy of the proposed sequential proximity based alarm clustering is demonstrated through a comparative study with existing similarity metrics.

Keywords: Alarm correlation, Sequential proximity, Clustering, Metrics.

1 Introduction

Intelligent network fault management via alarm correlation is a well-researched problem. There is a wide agreement that alarm correlation is one of the most effective solutions to network fault management. In general, alarm correlation provides a conceptual interpretation at a network level of multiple alarms to be used for effective fault diagnosis, isolation and identification, proactive maintenance, and trend analysis. In real-world operations, network alarms are streamed in sequences with the implication of temporal relation between a preceding alarm and a succeeding alarm. By dividing them into different subsequences, data mining algorithms can be applied to discover the correlations among these alarms and thus help locate the underlying causal relationships between different network faults.

A number of data mining algorithms have been applied to network alarm correlation, among which association rule learning, Bayesian belief networks, and clustering algorithms have become quite popular [1]. The challenges with associative learning lie in its inability to scale to large dataset due to the tradeoff

F. Sun et al. (Eds.): ISNN 2008, Part II, LNCS 5264, pp. 30–39, 2008.

problem between its efficiency and the completeness of mined patterns [2]. As a robust probabilistic model, although Bayesian learning has attracted increasing attention, the complexity and inefficiency associated with Bayesian learning is problematic and too costly for complex real-time systems [3]. Clustering algorithms based on certain properties such as frequencies of alarms and interval values have been proposed and applied as one way to efficiently classify network alarms for trend analysis and alarm storm detection [4]. However, they are rarely used for sequential pattern discovery due to the absence of a valid similarity metric that can precisely quantify the distance between alarms in a temporal space.

In this paper, we present our novel contribution by first showing that sparsely reported correlation patterns should be regarded as equally important as those frequent patterns. As a resolution, two new similarity metrics are proposed in an attempt to adequately capture the sequential proximity between alarm sequences. We first propose a new similarity metric that takes into account the implications of temporal relations between the binary reference vectors of alarm sequences. Based on this consideration, we further derive another distance metric using a numeric reference vector, in which the numeric values reflect the time of arrival of the alarms in a subsequence. By incorporating the temporal order, this metric provides a more accurate sequential proximity measurement for different classes of alarms that are distributed into the same interval. These two new metrics are evaluated using the classic agglomerative hierarchical clustering to prove their validity and improved efficacy against existing similarity metrics.

The remainder of this paper is organized as follows. Section 2 overviews the alarm correlation problem and alarm clustering. Section 3 introduces the new sequential similarity metrics as the basis for clustering the alarm sequences. Section 4 describes the experiments and results, which are further analyzed and discussed in comparison with other existing distance metrics used for clustering binary objects and numeric data. Section 5 concludes the paper with the description of some future work.

2 Alarm Correlation

As defined by Jakobson et.al. in [5], alarm correlation is a *"conceptual interpretation of multiple alarms such that new meanings are assigned to these alarms."* It is a *"generic process that underlies different network management tasks."* Alarms are indeed manifestation of a particular fault or faults reported by network devices in a standardized format. It is believed that the correlation between alarms can be used to infer the causal relations between their underlying faults. Thus, effective alarm correlation can not only lower the operational cost by reducing the number of actionable alarms, it can also aid the operators in fault diagnosis and isolation, as well as proactive maintenance.

Within a network fault management system, a typical alarm consists of multiple fields including device ID, time, alarm type, alarm, and Reason, etc. The following example shows two sample network alarms excerpted from an alarm log reported in Motorola's CDMA cellular networks.

BTSSPAN-654-1 05-10-20 01:58:56 bjomc4 XC-46 A000000.00000 137461/414273
** ALARM:35-18010 "SPAN Out-of-Service - Continuous Remote Faults (source)"
EVENT_TYPE=Communications Alarm EVENT_TIME=05-10-20 01:58:56
PROBABLE_CAUSE=Loss of Signal DETECTED_BY=MSIP-46-4-6

BTSLINK-654-1 05-10-20 01:59:34 bjomc4 MM-46 A000000.00000 761870/414293
** ALARM:12-51 "BTSLINK Out of Service"
OLD_STATE=INS NEW_STATE=OOS_AUTOMATIC PHY_STATE=OOS
REASON_CODE="Asynchronous OOS Received From Device"

In general forms, a correlation can be a statement about alarms reported by the network. In the above example, it is intuitive to speculate and state that "BTSLINK out of service" alarm reported at 01:58:56 by the network device BTSLINK-654-1 is likely related to the "SPAN Out-of-Service - Continuous Remote Faults (source)" reported by the network device BTSSPAN-654-1 at 01:59:34. In alarm correlation, a correlation is usually evidenced by statistical analyses and formalized by a correlation rule that defines the conditions under which correlations are asserted. Observations have been made while studying the network alarms, which provide an important basis for developing alarm correlation techniques.

2.1 Alarm Clustering

Given a data space, the aim of a clustering algorithm is to partition the data space into clusters so that the data objects within each cluster are "similar" to each other. The similarity is measured by certain distance metrics. Hierarchical clustering and k-means clustering are two main clustering methods. Hierarchical clustering algorithms use either "bottom-up" (agglomerative) or "top-down" (divisive) approach to build the hierarchy of clusters step by step, whereas k-means finds all clusters at once and repeats the partition until the solution converges. Both clustering techniques are well known for their efficiency in clustering large data set and has earned their reputation in many successful applications in various domains. However, the majority of existing distance metrics only works on numeric values and thus makes sequence clustering, which often represented by binary digits or categorical values, a challenging problem.

Alarm clustering has been proposed to mine for alarm patterns such as correlations, trend, and associations based on their behavior proximity. The proximity is usually defined by a set of significant attributes of the alarms [6]. Center to our clustering approach is the notion of similarity between alarms in terms of its sequential structure in a temporal space. Both associative learning and Bayesian networks rely on relative frequencies of the alarms to uncover and evaluate patterns, which becomes problematic when certain strongly correlated alarms only appear sparsely over time. Contrary to the idea of frequent patterns, the definition of similarity in our alarm clustering is independent of the absolute

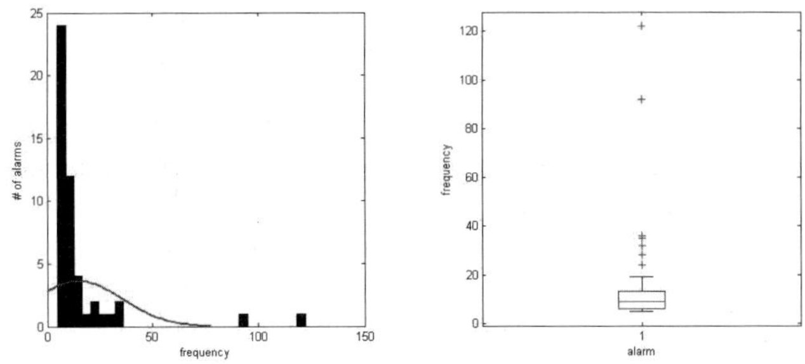

Fig. 1. Histogram and Boxplot of Sample Alarm Data

frequency of the alarms. Moreover, we believe similarity metrics should be custom defined and fine tuned given its domain-dependent nature.

The key observation in our study is that in a telecommunication network environment, a relatively small number of simultaneous alarms or alarms happening within a pre-defined time interval can be interpreted as a qualitatively important situation. In reality, there are times that such subtle correlation accounts for an alarm storm although the related alarms only appear together for a few times. Hence, often regarded as "transient noise" and ignored by other methods, these local clusters with a probable global impact can be discovered by clustering techniques using suitable similarity measurements. Figure 1 shows the histogram and boxplot of 49 alarms with their frequencies of sequential occurrence over 188 intervals using a 15 minutes division over a 72-hour period (some intervals do not contain any meaningful alarms). Near 86% alarms have a frequency lower than 0.10 and most of them only occur $5 - 10$ times in 188 intervals. Through closer examination, we discover more than 50% of correlations with high confidence actually exist between these alarms with low frequencies. Hence, it is our belief that for correlation patterns like this, models with high efficiency and high true positive ratio should be preferred over models with low efficiency and low false positive ratio. As clustering algorithms are well known for their efficiency, this further motivates us to use the clustering algorithms and investigate new similarity metrics with the aim of reducing the risks of false negatives while maintaining a reasonably high computational efficiency.

3 The New Similarity Metrics

When mining for sequential patterns from network alarms, one often looks for the correlation between certain alarms that implies "if alarm A occurs, then alarm B is likely to occur". A temporal relation is assumed with a time constraint imposed on the relationship. A binary vector can be used to represent the occurrence sequence of an alarm. Within the binary reference vector for a single alarm, each '0' or '1' represents whether this particular alarm is reported or not in a particular interval, respectively. Furthermore, numeric values can be assigned to each event that appears in one interval based on their ordering in terms of time of arrival. This numeric reference vector provides a basis for further separation of local correlation within the same interval, which can be particularly useful for long interval based alarm sequence divisions.

A fixed interval length is usually determined empirically by using the average length of the majority of alarms occurred in a certain period. The significant side effect of a fixed-length interval is that it is highly likely correlated alarms are separated into two neighboring intervals, which brings inaccuracy to the mined patterns. Flexible interval length could be an alternative solution to avoid the false separation problem. However, because different alarm patterns are often exhibited in various serendipitous forms, not only flexible length cannot solve the false separation problem, but also it adds extra computational burden and another layer of uncertainty to the pattern discovery process. In an attempt to solve this problem, we start with defining new similarity metrics suitable for fixed length interval-based alarm clustering.

3.1 Metric 1

The first new similarity metric we propose is for measuring the distance between the binary reference vectors derived from the alarm sequence. This new metric tackles the division bias problem introduced by interval-based division of alarms by defining a two-digit similarity metric based on weighted distance computation of two consecutive binary digits.

Given an alarm sequence is divided into n subsequences with m different alarms reported, the binary reference vector $S(i)$ for alarm $A_i, i = 1..m$ is composed of n binary digits. Each binary digit in $S(i)$, denoted by b_k, where $b_k = 1|0$ and $i = 1..n$, represents whether alarm A_i occurs in the kth interval or not, respectively. When two alarms are separated into two neighboring intervals, there are two scenarios of possible false separation as listed below.

- The first scenario involves a separation of two alarms into two intervals where each alarm is reported in only one interval. This can be caused by the incidence that one alarm is reported right before the "cut" and the next alarm is reported right after the"cut".
- In the second scenario, one alarm occurs in two consecutive intervals while the other occurs only in one of them. This could be explained by the case where the "cut" is made between two repeatedly reported alarms followed (or preceded) by the other possibly related alarm.

As shown in Table 1, these two scenarios can be revealed by using a double-digit comparison between two alarms. "01|10" and "10|01" reflect the first scenario listed above, whereas"01|11", "11|01", "10|11", and "11|10" reflect the second scenario of possible false separation.

Similar to Jaccard's similarity coefficient [7,8], which uses the size of the intersection divided by the size of the union of the sample sets, we use the following formula to calculate the similarity between alarm A_i and alarm A_j.

$$Sim(i,j) = 1 - \frac{3B_{11|11} + 3B_{01|01} + B_{01|10} + B_{01|11}}{B_{00|11} + B_{00|01} + 3B_{11|11} + 3B_{01|01} + B_{01|10} + B_{01|11}};$$

Table 1. Double-digit Binary Values

$b_k(i)b_{k+1}(i) \mid b_k(j)b_{k+1}(j)$	Count	False Separation
00|00	$B_{00\|00}$	No
01|01, 10|10	$B_{01\|01}$	No
00|11, 11|00	$B_{00\|11}$	No
01|10, 10|01	$B_{01\|10}$	Yes
11|11	$B_{11\|11}$	No
01|11, 11|01, 10|11, 11|10	$B_{01\|11}$	Yes
00|01, 00|10, 01|00, 10|00	$B_{00\|01}$	No

Clearly, pattern "11|11" and pattern "01|01" are given more weights than pattern "01|10" and "01|11" in this calculation.

3.2 Metric 2

The second new similarity metric we propose is for measuring the distance between the numeric reference vectors derived from the alarm sequence. A numeric value is first assigned to each alarm reported in one interval. Table 2 shows the example of how numeric values are assigned to alarms. In the example, the alarms are divided into 10-minute intervals. The first alarm occurs in an interval is given the value 1. Then, based on the time difference in minutes between this alarm and the succeeding alarm, the numeric value given to the next alarm is the addition of the time difference and the value of the preceding alarm. This gives a fair basis for quantifying the distances between alarms within one subsequence.

After the values are assigned to the alarms, the maximum distance between the alarms for every interval, denoted by $dmax_k, k = 1..n$, is used to compute the relative distance. The distance $D_k(i,j)$ between alarm A_i and alarm A_j that both happen in the kth interval is then computed as:

$$D_k(i,j) = \begin{cases} \frac{|S_k(i)-S_k(j)|}{|dmax_k|^2} & S_k(i) > 0 \wedge S_k(j) > 0 \\ 0 & otherwise; \end{cases}$$

Given that the number of intervals where alarm A_i or alarm A_j but not both appears is p_1 and the number of intervals where both alarms appear together is p_2, the similarity between these two alarms is then computed by the following formula.

$$Sim(i,j) = \begin{cases} 1 - \frac{\sum\limits_k D_k(i,j) + \frac{p_2}{n}}{p_1+(p_2/n)} & p_1 + p_2 > 0 \\ 0 & otherwise; \end{cases}$$

where $p_1 = |\{k|S_k(i) > 0 \wedge S_k(j) > 0\}|$ and $p_2 = |\{k|S_k(i) > 0 \vee S_k(j) > 0\}| - p_1$. It should be noted that by disregarding the "$S_k(i) = 0 \wedge S_k(j) = 0$" patterns,

Table 2. Assign Numeric Values to Alarms

Interval #	Alarm	Time of Arrival	Assigned Value
1	A_1	00:01:02	1
1	A_2	00:01:02	1
1	A_3	00:01:05	4
1	A_4	00:01:07	6
2	A_1	00:01:11	1
2	A_4	00:01:17	6
2	A_5	00:01:19	8

where both alarms are absent in an interval, the proximity measures are local to both alarms and thus do not necessarily reflect the absolute frequencies of both alarms or their correlations.

4 Experiments

There exist different types of clustering algorithms including partition-based k-means, hierarchical clustering, and some mixed models. Most of these algorithms have the significant advantage of computational efficiency at a bounded cost of polynomial time. For the purpose of demonstration, we choose the agglomerative clustering approach for its simplicity and relatively stable performance in this study. The alarm data we use to validate our approach are collected from a live cellular network through a distributed network management system that supports multiple network management interfaces for Motorola CDMA cellular networks. The input to the algorithm is an $m \times n$ matrix consisting of n vectors where every vector represents the sequential occurrences of an alarm over m intervals. The agglomerative hierarchical clustering algorithm treats every alarm sequence as a singleton cluster, and then successively merges these clusters until all alarms have been merged into a single remaining cluster. By proper thresholding, the output of the algorithm is the alarm clusters representing the probable correlations. The same alarm data as shown in Figure 1 is used for our experiments. It has 49 alarm sequences where each sequence is composed of 188 intervals. Therefore, the input of our clustering algorithm is a matrix of size 188×49. The performances of both metrics are evaluated using the following measures in a comparative examination against other existing metrics.

$$Sensitivity = \frac{\#\ of\ true\ correlations\ discovered}{Total\ \#\ of\ true\ correlations};$$

and

$$Precision = \frac{\#\ of\ true\ correlations\ discovered}{Total\ \#\ of\ correlations\ discovered}.$$

Sensitivity measures the true positive rate of the clustering approach, while precision measures the ability to discover only the correct correlations. Moreover, we use the Harmonic mean in our plots as a combined measure of both sensitivity and precision in order to incorporate both measures when we evaluate the clustering approaches. The Harmonic mean is defined as:

$$Harmonic\ Mean = \frac{2 \times Precision \times Sensitivity}{Sensitivity + Precision}.$$

4.1 Performance Evaluation

By assigning binary values to the alarm sequences, we first validate Metric 1 and compare it against existing distance metrics for clustering binary data. As our sequential proximity based approach mainly focuses on the relative frequency of '1"s, only metrics that do not value '00' counts are selected, which include Russel and Rao(RR), Tanimoto coefficient, Dice, Czekanowski, or Sorensen (DCS) distance, Kulczynski, and Jaccard's coefficient [9]. In applying agglomerative hierarchical clustering, we use Ward's clustering procedure for all metrics in order to minimize the "loss of information" during the cluster fusion process. The performance tuned over the threshold for cutting the clusters is measured by the Harmonic means as shown in the left plot of Figure 2. Evidently this plot shows that by using Metric 1 we obtain the best clustering performance over all thresholds. DCS and Kulczynski also produce performances that are close to that of Metric 1. While we examine the correlations uncovered by using these three metrics, we further identify that Metric 1 reveals more correlations at smaller thresholds. This verifies that the discovery of inherently strong patterns could be strengthened by accommodating sequential proximities across neighboring intervals.

Metric 2 is tested by first assigning numeric values to the sequence vectors of every alarm. We follow the scheme as shown in Table 2 by using the time of arrival for the value assignment. Then, we compare Metric 2 with other popular numeric value based distance metrics including Euclidean, cityblock, Minkowski, cosine, and correlation (standard Euclidean distance and Mahalanobis distance are not applicable to the alarm data set due to its singularity). For more information about these metrics, please refer to [10]. The right plot shown in Figure 2 displays the clustering performance over different thresholds. Clearly we can see the superior performance of Metric 2 over all other metrics. The second best performance is given by the correlation metric which also has a high true positive rate for strong correlation patterns.

4.2 Discussion

There are several interesting observations in this study that worth further investigation. We first observe that Metric 2 has outperformed Metric 1 using the 15 minute interval based division. As we are using a fairly long interval based division to collect alarm sequences, this further explains the effectiveness of Metric

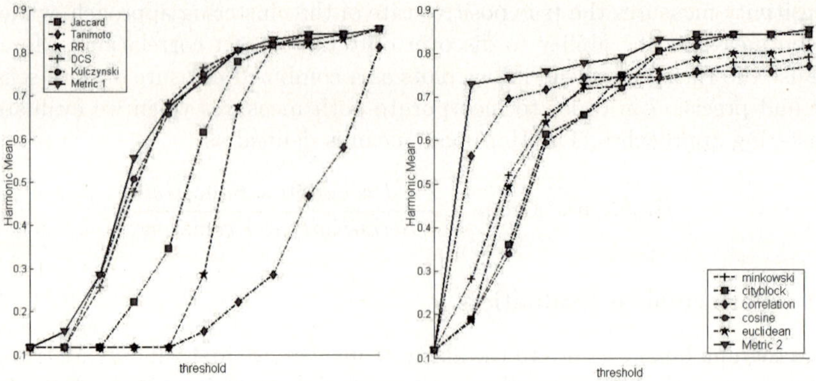

Fig. 2. Harmonic Mean Performance Evaluation of Metric 1(left) and Metric 2(right) against Existing Metrics

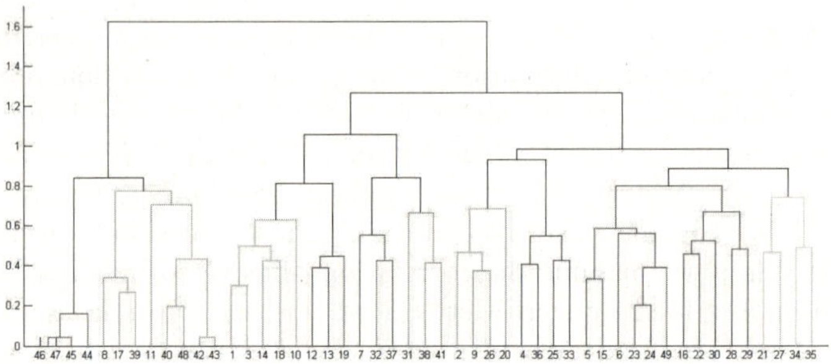

Fig. 3. The Dendrogram of Agglomerative Clustering using Metric 1

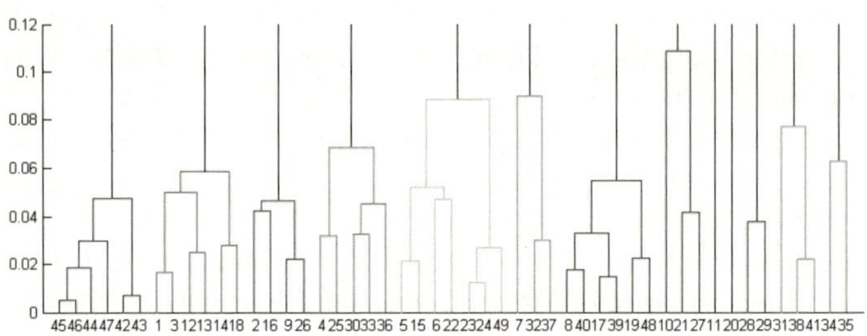

Fig. 4. The Dendrogram of Agglomerative Clustering using Metric 2

2 in capturing sequential proximity with long intervals. However, we also notice that Metric 1 reveals a few interesting correlations that has escaped from all other metrics including Metric 2 (e.g. alarm 19 is correlated with alarm 12 and alarm 13, see Figure 3 and Figure 4). After examination by network engineers, some of these correlations are verified to be correct correlations, which

confirms our assumption of missing correlations due to false separation of alarm sequences.

5 Conclusion

In this paper we tackle the alarm correlation problem in telecommunication networks by presenting a sequential proximity based clustering approach based on two new similarity metrics: one that tries to diminish the negative impact of false separation of correlated alarms, and the other that takes into account the temporal order within an interval to further improve the clustering performance. Both metrics are evaluated and demonstrated with superior performances in a comparative study against existing metrics. In the future, we will conduct more experiments with live network alarm data and investigate the applicability of sequential proximity based alarm clustering to online alarm correlation schemes.

References

1. Gary, M.W.: Data Mining in Telecommunications. The Data Mining and Knowledge Discovery Handbook, pp. 1189–1201 (2005)
2. Rakesh, A., Ramakrishnan, S.: Mining Sequential Patterns. In: Proceedings of Eleventh International Conference on Data Engineering, pp. 3–14. IEEE Computer Society Press, Taipei (1995)
3. Bowes, J., Neufeld, E., Greer, J.E., Cooke, J.: A Comparison of Association Rule Discovery and Bayesian Network Causal Inference Algorithms to Discover Relationships in Discrete Data. In: Hamilton, H.J. (ed.) Canadian AI 2000. LNCS (LNAI), vol. 1822. Springer, Heidelberg (2000)
4. Albaghdadi, M., Briley, B., Evens, M.W., Sukkar, R., Petiwala, M., Hamlen, M.: A Framework for Event Correlation in Communication Systems. In: MMNS 2001: Proceedings of the 4th IFIP/IEEE International Conference on Management of Multimedia Networks and Services, pp. 271–284. Springer, London (2001)
5. Jakobson, G., Weissman, M.: Alarm Correlation. IEEE Network 7, 52–59 (1993)
6. Bellec, J.H., Kechadi, M.T.: Feck: A New Efficient Clustering Algorithm for the Events Correlation Problem in Telecommunication Networks. In: Proceedings of the Future Generation Communication and Networking (FGCN 2007), pp. 469–475. IEEE Computer Society, Washington (2007)
7. Jaccard, P.: Ètude Comparative De La Distribution Florale Dans Une Portion Des alpes et des jura. Bulletin del la Socit Vaudoise des Sciences Naturelles 37, 1–3 (1901)
8. Tan, P.N., Steinbach, M., Kumar, V.: Introduction to Data Mining, 1st edn. Addison-Wesley Longman Publishing Co., Inc., Boston (2005)
9. Cha, S.H., Yoon, S., Tappert, C.C.: On Binary similarity Measures for Handwritten Character Recognition. In: Proceedings of Eighth International Conference on Document Analysis and Recognition, vol. 1, pp. 4–8 (2005)
10. Xu, R., Wunsch II, D.: Survey of Clustering Algorithms. IEEE Transactions on Neural Networks 16, 645–678 (2005)

A Fast Parallel Association Rules Mining Algorithm Based on FP-Forest

Jian Hu[1,2] and Xiang Yang-Li[1,2]

[1] School of Management, Harbin Institute of Technology, Harbin 150001
[2] Research Center of Technology, Policy and Management,
Harbin Institute of Technology, 150001
jianhu-hit@163.com

Abstract. Parallel association rules mining is a high performance mining method. Until now there are many parallel algorithms to mine association rules, this paper emphatically analyses existing parallel mining algorithms' realization skill and defects. On the basis, a new data structure, called FP-Forest, is designed with a multi-trees structure to store data. At the same time, a new parallel mining model is proposed according to the property of FP-Forest, which combines the advantage of data-parallel method and task-parallel method. First, database is reasonably divided to data processing nodes by core processor, and FP-Forest structure is built on data processing nodes for each sub-database. Secondly, core node perform a one-time synchronization merging for each FP-Forest, and every MFP-Tree on FP-Forest is dynamical assigned to corresponding mining node as sub-task by task-parallel technique. Furthermore, a fast parallel mining algorithm, namely F-FDPM, is presented to mine association rules according to above model, which mining process adopts frequent growth method basing on deepth-first searching strategy. From experimentation on real data sets, the algorithm has greatly enhanced association rules mining efficiency.

Keywords: Data mining; Association rules; Distributed and parallel algorithm; FP-Forest; MFP-Tree.

1 Introduction

Association rules mining is an important task in the data mining area and has a wide range of applications. The association rules mining problem has drawn much attention over the past decade. Despite all these efforts, association rules mining remains a time-consuming process due to its intrinsic characteristics: both I/O intensive and CPU intensive, which makes mining' efficiency becoming a very challenging task. However, high performance parallel algorithm is the key to solve the above problem.

Parallel association rules mining can be formally described as follows. Supposing $P_1, P_2, \ldots, P_n$ are n computers based on shared nothing architecture, that is information transmission only on network, and other resources are independent. D_i(i=1,2,…,n) is sub-database on P_i local disk, and transaction number is T_i. Then, the whole database is $D = \bigcup_{i=1}^{n} D_i$ and the entire transaction number is $T = \bigcup_{i=1}^{n} T_i$. Parallel association rules

F. Sun et al. (Eds.): ISNN 2008, Part II, LNCS 5264, pp. 40–49, 2008.

mining is how P_i only deals with private data D_i by n computers synchronously working and limited information transfers only on network. At last, association rules are mined on the whole database.

Many parallel mining algorithms have been proposed to mine association rules in recent years. However, most existing parallel mining algorithms can generate a lot of candidate set, and have poor expansibility and unbalanced computing load. In order to solve above problems, this paper proposes a new data structures, called FP-Forest, to store data using multi-trees. Furthermore, a new parallel mining model is proposed according to the property of FP-Forest, which combines the advantage of data-parallel method and task-parallel method. Then a new algorithm based on frequent growth method, namely F-FDPM, is designed to mine parallel association rules.

2 Related Work

In existing parallel association rules mining, most parallel data mining algorithms were developed from the Apriori-based method. Agrawal et al [1] first put forward three parallel algorithms, count distribution (CD), data distribution (DD) and candidate distribution (CaD). CD algorithm finished parallel mining by division database, if there are n processors, every processor got the 1/n of database, then mined association rules on every sub-database with Apriori-like algorithm. DD algorithm divided candidate sets of large sets between processors, every processor respectively calculated support of candidate sets. CaD algorithm combined CD and DD algorithms. In order to enhance performance, many researchers improved above three algorithms. Zaki et al [2] proposed the Common Candidate Partitioned Database (CCPD) and the Partition Candidate Common Database (PCCD) algorithms, which both are Apriori-like algorithms. Park et al [3] presented the parallel data mining (PDM) algorithm. This algorithm is a parallel implementation of the sequential DHP algorithm and it inherited its problems, which makes it impractical in some cases. Han et al [4] put forward intelligent data distribution (IDD) and hybrid distribution (HD) algorithms, which improved DD algorithm. Schuster et al [5] proposed DDM algorithm basing on Apriori. In spite of the significance of the association rule mining, above algorithms have few advances on parallelizing association rule mining because of transferring a lot of candidate sets and needing synchronization every iteration.

Cheung et al presented FDM [6] with the property local and global frequent itemsets reducing the transmission information, and hash method decreasing communication times. Schuster et al [5] improved FDM algorithm, and gave DDDM algorithm to reduce communication cost of iteration without depending on node number. Then, Cheung et al also gave FPM [7] algorithm. Furthermore, Cheung et al [8] gave a parallel association rules mining algorithm based on DIC algorithm. This implementation was sensitive to the skewness of the data and assumes that all data should be homogenous in order to get good results.

Another attempt was done on parallel the association rules mining based on FP-growth [9]. This algorithm only scanned database two times avoiding generating candidate sets, at the same time, adopted different division strategy in different mining phase. Pramudiono et al [10] reported results for a parallel FP-growth algorithm on a shared-nothing mining environment.

3 FP-Forest Construction and Related Operation

In order to conveniently realize parallel association rules mining, this paper put forward a new data structures, named FP-Forest.

3.1 FP-Forest Construction

FP-Forest is composed of some tree structure, which is called MFP-Tree. MFP-Tree is a mutation of FP-Tree. Fig. 1 shows FP-Forest construction process for the transaction data set in Table 1. Frequent 1-itemsets this data set are filled on Table 2.

Table 1. Transaction data set

Tid	Itemset	Tid	Itemset
100	$\{I_1, I_2, I_5\}$	600	$\{I_2, I_3\}$
200	$\{I_2, I_4\}$	700	$\{I_1, I_3\}$
300	$\{I_2, I_3\}$	800	$\{I_1, I_2, I_3, I_5\}$
400	$\{I_1, I_2, I_4\}$	900	$\{I_1, I_2, I_3\}$
500	$\{I_1, I_3\}$		

Table 2. Frequent 1-itemset

Frequent 1-itemset
I_5: 2
I_4: 2
I_1: 6
I_3: 6
I_2: 7

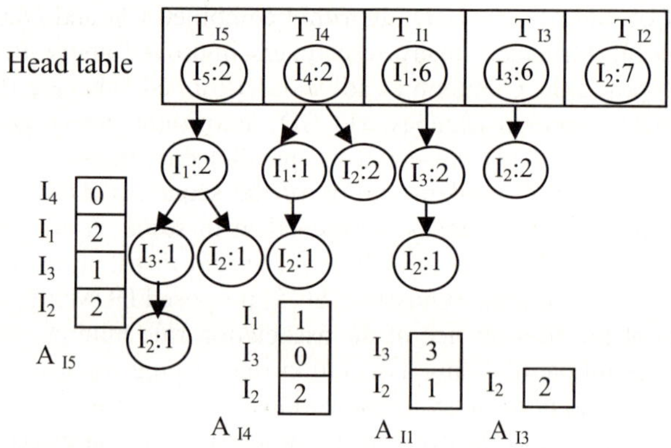

Fig. 1. FP-Forest construction

At first, frequency 1-itemsets, named $L = \{I_5, I_4, I_1, I_3, I_2\}$ are got by one-time scanning database, which are arranged according to support ascending order and stored in head-table H. The root of the MFP-Tree is an item in L which can identify this tree, not null, and seeks data with top-down method. So the length of L is the number of MFP-Tree. Then, database is scanned at second time, frequent itemset tables are obtained for every transaction, which are defined as F. For transaction $T100$, frequent itemset table which is arranged according to support ascending order, $F_1 = \{I_5, I_1, I_2\}$, and the first itemset I_5 is treated as root of MFP-Tree, defined this tree as T_{I5}, the other itemsets I_1, I_2 are insert T_{I5}. Also, $F_2 = \{I_4, I_2\}$ for transaction $T200$, the first itemset I_4 is the root T_{I4}, the itemset I_2 is inserted in T_{I4}. In the same way, multi-trees structure is built when every transaction is operated. At the same time, one-dimension array A_{Ti} is applied to store the counts of other nodes of MFP-Tree. When all the MFP-Trees have been completed, then FP-Forest is got.

The pseudocode of FP-Forest building algorithm is as following:

```
Algorithm : FP-Forest()
Input : transaction database D; Min_support;
Output: FP-Forest
  Scan D,get frequent 1-itemsets L,store in head table H;
  Build n Trees T_Ii using the items in L as root Node
and n one-dimension A_i;
  Scan D for second time;
      For every transaction t∈ D
        {
        Get frequent item table F acording the order of L;
          Find the T_Ii which root is the first item of F;
          Insert other nodes of F into T_Ii, and A_i stores
the count of node in tree T_Ii;
        }
```

3.2 Related Operation of FP-Forest

In this sub-section, we present two useful operations, called combination operation and join operation to merge two FP-Forests. At the same time, we also defined a peculiar MFP-Tree according to study the structure of MFP-Tree, named single frequent branch, which is found that it can get frequent itemsets by easy enumeration nodes on MFP-Tree.

Definition 1 Combination operation. When FP-Forest(y) is merged into FP-Forest(x), for the MPF-Trees which have the same items in their head table H, we will execute combination operation to the MFP-Trees corresponding to the same items, which is inserting the nodes of MFP-Trees(y) into MFP-Trees(x). At the same time, node number storing in array A of MFP-Trees(x) will correspondingly increase.

Dedinition 2 Insertion operation. When FP-Forest(y) is merged into FP-Forest(x), for the MPF-Trees which have not the same items in their head table H, then will execute insertion operation, that is inserting these MFP-Trees(y) into FP-Forest(x), and store roots of MFP-Trees(y) into head table H of FP-Forest(x).

Definition 3 Single frequent branch. When traversing an MFP-tree in top-down, If the count of a node of MFP-tree is less than the minimum threshold value of support and the count of father node is greater than or equal to the minimum threshold value of support, and every ancestor node only have one child node, then the MFP-tree is defined as single frequent branch.

4 F-FDPM Algorithm Describing

In this section, we present a new parallel model which employ data-parallel technique and task- parallel technique combination on a PC cluster. At the same time, a new parallel association rules mining algorithm, named F-FDPM, is designed for mining rules for FP-Forest structure. Each processor independently discovers corresponding association rules.

4.1 Parallel Mining Model

In most time, it can not solve performance problem by single using data-parallel method or task- parallel method. So we combine two techniques into the parallel mining model, as can be seen form Fig. 2. The cores of this parallel mining model are the unit of data management and the unit of data distribution on central node, also including mining algorithm on mining nodes.

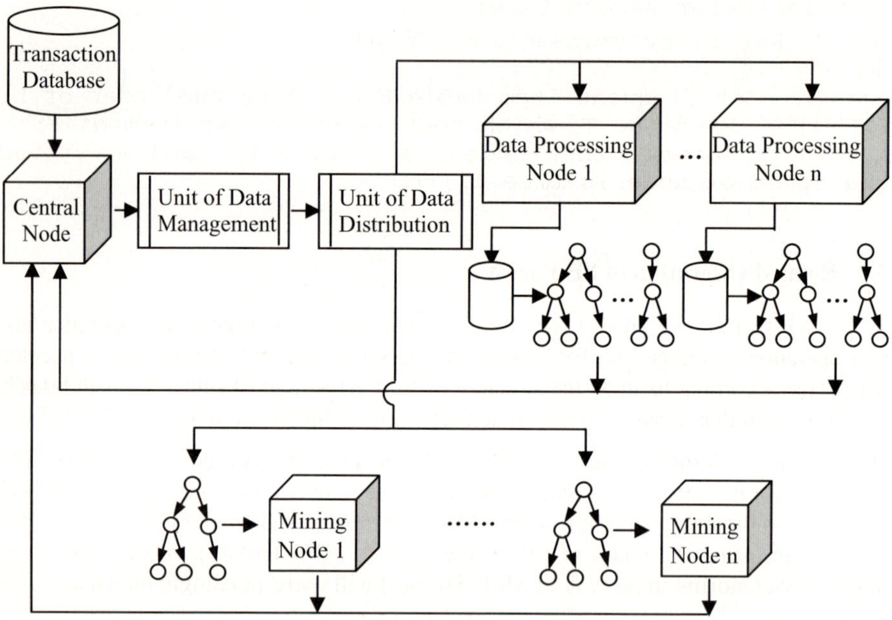

Fig. 2. Parallel mining model

Subordination node. We divide subordination nodes into two groups, one is used to receive sub-database from central node and generate FP-Forest, named data processing node, the other is mining association rules on MFP-Tree, called mining node, which is assigned by the former through central node control.

Central node. The main function is controlling the parallel mining flow. First, database is read and equally divided to data processing nodes. When FP-Forest is build for each sub-database on data processing nodes, central node receives the finishing information. Then, all data processing nodes need to perform a one-time synchronization to transfer table H to and FP-Forests to core code and execute combination operation or insertion operation. Furthermore, every MFP-Tree on FP-Forest is transferred to mining nodes to discover association rules. Last, central node takes over and gathers mining result form each mining node.

Unit of data management. The role of unit of data management is reading data of database, storing the information transferring by data processing nodes and gathering the mining result of each mining node.

Unit of data distribution. Unit of data distribution is the core of parallel mining model, which will determine the performance of parallel algorithm. According to the property of this model, we apply the dynamic data allocation method in this module. Central node do not allocate all MFP-Tree on FP-Forest to mining node, but transfer each MFP-Tree to mining node according to left-to-right on FP-Forest at first , until every node has one MFP-Tree. If some mining node have finished mining task, core node will assign the biggest capacity MFP-Tree to the node by comparing the array A'length. So the mining node completing in advance always gets the big capacity MFP-Tree, which will ensure the balance of load.

4.2 Mining Process on MTP-Tree

Mining process adopts frequent growth based on depth-first searching method to mine association rules on every MFP-Tree.

The pseudocode of mining algorithm is as following:

```
Algorithm: FMP( )
Input : MFP-tree structure; Min_support;
Output: Association rules ;
    For each MFP-tree T_Ii
       {
        Scan MFP-tree T_Ii;
        If T_Ii is single frequent branch
           Get combination of root node and T_Ii'nodes,
output combination as frequent itemsets;
          Else{
              Scan array A of T_Ii
              find frequent 1-itemsets L ;
                If the length L=1
                   Get combination of frequent 1-itemsets
and root node as frequent itemsets;
                Else
                  {
                   build new FP-Forest(L);
                       FMP(new FP-Forst);
                       Get subsets of frequent itemsets;
                   Calculate confidence of possible rules;
                       If confidence>Min_confidence
                       Output association rules;
                       }
              }
          }
```

Fig. 3, Fig. 4 and Table 3 show the mining association rules process on the T_{i5}. Therein, Min_support is equal to 0.22; Min_confidence is equal to 0.9.

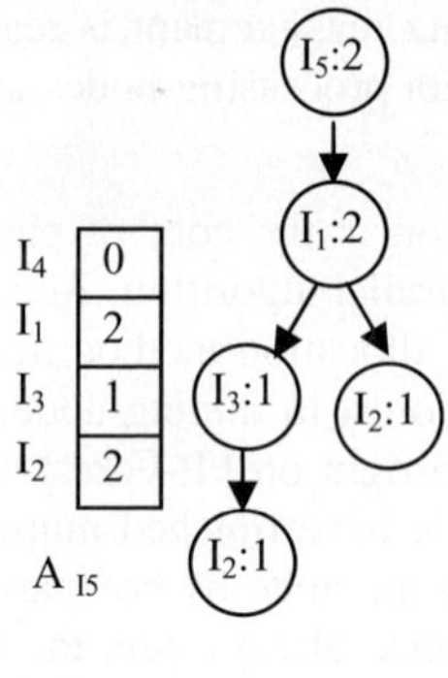

Fig. 3. MFP-tree construction on T_{i5}

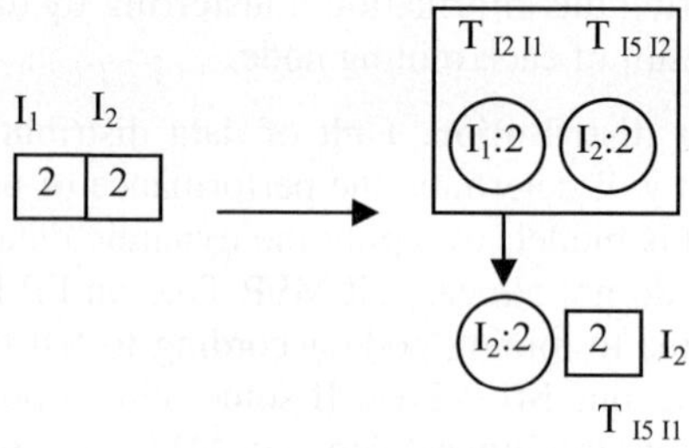

Fig. 4. Building new FP-Forest on T_{i5}

Table 3. Association rules mining result

Frequent itemsets	Confidence	Association rules
$\{I_5, I_1: 2\}$	1.0	$I_5 \Rightarrow I_1$
	1.0	$I_5 \Rightarrow I_1 \wedge I_2$
$\{I_5, I_1, I_2: 2\}$	1.0	$I_1 \wedge I_5 \Rightarrow I_2$
	1.0	$I_5 \wedge I_2 \Rightarrow I_1$
$\{I_5, I_2: 2\}$	1.0	$I_5 \Rightarrow I_2$

5 Algorithm Capability Analysis

We have tested the algorithm on a cluster of x86-64 PC. There are 20 IBM machines as subordinate nodes, each computer consists 3.0 GHZ CPU, 1 GB main memory and a 120 GB SATA disk. In addition, a dell workstation is as core node, which has a 3.7 GHZ Intel core processors, 4 GB main memory and 250G SATA disk. Each processor installs window XP operating system. The whole network is connected with ethernet switch, all machined are independent except for network connection. Message passing library adopts standard MPI, which is Mpich2 edition. We use C++ language implementation association rules mining algorithm and apply SQL server 2000 to store database.

In order to test parallel mining method capability, experiment adopts the real data. Pumsb* and connect data sets come from UCI [11]. T30I1.2D60k data set and T100I1.4D80K data set are generated by IBM synthetic data generator. Table 4 shows the characteristic of testing data set.

Table 4. Testing data sets

Data set	Item number	Record number
Pumsb*	7117	49046
Connect	129	67557
T30I1.2D60K	120	60000
T100I1.4D80K	140	80000

Running time is an important parameter of algorithm capability, so we select two serial algorithm Apriori and FP-Growth*, two parallel algorithm CD and DD to compare with algorithm in this paper. Speedup and sizeup are also two important measures to evaluate parallel mining algorithm, so we test the speedup and sizeup performance of F-FDPM.

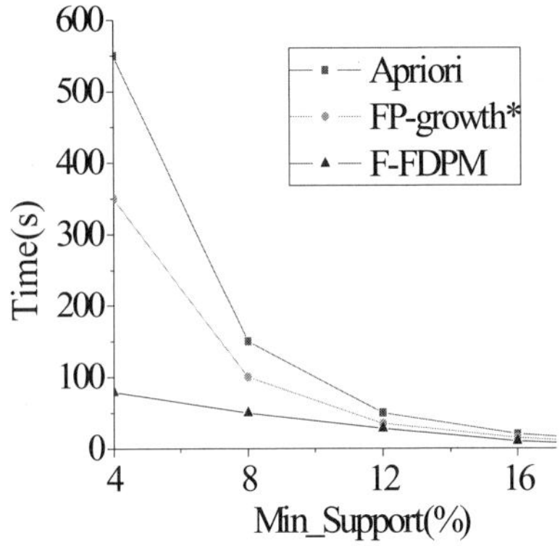

Fig. 5. Running time comparing on Pumsb* set

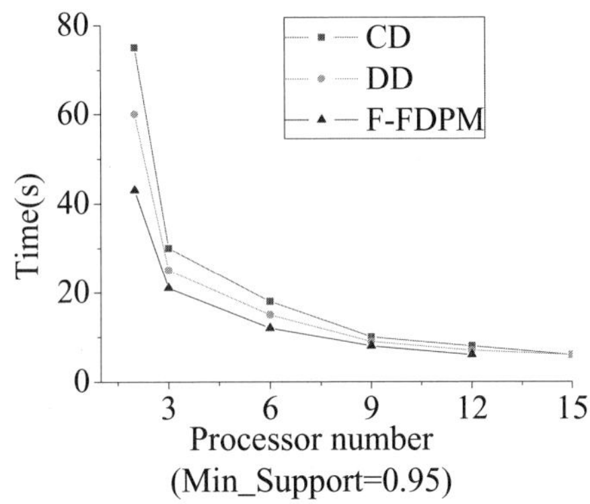

Fig. 6. Running time comparing on connect set

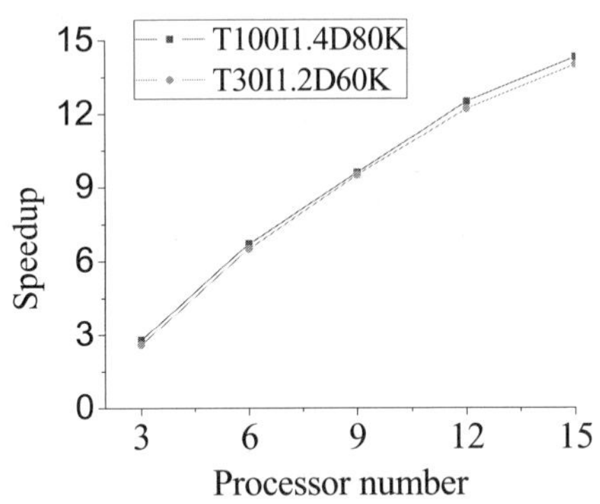

Fig. 7. Speedup of F-FDPM

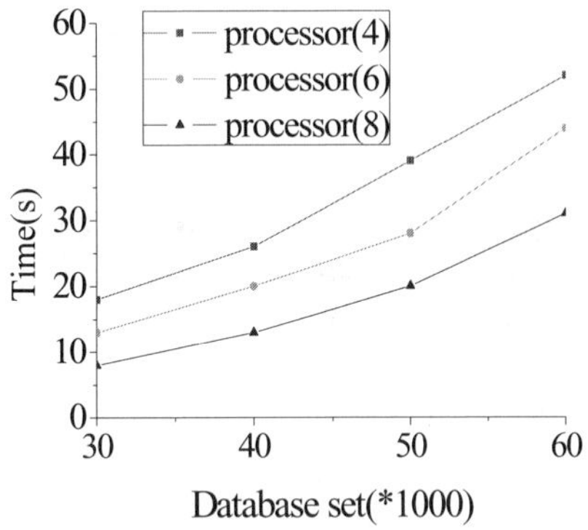

Fig. 8. Sizeup of F-FDPM

Fig. 5 illustrates the running times for the test data set mushroom. We ran F-FDPM algorithm with 8 processors. The results show that F-FDPM has better performance comparing to Apriori and FP-Growth* algorithm. Fig. 6 shows the times for the test data set using 2, 3, 6, 9, 12, and 15 processors comparing with count distribution algorithm and data distribution algorithm, Min_Support is equal to 0.95. The results show that run-time drops rapidly with the processors numbers increasing and is lower than CD and DD algorithms.

Fig. 7 shows the measured speedup on T30I1.2D60k data set and T100I1.4D80K data set with 3, 6, 9, 12, 15 processors. Therein, Min_support is equal to 0.8. It can be seen that speedup of F-FDPM is nearly linear with the number of processor increasing. Fig. 8 tests the sizeup capability of F-FDPM respectively using 4, 6, 8 processors on T30I1.2D60k data set, Min_support is equal to 0.85. We can know that time cost is slowly adding with sizeup rapidly increasing.

6 Conclusion

Rapid association rules extraction from large databases present challenges to data mining research in many areas, including run-time and memory requirements. In this paper, we propose a new data structure, namely F-Forest, which is suitable to parallel algorithm. On the basis, a parallel mining model and parallel association rules mining algorithm, namely F-FDPM, are designed to accelerate association rules mining. We present results of an implementation on a computer cluster which shows good performance between 3 and 15 processors and has better performance than some classical serial and parallel algorithms, this demonstrates that the F-Forest structure , parallel mining model and algorithm are suitable for parallel mine association rules on large database.

Acknowledgments. The authors thank the ISNN2008 anonymous referees for their substantive suggestions which have improved the paper. This work is partially supported by the National Natural Science Foundation of China (Grant No. 70571019) the Ph.D. Programs Foundation of Education Ministry of China (No. 20060213004) and the Research Center of Technology, Policy and Management, Harbin Institute of Technology.

References

1. Agrawal, R., Sharfer, J.: Parallel Mining of Association Rules. IEEE Trans on Knowledge and Dara Engineering 8(6), 962–969 (1996)
2. Zaki, M.J., Ogihara, M., Parthasarathy, S., Li, W.: Parallel Data Mining for Association Rules on Shared-memory Multi-processors. In: Super computing 1996, Pittsburg, PA, November 1996, pp. 88–91. IEEE Press, New York (2006)
3. Park, J.S., Chen, M.S., Yu, P.S.: Efficient Parallel Data Mining for Association Rules. In: ACM Int'l Conf on Information and Knowledge Management, pp. 31–36. ACM Press, New York (1995)
4. Han, E.H., Karpis, G., Kumar, V.: Scalable Parallel Data Mining for Association Rules. In: Proc of the ACM SIGMOD Conference on Management of Data 1997, pp. 277–288. IEEE Press, New York (1997)
5. Schuster, A., Wolff, R.: Communication Efficient Distributed Mining of Association Rules. In: Proc of the ACM SIGMOD Int'1 Conference on Management of Data, Santa Barbara, California, pp. 473–484. ACM Press, New York (2001)
6. Cheung, D., Han, J., Ng, V.: A Fast Distributed Algorithm for Mining Association rules. In: Proc of 1996 int'1 Conf on Parallel and Distributed Information System, Miami Beach Florida, pp. 31–44. IEEE Press, New York (1996)
7. Cheung, D., Xiao, Y.: Effect of Data Skewness in Parallel Mining of Association rules. In: 12th Pacic-Asia Conference on Knowledge Discovery and Data Mining, Melbourne, Australia, pp. 48–60. Springer, Heidelberg (1998)

8. Cheung, D., Hu, K., Xia, S.: A Synchronous Parallel Algorithm for Mining Association Rules on Shared-memory Multi-processors. In: 10th ACM Symp Parallel Algorithms and Architectures, pp. 219–228. ACM Press, New York (1998)
9. Zaiane, O.R., EI-Hajj, M., Lu, P.: Fast Parallel Association Rule Mining Without Candidacy Generation. In: Proceedings IEEE International Conference on Data Mining 2001, pp. 665–668. IEEE Press, New York (2001)
10. Pramudiono, I., Kitsuregawa, M.: Parallel FP-Growth on PC Cluster. In: Proceedings of the 7th Pacific-Asia Conference of Knowledge Discovery and Data Mining 2003, pp. 467–473. Springer, Heidelberg (2003)
11. Merz, C. J., Merphy, P.: UCI Repository of Machine Learning Databases (1996), http://www.ics.uci.edu/~mlearn/MLRRepository.html

Improved Algorithm for Image Processing in TCON of TFT-LCD

Feng Ran[1], Lian-zhou Wang[2], and Mei-hua Xu[2]

[1] Microelectronic Research and Development Center, Shanghai University
[2] School of Mechatronical Engineering and Automation, Shanghai University
Campus P.O.B. 110, 149 Yanchang Rd, Shanghai 200072, P.R. China
lianzhouwang@126.com

Abstract. In order to satisfy the special demand for image display in TFT-LCD, this paper presents a synthesis algorithm for image processing in TCON of TFT-LCD. The algorithm includes contrast adjustment, Gama correction and dithering technique which is improved to deal with the zoomed image, and at the same time, matches the real-time requirements. All the modules are simulated and verified using MATLAB, and then described in RTL. The experimental results show that the design achieves the anticipative effects.

Keywords: TCON, Contrast adjustment, Gama correction, Dithering technique, Error-diffusion.

1 Introduction

TFT-LCD takes the lead in entering into the display market due to its outstanding characteristics, such as low operation voltage, low power consumption, good display performance, convenient integration and carriage, and etc. Internal circuit of TFT-LCD is mainly composed of two parts. The first one is the LCD control module, including chips of SCALER, ADC (Analog-to-Digital Converter), OSD (On Screen Display) and MCU (Micro Controller Unit), which connects PC with LCD module. The other part is made up of driving IC and timing control IC integrated on the panel of TFT-LCD, and it forms the LCD panel module. As the connection between SCALER[1] and driving chips of SOURCE[2] and GATE[3], TCON[4,5] takes the responsibility in receiving and processing image data zoomed by SCALER, while providing correct control sequence and pixel data for the other driving chips. This paper presents an improved algorithm in dealing with the zoomed image after SCALER, including contrast adjustment, Gama correction and dithering technique.

2 Improved Algorithm

2.1 Contrast Adjustment

Contrast enhancement is used to highlight the contrast of different colors while hierarchical display and definition of images are improved. The images can also be softer

F. Sun et al. (Eds.): ISNN 2008, Part II, LNCS 5264, pp. 50–56, 2008.

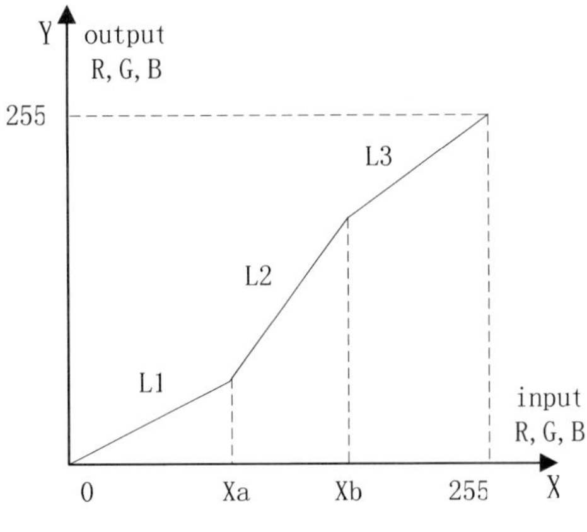

Fig. 1. Contrast adjustment diagram

with contrast reduction. Contrast adjustment function is shown in Fig.1, in which X-axis represents RGB input signal, Y-axis denotes RGB output signal and *L1*, *L2*, *L3* figure the slopes of respective functions.

The corresponding function formula is shown as follows:

$$Y=\begin{cases} L1*Xa & 0\le X<Xa \\ L2*(X-Xa)+Ya & Xa\le X\le Xb \\ L3*(X-Xb)+Yb & Xb<X\le 255 \end{cases} \tag{1}$$

Xa and *Xb* are the parameters of two registers, in which 0<*Xa*<*Xb*<255. Parameters *L*1, *L*2 and *L*3 in Fig.1 are restricted within limits. 5 bit-width is adopted for *L*1、*L*3 and *L*1、*L*3∈{1/4, 2/4, 3/4, 1, 5/4, 6/4, 7/4, 2, 4}, while *L*2 should be more accurate, because it connects line *L*1 and line *L*3. *L*2 with 8 bit ranges from 0.0000000 to 1.1111111, or 0 to 1.9921875. Completely independent and parallel processing structures are adopted for RGB pixels in data processing to improve the data processing speed and achieve the signal synchronization. Furthermore, the hardware structure is relatively simplified since the same logic units are adopted in the same way for general usage. The circuit framework of contrast adjustment is shown in Fig.2, in which the processing procedure of Red component is described. The same structures are adopted for Green and Blue components.

Furthermore, luminance adjustment is implemented by adding or reducing a certain constant ranging from 0 to 255 into the RGB weight of each pixel. Generally, the range of the certain constant is -128~+127. The increment of light, which can be allocated by SCM, is also stored into the registers bank of contrast adjustment circuit.

2.2 Gama Correction

There is a non-linear relationship between input voltage V of the driving ciruit and light intensity in TFT-LCD, such as, $L=v^c$ (c is 1~2.5, the range is given by the LCD

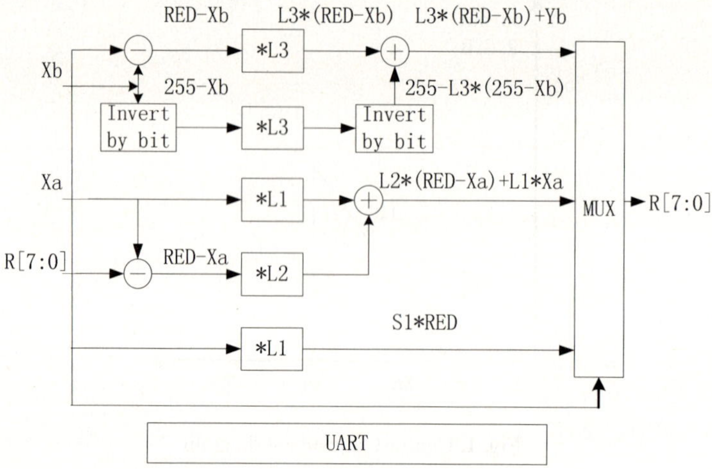

Fig. 2. Circuit framework of Contrast Adjustment

manufacturers according to different conditions). This non-linearity problem can be solved by the SOURCE driving chip with a set of resistances. However, digital technology is more accurate than analog technology in solving the non-linearity problem. Thus, digital Gama correction is used in the TFT-LCD display system, and as a result, it makes the light intensity linear with the input voltage, decreases image distortion and simplifies the design of SOURCE driving chip. Fig.3 shows the schematic diagram of Gama correction for TFT-LCD.

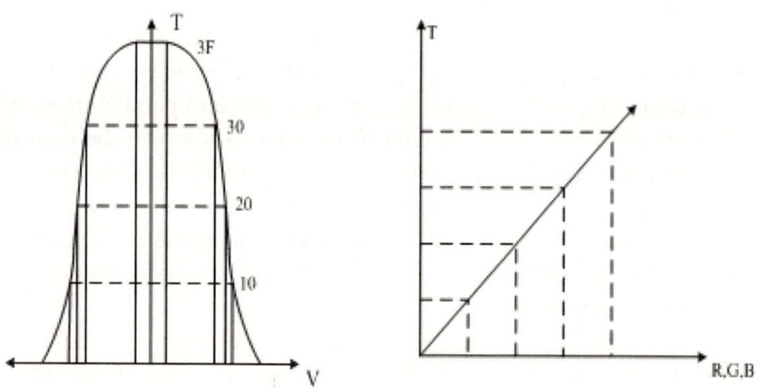

Fig. 3. Schematic diagram of Gama correction for TFT-LCD

The core algorithm of Gama correction focuses on the method of look-up table. The anticipate output values are written to a 256×8 bit RAM beforehand by Gama controller according to input signal values. Then, the output can be found by the address (input signals). In addition, the method of look-up table can be further optimized. The difference between Gama characteristic value and input value is stored in RAM instead of only Gama characteristic value. Compared with the former way of Gama characteristic value storage, the new method saves huge amounts of storage space. The image display color is improved after Gama correction. Fig.4 shows the difference storage circuit of Gama correction.

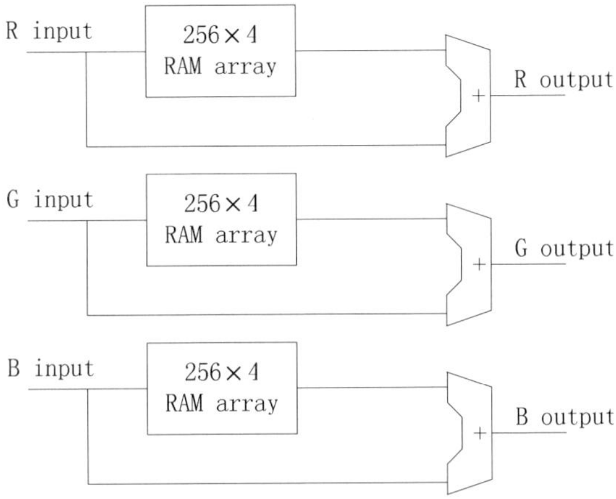

Fig. 4. Difference storage circuit of Gama correction

2.3 Dithering Technique

Since many image display equipments can only obtain finite number of output states, digital dithering is generally used to ensure the quality of obtained images. The SOURCE driving chip at medium-low end can only receive 6 bit-width RGB pixel data due to the manufacturing process and cost. However, the display data in computer are usually 8 bit-width. Therefore, 8 bit-width pixel data should be converted into 6 bit-width one first for advanced operation. The easiest way to solve this problem is to use the dithering matrix[6], such as the classical BAYER matrix. However, this is not an effective solution because it brings the effect of some squares on the edge of adjacency matrix in the image. In this paper, a better way called Floyd-Steinberg error diffusion algorithm[6] is used instead of BAYER. The principle of the error diffusion algorithm is not complex to obtain perfect result. It is actually a quantization process and also a neighborhood process which does not equal to the pattern dithering in function. It diffuses the quantization error of current pixel to adjacent pixel according to a certain proportion. Then, local quantization error becomes a compensation for adjacent pixel, and so, the error diffusion system has the ability of self-tuning. Delta-Sigma Modulation represents the foundation of Floyd-Steinberg error diffusion algorithm, and Noise-Shaping Feedback Coder (NSFC) which involves Floyd-Steinberg algorithm becomes a kind of DSM used for word-length compression. Fig.5 presents the framework of NSFC.

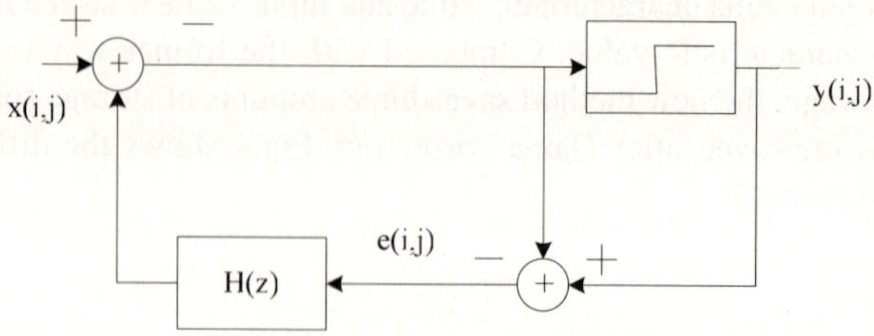

Fig. 5. Framework of NSFC

The two-dimensional signals x(i,j), y(i,j) and e(i,j) are respectively converted to X(Z), Y(Z), and E(Z) of frequency domain. At the same time, white noise N(Z) is introduced into the circuit. The corresponding equations are as follows.

$$Y(z) = X(z) - E(z) * H(z) + N(z) \tag{2}$$

$$E(z) = Y(z) - [X(z) - E(z) * H(z)] \tag{3}$$

E(z) can be eliminated by equation (2) and (3) , then the corresponding transfer function expression is:

$$Y(x) = X(z) + [1 - H(z)] * N(z) \tag{4}$$

The noise transfer function expression of Floyd-Steinberg algorithm and its corresponding expression of H(z) are:

$$NTF(Z_1, Z_2) = 1 - \frac{1}{16}(Z_1^{-1} Z_2^{-1} + 5Z_2^{-1} + 3Z_1 Z_2^{-1} + 7Z_1^{-1}) \tag{5}$$

$$H(z) = 1/16(z_1^{-1} z_2^{-1} + 5z_2^{-1} + 3z_1 z_2^{-1} + 7z_1^{-1}) \tag{6}$$

Fig.6 shows the frequency domain response of $NTF(z_1,z_2)$

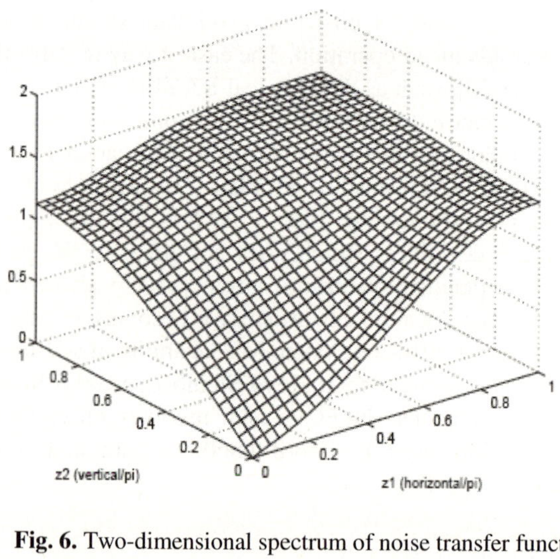

Fig. 6. Two-dimensional spectrum of noise transfer function

Floyd-Steinberg algorithm is simulated by MATLAB and the core code is listed as follows.

```
for i = 2 : Last_line
  for j = 2 : (Last_column-1)
       red_error_sum=int8(red_error(i-1,j-1)*1/16+red_error(i-1,j)*5/16+red_erro
    r(i-1,j+1)*3/16+red_error(i,j-1)*7/16);
    red(i,j)=red_error_sum+int8(bitshift(red(i,j),-2);
       red_error(i,j)=int8(bitshift(bitshift(red(i,j),6),-6))-red_error_sum;
  end
end;
```

Furthermore, the circuit of error diffusion algorithm mainly includes three parts: control module, data path and FIFO memory. The control module provides a read-write time sequence to FIFO. Data path mainly gets the weighted average of error diffusion coefficient to obtain the current pixel, and it also counts the corresponding error. FIFO

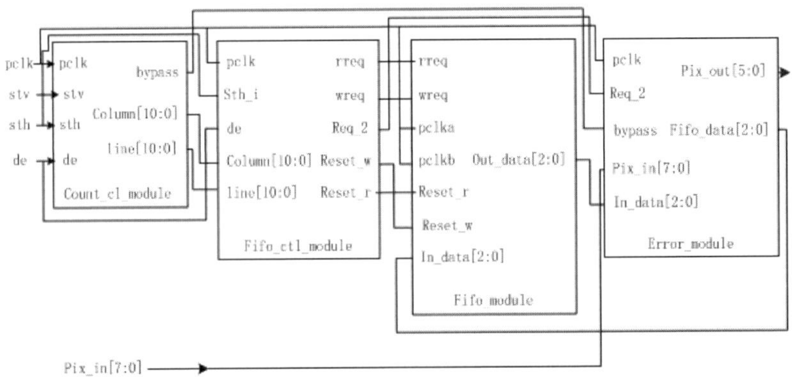

Fig. 7. Framework of error diffusion algorithm

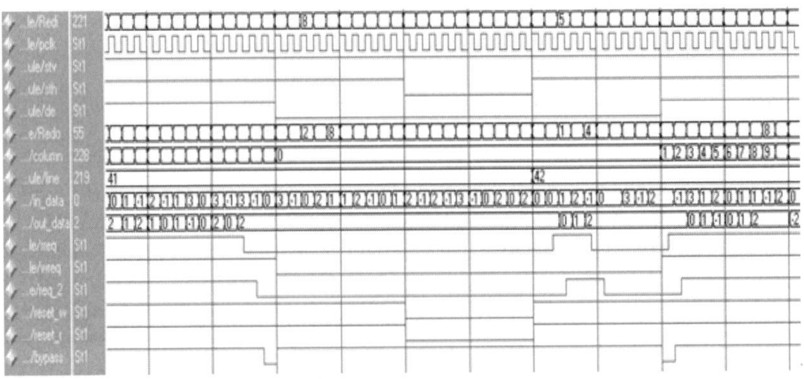

Fig. 8. Main controlling signal waveform

is used to store the errors of two neighbor lines. Considering that sometimes the error is a negative one, a 3 bit-width FIFO is used to store signed errors. The top framework of dithering algorithm and its control signal waveform are shown in Fig. 7 and Fig. 8.

3 Conclusion

A synthesis algorithm including contrast adjustment, luminance adjustment, Gama correction and dithering method for image processing in TCON of TFT-LCD is presented in this paper. Contrast adjustment improves hierarchical display and definition of images and also makes images to be softer. Gama correction solves the problem that nonlinear distortion of images is caused by non-linear relationship between penetration rate and external display voltage. Dithering algorithm converts 8 bit-width pixel data into 6 bit-width one to ensure the quality of obtained images. Finally, the hardware implementation of each module is successfully completed and then embedded in a TCON to test synthetically. The experimental results show that the proposed design can substitute the TCON in a commercial LCD, and transmit the video image with fine definition fluently.

Acknowledgments

The authors would like to acknowledge the National Natural Science Foundation of China for providing financial support for this work under Grant No. 60773081 and Grant No. 60777018, and also to acknowledge the financial support by the Shanghai Municipal Committee under Grant No.AZ028 and Grant No.06DZ22013.

References

1. Liu, Z.L., Zhao, H.B., Zhou, X.C.: Design of Picture enhancement module for scaler. TV engineering. 10, 56--61 (2004)
2. Deng, J.L., Wan, P., Zhou, M., Li, X.C.: Application of the LCD controller S1D13506 in the embedded system of AT91RM9200. Micro-computer information. 22, 9 (2006)
3. NT7167-00007 TFT-LCD Panel Timing Controller With BIST Function.
4. Liu, Z.L., Guo, X., Zhou, X.C., Xiao, J.P.: Image color enhancement technique based on improved Bayer dithering algorithm. J. Huazhong University of Science and Technology. 34, 5 (2006)
5. Lei, J.M., Zou, X.C., Zou, W.H., Liu, S.Q.: Dynamic Dithering Algorithm and Frame Rate Control Technique for LCD Controller. Microelectronics & Computer. 21, 5 (2004)
6. Floyd R., Steinberg L.: An Adaptive Algorithm for Spatial Grey Scale. SID 75 DIGEST. 75, 36--37 (1975)

Clustering Using Normalized Path-Based Metric

Jundi Ding[1,2], Runing Ma[3], Songcan Chen[2], and Jingyu Yang[1]

[1] School of Comp. Sci. & Tech., Nanjing University of Science and Technology
[2] Department of Computer Science & Engineering
[3]Department of Mathematics
[2,3] Nanjing University of Aeronautics & Astronautics
[1,2,3] Nanjing, Jiangsu, China
{dingjundi,mrning,s.chen}@nuaa.edu.cn, yangjy@mail.njust.edu.cn

Abstract. In this paper, we propose a normalized path-based metric based on an introduced neighborhood density index which can sufficiently exploit the local density "revealed" by data. The metric axioms (positive definite property, symmetry and triangular inequality) are strictly proved in theory. Using this idea of path, we further devise a heuristic clustering algorithm which can perform the elongated structure extraction, uneven lighting background isolation, grains of tiny objects segmentation and figure-ground separation. In particular, when the pairwise distances between data are given, the proposed algorithm has a computational complexity linear in the size of data. Extensive experiments are conducted to validate its effectiveness, efficiency and competitiveness in resistance to noise.

Keywords: Data Clustering, Metric Axioms, Elongated Structure, Linear Complexity, Image Segmentation.

1 Introduction

Clustering is a fundamental unsupervised learning problem. Researchers intend to partition a given set of objects $X = \{x_n\}_{n=1}^N$ into K clusters $\{C_k\}_{k=1}^K$ such that $\bigcup_{k=1}^K C_k = X$ and $C_i \bigcap C_j = \emptyset$ if $i \neq j$. With no priori information about the real distribution of the input data set, e.g., the number of clusters K, bounds on the number of points in the k'th cluster $|C_k|$, the task of tackling this problem is rather difficult in that there is often more than one plausible way to partition the data. In general, many researchers strive to seek a so-called optimal one from all possible partitions by solving a meaningful minimization or maximization problem, such as the pairwise data clustering [1], spectral clustering [2-3], path-based clustering [4-6]. Both the pairwise data clustering and normalized cut focus on the compactness of intra-cluster and fail to capture the essential elongated structures of data. The reason is that the (dis)similarity between two objects x_i and x_j is defined as a simple function of the Euclidean distance between them, considering no density information in the feature space. To deal with the non-compact clusters, Fischer et al. recently develop a path-based pairwise clustering approach [4]. The involved effective path-based dissimilarity between objects x_i

F. Sun et al. (Eds.): ISNN 2008, Part II, LNCS 5264, pp. 57–66, 2008.

and x_j is defined by $D^{\text{eff}}_{ij} = \min_{p \in P_{ij}} \{ \max_{1 \leq h \leq |p|} d_{p[h]p[h+1]} \}$, where P_{ij} denotes the set of all paths from x_i to x_j, $p[h]$ is the hth objects along the path p, $|p|$ denotes the number of objects that p goes through and $d_{p[h]p[h+1]}$ is the Euclidean distance between the two consecutive objects. The maximization term can be implicitly looked at as an approximation of the feature space density. Intuitively, when two points x_i and x_j reside in the same cluster, the dissimilarity between them should be small, otherwise large. However, if there exist some outliers between clusters, the dissimilarity between points falling into different clusters may become much smaller than it should be. That is, D^{eff}_{ij} is very sensitive to noise and outliers. To address the robust issue, Fisher et al. use a heuristic agglomerative strategy or bagging [5] to detect the outliers for reducing data variance. More recently, Chang and Yeung in [6] propose a robust path-based similarity measure based on the concept of M-estimation in robust statistics. This measure can reflect the genuine similarity between x_i and x_j even when outliers exist. Then, Chang and Yeung substitute the robust path-based similarity matrix for the affinity matrix commonly used in spectral clustering [3] to develop a robust path-based spectral clustering [6]. Omer and Werman in [7] point out that the two path-based (dis)similarity measures implicitly involve the density information of the data distribution which is insufficient. Because one point within a cluster may be connected with a far outlier point by a very long but dense path. Therefore, the bottleneck geodesic (BG) is introduced to search for a both short and dense path in a histogram domain [8]. Omer and Werman define the BG to be a function of the geodesic distance between points and of the minimal density (bottleneck) along this path. Moreover, they also choose the normalized cuts algorithm [2] to evaluate the defined BG. However, there is an underlying assumption that the feature points would be from convex or nearly convex clusters. In addition, all these mentioned path-based (dis)similarities could not obey the metric axioms except the symmetry.

On the other hand, the expensive computational cost is another unfavorable issue of these combinatorial optimization problems, like PC [4-6] normalized cuts [2] or Weiss's spectral clustering [3]. In fact, they are proven to be *NP*-hard. Spectral clustering methods utilize explicitly the eigenvector solvers to find clusters whose graph cut is minimal. In general, the computational complexity of all the eigenvector calculation is $O(N^3)$. The Nyström approximation [9] is used to decompose the similarity matrix efficiently by choosing a random sample of the data, so that the computational complexity is reduced to $O(m_1 N)$ where m_1 is the size of the sample. Fischer et al. suggest various ways to speed up their path-based clustering for many different applications in their papers, e.g., optimization by iterated conditional mode with an efficient implementation of update step takes $O(N^3)$ [4]; agglomerative optimization with automatic outlier detection takes $O(N^2 m_2 + N^3 logN)$ where m_2 is the number of used dissimilarities [5].

In this paper, the first consideration is to introduce a notion of the neighborhood density index (NDI) which falls within the range $[0, 1)$ and could provide a reliable and reasonable density information of the data. A small value for NDI(x_i) indicates that x_i is likely to be a dense point surrounded by many other

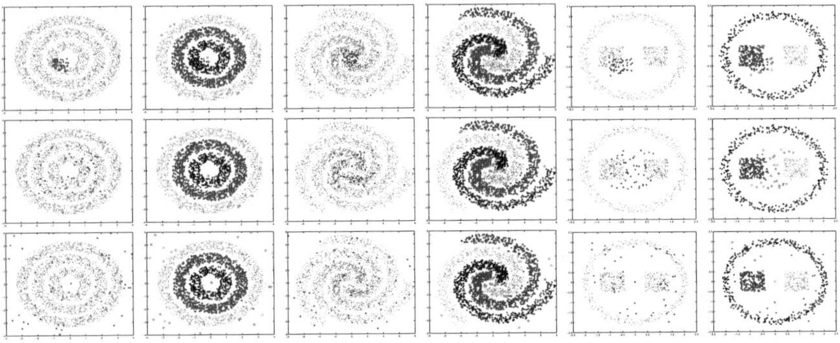

Fig. 1. Elongated non-convex structure extraction results with varying noise level: (a), (c) and (e) show the input noisy dataset; (b), (d) and (f) present the extraction results $k = 27, 27, 32$ respectively. But $k = 29$ for the bottom-right noisy data.

points, while a large value indicates that x_i is an even or sparse point with very few other points in its vicinity. Then, the defined NDI is used to derive a novel path-based normalized metric which explicitly captures the density information of the data point. Note that NDI has the different description and implication for the different specific data feature. Here, we focus on two distinct types of data, the artificial elongated data and the image data. Consequently, the normalized metric is proposed to respectively deal with the tasks of elongated structure extraction and image segmentation. Moreover, the mathematically strict proof of metric axioms, i.e., positive definite property, symmetry and triangular inequality, is also presented in detail, which is the main theoretical contribution of this paper and described in Section 2. In the transformed normalized metric space, the clusters can be interpreted as connected paths which traverse elements with a comparatively high density. To achieve these clusters, we finally devise a fast path-based clustering (FPC) algorithm in Section 3 which has a similar grouping principle to that of agglomerative methods, the single linkage algorithms [10], [11]. Apart from the calculation of the Euclidean distances between objects in the original data space, FPC has a nearly linear computational complexity in the number of the elements in the metric space. In addition to these theoretical results, we also demonstrate extensive experimental results in Section 4 to show that FPC is resistant enough to outliers, efficient and effective for several challenging tasks, such as the elongated structure extraction, uneven lighting background isolation, grains of tiny objects segmentation and figure-ground separation.

2 Normalized Metric and Metric Axioms

To access the description of the path-based normalized metric, we in this section first provide the definitions of NDI_d and NDI_i together with other useful related notations and notions. Note that NDI_d and NDI_i are applicable for different tasks, elongated data analysis and image segmentation respectively.

In [11, 12], the authors take both k-neighborhood (kNB) and reverse k-neighborhood (R-kNB) to expose the relationship between object x and its neighbors. $k\mathrm{NB}(x)$ describes who makes up of its own neighbors, while R-$k\mathrm{NB}(x)$ indicates whose neighborhood x belongs to. This two-way description of the relationship between x and its neighborhood would more clearly and precisely picture its position in the dataset than simply using only k nearest neighbor (kNN) or k-neighborhood (kNB). Coinciding with such a two-way description, a locally symmetric neighborhood of x ($\mathrm{SN}_d(x)$) is specified to be a set which contains elements shared by both $k\mathrm{NB}(x)$ and R-$k\mathrm{NB}(x)$. Namely, $\mathrm{SN}_d(x) = k\mathrm{NB}(x) \bigcap \text{R-}k\mathrm{NB}(x)$, which implies that $x \in \mathrm{SN}_d(y) \Leftrightarrow y \in \mathrm{SN}_d(x)$. This specification of symmetry is fundamental to our introduced path-based metric. In the case of image segmentation, $\Omega_s^v(p) = \{q \in \Omega_s(p) : |I(p) - I(q)| \leq v\}$ just is a locally symmetric neighborhood of p because $q \in \Omega_s^v(p) \Leftrightarrow p \in \Omega_s^v(q)$, where $p = (p_x, p_y), q = (q_x, q_y)$ are the arbitrary pixels in the image I and $I(p), I(q)$ are the corresponding pixel values (color, intensity or texture, etc.), while $\Omega_s(p) = \{q \in I : |p_x - q_x| \leq s, |p_y - q_y| \leq s\}$ is a square neighborhood of p. In what follows, we define NDI_d and NDI_i as:

$$\mathrm{NDI}_d(k;x) = \frac{|\text{R-}k\mathrm{NB}(x)|}{|k\mathrm{NB}(x)| + |\text{R-}k\mathrm{NB}(x)|}, \quad \mathrm{NDI}_i(s,v;p) = \frac{|\Omega_s^v(p)|}{|\Omega_s(p)|} \tag{1}$$

It is evident that $\mathrm{NDI}_d(k;x) \in [0,1), \mathrm{NDI}_i(s,v;p) \in [0,1)$. For simplicity, we uniformly formalize NDI_d and NDI_i by the term $\mathrm{NDI}(x)$, SN_d and $\Omega_s^v(q)$ by the term $\mathrm{SN}(x)$, whereas $x \in X$, X is the dataset (synthetic elongated data or image) to be tackled. Moreover, let $r(x) = 1-\mathrm{NDI}(x)$ and $r(x,y) = \max(r(x), r(y))$, then $r(x) \in (0,1]$, $r(x,y) \in (0,1]$. In practice, we are very interested in those objects with $r(x) \leq 1/2$, i.e., $\mathrm{NDI}(x) \geq 1/2$. On one hand, when $\mathrm{NDI}_d(k;x) \geq 1/2$ we can easily derive that $|\text{R-}k\mathrm{NB}(x)| \geq |k\mathrm{NB}(x)|$ and x is thus a dense point [11-12], which could be used to create a x's cluster. On the other hand, when $\mathrm{NDI}_i(s,v;p) \geq 1/2$ we could intuitively indicate that most of the p's neighboring pixels are similar to p and p is a dense pixel. Hence, these objects with $r(x) \leq 1/2$ can be taken as the intermediate objects of one path. Let $\mathrm{Path}(x,y)$ denotes the set of all paths from x to y, then it can be formally defined as follows:

$$\begin{aligned}\mathrm{Path}(x,y) = \{l = \{x = x_0, x_1, \cdots, x_t, x_{t+1} = y\} \mid x_i \neq x_j, \forall i \neq j \in \{0, \cdots, t+1\};\\ x_i \in \mathrm{SN}(x_{i+1}), i \in \{0, \cdots, t\}; r(x_i) \leq \tfrac{1}{2}, i \in \{1, \cdots, t\}, t \geq 1\}.\end{aligned}$$

Note that there are two strictly constrained conditions. One is that the intermediate objects must be dense ($\mathrm{NDI}(x) \geq 1/2, i.e., r(x) \leq 1/2$) while the start and end objects could be not dense, the other is that the two consecutive objects must be close enough ($x_i \in \mathrm{SN}(x_{i+1})$). In such a way, one path is ensured to traverse objects of high density and proximity. Besides, due to symmetry of $\mathrm{SN}(x)$, it is easy to discover that $l = \{x = x_0, x_1, \cdots, x_t, x_{t+1} = y\} \in \mathrm{Path}(x,y)$ indicates $l' = \{y = x_{t+1}, x_t, \cdots, x_1, x_0 = x\} \in \mathrm{Path}(y,x)$. Without regard to the direction of a path, we hence can say that $l = l'$ and $\mathrm{Path}(x,y) = \mathrm{Path}(y,x)$.

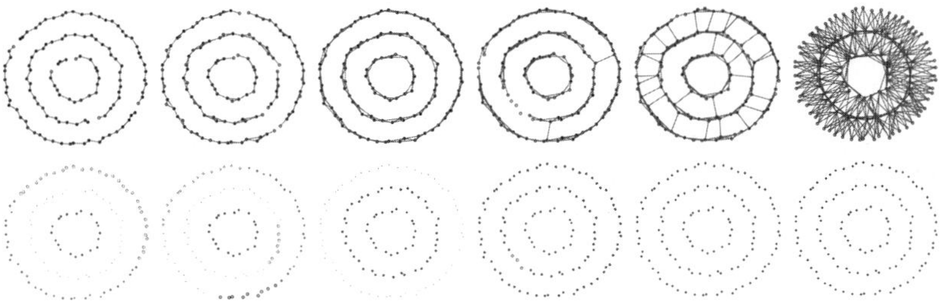

Fig. 2. Results of a simple 3-circle data on a range of different k values. From left to right, $k = 2, 3, 4, 5, 6, 15$ respectively.

Suppose $\text{Path}(x, y) \neq \emptyset$, then the length of each path $l = \{x = x_0, x_1, \cdots, x_t, x_{t+1} = y\} \in \text{Path}(x, y)$ could be intuitively determined by the density of objects and bounded by the minimum density of objects in the path, namely, $\text{Len}(l) \triangleq \max_{0 \leq i \leq t+1} \{r(x_i)\}$. The distance between two objects $d(x, y)$ can be subsequently defined as follows:

Definition 1. For $\forall x, y \in X$,
(a) $d(x, y) = 0$ if $x = y$;
(b) $d(x, y) = \min_{l \in \text{Path}(x,y)} \{\text{Len}(l)\}$ if $\text{Path}(x, y) \neq \emptyset$;
(c)$d(x, y) = r(x, y)$ if $x \neq y, \text{Path}(x, y) = \emptyset$ and $x \in \text{SN}(y)$; otherwise $d(x, y) = 1$.

This definition has several potential properties summarized in Lemma 1. From these properties, we conclude that the proposed $d(x, y)$ is a normalized metric satisfying the metric axioms. This will be formalized in Theorem 1.

Lemma 1
P1. $\forall y \neq x, d(x, y) = 1$ if $\text{SN}(x) = \emptyset$;
P2. $d(x, y) \leq \max\{1/2, r(x, y)\}$ if $\text{Path}(x, y) \neq \emptyset$;
P3. $r(x, y) \leq d(x, y) \leq 1$ for $\forall x \neq y$;
P4. $d(x, y) = 1 \Leftrightarrow \text{Path}(x, y) = \emptyset$ and $x \notin \text{NR}_k(y)$;
P5. $\forall x, y, z \in X, d(x, y) \leq d(x, z) + d(z, y)$ if $r(z) > 1/2$.

Theorem 1. For $\forall x \neq y \neq z \in X$, $d(x, y)$ is a normalized metric which satisfies
1) *Positiveness*: $0 \leq d(x, y) \leq 1$ and $d(x, y) = 0 \Leftrightarrow x = y$;
2) *Symmetry*: $d(x, y) = d(y, x)$;
3) *Triangular Inequality*: $d(x, y) \leq d(x, z) + d(z, y)$.

Proof: 1) If $d(x, y) = 0$ and $x \neq y$, we then can derive that $0 = d(x, y) \geq r(x, y) > 0$ from P3 in Lemma 1. This is a contradiction. So $d(x, y) = 0 \Rightarrow x = y$.
2) The symmetry will be checked for the case of $x \neq y$. On one hand, when $\text{Path}(x, y) \neq \emptyset$, from Definition 1(b) we can see $d(x, y) = d(y, x)$ because $\text{Path}(x, y) = \text{Path}(y, x)$. On the other hand, if $\text{Path}(x, y) = \emptyset, x \in \text{SN}(y) \Leftrightarrow y \in \text{SN}(x)$, so $d(x, y) = r(x, y) = d(y, x)$; In addition, $x \notin \text{SN}(y) \Leftrightarrow y \notin \text{SN}(x)$, so $d(x, y) = 1 = d(y, x)$;
3) From P5, triangular inequality holds true if $r(z) > 1/2$. Hence, we in the following assume that $r(z) \leq 1/2$. There are four cases possibly appearing.

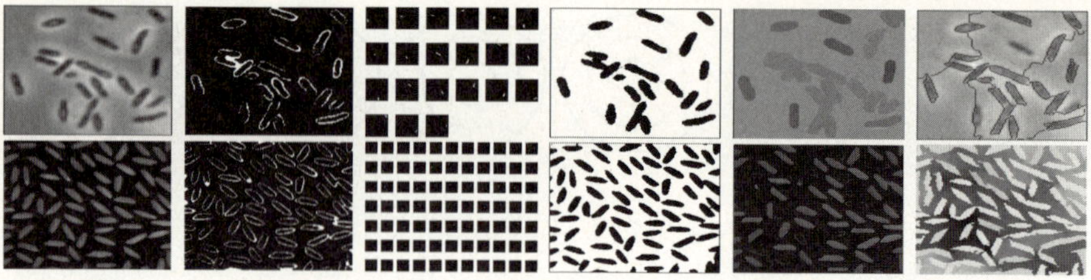

Fig. 3. Granules of objects segmentation results. From left to right, the six columns are input images, results segmented by FPC (2nd-4th columns, $s = 3; 3$ and $v = 19; 15$) and the methods in [10] and [2] respectively.

`Case one.` $\text{Path}(x, y) = \emptyset$: According to Definition 1 (c), in this case we know that if $x \in \text{SN}(y)$, then $d(x, y) = r(x, y) \le r(x) + r(y) \le r(x, z) + r(y, z) \le d(x, z) + d(y, z)$; if $x \notin \text{SN}(y)$, $d(x, y) = 1$. Consequently, we require to prove $d(x, z) + d(y, z) \ge 1$: i) when $\text{Path}(y, z) \ne \emptyset$, because $r(z) \le 1/2$ we have $\text{Path}(x, z) = \emptyset$ and $x \notin SN(z)$, then $d(x, z) = 1$, thereby $d(x, z) + d(y, z) \ge 1$ since $d(y, z) \ge 0$; ii) When $\text{Path}(y, z) = \emptyset$, if $y \in \text{SN}(z)$, then $\text{Path}(x, z) = \emptyset$ and $x \notin \text{SN}(z)$, thus $d(x, z) = 1$; if $y \notin \text{SN}(z)$, we have $d(y, z) = 1$. So, $d(x, z) + d(y, z) \ge 1$.
`Case two.` $\text{Path}(x, y) \ne \emptyset$, $\text{Path}(x, z) \ne \emptyset$, $\text{Path}(y, z) \ne \emptyset$: Let $l_1 = (x, x_1, \cdots, x_n, z) \in \text{Path}(x, z)$, $l_2 = (y, y_1, \cdots, y_n, z) \in \text{Path}(y, z)$, we have $l = (x, x_1, \cdots, x_n, z, y_n, \cdots, y_1, y) \in \text{Path}(x, y)$ due to $r(z) \le 1/2$. So $d(x, y) \le \text{Len}(l) \le \text{Len}(l_1) + \text{Len}(l_2)$ and $d(x, y) \le \min_{l_1}\{\text{Len}(l_1)\} + \min_{l_2}\{\text{Len}(l_2)\} = d(x, z) + d(y, z)$.
`Case three.` $\text{Path}(x, y) \ne \emptyset$, $\text{Path}(x, z) = \emptyset$: i) If $y \in \text{SN}(z)$ and $r(y) > 1/2$, then $P2 \Rightarrow d(x, y) \le \max\{1/2, r(x, y)\} = r(x, y) \le r(x) + r(y) \le d(x, z) + d(y, z)$; ii) If $y \in \text{SN}(z)$ and $r(y) \le 1/2$, then $\text{Path}(x, z) \ne \emptyset$. Let $l_1 = (x, x_1, \cdots, x_n, z) \in \text{Path}(x, z)$, since $r(z) \le 1/2$, we have $l = (x, x_1, \cdots, x_n, z, y) \in \text{Path}(x, y)$. Due to $d(x, y) \le \text{Len}(l) \le \text{Len}(l_1) + r(y, z) = \text{Len}(l_1) + d(y, z)$, we know $d(x, y) \le \min_{l_1}\{\text{Len}(l_1)\} + d(y, z) = d(x, z) + d(y, z)$; iii) If $y \notin \text{SN}(z)$, then $d(y, z) = 1$ and $d(x, y) \le 1 \le d(y, z) + d(x, z)$
`Case four.` $\text{Path}(x, y) \ne \emptyset$, $\text{Path}(y, z) = \emptyset$: See Case three. □

3 FPC Algorithm and Its Complexity

So far we have discussed a density-based normalized path metric which measures the dissimilarities between objects. The metric axioms ensure its reasonability: i) "positiveness" means the self-dissimilarity of objects vanishes; ii) "symmetry" illustrates the dissimilarities between two objects are unchangeable; iii) "triangular inequality" implies the dissimilarity between two objects is not larger than the sum of dissimilarities between other two objects in the same plane. Based on this sound metric, we devise a fast path-based clustering algorithm which is to heuristically make paths among objects closed enough, i.e., with very low dissimilarities.

FPC Algorithm: Let X be the input dataset,
(1) Find the dense set $X' = \{x \in X | r(x) \le 1/2\}$;

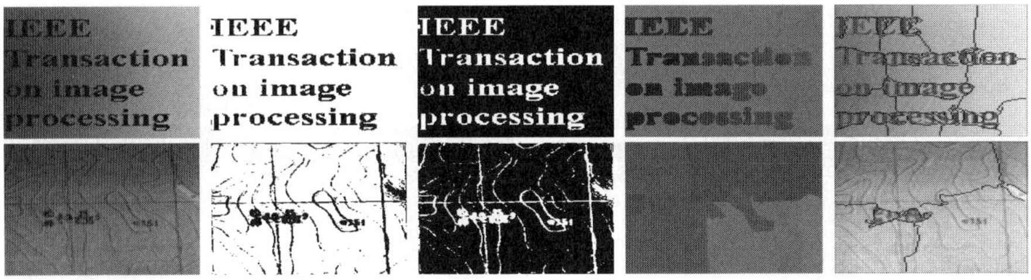

Fig. 4. Uneven lighting background isolation results. From left to right, the five columns are the input uneven images, results segmented by FPC (2nd-3rd columns, $s = 9; 9$ and $v = 19; 7$), the methods in [10] and [2] respectively.

(2) For $x \in X'$, let $\mathrm{CL}_x = \{x' \in X' | x' \in \mathrm{SN}(x)$ or Path $(x, x') \neq \emptyset\}$;
(3) For $y \notin X'$, if there exists $x \in X'$ such that $y \in \mathrm{SN}(x)$, y is merged into CL_x (when the number of such x is more than one, we randomly select one); if there is no $x \in X'$ such that $y \in \mathrm{SN}(x)$, all these y are pooled into one residual group.

In step (2), from the definition and properties of normalized metric $d(x, y)$, we have $\mathrm{CL}_x = \{y \in X | d(x, y) \leq 1/2\}$ which is a 1/2 closed-sphere, and $\mathrm{CL}_x \bigcap \mathrm{CL}_y = \emptyset$ if $x \neq y$. So, the dense set X' is partitioned into finite non-overlapping groups. Besides, if $y \in X' - \mathrm{CL}_x$, $x' \in \mathrm{CL}_x$, we have $\mathrm{Path}(x', y) = \emptyset$ and $y \notin \mathrm{SN}(x')$, thereby $d(x', y) = 1$. In fact, the maximal distance between objects in the transformed metric space is equal to 1. That is, our FPC yields the clusters such that the distances between objects belonging to the same cluster are as small as possible and the distances between objects belonging to the different clusters are as large as possible and even maximal. Therefore, CL_xs in step (2) are the solutions of the following maximum problem:

$$\max \sum_{i \neq j,\ i,j=1}^{K} d(\mathrm{CL}_i, \mathrm{CL}_j), \tag{2}$$

where $d(\mathrm{CL}_i, \mathrm{CL}_j) = \min\{d(x, y) : x \in \mathrm{CL}_i, y \in \mathrm{CL}_j\}$ and $\{\mathrm{CL}_i, i = 1, \cdots, K\}$ is a partition of X'. The number of clusters K is automatically determined by the adaptable selection of the parameters, e.g., k (the number of the nearest neighbors of the data) or s (the size of local neighborhood of the pixel) and v (the difference of the pixel values).

Complexity Analysis. The most time-consuming work of FPC is the search for all dense sets in Step 1. For data clustering, it is necessary to calculate the Euclidean distances between data points in the original input space to discover $k\mathrm{NB}(x)$ and R-$k\mathrm{NB}(x)$ which takes $\mathrm{O}(N^2)$. For image segmentation, the running time of computing all $\Omega_s^v(p)$ is $\mathrm{O}(MN)$, where $M = (2s + 1)^2, s \in \mathbb{Z}$. In general, $s \leq 12$ in our implementation and $M \ll N$, it takes nearly $\mathrm{O}(N)$ in proportion to the number of pixels. The serial procedure of generating clusters in Step 2-3 is similar to the method in [11]. For an arbitrary dense object x, all the

Fig. 5. Figure-ground separation results for three natural images coming from Berkeley database.From top to down, $s = 7, 4, 5$ and $v = 35, 20, 22$. The last two columns depict the results segmented by methods in [10] and [2] respectively.

objects within SN(x) are first connected with x to form the x' cluster, denoted as CL_x. And then the dense objects within SN(x) are found to extend CL_x. The process is continued until no dense object could be found and the CL_x is finally formed. In the same way, clusters of other rest dense objects could be yielded. Obviously, each object is scanned one time in such a serially recursive procedure, which results in a computation time O(N). So, the computational complexity of FPC is O(N), nearly linear in the number of elements in the transformed metric space, apart from the calculation of Euclidean distances in the original input space which is a compulsory work for most clustering methods.

4 Experimental Results

Although theoretically motivated, FPC is also outliers resistant and effective on several challenging clustering tasks. In this section, we first apply FPC with the metric derived from NDI_d and SN_d to perform some experiments on synthetic elongated non-convex data, three concentric circles (400, 800, 1200 points in respective circle), three spiral arms (each has 800 points) and two blocks (each has 200 points) within a circle (400 points). Our main concern in these experiments is the robustness against noise and outliers in the data. We intentionally add 50 outliers over the data in three varying noise levels as shown in Fig.1. From top to down, the three rows show the clustering results of the noisy data with 50 outliers located in the center, interior of the data and scattered all over the data. As expected, FPC is robust against the added outliers and gives the satisfactory results that appear to agree well with human judgement. The detected outliers are pooled into the residual group marked by azury roundlet. This shows that the normalized metric exploring the local density information is effective in reducing the influence of the outliers. In this application, the number of clusters is solely dependent on k which is chosen manually based on an intuitive expectation that which points would be in the same cluster or outliers. Fig.2 shows the results of a simple 3-circle data (20, 40, 60 points in respective circle) on a range of different

k values, from left to right $k = 2, 3, 4, 5, 6, 15$ respectively. As shown in the upper row, the points within the same cluster are connected by paths while outliers are isolated, where the dense points are floscular and the non-dense ones are hollow. The lower row demonstrates the corresponding clustering results with the different colorful marks.

Despite these promising results for elongated data clustering, we would like to test FPC with the metric derived from $\Omega_s^v(p)$ and $\Omega_s(p)$ on real-world image segmentation too. Although much effort has been done for image segmentation in the literature, there exist situations where those existing methods do not perform very well, e.g., the objects in the image are very tiny, long and thin, such as document words, contour map, grains of rice, granules of bacteria; the images are even severely degraded by uneven lighting, occlusion, poor illumination and shadow. In the interest of investigating whether our path-based normalized metric enables FPC to generate the accurate segmentation in these difficult problems as above, we thus compare FPC with spectral method [2] and minimum spanning tree algorithm [10] on such three types of images. One is the images containing granules of tiny objects, second is the uneven lighting images with long, thin lines and the last is the natural images coming from [13]. As shown in Fig.3-5, although different parameter (neighborhood size scale s and intensity difference scale v) values are required for different images, our segmented results indicate that FPC is effective in these problematic images: (1) each granule of bacteria and rice (except four smallest grains of rice located at the image boundary which vanish in the visual perspective) is correctly distinguished from the background; (2) the complex uneven lighting backgrounds are completely isolated from the long, thin contour map and document words; (3) the desired objects of interest in these natural images are extracted from the arbitrary scenes, realizing a figure-ground separation. In practice, the two scale parameters s and v are adjusted manually to achieve a set of coherent neighboring pixels which maximize the probability of being a single image content. In effect, an adaptive neighborhood size s reported in this application is in the range of 3-12, while v is adjusted nearby the mean intensity contrast of each pixel's given square neighborhood. While more desirable, dynamically finding optimal parameter values for each image to create the best segmentation results remains an open problem.

5 Conclusion

Utilizing the density information sufficiently and explicitly, we propose a novel normalized metric which reliably measures the dissimilarities between elements in the normalized metric space. The proposed metric obeys the metric axioms, i.e., positive definite property, symmetry and triangular inequality. The main theoretical contribution of this paper lies in the strict mathematical proof of the metric axioms. Based on this sound metric, we devise a fast path-based clustering (FPC) algorithm for elongated data analysis and image segmentation. Apart from the calculation of Euclidean distances between objects in the original input space, FPC has a nearly linear computational complexity in the number of elements in the transformed metric space. Extensive experimental results on the elongated structure extraction, uneven lighting background isolation, grains of tiny objects segmentation and figure-ground separation, have showed that FPC with the proposed metric often performs well for these difficult tasks and is outliers resistant enough, effective and efficient.

Acknowledgments. This work is supported by Nature Science Foundation of China 60632050.

References

1. Rattray, M.: A Model-Based Distance for Clustering. In: International Joint Conference on Neural Networks, pp. 13–16. IEEE Computer Society Press, Washington (2000)
2. Shi, J., Malik, J.: Normalized cuts and image segmentation. Pattern Analysis and Machine Intelligence 26, 888–905 (2000)
3. Andrew, Y., Ng, J., Weiss, Y.: On Spectral Clustering: Analysis and an algorithm. In: 15th Conference on Advances in Neural Information Processing Systems. MIT Press, Vancouver (2001)
4. Fischer, B., Zoller, T., Buhmann, J.M.: Path based pairwise data clustering with application to texture segmentation. Energy Minimization Methods in Computer Vision and Pattern Recognition 2134, 235–250 (2001)
5. Fischer, B., Buhmann, J.M.: Path-based clustering for grouping of smooth curves and texture segmentaion. IEEE Trans. Pattern Analysis and Machine Intelligence 25, 513–518 (2003)
6. Chang, H., Yeung, D.-Y.: Robust path-based spectral clustering with application to image segmentation. In: 10th IEEE International Conference on Computer Vision, pp. 278–285. IEEE Computer Society Press, Beijing (2005)
7. Omer, I., Werman, M.: The Bottleneck Geodesic: Computing Pixel Affinity. In: IEEE Computer Society Conference on Computer Vision and Pattern Recognition, pp. 1901–1907. IEEE Press, New York (2006)
8. Omer, I., Werman, M.: Image Specific Features Similarities. In: 9th European Conference on Computer Vision, pp. 321–333. Springer Press, Graz (2006)
9. Fowlkes, C., Belongie, S., Chung, F., Malik, J.: Spectral Grouping Using the Nystrom Method. IEEE Trans. Pattern Analysis and Machine Intelligence 26, 214–225 (2004)
10. Felzenszwalb, P., Huttenlocher, D.: Efficient Graph-Based Image Segmentation. Interantional Journal of Computer Vision 59, 167–181 (2004)
11. Ding, J., Chen, S., Ma, R., Wang, B.: A Fast Directed Tree Based Neighborhood Clustering for Image Segmentation. In: 13th International Conference on Neural Information Processing, pp. 369–378. Springer Press, Hong Kong (2006)
12. Zhou, S., Zhao, Y., Guan, J., Huang, J.: A Neighborhood-Based Clustering Algorithm. In: 9th Pacific-Asia Conference on Knowledge Discovery and Data Mining, pp. 361–371. Springer Press, Hanoi (2005)
13. Martin, D., Fowlkes, C., Tal, D., Malik, J.: A Database of Human Segmented Natural Images and Its Application to Evaluating Segmentation Algorithms and Measuring Ecological Statistics. In: 8th IEEE International Conference on Computer Vision, pp. 416–425. IEEE Computer Society Press, Vancouver (2001)

Association Rule Mining Based on the Semantic Categories of Tourism Information*

Yipeng Zhou[1,2], Junping Du[3], Guangping Zeng[1], and Xuyan Tu[1]

[1] Information Engineering School, University of Science and Technology Beijing, Beijing 100083, China
[2] School of Computer Science, Beijing Technology and Business University, Beijing 100037, China
[3] Beijing Key Lab of Intelligent Telecommunication Software and Multimedia, School of Computer Science and Technology, Beijing University of Posts and Telecommunications, Beijing 100876, China

Abstract. It is difficult for traditional data mining algorithms to mine semantic information from text set because of its complexity and high dimension. To solve this problem, the semantic categories of words appearing in tourism emergency reports are studied, and a semantic association rule mining algorithm is presented based on these categories. Association words are also gained from these rules, which can better describe the semantic contents of the texts. Quantum-inspired genetic algorithm is utilized to improve the effectiveness of rule-searching process. Experiments show the better results than traditional methods.

Keywords: Association rule, tourism emergency, genetic algorithm, text mining.

1 Introduction

As tourism is developing rapidly in our country, the load of service facilities is growing too and tourism emergencies happened frequently. Therefore, it is very necessary for us to gain tourism emergency information and analyze its patterns [1]. It is also an important basis of making decision and management. However, tourism emergency is a special concept, which is difficult to be described as a few topic words, so keyword based web searching algorithm can't be used directly[2]. We need a method to find semantic features of different kinds of tourism emergency information. An semantic association rule mining algorithm based on word's category is presented in this paper, which is used to find semantic relationships of characteristic words, which belong to five categories: object, environment, activity, event and result. To achieve a better performance and accuracy, quantum-inspired genetic algorithm is introduced into rule-searching process[3]. By using qubit coding and quantum gate transformation, the population of candidate rules is enlarged and the diversity of solutions is also ensured.

* This work is supported by National Natural Science Foundation of China (No. 60773112) and Beijing Natural Science Foundation (No.4082021, No.4072018).

F. Sun et al. (Eds.): ISNN 2008, Part II, LNCS 5264, pp. 67–73, 2008.

2 Semantic Association Rules of Tourism Emergency Information

Tourism Emergencies include many different types of events, such as traffic accident, security accident, and natural disaster and so on. Usually they are reported using different characteristic words, so it is difficult to describe them as a common keyword. By analyzing the reports, these words usually can be classified into five sets: object, environment, activity, event and result, which describe the contents of tourism emergencies from different aspects. In this paper, characteristic words are extracted from training text set and classified into five word sets as in table 1.

Table 1. Category of characteristic words

Category	Num. of words	Num. of concepts	Description
Object	123	51	Objects involved in the emergencies, such as 'tourist', 'vehicle'.
Environment	133	62	Geography or nature environment where the emergencies happened, such as 'strong wind', 'heavy rain'.
Aactivity	56	38	Activities of the objects, such as 'tour', 'taking photos'.
Event	209	146	Words which describe the emergency events, such as 'drowning'.
Result	41	27	The results of the events, such as 'injured', 'dead'.

In order to reduce the number of characteristic words, similar words in each word set are represented by a concept (sememe). Word similarity is calculated based on Hownet[4]. For two Chinese words w_1 and w_2, if w_1 has n concepts (sememes): s_{11}, s_{12}... s_{1n} , and w_2 has m concepts: s_{21}, s_{22}...s_{2m}, the similarity between w_1 and w_2 is defined as maximal sememe similarity [5] :

$$Sim(w_1, w_2) = \max_{i=1...n, j=1...m} Sim(s_{1i}, s_{2j}) \tag{1}$$

The similarity between two concepts is defined as follows:

$$Sim(s_1, s_2) = \frac{\alpha}{d + \alpha} \tag{2}$$

where d is the distance between s_1 and s_2 in hierarchy of Hownet, α is an adjustable parameter.

Mining association rules of the characteristic words in these five sets is useful to find their semantic relationships [6]. To avoid items appearing in association rules belong to a same category, association rules based on semantic categories is defined. We use W to denote the set of characteristic words, $W = \{w_1, w_2, \dots, w_n\}$, in which each word w_i belongs to a category K_{wi}, and T_i to denote the feature of text i. If the word appearing in text i is similar to a characteristic word w_j, the characteristic word will be entered into T_i, so $T_i \subseteq W$. D is denoted as the database of texts, $D=\{T_1, T_2, \dots, T_m\}$.

Define $A \rightarrow B$ as a semantic association rule based on category, where $A \subset W$, $B \subset W$, $A \cap B = \varnothing$, and for any two item u, v in A and B, $K_u \neq K_v$. For database D, we use s to denote the support of rule $A \rightarrow B$:

$$s = P(A \bullet B) \tag{3}$$

And c to denote confidence of the rule:

$$c = P(B|A). \tag{4}$$

According to this definition, semantic association rule based on category is better than traditional one which has same confidence. For example, rule 1 (*{vehicle, reverse} => tourism traffic accident*) is better than rule 2 (*{ tourist, vehicle} => tourism traffic accident*). Because "*vehicle*" and "*reverse*" belong to different semantic categories which include more semantic contents, but "*tourist*"' and "*vehicle*" both belong to "object" category with less information.

The association rules mining is also a search process [7]: data set is the search space, algorithm is the search policy. Quantum-inspired genetic algorithm (*QGA*) is a remarkable research area in recent years, and it is fit for solving search and optimization problems. In *QGA*, qubits are used to encode chromosomes, thus a more abundant population is obtained. Moreover, quantum transformation can easily make use of the best individual's information to control the variation of Q-chromosome, and make the population evolve towards the direction of excellent patterns.

3 Association Rules Mining Based on QGA

In QGA, a m qubits chromosome can be defined as: $\begin{bmatrix} \alpha_1 \\ \beta_1 \end{bmatrix} \begin{vmatrix} \alpha_2 \\ \beta_2 \end{vmatrix} \cdots \begin{vmatrix} \alpha_m \\ \beta_m \end{vmatrix}$

where α, β represent the probability amplitudes of the corresponding state, and $|\alpha|^2, |\beta|^2$ give the probability respectively that the qubit will be found in '0' state or in '1' state, $|\alpha|_i^2 + |\beta|_i^2 = 1$, $i = 1,\ldots,m$. The definition can represents any linear superposition of the states.

When generating the Q-chromosome of a candidate association rule ($A \rightarrow B$), encode each item in it as a qubit. If the observation state of a qubit is '1', that means the corresponding item will appear in the rule. The population of QGA is composed of Q-chromosomes, and the generation t, $Q(t)$, is defined as $\{q_1^t, q_2^t, \ldots, q_n^t\}$, where n is the population size; m is the length of chromosome; q_i^t is an individual chromosome in the population:

$$q_i^t = \begin{bmatrix} \alpha_{i1}^t \\ \beta_{i1}^t \end{bmatrix} \begin{vmatrix} \alpha_{i2}^t \\ \beta_{i2}^t \end{vmatrix} \cdots \begin{vmatrix} \alpha_{im}^t \\ \beta_{im}^t \end{vmatrix}, \qquad i = 1,2,\ldots,n$$

We use the following evolution procedure to obtain valid association rules:

(1) Initialization
Determine the size of population and number of qubits, and construct the initial population $Q = \{q_1, q_2, \ldots, q_n\}$, where the probability amplitudes of all qubits are set to $1/\sqrt{2}$, that means there are the same probabilities of superposition at the beginning of evolution.

(2) Generate Observation State P(t)
Create observation state *P(t)* for the population *Q(t)* from probability amplitudes of each individual in it: $P(t) = \{p_1^t, p_2^t, \cdots, p_n^t\}$, where p_i^t (i = 1 ,2 , ..., n) is the observation state of each individual, it is a m-bits binary group. The method of generating *P(t)* is as follows:

For each qubit $\begin{bmatrix} \alpha_{ij}^t \\ \beta_{ij}^t \end{bmatrix}$ of q_i^t, generate a random number r, $r\notin[0,1]$. If $r < |\alpha|_i^2$, the corresponding observation value is '0', otherwise it is '1'.

(3) Calculate Fitness
Calculate individual's fitness from its observation state:

$$f(A \rightarrow B) = s(A \rightarrow B) + c(A \rightarrow B) \tag{5}$$

Thus we can evaluate the rule's match with data set and get better individuals from the evaluation. During the evolution period, individual relies on its own fitness to compete with others. And individuals which fulfill the definition of semantic association rule based on category will have more chances to live.

(4) Selection
Select the best l individuals from current generation according to fitness. Then, compare them with the best ones from last generation and retain the better l individuals to consist of best solution set. At that time, if stop condition is met, the algorithm terminates; otherwise, go to next step.

(5) Variation of Q(t)
Apply quantum gate transformation to individuals of *Q(t)* and get new population *Q(t+1)*. In this paper, the quantum rotation gate G is applied [8].

$$G(\theta) = \begin{pmatrix} \cos(\theta) & -\sin(\theta) \\ \sin(\theta) & \cos(\theta) \end{pmatrix} \tag{6}$$

where θ is the rotation angle:

$$\theta = k \cdot h(\alpha, \beta) \tag{7}$$

k is a coefficient related to convergence, and it is 0.02π here. $h(\alpha, \beta)$ is a direction function pointing to the direction of the best solution. Table 2 shows the value of the direction function. In this table, p_{ij} is the j-th bit of current individual's observation state p_i . b_{kj} is the j-th bit of the best individual b_k's observation state. b_k is randomly

Table 2. Value of $h(\alpha, \beta)$

p_{ij}	b_{kj}	$f(p_i)>f(b)$	$h(\alpha_{ij}, \beta_{ij})$			
			$\alpha_{ij}=0$	$\beta_{ij}=0$	$\alpha_{ij}\cdot\beta_{ij}>0$	$\alpha_{ij}\cdot\beta_{ij}<0$
0	0	False	0	0	0	0
0	0	True	0	0	0	0
0	1	False	0	0	0	0
0	1	True	±1	0	-1	+1
1	0	False	±1	0	-1	+1
1	0	True	0	±1	+1	+1
1	1	False	0	±1	+1	-1
1	1	True	0	±1	+1	-1

selected from best solution set. f is the fitness function, and α_{ij}, β_{ij} are probability amplitudes of current individual.

By applying quantum rotation gate to each individual's probability amplitude[5], we can get a new population $Q(t+1)$ from $Q(t)$, as follows:

$$q_i^{t+1} = G(t) \cdot q_i^t \tag{8}$$

Where $G(t)$ is the rotation gate of the generation t, and q_i^t, q_i^{t+1} are respectively the probability amplitudes of each individual in generation t and t+1. The transformation allows a shift of each qubit in the direction of the corresponding bit value in the best solution. Thus, there is a great probability that the population will evolve towards multi optimal patterns with higher fitness.

(6) Migration

After g generations, a migration is performed. Firstly, generate a new population as initialization step. Then, choose some qubits randomly from the best individual of this population to substitute the best individual's of former population. In such way, a global optimal solution might be found.

(7) Count of generations increased, then go to step 2.

4 Experiment Results

In our experiments, the text set consists of 1500 tourism emergency reports and these reports belong to four types - traffic accident, security accident, public order event and natural disaster. Then these texts are divided into training set (including 1000 texts) and testing set (including 500 texts). Characteristic words (seen as table 1) are extracted from the training set. And the proposed algorithm is applied to mine association rules between characteristic words and emergency types. Then these rules are used to predict the types of the texts on the testing set and compute the correctness. Table 3 gives the results: the extraction rate (number of effective rules/ total number of mined rules) and the correctness. The results from classical method with genetic algorithm are also shown in table 3 for comparison. The results show that the algorithm presented in this paper can get semantic rules effectively with higher extraction rate and correctness. In an association rule with high confidence, the words appearing in it described the concepts of this type of emergency events. So several small association words sets are gained from these rules, and are used to search for tourism emergency reports. Some searching results are also shown in table 4. The results show that these association words can better describe the semantic contents of these events.

Table 3. Results of rules mining

Threshold of support	Threshold of confidence	Algorithm	Extraction rate	Correctness
0.1	0.8	QGA	0.61	0.86
		GA	0.49	0.78
0.2	0.8	QGA	0.79	0.91
		GA	0.64	0.85

Table 4. Results of web searching using association words sets

Association words set	Num. of searching results	Correctness
{ tourist, drowning}	110	0.96
{ tourist, surrounded}	369	0.95
{tourist, get lost}	134	0.86
{ tourist, failure, injured}	26	0.73

5 Conclusion

In this paper, a new algorithm for mining semantic association rules is proposed, which introduces quantum-inspired genetic algorithm into data mining. This algorithm divides characteristic words into several categories according to their semantic and tries to find association rules between them. *QGA* is used to search rules by coding the candidate rules to quantum-inspired chromosomes. Inducted by the best solution set, the population evolves towards multi-objective directions. Different types of association rules can be found. The experiments demonstrate that the semantic association rule mining based on category can obtain better results from complex text set of tourism emergences. Association words can also be gained from the rules for web searching or information extraction.

References

1. Guo, W.S., Du, J.P., Yin, Y.X.: Analysis Method for Holiday Tourism Information Based on Data Mining. In: The Third International Conference on Computational Intelligence, Robotics and Autonomous Systems, Singapore (2005)
2. Hulth, A., Megyesi, B.: A Study on Automatically Extracted Keywords in Text Categorization. In: International Conference of Association for Computational Linguistics, Sydney, Australia (2006)

3. Talbi, H., Draa, A., Batouche, M.: A New Quantum-Inspired Genetic Algorithm for Solving the Traveling Saleman Problem. In: IEEE International Conference on Industrial Technology, pp. 1192–1197 (2004)
4. Sun, J.G., Cai, D.F., Lv, D.X., Dong, Y.J.: HowNet Based Chinese Question Automatic Classification. Journal of Chinese Information Processing 21, 90–95 (2007)
5. Xia, T.: Study on Chinese Words Semantic Similarity Computation. Computer Engineering 6, 191–194 (2007)
6. Zhang, W.D., Yi, Y.H.: To Construct the Set of Synonyms and Association Words Using Latent Semantic Analysis and the Mining of Association Rules. Computer Engineering & Science 29, 103–104, 116 (2007)
7. Ren, J.: Study on Classification Association Rules Mining and Its Application in Complicated Industry Process. Dissertation Zhejiang University for the Degree of Doctor of Philosophy (submitted, 2006)
8. Zhou, Y.P., Du, J.P., Zuo, M., Tu, X.Y.: Application of Quantum-Inspired Genetic Algorithm for Mining Tourism Emergency Association Rules. In: China-Ireland International Conference on Information and Communications Technologies, Ireland (2007)

The Quality Monitoring Technology in the Process of the Pulping Papermaking Alkaline Steam Boiling Based on Neural Network*

Jianjun Su, Yanmei Meng, Chaolin Chen, Funing Lu, and Sijie Yan

Guangxi University 530004, Nanning, China
gxu_mengyun@163.com

Abstract. On the status quo that being lack of the testing equipment which gives reliable and direct parameters on measuring the quality of pulp in the cooking process, this article focus on the lignin value soft-measurement technology in the pulp and papermaking process. The pulp lignin value soft-measurement model is built basing on artificial neural network; It takes cooking process temperature, cooking time and the effective alkaline concentration as network input, and an improved BP algorithm to train the network for obtaining the predictive output value of the lignin value. Utilizing online measurement of cooking process temperature, cooking time and effective alkaline concentration, the soft-measurement model can monitor the quality of pulp.

Keywords: Neural network, Pulp and papermaking, Soft-measurement model.

1 Introduction

Pulp and papermaking cooking process is a complex and multi-reaction process, the main purpose of the automatic control in the cooking process is to produce pulp with certain quality (using Kappa Number or lignin as evaluation) and uniformity. Due to some technical reasons, it is difficult to compose the conventional quality control system by using Kappa Number or lignin value as adjustable variable. Aim at dealing with this technical problem of directly monitoring the quality parameters in the pulp and papermaking process, domestic and foreign research mainly concentrated on two aspects, one is to develop analytical instrument and online sensor which can obtain quality parameters directly from the pulp cooking pot; the other is to actively explore the pulp quality parameters soft-measurement model which can be used in the actual production. Because online sensor which developed abroad is expensive and difficult to maintain, it does not suit for China's pulp and papermaking industrial status quo. This paper bases on artificial neural network technology, and researches on estimating value of pulp lignin through testing cooking process temperature, cooking time and effective alkaline concentration to explore soft-measurement model that can play a role on practical industrial applications to monitor the quality of the pulp.

---

* The research is supported by Guangxi Province Nanning Key Technology Research Project (200501007A).

F. Sun et al. (Eds.): ISNN 2008, Part II, LNCS 5264, pp. 74–80, 2008.

2 Soft-Measurement Model of the Lignin Value in the Process of the Pulp and Papermaking Steam Boiling

2.1 Structural Design of the Lignin Value of Neural Network Model

Pulp and papermaking cooking process is complex and nonlinear, it is not only difficult to be described with mathematical model, but also hard to achieve the online direct measurement of Kappa Number. But a group of input and output data can be got from online measurement of cooking process pulp temperature, cooking time, effective alkaline concentration, and the pulp quality parameters got by laboratorial offline sampling measurements. This paper establishes the soft-measurement model through 60 groups of data provided by a paper mill about cooking process pulp lignin value, cooking process temperature, cooking time and the effective alkaline concentration. According to the experimental data, the temperature, time and effective alkaline concentration these three parameters are taken as input, and the value of lignin is taken as output. The neural network structure shows like Fig.1.

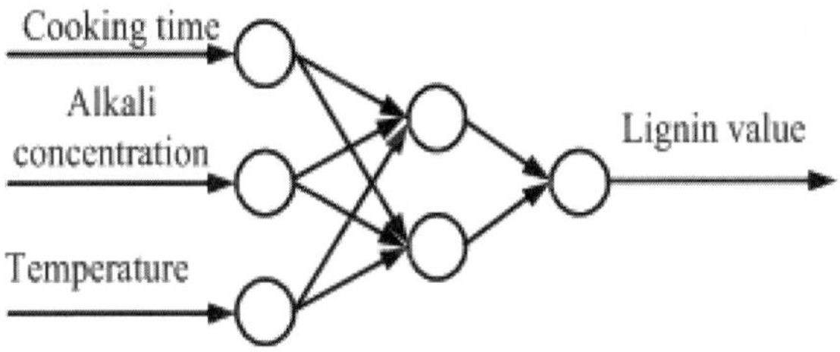

Fig. 1. Neural network model of cooking process

2.2 Data Fusion Process Based on Neural Network

When the neural network is being used in multi-sensors data fusion, firstly, the acquired parameters should be selected according to the specific requirement of the system, then sensors need to be chose suitably and the sensors' data should be dealt with appropriately. After that, a suitable neural network model should be chose according to the characteristic of the data, which including network topology, characteristic of neuron and rules of learning. At the same time, it is also necessary to establish the relationships between input and sensors' information, as well as output and the decision-making system. Finally, the obtaining sensors' information and the corresponding system's decision-making need to be studied; the distribution of weight value should be determined and the network's training needs to be completed. The trained neural network can participate in the actual fusion process, as showing in the Fig.2. The obtaining sensors' data firstly pass through the proper process 1, then the result is used as a neural network input. After that neural network would process the input, and the output is exported to the related structure. Finally, Process 2 would explain it as the material decision-making of the system.[1-4]

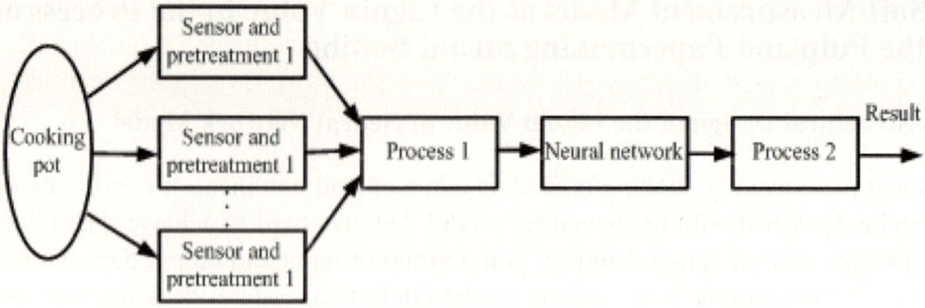

Fig. 2. Data fusion process based on neural network

2.3 Neural Network Parameters

Learning speed decides the change of weight value in each of circular training. High learning speed may lead to the system's less stability; low learning speed will result in a longer training time and a lower convergence speed, but it can guarantee the error small enough to jump out of the error table's bottom, eventually go to the smallest error. Therefore, in normal cases, a lower learning speed tends to be chose to ensure system's stability. Usually, the learning speed goes up and down between 0.01 and 0.8. The learning speed used in this article is 0.1.

The sample usually be divided into two separate set, one is training set and the other is testing set. The training set is used to estimate model, while the testing set is used to check out how capable the ultimate selecting optimal model is. All the sample set data will be handled by random method, a part of which will be used as training set data—training data, and the other part will be used as sample set data for detecting network performance—detecting data. The proportion of training set and sample set is 2:1. According to the experiential rule, the number of training sample is 5 to 10 times of the total number of connecting weight. There are all 60 samples in this subject, 40 of them are training set data, and 20 are detecting set data.

Kolmogorov theorem points out that arbitrary a continuous function can be approximated by only one hidden layer network in arbitrary precision, in other words, three-tiered neural network can approximate any continuous function. Therefore this article adopts the structure of three-tiered neural network, transfer function of the hidden layer and the output layer both are the S-type function.

The initial hidden nodes are decided by the fitting formula which is obtained from the least squares method.

$$S = \sqrt{0.43mn + 0.12n^2 + 2.54m + 0.77n + 0.35} + 0.51 \tag{1}$$

The calculation is rounded according to the rounding law (in the equation (1), S is hidden nodes, m is the nodes of input layer, n is the nodes of output layer). This paper uses three-input and single-output network structure, then m=3, n=1, S=3.6959, S is 4 after rounding, so the initial number of hidden layer neuron is 4. Hidden nodes which are obtained from the empirical formula accord with the most applied examples, they exceed slightly generally. In the preliminary stage of determinate network topology, the hidden nodes are decided by the empirical formula in advance, then they are estimated to see if there are more hidden nodes or not. According to the empirical formula [5-7]

$$m = \sqrt{n + l} + \alpha \tag{2}$$

or

$$m = \sqrt{nl} \tag{3}$$

(in the equation (2) and (3), m is hidden nodes, n is the nodes of input layer, ιis the nodes of output layer, α is an integer between 1 and 10) to determine the hidden nodes between 2 and 12. Because the initial hidden nodes are 4, the determined hidden nodes are between 2 and 4. Hidden nodes can be validated by the test method which firstly tests the hidden layer nodes from the smaller unit, then trains and tests the network performance, increases the hidden nodes slightly, lastly repeats training and testing.

After testing, the hidden nodes used in this article are 2, 0.1710 is taken as the target error, and 10,000 as the maximum training number. In order to objectively evaluate the network convergence, the subject uses the unitary error function as the network performance evaluation function.

$$E = \frac{1}{mn}\sqrt{\sum_{p=1}^{m}\sum_{j=1}^{n}(d_{pj} - y_{pj})^2} \tag{4}$$

When values of m and n are different, the values of E are also different.

2.4 Analysis of Simulation Results

This paper uses the neural network toolbox of MATLAB for testing. In addition, a procedure is compiled as the main program, and a neural network program which is called for comparing is compiled as the subroutine. Then the main program would transfer the subroutine 1,000 times, and then the testing error's mean value will be got base on the 1,000 times calculation. The result of running the comparing neural network program 1,000 times is quite objective, because both the neural network's weight value's initialization and the neural network's threshold value's initialization are random numbers. The neural network input data of testing sample will be put into the trained neural network, the result of neural network is obtained as show in the table 1.

Table 1. Test results of neural network toolbox function

Program	Traingd	Trainlm
Average testing error	0.0293	0.0334
Required time	7.1328e+004	9.0296e+004

Fig.3 is the error curve of traingd, testing error is 0.0293, the required testing time is 71.3280. Fig.4 is the error curve of trainlm, testing error is 0.0334, the required testing time is 90.2960.

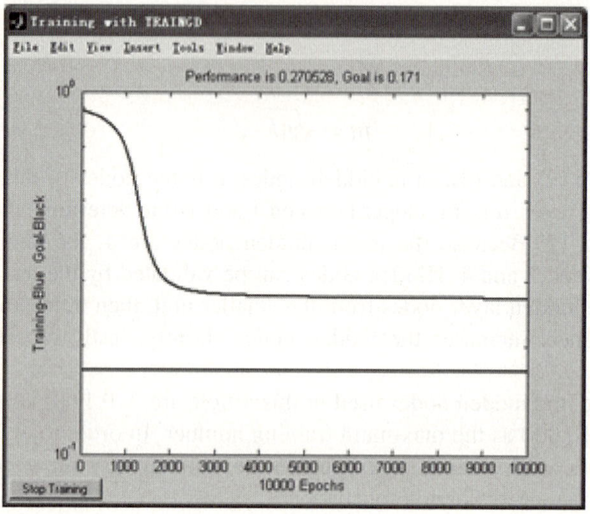

Fig. 3. Error curve of traingd

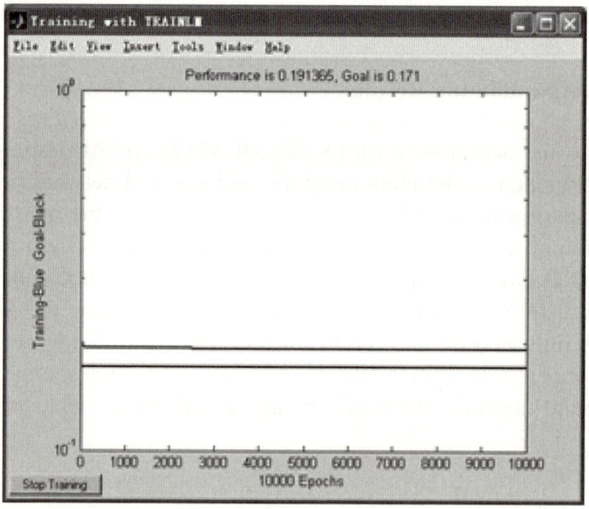

Fig. 4. Error curve of trainlm

The results of simulation show that the neural network model, which has been trained and tested, is feasible to forecast the lignin value in the process of the pulp and papermaking steam boiling.

2.5 Experimental Verification

In order to confirm whether the trained neural network can capture the relationships between the value of the pulp lignin and the parameters which concern about the temperature in the process of the pulping and papermaking, the cooking time and the alkali concentration, 8 groups of experimental data have been tested. The experimental results are showed in Table 2.

Table 2. Experimental results

Quantity value \ Measuring points			1	2	3	4	5	6	7	8
Input value	Heating or insulation (^{0}C)		100	125	150	175	175	175	175	175
	Heating Time (min)		98	132	170	228				
	Insulation time (min)					0	25	50	75	100
	Alkali concentration (Na_2O g/L)		47.20	44.15	41.23	37.51	33.72	32.86	33.16	31.89
Lignin value (%)	To raw material	Predictd value	27.31	24.82	20.33	2.61	1.30	0.73	0.65	0.62
		Observd value	27.43	24.95	20.43	2.72	1.40	0.79	0.68	0.65
	To sample without alcohol benzene	Predictd value	30.31	30.90	27.85	6.31	2.61	1.75	1.71	1.67
		Observd value	30.12	30.79	27.71	6.25	2.54	1.69	1.65	1.61

In the table, the temperature, the cooking time and the alkali concentration are measured online, the value of the pulp lignin is predicted by the trained neural network, and the observed value is tested by sample testing to the pulp in the laboratory, the time is the same as in the factory. The results show that the trained neural network can catch the rule between the value of the pulp lignin and the variational temperature, cooking time and alkali concentration in the process of the pulping and papermaking, and a better forecasted value of the pulp lignin can be got.

3 Conclusion

This article study the software measurement model, which is constructed by the BP neural network, to control the quality parameter of lignin value in the process of the pulping and papermaking steam boiling. The emulational and experimental results show that the established BP neural network model is feasible to forecast the lignin value in the process of the pulping and papermaking steam boiling. With high accuracy, good performance and better application prospect, the forecast lays the foundation for the soft measurement mode, which monitors the pulp quality, to play a role in the practical application of industry.

References

1. Gao, D.Q.: Research on the Structure of Prior to the Three-tiered Neural Network with Preceptorial Basic Linear Function. Computer Journal 21(1), 3, 80–86 (1998)
2. Hou, B.P., Lu, P.: Based on MATLAB BP Neural Network Modeling and Simulation System. Automation and Instrumentation 16(1), 34–36 (2001)
3. Du, Q.D., Xu, L.Y., Zhao, H.: Identification on Hydropower Plant Pressure Water System Based on Neural Fusion Algorithm. Control and Decision-making 16, 787–789 (2001)
4. Wang, X., Wang, H., Wang, W.H.: The Theory and Application of Artificial Neuron Network. Northeastern University Publisher, Shenyang (2000)
5. Carpenter, W.C., Hoffman, M.E.: Guidelines for the Selection of Network Structure. Artificial Intelligence for Engineering Design Analysis and Manufacturing 11(5), 395–408 (1997)
6. Hagan, M.T., Demuth, H.B., Beale, M.H.: Neural Network Design. Machinery Industry Publisher, Beijing (2005)
7. Li, J.C., Huang, H.X.: Research LMBP Algorithm to Improve the Convergence Speed in Neural Network. Computer Engineering and Application 16, 46–49 (2006)

A New Self-adjusting Immune Genetic Algorithm

Shaojie Qiao[1,2], Changjie Tang[1], Shucheng Dai[1,⋆], Mingfang Zhu[1], and Binglun Zheng[1]

[1] School of Computer Science, Sichuan University, Chengdu 610065, China
[2] School of Computing, National University of Singapore, 117590, Singapore
qiaoshaojie@cs.scu.edu.cn

Abstract. The genetic algorithm based on immunity has recently been an appealing research methodology in evolutionary computation. Aiming to cope with the problems of genetic algorithms, i.e., the solution is apt to trap into a local optimum and the convergence speed is slow, this paper proposes a new self-adjusting immune genetic algorithm, called SaiGa (**S**elf-**a**djusted **i**mmune **G**enetic **a**lgorithm), which seeks for an optimal solution with regard to complex problems such as the optimization of multidimensional functions by automatically tuning the crossover and the mutation probabilities, which can help avoid prematurity phenomena and maintain individual diversity. In particular, SaiGa introduces a variable optimization approach to improve the precision in terms of solving complex problems. The empirical results demonstrate that SaiGa can greatly accelerate convergence for finding an optimal solution compared with genetic algorithms and immune algorithms, achieve a better precision in function optimization, and avoid prematurity convergence.

Keywords: Immune genetic algorithms, Genetic algorithms, Self-adjusting.

1 Introduction

Most recently, there has been a growing interest in the investigation of natural phenomena such as evolution, heredity, and immunity [1]. Simple genetic algorithm (SGA) has already been extensively studied and widely used in a variety of practical applications such as machine learning [2], intelligent robot control [3], time series predication [4], and process planning [5]. The important genetic operators, i.e., crossover and mutation, provide opportunity for each individual to guarantee and optimize its evolutionary tendency upon the mechanism of *the survival of the fittest* [1]. However, these two genetic operators can make individuals evolve arbitrarily and finally lead to degeneracy. In addition, the generic crossover and the mutation operators lack the capability of meeting an actual situation and often neglect the assistant of prior knowledge [1], i.e., it cannot utilize the system information with a guarantee of finding a global optimum during evolutions.

⋆ Communication author.

F. Sun et al. (Eds.): ISNN 2008, Part II, LNCS 5264, pp. 81–90, 2008.

Artificial Immune System (AIS) is a computational system that is inspired by the processes of the vertebrate immune system [6]. It provides some evolutional learning mechanism such as noise tolerance, unsupervised learning, self-organizing [7]. In particular, it combines the advantages of classifiers, neural networks, machine reasoning systems, etc. In general, it has the potential capability of handling complex problems. Due to the inherent characteristics w.r.t. AIS, i.e., improving the capability of finding an optimal solution and maintain individual diversity, it has grown up to be a research hotspot in artificial intelligence area following artificial neural networks and evolutionary computation [7]. However, the existing immunity-based algorithms perform well only on coping with some specific problems. Immune Genetic Algorithm (IGA) [1] is a complement to genetic algorithms by combining the advantages of genetic algorithms and immune algorithms. Most specifically, IGA uses the immune regulation mechanism to achieve a dynamic balance between individual diversity and population convergence [8].

In this study, we make the following contributions.

1. We propose a new self-adjusting crossover and mutation operators, which can automatically adjust the crossover and the mutation probabilities that help avoid trapping into a local optimum and maintain individual diversity.
2. We analysis and optimize variables that affect the immune operators in order to adjust the suppressive threshold and the incentive value between antibodies and antigens.
3. A series of experiments are performed to demonstrate that SaiGa is a better solution to the function optimization problem.

2 Related Work

Immunity genetic algorithm is a new biology methodology which combines the principles of immunity in life sciences and genetic algorithms to improve the overall performance. In addition, IGA can utilize the information inherited from problems to suppress the degeneration phenomenon via variable optimization, and simulate immunity features to improve the performance of genetic algorithms. An influential IGA is proposed by Wang et al. [9]. The performance optimization of this algorithm is more powerful than SGA, but the mutation operator is lack of self-adaptive ability.

In general, the individuals defined in the IGA algorithm apply the binary encoding approach. For multi-dimensional function optimization issues, the binary encoding is faced with the problem of "dimension catastrophe". Mo et al. [10] encoded each individual in real number. However, this approach applies the mutation operator to entire antibodies, instead of reserving predecessors with high affinity via previous evolutions. In [11], the authors proposed an improved clonal selection algorithm, namely GLONALG. It explicitly takes into consideration the affinity maturation of the immune response and it is suitable for coping with optimization problems. But it does not consider the effect of relevant variables that play an essential role in immune operations. In order to optimize these important variables, Li et al. [12] analyzed the characteristics of two influential immune algorithms, i.e., aiNet [13] and RLAIS [14], and proposed a novel self-adaptive artificial immune algorithm. However, this algorithm does not take into account the multiple crossover operation, and it cannot guarantee the runtime efficiency as well.

3 A Self-adjusting Immune Genetic Algorithm

In this section, we first define the self-adjusting immune genetic algorithm and propose an IGA schema. Base on this schema, we introduce the self-adjusting crossover and mutation operations and analysis their effect in immune genetic algorithms.

The basic idea of immune genetic algorithms is to integrate the immune operation into genetic algorithms. The immune operation consists of two phases as: vaccination and immune selection [1]. In this study, we propose an IGA schema defined as follows, which includes the main components of an IGA.

Definition 1. *(Self-adjusting immune genetic algorithm—SaiGa). SaiGa is defined as a 9-tuple as follows:*

$$IGA\_Scheme = (C, F, P_0, S, \alpha, \beta, \varphi, \vartheta, T)$$

$$\begin{aligned} where: C &\longrightarrow encoding\ method;\\ F &\longrightarrow fitness\ function;\\ P_0 &\longrightarrow initial\ population;\\ S &\longrightarrow population\ cardinality;\\ \alpha &\longrightarrow selection\ operator;\\ \beta &\longrightarrow self-adjusting\ crossover\ operator;\\ \varphi &\longrightarrow self-adjusting\ mutation\ operator;\\ \vartheta &\longrightarrow immune\ operator;\\ T &\longrightarrow termination\ condition; \end{aligned}$$

SaiGa is defined beyond the IGA introduced in [1] and integrates self-adjusting crossover and mutation operators. In particular, the immune operator adopts the immune selection algorithm based on similarity and vector distance between individuals proposed in [15].

3.1 Algorithm Statement

In SaiGa, all individuals use a binary encoding, and apply one-point and two-point crossover operators simultaneously in order to maintain individual diversity without missing better individuals with high fitness values due to multi-crossover operations [16]. For easy understanding, we first describe some important symbols and their meanings. The crossover is achieved by our proposed self-adjusting crossover probability, denoted as P_c. While in mutation, we use the self-adjusting mutation probability, denoted as P_m. The state transition phases in SaiGa can be illustrated by a random process that is borrowed from [1].

$$\alpha(k) \overset{crossover}{\longrightarrow} \alpha^1(k) \overset{mutation}{\longrightarrow} \alpha^2(k) \overset{Vaccination}{\longrightarrow} \alpha^3(k) \overset{Immune\ selection}{\longrightarrow} \alpha(k+1)$$

The state transition from $\alpha(k)$ to $\alpha(k+1)$ composes a Markov chain. The main operations in SaiGa are presented as follows.

1. Initialize the population cardinality and the length of individuals, generate the first population, and encode each individual. Then, partition one population into sub-populations for separate evolutions.
2. Abstract vaccines according to the priori knowledge derived from the given problem.
3. Perform self-adjusting crossover and mutation in each sub-population from the first generation.
4. Perform concentration control and calculate the incentive value of each individual. The method for calculating incentive values is borrowed from [17] as follows.

$$A_i(n+1) = A_i(n) + ((\alpha \sum_{j=1}^{N} \gamma_{ij}\rho_j(n))/N + \beta m_i - k_i)\rho_i(n) \tag{1}$$

$$\rho_i(n) = (1 + \exp(0.5 - A_i(n)))^{-1} \tag{2}$$

   where $A_i(n)$ and $\rho_i(n)$ represent the incentive value and the concentration of the ith individual in the nth generation, respectively; γ_{ij} represents the affinity coefficient between individuals i and j; α and β are the ith antibody's effect intensity acting on other antibodies and antigens; m_i is the matching probability between the ith antibody and antigens; N is the number of antibodies. In this phase, we have to compute $\rho_i(n)$ by iteratively computing the above two equations. Then, employ the similar concentration control method used in [17] to calculate the clonal probability P_{clone}. Finally, perform the clonal operation.
5. Perform immune operations to duplicate individuals. Immune operators consist of the vaccination and the immune selection, introduced in [1]. The difference lies in that we perform immune selection by the similarity and vector distance based immune selection operator presented in [15].

3.2 Self-adjusting Crossover and Mutation

The crossover and the mutation are two important genetic operations in immune genetic algorithms, since it is essential for us to design crossover and mutation operators that have self-adjusting capability. In this section, we will briefly present our proposed crossover and mutation operators.

Self-adjusting Crossover. In order to maintain individual diversity, we uses one-point and two-point crossover simultaneously in SaiGa. In order to make the crossover operator automatically adjust, we propose the following crossover probability.

$$P_c^i = P_c^1 - \frac{(P_c^1 - P_c^{i-1}) * (f_i - \overline{f})}{\overline{f} - f_{min}} \quad f_i \geq \overline{f} \tag{3}$$

where P_c^i represents the crossover probability between two individuals in the ith generation, P_c^1 is the crossover probability of the first generation, f_{min} and $\overline{f}$ are the minimum and the average fitness value, and f_i is the bigger fitness value between two individuals in the ith generation.

As we can see from Equation 3, when $f_i \geq \overline{f}$, which implies that an antibody has achieved a relatively better evolutionary result from the previous crossover operation, we can safely decrease the crossover probability with a guarantee that SaiGa does not change into a random search. When $f_i < \overline{f}$, we use the original crossover probability P_c^1.

Self-adjusting Mutation. The self-adjusting mutation probability is defined as follows.

$$P_m^i = P_m^1 - \frac{(P_m^1 - P_m^{i-1}) * (f_i - f_{min})}{\overline{f} - f_{min}} \quad f_i \geq \overline{f} \tag{4}$$

where P_m^i represents the mutation probability of the ith generation, and f_i is the fitness value of the current individual.

By Equation 4, the idea behind the self-adjusting mutation operator is the same as the self-adjusting crossover operator. Similarly, when $f_i < \overline{f}$, we use the original mutation probability P_m^1.

There are three main advantages by combining the self-adjusting crossover and mutation operators into immune genetic algorithms: (1) complement the deficiency of missing important genetic information through crossover or mutation operations, (2) maintain individual diversity, and (3) prevent prematurity.

4 Variable Optimization

For IGA , the suppressive threshold can control antibody's diversity in order to avoid the occurrence of a large volume of similar antibodies [12]. Another importance variable is the incentive value that can determine the clone number of antibodies. Although these two parameters are very important, the evolutionary computation researchers often ignore them. In practice, they set them empirically that can cause experimental errors. In this study, we borrow the basic idea of variable optimization for the immune network proposed by Li et al. [12] and apply the variable optimization approaches to SaiGa. In Section 5, we will further evaluate the effect of variable optimization.

4.1 Optimization of Suppressive Threshold

Empirically, the setting of the suppressive threshold starts from a small value, e.g, $\sigma = 0.01$. Then, tune this parameter by gradually increasing this value until finally finding a better one that is suitable for a specified problem through several calculations. In general, the increment is set empirically. Aiming to avoid the arbitrariness, we use the following formula to define the suppressive threshold.

$$\sigma = \frac{2\omega \sum_{i=1}^{N-1} \sum_{j=i+1}^{N} d(X_i - X_j)}{n(n-1)} \tag{5}$$

where $d(\cdot)$ denotes the Manhattan distance between antibody X_i and X_j, and $\omega \in (0,1)$ is a tuning value that is employed to adjust the suppressive threshold based on the suggestion presented in [12]. We can see that σ is proportional to the average distance of antibodies. Therefore, as individuals evolve, the suppressive threshold can automatically adjust to avoid the occurrence of massive similar antibodies.

4.2 Optimization of Incentive Value

For SaiGa, the incentive value is denoted as $\lambda(a_b, a_g)$ where a_b and a_g represent antibody and antigen, respectively) and the Euclidean distance between antibody and antigen that is denoted as $d(a_b, a_g)$ can be summarized into two categories (see details, please refer to literature [12]). In this section, we first analysis the disadvantages of the traditional measure between antibodies and antigens, and then use a new function to calculate the incentive values.

In general, $\lambda(a_b, a_g) = 1/d(a_b, a_g)$. We can observe that when $d(a_b, a_g) < 0.1$, the small change of the d value can cause a great change to λ. Here, we give an example as follows.

Table 1. Effect of a small d value on incentive values

$d_1 = 0.03$	$d_2 = 0.08$	$\triangle d = 0.05$	$\left\lvert\frac{\triangle\lambda}{\triangle d}\right\rvert$	$= 416$
$\lambda_1 = 33.3$	$\lambda_2 = 12.5$	$\triangle\lambda = 20.8$		

Table 2. Effect of a big d value on incentive values

$d_1 = 0.3$	$d_2 = 0.8$	$\triangle d = 0.5$	$\left\lvert\frac{\triangle\lambda}{\triangle d}\right\rvert$	$= 4.1$
$\lambda_1 = 3.3$	$\lambda_2 = 1.25$	$\triangle\lambda = 2.05$		

As shown in Table 2, when $d(a_b, a_g) \geq 0.1$, the effect of λ on d is small. We can conclude that the incentive value is very sensitive to a small d value, and it is not helpful to generate a varying type of individuals. Based on the above analysis, it is important to define an appropriate function to measure the incentive value. Here, we present a new method to measure this variable as shown in Equation 6.

$$\lambda = \begin{cases} \frac{1}{\sqrt{d(a_b,a_g)}} & d(a_b, a_g) < 0.1 \\ \frac{1}{d(a_b,a_g)} & d(a_b, a_g) \geq 0.1 \end{cases} \tag{6}$$

Equation 6 can greatly reduce the effect of a smaller d value. Based on the example in Table 1, we can obtain that $|\Delta\lambda/\Delta d| = (5.77 - 3.54)/0.05 = 44.6$ according to Equation 6, this value is approximately 1/9 of the original value (i.e., 416). We can see that our proposed approach to calculating λ can help decrease the sensitivity of a small distance value between antibodies. In particular, the effect of a big incentive value is relatively small even d is big.

5 Experiments and Discussions

5.1 Experimental Setup

Our proposed immune genetic algorithm SaiGa is implemented on the platform of Microsoft Visual C++ 6.0. The experiments are performed on a PC with Pentium IV 2.4 GHz CPU, 512 Mb of RAM.

In order to evaluate the efficiency of SaiGa in handling complex problems, e.g., the function optimization problem, we compare SaiGa with other IGA algorithms, i.e., the adaptive chaos clonal evolutionary programming ACCEP [18], the simple genetic algorithm SGA [16]. First, we test the effectiveness of SaiGa in solving one-dimensional function optimization problem.

5.2 Application in Function Optimization

In this section, we compare SaiGa's capability of finding the optimal solution of a complex function with IGA proposed by Jiao et al. [1] and SGA [16]. To facilitate comparison, we use the similar testing function in [1] as shown in the following formula.

$$f(x) = 10 + \frac{sin(1/x)}{(x - 0.16)^2 + 0.1} \tag{7}$$

where the independent variable $x \in (0, 1)$. In terms of this problem, we aim at finding such independent variable x_{max} satisfying that: $f(x_{max}) \geq f(x), \forall x \in (0, 1)$.

The parameter settings are given as follows: the population cardinality is set to 100; the clonal rate is 5; the fitness function is $f(x)$; the termination condition is 100 iterative evolutions. We run each algorithm 30 times separately. Since the change of the fitness curves are nearly the same, we choose the result of one set of experiments to compare these three algorithms. The results are shown in Fig. 1.

The results show that both SaiGa and IGA can find the global optimal solution: $f(x)_{max} = 19.8949$, where $x_{optimal} = 0.1275$. However, SaiGa can converge in the ninth evolution. Whereas, IGA needs 12 iterations to find the best fitness

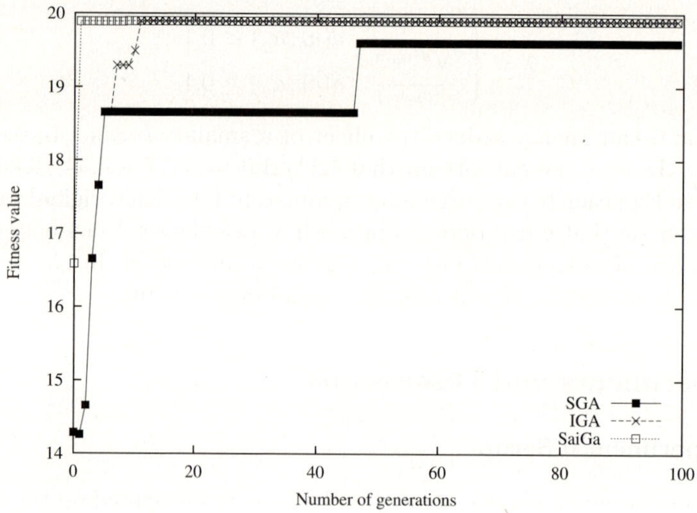

Fig. 1. Comparison of best fitness values across three algorithms

value. It shows that the convergence speed of SaiGa is faster than traditional immune genetic algorithms. This is because SaiGa uses self-adjusting crossover and mutation operators that can help accelerate the convergence speed. In addition, we can also see that the worst case is SGA which can only find a local optimal solution.

5.3 Effectiveness on Multi-dimensional Function Optimization

In order to further evaluate the effectiveness of SaiGa, we compare the optimal solution in terms of multi-dimensional function optimization among three algorithms: SaiGa, ACCEP, and SGA. To keep the consistency of algorithms, we choose 2-dimensional testing functions $f_1 \sim f_5$ provided in [18].

We use the possibly consistent parameter settings for distinct algorithms. The maximum number of evolutions is set as 100; for SGA, the population cardinality is 50; the crossover probability $P_c = 0.9$; the mutation probability $P_m = 0.08$; the clonal size is 50 for SaiGA and ACCEP. Table 5.3 shows the

Table 3. Optimization performance comparison of three algorithms

	SGA		ACCEP		SaiGa	
	Average	Error	Average	Error	Average	Error
$f1$	2.09	2×10^{-2}	2.12	4.1×10^{-8}	2.12	3.16×10^{-7}
$f2$	2.83×10^3	6.73×10^4	3.6×10^3	2.12×10^{-4}	3.6×10^3	0
$f3$	-5.4×10^{-9}	1.72×10^{-8}	-1.9×10^{-13}	3.8×10^{-13}	-1.27×10^{-9}	3.2×10^{-10}
$f4$	-2.02	5.05	-1.5×10^{-5}	1.76×10^{-5}	-1.7×10^{-3}	1.68×10^{-3}
$f5$	0.97	3.94×10^{-4}	1	2×10^{-15}	1	4.37×10^{-10}

results of 30 iterative evolutions. The results of SGA and ACCEP are borrowed from [18].

Note that, in Table 3, *Average* represents the average optimal fitness of 30 iterations, and *Error* represents the standard deviation. We can see that the accuracy and the errors in terms of SaiGa and ACCEP perform better than SGA. For SaiGa, this is due to the self-adjusting crossover and mutation operators. In addition, SaiGa can automatically tune immune parameters that helps improve the accuracy. For ACCEP, it use the logistic chaos approach to achieve the comparable performance as SaiGa.

6 Conclusions and Ongoing Work

In order to improve the self-adjusting capability of immune genetic algorithms, we integrate the advantages of genetic algorithms and immune algorithms and propose a new self-adjusting immune genetic algorithm, namely SaiGa. SaiGa applies the self-adjusting crossover and mutation operators. In particular, SaiGa can automatically tune immune parameters in order to reduce experimental errors.

Our future work includes: (1) rectifying immune operators in order to improve the global search ability; (2) improving SaiGa's convergence speed on global optimization of multi-dimensional functions; (3) applying SaiGa to other practical problems such as intrusion detection, pattern recognition and data mining.

Acknowledgments. This work is supported by the National Natural Science Foundation of China under Grant No. 60773169, the 11th Five Years Key Programs for Sci. and Tech. Development of China under Grant No. 2006BAI05A01, the Youth Software Innovation Project of Sichuan Province under Grant No. 2007AA0032 and 2007AA0028.

References

1. Jiao, L., Wang, L.: A Novel Genetic Algorithm Based on Immunity. IEEE Transactions on Systems, Man, and Cybernetics Part A 30(5), 552–561 (2000)
2. Riolo, R.L.: Modeling Simple Human Category Learning with a Classifier System. In: 4th International Conference on Genetic Algorithms, pp. 324–333. Morgan Kaufman, San Mateo (1991)
3. Qiao, S., Tang, C., Peng, J., Hu, J., Zhang, H.: BPGEP: Robot Path Planning Based on Backtracking Parallel-chromosome GEP. Dynamics of Continuous Discrete and Impulsive Systems-Series B-Applications and Algorithms 13E, 439–444 (2006)
4. Zuo, J., Tang, C., Li, C., Yuan, C., Chen, A.: Time Series Prediction Based on Gene Expression Programming. In: 5th International Conference on Web-Age Information Management, pp. 55–64. Springer, Dalian (2004)
5. Rocha, J., Ramos, C., Vale, Z.: Process Planning Using a Genetic Algorithm Approach. In: 1999 IEEE International Symposium on Assembly and Task Planning, pp. 338–343 (1999)

6. Azuaje, F.: Review of Artificial Immune Systems: a New Computational Intelligence Approach. In: de Castro, L.N., Timmis, J. (eds.) Neural Networks, vol. 16(8), pp. 1229–1229. Springer, London (2002)
7. Jiao, L., Du, H.: Development and Prospect of the Artificial Immune System. ACTA Electronica Sinica 31(10), 1540–1548 (2003)
8. Luo, W., Cao, X., Wang, X.: An Immune Genetic Algorithm Based on Immune Regulation. In: 2002 Congress on Evolutionary Computation, pp. 801–806. IEEE press, Hawaii (2002)
9. Wang, L., Jiao, L.: The Immune Genetic Algorithm and Its Convergence. In: 4th International Conference on Signal Processing, pp. 1347–1350. IEEE press, Beijing (1998)
10. Mo, H., Jin, H.: The Modified Immune Diversity Algorithm Used in Functionoptimization. Journal of Harbin Engineering University 25(1), 76–79 (2004)
11. de Castro, L.N., Von Zuben, F.J.: Learning and Optimization Using the Clonal Selection Principle. IEEE Transactions on Evolutionary Computation 6(3), 239–251 (2002)
12. Li, C., Zhu, Y., Mao, Z.: A Novel Adaptive Artificial Immune Algorithm. Computer Engineering and Application 22, 84–88 (2004)
13. de Castro, L.N., Von Zuben, F.J.: An Evolutionary Immune Network for Data Clustering. In: 6th Brazilian Symposium on Neural Networks, pp. 84–89. IEEE Computer Society, Washington (2000)
14. Timmis, J.: Artificial Immune Systems: a Novel Data Analysis Technique Inspired by the Immune Network Theory. PhD thesis, Department of Computer Science, University of Wales, Aberystwyth. Ceredigion. Wales (2000)
15. Duan, Y., Ren, W., Huo, F., Dong, H.: A Kind of New Immune Genetic Algorithm and Its Application. Control and Decision 20(10), 1185–1188 (2005)
16. Zhou, M., Sun, S.: Genetic Algorithms: Theory and Applications. National Defence Industry Press, Beijing (1999)
17. Luo, X., Wei, W.: General Discussion on Convergence of Immune Genetic Algorithm. Journal of Zhejiang University (Engineering Science) 39(12), 2006–2011 (2005)
18. Du, H., Gong, M., Liu, R., Jiao, L.: Adaptive Chaos Clonal Evolutionary Programming Algorithm. Science in China Ser. F Information Sciences 48(5), 579–595 (2005)

Calculation of Latent Semantic Weight Based on Fuzzy Membership

Jingtao Sun[1,2], Qiuyu Zhang[2], Zhanting Yuan[2],
Wenhan Huang[3], Xiaowen Yan[4], and Jianshe Dong[2]

[1] College of Electrical and Information Engineering, Lanzhou University of Technology, 730050 Lanzhou, China
[2] College of Computer and Communication, Lanzhou University of Technology, 730050 Lanzhou, China
[3] Department of Computer science and technology, Shaanxi University of Technology, 723003 Hanzhong, China
[4] Shaanxi Xiyu Highway Corporation Ltd. Hancheng, 715400 Shaanxi, China
sun2651@126.com

Abstract. One important process of Latent semantic analysis (LSA) is the weighting scheme to the term-document matrix by weight function, the weight function has directly affects the quality of LSA. This text leads to apriori information and global weighting of document on the traditional method and the modified weight function base on Fuzzy membership, Calculation of Latent Semantic Weight Based on Fuzzy Membership is proposed in this paper. By the last experiment, the results show that Latent Semantic Analysis based on modified weight function is better than that old one. The experiments show the expected results obtained, and the feasibility and advantage of the new spam filtering method is validated.

Keywords: Fuzzy membership, latent semantic analysis, weight function, entropy, spam.

1 Introduction

Fuzzy membership is proposed on the basis of representation of fuzzy concept using definite quantity and on objective laws of fuzzy objects [1]. Its basic thought is: to expand the membership in classic set so that the membership of elements in the "set" expands from 0 and 1 to any value in the interval (0, 1) to enable description of fuzzy object in a quantified way. This provides a more flexible method for it to solve problems with indefinite extension and contributes to extensive usage in such areas as fuzzy analysis, fuzzy identification and fuzzy statistics, etc [2].

Latent semantic analysis (LSA) is a computer technology that evolved to effectively extract information [3]. It analyzes a large amount for texts using statistical method to extract and quantify latent semantic information of characteristic word in documents, avoid influence of distorted characteristic word and improve accuracy of text identification. LSA was first used in searching for text information and it

F. Sun et al. (Eds.): ISNN 2008, Part II, LNCS 5264, pp. 91–99, 2008.

improved accuracy in information searching by 10% ~30% [4]. Yet LSA usually inherited vector space model in calculation of weight, failing to pay proper attention to its own characters. This results in the absence of priori information and global information implantation of documents as well as lack of flexibility in actual application. To solve the problem, it is necessary to introduce new weighting functions to improve existing ones. Therefore, it is an important direction of study how to introduce weighting functions to improve LSA more effectively.

Based on the above analysis, the article proposes a fuzzy membership-based latent semantic weight calculation method, defines priori information and global information of documents by constructing appropriate fuzzy membership functions so that the defined information concerts better with the fuzzy and indefinite characters of practical problems. It reduces the influence of the "1 or 0" assumption adopted when describing problems in a quantified way and improves accuracy in text identification with as little as possible influence on the training speed of LSA.

2 Analysis of Key Technologies

2. 1 Fuzzy Membership

Fuzzy member is the most basic concept for describing fuzzy sets as well as the most basic tool in the theory of fuzzy set and its application study [5]. Its definition is: assume U is the universe of discourse; a fuzzy set An in U is represented by a real-valued function in U $\mu_A : \begin{matrix} U \to [0,1] \\ u \mapsto \mu_A(u) \end{matrix}$. For $u \in U$, function value $\mu_A(u)$ is called membership of u to A, and function μ_A is called membership function of A. Please see Fig 1 for the graph [6].

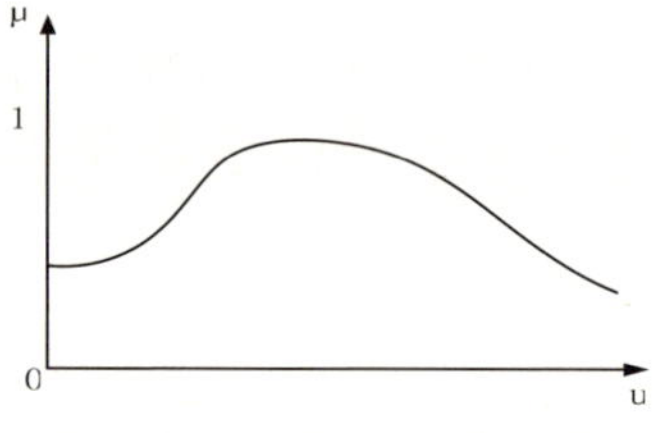

Fig. 1. Membership function

The value of $\mu_A(u)$ indicates membership of element u in discourse universe U to fuzzy set A. The closer the value of $\mu_A(u)$ is to 1, the higher the membership degree of u to A is; the closer the value of $\mu_A(u)$ is to 0, the lower the membership degree of u to A is. When $\mu_A(u)$ =1, u is considered to completely belong to A; When $\mu_A(u)$ =0, u is considered not to belong to A at all [7]. Therefore, it is the key to solve all kinds of practical problems how to construct membership function reasonably. Yet as formation of membership is people's subjective reflection of objective reality, membership functions may have more than one expression. So people can only construct membership functions using corresponding methods to address concrete problems.

2.2 LSA

The fundamental concept of LSA is to map a document represented by vector space model (VSM) of higher dimensions to latent semantic space of lower dimensions [8]. A dimension reduction matrix that contains *K* orthogonal factors is generated by performing Singular Value Decomposition (SVD) on the word-document matrix of the text collection to approximately represent the word-document matrix of the original text collection [9].

First, construct an m×n word-document matrix $X = [x_{ij}]$, in which x_{ij} is a nonnegative value, to indicate frequency of the number i word in the number j document. As there are a huge number of words and documents and the number of words in a single document is limited, X is usually a sparse matrix. To make the information that the word-document matrix X carries satisfy practical requirement, weighting processing should be conducted on x_{ij} to get a weighted m×n word-document matrix $X^* = [x_{ij}^*]$.

$$x_{ij}^* = x_{ij} \times M(i, j) \tag{1}$$

When performing SVD on X^* (assume m>n, rank(X)=r, K exists and K<<min(m,n), in the F-norm significance, X^*'s K-rank approximate matrix of $\boldsymbol{X}_k^*$ is: $\boldsymbol{X}^* \approx \boldsymbol{X}_k^* = \boldsymbol{U}_k \boldsymbol{\mathfrak{R}}_k \boldsymbol{V}_k^{\mathrm{T}}$, in which both $\boldsymbol{U}_k$ and $\boldsymbol{V}_k$'s column vectors are orthogonal vectors, and $\boldsymbol{I}_k$ is K-rank unit matrix, then:

$$\boldsymbol{U}_k^{\mathrm{T}} \boldsymbol{U}_k = \boldsymbol{V}_k^{\mathrm{T}} \boldsymbol{V}_k = \boldsymbol{I}_k \tag{2}$$

In the formula, columns of $\boldsymbol{U}_k$ and $\boldsymbol{V}_k$ are called left and right singular vector of matrix $\boldsymbol{X}_k^*$ respectively; row vectors of $\boldsymbol{U}_k$ and $\boldsymbol{V}_k$ server as the word vector and document vector respective; $\boldsymbol{\mathfrak{R}}_k$ is a diagonal matrix, and diagonal elements are called singular value of $\boldsymbol{X}_k^*$ [10].

Through SVD and K-rand approximate matrix, LSA diminishes the "noise" element in the original word-document matrix and clarifies the semantic relationship between words and documents one the one hand; on the other hand, some small semantic characteristic set that has rather great influence on text identification fails to be represented and thus the identification accuracy is lowed.

3 Calculation Method for Fuzzy Membership Based Latent Semantic Weight

3.1 Weight Calculation Method

At present, LSA generally adopts traditional weight calculation method, which divides weight into two parts [11, 12]: one is called local weight (marked as $L(i, j)$). It emphasizes the significance of a certain word in a certain document. The simplest form of definition is to use word frequency as its quantified expression. The other part is called global word weight (marked as $C(i)$). It emphasizes the significance of a certain word in the entire text collection and represents the significance of the role of a certain word in differentiating documents. Global word weight is usually calculated using statistical method. Latent relationship between two words with greater weights is more likely to be considered as important semantic relationship by LSA and thus retained. So the weighting function $M(i,j)$ is expressed in the following formula:

$$M(i,j) = L(i,j) \times C(i) \tag{3}$$

Nevertheless, the method only considers local weight and global word weight and overlooks the contribution of documents in differentiating words, which results in lowered accuracy in text identification. Therefore, it is an important direction of study on improving accuracy of LSA in text identification to define a more effective weight calculation method.

3.2 Expansion of LSA Weight Calculation Method

Given the above-mentioned problems and in combination of documents that provide more information to words, influence on basis vector of latent semantic space should be amplified; while the influence of documents that provide less information to words on basis vector of latent semantic space should be diminished. This information induction thought will be introduced in the definition of global weight of documents to expand calculation of LSA weight.

Semantics of a document is differentiated by the semantic of words it contains, while semantics of words is closely related to the theme of document, in which they appear. Fuzzy border exists between different themes. Different preference and interests lead to varied focuses of different themes, i.e. certain priori knowledge exists for different themes. Therefore, the author introduces the priori knowledge in the definition of global weight of document, thus further amplifying or diminishing the influence on basis vector of latent semantic space.

Global word weight $C(i)$ is an induction of horizontal information of matrix X and global document weight we defined $S(j)$ is an induction of vertical information of matrix X. Therefore, the weigh calculation formula can be expanded to:

$$M^*(i,j) = L(i,j) \times C(i) \times S(j) \tag{4}$$

If the effect of global document weight in the entire weight definition is not considered, take $S(j) = 1$.

We define global document weight and obtain new weight expression in the following way:

Define P themes $H_1, H_2, \ldots H_p$ in text collection U, each with certain priori weight Q_{H_i} $(i = 1, 2, \ldots, p)$. Each document in the collection can be expressed as a m -dimensional vector, which is marked as $u = (u_1, u_2, \ldots, u_m)$ and called characteristics index vector. It is widely believed that more information a certain document provides to the collection, the greater its role in text identification and the higher its global weight is. So we can define the global weight of a document using the product of priori weight Q_{H_i} and the membership of u to H_i $H_i(u)$. It is expressed as:

$$S(j) = H_i(u) \times Q_{H_i} \tag{5}$$

Here, $H_i(u)$ is constructed as below:

1) Select k_i samples from characteristic index vectors of class theme H_i, define $h_{ij} = (h_{ij_1}, h_{ij_2}, \ldots, h_{ij_m}), i = 1, 2, \ldots, p; j = 1, 2, \ldots, k_i$. In the formula, h_{ij} indicates the characteristic index vector of the number j sample in H_i. h_{ij_k} Indicates the measured data of the number k characteristic index of the number j sample in H_i, $k = 1, 2, \ldots, m$. According to practical problems involved in the article, we use word frequency as the measured data of characteristic index to perform calculation.
2) Calculate mean sample of k_i characteristic index vector h_{ij} $i = 1, 2, \ldots, p$; $j = 1, 2, \ldots, k_i$ selected from theme H_i using formula $h_i = (h_{i1}, h_{i2}, \ldots, h_{im})$, in which $h_{is} = \frac{1}{k_i} \sum_{j=1}^{k_i} h_{ij_k}$, $k = 1, 2, \ldots, m$.
3) Construct membership function of class theme H_i. Assume that $u = (u_1, u_2, \ldots, u_m) \in U$, calculate distance between u and h_i $d(u, h_i)$, and make $D = \max\{d_1(u, h_1), d_1(u, h_2), \ldots, d_1(u, h_p)\}$, membership function of H_i is calculated in the formula:

$$H_i(u) = 1 - \frac{d_i(u, h_i)}{D}, \quad i = 1, 2, \ldots, p \tag{6}$$

A new LSA weight calculation method is obtained through the above steps. Using the method to perform weighted conversion of word-document matrix, the matrix obtained can still been shrunk through Singular Value Decomposition and K-rank approximate matrix. Therefore, the expanded LSA weight calculation method is consistent with the traditional one in terms of format.

Compared with traditional LSA weight calculation method, the expanded method not only overcomes data sensitivity in LSA, but also implants priori information in basis vector of the latent semantic space to avoid lack of flexibility of SVD.

3.3 Update of Word's Identification History Vector

LSA is developed to quantify semantic information of words in the documents. Therefore, when every word has its vector representation, how to predict the semantic information of new words according to identification results (history) already obtained has become another key problem in the study of LSA. The article lessons from definition method of entropy to calculate weighted mean value of vectors of all words comprising the identification history to generate a history vector and merge it into long-distance semantic information. $\{X_1, X_2 \ldots, X_{i-1}\}$ and P_{i-1} represent vector of the word obtained before the moment $i-1$ and history vector at corresponding moment. Expand the result identified at moment i and add a new word W_i . $\{X_1, X_2 \ldots, X_{i-1}, X_i\}$ and P_i represent vector of new word at moment i and history vector at corresponding moment respectively. ω_j represents entropy of word W_j corresponding to the training language materials at moment j ($j = 1, 2, \ldots, i$).

According to: $P_{i-1} = \frac{1}{i-1}\sum_{j=1}^{i-1} X_j\left[1-\omega_j\right]$ and $P_i = \frac{1}{i}\sum_{j=1}^{i} X_j\left[1-\omega_j\right]$ (7)

There is:
$$P_i = \frac{1}{i-1} P_{i-1} + \frac{1}{i} X_i\left[1-\omega_i\right] \tag{8}$$

Formula (8) is the formula to update word identification history vector. When updating identification history vector, entropy is used for weighted calculation to differentiate contribution of each word to identification history.

3.4 Word Prediction Reliability

Standardize P_j and W_i upon obtaining representation of history vector (P_j) and new word vector (W_i) to make mode of the vectors 1. Then calculation similarity between identification history vector and new word vector using cosine similarity formula, as shown in formula (9).

$$Sim(W_i, P_j) = \cos\theta = \frac{\sum_{l=1}^{n} W_{il} \bullet P_{jl}}{\sqrt{\sum_{l=1}^{n} W_{il}^2} \bullet \sqrt{\sum_{i=1}^{n} P_{jl}^2}} \tag{9}$$

Calculate prediction reliability of new words through similarity normalization, as shown in formula (10).

$$C_{LSA}(W_i \mid p_j) = \frac{Sim(W_i, p_j)}{\sum_{w_j} Sim(W_j, p_{j-1})} \tag{10}$$

4 Experiment Design

To compare the latent semantic spaced calculated using traditional weight definition method and expanded weight definition method in solving the problem of text differentiation, we adopts $L(i,j)\times C(i)\times S(j)^*$ (weight not considered, i.e. $S(j)^*=1$)) and $L(i,j)\times C(i)\times S(j)$ weight definition method respectively to generate latent semantic spaces. To highlight practicability of the methods, we apply them in Chinese spam identification, using 100 spam mails of different types to form a training collection and generating latent semantic spaces based on the two above-mentioned weight definitions. The experiment platform is PM2.1G with 2GB memory.

First, perform Chinese word division, filtering of prohibited words, removal of words of extremely high or low frequency as well as other preprocessing tasks on the training collection to generate a 5672×1800 word-text matrix X . Weight matrix X using weight function $M(i,j)$ to get X^*. Then use SVD to perform basic conversion on the VSM space to generate latent semantic space X_K^*. In this process, selection of dimension reducing factor K has a direct influence on the efficiency of the latent semantic space model and similarity between X_K^* and X^* following dimension reduction. If the value of K is too small, useful information will be lost; if the value of K is too large, the calculation volume will increase. Therefore, an optimal K value should be selected in accordance with actual text collection and processing requirement. The article uses contribution rate δ as the criterion to assess the K value selected, i.e. $X^*=\mathrm{diag}(x_1^*,x_2^*,\cdots,x_n^*)$, and $x_1^*\geqslant x_2^*\geqslant\cdots\geqslant x_t^*= \ldots =x_n^*=0$, contribution rate δ:

$$\delta=\sum_{i=1}^{k}x_i^*\Big/\sum_{i=1}^{t}x_i^* \tag{11}$$

The contribution rate δ, proposed in reference of related factor analysis concept, indicates the degree, to which the K-dimensional space represents the entire space. Fig 2 shows that the closer the K value is to the rank of matrix A, the smaller $\|X^*-X_K^*\|_F$ is and the closer X_K^* is to X^*.Yet as the value of K continues to increase, its influence on δ will decrease or even disappear. Analysis indicates that when the value of K increases to a certain level, nearly all important characteristics of word-document matrix are represented. In this case, further increasing K value will only introduce noise. When K=400, the degree of representation is almost the same as when K=500. Yet when K=400, less time is consumed. So, we choose K=400.

In the results of spam identification, identifying spam is considered correct results, while identifying legal mail as spam is considered wrong result. Given the assumption, identification accuracy can be calculated. Table 1 lists identification accuracies using the two weight definitions. Experiment results shows that using the expanded weight model proposed herein and introducing priori information and global information of documents will remarkably improve identification accuracy.

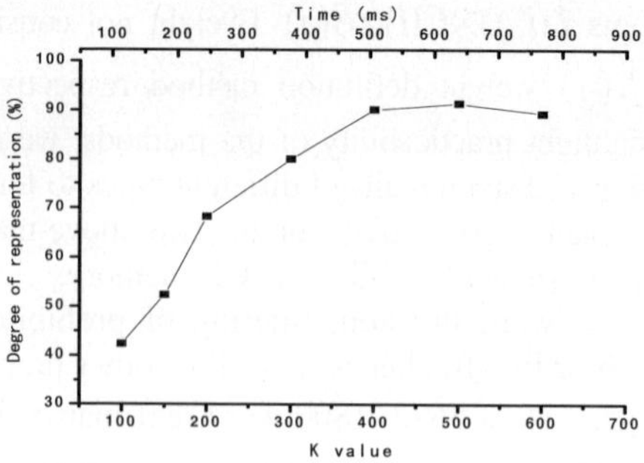

Fig. 2. Analysis of *K* value

Table 1. Identification accuracies using the two weight definitions

Identification accuracy	Accuracy
$L(i,j)\times C(i)\times S(j)^{*}$	71.1%
$L(i,j)\times C(i)\times S(j)$	80.07%

To compare identification results of the above two weight calculation methods in a more macroscopic way, Fig 3 shows F_1 value of the systems established by the latent semantic spaces generated on the basis of $L(i,j)\times C(i)\times S(j)^{*}$ (weight not considered, i.e. $S(j)^{*}=1$) and $L(i,j)\times C(i)\times S(j)$, the two weight definition methods.

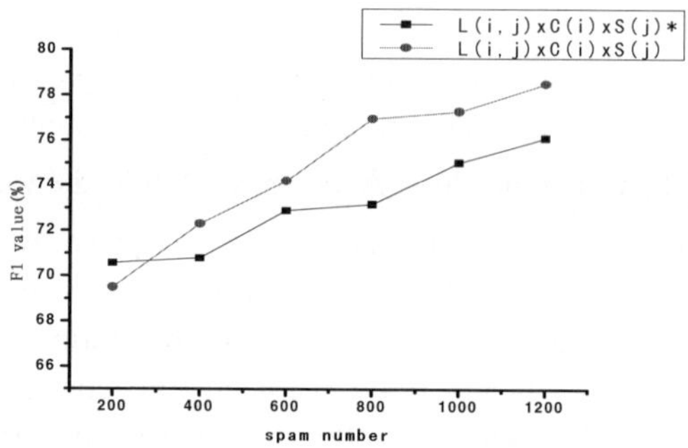

Fig. 3. F_1 value

5 Conclusion and Outlook

Ever improving accuracy in text identification accelerates improvement of LSA method. To address some problems in traditional method, we propose a definition method for global document weight using the definition of fuzzy membership as reference. Comparison between expanded weight definition method and traditional weight definition method proves that the LSA using expanded weight definition method is superior to the one using tradition method and greatly improves accuracy in text identification. Yet LSA fails to take into consideration the semantic relationship information at more in-depth level between words contained in sentence and grammar structures, which has an adverse effect on the ability of LSA to have an accurate understanding of text contents. Therefore, we need to study how to effectively combine the LSA thought and grammar information to improve its performance in text differentiation.

References

1. Wei, L.L., Long, W.J., Zhang, W.X.: Data Domain Description Based on Fuzzy Support Vector Machines. Computer Science 31(1), 108–109 (2004)
2. Liu, H., Huang, S.T.: A Fuzzy Method to Learn Text Classifier from Labeled and Unlabeled Examples. Journal of Harbin Institute of Technology 11(1), 98–102 (2004)
3. Miller, T., Wolf, E.: Word Completion with Latent Semantic Analysis. In: 18th IEEE International Conference on Pattern Recognition, pp. 2134–2137. IEEE Press, Hong Kong (2006)
4. Ishii, N., Murai, T., Yamada, T.: Text Classification by Combining Grouping, LSA and KNN. In: 5th IEEE/ACIS International Conference on Computer and Information Science, pp. 148–154. IEEE Press, USA (2006)
5. Zhang, X., Xiao, X.L., Xu, G.Y.: Fuzzy Support Vector Machine Based on Affinity Among Samples. Journal of Software 17(5), 951–958 (2006)
6. Li, S.L., Li, J.G., Wang, X.G.: Fuzzy Set Theory and Application. Science Press, China (2005)
7. Liu, S.Y., Du, Z.: An Improved Fuzzy Support Vector Machine Method. Caai Transactions on Intelligent Systems 2(3), 30–33 (2007)
8. Liu, Y.F., Qi, H., Dai, J.M., Wang, X.P.: Latent Semantic Analysis of Chinese Information. Journal of South China University of Technology (Natural Science) 32, 107–111 (2004)
9. Bellegarda, J.R.: Exploiting Latent Semantic Information in Statistical Language Odeling. Proceedings of the IEEE 88(8), 1279–1296 (2000)
10. Liu, Y.F., Qi, H., Hu, X., Cai, Z.Q., Dai, J.M.: Multi-hierarchy Documents Clustering Based on LSA Space Dimensionality Character. Journal of Tsinghua University (Science and Technology) 45(09), 1783–1786 (2005)
11. Liu, Y.F., Qi, H., Hu, X., Cai, Z.Q.: A Modified Weight Function in Latent Semantic Analysis. Journal of Chinese Information Processing 19(6), 64–69 (2005)
12. Landauer, T.K., Dumais, S.T.: A Solution to Plato's Problem: The Latent Semantic Analysis Theory of Acquisition, Induction, and Representation of Knowledge. Psychological Review 104(2), 211–240 (1997)

Research on Spatial Clustering Acetabuliform Model and Algorithm Based on Mathematical Morphology

Lichao Chen, Lihu Pan, and Yingjun Zhang

Institute of Computer Science and Technology, Taiyuan University of Science and Technology, Taiyuan, Shanxi 030024, China
chen_lc@263.net

Abstract. In this paper, a new spatial data analysis model is brought forward for the aid of analyzing to spatial terrain, which uses mathematical morphology method to carry through the research of 3-D spatial clustering analysis. The model algorithm is designed and implemented in this paper. Simulation results show that the model really solves the 3-D spatial clustering problems with high efficiency and practical features.

Keywords: Spatial clustering, Mathematical morphology, Acetabuliform model.

1 Introduction

In spatial terrain analysis, there are some problems about terrain chosen of a continuous region need to be solved that would meet many complex conditions and have suitable area [1]. The terrain is various and complex, perfectly flat terrain is difficult to be found. So few results of clustering on an absolutely ideal elevation value are available, perhaps not exist. The flatness is one of the mainly factors to affect and restrict the result of terrain chosen. So it needs to be considered to the spatial clustering analysis. The traditional methods of clustering analysis based 2D plane are unsuitable to the requirement of terrain chosen. If the analysis of flatness is directly imported into the analysis of spatial clustering, so 3D spatial clustering will be formed, and the problem could be solved. Kai-chang Di proposed an algorithm of Mathematical Morphology based Clustering algorithm (MMC) [1], which solves the problem of spatial data clustering based raster data structure preferably. The algorithm does closing operation by circularly utilizing rounded structure elements from small to big, and joins adjacent objects to cluster. However, the algorithm is the spatial clustering, which is based 2D plane, and it cannot solve 3D spatial clustering. On the basis of that, this paper puts forward a kind of acetabuliform model, constructs a creatable SSE, designs a new algorithm of 3DSCAMM,and solves the problem of non- protruding, complicated 3D spatial clustering.

2 Theory of Mathematical Morphology

Mathematical Morphology is earliest used at digital image manipulation [3], the morpha transform of which is a process of the manipulation to aggregate. On the basis of

F. Sun et al. (Eds.): ISNN 2008, Part II, LNCS 5264, pp. 100–109, 2008.

geometry, the geometry structure of the image is mainly to be researched. Its basic idea is to detect an image by a kind of structure element, to estimate whether the structure element is filled in the interior of the image or not, at the same time, to validate the way to fill is right or not. In Mathematical Morphology, the essential of morpha operator is to express the reciprocity between the aggregate of objects or shapes and the structure element. The shape of the structure element decides the information of the shape of the signal picked up by the operator. Morphology Image Manipulation is that moving a structure element on an image then does meet operator, join operator, erosion operator, and dilation operator between structure element and image aimed. In Morphology, structure element is the most important and elementary conception, whose function at the morpha transforms equals to a filer window of signal manipulation. The selection of size and shape of structure element is closely relevant to the structure and shape of image. The information of the structure of the image could be available through writing down the position of the image where the structure element is put in [4].

3 Ideas of Spatial Clustering Acetabuliforn Model

According to the principle of mathematical morphology, regarding the 2D grid spatial object as a binary image, closing operation is done to a binary image by choosing a circular structure element whose radius is R. According to the result of mathematical morphology, the grids whose distance is less than R are connected, and the objects are clustered. If this idea is extended to 3D, to use the 3D sphere of radius R of for closing operation, then the objects whose distance are less than R are connected in the 3D space, and these objects are clustered. According to this idea, considered the requirement of X-Y plane distance and flatness, an appropriate special structure element (SSE) is chosen, and an object O was selected as the origin of that. a plane S can be got by the X-Y plane intersects SSE at the origin O. The objects whose projections fall in plane S are probably in the same cluster with O. Granted that the elevation differences between these objects and O are smaller than a threshold, then they are in the same cluster, and the threshold is in direct proportion to the distance between objects and O on plane S.

3.1 Spatial Structure Element

The core of this algorithm is to construct SSE. Considering the factors of flatness, the sphere is used as SSE firstly, as shown in Figure 1.In the space taken up by SSE, on plane S, the further the object is difference from the center sphere of SSE, the smaller the elevation difference. Obviously, it is opposite to the actual situation, so it is unsuitable to be used as SSE of this algorithm.

In order to make the elevation difference of the objects that are far from the object O much bigger, and the elevation difference of the objects' point closer to O is relatively smaller. To use two-parabola rotator, which on the Z-axis cross the origin and cylinders top and bottom and use the spatial acetabuliform model entity as SSE. As shown in Figure 2, concrete definition as follows:

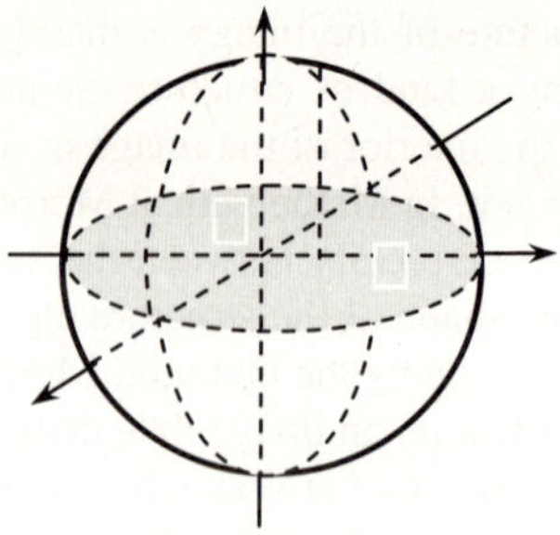

Fig. 1. Sphere SSE

X, Y-axis: The unit is the width of the DTM grid. Usually the width of DTM grid is the same as the length.

Z-axis: flatness and the actual length of X-Y axis decide the unit.

In clusters, the maximum elevation difference between the adjacent objects of two grids as follows:

$$H = L * tg\alpha \tag{1}$$

L is the actual width of a DTM grid; α is the largest angle met the requirement of the flatness; H is the actual length of a grid unit on z-axis.

The equation of the parabola rotator is:

$$z = \frac{1}{2L} tg\alpha(x^2 + y^2) \tag{2}$$

The formula (2) shows: when α enlarger, Z is also becoming bigger. It accords with the change request of the SSE chosen.

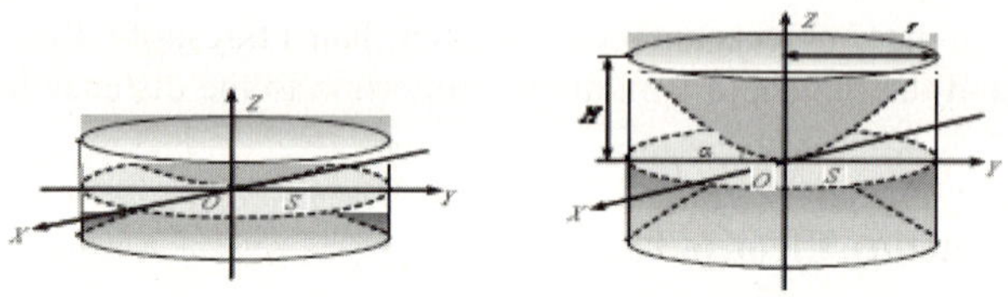

α=15° SSE template α=45° SSE template

Fig. 2. SSE template

In the 3D space taken this as a SSE, assuming that there are two objects, whose projection distance is a unit-1 on the plane S (value L), and the distance at Z direction (value H) as well, so the actual angle between the straight line which connects them and plane S is α, which is the largest angle the slope requires. The following description of all units in SSE is based this coordinate unit. Fig. 2. gives SSE model of different flatness, in which the directrix radius of cylinder in the podetium is R(R=L), height is 2H. When an object is selected as the structure element origin, all of the objects in the structure element are in the same clustering bunch with origin object. In 3D space, if the distance between two objects on plane S is smaller than the radius of structure element plane S, and their elevation difference is smaller than the corresponding maximum elevation difference, and they are connected. If α turns into zero gradually, the structure body turns into a plane, and 3D space clustering will degrade into 2-D plane clustering.

3.2 Algorithm Designing

This algorithm is suitable for space clustering questions that satisfy the following requirements:

1) Need to consider the topographical factors;
2) Have no request of the number of clustering bunch;

In order to find the region that is suitable for some particular request, taking output data of data preprocessing (considering the vegetation cover, transport facilities, etc.)

As input of space cluster, clustering bunch of cluster output need to meet flatness, conditions. Dimensionality condition is attained by abandon cluster bunch of unsatisfied conditions after clustering at the end of cluster. Flatness is considered directly in cluster of this algorithm.

The basic ideas of this algorithm is to connect with conjoint objects through closing operation for 3D objects by utilizing the structure element shown in Figure 2, and then get clusters in plane S with the pigmentation method of connecting region. The maximum distance between two adjacent points of cluster bunch in the plane S is the radius of SSE on the plane S.

Figure 3 is the input data example of spatial clustering algorithm after data preprocessing. The black spot in the figure shows clustering object. The background is the grid intersection without signed black spot. Each clustering object has elevation data. Usually there are two kinds of methods to deal with the questions that satisfy the requirement of flatness.

① The first method is to analyze flatness about the pretreated data at first, and then carry on 2D spatial clustering on the plane S to the data which is treated by the flatness .The shortcoming of this method is that it needs to analyze flatness about all the clustering objects .If there are so many spatial clustering objects, calculation of flatness is a very time-consuming job; if most objects are dissatisfied with the flatness, a large amount of flatness calculation will be a meaningless job .Even if clustering on the plane S at first ,the result of clustering may include the deserted region: A or B ,that is the region which is dissatisfied with the request of flatness ,then we use this result to analyze flatness ,although there will be less objects to do what we do like this ,there also exists similar problems.

② The second method is to carry on the spatial cluster on plane S according to the analytical method of the common 2D spatial clustering, and use the gained result to do the spatial clustering in the direction Z. This kind of method has smaller calculating amount than that of the first one, so it has higher efficiency. But, this kind of method involves multi-level, multi-value attribute of spatial clustering, and its algorithm complexity is relatively high, so the problem of algorithm parallel optimization should be considered.

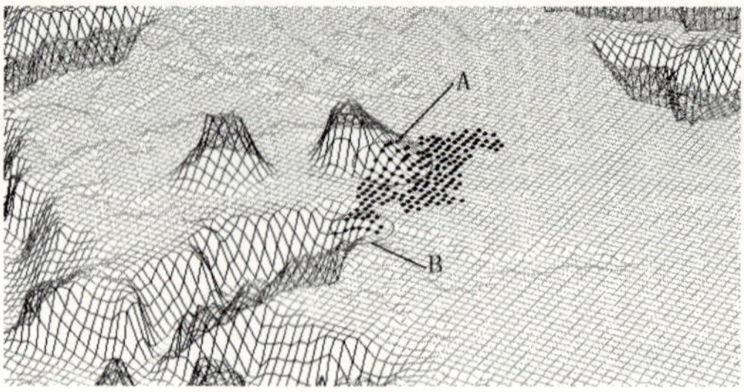

Fig. 3. Example of input data

To use the algorithm described in this paper, gets the result of the clusters which will not include region A or B which is dissatisfied with the requirement of flatness, just need to calculate boundary point of each clusters on the plane S, and then calculates the area dimensionality included by the boundary, and abandon clusters which is dissatisfied with the dimensionality terms, that is the spatial district which satisfies the requirement.

3.3 Algorithm Describing

1) Input and output

Input:

① Data X, which is an interested data though data preprocessing in the spatial database;

② SSE parameter r, namely the circle radius of the structure element on the plane S;

③ Standard quantity k, k is determined by flatness.

Output: Result of clustering Y. Clusters that satisfies the requirement of flatness and dimensionality.

2) Algorithm processing

① To allocate two 3D arrays named Array1 [V][T][U] and Array2 [V][T][U], they are separately used to preserve the original data and the temporary data .V、T and U are separately the size of each dimension. V is the difference between maximum and minimum of all objects with coordinate X; T is the subtraction between maximum and minimum of all objects with coordinate Y; U is (the maximum elevation value minus the minimum elevation value of all objects)/standard quantitative k, k is decided by the flatness. If k=5 means every 5 meters of elevation as a unit of direction Z, so a object whose elevation is 50 has the value 10 in the Z axis. Put the corresponding position of the object in Array1 for a unit-1, other positions for 0;

② To construct SSE B, whose circle radius is r, height is 2H on plane S;

③ Closing operation is Y=Array1*B, SSE in the discrete space is approached by the plane that parallels the planes of x-y、x-z、and y-z;

A) Dilation Operating

For the Points in each array {

If the point in the Array1 is the center of structure element B, and occupied position of B have destination points;

Then set the corresponding points of Array2 to unit- 1;}
End duplicates Array2 to Array1, set all elements of Array2 to 0;
B) Erosion Operating
For the points in each array {
If the point in the Array1 is the center of structure element B and occupied position of B are not all destination points;
Then set the corresponding points of Array2 to 0;}
End duplicates Array2 to Array1, set all element of Array2 to 0;

④ To get clusters on plane S with the pigmentation method of connected regions;

⑤ Each clusters C of the clustering results on plane S, is eroded by the circle structure element whose radius is the unit-1, then subtract the erosion results from C, and get the cluster bunch edge with single-wire width;

⑥ To calculate the area of each clusters through the function of GIS system, and abandon the clusters bunch which dissatisfied with the territory;

3) Explanation for parameter selection.

① The radius r of the circle that SSE projects on.

Plane S.

This parameter shows the maximum distance between two adjacent objects of clusters on plane S, that is, the width L among the grid points. When two objects in a circle whose radius is r on plane S, and the elevation difference also satisfies the request of flatness, then it can be considered that they are in the same clusters and connected.

② Standard quantity k which satisfies the flatness of adjacent objects

K is a standard quantity unit in Z axis, whose actual length is provided by formula (1), that is the H we discussed above .In the algorithm experiment examples, assume the slope of flatness≤15, the side of DTM grid L is 10m,after calculation it can be got that H=2.68,which is, k=2.68.

The algorithm described above reaches the basic requirement of solving problems, however, because of the 3-D spatial objects needed to deal with are sparse, and most space have no objects, though the speed of closing operation is far higher than common flatness analysis, the proportion of the objects in the data is too small, so the efficiency of the algorithm is not high enough .In order to improve the efficiency of the algorithm, we should decrease the 3D space to deal with.

3.4 Algorithm Performance Analyzing

Space complexity of the clustering algorithm mainly has relation to the size of 3D array Array1 (Size (Array1) =Size (Array2)), and the size of three dimensions is related to the distribution scope in direction X and Y and the compact degree of objects in direction Z (the range of Z is determined by the elevation scope and standard quantity k). If the distribution of objects on S plane is broader, the elevation distribution scope of the object is bigger, and flatness request is higher (namely smaller the maximum gradient leads to smaller standard quantity k), then 3-D array is bigger, and spatial complexity is also higher, there is not direct relation to the number of objects in input data. Considering the current computer internal memory capacity and the 3-D array entirely deposited in the memory may greatly enhance the speed of computer;

although the storage quantity that the algorithm uses is bigger, the existing equipment can versatility realizes the purpose.

Similar to spatial complexity, time complexity of the algorithm does not have not direct relations to the number of the objects in input data. Because closing operation in mathematical morphology is eroded after being dilated in the entire processing space, the size of processing space decides the processing time. Closing operation just compares and set data in spatial processing, so it is much simpler than flatness calculation, and it is also easy to realize the parallel computing.

4 Experiment Analyzing

4.1 Experimental Explanting

In order to explain the validity of (the 3D spatial clustering algorithm) 3DSCAMM of SSE, compare this algorithm with the traditional computational method. In order to facilitate description, makes the assumption as follows.

Algorithm 1: The data preprocessing is carried on flatness analysis first, then the objects set is carried on spatially clustering analysis on the plane S.

Algorithm 2: After importing SSE, the objects set are directly carried on spatially clustering analysis.

Its input parameters are r=2, α=15o, k=2.68.

The main hardware collocation of computer environment is CPU: P4/1.7G; Internal memory 512M/DDR. The test objects and data can be seen from Figure 4 to Figure 5, which are the test data not to be preprocessed.

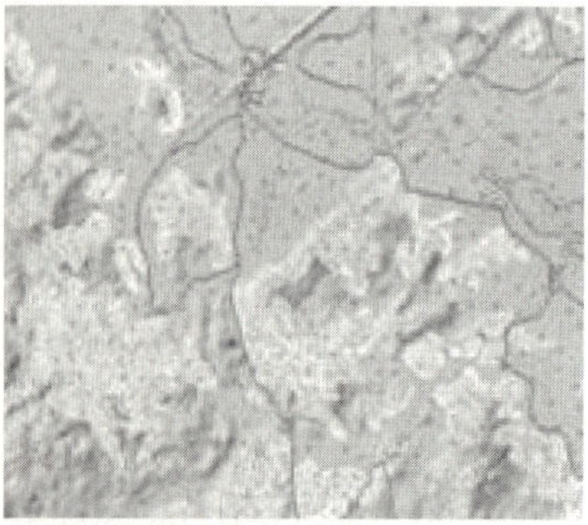

Fig. 4. Preprocess data 1. Width: 29.911km; High: 37.976km; Area: 1130.522km.

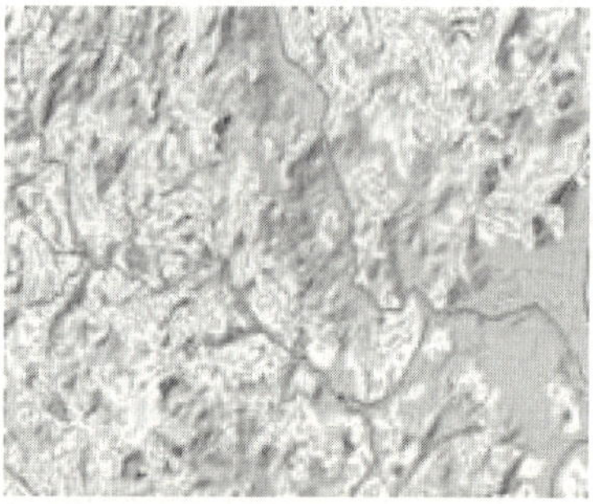

Fig. 5. Preprocess data 2. Width: 60.015km; High: 42.701km; Area: 2562.729km.

4.2 Experimental Results and Discussing

Clustering result as Figure 6 to Figure 9 shows. The algorithm 1 decides a object slope through calculating object slope in eight directions, then the object spots that satisfy the slope request will be carried on clustering; the algorithm 2 judges whether a goal is in clustering bunch by among neighboring goal angles, and carries on directly clustering to goal spots that the angle are smaller than the request. It is possible that the calculated slope not to satisfy the condition enter into clustering bunch, but considering the grid intervals lightly, moreover if the terrain has the suddenly changed, only is able to cause clustering butch edge to have a unit error, but it may be accepted for the major situation, the experimental result also showed shape of clustering bunch is basically same. More different place in clustering bunch shape is possible the reason that preserves clustering results with the straight line and the quadrangle ways in the spatial database.

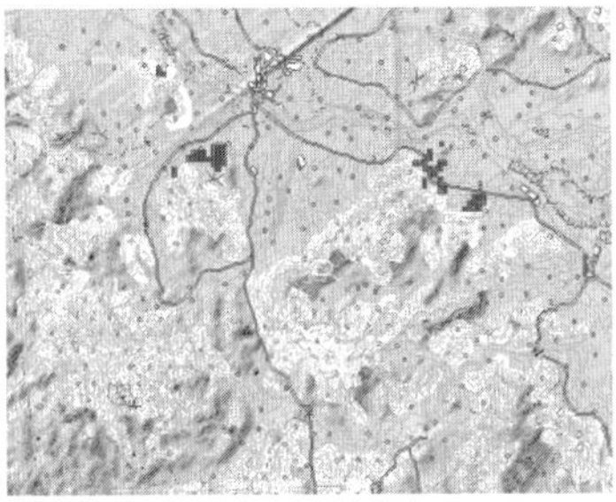

Fig. 6. Clustering result data 1 by the algorithm 1

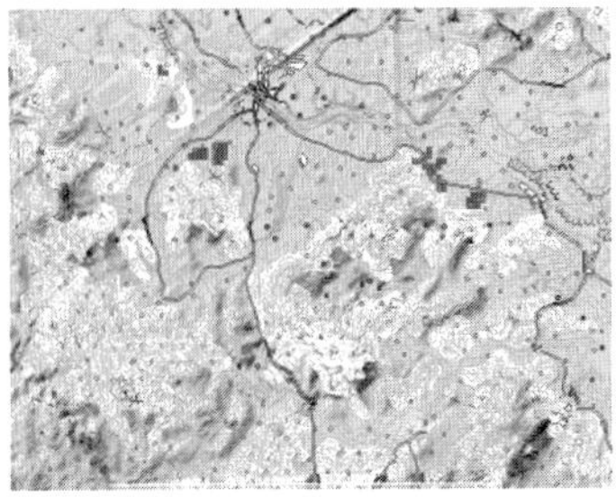

Fig. 7. Clustering result data 1 by the algorithm 1

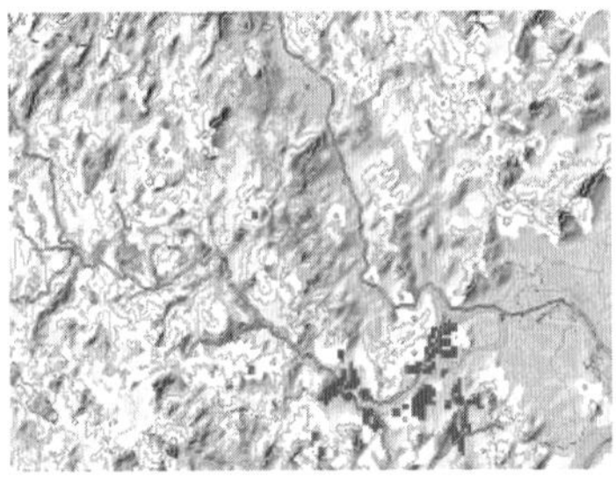

Fig. 8. Clustering result data 2 by the algorithm 1

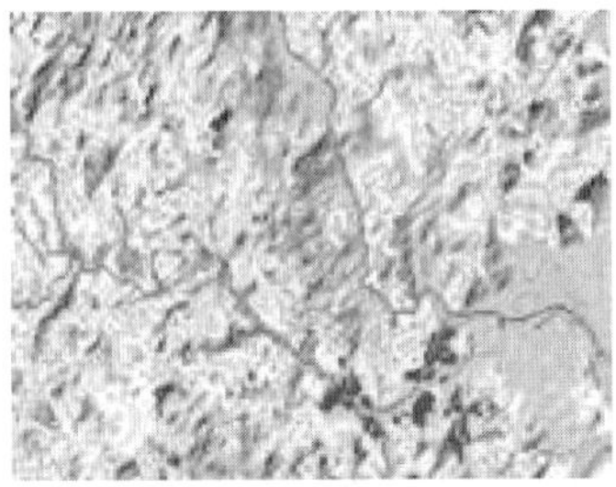

Fig. 9. Clustering result data 2 by the algorithm 1

Table 1. Test results of algorithm1 and algorithm2

	Using time of processing data 1 (s)	Using time of processing data 2 (s)
Algorithm 1	13	18
Algorithm 2	10	23

It can be seen from test data of table 1 that for the gentle terrain, the performance of algorithm 2 is slightly better than that of algorithm 1, but for the complex terrain, the performance of algorithm 2 is worse than that of algorithm 1. For gentle terrain, because the optimization parts of the algorithm 2 have abandoned much data not to satisfy the condition in the 3D array, thus cause to closing operation computation quantity greatly reducing; for complex terrain, because the goal in the Z direction distributes relative average, and closing operation computation quantity is still bigger, therefore, the algorithm efficiency is relatively lower. Simultaneously has two characteristics: (1) It is possible to further reduce the waited data on plane S to optimize the algorithm, so the space of the algorithm improvement is bigger; (2) Clustering algorithm based mathematical morphology is advantageous to realize fast parallel algorithm, so the algorithm performance can be more ideal.

5 Conclusions

On the theory of the mathematical morphology, this paper is in allusion to the spatial clustering of topography chosen, which has special request to research; puts forward a new kind of Spatial Clustering Acetabuliform Model, which is based mathematical morphology; constructs SSE, and designs 3DSCAMM.

This method can not only complete non-protruding, complicated 3D spatial clustering at one time, but also easily realize parallel algorithm .The experimental results show that, this algorithm has a good effect on spatial clustering. This algorithm has a certain commonality on the use of the spatial clustering of population and so on.

Acknowledgements. The author would like to acknowledge Natural Science Foundation of Shanxi Province (NO. 20051044), for its partial support to this work. The author would like to thank the anonymous reviewers for their valuable comments and suggestions.

References

1. Han, J., Kamber, M., Tung, A.K.H.: Spatial Clustering Methods in Data Mining. A Survey (2001)
2. Di, K.: Spatial Data Ming and Knowledge Discovery from Database. Wuhan University Press, Wuhan (2003)
3. Evans, A.N., Liu, X.: A Morphological Gradient Approach to Color Edge Detection. Image Processing 15(6), 1454–1463 (2006)
4. Chen, L.: Research on the Technology of the Battlefield Topography Analysis Based on the Spatial Data Mining. Beijing Institute of Technology, Beijing (2002)
5. Wang, Z.: Research on Certain Essential Technologies of SDM Based on GIS. Zhejiang University, Hangzhou (2005)
6. Di, K.: Spatial Data Mining and Knowledge Discovery in Database. Wuhan University Press, Wuhan (2000)
7. Han, J., Micheline, K.: Data Mining Concepts and Techniques. Morgan Kaufmann Publishers, US (2001)
8. Koperski, K., Adhikary, J., Han, J.: Spatial Data Mining: Progress and Challenges Survey Paper. In: ACM SIGMOD Workshop on Research Issues on Data Mining and Knowledge Discovery (1996)
9. Ganti, V., Ramakrishnan, R., GehLrke, J.: Clustering Large Datasets in Arbitrary Metric Spaces. In: 15th Int. Conf. on Data Engineering, pp. 502–511 (1998)
10. Tung, A.K.H., Hou, J., Han, J.: Spatial Clustering in the Presence of Obstacle. In: 2001 Int. Conf. on Data Engineering, pp. 359–367 (2001)

11. Yang, C., Zhang, Q., Tian, X., He, L., Su, Y.: Clustering Algorithm for Area Geographical Entities Based on Genetic Algorithm. Geography and Geography Information Science 20(3), 12–16 (2004)
12. Jing, S., Duan, S.: Application of Two-stage Fuzzy Classification Polymerization Method in Comprehensive Assessment of Stability of Slope Districts. Geotechnical Engineering Technique (4), 231–234 (2001)
13. He, B., Fang, T., Guo, D.: Uncertainty-Based Clustering Method for Spatial Data Mining. Computer Science 31(11), 196–198 (2004)
14. Zheng, H., Zhou, X., Wang, J.: Automatic Color Segmentation of Topographic Maps Based on Color Space Transformation and Fuzzy Restraint Clustering. Acta Geodaeticaet Cartographica Sinica 32(2), 183–187 (2003)
15. Liu, G.: Research and Implementation of SADBS Based on SADBS. Nanjing University of Aeronautics and Astronautics, Nanjing (2005)
16. Li, X., Zheng, X., Yan, H.: Research on Spatial Clustering Method Based on Integration of Coordinates and Attribute. Geography and Geography Information Science 20(2), 38–40 (2004)
17. Xiao, J., Zhuang, Y., Wu, F.: Recognition and Retrieval of 3D Terrain Based on Level of Detail and Minimum Spanning Tree. Journal of Software 14(11), 1955–1963 (2003)
18. Zhang, Y., Han, Y., Zhang, J.: Research and Realization of an Efficient Clustering Algorithm. Computer Applications 25(7), 1573–1576 (2005)
19. Yang, C.: Research on Clustering Algorithm in Spatial Data Mining. Geomatic Science and Engineering 25(2), 61–62 (2005)

Partner Selection and Evaluation in Virtual Research Center Based on Trapezoidal Fuzzy AHP

Zhimeng Luo, Jianzhong Zhou*, Qingqing Li, Li Liu, and Li Yang

College of Hydroelectricity and Digitalization Engineering, Huazhong University of Science and Technology, Wuhan 430074, China
lumhenghust@hotmail.com, jz.zhou@hust.edu.cn

Abstract. To the reduce the complexity and the uncertainty in selecting cooperation partner of virtual research center, a comprehensive evaluation model based on AHP and basic theory of the trapezoidal fuzzy number was proposed. In the model, the judgment matrix was first structured in trapezoid fuzzy number and the priority weights of criteria and sub-criteria were calculated, then the consistency of judgment matrix was tested in the necessary and sufficient condition of consistency. In the end, the alternatives are ranked with the integration of the fuzzy gravity method and the mean square difference method. Applying the trapezoidal fuzzy number makes the model combine qualitative analysis with quantitative analysis effectively, reduce influences of subjective factors and select the best cooperative partner objectively. In addition, the model can be easily programmed due to its simple, normative arithmetic formulas, thus can be widely used in practice. An example is presented where are four main criteria, sixteen sub-criteria and three alternatives.

Keywords: Analytic hierarchy process (AHP), Trapezoidal fuzzy number, Virtual research center, Partner selection.

1 Introduction

Virtual research center (VRC) is a new kind of organization that is formed by the penetration of virtual organization to science research organization with the development of computer science, network technology and other communication technique [1]. VRC has an extensive application in enterprise and research institute for its flexibility and dynamic characteristic. To maintain and improve the VRC's competitive power, as many researchers emphasized, it's a critical step to select agile, competent and compatible partners quickly and rationally during its formation phase. There are several approaches used to select the corporate partners: Fuzzy Analytical Approach [2], AHP/DEA [3], Rough Set Approach [4], etc. However, methods mentioned in those contributions are often unable to adequately represent the uncertainty of human judgments, and some are too complicated to be operated by programming. In addition, because the multiple attribute and the respective index weight are involved in VRC's partner selection, some indexes are quantified not in certain number but in fuzzy

*Corresponding author.

F. Sun et al. (Eds.): ISNN 2008, Part II, LNCS 5264, pp. 110–118, 2008.

number. Hence, a decision making method which combines qualitative analysis with quantitative analysis effectively is necessary for the core enterprise to select the best partner(s).

As one of the best multi-attribute decision methods, Fuzzy AHP has been widely applied in solving multiple criteria decision-making problems. Laarhoven, a Holland scholar, put forward a method to express the fuzzy judgment with trigonomertrical fuzzy numbers in 1983, and combined triangular fuzzy numbers operational laws with the logarithmic least square method to obtain the final ranking based on fuzzy weights [5].

At present, the research on the theory of fuzzy AHP and its application has made great progress, but most of them are based on trigonomertrical fuzzy number [6-8]. Since its membership function is more complex than trigonomertrical fuzzy number, trapezoidal fuzzy number is more suitable to describe the elements' uncertainty. In order to improve the ability to solving VRC partner selection problems, a model based on AHP and basic theory of the trapezoidal fuzzy number is proposed in this paper.

2 The Trapezoidal Fuzzy AHP Model

Assume that there are K hierarchies in the model structure of an evaluation system. Let $A=\{A_1, A_2, \ldots, A_m\}$ be the set of all alternatives, and $X=\{X_1, X_2, \ldots, X_n\}$ as the attribute of the objects. Let m_{ij} be the value that the experts are asked to give for each pairwise comparison between alternatives $A_1, A_2, \ldots, A_m$ for each criterion in a hierarchy, and also between the criteria. This value should be represented in trapezoidal fuzzy number, that is to say $m_{ij}=(a_{ij},b_{ij},c_{ij},d_{ij})$ is trapezoidal fuzzy number. When considering that there maybe several experts take part in the evaluation activity, the elements of the judgment matrix are expressed as below:

$$m_{ij} = \frac{1}{L}\sum_{k=1}^{L} m_{ij}^{k} . \tag{1}$$

There m_{ij}^{k} are the values that the experts give, and L present the number of experts who take part in this decision-making problem.

In the evaluation system, let $m_{ij}=(a_{ij},b_{ij},c_{ij},d_{ij})$ $(i\in N)$, and $m_{ii}=(1,1,1,1)$($i\in N$, $i\neq j$), then if

$$\begin{cases} a_i = \left(\prod_{j=1}^{n} a_{ij}\right)^{1/n}, & a = \sum_{i=1}^{n} a_i , \\ b_i = \left(\prod_{j=1}^{n} b_{ij}\right)^{1/n}, & b = \sum_{i=1}^{n} b_i , \\ c_i = \left(\prod_{j=1}^{n} c_{ij}\right)^{1/n}, & c = \sum_{i=1}^{n} c_i , \\ d_i = \left(\prod_{j=1}^{n} d_{ij}\right)^{1/n}, & d = \sum_{i=1}^{n} d_i . \end{cases} \tag{2}$$

the fuzzy priority weight of alternative A_i for criterion X_j is

$$x_{ij} = (\frac{a_i}{d}, \frac{b_i}{c}, \frac{c_i}{b}, \frac{d_i}{a}). \tag{3}$$

There, $i=1, 2, \ldots, n$;
And the membership function is

$$\mu_{x_{ij}}(x) = \begin{cases} 0, & x \leqslant \frac{a_i}{d}, \\ \frac{f_i(\alpha)}{g(\alpha)}, & \frac{a_i}{d} \leqslant x \leqslant \frac{b_i}{c}, \\ 1, & \frac{b_i}{c} \leqslant x \leqslant \frac{c_i}{b}, \\ \frac{g_i(\alpha)}{f(\alpha)}, & \frac{c_i}{b} \leqslant x \leqslant \frac{d_i}{a}, \\ 0, & x \geqslant \frac{d_i}{a}. \end{cases} \tag{4}$$

There

$$f_i(\alpha) = \left[\prod_{j=1}^{n} ((b_{ij} - a_{ij})\alpha + a_{ij}) \right]^{1/n}, \tag{5}$$

$$g_i(\alpha) = \left[\prod_{j=1}^{n} ((c_{ij} - d_{ij})\alpha + d_{ij}) \right]^{1/n}, \tag{6}$$

$$f(\alpha) = \sum_{i=1}^{n} f_i(\alpha), \tag{7}$$

$$g(\alpha) = \sum_{i=1}^{n} g_i(\alpha). \tag{8}$$

$\alpha(\alpha \in [0,1])$ presents the selected confidence level from equation (4) to equation (8).

In order to deduce the fuzzy weight vectors, a fuzzy reciprocal matrix is defined as follows

$$M_{n+1} = \begin{bmatrix} \omega_{11} & \omega_{12} & \cdots & \omega_{1n} \\ \omega_{21} & \omega_{22} & \cdots & \omega_{2n} \\ \vdots & \vdots & & \vdots \\ \omega_{n1} & \omega_{n2} & \cdots & \omega_{nn} \end{bmatrix}. \tag{9}$$

The elements of the matrix present the ratios for each pairwise comparison between attributes in the same hierarchy.

Because of the complexity of the objects and limitation of the subject's recognition, error often occurs when making decision. It is necessary to test the consistency of the judgment matrix to ensure the accuracy of the ranking results. The following theorem is introduced.

Theorem 1 [9]: Let $A=(m_{ij})_{n\times n}$ be a fuzzy, positive, reciprocal matrix with $m_{ij}=(a_{ij},b_{ij},c_{ij},d_{ij})$ $i, j=1, 2, \ldots, n$, if there are some real number elements $\tilde{m}_{ij} \in [b_{ij}, c_{ij}]$ $(i, j=1, 2, \ldots, n)$ which can assure the consistency of judgment matrix $\tilde{A}=(\tilde{m}_{ij})_{n\times n}$, that means the judgment matrix A also is a consistent fuzzy judgment matrix.

Generally, the method proposed by satty in the analytic hierarchy process is used to test the consistency of judgment matrix [10]. But it has obvious defects that saaty set the critical value of the consistent ratio as 0.1 by human experience. Another method known as necessary and sufficient condition of consistency[11], which can eliminate the defects, is selected to test the consistency of trapezoidal fuzzy number judgment matrix in this paper.

Assume $p_i(n)$ presents the column of the matrix $A=(m_{ij})_{n\times n}$, and q_i (n) is the row, thus

$$p_i(n)=\sum_{j=1}^{n} m_{ij} . \tag{10}$$

$$q_i(n)=\sum_{j=1}^{n} m_{ji} . \tag{11}$$

$$r=\sum_{i=1}^{n}\sum_{j=1}^{n} m_{ij} . \tag{12}$$

Let

$$\delta=\sum_{i=1}^{n}\max_{j}\{|\frac{1}{q_j(n)}-\frac{p_j(n)}{r}|\} . \tag{13}$$

Where, $i,j=1, 2, \ldots, n$ and $i\neq j$.

Then compute the equation (13) combined with equation (10), equation (11) and equation (12). We shall not demand all judgment matrices to be perfectly consistent. If $\delta\leq\varepsilon$ $(\varepsilon=0.1)$, the consistency of fuzzy number judgment matrix is acceptable. Else the feedback value needs to be adjusted until the matrix is reasonably consistent.

If the test of the consistency is qualified, then combine the equation (3) with equation (9), the fuzzy utility value of alternatives can be calculated by the following equation:

$$U_i=\sum_{j=1}^{n}\omega_j x_{ij} . \tag{14}$$

Where $i=1, 2, \ldots, m$, the ranking of alternatives is available, and the greater U_i is, the better the alternative is.

3 The Process of Partner Selection and Evaluation in VRC Based on Ttrapezoidal Fuzzy AHP Model

Generally, the application of the proposed trapezoidal fuzzy AHP model for partner selection and evaluation can be described as follows:

3.1 Structure the Hierarchical Model of Comprehensive Evaluation System in VRC

Firstly, analyze the hierarchical form and the practical needs of the research objects, then structure the hierarchical form with three or four hierarchies, containing the overall goal at the top with many levels of criteria and sub-criteria in the middle and the alternatives at the bottom.

Secondly, construct a set of pairwise comparison matrices by pairwise comparisons of decision elements from the top to down, and determine the priority weights of criteria and sub-criteria on the intermediate levels. Let $S=\{S_1, ..., S_n\}$ be the set of criteria in the second hierarchy. As shown in Figure 1, the model of comprehensive evaluation system for the virtual research center of Yalong river hydropower development is structured with a formulated four-hierarchy structure. The first hierarchy is the overall goal. Four main criteria and sixteen sub-criteria are in the second and the next hierarchy respectively. Three alternatives, namely A_1, A_2 and A_3, are at the bottom of the structure.

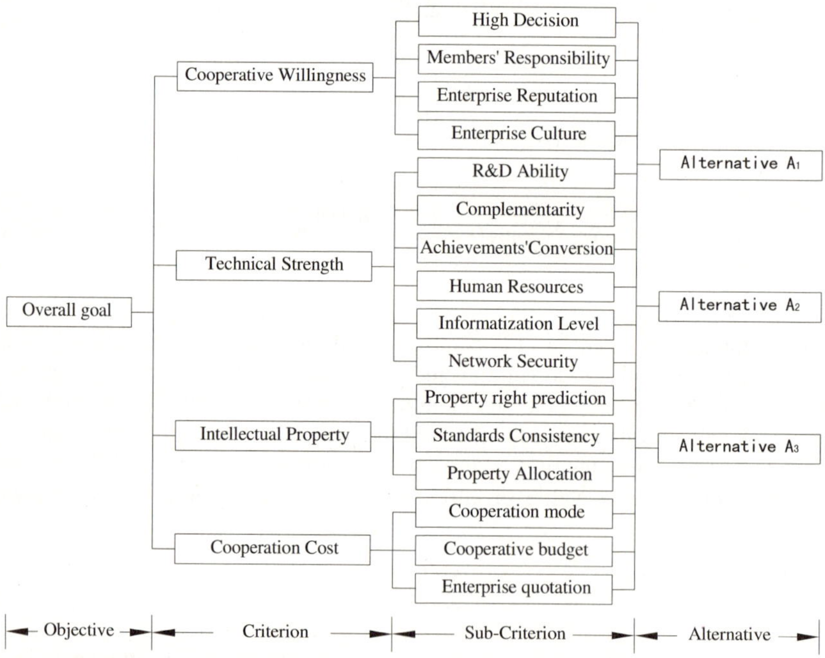

Fig. 1. Hierarchy model for partner selection of Yalong river VRC

3.2 Construct the Fuzzy Judgment Matrix, Test the Consistency and Calculate the Priority Weights

The evaluation teams are composed of professional experts, core enterprise leaders and department principals respectively. Based on the pairwise comparison between the four criteria, they give the judgment matrix C.

$$C=\begin{bmatrix} 1 & (7/8,1,13/9,9/5) & (4,5,6,7) & (2,3,7/2,4) \\ (5/9,9/13,1,8/7) & 1 & (2,3,3,4) & (4/3,1,6/5,2) \\ (1/7,1/6,1/5,1/4) & (1/4,1/3,1/3,1/2) & 1 & (1/2,2/3,1,2) \\ (1/4,2/7,1/3,1/2) & (1/2,5/6,1,4/3) & (1/2,1,3/2,2) & 1 \end{bmatrix}.$$

Then combine with the equation (13), test the consistency of the matrix $\overline{C}$ which is derived from matrix C.

$$\overline{C}=\begin{bmatrix} 1 & 13/9 & 6 & 7/2 \\ 13/9 & 1 & 3 & 1 \\ 1/6 & 1/3 & 1 & 2/3 \\ 2/7 & 1 & 3/2 & 1 \end{bmatrix}.$$

Because δ=0.0337<0.1, the consistency of matrix $\overline{C}$ is acceptable, and the trapezoidal fuzzy number judgment matrix C also is consistent matrix confirmed by theorem 1. The priority weights of the four criteria are obtained in Table 1 from matrix C.

Table 1. The priority weights of criteria

Criteria	The priority weight
Cooperative Willingness	(0.2630,0.3879,0.5451,0.7411)
Technical Strength	(0.1784,0.2366,0.3199,0.4836)
Intellectual Property	(0.0591,0.0865,0.1180,0.1967)
Cooperation Cost	(0.0808,0.1377,0.1953,0.2989)

Similarly, the priority weights of other sub-criteria under their corresponding criterion can be derived from Table 2. The pairwise comparisons between alternatives can be quantified precisely under some sub-criteria, such as cooperative budget, quoted price and achievements conversion. While under members responsibility, enterprise culture and high decision, the priority weights are obtained based upon the trapezoid fuzzy number AHP model where the exact comparative data are unavailable.

The complete processes about the calculation of fuzzy judgment matrix and the test of consistency have been omitted from the proceedings paper. The priority weights of the three alternatives under the four criteria are obtained and listed in Table 3.

Table 2. The priority weights of sub-criteria under its corresponding criterion

Sub-criterion	The priority weight
High Decision	(0.1769,0.2838,0.3828,0.6219)
Members' Responsibility	(0.1293,0.1875,0.2672,0.4996)
Enterprise Reputation	(0.1554,0.2589,0.3865,0.5551)
Enterprise Culture	(0.0716,0.1145,0.1474,0.1987)
R&D Ability	(0.2347,0.3466,0.4199,0.6213)
Complementarity	(0.0788,0.1355,0.1630,0.2674)
Achievements Conversion	(0.0261,0.0382,0.0462,0.0704)
Human Resources	(0.0346,0.0534,0.0661,0.1101)
Informatization Level	(0.0507,0.0818,0.0985,0.1647)
Network Security	(0.1541,0.2550,0.3045,0.4932)
Property Right Prediction	(0.1185,0.1917,0.2590,0.3831)
Standards Consistency	(0.1134,0.1675,0.2129,0.3831)
Property Allocation	(0.3599,0.5412,0.6387,0.9234)
Cooperation Mode	(0.2482,0.4291,0.5313,0.9022)
Cooperative Budget	(0.1319,0.2736,0.3524,0.6122)
Enterprise Quotation	(0.1198,0.1861,0.2414,0.4859)

Table 3. The priority weights of the three alternatives under four criteria accordingly

A	The priority weight			
	Cooperative Willingness	Technical Strength	Intellectual Property	Cooperation Cost
A_1	(0.0649,0.1905, 0.3558,0.9385)	(0.1368,0.3184, 0.5229,1.1466)	(0.1308,0.2427, 0.3459,0.7319)	(0.1387,0.2458, 0.3103,0.5488)
A_2	(0.0471,0.1233, 0.2156,0.6499)	(0.0885,0.1890, 0.3049,0.7272)	(0.1006,0.1736, 0.2338,0.4666)	(0.1293,0.2301, 0.2908,0.5143)
A_3	(0.0945,0.2478, 0.4576,1.1491)	(0.0679,0.1311, 0.2291,0.4947)	(0.1350,0.2427, 0.3266,0.6364)	(0.0981,0.1749, 0.2224,0.3992)

3.3 Determine the Global Priority Weights Vectors of All Alternatives

Finally, as shown in Table 4, the global priority weights are calculated by Table 2, Table 3 and equation (14).

Table 4. The global priority weights of alternatives

Alternative	The global priority weights
A_1	(0.0604,0.2041,0.4626,1.5580)
A_2	(0.0446,0.1392,0.2994,1.0788)
A_3	(0.0529,0.1722,0.4047,1.3353)

Calculated by the fuzzy gravity method under the comparative rules of trapezoid fuzzy number [12], the comprehensive priority weights of these alternatives are as follows: (0.6272, 0.4323, 0.5381). While by the mean square difference method, the comprehensive priority weights are: (0.2019, 0.1586, 0.1647).

Comprehensively considering these two methods, we can make the conclusion that alternative A_1 is the best one, A_1 ranks next to A_3 and is a little better than A_2 in the model of comprehensive evaluation system for the virtual research center of Yalong river hydropower development.

4 Conclusions

To solve the problem of partner selection and evaluation in the formation of a VRC, a comprehensive evaluation trapezoid fuzzy AHP model is proposed in this paper. The judgment matrix is structured in trapezoid fuzzy number and the priority weights of criteria and sub-criteria are calculated, then the consistency of judgment matrix is tested in the necessary and sufficient condition of consistency. At last, the alternatives are ranked with the integration of the fuzzy gravity method and the mean square difference method. The model makes experts judgment more accordant with human thought mode, and its arithmetic formulas can be programmed easily. No further human intervention is required after evaluators have filled the comparative data into the system, thus a quick selection of appropriate partner(s) is feasible. The example provided in the paper showed the feasibility, reliability and practicality of the model.

Acknowledgements

1.Thanks to the anonymous referees for their helpful comments and suggestion to improve the presentation of this paper.

2.This paper is supported by the National Natural Science Foundation of China (No. 50679098) and the Ph.D. Programs Foundation of Ministry of Education of China (20050487062).

References

1. Shen, M.B., Wu S.Y.: Discussion on The Virtual Research Center of Hydropower Development Enterprise. Journal of Hydroelectric Engineering, 26, 21--24 (2007)
2. Mikhailov, L.: Fuzzy Analytical Approach to Partnership Selection in Formation of Virtual Enterprise. The Information Journal of Management, 30, 393--401 (2002)
3. Lou, P., Chen, Y.P., Zhou, Z.D.: An AHP/DEA Method for Vendor Selection in the Agile Supply Chain. Journal of Huazhong University of Science and Technology. 30, 29--31 (2002)
4. Zhou, Q.M., Yin, B.B.: Rough Set Approach to Partnership Selection in Formation of Virtual Enterprises. Control and Decision. 20, 1047--1051 (2005)
5. Laarhoven, P.J.M., Pedrycz, W.: A fuzzy Extension of Saaty's Priority Theory. Fuzzy Sets and Systems. 11, 199--227 (1983)

6. Xu, R.N., Zhai, X.Y.: Fuzzy Logarithmic Least Squares Ranking Method in Analytic Hierarchy Process. Fuzzy Sets and Systems. 77, 175--190 (1996)
7. Cao, J., Yang, C.J., Li, P.: Partner Selection and Evaluation in Agile Virtual Enterprise Based Upon TFN-AHP Model. Journal of Zhejiang University (Engineering Science). 40, 1061--1065 (2006)
8. Ma, P.J., Zhu, D.B., Ding, Y.C.: Fuzzy-AHP Based Algorithm for Optimal Partner-Selecting in Agile Manufacturing. Journal of Xi'an Jiao Tong University. 33, 108--110 (1999)
9. Xu, R.N., Lui, K.: A Note on Consistency of Fuzzy Judgment Matrix. Journal of Systems Science and Mathematical Sciences. 20, 58--64 (2000)
10. Saaty, T L.: The Analytic Hierarchy Process. McGraw-Hill, New York (1980)
11. Li, X.Q., Li, S.R., Han, X.L.: Study of AHP Theory and Method: Test of Consistency and Calculation of Weight Vector. Journal of Systems Engineering. 12, 111--117 (1997)
12. Li, D.F.: Fuzzy Multiobjective Many-Person Decision Makings and Games. Defense Industry Publishing House, Beijing (2003)

A Nonlinear Hierarchical Multiple Models Neural Network Decoupling Controller

Xin Wang[1,2], Hui Yang[2], Shaoyuan Li[3], Wenxin Liu[4], Li Liu[4], and David A. Cartes[4]

[1] Center of Electrical & Electronic Technology, Shanghai Jiao Tong University, Shanghai, 200240, P.R. China
wangxin26@sjtu.edu.cn

[2] School of Electrical & Electronic Engineering, East China Jiao Tong University, Nanchang, Jiangxi, 330013, P.R. China

[3] Institute of Automation, Shanghai Jiao Tong University, Shanghai, 200240, P.R. China
syli@sjtu.edu.cn

[4] Center for Advanced Power Systems, Florida State University, Tallahassee, 32310, USA
{wliu,dave}@caps.fsu.edu

Abstract. For a nonlinear discrete-time Multi-Input Multi-Output (MIMO) system, a Hierarchical Multiple Models Neural Network Decoupling Controller (HMMNNDC) is designed in this paper. Firstly, the nonlinear system's working area is partitioned into several sub-regions by use of a Self-Organizing Map (SOM) Neural Network (NN). In each sub-region, around every equilibrium point, the nonlinear system can be expanded into a linear term and a nonlinear term. The linear term is identified by a BP NN trained offline while the nonlinear term by a BP NN trained online. So these two BP NNs compose one system model. At each instant, the best sub-region is selected out by the use of the SOM NN and the corresponding multiple models set is derived. According to the switching index, the best model in the above model set is chosen as the system model. Then the nonlinear term of the system are viewed as measurable disturbance and eliminated by the choice of the weighting polynomial matrices. The simulation example shows that the better system response can be got comparing with the conventional NN decoupling control method.

Keywords: Multiple models, hierarchical, neural network, decoupling.

1 Introduction

In recent years, for linear multivariable systems, researches on adaptive decoupling controller have made much success [1]. As for nonlinear Multi-Input Multi-Output (MIMO) systems, few works have been observed [2,3]. Ansari *et al.* simplified a nonlinear system into a linear system by using Taylor's expansion at the equilibrium point and controlled it using linear adaptive decoupling controller accordingly [4]. However, for a system with strong nonlinearity and high requirement, it can not get good performance [5]. In [6,7], an exact linear system can be produced utilizing a feedback linearization input-output decoupling approach and high dynamic performance was achieved. But accurate information, such as the parameters of the system, must be known precisely. Furthermore, a

F. Sun et al. (Eds.): ISNN 2008, Part II, LNCS 5264, pp. 119–127, 2008.

variable structure controller with sliding model was proposed [8] and industrial experiment in binary distillation columns was presented in [9], which required the system an affine system. Although the design methods above can realize nonlinear decoupling control, there were too many assumptions required on the system so that they can not be used in the industrial process directly. To solve this problem, Neural Network (NN) decoupling controller was proposed [10]. In [11], NN was used to identify the structure and the parameters of the nonlinear system. In [12], at the origin, the system was expanded into the linear term and the nonlinear term and two NNs were adopted to identify these two terms. Unfortunately, when the equilibrium point was far from the origin, the system lost its stability.

In this paper, for a nonlinear discrete-time MIMO system, a Hierarchical Multiple Models Neural Network Decoupling Controller (HMMNNDC) is designed. The nonlinear system's working area is partitioned into many sub-regions, which is described using a Self-Organizing Map (SOM) NN. In each sub-region, at every equilibrium point, the system is expanded into a linear term with a nonlinear term. Both terms are identified using BP NNs, which compose one system model. All models, which are got from all equilibrium points in this sub-region, compose a multiple models set. At each instant, the best sub-region is chosen out using the SOM NN. Then in the corresponding multiple models set, according to the switching index, the best model is selected out as the system model. To control the system, the nonlinear term the above model is viewed as measurable disturbance and eliminated by the choice of the weighting polynomial matrices.

2 Description of the System

The system is a nonlinear discrete-time MIMO system of the form

$$\boldsymbol{y}(t+1) = \boldsymbol{f}[\boldsymbol{y}(t),\cdots,\boldsymbol{u}(t),\cdots], \tag{1}$$

where $\boldsymbol{u}(t)$, $\boldsymbol{y}(t)$ are $n\times 1$ input, output vectors respectively and $\boldsymbol{f}[\cdot]$ is a vector-based nonlinear function which is continuously differentiable and Lipshitz.

Suppose that $(\boldsymbol{u}_1,\boldsymbol{y}_1),\cdots(\boldsymbol{u}_l,\boldsymbol{y}_l),\cdots(\boldsymbol{u}_m,\boldsymbol{y}_m)$ are all equilibrium points. At each equilibrium point $(\boldsymbol{u}_l,\boldsymbol{y}_l)$, using Taylor's formula, it obtains

$$\begin{aligned}\boldsymbol{y}(t+1) &= \boldsymbol{y}_l + \sum_{n_1=1}^{n_a} \boldsymbol{f}'_{n_1}\Big|_{\substack{u=u_l\\ y=y_l}} \cdot\left[\boldsymbol{y}(t-n_a+n_1)-\boldsymbol{y}_l\right]\\ &+\sum_{n_2=0}^{n_b} \boldsymbol{f}'_{n_2}\Big|_{\substack{u=u_l\\ y=y_l}} \cdot\left[\boldsymbol{u}(t-n_b+n_2)-\boldsymbol{u}_l\right] + \boldsymbol{o}\left[\boldsymbol{x}(t)\right],\end{aligned} \tag{2}$$

where $\boldsymbol{f}'_{n_1} = \dfrac{\partial \boldsymbol{f}}{\partial \boldsymbol{y}(t-n_a+n_1)}$, $n_1 = 1,\cdots,n_a$, $\boldsymbol{f}'_{n_2} = \dfrac{\partial \boldsymbol{f}}{\partial \boldsymbol{u}(t-n_b+n_2)}$, $n_2 = 0,\cdots,n_b$ and $\boldsymbol{x}(t) = \left[\boldsymbol{y}(t)-\boldsymbol{y}_l,\cdots;\boldsymbol{u}(t)-\boldsymbol{u}_l,\cdots\right]$ respectively. The nonlinear term $\boldsymbol{o}[\boldsymbol{x}(t)]$ satisfies $\lim_{\|\boldsymbol{x}(t)\|\to 0} \frac{\|\boldsymbol{o}[\boldsymbol{x}(t)]\|}{\|\boldsymbol{x}(t)\|} = 0$, where $\|\cdot\|$ is the Euclidean norm operator.

Define

$$\bar{\boldsymbol{y}}(t) = \boldsymbol{y}(t) - \boldsymbol{y}_l, \tag{3}$$

$$\bar{\boldsymbol{u}}(t) = \boldsymbol{u}(t) - \boldsymbol{u}_l, \tag{4}$$

$$\boldsymbol{v}(t) = \boldsymbol{o}[\boldsymbol{x}(t)], \tag{5}$$

$$\boldsymbol{A}_{n_l}^l = (-1)\cdot \boldsymbol{f}_{n_1}'\Big|_{\substack{u=u_l\\ y=y_l}}, n_1 = 1,\cdots,n_a, \tag{6}$$

$$\boldsymbol{B}_{n_2}^l = \boldsymbol{f}_{n_2}'\Big|_{\substack{u=u_l\\ y=y_l}}, n_2 = 0,\cdots,n_b, \tag{7}$$

$$\boldsymbol{A}^l(z^{-1}) = \boldsymbol{I} + \boldsymbol{A}_1^l z^{-1} + \cdots + \boldsymbol{A}_{n_a}^l z^{-n_a}, \tag{8}$$

$$\boldsymbol{B}^l(z^{-1}) = \boldsymbol{B}_0^l + \boldsymbol{B}_1^l z^{-1} + \cdots + \boldsymbol{B}_{n_b}^l z^{-n_b}. \tag{9}$$

Then system (2) can be rewritten as

$$\boldsymbol{A}^l(z^{-1})\bar{\boldsymbol{y}}(t+1) = \boldsymbol{B}^l(z^{-1})\bar{\boldsymbol{u}}(t) + \boldsymbol{v}(t). \tag{10}$$

Remark 1: Although the representation of the system (10) is linear, the term $\boldsymbol{v}(t)$ is a nonlinear term. Then it is viewed as measurable disturbance and eliminated by using feedforward method.

3 Design of HMMNNDC

In the industrial process, on the one hand, the environment, where the system runs, is too diverse to satisfy the strict requirement which the nonlinear controller needs. On the other hand, the engineers are willing to employ easy linear control theory because of less mathematical knowledge. So a nonlinear system would always be expanded around the equilibrium point. If better performance is required, more equilibrium points should be needed. However, too many equilibrium points mean too many models and too many computations. To solve this problem, a hierarchical structure is designed here.

polynomial matrices. Like the conventional optimal controller design, for the model j, the cost function is of the form

$$J_c = \left\| P(z^{-1})\bar{y}(t+k) - R(z^{-1})w(t) + Q(z^{-1})\bar{u}(t) + S(z^{-1})v(t) \right\|^2, \quad (15)$$

where $w(t)$ is the known reference signal, $P(z^{-1}), Q(z^{-1}), R(z^{-1}), S(z^{-1})$ are weighting polynomial matrices respectively. Introduce the identity as

$$P(z^{-1}) = F(z^{-1})A(z^{-1}) + z^{-1}G(z^{-1}). \quad (16)$$

Multiplying (14) by $F(z^{-1})$ from left and using (16), the optimal control law can be derived as follows

$$\begin{aligned} &G(z^{-1})\bar{y}(t) + [F(z^{-1})B(z^{-1}) + Q(z^{-1})]\bar{u}(t) \\ &+[F(z^{-1}) + S(z^{-1})]v(t) = Rw(t). \end{aligned} \quad (17)$$

Combing (17) with (14), the closed loop system equation is obtained as follows

$$\begin{aligned} &\left[P(z^{-1}) + Q(z^{-1})B^{-1}(z^{-1})A(z^{-1})\right]\bar{y}(t+1) = \\ &\left[Q(z^{-1})B(z^{-1})^{-1} - S(z^{-1})\right]v(t) + R(z^{-1})w(t). \end{aligned} \quad (18)$$

To eliminate the nonlinear form exactly, let

$$Q(z^{-1}) = R_1 B(z^{-1}), \quad (19)$$

$$S(z^{-1}) = R_1, \quad (20)$$

$$P(z^{-1}) + R_1 A(z^{-1}) = T(z^{-1}), \quad (21)$$

$$R(z^{-1}) = T(1), \quad (22)$$

where $T(z^{-1})$ is a stable diagonal polynomial matrix decided by the designer and R_1 is a constant matrix.. So the closed loop system is derived as

$$T(z^{-1})y(t+k) = T(1)w(t). \quad (23)$$

From the equation (23), it can be derived that, by the choice of weighting polynomial matrixes, the closed loop system not only decouples the system dynamically but also places poles arbitrarily.

Remark 2. Equation (17) is a nonlinear equation because $\bar{u}(t)$ is include into the nonlinear term $v(t)$. Considering $\bar{u}(t)$ will converge to a constant vector in steady state, then substitute $\bar{u}(t)$ in the nonlinear term $v(t)$ with $\bar{u}(t-1)$ and solve (17).

4 Simulation Studies

A discrete-time nonlinear multivariable system is described as follows

$$\begin{cases} y_1(t+1) = \dfrac{-0.2y_1(t)}{1+y_1^2(t)} + \sin[u_1(t)] - 0.5\sin[u_1(t-1)] \\ \qquad\qquad + 1.5u_2(t) + 0.2u_2(t-1) \\ y_2(t+1) = 0.6y_2(t) + 0.2u_1(t) + 1.3u_1(t-1) + u_2(t) \\ \qquad\qquad + u_2^2(t) + \dfrac{1.5u_2(t-1)}{1+u_2^2(t-1)} \end{cases}, \tag{24}$$

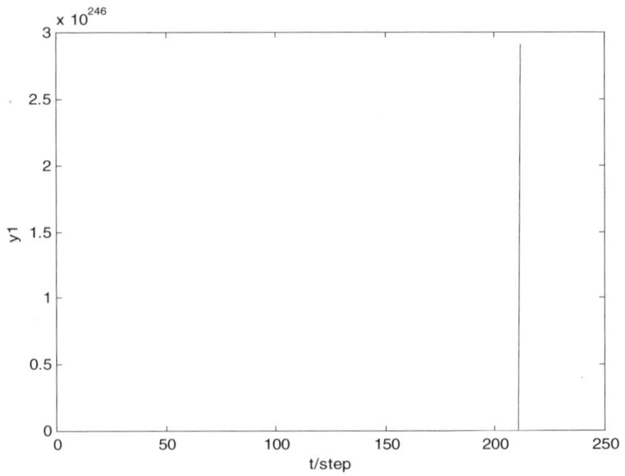

Fig. 2. The output $y_1(t)$ of NNDC

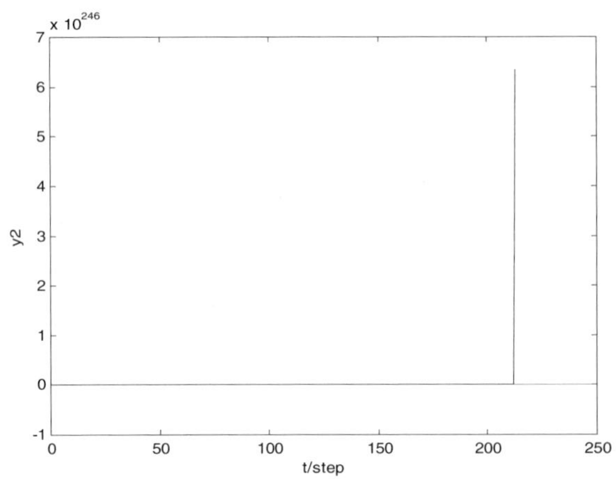

Fig. 3. The output $y_2(t)$ of NNDC

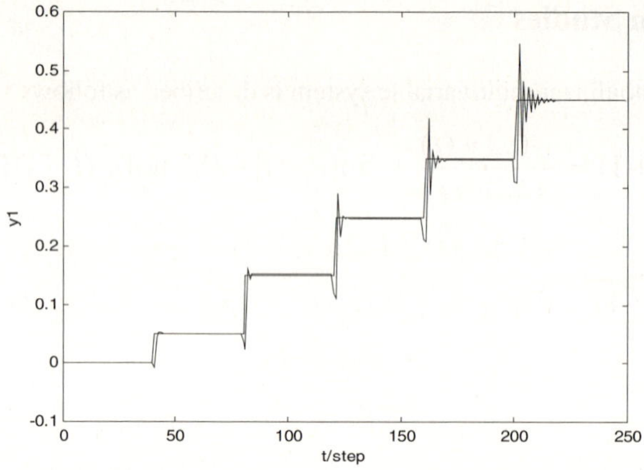

Fig. 4. The output $y_1(t)$ of HMMNNDC

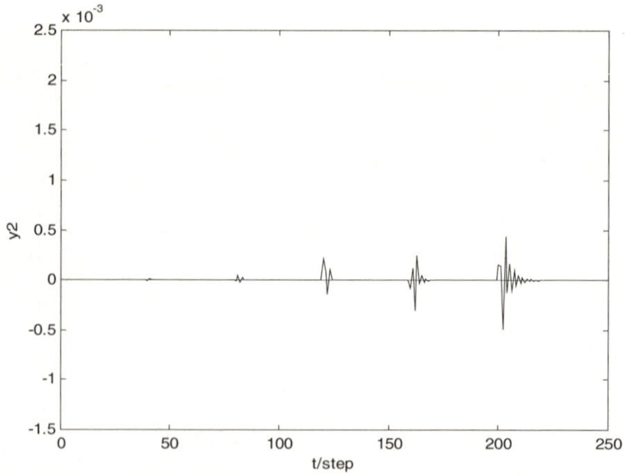

Fig. 5. The output $y_2(t)$ of HMMNNDC

which is the same as the simulation example in [12]. The known reference signal $\boldsymbol{w}$ is set to be a time-varying signal. When $t=0$, w_1 equals to 0 and when t is 40, 80, 120, 160, 200, it changed into 0.05, 0.15, 0.25, 0.35, 0.45 respectively, while w_2 equals to 0 all the time.

In Fig.2 and 3, the system (25) is expanded only at the original point $(0,0)$ and a Neural Network Decoupling Controller (NNDC) is used. In Fig.4 and 5, the system is expanded at six equilibrium points, *i.e.* $[0,0]^{\mathrm{T}}$, $[0.1,0]^{\mathrm{T}}$, $[0.2,0]^{\mathrm{T}}$, $[0.3,0]^{\mathrm{T}}$, $[0.4,0]^{\mathrm{T}}$ and. Note that the equilibrium points are far away from the set points. The results show that although the same NNDC method is adopted, the system using NNDC loses its stability (see Fig.2 and 3), while the system using HMMNNDC not only gets the good performance but also has good decoupling result (see Fig.4 and 5).

5 Conclusion

A HMMNNDC is designed to control the discrete-time nonlinear multivariable system. A SOM NN is employed to partition the whole working area into several sub-regions. In each sub-region, around each equilibrium point, one NN is trained offline to identify the linear term of the nonlinear system and the other NN is trained online to identify the nonlinear one. The multiple models set is composed of all models, which are got from all equilibrium points. According to the switching index, the best model is chosen as the system model. The nonlinear term of the system is viewed as measurable disturbance and eliminated by the use of the weighting polynomial matrices.

Acknowledgments. This work is supported by the National Natural Science Foundation of China (Grant: 60504010, 60774015), the High Technology Research and Development Program of China (Grant: 2008AA04Z129), the Specialized Research Fund for the Doctoral Program of Higher Education of China (Grant: 20060248001), and Natural Science Foundation of Jiangxi (Grant: 0611006).

References

1. Wang, X., Li, S.Y., et al.: Multi-model Direct Adaptive Decoupling Control with Application to the Wind Tunnel. ISA Transactions 44, 131–143 (2005)
2. Lin, Z.L.: Almost Disturbance Decoupling with Global Asymptotic Stability for Nonlinear Systems with Disturbance-affected Unstable Zero Dynamics. Systems & Control Letters 33, 163–169 (1998)
3. Lin, Z.L., Bao, X.Y., Chen, B.M.: Further Results on Almost Disturbance Decoupling with Global Asymptotic Stability for Nonlinear Systems. Automatica 35, 709–717 (1999)
4. Ansari, R.M., Tade, M.O.: Nonlinear Model-based Process Control: Applications in Petroleum Refining. Springer, London (2000)
5. Khail, H.K.: Nonlinear Systems. Prentice Hall Inc., New Jersey (2002)
6. Germani, A., Manes, C., Pepe, P.: Linearization and Decoupling of Nonlinear Delay Systems. In: Proceedings of the American Control Conference, pp. 1948–1952 (1998)
7. Wang, W.J., Wang, C.C.: Composite Adaptive Position Controller for Induction Motor Using Feedback Linearization. IEE Proceedings D Control Theory and Applications 45, 25–32 (1998)
8. Wai, R.J., Liu, W.K.: Nonlinear Decoupled Control for Linear Induction Motor Servo-Drive Using The Sliding-Mode Technique. IEE Proceedings D Control Theory and Applications 148, 217–231 (2001)
9. Balchen, J.G., Sandrib, B.: Elementary Nonlinear Decoupling Control of Composition in Binary Distillation Columns. Journal of Process Control 5, 241–247 (1995)
10. Haykin, S.S.: Neural Networks: A Comprehensive Foundation. Prentice Hall Inc., New Jersey (1999)
11. Ho, D.W.C., Ma, Z.: Multivariable Internal Model Adaptive Decoupling Controller with Neural Network for Nonlinear Plants. In: Proceedings of the American Control Conference, pp. 532–536 (1998)
12. Yue, H., Chai, T.Y.: Adaptive Decoupling Control of Multivariable Nonlinear Non-Minimum Phase Systems Using Neural Networks. In: Proceedings of the American Control Conference, pp. 513–514 (1998)
13. Kohonen, T.: Self-Organizing Feature Maps. Springer, New York (1995)
14. Hornik, K., Stinchcombe, M., White, H.: Universal Approximation of an Unknown Mapping and Its Derivatives using Multilayer Feedforward Networks. Neural Networks 3, 551–560 (1990)

Adaptive Dynamic Programming for a Class of Nonlinear Control Systems with General Separable Performance Index

Qinglai Wei[1], Derong Liu[2], and Huaguang Zhang[1]

[1] School of Information Science and Engineering, Northeastern University Shenyang, China
qinglaiwei@gmail.com, zhanghuaguang@mail.neu.edu.cn

[2] Department of Electrical and Computer Engineering, University of Illinois at Chicago, Chicago, USA
dliu@ece.uic.edu

Abstract. In this paper, an iterative scheme for a class of dynamic programming problem with general separable performance index has been studied. For the dynamic programming problem, the performance index function is time-varying and doesn't have the uniformed recurrent formulation. Noticing such prominent feature, adaptive dynamic programming (ADP) method is introduced. The proposed method aims to find out the efficient solution of the dynamic programming. Because of the approximation of the performance index, the optimal control can be computed forward-in-time. A proof is given to guarantee the convergence, and finally a case study shows the effectiveness of the proposed method.

Keywords: Dynamic programming, Adaptive dynamic programming, Separable performance index.

1 Introduction

This paper proposes an iterative algorithm to solve a class of dynamic programming problems with a general performance index that can be back forward separable. General separable performance index that are not stagewise additive which are common in control processes often result from optimization problems in areas such as minimax control, reliability optimization, multi-reservoir systems, chemical engineering process, and mathematical programming.

Consider the following performance index (P)

$$\begin{aligned} &\min_{U} J(X,U) \\ &\text{s.t. } x_{t+1} = f_t(x_t, u_t),\ t = 0, ..., T-1, \end{aligned} \tag{1}$$

where $U = \left(u_0^T, u_1^T, \ldots, u_{T-1}^T\right)^T$, $X = \left(x_0^T, x_1^T, \ldots, x_T^T\right)^T$, $x_t \in R^n$ is the state $u_t \in R^m$ is the control, and $f_t : R^n \times R^m \to R^n$, $t = 0, 1, \ldots, T-1$ T is terminal time and is system function.

F. Sun et al. (Eds.): ISNN 2008, Part II, LNCS 5264, pp. 128–137, 2008.

In this paper, we assume the performance index is general separable i.e. there exist functions φ_t: $R^n \times R^m \times R \to R$, $t = 0, 1, \ldots, T-1$, and φ_T: $R^n \to R$ such that

$$\begin{aligned} y_T &= \varphi_T(x_T) \\ y_t &= \varphi_t(x_t, u_t, y_{t+1}) \\ J(X,U) &= y_0. \end{aligned} \quad (2)$$

Here we make some assumptions:

(1) All the y_t, $t = 0, 1, \ldots, T$ is nonnegative,

(2) $\partial y_t / \partial y_{t+1} > 0$,this assumption ensures the satisfaction of the monotonicity condition.

(3) $J(X,U)$is convex.

(4) Both the performance index function and the system function have continuous second-order derivatives.

Problem formulation in (P) covers all types of optimal control problems with a separable structure in the sense of dynamic programming method. Dynamic programming with a general separable performance index was first studied in [1] and [2]. Li [3] applied the separable performance index dynamic programming into a class of nonseparable performance index dynamic programming problem whose performance index is of the form $J = \sum_i J_i$, while each J_i must be of a separable form. Liao [4] applied differential dynamic programming (DDP) method to the multipleobjective optimal control and obtained some good results. While DDP method is just quadratic approximate algorithm, it is difficult to compute the performance index accurately and the performance index function must be computed in backward time. Adaptive dynamic programming (ADP) is an effective approach to solve dynamic programming problems [5, 6]. Adaptive dynamic programming was proposed by Werbos [7], Barto *et. al.* [8], Watkins [9], Bertsekas [10], and others to solve optimal control problems in forward time. Werbos [5] classified adaptive dynamic programming approaches into four main schemes: Heuristic Dynamic Programming (HDP), Dual Heuristic Dynamic Programming (DHP), Action Dependent Heuristic Dynamic Programming (ADHDP), also known as Q-learning, and Action Dependent Dual Heuristic Dynamic Programming (ADDHP). In [11], Prokhorov and Wunsch developed new adaptive dynamic programming schemes known as Globalized Dual Heuristic Dynamic Programming (GDHP) and Action Dependent Globalized Dual Heuristic Dynamic Programming (ADGDHP). Landelius [12] applied HDP, DHP, ADHDP and ADDHP techniques to the discrete-time linear quadratic optimal control problem. Derong Liu [13] [14] applied ADHDP technique into cart-pole control problem. Huaguang Zhang [15] introduced ADACDs into system identification.

In this paper, we aim to develop ADP method into dynamic programming problems with general separable performance index function and propose an ADP for separable performance index function (ADPSPI) method. The importance of the paper is that using the ADPSPI method, we can approximate the performance index of next time step, and then we can iteratively compute the control and the performance index in forward-time.

2 Preliminaries

To solve the optimal control problem (P), we aim to find an efficient control sequence $U^* = ((u_0^*)^T, \ldots, (u_{T-1}^*)^T)^T$ so as to make the performance index minimum, and the corresponding state trajectory is formulated as $X^* = ((x_0^*)^T, \ldots, (x_T^*)^T)^T$.

Lemma 1. If problem (P) is satisfied with assumptions 1 – 4, then for all $t = 0, 1, \ldots, T-1$ there exist $\theta_t \geq 0$ to meet $(P_1(t))$

$$\begin{aligned} &\min_{\{u_t, u_{t+1}, \ldots, u_{T-1}\}} \quad \theta_t y_t \\ &x_{\tau+1} = f_\tau(x_\tau, u_\tau), \ \tau = t, \ldots, T-1, \end{aligned} \tag{3}$$

where θ_t, $t = 0, 1, \ldots, T-1$, satisfy the following equation $\theta_{t+1} = \theta_t \frac{\partial y_t}{\partial y_{t+1}}$ with all derivatives being evaluated at (X^*, U^*).

The proof is showed in [4].

As we can see that the performance index function is changed at different time stage, furthermore, the system function may also change at every time stage, thus it is impossible to build up a uniformed recurrent method for the whole time stage. Necessary development must be adopted based on the popular ADP method that can be applied to this environment.

3 Adaptive Dynamic Programming Method

Based on the basic properties discussed in the previous section, we will develop adaptive dynamic programming method to solve the dynamic programming problem with general separable performance index function. The elementary idea is based on [4]. While in [4] the performance index function is computed in backward-time and it is always difficult to apply in real world control systems. On the other hand, using DDP method, we should compute quadratic approximation of the performance index function every time stage, while the calculation of the second-order derivative is usually quite hard. So the computation complexity is very high. Using ADP, neural network is used to approximate the optimal performance index function. According to the character of neural network we can approximate the performance index function at any precision. Using ADPSPI method, the computation burden is much reduced, and the curse of dimensionality can also be avoided.

In the ADPSPI method, there are two procedures: feedback procedure and update procedure. During the feedback procedure we should compute the efficient control for $t = 0, 1, \ldots, T-1$, we start with initial performance index function$y^0(x_t) = 0$, which is not necessary the optimal performance index function. Then we find the control u^0as follows

$$u^0(x_t) = \arg\min_u \{\theta_t \phi_t(x_t, u, y^0(x_{t+1}))\}. \tag{4}$$

Next we can update the performance index function by

$$y^1(x_t) = \phi_t(x_t, u^0(x_t), y^0(x_{t+1})). \tag{5}$$

The HDP scheme therefore iterates between the following two equations

$$u^i(x_t) = \arg\min_u\{\theta_t\phi_t(x_t, u^i(x_t), y^i(x_{t+1}))\} \tag{6}$$

and

$$y^{i+1}(x_t) = \phi_t(x_t, u^i(x_t), y^i(x_{t+1})). \tag{7}$$

Then we obtain

$$\theta_{t+1} = \theta_t \frac{\partial y_t}{\partial y_{t+1}}. \tag{8}$$

Now we summarize the solution scheme discussed above, present the ADP algorithm for the general separable performance index function:

Step 1 Given θ_0 , x_0, $i_{\max}$ and ε.
Step 2 Let $y^i = 0$ when $i = 0$.
Step 3 Compute $u^i(x_k) = \arg\min_u\{\theta_t\phi_t(x_t, u, y^i(x_{t+1}))\}$.
Step 4 Compute $y^{i+1}(x_t) = \phi_t(x_t, u^i(x_t), y^i(x_{t+1}))$.
Step 5 If $\left|y^{i+1} - y^i\right| < \varepsilon$ then let $x^* = x$, $u^* = u$, and go to step 7.
Step 6 If $i > i_{\max}$stop, otherwise $i = i + 1$ go to step 3.
Step 7 Compute $\theta_{t+1} = \partial y_t / \partial y_{t+1}$.
Step 8 if $t = T$ stop, else $t = t + 1$ and $i = 0$ go to step 2.

Remark: As the additive performance index function is a special case to the general separable performance index function, so the above method is also effective for the additive performance index function, while in the iterative procedure, the coefficient θ_t is time-invariant which equals to θ_0.

4 Convergence Analysis

In the above section, we expatiate how to derive the feedback optimal control using the presented ADP method. In this section, a convergence analysis is carried out for the proposed method.

Theorem 1: If the system is controllable and assumptions 1 − 4 hold, then $u^i(x_t) - u^*(x_t)$ is quadratic bounded which can be expressed as $u^i(x_t) - u^*(x_t) = o(\sum_{k=1}^{t} (\bar{u}_k - u_k^*)^2)$, where $u^*(x_t)$ is efficient solution for(P) and $\bar{u}_k$ is a stabilized control.

Proof:As $\bar{u}_k$ is a stabilized control for $t = k$ that may not be optimal and $\bar{x}_k$ is the corresponding state strategy. According to assumption 4, quadratic Taylor expansion of the performance index function $\theta_t\phi_t(x_t, u, y^i(x_{t+1}))$ can be written as

$$QP\left(\theta_t\phi_t(x_t,u_t^i,y^i(x_{t+1}))\right)=\theta_t\phi_t\left(\bar{x}_t,\bar{u}_t,y^i(f_t(\bar{x}_t,\bar{u}_t))\right)+\alpha_t^i\Delta x_t+\beta_t^i\Delta u_t^i \quad (9)$$
$$+\frac{1}{2}(\Delta x_t)^T A_t^i\Delta x_t+(\Delta u_t^i)^T B_t^i\Delta x_t+\frac{1}{2}(\Delta u_t^i)^T C_t^i\Delta u_t^i$$

where $A_t^i\in R^{n\times n}$ and $C_t^i\in R^{m\times m}$, $t=0,1,\ldots,T-1$ are all positive definite matrices, B_t^i is an $m\times n$ matrix, $\alpha_t^i\in R^n$, $\beta_t^i\in R^m$ $\Delta x_t=x_t-\bar{x}_t$, $\Delta u_t^i=u_t^i-\bar{u}_t$.

The details of the matrices are formulated as

$$\alpha_t^i=\theta_t\left(\frac{\partial\phi_t}{\partial x_t}+\frac{\partial\phi_t}{\partial y_{t+1}^i}\frac{\partial y_{t+1}^i}{\partial x_{t+1}}\frac{\partial x_{t+1}}{\partial x_t}\right),\beta_t^i=\theta_t\left(\frac{\partial\phi_t}{\partial u_t}+\frac{\partial\phi_t}{\partial y_{t+1}^i}\frac{\partial y_{t+1}^i}{\partial x_{t+1}}\frac{\partial x_{t+1}}{\partial u_t}\right),$$
$$A_t^i=\theta_t\left(\frac{\partial^2\phi_t}{\partial x_t\partial x_t}+\frac{\partial^2\phi_t}{\partial x_t\partial y_{t+1}^i}\frac{\partial y_{t+1}^i}{\partial x_{t+1}}\frac{\partial x_{t+1}}{\partial x_t}\right.$$
$$+\left[\frac{\partial^2\phi_t}{\partial y_{t+1}^i\partial x_t}+\frac{\partial^2\phi_t}{\partial y_{t+1}^i\partial y_{t+1}^i}\frac{\partial y_{t+1}^i}{\partial x_{t+1}}\frac{\partial x_{t+1}}{\partial x_t}\right]\frac{\partial y_{t+1}^i}{\partial x_{t+1}}\frac{\partial x_{t+1}}{\partial x_t}$$
$$\left.+\frac{\partial\phi_t}{\partial y_{t+1}^i}\frac{\partial^2 y_{t+1}^i}{\partial x_{t+1}\partial x_{t+1}}\frac{\partial x_{t+1}}{\partial x_t}\frac{\partial x_{t+1}}{\partial x_t}+\frac{\partial\phi_t}{\partial y_{t+1}^i}\frac{\partial y_{t+1}^i}{\partial x_{t+1}}\frac{\partial^2 x_{t+1}}{\partial x_t\partial x_t}\right),$$
$$B_t^i=\theta_t\left(\frac{\partial^2\phi_t}{\partial u_t\partial x_t}+\frac{\partial^2\phi_t}{\partial u_t\partial y_{t+1}^i}\frac{\partial y_{t+1}^i}{\partial x_{t+1}}\frac{\partial x_{t+1}}{\partial x_t}\right.$$
$$+\left[\frac{\partial^2\phi_t}{\partial y_{t+1}^i\partial x_t}+\frac{\partial^2\phi_t}{\partial y_{t+1}^i\partial y_{t+1}^i}\frac{\partial y_{t+1}^i}{\partial x_{t+1}}\frac{\partial x_{t+1}}{\partial x_t}\right]\frac{\partial y_{t+1}^i}{\partial x_{t+1}}\frac{\partial x_{t+1}}{\partial u_t}$$
$$\left.+\frac{\partial\phi_t}{\partial y_{t+1}^i}\frac{\partial^2 y_{t+1}^i}{\partial x_{t+1}\partial x_{t+1}}\frac{\partial x_{t+1}}{\partial x_t}\frac{\partial x_{t+1}}{\partial u_t}+\frac{\partial\phi_t}{\partial y_{t+1}^i}\frac{\partial y_{t+1}^i}{\partial x_{t+1}}\frac{\partial^2 x_{t+1}}{\partial u_t\partial x_t}\right),$$
$$C_t^i=\theta_t\left(\frac{\partial^2\phi_t}{\partial u_t\partial u_t}+\frac{\partial^2\phi_t}{\partial u_t\partial y_{t+1}^i}\frac{\partial y_{t+1}^i}{\partial x_{t+1}}\frac{\partial x_{t+1}}{\partial u_t}\right.$$
$$+\left[\frac{\partial^2\phi_t}{\partial y_{t+1}^i\partial u_t}+\frac{\partial^2\phi_t}{\partial y_{t+1}^i\partial y_{t+1}^i}\frac{\partial y_{t+1}^i}{\partial x_{t+1}}\frac{\partial x_{t+1}}{\partial u_t}\right]\frac{\partial y_{t+1}^i}{\partial x_{t+1}}\frac{\partial x_{t+1}}{\partial u_t}$$
$$\left.+\frac{\partial\phi_t}{\partial y_{t+1}^i}\frac{\partial^2 y_{t+1}^i}{\partial x_{t+1}\partial x_{t+1}}\frac{\partial x_{t+1}}{\partial u_t}\frac{\partial x_{t+1}}{\partial u_t}+\frac{\partial\phi_t}{\partial y_{t+1}^i}\frac{\partial y_{t+1}^i}{\partial x_{t+1}}\frac{\partial^2 x_{t+1}}{\partial u_t\partial u_t}\right).$$

All the values of these coefficients are obtained at$(\bar{x}_t,\bar{u}_t)$. So we can calculate Δu_t^i to make the performance index function minimum. Following the idea of the principle of optimality, we can obtain $\Delta u_t^i=-(C_t^i)^{-1}\beta_t^i \quad -(C_t^i)^{-1}(B_t^i)^T\Delta x_t$.

In the following part, we prove $u^i(x_t)-u^*(x_t)=o(\sum\limits_{k=0}^{t}(\bar{u}_k-u_k^*)^2)$by mathematical induction. First, we prove it holds for $t=0$.

For $t=0$ we have

$$\theta_0\left(\frac{\partial\varphi_0}{\partial u_0}+\frac{\partial\varphi_0}{\partial y_1^i}\frac{\partial y_1^i}{\partial x_1}\frac{\partial x_1}{\partial u_0}\right)\bigg|_{(x_0^*,u_0^*)}-\theta_0\left(\frac{\partial\varphi_0}{\partial u_0}+\frac{\partial\varphi_0}{\partial y_1^i}\frac{\partial y_1^i}{\partial x_1}\frac{\partial x_1}{\partial u_0}\right)\bigg|_{(\bar{x}_0,\bar{u}_0)}$$
$$=\theta_0\frac{\partial^2\varphi_0}{\partial u_0\partial u_0}(u_0^*-\bar{u}_0)+\theta_0\frac{\partial^2\varphi_0}{\partial u_0\partial y_1^i}\frac{\partial y_1^i}{\partial x_1}\frac{\partial x_1}{\partial u_0}(u_0^*-\bar{u}_0)$$
$$+\theta_0\left[\frac{\partial^2\varphi_0}{\partial y_1^i\partial u_0}(u_0^*-\bar{u}_0)+\frac{\partial^2\varphi_0}{\partial y_1^i\partial y_1^i}\frac{\partial y_1^i}{\partial x_1}\frac{\partial x_1}{\partial u_0}(u_0^*-\bar{u}_0)\right]$$
$$\times\frac{\partial y_1}{\partial x_1}\frac{\partial x_1}{\partial u_0}$$
$$+\theta_0\frac{\partial\varphi_0}{\partial y_1^i}\frac{\partial^2 y_1^i}{\partial x_1\partial x_1}\frac{\partial x_1}{\partial u_0}(u_0^*-\bar{u}_0)\frac{\partial x_1}{\partial u_0}$$
$$+\theta_0\frac{\partial\varphi_0}{\partial y_1^i}\frac{\partial y_1^i}{\partial x_1}\frac{\partial^2 x_1}{\partial u_0\partial u_0}(u_0^*-\bar{u}_0)+o\left((u_0^*-\bar{u}_0)^2\right)$$
$$=\theta_0\left(\frac{\partial^2\varphi_0}{\partial u_0\partial u_0}+\frac{\partial^2\varphi_0}{\partial u_0\partial y_1^i}\frac{\partial y_1^i}{\partial x_1}\frac{\partial x_1}{\partial u_0}+\frac{\partial^2\varphi_0}{\partial y_1^i\partial u_0}\frac{\partial y_1}{\partial x_1}\frac{\partial x_1}{\partial u_0}\right.$$
$$+\frac{\partial^2\varphi_0}{\partial y_1^i\partial y_1^i}\frac{\partial y_1^i}{\partial x_1}\frac{\partial x_1}{\partial u_0}\frac{\partial y_2}{\partial x_2}\frac{\partial x_1}{\partial u_0}$$
$$\left.+\frac{\partial\varphi_0}{\partial y_1^i}\frac{\partial^2 y_1^i}{\partial x_1\partial x_1}\frac{\partial x_1}{\partial u_0}\frac{\partial x_1}{\partial u_0}+\frac{\partial\varphi_0}{\partial y_1^i}\frac{\partial y_1^i}{\partial x_1}\frac{\partial^2 x_1}{\partial u_0\partial u_0}\right)(u_0^*-\bar{u}_0)+o\left((u_0^*-\bar{u}_0)^2\right)$$

where the initial state $x_0 = \bar{x}_0 = x_0^*$. Because $(x_0^*, u_0^*)$is the efficient solution for $t = 0$, then we can derive that $\frac{\partial y^{i+1}(x_0)}{\partial u_0} = \frac{\partial \phi_0(x_0, u^i(x_0), y^i(x_1))}{\partial u_0}$ $= \left(\frac{\partial \phi_0}{\partial u_0} + \frac{\partial \phi_0}{\partial y_1^i}\frac{\partial y_1^i}{\partial x_1}\frac{\partial x_1}{\partial u_0}\right)\Big|_{(x_0^*, u_0^*)} = 0.$

As the optimal control increment is written as

$$\Delta u_0^i = -(C_0^i)^{-1}\beta_0^i - (C_0^i)^{-1}(B_0^i)^T\Delta x_0 \tag{10}$$

So according to (9) we have

$$-(C_0^i)^{-1}\beta_0^i = -(\bar{u}_0 - u_0^*) + o((\bar{u}_0 - u_0^*)^2) \tag{11}$$

while

$$-(C_0^i)^{-1}(B_0^i)^T\Delta x_0 = 0. \tag{12}$$

Then we have

$$u_0^i - u_0^* = o((\bar{u}_0 - u_0^*)^2). \tag{13}$$

Follow this idea, we also have

$$\begin{aligned} x_1 &= f_0(\bar{x}_0, \bar{u}_0) + \tfrac{\partial f_0}{\partial x_0}(x_0 - \bar{x}_0) + \tfrac{\partial f_0}{\partial u_0}(u_0^i - \bar{u}_0) + o((x_0 - \bar{x}_0)^2 + (u_0^i - \bar{u}_0)^2) \\ x_1^* &= f_0(\bar{x}_0, \bar{u}_0) + \tfrac{\partial f_0}{\partial x_0}(x_0^* - \bar{x}_0) + \tfrac{\partial f_0}{\partial u_0}(u_0^* - \bar{u}_0) + o((x_0^* - \bar{x}_0)^2 + (u_0^* - \bar{u}_0)^2) \\ x_1 - x_1^* &= \tfrac{\partial f_0}{\partial u_0}(u_0^i - u_0^*) + o((u_0^i - u_0^*)^2) = o((\bar{u}_0 - u_0^*)^2) \end{aligned} \tag{14}$$

Then it indicates $x_1 - x_1^*$ is also bounded. From assumption 4, we know that

$$\theta_1 = \theta_0 \frac{\partial \phi_t}{\partial y_{t+1}} \tag{15}$$

is also bounded.

Now we assume it is true for $t - 1$. Then for t, we have

$$\begin{aligned} &\theta_t\left(\frac{\partial \phi_t}{\partial u_t} + \frac{\partial \phi_t}{\partial y_{t+1}^i}\frac{\partial y_{t+1}^i}{\partial x_{t+1}}\frac{\partial x_{t+1}}{\partial u_t}\right)\Big|_{(\bar{x}_t, \bar{u}_t)} \\ &-\theta_t\left(\frac{\partial \phi_t}{\partial u_t} + \frac{\partial \phi_t}{\partial y_{t+1}^i}\frac{\partial y_{t+1}^i}{\partial x_{t+1}}\frac{\partial x_{t+1}}{\partial u_t}\right)\Big|_{(x_t^*, u_t^*)} \\ &= \theta_t\left\{\frac{\partial^2 \varphi_t}{\partial u_t \partial u_t} + \frac{\partial^2 \varphi_t}{\partial u_t \partial y_{t+1}^i}\frac{\partial y_{t+1}^i}{\partial x_{t+1}}\frac{\partial x_{t+1}}{\partial u_t}\right. \\ &+ \left[\frac{\partial^2 \varphi_t}{\partial y_{t+1}^i \partial u_t} + \frac{\partial^2 \varphi_t}{\partial y_{t+1}^i \partial y_{t+1}^i}\frac{\partial y_{t+1}^i}{\partial x_{t+1}}\frac{\partial x_{t+1}}{\partial u_t}\right]\frac{\partial y_{t+1}^i}{\partial x_{t+1}}\frac{\partial x_{t+1}}{\partial u_t} \\ &\left.+ \frac{\partial \varphi_t}{\partial y_{t+1}^i}\frac{\partial^2 y_{t+1}^i}{\partial x_{t+1}\partial x_{t+1}}\frac{\partial x_{t+1}}{\partial u_t}\frac{\partial x_{t+1}}{\partial u_t} + \frac{\partial \varphi_t}{\partial y_{t+1}^i}\frac{\partial y_{t+1}^i}{\partial x_{t+1}}\frac{\partial^2 x_{t+1}}{\partial u_t \partial u_t}\right\} \\ &\times(\bar{u}_t - u_t^*) \\ &+\theta_t\left\{\frac{\partial^2 \varphi_t}{\partial u_t \partial x_t} + \frac{\partial^2 \varphi_t}{\partial u_t \partial y_{t+1}^i}\frac{\partial y_{t+1}^i}{\partial x_{t+1}}\frac{\partial x_{t+1}}{\partial x_t}\right. \\ &+ \left[\frac{\partial^2 \varphi_t}{\partial y_{t+1}^i \partial x_t} + \frac{\partial^2 \varphi_t}{\partial y_{t+1}^i \partial y_{t+1}^i}\frac{\partial y_{t+1}^i}{\partial x_{t+1}}\frac{\partial x_{t+1}}{\partial x_t}\right]\frac{\partial y_{t+1}^i}{\partial x_{t+1}}\frac{\partial x_{t+1}}{\partial u_t} \\ &\left.+ \frac{\partial \varphi_t}{\partial y_{t+1}^i}\frac{\partial^2 y_{t+1}^i}{\partial x_{t+1}\partial x_{t+1}}\frac{\partial x_{t+1}}{\partial x_t}\frac{\partial x_{t+1}}{\partial u_t} + \frac{\partial \varphi_t}{\partial y_{t+1}^i}\frac{\partial y_{t+1}^i}{\partial x_{t+1}}\frac{\partial^2 x_{t+1}}{\partial u_t \partial x_t}\right\} \\ &\times(\bar{x}_t - x_t^*) + o\left((\bar{u}_t - u_t^*)^2 + (\bar{x}_t - x_t^*)^2\right) \end{aligned} \tag{16}$$

The optimal control increment is written as $\Delta u_t^i = -(C_t^i)^{-1}\beta_t^i - (C_t^i)^{-1}(B_t^i)^T\Delta x_t$.

We have

$$\begin{aligned} -(C_t^i)^{-1}\beta_t^i =& -(\bar{u}_t - u_t^*) - (C_t^i)^{-1}(B_t^i)^T(\bar{x}_t - x_t^*) \\ &+ o\left((\bar{u}_t - u_t^*)^2 + (\bar{x}_t - x_t^*)\right) \\ =& -(\bar{u}_t - u_t^*) - (C_t^i)^{-1}(B_t^i)^T \frac{\partial x_t}{\partial x_{t-1}}(\bar{x}_{t-1} - x_{t-1}^*) \\ &- (C_t^i)^{-1}(B_t^i)^T \frac{\partial x_t}{\partial u_{t-1}}(\bar{u}_{t-1} - u_{t-1}^*) \\ &+ o\left((\bar{u}_t - u_t^*)^2 + (\bar{u}_{t-1} - u_{t-1}^*)^2 + (\bar{x}_{t-1} - x_{t-1}^*)^2\right) \end{aligned} \tag{17}$$

and

$$\begin{aligned} &-(C^i)^{-1}(B_t^i)^T(x_t - \bar{x}_t) = -(C^i)^{-1}(B_t^i)^T \tfrac{\partial x_t}{\partial u_{t-1}}(u_{t-1}^i - \bar{u}_{t-1}) \\ &-(C^i)^{-1}(B_t^i)^T \tfrac{\partial x_t}{\partial x_{t-1}}(x_{t-1} - \bar{x}_{t-1}) \\ &+o\left((u_{t-1}^i - \bar{u}_{t-1})^2 + (x_{t-1} - \bar{x}_{t-1})^2\right) \end{aligned} \tag{18}$$

then we obtain

$$\begin{aligned} &\Delta u_t^i = -(\bar{u}_t - u_t^*) - (C_t^i)^{-1}(B_t^i)^T \tfrac{\partial x_t}{\partial x_{t-1}}(\bar{x}_{t-1} - x_{t-1}^*) \\ &-(C_t^i)^{-1}(B_t^i)^T \tfrac{\partial x_t}{\partial u_{t-1}}(\bar{u}_{t-1} - u_{t-1}^*) \\ &-(C_t^i)^{-1}(B_t^i)^T \tfrac{\partial x_t}{\partial u_{t-1}}(u_{t-1}^i - \bar{u}_{t-1}) \\ &-(C_t^i)^{-1}(B_t^i)^T \tfrac{\partial x_t}{\partial x_{t-1}}(x_{t-1} - \bar{x}_{t-1}) \\ &+o\left((u_{t-1}^i - \bar{u}_{t-1})^2 + (x_{t-1} - \bar{x}_{t-1})^2\right) \\ &+o\left((\bar{u}_t - u_t^*)^2 + (\bar{u}_{t-1} - u_{t-1}^*)^2 + (\bar{x}_{t-1} - x_{t-1}^*)^2\right). \end{aligned} \tag{19}$$

Then according to the assumption we get

$$u^i(x_t) - u^*(x_t) = o(\sum_{k=0}^{t}(\bar{u}_k - u_k^*)^2). \tag{20}$$

So $u_t^i - u_t^*$ is bounded. The proof is completed.

Theorem 2: If assumptions $1-4$ hold and the system is controllable, then$y^{i+1}(x_t) - y^i(x_t)$ is bounded.

Proof:

In the following part, we also use mathematical induction to prove the theorem.

For $t = 0$we have

$$y^{i+1}(x_0) - \bar{y}^{i+1}(x_0) = \frac{\partial \phi_0}{\partial u_0}(u_0^i - \bar{u}_0) + o((u_0^i - \bar{u}_0)^2) \tag{21}$$

and

$$y^*(x_0) - \bar{y}^{i+1}(x_0) = \frac{\partial \phi_0}{\partial u_0}(u_0^* - \bar{u}_0) + o((u_0^* - \bar{u}_0)^2) \tag{22}$$

where $\bar{y}^{i+1}(x_0) = \phi_0(\bar{x}_0, \bar{u}_0, y^i(f_0(\bar{x}_0, \bar{u}_0)))$. Then we can obtain

$$y^{i+1}(x_0) - y^*(x_0) = \frac{\partial \phi_0}{\partial u_0}(u_0^i - u_0^*) + o((u_0^i - \bar{u}_0)^2 - (u_0^* - \bar{u}_0)^2) \tag{23}$$

We can also derive that

$$\begin{array}{l} y^i(x_0) - y^*(x_0) = \frac{\partial \phi_0}{\partial u_0}(u_0^{i-1} - u_0^*) \\ +o((u_0^{i-1} - \bar{u}_0)^2 - (u_0^* - \bar{u}_0)^2). \end{array} \tag{24}$$

Then according to theorem 1, we can get

$$\begin{array}{l} y^{i+1}(x_0) - y^i(x_0) = (y^{i+1}(x_0) - y^*(x_0)) - (y^i(x_0) - y^*(x_0)) \\ = \frac{\partial \phi_0}{\partial u_0}(u_0^i - u_0^*) - \frac{\partial \phi_0}{\partial u_0}(u_0^{i-1} - u_0^*) \\ +o((u_0^i - \bar{u}_0)^2 - (u_0^* - \bar{u}_0)^2) - o((u_0^{i-1} - \bar{u}_0)^2 - (u_0^* - \bar{u}_0)^2) \\ = o((u_0^* - \bar{u}_0)^2). \end{array} \tag{25}$$

Then proposition holds for $t = 0$.

We assume it is true for $t-1$, then for t, we can derive

$$\begin{array}{l} y^{i+1}(x_t) - \bar{y}^{i+1}(x_t) = \frac{\partial \phi_t}{\partial x_t}(x_t - \bar{x}_t) + \frac{\partial \phi_t}{\partial u_t}(u_t^i - \bar{u}_t) \\ +o((x_t - \bar{x}_t)^2 + (u_t^i - \bar{u}_t)^2) \end{array} \tag{26}$$

and

$$\begin{array}{l} y^*(x_t) - \bar{y}^{i+1}(x_t) = \frac{\partial \phi_t}{\partial x_t}(x_t^* - \bar{x}_t) + \frac{\partial \phi_t}{\partial u_t}(u_t^* - \bar{u}_t) \\ +o((x_t^* - \bar{x}_t)^2 + (u_t^* - \bar{u}_t)^2) \end{array} \tag{27}$$

Then we can get

$$\begin{array}{l} y^{i+1}(x_t) - y^*(x_t) = \frac{\partial \phi_t}{\partial x_t}(x_t - x_t^*) + \frac{\partial \phi_t}{\partial u_t}(u_t^i - u_t^*) \\ +o((x_t - \bar{x}_t)^2 - (x_t^* - \bar{x}_t)^2) + o((u_t^i - \bar{u}_t)^2 - (u_t^* - \bar{u}_t)^2) \end{array} \tag{28}$$

We can also derive

$$\begin{array}{l} y^i(x_t) - y^*(x_t) = \frac{\partial \phi_t}{\partial x_t}(x_t - x_t^*) + \frac{\partial \phi_t}{\partial u_t}(u_t^{i-1} - u_t^*) \\ +o((x_t - \bar{x}_t)^2 - (x_t^* - \bar{x}_t)^2) + o((u_t^{i-1} - \bar{u}_t)^2 - (u_t^* - \bar{u}_t)^2). \end{array} \tag{29}$$

So

$$\begin{array}{l} y^{i+1}(x_t) - y^i(x_t) = \frac{\partial \phi_t}{\partial u_t}(u_t^i - u_t^*) + \frac{\partial \phi_t}{\partial u_t}(u_t^{i-1} - u_t^*) \\ +o((u_t^i - \bar{u}_t)^2 - (u_t^* - \bar{u}_t)^2) - o((u_t^{i-1} - \bar{u}_t)^2 - (u_t^* - \bar{u}_t)^2) \end{array} \tag{30}$$

According to theorem 1, we obtain

$$y^{i+1}(x_t) - y^i(x_t) = o(\sum_{k=0}^{t}(\bar{u}_k - u_k^*)^2) \tag{31}$$

The proof is completed.

Theorem 3: If assumptions 1–4 hold with$y^0(x_t) = 0$, we have $y^{i+1}(x_t) \geq y^i(x_t)$.

Proof: let $\Phi^{i+1}(x_t) = \phi_t(x_t, u_t^{i+1}, \Phi^i(x_{t+1}))$ where $\Phi^0(x_t) = \Phi^0(x_{t+1}) = 0$, then

$$y^1(x_t) - \Phi^0(x_t) = \phi_t(x_t, u_t^0, 0) \geq 0, \tag{32}$$

so we have $y^1(x_t) \geq \Phi^0(x_t)$.

Assume that $y^i(x_t) \geq \Phi^{i-1}(x_t)$. Since

$$\Phi^i(x_t) = \phi_t(x_t, u_t^i, \Phi^{i-1}(x_{t+1})) \tag{33}$$

and

$$y^{i+1}(x_t) = \phi_t(x_t, u_t^i, y^i(x_{t+1})) \tag{34}$$

then we have

$$y^{i+1}(x_t) - \Phi^i(x_t) = \frac{\partial y^{i+1}}{\partial y^i}(y^i(x_t) - \Phi^{i-1}(x_t)). \tag{35}$$

We know that $\frac{\partial y^{i+1}}{\partial y^i} > 0$ and according to the assumption above we have $y^{i+1}(x_t) \geq \Phi^i(x_t)$.

On the other hand, $y^{i+1} = \min_u\{\phi_t(x_t, u_t^i, y^i(x_{t+1}))\}$, It is straight forward from the fact thaty^i is a result of minimizing the right hand side of equation with respect to the control input u, so we have $\Phi^i(x_t) \geq y^i(x_t)$, finally we obtain $y^{i+1}(x_t) \geq y^i(x_t)$. The proof is completed.

Remark2: The basic idea of theorem 3 is appeared in reference [3], while reference [3] only analyzed the additive performance index. In this paper we expanded this result to the ADP problem with general separable performance index function and proved its effectiveness.

Corollary: In practice, we can choose$\phi_t > 0$. Therefore, we have$y^{i+1}(x_t) > y^i(x_t)$.

Proof: the process is the same as theorem 3 and so is omitted.

5 Conclusion

This paper discusses a new approach to solve a class of discrete-time nonlinear dynamic problem with general separable problem. The effectiveness is exhibited in the forward-time tuning the weights of the networks using the ADP technique, and more important giving an optimal control scheme when the system function is changed at different time stage. And the convergence proofs is also given to enhance the theory base.

Acknowledgment

This work was supported in part by the National Natural Science Foundation of China under Grants 60534010, 60572070, 60774048, and 60728307, in part by the Program for Changjiang Scholars and Innovative Research Groups of China under Grant No. 60521003, and in part by the National High Technology Research and Development Program of China under Grant No. 2006AA04Z183.

References

1. Nemhauser, G.L.: Introduction to Dynamic Programming. Wiley, New York (1966)
2. Furukawa, N., Iwamoto, S.: Dynamic Programming on Recursive Reward Systems. Bulletin of Mathematics Statistics 17, 103–126 (1976)
3. Li, D., Haimes, Y.Y.: New Approach for Nonseparable Dynamic Programmin Problems. Journal of Optimization Theory and Applications 64, 311–330 (1990)
4. Liao, L.Z., Li, D.: Adaptive Differential Dynamic Programming for Multiobjective Optimal Control. Automatica 38, 1003–1015 (2002)
5. Liu, D., Zhang, Y., Zhang, H.: A Self-learning Call Admission Control Scheme for CDMA Cellular Networks. IEEE Trans. Neural Networks 16, 1219–1228 (2005)
6. Li, B., Si, J.: Robust Dynamic Programming for Discounted Infinite-Horizon Markov Decision Processes with Uncertain Stationary Transition Matrices. In: Proceedings of the IEEE Symposium on Approximate Dynamic Programming and Reinforcement Learning, pp. 96–102 (2007)
7. Werbos, P.J.: A Menu of Designs for Reinforcement Learning Over Time. In: Miller, W.T., Sutton, R.S., Werbos, P.J. (eds.) Neural Networks for Control. MIT Press, Cambridge (1991)
8. Widrow, B., Gupta, N., Maitra, S.: Punish/reward: Learning with A Critic in Adaptive Threshold Systems. IEEE Trans. Syst., Man, Cybern. 3, 455–465 (1973)
9. Watkins, C.: Learning from Delayed Rewards. Ph.D. Thesis, Cambridge University, Cambridge, England (1989)
10. Bertsekas, D.P., Tsitsiklis, J.N.: Neuro-Dynamic Programming. Athena Scientific, MA (1996)
11. Prokhorov, D., Wunsch, D.: Adaptive Critic Designs. IEEE Trans. on Neural Networks 8, 997–1007 (1997)
12. Landelius, T.: Reinforcement Learning and Distributed Local Model Synthesis. PhD Dissertation, Linkoping University, Sweden (1997)
13. Liu, D., Zhang, H.: A Neural Dynamic Programming Approach for Learning Control of Failure Avoidance Problems. International Journal of Intelligence Control and Systems 10, 21–32 (2005)
14. Liu, D., Xu, X., Zhang, Y.: Action-Dependent Adaptive Critic Designs. In: IEEE Neural Networks Proceedings, pp. 990–995 (2001)
15. Zhang, H., Luo, Y., Liu, D.: A New Fuzzy Identification Method Based on Adaptive Critic Designs. In: Wang, J., Yi, Z., Żurada, J.M., Lu, B.-L., Yin, H. (eds.) ISNN 2006. LNCS, vol. 3971, pp. 804–809. Springer, Heidelberg (2006)

A General Fuzzified CMAC Controller with Eligibility

Zhipeng Shen, Ning Zhang, and Chen Guo

College of Information Science and Technology,
Dalian Maritime University, 116026, Dalian, China
s_z_p@263.net

Abstract. This paper presents an online neural network controller. Cerebellar Model Articulation Controller (CMAC) is suitable to online control due to its fast learning speed. By integrating the CMAC address scheme with fuzzy logic concept, a general fuzzified CMAC (GFAC) is proposed. Then by incorporating the concept of eligibility into the GFAC, a GFAC controller with eligibility is presented, named FACE. A learning algorithm for the FACE is derived to tune the model parameters. To achieve online control, an efficient implementation of the proposed FACE method is presented. As an example, the proposed FACE is applied to a ship steering control system. The simulation results show that the ship course can be properly controlled under the disturbances of wave, wind and current.

Keywords: General fuzzified CMAC (GFAC), receptive field function, eligibility, ship steering control.

1 Introduction

CMAC is an acronym for Cerebellar Model Articulation Controller,which was first described by Albus in 1975 as a simple model of the cortex of the cerebellum [1]. It has been successfully applied in the approximation of complicated functions [2], and it has been used in the field of system recognition and control [3]. However, because the original receptive field function of CMAC is zero rank (i.e., rectangle receptive field), it makes the network output discontinuous.In recent years, combining CMAC with fuzzy logic becomes a very popular direction [4-6]. However, until now there is no complete framework available. So it is necessary to have a systematic study on the fuzzy CMAC (FCMAC) structure and its learning algorithm. Thus a general FCMAC will be beneficial to further theoretical development and various applications.

Eligibility is a concept first described by Klopf [7], and it has been used for many years as part of the reinforcement learning paradigm [8]. Sutton had a systematic study of the eligibility in his doctoral thesis [9]. In this paper, by incorporating the eligibility in GFAC, a general fuzzy CMAC controller with Eligibility (FACE) is proposed. The basic idea is that each weight in the GFAC neural network is given an associated value called its "eligibility". When a weight is used, i.e. when its value makes a contribution to the network output, the eligibility is increased, or it decays toward zero. Training should change all eligible weights by an amount proportional to their eligibility, i.e., the input error signal is only effective at integrating the weight

F. Sun et al. (Eds.): ISNN 2008, Part II, LNCS 5264, pp. 138–147, 2008.

while the eligibility is nonzero. Eligibility can improve the system performance in two ways. First, it can increase the system stability by achieving a "braking" effect, and reduce the oscillations. The second improvement is that the cause and effect are better associated in the system, and the system state may be able to anticipate the change.

The organization of this paper is as follows: in section 2 a general fuzzified CMAC structure and its learning algorithm are proposed. In section 3, the FACE algorithm will be formally derived by considering how to optimize an error function of the controlled system states. To achieve online control, an efficient implementation of FACE method is developed. As an example the proposed FACE is finally applied to a ship steering control system in Section 4.

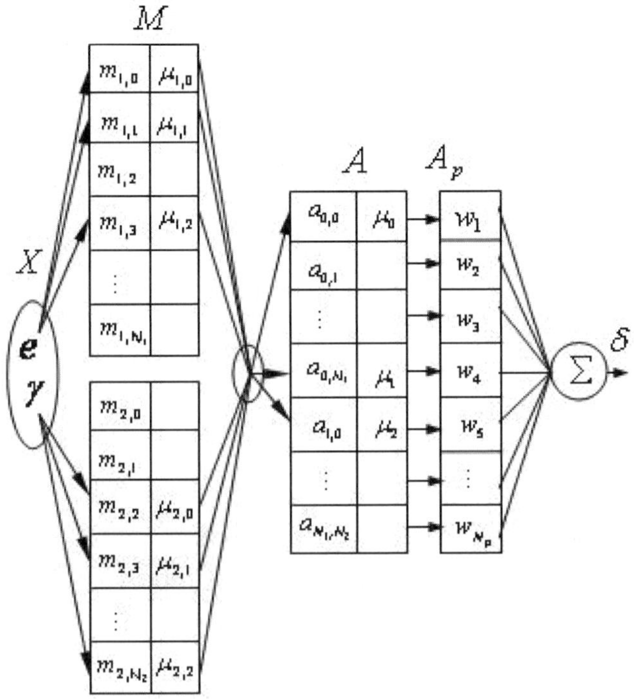

Fig. 1. The structure of GFAC

2 GFAC Structure and Its Learning Algorithm

The proposed network uses fuzzified language to define input variables, integrates fuzzy membership function $\mu(.)$ into association unit, so it has the fuzzy logic property. At the same time, it uses CMAC addressing method as mapping, so the input space can be demarcated better, which is different from conventional fuzzy CMAC [3-6]. Therefore, this network is called general fuzzified CMAC, abbreviated as GFAC.GFAC implementation is similar with traditional CMAC, converting the mapping $X \Rightarrow A$ into two sub-mapping. One is $R: X \Rightarrow M$, the other is $E: M \Rightarrow A$,where mapping $R(X)$ decides the excited unit position of middle variable *M* for input vector *X*, and calculates the value of the corresponding membership function. *E* is a synthesized function of excited input unit and association location unit *A*. Fig.1 shows the GFAC structure with double inputs, single output and generalization size 3. Where *A* and A_p are, respectively, conceptional memorizer and practical memorizer, $\hat{y}$ and y are, respectively, expected output and practical output, and $m_{i,j}$ is the mapping address in middle variable *M* of input vector x_i.

Suppose input vector is $X = [x_1 \quad x_2 \quad \dots \quad x_N]^T$, where $x_i \in I_i$, and I_i is a finite space and defined as:

$$I_i = \{x_i : x_{i,\min} \le x_i \le x_{i,\max}\} \tag{1}$$

The section is demarcated as

$$x_{i,\min} \le \lambda_{i,1} \le \lambda_{i,2} \le \cdots \le \lambda_{i,N_i} \le x_{i,\max} \tag{2}$$

Where $\lambda_{i,j}$ is the *j*-th inner node of x_i, corresponding to the fuzzy language positive big (PB), positive small (PS), negative big (NB), negative small (NS) etc. N_i is the demarcated number.

$D_i = \dfrac{x_{i,\max} - x_{i,\min}}{N_i}$ is the distinguish rate of *i*-th input variable. Use a linear quantization function:

$$q_i(x_j) = \left\lfloor \frac{x_{i,j} - x_{i,\min}}{D_i} \right\rfloor \tag{3}$$

where $q_i(x_j)$ is quantized value of *i*-th input x_i ,and $\lfloor \cdot \rfloor$ is a floor function.

Mapping $X \Rightarrow M$ adopts the mapping rule as followed.

$$m_{i,k} = \left\lfloor \frac{q_i + N_g - k}{N_g} \right\rfloor \cdot N_g + k \tag{4}$$

where $m_{i,k}$ is the address of mapping from input vector $q_i(x_j)$ to middle variable m, N_g is the number of excited units, k is the ordinal number of excited unit and $k = 0 \sim (N_g - 1)$. The membership function of input variable is defined as Gaussian function.

$$\mu_j(x) = \exp\left(\frac{\|x - c_j\|}{\sigma_j}\right) \tag{5}$$

where c_j and σ_j are, respectively, the center and the width of the j-th membership function. The width σ_j can be attained by the overlap algorithm of two near membership functions [3].

$$\mu_c = \exp\left(-\left(\frac{(c_j + c_{j+1})/2 - c_j}{\sigma_j}\right)^2\right) \tag{6}$$

$$\sigma_j = \sqrt{\frac{((c_{j+1} - c_j)/2)^2}{|\ln(\mu_c)|}} \tag{7}$$

where μ_c is selected in advance and $\mu_c \in (0,1]$.

Define error function as:

$$E = \sum_{i=1}^{P} \frac{1}{2}(\hat{y}_i - y_i)^2 = \sum_{i=1}^{P} \frac{1}{2}(\hat{y}_i - a^T h(x)w)^2 \tag{8}$$

The learning algorithm is

$$\Delta w_k = \frac{\beta}{a^T h(x) h^T(x) a}(\hat{y} - a^T h(x)w) a_k \mu_k \tag{9}$$

where $\rho = a^T h(x) h^T(x) a = \sum_{i=1}^{N_g} \mu_i^2$

Therefore, the GFAC differs from general basis function CMAC on weight value assignment. It determines the weight adjustment degree according to the membership function value of excited unit and the sum of all excited membership function square value, so it has adaptability.

3 FACE System Structure and Its Learning Algorithm

3.1 FACE Control System Structure

A FACE control system structure is shown in figure 2, where block F is the controlled system. The block F can also incorporate any additional feedback controller Q that the basic system might have. The GFAC has been split into two parts in the figure:

(1) The block A which represents the input layers (or association unit transformations). At each time step this block reads the system state Y_i (one expected value y_i^d or perhaps the current time t) and encodes it in the "sensor" vector S_i.

2) The output layer which multiplies the vector S with the weight matrix W to get the output X ($X = SW$).At each time step the "critic" block C computes a scalar error

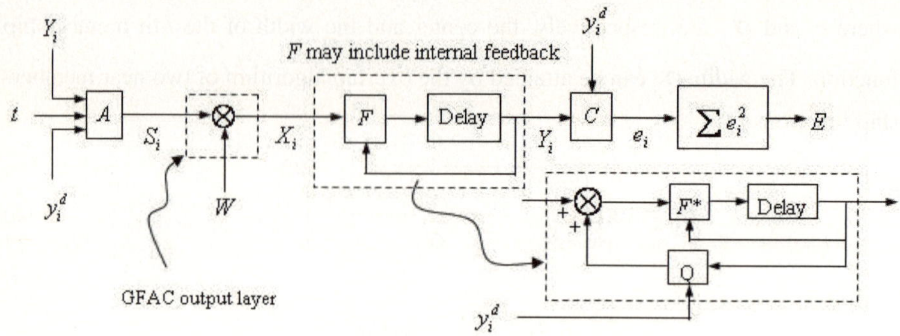

Fig. 2. FACE controller and controlled system

value e_i that depends on the current state. The squares of these error values are summed over time to get the global error value E.

If the system's desired behavior is to track a scalar reference position y_i^d, thus

$$e_i = y_i^d - CY_i \tag{10}$$

where C is a matrix that selects whatever element of y corresponds to a position. Then the total error E is given by:

$$E = \sum_{i=1}^{T} e_i^2 \tag{11}$$

The critic function is chosen so that when E is minimized the system achieves some desired behavior.

3.2 FACE Learning Algorithm

The purpose of the FACE learning algorithm is to modify W using gradient descent so that the error E is minimized. The gradient descent process for this system is:

$$w_j \leftarrow w_j + \alpha \frac{dE}{dw_j} \tag{12}$$

where w_j is an element of the matrix W and α is a scalar learning rate. This equation gives the modification made to weight w_j as a result of one iteration of learning. Now, from the chain rule:

$$\frac{dE}{dw_j} = \sum_{i=1}^{T-1} \left[2 \sum_{k=i+1}^{T} e_k \frac{de_k}{dX_i} \right] \cdot \frac{dX_i}{dw_j} \tag{13}$$

Now for a key step to determine the meaning of equation (12), and to derive a practical algorithm, F is approximated by a linear system F^*:

$$Y_{i+1} = AY_i + BX_i \tag{14}$$

$$e_i = CY_i - y_i^d \tag{15}$$

Combining equation (14) and (15) with equation (13):

$$\frac{de_{i+k}}{dX_i} = CA^{k-1}B \qquad (k > 0) \tag{16}$$

$$\frac{dE}{dw_j} = 2\sum_{k=2}^{T} e_k C\left[A^{k-i-1}B\hat{S}_i^j\right] = 2\sum_{k=2}^{T} e_k C\xi_k^j \tag{17}$$

where $\xi_k^j = \sum_{i=1}^{k-1} A^{k-i-1}B\hat{S}_i^j$, $\hat{S}_i^j = \frac{\partial X_i}{\partial w_j}$. $\hat{S}_i$ is all zero except for the element whose corresponding neural weight w_j is excited. And here ξ_k^j is called the *eligibility* signal. Based on the above equations, the FACE learning algorithm can be deduced:

$$\xi_1^j = 0 \tag{18}$$

$$\xi_{i+1}^j = A\xi_i^j + B\hat{S}_i^j \tag{19}$$

$$w_j \leftarrow w_j + \alpha\sum_{k=2}^{T} e_k C\xi_k^j \tag{20}$$

Note that a factor of 2 has been combined into α .Every FCMAC weight w_j requires an associated eligibility vector ξ^j .The order of the eligibility model is the size of the matrix *A*. There is a relationship between the two constants α and *C* : if the magnitude of *C* is adjusted then α can be changed to compensate. Because of this the convention will be adopted that the magnitude of *C* is always set to one ($|C| = 1$) and then the resulting α is the main FACE learning rate.

3.3 The Efficient Implementation of FACE Learning Algorithm

A naive implementation of the training equations is very simple: just update the eligibility state for every weight during each time step. Consider a GAFC with n_w weights and n_a association units. To compute the GAFC output without training (in the conventional way) requires one set of computations per association unit, so the computation required is $O(n_a)$ per time step. But if eligibilities must be updated as well then one set of computations per weight is needed, so the time rises by $O(n_w)$. A typical GAFC has $n_w >> n_a$ (e.g. n_a = 10 and n_w = 1000), so the naive approach usually requires too much computation to be practical in an online controller.

The algorithm described below requires the system F^* to have an impulse response that eventually decays to zero. This is equivalent to requiring that the eigenvalues of A all have a magnitude less than one. This will be called the "decay-to-zero" assumption. The next simulation part will explain how to get around this requirement in the ship steering system.

The weights are divided into three categories according to their values:

(1) *Active weights*: where the weight is one of the n_a currently being accessed by the GFAC. There are always n_a active weights.
(2) *Inactive weights*: where the weight was previously active and its eligibility has yet to decay to zero.
(3) *Retired weights*: where the weight's eligibility has decayed sufficiently close to zero, so no further weight change will be allowed to take place until this weight becomes active again.

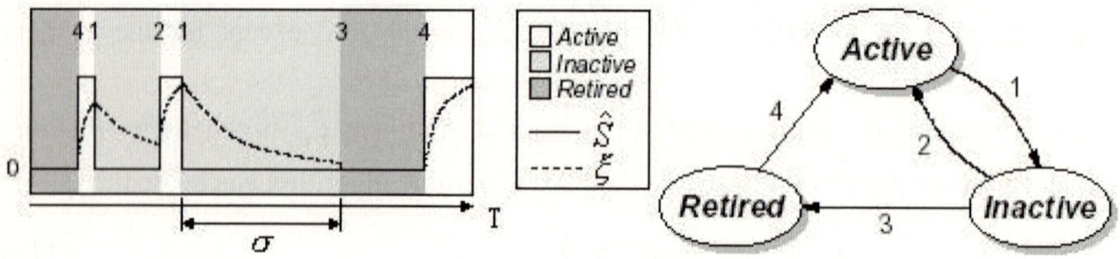

Fig. 3. The three states of FACE weight and the transition between them

Figure 3 shows how a weight makes the transition between these different states. FACE does not have to process the retired weights because their values do not change (their eligibilities are zero and will remain that way) and they do not affect the GFAC output. An active weight turns in to an inactive weight when the weight is no longer being accessed by the GFAC (transition 1 in figure 3). An inactive weight turns in to a retired weight after σ time steps have gone past (transition 3 in figure 3).The value of σ is chosen so that after σ time steps a decaying eligibility value is small enough to be set to zero. At each new time step a new set of weights are made active. Some of these would have been active on the previous time step, others are transferred from the inactive and retired states as necessary (transitions 2 and 4 respectively in figure 3).

4 FACE Control Simulation for Ship Steering

The linear ship Nomoto model has been widely accepted in designing ship course controller [10]. It omits the sway velocity but grasps the main characteristics of ship dynamics: $\delta \rightarrow \dot{\psi} \rightarrow \psi$. The disturbances of wind, waves can even be converted to a kind of equivalent disturbance rudder angle as an input signal. The second-order Nomoto model is

$$\ddot{\psi} + \frac{1}{T}\dot{\psi} = \frac{K}{T}\delta \tag{21}$$

where ψ is course, δ is ruder angle, T is time constant, K is rudder gain. For some unstalbe ship, $\dot{\psi}/T$ must be replaced with a nonlinear term and $H(\dot{\psi}) = a\dot{\psi} + \beta\dot{\psi}^3$. So the second-order nonlinear ship response model is expressed as

$$\ddot{\psi} + \frac{K}{T} H(\dot{\psi}) = \frac{K}{T}\delta \tag{22}$$

Parameters a, β and K, T is related to ship's velocity.

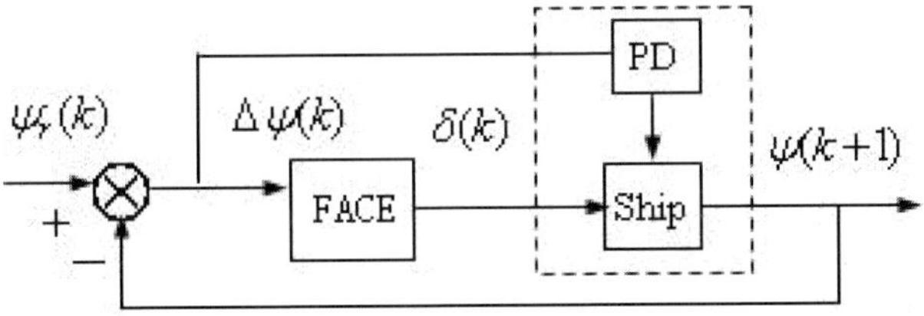

Fig. 4. Ship steering control system

Figure 4 shows the ship steering control system applying FACE controller. Its input are course error $\Delta\psi = \psi_r(k) - \psi(k)$ and fore turning angular velocity $r(k)$. Its output is the rudder angle $\delta(k)$, $\Delta\psi$ varies between $(-20^\circ, 20^\circ)$, r between $(-0.9^\circ/\sec, 0.9^\circ/\sec)$, and δ is $(-35^\circ, 35^\circ)$.

The FACE algorithm described above requires the system F^* to have an impulse response that eventually decays to zero. So a PD feedback control element is joined, then the ship state model is changed:

$$X = AX + Bu \tag{23}$$

$$Y = CX \tag{24}$$

where

$X = [\dot{\varphi}, \varphi]^T$, $A = \begin{bmatrix} -(1+Kkd)/T & -Kkp/T \\ 1 & 0 \end{bmatrix}$, $u = \delta$. Transfer the state matrix into discrete format, the eligibility curve can be attained as figure 5 shown. Here $K = 0.36, T = 230,\ k_p = 1.2$, $k_d = 15$. The eligibility decays to zero about 80s from figure 5, so the eligibility decay parameter can be selected as $\sigma = 100$.

Figure 6 shows the course angle and rudder angle curves result when set course is 10°, wind force is Beaufort 5 and wind direction is 30°.While figure 7 and figure 8 show the course angle and rudder angle curves respectively when set course is

5°~15°~20°. The dashed in figure 7 and figure 8 are control curves attained by conventional FCMAC control. From the compared curves, the proposed FACE control has better real-time quality and fast tracking speed. In term of course, it has no over-training results and has satisfied tracking effect; as to rudder angle, at beginning the bigger angle is accelerated to start up, then regained to stable angle needed. The curves indicate that the course tracking is fast, control action reasonable and meet the performance of ship steering. The control result is partial satisfied.

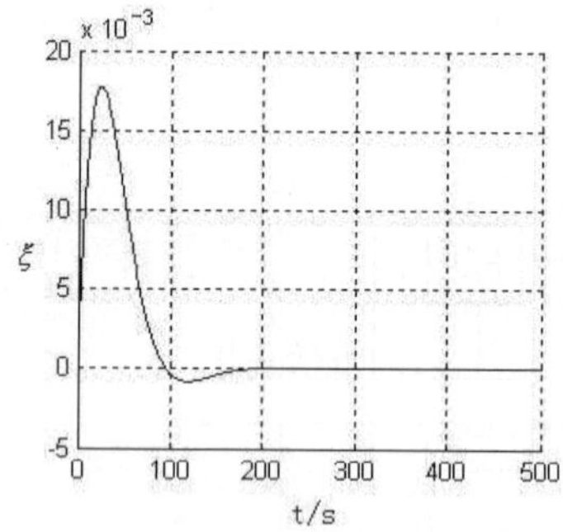

Fig. 5. Eligibility curve

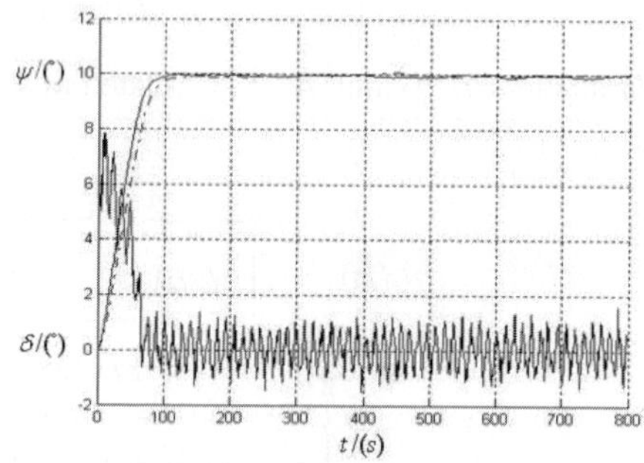

Fig. 6. Course and rudder response curves when set course is 10°

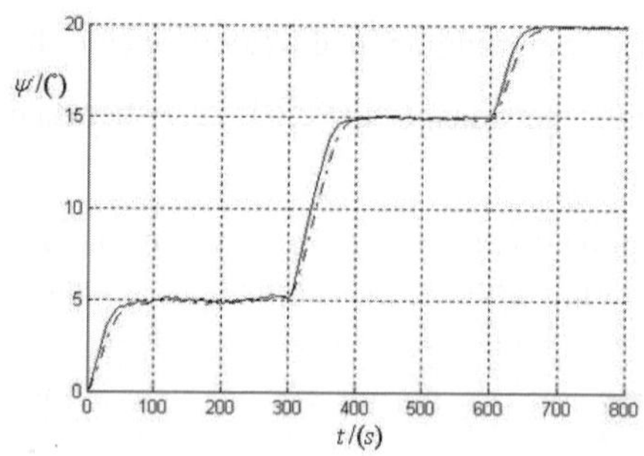

Fig. 7. Course response curve when set course is 5° ~15°~20°

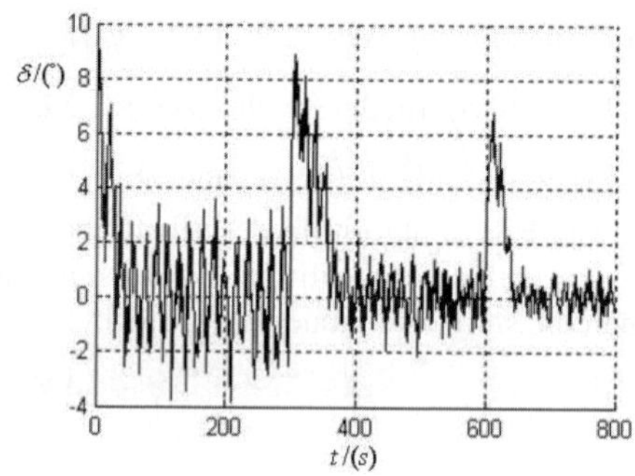

Fig. 8. Rudder response curve when set course is 5° ~15°~20°

5 Conclusion

Based on conventional Cerebellar Model Articulation Controller (CMAC), by preserving CMAC local learning and addressing schemes, as well as integrating fuzzy logic idea, a general fuzzified CMAC (GFAC) is proposed in this paper. The mapping of receptive field functions, the selection law of membership, and the learning algorithm are presented. By incorporating the eligibility into GFAC, a fuzzified CMAC controller with eligibility (FACE) is also proposed. The structure of FACE system is presented, and its learning algorithm is deduced. To make the algorithm fit to online control, an efficient implementation of FACE method is given. As an example, the proposed FACE controller is applied to a ship steering control system, and the simulation results show that the ship course can be properly controlled under disturbance of wind and wave.

Acknowledgments

This work was supported in part by the National Natural Science Foundation of China (No.60774046), in part by China Postdoctoral Science Foundation (No.20070421047).

References

1. Albus, J.S.: A New Approach to Manipulator Control: The Cerebellar Model Articulation Controller(CMAC). Trans. ASME-J Dyn. Syst. Meas. Control 97, 220–227 (1975)
2. Campagna, D.P.: Stability and Weight Smoothing in CMAC Neural Networks. Doctoral Dissertation, The University of New Hampshire, pp. 30–46 (1989)
3. Ker, J.S., Hsu, C.C., Kuo, Y.H.: A Fuzzy CMAC Model for Color Reproduction. Fuzzy Sets and Systems 91, 53–68 (1977)
4. Chan, L.C.Y., Asokanthan, S.F.: CMAC based Controller for Hydromechanical Systems. In: The American control conference, Arlington, VA, USA, pp. 25–27 (2001)
5. Zhou, X.D., Wang, G.D.: Fuzzy CMAC Neural Network. Acta Automatica Sinic 24, 173–177 (1998)
6. Nie, J.H., Linkens, D.A.: FCMAC: A Fuzzified Cerebellar Model Articulation Controller with Self-organizing Capacity. Automatica 30, 655–664 (1994)
7. Klopf, A.H.: A Comparison of Natural and Artificial Intelligence. Sigart Newsletter 53, 11–13 (1975)
8. Su, S.F., Hsieh, S.H.: Embedding Fuzzy Mechanisms and Knowledge in Box-Type Reinforcement Learning Controllers. IEEE Trans. Syst., Man, Cybern. part B: Cybern. 32, 645–653 (2002)
9. Sutton, R.S.: Temporal Credit Assignment in Reinforcement Learning. PhD thesis, University of Massachusetts, Amherst, MA (1984)
10. Jia, X.L., Yang, Y.S.: Ship Motion Mathematic Model, pp. 137–151. Dalian Maritime University Press (1998)

Case-Based Decision Making Model for Supervisory Control of Ore Roasting Process

Jinliang Ding, Changxin Liu, Ming Wen, and Tianyou Chai

Key Laboratory of Integrated Automation of Process Industry, Ministry of Education and Research Center of Automation, Northeastern University, Shenyang, China

Abstract. The shaft furnace roasting process is an important procedure in the mineral processing of the weakly magnetic iron ore. Its technical performance index is called the magnetic tube recovery rate(MTRR), which closely related to the overall performance of the mineral processing. In this paper, we mainly concern on the decision making of the supervisory control of the roasting process to control its MTRR into the target range. This model replaces the human operators to determine the set-points of the lower control loops. The experiment is given to evaluate the proposed model and the results show its validity and efficiency.

Keywords: Roasting Process, Supervisory Control, Decision-making.

1 Introduction

In the mineral processing industry, the magnetism of the weakly magnetic iron ore, such as, hematite, siderite, specularite, etc., needs to be enhanced through the reducing reaction in order to satisfy the technical requirement and to improve the efficiency of the magnetic separation. The shaft furnace is just a kind of thus devices used commonly for the magnetizing roasting process. With ever increasing competition in the globalized market environment, the process economics, efficiency and quality in the enterprises have attracted the attention of process industries. The distributed control systems (DCS) has been playing an important role in control of the iron ore roasting process operation and production. However, the decision of the set-points of this control loops has to rely on human operator's own experience. So new technologies are required to reduce the operator's cognitive load and achieve more consistent operations.

The roasting process is affected by many elements, such as, iron ore type, ore grade and size, furnace negative pressure, the caloricity of the gas, etc. All these factors vary in the roasting process, and are considered as boundary conditions of the shaft furnace. There are also many hidden variables, such as, the compact degree of the iron ore, affecting the process, which are very difficult or impossible to measure and thus impossible to control and affect the product quality.

Practically, the DCS-based process control has the local (lower) level and the higher level [1]. Since the control algorithms employed in local control loops, such as PID control, predictive control and fuzzy control, etc., are well developed, a satisfactory performance is always achieved in local loop. However, the overall

F. Sun et al. (Eds.): ISNN 2008, Part II, LNCS 5264, pp. 148–157, 2008.

performance of the roasting process may not satisfy the technical requirement due to the complexity of process. The human operator is often employed on the industrial site to select the proper set-points for local control loop. First, all boundary conditions are classified into different groups called operation modes in this paper. Through extensive experiments, a proper operating point is found for each operation mode. During the roasting process, the values of control variables of the relevant operating point will be taken as the set-points of DCS for the shaft furnace with the matched boundary conditions. Since each operation mode may have several different boundary conditions, the control accuracy will be limited by the operation mode classification. On the other hand, if the number of operation modes is more than what we can experiment, their appropriate operating points have to be found out via the interpolation or human experience. This human supervision is clearly a coarse control, which may not have a consistent performance and a well management of the energy waste. A better high-level supervision and control is required to maintain the satisfactory performance under variations of boundary conditions.

The PID control [2], fault diagnosis and tolerant control [3] , and multivariable control strategy [4] of the shaft furnace have been proposed. The two-layered strategy framework is proposed in [4], but the detailed realization doesn't be given. The intelligent control methods, like fuzzy control, neural network and knowledge engineering have been applied to control the complex industrial processes due to their loose requirement on the process model, and their capability to use human experience [5,6,7]. This paper presents a case-based decision making model for the supervisory control of the roasting process, which tries to replace the human operator by auto searching the proper operating points for the roasting process under variations of boundary conditions.

The paper is organized into four parts. Firstly, the roasting process and its DCS-based control is illuminated. Secondly, the case-based decision making model which replaced the human operators is described in detail. Thirdly, the industrial experiment result on a industrial shaft furnace is discussed. Finally, the conclusion is given.

2 Process Description

2.1 Shaft Furnace Roasting Process

The shaft furnace roasting process, which is to prepare ore for the magnetic separation, is composed of five subsections production process. The simplified cutaway view structure of the furnace main body is shown in Fig. 1, which is divided into three temperature zones according to the temperature profile of the solids material. These three zones are the preheating zone, the heat-up zone and the deoxidizing zone. The main purpose of the process is to convert Fe_2O_3 into Fe_3O_4 for improving the magnetism of the iron ore and to comminute the lump ores into smaller ones. The plant operation process is described as below:

The raw hematite ores are fed into the furnace from an ore-store slot and a square funnel at its top. In the preheating zone those ores contact the ascending

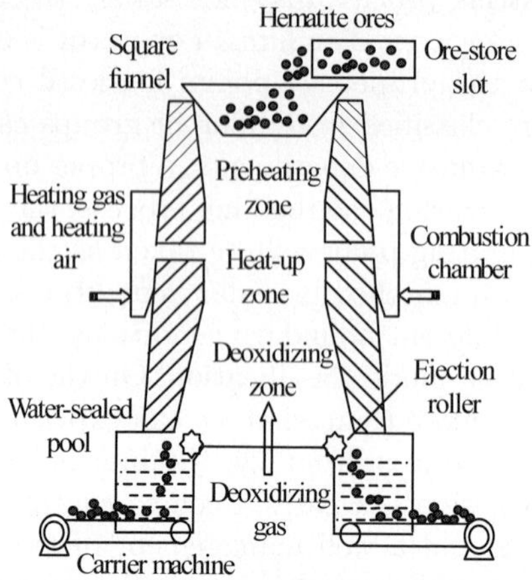

Fig. 1. The roasting process of shaft furnace

hot gas so that their temperature rises to 100~150°C. Then the ores fall into the heat-up zone, the ores temperature comes up to 700~850°C when they are attained the heat produced by the inflammation of air-mixed heat gas in the combustion chamber. In Deoxidizing phase, the hot low magnetic ores flow down into the deoxidizing zone and are deoxidized to high magnetic ones. The cooling zone is the final processing where the ores are laid down into the water-sealed pool by two ore ejection rollers and the ores are cooled and consequentially are moved out of the furnace by two carrier machines which operate synchronously with their corresponding rollers.

2.2 Process Analysis

The production quality of the shaft furnace is examined by an index called the magnetic tube recovery rate (MTRR), the bigger of its value (typically within the range of 0~1), and the higher of the concentrated ores grades may be attained.

During the process operation, a proper temperature range of the combustion chamber (i.e., 1000°C ~1200°C), plus the coordinated run-stop shift of the carrier machine and flow rate of deoxidization gas may offer a suitable temperature range (570°C±20°C) as a result of the following reactions:

$$\begin{array}{l} 3Fe_2O_3 + CO \xrightarrow{570^\circ C} 2Fe_3O_4 + CO_2 \\ 3Fe_2O_3 + H_2 \xrightarrow{570^\circ C} 2Fe_3O_4 + H_2O \end{array} . \tag{1}$$

The above produced Fe_3O_4 contains intensive magnetism, which is required in order to achieve high-grade concentrated ores after the final mineral process.

When the temperature of combustion chamber is relatively low or the flow rate of deoxidization gas is small, or the ores moving time quite long, the reactions are inadequate as they result in the production ores under deoxidization.

On the other hand, when the temperature is too high, and the flow rate of deoxidization gas is over abundant or the moving time is excessively short, it may lead to an over deoxidization reaction.

Under both of the above two circumstances, MTRR will be of a low value, showing the difficulty of gaining a high grade and concentrated ores.

As such, MTRR is influenced crucially by the temperature of combustion chamber, the flow rate of deoxidization gas, and the moving time in the run-stop shifts of the carrier machine (i.e., the moving duration). Only if MTRR is controlled within the target range (between 0.82 and 1), can a high-graded concentrated ores be obtained.

However, it is impossible to control MTRR through closed control loop directly. The reason is that: 1) MTRR can not be measured on line and only can be got from the laboratory assay; 2) the relation between MTRR and the manipulated variables is highly nonlinearity and it is difficult to describe using the accuracy dynamic mathematic model; 3) there is seriously interactive nature among the manipulated variables.

2.3 DCS-Based Process Control

The DCS-based process control system of the iron ore shaft furnace roasting process is shown in Fig. 2, which includes the sensors, actuators and control loops implemented in DCS (the distributed control system). The instrumentations of roasting process include the thermocouple and the resistance thermometer sensor for the temperature, the electromagnetic flow meter for the gas and air, the meter for the caloricity of the gas, etc. The actuators include values of the gas and the transducer for air and the motor of the carrier machines. The DCS provides three basic control loops: 1) a heating gas flow control loop; 2) a heating air flow control loop; and 3) a deoxidizing gas flow (F_{dg}) control loop; and one open control loop: the start and stop ($S_{\sum}$) of the carrier machine motors. There is a cascade control loop: a temperature ($\overline{T}_{cc}$) control loop of the combustion chamber whose control variables are the heating gas flow and the heating air flow. The relation of them is ratio control. The caloricity of gas is measured to control the combustion chamber temperature through the feedforward controller. The furnace negative pressure is monitored to ensure the production safety.

To maintain the satisfactory performance in the dynamical production environment, human experts are often required, as shown in Fig. 2, to process the data (γ_{target}, I, B, T_{cci},F_{dg}, S_j, i=1,2,...4, j=1,2) and determine the proper set-points for the control loops of combustion chamber temperature (T_{cc}^s), deoxidizing gas flow (F_{dg}^s) and the running duration of the carrier machine (S^s) through several typical experiments and a period of production operation.

In the next section, a case-based decision making model is developed to replace the human operators, which adopts the combination of the case-based reasoning, the intelligent optimization method and the prediction of MTRR.

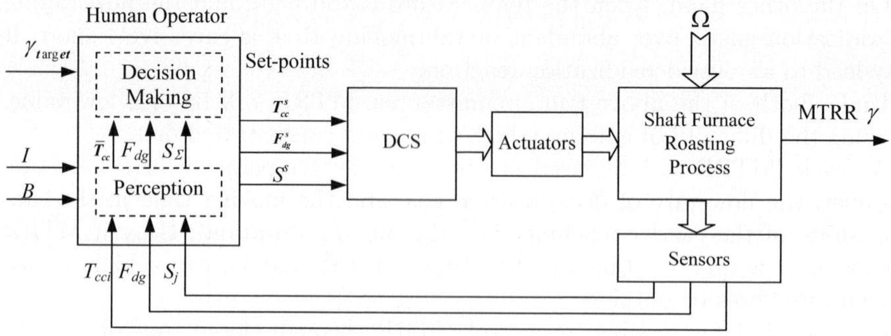

Fig. 2. DCS controlled roasting process and human supervisory

3 Case-Based Decision Making Model for the Supervisory Control

This model relies on an interactive approach of problem solving, case-based reasoning (CBR) [8], which is known in the area of artificial intelligence. The goal of CBR is to infer a solution for a current problem description in a special domain from solutions of a family of previously solved problems, named Case Base. The core idea of CBR is that "similar problems have similar solutions". The basic element of knowledge is a case, which is the solution to a specific understood problem. The CBR circle constitutes the following four processes: *retrieve*, *reuse*, *revise* and *retain* as introduced in reference [8]. Because of its successful applications in various fields, CBR is beginning to attract attention from the process industry [9].

In this paper, the developed case-base decision making model consists of the following modules mainly, as depicted in Fig. 3: 1) Production data process; 2) Case retrieval; 3) Case reuse; 4) Case evaluation; 5) Case revise; and 6) Case retain. The function of this model is to utilize the specific case information available as historical precedence for proposing solutions to current problem. The most important aspects of the existing cases are first stored and indexed. New problem situations are then presented and similar, existing cases are identified from the case base. Finally, the previous problem solutions are adapted and the revised solutions are proposed for the current situation. Next, every module of this model is introduced in detail as following.

3.1 Production Data Process

This module is to extract the data from the process (the work-condition I, the boundary condition B, including the ore type O_t, ore grade O_g and ore size O_s, the four points temperature of the combustion camber T_{cci}, i=1,2,. . . 4, the flow rate of the deoxidizing gas F_{dg}, the running duration of the two side carrier machines S_j, j=1,2) and the target range of MTRR γ_{target}, and to construct

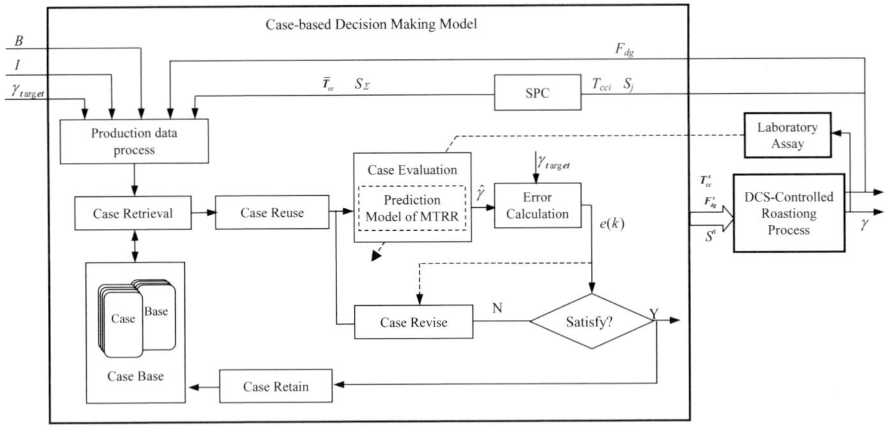

Fig. 3. Structure of the case-based decision making model

the case representation for the current work-condition. The statistical process control (SPC) module to process T_{cci} and S_j as following:

$$\overline{T}_{cc} = \frac{1}{4}\sum_{i=1}^{4} T_{cci} \quad (i = 1, 2, \ldots, 4); \quad S_\Sigma = \sum_{j=1}^{2} S_j \quad (J = 1, 2)\,. \tag{2}$$

The case structure is constructed in this paper, as shown in Table 1, where Q is the target value of the production yield of the furnace.

Table 1. Case structure of the decision model

Case Description F									*Case Solution FS*		
f_1	f_2	f_3	f_4	f_5	f_6	f_7	f_8	f_9	fs_1	fs_2	fs_3
I	O_t	O_g	O_s	$\overline{T}_{cc}$	F_{dg}	S_Σ	γ_{target}	Q	T^s_{cc}	F^s_{dg}	S^S

3.2 Case-Based Reasoning Procedure

1) *Case Retrieval*: In this procedure, the indexes of a new problem are used to retrieve similar cases from the case base. Let the current operating condition be C, define the case descriptors as F, solution as FS. Cases in the case base $C_l, (l = 1, \cdots, m)$, are expressed as case descriptors F_l and case solution FS_l. Similarity function between $F$and F_l is given by:

$$sim(f_i, f_{l,i}) = \begin{cases} 1 - \dfrac{f_i - f_{l,i}}{\mathrm{Max}(f_i, f_{l,i})}, & i = 1, \cdots, 4;\ l = 1, \cdots, m \\ 2 - \dfrac{f_i - f_{l,i}}{E}, & i = 5, \cdots, 9;\ l = 1, \cdots, m \end{cases} \,. \tag{3}$$

Similarity function between current operating condition C and C_l(l=1,$\cdots$,m) is described as follows:

$$SIM(C,C_l) = \sum_{i=1}^{9} \omega_i \times sim(f_i, f_{l,i}) \Big/ \sum_{i=1}^{9} \omega_i \ . \tag{4}$$

The r cases C_j, $(j = 1, \cdots, r)$, whose similarity is greater than a predefined threshold, are retrieved. ω_i is the weight of descriptors, which generally determined by the experience of the operators in industrial production.

2) *Case Reuse*: The similarity of C_j with current operating condition C is SIM_j, the case solution is $FS_j = (fs_{j,1}, fs_{j,2}, fs_{j,3})$.

The case solution of C:

$$fs_i = \sum_{j=1}^{r} w_j \times fs_{j,i} \Big/ \sum_{j=1}^{r} w_j \quad (i = 1,2,3) \ . \tag{5}$$

where w_j $(j = 1, 2, ...r)$ is determined by following:

$$if \quad SIM_r = 1 \quad then \quad w_j = \begin{cases} 1 & j = r \\ 0 & j \neq r \end{cases} \quad else \quad w_j = SIM_j \ . \tag{6}$$

3) *Case Evaluation and Retain*: The reused case solution should be evaluated before giving to the control system. The prediction model is used to predict MTRR γ with fs_1, fs_2, fs_3 as input (the prediction model will be introduced in the last of this section). Then calculates the error $e(k)$ between the prediction $\hat{\gamma}$ and the target range γ_{target}. If the error $e(k) = \hat{\gamma} - \gamma_{target} \geq 0$, the prediction is used as loop set-point and stored in the case base. Otherwise it is revised until satisfactory results are obtained.

4) *Case Revise*: The Fuzzy PI controller is used to revise the first two items of the solution, T_{cc}^s and F_{dg}^s, as shown in Fig. 4. The control strategy is shown as following:

$$\begin{bmatrix} \Delta U_T(k) \\ \Delta U_F(k) \end{bmatrix} = \begin{bmatrix} Kp_1 \\ Kp_2 \end{bmatrix} [e(k) - e(k-1)] + \begin{bmatrix} Ki_1 \\ Ki_2 \end{bmatrix} e(k) \ . \tag{7}$$

where $\Delta U = \begin{bmatrix} \Delta U_T(k) \\ \Delta U_F(k) \end{bmatrix}$ is the revisory value of the T_{cc}^s and $F_{dg}^s \cdot Kp(k) = \begin{bmatrix} Kp_1(k) \\ Kp_2(k) \end{bmatrix}$ is the proportion factor matrix and $Ki(k) = \begin{bmatrix} Ki_1(k) \\ Ki_2(k) \end{bmatrix}$ is the integral factor matrix.

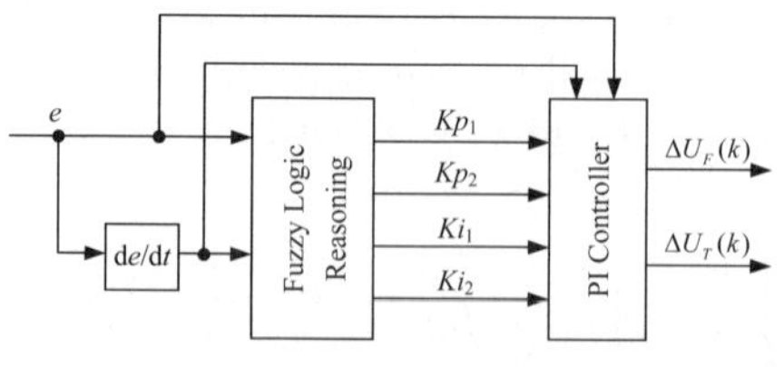

Fig. 4. Structure of the Fuzzy PI Controller

$K_p(k)$ and $K_i(k)$ are gained by the fuzzy logic reasoning. The universes of discourse of e, $\dot{e}$, k_{p1}, k_{p2}, k_{i1} and k_{i2} are [-3, 3], [-3, 3], [2, 10], [2, 10], [5, 30] and [10, 35], and their quantization grade are 7, 7, 11, 7, 11 and 7 respectively.

5) *Prediction model of MTRR*: MTRR is a key technical index to evaluate the quality of roasted iron ore. However it is difficult to measure online. Generally, MTRR is got from the laboratory assay with a large delay, so it cannot be obtained in time to guide the process operation. A modified Least Squares-Support Vector Machines(LS-SVM) is proposed to predict MTRR punctually. In this paper, the fuzzy weight coefficient of sample data is applied to avoid that the variation of the work situation influences the precision of the model.

According to Eqs. 1, MTRR γ is influenced dominantly by T_{cc}, F_{dg} and $S_{\sum}$. So them are taken as the input of the prediction model, whilst MTRR is the output, i.e. $x = (T_{cc}, F_{dg}, S_{\Sigma})$ and $y = \hat{\gamma}$.

To overcome the influence of the work-situation variation on the precision of model, a weighted LS-SVM is to be performed modeling task, whose optimization problem is shown as below:

$$\begin{array}{ll} \min & \mathrm{R}(w,\xi) = \frac{1}{2}||w||^2 + \frac{1}{2}\zeta \sum_{i=1}^{l} (p_i \xi_i)^2 \\ \text{s.t.} & y_i = w\Psi(x_i) + b + \xi_i \end{array} . \tag{8}$$

where $\mathrm{R}(w,\xi)$ is the structural risk; ξ_i is the allowable error; ζ is the regularization parameter; p_i is the fuzzy weight coefficient; i is the number of the sample data. p_i is determined as $p_i = f(t_i) = \gamma_1(t_i - \gamma_2)^2 + \gamma_3$, where t_i (i=1,2,…,l) is the time attribute of the sample data, and $t_1 \le t_2 \le \cdots \le t_l$. $\gamma_1, \gamma_2, \gamma_3$ are the correlation coefficients.

In this paper, define a lower limit of the weighting λ, λ>0. Let $p_1 = f(t_1) = \lambda$ and $p_l = f(t_l) = 1$. So we can get p_i, $p_i = f(t_i) = (1-\lambda)\left(\frac{t_i - t_1}{t_l - t_1}\right)^2 + \lambda$.

The kernel function is $K(x, x_i) = \exp(-\frac{\|x - x_i\|^2}{2\sigma^2}) = \Psi(x)^T\Psi(x_i)$, where σ is the width of the kernel function. σ and the above C are determined by the cross validation to get its optimal value.

The regression model of the LS-SVM becomes a following form:

$$f(x) = \sum_{i=1}^{l} \alpha_i K(x, x_i) + b . \tag{9}$$

where x is input vector; l is the number of samples.

3.3 Heuristic Rule Reasoning

In the case-based reasoning process, if there has no similar case be retrieved caused by the severe variation of the boundary condition, the heuristic rule reasoning is to be work to generate the setpoints of the lower control loops.

4 Industrial Experiment Results

The biggest hematite ore concentrator of China owns 22 shaft furnaces. The No. 12 shaft furnace is selected for the industrial experiment to evaluate the validity of the proposed model. The initialization case base is established from the actual production data and the expert knowledge. There are 100 initialization cases, and the weights of case descriptions ω_i, (i=1, 2, ..., 9), are 0.091, 0.083, 0.091, 0.108, 0.084, 0.078, 0.094, 0.195 and 0.176, respectively. The threshold θ is 0.98, which is determined by the expert knowledge.

The setpoints and responses of the three lower control loops are shown in Fig. 5, Fig. 6 and Fig. 7, respectively.

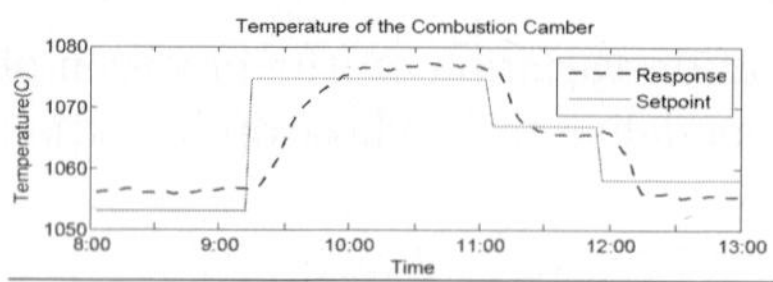

Fig. 5. Setpoint and response of the temperature of the combustion camber

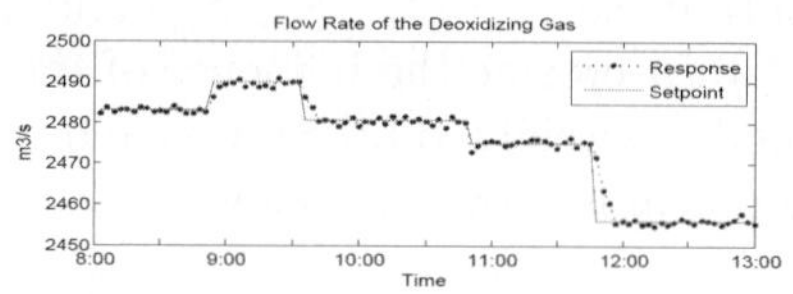

Fig. 6. Setpoint and response of the flow rate of the deoxidizing gas

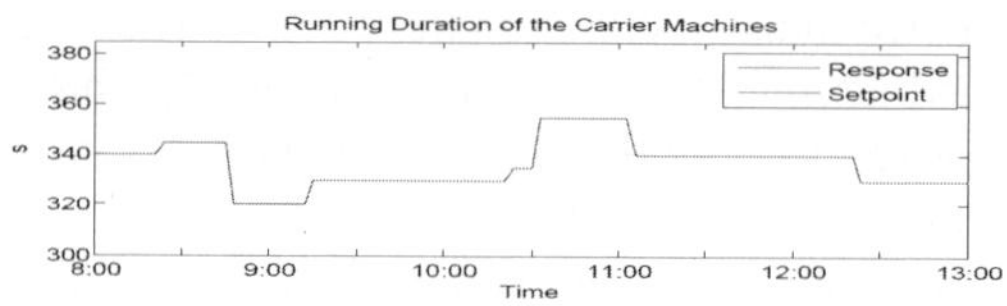

Fig. 7. Setpoint of running duration the carrier machines (summation)

From these figures, we can get that their setpoints can vary with the changes of the boundary conditions and they can track their setpoints very well.

In this five hours, five sample data of MTRR which come from the laboratory assay was got, as show in following Table 2:

From Table 2, we can get that the MTRRs are all greater than its low limit 82%, i.e. the quality of the production is satisfied the requirement.

Table 2. The laboratory assay of MTRR

No	1	2	3	4	5
MTRR(%)	83	82.4	82.6	83.1	82.9

5 Conclusion

A case-based decision making model for the supervisory control of the iron ore roasting process is proposed, which is mainly to determine the setpoints of the lower control loops which realized on the DCS. The procedure of the case-based reasoning and the prediction model of MTRR are illuminated in detail. The experiment is carried out the results show the validity and efficiency of the proposed model.

Acknowledgement

This work is supported by the projects with grant No.308007,B08015.

References

1. Li, H.X., Guan, S.: Hybrid Intelligent Control Strategy: Supervising a DCS-controlled Batch Process. IEEE Control System Magazine 21, 36–48 (2001)
2. Yan, A., Ding, J.L., Chai, T.Y.: Integrated Automation System for Shaft Furnace Roasting Process. Control Engineering of China 13, 120–122, 126 (2006)
3. Chai, T.Y., Wu, F.H., Ding, J.L., Su, C.Y.: Intelligent Work-situation Fault Diagnosis and Fault-tolerant System for Roasting Process of Shaft Furnace. In: Proc of the ImechE, Part I, Journal of Systems and Control Engineering, 9 (accepted for publication, 2007)
4. Yan, A., Chai, T.Y., Yue, H.: Multivariable Intelligent Optimizing Control Approach for Shaft Furnace Roasting Process. Acta Automation sinica 32, 636–640 (2006)
5. Lu, Y.Z., He, M., Xu, C.W.: Fuzzy Modeling and Expert Optimization Control for Industrial Processes. IEEE Transactions on control systems technology 5, 2–11 (1997)
6. Yao, L., Postlethwaite, I., Browne, W., Gu, D., Mar, M., Lowes, S.: Design, Implementation and Testing of an Intelligent Knowledge-based System for the Supervisory Control of a Hot Rolling Mill. Journal of Process Control 15, 615–628 (2005)
7. Frey, C.W., Kuntze, H.B.: A Neuro-Fuzzy Supervisory Control System for Industrial Batch Processes. IEEE Transactions on Fuzzy Systems 9, 570–577 (2001)
8. Kolodner, J.L.: An Introduction to Case-based Reasoning. Artif. Intell. Rev. 6, 3–34 (1992)
9. Ding, J.L., Zhou, P., Liu, C.X., Chai, T.Y.: Hybrid Intelligent System for Supervisory Control of Mineral Grinding Process. In: Conference Proceeding of 6th ISDA, Jinan, China, pp. 16–18 (2006)

An Affective Model Applied in Playmate Robot for Children

Jun Yu[1,2], Lun Xie[1], Zhiliang Wang[1], and Yongxiang Xia[2]

[1] School of Information Engineering, University of Science & Technology Beijing, 100083 Beijing, China

[2] Navy Flight Academy, 125001 Huludao, China

yujunfisher@163.com

Abstract. It is always the focus of researchers' attention to endow the robot with the emotion similar to human in human robot interaction. This paper present an artificial affective model based on Hidden Markov Model (HMM). It can achieve the transfer of several affective states under some basic hypothesis and restriction. The paper also shows some simulation results of affective states change. It is the basis for architecture in support of interactive robot. Then the paper explains the technical route of playmate robot for children in detail. The robot can behave like a human child and attempt daily communication with human supported by the affective model and these technologies.

Keywords: Artificial intelligence, Artificial Psychology, Affective model, Humanoid robot.

1 Introduction

With the development of the technology and economy and constant improvement of the living standards of the people, the robot is coming into family. It can provide all kinds of servers and entertainment for people, even can communicate with people. In the future, the robot will be in possession of intelligence and psychological activity such as emotion, character, will etc. Therefore it is the hotspot for researcher to endow the robot with the capability of affective interaction in these days. That is to say, if the robot were in possession of the genuine intelligence and could interact with human naturally, it would be endowed with the capabilities of emotion recognition, emotion comprehending, and emotion expression.

The paper builds a software and hardware platform of children playmate robot to imitate the change process of the human's emotion and corresponding action based on artificial intelligence theory and robotics. The paper puts forward an affective model based on artificial psychology and hidden markov models. It also proved the model can generate the emotion changes which are in correspondence with human's emotion changes. The model also is used in the playmate robot for children. This makes the robot possessing anthropoid emotion and can respond affectively to outside stimulation. Therefore the robot can interact with human sensibly. By incorporating the emotion to robotic architecture we hope the robot would have more behaviors similar to human.

F. Sun et al. (Eds.): ISNN 2008, Part II, LNCS 5264, pp. 158–164, 2008.

2 Affective Model of Playmate Robot for Children

In order to endow robot with emotion, first we should understand how the human's emotion is generated. Many theories and models on psychology are served for explaining it, such as simulation-response theory, physiological reaction theory, facial expression theory, motivation theory, subjective evaluation theory etc. Now it is cognitive evaluation theory of emotion that was widely accepted. According to the theory, emotion is generated when human estimates the important thing what he thinks is very important and obtains experience. This evaluation process is subjective and is influenced by special aim, faith and criterion of human. The different people have different internal mental structure, and then give the different explanation to the same external stimulus. Therefore, the emotion generated finally will depend on the people's cognition and subjective evaluation to stimulus.

2.1 Emotional Basis

There are many proposals for affective models. The paper will introduce the basic research of affective modeling in the way of application system. Ekman and Davidson[1] proposed six basic emotional states: happiness, anger, sadness, fear, surprise, and disgust. Elliott[2] proposed an OCC model. The model can generate 22 emotional states. The paper provides a classification of emotions and potential reasoning process between the emotions. Roseman [3] proposed cognitive evaluation model. The model includes five dimensionalities. Which emotion will be generated is inferred by the combination of them each other. Reilly and Bates [4] proposed EM module to generate true and credible social action. It is to decide which stimulus would induce a affective reaction by an accumulated threshold mechanism. Gratch and Marsella [5] proposed an EMA model. EMA discusses the effect of emotion on cognition and the action of cognition to emotion.

Professor Wang put forward a new theory of artificial psychology in 1999. The theory is based on artificial intelligence and deeply analyses human psychology in more comprehensives in the aspects of information scientific research methods, especially in the aspects of emotion, willingness, character, creativity and the realization of artificial machines [6]. It has broad application prospect such as developing the robot with emotion, consciousness and intelligence, research of humanoid machine in real meaning. It can make the control theory more similar to human brain's control mode.

It is the key and difficult point for playmate robot for children research to build the affective model using artificial psychology. Whether a robot can express the subhuman emotion lies on if or not there is a rational affective model. We take the change of affective states as a stochastic process, research the changing rules of affective states and build the mathematical model using the theory of stochastic process.

2.2 Precondition

The emotions of human are very complicated. They include low-grade emotion, high-grade emotion, basic emotion, composite emotion. Some of them are instinctive and related with human's sensitive power. The other are related with cognition and need learning process and experience to generate emotions. Now it is very difficult to endow robot with the profuse emotions as human. In order to simplify the problem and build the affective model used conveniently in robot, we must give some basic hypothesis and restriction.

Takes N as the total number of basic emotions, $S_i\ (i=1,2,...,N)$ the states variables, and then the states of emotion can be expressed by states set S.

$$S=\{S_1,S_2,...,S_N\}=\{1,2,...,N\},\quad S_i=i\ (i=1,2,...,N) \tag{1}$$

Takes $p_i(i=1,2,\dots N)$ as the probability of $S_i=i$ (No. i affective states), and satisfying affective states probability distribution equation:

$$\sum_{i=1}^{N} p_i = p_1 + p_2 + \cdots + p_N = 1,\quad 0 \le p_i \le 1\ (i=1,2,\cdots,N) \tag{2}$$

Such that the probability space model of affective emotion is as follow:

$$\begin{pmatrix} S \\ P \end{pmatrix} = \begin{pmatrix} S_1 & S_2 & \cdots & S_N \\ p_1 & p_2 & \cdots & p_N \end{pmatrix} \tag{3}$$

Hypothesis 1. The playmate robot for children possesses only several basic affective states. Commonly N = 3, 1 represents happy, 2 represents anger, and 3 represents sadness.

Hypothesis 2. With the effect of outside stimulus, any two kinds of affective states can transfer each other.

Hypothesis 3. One kind of stimulus can only induce one emotion, that is, stimulus V_i can only induce emotion i. Stimulus set can be expressed as follow:

$$V=\{V_1,V_2,\ldots,V_M\}=\{1,2,\cdots,M\},\quad V_m=m\ (m=1,2,\cdots,M) \tag{4}$$

Hypothesis 4. Every kind of affective state is mutual excluding. That is stimulus $m=i$ can increase the intension of affective states i, and reduce the intension of the others $j(j\neq i, j=1,2,...,N)$.

Hypothesis 5. The change of affective states includes two processes: spontaneous metastasis and stimulating metastasis.

Hypothesis 6. The tendency of spontaneous metastasis of affective states is always to quiet state.

2.3 The Transfer of Affective States

Due to different conditions, the transfer of the affective states has following cases:

- Stimulating transfer: On the effect of outside stimulation, a certain affective state or quiet state will transfer to stimulating state, as curve A, B showed.
- Stimulating transfer of affective states: On the effect of special outside stimulation, affective states will drift within a range of equilibrium point, as curve F showed.

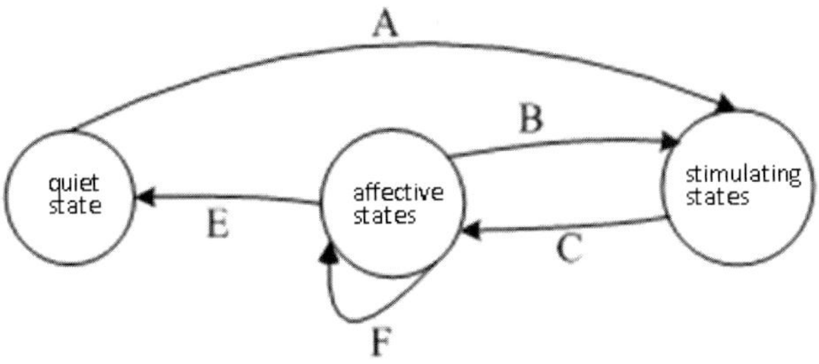

Fig. 1. Transfer picture of affective states

- Spontaneous metastasis of stimulating states: After the effect of outside stimulating finished, a certain affective state will spontaneously transfer from a certain stimulating state to a certain affective state in a certain time, as curve C showed.
- Spontaneous metastasis of affective states: In the case of without outside stimulation, a certain affective state will transfer to quiet state in a certain time, as curve E showed.

The transfer picture of affective states is showed as Fig.1.

2.4 Affective Model Based on HMM

Hidden Markov Model (HMM) is a double stochastic processes based on Markov chain. The model has following form [7]:

$$\lambda = (N, M, \pi, \hat{A}, B) \tag{5}$$

where,

N, the total number of basic affective states;

M = N, based on Hypothesis 3;

$\pi = [\pi_1, \pi_2, ..., \pi_N]$, probability distribution vector of initial states, and $\pi_i = P(S_i)$, $1 \le i \le N$

$$\hat{A} = (\hat{a}_{ij})_{N\times N} = \begin{bmatrix} \frac{\hat{\theta}\hat{\pi}_1^* - (N-1)}{\hat{\theta}\hat{\pi}_1^*} & \frac{1}{\hat{\theta}\hat{\pi}_1^*} & \cdots & \frac{1}{\hat{\theta}\hat{\pi}_1^*} \\ \frac{1}{\hat{\theta}\hat{\pi}_2^*} & \frac{\hat{\theta}\hat{\pi}_2^* - (N-1)}{\hat{\theta}\hat{\pi}_2^*} & \cdots & \frac{1}{\hat{\theta}\hat{\pi}_2^*} \\ \vdots & \vdots & \cdots & \vdots \\ \frac{1}{\hat{\theta}\hat{\pi}_N^*} & \frac{1}{\hat{\theta}\hat{\pi}_N^*} & \cdots & \frac{\hat{\theta}\hat{\pi}_N^* - (N-1)}{\hat{\theta}\hat{\pi}_N^*} \end{bmatrix}$$

$$=\begin{bmatrix} \frac{\hat{\theta}\hat{\pi}_1^* - (N-1)}{\hat{\theta}\hat{\pi}_1^*} & \frac{1}{\hat{\theta}\hat{\pi}_1^*} & \cdots & \frac{1}{\hat{\theta}\hat{\pi}_1^*} \\ \frac{1}{\hat{\theta}\hat{\pi}_2^*} & \frac{\hat{\theta}\hat{\pi}_2^* - (N-1)}{\hat{\theta}\hat{\pi}_2^*} & \cdots & \frac{1}{\hat{\theta}\hat{\pi}_2^*} \\ \vdots & \vdots & \cdots & \vdots \\ \frac{1}{\hat{\theta}\hat{\pi}_N^*} & \frac{1}{\hat{\theta}\hat{\pi}_N^*} & \cdots & \frac{\hat{\theta}\hat{\pi}_N^* - (N-1)}{\hat{\theta}\hat{\pi}_N^*} \end{bmatrix}$$

, stimulating transfer matrix of affective states, and $\overline{\pi}^* = [\overline{\pi}_1^*, \overline{\pi}_2^*, ..., \overline{\pi}_N^*]$, probability distribution of affective states in stable state;

$B = \{b_j(k)\}_{j \times k}$, stimulating matrix;

2.5 Simulation Results of Affective States Change

Fig.2 shows that after a certain stimulating state is generated by a certain outside stimulus, it will revert to a certain affective state when time goes by. Fig.3 shows that when there is a certain stimulus m, then the intensity of a certain affective state $i(i = m)$ will increase from initial probability intensity π_Δ to 1, and the intensity of the others will decrease to 0. The result of the matlab simulation is excellent agreement with **Hypothesis 3**, **Hypothesis 4** and human's psychological law.

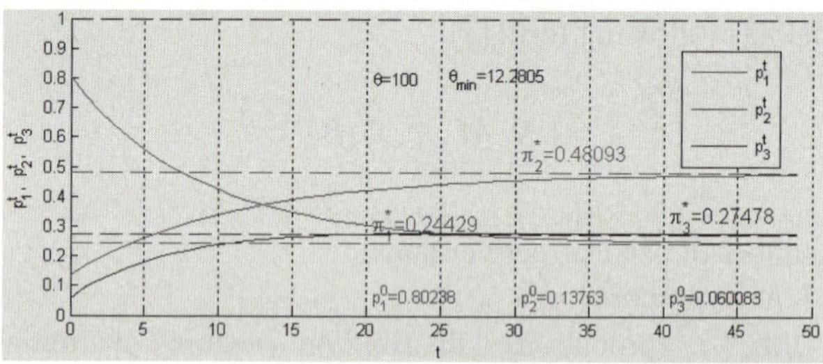

Fig. 2. Spontaneously transferring process of affective states

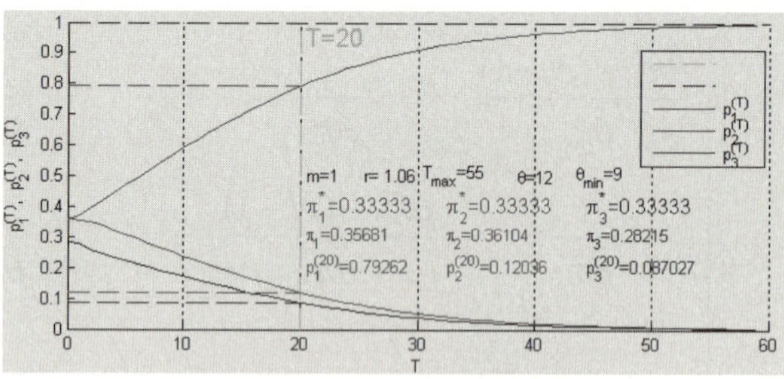

Fig. 3. Changing curve of affective intensity

3 Technical Route

The technical route of playmate robot for children is showed as Fig.4. First, we hope to obtain and deal with the environment information by kinds of sensor based on artificial psychology. The information apperceived by playmate robot for children is multimodal. Multimodal information means the different modes that can express the user's idea, execution action, or perception such as speech, eyesight, facial expression, gesture, posture, feeling, touch, or taste etc. The playmate robot for children uses the feature extracting tool to extract the primary characteristics of multimodal information. Then, it adopts the multimodal information fusion technology to deal with the information and to extract the information feature coinciding with human cognitive behaviour. The information expressed by each mode is complementary. If it was dealt with respectively, complete information will be lost. Therefore, the robot extracts the comprehensive information characteristic from not only each mode but also the combination of each mode. By this the robot can provide the essential information to complete the special interactive task only achieved by cooperative work of diversified channel models. And it is improved that the cognitive characteristics extraction capability and the expression ability of playmate robot for children. At the same time, user's information and behaviour intention is obtained by robot. So do the environment information. The information and the result of information fusion are used as the input of distributed cognitive information treatment. Third, the playmate robot for children adopts the cooperative perception interactive model of distributed cognitive

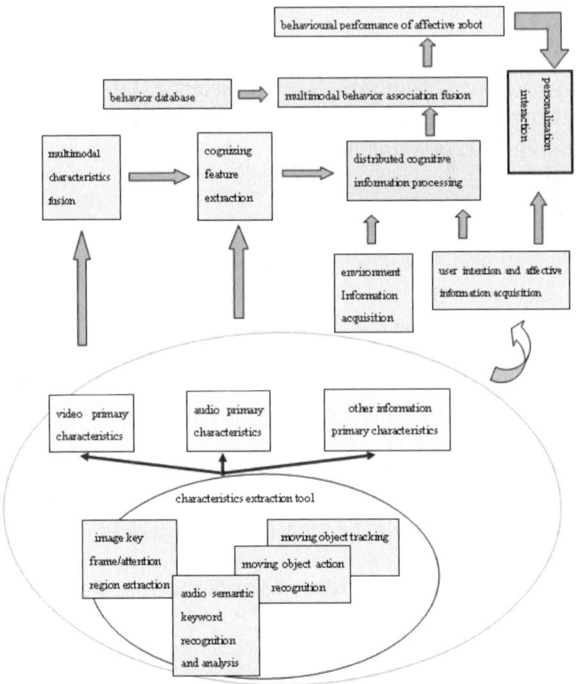

Fig. 4. The technical route of playmate robot for children platform

system to express and share the information and to set the environment. And affective model is used in distributed cognitive system to make the decision-making more similar to human. Therefore, by the distributed cognitive information processing module, the playmate robot for children can deal with the environment information, user information, and robot's cognitive feature and then can generate the all kinds of command to achieve the interactive activities between human and playmate robot for children. The module of distributed cognitive information treatment can output all kinds of behaviours. The playmate robot for children uses multimodal behaviour association fusion module to make the behaviours of the robot more natural and harmonious. Therefore, the natural and harmonious interaction between human and robot can be completed.

4 Conclusion

This paper reports the affective model and technical route for playmate robot for children. The robot has a human –like appearance and various sensors for interaction with human. In order to achieve the emotion similar to human, we build an affective model based on HMM and prove its validity by matlab stimulation. The robot can behave like a human child and attempt daily communication with human. We hope the research can be put to practical use for educational task, health tasks and family entertainment.

Acknowledgment

The paper is supported by National Natural Science Foundation of China (NO. 60573059), 863 Program (2007AA04Z218) and key programs of Natural Science Foundation of Beijing (KZ200810028016).

References

1. Ekman, P., Davidson, R.J.: The Nature of Emotion. Oxford University Press, Oxford (1994)
2. Elliott, C.: The Affective Reasoner: A Process Model of Emotions in a Multi-agent System. Ph.D. Dissertation, Northwestern University, The Institute for the Learning Sciences, Technical Report No.32 (1992)
3. Roseman, I.J.: Cognitive Aspects of Emotion and Emotional Behavior. The 87th Annual Convention (1979)
4. Reilly, W.S.: Believable Social and Emotional Agents. Technical Report CMU-CS-96-138, School of Computer Science, Carnegie Mellon University, Pittsburgh, PA (1996)
5. Gratch, J., Marsella, S.: Evaluating the Modeling and Use of Emotion in Virtual Humans. In: AAMAS 2004, vol. 1 (2004)
6. Wang, Z.L.: Artificial Psychology-a most Accessible Science Research to Human Brain. Journal of University of Science and Technology Beijing 22, 478–481 (2000)
7. Lu, C.G.: Stochastic Processes in Engineering System, pp. 32–36. Electronic Industry Publishing House, Beijing (2000)

The Application of Full Adaptive RBF NN to SMC Design of Missile Autopilot

Jinyong Yu[1], Chuanjin Cheng[2], and Shixing Wang[1]

[1] Department of Control Engineering,
Naval Aeronautical and Astronautical University, Yantai, China
[2] Department of Aeronautical Technology Support, BeiJing, China
tothelast@163.com

Abstract. A new adaptive sliding-mode control (SMC) scheme was proposed, which incorporated Full Adaptive RBF NN into sliding-mode control using Full Adaptive RBF NN to approximate the equivalent control and the upper bound of uncertainty which involved the disturbance and approximation error, thus the influence of modeling error was reduced and the gain of sliding-mode control part was more fitting, such that the chattering effects could be alleviated . Lyapunov stability theorem was used to prove the stability of the system and the adaptive laws were deduced. Finally, simulation results of some BTT missile were included to illustrate the effectiveness of the adaptive sliding-mode control scheme.

Keywords: RBF, adaptive, SMC, missile.

1 Introduction

A Radial Basis Function (RBF) NN differs in structure from the SHL NN in that there are no weight parameters associated with any of the input layer interconnections. Inaddition, the activation function is a bell shaped Gaussian function[1]. This kind of NN is usually considered linear parameterized, but if the centers and the widths are adjusted, this NN structure becomes nonlinearly parameterized. This structure can uniformly approximate continuous functions to arbitrary accuracy on compact sets provided that when a sufficient number of Gaussian functions is employed [2, 3]. When the function approximation is over a large domain , the local characteristic of RBF networks is considered an unattractive feature, The choice of a sufficient number of Gaussian functions can quickly lead to an intractable problem due to the curse of dimensionality [4]. In order to deal with highly uncertain nonlinear systems, approximator-based control schemes have been extensively studied [5-6], a full adaptive RBF control scheme is proposed, which allow not only the weights but also the centers and widths of the Gaussian functions to adapt online based on a Lyapunov derived update law that yield boundedness of signals in the closed loop. In this note, by combining sliding mode control with NN technologies, we present a novel sliding mode NN control scheme, the full adaptive RBF NN will be used to approximate the uncertain nonlinear term in the control scheme.

The paper is organized as follows. A brief description of RBF NN is made in Section II. The problem formulation and the design of SMC controller is made based on full

F. Sun et al. (Eds.): ISNN 2008, Part II, LNCS 5264, pp. 165–170, 2008.

adaptive RBF NN in Section III. The correctness and effectiveness of the proposed scheme have been verified by mathematical simulations in Section IV. A Conclusion has been drawn in Section V.

2 Description of RBF NN

Fig.1 shows the typical structure of a RBF NN.

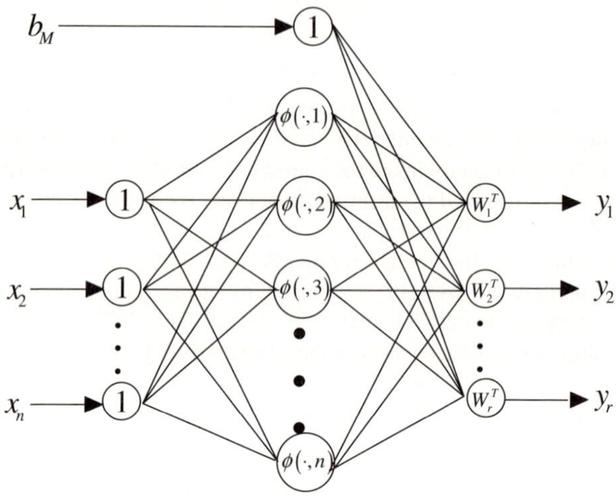

Fig. 1. RBF NN Structure

The activation functions ϕ are defined as follows:

$$\phi(\bar{x}, c, \sigma) = \exp(\|\bar{x} - c\|^2 / \sigma^2) \tag{1}$$

where $\bar{x}$ is a vector of input variables, σ is the width of the Gaussian function, c is a vector of Gaussian center positions.

The argument of the activation function of the hidden layer units represents the Euclidean norm between the input vector and the units' center position. This operation characterizes the exponentially decaying globalized nonlinearity of Gaussian functions. The output of an RBF network can hence be written as

$$y = M^T \phi(\bar{x}, c, \sigma) \tag{2}$$

3 Problem Formulation and Controller Design

Consider a n-th SISO nonlinear system in the following form:

$$x^{(n)} = f(x, \dot{x}, ..., x^{(n-1)}) + g(x, \dot{x}, ..., x^{(n-1)})u + d \tag{3}$$

$$y = x \tag{4}$$

Where $u \in R$ is the control input, $y \in R$ is system output, $f(.)$ and $g(.)$ are unknown smooth functions and $g(.) \neq 0$. First, define the tracking error

$$e = y - y_d \tag{5}$$

Where y_d is the desired output of the system that is bounded. Then an ideal feedback linearization control law can be obtained, and the total control can be written in the following form

$$u = u_{eq} + u_s \tag{6}$$

$$u_{eq} = \frac{1}{g(.)}[-f(.) + y_d^{(n)} - K_1^T E] \tag{7}$$

where $[K_1 = (k_n, k_{n-1}, ..., k_1)^T$, $E = (e, \dot{e}, ..., e^{(n-1)})^T$, $u_s = -Ksign(S)$ is sliding-mode control part which aims to offset disturbance, substituting (6) and (7) into (5) gives

$$\begin{aligned} &e^{(n)} + k_1 e^{(n-1)} + k_2 e^{(n-2)} + ... + k_{n-1}\dot{e} + k_n e \\ &= g u_s + d = \bar{g} u_s + d' \end{aligned} \tag{8}$$

Where $d' = d + \Delta g u_s$, $\bar{g}$ is the nominal value of $g(.)$, Δg is the uncertain part of $g(.)$, k_i makes the left part of (7) a Hurwitz polynomial. Because $f(.)$ and $g(.)$ are unknown, the ideal control law (6) can not be realized, now we use the following RBF NN to approximate it.

$$\hat{u} = \hat{M}^T \hat{\phi} \tag{9}$$

Define the error variables $\tilde{M} = \hat{M} - M^*$, $\tilde{\phi} = \hat{\phi} - \phi^*$, $\tilde{c} = \hat{c} - c^*$, $\tilde{\sigma} = \hat{\sigma} - \sigma^*$ then we have

$$\begin{aligned} \tilde{u} &= \hat{u} - u^* = \hat{M}^T \hat{\phi} - M^{*T} \phi^* \\ &= \hat{M}^T \hat{\phi} + \hat{M}^T \tilde{\phi} - \tilde{M}^T \tilde{\phi} \\ &= \tilde{M}^T \left(\hat{\phi} - \hat{\phi}'_c . * \hat{c} - \hat{\phi}'_\sigma . * \hat{\sigma}\right) \\ &+ \hat{M}^T \left(\hat{\phi}'_c . * \tilde{c} + \hat{\phi}'_\sigma . * \tilde{\sigma}\right) + d_u \end{aligned} \tag{10}$$

where $\begin{aligned} &\|d_u\| \le \|c^*\| \left\|\hat{M}^T \hat{\phi}'_c\right\| + \|\sigma^*\| \left\|\hat{M}^T \hat{\phi}'_\sigma\right\| + \left\|M^{*T}\right\| \left\|\hat{\phi}'_c . * \hat{c}\right\| \\ &+ \left\|M^{*T}\right\| \left\|\hat{\phi}'_\sigma . * \hat{\sigma}\right\| + \sqrt{R} \left\|M^{*T}\right\| \end{aligned}$ [5], $\hat{\phi}'_c$ and $\hat{\phi}'_\sigma$ are the partially differentials with respect to c and σ.

If we choose the sliding manifold as

$$S = -y_d^{(n-1)} + \int \begin{pmatrix} y^{(n)} + k_1 e^{(n-1)} + k_2 e^{(n-2)} \\ + ... + k_{n-1}\dot{e} + k_n e \end{pmatrix} dt \tag{11}$$

Substituting (10) into (11) after some manipulations, it can obtain that

$$\begin{aligned} \dot{S} &= e^{(n)} + k_1 e^{(n-1)} + k_2 e^{(n-2)} + ... + k_{n-1}\dot{e} + k_n e \\ &= g\tilde{u} + \bar{g} u_s + d' = \bar{g}\tilde{u} + \bar{g} u_s + d'' \end{aligned} \tag{12}$$

Where $d'' = d' + \Delta g \tilde{u}$, define the following Lyapunov function:

$$V_1 = \frac{1}{2} S^2 + \frac{1}{2a_1} \tilde{M}^T \tilde{M} + \frac{1}{2a_2} \tilde{c}^T \tilde{c} + \frac{1}{2a_3} \tilde{\sigma}^T \tilde{\sigma} \tag{13}$$

where a_1, a_2, a_3, are positive constants, differentiating it with respect to time, it can obtain that

$$\begin{aligned}
\dot{V}_1 &= S\dot{S} + \tfrac{1}{a_1}\tilde{M}^T\dot{\tilde{M}} + \tfrac{1}{a_2}\tilde{c}^T\dot{\tilde{c}} + \tfrac{1}{a_3}\tilde{\sigma}^T\dot{\tilde{\sigma}} \\
&= \mathrm{S}[\bar{g}\tilde{u} + \bar{g}u_s + d''] + \tfrac{1}{a_1}\tilde{M}^T\dot{\tilde{M}} + \tfrac{1}{a_2}\tilde{c}^T\dot{\tilde{c}} + \tfrac{1}{a_3}\tilde{\sigma}^T\dot{\tilde{\sigma}} \\
&= S\{\bar{g}[\tilde{M}^T\left(\hat{\phi} - \hat{\phi}'_{c\cdot} * \hat{c} - \hat{\phi}'_{\sigma\cdot} * \hat{\sigma}\right) \\
&+ \hat{M}^T\left(\hat{\phi}'_{c\cdot} * \tilde{c} + \hat{\phi}'_{\sigma\cdot} * \tilde{\sigma}\right) + d_u] + \bar{g}u_s + d''\} \\
&+ \tfrac{1}{a_1}\tilde{M}^T\dot{\tilde{M}} + \tfrac{1}{a_2}\tilde{c}^T\dot{\tilde{c}} + \tfrac{1}{a_3}\tilde{\sigma}^T\dot{\tilde{\sigma}} \\
&= \tilde{M}^T[S\bar{g}\left(\hat{\phi} - \hat{\phi}'_{c\cdot} * \hat{c} - \hat{\phi}'_{\sigma\cdot} * \hat{\sigma}\right) + \tfrac{1}{a_1}\dot{\tilde{M}}] \\
&+ \tilde{c}^T(S\bar{g}\hat{M}. * \hat{\phi}'_c + \tfrac{1}{a_2}\dot{\tilde{c}}) + \tilde{\sigma}^T(S\bar{g}\hat{M}. * \hat{\phi}'_\sigma + \tfrac{1}{a_3}\dot{\tilde{\sigma}}) \\
&+ S\bar{g}d_u - S\bar{g}K\,sign(S) + S d''
\end{aligned} \tag{14}$$

If the parameter adaptive laws are chosen as(16),(17),(18) and and K satisfies the inequality shown in (19)

$$V_1 = \frac{1}{2}S^2 + \frac{1}{2a_1}\tilde{M}^T\tilde{M} + \frac{1}{2a_2}\tilde{c}^T\tilde{c} + \frac{1}{2a_3}\tilde{\sigma}^T\tilde{\sigma} \tag{15}$$

$$\dot{\hat{\xi}} = -a_1 S\bar{g}\left(\hat{\phi} - \hat{\phi}'_{c\cdot} * \hat{c} - \hat{\phi}'_{\sigma\cdot} * \hat{\sigma}\right) \tag{16}$$

$$\dot{\hat{c}} = -a_2 S\bar{g}\hat{M}. * \hat{\phi}'_c \tag{17}$$

$$\dot{\hat{\sigma}} = -a_3 S\bar{g}\hat{M}. * \hat{\phi}'_\sigma \tag{18}$$

$$K > \|d_u\| + \left\|d''/\bar{g}\right\| \tag{19}$$

It can obtain that

$$\begin{aligned}
\dot{V}_1 &= S\bar{g}d_u - S\bar{g}K\,sign(S) + S d'' \\
&\leq -\|S\|(\bar{g}K - \|\bar{g}d_u\| - \|d''\|) \\
&\leq 0
\end{aligned} \tag{20}$$

According to the Barbalat Lemma, we can easily know that $S \to 0$, $\tilde{u} \to 0$, when $t \to \infty$, then a conclusion can be made that the adaptive sliding-mode control scheme with the conditions of (16), (17),(18),(19)to be satisfied can guarantee the stability of the system.

4 Simulation

In order to verify the rightness and effectiveness of the control scheme, mathematical simulation is conducted for the overload model of the pitch channel of some BTT missile. The mathematical model is shown in (21). A series of step-commands are inputted to the autopilot continuously (overload command n_{y_c}), in this way, the performance of the designed autopilot is tested. Simulation results with 50% aerodynamic coefficients uncertainty and the disturbance $d = 0.5\sin(5t)$ are shown in Fig.2 and Fig.3.

$$\begin{aligned}
\ddot{n}_y &= f(.) + g(.)\delta_z + d \\
y &= n_y
\end{aligned} \tag{21}$$

where $f(.) = -(a_1 + a' + a_4)\dot{n}_y - (a_2 + a_1 a_4)n_y$, $g(.) = -\frac{Va_4}{57.3g}(a_1 a_5 + a_3)$, $d = (a' + a_1)f_1 + \frac{Va_4}{57.3g}f_2 + \dot{f}_1 - \frac{Va_4 a_5}{57.3g}\dot{\delta}_z$, $f_1 = -\frac{Va_4}{57.3g}\cdot\frac{\omega_x}{57.3}\beta$, $f_2 = \frac{a'_1\omega_x}{57.3}\beta + \frac{J_x - J_y}{57.3J_z}\omega_x\omega_y$.

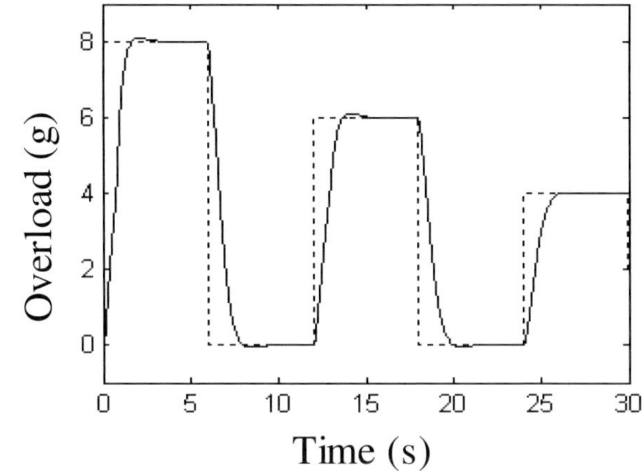

Fig. 2. Trajectories of the output

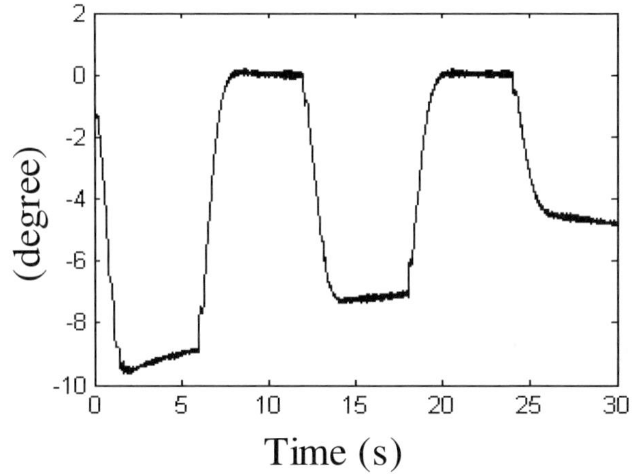

Fig. 3. Trajectories of the fin deflection

5 Conclusion

In this paper, a new adaptive sliding mode control scheme is proposed, which incorporated Full Adaptive RBF NN into sliding-mode control using Full Adaptive RBF NN to approximate the equivalent control and the upper bound of uncertainty which involved the disturbance and approximation error, thus the influence of modeling error was reduced and the gain of sliding-mode control part was more fitting, such that the chattering effects could be alleviated. Lyapunov stability theorem was used to prove the stability of the system and the adaptive laws were deduced. Finally, simulation results of some BTT missile were included to illustrate the effectiveness of the adaptive sliding-mode control scheme.

References

1. Girosi, F., Poggio, T.: Networks and the best Approximation Property. Articial Intelligence Lab. Memo, 1164 (1989)
2. Poggio, T., Girosi, F.: Networks for Approximation and Learning. Proc. of the IEEE 78, 1481–1497 (1990)
3. Sanner, R., Slotine, J.J.: Gaussian Networks for Direct Adaptive control. IEEE Transactions on Neural Networks 3, 837–864 (1992)
4. Zhang, Y.A., Hu, Y.A.: Nonlinear Design Approaches for Missile Control and Guidance, pp. 78–81. Defense Industry Press (2003)
5. Ge, S.S., Ren, B., Tee, K.P.: Adaptive Neural Network Control of Helicopters with Unknown Dynamics. In: Proceedings of the 45th IEEE Conference on Decision and Control, San Diego, USA (2006)
6. Chen, G.: Sliding Mode Neural Network Control for Nonlinear Systems. In: Proceedings of the 2006 IEEE International Conference on Mechatronics and Automation, Luoyang, China (2006)

Multi-Objective Optimal Trajectory Planning of Space Robot Using Particle Swarm Optimization

Panfeng Huang[1], Gang Liu[2], Jianping Yuan[1], and Yangsheng Xu[3]

[1] College of Astronautics
Northwestern Polytechnical University
Xi'an, China
pfhuang@nwpu.edu.cn, jyuan@nwpu.edu.cn
[2] Infineon Technologies(Xi'an) Ltd
Xi'an, China
[3] Department of Mechanical and Automation Engineering
The Chinese University of Hong Kong
Hong Kong, China

Abstract. Space robots are playing significant roles in the maintenance and repair of space station and satellites and other future space services. The motion trajectory planning is a key problem for accomplishing above missions. In order to obtain the high efficiency, safety motion trajectory of space robot, the motion trajectory should be optimized in advance. This paper describes the multi-objective optimization for optimizing the motion trajectory of space robot using a multi-objective particle swarm optimization (MOPSO). In this formulation, the multi-objective function is generated which includes some parameters such as motion time, dynamic disturbance, and jerk, and so on. Then a number of relative parameters can be simultaneously optimized through searching in the parameter space using MOPSO algorithms. The simulation results attest that MOPSO algorithm has satisfactory performance and real value in fact.

1 Introduction

Space-base robot is unlike ground-base robot, it has some special characteristics such as kinematic nonholonomic, dynamic coupling which make its planning and control complicated. When the space robot is in free-floating situation, The longer the motion time of space manipulator is, the greater the disturbance to the base will be. Hence, the operation precise of the end-effector will be affected severely. Fortunately, the kinematic and dynamic model of space robot can be obtained accurately in space environment. Moreover, the interactive disturbance between the manipulator and its base can be estimated and calculated according to Yangsheng Xu's papers [1]. Therefore, we can realize the accurate operation of space robot by optimizing its motion trajectory with multi-objective parameters.

F. Sun et al. (Eds.): ISNN 2008, Part II, LNCS 5264, pp. 171–179, 2008.

Several researchers focused on the motion path or trajectory planning of space robot, Agrawal and Xu[2] addressed the global optimum path planing for redundant space manipulator. They considered the linear and angular momentum as constraint conditions, then using Lagrange multiplier technique to change the optimum problems subject to constraint conditions to non-constraint problems. Lastly, using differential and algebraic equations to solve the objective functions. Dubowsky and Torres[3] proposed a method called the Enhanced Disturbance Map (EDM) to plan the space manipulator motions so that the disturbance to the space base is relatively minimized. Their technique also aided to understand the complicated problem and developed the algorithm to reduce the disturbance. Evangelos G. Papadopoulos[4] exhibited the nonholonomic behavior of free-floating space manipulator by planing the path of space manipulator. They proposed a path planning method in Cartesian space to avoid dynamically singular configurations. Yoshida and Hashizume [5] utilized the ETS-VII as an example to address a new concept called Zero Reaction Maneuver (ZRM) that is proven particularly useful to remove the velocity limit of manipulation due to the reaction constraint and the time loss due to wait for the attitude recovery. Moreover, they found that the existence of ZRM is very limited for a 6 DOF manipulator, besides the kinematically redundant arm. However, the redundant manipulator give rise to the computational cost and complicated kinematics problems. Panfeng Huang et,al[6] proposed a novel motion trajectory planning of space manipulator using genetic algorithms, they developed a single objective genetic algorithms for minimum-torque trajectory planning of space manipulator.

In this paper, we propose a multi-objective particle swarm optimization (MOPSO) to obtain optimal trajectory of space robots with minimum disturbance and minimum time. PSO is a swarm intelligence method for global optimization. It differs from other well known evolutionary algorithms (EA) [7], such as genetic algorithm, no operators inspired by evolutionary procedures are applied on the population to generate new promising solutions in the PSO. However, each individual, named particle, of the population, called swarm, adjusts its trajectory toward its own previous best position, and toward the previous best position attained by any member of its topological neighborhood [8]. In the global variant of PSO, the whole swarm is considered as the neighborhood. Thus, global sharing of information takes place and the particles profit from the discoveries and previous experience of all other companions during the search for promising regions of the landscape.

This paper is organized as follows. In section 2, we mainly address Multi-objective optimization problem and multi-objective trajectory planning of space manipulator. Section 3 presents the approved multi-objective optimal algorithm to solve the multi criteria trajectory planning problem based on the MOPSO. In section 4, we use an illustrative example to illustrate the effectiveness of the proposed method. Final section summarizes the whole paper and give some conclusions.

2 Multi-Objective Optimal Trajectory Planning

2.1 Multi-Objective Optimization Problem

The general multi-objective optimization problem can be mathematically stated as follows:

Find the vector $\tilde{x} = [x_1, x_2, ...x_n]^T$, which satisfies the m inequality constraints:

$$g_i(x) \leq 0, i = 1, 2, ..., m \tag{1}$$

and the p equality constraints:

$$h_i(x) = 0, i = 1, 2, ..., p \tag{2}$$

and optimizes the vector function:

$$f(\tilde{x}) = [f_1(\tilde{x}), f_2(\tilde{x}), ...f_k(\tilde{x})]^T \tag{3}$$

Without loss of generality, we consider the minimization case for objective function, $f(\tilde{x})$. However, these objective functions such as $f_1(\tilde{x})$,$f_2(\tilde{x})$,...$f_k(\tilde{x})$ may be in conflict, thus, it is impossible to obtain the global minimum for all objectives at the same point. The goal of Multi-Objective Optimization is to provide a set of Pareto Optimal solutions to aforementioned problem using the concept of Pareto dominance. This concept formulated by Vilfredo Pareto is defined as [10]:

Define $u = (u_1, ..., u_k)$ and $v = (v_1, ..., v_k)$ as two vectors, Thus, if and only if $u_i \leq v_i$,$i = 1, ..., k$, u dominates v, and $u_i \leq v_i$ for at least one component,which property is known as Pareto Dominance and it is define the Pareto Optimal points. So the solution x of the Multi-Objective Problem is said to be Pareto Optimal if and only if there does not exist another solution there does not exist another solution y such that $f(y)$ dominates $f(x)$. The set of all Pareto Optimal solutions of an Multi-Objective problem is called Pareto Optimal Set and we denote it with $\Re^*$. The set $\Re F^* = \{(f_1(x), ..., f_k(x)) \mid x \in \Re^*\}$ is called Pareto Front. A Pareto Front $\Re F^*$ is called convex if and only if $\forall u, v \in \Re F^*$, $\forall \lambda \in (0, 1)$, $\exists \omega \in \Re F^* : \lambda \parallel u \parallel +(1-\lambda) \parallel v \parallel \geq \parallel \omega \parallel$. Respectively, it is called concave if and only if $\forall u, v \in \Re F^*$, $\forall \lambda \in (0, 1)$,$\exists \omega \in \Re F^* : \lambda \parallel u \parallel +(1-\lambda) \parallel v \parallel \leq \parallel \omega \parallel$. A Pareto Front can be convex, concave or partially convex and/or concave and/or discontinuous. The last three cases present the greatest difficulty for most MO techniques.

2.2 Multi-Objective Optimal Trajectory Planning of Space Manipulator

The trajectory planning problem is generally defined here as the point to point problem, i.e, that of determining the time history of the robot joints and spacecraft state (position and orientation) in order to move the end-effector of the robot form a given initial state to a given final state in inertial space. However, such planning path only ensures that the end-effector of robot move to the desired state. Whereas, this trajectory must be optimized in order to satisfy kinematic and dynamic constraints and fully use the capability of robotic manipulator. Especially, for space robot system, optimizing the motion path becomes a more and more important problem in order to minimize multiple objectives simultaneously, such as minimizing the disturbances, minimizing the mechanical energy of the actuators and minimizing the traveling time, and so on. Therefore, all these optimal objectives are considered together to build a multi-objective function and the results depend on the associated weighting factors.

According to the multi-objective optimization problem mentioned in section 2.1, we can define the objective function for trajectory planning of space manipulator as follows.

$$MinF(x) = \omega_1 f_1(x) + \omega_2 f_2(x)+, \ldots, +\omega_n f_n(x) \tag{4}$$

where, $f_i(x)$ is $i-th$ objective function,and ω_i is a constant weight for $f_i(x)$, n is the number of objectives. In Equation (3), if we define ω as constant weights, the search direction in MPSO is fixed. Therefore we propose a selection procedure with random weights to search for Pareto-optimal solutions by utilizing various search directions. Therefore, we assign a random real number to each weight as follows when a pair of strings are selected for a crossover operation.

$$\omega_i = random_i(\cdot)/\sum_{k=1}^{m} random(\cdot), i = 1, 2, \ldots, n. \tag{5}$$

where $random(\cdot)$ is a non-negative random number. From Equation (4), we can see that ω_i is a real number in the closed interval [0,1]. Next pair of strings are selected with different weight values newly given by Equation (4), and so on.

In this paper, we consider two objectives which are conflicting, these two objectives are minimization disturbance to the space base and minimization traveling time of end-effector of space manipulator for simplification. According to our previous works in [11,12], we can obtain the two objective functions respectively as follows.

The objective function for minimum disturbance trajectory planning can be defined as following constraint optimization problems.

$$min\Gamma = \frac{1}{N}\sum_{j=0}^{N-1} max(F_b(t_j))^2 \tag{6}$$

subject to

$$\begin{aligned} &|\theta_i(t_j)| \le \theta_{max}, 1 \le j \le N \\ &|\dot{\theta}_i(t_j)| \le \omega_{max} \\ &|\ddot{\theta}_i(t_j)| \le a_{max} \end{aligned} \tag{7}$$

where, $F_b(t_j)$ represents the disturbance at time t_j, which can be computed by calculating three derivative of position of desired path.

On the other hand, the aim of time-optimal trajectory planning is the determination of the maximum velocity profile along a given path that complies with all given dynamic and kinematic robotic constraints. The motion of manipulator can be denoted as a position vector, $\boldsymbol{l}$, which starts from the starting point, p_0, to the end point, p_f. Thus, the path length can be defined as following function.

$$s = \int_{t_0}^{t} \parallel \frac{d\boldsymbol{l}}{dt} \parallel dt, t \in [t_0, t_f] \tag{8}$$

Therefore, the objective function for the time-optimal trajectory can be defined in terms of s as follows.

$$T = \int_{t_0}^{t_f} dt = \int_{t_0}^{t_f} \frac{1}{v} ds, s \in [p_0, p_f] \tag{9}$$

Subject to the constraints:

$$\tau_{min}^i \leq \tau_i \leq \tau_{max}^i, i = 1, 2, ..., n \tag{10}$$

$$f_{bmin} \leq f_b \leq f_{bmax} \tag{11}$$

where, v represents the velocities of manipulator. Obviously, the objective function becomes minimum when v is maximized while it is kept under dynamic constraints for keeping the safe of space robot system.

In two objective functions above mentioned, the trajectory of manipulator is in joint space. Therefore, the motion of manipulator must be subject to physical constraints of manipulator, such as, angle joint, joint velocities and joint acceleration, and so on. Thus, we can obtain the final multi-objective function as follows.

$$MinF(x) = \omega_1 \Gamma + \omega_2 T \tag{12}$$

subject to

$$\begin{array}{l} |\theta_i(t_j)| \leq \theta_{max}, 1 \leq j \leq N \\ |\dot{\theta}_i(t_j)| \leq \omega_{max} \\ |\ddot{\theta}_i(t_j)| \leq a_{max} \\ \tau_{min}^i \leq \tau_i \leq \tau_{max}^i, i = 1, 2, ..., n \end{array} \tag{13}$$

In Equation (12), the ω can be generated by Equation (4), and one objective function for minimization disturbance can be measured by the dynamics factors [12]. the other can be gotten by the Equation (8). We use the linearization method to combine these two objective functions and generate a multi-objective cost function, which can get the Pareto-optimal results using multi-objective particle swarm optimizer.

3 Multi-Objective Particle Swarm Optimizer

The basic concept and algorithms of Particle Swarm Optimizer can be obtained from [10]. In order to adapt PSO for multi-objective optimization, the p-vector in the update function [12] was modified to keep track of all

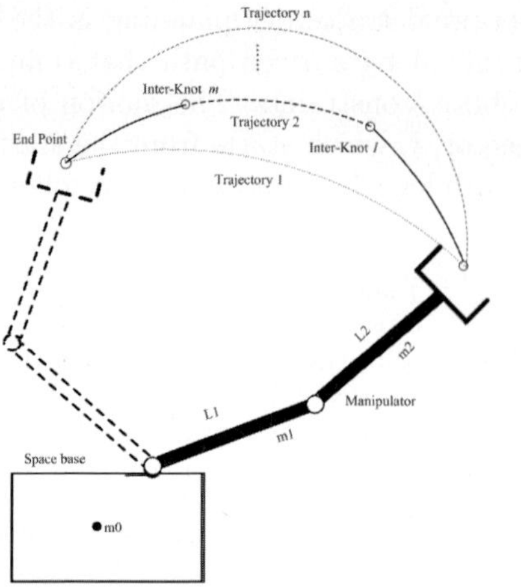

Fig. 1. Model of 2 DOF planar space robot system

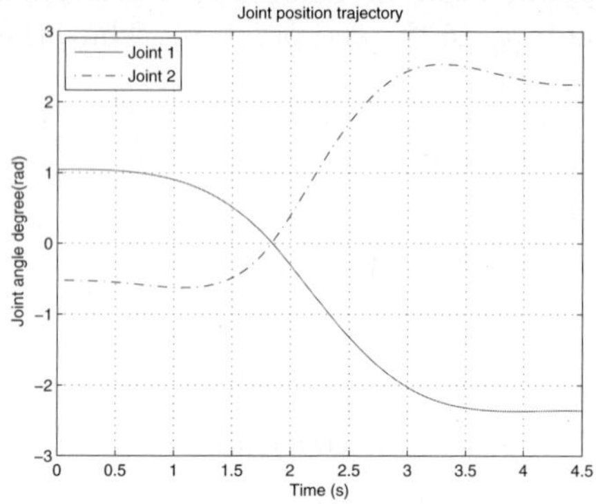

Fig. 2. Joint position trajectory

nondominated solutions (according to Pareto preference) that a particle encountered as it explored the search space. The p-vector has now become a list of solutions.

MPSO begins by randomly initializing the x and v vectors[12]. Each time the x-vector of a particle is updated, it is compared with the solutions in the p-list to determine if it is a nondominated solution. If it is a nondominated solution, it is added to the p-list. The p-list is also constantly updated to ensure that it contains only nondominated solutions. The exploration of the search space is guided by at most two pieces of information: the best potential solution discovered by an individual and the best potential solution discovered by its neighborhood. Since the p-list, formally the p-vector, can now contain numerous solutions, the best potential solution discovered by an individual is randomly selected from its list of nondominated solutions. To determine the best potential solution in the neighborhood, we compare the nondominated solutions found in the p-lists to find one that is not dominated within the neighborhood.

Therefore, we can use the MPSO to solve the optimal trajectory planning described by the cost function (12) with dynamic model under initial and final conditions and constraints. According to the trajectory planning strategies, we divide whole trajectory into several trajectory segments, the path point connecting with two segments is called knot point, our proposed method is to search optimum parameters of each knot point, such as, joint angle, joint angular velocity, and joint acceleration, then to realize optimal time trajectory planning. Therefore, we can obtain new algorithm procedure to optimize trajectory as follows:

Step 1. Define the control points (inter-knots) n and maximum iterative number N_{max}, then generate randomly particles $P_s - list$, $P_a^{(i)} - list, i = 1, ..., P_s$. Define the initial parameters χ, c_1, c_2, and w[12].

Step 2. Implement the MPSO algorithm from $P_a^{(i)} - list$ to calculate the maximum velocity v and acceleration $\dot{v}$ of manipulator using Equation (8). When iterative number reaches at N_{max},stop MPSO algorithm and record the best as $P_a^* - list$.

Step 3. Stop the algorithm. if $n > N_{max}$.

Step 4. Redefine the control points as $T_{midst} = (\frac{s_1+s_2}{2}, \frac{s_3+s_4}{2}, ..., \frac{s_n+s_{n+1}}{2})$ and insert these knots T_{midst} into T one by one, and $P_a^* - list$ and n.

Step 5. Update the velocity and position of $P_s - list$ particles. Then go to Step 2.

Step 6. Obtain the minimum fitness function value, thus, the parameters in this situation is optimal values. Get the optimum knot point. Then get the optimal trajectory.

4 Simulation Result

In order to verify the performance of our proposed optimal algorithm. Let's consider an example to better understand the optimum algorithms. A model of a planar 2 DOFs free-flying space robot is shown in Fig. 1. The parameters of the space robot are shown here: $m_0 = 40kg$, $m_1 = 4kg$, $m_2 = 3kg$, $L = L_0 = L_1 = L_2 = 1m$, $I_0 = 6.67kg \cdot m^2$, $I_1 = 0.33kg \cdot m^2$, $I_2 = 0.25kg \cdot m^2$. For a real space robot system, the joint angle, angular velocity, acceleration, and torque of the manipulator should have constraint values. We can define the constraint conditions of the model of space robot as follows.

$-pi \leq \theta_j \leq pi, i = 1, ..., 2$
$v_{jmax} = 5rad/s, j = 1, ..., 2$
$a_{jmax} = 20rad/s^2, j = 1, ..., 2$
$\tau_{1max} = 100Nm$, $\tau_{2max} = 50Nm$.

We plan a point to point trajectory in joint space. The manipulator starts from $\theta_{1s} = pi/3$, $\theta_{2s} = -pi/6$ in joint space, and the end point, $\theta_{1e} = -3 * pi/4$, $\theta_{2e} = 5 * pi/7$. The initial and final velocities, and accelerations are taken to be zero. According to the optimal objective, we need use the MPSO to search the best inter-knot points in the constraint conditions. To simplify the complicated computation, one inter-knots and two second execution time for each segment are chosen. Thus, there are six parameters: the position and velocities of inter-knots, θ_{11}, v_{11}, θ_{12}, v_{12}, and the traveling time of the first and second trajectory segments, t_0, t_1. We use the proposed algorithms to optimize these six parameters.

The goal of simulation is to verify the performance of MPSO. From the simulation result, the parameters of one inter-knot points can be obtained as follows.

$\theta_{1mp1} = 0.6779$, $\theta_{2mp1} = -0.8802$
$\dot{\theta}_{1mp1} = -1.5203$, $\dot{\theta}_{2mp1} = 1.7563$
$t_0 = 1.8s$, $t_1 = 2.7s$

Thus, the total optimal time: $t_f = t_0 + t_1 = 4.5s$.

According to the simulation results, Fig. 2 shows the joint position path of joint θ_1 and θ_2. The plot shows that both θ_1 and θ_2 start from initial

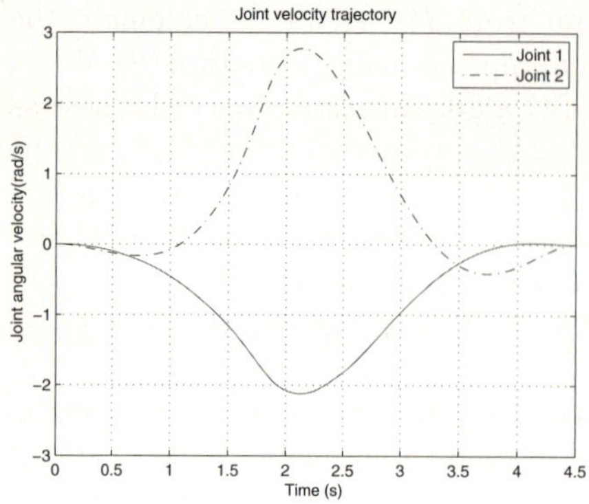

Fig. 3. Joint velocity trajectory

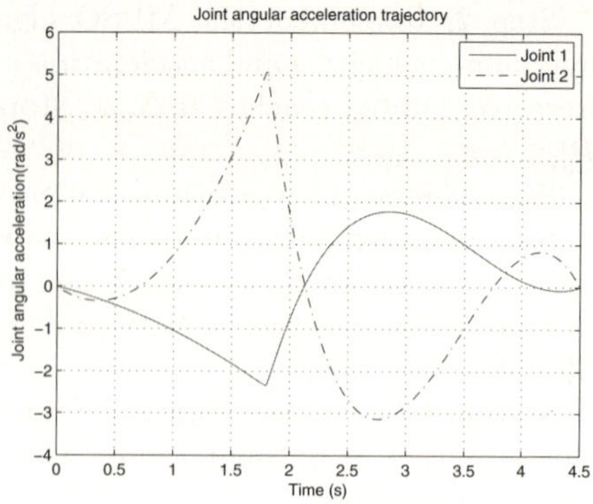

Fig. 4. Joint acceleration trajectory

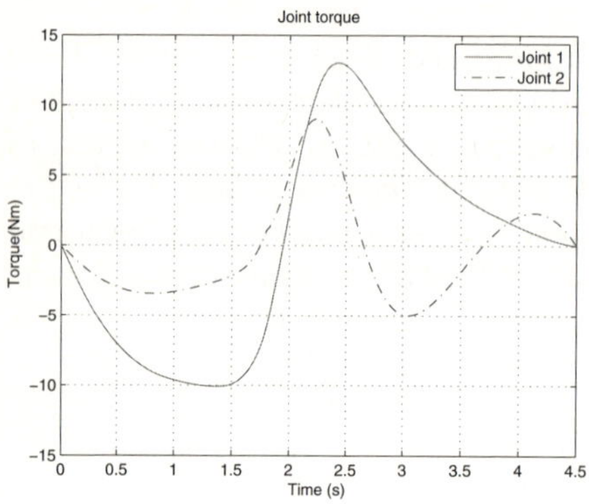

Fig. 5. Joint torque trajectory

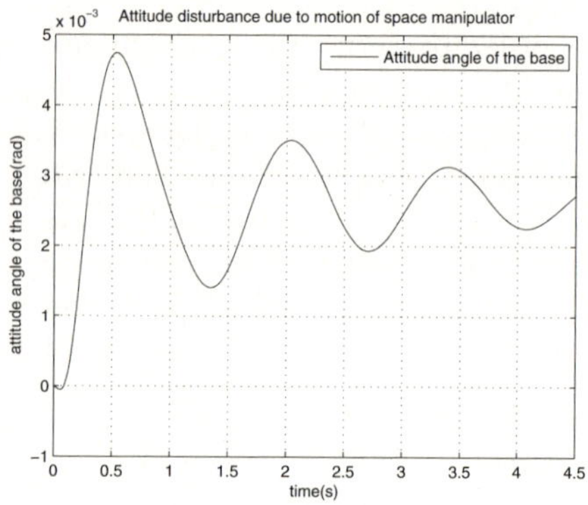

Fig. 6. Attitude disturbance to the base after optimization within optimal traveling time

position at $t = 0sec$, and reach the optimal time at $t = 4.5sec$. Fig. 3 shows the joint angular velocities whose values are limited in the constraints of joint angular velocity. Fig. 4 shows the joint angular acceleration after optimization. Moreover, the acceleration values are limited in the constraint conditions. Fig. 5 shows the joint torque when the manipulator motions along the optimal path. Fig.6 shows the attitude disturbance to the base of space robot after optimizing the motion trajectory of space manipulator. Therefore, the proposed method is useful and valid to reduce the disturbance within optimal traveling time.

Because the number of inter-knots is chosen manually, it is necessary to study how many inter-knots are optimal for optimization. Obviously, the more inter-knots, the more problem complicated, which certainly cost more computation time. Thus, choosing the smallest number of inter-knot is optimal. However, optimal precision may increase when adding the number of inter-knot. more inter-knots will be verified in the future work.

5 Conclusions

This paper presents a multi-objective particle swarm optimization for trajectory optimization of space manipulator. The proposed algorithm can globally search most satisfactory parameters of inter-knots to generate the optimal motion trajectory based on multi-objective functions. The optimal trajectory obtained is fitful for high velocity and high precision dynamic control. We use an example to verify the performance of proposed MPSO and suggest its potential application to real space robot system. Multi-objective trajectory optimization of space manipulator is a complicated and nonlinear problem, which will become an key point for improving and enhancing the productivity and save the fuel and energy.

References

1. Xu, Y.H.: The Measure of Dynamic Coupling of Space Robot System. In: Proc. of IEEE Int. Conf. on Robotics and Automation, pp. 615–620 (1993)
2. Agrawal, O.P., Xu, Y.S.: On the Global Optimum Path Planning for Redundant Space Manipulator. IEEE Trans. System, Man, and Cybernetics 24(9), 1306–1316 (1994)
3. Dubowsky, S., Torres, M.A.: Path Planning for Space Manipulator to Minimize Spacecraft Attitude Disturbances. In: Proc. of IEEE Int. Conf. on Robotics and Automation, pp. 2522–2528 (1991)
4. Papadopouls, E.: Path Planning for Space Manipulators Exhibiting Nonholonomic Behavior. In: Proc. of IEEE Int. Conf. on Intelligent Robots and Systems, pp. 669–675 (1992)
5. Yoshida, K., Hashizume, K.: Zero Reaction Maneuver: Flight Velification with ETS-VII Space Robot and Extention to Kinematically Redundant Arm. In: Proc. of IEEE Int. Conf. on Robotics and Automation, pp. 441–446 (2001)
6. Huang, P.F., Xu, Y.S., Liang, B.: Global Minimum-Jerk Trajectory Planning of Space Manipulator Using Genetic Algorithms. Int. J. Robotics and Automation 21(3), 229–236 (2006)
7. Banzhaf, W., Nordin, P., Keller, R.E., Francone, F.D.: Genetic Programming - An Introduction. Morgan Kaufmann, San Francisco (1998)
8. Kennedy, J.: The Behavior of Particles, Evolutionary Programming VII, pp. 581–587 (1998)
9. Coello, C.A.C., Veldhuizen, D.A.V., Lamount, G.B.: Evolutionary Algorithms for Solving Multi-objective Problems. Kluwer Academic Publishers, Dordrecht (2001)
10. Kennedy, J., Eberhart, R.: Particle Swarm Optimization. In: Proceeding of IEEE International Conference on Neural Networks, pp. 1942–1948 (1995)
11. Huang, P.F., Chen, K., Xu, Y.S.: Optimal Path Planning for Minimizing Disturbance of Space Robot. In: Proc. of IEEE Ninth Int. Conf. on Control, Automation, Robotics and Vision, pp. 139–144 (2006)
12. Huang, P.F., Liang, B., Xu, Y.S.: PSO-based Time-optimal Trajectory Planning for Space Robot with Dynamic Constraints. In: Proc. of IEEE Int. Conf. on Robotics and Biomimetics, pp. 1402–1407 (2006)

The Direct Neural Control Applied to the Position Control in Hydraulic Servo System

Yuan Kang[2], Yi-Wei Chen[1], Yeon-Pun Chang[2], and Ming-Huei Chu[1]

[1] Department of Mechatronic Technology,
Tung Nan University, Taipei 222, Taiwan, R.O.C.
[2] Department of Mechanical Engineering,
Chung Yuan Christian University, Chung Li 320, Taiwan, R.O.C.
mhchu48@yahoo.com.tw

Abstract. This study utilizes the direct neural control (DNC) based on back propagation neural networks (BPN) with specialized learning architecture applied to control the position of a cylinder rod in an electro-hydraulic servo system (EHSS). The proposed neural controls without the specified reference model use a tangent hyperbolic function as the activation function, and the back propagation error is approximated by a linear combination of error and error's differential. The hydraulic cylinder subjected to varied load is also proposed. The simulation and experiment results reveal that the proposed neural controller is available to position control with high convergent speed, and enhances the adaptability and stability in varied load condition.

Keywords: Electro-hydraulic servo system, Position control, Neural networks, Back propagation.

1 Introduction

The electro-hydraulic servo systems are used in aircraft, industrial and precision mechanisms [1]. They are always used for servomechanism to transmit large specific powers with low control current and high precision. The electro-hydraulic servo system (EHSS) consists of hydraulic supply units, actuators and an electro-hydraulic servo valve (EHSV) with its servo driver. The EHSS is inherently nonlinear, time variant and usually operated with load disturbance. It is difficult to determine the parameters of dynamic model for an EHSS. Furthermore, the parameters are varied with temperature, external load and properties of oil etc. The modern precise hydraulic servo systems need to overcome the unknown nonlinear friction, parameters variations and load variations. It is reasonable for the EHSS to use a neural network based adaptive control to enhance the adaptability and achieve the specified performance.

In recent years, the neural network controls have been used in various fields owing to their capability of on-line learning and adaptability. Tremendous studies for neural network controllers have been conducted to dynamic systems. Psaltis et al. [2] discussed the general learning and specialized learning architectures, populated the input space of the plant with training samples so that the network can interpolate for intermediate points. The specialized learning architecture doesn't need off-line training the

F. Sun et al. (Eds.): ISNN 2008, Part II, LNCS 5264, pp. 180–189, 2008.

connective weights with all data pairs of working region, and can be easily implemented. The error between the actual and desired outputs of the plant is used to update the connective weights. In this sense, the controller learns continuously, and hence it can control plants with time-varying characteristics.

There are two strategies to facilitate the specialized learning, one being direct control and the other indirect control. In the former, the plant can be viewed as an additional but no modifiable layer of the neural network. The latter, which has been used in many applications, is a two-step process including identification of plant dynamics and control. In the indirect control strategy, a sub-network (called "emulator") is required to be trained before the control phase, and the quality of the trained emulator is crucial to the controlling performance. It is therefore very important that the data sets for training the emulator must cover a sufficiently large range input and output pairs, but it is very possible that the future behaviors in on-line control may outside the range that was used during the emulator's training, the back propagation through the emulator fails, causing poor or even unstable control performance.

The direct control strategy can overcome this problem if a priori qualitative knowledge or Jacobian of the plant is available. But it is usually difficult to approximate the Jacobian of a dynamic plant. Zhang and Sen [3] presented a direct neural controller for on-line industrial tracking control application, and a simple sign function applied to approximate the Jacobian of a ship track keeping dynamics. The results of a nonlinear ship course-keeping simulation were presented, and the on-line adaptive control was available. But their schematic is not feasible for high performance motion controls. A motion control system needs a neural controller with faster convergent speed.. Chu et al. [4] proposed a linear combination of error and error's differential to approximate the back propagation error, by this way the convergent speed can be increased. However the neural based adaptive control for the EHSS has rarely been proposed. Gao and Wu [5] performed stable position control of an EHSS with a specified fuzzy neural control. They could tune the fuzzy membership function on-line by neural networks, and the stability of the position control is proven by experiment.

In this paper, the proposed neural controls without the specified reference model use a tangent hyperbolic function as the activation function, and the back propagation error is approximated by a linear combination of error and error's differential [4]. The simulation and experiment results show the proposed direct neural control (DNC) is available to hydraulic position control with external force load.

2 Description of the Electro-Hydraulic Servo Control System

The EHSS is shown in Fig.1 consists of hydraulic supply units, actuators and an electro-hydraulic servo valve (EHSV) with its servo driver. The EHSV is a two-stage electro hydraulic servo valve with force feedback. The actuators are hydraulic cylinders with double rods.

2.1 The Simplified Servo Valve Model

The EHSV is a two-stage electro hydraulic servo valve with force feedback. The dynamic of EHSV consists of inductance dynamic, torque motor dynamic and spool

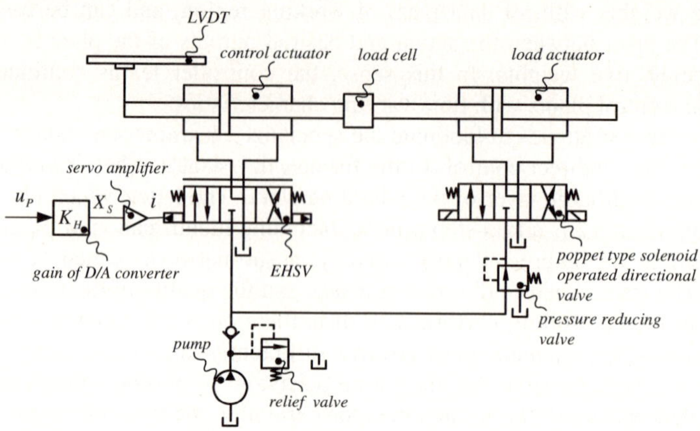

Fig. 1. The hydraulic circuit of EHSS

dynamic. The inductance and torque motor dynamics are much faster than spool dynamic, it means the major dynamic of EHSV determined by spool dynamic, so that the dynamic model of servo valve can be expressed as:

$$\frac{\Delta x_v}{\Delta e} = \frac{K_t}{1 + S/K_{vf}} . \tag{1}$$

Δx_v : The displacement of spool

Δe : The input voltage.

2.2 The Dynamic Model of Hydraulic Cylinder

The EHSV is 4 ports with critical center, and used to drive the double rods hydraulic cylinder. The leakages of oil seals are omitted and the valve control cylinder dynamic model [6] can be expressed as:

$$X_P = \frac{\frac{kq}{A_P} x_v - \frac{k_c}{A_P^2}(1 + \frac{V_t}{4\beta_e k_c} s) F_L}{s(\frac{V_t M_t}{4\beta_e A_P^2} s^2 + (\frac{k_c M_t}{A_P^2} + \frac{B_P V_t}{4\beta_e A_P^2}) s + (1 + \frac{B_P k_c}{A_P^2}))} . \tag{2}$$

x_v : The displacement of spool

F_L : The load force

X_P :The piston displacement.

2.3 Direct Neural Control System

The application of the direct neural controller for EHSS is shown in Fig.2, where y_r is the position command and y_p is the actual position response.The difference between command y_r and the actual output position response y_p is defined as error e. The error e and its differential $\dot{e}$ are normalized between –1 and +1 by multiplied by parameters K_1 and K_2 respectively in the input neurons. In this study, the back propagation error term is approximated by the linear combination of e and $\dot{e}$. A tangent hyperbolic function is designed as the activation function of the nodes in the output and hidden layers. So that the output signal of the output neuron is bounded between – 1 and +1, and converted into a bipolar analogous voltage signal through a D/A converter, then amplified by a servo-amplifier for enough current to drive the EHSV.

De Villiers et al. [7] has shown that one hidden layer with sigmoidal function is sufficient to compute arbitrary decision boundaries for the outputs. Although a network with two hidden layers may give better approximation for some specific problems, but De Villiers has demonstrated that networks with two hidden layers are more prone to fall into local minima and take more CPU time. In this study, a network with single hidden layer is applied to the position controller. Another consideration is the right number of units in a hidden layer. Lippmann [8] has provided comprehensive geometrical arguments and reasoning to justify why the maximum number of units in a single hidden layer should equal to M(N+1), where M is the number of output units and N is the number of input units. Zhang and Sen. [3] have tested different numbers units of the single hidden layer. It was found that a network with three to five hidden units is often enough to give good results. There are 5 hidden neurons in the proposed neural controller. The proposed DNC is shown in Fig 3 with a three layers neural network.

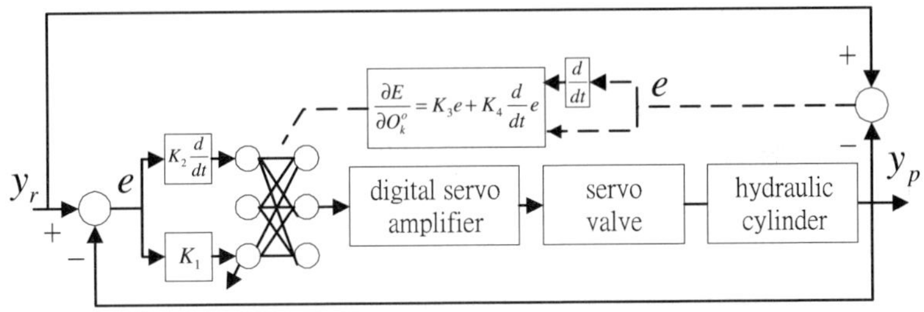

Fig. 2. The block diagram of EHSS control system

The proposed three layers neural network, including the hidden layer (j), output layer (k) and input layer (i) as illustrated in Fig.3. The input signals e and $\dot{e}$ are normalized between – 1 and +1, and defined as signals O_i feed to hidden neurons. A tangent hyperbolic function is used as the activation function of the nodes in the hidden and output layers. The net input to node j in the hidden layer is

$$net_j = \sum (W_{ji} \cdot O_i) + \theta_j \quad i = 1,2,...I, \quad j = 1,2,...J \,. \tag{3}$$

the output of node j is

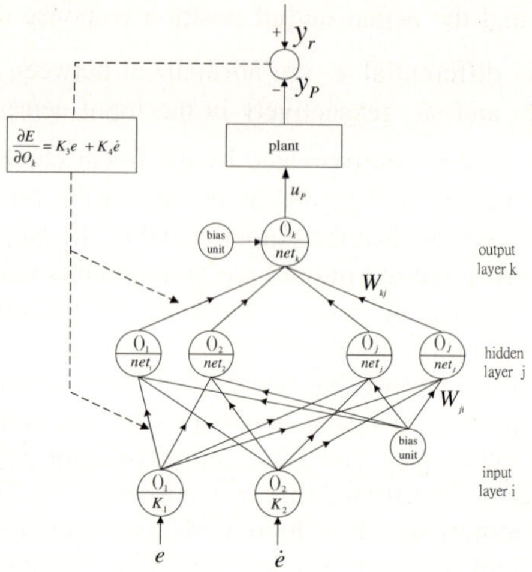

Fig. 3. The structure of proposed neural controller

$$O_j = f(net_j) = \tanh(\beta \cdot net_j) \cdot \tag{4}$$

where $\beta > 0$, the net input to node k in the output layer is

$$net_k = \Sigma(W_{kj} \cdot O_j) + \theta_k \quad j = 1,2,...J, \quad k = 1,2,...K \,. \tag{5}$$

the output of node k is

$$O_k = f(net_k) = \tanh(\beta \cdot net_k) \cdot \tag{6}$$

The output O_k of node k in the output layer is treated as the control input u_p of the system for a single-input and single-output system. As expressed equations, W_{ji} represent the connective weights between the input and hidden layers and W_{ij} represent the connective weights between the hidden and output layers. θ_j and θ_k denote the bias of the hidden and output layers, respectively.

The error energy function at the Nth sampling time is defined as

$$E_N = \frac{1}{2}(y_{rN} - y_{PN})^2 = \frac{1}{2}e_N^2 \,. \tag{7}$$

where y_{rN}, y_{PN} and e_N denote the reference command, the output of the plant and the error term at the Nth sampling time, respectively. The weights matrix is then updated during the time interval from N to N+1.

$$\Delta W_N = W_{N+1} - W_N = -\eta \frac{\partial E_N}{\partial W_N} + \alpha \cdot \Delta W_{N-1} \,. \tag{8}$$

where η is denoted as learning rate and α is the momentum parameter. The gradient of E_N with respect to the weights W_{ij} is determined by

$$\frac{\partial E_N}{\partial W_{kj}} = \frac{\partial E_N}{\partial net_k}\frac{\partial net_k}{\partial W_{kj}} = \delta_k O_j \quad (9)$$

and δ_k is defined as

$$\begin{aligned}\delta_k &= \frac{\partial E_N}{\partial net_k} = \sum_n \frac{\partial E_N}{\partial X_P}\frac{\partial X_P}{\partial u_P}\frac{\partial u_P}{\partial O_n}\frac{\partial O_n}{\partial net_k} = \sum_n \frac{\partial E_N}{\partial O_n}\frac{\partial O_n}{\partial net_k} \\ &= \sum_n \frac{\partial E_N}{\partial O_N}\beta(1-O_k^2) \qquad n=1,2,\cdots,K\end{aligned} \quad (10)$$

where $\partial X_P/\partial u_P$ is difficult to be evaluated. The EHSS is a single-input and single-output control system (i.e., $n=1$), in this study, the sensitivity of E_N with respect to the network output O_K is approximated by a linear combination of the error and its differential shown as：

$$\frac{\partial E_N}{\partial O_k} = K_3 e + K_4 \frac{de}{dt} \quad (11)$$

where K_3 and K_4 are positive constants . Similarly, the gradient of E_N with respect to the weights, W_{ji} is determined by

$$\frac{\partial E_N}{\partial W_{ji}} = \frac{\partial E_N}{\partial net_j}\frac{\partial net_j}{\partial W_{ji}} = \delta_j O_i \quad (12)$$

where

$$\begin{aligned}\delta_j &= \frac{\partial E_N}{\partial net_j} = \sum_m \frac{\partial E_N}{\partial net_k}\frac{\partial net_k}{\partial O_m}\frac{\partial O_m}{\partial net_j} \\ &= \sum \delta_k W_{km}\beta\left(1-O_j^2\right) \qquad m = 1,2,\cdots,J\end{aligned} \quad (13)$$

The weight-change equations on the output layer and the hidden layer are

$$\begin{aligned}\Delta W_{kj,N} &= -\eta\frac{\partial E_N}{\partial W_{kj,N}} + \alpha\cdot\Delta W_{kj,N-1} \\ &= -\eta\delta_k O_j + \alpha\cdot\Delta W_{kj,N-1}\end{aligned} \quad (14)$$

$$\begin{aligned}\Delta W_{ji,N} &= -\eta\frac{\partial E_N}{\partial W_{ji,N}} + \alpha\cdot\Delta W_{ji,N-1} \\ &= -\eta\delta_j O_i + \alpha\cdot\Delta W_{ji,N-1}\end{aligned} \quad (15)$$

where η is denoted as learning rate and α is the momentum parameter， δ_j and δ_k can be evaluated from Eq.(11) and (8)， The weights matrix are updated during the time interval from N to N+1 :

$$W_{kj,N+1} = W_{kj,N} + \Delta W_{kj,N}. \quad (16)$$

$$W_{ji,N+1} = W_{ji,N} + \Delta W_{ji,N}. \quad (17)$$

3 Numerical Simulation

An EHSS shown as Fig.1 with a hydraulic double rod cylinder controlled by an EHSV is simulated. A LVDT of 1 V/m measured the position response of EHSS. The numerical simulations assume the supplied pressure $P_s = 70\,Kg_f/cm^2$, the servo amplifier voltage gain of 5, the maximum output voltage of 5V, servo valve coil resistance of 250 ohms, the current to voltage gain of servo valve coil of 4 mA/V (250 ohms load resistance), servo valve settling time$\approx$20ms, the serve valve provides maximum output flow rate = 19.25 $\ell/\min$ under coil current of 20mA and ΔP of 70 Kg_f/cm^2 condition. The spool displacement can be expressed by percentage (%), and then the model of servo valve can be built as

$$\frac{x_v(100\%)}{i\ (mA)} = \frac{0.05}{S/200+1}. \tag{18}$$

or

$$\frac{x_v(100\%)}{v\ (V)} = \frac{0.2}{S/200+1}. \tag{19}$$

The cylinder diameter = $40mm$, rod diameter= $20mm$, stroke=200 mm, and the parameters of the EHSS listed as following:

$A_p = 9.4248cm^2 = 0.00094248m^2$, $V_t = 188.5cm^3 = 0.0001885m^3$, $B_p = 40\ N \cdot s/m$, $k_c = 3.727(10^{-5})\ m^3/MPa \cdot s$, $M_t = 1kgm$, $k = 0\ N/m$, $\beta_e = 1000MPa$ and $kq = 19.25\ \ell/\min$ (at $\Delta P = 70.3\ Kg_f/cm^2$)$= 320.833m^3/s = 0.000320833\ m^3/s$

According to Eq.(2), the no load transfer function is shown as

$$\frac{y_P}{x_v} = \frac{340414}{s(0.053s^2 + 44.122s + 1001678)}. \tag{20}$$

The direct neural controller without reference model is applied to control the EHSS shown as Fig.2, and the time responses for piston position are simulated. The appropriate parameters K_1 and K_2 can be assigned to normalize the input signals, and $K_3 \approx K_1, K_4 \approx K_2$ also can be available because the proposed servo control no reference model. According to Eq.(11), The constants K_3 and K_4 are defined as the appropriate parameters for the linear combination of error and its differential. A tangent hyperbolic function is used as the activation function, so that the neural controller output is between ± 1, and converted to be the analog voltage between $\pm 5\ Volt$ by a D/A converter and amplified with sufficient current by a servo amplifier to drive the EHSV. A conventional PD controller with well-tuned parameters is also applied to the simulation stage as a comparison performance. The square signal with a period of 5 sec and amplitude of 0.1m is used as the command input. The simulation results for PD control is shown in Fig.4 and for DNC is shown in Fig.5. Fig. 5 reveals that the EHSS with DNC track the square command with sufficient convergent speed, and the tracking performance will become better and better by on-line trained. Fig.6 shows the time response of piston displacement with 1200N force disturbance. Fig.6 (a) shows the EHSS with PD controller is induced obvious overshoot by the external force disturbance, and Fig.6 (b) shows the EHSS with the DNC can against the force disturbance with few overshoot. The simulation results show that the proposed DNC can provide favorable tracking characteristics, because of the neural controller is on-line trained with sufficient convergent speed by the proposed algorithms.

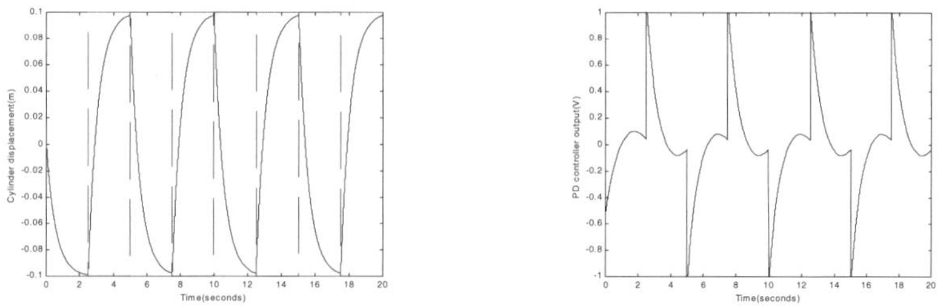

(a) Time response for piston displacement (b) Controller output

Fig. 4. The simulation results for EHSS with PD controller (Kp=7, Kd=1, Amplitude=0.1m and period=5 sec)

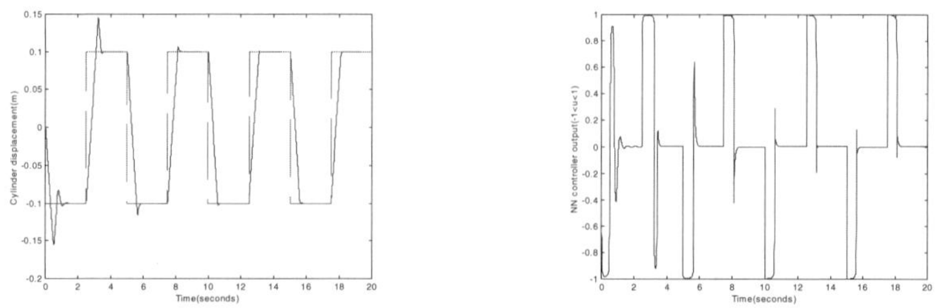

(a) Time response for piston displacement (b) Controller output

Fig. 5. The simulation results for EHSS with DNC (Amplitude=0.1m and period=5 sec)

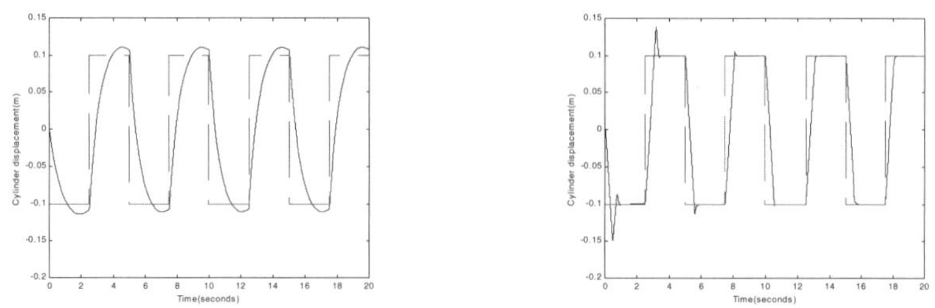

(a) EHSS with PD controller (b) EHSS with DNC

Fig. 6. Simulation results of position response with 1200N force disturbance

4 Experiment

The EHSS shown in Fig. 1 is established for our experiment. A hydraulic cylinder with 200 mm stroke, 20 mm rod diameter and a 40 mm cylinder diameter is used as the system actuator. The Atchley JET-PIPE-206 servo valve is applied to control the piston position of hydraulic cylinder.The output range of the neural controller is between ± 1, and converted to be the analog voltage between ± 5 $Volt$ by a 12 bits bipolar DA /AD servo control interface, It is amplified in current by a servo amplifier to drive the EHSV. A crystal oscillation interrupt control interface provides an accurate 0.001 sec sample rate for real time control.

A square signal with amplitude of 10 mm and period of 4 sec is used as reference command. Fig.7 (a) shows the EHSS with PD controller has obvious overshoot, which is induced by the external force disturbance created by load actuator with operation pressure 9 Kg_f/cm^2 . Fig.7 (b) shows the EHSS with the DNC can against the force disturbance with few overshoot. The experiment results show the proposed DNC is available for position control of EHSS with favorable performance, and can decrease the effect of force disturbance.

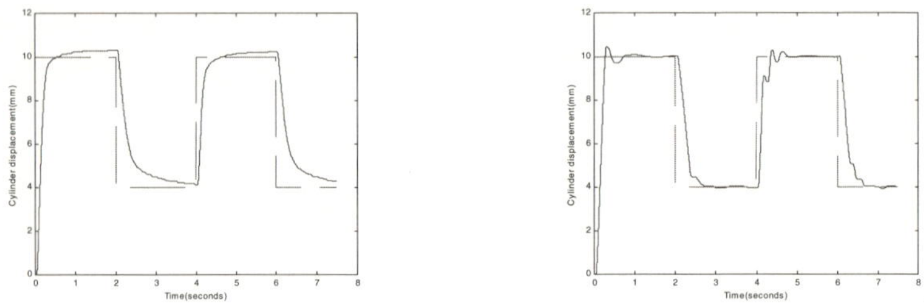

(a) EHSS with PD controller (b) EHSS with DNC

Fig. 7. Experiment results of position response with the load actuator pressure of 9 Kg_f/cm^2

5 Conclusion

The proposed DNC is applied to control the piston position of a hydraulic cylinder in an EHSS. The time responses of the proposed neural control and the conventional PD control systems with force disturbance are analyzed by simulation and experiment. The results show that the DNC has favorable tracking characteristic, and better performances than conventional PD control under external force load condition. The proposed neural control also improves the adaptability and stability of the EHSS.

Reference

1. Li, X., Ou, Y., Guan, X.P., Du, R.: Ram Velocity Control in Plastic Injection Molding Machines with Higher Order Iterative Learning. Control and Intelligent Systems 34(1), 64–71 (2006)
2. Psaltis, D., Sideris, A., Yamamura, A.A.: A Multilayered Neural Network Controller. IEEE Control Systems Magazine 8(2), 17–21 (1988)

3. Zhang, Y., Sen, P., Hearn, G.E.: An On-line Trained Adaptive Neural Controller. IEEE Control Systems Magazine 15(5), 67–75 (1995)
4. Chu, M.H., Kang, Y., Chang, Y.F., Liu, Y.L., Chang, C.W.: Model-Following Controller Based on Neural Network for Variable Displacement Pump. JSME International Journal, Series C 46(1), 176–187 (2003)
5. Gao, J.C., Wu, P.: A Fuzzy Neural Network Controller in the Electrohydraulic Position Control System. In: IEEE International Conference on Intelligent Processing Systems, vol. 1, pp. 58–63. IEEE Press, New York (1997)
6. Merritt, H.E.: Hydraulic Control Systems. Wiley, New York (1967)
7. Villiers, J., Barnard, E.: Backpropagation Neural Nets with One and Two Hidden layers. IEEE Trans. Neural Networks 4(1), 136–141 (1993)
8. Lippmann, R.P.: An Introduction to Computing with Neural Nets. IEEE Acoustics,

An Application of Wavelet Networks in the Carrying Robot Walking

Xiuxia Yang*, Yi Zhang, Changjun Xia, Zhiyong Yang, and Wenjin Gu

Department of Control Engineering,
Naval Aeronautical and Astronautical University, Yantai 264001, China
yangxiuxia@21cn.com

Abstract. We found that neuron model is inadequate owing to its defects such as those inherent in its structure and in its capability of information storage. So we propose an intelligent neurons assemblage model with generalized wavelet basis function network as its excited function. Not only the wavelet neural networks' convergence rate is much faster and its nonlinear approach capability is much better but also its intelligent characteristics, such as the variable-scale adaptive adjustment of structure and the generalized information storage, make it reflect much more faith fully the biological original. Static learning of the inverse dynamics model and adaptive virtual torque control based on Lyapunov stability of the carrying robot walking are demonstrated to prove that the proposed mechanism is valid.

Keywords: Carrying robot, Wavelet neural networks, Dynamics model learning, Adaptive control.

1 Introduction

Lower extremity exoskeleton intelligent carrying system is a new concept human-machine intelligent robot system. It has two mechanic legs similar to humans, and integrates with human through connection at operator's waist, foot or lower limb. In the human-machine system, the mechanical legs carry the entire load and the operator serves as the control centre to determine the walking direction and speed. There are many sensors installed in the exoskeleton legs which can measure the motion information of human. Then the control algorithm will judge human's motion consciousness and control the exoskeleton leg to move. The greatest difference between exoskeleton robot and other robot is that the CPU of exoskeleton robot is human but not the machine itself.

The unique human-machine integral system has cause widely attention of international scholars [1-8]. At present the mostly successful exoskeleton are HAL (Hybrid Assistive Leg) which was developed in the University of Tsukuba [1-3] and BLEEX (Berkeley Lower Extremity Exoskeleton) [5-8]. HAL use the s-EMG (surface ElectroMyogram) to sense the neuromuscular signal which are generated when motion,

* This work was supported by National Natural Science Foundation of China under Grant No. 60705030 and postdoctoral science Foundation of China under Grant No 2006040029.

F. Sun et al. (Eds.): ISNN 2008, Part II, LNCS 5264, pp. 190–199, 2008.

but there are many disadvantages when using s-EMG, for example, surface electrode is prone to fall off and the accuracy of the electrode will be influenced when people perspiring after a long time locomotion. Virtual joint torque control is used to BLEEX, which needs no direct measurements from the pilot or the human-machine interface (e.g. no force or EMG sensors between the two); instead, the controller estimates, based on measurements from the exoskeleton suits only, how to move so the pilot feels very little force [8]. This control scheme is an effective method of generating locomotion when the contact location between the pilot and the exoskeleton is unknown and unpredictable, but the control method need the mathematic model of the exoskeleton dynamic equation, while for the lower extreme carrying exoskeleton robot has the complex elements as nonlinear, uncertain and parameters time-varying and the accurate mathematics model is difficult to built, which bring difficult for the system design and affect the control results. BP neural network has the ability of approximating nonlinear functions, which has attracted people many attention and interest, but the BP network has some defects, such as the primary functions are not orthogonal, the convergence velocity is slow and difficult to determine the resolution scale. The wavelet transform has the good advantage of time-frequency localization properties and can construct a set of orthogonal primary functions, which can make up for the defects of BP network [9-11].

Aiming to the characteristics and the demands of the carrying system, in this paper, the virtual joint torque control is used. The wavelet network is used to approach the dynamics inverse model and calculate the input virtual input torque of the carrying robot and the human-machine interaction. Based on Lyapunov stability theory, the adaptive control is applied to lower extreme exoskeleton. Theoretical analyse and simulation results test the feasibility and validity of this control method.

2 Description of Virtual Torque Control

Virtual torque control selects a generalized force vector such that the control law is constructed in the machine's joint space rather than a set of forces and torques applied at a point on the body. The block diagram of the virtual torque control law is shown in Fig. 1. Where G_a represents the system transfer function. $G_a^{'}$ is an estimate of the machine forward dynamics. $K(s)$ is the controller.

T_{hm} denote the torque exerted on the plant by human. T_a denote the torque exerted by actuator. T denote all the external torque exerted on the exoskeleton. The human-machine torque can be modeled as:

$$T_{hm} = K_h(q_h - q) . \tag{1}$$

K_h is the impedance between the human and the machine, q_h is the human's position, q is the machine's position.

The system dynamics model can be built using Lagrange equation:

$$\boldsymbol{T} = \boldsymbol{J}(\boldsymbol{q})\ddot{\boldsymbol{q}} + \boldsymbol{B}(\boldsymbol{q},\dot{\boldsymbol{q}})\dot{\boldsymbol{q}} + \boldsymbol{G}(\boldsymbol{q}) . \tag{2}$$

Where $\boldsymbol{J}$ is the inertia matrix and is a function of $\boldsymbol{q}$, $\boldsymbol{B}$ is the centripetal and Coriolis matrix and is a function of $\boldsymbol{q}$ and $\dot{\boldsymbol{q}}$, $\boldsymbol{G}$ is a vector of gravitational torques, is a function of $\boldsymbol{q}$ only.

The tracking objective of $T_{hm} \to 0$ is identical to the tracking objective $q \to q_h$.

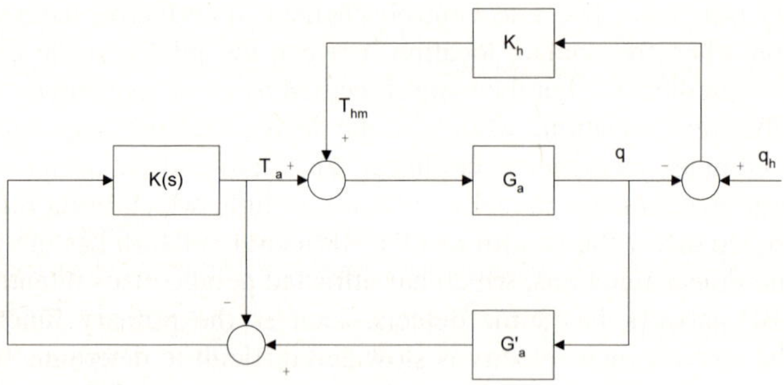

Fig. 1. Block diagram of the virtual torque control law

3 Virtual Prototyping Model Building

Using SimMechanics toolbox in Matlab, the virtual prototyping model of the control object is built. The inputs are three joint torque signals, the outputs are three joint angular signal, three joint angular velocity signals and three joint angular acceleration signals. The SmiMechanics toolbox has a set of visual tools, which can be used to display the simulation result dynamically, as presented in figure2.

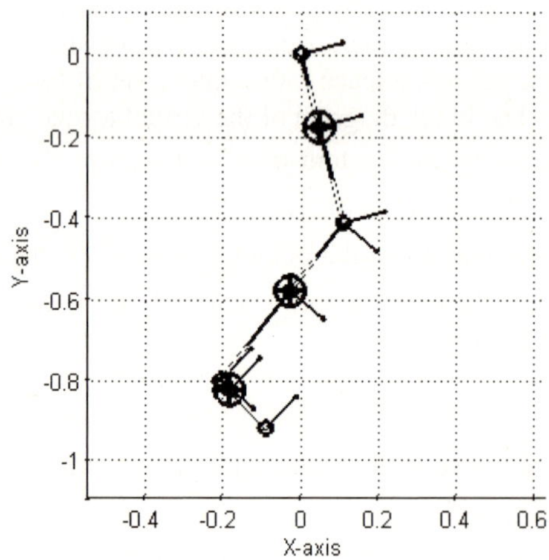

Fig. 2. Demo model of swing leg

4 Dynamics Model Identification Using Wavelet Network

For the dynamics mathematics model $G_a^{'}$ used in virtual torque controller can't be gotten accurately, such as $\boldsymbol{J}$, $\boldsymbol{B}$, $\boldsymbol{G}$, the wavelet neural network(WNN) is used to approximate the dynamics mathematics model.

4.1 Wavelet Neural Network Structure

The wavelet series denote preset function by the summation of a series of function $\psi_{j,k}(x)=2^{j/2}\psi(2^j x-k)$ gotten by binary system expand and integer translation, which is compact in $L^2(R)$. If $\psi(x)$ is a orthogonal wavelet, it is self dual and $\psi_{j,k}(x)$ is a orthogonal basis, which can denote any function $f(x)\in L^2(R)$. But for the actual system, infinity series summation is insignificant and is not necessary, the function can be signified approximately as followings:

$$f(x)\approx\sum_{j,k=-N}^{N}W_{j,k}\psi_{j,k}(x),\forall f(x)\in L^2(R). \tag{3}$$

In this study, the Morlet wavelet $\psi(\mathrm{t})$ is chosen as the mother wavelet, which has good finity support both in time-domain and frequency-domain:

$$\psi(x)=-xe^{-\frac{1}{2}x^2}. \tag{4}$$

Transform equation (3) to equation (5), the wavelet neural network is constructed

$$f(x)\approx\sum_{j=1}^{N}W_j\psi\left(\frac{x-t_j}{s_j}\right) \tag{5}$$
$$\forall f(x)\in L^2(R)$$

The input vector of the wavelet network is $\mathrm{x}=[\mathrm{x}_1,\mathrm{x}_2,\cdots,\mathrm{x}_\mathrm{n}]$, the wavelet has N nodes, w_{ij} is coefficient of the *j*th wavelet node to the *k*th output variable, and the kth output variable is

$$y_k\approx\sum_{j=1}^{N}W_{jk}\psi\left(\frac{x-t_j}{s_j}\right). \tag{6}$$

The less s , that is, the higher frequency, the time precision is higher. On the other hand, the higher s , the less frequency, the frequency precision is higher. If the flex factor s_i and the translation factor t_i are not selected appropriately, the preset function can not be approximated exactly.

4.2 Initialization of Wavelet Network Parameters

In general, the range of system output can be gotten. If the maximum of the output is $f_{\max}$, the minimum is $f_{\min}$, choose the first partition t_1 in the interval $[f_{\min}, f_{\max}]$, that is, $t_1 = f_{\min} + \xi(f_{\max} - f_{\min})$, s_1 is the interval contract, that is, $s_1 = \xi(f_{\max} - f_{\min})$, the typical value of ξ is 0.5. The interval $[f_{\min}, f_{\max}]$ is separated into two subinterval; In every subinterval, the separation is repeated and $s_2, t_2, s_3, t_3 \ldots$ are selected, similarly, all the wavelets are initiated. The number of wavelet cell used is $N = 2^0 + 2^1 + \cdots + 2^{m-1}$, where m is the number of interval separation.

4.3 Dynamics Model Identification

Taking the wavelet neural network as the inverse model in virtual joint torque control, we must get the input and output data of the system to train the neural network so that the network has the same properties with the inverse model. The exoskeleton joint

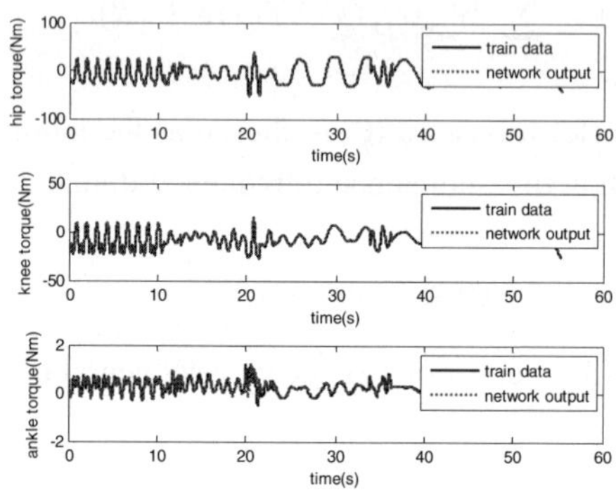

Fig. 3. Train data and the wavelet network output data

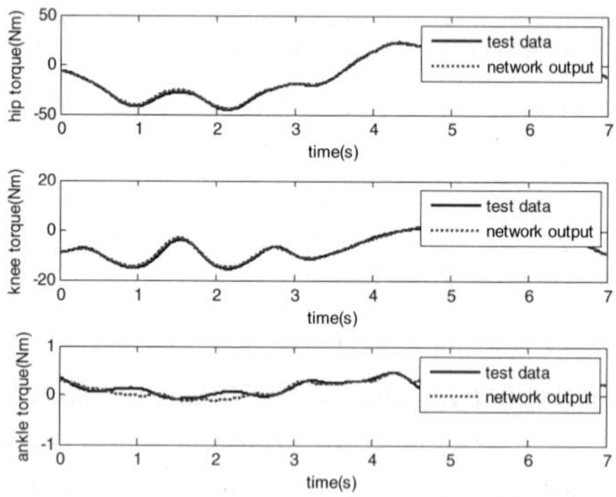

Fig. 4. Test data and the wavelet network output data

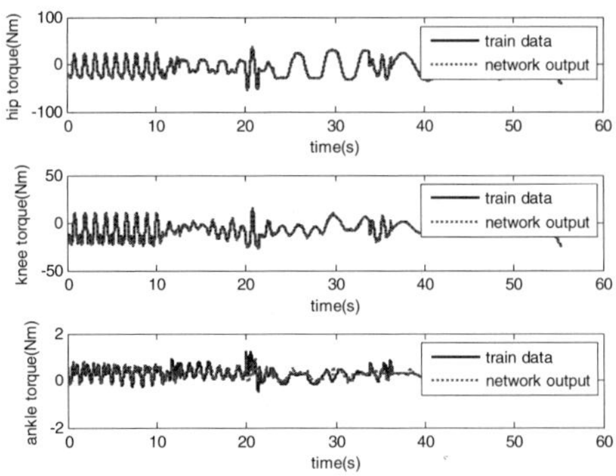

Fig. 5. Train data and the neural network output data

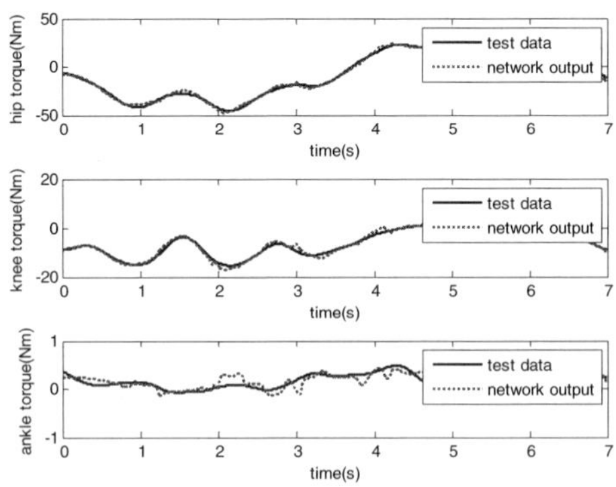

Fig. 6. Test data and the neural network output data

angular, joint angle velocity and joint angle acceleration are taken as wavelet network input, while the virtual prototyping model (shown in Fig. 2) joint torque T is taken as output. Train the network using BP method, the adaptive law of the weight is selected according to the paper [9], the number of the wavelet cell is selected as 7. In Figs. 3 and 4, the training and test results are given, where the error square summation of the network output and the train data is 0.005, which indicates the effectiveness of the wavelet network. On the same condition, the neural network training and test results are given in Figs. 5 and 6. From the simulation results we can draw a conclusion that the map ability of wavelet network is stronger than the neural network.

5 Adaptive Control Based on Lyapunov Function Using WNN

In the virtual torque control, the wavelet network is used to approach the dynamics inverse model $G_a^{'}$, which are adopted to calculate the virtual input torque $T^{'}$ of the carrying robot and the human-machine interaction to present the human's intelligence,

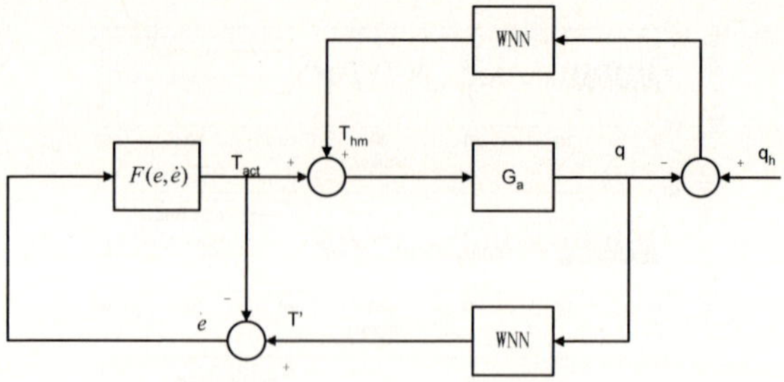

Fig. 7. Adaptive virtual torque control law based on WNN

that is, $K_h = G_a{}'$, which is shown in Fig. 7. To improve the performance of the system, using the nonlinear adaptive controller to replace the PD controller in Fig. 1, supposing the nonlinear controller is $F(e,\dot{e})$, which is also shown in Fig. 7.

The virtual human-machine force T_{hm} can be gotten from the wavelet network evaluation of the input torque of the exoskeleton inverse model and the output torque of the actuator. The reference signal T_{hm_ref} of T_{hm} is 0, the error signal of the control system is e, then

$$e \approx T_{hm} - T_{hm\_ref} = T_{hm} . \tag{7}$$

From the system we can get

$$\boldsymbol{F}(\boldsymbol{e},\dot{\boldsymbol{e}}) + \boldsymbol{e} = \boldsymbol{J}(\boldsymbol{q})\ddot{\boldsymbol{q}} + \boldsymbol{B}(\boldsymbol{q},\dot{\boldsymbol{q}})\dot{\boldsymbol{q}} + \boldsymbol{G}(\boldsymbol{q}) . \tag{8}$$

On the following, based on Lyapunov stability theory, the adaptive control law is driven.

Let

$$\boldsymbol{F}(\boldsymbol{e},\dot{\boldsymbol{e}}) + \boldsymbol{e} = \dot{\boldsymbol{e}} + \boldsymbol{F}_1(\boldsymbol{e}) . \tag{9}$$

Then

$$\dot{\boldsymbol{e}} = \boldsymbol{J}(\boldsymbol{q})\ddot{\boldsymbol{q}} + \boldsymbol{B}(\boldsymbol{q},\dot{\boldsymbol{q}})\dot{\boldsymbol{q}} + \boldsymbol{G}(\boldsymbol{q}) - \boldsymbol{F}_1(\boldsymbol{e}) . \tag{10}$$

Defining the Lyapunov function

$$V = \frac{1}{2}\boldsymbol{e}^T \boldsymbol{\Gamma} \boldsymbol{e} . \tag{11}$$

Where, $\boldsymbol{\Gamma}$ is symmetry positive definite matrix, here the unit matrix is selected. The differential of V is

$$\dot{V} = \boldsymbol{e}^T \cdot \dot{\boldsymbol{e}} = \boldsymbol{e}^T \cdot [\boldsymbol{J}(\boldsymbol{q})\ddot{\boldsymbol{q}} + \boldsymbol{B}(\boldsymbol{q},\dot{\boldsymbol{q}})\dot{\boldsymbol{q}} + \boldsymbol{G}(\boldsymbol{q}) - \boldsymbol{F}_1(\boldsymbol{e})]s . \tag{12}$$

Let

$$\boldsymbol{F}_1(\boldsymbol{e}) = \boldsymbol{J}(\boldsymbol{q})\ddot{\boldsymbol{q}} + \boldsymbol{B}(\boldsymbol{q},\dot{\boldsymbol{q}})\dot{\boldsymbol{q}} + \boldsymbol{G}(\boldsymbol{q}) + \boldsymbol{K}\,\mathrm{sgn}(\boldsymbol{e}) . \tag{13}$$

Where, $\boldsymbol{K}$ is constant matrix and the element of $\boldsymbol{K}$ is positive, sgn($\boldsymbol{e}$) is symbol function, then

$$\dot{V} \le 0 . \tag{14}$$

From above the nonlinear adaptive controller $\boldsymbol{F}(\boldsymbol{e},\dot{\boldsymbol{e}})$ in Fig. 9 can be gotten:

$$\boldsymbol{F}(\boldsymbol{e},\dot{\boldsymbol{e}}) = \dot{\boldsymbol{e}} + \boldsymbol{F}_1(\boldsymbol{e}) - \boldsymbol{e} = \boldsymbol{J}(\boldsymbol{q})\ddot{\boldsymbol{q}} + \boldsymbol{B}(\boldsymbol{q},\dot{\boldsymbol{q}})\dot{\boldsymbol{q}} + \boldsymbol{G}(\boldsymbol{q}) + \boldsymbol{K}\,\mathrm{sgn}(\boldsymbol{e}) + \dot{\boldsymbol{e}} - \boldsymbol{e} . \tag{15}$$

6 Simulation Results

Using the anthropometric data computed from Winter D. A. as the parameters of the exoskeleton leg [12] and choosing the swing phase data from the Clinical Gait Analysis (CGA) data [13] as the desired motion of the human, assuming the pilot tied together with the exoskeleton at the hip joint and foot, making the SmiMechanics model of swing leg as G_a and applying the adaptive controller as shown in Fig. 7 , where the

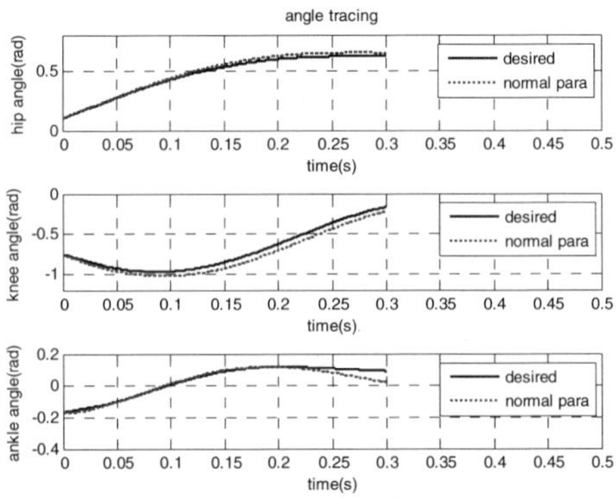

Fig. 8. Trajectory of joint angle

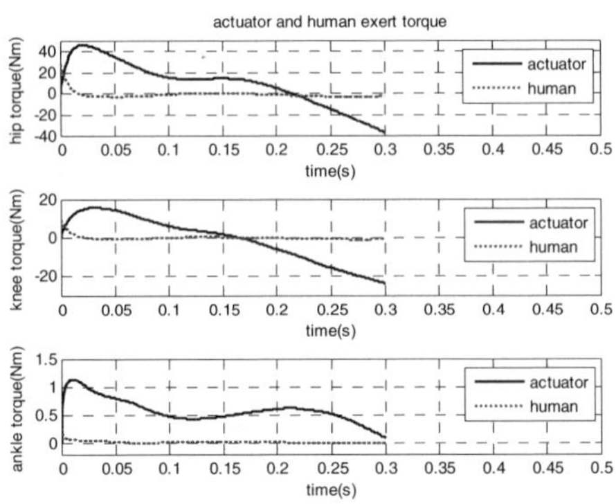

Fig. 9. Torque exerted by human and actuator with WNN controller

gain matrix is selected as $\boldsymbol{K} = [3,3,2]^T$, using the wavelet neural network as the identifier,, the simulation results are presented in Fig.8 and Fig. 9, which illustrate that the exoskeleton tracks the motion of the human very well and the torque exerted by the human is very small and the actuator exert the most which means the pilot (human) can swing the exoskeleton easily and only need to exert a little torque of the actuator. To analysis the anti-interference performance, add perturbation to the parameters. The simulation results are given in figure 10 for the exoskeleton mass parameters are all added, decreased 20% and in the normal condition, which show that the exoskeleton can track the motion of the human and the robust of the system is well.

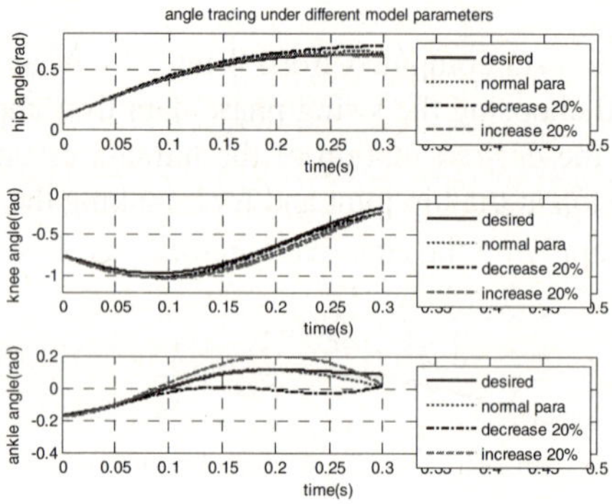

Fig. 10. Angle tracing under three model parameters

7 Conclusion

In this paper, the wavelet network is used to approach the dynamics inverse model, an adaptive control system based on Lyapunov stability theory, with two wavelet neural networks as virtual torque evaluator and human-machine interaction force calculation respectively, is developed for lower extreme carrying exoskeleton robot. Simulation results show that this control system is more effective than those based on neural network identifier, where the exoskeleton tracking precision is high, the operator feels very little torque and the system is robust.

References

1. Okamura, J., Tanaka, H., Sankai, Y.: EMG-based Prototype Powered Assistive System for Walking Aid. In: Asian Symposium on Industrial Automation and Robotics (ASIAR 1999), Bangkok, Thailand, pp. 229–234 (1999)
2. Lee, S., Sankai, Y.: Power Assist Control for Walking Aid with HAL-3 Based on EMG and Impedance Adjustment around Knee Joint. In: 2002 IEEE/RSJ International Conf on Intelligent Robots and Systems (IROS 2002), EPFL, Switzerland, pp. 1499–1504 (2002)
3. Hiroaki, K., Yoshiyuki, S.: Power Assist System HAL-3 for Gait Disorder Person. In: Miesenberger, K., Klaus, J., Zagler, W. (eds.) ICCHP 2002. LNCS, vol. 2398, pp. 196–203. Springer, Heidelberg (2002)

4. Eppinger, S.D., Seering, W.P.: Understanding Bandwidth Limitations in Robot Force Control. In: 1987 IEEE International Conference on Robotics and Automation, pp. 904–909. IEEE Press, New York (1987)
5. Zoss, A.B., Kazerooni, H., Chu, A.: Biomechanical Design of the Berkeley Lower Extremity Exoskeleton (BLEEX). IEEE/ASME Transactions on Mechatronics 11, 128–138 (2006)
6. Kazerooni, H., Racine, J.-L., Huang, L., Steger, R.: On the Control of the Berkeley Lower Extremity Exoskeleton (BLEEX). In: 2005 IEEE International Conference on Robotics and Automation, pp. 4353–4360. IEEE Press, New York (2005)
7. Racine, J.L.: Control of a Lower Extremity Exoskeleton for Human Performance Amplification. Ph. D. dissertation, University of California, Berkeley (2003)
8. Steger, R.: A Design and Control Methodology for Human Exoskeletons. Ph. D. dissertation, University of California, Berkeley (2006)
9. Lin, C.-M., Hung, K.-N., Hsu, C.-F.: Adaptive Neuro-Wavelet Control for Switching Power Supplies. IEEE Transactions on Power Electronics 22, 87–95 (2007)
10. Yacine, O., Gérard, D.: Initialization by Selection for Wavelet Network Training. Neurocomputing 34, 131–143 (2000)
11. Chen, H., Chen, W., Xie, T.: Wavelet Network Solution for the Inverse Kinematics Problem in Robotic Manipulator. Journal of Zhejiang University Science A 7, 525–529 (2006)
12. Winter, D.A.: Biomechanics of Human Movement. John Wiley and Sons, New York (1979)
13. Hong Kong Polytechnic University, `http://guardian.curtin.edu.au:16080/cga/data/HKfyp98/All.gcd`

TOPN Based Temporal Performance Evaluation Method of Neural Network Based Robot Controller*

Hua Xu and Peifa Jia

State Key Laboratory on Intelligent Technology and Systems,
Tsinghua National Laboratory for Information Science and Technology,
Department of Computer Science and Technology,
Tsinghua University, Beijing 100084, China
{xuhua,dcsjpf}@mail.tsinghua.edu.cn

Abstract. These years, for some neural network (NN) controller based time critical systems, temporal performance is always required to be evaluated. In order to model complex time critical systems, timed hierarchical object-oriented Petri net (TOPN) has been proposed. On the base of TOPN method, this paper has proposed worst case execution time (WCET) calculation method called time accumulation effect (TAE) calculation, whose goal is to evaluate the performance of TOPN models such as NN based robot controller systems etc al. TAE method can be used to calculate the worst case execution time interval directly on the base of integral linear programming (ILP) method. The use and benefits of TAE calculation for TOPN models have also been illustrated by analyzing one robot controller system model.

Keywords: Neural network, Petri nets, Robot controller.

1 Introduction

Recently, performance evaluation of NN based systems, such as robot controller systems or multi-robot systems [1,2], has become one of the critical problems. Several experiment methods [3] and analysis methods such as threshold functions based NN evaluation method [4] have been proposed. Although many kinds of evaluation methods for NN performance have been proposed, most of them focus on evaluating robustness and generalization. There is still no temporal performance evaluation research of NN based robot controller published.

These years, in order to model complex temporal systems, high level Petri net–timed object-oriented Petri net (TOPN) [5] has been proposed on the base of object Petri net-HOONet[6] and Time Petri net (TPN) [7]. When TOPN is used to analyze time critical system models, the temporal knowledge of the basic function modules can be evaluated on the base of the existing performance

* This work is jointly supported by National Natural Science Foundation of China (Grant No: 60405011, 60575057) and the China Postdoctoral Foundation for China Postdoctoral Science Fund (Grant No: 20040350078).

F. Sun et al. (Eds.): ISNN 2008, Part II, LNCS 5264, pp. 200–209, 2008.

evaluation method [8] beforehand. But it needs a methodology to calculate the temporal knowledge of the whole TOPN model. That's to say, the worst case execution time (WCET) of the whole TOPN model needs to be evaluated. Because the execution time of the whole TOPN model is accumulated by the execution time of every TOPN object in the execution path, it can also be called time accumulation effect (TAE) in TOPN.

This paper proposes an integral linear planning (ILP) based WCET or TAE calculation method to evaluate the NN controller model. In the proposed TAE calculation method, the structure of TOPN is firstly analyzed. Then, the conditional inequalities are used to describe the structure constraint relation. Similar to the structure constraints, the function constraint relation can also be got at the same time. These two kinds of constraint conditions are used as the constraint relation in ILP. Finally, the TEA calculation has been transformed to two ILP problems, whose aim is to calculate the upper limit (LFT) and the lower one (EFT) of the WCET respectively. That's to say, the performance of the corresponding model can be evaluated.

This paper is organized as the following. In Section 2, basic formal definitions are reviewed and presented firstly. Then, ILP based TAE calculation method is proposed in Section 3. In section 4, TAE calculation method is used to analyze one NN based controller model of one cooperative multiple robot system. Section 5 concludes our paper and suggests further research issues.

2 Basic Concepts

The work in this paper is mainly on the base of TOPN [5] and TAE concepts. The details of TOPN can be referred to [5] and the concepts of TAE are proposed as the following.

Definition 1: If the initial state of one TOPN model N is M_0, then there exists a set of all possible execution paths path_set, where path_set = { path$_1$, path$_2$, ..., path$_n$ }. The execution path set path_set is called the execution path set of TOPN model N under the initial state M_0. □

Definition 2: Suppose $[a, b]$ and $[c, d]$ are two time intervals, where $b \geq a \geq 0$ and $d \geq c \geq 0$. The time interval operation can be defined as the following:

The addition of two time intervals: $[a, b] + [c, d] = [a + c, b + d]$.

The addition of multi-time intervals: $[a_1, b_1]+[a_2, b_2]+\ldots+[a_n, b_n]=\sum_{i=1}^{n}[a_i, b_i]$, where $n \in \mathbb{Z}^+$ and $\mathbb{Z}^+$ is the positive integer set.

The multiplication between one positive integer and one time interval: $[a, b] \times f = [a \times f, b \times f]$, where $f \in \mathbb{Z}^+$. □

Just like the object-oriented model, TOPN is also a kind of hierarchical Petri net model. The higher layer of one TOPN model defines and contains abstract TOPN objects, so it always manifests simple architecture. However, the realization details of these objects are depicted in the lower layers of TOPN model. For example, Fig.1 is a typical TOPN model sketch figure. In the higher layer, there are three child TOPN objects: A, B and C. In the lower layer, there are three

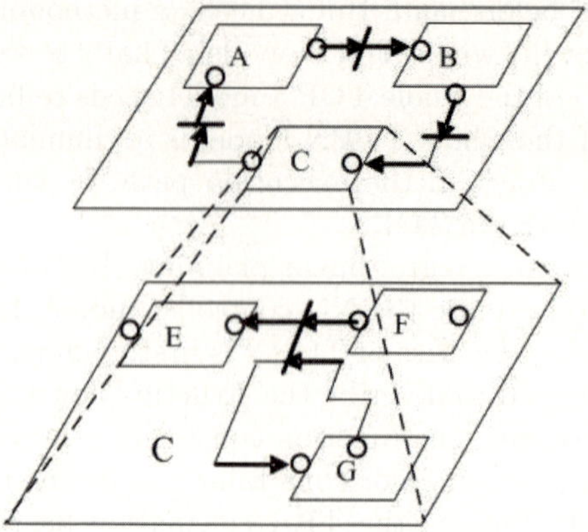

Fig. 1. A typical TOPN model sketch

parent TOPN objects: G, E and F. Suppose G, E, and F are the parent objects of C. In Fig.1, it can be known that the temporal knowledge of C can be got on the base of that of G, E and F. So TAE calculation essentially is to accumulate the temporal knowledge of the objects in the lower layer.

Similar to those in the WCET calculation, in TAE calculation of TOPN models, the execution time of the same type of objects or transitions is supposed to be similar.

Definition 3: The execution time of one system or one model can be defined as the following:

$$TI = \sum_{i=1}^{n} f_i c_i \tag{1}$$

where f_i is the executing frequency (or executing times) of one command on the execution path. The constant c_i is the execution time of this command. n is the number of this command on the execution path. TI is the execution time of the whole path. □

Definition 4: In the TOPN model N, $N = (\text{OIP}, \text{ION}, \text{DD}, \text{SI}_P)$. If

$$X = \text{ION.T} \bigcup \text{ION.TABP} \tag{2}$$

Suppose the path set "path_set" is the set of all possible execution path of N under the initial state M_0. SI() is the static mapping function between the objects and the temporal knowledge. The execution time interval TI_i is the execution time interval of the path "path$_i$" in the path set path_set. TI_i can be represented as the following forms:

Form 1:

$$TI_i = \sum_{j=1}^{|\text{path}_i|} SI(t_{ij}) = \sum_{j=1}^{|\text{path}_i|} [a_{ij}, b_{ij}] = [a_i, b_i] \tag{3}$$

where $\text{path}_i \in \text{path\_set}$. $|\text{path}_i|$ represents the number of objects on the path. t_{ij} is the jth object on the path path_i and $t_{ij} \in X$. The temporal knowledge of t_{ij} is $[a_{ij}, b_{ij}]$.

Form 2:

$$TI_i = \sum_{j=1}^{|N.X|} f_{ij} SI(t_j) = \sum_{j=1}^{|N.X|} f_{ij}[a_j, b_j] = [a_i, b_i] \quad (4)$$

where $t_j \in N.X$, f_{ij} is the execution frequency of the object t_j on path_i. $[a_j,b_j]$ is the mapping results of $SI(t_j)$.

So the WCET of the TOPN model can be calculated as the following:

$$\text{WCET} = [\min(a_i), \max(b_i)] \quad (5)$$

where $i = 1, \ldots, |\text{path\_set}|$.

The temporal knowledge of ION can be got from the WCET. The reason is that the temporal knowledge of TOPN model is the accumulated result of the temporal knowledge of objects on the execution path. It is also called time accumulation effect (TAE) in TOPN model. The corresponding calculation is also called the TAE calculation. □

3 TAE Calculation

TAE calculation is proposed on the base of ILP method [8, 9]. In the corresponding ILP, there are two types of constraint conditions: structure constraint condition and logical constraint condition. In this section, several calculating hypotheses are presented. Then, how to transform TAE object function to ILP object function is discussed. Finally, ILP based TAE calculation method and constraint condition analysis methodology are proposed.

3.1 Several Hypotheses

Before the TAE calculation is discussed, several hypotheses are made.

- Similar to those of OPN model, it is supposed that every TOPN model owns single-entry and single-exit.
- State changes in TOPN can be represented as an execution path, which is made up of the corresponding fired transitions and abstract place objects.
- Every object with behaviors in TOPN models should be on one or more execution path.
- The temporal knowledge of the objects on lower layer can be got before the TAE calculation.
- The time interval of TOPN model is a kind of relative time interval, which is relative to the actual enabled time of the whole TOPN model.

3.2 ILP Based TAE Calculation

The object of TAE calculation is to get the EFT and LFT of the whole TOPN model. Similar to those of WCET in TOPN, the execution time of TOPN model can be represented as the following:

$$Z = \sum_{j=1, t_j \in \text{path}_i}^{|\text{path}_i.X|} f_j TI_{t_j} \tag{6}$$

where the decision variable f_j represents the execution frequency of the object t_j and the constant TI_{t_j} represents the execution time of the object t_j on the path path$_i$. The aim of ILP is to find the set of $f_j (j = 1, \ldots, |\text{path}_i.X|)$, which can get the extremums of the object function Z.

To calculate EFT is to calculate the minimum of the object function in TEA calculation. Correspondingly, to calculate LFT is also to calculate the maximum of the object function in TEA calculation.

$$\text{EFT} = \min(Z) \tag{7}$$

$$\text{LFT} = \max(Z) \tag{8}$$

According to the definition of EFT and LFT, the temporal knowledge of TOPN model N has also been represented as the following.

$$SI(N) = [\text{EFT}, \text{LFT}] \tag{9}$$

3.3 Constraints of the ILP

The constraints of the ILP can be divided into two types: structure constraints and logical constraints. The structure constraints can be got from the analysis results of TOPN model structure. However, the logical constraints can be got from the data dictionary and bool condition of transition firing.

(1) Structure Constraints

Structure constraints include *entry/exit constraint condition, connection constraint condition, branch constraint condition* and *combination constraint condition.* To begin with discussing the constraint condition, it hypothesizes that f_i is the execution frequency of the corresponding transition t_i in Fig.2, where $i = 1, \ldots, 10$.

Entry/exit constraint condition: Single entry and single exit is one of the TOPN characters, which is illustrated in Fig.2 (a). According to this structure feature, the following constraint can be got.

$$f_1 = f_2 = 1 \tag{10}$$

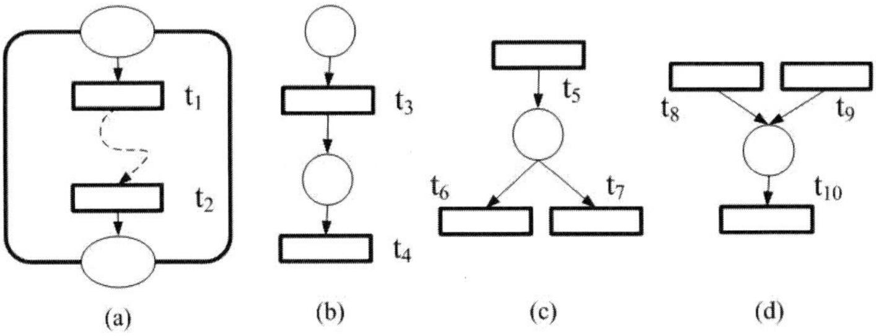

Fig. 2. The illustration of structure constraints

Connection constraint condition: If two transitions are connected as that in Fig.2 (b), the execution frequency of these two transitions is similar. So the connection constraint is like the following.

$$f_3 = f_4 \tag{11}$$

Branch constraint condition: If the transitions in Fig.2 (c), t_6 and t_7 are on the branch of the execution path of the transition t_5. So the execution frequency of these transitions fulfills the following constraints.

$$f_5 = f_6 + f_7 \tag{12}$$

Combination constraint condition: Similar to those in the branch constraint, there also exists a combination constraint in TOPN models, which is shown in Fig.2 (d). So the execution frequency of these transitions fulfills the following constraints.

$$f_8 + f_9 = f_{10} \tag{13}$$

(2) Logical Constraints

In most circumstances, actual logical constraints depend on the input data and the value of bool variables. In TOPN models, typical logical constraints include loop times and the information of the execution path. For example, one bool variable decides the following execution branch. The aim of logical constraints is not to make the maximum or minimum execution number accurate, but to constrain the frequency of all possible execution paths.

The logical constraints in TOPN can be represented as the following form.

$$\sum_{p_j \in \text{path}_i} a_{ij} f_j \circ \sum_{p_j \in \text{path}_i} a'_{ij} f_j + k \tag{14}$$

where $a_{ij}, a'_{ij} \in Z_0, k \in N_0$ and $\circ \in \{\leq, <, =\}$. In the above inequality, a_{ij} is the coefficient of the transition t_j on the execution path path_i. f_j is the execution frequency of the transition t_j. This constraint can be used to describe the loop constraint (maximum loop times) and the dependency between different modules in TOPN model.

(3) Positive Constraint
All the execution frequency f_i should be positive, so $f_i \geq 0$.

3.4 Reducing the Calculation Complexity of ILP

In order to reduce the calculation complexity of ILP, the following methods can be adopted on the base of the work in [9].

1. Delete the equation whose form is like "A=B" and substitute A with B in all the other equations and inequalities.
2. If there are constraints like "A=X and B=X", then substitute "A=X" with "B=X". And substitute B with X to simplify the constraint conditions.

4 A Simple Example

In cooperative multi-robot systems (CMRS), NN is always used to control every autonomous robot. That's to say, every robot uses one NN based controller. In the control of every robot of CMRS, NN controller needs the information of other robots and itself state information as the input. And then NN model is used to give the control commands and parameters. The execution sequence can be divided into 5 phases. In the first phase, the robot controller will get the information of other robots by itself and input the information into the information buffer. The controller will spend about 50 unit times on this procedure. (One unit time equals one millisecond.). In the second phase, numeric controller (NC) will know that parameters in the buffer are ready after receiving the synchronous signal. Then controller will get the parameters from the DPR. The second phase will last for about 50 unit times. In the next, the controller spends 50 unit times on giving the control commands and parameters by means of NN control model in the third phase. In the forth phase, the control parameters will be written back into the buffer. The time is the same as that in the first phase. At last, the new control parameters or results in the buffer will be read and transferred to other robots. It also needs about 50 unit time. When the buffer receives the control parameter "AT", it maybe exists about 50 unit time delay. In this control procedure, the hardware resources and temporal constraints need to be considered simultaneously, so TOPN is chosen to model this control module and to evaluate its performance.

1. *Definition and Modeling on High Level*: According to the introduction of the robot control procedure, TOPN has been used to model the NN based controller module in CMRS. The model is depicted in Fig.3.
2. *Data Dictionary*: The data dictionary defines the token type, variables and functions (object behavior). They are described by Hong's modified CPN ML [6] in the data dictionary.

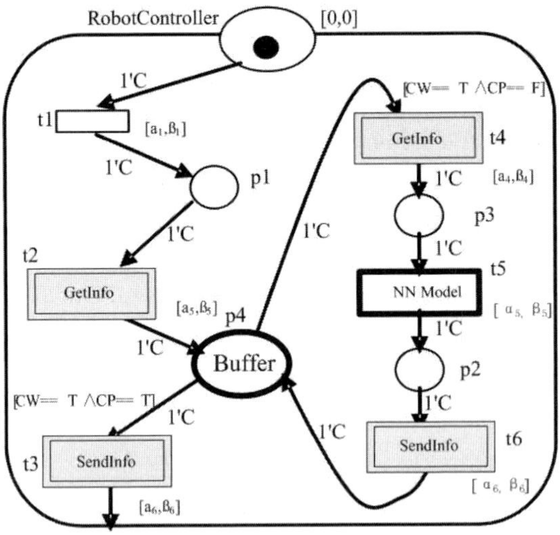

Fig. 3. The TOPN model of example

Table 1. The data dictionary for the example

```
Var +CW = boolean; /*Writing buffer tag*/
Var +CP = boolean; /*Control parameter processing Tag*/
/* CP is set to T in the transition "ProcessData" */
Var +Time = Integer; /*Current relative time*/
TT C = with hollow | solid;
TCOT (GetInfo)={
Fun(CW= =F ∧ CP= =F ∧ (α2 ≤ Time ≤ β2)):
  GetInfo(p4) ∧ CW=T ∧ Mark(p4,C);
Fun(CW= =F ∧ CP= =T ∧ (α6 ≤ Time ≤ β6)):
  GetInfo(p4) ∧ CW=T ∧ Mark(p4,C);
};
/* Mark(P,C): set the corresponding mark C according to the type of P */
TCOT (SendInfo)={
Fun(CW= =T ∧ CP= =F ∧ (α4 ≤ Time ≤ β4)):
  SendInfo(p4) ∧ CW=F ∧ CP=T ∧ M(p3,C);
Fun(CW= =T ∧ CP= =T ∧ (α3 ≤ Time ≤ β3)):
  SendInfo(p4) ∧ CW=F ∧ CP=F ∧ M(oip,C);
};
/* M(P,C): set the place P with the mark C.*/
A(Arc)={
  Fun((C= =solid) ∧ ((Arc= =Arc1) ∨ (Arc= =Arc2))): C=hollow;
}; /*When TABP has been marked, its tokens are set to be hollow. */
TABP(DPR)={
Fun(C= =hollow): C=solid;
}; /*After the ION of TABP has been conducted, the tokens change to solid ones. */
Mark(Place,C)={
  Fun(Place is a TABP ∧ (αp ≤ Time ≤ βp)):
    OIP(Place) ∧ M(Place,C);
  Fun(Place ∉ N.TABP ): M(Place,C);
};/*Mark different places, which include common places and abstract places */
```

3. *TAE calculation*: The TAE calculation mentioned above is used to evaluate the performance of the whole TOPN model. According to the TOPN model, the temporal constraint of every transition and every TABP is [0, 50], which is a kind of relative temporal knowledge.
   - Get the Object Function
     On the base of TOPN analysis method [5], the execution path can be got as the following.
     $$\text{Path\_set} = \{\text{path}_1\} \tag{15}$$
     $$\text{Path}_1 = \{t_1, t_2, P_4, t_4, t_5, t_6, P_4, t_3\} \tag{16}$$
     $$\text{Path}_1.X = \{t_1, t_2, t_3, t_4, t_5, t_6, P_4\} \tag{17}$$
     So the object function is like the following
     $$Z = \sum_{j=1}^{7} f_{ij} TI_{t_j} \tag{18}$$
     where $t_j \in \text{Path}_i.X$ and $\text{Path}_i \in \text{Path\_set}$. TI_{t_j} is the execution time of the object t_j on the path path_i. f_j is the execution frequency of the object t_j on the path path_i $(j = 1, \ldots, 6)$. f_7 is the execution frequency of the TABP P_4 on the path.
   - Get the structure constraints
     Single-entry and single-exit constraints:
     $$\text{Single Entry:} \qquad f_1 = 1 \tag{19}$$
     $$\text{Single Exit:} \qquad f_3 = 1 \tag{20}$$
     Connection Constraint:
     $$f_1 = f_2 \tag{21}$$
     Branch Constraint:
     $$f_7 = f_3 + f_4 \tag{22}$$
     Combination Constraint:
     $$f_2 + f_6 = f_7 \tag{23}$$
   - Get the logical constraint
     According to the data dictionary, after the work flow of this TOPN model has been analyzed, the following logical constraint can be got.
     $$f_4 = f_5 = f_6 = 1 \tag{24}$$
   - Positive constraint
     $$f_i \geq 0 \qquad i = 1, \ldots, 7 \tag{25}$$
4. Get the temporal knowledge of the whole TOPN model.
   On the base of the object function and the corresponding constraints, the temporal knowledge or the system performance–the calculating results are just like the following:
   $$\text{EFT} = \min(Z) = 0 \tag{26}$$

$$\text{LFT} = \max(Z) = 400 \tag{27}$$

So the temporal knowledge of the whole example is [0,400]. □

That's to say, the temporal performance of NN based robot controller is [0,400]. The controller will take 400 unit time at most to complete the controlling task.

5 Conclusions

Neural network based controllers have been widely used in robot systems or CMRS. However, for time critical systems such as NN based robot controllers etc al, the temporal performance of these systems requires to be evaluated. TOPN is a powerful modeling and analysis tool to model and analyze complex time critical systems, which is also suitable for model NN based controllers. On the base of the existing TOPN method, Time Accumulation Effect (TAE) Calculation Method is proposed in this paper. Its aim is to calculate the WCET of the whole TOPN model or abstract TOPN objects which may represent as NN based controller models.

The automatic mapping method from NN model to TOPN model will be studied in the future. It can help to realize automatic performance evaluation for NN based models.

References

1. Yıldırım, S.: Adaptive Robust Neural Controller for Robots. Robotics and Autonomous Systems 46(3), 175–184 (2004)
2. Lewis, F.L., Liu, K., Yesildirek, A.: Neural Net Robot Controller with Guaranteed Tracking Performance. IEEE Transactions on Neural Networks 6(3), 703–715 (1995)
3. Sato, M., Kanda, A., Ishii, K.: ESEC 1997 and ESEC-FSE 1997. International Congress Series, vol. 1301, pp. 160–163 (2007)
4. Bolouri, H., Morgan, P., Gurney, K.: Design, Manufacture and Evaluation of a Scalable High-Performance Neural System. Electronics Letters, 3rd 30(5), 426–427 (1994)
5. Xu, H., Jia, P.F.: Timed Hierarchical Object-Oriented Petri Net-Part I: Basic Concepts and Reachability Analysis. In: Wang, G.-Y., Peters, J.F., Skowron, A., Yao, Y. (eds.) RSKT 2006. LNCS (LNAI), vol. 4062, pp. 727–734. Springer, Heidelberg (2006)
6. Hong, J.E., Bae, D.H.: Software Modeling and Analysis Using a Hierarchical Object-Oriented Petri Net. Information Sciences 130, 133–164 (2000)
7. Yao, Y.L.: A Petri Net Model for Temporal Knowledge Representation and Reasoning. IEEE Transactions on Systems, Man and Cybernetics 24, 1374–1382 (1994)
8. Peter, P., Alan, B. Guest Editorial: A Review of Worst Case Execution-Time Analysis. Real-Time Systems 18(2-3), 115–128 (2000)
9. Theiling, H., Ferdinand, C., Wilhelm, R.: Fast and Precise WCET Prediction by Separated Cache and Path Analyses. Real-Time Systems 18(2-3), 157–179 (2000)

A Fuzzy Timed Object-Oriented Petri Net for Multi-Agent Systems*

Hua Xu and Peifa Jia

State Key Laboratory on Intelligent Technology and Systems,
Tsinghua National Laboratory for Information Science and Technology,
Department of Computer Science and Technology,
Tsinghua University, Beijing 100084, China
{xuhua,dcsjpf}@mail.tsinghua.edu.cn

Abstract. In this paper, a multi-agent system (MAS) modeling method called fuzzy timed object-oriented Petri nets (FTOPN) is proposed. FTO-PN has extended Petri nets (PN) supporting object-oriented modeling and temporal fuzzy learning based on timed hierarchical object-oriented Petri net (TOPN) and fuzzy timed Petri net (FTPN). Our focus is the adaptation according to TOPN concepts of cooperation objects for supporting synchronous and asynchronous communications and the temporal fuzzy learning proposed in FTPN. These two diagrams have been chosen because they are the most commonly used in modeling MAS and describing agent learning and reasoning ability. That is to say, they can be used to model and illustrate both the structural and dynamic aspects of MAS. Not only the proposed FTOPN can be used to model complex MAS, but also FTOPN model can be refined into the object-oriented implementation easily. It has bridged the gap between the formal modeling and the system refinement, which can overcome the development problems in agent-oriented software engineering. At the same time, it also can be regarded as a conceptual and practical artificial intelligence (AI) tool for the integration of MAS into the mainstream practice of software development.

1 Introduction

Characterized as autonomy, situatedness and sociality, multi-agent systems (MAS) have recently emerged as a powerful paradigm for designing and developing software systems [1, 2] in both industry and academia. Currently, most of MAS are described by logical models [1], but they are difficult to be refined into exact implementations [3]. These years, several methods have been proposed to describe the MAS architecture, such as Shoham's AGENT0 [4], METATEM [5] and TAO [6]. On the base of Petri nets and objects, Object Control Structure (OBCS) [3] and Object Petri net (OPN) [7] as new formalism methods have been

* This work is jointly supported by National Natural Science Foundation of China (Grant No: 60405011, 60575057) and China Postdoctoral Foundation for China Postdoctoral Science Fund (Grant No: 20040350078).

F. Sun et al. (Eds.): ISNN 2008, Part II, LNCS 5264, pp. 210–219, 2008.

proposed. But in OBCS models, only the static knowledge can be represented and OPN is difficult to describe actual large-scale complex MAS. Recently, fuzzy timed Petri net (FTPN) [8] has been presented to model and analyze different systems, which can also be considered as supporting autonomous judging or reasoning ability in MAS.

In this paper, fuzzy timed object-oriented Petri net (FTOPN) is proposed on the base of TOPN [10] and FTPN [9], whose aim to solve the reasoning ability and other modeling problems in large-scale MAS and construct a bridge between MAS models and their implementations. It has provided a methodology to overcome the development problems in agent-oriented software engineering. At the same time, it can also be regarded as a conceptual and practical artificial intelligence (AI) tool for integration of MAS into the mainstream practice of software development.

This paper is organized as the following. Section 2 presents our FTOPN and discusses the learning of FTOPN and the relation between FTOPN and MAS. In section 3, FTOPN is used to model one cooperation robot system in the wafer etching procedure of circuit industry to demonstrate its benefits in modeling MAS. Finally, the conclusion and future work can be found in section 4.

2 Fuzzy Timed Object-Oriented Petri Net

In this section, on the base of TOPN [10] and FTPN [9], FTOPN is proposed and its concepts are discussed. Then, the learning in FTOPN is also presented according to that in FTPN [9]. Finally, the relation between FTOPN representation and MAS is discussed.

2.1 Basic Concepts

Similar to FTPN [9], fuzzy set concepts are introduced into TOPN [10]. Then FTOPN is proposed.

Definition 1: FTOPN is a six-tuple, FTOPN= (OIP, ION, DD, SI, R, I) where

1. Suppose OIP=(oip, pid, M_0, status), where oip, pid, M_0 and status are the same as those in HOONet [7] and TOPN [10].
   - oip is a variable for the unique name of a FTOPN.
   - pid is a unique process identifier to distinguish multiple instances of a class, which contains return address.
   - M_0 is the function that gives initial token distributions of this specific value to OIP.
   - status is a flag variable to specify the state of OIP.
2. ION is the internal net structure of FTOPN to be defined in the following. It is a variant CPN [8] that describes the changes in the values of attributes and the behaviors of methods in FTOPN.
3. DD formally defines the variables, token types and functions (methods) just like those in HOONet [7] and TOPN [10].

4. SI is a static time interval binding function, SI: {OIP}$\rightarrow Q^*$, where Q^* is a set of time intervals.
5. R: {OIP} $\rightarrow r$, where r is a specific threshold.
6. I is a function of the time v. It evaluates the resulting degree of the abstract object firing. □

Definition 2: An internal object net structure of TOPN, ION = $(P, T, A, K, N, G, E, F, M_0)$, where

1. P and T are finite sets of places and transitions with time restricting conditions attached respectively.
2. A is a finite set of arcs such that $P \bigcap T = P \bigcap A = T \bigcap A = \Phi$.
3. K is a function mapping from P to a set of token types declared in DD.
4. N, G, and E mean the functions of nodes, guards, and arc expressions, respectively. The results of these functions are the additional condition to restrict the firing of transitions. So they are also called *additional restricting conditions.*
5. F is a special arc from any transitions to OIP, and notated as a body frame of ION.
6. M_0 is a function giving an initial marking to any place the same as those in HOONet [7] and TOPN [10]. □

Definition 3: A set of places in TOPN is defined as P=PIP $\bigcup$ TABP, where

1. Primary place PIP is a three-tuple: PIP $=(P, R, I)$, where
   - P is the set of common places similar to those in Petri Nets.
2. Timed abstract place (TABP) is a six-tuple: TABP= TABP(pn, refine state, action, SI, R, I), where
   - pn is the identifier of the abstract timed place.
   - refine state is a flag variable denoting whether this abstract place has been refined or not.
   - action is the static reaction imitating the internal behavior of this abstract place.
3. SI, R and I are the same as those in *Definition 1.* □

Definition 4: A set of transitions in TOPN can be defined as T= TPIT $\bigcup$ TABT $\bigcup$ TCOT, where

1. Timed primitive transition TPIT = TPIT (BAT, SI), where
   - BAT is the set of common transitions.
2. Timed abstract transition TABT= TABT (tn, refine state, action, SI), where
   - tn is the name of this TABT.
3. Timed communication transition TCOT=TCOT (tn, target, comm type, action, SI), where
   - tn is the name of TCOT.

   - target is a flag variable denoting whether the behavior of this TCOT has been modeled or not. If target="Yes", it has been modeled. Otherwise, if target="No", it has not been modeled yet.
   - comm type is a flag variable denoting the communication type. If comm type ="SYNC", then the communication transition is synchronous one. Otherwise, if comm type="ASYN", it is an asynchronous communication transition.
4. SI is the same as that in *Definition 1.*
5. *refine state* and *action* are the same as those in *Definition 3.* □

Similar to those in FTPN [9], the object t fires if the foregoing objects come with a nonzero marking of the tokens; the level of firing is inherently continuous. The level of firing ($z(v)$) assuming values in the unit interval is governed by the following expression (1):

$$z(v) = (\mathop{\mathrm{T}}_{i=1}^{n}(r_i \rightarrow x_i(v'))sw_i)t\mathrm{I}(v) \tag{1}$$

where T (or t) denotes a t-norm, while "s" stands for any s-norm. "v" is the time instant immediately following v'. More specifically, $x_i(v)$ denotes a level of marking of the ith place. The weight w_i is used to quantify an input coming from the ith place. The threshold r_i expresses an extent to which the corresponding place's marking contributes to the firing of the transition. The implication operator ($\rightarrow$) expresses a requirement that a transition fires if the level of tokens exceeds a specific threshold (quantified here by r_i).

Once the transition has been fired, the input places involved in this firing modify their markings that are governed by the following expression (2):

$$x_i(v) = x_i(v')\boldsymbol{t}(1 - z(v)) \tag{2}$$

(Note that the reduction in the level of marking depends upon the intensity of the firing of the corresponding transition, $z(v)$.) Owing to the t-norm being used in the above expression, the marking of the input place gets lowered. The output place increases its level of tokens following the expression (3):

$$y(v) = y(v')sz(v) \tag{3}$$

The s-norm is used to aggregate the level of firing of the transition with the actual level of tokens at this output place. This way of aggregation makes the marking of the output place increase.

The FTOPN model directly generalizes the Boolean case of TOPN and OPN. In other words, if $x_i(v)$ and w_i assume values in $\{0, 1\}$ then the rules governing the behavior of the net are the same as those encountered in TOPN.

2.2 Learning in FTOPN

The parameters of FTOPN are always given beforehand. In general, however, these parameters may not be available and need to be estimated just like those in

FTPN [9]. The estimation is conducted on the base of some experimental data concerning marking of input and output places. The marking of the places is provided as a discrete time series. More specifically we consider that the marking of the output place(s) is treated as a collection of target values to be followed during the training process. As a matter of fact, the learning is carried in a supervised mode returning to these *target* data.

The connections of the FTOPN (namely weights w_i and thresholds r_i) as well as the time decay factors α_i are optimized (or trained), so that a given performance index Q becomes minimized. The training data set consists of (a) initial marking of the input places $x_i(0), \ldots, x_n(0)$ and (b) target values-markings of the output place that are given in a sequence of discrete time moments, that is target(0), target(1), ..., target(K).

In our FTOPN, the performance index Q under discussion assumes the form of the following sum:

$$Q = \sum_{k=1}^{K} (\text{target}(k) - y(k))^2 \tag{4}$$

where the summation is taken over all time instants ($k = 1, 2, \ldots, K$).

The crux of the training in FTOPN models follows the general update formula being applied to the parameters:

$$\mathbf{param}(\text{iter} + 1) = \mathbf{param}(\text{iter}) - \gamma \nabla_{\mathbf{param}} Q \tag{5}$$

where γ is a learning rate and $\nabla_{\mathbf{param}} Q$ denotes a gradient of the performance index taken with respect to all parameters of the net (here we use a notation **param** to embrace all parameters in FTOPN to be trained).

In the training of FTOPN models, marking of the input places is updated according to the following form:

$$\tilde{x}_i = x_i(0) T_i(k) \tag{6}$$

where $T_i(k)$ is the temporal decay. And $T_i(k)$ complies with the following form. In what follows, the temporal decay is modeled by an exponential function,

$$T_i(k) = \begin{cases} \exp(-\alpha_i(k - k_i)) & \text{if } k > k_i, \\ 0 & \text{others} \end{cases} \tag{7}$$

The level of firing of the place can be computed as the following:

$$z = (\mathop{\mathrm{T}}_{i=1}^{n} ((r_i \to \tilde{x}_i) s w_i) \tag{8}$$

The successive level of tokens at the output place and input places can be calculated as:

$$y(k) = y(k-1) \boldsymbol{s} z, x_i(k) = x_i(k-1) \boldsymbol{t} (1 - z) \tag{9}$$

We assume that the initial marking of the output place $y(0)$ is equal to zero, $y(0) = 0$. The derivatives of the weights w_i are computed as follows:

$$\frac{\partial}{\partial w_i}(\text{target}(k) - y(k))^2 = -2(\text{target}(k) - y(k)\frac{\partial y(k)}{\partial w_i}) \quad (10)$$

where $i = 1, 2, \ldots, n$. Note that $y(k+1) = y(k)\boldsymbol{s}z(k)$.

2.3 FTOPN and MAS

Objects in FTOPN have integrated concepts from object-oriented approaches, agents and Petri nets. The system modeled by FTOPN can be characterized by the following equations:

$$\text{System} = \text{FTOPN Objects(or Agents)} + \text{Cooperation} \quad (11)$$

$$\text{FTOPN Objects} = \text{Objects} + \text{Mental States} \quad (12)$$

One object in FTOPN model can be regarded as an agent. The reason is that: (1) FTOPN can describe the mental states of agents for having integrated object-oriented concepts and Petri nets; (2) Not only the interaction but also the autonomy can be represented in FTOPN for the integration of fuzzy reasoning with temporal knowledge.

The features of FTOPN decide its representing ability of agent mental states characterized by the set of *goals*, *beliefs*, *actions* and *plans* defined by the agent [2]. The definition of state for objects in FTOPN has been extended and its behaviors have been added. From the concepts of FTOPN, one FTOPN object can not only be transformed to state graph [10] according to its marking change, but also represent its behavior by its FTOPN structure. Thus, one state of objects in FTOPN consists of both its state and behavior.

On the other hand, fuzzy reasoning in FTOPN support every FTOPN object owns a kind of autonomous decision ability. By this reasoning ability, every FTOPN object can make decisions according to the environments (also represented as objects in FTOPN model) change and its own knowledge (learning results or FTOPN structure). An object in FTOPN is an autonomous and interactive element that sends and receives messages similar to an agent, which leads to cooperate with each other.

Cooperation between different objects is also strongly supported in FTOPN. On the base of communication transactions (synchronous ones or asynchronous ones), objects in one FTOPN model can cooperate to complete tasks. One object can request the services presented in other objects by the communication. Cooperative objects in FTOPN support multi-tasking inside objects and asynchronous communications. Accordingly, a FTOPN object enjoys from a high level of autonomy and FTOPN can be used to describe MAS.

From the definitions of FTOPN, it is clear that FTOPN model is a kind of hierarchical object structure. For supporting the encapsulation in OO, the state analysis can be conducted in each encapsulation independently, even if the model has not been completed in other hierarchies [7, 10]. This has given an easy way to analyze the state change for a complex agent-oriented system. For not only an agent but also MAS, the state changes can be analyzed in various hierarchies considering temporal knowledge.

3 A Modeling Example

Cooperative multiple robot systems (CMRS) such as cooperative industrial robots, soccer robots, etc al, are typical multi-agent systems (MAS). Every robot in CMRS is controlled according to the environments, its own goals and its own states. Information about environments comes from sensors or other robots. As the information may not be available from all sensors or sources at the same time moment, the one that occurs earlier needs to be discounted over time as becoming less relevant or to be mounted up over time as becoming more relevant. That is to say, information timing effects exist in this kind of dynamic systems. However, in the control of every robot system, every kind of information is required simultaneously. As the information readings could come at different time instants and be collected at different sampling frequency, we encounter an inevitable timing effect of information collected by the system and sensors. So FTOPN is used to model our CMRS. At the same time, FTOPN can reduce the model complexity and can model complex decision making processes in different levels, because of the OO abstraction concept supported in FTOPN.

3.1 CMRS Example

In the wafer etching procedure of circuit industries, two industrial robots always need to cooperate to transfer wafers from the cassette to the chambers. In the transformation, one robot (RA) firstly fetch one wafer from input cassette to the input load lock, then another robot (RB) fetch the wafer to different chambers in turn. At last, RB will fetch the processed wafer from cooling chamber to the output load lock. And then RA will fetch the wafer from the output load lock to the output cassette. In actual industrial applications, productivity needs to be improved to the highest. So in the wafer etching procedure, robots need to cooperate seamlessly. Waiting for another robot means that productivity has been reduced. In reality, the waiting time always happens, when the wafer is transferred from one robot to another. In order to avoid the waiting time, the accepting robot always need to judge the wafer arrival time, when the transferring robot is sending it to the goal location. At this time, not all of the information required in the judging procedure can be got at the same time. So in the control of our CMRS systems accepting robots, FTOPN is used to model the system. Because the model is hierarchical, only the highest level of the model is depicted in Fig. 1.

In the model of Fig. 1, three communication transition objects are used to represent the service requirements for getting different kinds of system states. These states include the state of the other robot, its own goal and its current state, which can be required by the conductions of the communication transactions t1, t2 and t3. When one condition has been got, the following place will be marked. In order to make control decision (transition object t4) in time, all of these state parameters are required in the prescriptive time interval. However, the parameter arrival times are different. The timing effect on the control decision is depicted in Fig. 2. The information "its current state" complies with the

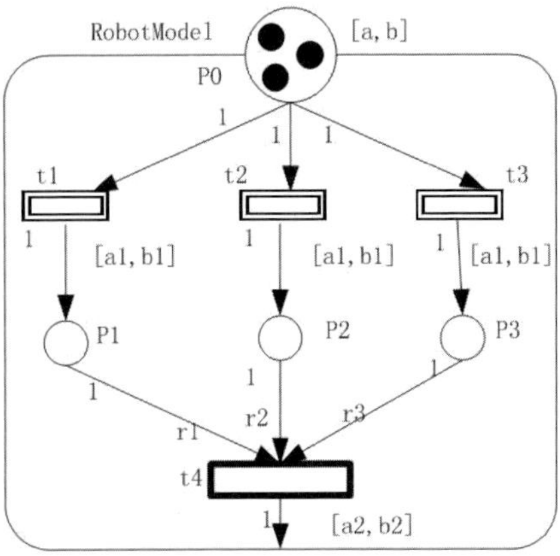

Fig. 1. The FTOPN Model

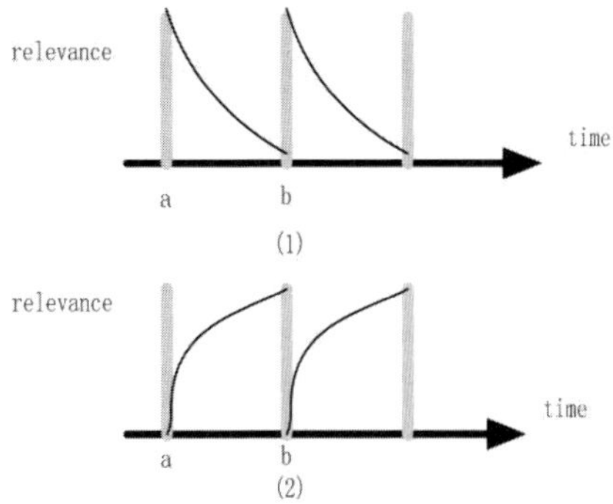

Fig. 2. The Relevance

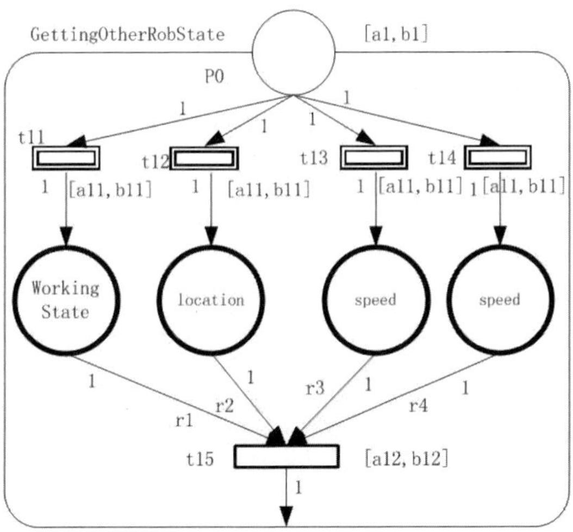

Fig. 3. The Transaction Object t1-"Getting Other Robot State" Model

rule in Fig. 2(1). The other two kinds of information comply with that in Fig. 2(2). After the decision, a new command will be sent in this relative interval.

What's more, all the objects in Fig. 1 can also be depicted in details by FTOPN. For example, the object-"Other Robot State" in Fig. 1 can also be modeled concretely with FTOPN. The detailed model of the object is depicted in Fig. 1. It is also an independent fuzzy reduction process. According to the modeling and analysis requirements, the detailed model can be unfolded directly in the model of Fig. 1 according to the TOPN reduction rules in [11]. At the same time, its training can be conducted independently. It can also be reduced independently and the reduction results will be used as the believing effect of the corresponding object in the higher level of the FTOPN model in Fig. 1.

After completing the FTOPN model, the learning algorithm of FTOPN can be used to train our model and adjust it to fulfill the practical requirements.

4 Conclusions and Future Works

In this paper, an enhanced high-level Petri net called Fuzzy Timed Object-oriented Petri net is proposed. Firstly, it aims to bridge the gap between the world of formal theory and infrastructure and the world of practical OO, agent or MAS system development. On the other hand, it also tries to conduct on the identification and definition of suitable models and techniques to support the development of complex software systems in terms of MAS. For artificial intelligence, it is also a kind of effort in producing a reasonable set of conceptual and practical tools, which could promote the integration of such a vast amount of research findings into the mainstream practice of software development. Previously, OBCS [3] have been presented to be easily refined into implementation. But its models can not support all the OO concepts and can not be analyzed like those in Petri nets.

The proposed FTOPN in this paper is a formal method on the base of TOPN [10] and FTPN [9]. It is for the design, the analysis, the validation and the implementation of MAS using the concurrent dynamic system modeling and analyzing benefits of Petri nets. In FTOPN, all OO concepts and fuzzy reasoning with temporal knowledge have been indulged into Petri nets. Mental states can be represented in FTOPN objects. High-level autonomy has been provided to FTOPN objects. So objects in FTOPN can be modeled as agents. The behavior of each object and its cooperation with other objects is defined by FTOPN. Thus, MAS can be easily modeled and analyzed, because of the Petri net advantage in modeling and analyzing concurrent dynamic systems.

State analysis needs to be studied in the future. Xu [10, 11] has proposed an extended State Graph to analyze the state change of TOPN models. With the temporal fuzzy sets introduced into TOPN, the certainty factor about object firing (state changing) needs to be considered in the state analysis.

References

1. Jennings, N.R., Sycara, K., Wooldridge, M.: A Roadmap of Agent Research and Development. Autonomous Agents and Multi-Agent Systems 1, 7–38 (1998)
2. Luck, M., McBurney, P., Preist, C.: A Manifesto for Agent Technology: Towards Next Generation Computing. Autonomous Agents and Multi-Agent Sytems 9, 203–252 (2004)
3. Chainbi, W.: Multi-agent systems: a Petri net with objects based approach, Intelligent Agent Technology. In: Proceedings of IEEE/WIC/ACM International Conference on Intelligent Agent Technology (IAT 2004), pp. 429–432 (2004)
4. Shoham, Y.: Agent-oriented programming. Artificial Intelligence 60(1), 51–92 (1993)
5. Fisher, M.: Representing and executing agent-based systems. In: Wooldridge, M., Jennings, N.R. (eds.) Intelligent Agents: Theories, Architectures, and Languages. LNCS (LNAI), vol. 890, pp. 307–323. Springer, Berlin (1995)

6. Silva, V., Garcia, A., Brandao, A., Chavez, C., Lucena, C., Alencar, P.: Taming Agents and Objects in Software Engineering. In: Garcia, A., Lucena, C., Zamboneli, F., Omicini, A., Castro, J. (eds.) Software Engineering for Large-Scale Multi-Agent System. LNCS. Springer, Heidelberg (2003)
7. Hong, J.E., Bae, D.H.: Software Modeling And Analysis Using a Hierarchical Object-oriented Petri net. Information Sciences 130, 133–164 (2000)
8. Jensen, K.: Coloured Petri Nets: Basic Concepts, Analysis methods and Practical Use, vol. 1, pp. 65–85. Springer, Berlin (1992)
9. Pedrycz, W., Camargo, H.: Fuzzy timed Petri nets. Fuzzy Sets and Systems 140(2), 301–330 (2003)
10. Xu, H., Jia, P.F.: Timed Hierarchical Object-Oriented Petri Net-Part I: Basic Concepts and Reachability Analysis. In: Wang, G.-Y., Peters, J.F., Skowron, A., Yao, Y. (eds.) RSKT 2006. LNCS (LNAI), vol. 4062, pp. 727–734. Springer, Heidelberg (2006)
11. Xu, H.: Studies on Timed Hierarchical Object-oriented Petri Net and Its Application, Doctor Thesis, Tsinghua University (2003)

Fuzzy Reasoning Approach for Conceptual Design

Hailin Feng[1], Chenxi Shao[2], and Yi Xu[2]

[1] School of Information Science and Technology, Zhejiang Forestry University
311300 Hangzhou, China
[2] Computer Science Depart., University of Science and Technology of China
230027 Hefei, China
hlfeng@mail.ustc.edu.cn

Abstract. To deal with incomplete quantitative information in configuration design stage, a fuzzy approach is presented for conceptual design. Fuzzy information vector and fuzzy constraint matrices were given for the description of components selected. Fuzzy sign algebra was used to build the reasoning rules and create the configuration branches. The possible reasonable spatial configurations of mechanisms were predicted with the constraint spread rules. Reasoning structure for space configuration was given to determine the configuration matrices and output. The result of reasoning process applied to the spatial configuration inside a mechanism shows that the fuzzy approach is effective for conceptual design with little quantitative information.

Keywords: Conceptual design, Fuzzy reasoning, Spatial reasoning, Mechanical configuration.

1 Introduction

To be compatible with the spatial configuration, the numerical information of the arrangement of connected parts in a machine needs to be known in the conceptual design stage[1,2]. However, it is not easy to gain the quantitative information necessary to reason spatial configuration of a mechanism, so the traditional quantitative methods to reason possible spatial solutions of a mechanism are less effective[3].

Therefore, the fuzzy and qualitative spatial reasoning can be used as an alternative approach to be used in the stage of conceptual design[4,5]. Qualitative and fuzzy reasoning and simulation is an effective strategy in studying systems with incomplete knowledge[6]. Qualitative diagram and spatial reasoning, as similar to other cognizing ways of human being, is one of the focuses in this field in recent years and has wide application requirement[7,8,9,10]. There are many possible applications of qualitative spatial reasoning [11], such as physical systems, the traditional domain of QR systems[12], Geographical Information Systems [13], Computer aided design [14], and so on. The challenge of qualitative spatial reasoning is to provide calculi which allow a machine to represent and reason with spatial entities of higher dimension without resorting to the traditional quantitative techniques.

In this paper, we reason about the spatial solutions of a mechanism by using fuzzy spatial reasoning, which means the reasoning process is based mostly on incompletely

F. Sun et al. (Eds.): ISNN 2008, Part II, LNCS 5264, pp. 220–226, 2008.

qualitative information. Here we introduce the method based on fuzzy reasoning for the spatial configuration of mechanisms that could be applied earlier in mechanism design. We present the reasoning structure and algorithm to reason automatically the qualitative configuration of mechanisms composed of functional parts, based on some qualitative information vectors or matrices. The result of reasoning process applied to the spatial configuration inside a mechanism shows that the fuzzy approach is effective for conceptual design with little quantitative information.

2 Representation of Components

Though qualitative reasoning approach is introduced to the stage of conceptual design of mechanisms, there are still several steps we have to follow in traditional ways. Firstly, appropriate types of mechanisms are chosen for a specific task through conceptual synthesis, and then a further evaluation process is carried out to select the most suitable mechanism for the subsequent design process.

2.1 Fuzzy Information Vector and Position Vector

The fuzzy information vector (FIV), composed of two sub-vectors, is introduced to represent the components in 3-D space. The Position Vector (PV) is defined to represent the position of the components.

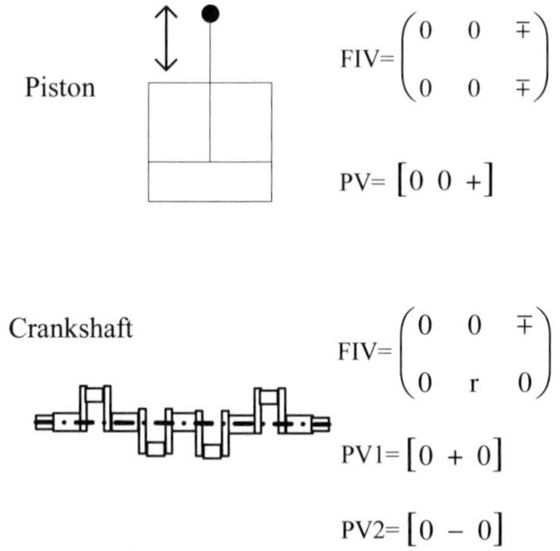

Fig. 1. FIV and PV of primitive mechanism components

One row of PIV is to describe the dynamic function of the component S (source) and the other is to describe the output function of the component O (output). S and O are both represented as [x, y, z] (x, y, z $\in$ {+,0,-,r}). The new PV of a complex mechanism, which is composed by its components, can be qualitatively reasoned based on their PV. Given the input position of the component as the source, the PV presents the output position in the space. In order to reason logically, PV also follows the format [x, y, z] (x, y, z∈{+,0,-,r}). In addition, some components provide more than one output, and these outputs may have different positions in space. In such cases, more PVs are necessary to describe the position information of the multi-output. Further more, we introduce r to describe rotation in 3D space, which means the rotation around certain axis.

2.2 Fuzzy Constraints Matrix

Fuzzy Constraints Matrix contains the dynamic constraints information, which is used to eliminate the redundant reasoning results and delete the schemes that don't match the reasoning rules. This matrix is composed by the three 3D vectors as follows: (*tc, rc, pc* = (a b c); a,b,c∈{0,+,−}). For reasoning, the definition of *tc*, *rc* and *pc* is similar to the format of S, O and PV. The *tc* is defined as the transferring constraint, which implies the transferring direction of mechanism component in 3D space. The *rc* is defined as the rotating constraint, which implies the rotating style of mechanism component in 3D space. Obviously there would be definitely no such situation that a clockwise rotating component is linked to an anticlockwise rotating component. Finally, we define *pc* as the position constraints, which imply the output position in 3D space of the mechanism component relative to the input position. The three elements in this vector present the different position as X, Y, Z axis in space.

We present a piston and a crankshaft with the matrixes and vectors defined above, as shown in Fig. 1. The crankshaft moves in direction of Z-axis, and has no change in X, Y-axis, so the source vector S = [0 0 ∓]. At the same time, the crankshaft provides an output which is a rotation around the Y axis, the output in X and Z are both zero, so we have O = [0 *r* 0]. The crankshaft provides outputs in two different positions in the space; one is on the left of the crankshaft while the other is on the right, so the position vectors of the crankshaft is PV1 = [0 + 0], PV2 = [0 − 0]. As to constraints, in this situation, it is supposed that we need no transferring constraints in crankshaft and the clockwise rotation is permitted to be the output, and we only need the output in the positive direction of the Y axis. So according to the definition above, we get the 3D Fuzzy Constraints Matrix.

3 Fuzzy Sign Algebra

To reason useful information with the defined vectors of mechanism components, the fuzzy sign algebra reasoning rules must be specified firstly. The reasoning rules are shown in Fig 2. The ⊕ is defined as the ADD operator for fuzzy sign algebra. In the advanced reasoning step, if (+ +) ADD a movement (+−), the result would be unspecific:(+ +)⊕(+−)= (+ *). Obviously, the result has three possible styles: (+ +), (+−) or (+ 0). It would not be determinate until more quantitative information is provided.

With these rules, we can extend the reasoning process to 2D or 3D space easily. The rules are defined based on fuzzy sign operations, *r* means it rotates around certain axis while '+' means it is in the positive direction of an axis, '−' means negative direction of an axis, '0' means there has no movement in this direction. The sign ' * ' means that we can not specify the result, it might be one of the {+, −, 0}. The sign

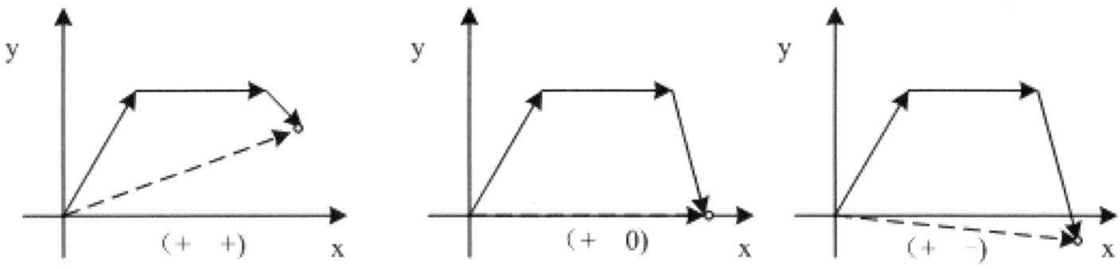

Fig. 2. Sign algebra reasoning rules

' $\neq$ ' presents that there is no meaning for this reasoning. For example, we can catch the reasoning process quickly in 2D space, and the ADD of two movements in 2D space is as: (+ +) $\oplus$ (+ 0) = (+ +). The result (+ +) means the final movement locates in the positive direction of X and Y axis.

4 Fuzzy Reasoning Process

As mentioned above, a complicated mechanism is composed of the arrangement of connected parts. The primary problem is to judge whether the designed parts match the designed requirements imposed by spatial constraints and movement character in the conceptual design stage. In the description of information vector, it is possible that there have several reasoning outcomes. The mechanical configuration matrix and output vector can be determined with the fuzzy information vector and position vector of every functional component. The design plan of machine cannot be determined entirely before the quantitative information is given, so we need to predict possible solutions of spatial configurations. The reasoning process can be expressed clearly by using reasoning tree pattern.

For example, if there is an output movement in direction Z for certain functional part which has been linked with another functional part, there can have positive or negative changes in direction Y and bring two branches. In the following design process there can also have positive or negative changes in direction Z and it can bring out four branches that indicate there are four possible output manners. Correspondingly, it is possible that there are branches in the vector reasoning process. The fuzzy configuration matrix and output position vector are derived on the leaf nodes of reasoning tree through the path from root to leaf. The definition of fuzzy configuration matrix is [S O]. S denotes initial movement character of functional component that can support the original power for the mechanism. O means output movement for the whole machine, and output position vector means the output for the whole machine, which is the space information of output relative to input. So the space configuration information of output movement character and output position can be reasoned based on description of sign algebra.

The space configuration information and output position information will spread in the design reasoning process. Some branches that do not suit fuzzy constraint matrix will be eliminated from the reasoning tree. The reasoning machine aims at reasoning qualitative configuration of every functional part, space position vector and fuzzy constraint matrix. Though it is possible that there have several inputs and outputs for a complicated mechanism, the reasoning machine defined here is only limited in single input scope for simplification. If an input can produce multiple outputs, the machine can be viewed as output chain caused by multiple inputs in order to describe the system of one-input and multiple-outputs, and we need to reason every chain of one-input and one-output.

5 Application Example and Discussion

The mechanism inside a truck has two pieces of mechanical chains belt and we can consider it as multi-input and output system. The gears of two sides are driven through main transducer to reduce speed and enhance torque. When the engine starts, movement of engine piston leads to the running of crank, and then links to clutch to output main power shaft by ratio gear. However, regarding lifting device of track case, the principle of lift case is to link with oil pump by footswitch, and to link with case by pulley.

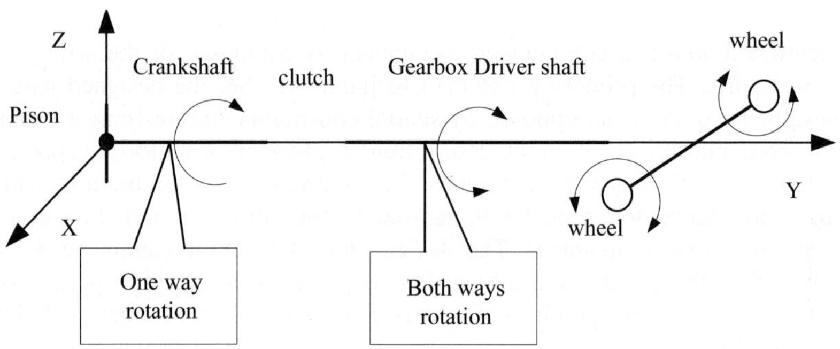

Fig. 3. Reasoning result of mechanism inside truck

Unloading track case has to be in the position of direction positive of YZ plane of footswitch; the block has to be at the same side of footswitch and unloading box; engine crank is only allowed to rotate in unilateral way and is only acceptable for output in positive direction of Y; wheel axle must be in position of positive direction Y of engine and rotate in bilateral way. Therefore, according to the requirement of design, we need to predefine Fuzzy Constraint Matrix of unloading case, block, crankshaft and wheel axle. The mechanism chain of unloading truck is shown in Fig.3. One arrowhead means the component rotating in one direction.

Footswitch: $\text{FIV} = \begin{pmatrix} 0 & - & - \\ 0 & \mp & 0 \end{pmatrix}$ $\quad$ $\text{PV} = [\,0\ -\ 0\,]$

Piston: $\text{FIVr} = \begin{pmatrix} 0 & \mp & 0 \\ 0 & \mp & 0 \end{pmatrix}$ $\quad$ $\text{PV} = [\,0\ 0\ -\,]$

Pump: Virtual FIVr $=\begin{pmatrix} 0 & \mp & 0 \\ 0 & - & + \end{pmatrix}$ PV = [0 − 0]

Block: FIV $=\begin{pmatrix} 0 & - & + \\ 0 & 0 & + \end{pmatrix}$ PV = [0 + +] FCM $=\begin{pmatrix} 0 & 0 & 0 \\ 0 & 0 & 0 \\ 0 & + & 0 \end{pmatrix}$

Case: FIV $=\begin{pmatrix} 0 & 0 & + \\ 0 & + & + \end{pmatrix}$ PV = [0 + +] FCM $=\begin{pmatrix} 0 & 0 & 0 \\ 0 & 0 & 0 \\ 0 & + & + \end{pmatrix}$

Chain: FCM = $FIV_1 \oplus FIV_2 \oplus FIV_3 \oplus FIV_4 \oplus FIV_5$

$$=\begin{pmatrix} 0 & - & - \\ 0 & \mp & 0 \end{pmatrix} \oplus \begin{pmatrix} 0 & \mp & 0 \\ 0 & \mp & 0 \end{pmatrix} \oplus \begin{pmatrix} 0 & \mp & 0 \\ 0 & - & + \end{pmatrix} \oplus \begin{pmatrix} 0 & - & + \\ 0 & 0 & + \end{pmatrix} \oplus \begin{pmatrix} 0 & 0 & + \\ 0 & + & + \end{pmatrix}$$

$$=\begin{pmatrix} 0 & - & - \\ 0 & + & + \end{pmatrix}$$

OPV= $PV_1 \oplus PV_2 \oplus PV_3 \oplus PV_4 \oplus PV_5$

= [0 − 0] ⊕ [0 0 −] ⊕ [0 − 0] ⊕ [0 + +] ⊕ [0 + +]

= [0 − −] ⊕ [0 + +] ⊕ [0 + +]

= [0 + *] ⊕ [0 + +]

= [0 + +];

According to the result, the unloading box has a movement towards negative direction of YZ flat while the footswitch has a movement towards positive direction of YZ flat. There are some branches that we can't specify in the reasoning early stage. If there is no FCM, it is obvious that the reasoning tree has 9 branches. The ADD operation [0 − −] ⊕ [0 + +] should have gotten [0 * *], but the pc of wheel's FCM limits that the wheel must locate in the right flat of the footswitch, so the branches [0 − *] and [0 0 *] are deleted and only [0 + *] is reasonable. In the next step, follow the same way, branches [0 + −] and [0 + 0] are deleted from the reasoning tree.

6 Conclusion

In this paper, the spatial solutions of a mechanism were studied by using fuzzy spatial reasoning. The reasoning process is based mostly on incompletely qualitative information. The method based on fuzzy reasoning was introduced for the spatial configuration of mechanisms that could be applied earlier in mechanism design. We present the reasoning structure to reason automatically the qualitative configuration of mechanisms composed of functional parts, based on some fuzzy information vectors or matrices. Sign algebra is used to express these information vectors, and with the special algebra rules, these vectors can be reasoned to further result based on the reasoning structure.

We applied the mechanical reasoning structure to the spatial configuration inside a mechanism, and reasoned statically in the stage of conceptual design. As a result, the static spatial configuration and positional information of the mechanisms are gained qualitatively. We can retain the parameters that we need to the future to be determined with quantitative knowledge, and at the same time describe other parameters qualitatively.

Acknowledgments. This work was supported by Scientific Research Foundation of Zhejiang Forestry University, 2351000848; Zhejiang Science and Technology Project, 2007C21045.

References

1. Chakrabarti, A., Bligh, T.P.: An Approach to Functional Synthesis of Solutions in Mechanical Conceptual Design. Part III: Spatial Configuration. Research in Engineering Design 8(2), 116–124 (1996)
2. Erdman, A.G., Sandor, G.N.: Mechanism Design. Prentice–Hall, New Jersey (1997)
3. Younghyun, H., Kunwoo, L.: Using Sign Algebra for Qualitative Spatial Reasoning about the Configuration of Mechanisms. Computer-Aided Design 34(11), 835–848 (2002)
4. Jiming, L.: Method of Spatial Reasoning Based on Qualitative Trigonometry. Artificial Intelligence 98(1-2), 137–168 (1998)
5. Jochen, R., Bernhard, N.: On the Complexity of Qualitative Spatial Reasoning: a Maximal Tractable Fragment of the Region Connection Calculus. Artificial Intelligence 108(1), 69–123 (1999)
6. Benjamin, K.: Qualitative Reasoning. MIT Press, Cambridge (1994)
7. Fangzhou, B., Lei, Z.: Introduction of Qualitative Simulation. Hefei Publishing Company of University of Sci. & Tech. of China (1998)
8. Oliver, L., Ian, P.: Complete Logics for QSR: A Guide to Plane Mereotopology. International Journal of Visual Languages and computing 9(5), 5–21 (1998)
9. De, K.J., Brown, J.S.: A Qualitative Physics Based on Confluence. Artif. Intell. 59, 7–15 (1983)
10. Lunze, J.: Qualitative Modeling of Linear Dynamical Systems with Quantized State Measurements. Automatica 30(3), 417–431 (1994)
11. Anthony, G.C., Brandon, B., John, G., Nicholas, G.: Qualitative Spatial Representation and Reasoning with the Region Connection Calculus. GeoInformatica 1, 275–316 (1997)
12. Forbus, K.D., Nielsen, P., Faltings, B.: Qualitative Spatial Reasoning: the Clock Project. Artificial Intelligence 51(1-3), 417–471 (1991)
13. Clementini, E., Sharma, J., Egenhofer, M.J.: Modeling Topological Spatial Relations: Strategies for Query Processing. Computers and Graphics 18(6), 815–822 (1994)
14. Han, Y.H., Lee, K.: A Case-based Framework for Reuse Design Concepts in Conceptual Synthesis of Mechanisms. Computers in Industry 57(4), 305–318 (2006)

Extension Robust Control of a Three-Level Converter for High-Speed Railway Tractions

Kuei-Hsiang Chao

National Chin-Yi University of Technology, Department of Electrical Engineering, 35, 215 Lane, Sec. 1, Chung Shan Road, Taiping, Taichung, Taiwan, R.O.C.
chaokh@ncut.edu.tw

Abstract. In this paper, the dynamic model of a three-level neutral point clamped (NPC) converter with neutral-point voltage-balance control (NPVBC) and input power factor correction for high-speed railway tractions is derived. And accordingly, a feedback controller is designed to meet the given voltage regulation control specifications. As the variation of converter parameters occurs, a compensation signal is yielded by an extension robust controller (ERC) to preserve the prescribed response. The compensation signal is adaptively tuned by a model error driven extension weighting controller. Some simulation results are presented to demonstrate the effectiveness of the proposed controller.

Keywords: Three-level converter, power factor correction, extension robust controller, high-speed railway tractions.

1 Introduction

Several numerous three-level converters circuit topologies have been developed [1,2] to reduce the voltage stress of power semiconductors, voltage harmonics, and EMI in medium and high power applications. Although the input current of three-level NPC converters can be regulated to be sinusoidal and maintained almost in phase with the input voltage [3,4], the dynamic modeling and quantitative controller design to obtain well-regulated dc output voltage under good input power factor correction have not get been performed. In addition, it is well known that robust control is one of the most effective techniques for dealing with parameter variations. Although a robust control technique has been applied to many processes [5,6], it is seldom used for control of three-level converters. Moreover, most of the existing robust control techniques are too theoretically complex for use by practical engineers. It follows that good control performance generally can not be achieved.

In this paper, the dynamic model of the proposed three-level converter is firstly derived from the system parameters and measured data for a nominal case by averaged power method and circuit theory. Then, a quantitative design procedure is developed to find the parameters of the voltage controller according to prescribed control specifications. Finally, an extension robust controller is proposed to overcome the control performance degradation due to system parameter and load current changes. The key feature of the proposed ERC is that its weighting factor is adaptively set by an extension

F. Sun et al. (Eds.): ISNN 2008, Part II, LNCS 5264, pp. 227–236, 2008.

weighting controller. In addition, the compromise between control effort and response is considered through tuning the weighting factor automatically. Since the model error is used as the input of the extension controller and the linguistic algorithms for tuning the weighting factor are properly set, more robust and better voltage control performance than those of the conventional robust control [7,8] are obtained by the proposed controller. The performance of the developed quantitative robust controller is demonstrated by some simulation results with the PSIM software package [9].

2 Small Signal Model of Three-Level Converter

The system configuration of the proposed three-level NPC converter with NPVBC and hysteresis current control for high-speed railway tractions is shown in Fig.1. The dynamic model of the proposed three-level NPC converter can be derived at the nominal case as follows [10]:

$$\Delta V_{dc} = \frac{\hat{V}^*_{ac}/(2K_s\bar{V}_{dc})}{(1/2)Cs+1/(Z_a+Z_b)}\Delta\hat{I}^*_{ac} - \frac{1}{(1/2)Cs+1/(Z_a+Z_b)}\Delta i_o \overset{\Delta}{=} (\Delta\hat{I}^*_{ac} - \frac{1}{K_x}\Delta i_o)\frac{\beta}{s+\alpha} \tag{1}$$

where

$$K_x \overset{\Delta}{=} \frac{\hat{V}^*_{ac}}{2K_s\bar{V}_{dc}},\ \alpha \overset{\Delta}{=} \frac{2}{Z_T C},\ \beta \overset{\Delta}{=} \frac{\hat{V}^*_{ac}K_v}{C\bar{V}_{dc}K_s},\ Z_T \overset{\Delta}{=} Z_a + Z_b$$

and K_v is the conversion constant of voltage sensor.

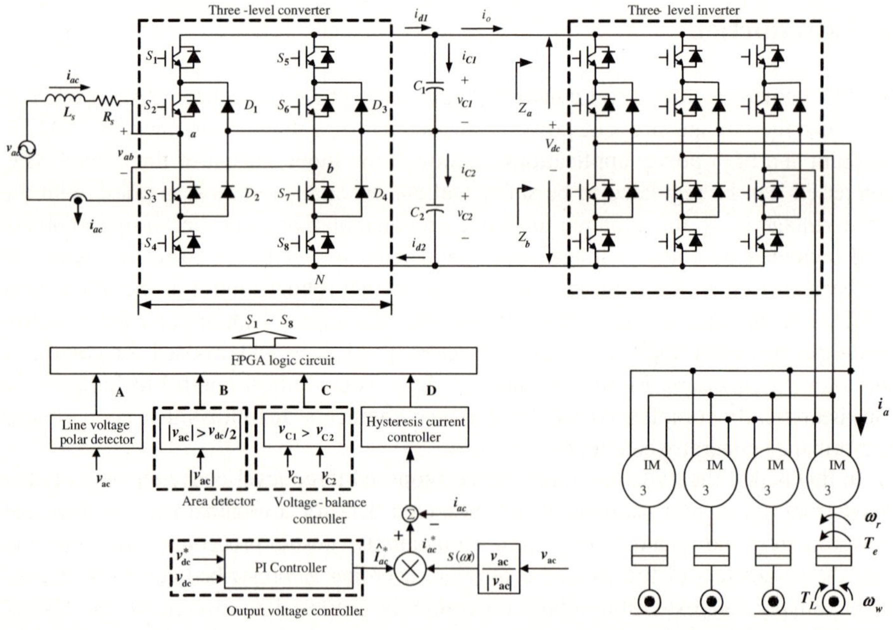

Fig. 1. System configuration of the three-level NPC converter for high-speed railway tractions

The relative parameters of the proposed three-level NPC converter at nominal case are listed in Table 1. Then the parameters of the small-signal model at nominal case can be determined as: $K_x = 73.26$, $\alpha = 12.76$ and $\beta = 26.16$.

Table 1. Some parameters of the proposed three-level NPC converter at nominal case

P_{out}	500kW	$C = C_1 = C_2$	0.01 F
$\hat{V}_{ac}$	2050.6V	K_s	$1/200$
R_s	13.5$m\Omega$	K_v	$1/560$
L_s	1.75mH	$Z_a = Z_b$	7.84Ω
$\overline{V}_{dc}$	2800V	Z_T	15.68Ω

3 Quantitative PI Type Voltage Controller Design

To achieve the desired control requirements with easy implementation, the following PI controller $G_{cv}(s)$ is chosen:

$$G_{cv}(s) = K_{Pv} + \frac{K_{Iv}}{s}. \tag{2}$$

The following control requirements for the response of ΔV_{dc} due to step load current change at nominal case ($\overline{V}_{dc} = 2800V$, $P_{out} = 500kW$) are specified:

(i) Steady state error=0;
(ii) Overshoot=0;
(iii) The maximum voltage dip due to step load current change Δi_o =20A is $\Delta\hat{v}_{dc,\max} = 115V$;
(iv) The restore time is $t_r = 0.3\sec$, which is defined as the time at which $\Delta v_{dc}(t = t_r) = 0.05\hat{v}_{dc,\max}$.

Following the design procedure developed in [10], one can find the parameters of voltage controller $G_{cv}(s)$ as follows:

$$K_{Pv} = 0.58, K_{Iv} = 7.4. \tag{3}$$

4 The Proposed Robust Controller Based on Extension Theory

The robust control technique based on direct cancellation of uncertainties presented in [10] is easy to apply and effective in reducing the effects of system parameter variations. However, since the weighting factor set to determine the extent of disturbance compensation is fixed, it lacks control adaptability. This will lead to the performance degradation and even the stability problem during wide operation range, especially for the system having some kinds of nonlinearities. Before introducing the proposed ERC, the conventional robust control is briefly described.

4.1 Robust Controller with Fixed Weighting Factor

When system configuration and plant parameter variations occur, the PI-type voltage controller designed for the nominal case can no longer satisfy the prescribed control requirements. To overcome this problem, a robust voltage controller based on direct cancellation of uncertainties is proposed in Fig. 2. A model error, denoted by e, is extracted using an inverse nominal plant model $G_I(s) \overset{\Delta}{=} (s+\bar{\alpha})/(\bar{\beta}K_v)$, and then a compensation control signal, $\Delta I = we$, $(0 < w \le 1)$, is generated for disturbance cancellation.

The transfer function of load disturbance Δi_o to output voltage ΔV_{dc} is derived as:

$$\Delta V_{dc} = \frac{\Delta \hat{I}_{ac} - (1-w)/K_x \Delta i_o}{\left[\bar{\alpha} + (1-w)\Delta\alpha\right]s + \left[\bar{\beta} + (1-w)\Delta\beta\right]} K_v. \tag{4}$$

where $\bar{\alpha}, \bar{\beta}$ are plant parameters for the nominal case and $\Delta\alpha, \Delta\beta$ are system uncertainties.

For the ideal case ($w = 1$), one can find from (4) that

$$\Delta V_{dc} = \frac{\Delta \hat{I}_{ac}}{\bar{\alpha}s + \bar{\beta}} K_v. \tag{5}$$

That is, all the load disturbances and uncertainties have completely eliminated by the compensation control signal ΔI. However, this ideal case is practically unrealizable, and so suitable compromise between control performance and operating stability should be made. Hence for obtaining good performance without overshoot and taking into account the maximum control effort, the value w must be regulated automatically.

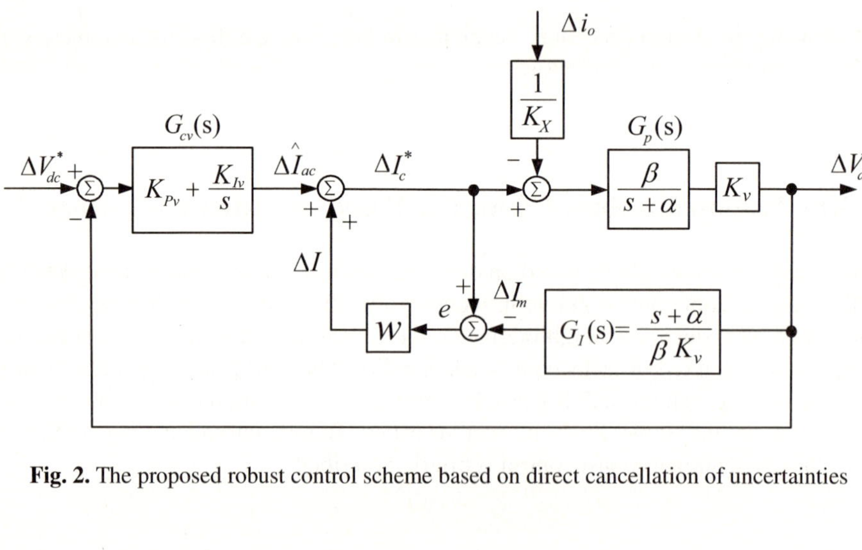

Fig. 2. The proposed robust control scheme based on direct cancellation of uncertainties

4.2 The Proposed Extension Robust Controller

4.2.1 Matter-Element Theory

In extension theory, a matter-element (***R***) contains three fundamental elements: matter name (*N*), matter characteristics (***C***) and matter characteristics of values (***V***). The matter-element can be described as follows [11]:

$$R = (N, C, V) \tag{6}$$

Where ***C*** is a matter characteristic or a characteristic vector, ex: $\boldsymbol{C}=[c_1,c_2,\ldots,c_n]$, and ***V*** the same as ***C*** is a value or a vector, ex: $\boldsymbol{V}=[v_1,v_2,\ldots,v_n]$.

4.2.2 Application of Correlation Function

The correlation functions have many forms dependent on application. If we set $X_o = < k_1, k_2 >$, $X = < n_1, n_2 >$, and $X_o \in X$, then the extended correlation function can be defined as follows [11]:

$$K(x) = \frac{\rho(x, X_o)}{D(x, X_o, X)} \tag{7}$$

If one wants to set K(x) = 1, then

$$\rho(x, X_o) = \left| x - \frac{k_1 + k_2}{2} \right| - \frac{k_2 - k_1}{2} \tag{8}$$

$$D(x, X_o, X) = \begin{cases} \rho(x, X) - \rho(x, X_o) & x \notin X_o \\ -\dfrac{|(k_2 - k_1)|}{2} & x \in X_o \end{cases} \tag{9}$$

where

$$\rho(x, X) = \left| x - \frac{n_1 + n_2}{2} \right| - \frac{n_2 - n_1}{2} \tag{10}$$

The correlation function can be used to calculate the membership grade between x and X_o. The extended correlation function is shown in Fig. 3. When K(x) = 0, this indicates the degrees to which x belongs to X_o. When K(x) < 0 it describes the degree to which x does not belong to X_o. When -1 < K(x) < 0, it is called the extension domain, which means the element x still has a chance to become part of the set if conditions change.

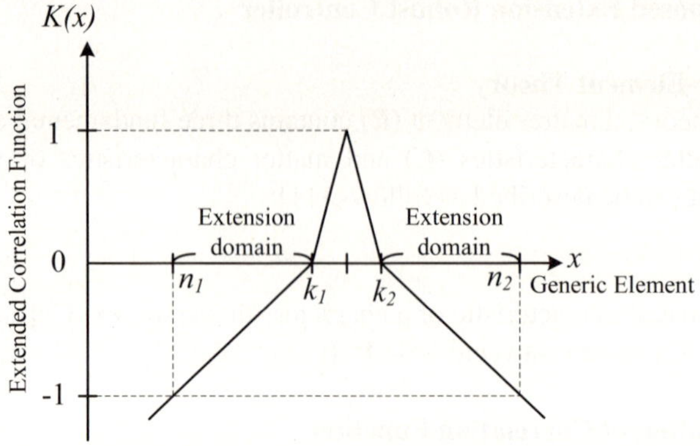

Fig. 3. The extension correlation function

4.2.3 Extension Weighting Controller

To let the robust controller (RC) possess adaptive capability, it is proposed that the weighting factor of the RC is adaptively tuned by the extension error tuning scheme, which is driven by a model error and its change defined as $e(k) \overset{\Delta}{=} \Delta I_c^*(k) - \Delta I_m(k)$ and $e(k) \overset{\Delta}{=} (1-B)e(k) = e(k) - e(k-1)$ with I_m and I_c^* being the output of the inverse model and the plant model input at k-th sampling interval, respectively. The major purpose of the proposed controller is to let the resulted output voltage tracking response closely follow those of reference. Thus the general model error trajectory can be predicted and plotted in Fig. 4. Incorporating with the extension matter-element, the numbers of quantization levels of the input variables $e(k)$ and $\Delta e(k)$ are chosen to be 13 and listed in Table 2 (The scaling is set as 1V to 10A). Based on the experience about the three-level converter to be controlled and the

Table 2. Quantized error and error change

error e(V)	error change Δe(V)	quantized level
$-3.2 < e \le -1.6$	$-3.2 < \Delta e \le -1.6$	-6
$-1.6 < e \le -0.8$	$-1.6 < \Delta e \le -0.8$	-5
$-0.8 < e \le -0.4$	$-0.8 < \Delta e \le -0.4$	-4
$-0.4 < e \le -0.2$	$-0.4 < \Delta e \le -0.2$	-3
$-0.2 < e \le -0.1$	$-0.2 < \Delta e \le -0.1$	-2
$-0.1 < e \le -0.05$	$-0.1 < \Delta e \le -0.05$	-1
$-0.05 < e \le 0.05$	$-0.05 < \Delta e \le 0.05$	0
$0.05 < e \le 0.1$	$0.05 < \Delta e \le 0.1$	1
$0.1 < e \le 0.2$	$0.1 < \Delta e \le 0.2$	2
$0.2 < e \le 0.4$	$0.2 < \Delta e \le 0.4$	3
$0.4 < e \le 0.8$	$0.4 < \Delta e \le 0.8$	4
$0.8 < e \le 1.6$	$0.8 < \Delta e \le 1.6$	5
$1.6 < e \le 3.2$	$1.6 < \Delta e \le 3.2$	6

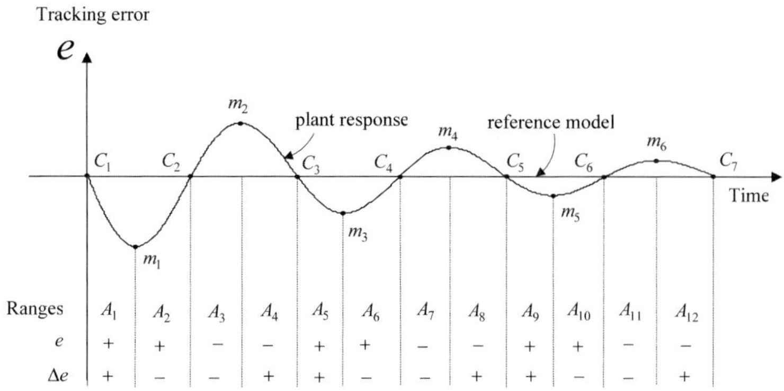

Fig. 4. General model reference tracking error dynamic behavior

Table 3. The decision weight of the proposed extension robust controller

Δe \ e (w_Δ)	-6	-5	-4	-3	-2	-1	0	1	2	3	4	5	6
-6	1	1	1	1	1	1	1	-5/6	-4/6	-3/6	-2/6	-1/6	0
-5	1	1	1	1	1	1	-5/6	-4/6	-3/6	-2/6	-1/6	0	1/6
-4	1	1	1	1	1	-5/6	-4/6	-3/6	-2/6	-1/6	0	1/6	2/6
-3	1	1	1	1	-5/6	-4/6	-3/6	-2/6	-1/6	0	1/6	2/6	3/6
-2	1	1	1	-5/6	-4/6	-3/6	-2/6	-1/6	0	1/6	2/6	3/6	4/6
-1	1	1	-5/6	-4/6	-3/6	-2/6	-1/6	0	1/6	2/6	3/6	4/6	5/6
0	1	-5/6	-4/6	-3/6	-2/6	-1/6	0	1/6	2/6	3/6	4/6	5/6	1
1	-5/6	-4/6	-3/6	-2/6	-1/6	0	1/6	2/6	3/6	4/6	5/6	1	1
2	-4/6	-3/6	-2/6	-1/6	0	1/6	2/6	3/6	4/6	5/6	1	1	1
3	-3/6	-2/6	-1/6	0	1/6	2/6	3/6	4/6	5/6	1	1	1	1
4	-2/6	-1/6	0	1/6	2/6	3/6	4/6	5/6	1	1	1	1	1
5	-1/6	0	1/6	2/6	3/6	4/6	5/6	1	1	1	1	1	1
6	0	1/6	2/6	3/6	4/6	5/6	1	1	1	1	1	1	1

properties of dynamic signal analyses made in [10], the linguistic rules of the extension error tuning scheme are decided and listed in the Table 3.

According to the measured model error and error change of the three-level converter, the matter-elements have been summarized in Table 2. The value ranges <k1, k2> of classical regions for each characteristic are assigned by the lower and upper boundary of model errors and error changes. In addition, one can set a matter-element model to express the neighborhood domain of every characteristic for describing the possible range of all model errors and error changes. The value range < n1, n2 > of the neighborhood domain could be determined from the maximum and minimum values of every characteristic in the measured records. For the controlled converter, it can be represented as:

$$R_s = (N_s, C_s, V_s) = \begin{bmatrix} N_s, & c_1, & <-3.2,\ 3.2> \\ & c_2, & <-3.2,\ 3.2> \end{bmatrix} \tag{11}$$

where matter name (N_s) is three-level converter, matter characteristics c_1 and c_2 represent the model error and error change, respectively.

The process of the proposed control method is shown below:

Step 1) Establish the matter-element model of model error and error changes category, which is performed as follows:

$$R_j = (N_j, C_j, V_j) = \begin{bmatrix} N_j, & c_1, & V_{j1} \\ & c_2, & V_{j2} \end{bmatrix} \quad j = 1, 2, \ldots, 13 \tag{12}$$

where $V_{jk} = \langle a_{jk}, b_{jk} \rangle$ is the classical region of every characteristic sets. In this paper, the classical region of each matter-element is assigned by the maximum and minimum values of model error and model error change at any instant.

Step 2) Set the matter-element of the input model error and error change as (13):

$$R_{new} = \begin{bmatrix} N_{new}, & c_1, & V_{new1} \\ & c_2, & V_{new2} \end{bmatrix} \tag{13}$$

Step 3) Calculate the correlation degrees of the input model errors and error changes with the characteristic of each matter-element by the proposed extended correlation function as follows:

$$K(v_{new,k}) = \frac{\rho(v_{new,k}, V_j)}{D(v_{new,k}, V_j, V_s)}, \quad k = 1, 2 \tag{14}$$

Step 4) Assign weights to the matter characteristic such as W_{j1}, W_{j2} denoting the significance of every characteristic. In this paper, W_{j1}, W_{j2} are set as $W_{j1} = W_{j2} = 0.5$.

Step 5) Calculate the correlation degrees of every category:

$$\lambda_j = \sum_{k=1}^{2} W_{jk} K_{jk}, \quad (j = 1,\ 2,\ \ldots,13) \tag{15}$$

Step 6) Select the maximum value from the normal correlation degrees to recognize the reference range of the input model error and error change and determine the weighting factor w_Δ from Table 3. To increase the sensitivity and adaptive capability, the weighting factor w_Δ of the extension robust controller at the instant is determined as follows:

$$w = w_\Delta * \lambda_j . \tag{16}$$

5 Simulation Results

In order to demonstrate the effectiveness of the proposed quantitative designed voltage controller ($K_{Pv} = 0.58$, $K_{Iv} = 7.4$) for the proposed three-level NPC converter, some simulations are made using the PSIM software package. The simulated voltage response due to step load current change Δi_o =20A by the quantitative designed PI controller at nominal case is shown in Fig. 5. It can be seen from the results that the given specifications $\hat{v}_{dc,\max} = 115V$, $t_r = 0.3$ sec are fully satisfied. For comparison, the simulated dynamic output voltage responses of PI controller without and with the proposed extension robust controllers under the load current changes Δi_o =30A are shown in Fig. 6. The results clearly show that better control performance is obtained by adding the proposed extension robust controller when load current change occurs.

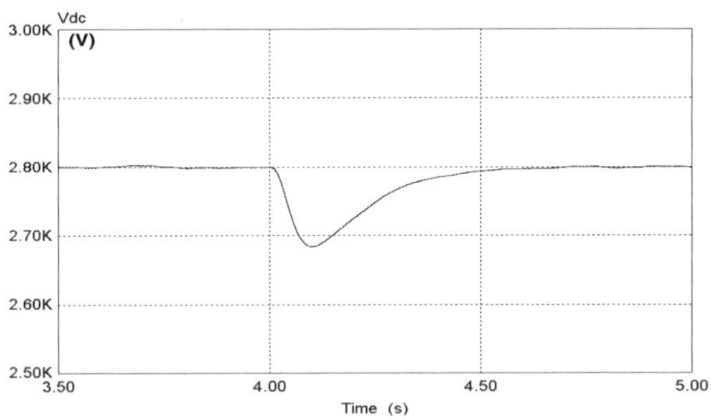

Fig. 5. The simulated result of output voltage Δv_{dc} due to step load current change of $\Delta i_o = 20A$ with the proposed PI-type voltage controller

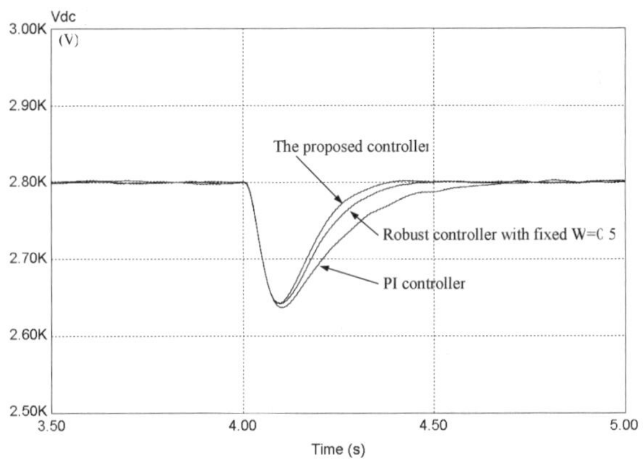

Fig. 6. The simulated result of output voltage Δv_{dc} due to step load current change of $\Delta i_o = 30A$ with PI-type, robust controller with fixed weighting factor $w = 05$ and the proposed extension robust controller

6 Conclusions

An extension robust controller for a three-level converter considering the parameter variation is proposed. First, the dynamic modeling and quantitative design of an output voltage controller for a three-level NPC converter have been presented. Voltage regulation performance can be achieved according to the prescribed specifications. In addition, the dynamic responses of the proposed three-level NPC converter are insensitive to operating conditions and parameter changes, as the PI controller is augmented with the extension robust controller. The simulation results indicate that good control performance in load regulation are achieved.

Acknowledgment. This work was supported by the National Science Council, Taiwan, Republic of China, under the Grant NSC 94-2213-E-167-016.

References

1. Lai, J.S., Peng, F.Z.: Multilevel converters - a new breed of power converters. IEEE Trans. Industry Applications 21, 509–517 (1996)
2. Osawa, C., Matsumoto, T., Mizukami, T., Ozaki, S.: A state-space modeling and a neutral point voltage control for an NPC power converter. In: IEEE Power Conversion Conf., pp. 225–230. IEEE Press, New York (1997)
3. Bendre, A., Venkataramanan, G.: Modeling and design of a neutral point regulator for a three level diode clamped rectifier. In: IEEE Industry Applications Conf., pp. 1758–1765. IEEE Press, New York (2003)
4. Lin, B.R., Chen, D.J.: Power factor correction based on diode clamped rectifier. International Journal of Electronics 88, 595–614 (2001)
5. Hsia, T.C.: A new technique for robust control of servo systems. IEEE Trans. Industrial Electronics 36, 1–7 (1989)
6. Iftar, A., Ozgune, U.: Techniques in modeling uncertain dynamics for robust control system design. Control Dynamic System 50, 255–296 (1992)
7. Liaw, C.M., Lin, F.J.: Control of indirect field-oriented induction motor drives considering the effects of dead-time and parameter variations. IEEE Trans. Ind. Electron 40, 486–495 (1993)
8. Hong, K., Nam, K.: A load torque compensation scheme under the speed measurement delay. IEEE Trans. Ind. Electron 45, 283–290 (1998)
9. PSIM User's Guide: Powersim Inc. (2001-2003)
10. Chao, K.H., Chen, P.Y., Cheng, C.H.: A three-level converter with output voltage control for high-speed railway tractions. In: 33rd IEEE Annual Conference on Industrial Electronics, pp. 1793–1798. IEEE Press, New York (2007)
11. Cai, W.: Extension set and incompatible problems. Science Exploration 3(1), 83–97 (1983)

Blind Image Watermark Analysis Using Feature Fusion and Neural Network Classifier

Wei Lu[1], Wei Sun[2], and Hongtao Lu[3]

[1] School of Information Science and Technology and Guangdong Key Laboratory of Information Security Technology, Sun Yat-sen University, Guangzhou 510275, China
[2] School of Software and Guangdong Key Laboratory of Information Security Technology, Sun Yat-sen University, Guangzhou 510275, China
[3] Department of Computer Science and Engineering, Shanghai Jiao Tong University, Shanghai 200240, China
{luwei3,sunwei}@mail.sysu.edu.cn, lu-ht@cs.sjtu.edu.cn

Abstract. Over the past two decades, great efforts have been made to develop digital watermarking techniques for multimedia copyright protection and authentication. However, most of watermark detection methods are designed based on the corresponding specific watermark embedding procedures. In this paper, we propose a general blind watermarking analysis scheme to recognize whether images are watermarked no matter what kind of watermark embedding schemes are used. In the proposed method, multiscale feature fusion are used to construct statistical characteristics between non-watermarked images and watermarked images. Then, RBF neural networks are used to classify these characteristics. Numerical simulations show that the proposed scheme describes intrinsic statistical characteristics and the proposed blind watermark analysis method is effective.

Keywords: Digital watermark analysis, Feature fusion, Neural network classifier.

1 Introduction

Steganography, as a field of information hiding, focuses on establishing a secret communication channel for transferring hidden information. Digital watermarking, as a branch of steganography, aims to develop effective methods for protecting digital copyright and data authentication. Generally speaking, digital watermarking involves two aspects. One is to design a good watermark embedding scheme, which should achieve good trade-off between the quality of cover data and the robustness of watermarks. The other is to design watermark detection scheme. A limitation here is that most existing detection methods are based on the specific watermarking embedding process. In most cases, watermark detection is simply partial repetition of the embedding process, and this system is called symmetric watermarking. In this paper we presents a new watermarking

F. Sun et al. (Eds.): ISNN 2008, Part II, LNCS 5264, pp. 237–242, 2008.

method, which can recognize non-watermarked images and watermarked images without any other assistant process, including watermark embedding process. which is a blind watermark analysis.

In recent years, blind watermark analysis has been studied [1,2,3,4,5]. In [3], Lie and Lin proposed a steganalysis method based on a feature classification technique that analyzes two statistical properties in the spatial and discrete cosine transform (DCT) domains, then they use a nonlinear neural classifier in class separation. In [4], a universal approach to steganalysis is developed for detecting the presence of hidden messages embedded within digital images. The experiments there demonstrate that the scheme was effective.

In this paper, we propose a blind watermark analysis scheme based on multiscale feature fusion and RBF neural network classifier. The rest of this paper is organized as follows. In section 2, we discuss where watermarks are inserted in most watermarking applications. In section 3, we describe multiscale feature fusion based on DWT in detail. In Section 4, we introduced feature dimension reduction and RBF neural networks classifier. The experimental results are given in section 5. Finally, conclusions are given in section 6.

2 Where Watermarks Are

In order to expose the intrinsic characteristics of non-watermarked images and watermarked images, first of all, we need to know where watermarks are in cover images, i.e., what is the difference between non-watermarked images and their watermarked versions. In Fig. 1, we show a test image and its watermarked version. Here we make use of the classical DCT based watermarking scheme proposed by Cox [6] to produce the watermarked image. We can see that the fidelity is very good for the watermarked version compared with the cover image. For this scheme, watermarks are general embedded in the middle frequency range of the cover images, which can achieve better trade off between the fidelity and the robustness. Fig. 1 also shows the pixel difference between the cover image and the watermarked image, and its 3 level 2D DWT. We can see that large coefficients in DWT are covered in boundary of the texture and flat domain. So, we think that the range of the middle frequency carries more watermark information.

Two dimensional discrete wavelet transformation (2D-DWT) is one of the most useful tools in many image processing applications, such as image compression, image coding and image analysis. Given an image I, 2D-DWT splits the spatial and frequency domains into 3 levels and orientations, denoted as $H_i(x,y)$, $V_i(x,y)$ and $D_i(x,y)$, where (x,y) is the two dimensional coordinate, and $i = 1, 2, \cdots, l$ denotes the decomposition level. Through 2D-DWT, an image can be analyzed in different detail levels composed of frequency and spatial domains. We think DWT is a better analyzing tool to expose the intrinsic details in images, and some DWT coefficients can convey the information for blind watermark analyzable features.

Fig. 1. The example image F16 (first) and its watermarked version (second). Their difference image (third) and its 3 level 2D DWT (forth).

3 Multiscale Feature Fusion

From an image analysis point of view, the most intrinsic characteristics for revealing the differences between non-watermarked images and watermarked are perhaps local one, such as noise distribution, etc. Hence, we consider that well constructed local statistics can depict the unique features of non-watermarked images and watermarked images, since they can describe the basic feature correlations. Without loss of generality, consider the horizontal band $H_i(x, y)$, $i = 1, 2, \cdots, l$, we construct the statistical relation of the neighborhood characteristics as follows:

$$\begin{bmatrix} H_i(x,y) & H_i(x,y+1) \\ H_i(x+1,y) & H_i(x+1,y+1) \end{bmatrix} = \begin{bmatrix} H_{i-1}(2x-1,2y-1) & H_{i-1}(2x-1,2y) \\ H_{i-1}(2x,2y-1) & H_{i-1}(2x,2y) \end{bmatrix} T_i(x,y). \quad (1)$$

Through resolving the matrix equation, we get the solution $T_i(x, y)$ as a 2×2 matrix. Based on the property of DWT, if there is a large coefficient $H_i(x, y)$ at scale i, it is more likely that it is also large for $H_{i-1}(2x-1, 2y-1)$ at scale $i+1$. Thus, the two matrix

$$\begin{bmatrix} H_i(x,y) & H_i(x,y+1) \\ H_i(x+1,y) & H_i(x+1,y+1) \end{bmatrix} \text{ and } \begin{bmatrix} H_{i-1}(2x-1,2y-1) & H_{i-1}(2x-1,2y) \\ H_{i-1}(2x,2y-1) & H_{i-1}(2x,2y) \end{bmatrix}$$

are more likely similar, and $T_i(x, y)$ expresses the similarity of wavelet coefficients between neighborhood scales. Then, we update the coefficient $H_i(x, y)$ as follows:

$$H_i(x,y) \longleftarrow \frac{H_i(x,y) \cdot |\det(T_i(x,y))|}{1 - \dfrac{\sigma(H_i(x,y))}{\sigma(H_{i-1}(2x-1,2y-1))}}, \quad (2)$$

where $\det(\cdot)$ denotes the determinant of matrix, and $\sigma(\cdot)$ denotes the variance in a 3×3 neighborhood. In Eq. (2), the denominator item $|\det(T_i(x, y))|$ imports $(i-1)$-th scale detail coefficient feature into i-th scale, and the numerator item

$$\frac{\sigma(H_i(x,y))}{\sigma(H_{i-1}(2x-1,2y-1))}$$

imports the neighborhood texture or edge features from $(i-1)$-th scale to i-th scale. Thus, we obtained the fused feature coefficients at scale i. Through repeating Eq. (2) from the decomposition scale 2 to l, we update the coefficients through fusing the lower scale coefficients into the higher scale coefficients, i.e., $H_1, H_2, \cdots, H_l$. Thus we obtain the final updated l-th scale coefficients H_l. Then, though repeating the same operation in the vertical and diagonal bands V_i and D_i, $i = 1, 2, \cdots, l$, we can get three fused feature matrices H_l, V_l and D_l.

4 Feature Dimension Reduction and Classification

Dimension reduction is a very common technique in pattern recognition, since when the feature dimension increasing, the computation complexity is unavoidable, that is "Dimension Curse". In our methods, we constructed three fused feature matrix. If we use them as the features directly, the dimension will be very large. In order to avoid this problem, dimension reduction is applied before classification. Here we use principal components analysis (PCA). Firstly, the three fused feature matrices H_l, V_l and D_l are spread into three column vectors in zigzag order, and then joint them to a long column vector end to end. Thus, for each image, a column vector is constructed as the initial features. For all the training images, we can obtain a matrix by arranging their initial features column by column, and thus form a matrix F with size $m \times n$, where m and n denote the length of initial features and the quantity of training images respectively. Through PCA, the dimension of F is reduced to $r \times n$, where $r < n$.

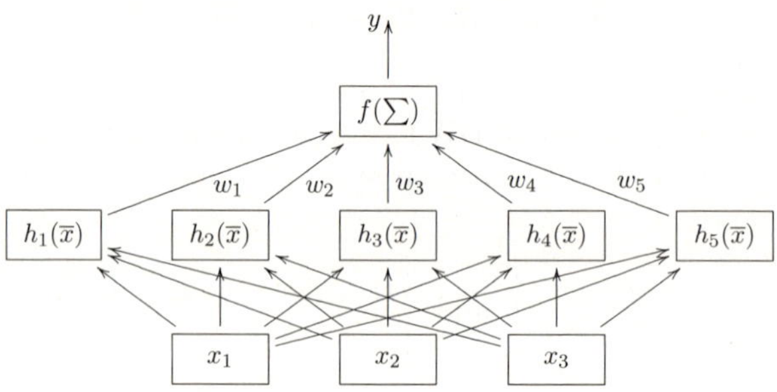

Fig. 2. The 3-5-1 structure of RBF neural network

The extracted features are not proper for the classification and detection, although they contain many information for recognized objects. Here, we use a 3-layer RBF neural network as the classifier to detect watermarked images, as it recombines the features using trained nonlinear mapping before classification. We use the reduced features above using PCA as the input features. To the training stage, if the tested image is watermarked, the output is set to -1, and if the tested image is not watermarked, the output is set to 1. Then, to the classification, if the output of the neural network classifier is less than 0, then the test image is labeled as a watermarked one, otherwise a un-watermarked one.

Table 1. Classification accuracy (percent) using RBF neural network classifier

database	training	testing
un-watermarked images	51.2	46.3
watermarked images	96.6	95.4

Table 2. Classification accuracy (percent) for the proposed schemes using RBF neural network classifier under random assignments

database	training	testing
un-watermarked images	32.1	29.4
watermarked images	96.3	93.8

5 Simulations and Discussions

In our experiments, a image database of 1000 images and 1000 watermarked images is used to train and test the proposed scheme, where half of these images are used to train the RBF neural network classifier, and the others for test. Two examples taken from the database are shown in Fig. 1. The decomposition parameter l is set to 3. Table 1 shows the classification accuracy using the proposed classifier. Under the false negative rate 3.4%, about 51.2% watermarked images are classified correctly, where false negative refers that a un-watermarked image is classified as a watermarked image. Note that the testing accuracy is close to the training accuracy, which shows that the classifiers are general.

In Fig. 3, the ROC curves between the false positive rate and the true positive rate for the three classifiers are shown. Again, the false positive rate is the percentage of un-watermarked images that are incorrectly detected as watermarked images, and the true positive rate is the percentage of watermarked images that

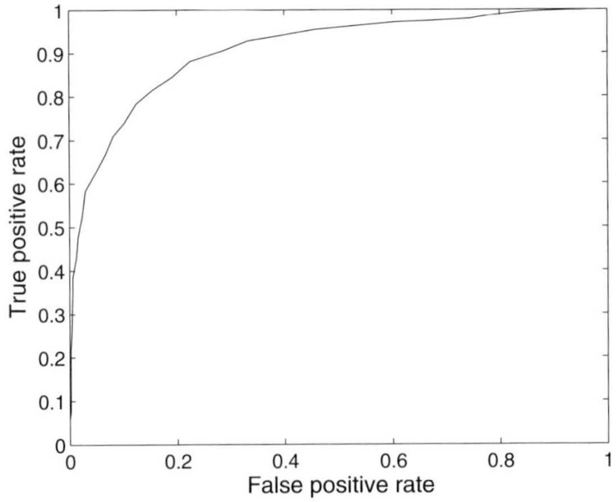

Fig. 3. The ROC curves between the false positive rate and the true positive rate

are correctly classified as watermarked images. It can be concluded that the performances is good for the proposed classification scheme based on multiscale feature fusion and neural network classification.

We also assigned the training images with random outputs of 1 and -1, where half of the images are randomly assigned to the watermarked images and the others are the un-watermarked images. Then we trained the RBF neural network classifier using the data set and then tested it. We found that the classification accuracy is 29.4% for the un-watermarked images with the false negative rate 6.2%, which is worse than the case when the correct outputs are assigned. This indicates that the proposed features and RBF neural network classifiers are based on rational fused features for un-watermarked and watermarked images.

6 Conclusions

Blind watermark analysis has not been research widely in recent years. In this paper, we have proposed a blind digital image watermarking analysis scheme using multiscale feature fusion in DWT domain, which uses RBF neural networks to classify the watermarked and un-watermarked images. Simulation results demonstrate that the proposed detection scheme is effective.

Acknowledgments

This work is supported by the Scientific Research Foundation for the Young Teachers in Sun Yat-sen University, NSFC under project no. 60573033, and Program for New Century Excellent Talents in University (no. NCET-05-0397).

References

1. Dumitrescu, S., Wu, X.: A New Framework of LSB Steganalysis of Digital Media. IEEE Trans. Signal Processing 53, 3936–3947 (2005)
2. Ker, A.D.: Steganalysis of LSB Matching in Grayscale Images. IEEE Signal Processing Letters 12, 441–444 (2005)
3. Lie, W.N., Lin, G.S.: A Feature-based Classification Technique for Blind Image Steganalysis. IEEE Trans. Multimedia 7, 1007–1020 (2005)
4. Lyu, S., Farid, H.: Steganalysis using Higher-Order Image Statistics. IEEE Trans. Information Forensics and Security 1, 111–119 (2006)
5. Lyu, S., Farid, H.: Steganalysis using Color Wavelet Statistics and One-Class Support Vector Machines. In: Proc. SPIE, San Jose, CA, vol. 5306, pp. 35–45 (2004)
6. Cox, I.J., Kilian, J., Leighton, F.T., Shamoon, T.: Secure Spread Spectrum Watermarking for Multimedia. IEEE Trans. Image Processing 6, 1673–1687 (1997)

Gene Expression Data Classification Using Independent Variable Group Analysis

Chunhou Zheng[1,2], Lei Zhang[3,*], Bo Li[2], and Min Xu[1]

[1] College of Information and Communication Technology, Qufu Normal University, Rizhao, Shandong, 276826 China
Zhengch99@126.com

[2] Intelligent Computing Lab, Institute of Intelligent Machines, Chinese Academy of Sciences, Hefei, Anhui 230031, China

[3] Biometric Research Center, Dept. of Computing, The Hong Kong Polytechnic University, Hong Kong, China
cslzhang@comp.polyu.edu.hk

Abstract. Microarrays are capable of detecting the expression levels of thousands of genes simultaneously. In this paper, a new method for gene selection based on independent variable group analysis is proposed. In this method, we first used *t*-statistics method to select a part of genes from the original data. Then we selected the key genes from the selected genes by *t*-statistics for tumor classification using IVGA. Finally, we used SVM to classify tumors based on the key genes selected using IVGA. To validate the efficiency, the proposed method is applied to classify three different DNA microarray data sets. The prediction results show that our method is efficient and feasible.

Keywords: Gene expression data, Independent variable group analysis, Gene selection, Classification.

1 Introduction

With the wealth of gene expression data from microarrays, many information processing techniques, such as prediction, classification and clustering are used to analyze and interpret the data. The analysis of gene expression data can be motivated by the problem of distinguishing between cancer classes or identifying and discovering various subclasses of cancers [2,5,6,15]. For this problem, there are two distinct methods, supervised and unsupervised classification respectively, which can be addressed by discriminant analysis and clustering techniques.

In statistical terms, the very large number of predictors or variables (genes) compared to small number of samples or observations (experimnets) make most of classical "class prediction" methods unemployable. Fortunately, this problem can be avoided by selecting only the relevant features or extracting new features containing the maximal information about the class label from the original data. The former methodology is called feature selection or subset selection, while the latter is named

[*] Corresponding author.

F. Sun et al. (Eds.): ISNN 2008, Part II, LNCS 5264, pp. 243–252, 2008.

feature extraction. In this paper, we will focus on the feature selection (gene selection) problem.Generally, gene selection is treated as a variable selection problem in statistics and a dimension reduction problem in machine learning. For gene selection, there are a vast amount of literatures focused on how to use it for tumor classification[3,4,10,14]. Although being useful in practice, all these methods that select important genes based on individual gene information thus fail to take into account mutual information among genes.

Independent variable group analysis (IVGA) [9] is a principle for grouping variables that are mutually dependent together so that independent or only weakly dependent variables are placed to different groups. In this paper, we proposed a new method for gene selection using independent variable group analysis. Other than the feature selection method proposed in the literature [1], we first used t-statistics method to select a part of genes from the original data. Then we chose the independent key genes using IVGA from the selected genes for tumor classification. Finally, we used SVM to classify tumors based on the key genes chose by IVGA. To validate the efficiency, the proposed method is applied to classify two different DNA microarray data sets including colon cancer data [2], and acute leukemia data [7]. The prediction results show that our method is efficient and feasible.

2 Independent Variable Group Analysis

2.1 Principle of Independent Variable Group Analysis

In more conventional terms, IVGA can be regarded as a clustering method where samples are taken as random variables and the criterion is to minimize the mutual information between the groups. A similar criterion has been used for hierarchical clustering in [8]. What should be emphasized is that IVGA should be regarded as a principle, not an algorithm. The IVGA principle is depicted in Fig. 1. In fact, the IVGA can be viewed in many different ways. Firstly, it can be regarded as a method

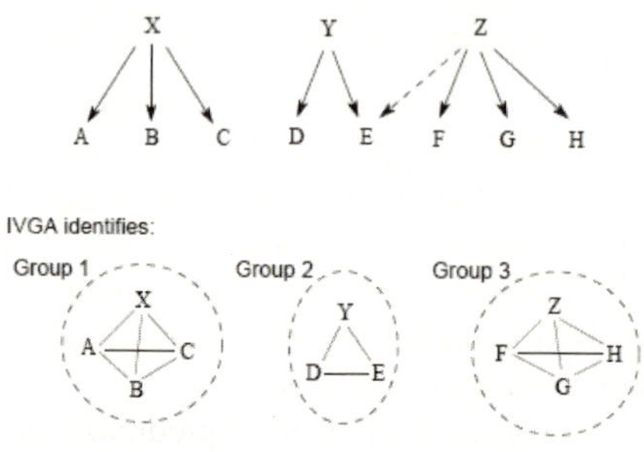

Fig. 1. An illustration of the IVGA principle. The upper part of the figure shows the actual dependencies between the observed variables. The arrows that connect variables indicate causal dependencies. The lower part depicts the variable groups that IVGA might find here. One actual dependency is left unmodeled, namely the one between Z and E. Note that the IVGA does not reveal causalities, but dependencies between the variables only [1].

for finding compact representation of data using multiple independent models. Secondly, IVGA can also be regarded as a method of clustering variables. Note that it is not equivalent to taking the transpose of the data matrix and performing ordinary clustering, since dependent variables need not be close to each other in the Euclidean or any other common metric. Thirdly, IVGA can also be used as a dimensionality reduction or feature selection method [1].

2.2 Algorithm for IVGA

As described above, the goal of the IVGA algorithm is to find such a variable grouping and such models for the groups that the total cost over all the models is minimized. In this method, a natural criterion for solving the IVGA problem is to minimize the mutual information or in general multi-information, within the grouping evaluated by considering each group a separate random variable. The actual objective function for IVGA can be derived by the following process: Assuming that the data set D consists of vectors, $\mathbf{x}(t)$, t = 1, . . . , T . The vectors are N-dimensional with the individual components denoted by x_j , j = 1, . . . , N, and all observed x_j by $X_j = (x_j(1), \cdots, x_j(T))$. The aim here is to find a partition of {1, . . . ,N} to M disjoint sets $\Re = \{\Re_i \mid i = 1, \cdots M\}$ such that the mutual information

$$I_{\Re}(\mathbf{x}) = \sum_i H(\{x_j \mid j \in \Re_i\}) - H(\mathbf{x}), \tag{1}$$

between the sets is minimized. In case, $M > 2$, this is actually a generalization of mutual information commonly known as multi-information [13]. As the entropy $H(x)$ is constant, this can be achieved by minimizing the first sum. The entropies of that sum can be approximated through

$$\begin{aligned} H(x) &= -\int p(x) \log p(x) dx \approx -\frac{1}{T}\sum_{t=1}^{T} \log p(x(t)) \\ &\approx -\frac{1}{T}\sum_{t=1}^{T} \log p(x(t) \mid x(1), \cdots x(t-1), \Phi) \\ &= -\frac{1}{T} \log p(D \mid \Phi), \end{aligned} \tag{2}$$

where Φ denotes the model of the data.

Two approximations were made in this derivation. First, the expectation over the data distribution was replaced by a discrete sum using the data set as a sample of points from the distribution. Next, the data distribution was replaced by the posterior predictive distribution of the data sample given the past observations. The sequential approximation is necessary to avoid the bias caused by using the same data twice, both for sampling and for fitting the model for the same point. A somewhat similar approximation based on using the probability density estimate implied by a model has been applied for evaluating mutual information also in [11].

Using the result of Eq. (2), minimizing the criterion of Eq. (1) is equivalent to maximizing

$$L = \sum_i \log p(\{D_i \mid j \in \Re_i\} \mid \Phi_i). \tag{3}$$

This reduces the problem to a standard Bayesian model selection problem. The two problems are, however, not exactly equivalent. The mutual information cost in Eq. (1) is always minimized when all the variables are in a single group, or multiple statistically independent groups. In case of the Bayesian formulation in Eq. (3), the global minimum may actually be reached for a nontrivial grouping even if the variables are not exactly independent. This allows to determine a suitable number of groups even in realistic situations when there are weak residual dependencies between the groups. More details for the algorithm can be found in [1].

3 Gene Selection through IVGA

In literature [1], the authors proposed a method using IVGA to select features for classification. In this method, the variables which grouped together with the class variable were selected out, and only some of them were used in the actual classifier [1]. In this study, we have directly used this method for tumor classification, yet the experimental result shows that the accuracy is not steady. In other words, we may get high accuracy when using it to classify one tumor data set. Also, we may get bad results. The detailed results for this method can be found in Results Section of this paper.

3.1 IVGA Based Gene Selection

In this paper, we proposed another method to select key genes for tumor classification. We first used IVGA to group the genes. In other words, we cluster the genes using the IVGA algorithm. Because the statistical dependencies of the genes within every group are strong, so the classification information contained in the genes of one group is redundant. On the contrary, the information contained in the genes of different groups should be complementary, since the gene expression profiles in different group are independent. According to the analysis above, we select one gene from every group to form the features for tumor classification. Since every group may contain several genes, then which gene should be selected out for classification from every group is another problem to be solved. We will give out the detail method in Preliminary gene selection in next section.

3.2 Preliminary Gene Selection

In our method, we will select one gene from each cluster, so we must figure out which gene, i.e. the key gene, should be chosen from the cluster. To solve this problem, we first rank all of the genes before they are grouped using IVGA. Then, after clustering, the genes in every cluster are also ranked. Finally we can simply choose the first gene of every cluster as the key gene.

For two-class prediction problem, the ranking of the genes can be made based on the simple t-statistics introduced by Golub et al. [7], as also used in Furey et al. [6]:

$$f(g_j) = \left| \frac{\mu_j^1 - \mu_j^2}{\sigma_j^1 + \sigma_j^2} \right|. \tag{4}$$

This method allows to find the individual gene expression profiles that help to discriminate between two classes by calculating for each gene expression profile $\mathbf{g}_j$ a score based on the mean μ_j^1 (respectively μ_j^2) and the standard deviation σ_j^1 (respectively σ_j^2) of each class of samples. In this study, considering the power of our computer and the quantity of the gene expression data, we first ranked the genes by their scores and retained a set of the top 200, and 500 genes of the two data sets for IVGA, respectively.

4 Results

To test the effectiveness of the proposed methodology, in this section, we applied it to classify two data sets with various human tumor samples, which are colon cancer data [2], and acute leukemia data [7] respectively. In these datasets, all data samples have already been assigned to a training set or test set.

In this study, we used the SVM with RBF kernel as the classifier. Since building a prediction model requires good generalization towards making predictions for previously unseen test samples, tuning the parameters is an important issue. So it requires the optimization of the regularization parameter as well as the kernel parameter of SVM. This is done by searching a two dimensional grid of different values for both parameters. Moreover, the small sample size characterizing microarray data restricts the choice of an estimator for the generalization performance. To solve these problems, the optimization criterion used in this study is the leave-one-out cross-validation (LOO-CV) performance. Though Ambroise et al [3] had shown that LOO-CV may bring in selection bias when used for gene selection, yet considering that we only using LOO-CV to optimize the parameter of SVM and the case of litter samples of tumor, we used it as Pochet et al [12] had done. In each LOO-CV iteration (the number of iterations equals the sample size), one sample is left out of the data, a classification model is trained on the rest of the data and this model is then evaluated on the left out data point. As an evaluation measure, the LOO-CV performance [(No. of correctly classified samples)/(No. of samples in the data) · 100]% is used. The value of the regularization parameter corresponding to the largest LOO-CV performance is then selected as the optimal value.

To obtain reliable results allowing comparability and repeatability of the different numerical experiments, this study does not use the original division of each data set in training and test set, but reshuffles (randomizes) all data sets. Consequently, all numerical experiments were performed with 20 randomizations of the original data set. In other words, we considered an equal random splitting of all of the N samples: N/2 training and N/2 test samples, e.g., for Colon cancer data set, N =62, and the cancerous (and noncancerous) samples in training and test set are also equivalent.

For the results, two kinds of measures are used. The first one is the LOO-CV performance. This is estimated by only making use of the training data sets for tuning the parameters. The second measure is the accuracy, which gives an idea of the classification performance by reflecting the percentage of correctly classified samples. When measured on independent test sets, this gives an idea of the generalization performance. But when measured on the training set, one can get an idea of the degree of overfitting.

4.1 Colon Cancer Data Set

Colon adenocarcinoma tissues were collected from patients and from some of these patients, paired normal colon tissue also was obtained [2]. Gene expression in 40 tumor and 22 normal colon tissue samples was analyzed with an Affymetrix oligonucleotide array complementary to more than 6500 human genes. The data set contains the expression of the 2000 genes with highest minimal intensity across the 62 tissues.

The training set consists of 40 colon tissues, of which 14 are normal and 26 tumor samples. The test set consists of 22 tissues, of which 8 are normal and 14 tumor samples. The number of gene expression levels is 2000. The goal here is how to classify the tissues as being cancerous or noncancerous.

In this experiment, we first rank the genes by the t-statistics method listed above using the original training data, and retained a set of the top 200 genes. Then we group the 200 genes using IVGA. Since every gene has been ranked, the genes in every group are also ranked. Finaly, we selected the first gene from every cluster, which forms the key genes. During the experiment, we also found that the key genes are not coincident in different IVGA runs. To solve this problem, we run IVGA several times (say, 100 times) and rank the key genes selected from all of the several times according to their frequencies of appearance. In the end, the first P key genes are selected for tumor classification.

We choose P= 3, 5, 10, 15, 20, 40, 100, 200, respectively. The detailed results are listed in Table 1. The results of all numerical experiments in the table represent the mean and standard deviation of the results on 20 randomized splitting of the original data. From Table 1 it can be seen that the classification results of our method (denoted by IVGA_I) is fare well. To farther show the advantage of our gene selection method, in this experiment, we also used other two gene selection methods to choose key genes for classification, i.e. the frequently used t-statistics [7] and the IVGA based method listed in [1] (denoted by IVGA_II). For IVGA_II method, we also rank the selected key genes according to their frequencies of appearance as we have done in IVGA_I. The classification results are also listed in Table 1.

From Table1 it can be seen that, for the accuracy on test set, our method achieves the best and most stable classification results. The highest accuracy is achieved when P=20. And when the selected number is small, our method can even achieve a famous result. For the LOO-CV performance and accuracy (ACC) on training set, the general trend is that they are gradually ascend with the increase of P, yet they have no obvious relation with the accuracy on test data.

4.2 Acute Leukemia Data Set

The initial leukemia data set consisted of 38 bone marrow samples obtained from adult acute leukemia patients at the time of diagnosis, before chemotherapy [7]. RNA

Table 1. Summary of the results of the numerical experiments on Colon cancer classification problems, comprising the LOO-CV performance and the accuracy (ACC) on test and training set

Selected gene number	Method	LOO-CV performance	ACC on test set	ACC on training set
P=3	*t*-statistics	75.48±6.31	79.68±3.74	86.77±3.86
	IVGA_I	84.19±5.37	87.42±3.86	88.71±4.09
	IVGA_II	83.23±8.44	80.97±5.14	88.71±5.09
P=5	*t*-statistics	83.23±6.42	82.90±5.49	92.90±4.99
	IVGA_I	84.52±6.04	86.45±3.33	89.35±5.05
	IVGA_II	78.39±3.06	80.32±1.83	90.00±5.98
P=10	*t*-statistics	85.16±5.31	84.52±4.25	91.92±3.13
	IVGA_I	87.39±3.37	86.45±2.96	91.61±2.25
	IVGA_II	80.97±6.88	74.89±4.82	93.55±3.72
P=15	*t*-statistics	85.81±7.63	85.16±7.00	92.90±5.00
	IVGA_I	88.06±4.57	89.68±4.24	93.22±4.15
	IVGA_II	79.36±7.78	79.68±4.32	92.90±3.66
P=20	*t*-statistics	86.13±4.57	92.58±5.05	91.61±4.07
	IVGA_I	85.81±3.47	93.22±3.54	91.29±3.42
	IVGA_II	80.97±6.71	84.50±6.46	93.55±3.04
P=40	*t*-statistics	91.29±4.57	89.35±4.57	91.93±4.87
	IVGA_I	90.32±5.04	89.35±4.57	93.87±3.55
	IVGA_II	88.39±6.66	84.19±5.98	91.93±3.80
P=100	*t*-statistics	88.38±5.09	91.93±4.87	91.61±5.52
	IVGA_I	88.39±5.09	91.93±4.87	91.61±5.52
	IVGA_II	84.84±5.28	89.36±5.05	90.32±6.80
P=200	*t*-statistics	91.61±3.79	87.10±4.30	92.26±3.11
	IVGA_I	91.61±3.79	87.10±4.30	92.26±3.11
	IVGA_II	91.61±3.79	87.10±4.30	92.26±3.11

prepared from bone marrow mononuclear cells was hybridized to high-density oligonucleotide microarrays, produced by Affymetrix and containing 6817 human genes. The training set consists of 38 leukemia patients, of which 11 suffer from acute myeloid leukemia (AML) and 27 from acute lymphoblastic leukemia (ALL). The test set consists of 34 patients, of which 14 suffer from AML and 20 from ALL. The number of genes is 7129. Separating the AML samples from the ALL samples is the issue here.

In this experiment, we first ranked the genes and retained a set of the top 500 genes, then chose P= 3, 5, 10, 15, 20, 40, 150, 500, respectively. The detailed results are listed in Table 2. From Table2 we can see that, different from the first experiment, for the accuracy on test set, the IVGA based method IVGA_II achieves the best classification results, though it is not stable. Particularly, when P takes 15 and 40, it can get very high accuracy. Of course, our method is also better than the t-statistics when the key genes are small. For the LOO-CV performance and accuracy (ACC) on training set, the results are similar with the first experiment.

Table 2. Summary of the results of the numerical experiments on Acute leukemia cancer classification problems, comprising the LOO-CV performance and the accuracy (ACC) on test and training set

Selected gene number	Method	LOO-CV performance	ACC on test set	ACC on training set
P=3	*t*-statistics	83.06±4.43	85.83±2.43	84.72±3.27
	IVGA_I	86.39±4.62	87.78±4.93	89.72±3.48
	IVGA_II	79.17±5.43	82.22±3.74	81.94±4.58
P=5	*t*-statistics	78.47±7.08	82.99±4.04	92.71±2.94
	IVGA_I	82.50±5.85	84.37±5.34	94.79±1.77
	IVGA_II	90.97±5.30	86.46±3.12	96.53±2.87
P=10	*t*-statistics	91.36v2.76	89.44±2.86	95.55±2.98
	IVGA_I	88.61±5.62	90.01±5.08	99.17±1.87
	IVGA_II	97.49±2.05	96.67±2.86	99.17±1.34
P=15	*t*-statistics	88.61±4.97	92.22±3.99	94.72±3.05
	IVGA_I	91.39±3.05	92.78±3.97	93.89±2.86
	IVGA_II	98.99±2.56	99.72±0.87	100.0±0.00
P=20	*t*-statistics	91.11±3.15	87.49±3.27	95.55±2.98
	IVGA_I	92.22±4.09	91.39±2.76	96.39±2.63
	IVGA_II	90.28±4.76	92.22±5.97	98.61±1.96
P=40	*t*-statistics	93.61±4.90	94.72±3.22	97.77±7.55
	IVGA_I	88.33±3.65	92.50±4.54	95.55±3.97
	IVGA_II	98.05±2.28	99.17±1.34	100.0±0.0
P=150	*t*-statistics	95.27±2.28	98.05±1.87	97.50±1.57
	IVGA_I	92.57±4.57	95.83±2.69	98.05v1.34
	IVGA_II	93.88±3.41	97.22±2.92	99.72±0.87
P=500	*t*-statistics	97.22±2.26	96.67±2.86	100.0±0.0
	IVGA_I	97.22±2.26	96.67±2.86	100.0±0.0
	IVGA_II	97.22±2.26	96.67±2.86	100.0±0.0

4.3 General Comments

From the two experiments it can be seen that, for all the two data sets, the mean accuracy (ACC) on the test set of our method is high and stable. The smaller the key genes selected, the more advantage of our method embodied. For Colon data, our method is the best one, IVGA_II method is the worst, yet it gets the best results when using Acute leukemia data set.

One more thing should be explained is that, for colon data, when P=200, the key genes selected by the three methods are actually the same. So the results of the three methods are also coincident. In other words, for VGA_I and IVGA_II, when P=200, the genes with zero frequencies of appearance are also comprised in the key genes set. We listed this result only for the integrality of the experiment. This is the same for the other data sets when P=500.

5 Conclusions

In this paper, we proposed an independent variable group analysis based method for gene selection. The methodology is involved in selecting key genes using IVGA, followed by the classification applying SVM. We have compared our method with other two methods such as t-statistics and the IVGA based method listed in [1], the experimental results show that our method is effective and efficient in predicting normal and tumor samples from four human tissues. Furthermore, these results hold under re-randomization of the samples.

In this method, we only selected one gene from every cluster, this may discard useful information for tumor classification. In future works, we will deeply study the IVGA model of gene expression data, how to apply the method proposed in this paper to solving multiclass problems of tumor classification, and also study how to make full use of the information contained in every gene cluster to construct more effective and efficient gene selection method so that more exact prediction results of tumor class can be achieved.

Acknowledgements

This work was supported by the grants of the National Science Foundation of China, No. 30700161, China Postdoctoral Science Foundation, No. 20070410223.

References

1. Alhoniemi, E., Honkela, A., Lagus, K., Seppä, J., Wagner, P., Valpola, H.: Compact Modeling of Data Using Independent Variable Group Analysis. Technical Report E3, Helsinki University of Technology. Publications in Computer and Information Science, Espoo, Finland (2006)
2. Alon, U., Barkai, N., Notterman, D.A., Gish, K., Ybarra, S., Mack, D., Levine, A.J.: Broad Patterns of Gene Expression Revealed by Clustering Analysis of Tumor and Normal Colon Tissues Probed by Oligonucleotide Arrays. Proc. Natl. Acad. Sci. USA 96, 6745–6750 (1999)
3. Ambroise, C., McLachlan, G.J.: Selection Bias in Gene Extraction on the Basis of Microarray Gene-Expression Data. Proc. Natl. Acad. Sci. USA 99, 6562–6566 (2002)
4. Bae, K., Mallick, B.K.: Gene Selection Using a Two-Level Hierarchical Bayesian Model. Bioinformatics 20, 3423–3430 (2004)
5. Bittner, M., Meltzer, P., Chen, Y., Jiang, Y., Seftor, E., Hendrix, M., Radmacher, M., Simon, R., Yakhini, Z., Ben-Dor, A., Sampas, N., Dougherty, E., Wang, E., Marincola, F., Gooden, C., Lueders, J., Glatfelter, A., Pollock, P., Carpten, J., Gillanders, E., Leja, D., Dietrich, K., Beaudry, C., Berens, M., Alberts, D., Sondak, V., Hayward, N., Trent, J.: Molecular Classification of Cutaneous Malignant Melanoma by Gene Expression Profiling. Nature 406, 536–540 (2000)
6. Furey, T.S., Cristianini, N., Duffy, N., Bednarski, D.W., Schummer, M., Haussler, D.: Support Vector Machines Classification and Validation of Cancer Tissue Samples Using Microarray Expression Data. Bioinformatics 16, 906–914 (2000)

7. Golub, T.R., Slonim, D.K., Tamayo, P., Huard, C., Gaasenbeek, M., Mesirov, J.P., Coller, H., Loh, M.L., Downing, J.R., Caligiuri, M.A., Bloomfield, C.D., Lander, E.S.: Molecular Classification of Cancer: Class Discovery and Class Prediction by Gene Expression Monitoring. Science 286, 531–537 (1999)
8. Kraskov, A., Stögbauer, H., Andrzejak, R.G., Grassberger, P.: Hierarchical Clustering Using Mutual Information. Europhysics Letters 70, 278–284 (2005)
9. Lagus, K., Alhoniemi, E., Valpola, H.: Independent Variable Group Analysis. In: Dorffner, G., Bischof, H., Hornik, K. (eds.) ICANN 2001. LNCS, vol. 2130, pp. 203–210. Springer, Heidelberg (2001)
10. Li, W., Sun, F., Grosse, I.: Extreme Value Distribution Based Gene Selection Criteria for Discriminant Microarray Data Analysis Using Logistic Regression. Journal of Computational Biology 1, 215–226 (2004)
11. Nilsson, M., Gustafsson, H., Andersen, S.V., Kleijn, W.B.: Gaussian Mixture Model Based Mutual Information Estimation between Frequency Bands in Speech. In: Proceedings of the IEEE International Conference on Acoustics, Speech and Signal Processing, vol. 1, pp. I–525–I–528 (2002)
12. Pochet, N., De Smet, F., Suykens, J.A.K., De Moor, B.L.R.: Systematic Benchmarking of Microarray Data Classification: Assessing the Role of Non-Linearity and Dimensionality Reduction. Bioinformatics 20, 3185–3195 (2004)
13. Studený, M., Vejnarová, J.: The Multiinformation Function as a Tool for Measuring Stochastic Dependence. In: Learning in Graphical Models, pp. 261–297. MIT Press, Cambridge (1999)
14. Zhang, H.H., Ahn, J., Lin, X., Park, C.: Gene Selection Using Support Vector Machines with Non-Convex Penalty. Bioinformatics 22, 88–95 (2006)
15. Huang, D.S., Zheng, C.H.: Independent Component Analysis Based Penalized Discriminant Method for Tumor Classification Using Gene Expression Data. Bioinformatics 22, 1855–1862 (2006)

The Average Radius of Attraction Basin of Hopfield Neural Networks

Fan Zhang[1,2] and Xinhong Zhang[3]

[1] Institute of Advanced Control and Intelligent Information Processing, Henan University, Kaifeng 475001, China
zhangfan@vip.sohu.com
[2] College of Electronic and Information Engineering, Tianjin University, Tianjin 300072, China
[3] Computing Center, Henan University, Kaifeng 475001, China
hnkfzxh@163.com

Abstract. This paper introduces a derivation of the attraction basin to the Hopfield neural networks and obtains an average radius of the attraction basin, which is a expression of Hamming distance. The average radius of the attraction basin is $(N-1)/2P$. If the average of Hamming distance between the probe pattern and a stored pattern is less than $(N-1)/2P$, the neural network will converge to the stored pattern.

Keywords: Hopfield neural networks, Attraction basin, Radius.

1 Introduction

The Hopfield neural network is a recurrent neural network that stores information in a dynamically stable configuration [1]. An interesting property of recurrent type networks is that their state can be described by an energy function. The energy function is used to prove the stability of recurrent type networks. The local minima of the energy function correspond to the energy of the stored patterns. An energy function is used to evaluate the stability property, and the energy function always decreases to a state of the lowest energy. Hopfield has shown that the energy of the discrete Hopfield model decreases or remains the same after each unit update. Therefore, the network will eventually converge to a local minimum that corresponds to a stored pattern. The stored pattern to which the network converges depends on the input pattern and the connection weight matrix.

An important question in the study of neural networks is on the "inabilit" of the system, and the conditions under which the system is trainable is a critical issue. The trainability is known to be directly related to the type of the equilibrium points of the set of nonlinear differential equations describing the system. In a neural network, the basins of attraction in state space are related

F. Sun et al. (Eds.): ISNN 2008, Part II, LNCS 5264, pp. 253–258, 2008.

to the specific configurations that have been stored, furthermore, uncertain equilibrium points in regions of unpredictability cause noisy patterns in state space with interacting and eroded basins of attraction. Attractors of Hopfield network represent the stored patterns. The basin of attraction is the set of states in the system within which almost all states flow to one attractor. For a trained neural network, the attraction basin gives a measure of neural network error-correcting capability. Once a pattern is stored, the Hopfield network can reconstructs the original pattern from the degraded or incomplete pattern.

Attraction basins and capacity are not independent and there have been methods to try to use the attraction basin size in a capacity like definition [2]. McEliece discussed the capacity of the Hopfield associative memory and the attraction basin size [3]. Storkey established some initial results about the size and shapes of the attraction basins [4]. Empirical methods have been used to gain knowledge of the sizes and shapes of the basins for independent unbiased storage patterns. This paper has shown that the new rule has larger, more evenly distributed and more rounded attraction basins than those of the Hebb rule. Other approaches include the introduction of dynamic or noisy external fields [5] and the minimum overlap method [6]. Li discussed the binary orthogonal memory patterns in Hopfield neural networks [7].

In this work, we introduce a derivation of the attraction basin to the Hopfield neural network. We obtain an average radius of the attraction basin, which is a expression of Hamming distance. In Section 2, we introduce the Hopfield model of associative memory and explain the function of the neural network. The average radius of the attraction basin is derived in Section 3. In Section 4, we discuss the result of our derivation about the attraction basin. The conclusion is drawn in Section 5.

2 The Hopfield Network

The dynamics of the Hopfield model is different from that of the linear association model in that it computes its output recursively in time until the system becomes stable. Unlike the linear association model which consists of two layers of processing units, one serving as the input layer while the other as the output layer, the Hopfield model consists of a single layer of processing elements where each unit is connected to every other unit in the network other than itself. Each unit has an extra external input. This extra input leads to a modification in the computation of the net input to the units.

Hopfield found that a set of asynchronously operating nonlinear neurons can store information with stability and efficiency, recall it with some error-correcting capability, and exhibit a sense of time order [8]. Also, his model is quite robust and should work even when more neurological details are added.

The Hopfield model neurons we consider are simple bistable elements each being capable of assuming two values: -1 (off) and $+1$ (on). The state of each neuron then represents one bit of information, and the state of the system as a whole is described by a binary N-tuple if there are N neurons in the system. We assume that the neural net is (possibly) densely interconnected, with neuron i transmitting information to neuron j through a linear synaptic connection W_{ij}. The neural interconnection weights W_{ij} are throughout considered to be fixed; i.e., learning of associations has already taken place, and no further synaptic modifications are made in the neurobiological interpretation. The connection matrix is also assumed to be symmetric with zero diagonal in this paper.

Logical computation in the network takes place at each neural site by means of a simple threshold decision rule. Each neuron evaluates the weighted sum of the binary states of all the neurons in the system; the new state of the neuron is -1 if the sum is negative, and $+1$ if the sum (equals or) exceeds zero. (In this and what follows we almost always assume a threshold of zero.) Specifically, if $x = (x_1, x_2, \cdots, x_n)$ is the present state of the system (with $x_j = +1$ being the state of the jth neuron), the new state x_i' of the ith neuron is determined by the rule

$$x_i' = sgn\left[\sum_{\substack{j=1\\ j\neq i}}^{N} w_{ij}x_j\right] = \begin{cases} +1, & w_{ij}x_j \geq 0 \\ -1, & w_{ij}x_j < 0 \end{cases}. \tag{1}$$

3 The Attraction Basin

Information in the Hopfield model is stored as stable states. A stable state is a state that is a fixed point of the neural network. Each of the N neurons randomly and repeatedly looks at the weighted sum of all its inputs and then decides not to change from its previous state. Attractors represent the stored patterns. The basin of Attractor is the set of states in the system within which almost all states flow to one attractor.

Usually the basin of attraction is measured by Hamming distance. Hamming distance is the number of components that differ between two vectors. The distance between two vectors S^1 and S^2 is d_h (S^1, S^2). The Hamming distance can be used to measure the basin of attraction [9].

Let P denotes the number of stored patterns and N denotes the number of neurons. The Hopfield network model can be expressed as,

$$x_i^{t+1} = sgn\left[\sum_{\substack{j=1\\ j\neq i}}^{N} w_{ij}x_j^t\right], \tag{2}$$

where x_i^t is the state of neurons at time t; *sgn* is the sign function. The weight matrix W be an $N \times N$ real-valued, zero-diagonal symmetric matrix,

$$w_{ij} = \frac{1}{N}\sum_{\mu=1}^{P} \xi_i^{\mu}\xi_j^{\mu}. \tag{3}$$

The entries of W are the W_{ij}, which is the strength of the synaptic connection from neuron j to neuron i. Each choice of W defines a specific neural network of N neurons with specific values for the strengths of the synaptic connections of the neurons. The network starts in an initial state and runs with each neuron randomly and independently reevaluating itself. Often, the network enters a stable point in the state space in which all neurons remain in their current state after evaluating their inputs. This stable vector of states constitutes a stored word in the memory, and the basic operation of the network is to converge to a stable state if we initialize it with a nearby state vector (in the Hamming sense).

Let $\xi = (\xi^1, \xi^2, \cdots \xi^P)$ be an P-set of N-dimensional binary (± 1) column vectors, which are to be stored. Hebb rule determines the change in the weight connection. Let $X^0 = \{x_1^0, x_2^0, \cdots, x_N^0\}^T$ denotes the initial state of neural network, $X^t = \{x_1^t, x_2^t, \ldots, x_N^t\}^T$ denotes the state at a time t.

$$\sum_{\substack{j=1\\j\neq i}}^{N} w_{ij}x_j^0 = \sum_{\substack{j=1\\j\neq i}}^{N} \frac{1}{N}\sum_{\mu=1}^{P} \xi_i^\mu \xi_j^\mu x_j^0$$

$$= \frac{1}{N}\sum_{\mu=1}^{P}\sum_{j=1}^{N} \xi_i^\mu \xi_j^\mu x_j^0 - \frac{1}{N}\sum_{\mu=1}^{P} \xi_i^\mu \xi_i^\mu x_i^0$$

$$= \frac{1}{N}\left[\sum_{\mu=1}^{P} \xi_i^\mu (\xi^\mu)^T X^0 - P x_i^0\right]. \tag{4}$$

$\xi^\mu = \{\xi_1^\mu, \xi_2^\mu, \cdots, \xi_N^\mu\}^T$ and $X^0 = \{x_1^0, x_2^0, \cdots, x_N^0\}^T$ are the N-dimensional binary (± 1) column vectors. The Hamming distance between the ξ^μ and the X^0 is $d_{h\mu}(X^0, \xi^\mu)$, so,

$$(\xi^\mu)^T X^0 = N - 2d_{h\mu}(X^0, \xi^\mu). \tag{5}$$

Then,

$$\sum_{\substack{j=1\\j\neq i}}^{N} w_{ij}x_j^0 = \frac{1}{N}\left[\sum_{\mu=1}^{P} \xi_i^\mu (N - 2d_{h\mu}(X^0, \xi^\mu)) - P x_i^0\right]. \tag{6}$$

The average of Hamming distance between the ξ^μ and the X^0 is,

$$\widetilde{d_h} = \frac{d_{h1}(X^0, \xi^1) + d_{h2}(X^0, \xi^2) +, \cdots, d_{hP}(X^0, \xi^P)}{P}. \tag{7}$$

Then,

$$\sum_{\substack{j=1\\j\neq i}}^{N} w_{ij}x_j^0 = \frac{1}{N}\left[N\sum_{\mu=1}^{P} \xi_i^\mu - 2\sum_{\mu=1}^{P} \xi_i^\mu d_{h\mu}(X^0, \xi^\mu) - P x_i^0\right]$$

$$= \frac{1}{N}\left[N\sum_{\mu=1}^{P} \xi_i^\mu - 2P\sum_{\mu=1}^{P} \xi_i^\mu \widetilde{d_h} - P x_i^0\right]. \tag{8}$$

We assume that the probe pattern is one of the stored patterns.

When $\xi_i^\mu = +1$,

$$\sum_{\substack{j=1\\j\neq i}}^{N} w_{ij}x_j^0 = \frac{1}{N}\left[NP - 2P^2\widetilde{d}_h - P\right]. \tag{9}$$

IF $\widetilde{d}_h < (N-1)/2P$, then $\sum_{\substack{j=1\\j\neq i}}^{N} w_{ij}x_j^0 > 0$.

When $\xi_i^\mu = -1$,

$$\sum_{\substack{j=1\\j\neq i}}^{N} w_{ij}x_j^0 = \frac{1}{N}\left[-NP + 2P^2\widetilde{d}_h + P\right]. \tag{10}$$

IF $\widetilde{d}_h < (N-1)/2P$, then $\sum_{\substack{j=1\\j\neq i}}^{N} w_{ij}x_j^0 < 0$.

So, $\xi_i^\mu = sgn\left[\sum_{\substack{j=1\\j\neq i}}^{N} w_{ij}x_j^0\right]$, and then,

$$x_i^v = sgn\left[\sum_{\substack{j=1\\j\neq i}}^{N} w_{ij}x_j^0\right] = \xi_i^\mu. \tag{11}$$

According to Eq. 11, if the average of Hamming distance between the probe pattern and a stored pattern,

$$\widetilde{d}_h < (N-1)/2P, \tag{12}$$

the neural network will converge to this stored pattern. The attraction basin of each stored pattern can be expressed as the Hamming distance shown in Eq. 12.

4 Conclusion

The basin of attraction is the set of states in the system within which almost all states flow to one attractor. In this work, we introduce a derivation of the attraction basin to the Hopfield neural network. We obtain an average radius of the attraction basin, which is a expression of Hamming distance. The average radius of the attraction basin is $(N-1)/2P$. If the average of Hamming distance between the probe pattern and a stored pattern is less than $(N-1)/2P$, the neural network will converge to this stored pattern. Although only the average radius of attraction basin in Hamming distance is discussed, but the estimation of the attraction basin is surely a useful work to the analysis of capacity of Hopfield neural networks.

Acknowledgements. This work was supported by the Natural Science Foundation of Education Bureau of Hunan Province, China (2008A520003) and the Post-doctoral Science Foundation of China (20070420707).

References

1. Davey, N., Hunt, S.: The Capacity and Attractor Basins of Associative Memory Models. In: Proceedings 5th International Conference on Artificial and Natural Neural Networks. LNCS, pp. 340–357. Springer, Heidelberg (1999)
2. Schwenker, F., Sommer, F., Palm, G.: Iterative Retrieval of Sparsely Coded Associative Memory Patterns. Neural Networks 9, 445–455 (1996)
3. McEliece, R., Posner, C., Rodemich, R., Santosh, R.: The Capacity of the Hopfield Associative Memory. IEEE Transactions on Information Theory 33, 461–482 (1987)
4. Storkey, A., Valabregue, R.: The Basins of Attraction of a New Hopfield Learning Rule. Neural Networks 12, 869–876 (1999)
5. Wang, T.: Improving Recall in Associative Memories by Dynamic Threshold. Neural Networks 7, 1379–1385 (1994)
6. Chang, J., Wu, C.: Desing of Hop Eld Type Associative Memory with Maximal Basin of Attraction. Electronics Letters 29, 2128–2130 (1993)
7. Li, Y.: Analysis of Binary Orthogonal Memory Patterns in Hopfield Neural Networks. Chinese Journal of Computers 24, 1334–1336 (2001)
8. Hopfield, J.: Neural Networks and Physical Systems with Emergent Collective Computational Abilities. Proceedings of the National Academy of Sciences 79, 2554–2558 (1982)
9. Castillo, P., Merelo, J., Arenas, M., Romero, G.: Comparing Evolutionary Hybrid Systems for Design and Optimization of Multilayer Perception Structure along Training Parameters. Information Sciences 177, 2884–2905 (2007)
10. Floréen, P., Orponen, P.: Attraction Radii in Binary Hopfield Nets are Hard to Compute. Neural Computation 5, 812–821 (1993)

A Fuzzy Cluster Algorithm Based on Mutative Scale Chaos Optimization

Chaoshun Li, Jianzhong Zhou*, Qingqing Li, and Xiuqiao Xiang

College of Hydroelectric Digitization Engineering , Huazhong University of Science and Technology,
Wuhan 430074, China
lichaoshun@yahoo.com.cn, jz.zhou@hust.edu.cn

Abstract. The traditional Fuzzy C-Means (FCM) algorithm has some disadvantages in optimization method, which makes the algorithm liable to fall into local optimum, thus failing to get the optimal clustering results. According to the defect of FCM algorithm, a new Fuzzy Clustering algorithm based on Chaos Optimization (FCCO) is proposed in this paper, which combines mutative scale chaos optimization strategy and gradient method together. Moreover, a fuzzy cluster validity index (PBMF) is introduced to make the FCCO algorithm capable of clustering automatically. Three other fuzzy cluster validity indices, namely XB, PC and PE, are utilized to compare the performances of FCCO, FCM and another algorithm, when applied to artificial and real data sets classification. Experiment results show FCCO algorithm is more likely to obtain the global optimum and achieve better performances on validity indices than other algorithms.

Keywords: Clustering,Fuzzy c-means algorithm, Chaos optimization, Cluster validity indices.

1 Introduction

The cluster analysis is one kind of multivariate statistical analysis, and a very important branch of non-supervision pattern recognition as well. As a type of non-supervision classification method, cluster analysis is widely used in fields such as pattern recognition, image processing and data mining, etc [1,2,3]. The conventional clustering methods regard clustering issue as a kind of issues of function optimization according to similarity principle, and try to categorize the samples into different kinds based on achieving function extremum. Classical clustering method include conventional c-means method and Fuzzy C-Means(FCM) algorithm, etc.

Although FCM algorithm is used in various fields, it still has its disadvantages. They are mainly manifested in the following aspects: 1) the classification results are so dependent upon sample distribution that when sample distribution is uneven, the clustering results is not satisfactory. 2)the existing cluster validity

* Corresponding author.

F. Sun et al. (Eds.): ISNN 2008, Part II, LNCS 5264, pp. 259–267, 2008.

function has its limits when evaluating fuzzy clustering effects. 3) when seeking for clustering objective function extremum through gradient method, it is likely to get local minimum and fail to get optimal classification. The disadvantage of FCM algorithm at optimization will lead to the result that the algorithm is sensitive to initial values, and is liable to fall into the local optimum, which means different initial values will generate different results or even lead to non-solution. In order to overcome those disadvantages manifested above, Wang proposed kernel function to optimize the original samples in [4], which reduced the effect that the sample distribution exerted upon FCM algorithm; Li assigned weights to every dimension of samples in [5], considering the different contributions each sample dimension does for the fuzzy clustering results; Pakhira put forward a new fuzzy cluster validity measure in [6].

Considering the shortage of FCM algorithm in function optimization aspect, a novel fuzzy Clustering algorithm based on Chaos Optimization (FCCO) is proposed in this paper, which optimizes objective function with mixed optimization method consisting of mutative scale chaos optimization strategy and gradient method and can classify automatically with the help of PBMF index. We design a group of experiments to verify the performance of FCCO with objective function optimum and three cluster validity indices when dealing with artificial and real data sets. Finally FCCO algorithm is used in image segmentation, which proves the validity and feasibility of the proposed method.

2 FCM Algorithm and Fuzzy Cluster Validity Indices

2.1 Fuzzy C-Mean (FCM) Algorithm

When dealing with problem of classify n samples into c classes, the cluster objective function of the well known Fuzzy C-Mean (FCM) model is defined as:

$$J_m(U,V)=\sum_{k=1}^{n}\sum_{i=1}^{c}(\mu_{ik})^m(d_{ik})^2 \ , \tag{1}$$

where d_{ik} is the distance between sample x_k and center v_i , which is usually Euclidean distance and can be defined as $(d_{ik})^2=\| x_k-v_i \|^2$. FCM assumes that the sum of the fuzzy membership grades to each cluster is equal to 1, which can be described as:

$$\sum_{i=1}^{c}\mu_{ik}=1, k=1,...,n \ . \tag{2}$$

It's expected that the optimal cluster structure will be achieved if the clustering objective function J_m reaches its minimization. Considering the constraint equation (2), the extreme of J_m could be found using Lagrange Multiplier Rule only when fuzzy membership grades and cluster center are in accordance with:

$$\mu_{ik}=\frac{1}{\sum_{j=1}^{c}(d_{ik}/d_{jk})^{\frac{2}{m-1}}} \ , \tag{3}$$

$$v_i = \sum_{k=1}^{n} (\mu_{ik})^m x_k / \sum_{k=1}^{n} (\mu_{ik})^m . \tag{4}$$

The FCM algorithm calculates the fuzzy membership grades matrix U and the fuzzy cluster center matrix V by using gradient method. In every iteration of this algorithm J_m will be reduced, and meanwhile U and V will be refreshed. When the difference in the cluster centers are within a given threshold, the algorithm will be stopped and the final U and V be obtained.

2.2 Fuzzy Cluster Validity Measure Indices

Several validity measures of fuzzy clusters have been proposed in literatures, which provide us methods to prove the validity of new fuzzy cluster algorithm. In this article, four validity measures are introduced to verify and compare the effects of FCCO with another two fuzzy cluster algorithms.

Pakhira developed PBMF index in [6], which is defined as:

$$PBMF = \frac{1}{c} \times \frac{E_1 \max_{i,j=1}^{c} \|v_i - v_j\|}{E_c} , \tag{5}$$

where c is the number of clusters, v presents a cluster center vector, and E_c is defined as:

$$E_c = \sum_{k=1}^{n} \sum_{i=1}^{c} (\mu_{ik})^m \|x_k - v_i\| , \tag{6}$$

where μ_{ik} is an element of fuzzy membership grades matrix U,x_k is a sample vector. When $c = 1$, E_c equals to E_1 , which means E_1 is a constant term for a particular data set. PBMF was proposed to find the exact number of fuzzy clusters for a data set, while the better the value of PBMF is, the more excellent the cluster validity is. The optimal number of clusters can be obtained when the best PBMF value is achieved with changed c.

Xie-Beni (XB) index [7] is defined as:

$$XB = \frac{\sum_{i=1}^{c} \sum_{k=1}^{n} \mu_{ik}^2 \|x_k - v_i\|^2}{n \times min_{i \neq j} \|v_i - v_j\|} . \tag{7}$$

The smaller the value of XB index is, the better the cluster validity is.

Bezdek's PC and PE index, co-efficient (PC) and partition entropy (PE), were defined in [8] for any fuzzy clusters, which have the form as:

$$PC = \frac{\sum_{i=1}^{c} \sum_{k=1}^{n} (\mu_{ik})^2}{n} , \tag{8}$$

$$PE = \frac{1}{c} \sum_{i=1}^{c} \sum_{k=1}^{n} (\mu_{ik})^2 \log_b(\mu_{ik}) , \tag{9}$$

where b is the logarithmic base. When the cluster structure is optimal, the PC has its maximum value and PE takes its minimum value.

3 FCCO Algorithm

3.1 Chaos Optimization Strategy

Chaos is a kind of seemingly random or irregular movement, which appears in a deterministic system, and is a kind of complex movement and natural phenomenon existing universally. Chaos variables are random, ergodic and regular to some extent. The basic idea of searching optimum using chaos variables is: producing chaos variables with a kind of chaotic map, projecting chaos variables to optimization variables interval and then searching optimal solution with chaos variable. Randomness and ergodicity of chaos variables make chaos optimization possible to achieve global optimum quickly.

We choose the famous Logistic map as chaotic map, which is a one-dimensional quadratic map defined by

$$x^{(i+1)} = \mu x^{(i)}(1 - x^{(i)}) \ , x^{(i)}) \in [0, 1] \ , \tag{10}$$

where μ is a control parameter and when $\mu = 4$, equation (10) generates chaotic evolutions, and the sequence of exhibits chaotic state completely, which is called chaos sequence. The values in chaos sequence can not repeated, which means every value in the given optimization variable's interval can be reached by projecting the chaos variables to optimization variables, and thus the global optimum of the objective function could be achieved.

3.2 Implementation of FCCO Algorithm

When solving the problem of non-convex function's optimization, chaos optimization strategy can avoid sinking into local minimum while chaos variable has excellent property, consequently satisfactory results can be achieved. Fuzzy clustering objective function J_m is a typical non-convex function, when optimizing it as FCM does by using gradient method, algorithm is liable to sink into local minimum, so we use chaos optimization strategy to optimize , local minimum is expected to be avoided and global minimum is expected to be achieved.

The main idea of FCCO is to optimize the clustering objective function by using chaos optimization strategy and get the exact number of clusters automatically with the PBMF index. We start a circulation, while the number of clusters c increases from 2 to $\sqrt{n}$ and search the optimal clustering result at each circulation. The best PBMF index value is believed to in accordance with the exact number of clusters and the optimal clustering result. The searching method of FCCO can be described as follows: using Logistic map generates chaos sequence, then projecting the chaos variables to cluster center matrix's elements v_{ij} and refresh fuzzy membership grades matrix U accordantly, finally calculating the function value J_m and judging whether the current value is the optimal or not. In order to improve efficiency of FCCO, we use a new chaos optimization strategy listed in [9], which reduces the searching range gradually, and combine the gradient method with FCCO, which means when obtaining current optimal solution, namely cluster center matrix V, we get new V and U by calculating equation (3) and (4) once. FCCO combines mutative scale chaos optimization strategy with gradient method, and thus it could search the global optimum quickly and effectively.

The specific steps of FCCO are described as:

Step 1: Initialize the number of clusters c, $c = 2$, the best PBMF index value b_{pbmf} , the exact number of clusters c_b . Set the threshold θ, which determines when to stop once circulation.

Step 2: Circulate the searching of best clustering structure and results as the number of clusters increase form 2 to $\sqrt{n}$, meanwhile calculate the PBMF value and refresh b_{pbmf} along with c_b. If start a optimization process, use the following searching strategy ,which mainly make up of the following key points.

a) Initialize the best partition matrix U_b,the best cluster center matrix V_b and the optimal objective function value J_{mb}.

b) Generate chaos sequence and project chaos variables to elements of cluster center matrix V, calculate the corresponding partition matrix U and the objective function value J_m.Refresh V_b,U_b and J_{mb} if necessary.

c) Run calculating equation (3) and (4) once when J_{mb} don't refresh for a given iterations which can accelerate convergence.

d) Reduce the scale of cluster center variables when projecting chaos variables if J_{mb} stay unchanged for a given iterations.

e) Stop the searching when the distance between the current U_b and the last U_b is within the given threshold,which can be expressed as $\|U_b - U_{bl}\| < \theta$.

Step 3 $c = c + 1$, if $c > \sqrt{n}$, stop the algorithm. The auto determined best number of clusters c_b can be obtained as well as the corresponding optimal fuzzy cluster center matrix U_b and objective function value J_{mb}.

4 Experiment Results

4.1 Data Sets

In order to test the performance of FCCO, we design a series of experiments based on artificial and real data sets. These data sets are described as below.

Data1-Data4 are artificial data sets, all of which are uniform distributed. The sizes of these data sets become larger gradually from 300 to 600, while the dimensions stay as 2.

IRIS is a data set which represents the categories of iris flower with four features. The four feature values represent the sepal length, sepal width, petal length and the petal width in centimeters [10]. This data set can be partitioned into three clusters, the first, second and the last 50 samples of that belong to one cluster, while the first category is easy to be separated from the other two, and the second and the third categories are difficult to be differentiated from each other.

VOWEL consists of 871 Indian Telugu vowel sounds [11], each pattern of which has 3 features, namely F1, F2 and F3, indicating the first, second and third vowel formant frequencies. The detailed information of all data sets are

Table 1. Information of data set

Data set	Number of clusters	Number of samples	Number of dimensions
Data1	3	300	2
Data2	4	400	2
Data3	5	500	2
Data4	6	600	2
IRIS	3	150	4
VOWEL	6	871	3

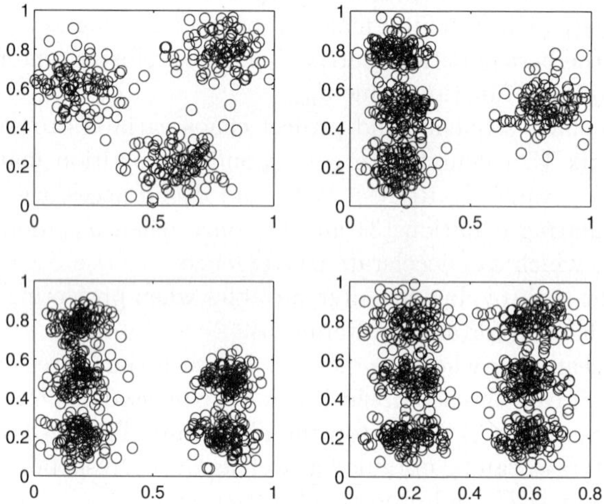

Fig. 1. Distributions of Data1 to Data4

listed in Table 1. The distributions of artificial data sets are presented in Fig. 1, where (a), (b), (c) and (d) represent Data1 to Data4 respectively.

4.2 Comprehensive Performance of the Cluster Algorithms

In FCM and FCCO algorithms, we set fuzzy factor m as 1.5. The threshold of FCM is 1e-6. The threshold of FCCO is also 1e-6. We have executed every algorithm for 50 times on all data sets, and the average results are presented in Tables 2-5. The average values of fuzzy cluster objective function of FCM and FCCO are listed in Table 2, while the average values of fuzzy cluster validity index are presented in Tables 3-5, that is, the values of XB, PC and PE index. We choose the PBMF index values of data set Data1 with changed number of clusters presented in Fig. 2, which shows the best number of clusters of data set Data1.

It's clear that the PBMF index value changes according to different number of clusters, and achieves the best value when number of clusters is equal to 3, which conforms with the reality. We have verified that FCCO algorithm can find out the right number of clusters correctly, but just show only one figure because

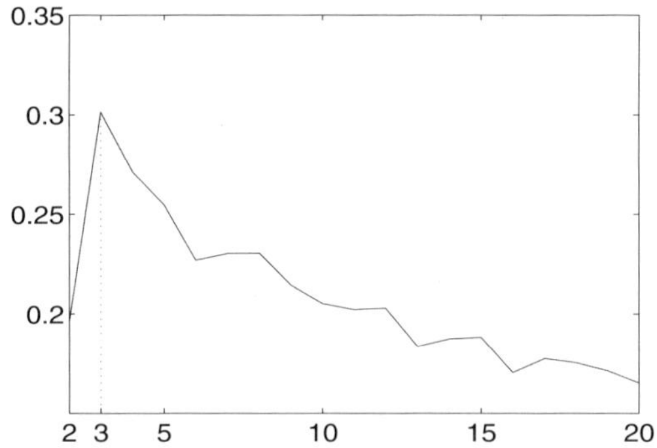

Fig. 2. Distributions of Data1 to Data4

Table 2. Comparison of FCM and FCCO in term of optimum

Argorithms	Data1	Data2	Data3	Data4	IRIS	VOWEL
FCM	5.2751	5.0577	4.7734	5.4622	78.019	26.453
FCCO	5.2751	4.0853	3.8657	4.5981	74.412	26.439

of the limitation of paper length. In Table 2,we take fuzzy clustering function value into consideration to check the performance of algorithms. It's clear that FCCO algorithm get better value than FCM, apart from Data1. The result shows that the traditional FCM algorithm with gradient method is more likely to get trapped in local optimal if the data set scale is large or the dimension of the data set is high, while FCCO algorithm has wonderful performance since we improved the optimization method with chaos searching strategy.

In Tables 3-5, we test algorithms with three validity measures. The XB values of FCCO algorithm are smaller compared with FCM on all data sets apart from Data1, clearly indicated in Table 3, which means FCCO algorithm has better performance. The PC values of FCCO algorithm are larger than those of FCM on all data sets besides VOWEL in Table 4, which demonstrates that FCCO algorithm is more effective under this index. In Table 5, FCCO algorithm achieves better performance under the PE index on all data sets apart from Data1 and VOWEL.

From the results of experiments, FCCO algorithm has been proved to be a more excellent algorithm compared with FCM, while FCCO has taken a new optimization strategy, which will make the algorithm avoid sinking into local optimum, and achieve global optimum. The results in Table 2 have proved the validity of the chaos optimization strategy embedded in FCCO algorithm, and the results in Tables 3-5 verify FCCO algorithm have wonderful performance under the fuzzy cluster validity measures.

Table 3. Comparison of FCM and FCCO in term of XB index

Argorithms	Data1	Data2	Data3	Data4	IRIS	VOWEL
FCM	0.0538	0.3077	0.2135	0.2835	0.1566	0.2734
FCCO	0.0538	0.1172	0.0949	0.0882	0.1566	0.2729

Table 4. Comparison of FCM and FCCO in term of PC index

Argorithms	Data1	Data2	Data3	Data4	IRIS	VOWEL
FCM	0.9693	0.9267	0.9417	0.9127	0.9188	0.7957
FCCO	0.9694	0.9511	0.9484	0.9357	0.9190	0.7956

Table 5. Comparison of FCM and FCCO in term of PE index

Argorithms	Data1	Data2	Data3	Data4	IRIS	VOWEL
FCM	0.0691	0.1407	0.1185	0.1876	0.1462	0.3908
FCCO	0.0691	0.1022	0.1071	0.1435	0.1460	0.3908

5 Discussions and Conclusions

A new fuzzy cluster algorithm, called FCCO, is proposed in this paper. FCCO algorithm is based on the FCM algorithm and improves mainly in two aspects to overcome the disadvantages of FCM algorithm, namely, 1) putting forward a new optimization strategy to optimize the clustering objective function, which combines a kind of mutative scale chaos optimization strategy with gradient method; 2) introducing a fuzzy cluster validity measure index to realize clustering automatically without requiring the number of clusters as a known priori. Simulation results and experiment show that FCCO algorithm achieve better performance than compared algorithms, which proves the validity and efficiency of FCCO algorithm.

Acknowledgments. This paper is supported by the Ph.D. Programs Foundation of Ministry of Education of China (20050487062),Project supported by the State Key Development Program for Basic Research of China (Grant No. 2007CB714107) and National Natural Science Foundation of China (50579022).

References

1. Ruan, X.G.: A Pattern Recognition Machine with Fuzzy Clustering Analysis. Intelligent Control and Automation 4, 2530–2534 (2000)
2. Xia, Y., Feng, D.G., Wang, T.J., Zhao, R.C., Zhang, Y.N.: Image Segmentation by Clustering of Spatial Patterns. Pattern Recognition Letters 28, 1548–1555 (2007)
3. Wang, S.Y., Zhou, M.Q., Geng, G.H.: Application of Fuzzy Cluster Analysis for Medical Image Data Mining. Mechatronics and Automation 2, 631–636 (2005)
4. Wang, J.H., Lee, W.J., Lee, S.J.: A Kernel-Based Fuzzy Clustering Algorithm. In: Innovative Computing, Information and Control 2006, pp. 550–553 (2006)

5. Li, J., Gao, X.B., Jiao, L.C.: A New Feature Weighted Fuzzy Clustering Algorithm. Acta Electronica Sinica 1, 89–92 (2006)
6. PakhiraMalayK, B.S., Maulik, U.: Study of Some Fuzzy Cluster Validity Indices,Genetic Clustering and Application to Pixel Classification. Fuzzy Sets and Systems 155, 191–214 (2005)
7. Xie, X.L., Beni, G.: A Validity Measure for Fuzzy Clustering. IEEE Trans PAMI 13, 841–847 (1991)
8. Bezdek, J.C.: Mathematical Models for Systematics and Taxonomy. In: Eighth International Conference on Numerical Taxonomy, pp. 143–165 (1971)
9. Zhang, T., Wang, H.W., Wang, Z.C.: Mutative Scale Chaos Optimization Algorithm and Its Application. Control and Decision 14, 285–287 (1999)
10. Fisher, R.A.: The Use of Multiple Measurements in Taxonomic Problems. Ann. Eugen. 3, 179–188 (1936)
11. Pal, S.K., Majumder, D.D.: Fuzzy Sets and Decision Makingapproaches in Vowel and Speaker Recognition. IEEE Trans Syst. Man Cybern. 7, 625–629 (1977)

A Sparse Sampling Method for Classification Based on Likelihood Factor

Linge Ding, Fuchun Sun, Hongqiao Wang, and Ning Chen

Department of Computer Science and Technology, Tsinghua University,
State Key Lab of Intelligent Technology and Systems, 100084 Beijing , China
dlg06@mails.tsinghua.edu.cn

Abstract. The disadvantages of large computing and complex discriminant function involved in classical SVM emerged when the scale of training data was larger. In this paper, a method for classification based on sparse sampling is proposed. A likelihood factor which can indicate the importance of sample is defined. According to the likelihood factor, non-important samples are cliped and misjudged samples are revised, this is called sparse sampling. Sparse sampling can reduce the number of the training samples and the number of the support vectors. So the improved classification method has advantages in reducing computational complexity and simplifying discriminant function.

Keywords: Sparse Sampling, Non-important Sample Cliping, Misjudged Sample Revising.

1 Introduction

Statistical learning theory is proposed by V. Vapnik specially for the learning problem of small samples. The VC dimension is defined as a scalar indicator representing the capacity of a set of functions. Larger the VC dimension is, better the discriminant capacity of the functions can be. The rule of the decision should follow structural risk minimization(SRM) rather than empirical risk minimization(ERM) [1,2].

Support vector machine(SVM) is a machine learning model based on SRM. The model not only minimizes the ERM, but also minimizes a regularization parameter which indicates the capacity of classification. Compared with neural networks, fuzzy learning machine and genetic algorithm, SVM has better generalization ability in case of small scale application[3], outstanding performance for nonlinear and high dimensional processing and no suboptimization problem. So SVM has been applied in pattern classification, regression and density estimation popularly[4,5,6].

However, there are some disadvantages of SVM. As the optimizational model of SVM is a quqdratic programming(QP) problem. It is necessary to calculate and store kernal matrix which size is n^2 . Solving QP problem needs to deal with much matrix operations[7]. The memory and time used in solving SVM increased rapidly when the problem scale was large.

In order to make a progress in solving SVM for large scale application, the Clip-SVM, Mean-SVM, Sparse-SVC are proposed[8,9]. These methods applied sample preprocessing technology to reduce the number of samples or support vectors. In this

F. Sun et al. (Eds.): ISNN 2008, Part II, LNCS 5264, pp. 268–275, 2008.

paper, a development of SVM based on sparse sampling is presented, which can make the number of the samples and vectors involved reducing observably.

2 Sample Importance Analyses

2.1 Discriminant Function for SVM

Classical SVM is applied to looking for a maximal margin optimal classification hyperplane through mapping the sample vector to high dimension space by kernal function. The optimal classification hyperplane formulation is $\omega^T \Phi(\boldsymbol{x}) + b = 0$,the discriminant function is $y(x) = \text{sign}\left(\omega^T \Phi(\boldsymbol{x}) + b\right)$,where $\Phi(\boldsymbol{x})$is a nonlinear mapping. The optimization model of SVM is formulized as follow

$$\begin{aligned} \min \quad & \tfrac{1}{2}\|\omega\|^2 + C\sum_{i=1}^{l} \xi_i \\ \text{s.t.} \quad & y_i\left(\omega^{\mathrm{T}}\Phi(\boldsymbol{x}_i) + b\right) \geq 1 - \xi_i, \\ & \xi_i \geq 0 \quad i = 1, 2, \ldots, l. \end{aligned} \tag{1}$$

it's dualty problem is

$$\begin{aligned} \min \quad & Q(\alpha) = \tfrac{1}{2}\sum_{i=1}^{l}\sum_{j=1}^{l} \alpha_i\alpha_j y_i y_j K\left(\boldsymbol{x}_i, \boldsymbol{x}_j\right) - \sum_{i=1}^{l} \alpha_i \\ \text{s.t.} \quad & \sum_{i=1}^{l} y_i\alpha_i = 0 \\ & 0 \leq \alpha_i \leq C \quad i = 1, 2, \ldots, l. \end{aligned} \tag{2}$$

where $K(\boldsymbol{x}_i, \boldsymbol{x}_j) = \boldsymbol{\Phi}(\boldsymbol{x}_i) \cdot \boldsymbol{\Phi}(\boldsymbol{x}_j)$ is a kernel function which can map the sample vector to high dimension Hilbert space. Discriminant functin is $y(\boldsymbol{x}) = \text{sign}\left(\sum_{\boldsymbol{x}_i \in SV} \alpha_i y_i K(\boldsymbol{x}_i, \boldsymbol{x}) + b\right)$.

Considering problem (1) and (2), although SVM maps the sample vector to high dismension space and looks for optimal hyperplane in feature space, but the computation increases little. What increased only is the computation of kernel innerproduct matrix. This advoid dimensionality curse.

According to Wolfe dualty theorem

$$\alpha_i\left(y_i\left(\omega^T \Phi(\boldsymbol{x}_i) + b\right) - 1 + \xi_i\right) = 0 \tag{3}$$

If sample vector $\boldsymbol{x}_s$ is support vector (namely $\alpha_s > 0$), then $y_i\left(\omega^T \Phi(\boldsymbol{x}_i) + b\right) = 1 - \xi_i$. Because $\xi_i \geq 0$, so support vectors (the A class vectors outer red line and B class vectors outer blue line in figure 1) are the vectors satisfy $y_i\left(\omega^T \Phi(\boldsymbol{x}_i) + b\right) \leq 1$. Perspicuously, most support vectors lie aside the overlaped area of A class sample and B class sample.

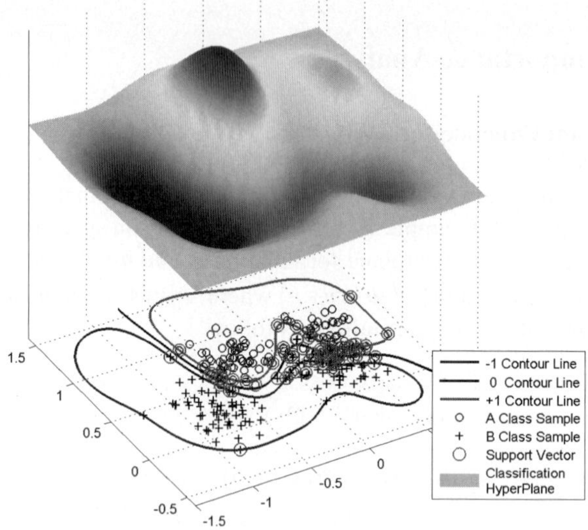

Fig. 1. Classical SVM Classification

2.2 Sample Likelihood Factor

Definition 1. *Suppose we have two class sample A and B, N is the total of sample vectors, $\boldsymbol{x}_i$ is the ith sample vector, $y_i$is the lable of $\boldsymbol{x}_i$, +1 denotes A class sample, -1 denotes B class sample. We define a likelihood factor of $\boldsymbol{x}_i$ as follow*

$$\eta_i = y_i \ln \frac{P(A|\boldsymbol{x}_i)}{P(B|\boldsymbol{x}_i)} = y_i \ln \frac{P(\boldsymbol{x}_i|A)P(A)P(\boldsymbol{x}_i)}{P(\boldsymbol{x}_i)P(\boldsymbol{x}_i|B)P(B)} = y_i \ln \frac{P(\boldsymbol{x}_i|A)P(A)}{P(\boldsymbol{x}_i|B)P(B)} \tag{4}$$

At the same time, we suppose the density functions of A and B are fixed ($P(A)$ and $P(B)$ are fixed) , then η_i only depends on $P(\boldsymbol{x}_i|A)$ and $P(\boldsymbol{x}_i|B)$.

Following six cases are discussed

a. $\boldsymbol{x}_i \in A$,$y_i = +1$, if $y_i\left(\omega^T \Phi(\boldsymbol{x}_i) + b\right) > 1$ is satisfied , then $\boldsymbol{x}_i$ is not a support vector ($\alpha_i = 0$), η_i is a plus lage numeral.
b. $\boldsymbol{x}_i \in A$,$y_i = +1$, if $-1 \leq y_i\left(\omega^T \Phi(\boldsymbol{x}_i) + b\right) \leq 1$ is satisfied , then $\boldsymbol{x}_i$ is a support vector ($\alpha_i > 0$), η_i is a small numeral.
c. $\boldsymbol{x}_i \in A$,$y_i = +1$, if $y_i\left(\omega^T \Phi(\boldsymbol{x}_i) + b\right) \leq -1$ is satisfied , then $\boldsymbol{x}_i$ is support vector ($\alpha_i > 0$), η_i is a negative large numeral.
d. $\boldsymbol{x}_i \in B$,$y_i = -1$, if $y_i\left(\omega^T \Phi(\boldsymbol{x}_i) + b\right) > 1$ is satisfied , then $\boldsymbol{x}_i$ is not a support vector ($\alpha_i = 0$), η_i is a plus lage numeral.
e. $\boldsymbol{x}_i \in B$,$y_i = -1$,if $-1 \leq y_i\left(\omega^T \Phi(\boldsymbol{x}_i) + b\right) \leq 1$ is satisfied , then $\boldsymbol{x}_i$ is a support vector($\alpha_i > 0$), η_i is a small numeral.
f. $\boldsymbol{x}_i \in B$,$y_i = -1$,if $y_i\left(\omega^T \Phi(\boldsymbol{x}_i) + b\right) \leq -1$ is satisfied , then $\boldsymbol{x}_i$ is support vector ($\alpha_i > 0$), η_i is a negative large numeral.

Just like mentioned above, if η is larger, the possibility of that the corresponding sample vector is support vector is smaller. Reversely, if η is smaller, the possibility of that the corresponding sample vector is support vector is larger. So we can regard η as a indicator of the influence of sample vector to optimal classification hyperlane. $\eta < 0$ means that the sample will be misjudged in Bayesian Classifier.

3 Sparse Sampling

3.1 Non-important Sample Cliping

Theorem 1. *Suppose we have N samples of two class A and B. If* $\mathbf{x}_i$ *is not a support vector of SVM, then after* $\mathbf{x}_i$ *is removed, the optimal classification hyperlane for the* $N-1$ *samples is invariable.*

Proof. Let $\boldsymbol{\alpha}^*$ be the solution of the dualty problem (2) for the N samples.Since $\boldsymbol{x}_i$ is not a support vector,we can know $\alpha_i = 0$ from Wolfe Dualty Theorem.If $\boldsymbol{x}_i$ is removed, because $\boldsymbol{\alpha}^{**}$ is the solution of the dualty problem (2) for the $N-1$ samples. It is obviously that $[\boldsymbol{\alpha}^{**}, \alpha_i = 0]$ is a feasible solution of the dualty problem (2) for the N samples,then

$$Q_N(\boldsymbol{\alpha}^*) \leq Q_N\left([\boldsymbol{\alpha}^{**}, \alpha_i = 0]\right) \tag{5}$$

similarly, the vector $\bar{\boldsymbol{\alpha}}^*$ which gets from $\boldsymbol{\alpha}^*$ by removing $\boldsymbol{\alpha}_i^* = 0$ is a feasible solution of the dualty problem (2) for the $N-1$ samples,then

$$Q_{N-1}(\boldsymbol{\alpha}^{**}) \leq Q_{N-1}(\bar{\boldsymbol{\alpha}}^*) \tag{6}$$

observing the dualty problem (2), the following equations can be gained.

$$\begin{aligned} Q_N(\boldsymbol{\alpha}^*) &= Q_{N-1}(\bar{\boldsymbol{\alpha}}^*) \\ Q_{N-1}(\boldsymbol{\alpha}^{**}) &= Q_N\left([\boldsymbol{\alpha}^{**}, \alpha_i = 0]\right) \end{aligned} \tag{7}$$

from (5),(6)and (7),we can get

$$Q_N(\boldsymbol{\alpha}^*) = Q_N\left([\boldsymbol{\alpha}^{**}, \alpha_i = 0]\right) \tag{8}$$

since,the dualty problem (2) is a convex optimization problem, so Problem (2) has only one optimal solution,then we can get $\boldsymbol{\alpha}^* = [\boldsymbol{\alpha}^{**}, \alpha_i = 0]$. This show that the optimal classification hyperlane for the $N-1$ samples is same to that for the N samples.

According to **Theorem 1**, we can know the discriminant function for SVM only depends on the support vectors, non-support vectors are redundant. So we can remove those sample vectors which likelihood factors η are larger, this is called *non-important sample cliping*.

3.2 Misjudged Sample Revising

Theorem 2. *Suppose we have two class sample A and B, if the likelihood factor of sample vector* $\mathbf{x}_i$ *satifies* $\eta_i < 0$ *, then after the label* y_i *of* $\mathbf{x}_i$ *is reversed, the convergence property of the discriminant function for SVM is invariable.*

Proof. The decision rule of Binary Bayesian Classifier is that if $g(\boldsymbol{x}) > 0$, then $\boldsymbol{x}$ is judged as A, otherwise , $\boldsymbol{x}$ is judged as B, where $g(\boldsymbol{x}) = \ln \frac{P(x|A)}{P(x|B)} + \ln \frac{P(A)}{P(B)}$. Since the likelihood factor η_i of $\boldsymbol{x}_i$ satifies inequation $\eta_i < 0$, then

$$\eta_i = y_i \left(\ln \frac{P(\boldsymbol{x}_i|A)}{P(\boldsymbol{x}_i|B)} + \ln \frac{P(P(A)}{P(P(B)} \right) < 0 \tag{9}$$

so the sample $\boldsymbol{x}_i$ is misjudged by the Binary Bayesian Classifier. It is obvious that if the lable y_i of sample $\boldsymbol{x}_i$ is reversed, the discriminant function of Binary Bayesian Classifier is not changed. Because the discriminant function of SVM converges to that of Binary Bayesian Classifier [10,11,12]. So the label y_i of $\boldsymbol{x}_i$ is reversed, the convergence property of the discriminant function for SVM is invariable.

According **Theorem 2**, if we reverse the label y_i which $\eta_i < 0$, the solution of SVM is also available. Furthermore, the number of support vectors become little. This is called *misjudged sample revising*.

4 Sparse Sampling Classification

4.1 Algorithm

According to the content above , we can clip and revise the sample through the likelihood factor η . But (4) tell us if we want to compute η_i , we must know prior probability $P(A)$,$P(B)$ and posterior probability $P(\boldsymbol{x}_i|A)$,$P(\boldsymbol{x}_i|B)$. $P(A)$ and $P(B)$ can be estimated by the frequency of A and B. $P(\boldsymbol{x}_i|A)$ and $P(\boldsymbol{x}_i|B)$ can be estimated by Parzen window method [13],

$$P(\boldsymbol{x}_i|A) = \frac{1}{|A|} \sum_{\boldsymbol{x}_j \in A} \frac{1}{V} \varphi(\boldsymbol{x}_i, \boldsymbol{x}_j, \sigma^2) \tag{10}$$

$$P(\boldsymbol{x}_i|B) = \frac{1}{|B|} \sum_{\boldsymbol{x}_j \in B} \frac{1}{V} \varphi(\boldsymbol{x}_i, \boldsymbol{x}_j, \sigma^2) \tag{11}$$

$|A|$ and $|B|$ are the amounts of A and B respectively. V is the volume of the window function. $\varphi(\boldsymbol{x}_i, \boldsymbol{x}_j, \sigma^2)$ is a normal density function with mean $\boldsymbol{x}_j$ and variance σ^2 .

If we have the likelihood factors of the samples, we can make a progress in classical SVM. Below is the detail of the algorithm.

Step 1. Likelihood factors computing
Select a appropriate variance σ^2 to estimate every sample's likelihood factor. And sort the samples in ascending order by likelihood factor.

Step 2. Non-important sample cliping
Select necessary frontal M samples for using.

Step 3. Misjudged sample revising
Reversed those samples lables whose likelihood factors satify qualification $\eta < 0$.

Step 4. SVM solving
Use classical SVM algorithm to solve the problem2.

Step 5. Algorithm over

This improved algorithm increases only one time computation of kernel matrix, but after cliping and revising, the total samples used in SVM decreases obviously, so the computation of the problem (2) is sharply reduced. And at the same time, the number of support vectors also becomes small, this is beneficial for sample testing.

4.2 Experiment

Apply the sparse sampling SVM(SS-SVM) algorithm stated above to make a classification on Ripley data set[1] . The data set has two class synthetical data which has overlap area shown in figure2. The training data number is 250 and the test data number is 1000. We select the Gauss function with variance $\sigma^2 = 0.04$ as window function to estimate the likelihood factor of samples and select the Gauss function with $\sigma^2 = 0.2$ as kernel function to solve the dualty problem (2). The parameter C in problem (2) is seted as $C = 30$. Compared with with Classical SVM and Sparse SVC [9], the experiment results are shown in figure3 and table 1.

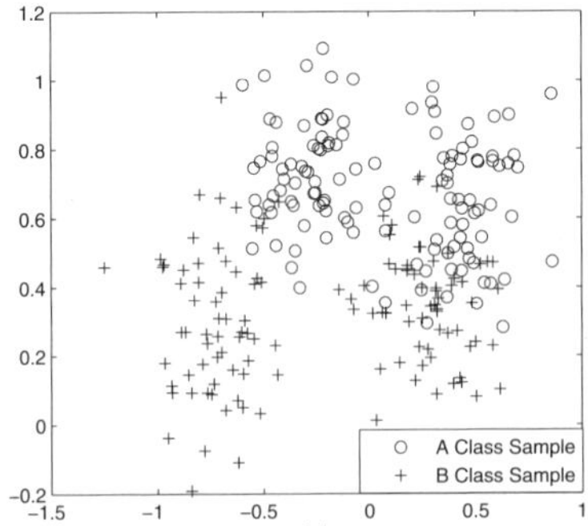

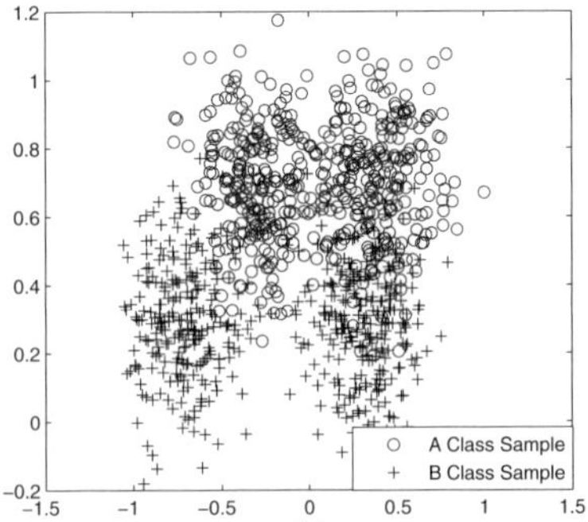

Fig. 2. Ripley data set (a) train data, (b) test data

Table 1. The result of SS-SVM, Classical SVM, Sparse SVM

Method	Train Samples	Cliped	Revised	SVs	Test Error(%)	Elapsed time(s)
Classical SVM	250	No	No	74	9.9	2.586625
Sparse SVC	250	No	No	5	10.2	1073
SS-SVM	250	No	Yes	21	8.7	2.754693
SS-SVM	150	Yes	Yes	20	8.4	0.678751

From **Table** 1 and **Fig** 3, we can see SS-SVM does better in computation over other classification methods through *non-important sample cliping* and *misjudged sample revising*. As the amount of train samples becomes small, the computation in step 4 of the SS-SVM algorithm is sharply reduced. So the speed of SS-SVM is higher than Classical SVM and sparse SVC[9]. Although the SVs of SS-SVM is not lessthan sparse SVC, but because the sparse SVC is a iterative algorithm which will elapse many time[9]. So generally speaking, the SS-SVM is better than classical SVM and sparse SVM.

[1] Available from "http://www.stats.ox.ac.uk/pub/PRNN/"

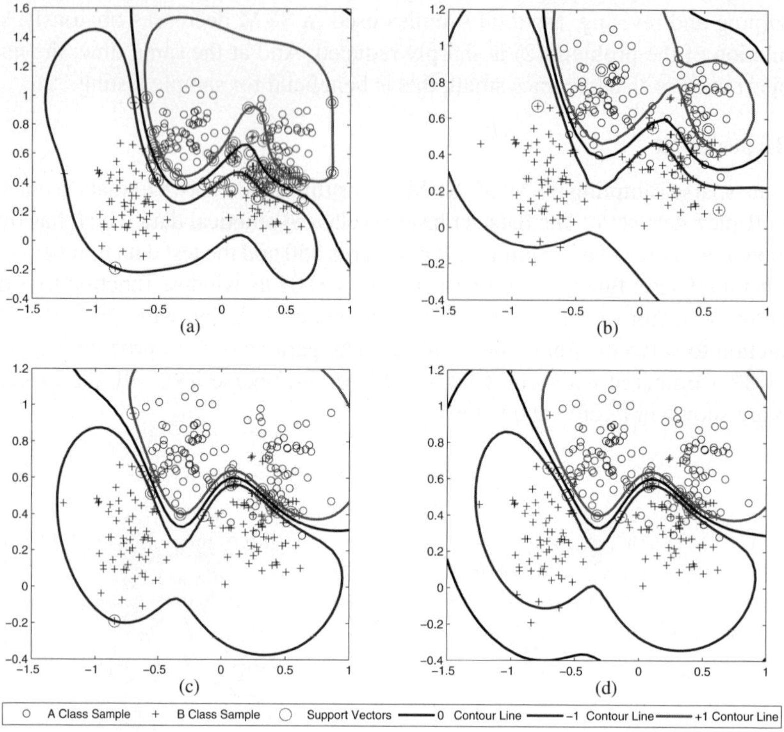

Fig. 3. Classification results: (a) Classical SVM, (b) Sparse SVM, (c) SS-SVM with M=250, (d) SS-SVM with M=150

5 Conclusion

In this paper, we start with the influence of sample to optimal classification hyperplane and then analyse the property of non-important sample and misjudged sample. The likelihood factor is defined to indicate the importance of sample. Finally, we design a classification method SS-SVM. The proposed algorithm has three merits:(a) Simplified algorithm.(b)Small computation.(c)Less support vectors. The results of the application on Ripley data set show that SS-SVM has an advantage over other methods.

Acknowledgements

This work was jointly supported by the National Science Fund for Distinguished Young Scholars (Grant No: 60625304), the National Natural Science Foundation of China (Grants No: 60621062, 60504003, 60474025, 90405017), the National Key Project for Basic Research of China (Grant No: G2007CB311003, G2002CB312205) and the Specialized Research Fund for the Doctoral Program of Higher Education (Grant No: 20050003049).

References

1. Vapnik, V.N.: The Nature of Statistical Learning Theorem. Springer, Berlin (1995)
2. Vapnik, V.N.: Statistical Learning Theory. John Wiley and Sons, New York (1998)
3. Zhang, X.G.: Introduction to Statistical Learning Theory and Support Vector Machines. Acta Automatica Sinica 26, 32–41 (2000)
4. Christopher, J.C.B.: A Tutorial on Support Vector Machines for Pattern Recognition. Data Mining and Knowledge Discovery 2, 121–167 (1998)
5. Alex, J.S., Bernhard, S.: A Tutorial on Support Vector Regression. Statistics and Computing 14, 199–222 (2004)
6. Vapnik, V.N., Mukherjee, S.: Support Vector Method for Multivariate Density Estimation. In: Advances in Neural Information Processing Systems. MIT Press, Cambridge (1999)
7. Li, X.Y., Zhang, X.F., Shen, L.S.: Some Developments on Support Vector Machine. Chinese Journal of Electronics 25, 7–12 (2006)
8. Zhang, X.G.: Using Class-center Vectors to Build Support Vector Machines. In: Neural Networks for Signal Processing IX-Proceedings of the 1999 IEEE Workshop, pp. 3–11. IEEE, Wisconsin (1999)
9. Zheng, D.N.: Research on Kernel Methods in Machine Learning. PH.D thesis Tsinghua University vol. 17–67 (2006)
10. Lin, Y.: Support Vector Machines and the Bayes Rule in Classification. Data Mining and Knowledge Discovery 6, 259–275 (2002)
11. Steinwart, I.: Support Vector Machines Are Universally Consistent. J. Complexity 18, 768–791 (2002)
12. Wu, Q., Zhou, D.X.: Analysis of Support Vector Machine Classification (preprint, 2004)
13. Richard, O.D., Peter, E.H., David, G.S.: Pattern Classification, 2nd edn. ISBN:0-471-05669-3

Estimation of Nitrogen Removal Effect in Groundwater Using Artificial Neural Network

Jinlong Zuo

Department of Environmental Engineering, Harbin University of Commerce, 50# box, Tongda Road 138#, Daoli district, Harbin, 150076, China
mdjzjl@163.com

Abstract. Groundwater contamination by nitrate is a globally growing problem. Biological denitrification is a simple and cost effective method. However, this process is non-linear, complex and multivariable. This paper presents the application of artificial neural network (ANN) in denitrification process in ground water. Experimental results showed that the ANN was able to predict the output water quality parameters—including nitrate as well as nitrite and COD. Most of relative error of NO_3^--N and COD were in the range of ±10% and ±5% respectively. The ANN model of nitrate removal in ground water prediction results produced good agreement with experimental data.

Keywords: groundwater; nitrogen removal; artificial neural networks (ANN).

1 Introduction

Groundwater serves as an important source of drinking water in various parts of the world [1]. Groundwater contamination by nitrate is a globally growing problem due to the population growth and increase of demand for food supplies [2]. Nitrate is identified as one of the hazardous contaminants in potable water that may reduce to nitrosamines in the stomach which is suspected to cause gastric cancer. In addition, the nitrite reacts with the hemoglobin in blood and converts the hemoglobin into methaemoglobin, which does not carry oxygen to cell tissues. This phenomenon results in a bluish color of infant's skin so called methaemoglobinemia or the blue baby syndrome [3-4].

So nitrate problem prohibited the direct use of the groundwater resources for human consumption in some parts of the world including India, Japan, China, Saudi Arabia, USA, UK and several parts of Europe [5]. With the aim to protect the consumers from adverse effects of high nitrate intake, the United States Environmental Protection Agency (USEPA) has regulated the maximum contaminant levels (MCLs) that nitrate and nitrite concentration in drinking water not greater than 10 mg NO_3^--N/L and 1.0 mg NO_2^--N/L [6]. The World Health Organization and the European Economic Community have set standards of 10 mg NO_3^--N/L and 0.03mg NO_2^--N/L [7]. The European standard for nitrite is stricter to account for the direct toxic effects from nitrite [8-9].

F. Sun et al. (Eds.): ISNN 2008, Part II, LNCS 5264, pp. 276–283, 2008.

In order to comply with this regulation, nitrate must be removed from groundwater effectively. The conventional processes of water treatment method which are applied for water portability are not efficient for the elimination of nitrate ion from the water. Existing remediation technologies include ion exchange, reverse osmosis, electro dialysis, and biological denitrification [10-11]. Biological denitrification, the microbial reduction of NO_3^- and NO_2^- to gaseous dinitrogen (N_2) holds an environmental and economic advantage over other methods because it is simple, selective, and cost effective. However, it needs improving effluent water quality for potable water purpose when carbon source is added as electron donors to facilitate denitrifying nitrate. In addition, water treatment systems are non-linear and multivariable, and they consist of a great number of complex processes. How to control nitrogen removal efficiently and meet stricter effluent quality standard at a minimum cost has attracted lots of attentions.

The ANN can imitate such basic characteristics of the human brain as self-adaptability, self-organizing, and error tolerant and has been widely adopted for mode identification, analysis and forecast, system recognition and design optimization [12-13]. MATLAB is mathematics software with high-level numerical computation and data visualization ability. It provides users with ANN design and simulation conveniently. Recently, the artificial neural network (ANN) has been increasingly applied in the area of environmental and water resources engineering [14-17]. In this paper, ANN is used to estimate the biological denitrification effect in groundwater. At First, architecture of ANN is constructed for estimate the nitrate removal effect. And then the input and output of the neurons are obtained by the ANN model and MATLAB GUI function. At last, the values estimated by the ANN compared with the experiment results.

2 Material and Methods

The experiments were carried out in a modified configuration lab-scale plant (Fig.1); this continuous-flow system was set up as a rectangular tank with an effective volume of approximately 60 liters. It consisted of a bioreactor and a secondary settler (20 liters). The bioreactor was composed of four compartments (15 liters each). The first two compartments were non-aerated and the last two were aerated. The inflow, nitrate recirculation flow and sludge recycle flow were controlled by peristalsis pump. Influent flow was 750mL/h. The MLSS was controlled at 2500~3000mg/L and SRT was 20 days. The sludge recycle ratio was 100%. All experiments were conducted at temperature (20~25℃).

Synthetic water was used to simulate groundwater and fed continuously at a constant flow rate of 750 mL/h. The supported aeration flow rate was fixed at 10 L/h. With an overall hydraulic retention time (HRT) of 8 hours, the time intervals for different serial operation phases were fixed as followings: 4 hours for anoxic, 4 hours for aerobic, along with 2 hours for settling.

The sludge was obtained from an A/O process municipal water treatment plant. The synthetic wastewater was prepared with distilled water and the composition is as follows: KNO_3 (0.04~0.06 g/L), KH_2PO_4 (0.01~0.03g/L), $NaHCO_3$ (0.1~ 0.2g/L), $MgSO_4{\cdot}7H_2O$ (0.05g/L), $CaCl_2{\cdot}2H_2O$ (0.02g/L), $FeSO_4{\cdot}2H_2O$ (0.001g/L).

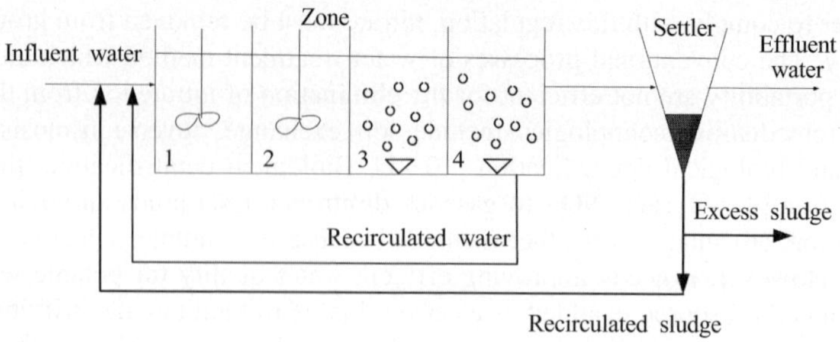

Fig. 1. The schematic graph of the test layout

Measurement of pH, DO (dissolved oxygen) and mixed liquid suspend solid (MLSS), the variations in chemical oxygen demand (COD), NO_3^--N (nitrate), NO_2^--N (nitrite), HRT (Hydraulic retention time), and SRT (sludge retention time) are also analyzed following Standard Methods [18]. Aerobic zone DO and pH are measured by WTW-340i inolab analyzer.

3 Artificial Neural Networks

3.1 Architecture of Neural Networks

The ANN model was done via MATLAB, mathematical software introduced by Mathworks of USA in 1982 which has high level numerical computation and data visualization capacity. MATLAB Neural Network Toolbox 4.0 is an integral part of MATLAB6.x high-performance visualized numerical computation software. Aimed for analysis and design of ANN, Toolbox 4.0 offers many toolbox functions that can be called directly. GUI and Simulink, the simulation tool, has become an ideal tool for analysis and design of ANN. The model can be modified subject to actual needs to forecast water quality under various conditions.

The model created in this paper is a ANN with a single hidden layer (Fig.2), with R as the input layer, S^1 the hidden layer, S^2 the output layer, $IW^{1.1}$ the weight matrix of the input layer, $LW^{2.1}$ the weight matrix from the hidden layer to the output layer, b^1 and b^2 threshold values of the hidden and output layer respectively, f^1 and f^2 the neuron transfer function of the hidden and output layer respectively. The ANN model shown as in Fig.2, the input and output variables were established for evaluation of water quality.

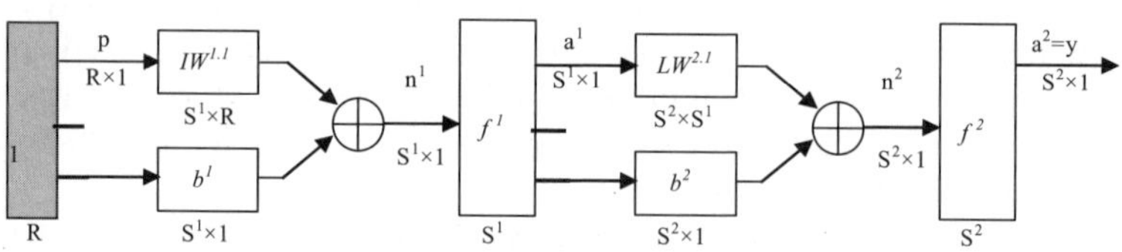

Fig. 2. ANN model with a single hidden layer

3.2 ANN Forecast Model with MATLAB

By keying in nntool in command window of MATLAB, the user will enter the main page of the neural network GUI (Fig.3), the Network/Data Manager.

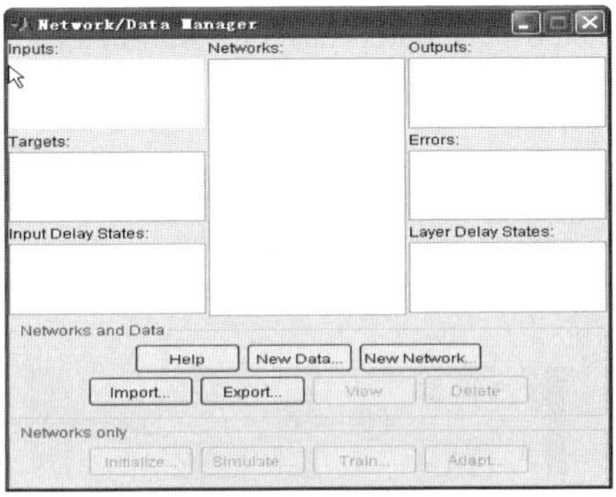

Fig. 3. GUI main page

First, upload respectively under Input and Targets in GUI main page the input and output data that have been previously written into Excel worksheet. The input variables are set at 9. Next, click New Network to create a new network model as shown in Fig.4.

Select Feed-forward backdrop as the Network Type. Based on experiences, the *LOGSIG* or *TANSIG* can be chosen as the neuron transfer function of the hidden layer. The output characteristics of the entire neural network will be decided by the characteristics of the last layer of the ANN. When *Sigmoid* functions are applied to the last layer, the output of the entire network will be limited to a smaller range; and if *Purelin* functions is applied to the last layer, the output could be an arbitrary value. As a result, *Purelin* is chosen as the transfer function for the neurons of the output layer.

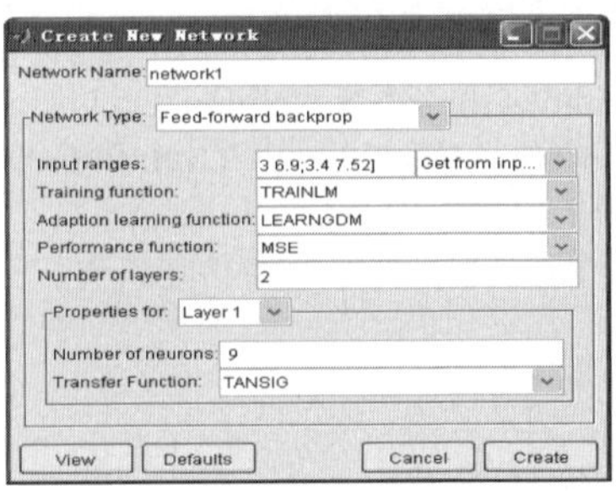

Fig. 4. Create new network

3.3 Inputs and Output Neurons

Many water quality parameters are monitored in treated water. For the purpose of estimate Nitrogen removal effect in groundwater, we are able to get daily raw water quality parameters such as COD、NO_3^--N、NO_2^--N、pH、Temperature、 MLSS、DO、SRT and HRT. These parameters could be used for input nodes. The experimental effluent water quality parameters, which are selected as COD、NO_3^--N、NO_2^--N, could be used for output nodes. Because the European standard for nitrite is very strict (0.03mg NO_2^--N/L) account for the direct toxic effects to human health. Therefore, the study of the nitrite concentration in the effluent water must be controlled zero by adopting two stages aerobic in the experiment.

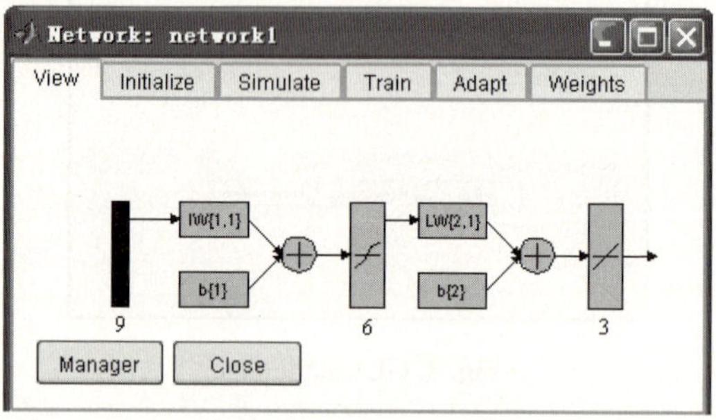

Fig. 5. The ANN structure

A neural network used here has one hidden layer composed of 6 neurons, with IW{1,1}as the weight matrix of the input layer, LW{2,1}as the weight matrix from hidden layer to output layer, b{1}and b{2} threshold values of the hidden and output layer respectively. As it can be seen from Fig. 5, the ANN has three layers, an input layer consisting of nine nodes, a hidden layer consisting of six nodes, and an output layer consisting of three nodes.

4 Results and Discussion

4.1 ANN Training Output Error Variation

The training process determines the ANN weights and is similar to the calibration of a mathematical model. The ANN is trained with a training set of experimental input and

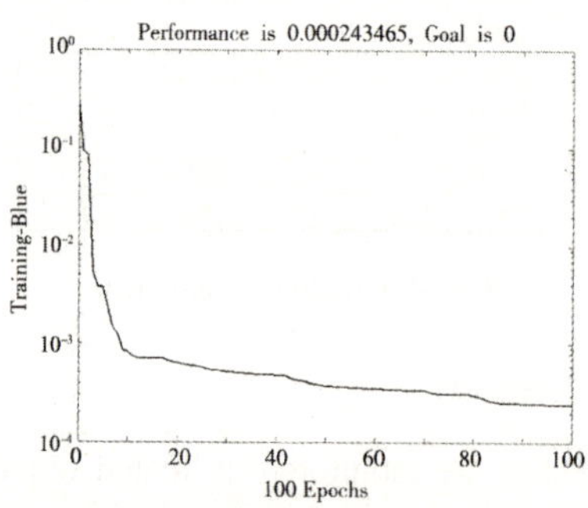

Fig. 6. The output error variability of ANN with epochs

output data. In this paper, all 97 data are used for study, 54 sets of data for training and remaining 43 data sets for verification.

At the beginning of training, the weights are initialized with a set of random values. The goal of learning is to determine a set of weights that will minimize the error function. As training proceeds, the weights are systematically updated according to a training rule. The process is terminated when the difference between measured and estimated value is less than a specified value. As shown in Fig. 6, the difference (error) between experimental and ANN output value is less than 10^{-3} after 100 epochs.

4.2 Verification by ANN

The values of COD, NO_3^--N, NO_2^--N estimated by the ANN are compared against the respective measured values of the experiments. The output NO_2^--N concentration of ANN is the same as the experimental result because they are zero.

As can be seen from Fig. 7, the COD results of training and verification of ANN are quite similar to those of the experimental data. Fig. 8 shows the NO_3^--N of ANN output difference from the measured data. The NO_3^--N maximum and minimum values of ANN output are 10.36427 mg/L and 1.93677mg/L, respectively. Compared with corresponding ANN input value, the removal efficiency of NO_3^--N could reach 92.3% and 94.5%, respectively.

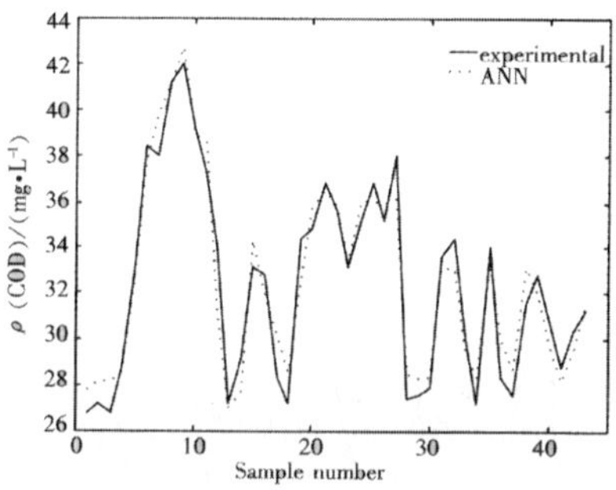

Fig. 7. The ANN verification and experimental data of COD

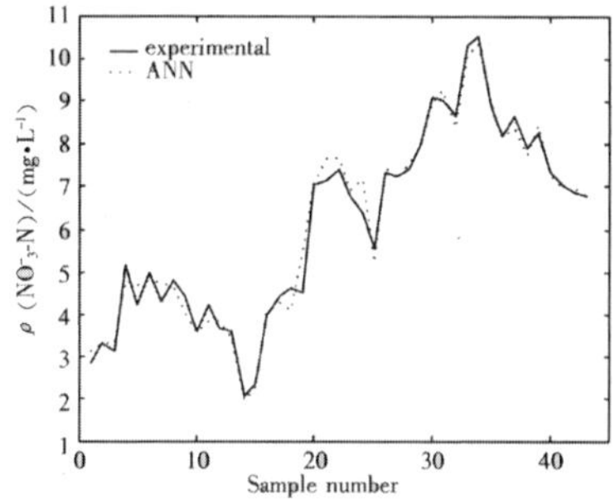

Fig. 8. The ANN verification and experimental data of NO_3^---N

4.3 Relative Error Distributions

The relative error distribution of NO_3^--N and COD output between experiment and ANN is showed in Fig. 9 and Fig. 10. The maximum relative error of NO_3^--N and COD are 21.45% and 33.42%, respectively.

However, most of relative error of NO_3^--N and COD are in the range of ±10% and ±5% respectively. Thus the average forecast error rate indicates that the overall forecast results are fairly good with error controlled within an acceptable range, proving the viability of the forecast model.

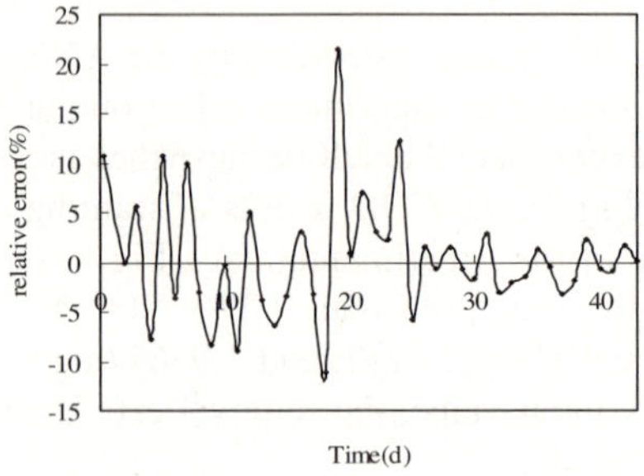

Fig. 9. The relative error distribution of NO_3^--N output between experiment and ANN

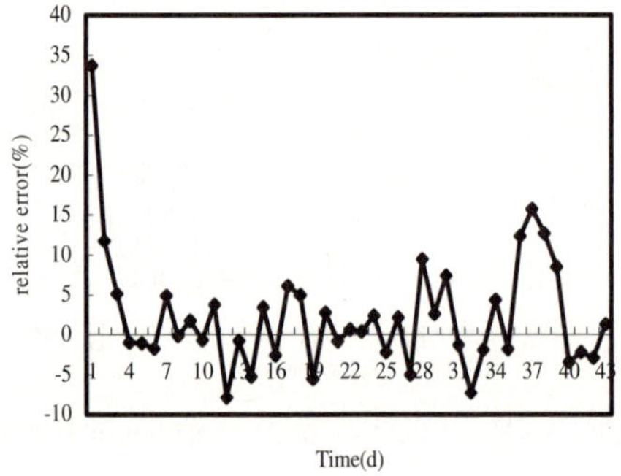

Fig. 10. The relative error distributions of COD output between experiment and ANN

5 Conclusions

In solving the present problems in ground water, ANN applications appeared simple and robust. This method to estimate an unknown parameter could use the correlation among water quality parameters. This paper has used the ANN to predict the output water quality parameters—including nitrate as well as nitrite and COD. Most of relative error of NO_3^--N and COD are in the range of ±10% and ±5% respectively. It showed that the ANN model of nitrate removal in ground water prediction results produced good agreement with experimental data.

Reference

1. Schubert, C., Kanarek, M.S.: Public Response to Elevated Nitrate in Drinking Water Wells in Wisconsin. Archi. Environ. Health 4, 242–247 (1999)
2. Insaf, S., Babiker, B., Mohamed, A.A., Terao, H., Kato, K., Keiichi, O.: Assessment of Groundwater Contamination by Nitrate Leaching from Intensive Vegetable Cultivation Using Geographical Information System. Environ. Inter. 29, 1009–1017 (2004)
3. Galvez, J.M., Gomez, M.A., Hontoria, E., Gonzalez, L.J.: Influence of hydraulic loading and air flowrate on urban wastewater nitrogen removal with a submerged fixed-film reactor. J. Hazard. Mater 101, 219–229 (2003)
4. Shrimali, M., Singh, K.P.: New Methods of Nitrate Removal from Water. Environ. Pollut. 112, 351–359 (2001)
5. Nolan, B.T., Ruddy, B.C., Hitt, K.J.: A National Look at Nitrate Contamination of Ground Water. Wat. Con. Puri. 39, 76–79 (1998)
6. Drinking Water Health Advisories, `http://www.epa.gov`
7. Water Sanitation and Health, `http://www.who.int`
8. Urbain, V., Benoit, R., Manem, J.: Membrane Bioreactor: A New Treatment Tool. J. AWWA 88, 75–86 (1996)
9. Kapoor, A., Viraraghavan, T.: Nitrate Removal from Drinking Water Review. J. Environ. Eng. 123, 371–380 (1997)
10. Haugen, K.S., Semmens, M.T., Novak, P.J.: A Novel in Situ Technology for the Treatment of Nitrate Contaminated Groundwater. Wat. Res. 36, 3497–3506 (2002)
11. Kuo, Y.M., Liu, C.W., Lin, K.H.: Evaluation of the Ability of an Artificial Neural Network Model to Assess the Variation of Groundwater Quality in an Area of Blackfoot Disease in Taiwan. Wat. Res. 38, 148–158 (2004)
12. Lek, S., Guiresse, M., Giraudel, J.L.: Predicting Stream Nitrogen Concentration from Watershed Features Using Neural Networks. Wat. Res. 33, 3469–3478 (1999)
13. Wen, C.W., Lee, C.S.: A Neural Network Approach to Multiobjective Optimization for Water Quality Management in a River Basin. Wat. Resource Res. 34, 427–436 (1998)
14. Gail, M.B., Neelakantan, T.R., Srinivasa, L.: A Neural-network-based Classification Scheme for Sorting Sources and Ages of Fecal Contamination in Water. Wat. Res. 36, 3765–3774 (2002)
15. Chang, T.C., Chao, R.J.: Application of Backpropagation Networks in Debris Flow Prediction. Eng. Geology 85, 270–280 (2006)
16. Chen, L.H., Chang, Q.C., Chen, X.G.: Using BP Neural Network to Predict the Water Quality of Yellow River. J. Lanzhou Univ (Natural Sciences) 39, 53–56 (2003) (in Chinese)
17. Vandenberghe, V., Bauwens, W., Vanrolleghem, P.A.: Evaluation of Uncertainty Propagation into River Water Quality Predictions to Guide Future Monitoring Campaigns. Environ. Mod. Soft 22, 725–732 (2007)
18. Chinese EPA. Water and Wastewater Monitoring Methods, 3rd edn. Chinese Environmental Science Publishing House, Beijing (1997) (in Chinese)

Sequential Fuzzy Diagnosis for Condition Monitoring of Rolling Bearing Based on Neural Network

Huaqing Wang[1,2] and Peng Chen[1]

[1] Graduate School of Bioresources, Mie University
1577 Kurimamachiya-cho, Tsu, 514-8507 Mie, Japan
chen@bio.mie-u.ac.jp

[2] School of Mech. & Elec. Engineering, Beijing University of Chemical Technology
ChaoYang District, 100029 Beijing, China
wanghq_buct@hotmail.com

Abstract. In the case of fault diagnosis of the plant machinery, diagnostic knowledge for distinguishing faults is ambiguous because definite relationships between symptoms and fault types cannot be easily identified. This paper propose a sequential fuzzy diagnosis method for condition monitoring of a rolling bearing used in a centrifugal blower by the possibility theory and a neural network. The possibility theory is used for solving the ambiguous problem of the fault diagnosis. The neural network is realized with a developed back propagation neural network. As input data for a neural network, the non-dimensional symptom parameters are also defined in time domain. Fault types of a rolling bearing can be effectively, sequentially distinguished on the basis of the possibilities of the normal state and abnormal states at early stage by the fuzzy diagnosis approach. Practical examples of diagnosis are shown in order to verify the efficiency of the method.

Keywords: Sequential fuzzy diagnosis, Neural network, Possibility theory, Condition monitoring, Rolling bearing.

1 Introduction

Rolling bearings are an important part of and widely used in rotating machinery. The fault of a rolling bearing may cause the breakdown of a rotating machine, and furthermore, serious consequences may arise due to the fault. Therefore, condition monitoring and fault diagnosis of rolling bearings is most important for guaranteeing production efficiency and plant safety [1] [2]. In the field of machinery diagnosis, utilization of vibration signals is effective in the detection of faults and the discrimination of fault type, because the signals carry dynamic information about the machine state [3] [4]. However, the values of symptom parameters calculated from vibration signals for fault diagnosis are ambiguous. Although fault diagnosis of rolling bearings is often artificially carried out using time or frequency analysis of vibration signals, there is a need for a reliable, fast automated diagnosis method thereof. Neural networks (NN) have potential applications in automated detection and diagnosis of machine failures [5]-[8]. However a conventional neural network cannot adequately

F. Sun et al. (Eds.): ISNN 2008, Part II, LNCS 5264, pp. 284–293, 2008.

reflect the possibility of ambiguous diagnosis problems, and will never converge, when the symptom parameters, input to the first layer of the NN, have the same values in different states [9].

For the above reasons, in order to solve these problems and improve the efficiency of the fault diagnosis, this paper proposes a sequential fuzzy diagnosis method for condition monitoring of a rolling bearing using a neural network and the possibility theory. A neural network is realized with the partially-linearized neural network, by which the condition of a bearing can be automatically judged on the basis of the possibility distribution of symptom parameters. The non-dimensional symptom parameters (NSP) in time domain are also defined for the condition diagnosis of a bearing. The detection index (DI) is used to evaluate the sensitivity of the NSP for distinguishing faults. Practical examples of fault diagnosis for a rolling bearing used in a centrifugal blower are shown to verify the efficiency of this method.

2 Experimental System for Bearing Diagnosis

The experimental system is shown in Fig.1, including the rotating machine (a centrifugal blower TERAL CLF3), the rolling bearing and the accelerometers. A 2.2 kW induction motor with three-phases and a maximum revolution of 1420 rpm is employed to drive a blower through two V-belts. Two Accelerometers are used to measure the vibration signals for the bearing diagnosis. The faults often occurring in a rolling bearing at an early stage, such as the outer race flaw and inner race flaw are considered in the present work. We artificially made those flaws as shown in Fig. 1 for the tests of condition diagnosis.

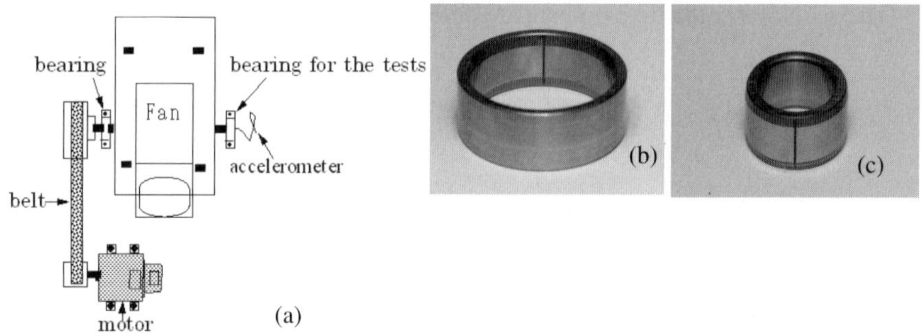

Fig. 1. Experiment system and rolling bearing flaws (a) Blower system, (b) Outer race flaw, (c) Inner race flaw

The vibration signals are measured at a rotational speed of 800 rpm. The sampling frequency is 100 kHz, and the sampling time is 5s. A high-pass filter with a 5 kHz cut-off frequency is used to cancel noise in these vibration signals. After preprocessing, the time-domain signal is divided into 20 parts each containing 25,000 (3.75 cycles) samples to calculate the symptom parameters. As an example, Fig. 2 shows the normalized vibration signals of each state.

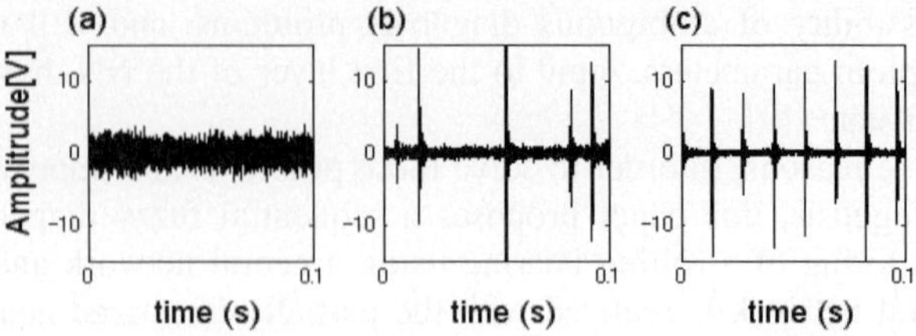

Fig. 2. Vibration signals in each state (a) Normal, (b) Inner flaw, (c) Outer flaw

3 Symptom Parameters for Fault Diagnosis

For automatic diagnosis, symptom parameters (SP) are needed that can sensitively distinguish the fault types. A large set of non-symptom parameters (NSP) has been defined in the pattern recognition field [10]. In this paper, six of those NSPs, are considered.

3.1 Non-dimensional Symptom Parameter (NSP)

To make the signals comparable regardless of differences in magnitude, the signals of each state are normalized by the following formula before calculating the NSPs.

$$x_i = (x'_i - \mu')/\sigma' \tag{1}$$

where, μ' and σ' are the mean and standard deviation of the original signals $x_i'(i=1\sim N)$, respectively. x_i is the ith element of the signal series after normalization.

The six NSPs in the time domain are described as follows.

$$p_1 = \sigma / \mu_{abs} \tag{2}$$

$$p_2 = \sum_{i=1}^{N} (x_i - \mu)^3 \Big/ N\sigma^3 \tag{3}$$

$$p_3 = \sum_{i=1}^{N} (x_i - \mu)^4 \Big/ N\sigma^4 \tag{4}$$

$$p_4 = \mu_p / \mu_{abs} \tag{5}$$

$$p_5 = \bar{x}_{\max} / \mu_p \tag{6}$$

$$p_6 = \sigma_p / \mu_p \tag{7}$$

where μ and σ is the average and standard deviation of x_i, σ_p and μ_p are the standard deviation and average of peak values of $|x_i|$, $\bar{x}_{\max}$ is the average of ten peak values (from top peak value to the tenth value) of $|x_i|$, and $\mu_{abs} = \sum_{i=1}^{N} |x_i| \Big/ N$, respectively.

3.2 Sensitivity Evaluation of Symptom Parameter

The sensitivity of a SP, which will be used to distinguish two states, is derived in the following way. Supposing that x_1 and x_2 are the SP values calculated from the signals measured in state 1 and state 2 respectively, and they conform respectively to the normal distributions $N(\mu_1,\sigma_1)$ and $N(\mu_2,\sigma_2)$. Here, μ and σ are the average and the standard deviation of the SP respectively. The larger the value of $|x_2-x_1|$ is, the higher the sensitivity of distinguishing the two states by the SP is. $z=x_2-x_1$ is also conform to the normal distribution $N(\mu_2-\mu_1,\sigma_2^2+\sigma_1^2)$ [11].

The probability of $x_2<x_1$ can be calculated as following:

$$P_0 = \int_{-\infty}^{0} \frac{1}{\sqrt{2\pi(\sigma_1^2+\sigma_2^2)}} \exp(-\frac{\{z-(\mu_2-\mu_1)\}^2}{2(\sigma_1^2+\sigma_2^2)}dz \tag{8}$$

where, $\mu_2 \geq \mu_1$ (we can obtain the same conclusion when $\mu_1 \geq \mu_2$).

With the substitution $u=\{z-(\mu_2-\mu_1)\}/\sqrt{\sigma_1^2+\sigma_2^2}$ to (7), the P_0 can be obtained as following:

$$P_0 = \frac{1}{\sqrt{2\pi}} \int_{-\infty}^{-DI} \exp(-\frac{u^2}{2})du \tag{9}$$

where, Distinction Index (DI) is calculated by

$$DI = \frac{|\mu_2-\mu_1|}{\sqrt{\sigma_1^2+\sigma_2^2}} \tag{10}$$

Distinction Rate (DR) is defined as following:

$$DR = 1 - P_0 \tag{11}$$

It is obvious that the larger the value of DI, the larger the value of DR will be, and therefore, the better the SP will be. So the DI can be used as the index of the quality to evaluate the distinguishing sensitivity of SP.

4 Sequential Fuzzy Diagnosis by Possibility Theory

In most practical cases of plant, it is difficult to establish the exacting accurate mathematical models for machinery because the fault mechanisms of machinery and the features of fault types cannot be perfectly clarified by a theoretical approach. In addition, due to the complexity of plant machinery conditions, it is very hard to find out one or a few symptom parameters that can identify all faults simultaneity.

In order to solve these problems, a sequential fuzzy diagnosis method is proposed. An example of the sequential diagnosis is shown in Fig. 3. In this case, a bearing in the normal state, the inner race flaw state and outer race flaw state should be identified sequentially.

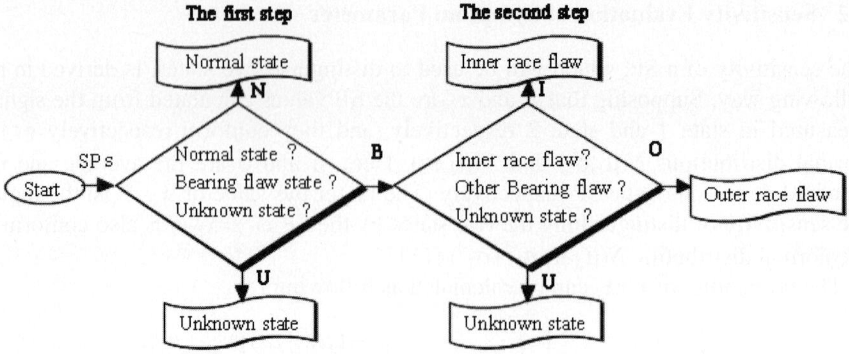

Fig. 3. Flowchart of sequential diagnosis for bearing diagnosis

The first step of the sequential diagnosis can be used to distinguish the normal state (N) from bearing faults (B) and the unknown state (U) with the relevant possibility functions of symptom parameters. The second step can be used to distinguish the inner race flaw (I) from the outer race flaw (O) and the unknown state.

In the present work, we used the *DI*s calculated by (10) to select the two best symptom parameters (P_i and P_j) for each step of the sequential diagnosis, respectively. The selection results of the SPs for the sequential classification are, P_1 and P_5 for the fist step, and P_1 and P_4 for the second step.

Table 1 shows the *DI*s of the NSPs (P_i and P_j) for each step of the sequential diagnosis respectively. Since all of those *DI*s are larger than 3.0, all of relevant detection rates (*DR*) are larger than 99.9% [12].

Table 1. The *DI*s of the NSPs for sequential diagnosis

	For the first step			For the second step
	N:I	*N:O*		*I:O*
P_1	9.66	22.22	P_1	5.86
P_5	8.08	12.39	P_4	8.93

The possibility theory is used to solve the ambiguous problem of the fault diagnosis in this work. The possibility function of the NSP can be obtained from its probability density function [11]. When the probability density function of the NSP conforms to the normal distribution, it can be changed to possibility distribution functions $P(x_i)$ by the following formula:

$$P(x_i) = \sum_{k=1}^{N} \min \{\lambda_i , \lambda_k\} \tag{12}$$

where, $$\lambda_i = \int_{x_{i-1}}^{x_i} \frac{1}{\sigma\sqrt{2\pi}} \exp\left\{-\frac{(x-\mu)^2}{2\sigma^2}\right\} dx \tag{13}$$

$$\lambda_k = \int_{x_{k-1}}^{x_k} \frac{1}{\sigma\sqrt{2\pi}} \exp\left\{-\frac{(x-\mu)^2}{2\sigma^2}\right\} dx \tag{14}$$

where, μ and σ are the average and the standard deviation of the SP respectively, and $x=\mu\text{-}3\sigma\sim\mu+3\sigma$.

In the case of sequential diagnosis, the possibility functions and the probability density functions of SP (P_i) used for each step, as examples, are shown in Fig. 4 respectively.

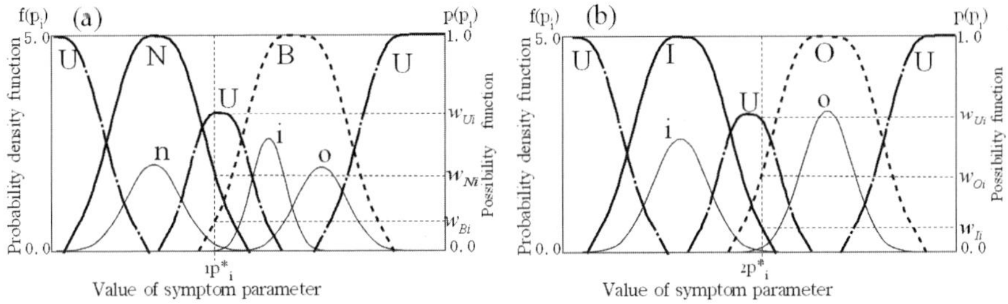

Fig. 4. Possibility function for fuzzy diagnosis (a) for the first step, (b) for the second step

In Fig. 4 (a), *N*, *B*, and *U* are the possibility functions of the normal state, bearing flaw, and the unknown state, respectively; *n*, *o* and *i* are the probability functions of the normal state, outer race flaw and inner race flaw, respectively. The w_{Ni}, w_{Bi} and w_{Ui} are the possibilities of p_i in the normal state, bearing faults and unknown states, respectively. w_{Ni} and w_{Bi} can be calculated by (12) from the probability density functions of p_i, and w_{Ui} can be calculated by the following:

$$w_{Ui} = 1 - (w_{Ni} + w_{Bi}) \tag{15}$$

In Fig. 4 (b), *I*, *O*, and *U* are the possibility functions of the inner race flaw, the outer race flaw and the unknown state, respectively; *i* and *o* are the probability functions of the inner race flaw and the outer race flaw, respectively.

In order to process the ambiguous relationship between symptom parameters and fault types, the combining possibility function of each state to be distinguished can be obtained by the Mycin certainty factor [13].

In the first step of the sequential diagnosis, the combining possibility functions of the normal state (w_N), bearing fault state (w_B), and unknown state (w_U) can be obtained through the possibilities of the two selected symptom parameters p_i and p_j (here i=1, and j=5), respectively, as follows:

$$w_K = w_{Ki} + w_{Kj} - w_{Ki}w_{Kj}, \quad K = N, B, U \tag{16}$$

The combining possibility functions are normalized as following:

$$w_K{}' = w_K \Big/ \sum_K w_K, \quad K = N, B, U \tag{17}$$

Similarly, the combining possibility functions for the second step can be obtained.

As teacher data of the neural network, the combining possibilities are used for the learning of the NN. The fuzzy NN should be described in the next chapter.

5 Fuzzy Neural Network

The fuzzy neural network is applied to diagnose the fault types of a rolling bearing by the sequential diagnosis algorithm, and realized with a developed back propagation neural network called as "the partially-linearized neural network (PNN)" [2]. A back propagation neural network is only used for training the data, and the PNN is used for testing the learned NN. Here, the basic principle of the PNN for the fault diagnosis is described as follows.

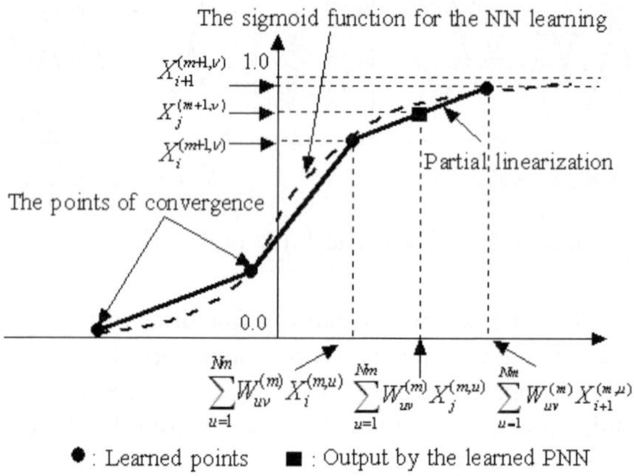

Fig. 5. The partial linearization of the sigmoid function

The neuron number of the mth layer of the NN is N_m. The set $X^{(1)}=\{X_i^{(1,j)}\}$ represents the pattern inputted to the 1st layer and the set $X^{(M)}=\{X_i^{(M,k)}\}$ is the training data for the last layer (Mth layer). Where, $X_i^{(1,j)}$: the value inputted to the jth neuron in the input (1st) layer; $X_i^{(M,k)}$: the output value of the kth neuron in the output (Mth) layer, $i=1\sim P$, $j=1\sim N_1$, $k=1\sim N_M$.

Even if the NN converges by learning $X^{(1)}$ and $X^{(M)}$, it cannot deal well with the ambiguous relationship between the new $X^{(1)*}$ and $X^{(M)*}$, which had not been learned. In order to predict $X^{(M)*}$ according to the probability distribution of $X^{(1)*}$, a partially linear interpolation of the NN is introduced in Fig. 5.

In the PNN that has converged by the training data $X^{(1)}$ and $X^{(M)}$, the symbols are used as follows.

$X_i^{(m,t)}$: The value of the tth neuron in the hidden (mth) layer; $t=1\sim N_m$.

$W_{uv}^{(m)}$: The weight between the uth neuron in the mth layer and the vth neuron in the (m+1)th layer; $m=1\sim M$, $u=1\sim N_m$, $v=1\sim N_{m+1}$.

If these values are all remembered by the computer, then when new values $X_j^{(1,u)*}$ ($X_i^{(1,u)} < X_j^{(1,u)*} < X_{i+1}^{(1,u)}$) are inputted to the first layer, the predicted value of the vth neuron ($v=1$ to N_m) in the $(m+1)$th layer ($m=1$ to $M-1$) will be estimated by

$$X_j^{(m+1,v)} = X_{i+1}^{(m+1,v)} - \frac{\{\sum_{u=1}^{Nm} W_{uv}^{(m)}(X_{i+1}^{(m,u)} - X_j^{(m,u)})\}(X_{i+1}^{(m+1,v)} - X_i^{(m+1,v)})}{\sum_{u=1}^{Nm} W_{uv}^{(m)}(X_{i+1}^{(m,u)} - X_i^{(m,u)})} \quad (18)$$

By using the operation above, the sigmoid function is partially linearized, as shown in Fig. 5. If a function must be learned, the PNN will learn the points indicated by the symbols (●) shown in Fig. 6. When new data (s_1', s_2') are inputted into the converged PNN, the value indicated by the symbols (■) corresponding to the data (s_1', s_2') will be quickly identified as P_e. Thus, the PNN can deal with ambiguous diagnosis problems.

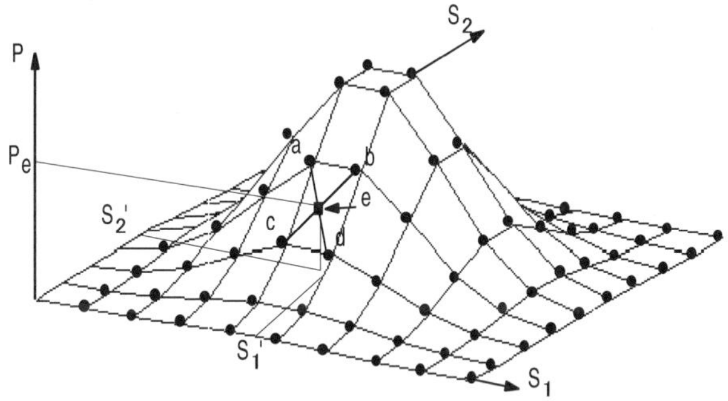

Fig. 6. Interpolation by the PNN

The new data (s_1', s_2') inputted into the converged PNN, which are not learned by the PNN to identify, must satisfy the following condition.

$$s_{1(\min)} < s_1' < s_{1(\max)}, s_{2(\min)} < s_2' < s_{2(\max)} \quad (19)$$

where, $s_{1(\min)}$ and $s_{2(\min)}$, $s_{1(\max)}$ and $s_{2(\max)}$ are the minimum and maximum value of s_1 and s_2, respectively, which have been learned by the PNN.

6 Diagnosis and Verification

Fig. 7 shows PNNs built for condition diagnosis of a bearing by a sequential diagnosis algorithm. In an example of the first step, the PNN consists of the first layer, the hidden layer and the last layer. The neurons in the first layer are inputted symptom parameters (P_1 and P_5), and the outputs in the last layer are w_N', w_B' and w_U' which are the possibilities of normal state, bearing flaw and unknown state respectively.

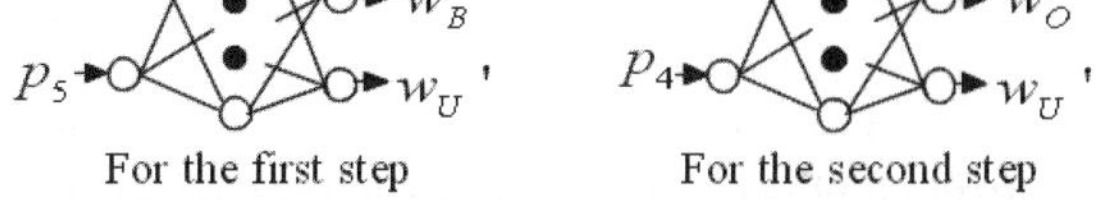

Fig. 7. PNNs built for the sequential fuzzy diagnosis

The knowledge for training of the PNN is acquired by the possibility theory and the Mycin certainty factor, and it can deal with the vagueness and uncertainty relationships between the symptoms and the fault types. Therefore, the PNN can obtain good convergence when learning the acquired knowledge.

We used the data measured in each state that had not been learned by the PNN to verify the diagnosis capability of the PNN. When the value of the SPs is input into the learned PNN, it can quickly judge conditions according to the possibilities of the relevant states. The diagnosis results are shown in Table 2 and Table 3.

Table 2. Verification Results for the first step

P_1	P_5	w_N'	w_B'	w_U'	Judge
1.259	3.613	0.789	0.005	0.217	N
1.26	3.75	0.854	0.018	0.118	N
2.95	28.15	0.004	0.919	0.095	B
2.237	23.1	0.032	0.765	0.238	B
3.67	34.56	0.004	0.312	0.694	U

Table 3. Verification Results for the second step

P_1	P_4	w_I'	w_O'	w_U'	Judge
2.147	1.39	0.847	0.005	0.176	I
2.237	1.40	0.827	0.011	0.148	I
2.95	1.58	0.003	0.835	0.158	O
3.03	1.60	0.008	0.808	0.192	O
3.67	1.83	0.000	0.226	0.796	U

According to the verification results, the possibilities output by the PNN show correct judgments for each state. Therefore, the PNN can precisely distinguish the types of bearing faults on the basis of the possibilities of the symptom parameters.

7 Conclusions

To process the ambiguous information in the condition diagnosis, and effectively identify fault types, this paper proposed a sequential diagnosis method for a rolling bearing by using the possibility theory and the fuzzy neural network. The possibility theory was used for solving the ambiguous problems of the fault diagnosis and the obtaining method of possibility function of a bearing state was shown for the condition diagnosis. The fuzzy neural network is realized with a developed back propagation neural network by which the conditions of a bearing can be effectively identified on the basis of the possibilities of normal and each abnormal state. The non-dimensional symptom parameters in time domain were also defined, which can reflect the characteristics of time signal of a bearing. Practical examples of diagnosis for a bearing used in a centrifugal blower verified the efficiency of this method.

References

1. Pusey, H.C.: Machinery Condition Monitoring. Journal of Sound and Vibration 34(5), 6–7 (2000)
2. Mitoma, T., Wang, H., Chen, P.: Fault Diagnosis and Condition Surveillance for Plant Rotating Machinery Using Partially-linearized Neural Network. Computers & Industrial Engineering (2008), doi:10.1016/j.cie.2008.03.002H
3. Wang, H., Chen, P.: Fault Diagnosis for a Rolling Bearing Used in a Reciprocating Machine by Adaptive Filtering Technique and Fuzzy Neural Network. WSEAS Transactions on Systems 7, 1–6 (2008)
4. Williams, T., Ribadeneira, X., Billington, S., Kurfess, T.: Rolling Element Bearing Diagnostics in Run-to-failure Lifetime Testing. Mechanical Systems and Signal Processing 15, 979–993 (2001)
5. Samanta, B., Al-Balushi, K.R., Al-Araimi, S.A.: Artificial Neural Networks and Genetic Algorithm for Bearing Fault Detection. Soft Computing 10, 264–271 (2006)
6. Li, R.Q., Chen, J., Wu, X.: Fault Diagnosis of Rotating Machinery Using Knowledge-based Fuzzy Neural Network. Appl. Math. Mech-Eng. 27, 99–108 (2006)
7. Blowerg, R.M.: Fault Diagnosis of Induction Machine Using Artificial Neural Network and Support Vector Machine. Dynamics of Continuous Discrete and Impulsive Systems-series A-Mathematical Analysis 13 (Part 2 Suppl. S) 658–661 (2006)
8. Saxena, A., Saad, A.: Evolving an Artificial Neural Network Classifier for Condition Monitoring of Rotating Mechanical Systems. Applied Soft Computing 7, 441–454 (2007)
9. Bishop, M.C.: Neural Networks for Pattern Recognition. Oxford Univ. Press, Oxford (1995)
10. Fukunaga, K.: Introduction to Statistical Pattern Recognition. Academic Press, London (1972)
11. Bendat, J.S.: Probability Function for Random Processes: Prediction of Peak, Fatigue Damage, and Catastrophic Failure. NASA Report CR-33 (1969)
12. Chen, P., Taniguchi, M., Toyota, T., He, Z.: Fault Diagnosis Method for Machinery in Unsteady Operating Condition by Instantaneous Power Spectrum and Genetic Programming. Mechanical Systems and Signal Processing 19, 175–194 (2005)
13. Shafer, G.: A Mathematical Theory of Evidence. Princeton Univ. Press, Princeton (1976)

Evolving Neural Network Using Genetic Simulated Annealing Algorithms for Multi-spectral Image Classification

Xiao Yang Fu[1] and Chen Guo[2]

[1] Institute of Computer Science and Technology, Jilin University, Zhuhai 519041, China
dvndavidfu@vip.163.com

[2] College of Information Science and Technology, Dalian Maritime University, Dalian 116026, China
guoc@dlmu.edu.cn

Abstract. In this paper, an evolving neural network classifier using genetic simulated annealing algorithms (GSA) and its application to multi-spectral image classification is investigated. By means of GSA, the classifier presented is available to automatically evolve the appropriate architecture of neural network and find a near-optimal set of connection weights globally. Then, with Back-Propagation (BP) algorithm, the conformable connection weights for multi-spectral image classification can be found. The GSA-BP classifier, which is derived from hybrid algorithm mentioned above, is demonstrated on SPOT multi-spectral image data effectively. The simulation results demonstrated that GSA-BP classifier possesses better performance on multi-spectral image classification. Its overall accuracy is improved by 4%~6% than conventional classifiers.

Keywords: Genetic simulated annealing algorithms, Multi-spectral image, Neural network, Classification.

1 Introduction

Classification of pixels for partitioning different land cover regions is an important investigation in remote sensing imagery classification research. So far, besides conventional classifiers such as Bayes maximum-likelihood classifier [11], several image recognition classifiers have been adopted, for example, the Fuzzy Classifier [4], Genetic Classifier [1, 9] and Neural Network Classifier [2, 12]. Within these methods, a neural network classifier with Evolutionary Algorithms (EA's) has been developed rapidly [8, 15-16].

The most widely used neural network model is the multi-layer perceptron (MLP), in which the connection weight training is normally completed by a BP learning algorithm [10]. The essential character of the BP algorithm is gradient descent. Because of the gradient descent algorithm is strictly dependent on the shape of the error surface, and the error surface may have some local minimum and multimodal. This results in falling into some local minimum and appearing premature convergence [7].

On the other side, GA [5, 6] is optimization technique guided by the principles of evolution and natural genetics. They are efficient, adaptive and robust searching processes,

F. Sun et al. (Eds.): ISNN 2008, Part II, LNCS 5264, pp. 294–303, 2008.

producing near-optimal solutions and can handle large, highly complex and multimode spaces. Of course, it is not surprise to evolve the neural network with GA [13, 14], and to study the remote sensing land cover classification using GA evolving neural network [8]. By searching a near-optimal set of initial connection weights of BP network with GA, Liu et al have showed that the hybrid GA-BP approach was more efficient than BP algorithm used alone for multi-spectral image classification. But their works assumed that the architecture of a neural network was predefined and fixed during the evolution of connection weights. So, it is necessary to find a method of evolving the architecture and connection weights of neural network simultaneously. On the other hand, Yao (1999) showed that the neural network architecture could be destroyed by using crossover operator.

In order to overcome the shortcomings mentioned above, in this paper, we proposed a hybrid genetic simulated annealing algorithm (GSA) for evolving the three neural network architecture and connection weights. With GSA, we can evolve simultaneously the appropriate number of hidden nodes and their connections with input and output nodes and find a near-optimal set of connection weights globally. Then, with BP algorithm, find the best connection weights for multi-spectral image classification.

The paper is organized as follows: Section 2 introduces the three-layer neural network and a hybrid GSA and BP algorithm based neural network classifier (GSA-BP). Section 3 provides the results of the classification using our GSA-BP classifier on SPOT (Systems Probataire d'Observation de la Terre) image data and compared with other classifiers. Finally, in Section 4, some conclusions are proposed.

2 Soft Computing Based Algorithms

2.1 The Three-Layer Neural Network

Assuming a three-layer neural network with $\boldsymbol{m}$ inputs (spectral bands), $\boldsymbol{q}$ outputs (categories), and $\boldsymbol{l}$ hidden nodes, see Fig 1.

The relations of input and output can be formulated as follows:

$$net_j = \sum_{i=1}^{m} W_{ji} x_i + w_{j0} \tag{1}$$

$$y_j = f(net_j) \tag{2}$$

$$net_k = \sum_{j=1}^{l} V_{kj} y_j + v_{k0} \tag{3}$$

$$O_k = f(net_k) \tag{4}$$

where W_{ji} is the connection weight between the jth hidden node and the ith input node, w_{j0} is its bias; the net_j is the input of the jth hidden node ; V_{kj} is the connection weight between the jth hidden node and the kth output node, v_{k0} is its bias. net_k is the input of the k output node. The $f(net)$ is a sigmoid activation function, it is defined as:

$$f(net) = \frac{1}{1 + e^{(-net)}} \tag{5}$$

where $net \in [-\infty, +\infty]$.

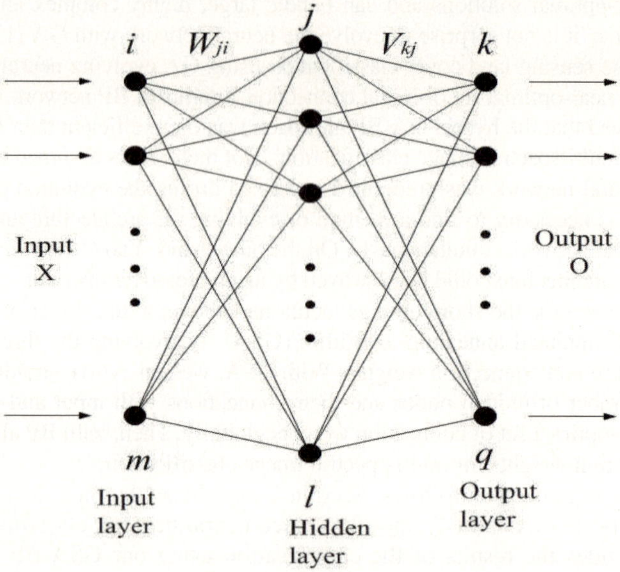

Fig. 1. Three layers neural network structure

Suppose we have a set of training patterns $X=\{X_1, X_2, \ldots, X_n\}$, where n is the number of training patterns, each training pattern X_i in set X is an m-dimensional feature vector. Let $T=\{T_1, T_2, \ldots, T_n\}$ as set X's corresponding output classes, $T_i=\{t_1, t_2, \ldots, t_q\}$ is a q-dimensional class vector. If the target class for a specific pattern is k $(1 \le k \le q)$, then $t_k=1$, otherwise, $t_k=0$. Let's denote O_{ik} as the ith actual neuron output for input training pattern x_i at the output node k while t_{ik} as its desired response. The mean square error function (MSE) for this neural network is defined as:

$$MSE(W) = \frac{1}{nq}\sum_{i=1}^{n}\sum_{k=1}^{q}(t_{ik} - o_{ik})^2 \tag{6}$$

where W represents all the weights in the network. Thus, this error is some scalar function of the weights and is minimized when the network output match the desired outputs.

2.2 Genetic Algorithms Based Simulated Annealing (GSA)

GA is an optimization searching technique by means of crossover, mutation, and selection.

Simulated Annealing Algorithm (SA) is a searching and optimal techniques based on the annealing principle of physics. In annealing process, first, the temperature is raised, then, decreased gradually to a very low value, while ensuring that one spends sufficient time at each temperature value, finally, the system reaches a stability condition.

GSA is a hybrid algorithm of GA and SA mentioned above.

The basic steps of GSA are described as follows:

1. Randomly construct an initial population.
2. Do SA operation for each individual in population.
3. Select parents from the current generation according to their fitness.
4. Apply mutation operator to parents to generate offspring which form the new generation.
5. Do step 2 to 4 repeatedly until some stopping criterion is met.

2.3 GSA-BP Classifier

2.3.1 Description of the GSA-BP Algorithm

GSA-BP algorithm is composed of GSA and BP algorithms, as follows in Fig. 2.

GSA Algorithm adopts two-layer architecture, out layer is used for evolving the architecture by GA and inner layer is used for evolving the connection weights by SA.

Thus the GSA-BP evolution procedure can be described as follows:

1. Construct a set of neural networks with randomly generated architectures and initial connection weights; decode each individual in the current generation into architectures.
2. Train their connection weights using SA for each individual.

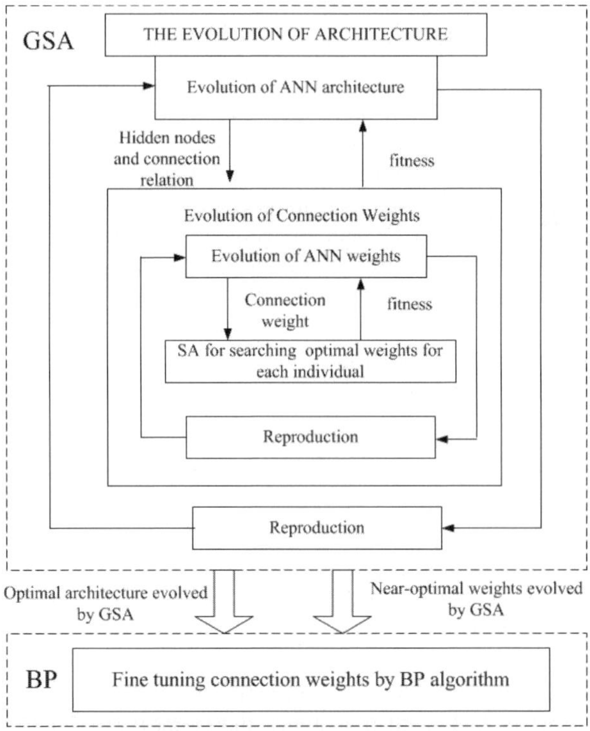

Fig. 2. Two-layer framework for GSA-BP

3. Calculate the fitness of each individual (chromosome) and select parents from the current generation according to their fitness.
4. Apply mutation operator to parents to generate offspring which form the new generation.
5. Do the step 2 to 4 repeatedly until some stopping criterion is met.
6. Continue BP training based on the architecture and the connection weights of the best individual in order to find the best solution.

2.3.2 Architecture Encoding

Here, the architecture of neural network is adopted indirect encoding, and only the hidden nodes and their connection relations with input and output nodes can be specified by a string (chromosome) *StrA*. Let:

$$StrA=(l_1,\cdots l_i,\cdots l_b;IN_{11},\cdots IN_{m1},\cdots IN_{ij},\cdots IN_{1l},\cdots IN_{ml};\cdots ON_{11},\cdots ON_{q1},\cdots ON_{jk},\cdots ON_{1l},\cdots ON_{ql}), \tag{7}$$

$$l = \sum_{i=1}^{b} l_i \cdot 2^{i-1} \tag{8}$$

In formulation (7), the first set of parameters $l_1,l_2,\ldots,l_i, \ldots l_b$ represents the binary bits of the hidden nodes, its decimal value is get by formulation (8) . The second set of parameters $IN_{11}\ldots IN_{ij}\ldots IN_{ml}$ represents the connection relations between the input and hidden nodes. If IN_{ij}=1, then the ith input node is connected with the jth hidden node; if IN_{ij}=0, then no connection. The third set parameters $ON_{11}, \ldots ON_{kj}\ldots ON_{ql}$ represents the connection relations between the output and hidden nodes. If ON_{jk}=1, then the jth hidden node is connected with the kth output node; if ON_{jk}=0, then no connection.

By convenient rule of thumbs [3], the number of hidden nodes is chosen such that the total number of weights in the network is roughly $n/10$, or more than this number, but it should not be more than the total number of training points, n. For the three-layer neural network, the total number of weights (except bias) is $l*(m+q)$, so the maximum number of hidden nodes $l_{max} \leq n/(m+q)$, the hidden node l can be chosen number from range$[l_{max}/10,l_{max}]$.

2.3.3 Connection Weights Encoding

Assuming the connection weight chromosomes represented by two strings which have hidden nodes l.

$$StrV=(v_{10},v_{20},\cdots v_{q0},v_{11},v_{21}\cdots v_{q1},\cdots v_{1l},v_{2l},\cdots v_{ql}), \tag{9}$$

$$StrW=(w_{10},w_{20},\cdots w_{m0},w_{11},w_{21},\cdots w_{m1},\cdots w_{1l},w_{2l},\cdots w_{ml}). \tag{10}$$

The connection weights are represented by using a real number form. In initialization, the number of hidden nodes can be got from the first set parameters of *StrA*, and what connection items is true or not can be got from the second and third set of parameters in *StrA*. If the connection item is not true, then connection weight corresponding to it set 0. For all true items, the connection weight values are randomly chosen number from range [-1, +1].

2.3.4 Fitness Computation

For a pair of strings ($StrV_i$ and $StrW_i$) with the same length l_i, we can calculate the MSE for a set of training patterns through formulation (1)—(6). The fitness function is defined as:

$$f_i = e^{-(MSE_i + \alpha \cdot l_i / l_{\max})} . \tag{11}$$

Where f_i is the fitness of the ith string, l_i is its number of hidden nodes, α is a positive architecture constant. Therefore, maximization of the fitness ensures the minimization of the MSE, and the term l_i/l_{max} will force the minimization of the number of hidden nodes. The larger α is, the stronger the effectiveness of l_i/l_{max} is.

3 Simulation Results and Discussion

The data extracted from SPOT image of Calcutta in India is used for classification. The set has three bands, i.e. green band (0.50—0.59μm), red band (0.61—0.68μm) and infrared band (0.79—0.89μm), and comprises 932 points belonging to seven classes: turbid water (TW), pond water (PW), concrete (concr), vegetation (Veg), habitation (Hab), open space (OS), and roads (including bridges) (B/R). Some points are extracted randomly for training, the remaining extracted for testing.

In GA algorithm, the population size is 20, the maximum number of hidden nodes is $n/(m+q)$, and μ_m is variable within the range [0.01, 0.15], α is variable within the range [0.3, 2.0], the GA algorithm is executed for 300 generations. In SA algorithm, the maximum temperature, T_{max} =100; the minimum temperature, T_{min} = 0.01; the temperature descend rate, r=0.8; the repeat annealing counts at the same temperature, k = 50. In BP algorithm, the learning rate is 0.02, the learning rate increment is set to 1.001, the momentum is 0.9, and the algorithm is executed for 3000 epochs. Table 1 shows the performance comparison for different proportion training data.

Table 1. The performance comparation of different proportion training dataset

Training percent in dataset	allScore (%)	
	Training	testing
10%	90.3	81.4
30%	90.7	85.7
50%	91.4	86.5
70%	90.6	86.8

allScore=(n-miss)/n, n is total number of dataset. miss is total number of misclassified points.

Seen as Table 1, for different proportion training dataset, when the proportion of training dataset is larger than 30%, the testing allScore of GSA-BP is consistent considerably. It is showed that the classifier has better performance of generalization. But when the training data are less than 10%, the testing performance will decline.

The classification performance of GSA-BP classifier has compared with Bayes classifier, GA classifier. Table 2 shows the testing results of classification accuracy when 30% data set for training and 70% for testing.

Table 2. Testing Results of Classification Accuracy for Different Classifiers

Class	User's Accuracy (%)			Kappa Coefficient		
	Bayes	GA	GSA-BP	Bayes	GA	GSA-BP
TW	100.00	96.70	100.00	1.000	0.962	1.000
PW	85.00	68.33	83.33	0.809	0.632	0.796
Concr	84.96	94.69	84.07	0.818	0.934	0.809
Veg	84.57	88.00	92.57	0.794	0.839	0.898
Hab	66.67	68.42	70.18	0.630	0.652	0.676
OS	81.48	86.21	89.46	0.829	0.843	0.897
B/R	28.95	18.42	39.47	0.266	0.172	0.362
overall	75.95	74.40	79.87	0.735	0.719	0.777

As seen from Table 2, the overall classification accuracy of the GSA-BP classifier is better than other classifiers, about 4%~ 6%. GSA-BP classifier recognizes the different classes consistently with a high degree of accuracy. On the contrary, the other classifiers can recognize some classes very nicely, however, much poorer for other classes. For example, the Bayes classifier provides User's accuracy of 100.00% and 85.00% for TW and PW, respectively, but its User's accuracy for B/R is 28.95%.

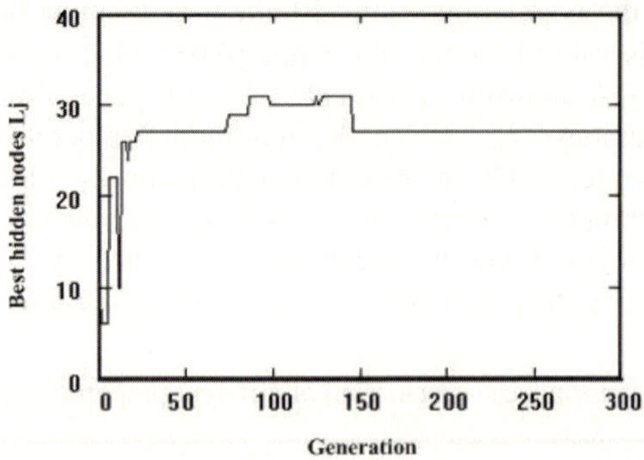

Fig. 3. The Neural Network Hidden Nodes changed by GSA Evolutionary Training

Fig. 3 is the number of hidden nodes (L_j) corresponding to the best MSE during the evolutionary training (GSA). From this Figure, we can see that the best number of hidden nodes (27) is obtained just after the 142 generations. Because of the mutation probability μ_m and architecture coefficient α are varied with the number of generations. Initially the μ_m has a high value, thus ensuring a lot of diversity in the population, at this stage, the architecture (L_j) of the best individual has much more changed. As generations pass and evolving progress reaches the vicinity of an optimal solution, the μ_m should be decreased for fine tuning the architecture and α should be increased for obtaining a more simple architecture.

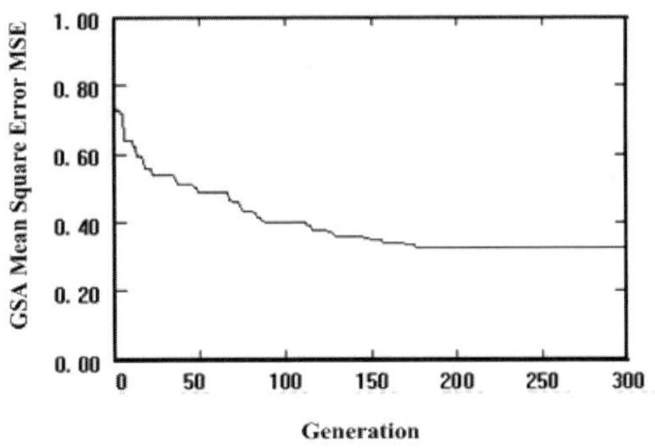

Fig. 4. The Best MSE changed by GSA Evolutionary Training.

Fig. 4 shows the best MSE corresponding to each generation during the evolutionary training (GSA).

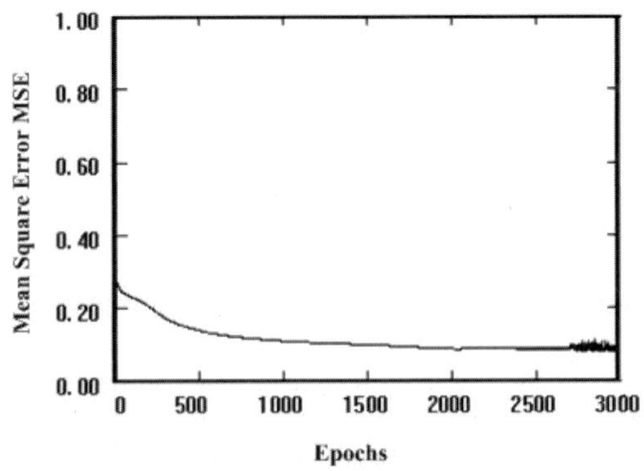

Fig. 5. The MSE changed by BP Training

Fig. 5 is the MSE corresponding to each epoch during BP training. It is found that although GSA algorithm has greatly reduced the total MSE of the neural network, a more improvement of training performance is achieved by applying a back propagation weight adjustment procedure.

4 Conclusions

In this paper, an evolving neural network classifier using genetic simulated annealing algorithms (GSA) and its application to multi-spectral image classification is proposed. GSA made it feasible to automatically evolve the appropriate architecture of neural network and to globally find near-optimal connection weights. Then, with BP algorithm, the conformable connection weights for multi-spectral image classification can be found. The GSA-BP classifier, which is derived from hybrid algorithm mentioned above, is demonstrated on SPOT multi-spectral image data effectively. The simulation results show that GSA-BP classifier possesses better performance on multi-spectral image classification.

The effectiveness of the algorithm is also demonstrated on SPOT multi-spectral image data. Compared with standard classifiers, such as Bayes classifier and GA classifier, it has shown that the hybrid GSA algorithm based neural network classifier giving better performance on multi-spectral image classification.

The notable feature of the proposed soft computing algorithms is that the neural network structure is evolved automatically while connection weights are being evolved. But proper selection of control parameters for GSA algorithm is still an open issue, which will be a part of our further work.

Acknowledgments

The authors acknowledge the help of Ph.D. Bandyopahyay (Machine Intelligence Unit, Indian Statistical Institute, India) for providing the SPOT image data, and also acknowledge the suggestions of Prof. Shuqing Zhang (Northeast institute of Geography and Agricultural Ecology, Chinese Academy of Sciences, Changchun, China) for meaningful discussion of the paper. [1] This work was also supported by the National Science Foundation of China (No. 60774046)

References

1. Bandyopadhyay, S., Murthy, C.A., Pal, S.K.: Pattern Classification Using Genetic Algorithms: Determination of H. Pattern Recognition Letters 19, 1171–1181 (1998)
2. Benediktsson, J.A., Swain, P.H., Ersoy, O.K.: Neural Network Approaches versus Statistical Methods in Classification of Multi Source Remote Sensing Data. IEEE Transaction on Geo-science and Remote Sensing 28, 540–552 (1990)
3. Duda, R.O., Hart, P.E., Stork, D.G.: Pattern Classification, 2nd edn. China Machine Press, Beijing (2001)
4. Filippi, A.M., Jensen, J.R.: Fuzzy Learning Vector Quantization for Hpyerspectral Coastal Vegetation Classification. Remote Sensing of Environment 100, 512–530 (2006)
5. Goldberg, D.E.: Genetic Algorithms in Search, Optimization and Machine Learning. Addison-Wesley, New York (1989)
6. Holland, J.H.: Adaptation in Natural and Artificial Systems. University of Michigan Press, Ann Arbor (1975)
7. Hertz, J., Krogh, A., Palmer, R.: An Introduction to the Theory of Neural Computation. Addison Wesley Publ. Comp., Redwood City (1991)
8. Liu, Z.J., Wang, C.Y., Liu, A.X., Niu, Z.: Evolving Neural Network Using Real Coded Genetic Algorithm (GA) for Multi-spectral Image Classification. Future Generation Computer Systems 20, 1119–1129 (2004)
9. Pal, S.K., Bandyopadhyay, S., Murthy, C.A.: Genetic Classifiers for Remotely Sensed Images: Comparison with Standard Methods. International Journal of Remote Sensing 22, 2545–2569 (2001)
10. Rumelhart, D.E., Hinton, G.E., Williams, R.J.: Learning Representations by Back-propagating Errors. Nature 323, 533–536 (1986)

11. Tou, T.T., Gonzalez, R.C.: Pattern Recognition Principles. Addison-Wesley, New York (1974)
12. Van Coillie, F.M.B., Verbeke, L.P.C., De Wulf, R.R.: Previously Trained Neural Networks as Ensemble Members: Knowledge Extraction and Transfer. International Journal of Remote Sensing 25, 4843–4850 (2004)
13. Van Coillie, F.M.B., Verbeke, L.P.C., De Wulf, R.R.: Feature Selection by Genetic Algorithms in Object-Based Classification of IKONOS Imagery for Forest Mapping in Flanders. Belgium, Remote Sensing of Environment 110, 476–487 (2007)
14. Van Rooij, A.J.F., Jain, L.C., Johnson, R.P.: Neural Network Training Using Genetic Algorithm. World Scientific Publishing, Co., Inc., River Edge (1996)
15. Yao, X.: Evolving Artificial Neural Networks. Proceeding of the IEEE 87, 1423–1447 (1999)
16. Yao, X., Xu, Y.: Recent Advances in Evolutionary Computation. Journal of Computer Science and Technology 21, 1–18 (2006)

Detecting Moving Targets in Ground Clutter Using RBF Neural Network

Jian Lao, Bo Ning, Xinchun Zhang, and Jianye Zhao*

School of Electronics Engineering and Computer Science
Peking University, Beijing,100871, China
laojian1982@gmail.com

Abstract. In this paper, a new structure for moving targets detection and characteristics extraction in ground clutter is proposed. This structure combines Radial Basis Function (RBF) neural network, Burg algorithm, and notch filter. After dynamical reconstruction, the RBF network is used to predict the ground clutter. Spectral characteristics of the ground clutter are estimated using the Burg algorithm. We apply notch filter to cancel the interference caused by the ground clutter. Moreover, a hardware platform based on FPGA is also realized for this paper to demonstrate this proposed structure and sufficient details of the hardware platform are discussed. The results of simulation and hardware implementation show that the presented structure has a good performance in processing target signals mixed with the ground clutter.

Keywords: Ground clutter, Reconstruction, RBF, Burg algorithm, Notch filter, FPGA.

1 Introduction

Detection and characteristic extraction of moving target from ground clutter is an important aspect of Neutral Network applications. On this issue, target signals should be detected in environment with uncertain interference. Considering the ground clutter shows certain statistic correlation properties which are different from that of the moving target signals, we can analyze the spectral characteristics of received data to determine whether the target signal exists.

Many researches have been introduced into this area. In [1-4], reconstruction in Time Domain can be regarded as a more suitable approach for clutter modeling. This concept is the foundation of signal detecting methods based on single step prediction. Among all the neural networks used for prediction, Radial Basis Function (RBF) neutral network is found to have the best performance owing to its high accuracy [5]. In [6], by adopting Genetic Algorithm (GA), Henry Leung developed the RBF prediction algorithm with regard to the prediction error in environment of sea clutter. But this proposal ignores the spectral characteristics of the clutter, which is very important for extracting useful spectral information of moving target signals.

---

* The paper is supported by NSF China, Grant No. 60704040.

F. Sun et al. (Eds.): ISNN 2008, Part II, LNCS 5264, pp. 304–312, 2008.

In order to cancel the interference caused by clutter, the dominating method till now is to employ Adaptive Moving Target Indication (AMTI) filter, as listed in [7-8]. A typical function of AMTI filter is to remove the low frequency interference. But in some situations, the frequency of interference is not so low compared with the target signal frequency. Take the experimental input data for this paper as an example. After the power spectrum estimation using Burg algorithm, we find that the central frequencies of ground clutter are mainly located between 15 KHz and 18 KHz, and the target signal frequency is about 25 KHz. Furthermore, the power spectrum of pure ground clutter has a sharp attenuation when the frequencies deviate from 15 KHz and 18 KHz. So the ground clutter can be considered to be a mixed signal composed by two single-frequency signals here. Obviously, AMTI filter is not the best choice under these conditions. In this paper, it is better to employ the notch filter to remove ground clutter interference because the notch filter just has the frequency property of sharp attenuation to some given frequencies. This property can be adopted to remove the ground clutter which concentrates its power at some single frequencies. The structure of the notch filter has been discussed in [9-10].

Since there are few researches published on hardware implementation of the issue discussed above, a hardware platform based on FPGA is also designed for this paper. We have realized the whole structure proposed in this paper on the hardware platform. The validity of our structure is confirmed by the hardware implementation results.

The rest of the paper is organized as follows. After the description of the proposed system model in Section 2, we analyze the theories which the designed structure is based on in Section 3. The results of simulation and hardware implementation are provided in Section 4 and conclusions are shown in Section 5.

2 System Model Description

The proposed structure is based on the design idea that the ground clutter prediction error can be an indication of the moving target signal's existence. After the training process to predict the ground clutter accurately, the prediction error can be very low. But when the moving target signals come, the error will increase significantly. Thus we can detect the targets by the abrupt change of the prediction error.

Fig.1 shows the diagram of the proposed target detection and characteristic extraction system. This diagram indicates that the system processes in two key steps. One is the preprocessing step, and the other is the processing step.

2.1 Preprocessing Step

This step focuses mainly on the ground clutter, as pointed by dashed line in the diagram of Fig. 1. In practical application, this step works when the moving target is far from the detection system or there are no moving target signals at all. Since the input data is pure ground clutter without the existence of target signals in this stage, characteristics of ground clutter both in Time Domain and Frequency Domain can be extracted and analyzed.

In Time Domain, in order to make the RBF neural network predict the ground clutter accurately, reconstruction based on Takens Embedding Theorem is employed. On the

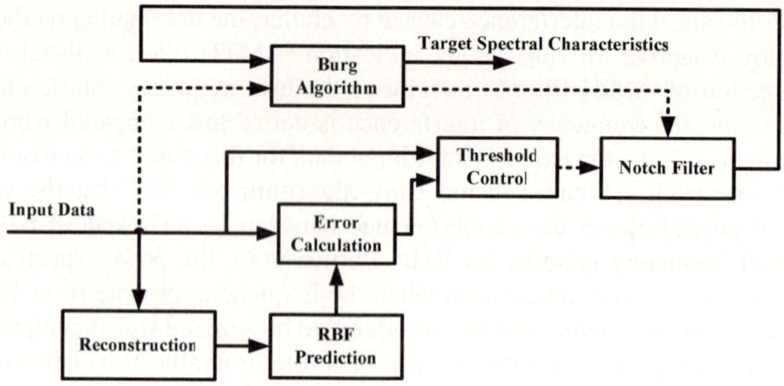

Fig. 1. The proposed system diagram

other side, in Frequency Domain, Burg algorithm is brought in for the purpose of getting the power spectrum information of ground clutter. According to the spectrum properties obtained by Burg algorithm, the notch filter can adjust its notch frequency automatically to cancel the interference the ground clutter caused.

2.2 Processing Step

This step deals with the mixed-signal composed by the target signal and the ground clutter. Due to the training process in the preprocessing step, the RBF neural network has already been able to predict the ground clutter with minor error. Thus because of the existence of the target signal with different statistic characteristics in this processing stage, the prediction error of RBF neural network can increase significantly. If the prediction error here is larger than a given threshold, it indicates that the target signal exists in the ground clutter and has been detected. Then the notch filter should start up to process the mix-signal and only the target signal will be retained. Then, the spectral characteristics of the moving target, such as distance and velocity, can be obtained by Burg algorithm.

3 Structure Analysis

3.1 Radial Basis Function Neutral Network as Predicting Module

According to Takens Embedding Theorem, a time series $\overline{x} = (x_{i-1}, x_{i-2}, ..., x_{i-d})$ generated by d-dimension dynamical system can be reconstructed by a m-dimension serial, provided m>2d+1.That is, there exists such a projection function which can make the reconstruction problem become a single step prediction problem. The projection function can be expressed by

$$\hat{x}_i = f(x_{i-1}, x_{i-2}, ..., x_{i-m}) \quad \text{if m>2d+1.} \tag{1}$$

From (1) we can see that the main problem is to find a hidden mapping function *f* to minimize the error or Mean-Square Error (MSE) between the true value x_i and the predicted value $\hat{x}$. These two errors can be given by

$$e = \sum_{i=1}^{N} \| x_i - f(x_{i-1}, x_{i-2}, \ldots, x_{i-m}) \| \tag{2}$$

$$e_{MSE} = \frac{\sum_{i=1}^{N} \| x_i - f(x_{i-1}, x_{i-2}, \ldots, x_{i-m}) \|^2}{N} . \tag{3}$$

Here, we choose the Radial Basis Function Neutral Network as the prediction approach because of its quick computing and high accuracy in function approximation.

A typical diagram of RBF network is shown in Fig. 2. The network consists of three parts: the input layer, the hidden layer, and the output layer. The synaptic weights from input layer to hidden layer are all chosen to be constant 1.

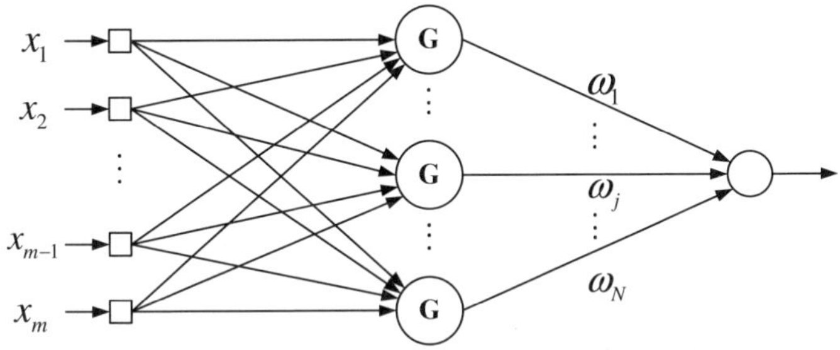

Fig. 2. Typical diagram of RBF network

The hidden radial basis function is given by

$$G(\overline{x}, \overline{c}_i) = \exp(-\frac{1}{2\sigma_i^2} \| \overline{x} - \overline{c}_i \|^2) , \tag{4}$$

where $\overline{x}$ is the m-dimension series $(x_{n-1}, x_{n-2}, \ldots, x_{n-m})$ input to neuron, $\overline{c}_i$ is the center, σ_i^2 is the variance and G is the output of hidden layer. The response of the output neuron can be given by

$$F = \sum_{i=1}^{N} w_i G(\| \overline{x} - \overline{c}_i \|) , \tag{5}$$

where $\overline{w} = (w_1, w_2, \ldots, w_N)$ is the weights connecting hidden layer and output layer. These parameters can be obtained by training procedure according to MSE criterion.

3.2 Burg Algorithm

According to the theory of Simon Haykin, most of the clutter, including the ground clutter, can be modeled by a low rank Auto Regressive (AR) sequence. The coefficients of AR model vary with the statistic characteristics of the clutter. The following is the power spectrum based on Burg algorithm

$$P(w) = \frac{\sigma^2}{|A(e^{jw})|^2} = \frac{\sigma_p^2}{|1 + \sum_{k=1}^{p} a_k e^{-jw}|^2}. \tag{6}$$

Using the coefficients calculated by (6), we can obtain the power spectrum characteristics of clutter, such as the central frequency and the bandwidth. These characteristics in Frequency Domain are important for the notch filter design.

3.3 Notch Filter

Notch Filter is a particular kind of band-stop filter, which can be used to cancel the interference caused by signals with a certain frequency. The frequency response of notch filter can be expressed by

$$H_d(e^{j\omega}) = \begin{cases} 1, \omega = \omega_N \\ 0, otherwise \end{cases}, \tag{7}$$

where ω_N is the notch frequency. According to [9-10], the frequency response for notch filter is 1 for all frequencies except the notch frequency, where the frequency response is 0.

Using Inverse Fourier Transformation to (7), we get the time-domain impulse respond function as

$$h(n) = \delta(n) - 2\cos(\omega_N n). \tag{8}$$

If there is more than one frequency in the interference, the total impulse response can be obtained by employing convolution computation between $h_1(n)$ and $h_2(n)$, which can be expressed by

$$h(n) = h_1(n) * h_2(n). \tag{9}$$

4 Simulation and Hardware Implementation

In this section, we give the results of simulations and hardware implementation to evaluate the performance of this proposed structure. Experimental data used here is divided into two groups. One is pure clutter ground clutter and the other is the mixed signal composed by the ground clutter and the moving target signal. The frequency of the target signal is 25 KHz.

4.1 Simulation Results of RBF Prediction Module

Fig.3.a shows the prediction curve of RBF network when the input data is pure ground clutter, and Fig.3.b is the prediction error between the actual value and the predicted

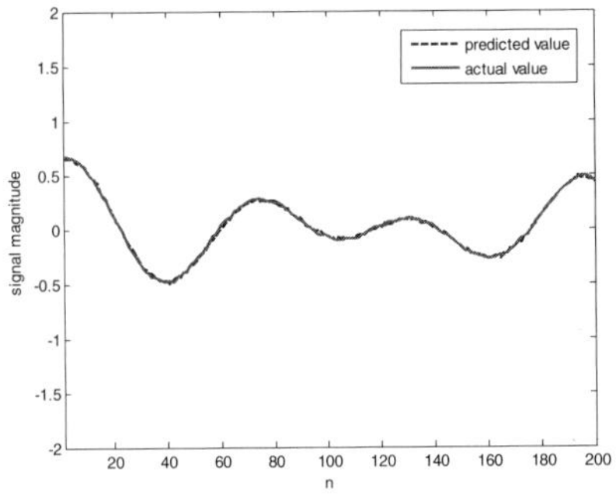

Fig. 3(a). Magnitude of ground clutter

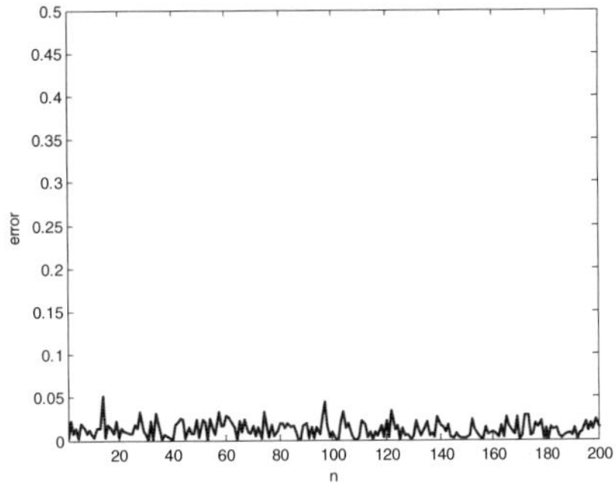

Fig. 3(b). Prediction error of ground clutter

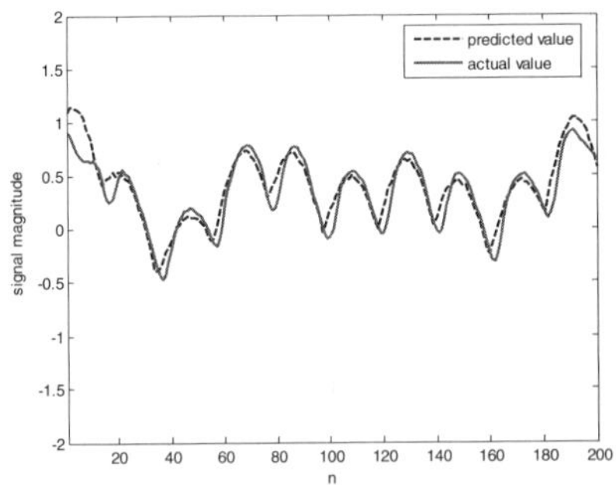

Fig. 4(a). Magnitude of mixed signal

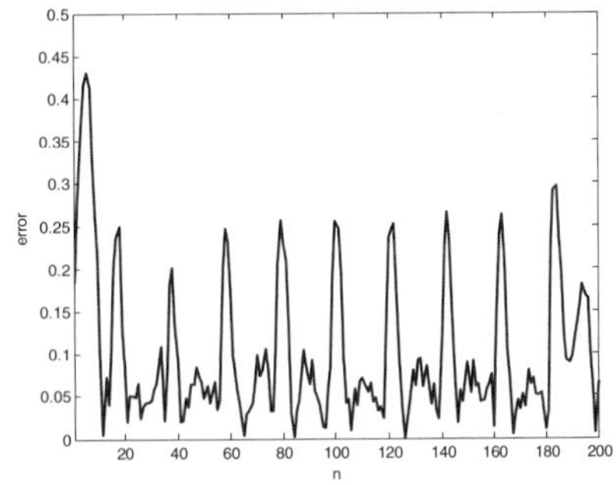

Fig. 4(b). Prediction error of mixed signal

value. Fig.4.a gives the prediction result when the input clutter is mixed with target signal at the frequency of 25 KHz, and the prediction error under this condition is shown in Fig.4.b.

From Fig.3.a and Fig.3.b, we can see that the RBF prediction module has a good performance without the target signal, since the prediction error is very low. But if the target signal is added into the input data, the prediction error will increase significantly, and the predicted value will deviate from the actual value obviously, as illustrated in Fig.4.a and Fig.4.b. The abrupt change of prediction error indicates that the target signal is mixed in the ground clutter. It is shown that by employing the RBF predicting module and recognizing the larger prediction error, we can successfully predict the ground clutter and detect the target signal.

4.2 Simulation Results of Burg Algorithm and Notch Filter

The spectral characteristics estimation results illustrated in Fig.5.a show that the power spectrum of ground clutter has two peaks at the frequency point of 15 KHz and 18 KHz. So, in order to remove the interference caused by these two frequencies, the notch filter should be able to adjust its notch frequency to these two frequencies automatically. This adjustment can be realized by scanning the data obtained with Burg algorithm for the local maximum values. After getting the index numbers of the maximum value, notch filter can convert these indexes to corresponding frequencies.

Here, we site the computation code for this paper as an example. The frequency parameter w in (6) is initialized between 0 and 50 KHz. The step length is 100 Hz. So $P(w)$ is an array with 500 data. After scanning the $P(w)$, notch filter find that the 144th and 175th data are the local maximum values. That means these two frequencies, 14.4 KHz and 17.5 KHz, are the peak values of ground clutter power spectrum. Then, notch filter can adjust its notch frequency to 14.4 KHz and 17.5 KHz.

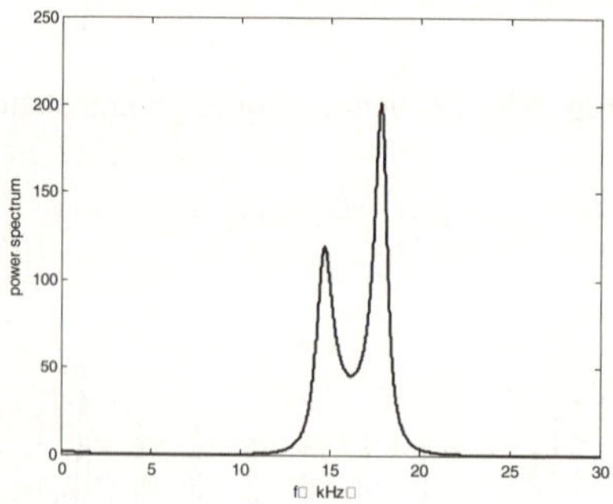

Fig. 5(a). Power spectrum of pure ground clutter

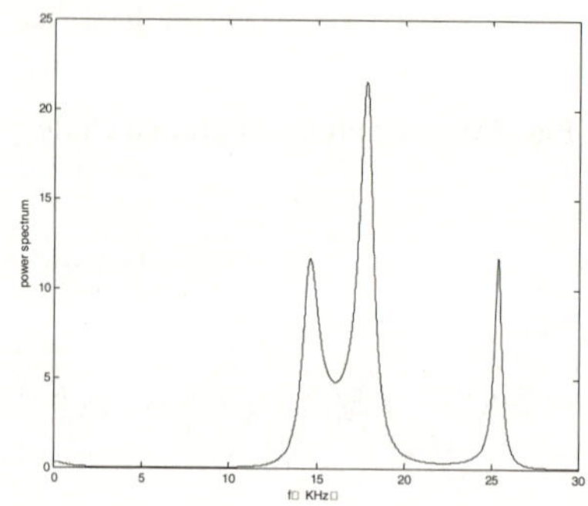

Fig. 5(b). Power spectrum of mixed signal

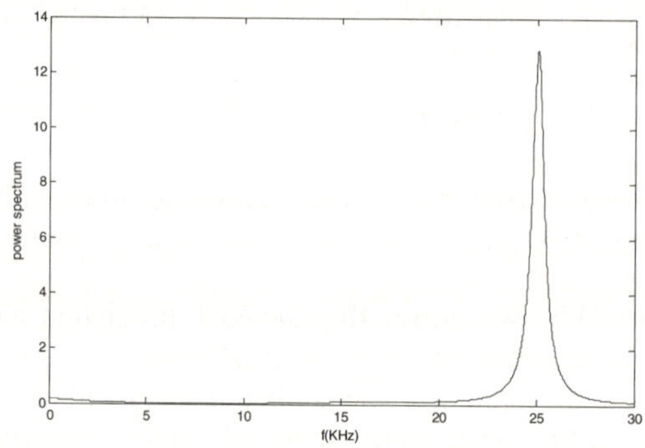

Fig. 5(c). Power spectrum of mixed signal filtered with notch filter

4.3 Hardware Implementation

We have accomplished the proposed structure on a hardware platform using FPGA. The conclusions declared in simulation sections above have been confirmed by the implementation results. The structure is listed below.

If the prediction error of RBF neural network is reduced below the given threshold by training process, parameters of the network can be extracted for FPGA hardware design. As to Burg algorithm module, there are only two parameters needed: the order number and the length of input data. The coefficients of notch filter can be obtained by calculation based on the notch frequency.

Here, we choose the FPGA chip EPS290 of StratixII family as the hardware platform. Single neuron unit designed of our scheme is illustrated in Fig. 6.

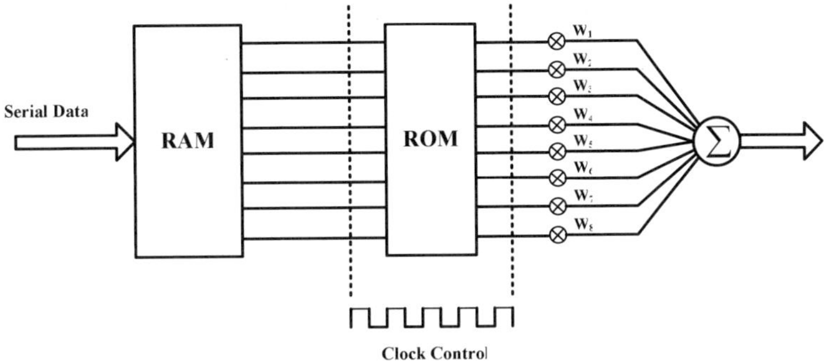

Fig. 6. FPGA structure of RBF module

The computation of function G can be implemented by searching a table stored in ROM, and RAM serves as data cache to avoid data overflow. By the method of time division multiplexing, this module can be adopted by all neuron units in the hidden layer.

As to Burg algorithm module, considering the designed frequency of sample clock is 1.25 MHz, and the length of data input each time for computation is 512 samples, so the cache time is $T = 0.8 * 512 = 409.6us$. The computation time assumed by Burg algorithm module must be lees than *T*. We take the strategy that RAM caches the input data at the same time of ROM computation to improve efficiency. The process time provided by accurate test of Burg algorithm module here is 333 us.

The order number of FIR Notch filter here is chosen to be 32. In order to improve the computation speed, we apply convert structure and two-level pipeline to the filter design. Convert structure illustrated in Fig. 7 can reduce by half the amount of computation when the filter is carrying out convolution. And pipeline can reduce the waiting time of input data due to its parallel computation characteristic.

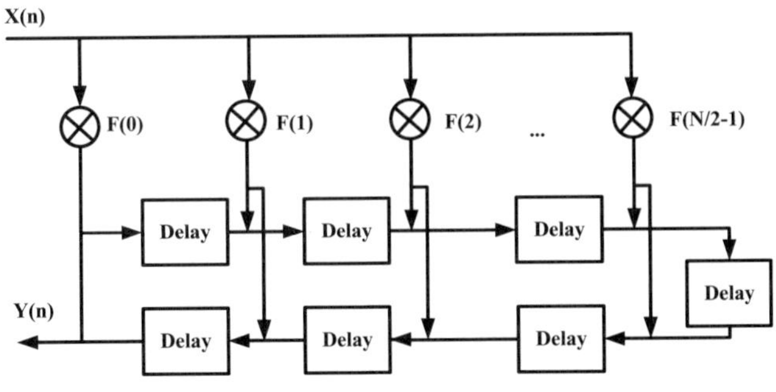

Fig. 7. FPGA structure of notch filter module

5 Conclusion

In this paper, a new structure for moving target detection and characteristics extraction is proposed. We accomplish the system structure design by combining RBF network, Burg algorithm, and notch filter. After reconstruction, RBF network is used to predict the ground clutter with high accuracy. Burg algorithm is applied here to estimate the spectral information of input data. In order to remove the interference caused by the ground clutter, notch filter is employed here corresponding to the central frequency points of the ground clutter. A hardware platform based on FPGA is realized. Simulation results have confirmed the validity of the designed structure, and the hardware implementation results have also proved this point.

References

1. Leung, H.: Experimental Modeling of Electromagnetic Wave Scattering from an Ocean Surface Based on Chaotic Theory. Chaos, Fractals and Solitons 2, 25–43 (1992)
2. Palmer, A.J., Kropfli, R.A., Fairall, C.W.: Signatures of Deterministic Chaos in Radar Sea Clutter and Ocean Surface Winds. Chaos 5, 613–616 (1995)
3. Abarbanel, H.D.I.: Analysis of Observed Chaotic Data. Springer, New York (1996)
4. Haykin, S., Puthusserypady, S.: Chaotic Dynamics of Sea Clutter. Wiley, New York (1999)
5. Haykin, S.: Neural Networks: A Comprehensive Foundation, 2nd edn. China Machine Press (1998)
6. Leung, H., Dubash, N., Xie, N.: Detection of Small Objects in Clutter Using a GA-RBF Neural Network. IEEE Trans. Aerosp. Electron. Syst. 38, 98–118 (2002)
7. Huang, Y., Peng, Y.N.: Design of Airborne Adaptive Recursive MTI Filter for Detecting Targets of Slow Speed. In: 2000 IEEE National Radar Conference-Proceedings, pp. 215–218. IEEE Press, New York (2000)
8. Xiang, Y., Ma, X.Y.: AR Model Approaching-based Method for AMTI Filter Design. Systems Engineering and Electronics 27, 1826–1830 (2005)
9. Tseng, C.C., Pei, S.-C.: Sparse FIR Notch Filter Design and Its Application. Electronics Letters 33(13), 1131–1133 (1997)
10. Er, M.H.: Designing Notch Filter with Controlled Null Width. Signal Process 24, 319–329 (1991)

Application of Wavelet Neural Networks on Vibration Fault Diagnosis for Wind Turbine Gearbox

Qian Huang, Dongxiang Jiang, Liangyou Hong, and Yongshan Ding

Department of Thermal Engineering, Tsinghua University, Beijing 100084, China
hg03@mails.tsinghua.edu.cn, jiangdx@tsinghua.edu.cn

Abstract. This paper applies an Artificial Neural Networks (ANN) method－Wavelet Neural Networks (WNN) on fault diagnosis for a wind turbine gearbox. A gearbox is one of the most important units in a wind turbine drive train. It is significant to study fault diagnosis of gearbox conditions. First this paper presents the principles and advantages of Wavelet Neural Networks. Second this paper specifies the vibration mechanism of the gearbox and the feature parameter group reflecting fault feature, and then the standard fault samples (training samples) and simulation samples (testing samples) are obtained. Third this paper applies the WNN method to perform diagnosing. The accurate diagnostic results have proved the effectiveness of the method for vibration fault diagnosis of gearbox. Finally, the relative advantages of the WNN method are contrasted with those of BPNN method.

Keywords: Gearbox, wavelet neural networks, fault diagnosis.

1 Introduction

Gearbox is one of the most important units in a wind turbine drive. An unexpected failure of the gearbox may cause significant economic losses and wide-ranging social influence. It is necessary to distinguish the status of gearbox. Usually, vibration signal are acquired from accelerometers mounted on the outer surface of a bearing case. The signals include vibrations from the meshing gears, shafts, bearings, and other parts. Thus, the signals are always complex and it is difficult to diagnose a gearbox from such vibration signals. Recently, the research on ANN has great development, ANN has applied to fault diagnosis field widely due to its capacity for associative memory, self-organization, self-learning and strong nonlinear mapping capacity to solve complex nonlinear problem.[1] But conventional back-propagation neural networks (BPNN) most frequently used in practical applications has low learning speed, difficulty in choosing the proper size of network, and easy to fall into local minima. This paper uses WNN to diagnose gearbox. WNN combines the time-frequency characteristic of wavelet transformation with the self-learning of conventional neural network. The basic of WNN is using wavelet function as the activation function of neurons and combining wavelet with neural network directly for fault diagnosis.[2] As wavelet analysis employs mainly the expansion and contraction of basis function to detect simultaneously the characteristics of global and local of the measured signal[3], WNN inherits these characters form wavelet analysis, and has stronger approximating,

F. Sun et al. (Eds.): ISNN 2008, Part II, LNCS 5264, pp. 313–320, 2008.

tolerance and classification capacity than conventional neural network, which makes it have strong advantages in dealing with nonlinear mapping and on-line estimates.[4] According to the nonlinear relationship between the failure mechanism and fault, this paper selects nonlinear Morlet wavelet as the basis function, use error BP algorithm to train network in wind turbine gearbox fault diagnosis, which can overcome the defects of conventional BPNN.

2 Principles of Wavelet Neural Networks

2.1 Wavelet Transform

The wavelet transform allows a multi-resolution decomposition which permits a scale-invariant representation.[5] This transform, introduced by Morlet to overcome the time/frequency uncertainty shortcomings of the windowed Fourier transform, can be viewed as the decomposition of a signal into a set of frequency channels having the same hand-width on a logarithmic scale. The basic idea of Wavelet transform is to hierarchical decomposition of a function into a set of Basis Functions and Wavelet Functions. The so-called wavelet is a square-integrable function with zero mean and that is localized in both frequency and time. It oscillates in amplitude and decays to zero quickly on both sides of the central position of the waveform. The wavelet is derived form the mother wavelet by dilation and translation. If a mother wavelet is defined as $\Psi(t)\square L^2(R)$ ($L^2(R)$ denote the square integrate real space). Its Fourier transform is $\hat{\psi}(\omega)$, when $\hat{\psi}(\omega)$ is satisfied by admissible condition:

$$C_\psi = \int_R \frac{|\hat{\psi}(\omega)|^2}{|\omega|} d\omega < \infty \quad (1)$$

R is the set of real numbers. The daughter wavelet can be described as

$$\psi_{a,b}(t) = \frac{1}{\sqrt{a}} \psi\left(\frac{t-b}{a}\right) \quad (2)$$

where the scaling parameter a determines the amplitude and the duration of $\psi_{a,b}(t)$, and the time translation b shifts the wavelet on the time axis. Therefore, localization properties in both the time and frequency can be achieved simultaneously when the signals under analysis are examined using such wavelets. By varying the parameters a and b, we can obtain different daughter wavelets that constitute a wavelet family. Wavelet transform is to perform the following operation:

$$W(a,b) = \frac{1}{\sqrt{a}} \int_c \psi^*_{a,b}(t) x(t) dt \quad (3)$$

where $a,b \in R, a \neq 0$, * stands for complex conjugation. The Wavelet transform calculates the wavelet coefficients at every possible scale and along every time instant. The value of $W(a,b)$ represents the similarity extent between the examined section of $x(t)$ and the scaled and shifted wavelets. The greater the $W(a,b)$, the more energetic the $W(a,b)$, and the greater the similarity between the wavelets and the original signal.

2.2 Wavelet Neural Networks Architecture

WNN have been successfully used for various classification tasks.[6] WNN replaces the global sigmoid activation units of the classical feedforward NN with wavelets, while preserving the network's universal approximation property.[7] During the training phase, the wavelet network learns the decision functions and regions in the feature space defined by the network coefficients to reliably classify the input signals.

One method of integrating wavelets with neural network is to use the wavelet transform as a preprocessor, whereby feature (i.e. wavelet coefficients) extracted from the input signal are fed to a classical ANN for nonlinear classification. Another method is to use wavelet networks (Fig. 1), where wavelets are used to replace the sigmoid hidden nodes of the ANN. A third method is to update both the feature set and the neural network. The second method is chosen in this paper.

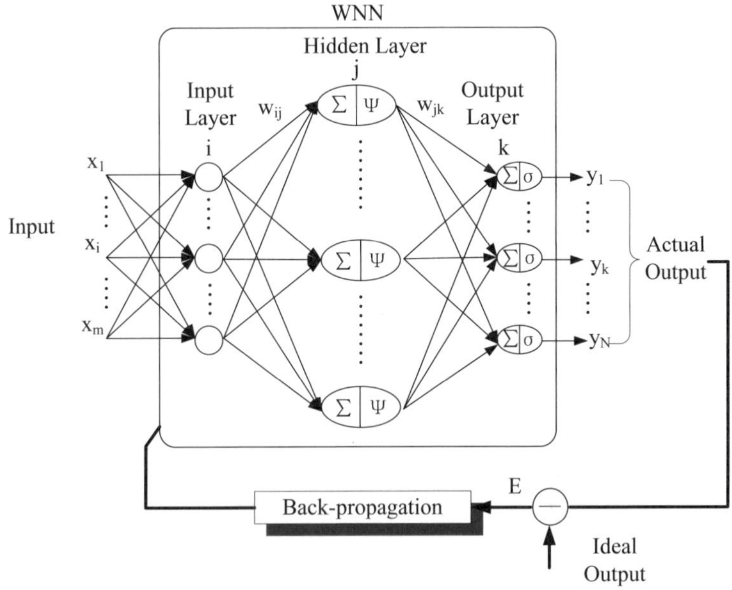

Fig. 1. Structure of WNN Architecture

$X_P = [x_1^P, x_2^P, \cdots, x_m^P]$ represents the input mode vector, and $Y_P = [y_1^P, y_2^P, \cdots, y_N^P]$ is the output mode vector of the network; w_{ij} represents the connection weight between the number i node of the input layer and the number j neuron of the hidden layer; w_{jk} represents the connection weight between the number j neuron of the hidden layer and the number k node of the output layer; a_j and b_j are the scale factor and translation factor of the number j neuron node of the hidden layer, respectively; P (p=1,2,⋯,P) represents the number of input mode; m (i=1,2,⋯,m) represents the number of nodes of the input layer; n (j=1,2,⋯,n) represents the number of neurons of the hidden layer; N (k=1,2,⋯,N) represents the number of nodes of the output layer. The output of the WNN is shown as follows:

$$y_k(t) = \sigma\left[\sum_{j=1}^{n} w_{jk}\psi_{a,b}\left(\sum_{i=1}^{m} w_{ij}x_i(t)\right)\right] \quad (k=1,2,\cdots,N)\ . \tag{4}$$

The function σ is a non-linear activation function which transforms the sum of the weighted inputs into the output node. The type of activation function adopted for the present work discussed above is the sigmoid function: $f(x)=1/[1+\exp(-x)]$.

2.3 Wavelet Neural Network Training Algorithm

A nonlinear optimization algorithm, such as gradient descent, conjugate gradients or Byden-Fletcher-Goldfarb Shanno (BFGS), could be applied to training a wavelet neural network. However, an advantage of the wavelet neural network architecture is that it can be trained in stages using linear optimization algorithms, which allows for faster training and improved convergence compared with nonlinear alternatives.

One method often used to vary the weights and biases is known as back-propagation algorithm, in which the weights and biases are modified so as to minimize an average quadratic error function of the form:

$$E = \frac{1}{2}\sum_{p=1}^{P}\sum_{k=1}^{N}\left[d_k^p - y_k^p\right]^2 \ . \tag{5}$$

where d_k^p is the expected output of WNN. The back-propagation algorithm actually adopts gradient descent to minimize E and the corresponding iterative formulas are presented as following:

$$\frac{\partial E}{\partial w_{jk}} = -\sum_{p=1}^{P} y_k(1-y_k)(d_k - y_k)\psi_{a,b}\left(\sum_{i=1}^{m} w_{ij}x_i(t)\right) \ . \tag{6}$$

$$\frac{\partial E}{\partial w_{ij}} = -\sum_{p=1}^{P}\sum_{k=1}^{N} y_k(1-y_k)(d_k - y_k)w_{jk}\psi'_{a,b}\left(\sum_{i=1}^{m} w_{ij}x_i(t)\right)x_i^p / a_j \ . \tag{7}$$

$$\frac{\partial E}{\partial b_j} = \sum_{p=1}^{P}\sum_{k=1}^{N} y_k(1-y_k)(d_k - y_k)w_{jk}\psi'_{a,b}\left(\sum_{i=1}^{m} w_{ij}x_i(t)\right)/ a_j \ . \tag{8}$$

$$\frac{\partial E}{\partial a_j} = \sum_{p=1}^{P}\sum_{k=1}^{N} y_k(1-y_k)(d_k - y_k)w_{jk}\psi'_{a,b}\left(\sum_{i=1}^{m} w_{ij}x_i(t)\right)\left(\frac{\left(\sum_{i=1}^{m} w_{ij}x_i(t)\right)-b_j}{a_j}\right)/ a_j^2 \ . \tag{9}$$

$$w_{ij}(t+1) = w_{ij}(t) - \eta\frac{\partial E}{\partial w_{ij}} + \mu\Delta w_{ij}(t) \ . \tag{10}$$

$$w_{jk}(t+1) = w_{jk}(t) - \eta\frac{\partial E}{\partial w_{jk}} + \mu\Delta w_{jk}(t) \ . \tag{11}$$

$$a_j(t+1) = a_j(t) - \eta\frac{\partial E}{\partial a_j} + \mu\Delta a_j(t) \ . \tag{12}$$

$$b_j(t+1) = b_j(t) - \eta\frac{\partial E}{\partial b_j} + \mu\Delta b_j(t) \ . \tag{13}$$

where η refers to w_{ij}, *wjk*, *aj* and *bj* learning rate parameter; μ refers to momentum their own factor.

The selection of the mother wavelet is very important and depends on particular application. Morlet wavelet has been adopted for the present work discussed above, and is shown as follows:

$$\psi(t) = \cos(1.75t)\exp(-t^2/2) \ . \tag{14}$$

Its first-order derivative is in the form of:

$$\psi'(t) = -1.75\sin(1.75t)\exp(-t^2/2) - t\cos(1.75t)\exp(-t^2/2) \ . \tag{15}$$

and the reason of such a basic function application is a possibility of its analytical expression leading to an easy calculation of the error function partial derivatives, what is necessary for WNN parameters modification during training process.

3 Fundamental of Gearbox Fault Diagnosis

A gearbox is a typical rotating machine (shown in Fig.2). The signals from a gearbox are often complicated since there are many components in a gearbox. According to field experts' experience and vibration research outcomes. Table 1 shows the vibration faults classification of gearbox for wind turbine. The table includes typical sixteen conditions numbered F0 to F15. Gear tooth abrasion is the main type of fault in a gearbox.

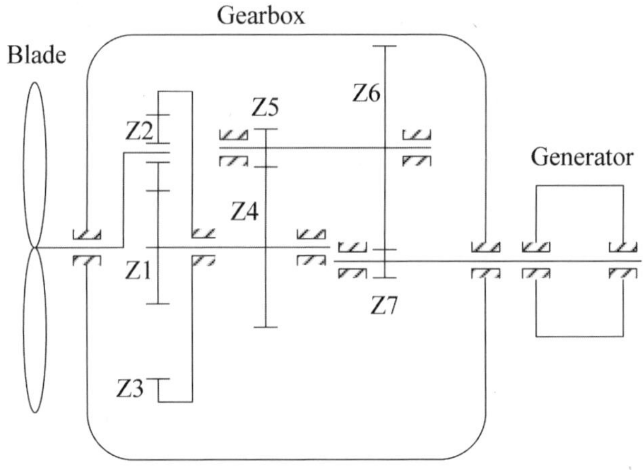

Fig. 2. Structure of Gearbox

Where Z1 is the sun gear, Z2 is the planet gear, Z3 is the annulus, Z4 is the large gear of the 2nd transmission stage, Z5 is the small gear of the 2nd transmission stage, Z6 is the large gear of the 3rd transmission stage and Z7 is the small gear of the 3rd transmission stage.

Table 1. Faults classification of gearbox

Faults No.	Faults name	Faults No.	Faults name
F0	normal	F8	the 7th Gear eccentricity
F1	Sun Gear abrasion	F9	Sun Gear profile error
F2	the 4th and 5th Gears abrasion	F10	the 4th Gear profile error
F3	the 6th and 7th Gears abrasion	F11	the 5th Gear profile error
F4	Sun Gear eccentricity	F12	the 6th Gear profile error
F5	the 4th Gear eccentricity	F13	the 7th Gear profile error
F6	the 5th Gear eccentricity	F14	Sun axle and the middle axle misalignment
F7	the 6th Gear eccentricity	F15	the middle axle and the high speed axle misalignment

Nowadays spectrum analysis is well proven as a practical and powerful tool for fault diagnosis of rotating machinery because it results from a great deal of engineering experience. However, the relationship of faults and spectrum data is complex. We can not establish mathematical model to describe it. We can get the relation through establishing the coefficient matrices of gearbox spectrum parameters.[8] Table 2 shows the spectrum symptoms description.

Table 2. Spectrum description

Spectrum	Description	Spectrum	Description
s1	f_{r1}	s10	$2f_{r2}$~$3f_{r2}$
s2	$2f_{r1}$~$3f_{r1}$	s11	$f_{m2} \pm f_{r2}$
s3	f_{m1}	s12	f_{m3}
s4	$f_{m1} \pm f_{r1}$	s13	$f_{m3} \pm f_{r2}$
s5	$2f_{m1}$~$3f_{m1}$	s14	$2f_{m3}$~$3f_{m3}$
s6	f_{m2}	s15	f_{r3}
s7	$f_{m2} \pm fr1$	s16	$2f_{r3}$~$3f_{r3}$
s8	$2f_{m2}$~$3f_{m2}$	s17	$f_{m3} \pm f_{r3}$
s9	f_{r2}		

where f_{r1} refers to the frequency of sun axle, f_{r2} refers to the frequency of middle axle, f_{r3} refers to the frequency of the high speed axle, f_{m1} refers to the meshing frequency of the first stage transmission, f_{m2} refers to the meshing frequency of the second stage transmission, f_{m3} refers to the meshing frequency of the third stage transmission.

Then establish the gearbox fault cause and sign normalization data table, the data is partly shown in Table 3.

Table 3. Part of normalization data of gearbox fault-symptom matrix

Fault	Spectrum symtoms								
types	s1	s2	s3	s4	s5	s6	s7	s8	s9
F0	0	0	0	0	0	0	0	0	0
F1	0.5	0.5	0	1	0	0	0	0	0.2
F2	0.5	0.5	0	0	0	0	1	0	0.2
F3	0.2	0.2	0	0	0	0	0	0	0.5

4 Application of WNN Method on Fault Diagnosis

In section 3 we get standard fault samples—sixteen standard fault samples (F0-F15). In order to prove the practicability of this method, we provide five testing samples (simulation fault samples) T0—T4, which are simulated from F0, F1, F6, F9 and F14,. Now we have both the training samples and testing samples to continue our research.

The WNN is constructed according to the symptom-fault relationship matrix and the principle of WNN discussed above. For there are seventeen spectrum symptoms, the number of input nodes is seventeen. For there are sixteen standard conditions, the number of output nodes is sixteen. Simultaneously the number of hidden neurons is seventeen. Other than input the standard training samples, we also need to input the prospective output. The prospective output is a 16×16 unit matrix "A", and "A_{ii}=1" means the "i"th fault has strictly occurred. Now we input the training samples and prospective output, construct the network, and diagnose the testing samples using the WNN. Table 4 showed the part of the diagnostic output results.

Table 4. Part of the diagnosis output by WNN network

Test	Output nodes									
samples	F0	F1	F2	F3	F4	F5	F6	F7	F8	F9
T0	0.96	0	0	0	0	0	0.02	0.02	0.02	0
T1	0	0.97	0.01	0	0	0	0	0	0	0.02
T2	0.01	0	0	0	0.01	0	0.97	0.01	0	0.01
T3	0	0.02	0.01	0	0.01	0.01	0.01	0.01	0	0.97

The result matrix demonstrates that the wavelet neural network is effective for gearbox fault diagnosis.

Also we compared the error curves of WNN with BPNN during the training processes (shown in Fig. 3 and Fig. 4).

According to the figure, the convergence speed and error accuracy are better than BPNN with the same structure.

WNN method achieves the same quality of classification with a network of reduced size. Reported results would favor this claim. It is worth noticing that the number of neurons is generally smaller in WNN than in BPNN, especially for problems of higher dimension. And it turns out that, for the particular wavelet we have chosen, the cost of implementing the nonlinearity is mainly proportional to the number of neurons.

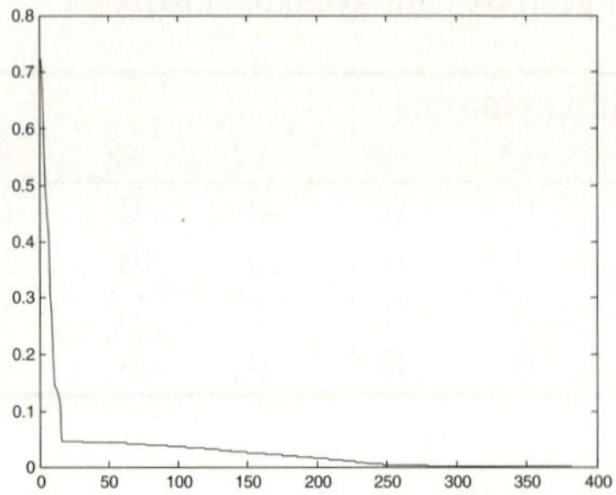

Fig. 3. The error curve of WNN

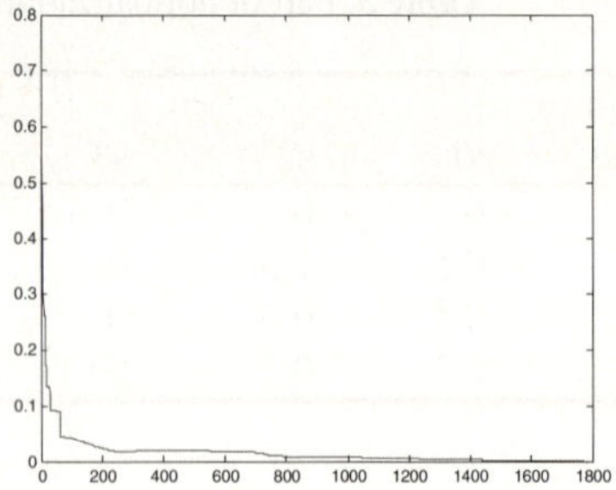

Fig. 4. The error curve of conventional BPNN

5 Conclusions

In this paper, Wavelet Neural Network method has been used to diagnose vibration faults of a wind turbine gearbox. First, both the principles and advantages of WNN were presented. Second, the fundamentals of vibration fault diagnosis were described. The standard fault samples are symptom-fault matrix including sixteen conditions, each is characterized by seventeen symptoms. Third we applied WNN method to perform diagnosing. The research diagnostic results indicate that WNN can solve complicated condition identification problems in the gearbox fault diagnosis, due to its multi-scale, multi-resolution. Finally, the performances of the WNN method were compared with BPNN. The test results show that the proposed WNN method improves the diagnosis accuracy and learning speed, as well as reduces the size of network compared with conventional method.

Acknowledgment. This work is supported by National Basic Research (973) Program of China (No.2007CB210304).

References

1. Chinese Journal of Stereology and Image Analysis 4, 239–245 (2001)
2. Zhang, Q., Benveniste, A.T.: Wavelet networks. IEEE Trans. On Neural Networks 3(6), 889–898 (1992)
3. Xu, Q., Meng, X., Han, X., Meng, S.: Gas turbine fault diagnosis based on wavelet neural network. In: Proceedings of the 2007 International Conference on Wavelet Analysis and Pattern Recognition, Beijing, China, 2-4, November, pp. 738–741 (2007)
4. Rui, Z., Xu, L., Feng, R.: Gear faults diagnosis based on wavelet neural network. Journal of Mechanical Transmission 01, 80–83 (2008)
5. Martin, V., Cormac, H.: Wavelets and filter banks: theory and design. IEEE Trans. On Signal Processing 40, 2207–2231 (1992)
6. Leonard, J.A., Kramer, M.A.: Radial basis function networks for classifying process faults. IEEE Control Systems, 31–38 (1991)
7. Dickhaus, H., Kohonen, H.H.: Classifying biosignals with wavelet networks. IEEE Engineering in Medicine and Biology, 103–111 (1996)
8. Tang, X., Xie, Z., Wang, Z., Wu, J.: Fault diagnosis of gearbox for wind turbine. Noise and Vibration Control 01, 120–124 (2007)

Dynamical Pattern Classification of Lorenz System and Chen System

Hao Cheng and Cong Wang

College of Automation and the Center for Control and Optimization,
South China University of Technology,
Guangzhou 510641, P.R. China
cheng.hao@mail.scut.edu.cn,
wangcong@scut.edu.cn

Abstract. Recently, an approach for rapid dynamical pattern recognition was proposed, by which a dynamical pattern can be locally accurately identified and rapidly recognized using localized radial basis function (RBF) networks. Further, a scheme for classification of dynamical patterns was presented. In this paper, we investigate the construction of the recognition system for classification of Lorenz system and Chen system, both of which can generate various types of dynamical patterns. Simulation studies are included to demonstrate the effectiveness of this method.

Keywords: Deterministic learning, Dynamical pattern, Recognition, Classification, Lorenz system, Chen system.

1 Introduction

In the last few years, a great deal of progress has been made for recognition of static patterns [1], however, the problem of recognition of temporal patterns is still a difficult problem in the pattern recognition area. For temporal or dynamical patterns, a very difficult and fundamental problem is how time-varying patterns can be represented in a time-independent way appropriately [2].

Recently, a dynamical and deterministic frame for identification, representation and recognition has been developed in [3][4]. Through deterministic learning, a time-varying dynamical pattern can be effectively represented in a time-invariant and spatially distributed manner by using constant neural weights [3]. Similarity definition for dynamical patterns based on system dynamics was given in [4]. With the time-invariant representation and similarity definition, an mechanism was proposed in which rapid recognition of dynamical patterns can be implemented by the achievement of state synchronization with a recognition system according to a kind of internal and dynamical matching on system dynamics [4]. The synchronization errors are proven to be proportional to the difference on system dynamics.

With these results, a recognition system for classification of dynamical patterns was constructed in [11]. Since the synchronization errors being taken as

F. Sun et al. (Eds.): ISNN 2008, Part II, LNCS 5264, pp. 321–330, 2008.

the similarity measures between the test and the training dynamical patterns, the recognition system can be constructed by using the principle of minimal distance or nearest-neighbor classification. With the help of the qualitative analysis of dynamical systems, the dynamical template models are arranged in some specific order and a hierarchical structured knowledge representation can be set up based on the similarity of system dynamics [11].

In this paper, we further investigate the construction of hierarchical structure recognition system for classification of Lorenz system [9] and Chen system [10]. Since patterns of Lorenz system and Chen system represent two large classes of dynamical patterns generated from dynamical system and include rich dynamical behaviors, it is meaningful for the identification, recognition and classification of these dynamical patterns. First, through the qualitative analysis of nonlinear dynamical systems [5], in which the concepts of topological equivalence, structural stability, bifurcation and chaos altogether provide an inclusive classification of various types of dynamical patterns, we can divide various patterns of Lorenz system and Chen system into classes of dynamical patterns with different dynamical behaviors. Second, we choose the most representative patterns as appropriate template models and arrange these dynamical template models according to a hierarchical structured knowledge representation based on the similarity of system dynamics. With the simulation results, we verify the effectiveness of the method for classification of various patterns generated from Lorenz system and Chen system.

2 Preliminaries

In this section, we briefly review the deterministic learning theory, the mechanism for recognition of dynamical patterns and the scheme for dynamical pattern classification [3][4][11].

2.1 Deterministic Learning

Consider the following parameter-dependent nonlinear dynamical system:

$$\dot{x} = F(x;p), x(t_0) = x_0 \tag{1}$$

where $x = [x_1, \cdots, x_n]^T \in R^n$ is the state of the system, p is a system parameter vector, $F(x;p) = [f_1(x;p), \cdots, f_n(x;p)]^T$ is a smooth but unknown nonlinear vector field. The objective of deterministic learning is to achieve accurate NN identification (or learning) of the dynamics $F(x;p) = [f_1(x;p), \cdots, f_n(x;p)]^T$ via state measurements. The main elements of deterministic learning include:

1) Dynamical Localized RBF Networks: In deterministic learning, the following dynamical RBF network is employed:

$$\dot{\hat{x}} = -A(\hat{x} - x) + \widehat{W}^T S(x) \tag{2}$$

where $\hat{x} = [\hat{x}_1, \cdots, \hat{x}_n]^T$ is the state of the dynamical RBF network, x is the state of system (1), $A = diag\{a_1, \cdots, a_n\}$ is a diagonal matrix, with $a_i > 0$ being designed constants, and $\widehat{W}^T S(x) = [\widehat{W}_1^T S_1(x), \cdots, \widehat{W}_n^T S_n(x)]^T$ are localized RBF [6] networks.

2) Accurate NN Approximation: With the partial PE condition satisfied, and with the following NN weight adaptation law:

$$\dot{\widehat{W}}_i = \dot{\widetilde{W}}_i = -\Gamma_i S_i(x)\tilde{x}_i - \sigma_i \widehat{W}_i, i = 1, \cdots, n \tag{3}$$

where $\widetilde{W}_i = \widehat{W}_i - W_i^*$, $\widehat{W}_i$ is the estimate of W_i^*, $\Gamma_i = \Gamma_i^T > 0$, and $\sigma_i > 0$ is a small value. The exponential stability of a closed-loop identification system, and consequently the exponential convergence of some neural weights to small neighborhoods of zero could be achieved [4]. Accordingly, the RBF network $W_i^T S_i(x)$ can approximate the unknown dynamics $f_i(x;p)$ along the trajectory $\varphi_\zeta(x_0)$ as:

$$f_i(\varphi_\zeta;p) = \widehat{W}_i^T S_i(\varphi_\zeta) + \varepsilon_{i_1} = \overline{W}_i^T S_i(\varphi_\zeta) + \varepsilon_{i_2} \tag{4}$$

where ε_{i_1} is the NN approximation error, which is small in the local region along $\varphi_\zeta(x_0)$. $\overline{W}_i$ is a constant neural weights obtained from $\widehat{W}_i$, and $|\varepsilon_{i_2}|$ is close to $|\varepsilon_{i_1}|$.

3) Representation: From (4), it is seen that by using the constant $\overline{W}^T S(x)$, corresponding to a training dynamical pattern we construct a dynamical model as:

$$\dot{\bar{x}} = -B(\bar{x} - x) + \overline{W}^T S(x) \tag{5}$$

where $\bar{x} = [\bar{x}_1, \cdots, \bar{x}_n]^T$ is the state of the dynamical model, x is the state of an input pattern generated from system (1), $B = diag\{b_1, \cdots, b_n\}$ is a diagonal matrix, with $b_i > 0$ normally smaller than a_i (a_i is given in (2)).

4) A Fundamental Similarity Measure: A similarity definition of dynamical patterns was proposed in [4], which states that: A test pattern φ_ς is said to be similar with a training dynamical pattern φ_ζ^k, if the state of the test pattern stays within a local region of the state of the training pattern, i.e.:

$$|f_i'(x;p') - f_i^k(x;p^k)| < \varepsilon_i^{k*}, \forall x \in \varphi_\varsigma(x_{\varsigma 0};p') \tag{6}$$

where $\varepsilon_i^{k*} > 0$ is a small constant.

2.2 The Recognition Mechanism

The recognition problem is to search from the training dynamical patterns $\varphi_\zeta^k (k = 1, \cdots, M)$ for those similar to the given test pattern φ_ς. Rapid recognition can be implemented according to a kind of indirect and dynamical matching of the system dynamics [4]. Specifically, for the kth training pattern, a dynamical model is constructed as

$$\dot{\bar{x}}^k = -B(\bar{x}^k - x) + \overline{W}^{k^T} S(x) \tag{7}$$

where $\bar{x}^k = [\bar{x}_1^k, \cdots, \bar{x}_n^k]^T$ is the state of the dynamical model, x is the state of an in put test pattern. Then, corresponding to the test pattern φ_ς and the training pattern φ_ζ^k, we obtain the following closed-loop recognition system:

$$\dot{\tilde{x}}_i^k = -b_i \tilde{x}_i^k + \overline{W}_i^{k^T} S_i(x) - f_i(x;p'), i = 1, \cdots, n \tag{8}$$

where $\tilde{x}_i^k = \bar{x}_i^k - x_i$ is the state tracking error. When pattern φ_ς is similar to pattern φ_ζ^k, the origin $\tilde{x} = 0$ of the closed-loop recognition system (8) will converge exponentially to a small neighborhood of zero, and the state tracking (or synchronization) $\bar{x}^k \to x$ will be achieved.

2.3 Construction of Recognition System for Classification

In [11], a scheme was presented for the construction of recognition system for which classification is to assign an input dynamical pattern φ_ς to one of K classes $\Psi_1, \cdots, \Psi_K$ based on the predefined similarity measure on system dynamics.

1) Nearest-Neighbor Decision: The nearest-neighbor decision is a commonly used classification algorithm in pattern recognition [1], in which each class is represented by a set of chosen templates. When an unknown pattern is to be classified, its closest neighbor is found from among all the templates, and the class label is decided accordingly. If the number of pre-classified templates is large, it makes good sense to use, instead of the single nearest neighbor, the majority vote of the nearest k neighbors. This method is referred to as the k-nearest neighbor rule [1].

2) Qualitative Analysis of Dynamical Patterns: The recognition system is constructed with dynamical models being arranged in some specific order. This specific order can be designed according to the qualitative analysis of nonlinear dynamical systems [5], in which the concepts of topological equivalence, structural stability, bifurcation and chaos altogether provide an inclusive classification of various types of dynamical patterns.

3) A Hierarchical Structure: To save the space of memory, it is desirable not to store all the identified training patterns as templates. Compared with the periodic patterns, quasi-periodic and chaotic patterns are more spatially expanded, and usually occur under a slight parameter variation. This means that the quasi-periodic and chaotic patterns are very suitable being taken as template models in the recognition system. Specifically, at the first level of the hierarchical structure, a few chaotic patterns are chosen as templates to represent classes of dynamical patterns in a large. In the subsequent levels, quasi-periodic and chaotic patterns are used to represent classes and subclasses of dynamical patterns. In this way, the recognition system is constructed with the dynamical template models being arranged according to a hierarchical structured knowledge representation based on the similarity of system dynamics [11].

3 Construction of Recognition System for Classification of Lorenz System and Chen System

In this section, we study the construction of the recognition system for classification of Lorenz system and Chen system.

The Lorenz system is described by [9]:

$$\begin{cases} \dot{x} = -ax + ay \\ \dot{y} = cx - y - xz \\ \dot{z} = -bz + xy \end{cases} \tag{9}$$

where $a = 10, b = 8/3$ and $c \in [24, 340]$ is a variable parameter of the system.
The Chen system is described by [10]:

$$\begin{cases} \dot{x} = -ax + ay \\ \dot{y} = (c-a)x + cy - xz \\ \dot{z} = -bz + xy \end{cases} \tag{10}$$

where $a = 35, b = 8/3$ and $c \in [24, 44]$ is a variable parameter of the system.
A generalized Lorenz systems was presented in [8]:

$$\begin{cases} \dot{x} = a_{11}x + a_{12}y \\ \dot{y} = a_{21}x + a_{22}y - xz \\ \dot{z} = a_{33}z + xy \end{cases} \tag{11}$$

According to the classification given in [8], the Lorenz system (9) satisfies the condition $a_{12}a_{21} > 0, a_{12} = -a_{11} = a, a_{21} = c, a_{22} = -1$, and $a_{33} = -b$. Chen system (10) satisfies the condition $a_{12}a_{21} < 0, a_{12} = -a_{11} = a, a_{21} = c - a, a_{22} = c$, and $a_{33} = -b$.

3.1 Qualitative Analysis of Lorenz System and Chen System Patterns

A bifurcation is a change of the topological types of dynamical behaviors as a parameter-dependent dynamical system varies its parameters across critical values, which is referred to as a bifurcation point. Bifurcation means structurally unstable, and bifurcation points actually form the boundaries between different subclasses of a set of dynamical patterns [5]. Eckmann studied varied possible bifurcation phenomena and included three accesses to chaos: Feigenbaum access (chaos generated through pitchfork bifurcation), Ruelle-Takens-Newhouse access (chaos generated through Hopf Bifurcation), Pomeau-Manneville access (chaos generated through paroxysm) [7].

A. Qualitative Analysis of Lorenz System
Observe the variation of the dynamics of Lorenz system when the value of variable parameter c has been changed. When $229 < c \leq 340$, system dynamics is a limit cycle(Fig.1(a)). Decrease c, when $216 < c \leq 229, 215 < c \leq 216$, double-periodic bifurcation of the system takes place and there are two cycles, four cycles, eight cycles appear in the phase space in turn(Fig.1(b),(c)). When $166 < c \leq 214$, there is a strange attractor in the phase space, system dynamic is chaotic(Fig.1(d)). Continue to decrease c, when $148 < c \leq 146$, limit cycle appears in the phase space again(Fig.1(e)). When $146 < c \leq 148, 145.5 < c \leq 146$,

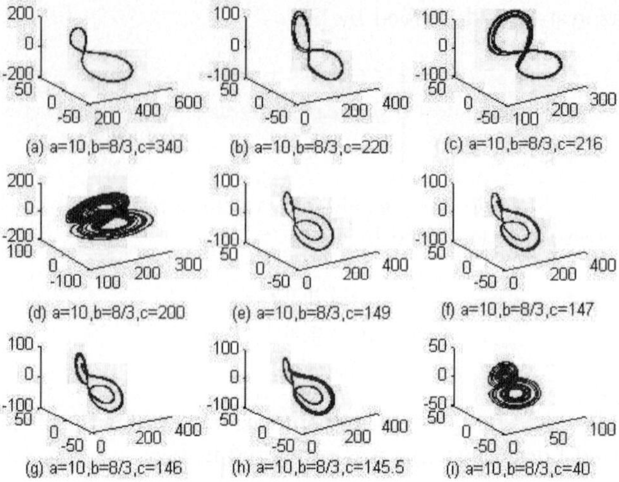

Fig. 1. Variations of Lorenz system dynamics according to the decrement of the value of variable parameter c

double-periodic bifurcation takes place again and there are two cycles, four cycles, eight cycles appear again in the phase space in turn(Fig.1(f),(g),(h)). Continue decreasing c, when $24 < c \leq 145.5$, strange attractor appears again in the phase space(Fig.1(i)). As shown in Fig.1, the dynamics of Lorenz system is periodic motion ,double-periodic motion and chaotic motion which appeared alternately according to the decrement of c and belongs to Pomean-Manneville access which generate chaos through paroxysm. This paroxysm is related to Hopf bifurcation and double-periodic bifurcation.

B. Qualitative Analysis of Chen System

Similar to Lorenz system, following the decrement of variable parameter c, Chen system dynamics undergoing chaotic motion, periodic motion and chaotic motion in turn. When $33.5 < c \leq 34$, the dynamics of Chen system is chaotic motion, when $29 < c \leq 33.5$, the dynamics of Chen system is a limit cycle, when $24 < c \leq 29$, Chen system dynamics is chaotic motion again.

3.2 Classification and Hierarchical Structure

Based on the analysis of dynamics of Lorenz system and Chen system above, we could classify Lorenz system and Chen system according to the concept of bifurcation and chaos. Following the variation of parameter c, the range of the state variation of Lorenz system and Chen system has been changed greatly. In order to restrict the state variation in a certain range such that unified state transform could be used in the simulation analysis, the variable parameter c of Lorenz system supposed vary only between interval [140, 180]. This interval include periodic, double-periodic, quasi-periodic and chaotic dynamics of Lorenz system, so Lorenz system in this interval is representative of the whole Lorenz system. When $140 \leq c < 145$, the dynamics of Lorenz system is chaotic, we define this class of pattern belongs to pattern Ψ_1; when $145 \leq c < 148$, the dynamics of Lorenz system is double-periodic and we define this class of pattern belongs to pattern Ψ_2; when $148 \leq c < 166$, the dynamics of Lorenz system is periodic and we define this class of pattern belongs to pattern Ψ_3; when $166 \leq c < 180$, the dynamics of Lorenz system is chaotic again and we define this class of pattern belongs to pattern Ψ_4.

Accordingly, in order to restrict the variation of the state of Chen system in certain range, we suppose that the variable parameter c vary only between interval [24, 33.5]. When $24 \leq c < 29$, the dynamics of Chen system is chaotic, so we define this class of pattern belongs to pattern Ψ_5; when $29 \leq c \leq 33.5$, the dynamics of Chen system is periodic and we define this class of pattern belongs to pattern Ψ_6.

The first level of the recognition system: Choose chaotic patterns generated from Lorenz system when $c = 145$ and $c = 166$ (we write them as M_{L145}, M_{L166} for concise, where M refers to model, subscript L refers to Lorenz system and 145 refers to $c = 145$) as template models to represent Lorenz system in a large. Choose chaotic pattern generated from Chen system when $c = 26(M_{C26})$ as template model to represent Chen system in a large.

The subsequent level: Choose quasi-periodic and periodic patterns even distributed in each subclass to represent subclasses of Lorenz system and Chen system.

Choose M_{L140} and M_{L145} to represent pattern Ψ_1; choose M_{L146} and M_{L148} to represent pattern Ψ_2; choose M_{L156} and M_{L160} to represent pattern Ψ_3; choose M_{L166} and M_{L170} to represent pattern Ψ_4; choose M_{C24} and M_{C26} to represent pattern Ψ_5; choose M_{C29} and M_{C33} to represent pattern Ψ_6.

4 Simulations

Since the neural networks used in the simulation for deterministic learning and rapid dynamical pattern recognition only have limited number of nodes [3][4], as for all the dynamical patterns generated from Lorenz system and Chen system mentioned above, we need to introduce a kind of state transform in order to restrict the states of all dynamical patterns vary in the small interval [-3,3]. The state transform used here just change the value of the states without affecting the inherent system dynamics of the patterns.

We introduce the following state transform: $x_1 = \frac{x}{20}, x_2 = \frac{y}{50}, x_3 = \frac{z-170}{45}$, and then with a little abuse of notations, we assume that: $x_1 = x, x_2 = y, x_3 = z$, Lorenz system can be rewrite as:

$$\begin{cases} \dot{x} = a(2.5y - x) \\ \dot{y} = 0.4cx - y - 18x(z + 3.4) \\ \dot{z} = -b(z + 3.4) + 20xy \end{cases} \tag{12}$$

The corresponding state transform is needed for Chen system to become suitable for simulation analysis. Transform process is omitted here since it is similar to the process of Lorenz system. Transformed Chen system is described as follow:

$$\begin{cases} \dot{x} = a(2y - x) \\ \dot{y} = 0.5(c - a)x - 8x(z + 3.125) + cy \\ \dot{z} = 8xy - b(z + 3.125) \end{cases} \tag{13}$$

All the dynamical patterns used in the simulation analysis have been transformed and without changing their inherent dynamical structure, just altered the magnitude of the states. Through deterministic learning, for various unknown parameter c, the unknown system dynamics $f_L(x, y, z) = 0.4cx - y - 18x(z+3.4)$ and $f_C(x, y, z) = 0.5(c - a)x - 8x(z + 3.125) + cy$ of various patterns can be accurately identified along their system orbits, and the dynamical patterns can be effectively represented in a time-independent and spatially-distributed manner [4].

To verify the validity of the recognition system in this paper, we take four patterns p_1, p_2, p_3, p_4 as test patterns, these patterns are generated from Lorenz system when $c = 144, c = 148, c = 155$ and Chen system when $c = 27$, respectively. We take the synchronization errors of state y between training pattern and test pattern as similarity measure. The recognition process is as follow:

From Fig.2, it is seen that test patterns p_1, p_2, p_3 are similar to the training pattern M_{L145} since the synchronization errors between themselves and the

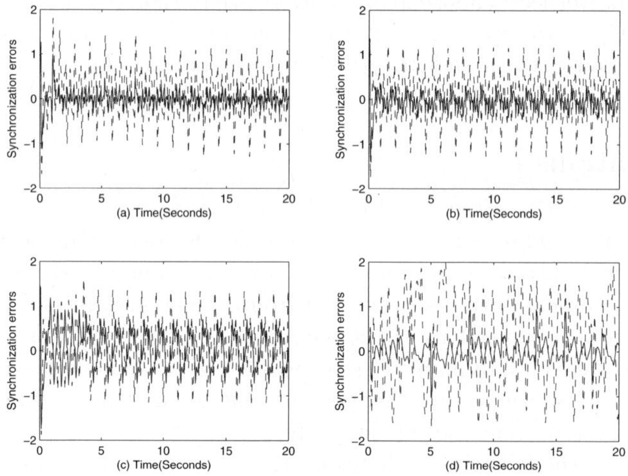

Fig. 2. Synchronization errors: (a) test pattern p_1 and training pattern M_{L145} "—", test pattern p_1 and training pattern M_{C26} "- -" (b) test pattern p_2 and training pattern M_{L145} "—", test pattern p_2 and training pattern M_{C26} "- -" (c) test pattern p_3 and training pattern M_{L145} "—", test pattern p_3 and training pattern M_{C26} "- -" (d) test pattern p_4 and training pattern M_{L145} "- -", test pattern p_4 and training pattern M_{C26} "—" .

training pattern M_{L145} are relative small, so test patterns p_1, p_2, p_3 are being recognized as patterns generated from Lorenz system. Test pattern p_4 similar to training pattern M_{C26} since the synchronization errors between itself and the training pattern M_{C26} are relative small, so test pattern p_4 is being recognized as pattern generated from Chen system.

Subsequently we compare the synchronization errors between each of test patterns and each of training model patterns which represent subclasses of patterns in which system the test pattern lies. If all the synchronization errors are relative small, from the figures it is not suitable to compare the synchronization errors quantitatively, so we take the norm of synchronization errors to represent the magnitude of synchronization errors. First, we solve the synchronization errors using the fourth-order Runge-Kutta method (step size is 0.01 second), and ignore the errors generated from the first 500 operations, then calculate the L_2 norm of the synchronization errors obtained from subsequent 500 operations. Results are described as follow:

Table 1. Synchronization errors between test patterns and training patterns in Lorenz system

	M_{L140}	M_{L145}	M_{L146}	M_{L148}	M_{L156}	M_{L160}	M_{L166}	M_{L170}
p_1	4.6110	3.0219	3.2228	4.4770	10.3349	13.3941	17.0760	17.0302
p_2	6.7126	3.5233	2.7023	2.5188	7.6100	10.7878	14.6080	14.8736
p_3	10.8295	8.3961	8.4623	6.9609	2.9823	5.7579	9.9886	11.2632

Table 2. Synchronization errors between test patterns and training patterns in Chen system

	M_{C24}	M_{C26}	M_{C29}	M_{C33}
p_4	10.3105	5.3057	9.4375	9.3915

From Table 1, it is clear that the norm of synchronization errors between test pattern p_1 and training model pattern M_{L145} is the smallest, so test pattern p_1 belongs to pattern Ψ_1, variable parameter c is close to 145. For the same reason, test pattern p_2 belongs to pattern Ψ_2, variable parameter c is close to 148; test pattern p_3 belongs to pattern Ψ_3, variable parameter c is close to 156. From Table 2, we could conclude that test pattern p_4 belongs to pattern Ψ_5, variable parameter c is close to 26. Simulation results verified the validity of this recognition system.

5 Conclusions

In this paper, a recognition system has been presented which not only can classify different classes of dynamical patterns, but also can distinguish a set of dynamical patterns generated from Lorenz system and Chen system. It is also be able to apply to practical industry such as power system.

Acknowledgments. The authors acknowledge support by the Natural Science Foundation of China under Grant No. 60743011, the program of New Century Excellent Talents in Universities (NCET), and National 973 Project.

References

1. Jain, A.K., Duin, R.P.W., Mao, J.: Statistical Pattern Recognition: a Review. IEEE Transactions on Pattern Analysis and Machine Intelligence 22, 4–37 (2000)
2. Covey, E., Hawking, H.L., Port, R.F. (eds.): Neural Representation of Temporal Patterns. Plenum Press, New York (1995)
3. Wang, C., Hill, D.J.: Learning from Neural Control. IEEE Transactions on Neural Networks 17, 130–146 (2006)
4. Wang, C., Hill, D.J.: Deterministic Learning and Rapid Dynamical Pattern Recognition. IEEE Transactions on Neural Networks 18, 617–630 (2007)
5. Shilnikov, L.P., et al.: Methods of Qualitative Theory in Nonlinear Dynamics. World Scientific, Singapore (2001)
6. Powell, M.J.D.: The Theory of Radial Basis Function Approximiation in 1990. In: Light, W.A. (ed.) Advances in Numerical Analysis II: Wavelets, Subdivisions, Algorithms, and Radial Basis Functions, pp. 105–210. Oxford University Press, Oxford (1992)
7. Eckmann, J.P.: Roads to Turbulence in Dissipative Dynamics System. Rev. Mod. Phys. 53, 643–649 (1981)
8. Celikovsky, S., Chen, G.: On a Generalized Lorenz Canonical Form of Chaotic Systems. Int. J. of Bifurcation Chaos 12, 1789–1812 (2002)
9. Lorenz, E.N.: Deterministic Nonperiodic Flow. J. Atmos. Sci. 20, 130–141 (1963)
10. Chen, G., Ueta, T.: Yet Another Chaotic Attractor. Int. J. of Bifurcation Chaos 9, 1465–1466 (1999)
11. Wang, C., Hill, D.J.: Dynamical Pattern Classification. In: IEEE Conference on Intelligent Automation, Hong Kong (2003)

Research of Spam Filtering System Based on LSA and SHA

Jingtao Sun[1,2], Qiuyu Zhang[2], Zhanting Yuan[2],
Wenhan Huang[3], Xiaowen Yan[4], and Jianshe Dong[2]

[1] College of Electrical and Information Engineering, Lanzhou University of Technology, 730050 Lanzhou, China
[2] College of Computer and Communication, Lanzhou University of Technology, 730050 Lanzhou, China
[3] Department of Computer science and technology, Shaanxi University of Technology, 723003 Hanzhong, China
[4] Shaanxi Xiyu Highway Corporation Ltd. Hancheng, 715400 Shaanxi, China
sun2651@126.com

Abstract. Along with the widespread concern of spam problem, at present, there are spam filtering system nowadays about the problem of semantic imperfection and spam filter low effect in the multi-send spam. This paper proposes a model of spam filtering which based on latent semantic analysis (LSA) and message-digest algorithm 5 (SHA). Making use of the LSA marks the latent feature phrase in the spam, semantic analysis is led into the spam filtering technique; the "e-mail fingerprint" of multi-send spam is born with SHA on the LSA analytical foundation, the problem of filtering technique's low effect in the multi-send spam is resolved with this kind of method. We have designed a spam filtering system based on this model. Our designed system was evaluated with an optional dataset. The results obtained were compared with KNN algorithm filter experiment results show that system based on Latent Semantic Analysis and SHA performs KNN. The experiments show the expected results obtained, and the feasibility and advantage of the new spam filtering method is validated.

Keywords: Latent Semantic Analysis, Secure Hash Algorithm, Mail Characteristic ID, Slipping Windows, Spam Filtering.

1 Introduction

With rapid popularization of Internet, Email is widely used in companies, government organs, colleges and universities, middle schools and families, etc [1]. While serving as a convenient communication approach to facilitate people's work, study and life, Email also provides an important carrier for spreading virus, hacker programs, porn, reactionary and superstitious information [2, 3]. There is now a couple of filtering software programs available on the market, yet we found by comparing their theories that current filtering software programs are more or less having the problem of absence of semantics. Therefore, when the spam develops to a certain degree, these Email filtering algorithms or

F. Sun et al. (Eds.): ISNN 2008, Part II, LNCS 5264, pp. 331–340, 2008.

filtering systems may not be able to handle them. Besides, the sender address of most spam today varies in a dynamic way, yet contents of the text or attachment are the same. In large-scale LAN that includes tens of thousands of users, spam usually spreads itself across the network by means of mass mailing. In consideration of these characteristics, it is necessary to introduce new theory to improve existing solutions.

2 Overview of Key Technologies

LSA is a knowledge representation-related algorithm established on the basis of a semantic space that includes a vast amount of texts [3, 4]. Its ability to clarify the hidden semantic relationship between words and text provides an extremely important and significant direction for studies on fighting spam that contains hidden information. Yet a number of difficulties are to be overcome for LSA that are related to characteristics of Chinese language. Secure Hash Algorithm (SHA) is a common data encryption algorithm published in 1993 by the National Institute of Standards and Technology of the United States as the national standard for information processing (i.e. the first generation of SHA algorithm SHA-0)[4, 5]. SHA algorithm can process input with 2-bid data block as the unit and generate 160-bit information abstract. The algorithm (SHA-1) has so far been widely used in digital signature files and authentication in E-commerce transactions [6, 7].

3 Analysis of Key Technologies

3.1 Basic Method Based on Latent Semantic Analysis

The fundamental concept of LSA is to map a document represented by vector space model (VSM) of higher dimensions to latent semantic space of lower dimensions [8, 9]. A dimension reduction matrix that contains *K* orthogonal factors is generated by performing Singular Value Decomposition (SVD) [10, 11] on the word-document matrix of the text collection to approximately represent the word-document matrix of the original text collection.

First, an m×n matrix of terms and documents, $X=[x_{ij}]$, is structured by a large amount of regulations, where $x_{ij} \geq 0$, x_{ij} is the ith term frequency which appears in the j-th document. The amount of terms and documents is so larger and term appearing in individual document is so few. Thus X is generally a sparse matrix, and x_{ij} is usually involved in two aspect factors, which are terms partial weight L (i, j) and terms whole weight C (i), where L (i, j) is the i-th term weight which appears in the j-th documents, and C (i) is the i-th term which appears in the entire documents library. Weighting processing should be conducted on x_{ij} to get a weighted m×n word-document matrix $X^{'}=[x_{ij}^{'}]$.

$$x_{ij}^{'} = x_{ij} \times L(i,j) \times C(i) \tag{1}$$

$X^{'}$ is analyzed by singular value decomposition (assumption: m > n, rank(X) = r, $\exists$ K, K< r and K< < min (m, n)). Under the F-norm, $X_k^{'}$ can be decomposed into the product of three other matrices: $X^{'} \approx X_k^{'} = U_k \mathfrak{R}_k V_k^{\mathrm{T}}$, where $X_k^{'}$ is a k rank approximate matrix; U_k 's column vector and V_k 's column vector are Orthogonal vectors; I_k is k-rank eye matrix.

$$U_k^T U_k = V_k^T V_k = I_k \tag{2}$$

Where U_k and V_k is the matrices of left and right singular vectors, and $\Re_k$ is the diagonal matrix of singular values. Where U_k and V_k is terms and documents vectors.

3.2 SHA Algorithm

Secure Hash Algorithm (SHA) is a popular data encryption algorithm. Message *m* is converted by SHA into a 160-bit characteristic string consisting solely of 0 and 1. Steps of SHA algorithm [12, 13]:

(1) Filling position: fill positions of message *m* in SHA algorithm so that the reminder when dividing final number of bits of message *m* by 512 is 448. That is to say, the number of bits filled makes the total number of bits 64 bits less than a multiple of 512. To fill the positions, add a 1 first, and then add 0 until the above requirement is satisfied.

(2) Expanding length: after the position is filled, affix a 64-bit segment to the end, which is regarded as a 64-bit integer.

(3) Initializing variables: 160-bit buffer is used to store intermediate results and final Hash value. The buffer consists of five registers, namely A, B, C, D and E, each being 32 bits long. After initialization, they are (hexadecimal system): A = 67 45 23 01, B = EF CD AB 89, C = 98 BA DC FE, D = 10 32 54 76, E = C3 D2 E1 F0.

(4) Processing information: process 512-bit information groups. The core of the algorithm includes four rounds of operations, each of which includes 20 steps. Four logical functions will be defined firstly:

$f_1(X,Y,Z)=(X\wedge Y)\vee(\bar{X}\wedge Z)$, $f_2(X,Y,Z)=X\oplus Y\oplus Z$, $f_3(X,Y,Z)=(X\wedge Y)\vee(X\wedge Z)\vee(Y\wedge Z)$, $f_4(X,Y,Z)=X\oplus Y\oplus Z$. In the functions, X, Y and Z are all 32 bits long, $(\wedge, \vee, -, \oplus)$ indicates logical operations (AND, OR, NOT, XOR) respectively. If the corresponding bits of X, Y and Z are independent and even, all bits in the result will also be independent and even.

Input of each round of data processing is a 512-bit variable and the output is a 160-bit variable (ABCDE). Constant K_t is use in each round, in which $0 \le t \le 79$. There are in fact a total of only 4 different constants.

$$K_t = \begin{cases} 5A \quad 82 \quad 79 \quad 99 & 0 \le t \le 19 \\ 6E \quad D9 \quad EB \quad A1 & 20 \le t \le 39 \\ 8F \quad 1B \quad BC \quad DC & 40 \le t \le 59 \\ CA \quad 62 \quad C1 \quad D9 & 60 \le t \le 79 \end{cases}$$

They are $\lfloor 2^{30}\sqrt{2} \rfloor, \lfloor 2^{30}\sqrt{3} \rfloor, \lfloor 2^{30}\sqrt{5} \rfloor, \lfloor 2^{30}\sqrt{10} \rfloor$.

(5) Output: ABCDE obtained after the above steps are the output results. They are stored in an uninterrupted sequence and occupy a total of 20B and 160 bits. A is the lowest bit and E is the highest bit.

4 Model of the Spam Filtering System

LSA and SHA-based spam filtering system is designed to filter similar mass mailing spam efficiently and accurately. Semantic analysis, generation of mail characteristic ID and other technologies will be introduced to enable the system to have fairly high flexibility and good adaptability. Please see Fig 1 for model of the spam filtering system.

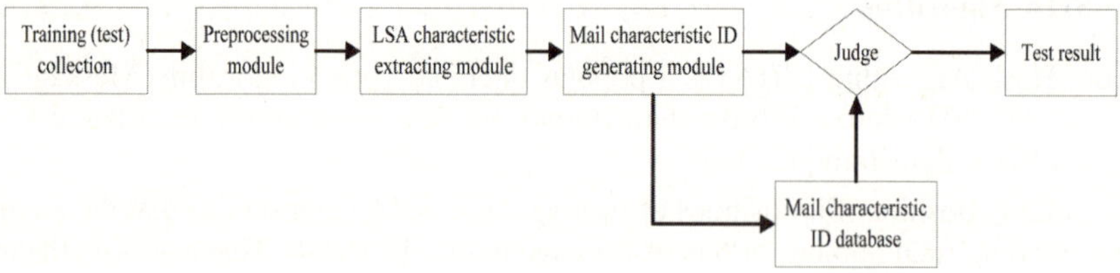

Fig. 1. Model of spam filtering figure

First and foremost, the system performs training using given mail collection so that LSA characteristic extracting module can extract characteristics of spam and legal Emails from already known legal Emails and spam and save these characteristics in an appropriate word-document matrix. Then the inspected Email, which has been preprocessed by the preprocessing module, is sent to the LSA characteristic extracting module for information extracting. The information obtained fully reflects the capability of LSA method in extracting information. The extracted information will be used as the "anchor" value of the inspected mail. This mail, which contains anchor value, will be sent to the mail characteristic ID generating module, which uses sliding windows and SHA algorithm to get a characteristic ID of certain length. The method solves the problem of inaccurate representation of documents by mail characteristic ID generated on the basis of independent characteristic word. Therefore, the spam is identified by comparing the mail characteristic ID generated with information in the mail characteristic ID database.

5 Module Design of the Spam Filtering System

5.1 Preprocessing Module

Different from structured data in traditional database, Emil's header information has certain structure, while its content is not structured. To process such semi-structured data as Email, preprocessing must be conducted that consists of the following steps: analysis of document characteristic and format, Chinese word division and word frequency weighting, etc. The article focuses on Chinese lexical analysis system

ICTCLAS (Institute of Computing Technology, Chinese Lexical Analysis System) using an approach based on multi-layer HMM, which features Chinese word division, mark of syntactical function and identification of unregistered words, etc. The system will be used together with filtering of prohibited words, removal of words of extremely high or low frequency as well as other preprocessing methods to remove words and expressions of little significance and meet the requirements for subsequent processing of mails.

5.2 LSA Characteristic Extracting Module

LSA characteristic extracting module adopts LSA technology and reveals the semantic relationship between words and documents through SVD and K-rank approximate matrix. When every word has its vector representation, latent characteristic words (phases) can be predicted on the basis of already obtained identification results (history) in identifying latent characteristic words (phases). $\{X_1, X_2, \cdots, X_{i-1}\}$ and P_{i-1} indicate respectively vector of the word obtained before moment i–1 and the historical vector obtained at corresponding moments. Expanding the results identified at moment i and add a new word W_i, $\{X_1, X_2, \cdots, X_{i-1}, X_i\}$ and P_i indicate respectively vector of the word obtained at moment i and the historical vector obtained at corresponding moments, ω_j indicates the entropy of word W_j at moment $j(j=1,2,\cdots,i)$ in relation to the training language materials. According to:

$$P_{i-1} = \frac{1}{i-1}\sum_{j=1}^{i-1} X_j\left[1-\omega_j\right] , \quad P_i = \frac{1}{i}\sum_{j=1}^{i} X_j\left[1-\omega_j\right] \tag{3}$$

We get

$$P_i = \frac{1}{i-1}P_{i-1} + \frac{1}{i}X_i\left[1-\omega_i\right] \tag{4}$$

Formula (4) is the historical vector update formula in identifying latent characteristic words (phases). When updating and identifying historical vector, use the entropy to conduct weighting calculation so as to differentiate the contribution of each word to identification history.

5.3 Email Characteristic ID Generating Module

The major function of Email characteristic ID generating module in the spam filtering system is to generate the mail characteristic ID of spam. Please see Fig 2 for the realization process.

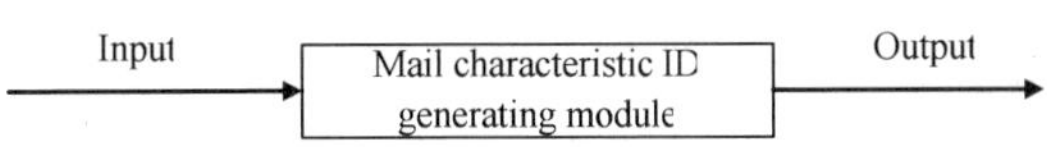

Fig. 2. Generation of Email characteristic ID

(1) Input: in order to get characteristic ID of inspected mails in a rapid, highly efficiently and accurately way using SHA algorithm, sampling points need to be set up in mail text. Mail text containing "anchor" values are produced following processing by modules of upper levels. These "anchor" values serve as the sampling points of the mails. To reflect document information more effectively and in order not to damage relationship between words, sliding window characteristic extracting algorithm is introduced to restructure words in the nearby area of the "anchor" values. This further expands the scope of characters to be chosen so that the characteristic word extract can reflect characteristic of the document more accurately and be used as input of the module at upper level.

(2) Output: the characteristic value obtained using sliding window characteristic extracting algorithm outputs 160-bit mail characteristic ID through SHA algorithm and store the ID in the mail characteristic ID database at the background.

5.4 Mail Characteristic ID Database

The database module adopts MySQL database and is a high-speed, multi-thread, multi-user and robust SQL database server. Compared with existing database systems, it features quick response and can be used across different platforms [14]. The database mainly stores tab-files table and includes "files" and "characteristic" fields. The "files" field is used store file information and the "characteristic" field is used to store digital fingerprint information generated to avoid file repetition from occurring.

Data is imported into tab-files table in the following steps:

(1) Conduct processing to get mail document *M*, and calculate mail characteristic of *M* ID-CT_M. (2) Check whether there is identical mail characteristic ID as CT_M in the database. (3) Skip the document if there is and go to (1) to process the next mail until all mails are imported in the database. (4) If there is not, save the mail document and corresponding mail characteristic ID in tab-files table and go to (1) to process the next mail until all mails are imported in the database.

6 System Test and Analysis

Selection of language material database is critical for system test. There are some benchmark and widely recognized language material database overseas, such as PUI language material database [15, 16]. Yet in the field of Chinese spam classification, there is not a widely recognized Chinese material database. Given this, the article collects 1800 spam mails of different types from extensive sources to form a 15MB training collection. The test platform is PM2.1G with 2GB memory.

First, extract and process texts of these mails. Conduct Chinese word division, filtering of prohibited words, removal of words of extremely high or low frequency as well as other preprocessing measures to generate a 5672×1800 word-document matrix A. Then perform SVD to generate latent semantic space A_k. In the process, selection of dimension reducing factor *K* has a direct influence on the efficiency of the latent semantic space model and similarity between A_k and A following dimension reduction. If the value of *K* is too small, useful information will be lost; if the value of *K* is too large, the calculation volume will increase. The article uses contribution rate δ as the criterion to assess the K value selected, i.e. $A=\mathrm{diag}(a_1,a_2,\cdots,a_n)$, and $a_1 \geqslant a_2 \geqslant \cdots \geqslant a_t = \cdots = a_n = 0$, contribution rate δ:

$$\delta = \sum_{i=1}^{k} a_i \Big/ \sum_{i=1}^{t} a_i \tag{5}$$

The contribution rate δ, proposed in reference of related factor analysis concept, indicates the degree, to which the K-dimensional space represents the entire space. Fig 3 shows that the closer the K value is to the rank of matrix A, the smaller $\| A - A_K \|_F$ is and the closer A_k is to A. Yet as the value of K continues to increase, its influence on δ will decrease or even disappear. Analysis indicates that when the value of K increases to a certain level, nearly all important characteristics of word-document matrix are represented. In this case, further increasing K value will only introduce noise. When K=900, the degree of representation is almost the same as when K=1000. Yet when K=900, less time is consumed. So, we choose K=900.

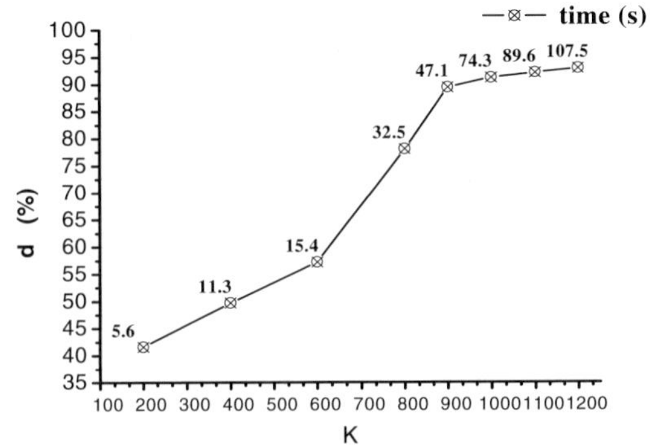

Fig. 3. Analysis of *K* value

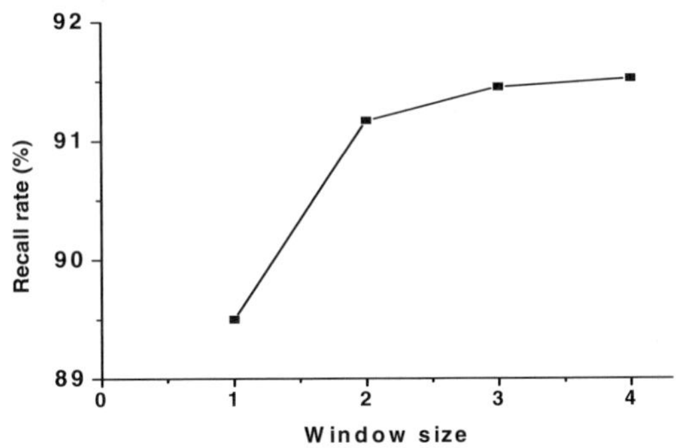

Fig. 4. Analysis of window size

When generating mail characteristic ID, configuration of size of the sliding window will also affect performance and efficiency of the entire filter. As shown in Fig 4, the larger the sliding window is, the better the entire filtering system performs [17]. This is because when the window becomes larger, more characteristics will be selected and more document characteristics will be represented. In the mean time, both recall rate and correctness rate will improve. Yet the large the window is, the slow the operation speed and the longer the operation time will be. When the window size is 2, an optimal balance will be stricken between performance and operation speed.

To explain actual effect of the system in filtering spam, an experiment was performed to compare it with KNN method [18, 19]. See Table 1 for the result.

Table 1. Experiment results of LSA and SHA algorithm and KNN

	Recall rate (%)	Correctness rate (%)	F_1 value (%)
KNN	82.71	85.21	83.94
LSA and SHA	89.17	91.37	90.26

Table 1 shows results of the experiment on 900 mails. According to the results, the method specified herein increases the recall rate of the mail system by 6.46%, accuracy of identification by 6.16% and the F_1 value by 6.32%.

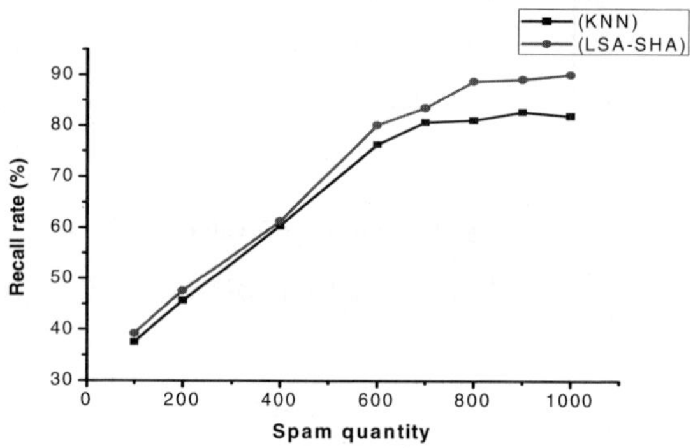

Fig. 5. Recall rate

Fig 5 and 6 is the curve chart of performance of the LSA and SHA-based spam filtering system and the KNN algorithm [20] based system given different quantity of Emails. Analysis of the data shows that the LSA and SHA-based spam filtering system performs better than the KNN algorithm-based system in spam filtering and its system design measures up to expected requirement. So the system has a promising future of application.

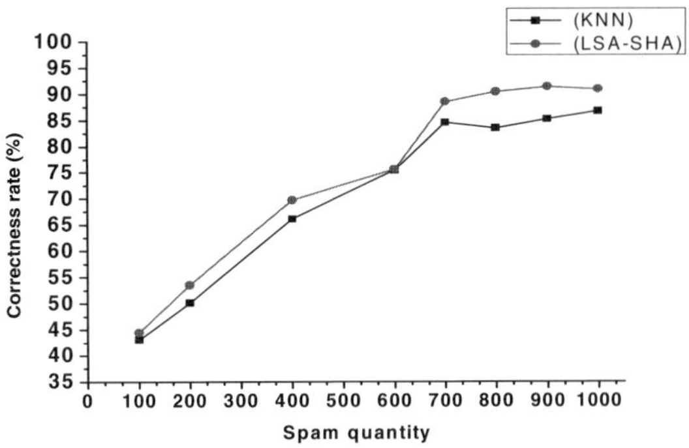

Fig. 6. Correctness rate

7 Conclusion

The article proposes and realizes the LSA and SHA-based spam filtering system. It combines LSA and SHA algorithms as well as database technology to introduce Latent Semantic Analysis into filtering technologies. In consideration of characteristics of mass mailing spam, it generates mail characteristic ID using SHA algorithm to enable highly efficient and accurate filtering of mass mailing spam and invent a new theory for spam filtering.

References

1. Anti-spam Alliance in China, `http://www.anti-spam.org.cn`
2. Hoanca, B.: How Good are Our Weapons in the Spam Wars? Technology and Society Magazine 25(1), 22–30 (2006)
3. Whitworth, B., Whitworth, E.: Spam and the Social Technical Gap. Computer & Graphics 37(10), 38–45 (2004)
4. Tang, P.Z., Li, L.Q., Zuo, L.M.: A New Verification Technology Based on SHA and OTP. Journal of East China Jiao Tong University 22(2), 55–59 (2005)
5. Wang, G.P.: An Efficient Implementation of SHA-1 Hash Function. In: The 2006 IEEE International Conference on Information Technology, pp. 575–579. IEEE Press, China (2006)
6. Chen, H., Zhou, J.L., Feng, S.: Double Figure Authentication System Based on SHA and RSA. Network & Computer Security 4, 6–8 (2006)
7. Burr, W.E.: Cryptographic Hash Standards: Where Do We Go From Here? Security & Privacy Magazine 4(2), 88–91 (2006)
8. Zhu, W.Z., Chen, C.M.: Storylines: Visual Exploration and Analysis in Latent Semantic Spaces. Computers & Graphics 31(3), 78–79 (2007)
9. Maletic, J.I., Marcus, A.: Using Latent Semantic Analysis to Identify Similarities in Source Code to Support Program Understanding. In: 12th IEEE International Conference on Tools with Artificial Intelligence, pp. 46–53. IEEE Press, New York (2000)
10. Martin, D.I., Martin, J.C., Berry, M.W.: Out-of-core SVD Performance for Document Indexing. Applied Numerical Mathematics 57(11-12), 224–226 (1994)

11. Gai, J., Wang, Y., Wu, G.S.: The Theory and Application of Latent Semantic Analysis. Application Research of Computers 21(3), 161–164 (2004)
12. Michail, H., Kakarountas, A.P.: A Low-power and High-throughput Implementation of the SHA-1 Hash Function. In: The 2005 IEEE International Symposium on Circuits and Systems, vol. 4, pp. 4086–4089. IEEE Press, Kobe Japan (2005)
13. Wang, M.Y., Su, C.P., Huang, C.T., Wu, C.W.: An HMAC Processor with Integrated SHA-1 and MD5 Algorithms. In: Design Automation Conference, Proceedings of the ASP-DAC 2004, Japan, pp. 456–458 (2004)
14. Paul, D.B.: MySQL: The Definitive Guide to Using, Programming, and Administering MySQL 4, 2nd edn. China Machine Press, China (2004)
15. Learning to Filter Unsolicited Commercial E-mail, `http://www.aueb.gr/users/ion/docs/TR2004_updated.pdf`
16. Deshpande, V.P., Erbacher, R.F., Harris, C.: An Evaluation of Naïve Bayesian Anti-Spam Filtering. In: Information Assurance and Security Workshop, pp. 333–340. IEEE SMC Press, Spain (2007)
17. Li, J.Z., Zhang, D.D.: Algorithms for Dynamically Adjusting the Sizes of Sliding Windows. Journal of Software 15(12), 13–16 (2004)
18. Parthasarathy, G., Chatterji, B.N.: A Class of New KNN Methods for Low Sample Problems. Systems, Man and Cybernetics 20(3), 715–718 (1990)
19. Yuan, W., Liu, J., Zhou, H.B.: An Improved KNN Method and Its Application to Tumor Diagnosis. In: The 2004 IEEE International Conference on Machine Learning and Cybernetics, vol. 5, pp. 2836–2841. IEEE Press, Shanghai (2004)
20. Soucy, P., Mineau, G.W.: A Simple KNN Algorithm for Text Categorization. In: Data Mining. The 2001 IEEE International Conference on Data Mining, pp. 647–648. IEEE Press, USA (2001)

Voice Translator Based on Associative Memories

Roberto A. Vázquez and Humberto Sossa

Centro de Investigación en Computación – IPN
Av. Juan de Dios Batíz, esquina con Miguel Othón de Mendizábal
Ciudad de México, 07738, México
ravem@ipn.mx, hsossa@cic.ipn.mx

Abstract. An associative memory is a particular type of neural network for recalling output patterns from input patterns that might be altered by noise. During the last 50 years, several associative models have emerged and they only have been applied to solve problems where input patterns are images. Most of these models have several constraints that limit their applicability in complex problems. Recently in [13] it was introduced a new associative model based on some aspects of the human brain. This model is robust under different type of noises and image transformations, and useful in complex problems such as face and 3d object recognition. In this paper we adopt this model and apply it to problems that not involve images patterns, we applied to speech recognition problems. In this paper it is described a novel application where an associative memory works as a voice translator device performing a speech recognition process. In order to achieve this, the associative memory is trained using a corpus of 40 English words with their corresponding translation to Spanish. Each association used during training phase is composed by a voice signal in English and a voice signal in Spanish. Once trained our English-Spanish translator, when a voice signal in English is used to stimulate the associative memory we expect that the memory recalls the corresponding voice signal in Spanish. In order to test the accuracy of the proposal, a benchmark of 14500 altered versions of the original voice signals were used.

1 Introduction

An associative memory AM can be seen as a particular type of neural network specially designed to recall output patterns in terms of input patterns that can appear distorted by some kind of noise. Several associative models have been proposed in the last 50 years. Refer for example [1-9].

Let $\mathbf{x} \in \mathbf{R}^n$ and $\mathbf{y} \in \mathbf{R}^m$ an input and output pattern, respectively. An association between input pattern $\mathbf{x}$ and output pattern $\mathbf{y}$ is denoted as $\left(\mathbf{x}^k, \mathbf{y}^k\right)$, where k is the corresponding association. Associative memory $\mathbf{W}$ is represented by a matrix whose components w_{ij} can be seen as the synapses of the neural network. If $\mathbf{x}^k = \mathbf{y}^k \forall k = 1, \ldots, p$ then $\mathbf{W}$ is auto-associative, otherwise it is hetero-associative. A distorted version of a pattern $\mathbf{x}$ to be recuperated will be denoted as $\tilde{\mathbf{x}}$. If an associative

F. Sun et al. (Eds.): ISNN 2008, Part II, LNCS 5264, pp. 341–350, 2008.

memory $\mathbf{W}$ is fed with a distorted version of $\mathbf{x}^k$ and the output obtained is exactly $\mathbf{y}^k$, we say that recalling is robust.

Most of these AMs have several constraints that limit their applicability in the solution of problems from real life. Among these constraints we could mention their capacity of storage (limited), the type of patterns (only binary, bipolar, integer or real patterns), robustness to noise (additive, subtractive, mixed, Gaussian noise, deformations, etc). The most common application of an AM is as a filter, where the input stimulus is an image; refer for example to [5-9]. Recently in [13] it was introduced a new associative model based on some aspects of human brain. Although authors show the robustness of the model applied to face and 3D object recognition [11] and [12], even if patterns are contaminated by different type of noises and transformations, they do not report results using other type of stimulus patterns such as voice signal patterns.

The concept of AM emerges from psychological theories of human and animals learning [26]. These memories store information by learning correlations among different stimuli. When a stimulus is presented as a memory cue, the other is retrieval as a consequence; this means that the two stimuli have become associated each other in the memory.

The storage of voice signal or other type patterns have sense because human memory not only stores patterns acquired from the vision system such as objects, faces, letters, cars, but also stores patterns acquired from the auditory system, olfactory system. In this paper we adopt the model described in [13] and apply it to problems that not involve images patterns. In this paper it is described a novel application where an AM work as a voice translator device. In order to achieve this, the AM is trained using a corpus of 40 English words with their corresponding translation in Spanish. Each association used during training phase is composed by a voice signal in English and a voice signal in Spanish. Once trained our English-Spanish translator, when a voice signal in English is used to stimulate the AM we expect that the memory recall the corresponding voice signal in Spanish. In order to test the accuracy of the proposal, a benchmark of 14500 altered versions of the original voice signals were used.

2 The Associative Model

The dynamic associative model described in [13] is not an iterative model as Hopfield's model [4]. The principal difference of this model against other classic models is that once trained, during recalling phase the synapses' values could change as a respond to an input stimulus. The formal set of propositions that support the correct functioning of this model and the main advantages against other classical models can be found in [13]. This model defines several interacting areas, one per association we would like the memory to learn. It also integrates the capability to adjust synapses in response to an input stimulus. Before an input pattern is learned or processed by the brain, it is hypothesized that it is transformed and codified by the brain. This process is simulated using the procedure introduced in [7]. This procedure allows computing *codified patterns* from input and output patterns denoted by $\bar{\mathbf{x}}$ and $\bar{\mathbf{y}}$ respectively; $\hat{\mathbf{x}}$ and $\hat{\mathbf{y}}$ are *de-codifying patterns*.

Codified and de-codifying patterns are allocated in different interacting areas and d defines how much these areas are separated. On the other hand, d determines the noise supported by the model. In addition a simplified version of $\mathbf{x}^k$ denoted by s_k is obtained as:

$$s_k = s\left(\mathbf{x}^k\right) = \mathbf{mid}\ \mathbf{x}^k \tag{1}$$

where **mid** operator is defined as $\mathbf{mid}\ \mathbf{x} = x_{(n+1)/2}$.

In this model, the most excited interacting area is call *active region* (AR) and could be estimated as follows:

$$ar = r(\mathbf{x}) = \arg\left(\min_{i=1}^{p}\left|s(\mathbf{x}) - s_i\right|\right) \tag{2}$$

Once computed the *codified patterns*, the *de-codifying patterns* and s_k we can compute the synapses of the associative memory as follows:

Let $\left\{\left(\overline{\mathbf{x}}^k, \overline{\mathbf{y}}^k\right) \middle| k = 1,\ldots,p\right\}$, $\overline{\mathbf{x}}^k \in \mathbf{R}^n$, $\overline{\mathbf{y}}^k \in \mathbf{R}^m$ a fundamental set of associations (codified patterns). Synapses of associative memory $\mathbf{W}$ are defined as:

$$w_{ij} = \overline{y}_i - \overline{x}_j \tag{3}$$

In short, building of the associative memory can be performed in three stages as:

1. Transform the fundamental set of association into codified and de-codifying patterns by means of Procedure 1 described in [7].
2. Compute simplified versions of input patterns by using equation 1.
3. Build $\mathbf{W}$ in terms of codified patterns by using equation 3.

Synapses could change in response to an input stimulus. There are synapses that can be drastically modified and they do not alter the behavior of the associative memory. In the contrary, there are synapses that only can be slightly modified to do not alter the behavior of the associative memory; this set of synapses is call *the kernel* of the associative memory and it is denoted by $\mathbf{K}_\mathbf{W}$. Let $\mathbf{K}_\mathbf{W} \in \mathbf{R}^n$ the kernel of an associative memory $\mathbf{W}$. A component of vector $\mathbf{K}_\mathbf{W}$ is defined as:

$$kw_i = \mathbf{mid}\left(w_{ij}\right), j = 1,\ldots,m \tag{4}$$

Synapses that belong to $\mathbf{K}_\mathbf{W}$ are modified as a response to an input stimulus. Input patterns stimulate some ARs, interact with these regions and then, according to those interactions, the corresponding synapses are modified. Synapses belonging to $\mathbf{K}_\mathbf{W}$ are modified according to the stimulus generated by the input pattern. This adjusting factor is denoted by Δw and can be computed as:

$$\Delta w = \Delta(\mathbf{x}) = s\left(\overline{\mathbf{x}}^{ar}\right) - s\left(\widehat{\mathbf{x}}\right) \tag{5}$$

where ar is the index of the AR and $\widehat{\mathbf{x}} = \mathbf{x} + \widehat{\mathbf{x}}^{ar}$.

Finally, synapses belonging to $\mathbf{K}_{\mathbf{W}}$ are modified as:

$$\mathbf{K}_{\mathbf{W}} = \mathbf{K}_{\mathbf{W}} \oplus (\Delta w - \Delta w_{old}) \tag{6}$$

where operator $\oplus$ is defined as $\mathbf{x} \oplus e = x_i + e \ \forall i = 1,\ldots,m$. As you can appreciate, modification of $\mathbf{K}_{\mathbf{W}}$ in equation 6 depends on the previous value of Δw denoted by Δw_{old} obtained with the previous input pattern. Once trained the **DAM**, when it is used by first time, the value of Δw_{old} is set to zero.

Once synapses of the associative memory have been modified in response to an input pattern, every component of vector $\overline{\mathbf{y}}$ can be recalled by using its corresponding input vector $\overline{\mathbf{x}}$ as:

$$\overline{y}_i = \mathbf{mid}(w_{ij} + \overline{x}_j),\ j = 1,\ldots,n \tag{7}$$

In short, pattern $\overline{\mathbf{y}}$ can be recalled by using its corresponding key vector $\overline{\mathbf{x}}$ or $\tilde{\mathbf{x}}$ in six stages as follows:

1. Obtain index of the active region ar by using equation 2.
2. Transform $\mathbf{x}^k$ using de-codifying pattern $\hat{\mathbf{x}}^{ar}$ by applying the following transformation: $\widehat{\mathbf{x}}^k = \mathbf{x}^k + \hat{\mathbf{x}}^{ar}$.
3. Compute adjust factor $\Delta w = \Delta(\widehat{\mathbf{x}})$ by using equation 5.
4. Modify synapses of associative memory $\mathbf{W}$ that belong to $\mathbf{K}_{\mathbf{W}}$ by using equation 6.
5. Recall pattern $\widehat{\mathbf{y}}^k$ by using equation 7.
6. Obtain $\mathbf{y}^k$ by transforming $\widehat{\mathbf{y}}^k$ using de-codifying pattern $\hat{\mathbf{y}}^{ar}$ by applying transformation: $\mathbf{y}^k = \widehat{\mathbf{y}}^k - \hat{\mathbf{y}}^{ar}$.

3 Implementation of the Voice Translator

The proposal consists of a dynamic associative memory DAM. The DAM is trained using a selected corpus. Each association $(\mathbf{x}^k, \mathbf{y}^k)$ is composed by two voice signals where $\mathbf{x}^k$ is the k-th voice signal in a language (for example English) and $\mathbf{y}^k$ is its corresponding version in other language (for example Spanish).

As was show in [11] and [12], the original DAM performs a low accuracy in complex problems such as face or 3d object recognition. In order to increase the accuracy of the DAM the authors suggest computing a simplified version of the DAM model by using a random selection of stimulating points. Some pixels (stimulating points) of pattern $\mathbf{x}^k$ are random selected, where k defines the class of the pattern. These stimulating points **SP** are used by the DAM to determine an active region and are given by $\mathbf{sp} \in \{\mathbb{Z}^+\}^c$ where c is the number of used **SP**. $sp_i = random(n), i = 1,\ldots,c$ where n is the size of the pattern.

To determine the active region, the DAM stores during training phase an alternative simplified version of each pattern $\mathbf{x}^k$ given by:

$$\mathbf{ss}^k = ss\left(\mathbf{x}^k\right) = \mathbf{x}^k\Big|_{\mathbf{sp}} = \left\{x^k_{sp_1},\ldots,x^k_{sp_c}\right\} \tag{8}$$

During recalling phase, each element of an input simplified pattern $\tilde{\mathbf{x}}^k\Big|_{\mathbf{sp}}$ excites some of these regions and the most excited region will be the active region. To determine which region is excited by an input pattern we use:

$$b = \arg\min_{k=1}^{p}\left|\left[ss(\mathbf{x})\right]_i - \mathbf{ss}^k_i\right| \tag{9}$$

For each element of $\tilde{\mathbf{x}}^k\Big|_{\mathbf{sp}}$ we apply equation 9 and the most excited region (the region that more times was obtained) will be the active region.

Building of the DAM is done as follows: Let $\mathbf{S}^k_x$ and $\mathbf{S}^k_y$ an association of voice signals and c be the number of stimulating points. First, take at random c stimulating point sp_i. Then, for each association transform the voice signals into a raw vector ($\mathbf{x}^k$, $\mathbf{y}^k$) and finally, train the DAM.

Pattern $\mathbf{S}^k_y$ can be recalled by using its corresponding voice signal $\mathbf{S}^k_x$ or distorted version $\tilde{\mathbf{S}}^k_x$ as follows: first, use the same c stimulating point sp_i. Then, transform the voice signal into a raw vector and finally, operate the DAM.

A schematic representation of the proposal is show in Fig. 1.

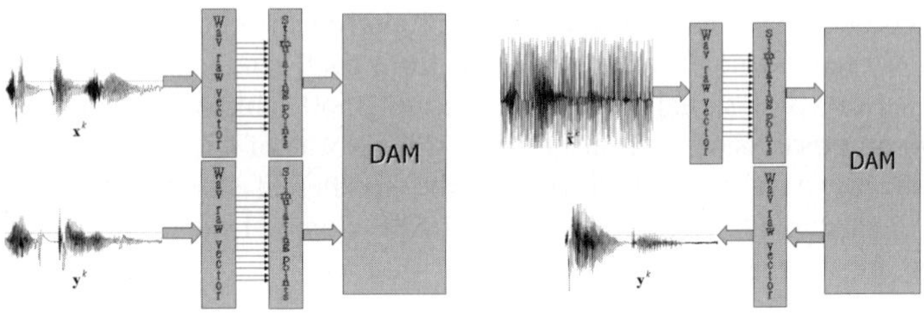

(a) (b)

Fig. 1. Schematic representation of a voice translator based on associative memory. (a)Training phase. (b) Recalling phase.

4 Behavior of the Proposal

To corroborate the behavior and accuracy of the associative memory based voice translator we have performed several experiments divided into two cases. In the first case we have verified the behavior and accuracy of the model with voice signal patterns altered with additive, subtractive, mixed and Gaussian noise. Different sets of voice signal used in the first kind of experiments are shown in Fig. 2(a-d). In the second kind of experiments we verified the behavior and accuracy of the proposal with slightly distorted voice signal patterns such as voice signals recorded at different tempo, volume, velocity and tone. Different sets of voice signals used in the second kind of experiments are shown in Fig. 2(e). For both kinds of experiments, each voice signal in English was associated with its corresponding voice signal in Spanish.

Each voice signal was recorded in a WAV file (PCM format, 44.1 KHz, 16 bits and mono). Before training the associative memory, each voice signal has to be transformed into a voice signal pattern. In order to build a voice signal pattern from the wav file, we only read the wav information chunk of the file and then we stored it into an array. It is important to remark that no preprocessing technique or transformation was applied to the wav information; we only used the raw information combined with the random selection of stimulation points.

4.1 First Kind of Experiments

EXPERIMENT 1: Recalling the fundamental set of associations: In this experiment we firstly trained the associative memory with the set of voice signals in English and Spanish. In this case, each voice signal in English was associated its corresponding voice signal in Spanish (40 associations). Once trained the associative memory, as described in section 3, we proceeded to test the accuracy of the proposal. First we verified if the DAM was able to recall the fundamental set of association using set of voice signals. In this experiment the DAM provided a 100% of accuracy using only 1 stimulating point. Whole associations used to train the DAM were perfectly recalled.

EXPERIMENT 2: In this experiment, we verified if the DAM was able to recall the voice signal in Spanish associated to the voice signal in English (input pattern), even if the voice signal is altered by additive noise (AN). To do this, each voice signal previously recorded was contaminated with AN altering from 2% until 90% of the information. 89 new samples were generated from each voice signal already recorded. This new set of voice signals was composed of 3560 samples. In average, the accuracy of the proposal using this set of voice signals was of 42.5% using only 1 stimulating point; however when we increased the number of stimulation points to more than 100, the accuracy increased to almost 100%. Some of the results obtained in this experiment are shown in Fig.2 (a).

EXPERIMENT 3: In this experiment, we verified if the DAM was able to recall the voice signal in Spanish associated to the voice signal in English (input pattern), even if the voice signal is altered by subtractive noise (SN). To do this, each voice signal previously recorded was contaminated with SN altering from 2% until 90% of the information. 89 new samples were generated from each voice signal already recorded. This new set of voice signals was composed of 3560 samples. In average, the accuracy of the proposal using this set of voice signals was of 44.4% using only 1 stimulating point; however when we increased the number of stimulation points to more than 100, the accuracy was of almost 100%. Some of the results obtained in this experiment are shown in Fig. 2 (b).

EXPERIMENT 4: In this experiment, we verified if the DAM was able to recall the voice signal in Spanish associated to the voice signal in English (input pattern), even if the voice signal is altered by mixed noise (MN). To do this, each voice signal previously recorded was contaminated with MN altering from 2% until 90% of the information. 89 new samples were generated from each voice signal already recorded. This new set of voice signals was composed of 3560 samples. In average, the accuracy of the proposal using this set of voice signals was of 36.6% using only 1 stimulating point; however when we increase the number of stimulation points to more than 100, the accuracy was of almost 100%. Some of the results obtained in this experiment are shown in Fig. 2 (c).

EXPERIMENT 5: In this experiment, we verified if the DAM was able to recall the voice signal in Spanish associated to the voice signal in English (input pattern), even if the voice signal is altered by Gaussian noise (GN). To do this, each voice signal previously recorded was contaminated with GN altering from 2% until 90% of the information. 89 new samples were generated from each voice signal already recorded. This new set of voice signals was composed of 3560 samples. In average, the accuracy of the proposal using this set of voice signals was of 46.5% using only 1 stimulating point; however when we increased the number of stimulation points to more than 100, the accuracy increased to almost 100%. Some of the results obtained in this experiment are shown in Fig. 2 (d).

4.2 Second Kind of Experiments

EXPERIMENT 6: In this experiment, we verified if the DAM was able to recall the voice signal in Spanish associated to the voice signal in English used as input pattern, even if the voice signal experimented slightly deformations such as voice signals recorded at different tempo, volume, velocity and tone. To do this, each voice signal previously recorded was recorded 10 times. Ten new deformed samples (DEF) were recorded from each voice signal already recorded. This new set of voice signals was composed of 400 samples, some examples are shown in Fig. 2 (e). In average, the accuracy of the proposal using this set of voice signals was of 10.5% using only 1 stimulating point; compared with the previous experiments when we increased the number of stimulation points to more than 100, the accuracy slightly increases to 20%.

Despite of the low accuracy obtained, the results are encouraging. First to all, we have demonstrated the applicability of the associative models in a complete different domain.

For the first kind of experiments, we realized that a human was unable to perceive the voice signal if the voice signal was contaminated with noise in more than the 20%; therefore unable to translate the voice signal. Using this DAM, we obtained a 100% of accuracy even if the voice signal was contaminated with noise until 90%. For the second kind of experiments, we realized that a human was able to perceive the voice signal even if the voice signal was reproduced in different tempo and tone. Using this DAM, we obtained a 17.5% of accuracy. Although the percentage of recalling is low, it is one of the first results reported in literature for recalling voice signal patterns based on AMS.

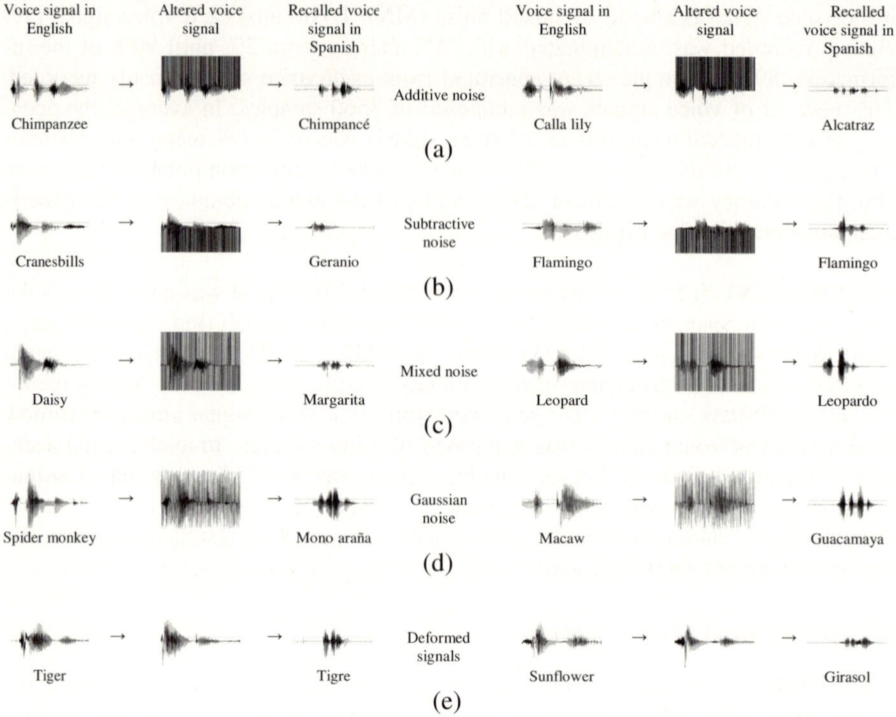

Fig. 2. Some voice signals recalled using voice signals altered by different type of noises

It is worthy mentioning, that to our knowledge nobody in this field had before reported results of this type. Authors only report results when images patterns are distorted by additive, subtractive or both noises or when images under image orientations but not when the associative memory is trained with other type of patterns such as voice signals.

Furthermore, this model is capable to associate patterns from different domains, suggesting its applicability in a large domain of complex problems such as image retrieval using voice signal queries, control of robots using voice commands and associative memories, speech recognition using associative memories, etc.

No comparison with other AM was performed because constrains of these models limits their applicability in this problem and the accuracy would be too low. The general behavior of the voice translator is shown in Fig. 3. Note that using a small number of stimulating points, the accuracy of the proposal is low. However, if the number of stimulating points is increased, the accuracy of the proposal also increases. After using a number of stimulating points greater than 100, the obtained accuracy was of 100%. In addition, as you can appreciate from Fig. 3, no matter the type of noise or amount of noises added to the patterns, the behavior of the proposal for each typo of noise was almost the same.

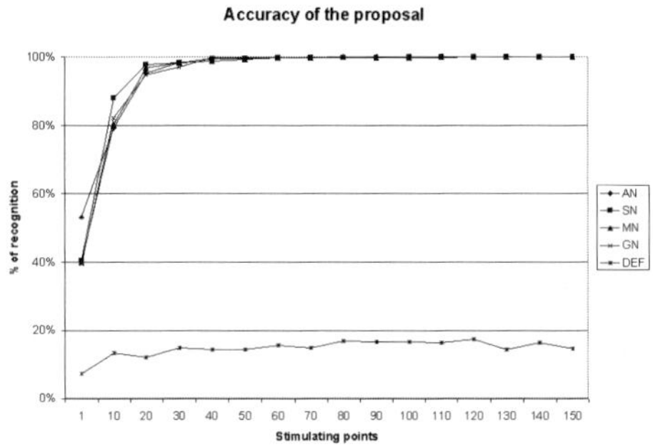

Fig. 3. Accuracy of the proposal using different number of stimulating points

Finally, voice translators are used in real-time systems; therefore the translation speed is very important. Due to this approach do not use complex and expensive technique for training and testing the associative model, the recall of an associated pattern is performed in a very short time. For example, in PC with a Pentium 4 CPU 2.80GHz and 500 Mb of RAM, a voice signal of a word is translated in less than 1ms using 100 stimulating points. This result also supports the applicability of the proposal in real time systems.

5 Conclusions

In this paper we have described a novel voice translator based on associative memories. We have shown the robustness of the dynamic associative model. The results obtained, through the different experiments, using a benchmark composed by 14440 voice signal samples, support the applicability of this model in different complex problems that not involve only computer vision but also voice processing such as voice translator devices.

It is worth mentioning that, even without applying preprocessing methods for adequate the voice signals, the accuracy was highly acceptable, 100% for the first kind of experiments; in the last experiment the model presented an accuracy of 17.5%.

Some important to remark is that the associative model only uses less than the 1% of whole information of the voice signal for recalling its corresponding translated version.

The results reported in this paper support the robustness of the associative model, show the versatility of the model in different environments and make it an excellent associate model for solve the voice translator problem.

This was just a first step, nowadays we are working with some voice preprocessing techniques in order to increase the accuracy of the proposal with the type of voice signals used in the second type of experiments even if the voice signal is produced by different people. Of course, we are working on integrating some natural language techniques to translate speech-to-speech not only words but also phrases in real time.

Acknowledgment. This work was economically supported by SIP-IPN under grant 20082948 and CONACYT under grant 46805.

References

1. Steinbuch, K.: Die Lernmatrix. Kybernetik 1, 26–45 (1961)
2. Anderson, J.A.: A simple neural network generating an interactive memory. Math. Biosci. 14, 197–220 (1972)
3. Kohonen, T.: Correlation matrix memories. IEEE Trans. on Comp. 21, 353–359 (1972)
4. Hopfield, J.J.: Neural networks and physical systems with emergent collective computational abilities. Proc. Natl. Acad. Sci. 79, 2554–2558 (1982)
5. Sussner, P.: Generalizing operations of binary auto-associative morphological memories using fuzzy set theory. J. Math. Imaging Vis. 19, 81–93 (2003)
6. Ritter, G.X., et al.: Reconstruction of patterns from noisy inputs using morphological associative memories. J. Math. Imaging Vis. 19, 95–111 (2003)
7. Sossa, H., Barron, R., Vazquez, R.A.: Transforming Fundamental set of Patterns to a Canonical Form to Improve Pattern Recall. In: Lemaître, C., Reyes, C.A., González, J.A. (eds.) IBERAMIA 2004. LNCS (LNAI), vol. 3315, pp. 687–696. Springer, Heidelberg (2004)
8. Ritter, G.X., Sussner, P., Diaz de Leon, J.L.: Morphological associative memories. IEEE Trans. Neural Networks 9, 281–293 (1998)
9. Sussner, P., Valle, M.: Gray-Scale Morphological Associative Memories. IEEE Trans. on Neural Netw. 17, 559–570 (2006)
10. James, W.: Principles of Psychology, Holt, New York (1980)
11. Vazquez, R.A., Sossa, H., Garro, B.A.: 3D Object recognition based on low frequencies response and random feature selections. In: Gelbukh, A., Kuri, A.F. (eds.) MICAI 2007. LNCS (LNAI), vol. 4827, pp. 694–704. Springer, Heidelberg (2007)
12. Vazquez, R.A., Sossa, H., Garro, B.A.: Low frequency responses and random feature selection applied to face recognition. In: Kamel, M., Campilho, A. (eds.) ICIAR 2007. LNCS, vol. 4633, pp. 818–830. Springer, Heidelberg (2007)
13. Vazquez, R.A., Sossa, H.: A new associative memory with dynamical synapses (submitted to Neural Processing Letters, 2007)

Denoising Natural Images Using Sparse Coding Algorithm Based on the Kurtosis Measurement

Li Shang, Fengwen Cao, and Jie Chen

Department of Electronic Information Engineering,
Suzhou Vocational University, Suzhou, Jiangsu 215104, China
{sl0930,cfw,cj}@jssvc.edu.cn

Abstract. A new natural image denoising method using a modified sparse coding (SC) algorithm proposed by us was discussed in this paper. This SC algorithm exploited the maximum Kurtosis as the maximizing sparse measure criterion at one time, a fixed variance term of sparse coefficients is used to yield a fixed information capacity. On the other hand, in order to improve the convergence speed, we use a determinative basis function as the initialization feature basis function of our sparse coding algorithm instead of using a random initialization matrix. This denoising method is evaluated by values of the normalized mean squared error (NMSE) and signal to noise ratio (NSNR). Compared with other denoising methods, the simulation results show that our SC shrinkage technique is indeed effective.

Keywords: Sparse coding; Kurtosis; Fixed variance; Image feature extraction; Denoising.

1 Introduction

Image Reconstruction is generally an inverse problem, which intends to recover the original ideal image from its given bad version [1]. In this paper, we only consider the contaminated source, i.e., noise, of natural images. In other words, the purpose of image denoising is to restore the original image with noise-free. Classical image denoising techniques are based on filtering method [2]. More recently, there are more and more new denoising techniques explored, such as wavelet-based approach [5], principal components analysis (PCA) approach [3], and standard sparse coding (SC) shrinkage proposed by Alpo Hyvärinen in 1997 [4], etc. These methods can successfully denoise images by using different skills and strategies. Moreover, literature [5] gave an important conclusion: when ICA is applied to natural data, ICA is equivalent to SC. However, ICA emphasizes independence over sparsity in the output coefficients, while SC requires that the output coefficients must be sparse and as independent as possible. Because of the sparse structures of natural images, SC is more suitable to process natural images than ICA. Hence, SC method has been widely used in natural image processing [6].

In this paper, we propose a modified SC algorithm, which exploits the maximum Kurtosis as the maximizing sparse measure criterion, so the natural image structure

F. Sun et al. (Eds.): ISNN 2008, Part II, LNCS 5264, pp. 351–358, 2008.

captured by the Kurtosis not only is indeed sparse, but also is surely independent. At the same time, a fixed variance term of coefficients is used to yield a fixed information capacity. This term can well balance the reconstructed error and sparsity. On the other hand, we use a determinative basis function, which is obtained by a fast fixed-point independent component analysis (FastICA) algorithm [7], as the initialization feature basis function of our SC algorithm instead of using a random initialization matrix, so that the convergent speed of SC is further speeded up. The experimental results also showed that utilizing our SC algorithm, the edge features of natural images can be extracted successfully. Further, applying the features extracted, the images contaminated by additive Gaussian white-noise can be reconstructed clearly.

2 The Early Sparse Coding Algorithm

The SC algorithm proposed by B. A. Olshausen et al in 1996 is deemed to be the classical SC algorithm [6], since it explains the characteristics of the respective fields of the simple cells in mammalian primary visual cortex (i.e., V1 field) for the first time, which are spatially localized, oriented and bandpass (selective to the structure at different spatial scales) comparable to the basis functions of wavelet transforms. B. A. Olshausen and D. J. Field combined the preserve informational and sparseness of coefficients to construct the following cost function:

$$E=\sum_{x,y}[\dot{I}(x,y)-\sum_{i} a_i \varphi_i(x,y)]^2+\lambda\sum_{i} S(a_i/\sigma) \tag{1}$$

where a_i are the coefficients, φ_i are the basis vectors, σ is a scaling constant, λ is a positive constant that determines the importance of the second term relative to the first, and $S(\cdot)$ is a nonlinear function. The first term, which measures how well the code describes the images, is the mean square of the error between the actual image and the reconstructed one. The second term assesses the sparseness of the code for a given image by assigning a cost depending on how activity is distributed among the coefficients. The choices for $S(\cdot)$ used by Olshausen, et al, are the forms of $-e^{-x^2}$, $\log(1+x^2)$ and $|x|$. The reason for these choices is that they will favor those with the fewest number of non-zero coefficients among activity states with equal variance.

Using gradient descent method to minimizing Eqn. (1), the optimization problem can be solved. Thus for a given image, the a_i can be determined from the equilibrium solution to the differential equation:

$$\dot{a}_i = b_i - \sum_j C_{ij} a_j - \frac{\lambda}{\sigma} S'\left(\frac{a_i}{\sigma}\right). \tag{2}$$

where $b_i = \sum_{x,y} \varphi_i(x,y) I(x,y)$ and $C_{ij} = \sum_{x,y} \varphi_i(x,y) \varphi_j(x,y)$. The learning rule for updating the φ_i is then:

$$\Delta\varphi_i(x_m, y_n) = \eta \left\langle a_i\left[I(x_m, y_n) - \hat{I}(x_m, y_n)\right]\right\rangle . \tag{3}$$

where $\hat{I}$ is the reconstructed image, $\hat{I}(x_m, y_n) = \sum_i a_i \varphi_i(x_m, y_n)$, and η is the learning rate. Olshausen and Field applied their algorithm to natural images data processing and verified its efficiency [6]. In their experiments, the sparse measure function is selected as $\log(1+x^2)$, σ^2 is set to the variance of test images, and the parameter λ was set so that $\lambda/\sigma = 0.14$. Experimental results obtained by Olshausen and Field demonstrated successfully that the localized, oriented, bandpass receptive fields emerge when only two global objectives (see Eqn. (1)) are placed on a linear coding of natural images (please see literature [6]).

3 Our Modified Sparse Coding Algorithm

3.1 Modeling NNSC of Natural Images

Referring to the classical SC algorithm [6], and combining the minimum image reconstruction error with Kurtosis and fixed variance, we construct the following cost function of the minimization problem:

$$J(A,S) = \frac{1}{2}\sum_{x,y}\left[X(x,y) - \sum_i a_i(x,y)s_i\right]^2 - \lambda_1 \sum_i \left|kurt(s_i)\right| + \lambda_2 \sum_i \left[\frac{\langle s_i \rangle}{\sigma_t}\right]^2 \tag{4}$$

where the symbol $\langle\cdot\rangle$ denotes the mean, $X=(x_1,x_2,\ldots,x_n)^T$ denotes the n-dimensional natural image data, $A=(a_1,a_2,\ldots,a_m)$ denotes the feature basis vectors, $S=(s_1,s_2,\ldots,s_m)^T$ denotes the m-dimensional sparse coefficients. In this paper, note that only the case of $m=n$ is considered, i.e., A is a square matrix. Parameters λ_1 and λ_2 are positive constant, σ_t^2 is the scale of coefficient variance. Generally, σ_t^2 is set to be the variance of an image. In Eqn. (4), the first term is the image reconstruction error and it ensures a good representation for a given image, and the second term is the sparseness measure based on the absolute value of Kurtosis, which is defined as:

$$\left|kurt(s_i)\right| = \left|E\{s_i^4\} - 3\left(E\{s_i^2\}\right)^2\right| \tag{5}$$

and maximizing $\left|kurt(s_i)\right|$ is equivalent to maximizing the sparseness of coefficient vectors; The last term can penalize the case in which the coefficient variance of the *ith* vector $\langle s_i^2 \rangle$ deviates from its target value σ_t^2. Without this term, the variance becomes so small that the sparseness constraint can only be satisfied, and the image reconstruction error would become large, which is not desirable either.

3.2 Learning Rules

Using the simple gradient descent algorithm to minimize the objective function, this differential equation of a_i is defined as follows:

$$\frac{\partial J(a_i, s_i)}{\partial a_i} = -\left[X - \sum_i a_i s_i\right] s_i^T \tag{6}$$

and further, the Eqn. (6) can be rewritten as follows:

$$a_i(k+1) = a_i(k) + \left[X - \sum_i a_i(k) s_i(k) \right] (s_i(k))^T \cdot \tag{7}$$

In a similar manner, the differential equation of s_i can be obtained as the following equation:

$$\frac{\partial J(a_i, s_i)}{\partial s_i} = -a_i^T \left[X - \sum_i a_i s_i \right] - \lambda_1 f_1(s_i) + \lambda_3 \frac{\ln\left(\langle s_i^2 \rangle / \sigma_i^2 \right)}{\langle s_i^2 \rangle} \langle s_i \rangle \tag{8}$$

where $\lambda_3 = 4\lambda_2$ (a positive constant), $f_1(s_i) = \dfrac{\partial |kurt(s_i)|}{\partial s_i}$. According to Eqn.(4), the function $f_1(s_i)$ can be deduced as follows:

$$f_1(s_i) = \frac{\partial |kurt(s_i)|}{\partial s_i} = \beta \left[\langle s_i^3 \rangle - 3 \langle s_i^2 \rangle \langle s_i \rangle \right] \tag{9}$$

where $\beta = sign(kurt(s_i))$, and for super-Gaussian signals, $\beta = 1$, and for sub-Gaussian signals, $\beta = -1$. Because of natural image data belonging to super-Gaussian, β is equal to 1. Thus, combined Eqn. (9) into Eqn. (8), then the updating rule of the coefficient variables can be obtained as follows:

$$s_i(k+1) = s_i(k) + (a_i(k))^T \left[X - \sum_i a_i(k) s_i(k) \right] + \lambda_1 \beta \left[\left\langle (s_i(k))^3 \right\rangle - 3 \left\langle (s_i(k))^2 \right\rangle \langle s_i(k) \rangle \right] - \lambda_4 \langle s_i(k) \rangle \tag{10}$$

where $\lambda_4 = 4\lambda_2 \left[\ln\left(\langle s_i^2 \rangle / \sigma_i^2 \right) \right] / \langle s_i^2 \rangle$.

In performing loop, we update S and A in turn. First, holding A fixed, S is updated, then, holding S fixed, A is updated. To speed up the search process for the optimal basis vectors, the initialization values of A and S are determined by the ICA basis, which are computed by using FastICA algorithm [3]. Otherwise, for the convenience of computation, A is scaled in programming. Using the learning rules of A and S, the obtained results for 64 basis functions extracted from natural scenes are

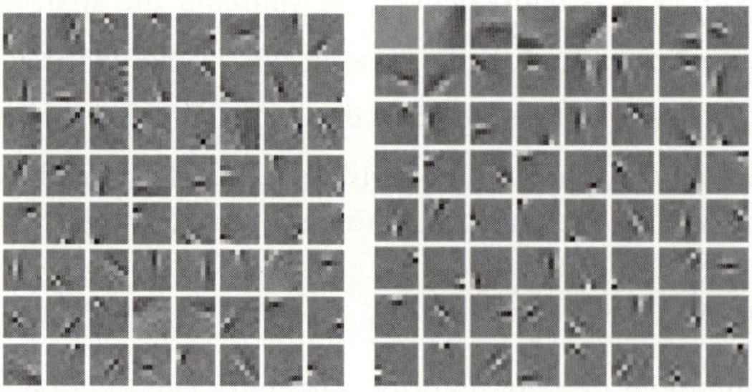

(a) (b)

Fig. 1. Basis vectors obtained by applying our sparse coding algorithm to natural scenes. (a) Bases obtained by our algorithm; (b) Bases obtained by orthogonalized ICA.

shown in Fig. 1 (a). Moreover, it should be noted that, for the same training set of natural image, these basis vectors are very similar to those orthogonal basis vectors shown in Fig.1 (b) obtained by the orthogonalized ICA method used in the Ref. [3]. This experimental result also testifies that our SC algorithm is indeed efficient in natural image feature extraction.

4 The Sparse Coding Shrinkage Function

In this section, the estimators of sparse components are presented based on the statistical distributions. First considering a single noisy component y, the original (noise-free) non-Gaussian random variable s, and the Gaussian noise n with zero mean and variance, then the observed random variable, y, can be represented as:

$$y=s+n. \tag{11}$$

Then given y, we need to estimate the original s using $\hat{s}=g(y)$. Here, the estimation equation takes the form [4]:

$$\hat{s}=g(y)=sign(y)\max(0,|y|-\sigma^2|f'(y)|) \tag{12}$$

where $f'(\cdot)$ is the derivative of the sparse punitive function $f(\cdot)$, moreover, $f(\cdot)$ is the negative log-density (i.e., $f(\cdot)=-\log[p(\cdot)]$). Here, the sparse density model $p(\cdot)$ is defined as the classical Laplace distribution:

$$p(s)=\frac{1}{\sqrt{2d}}\exp\left(-\frac{\sqrt{2}}{d}|s|\right) \tag{13}$$

and considering $f(\cdot)'=(-\log[p(\cdot)])'$, then the sparse coding shrinkage function $g(\cdot)$ is written as [16]:

$$g(y)=sign(y)\max(0,|y|-\sqrt{2}\sigma^2/d) \tag{14}$$

where d is the scale parameter, which is a shrinkage function that has a certain thresholding flavor.

5 Experimental Results

All test images used in experiments are available on the following Internet web: http:// *www.cns.nyu.edu/lcv/denoise*. Firstly, selecting randomly 10 noise-free natural images with 512×512 pixels. Then, we sampled patches of 8×8 pixels 5000 times from each original image, and converted every patch into one column. Thus, the input data set X with the size of 64×50000 is acquired. Further, the data set X was centered and whiten by principal component analysis (PCA), and the preprocessed data set was denoted by $\hat{\mathrm{X}}$. Then, using the updating rules of A and S in turn defined in

Eqns. (7) and (10), we minimized the objective function given in Eqn. (4). The 64 feature basis of natural scenes are shown in Fig. 1, as described in subsection 3.2.

In denoising, the test image was chosen as Lena image. The noise versions were shown in the topmost of Fig. 2. And The denoised results were showed in the bottommost of Fig. 2. Here, the quality of denoised images was evaluated by objective measures of normalized mean square error (NMSE) and normalized signal to noise ratio (NSNR), which are respectively defined as follows [4]:

$$NMSE = \sum_{i=1}^{M} \sum_{j=1}^{N} \left[X(i,j) - \hat{X}(i,j) \right]^2 \Big/ \sum_{i=1}^{M} \sum_{j=1}^{N} X(i,j)^2 \tag{15}$$

$$NSNR = 10 \log_{10} \left\{ \sum_{i=1}^{M} \sum_{j=1}^{N} \left(X_{ij} - \bar{X}_{ij} \right)^2 \Big/ \sum_{i=1}^{M} \sum_{j=1}^{N} \left(X_{ij} - \hat{X}_{ij} \right)^2 \right\} \quad (dB) \tag{16}$$

where M and N denote the original image's size. The pixel coordinate in an image is (i, j), $x_{i,j}$ and $\hat{x}_{i,j}$ denotes respectively the pixel values of X and $\hat{X}$, and $\bar{X}$ denotes the mean value of X. Using Eqns. (15) and (16), the values of NMSE and NSNR under different noise levels were listed in Table 1. It is clear to see that the noise has been effectively reduced and the visual effect has been highly enhanced.

We also compared this denoising technique to other three denoising algorithms: the Wiener filter, the wavelet-based soft shrinkage [3] and standard ICA or SC shrinkage [4]. The denoised results corresponding to the noise level 0.3 were shown in Fig. 3, and the corresponding NMSE and NSNR were also listed in Table 1. It can be concluded that, under the same noise level, our SC shrinkage method is the best denoiser than other methods considered here, since it yielded the minimum NMSE and the maximum NSNR values. Wiener filter is the worst denoiser, it can hardly reduce the noise, but it is worse than wavelet-based shrinkage. Moreover, it was clear that the visual effectiveness of denoising results obtained by our SC shrinkage excelled the ones obtained using the other methods. Furthermore, from Table 1, the larger the noise level is, the more notable the advantage of denoising images using our algorithm is. So, it can be easily concluded that our denoising method is indeed successful and efficient in application.

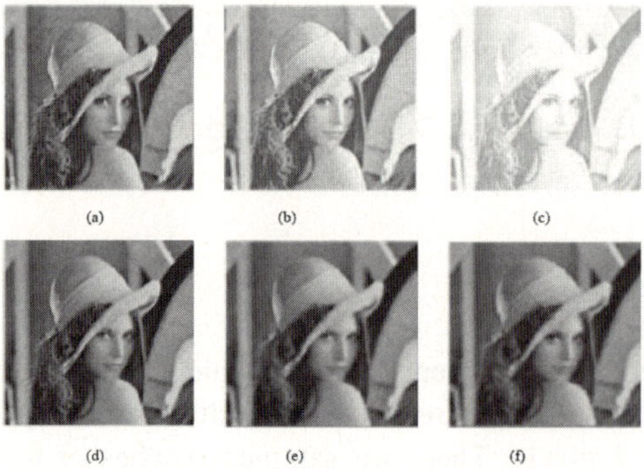

Fig. 2. Denoised results of Lenna image with different noise levels using shrinkage rules based on our SC algorithm. Topmost: Noise versions, from (a) to (d), the noise level is orderly 0.05, 0.2 and 0.5 ; Bottommost: from (e) to (h), denoised results corresponding to the noise level 0.05, 0.2, 0.3 and 0.5.

Table 1. Values of normalized MSE and SNR obtained by different denoising methods. Image: Lena.

Noisy levels	Noise images		Our SC shrinkage		ICA shrinkage		Wavelet-based shrinkage	
	NMSE	NSNR	NMSE	NSNR	NMSE	NSNR	NMSE	NSNR
0.01	0.0369	5.4826	0.0170	11.2935	0.0181	8.5705	0.0312	6.2072
0.05	0.0457	4.5526	0.0253	10.8635	0.0263	6.9542	0.0398	5.1563
0.1	0.0722	2.5714	0.0523	9.6698	0.0539	3.8383	0.0664	2.9320
0.2	0.1746	2.2591	0.0945	6.872	0.1584	1.4246	0.1697	2.2925
0.3	0.3325	1.1205	0.1207	4.5223	0.3171	1.1280	0.3873	0.7437
0.4	0.5189	0.4860	0.2688	2.8186	0.5165	0.4806	0.5055	0.4319
0.5	0.7048	0.0831	0.4057	1.6241	0.6878	0.2563	0.7029	0.0779

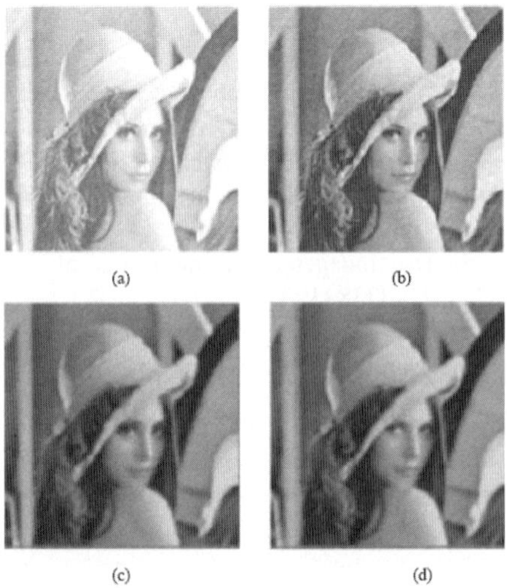

Fig. 3. Denoising results obtained by different techniques corresponding to the noise level 0.3. (a) Wiener filter; (b) Wavelet-based shrinkage; (c) Standard ICA/SC shrinkage; (d) Our SC shrinkage.

6 Conclusions

In this paper, a novel natural image reconstruction method based on a modified sparse coding (SC) algorithm developed by us is proposed. This modified SC algorithm exploited the maximum Kurtosis as the maximizing sparse measure criterion, so the natural image structure captured by the Kurtosis not only is surely sparse, but also is surely independent. At the same time, a fixed variance term of coefficients is used to yield a fixed information capacity. Edge features of natural images can be extracted successfully by exploiting our SC algorithm. Utilizing these features, the natural images corrupted with additive Gaussian noise can be reconstructed efficiently. Compared with other denoising methods of Wiener filter, wavelet-based soft shrinkage and standard ICA/SC shrinkage, our method is very effective in denoising based on statistics of NMSE and NSNR. Moreover, for our SC shrinkage technique, the larger the noise level is, the better the effect on the denoising results is.

Acknowledgments

The work is supported by the National Natural Science Foundation of China (No. 60472111 and No. 60405002).

References

1. Jähne, B.: Digital Image Processing: Concepts, Algorithms and Scientific Applications. Springer, Berlin (1991)
2. Alan, C.: Bovik: Handbook of Image and Video Processing. Academic Press, San Diego (2000)
3. Diamantaras, K.I., Kung, S.Y.: Principal Component Neural Networks: Theory and Applications. John Wiley & Sons, New York (1996)
4. Hyvärinen, A., Hoyer, P., Oja, E.: Image Denoising by Sparse Code Shrinkage. In: Haykin, S., Kosko, B. (eds.) Intelligent Signal Processing, pp. 554–568. IEEE Press, New York (2001)
5. Bell, A.J., Sejnowski, T.J.: The 'Independent Components' of Natural Scenes are Edge Filters. Vision Research 37, 3327–3338 (1997)
6. Olshausen, B.A., Field, D.J.: Emergence of Simple-cell Receptive Field Properties by Learning A Sparse Code for Natural Images. Nature 381, 607–609 (1996)

A New Denoising Approach for Sound Signals Based on Non-negative Sparse Coding of Power Spectra

Li Shang, Fengwen Cao, and Jinfeng Zhang

Department of Electronic Information Engineering, Suzhou Vocational University, Suzhou, Jiangsu 215104, China
{sl0930,cfw,zjinfeng}@jssvc.edu.cn

Abstract. In this paper, a novel sound denoising approach based on a statistical model of the power spectrogram of a sound signal is proposed by using an extended non-negative sparse coding (NNSC) algorithm for power spectra. This approach is self-adaptive to the statistic property of spectrograms of sounds. The basic idea for denoising is to exploit a shrinkage function to reduce noises in spectrogram patches. Experimental results show that our approach is indeed effective and efficient in spectrogram denoising. Compared with other denoising methods, the simulation results show that the NNSC shrinkage technique is indeed effective and efficient.

Keywords: Non-negative sparse coding; Power spectra; Spectrograms; Denoising sound signals.

1 Introduction

Recently, it has been shown that the characteristics of the auditory system can also be understood in terms of sparse activity in response to speech data that are represented by spectrograms [1]. Therefore, we can say that sound data belong to super-Gaussian distribution, which causes the sound spectral coefficients to be sparse as well. Therefore, one can use feasible methods of sparse representations, such as independent component analysis (ICA) [2], sparse coding (SC) [3], and non-negative sparse coding (NNSC) [4], to process sound data. Just so, this paper focuses on using an extended NNSC algorithm developed by us to denoise sound signals. The magnitude spectrogram representing time-dependent spectral energies of sounds is exploited as the observed input data. Spectrograms are segmented into a series of image patches through time. Therefore, each image patch in spectrogram can be modeled in terms of linear superposition of localized basis images with non-negative sparse encoding variables. Then, by referring to sparse coding shrinkage rule [5], we utilize the shrinkage function selected to perform the spectrogram denoising process. The simulation results showed that our approach is indeed effective and efficient in denoising sounds. Compared with the methods of DFT, wavelet shrinkage and standard SC shrinkage, our method outperforms them in denoising sounds.

F. Sun et al. (Eds.): ISNN 2008, Part II, LNCS 5264, pp. 359–366, 2008.

2 The Extended NNSC Algorithm for Spectrograms

2.1 Modeling Spectrograms

A sound source can be characterized by its magnitude spectrogram and the time-varying gain [6]. The spectrogram of any audio signal in nature can be modeled as a linear superposition of spectrum basis functions $a_{i,j}$:

$$x_i(t,f) = \sum_{j=1}^{m} a_{i,j}(t,f) s_j + \varepsilon \cdot \tag{1}$$

where $X = (x_1, x_2, \cdots, x_i, \cdots, x_n)^T \ (X > 0)$ and $S = (s_1, s_2, \cdots, s_j, \cdots, s_m)^T \ (S > 0)$ denote the n-dimensional multivariate observation of spectrogram data and the m-dimensional ($m \leq n$) weight coefficients (i.e., sparse sources), respectively. (t,f) is the time-frequency coordinate of power spectrum, $a_{i,j}$ is the time-varying gain of the $j-th$ sparse source in the $i-th$ observation, and ε is the Gaussian additive noise independent of the clean source signal.

2.2 The Cost Function

In this paper, the cost function is constructed as follows:

$$J(A,S) = \frac{1}{2}\sum_{i,j}\left[X_{i,j}(t,f) - \sum_{i=1}^{n}\sum_{j=1}^{m} a_{i,j}(t,f) s_j\right]^2 + \lambda \sum_{j=1}^{m} F\left(\frac{s_j}{\sigma_j}\right) + \gamma \sum_{i=1}^{n}\sum_{j=1}^{m} \left|a_{i-1,j} - a_{i,j}\right| \cdot \tag{2}$$

subject to the constraints: $X_{i,j}(t,f) \geq 0$, $\lambda > 0$, $\gamma > 0$, $\forall_{i,j}$: $a_{i,j} \geq 0$, $s_j \geq 0$, and $\|a_j\| = 1$. Where $\sigma_j = \sqrt{\langle s_j^2 \rangle}$; $X_{i,j}(t,f)$ denotes an element of the spectrum input matrix $X(t,f)$; a_j and s_j denote respectively the $j-th$ column of A and the $j-th$ row of coefficients S ; λ and γ are scalar parameters; $F(\cdot)$ is the sparseness measure function, and is determined by the negative logarithm of sparse coefficients.

This constrained optimization problem could be solved with a classical gradient-based descent method. According to Eqn. (3), the partial derivatives of $\partial J / \partial s_j$ and $\partial J / \partial a_j$ can be written as follows:

$$\frac{\partial J}{\partial s_j} = -a_j^T \left[X(t,f) - \sum_{j=1}^{m} a_j(t,f) s_j\right] + \lambda F'\left(\frac{s_j}{\sigma_j}\right) \cdot \tag{3}$$

$$\frac{\partial J}{\partial a_j} = -\left[X(t,f) - \sum_{j=1}^{m} a_j(t,f) s_j\right] s_j^T + \gamma\varphi \tag{4}$$

where $\sigma_j = \sqrt{\langle s_j^2 \rangle}$, φ is the partial derivative of the temporal continuity term, which is defined as:

$$\varphi = \begin{cases} -1, & a_{i,j} < a_{i-1,j} \wedge a_{i,j} < a_{i+1,j} \\ +1, & a_{i,j} > a_{i-1,j} \wedge a_{i,j} > a_{i+1,j} \\ 0, & otherwise \end{cases} . \tag{5}$$

For the convenience of computation, the matrix forms of the updating rules of A and S are respectively rewritten as:

$$S^{k+1} = S^k + A^T(X - AS^k) - \lambda F'\left(S^k / \sqrt{\langle (S^k)^2 \rangle}\right) . \tag{6}$$

$$A^{k+1} = A^k + \eta^k\left((X - A^k S)S^T - \gamma\varphi\right) . \tag{7}$$

where k is the iteration index, and $\eta^k > 0$ is the optimal step size estimated by using a line search, which is determined by tentative methods. Importantly, in the process of iteration, A and S must be ensured to be non-negative, and the scaling of A is compensated by rescaling S .

3 The Shrinkage Function of Spectrogram Components

In this section, the estimators of sparse components of power spectrograms are obtained by using the Maximum Likelihood (ML) rule. For a noisy component y , the shrinkage function given by the ML rule is written as:

$$\hat{y} = g(y) = sign(y) \max\left(0, |y| - \sigma^2 |F'(y)|\right) . \tag{8}$$

where $F'(y)$ is the derivative of $F(y)$, and $F(y)$ is equal to the negative log-density of y , i.e., $F(y) = -\log p(y)$. Here, $p(\cdot)$ denotes the normal inverse Gaussian (NIG) density of any random variable, which is defined as follows [7]:

$$p(u) = C \cdot \exp\left[\beta(u-\mu) - \alpha\sqrt{(u-\mu)^2 + \delta^2}\right] \cdot \left[(u-\mu)^2 + \delta^2\right]^{-\frac{3}{4}} . \tag{9}$$

where $C = \delta \cdot \sqrt{\alpha/2\pi} \cdot \exp\left(\delta\sqrt{\alpha^2 - \beta^2}\right)$, and subject to the constraints of $0 \le |\beta| < \alpha$, $\delta > 0$, and $-\infty < \mu < \infty$. Clearly, the shape of the NIG density is specified by the parameter vector $[\alpha, \beta, \mu, \delta]$. Parameter α controls the steepness of the NIG density. Parameter β controls the skewness. Parameters of μ and δ are scale-like constants.

By estimating the first four lowest cumulants from the sample data, denoted by $k^{(1)}$, $k^{(2)}$, $k^{(3)}$, and $k^{(4)}$, and using them to estimate the skewness $r_3 = k^{(3)} / \left[k^{(2)}\right]^{3/2}$ and normalized kurtosis $r_4 = k^{(4)} / \left[k^{(2)}\right]^2$. Then, two auxiliary variables can be readily computed as [7]:

$$\zeta = 3\left(r_4 - \frac{4r_3^2}{3}\right)^{-1}, \rho = \frac{r_3}{3}\sqrt{\zeta} . \tag{10}$$

Then the four-parameter vector estimators can be derived as follows:

$$\delta = \sqrt{k^{(2)}\zeta\left(1-\rho^{2}\right)}, \alpha = \frac{\zeta}{\delta\sqrt{1-\rho^{2}}}, \beta = \alpha\rho, \mu = k^{(1)} - \rho\sqrt{k^{(2)}\zeta} \cdot \tag{11}$$

Thus, according to the Eqns. (9) to (11), the score function $F'(\cdot)$ of the NIG density is given:

$$F'_{NIG}(y) = \frac{\alpha(y-\mu)}{\sqrt{(y-\mu)^2+\delta^2}}\left(\frac{K_0\left[\alpha\sqrt{(y-\mu)^2+\delta^2}\right]}{K_1\left[\alpha\sqrt{(y-\mu)^2+\delta^2}\right]}+\frac{2}{\alpha\sqrt{(y-\mu)^2+\delta^2}}\right)-\beta \cdot \tag{12}$$

where $K_0(\cdot)$ and $K_1(\cdot)$ is the modified Bessel function of the second kind with index 0 and 1, respectively. It is obvious that $|F'_{NIG}(y)|$ depends on the parameter vector $[\alpha,\beta,\delta,\mu]^T$ estimated by the spectrum sparse components.

4 Experimental Results

4.1 Preprocessing Spectra Data

All test sound signals were obtained at *http://www2.arts.gla.ac.uk/IPA/sounds*. Firstly, 5 sound signals with the same statistical properties were randomly selected. Then, each signal was transformed into its corresponding power spectrogram. The frequency range of sounds was limited to 8KHz, so that only 257 frequency lines of the power spectrogram were used. Next, each power spectrogram were randomly sampled 10, 000 patches through time (see Fig.1), and the time points of each patch were set as 25. So, each patch was 257×25 pixels (i.e., 6425). Then, each patch was converted into one column vector. Thus, the input data set with the size of 6425×50, 000 was acquired and denoted as matrix X_1. Using PCA technique, the dimension of X_1 was reduced to 256, and the processed set was denoted by X_2. Considering the non-negativity of the input data, we separated X_2 into two non-negative matrixes Y and Z. Thus, the nonnegative input data matrix $X = (Y;Z)$ with the size of (2×256)×50,000 was obtained. And then, using the updating rules of A and S, the

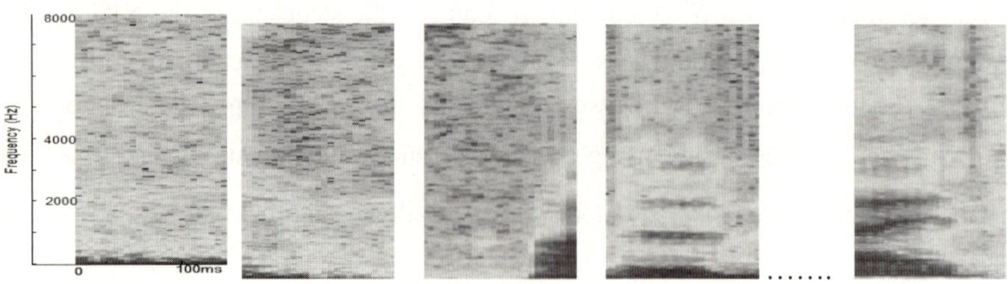

Fig. 1. Segmenting a spectrogram into a series of image patches through time. Each image patch covers 100ms.

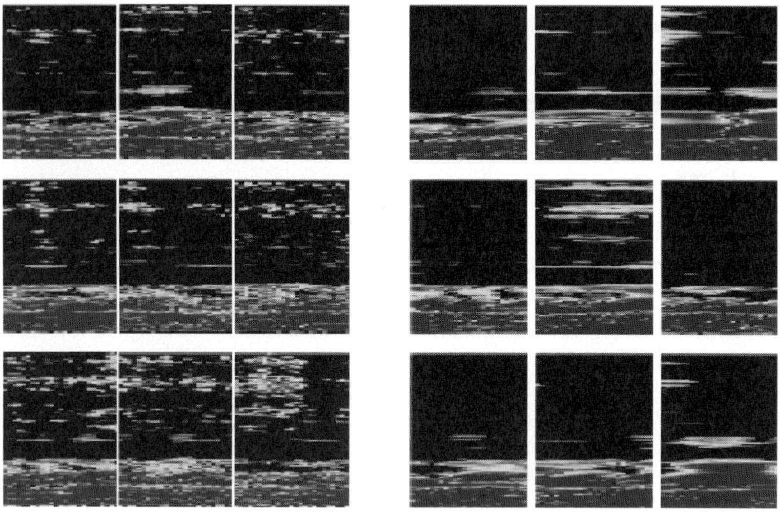

(a) (b)

Fig. 2. Basis images of the extended NNSC and ICA. (a) The NNSC basis images; (b)The ICA basis images .

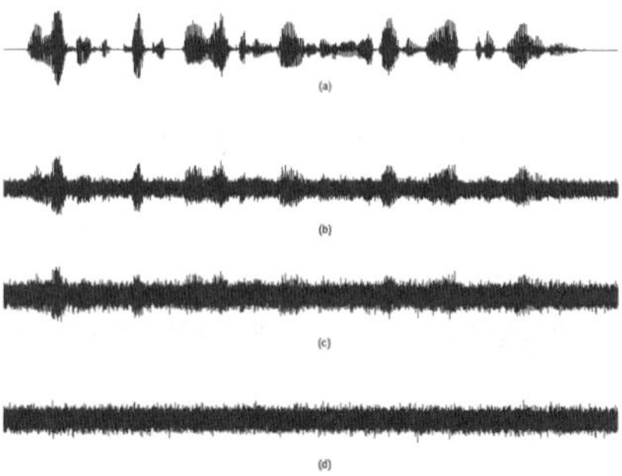

(a) The clean sound signal; (b), (c) and (d) are the noisy signal corresponding to the noise level $\sigma = 0.05$, $\sigma = 0.1$ and $\sigma = 0.5$.

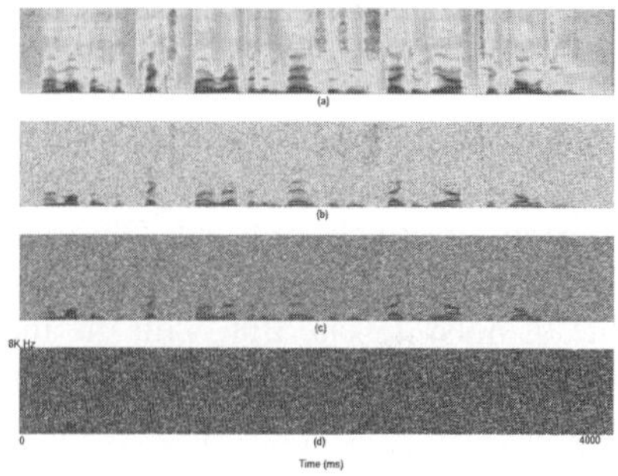

(a) The clean signal's spectrogram; (b), (c) and (d) are the noisy signal spectrogram corresponding to the noise level $\sigma = 0.05$, $\sigma = 0.1$ and $\sigma = 0.5$.

Fig. 3. The original sound signal and noise versions, as well as the corresponding spectrograms with the different noise level

objective function can be minimized. Figure 2a is an illustration of the first 9 feature basis images estimated by our NNSC algorithm. For comparison, the first 9 basis images of ICA were also given, as shown in Fig. 2b. These basis images are color-coded in a scale, where red represents larger values while blue represents smaller values. Clearly, NNSC basis images exhibit much more localized characteristics than those of ICA.

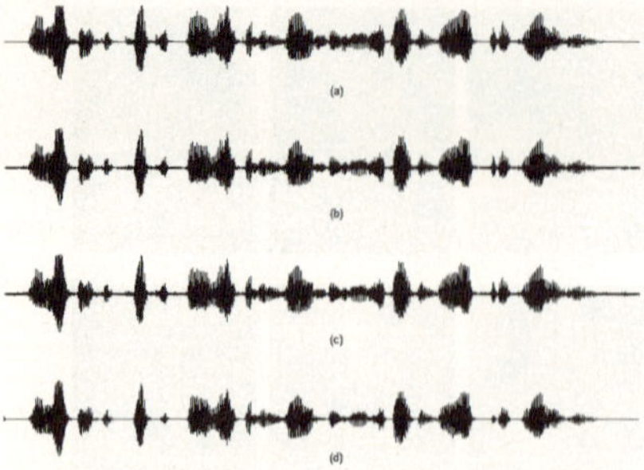

(a) The original sound signal; (b), (c) and (d) are the reconstructed sound signal corresponding to $\sigma = 0.05$, $\sigma = 0.1$ and $\sigma = 0.5$.

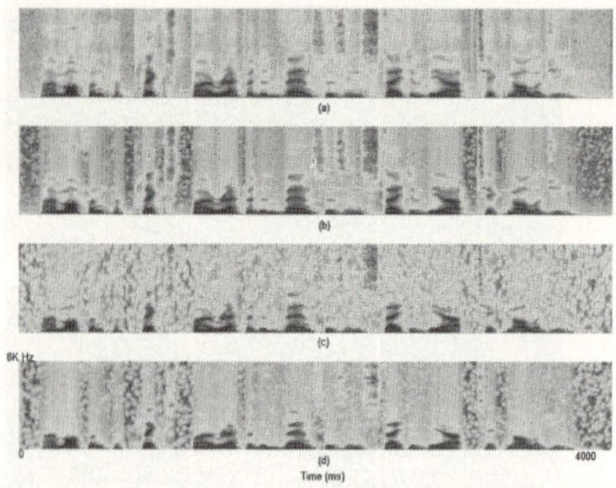

(a) The original sound signal; (b), (c) and (d) are the reconstructed spoectrograms corresponding to $\sigma = 0.05$, $\sigma = 0.1$

Fig. 4. Reconstructed signals and spectrograms with different noisy levels using our SC shrinkage method

4.2 Denoising Results

In this section, we performed the denoising procedure using the proposed NIG-based NNSC shrinkage function given in Eqn. (8). The clean sound signal used is shown in Fig.3(a). The noise versions with different Gaussian additive noise variance $\sigma = 0.05$, 0.1, 0.5 are also given. And the corresponding spectrograms of this clean signal and its noisy versions are also shown in Fig.3. The reconstructed signal waveforms and power spectrograms were respectively shown in Fig. 4. It can be seen that the noise has been effectively reduced. Observing the reconstructed power spectrograms, one can find that the energy existing in the low frequency field was retained mostly in spite of noisy levels. But, with the increasing of noisy levels, more and more little energy in the high frequency field was retained.

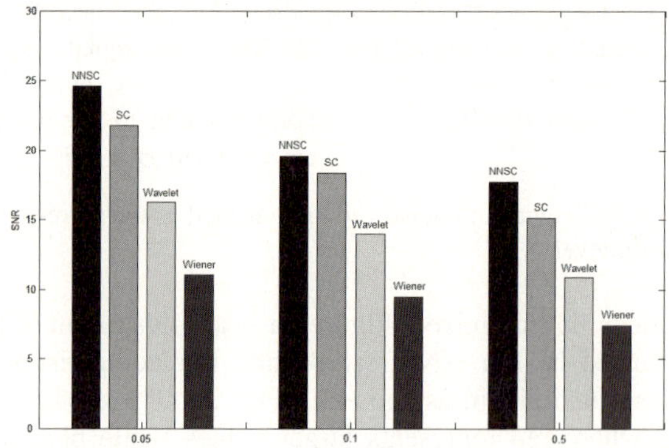

Fig. 5. Values of SNR corresponding to reconstruction spectrograms obtained by different denoising algorithms with different noisy levels

We also compared this technique to other denoising algorithms: the standard sparse coding shrinkage, the wavelet soft shrinkage, and the usual Wiener filter. As a result, the denoised spectrograms and the values of the corresponding normalized MSE and SNR obtained by different algorithms were respectively shown in Fig. 5. It can be found that the NNSC shrinkage method was the best denoiser, while the Wiener filter was the worst denoiser since the former yielded the minimized normalized MSE and the maximized SNR values under the condition of the same noise level. Moreover, with the increase of the noisy level, the visual differences of the denoised power spectrograms using different methods became more and more distinct. From Fig. 5, it was also clearly seen that the NNSC shrinkage method retains much more energy of the sound signal.

5 Conclusions

This paper proposed a new sound denoising technique by using the normal inverse Gaussian (NIG) model and an extended non-negative sparse coding (NNSC) algorithm for power spectra. This denoising method was proposed on the basis of the statistical model of the sound power spectra. In performing denoising, the original spectrogram is contaminated by Gaussian additive white noise with different noisy levels. The basic principle for denoising is to utilize the shrinkage function selected in advance to deal with the sparse components of power spectra learned by our NIG-based NNSC algorithm. The shrinkage function depends on the NIG density model determined by the given sparse spectral data. The simulation results showed that it was successful in denoising power spectrograms using our algorithm. Compared with the other denoising methods of the Wiener filter, the SC shrinkage and wavelet-based soft threshold on test spectrograms with known noise characteristics, the experimental results showed that based on the two statistics of MSE and SNR in the case of the same noisy level, the NIG-based NNSC shrinkage method outperforms other three types of denoising methods considered here. However, the responses of neurons in the central auditory system share similar properties with the spatiotemporal processing of the visual system, therefore, new theories and methods developed in visual neuro-system can be discussed in the research field of sound enhancement.

Acknowledgments

This work was supported by Natural Science Foundation of China (No. 60472111 and No. 60405002).

References

1. Klein, D.J., Peter, K., Körding, K.P.: Sparse Spectrotemporal Coding of Sounds. EURASIP Journal on Applied Signal Processing 7, 659–667 (2003)
2. Hanuch, L.A., Yariv, E.: Extension of the Signal Subspace Speech Enhancement Approach to Colored Noise. IEEE Signal Processing Letters 10, 104–106 (2003)

3. Mahmoudi, D., Drygajlo, A.: Combined Wiener and Coherence Filtering Array Speech Enhancement. In: International Conference on Acoustics, Speech, and Signal Processing (ICASSP 1998), pp. 385–388. Seattle Press, Washington (1998)
4. Wan, E., Vander, M.R.: Noise-regularized Adaptive Filtering for Speech Enhancement. In: 6th European Conference on Speech Communication and Technology (EUROSPEECH 1999), pp. 156–163. Budapest Press, Hungary (1999)
5. Hyvärinen, A.: Sparse Coding Shrinkage: Denoising of Nongaussian Data by Maximum Likelihood Estimation. Neural Computation 11, 1739–1768 (1997)
6. Gazor, S., Zhang, W.: Speech Enhancement Employing Laplacian-Gaussian Mixture. IEEE Transactions on Speech Audio processing 13(5), 896–904 (2005)
7. Hanssen, Ø.T.A.: The Normal Inverse Gaussian Distributions as A Flexible Model for Heavy Tailed Processes. In: Proc. IEEE-EURASIP Workshop on Nonlinear Signal and Image Processing. Baltimore Press, Maryland (2001)

Building Extraction Using Fast Graph Search

Dong-Min Woo, Dong-Chul Park, Seung-Soo Han, and Quoc-Dat Nguyen

Information Engineering Department, Myongji University
Gyeonggido, Korea 449-728
{dmwoo,parkd,shan}@mju.ac.kr, iavmm@yahoo.com

Abstract. This paper presents a new building rooftop extraction method from aerial images. In our approach, we extract the useful building location information from the generated disparity map to segment the interested objects and consequently reduce unnecessary line segments extracted in low level feature extraction step. Hypothesis selection is carried out by using undirected graph, in which close cycles represent complete rooftops hypotheses. We test the proposed method with the synthetic images generated from Avenches dataset of Ascona aerial images. The experiment result shows that the extracted 3D line segments of the reconstructed buildings have an average error of 1.69m and our method can be efficiently used for the task of building detection and reconstruction from aerial images.

Keywords: Perceptual grouping, Building detection, Building reconstruction, Aerial images.

1 Introduction

The building detection and reconstruction from aerial images is one of the challenging tasks in computer vision. It has been used widely in various applications including traditional applications such as cartography and photo-interpretation and recent application including mission planning, urban planning, computer graphics and virtual reality.

Early approaches used 3D interpretation of a single 2D image [1,2]. This direction has some restrictions such as inferring 3D information from one image is very difficult and there are still some ambiguities in the detected buildings that can be only resolved by feature matching in multiple images. Since multiple aerial images can be obtained with only small extra cost, most recent works have focused on the multiple-view analysis [3,4,5].

Mohan and Nevatia [6] proposed an approach for detecting and describing buildings in aerial image using perceptual grouping. They demonstrated the usefulness of the structural relationships called collated features which can be explored by perceptual organization in complex image analysis. Huertas [7] suggested using extracted cues from the IFSAR data, while Kim [8] gets from commercial DEM image to solve the segmentation of interested objects problem. The extracted cues do not give us the shape of the buildings. However they can give us the idea where the buildings are located in the image. Unfortunately, it is not easy to have IFSAR data or DEM image

F. Sun et al. (Eds.): ISNN 2008, Part II, LNCS 5264, pp. 367–375, 2008.

in all cases. Jaynes [9] proposed task driven perceptual organization. Features such as corner and line segment are first extracted and assigned into certainty values. Features and their grouping are stored in a feature relation graph. Close cycles in the graph represent the grouped polygon hypotheses. The independent set of closed groups that have maximum sum of certainty values of its parts is the final grouping choice. This approach is limited on rectangular buildings and tends to have false hypotheses in complexity images.

In this context, we propose a new method based on hypothesis generation and selection in terms of perceptual organization strategy to solve the building detection task. The key idea is that we use the proposed suspected building regions extracted from the disparity map for obtaining locations of interested objects in the image. This building location information can support the removal of unnecessary line segments, and reduces computational complexity and false hypotheses in later steps. Additionally, hypothesis selection is carried out by graph searching for close cycles in an undirected graph.

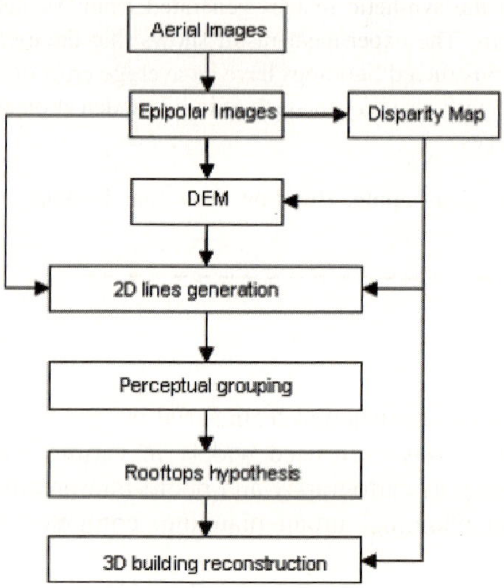

Fig. 1. System overview

Fig. 1 shows the main components in our system. The epipolar images are generated from the aerial images by epipolar resampling process. The disparity map is obtained by area-based stereo matching. From the disparity map, we generate the DEM as a 3D terrain model. The building location information extracted from disparity map can be used to remove the unnecessary line segments. Next, we apply perceptual grouping to the filtered line segments in order to obtain the structural relationship features such as parallel line segment pairs and U-shapes, which can be used to generate rooftop hypotheses. Then, hypothesis selection is carried out by searching close cycles in the undirected graph. Finally, we retrieve 3D buildings by using 3D triangulation for each line segment of detected rooftops.

2 Low-Level Feature Extraction

2.1 Grouping and Filtering of 2D Lines

To detect 2D lines from epipolar image, edge detection is carried out and then 2D lines are detected from edges. We employed Canny edge detector, since it is optimal in a sense of "edge definition". To obtain 2D line segment, we use Boldt algorithm [10] based on token grouping. A basic line element is extracted as a token in terms of the properties of a line, and 2D lines are formed by grouping process. Suspected building regions are used to remove line segment that outside or far from interested object boundaries.

The closely parallel linear segments are grouped into one line, since they usually represent a linear structure of objects in image, like the border of a roof or the divider between ground terrain and building. For this grouping process, we utilize "folding space" between two line segments. If both line segments are inside the folding space, two line segments can be replaced by a single line. Its orientation is determined as that of the longer line segment, and its length can be calculated as the total length of two segments. This process can make the closely overlapping and parallel line segments represented only by one single line.

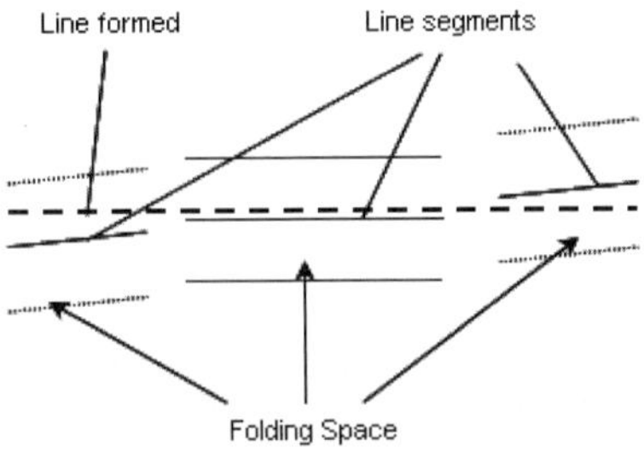

Fig. 2. Folding space

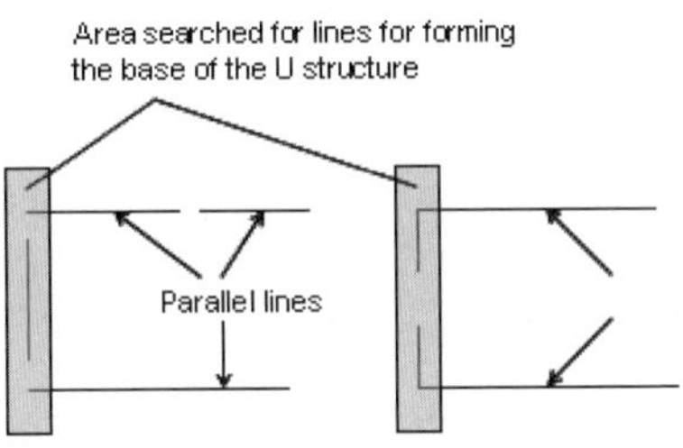

Fig. 3. U-structure

Fig. 2 showed the typical example of near collinear segments grouping. the first condition is the angle between them should be from 0^0 to 10^0. If two line segments are fragmented lines from one edge, these line segments must be close and should be inside a folding space created by them.

The U shaped structure in Fig. 3 is used to detect candidates for rooftop hypothesis generation. Any line segment in a set of parallel lines with aligned end is a U shaped structure candidate which is kept as an input for hypothesis generation. Otherwise the line segment will be removed.

2.2 Corner Detection

Corner can be calculated as intersection of two line segments which their angle is from 80^0 to 100^0 and one of them has nearest distance to another one. We define four types of corner. As shown in Fig. 4, they are labeled as I, II, III and IV. Each corner has an attribute to indicate whether it is L-junction or T-junction. This attribute is used to decide whether two different corners have a connection or not. For example, if a corner's label is I and type is L-junction, it connects to any type of corner. However, it prefers connecting to a corner which label is II or IV. If that corner is T-junction, it can only connect to a corner which label is II or IV. This rule is used in hypothesis generation to build collated features.

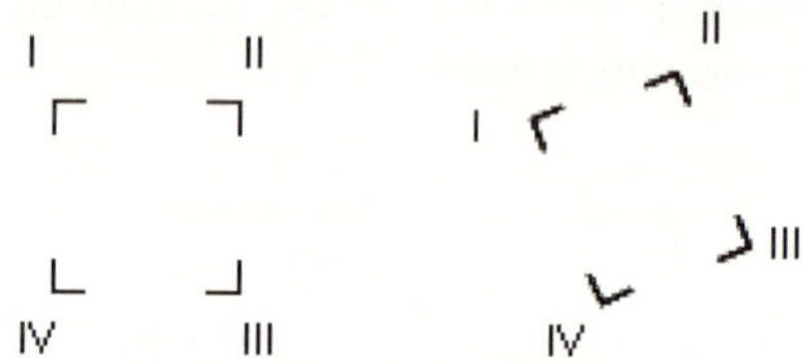

Fig. 4. Corner labels

With the flexible connection between corners, our method is able to detect rectilinear rooftops. Fig. 5 show some examples of corner detection, A, B, E, F, G are L-junctions while C, D are T-junctions.

3 3D Rooftop Detection and Reconstruction

3.1 Rooftop Hypothesis Generation

A collated feature is a sequence of perceptually grouped corners and line segments. Here, collated features are constructed from filtered line segments and corners obtained from the filtering and grouping process. That reduces computational effort and false hypotheses.

Hypotheses are formed by alternation of corners and line segments that form collated features. In a collated feature, two corners have connectivity only if they satisfy the corner relation condition and they are the nearest appropriate corner to each other. Beside, every corner connects to only one corner on each its line segment direction. Hypothesis generation is performed by constructing the feature graph. Construction of the graph can be seen as placing corners as nodes and edges between nodes if there is the relation between the corresponding corners in the collated features. When a node is inserted into the graph, the system looks into the remaining nodes whether any node has the relation with the inserted node. If some nodes satisfy the connectivity relation rules, those nodes are inserted into the graph and the system creates an edge between them. As shown in Fig. 5, C is T-junction, so it can connect to D, A and E. Meanwhile, A can connect to B, C and E but C is nearer than E towards A on the line segment AE so that A only connects to B and C. So there will be two collated features ACGB and CEFD in the Fig. 5.

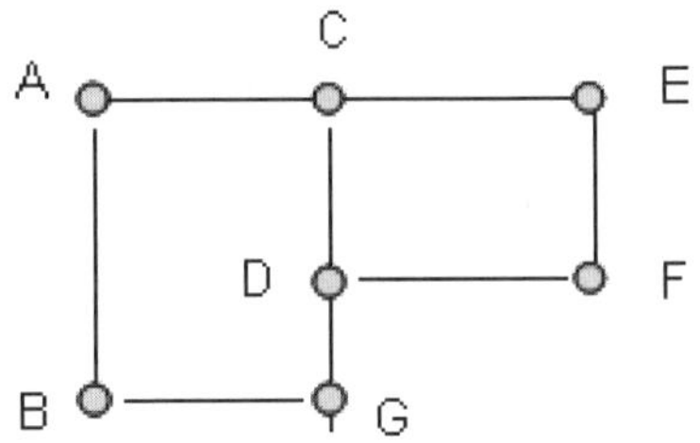

Fig. 5. Corner detection

3.2 Rooftop Hypothesis Selection

The graph is the place to store features and their groupings. Feature as corner is node in the graph and relations between corners are represented with an edge between the corresponding nodes. Closed cycles in the graph represent the rooftop candidates. The hypothesis selection can be seen as a simple graph search problem. The close cycles in the graph are rooftops that we need to detect. Fig. 6 show a graph constructed from the example in Fig. 5. Corner C and corner D are T-junctions so that there are two nodes in the graph for each corner. Node C1, C2 for corner C and node D1, D2 for corner D. There are two close cycles C1 and C2 as shown in Fig. 6.

3.3 3D Building Reconstructions

3D triangulation is used to generate 3D line segments. The relationship between a point k located at $X_k = (X_k, Y_k, Z_k)$ in model/objects space and the projection of the point k located at $x_{Lk} = (x_{Lk}, y_{Lk}, f_L)$ in the image of camera L is

$$\begin{bmatrix} X_k \\ Y_k \\ Z_k \end{bmatrix} = \begin{bmatrix} X_L^0 \\ Y_L^0 \\ Z_L^0 \end{bmatrix} + \lambda_{Lk} \begin{bmatrix} m_{L11} & m_{L21} & m_{L31} \\ m_{L12} & m_{L22} & m_{L32} \\ m_{L13} & m_{L23} & m_{L33} \end{bmatrix} \begin{bmatrix} x_{Lk} \\ y_{Lk} \\ -f_L \end{bmatrix} \quad (1)$$

where $X_L^0 = (X_L^0, Y_L^0, Z_L^0)$ is the model space coordinates of the focal point of camera L, f_L is the focal length of camera L, λ_{Lk} is the scale factor for point k projected on the focal plane of camera L and m_L is the rotation matrix between the image space coordinate system and the model space coordinate system.

We have a system of equations for five variables from each pair of points in two images. Solving that system of equations we have the real 3D coordinates of the selected points in two images. As a result, we have 3D line segments from the corresponding 2D line segments.

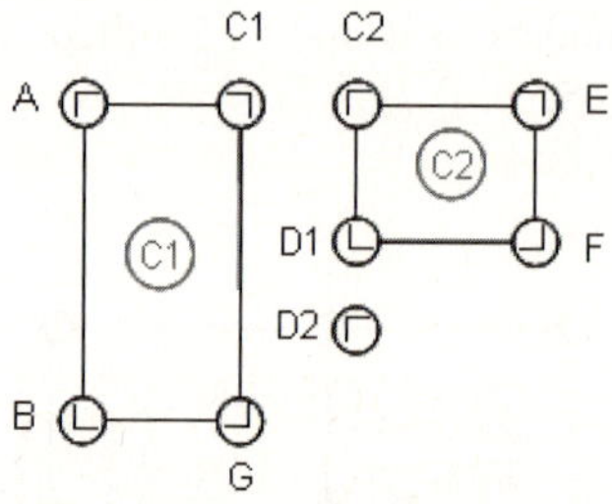

Fig. 6. Feature graph

4 Experimental Results

The experimental environment was set up based on Ascona aerial images of the Avenches area. There are two aerial images as shown in Fig. 7.

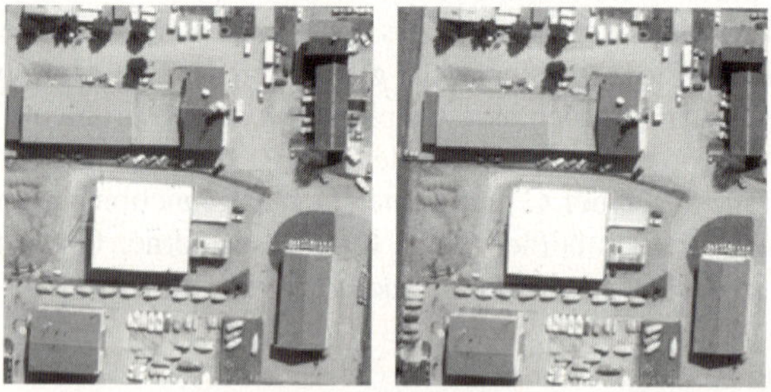

Fig. 7. Ascona aerial images

Fig. 8 shows the line segments obtained from low level feature extraction process. By removing unnecessary line segments, we obtain the suspected building regions extracted from the disparity map as shown in Fig. 9.

After removing unnecessary line segments, we carry out perceptual filtering and grouping process to obtain line segments which can be part of any U-structure group. The close parallel line segments which are inside their folding space of each other will be grouped into one representation line. The line segments which are part of a collection of line segments forming U-structure will be used to generate hypotheses in the next step. Fig. 10 shows the line segments forming U-structures in a collection of line segments.

The corners are calculated form the intersection of the line segments which satisfy two conditions: their angle is from 85^0 to 95^0 and one of them has nearest distance to another one. Fig. 10 shows extracted corners from the line segments collection.

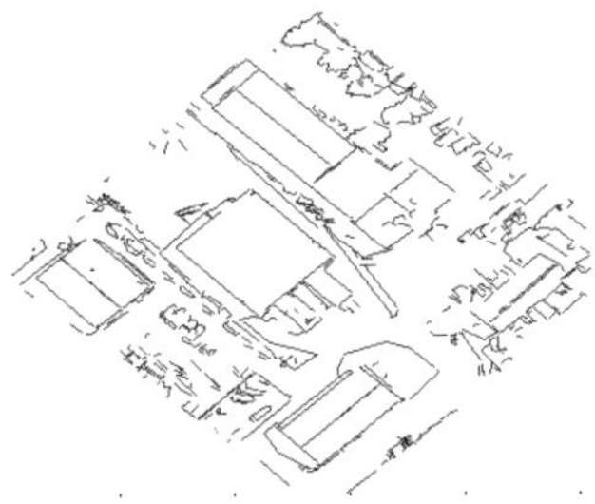

Fig. 8. Example of low level extraction result

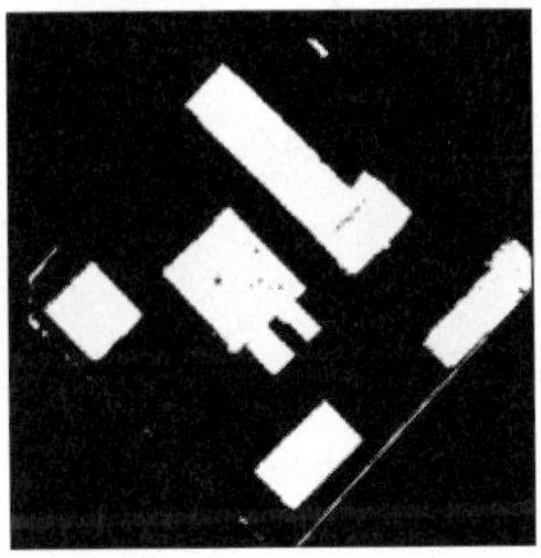

Fig. 9. Example of suspected building regions

Using the obtained corners and line segments from the previous steps, we can build the collated features. In order to have a link between each other, two corners must satisfy the connecting relation of corner type and the required condition of their distance. Another important rule that help to define the corner connectivity is on each line segment of a corner, there is only one corner has connection with it. Fig. 10 shows the collated features obtained from the line segments collection.

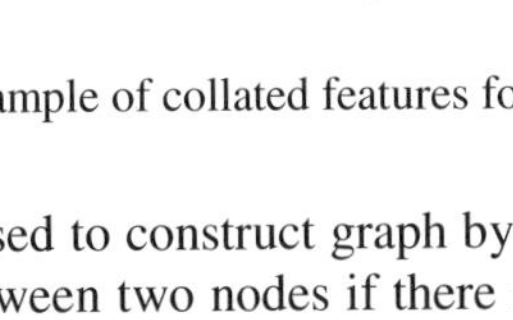

Fig. 10. Example of collated features for U-structure

The collated features are used to construct graph by placing a corner as a node and a line segment as an edge between two nodes if there is the relation between the corresponding corners in the collated features. Closed cycles in the graph represent the possible rooftops. Hypothesis selection becomes the searching of close cycles in the graph.

Fig. 11 shows rooftop detection result of the entire area. There is a building located near the border line of the epipolar image that the system can not detect correctly due to missing line segments in low level extraction step. The result of remaining building is very good. From the detected rooftop and the known geometric parameters of image acquisition, we reconstruct 3D building using 3D triangulation as shown in Fig. 12.

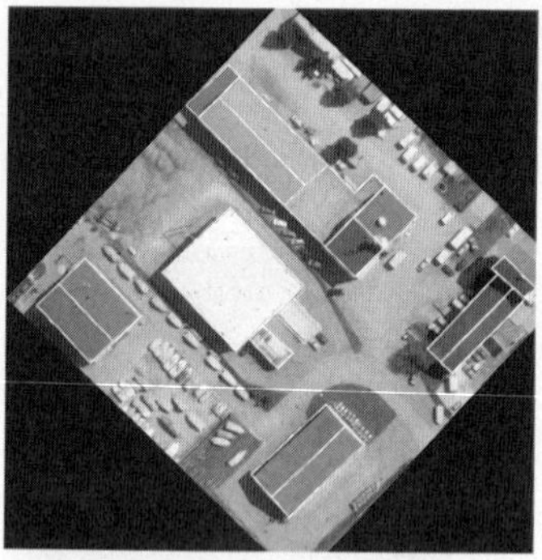

Fig. 11. Example of detected rooftops

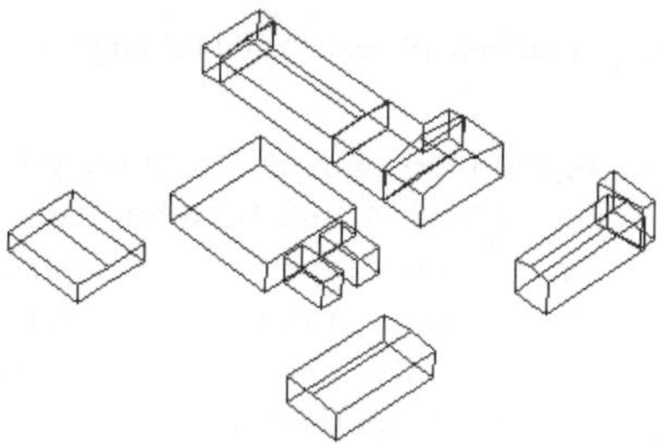

Fig. 12. Example of 3D building reconstruction

To represent the quantitative accuracy of 3D building reconstructed by our approach, we obtain the error by calculating the average distance between the extracted 3D line segments and the ground truth line segments as the following equation.

$$E = \frac{\sum \frac{e_{1i} + e_{2i}}{2} \times d_i}{\sum d_i} \tag{2}$$

In Eq.(2), e_{1i} is the distance from the starting point of line segment i to the ground truth 3D line, while e_{2i} is the distance from the end point of line segment i to the ground truth 3D line and d_i is the length of line segment i. Error calculation shows that the average error of the reconstructed buildings is 1.65m while the error of corresponding digital elevation model is 1.93m.

5 Conclusion

A new technique to detect and reconstruct buildings from two aerial images has been suggested. The suspected building regions are used to remove the unnecessary line segments before generating rooftop hypotheses. This approach enables to reduce computational complexity and false hypotheses. Using undirected feature graph, the selection of rooftop hypotheses is reduced to a simple graph searching for close cycles. Experimental result shows that the proposed technique can be very effectively utilized for detecting rectilinear building structures in urban area.

Acknowledgement

This work was supported in part by the Korea Science and Engineering Foundation (KOSEF) grant funded by the Korean government (MOST)(Grant No.: R01-2007-000-20330-0), and in part by the ERC program of MOST/KOSEF (Next-generation Power Technology Center).

References

1. Huertas, A., Nevatia, R.: Detecting Buildings in Aerial Images. Computer Vision, Graphics and Image Processing 41, 131–152 (1988)
2. Lin, C., Nevatia, R.: Building Detection and Description from a Single Intensity Image. Computer Vision and Image Understanding 72, 101–121 (1998)
3. Fischer, A., Kolbe, T., Lang, F., Cremers, A., Forstner, W., Plumer, L., Steinhage, V.: Extracting Buildings from Aerial Images Using Hierarchical Aggregation in 2D and 3D. Computer Vision and Image Understanding 72, 185–203 (1998)
4. Noronha, S., Nevatia, R.: Detection and Modeling of Buildings from Multiple Aerial Images. IEEE Transaction on Pattern Analysis and Machine Intelligence 23, 501–518 (2001)
5. Collins, R., Jaynes, C., Cheng, Y., Wang, X., Stolle, F., Riseman, E., Hanson, A.: The Ascender System: Automated Site Modeling from Multiple Aerial Images. Computer Vision and Image Understanding 72, 143–162 (1998)
6. Mohan, R., Nevatia, R.: Using Perceptual Organization to Extract 3D Structure. Trans. Pattern Analysis and Machine Intelligence 11, 1121–1139 (1989)
7. Huertas, A., Kim, Z., Nevatia, R.: Use of Cues from Range Data for Building Modeling. In: Proc. DARPA Image Understanding Workshop (1998)
8. Kim, Z., Nevatia, R.: Automatic Description of Complex Buildings from Multiple Images. Computer Vision and Image Understanding 96, 60–95 (2004)
9. Jaynes, C., Stolle, F., Collin, R.: Task Driven Perceptual Organization for Extraction of Rooftop Polygons. In: IEEE Workshop on Application of Computer Vision (1994)
10. Boldt, M., Weiss, R., Riseman, E.: Token-based Extraction of Straight Lines. IEEE Trans. Systems Man Cybernetics 19, 1581–1594 (1989)

Image Denoising Using Three Scales of Wavelet Coefficients

Guangyi Chen and Wei-Ping Zhu

Department of Electrical and Computer Engineering,
Concordia University,
Montreal, Quebec, Canada H3G 1M8
guang_c@encs.concordia.ca, weiping@encs.concordia.ca

Abstract. The denoising of a natural image corrupted by the Gaussian white noise is a classical problem in image processing. In this paper, a new image denoising method is proposed by using three scales of dual-tree complex wavelet coefficients. The dual-tree complex wavelet transform is well known for its approximate shift invariance and better directional selectivity, which are very important in image denoising. Experiments show that the proposed method is very competitive when compared with other existing denoising methods in the literature.

Keywords: Image denoising, dual-tree complex wavelets, wavelet transforms, thresholding.

1 Introduction

Wavelet denoising for two-dimensional (2D) images has been a popular research topic in the past decade. The denoising problem to be solved in this paper can be defined as follows. Let $g(t)$ be a noise-free image and $f(t)$ the image corrupted with Gaussian white noise $z(t)$, i.e.,

$$f(t) = g(t) + \sigma_n z(t),$$

where $z(t)$ has a normal distribution $N(0,1)$ and σ_n is the noise variance. Our goal is to remove the Gaussian noise and recover the noise-free image $g(t)$. The basic procedure of wavelet denoising is to transform the noisy signal into the wavelet domain, threshold the wavelet coefficients, and then perform the inverse wavelet transform to obtain the denoised image. The thresholding may be undertaken on a term-by-term basis or by considering the influence of other wavelet coefficients on the wavelet coefficient to be thresholded. For term-by-term denoising, the readers can be referred to [1], [2], [3].

Here we briefly review the most popular wavelet denoising methods that consider the influence of other wavelet coefficients on the current coefficients to be thresholded. Cai and Silverman [4] proposed a thresholding scheme for signal denoising by taking the immediate neighbour coefficients into account. They claimed that this approach gives better results over the traditional term-by-term approach for both translation

F. Sun et al. (Eds.): ISNN 2008, Part II, LNCS 5264, pp. 376–383, 2008.

invariant (TI) and non-TI single wavelet denoising. Chen and Bui [5] extended this neighbouring wavelet thresholding idea to the multiwavelet case. They claimed that neighbour multiwavelet denoising outperforms neighbour single wavelet denoising and the term-by-term multiwavelet denoising [6] for some standard test signals and real-life signals. Chen et al. [7] proposed an image denoising scheme by considering a square neighbourhood window in the wavelet domain. Chen et al. [8] also considered a square neighbourhood window, and tried to customize the wavelet filter and the threshold for image denoising. Experimental results show that both methods produce better denoising results. Mihcak et al. [9] performed an approximate maximum a posteriori (MAP) estimation of the variance for each coefficient, using the observed noisy data in a local neighbourhood. Then an approximate minimum mean squared error estimation procedure is used to denoise the noisy image coefficients. Sendur and Selesnick [10], [11] developed a bivariate shrinkage function for image denoising. Their results showed that the estimated wavelet coefficients depend on the parent coefficients. The smaller the parent coefficients, the greater the shrinkage is. Crouse et al. [12] developed a framework for statistical signal processing based on wavelet-domain hidden markov models (HMM). This framework enables us to concisely model the non-Gaussian statistics of individual wavelet coefficients and capture statistical dependencies between coefficients. Simoncelli and Adelson [13] proposed a Bayesian wavelet coring approach by incorporating the higher-order statistical regularity present in the point statistics of subband representation.

It is well known that the ordinary discrete wavelet transform is not shift invariant because of the decimation operation during the transform. A small shift in the input signal can cause very different output wavelet coefficients. One way of overcoming this is to do the wavelet transform without decimation. The drawback of this approach is that it is computationally inefficient, especially in multiple dimensions. Kingsbury [14], [15], [16] introduced a new kind of wavelet transform, called the dual-tree complex wavelet transform, which exhibits approximate shift invariant property and improved angular resolution. The success of the transform is attributed to the use of filters in two trees, a and b. He proposed to use a simple one-sample delay between the level *1* filters in each tree, in conjunction with alternate odd-length and even-length linear-phase filters. As he pointed out, there are some difficulties in the odd/even filter approach. Therefore, he turned to a new *Q-shift* dual-tree [17] where all the filters beyond level *1* have even length. The filters in the two trees are just the time-reversal of each other, as are the analysis and reconstruction filters. The new filters are shorter than before, and the new transform still satisfies the shift invariant property and good directional selectivity in multiple dimensions. Recently, Chen et al. have successfully applied the dual-tree complex wavelets to image denoising [18] and pattern recognition [19].

In this paper, we proposed a new image denoising method by considering three scales of complex wavelet coefficients during the thresholding process. A simple thresholding formula is developed by exploiting the statistical dependency between a complex wavelet coefficient and its parent and its children. It maintains the simplicity, efficiency, and intuition of soft thresholding. However, it is different from previous developed methods published in the literature. Experiments conducted in this paper confirm the superiority of the proposed image denoising method.

2 Wavelet Thresholding Using Three Scales of Wavelet Coefficients

In this section, we focus on the dependency among a wavelet coefficient, its parent and its children. By considering these three scales, we will derive the wavelet thresholding formula. For any given wavelet coefficient w_1, it has four children at the next detail scale. Let w_2 be the parent of w_1, and w_3 the average of the four children of w_1. Define

$$y = w + n$$

where $w = (w_1, w_2, w_3)$ is the noise-free wavelet coefficients, $y = (y_1, y_2, y_3)$ the noisy coefficients, and $n = (n_1, n_2, n_3)$ the noise. The maximum *a posteriori* (MAP) estimator for w is given by

$$\hat{w}(y) = \arg\max_{w} p_{w|y}(w \mid y).$$

which can be rewritten as

$$\hat{w}(y) = \arg\max_{w} [p_{y|w}(y \mid w) \cdot p_w(w)].$$

Namely,

$$\hat{w}(y) = \arg\max_{w} [p_n(y - w) \cdot p_w(w)].$$

Or equivalently,

$$\hat{w}(y) = \arg\max_{w} [\log(p_n(y - w)) + \log(p_w(w))].$$

By assuming that the noise is i.i.d. Gaussian with

$$p_n(n) = \frac{1}{(\sqrt{2\pi\sigma_n})^3} e^{-\frac{n_1^2 + n_2^2 + n_3^2}{2\sigma_n^2}}$$

and defining $f(w) = \log(p_w(w))]$, we have

$$\hat{w}(y) = \arg\max_{w} [-\frac{(y_1 - w_1)^2}{2\sigma_n^2} - \frac{(y_2 - w_2)^2}{2\sigma_n^2} - \frac{(y_3 - w_3)^2}{2\sigma_n^2} + f(w)].$$

Letting its 1st-order derivative with respect to w_i be zero, (i=1,2,3), we have

$$\frac{y_i - w_i}{\sigma_n^2} + \frac{\partial f(w)}{\partial w_i} = 0.$$

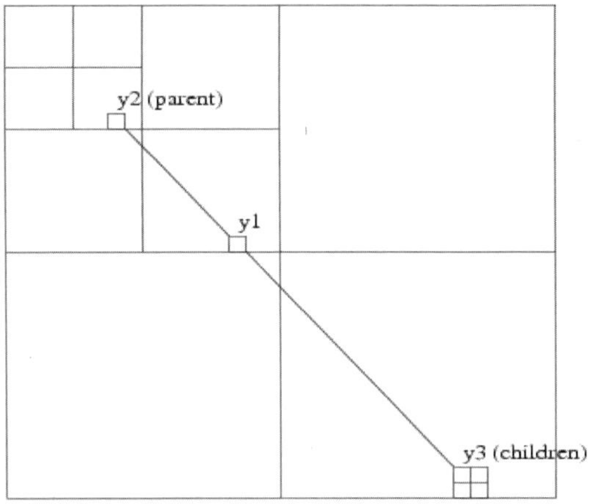

Fig. 1. An illustration of the wavelet coefficient y_1, its parent y_2 and its children y_3

We propose a non-Gaussian probability density function (pdf) for the wavelet coefficient and its parent and its children as

$$p_w(w) = \left(\frac{4}{\sqrt{2\pi}\sigma}\right)^3 e^{-\frac{2}{\sigma}\sqrt{w_1^2+w_2^2+w_3^2}}.$$

Therefore,

$$f(w) = 3\log\left(\frac{4}{\sqrt{2\pi}\sigma}\right) - \frac{2}{\sigma}\sqrt{w_1^2 + w_2^2 + w_3^2}.$$

$$\frac{\partial f(w)}{\partial w_i} = -\frac{2}{\sigma} \cdot \frac{w_i}{\sqrt{w_1^2 + w_2^2 + w_3^2}}.$$

$$y_i = w_i + \frac{2\sigma_n^2}{\sigma} \cdot \frac{w_i}{\sqrt{w_1^2 + w_2^2 + w_3^2}}.$$

After some derivation, we obtain the following thresholding formula

$$w_1 = y_1 \cdot \left(1 - \frac{\frac{2}{\sigma}\sigma_n^2}{\sqrt{y_1^2 + y_2^2 + y_3^2}}\right)_+.$$

where $(x)_+ = \max(x,0)$. The noise variance σ_n can be approximated as [20]

$$\sigma_n = \frac{median(|y_{1i}|)}{0.6745}, \quad y_{1i} \in \text{subband } HH_1.$$

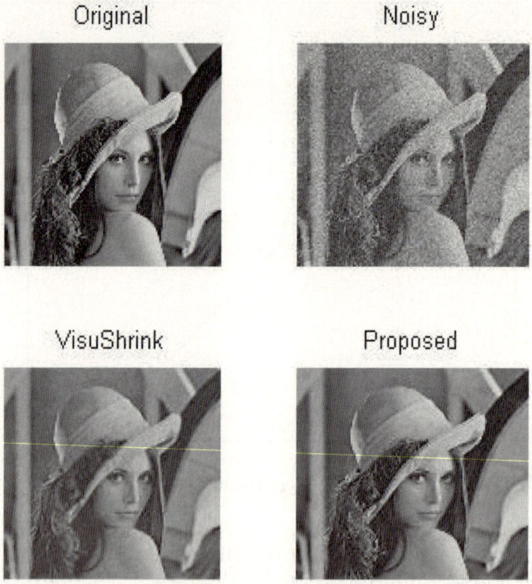

Fig. 2. Original noise-free image, the noisy image with σ_n=20, and the denoised image with VisuShrink and the proposed method, respectively

and

$$\sigma = \sqrt{\left(\frac{1}{M}\sum_{y_{1i}\in S} y_{1i}^2 - \sigma_n^2\right)_+}$$

where M is the number of pixels in the neighborhood S.

For the wavelet coefficients in the first decomposition scale, we use the bivariate thresholding formula since this decomposition scale does not have children. The bivariate thresholding formula is given by [11]

$$w_1 = y_1 \cdot \left(1 - \frac{\frac{\sqrt{3}}{\sigma}\sigma_n^2}{\sqrt{y_1^2 + y_2^2}}\right)_+.$$

The proposed three-scale wavelet-denoising algorithm can be summarized as follows.

1. Perform the forward 2D dual-tree complex wavelet transform on the noisy image until certain specified decomposition scales.
2. Estimate the noise variance σ_n.
3. Threshold the dual-tree complex wavelet coefficients in the first scale by using the bivariate thresholding formula.

4. Threshold the dual-tree complex wavelet coefficients in other scales by using the proposed three scale thresholding formula.
5. Conduct the inverse 2D dual-tree complex wavelet transform to obtain the denoised image.

The above thresholding formula uses the magnitude of the complex wavelet coefficients, since it is shift invariant even though the real and imaginary parts are not individually so. The experiments conducted in this paper show that the proposed method outperforms other existing denoising methods published in the literature.

3 Experimental Results

In this section, we conduct some experiments to denoise the noisy images of 512×512 pixels, and compare the proposed method with a number of existing denoising methods, including BayesShrink [21], locally adaptively window-based denoising using MAP (LAWMAP estimator [22], and the hidden Markov tree (HMT) model [23]. The noisy images are obtained by adding Gaussian white noise on the noise-free image. The noise variance σ_n goes from 10 to 30 in the experiments conducted in this paper. The Daubechies-8 wavelet filter is used for the existing denoising methods. The neighbourhood window size is chosen as 7×7 pixels and the dual-tree complex wavelet transform is performed for 6 decomposition levels. Tables 1-3 tabulate the PSNR values of the denoised images resulting from three existing denoising methods as well as the proposed method for Lena, Barbara and Boat images at different levels of noise variance. The peak signal to noise ratio (PSNR) is defined as

$$PSNR = 10\log_{10}\left(\frac{N \times 255^2}{\sum_{i,j}(B(i,j) - A(i,j))^2}\right)$$

where N is the number of pixels in the image, and B and A are the denoised and noise-free images. Fig. 2 shows the noise-free image, the noisy image with noise added ($\sigma_n = 20$), and the denoised ones with VisuShrink and the proposed method. From the experiments conducted in this paper we find that the proposed method is competitive with other existing methods. Therefore, it is preferred in denoising real life noisy images.

Table 1. PSNR values of different denosing methods for lena

σ_n	Noisy	BayesShrink	HMT	LAWMAP	Proposed
10	28.15	33.32	33.84	34.10	**35.32**
15	24.63	31.41	31.76	32.23	**33.60**
20	22.13	30.17	30.39	30.89	**32.36**
25	20.19	29.22	29.24	29.89	**31.38**
30	18.61	28.48	28.35	29.05	**30.56**

Table 2. PSNR values of different denosing methods for barbara

σ_n	Noisy	BayesShrink	HMT	LAWMAP	Proposed
10	28.15	30.86	31.36	31.99	**33.66**
15	24.63	28.51	29.23	29.60	**31.49**
20	22.13	27.13	27.80	27.94	**29.97**
25	20.19	26.01	25.99	26.75	**28.78**
30	18.61	25.16	25.11	25.80	**27.84**

Table 3. PSNR values of different denosing methods for boat

σ_n	Noisy	BayesShrink	HMT	LAWMAP	Proposed
10	28.15	31.80	32.28	32.25	**33.23**
15	24.63	29.87	30.31	30.40	**31.35**
20	22.13	28.48	28.84	29.00	**30.01**
25	20.19	27.40	27.68	27.91	**28.98**
30	18.61	26.60	26.83	27.06	**28.16**

4 Conclusion

In this paper, we have proposed a new method for image denoising by using the dual-tree complex wavelet transform, which has an approximate shift invariant property and a good directional selectivity. The thresholding formula uses three scales of complex wavelet coefficients for image denoising. Experimental results show that the proposed method is competitive in comparison with other state-of-the-art methods published in the literature. Further investigation will be carried out by exploiting both inter-scale and intra-scale relationships in the dual-tree complex wavelet coefficients. The parent-child relations in the multiwavelet coefficients could also be investigated to achieve better denoising results.

Acknowledgments. This work was supported by the Natural Sciences and Engineering Research Council of Canada (NSERC).

References

1. Strela, V., Heller, P.N., Strang, G., Topiwala, P., Heil, C.: The Application of Multiwavelet Filter Banks to Image Processing. IEEE Transactions on Image Processing 8, 548–563 (1999)
2. Downie, T.R., Silverman, B.W.: The Discrete Multiple Wavelet Transform and Thresholding Methods. IEEE Transactions on Signal Processing 46, 2558–2561 (1998)

3. Coifman, R.R., Donoho, D.L.: Translation Invariant Denoising. In: Wavelets and Statistics. Springer Lecture Notes in Statistics, vol. 103. Springer, New York (1994)
4. Cai, T.T., Silverman, B.W.: Incorporating Information on Neighbouring Coefficients into Wavelet Estimation. Sankhya: The Indian Journal of Statistics 63(B), pt. 2, 127–148 (2001)
5. Chen, G.Y., Bui, T.D.: Multiwavelet Denoising Using Neighbouring Coefficients. IEEE Signal Processing Letters 10, 211–214 (2003)
6. Bui, T.D., Chen, G.Y.: Translation-invariant Denoising Using Multiwavelets. IEEE Transactions on Signal Processing 46, 3414–3420 (1998)
7. Chen, G.Y., Bui, T.D., Krzyzak., A.: Image Denoising Using Neighbouring Wavelet Coefficients. Integrated Computer-Aided Engineering 12, 99–107 (2005)
8. Chen, G.Y., Bui, T.D., Krzyzak., A.: Image Denoising with Neighbour Dependency and Customized Wavelet and Threshold. Pattern Recognition 38, 115–124 (2005)
9. Mihcak, M.K., Kozintsev, I., Ramchandran, K., Moulin., P.: Low-Complexity Image Denoising Based on Statistical Modeling of Wavelet Coefficients. IEEE Signal Processing Letters 6, 300–303 (1999)
10. Sendur, L., Selesnick, I.W.: Bivariate Shrinkage Functions for Wavelet-Based Denoising Exploiting Interscale Dependency. IEEE Transactions on Signal Processing 50, 2744–2756 (2002)
11. Sendur, L., Selesnick, I.W.: Bivariate Shrinkage with Local Variance Estimation. IEEE Signal Processing Letters 9, 438–441 (2002)
12. Crouse, M.S., Nowak, R.D., Baraniuk, R.G.: Wavelet-Based Signal Processing Using Hidden Markov Models. IEEE Transactions on Signal Processing 46, 886–902 (1998)
13. Simoncelli, E.P., Adelson, E.H.: Noise Removal via Bayesian Wavelet Coring. In: The 3rd International Conference on Image Processing, Lausanne, Switzerland, pp. 379–382 (1996)
14. Kingsbury, N.G.: The Dual-Tree Complex Wavelet Transform: A New Efficient Tool for Image Restoration and Enhancement. In: Proceedings European Signal Processing Conference, Rhodes, pp. 319–322 (1998)
15. Kingsbury, N.G.: Shift Invariant Properties of the Dual-Tree Complex Wavelet Transform. In: Proceedings of the IEEE International conference on Acoustics, Speech, and Signal Processing, Phoenix, AZ, pp. 1221–1224 (1999)
16. Romberg, J., Choi, H., Baraniuk, R., Kingsbury, N.G.: Multiscale Classification Using Complex Wavelets and Hidden Markov Tree Models. In: Proceedings of International Conference on Image Processing, Vancouver, pp. 371–374 (2000)
17. Kingsbury, N.G.: A Dual-Tree Complex Wavelet Transform with Improved Orthogonality and Symmetry Properties. In: Proceedings of International Conference on Image Processing, Vancouver, pp. 375–378 (2000)
18. Chen, G.Y., Kegl, B.: Image Denoising with Complex Ridgelets. Pattern Recognition 40, 578–585 (2007)
19. Chen, G.Y., Xie, W.F.: Pattern Recognition with SVM and Dual-tree Complex Wavelets. Image and Vision Computing 25, 960–966 (2007)
20. Donoho, D.L., Johnstone, I.M.: Ideal Spatial Adaptation by Wavelet Shrinkage. Biometrika 81, 425–455 (1994)
21. Chang, S., Yu, B., Vetterli, M.: Adaptive Wavelet Thresholding for Image Denoising and Compression. IEEE Transactions on Image Processing 9, 1532–1546 (2000)
22. Crouse, M.S., Nowak, R.D., Baraniuk, R.G.: Wavelet-based Signal Processing Using Hidden Markov Models. IEEE Transactions on Signal Processing 46, 886–902 (1998)
23. Mihcak, M.K., Kozintsev, I., Ramchandran, K., Moulin, P.: Low-complexity Image Denoising Based on Statistical Modeling of Wavelet Coefficients. IEEE Signal Processing Letters 6, 300–303 (1999)

Image Denoising Using Neighbouring Contourlet Coefficients

Guangyi Chen and Wei-Ping Zhu

Department of Electrical and Computer Engineering,
Concordia University,
Montreal, Quebec, Canada H3G 1M8
guang_c@encs.concordia.ca, weiping@encs.concordia.ca

Abstract. The denoising of a natural image corrupted by Gaussian white noise is a classical problem in image processing. In this paper, a new image denoising method is proposed by using the contourlet transform. The thresholding process employs a small neighbourhood for the current contourlet coefficient to be thresholded. This is because the contourlet coefficients are correlated, and large contourlet coefficients will normally have large coefficients at its neighbour locations. Experiments show that the proposed method is better than the standard contourlet denoising and the wavelet denoising.

Keywords: Image denoising, the contourlet transform, thresholding.

1 Introduction

Wavelet denoising for two-dimensional (2D) images has been a popular research topic in the past decade. Let $g(t)$ be a noise-free image and $f(t)$ the image corrupted with Gaussian white noise $z(t)$, i.e.,

$$f(t) = g(t) + \sigma_n z(t),$$

where $z(t)$ has a normal distribution $N(0,1)$ and σ_n is the noise variance. Our aim is to remove the Gaussian white noise and recover the noise-free image $g(t)$. The basic procedure of wavelet denoising is to transform the noisy image into the wavelet domain, threshold the wavelet coefficients, and then perform the inverse wavelet transform. The thresholding processing may be on a term-by-term basis or consider the intra- or inter-scale dependency. For the term-by-term wavelet denoising, the readers are referred to [1], [2], [3].

We briefly review the most popular wavelet denoising methods that consider the intra- or inter-scale dependency. Cai and Silverman [4] proposed a thresholding scheme for signal denoising by taking the immediate neighbour coefficients into account. They claimed that this approach gives better results than the traditional term-by-term approach for both translation invariant (TI) and non-TI single wavelet denoising. Chen and Bui [5] extended this neighbouring wavelet thresholding idea to the multiwavelet case, and they found that neighbor multiwavelet denoising outperforms neighbour

F. Sun et al. (Eds.): ISNN 2008, Part II, LNCS 5264, pp. 384–391, 2008.

single wavelet denoising and the term-by-term multiwavelet denoising [6] for some standard test signals and real-life noisy signals. Chen et al. [7] proposed an image-denoising scheme by considering the intra-scale dependency in the wavelet domain. Chen et al. also considered intra-scale dependency in [8], and tried to customize the wavelet filter and the threshold for image denoising. Experimental results show that both methods produce promising denoising results. Mihcak et al. [9] performed an approximate maximum a posteriori (MAP) estimation of the variance for each coefficient, using the observed noisy data in a local neighbourhood. Then an approximate minimum mean squared error estimation procedure is used to denoise the noisy image coefficients. Sendur and Selesnick [10], [11] developed several bivariate shrinkage functions for image denoising. Their results showed that the estimated wavelet coefficients depend on the parent coefficients. The smaller the parent coefficients, the greater the shrinkage. Crouse et al. [12] proposed a framework for statistical signal processing based on wavelet-domain hidden markov models (HMM). The framework enables us to concisely model the non-Gaussian statistics of individual wavelet coefficients and capture statistical dependencies between coefficients. Simoncelli and Adelson [13] developed a Bayesian wavelet coring approach by incorporating the higher-order statistical regularity present in the point statistics of subband representation.

The contourlet transform was recently proposed by Do and Vetterli [14] to overcome the limitations of wavelets. They constructed a double filter bank structure in which at first the Laplacian pyramid (LP) is used to capture the point discontinuities into linear structure. The overall result is an image expansion with basis images as contour segments, and thus it is named the contourlet transform. The contourlet transform is an extension to the wavelet transform in 2D by using nonseparable and directional filter banks. It uses basic images oriented at varying directions in multiple scales with flexible aspect ratios. The contourlet construction allows for any number of directional filter bank (DFB) decomposition levels to be applied at each LP level. For the contourlet transform to satisfy the anisotropy scaling relation, we impose that in the pyramid DFB, the number of directions is doubled at every other finer scale of the pyramid. Also, the support size of the LP is reduced by four times while the number of directions of the DFB is doubled. The contourlet transform involves basis functions that are oriented at any power of two's number of directions with flexible aspect ratios. Therefore, it can represent smooth edges with close to optimal efficiency. More recent developments on contourlets include [15], [16], [17] and [18].

In this paper, we proposed a new image denoising method by considering a small neighbourhood of the contourlet coefficients to be thresholded during the thresholding process. The reason why we consider a small neighborhood is because the contourlet coefficients are correlated, just like the wavelet coefficients. A large contourlet coefficient will probably have large contourlet coefficients at its neighbour locations. Experiments conducted in this paper confirm the superiority of the proposed image denoising method.

2 Contourlet Thresholding Using Neighbouring Coefficients

The contourlet transform was recently developed by Do and Vetterli [14] to overcome the limitations of wavelets. It is based on an efficient 2D multiscale and directional filter bank that can deal effectively with images having smooth contours. It uses a combination of a Laplacian Pyramid (LP) that decomposes an image into a number of radial subbands, and a Directional Filter Bank (DFB), where each LP detail subband is fed to this stage to be decomposed into a number of directional subbands. Contourlets have elongated supports at various scales, directions, and aspect ratios. Therefore, they are good at capturing directional features in images in a multiresolutional way. Moreover, the discrete contourlet transform has a fast-iterated filter bank algorithm that requires an order N operations for N-pixel images. However, the contourlet transform is up to 33% overcomplete, which comes from the Laplacian pyramid.

In this section, we propose a new image denoising method by using the subsampled contourlet transform. The thresholding formula employs the intra-scale dependency in the contourlet subbands. For every contourlet coefficient $c_{i,j}$ to be thresholded, we consider a neighborhood window $N_{i,j}$ around it. We choose the window by having the same number of pixels above, below, and on the left or right of the pixel to be thresholded. This means the neighbourhood window size should be $(2L+1)\times(2L+1)$, where L is a non-negative integer. Fig. 1 illustrates a 3×3 neighbourhood window centered at the contourlet coefficient to be thresholded. We threshold different contourlet subbands independently. Therefore, when the small neighborhood window surrounding the contourlet coefficient to be thresholded touches the coefficients in other subbands, we do not include those coefficients in our calculation. Let

$$S_{i,j} = \frac{1}{(2L+1)^2} \sum_{m=i-L}^{i+L} \sum_{n=j-L}^{j+L} | c_{m,n} |$$

be the average of the magnitude of the contourlet coefficients in a neighborhood window $N_{i,j}$ centered at $c_{i,j}$, then the thresholding formula can be defined as

$$c_{i,j} = c_{i,j} \cdot (1 - \alpha \frac{\lambda}{S_{i,j}})_+$$

where $\alpha = 0.45$ is a scaling factor, $(x)_+ = \max(0, x)$, and λ is the threshold defined as in [19]. We set $\lambda = 4\sigma_n \sigma_{i,j}$ for the finest scale, and $\lambda = 3\sigma_n \sigma_{i,j}$ for other scales, where σ_n is the noise variance and $\sigma_{i,j}$ is the individual variance of the contourlet coefficient $c_{i,j}$. Similar thresholding approach was also used in [20] for image denoising by means of the complex ridgelet transform.

The neighbour contourlet-denoising algorithm can be given as follows.

1. Perform the forward contourlet transform on the noisy image.
2. Threshold the contourlet coefficients by using the proposed thresholding formula.
3. Conduct the inverse contourlet transform in order to obtain the denoised image.

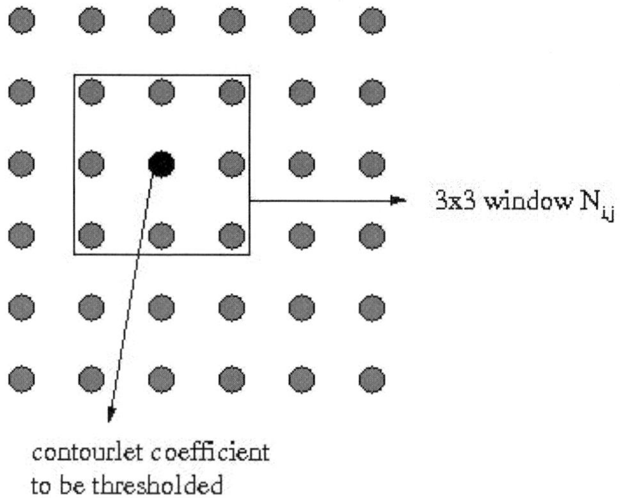

Fig. 1. An illustration of the neighbourhood of 3x3 pixels

The reason why we consider a neighbourhood in the thresholding process is because the contourlet coefficients are correlated. A large contourlet coefficient will probably have large coefficients at its neighbour locations. This is also true for the wavelet coefficients. Experiments conducted in this paper show that the proposed method outperforms the standard contourlet denoising and the wavelet denoising.

3 Experimental Results

In this section, we conducted some experiments to denoise three noisy images with 512×512 pixels, namely Lena, Barbara and Boat. We compared our proposed method with the wavelet denoising and the standard contourlet denoising. The noisy images are obtained by adding Gaussian white noise to the noise-free image. The noise variance σ_n goes from 10 to 30 in the experiments conducted in this paper. The Daubechies-8 wavelet filter is used for the existing wavelet denoising. The neighbourhood window size is chosen as 3×3 pixels in the proposed denoising method. The neighbourhood sizes of 5×5 and 7×7 are also considered in this section, but they are not as good as 3×3. We have chosen the optimal scaling factor as $\alpha = 0.45$ because it generates the best denoising results for all three images tested in this section. Tables 1-3 tabulate the PSNR values of the denoised images using the wavelet denoising, the standard contourlet denoising, and the proposed method for different levels of noise variance for images Lena, Barbara and Boat, respectively. The peak signal to noise ratio (PSNR) is defined as

$$PSNR = 10\log_{10}(\frac{M \times 255^2}{\sum_{i,j}(B(i,j) - A(i,j))^2})$$

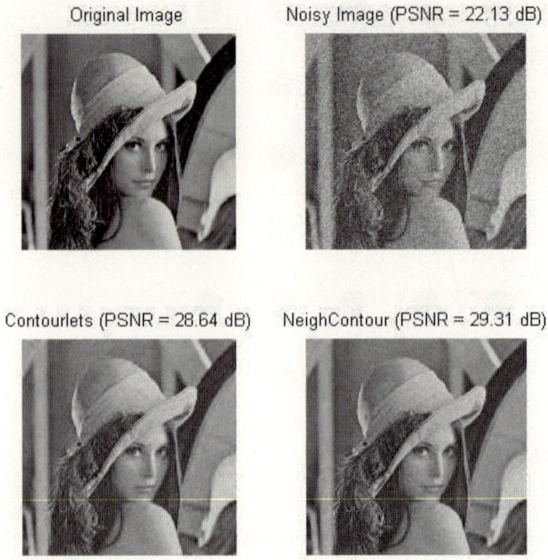

Fig. 2. Original noise-free image Lena, the noisy image with σ_n=20, and the denoised image with the standard contourlets and the proposed method, respectively

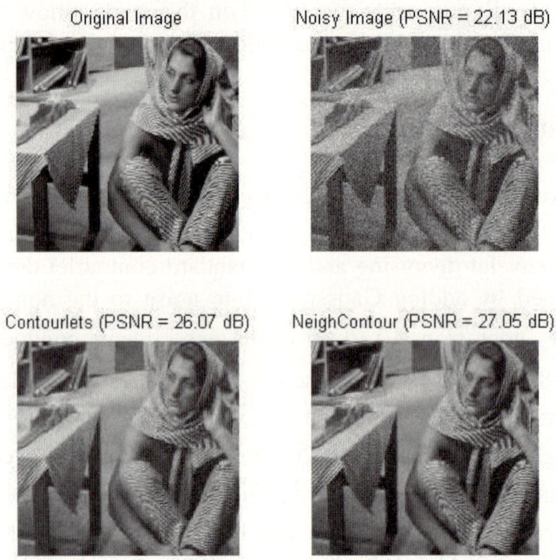

Fig. 3. Original noise-free image Barbara, the noisy image with σ_n=20, and the denoised image with the standard contourlets and the proposed method, respectively

Fig. 4. Original noise-free image Boat, the noisy image with σ_n=20, and the denoised image with the standard contourlets and the proposed method, respectively

Table 1. PSNR of lena image from different denosing methods at various noise levels

σ_n	Noisy	Wavelets	Contourlets	NeighContour
10	28.15	31.98	31.69	**32.59**
15	24.63	29.95	29.89	**30.65**
20	22.13	28.51	28.64	**29.31**
25	20.19	27.31	27.65	**28.27**
30	18.61	26.26	26.84	**27.33**

Table 2. PSNR of barbara from different denosing methods at various noise levels

σ_n	Noisy	Wavelets	Contourlets	NeighContour
10	28.15	29.69	29.28	**30.71**
15	24.63	27.18	27.37	**28.54**
20	22.13	25.57	26.07	**27.05**
25	20.19	24.31	25.09	**26.06**
30	18.61	23.38	24.28	**25.22**

Table 3. PSNR of boat image from different denosing methods at various noise levels

σ_n	Noisy	Wavelets	Contourlets	NeighContour
10	28.15	30.21	29.72	**30.70**
15	24.63	28.21	27.96	**28.84**
20	22.14	26.73	26.72	**27.54**
25	20.19	25.69	25.87	**26.46**
30	18.61	24.78	25.13	**25.71**

where M is the number of pixels in the image, and B and A are the denoised and noise-free images. Figs. 2-4 show the noise-free image, the noisy image with noise added ($\sigma_n = 20$), and the denoised images with the standard contourlet denoising and the proposed method for images Lena, Barbara and Boat, respectively. From the experiments conducted in this paper we find that the proposed method is better than the standard contourlet denoising and wavelet denoising. Therefore, it is preferred in denoising real life noisy images.

4 Conclusion

In this paper, we proposed a new method for image denoising by using the contourlet transform. The thresholding formula considers a small neighbourhood of the current contourlet coefficient to be thresholded. This is because the contourlet coefficients are correlated. A large contourlet coefficient will probably have large coefficients at its neighbour locations. Experimental results show that the proposed method is better than the standard contourlet denoising and the wavelet denoising.

Even though the neighbourhood strategy is applied to threshold the critically sampled contourlet coefficients in this paper, it can also be applied to the nonsubsampled contourlet coefficients. It is expected that much better denoised results can be obtained by combining the neighbourhood strategy with the nonsubsampled contourlet transform [15].

Acknowledgments. The authors would like to thank M. N. Do for making his contourlet software available on his website. This work was supported by the Natural Sciences and Engineering Research Council of Canada (NSERC).

References

1. Strela, V., Heller, P.N., Strang, G., Topiwala, P., Heil, C.: The Application of Multiwavelet Filter Banks to Image Processing. IEEE Transactions on Image Processing 8, 548–563 (1999)
2. Downie, T.R., Silverman, B.W.: The Discrete Multiple Wavelet Transform and Thresholding Methods. IEEE Transactions on Signal Processing 46, 2558–2561 (1998)

3. Coifman, R.R., Donoho, D.L.: Translation Invariant Denoising. In: Wavelets and Statistics. Springer Lecture Notes in Statistics, vol. 103, pp. 125–150. Springer, New York (1994)
4. Cai, T.T., Silverman, B.W.: Incorporating Information on Neighbouring Coefficients into Wavelet Estimation. Sankhya: The Indian Journal of Statistics 63(B), pt. 2, 127–148 (2001)
5. Chen, G.Y., Bui, T.D.: Multiwavelet Denoising Using Neighbouring Coefficients. IEEE Signal Processing Letters 10, 211–214 (2003)
6. Bui, T.D., Chen, G.Y.: Translation-invariant Denoising Using Multiwavelets. IEEE Transactions on Signal Processing 46, 3414–3420 (1998)
7. Chen, G.Y., Bui, T.D., Krzyzak., A.: Image Denoising Using Neighbouring Wavelet Coefficients. Integrated Computer-Aided Engineering 12, 99–107 (2005)
8. Chen, G.Y., Bui, T.D., Krzyzak., A.: Image Denoising with Neighbour Dependency and Customized Wavelet and Threshold. Pattern Recognition 38, 115–124 (2005)
9. Mihcak, M.K., Kozintsev, I., Ramchandran, K., Moulin., P.: Low-Complexity Image Denoising Based on Statistical Modeling of Wavelet Coefficients. IEEE Signal Processing Letters 6, 300–303 (1999)
10. Sendur, L., Selesnick, I.W.: Bivariate Shrinkage Functions for Wavelet-Based Denoising Exploiting Interscale Dependency. IEEE Transactions on Signal Processing 50, 2744–2756 (2002)
11. Sendur, L., Selesnick, I.W.: Bivariate Shrinkage with Local Variance Estimation. IEEE Signal Processing Letters 9, 438–441 (2002)
12. Crouse, M.S., Nowak, R.D., Baraniuk, R.G.: Wavelet-Based Signal Processing Using Hidden Markov Models. IEEE Transactions on Signal Processing 46, 886–902 (1998)
13. Simoncelli, E.P., Adelson, E.H.: Noise Removal via Bayesian Wavelet Coring. In: The 3rd International Conference on Image Processing, Switzerland, pp. 379–382 (1996)
14. Do, M.N., Vetterli, M.: The Contourlet Transform: An Efficient Directional Multiresolution Image Representation. IEEE Transactions Image on Processing 14, 2091–2106 (2005)
15. Cunha, A.L., Zhou, J., Do, M.N.: The Nonsubsampled Contourlet Transform: Theory, Design, and Applications. IEEE Transactions on Image Processing 15, 3089–3101 (2006)
16. Eslami, R., Radha, H.: Translation-invariant Contourlet Transform and Its Application to Image Denoising. IEEE Transactions on Image Processing 15, 3362–3374 (2006)
17. Matalon, B., Zibulevsky, M., Elad, M.: Improved Denoising of Images Using Modeling of the Redundant Contourlet Transform. In: Proc. of the SPIE conference wavelets, vol. 5914 (2005)
18. Chappelier, V., Guillemot, C., Marinkovic, S.: Image Coding with Iterated Contourlet and Wavelet Transforms. In: Proc. of International Conference on Image Processing, Singapore, pp. 3157–3160 (2004)
19. Starck, J.L., Candes, E.J., Donoho, D.L.: The Curvelet Transform for Image Denoising. IEEE Transactions on Image Processing 11, 670–684 (2002)
20. Chen, G.Y., Kegl, B.: Image Denoising with Complex Ridgelets. Pattern Recognition 40, 578–585 (2007)

Robust Watermark Algorithm Based on the Wavelet Moment Modulation and Neural Network Detection

Dianhong Wang , Dongming Li, and Jun Yan

Institute of Mechanical & Electronic Engineering, China University of Geosciences
430074 Wuhan, China
{universelidongming,universelister,universeli}@gmail.com

Abstract. Moment-domain based watermark can resist geometric attacks but can not be detected blindly. The purpose of this paper is to outline the state of the research of wavelet moment modulation-based watermark and to propose a neural network detection algorithm towards it. With regard to the later we first analyze the computation of the wavelet moment and inverse wavelet moment. Then we focus on watermark added with template embedding and detection based on neural network. Results of the experiments revealed that our watermark detection algorithm is more robust comparing with conventional wavelet-based algorithm. In addition, it detects the watermark blindly.

Keywords: Digital watermarking, Wavelet moment, Neural network, Information hiding.

1 Introduction

Digital watermarking was a new technology that could be used in copyright protection, source authentication and integrity in network. Two types of commonly used watermark embedding algorithms works in spatial domain and in transform domain respectively. The shortcoming of spatial watermarking algorithm, which directly embeds information into the spatial domain of digital media, is that it is not robust enough to image processing. On the other hand, transform domain watermarking algorithms, which hide information in the transform domain, attract more attention of recent study for its excellent stability and robustness. As one method work in the moment domain The literature [1] embedded the watermark in the Fourier-Mellin (FMT) transform amplitude domain, which is invariant to rotation, scaling and translation, but the log-polar mapping in the computation of the FMT makes reconstruction of the image contains greater error. In literature [2], another watermarking method based in the RST domain was proposed. However, this kind of algorithm has lots of computation complexity and can not reconstruct the image with high quality; meanwhile watermarking method base on this kind of algorithm can not resist attacks such as compression and filtering. In literature [3, 4, 5] Masoud A proposed an algorithm to estimate the rotate angle and the scale changes through the wavelet coefficients, then apply geometric correction to the attacked watermarked image with moment information of the host image before watermarking detection. However, the experimental results show that the geometric correction has greater error except four integer

F. Sun et al. (Eds.): ISNN 2008, Part II, LNCS 5264, pp. 392–401, 2008.

angles 90^0, 180^0, 270^0, 360^0, thus easily lead to failure of the watermark detection, not to mention the detection can not be blind. On the other hand, the watermarked image attacked by the unpredictable cropping can not be reconstructed. Literature [6] use Zernik moments of the host image as watermark carrier, but the Zernik moment has greater error in the inverse moment computation.

Moment information of the host image is of great importance of the copyright protection, and they have RST Invariance [7-8]. To enhance the anti-attack capability of the watermark, an algorithm based on neural network detection and wavelet moment is proposed in this paper. First, we analyze the computation of the wavelet moment and inverse wavelet moment. Then we add watermark and template into wavelet moment domain. Last we detect the watermark in wavelet moment domain by the trained neural network. Simulation experiments indicate that our watermarking algorithm has effectively improved the anti-attack capability of the watermark.

2 Wavelet Moment Feature Extraction and Image Reconstruction

The image moment can be defined as follows:

$$F_{p,q} = \int s_q(r) g_p(r) r dr \tag{1}$$

Where $s_q(r) = \int f(r,\theta) e^{jq\theta} d\theta$, $f(r,\theta)$ is the image which was mapped into log-polar coordinates system.

$g_p(r)$ is a function with variable r, if $g_p(r) = \psi_{m,n}(r)$,we have wavelet moment:

$$F_{m,n,q} = \int F_q(r) \psi_{m,n}(r) dr \tag{2}$$

Where $\quad F_q(r) = s_q(r) r$

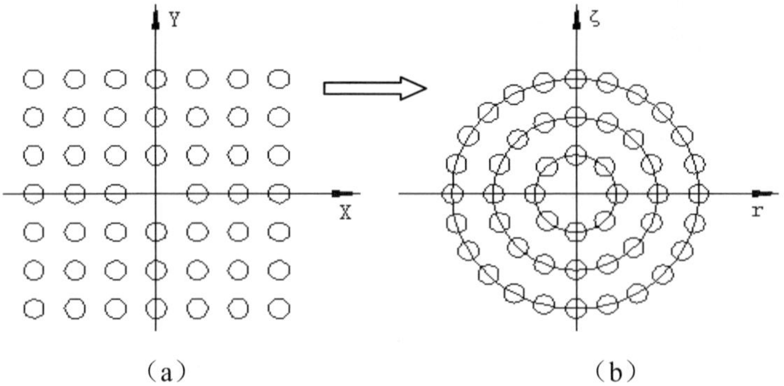

Fig. 1. Square-to-Circle Transform. Coordinates (x, y) in figure (a) is translated to Circle coordinates (γ, ξ) in figure(b).

(x, y) is coordinates of Square map, while (γ, ξ) is coordinates of Circle map, and the Integer γ, ξ could be calculated by following expressions:

$$\gamma = \max\{|x|, |y|\} \tag{3}$$

If $|x| = r$, then $\xi = 2(r - x)\frac{y}{|y|} + \frac{xy}{r}$, If $|y| = r$, then $\xi = 2y - \frac{xy}{r}$, Square-to-Circle Transform is an one-to-one mapping, Assume that the original image gray value remain unchanged in the transformation, we get $f(x, y) = f(r, \xi)$, if the image size is $N \times N$, so we have : $-\frac{N-1}{2} \le x, y \le \frac{N-1}{2}$, $0 \le r \le \frac{N-1}{2}$, $0 \le \xi \le 8r - 1$, $\theta = \frac{\pi\xi}{4r}$, we get equations

$$rd\theta = r\frac{\pi}{4r} = \frac{\pi}{4}, \tag{4}$$

and

$$F_q(r) = s_q(r)r = \int f(r,\theta)e^{jq\theta} rd\theta$$

$$= \frac{\pi}{4}\sum_{\xi=0}^{8r-1} f(r,\theta)e^{jq\frac{\pi\xi}{4r}} \quad . \tag{5}$$

From equation (5) and (2) we have $F_{m,n,q} = \int F_q(r)\psi_{m,n}(r)dr$

$$= \int [\frac{\pi}{4}\sum_{\xi=0}^{8r-1} f(r,\theta)e^{jq\frac{\pi\xi}{4r}}]\psi_{m,n}(r)dr$$

$$= <\frac{\pi}{4}\sum_{\xi=0}^{8r-1} f(r,\theta)e^{jq\frac{\pi\xi}{4r}}, \psi_{m,n}(r)> \tag{6}$$

Equation (6) is the Discrete Wavelet Transform (DWT) of $F_q(r)$.

Now we use wavelet filter $\{h\},\{g\}$ to calculate the Details wavelet-moment $F^H_{m,k,q}$ and Rough wavelet-moment $F^L_{m,k,q}$ with q^{th} order:

$$\begin{cases} F^L_{m,k,q} = \sum_n F_q(n)\bar{h}_{n-2k} \\ F^H_{m,k,q} = \sum_n F_q(n)\bar{g}_{n-2k} \end{cases}$$

$F^L_{m,k,0}$ can not be changed even if the host image was rotated from angle θ to $\theta + \alpha$, so the wavelet-moment could be used as information carrier.

Now we analyze the computation of inverse wavelet moment. In Equation (6), the wavelet moment is in fact a Discrete Wavelet Transform (DWT) of $s_q(r)r$:

$$F_{m,n,q} =< s_q(r)r, \Psi_{m,n}(r) > \tag{7}$$

If the wavelet $\{\psi_{m,n}\}$ is a standard Orthogonal base of $L^2(2)$, we could use $\{\tilde{\psi}_{m,n}\}$ to reconstruct the signal $s_q(r)r$

$$\begin{aligned} F_q(r) &= \int f(r,\theta) r e^{jq\theta} d\theta \\ &= \sum_m \sum_n < F_q(r), \psi_{m,n}(r) > \tilde{\psi}_{m,n}(r) \\ &= \sum_m \sum_n F_{m,n,q} \tilde{\psi}_{m,n}(r) \end{aligned} \tag{8}$$

With the reconstruction wavelet filter $\{h'\},\{g'\}$, the original host image could be reconstructed by rough wavelet moment $F^L_{m,k,q}$ and detailed wavelet moment

$$F_q(r) = \sum_l F^L_{m,r,q} h'_{r-2l} + \sum_l F^H_{m,r,q} g'_{r-2l} \tag{9}$$

$F_q(r) = \frac{\pi}{4}\sum_{\xi=0}^{8r-1} f(r,\theta) e^{jq\frac{\pi\xi}{4r}}$ is in fact the Fourier Transform of circle serial, so we have:

$$f(r, \frac{\pi\xi}{4r}) = \frac{4}{N\pi}\sum_{q=0}^{8r-1} F_q(r) e^{-jq\frac{\pi\xi}{4r}} \tag{10}$$

From (9) and (10),we reconstruct image $f(r,\theta)$ in Polar coordinates system.

$$\begin{aligned} f(r,\theta) = f(r, \frac{\pi\xi}{4r}) &= \frac{4}{N\pi}\sum_{q=0}^{8r-1}[\sum_l F^L_{m,r,q} h'_{r-2l} + \sum_l F^H_{m,r,q} g'_{r-2l}] e^{-jq\frac{\pi\xi}{4r}} \\ &= \frac{4}{N\pi}\sum_{q=0}^{8r-1}\sum_l F^L_{m,r,q} h'_{r-2l} + \frac{4}{N\pi}\sum_{q=0}^{8r-1}\sum_l F^H_{m,r,q} g'_{r-2l} e^{-jq\frac{\pi\xi}{4r}} \end{aligned} \tag{11}$$

Through an inverse square-to-circle transform, we can get the reconstructed host image $f(x, y)$.

3 Watermark Embedding and Detection

3.1 To Produce Watermark Sequence

We choose one-dimensional Logistic mapping to get chaotic sequence. Logistic mapping $x_{k+1} = 1 - \mu x_k^2$, $x_k \in (-1,1)$,$\mu \in (0,2]$ is defined in interval (-1, 1), and the μ is a chaotic parameter. If $\mu = (1.40115,2]$, the system goes into chaotic state; if $\mu = 2$, the output sequence of the system is similar to white noise, which could be used as watermark. In this paper, the binary watermark was produced from $\{x_k, k=0,1,2,3\cdots\}$ as follows:

$$W=\{\Gamma(x_k)=\begin{cases}-1, -1\le x_k<0\\ 1, 0\le x_k\le 1\end{cases}.$$

3.2 Watermark Embedding

Step1. To generated $t=M$ pseudo-random numbers with key k_0, which will be used as watermark embedding position $t\in\{0,1,\cdots,M+L-1\}$.

Step2. Adding a piece of binary data as template which length is L into $W(t)$, we get watermark $S(t)=\{h_L \sim W_k\}, L=0,1\cdots M-1, W_k=\Gamma(x_k), K=M, M+1,\cdots, L+M-1$

Step3. To calculate the wavelet moment $F^L_{m,n,q}=\int s_q(r)\Psi_{m,n}(r)rdr$, and we embedded watermark into $F^L_{m,k,q}$ as follows $F^L_{m,k,q}=F^L_{m,k,q}+\alpha S(t)$, The watermarking embedding was shows as figure 2.

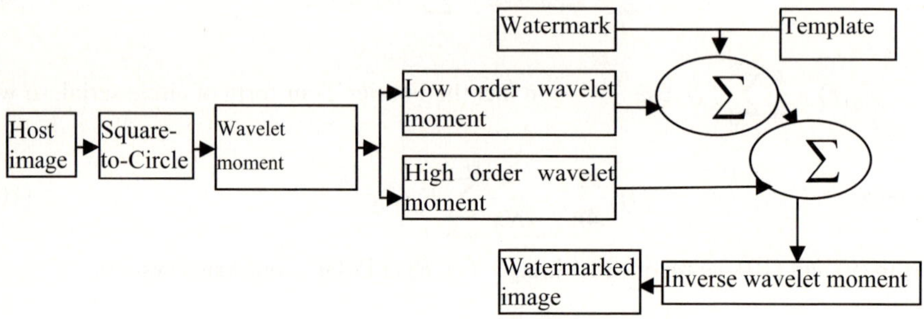

Fig. 2. Watermarking embedding

3.3 Watermark Detection

The information processed in the BP network is transmitted forward; meanwhile the error information is transmitted backward to revised weights of the network.

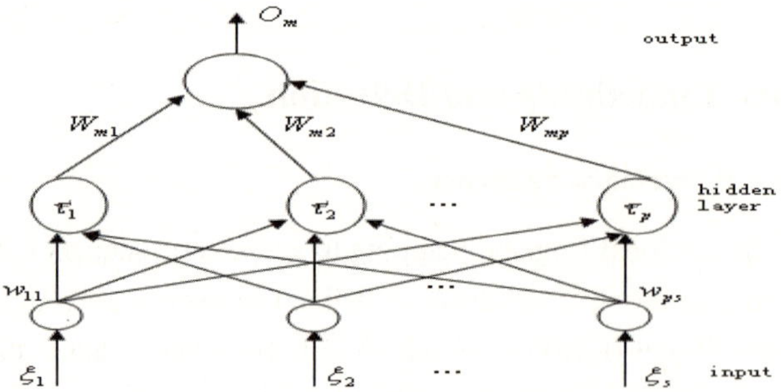

Fig. 3. A BP network with only one hidden layer

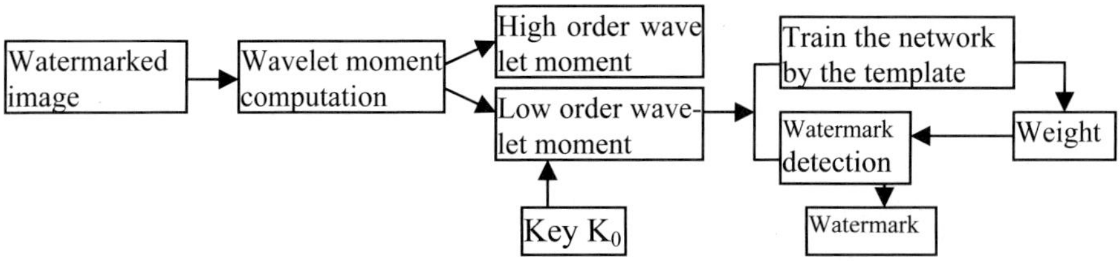

Fig. 4. Watermarking detection by BP network

In the watermark detection, first the network learns the relation between template and wavelet moment of host image, then the learned network detect the watermark. The watermark detection was as follows:

Step1. Calculate the wavelet moment of the host image.

Step2. Calculate the embedding position t and the wavelet moment $F^{L}_{m,k,q}$.

Step3. Set a data block which center is $F^{L,t}_{m,k,q}$, the relation between watermark s_t and the wavelet moment $F^{L,t}_{m,k,q}$ is as follows, the pattern vector P:

$$P = \{(\delta_{t-c}, \delta_{t-c+1}, ..., \delta_t, ...\delta_{t+c-1}, \delta_{t+c}), (d_t)\}_{t=0}^{L+M-1} \tag{12}$$

Where $\delta_t = \left\|F^t_{m,n,q}\right\|_t - \frac{1}{2c}(\sum_{r=-c}^{r=c} \left\|F^t_{m,n,q}\right\|_{t+r} - \left\|F^t_{m,n,q}\right\|_t)$, and d_t is the tth output of the tth input pattern of vector P.

$$d_t = \begin{cases} \frac{1}{255}(\left\|F^t_{m,n,q}\right\|_t - \delta_t), if (S_t = 1) \\ -\frac{1}{255}(\left\|F^t_{m,n,q}\right\|_t - \delta_t), if (S_t = 0) \end{cases} \tag{13}$$

step4. Use the first L template patterns $\{P_L, T_L\}$ to train the network

$$\{P_L, T_L\} = \{(\delta_{t-c}, \delta_{t-c+1}, ..., \delta_t, ...\delta_{t+c-1}, \delta_{t+c}), (d_t)\}_{t=0}^{L-1} \tag{14}$$

The neural network learned the relation between the P_L and the templates, and predict the rest binary $d_t \in (-1,1)$, thus we have binary watermark $W = W_M, W_{M+1} ... W_{L+M-1}$

$$S_{L+t} = W_t = \begin{cases} 1, if (d_t > 0) \\ 0, other \end{cases} \quad (M \le t \le L + M - 1) \tag{15}$$

4 Experiment Results

The original image is 512×512×8bit Wbarb, M = 64, the wavelet we used is db10, the embedding intensity is 0.005. The experiments show that the PSNR of our algorithm is 11.2dB greater than literature [7] and 13.6dB greater than literature [8].

Anti-attack test contains of three parts: (1)to test the anti-attack ability of the watermark with different embedding intensity;(2)to test the anti-attack ability of the watermark with different wavelet moment;(3)to test the anti-attack ability of the watermark

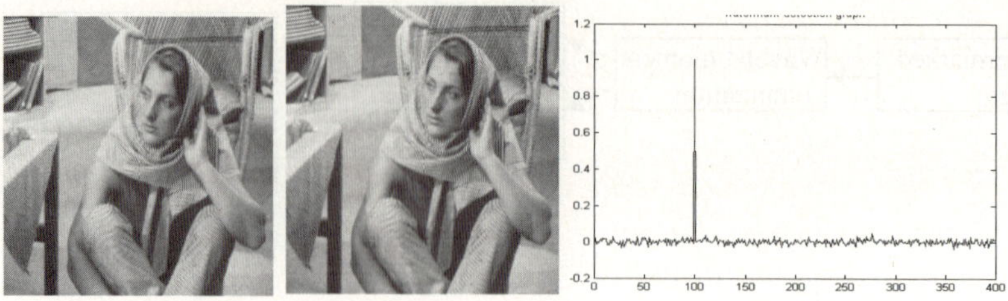

(a) (b) (c)

Fig. 5. (a)host image(b)watermarked image(PSNR=48.6440dB)and(c)detection graph

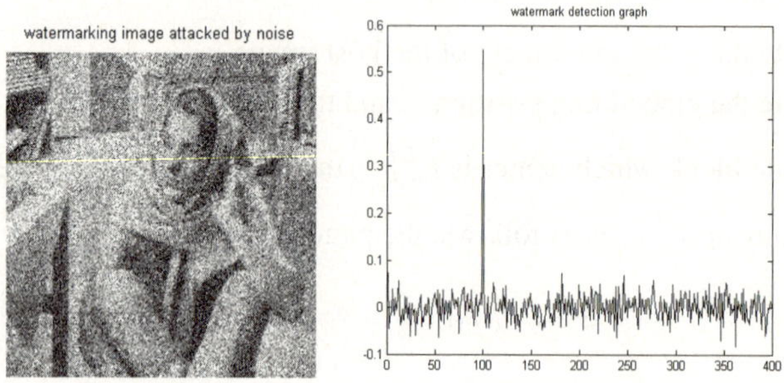

(a) (b)

Fig. 6. (a)white noise attack with mean 0, variance 0.05, (b)detection graph

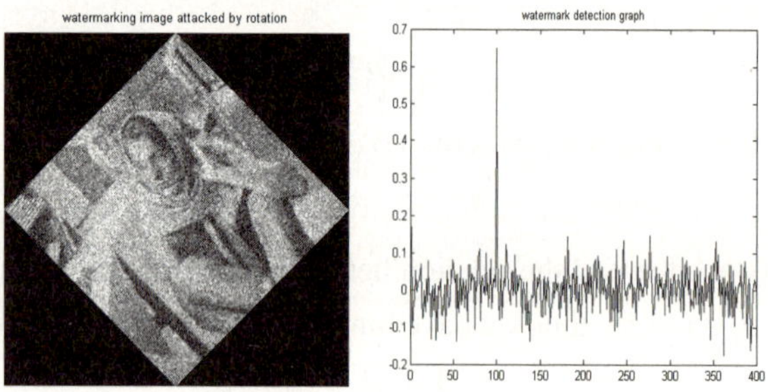

(a) (b)

Fig. 7. (a)rotate 45^0 attack, and add noise with mean 0,variance0.05,(b)detection graph

in different scale of wavelet moment domain; we use Error Bit Rate (EBR, the error bits of detection divided by the total number of watermark bits) to measure the performance of the watermark detector, and the smaller of EBR the better detection performance of the watermark.

The experiments show that our algorithm can detect the watermark with neural network blindly and can resist geometric attack or the combination of geometric and noise attack effectively.

Table 1. Test the anti-attack ability of the watermark with different embedding intensity

Intensity α	no attack	Gaissian noise	compression (32:1)	zoom out （1/2）	rotate 80^0	rotate 120^0	rotate+cropping 45^0+50×50+10×10	3×3median filter
	PSNR/EB	EB	EB	EB	EB	EB	EB	EB
0.001	62.6/0	0.064（4/64）	0.080	0.048	0.032	0.064	0.064	0.016
0.002	58.3/0	0.016（1/64）	0.032	0.016	0.016	0.016	0.016	0
0.004	52.0/0	0.016	0	0.016	0	0.016	0.016	0
0.005	48.6/0	0	0.016	0	0	0	0.016	0
0.007	46.5/0	0	0	0	0	0	0	0
0.009	44.6/0	0	0	0	0	0	0	0
0.01	42.7/0	0	0	0	0	0	0	0

Table 2. Test the anti-attack ability of the watermark with different wavelet moment

α=0.005	no attack	Gaissian noise	compression (32:1)	zoom out （1/2）	rotate 80^0	rotate 120^0	rotate+cropping 45^0+50×50+10×10	3×3median filter
	PSNR/EB	EB	EB	EB	EB	EB	EB	EB
db4	46.4/0	0	0.048	0.032	0	0.016	0.032	0.016
db10	48.6/0	0	0.016	0	0	0	0.016	0
cubic B-spline	46.6/0	0	0	0	0	0	0.032	0
Biorthogonal 9/3	48.2/0	0	0.032	0	0	0	0.032	0
Biorthogonal 9/7	48.8/0	0	0	0	0	0	0.016	0
Biorthogonal 12/4	46.2/0	0.016	0.032	0.016	0	0	0.032	0

Table 3. Anti-attack ability test of the watermark in different scale of wavelet moment

α=0.005	no attack	Gaissian noise	compression (32:1)	zoom out （1/2）	rotate 80^0	rotate 120^0	rotate+cropping 45^0+50×50+10×10	3×3median filter
m	PSNR/EB 值	EB	EB	EB	EB	EB	EB	EB
1	48.6/0	0	0	0	0	0	0.016	0
2	46.4/0	0.032	0.016	0.016	0	0	0.032	0
3	42.8/0	0.032	0.016	0.016	0	0	0.048	0

To compare our method with the literature[9].

Literature [9] embedded zero-mean pseudorandom sequence into the local features detected by the multi-scale Harris corner detector, but if the host image was largely cropped and the local features will be destroyed as shows in Figure 9, the watermark could be removed and will not be detected. On the other hand, in our method the watermark covers the entire host image in spatial domain, because the watermark is embedded into the wavelet-moment transform domain of the host image, figure 9 shows that our method can detected the watermark even the local features were almost cropped.

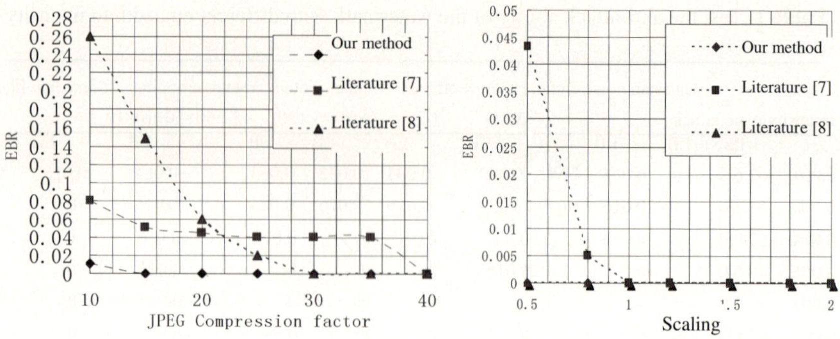

(a) (b)

Fig. 8. (a)Anti-attack ability under different compression ratio, (b) anti-attack ability under different scaling

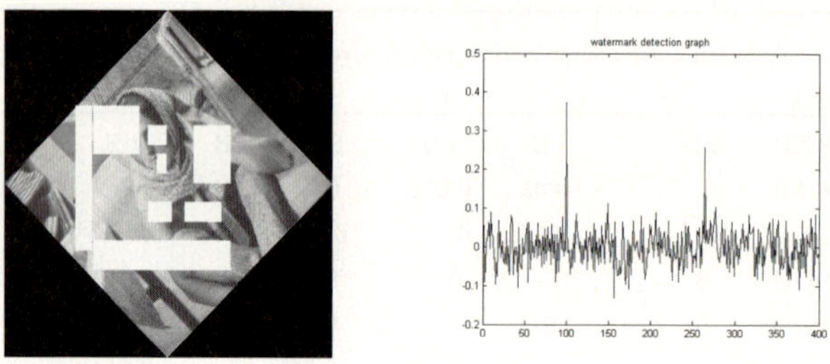

(a) (b)

Fig. 9. (a) rotate 45^0attack, and crop local features, (b) detection graph

5 Conclusions

We have researched the wavelet moment and inverse wavelet moment computation of the host image, and hide watermark into the wavelet moment domain, which performed better than the conventional wavelet domain method. We establish a BP network to learn the relation between the template and the watermark, and then use the trained network to detect the watermark blindly. The next step of research should focus on the location tamper and the design of special meaning watermark.

References

1. Tirkel, A.Z.: Electronic Watermark, Digital Image Computing Technology and Applications (DICTA 1993) [M], pp. 666–673. Macquarie Universeity (1993)
2. Matsui, K.T.: Video-Steganography: How to Secretly Embed a Signature in a Picture. Proceeding of Technological Strategies for Protecting Intellectual Property in the Networked Multimedia Environment [J]. Journal of the Interactive Multi-media Association Intellectual Property Project 1, 187–205 (1994)

3. Pereira, S., Voloshynovskiy, S., Pun, T.: Optimized Wavelet Domain Watermark Embedding Strategy Using Linear Programming[J]. In: proceeding of SPIE AeroSence, Wavelet Applications VII, Orlando, USA, pp. 26–28 (2000)
4. Yu, P.T., Tsai, H.H., Lin, J.S.: Digital Watermarking Based on Neural Networks for Color Images[J]. Signal Processing 81, 663–671 (2001)
5. Li, D.M., Wang, D.H., Chen, F.X.: Robust Watermark Algorithm Based on the Wavelet Moment and Neural Network Detection [J]. Journal of Computer Applications 26, 1833–1835 (2006)
6. Li, D.M., Wang, D.H., Yan, J.: Digital Watermarking Algorithm Based on Wavelet Moment Modulating [J]. Journal of Computer Applications 27, 1599–1602 (2007)
7. Yang, W.X., Zhao, Y.: Multi-bits Image Watermarking Resistant to Affine Transformations Based on Normalization. Signal Processing 20, 245–250 (2004)
8. Ping, D., Galatsanos, N.P.: Affine Transformation Resistant Watermarking Based on Image Normalization. In: Proceedings of International Conference on Image Processing, pp. 489–492 (2002)
9. Deng, C., Gao, X., Tao, D., Li, X.: Digital Watermarking in Image Affine Co-Variant Regions. In: IEEE International Conference on Machine Learning and Cybernetics, vol. 4, pp. 2125–2130 (2007)

Manifold Training Technique to Reconstruct High Dynamic Range Image

Cheng-Yuan Liou* and Wei-Chen Cheng

Department of Computer Science and Information Engineering
National Taiwan University
Republic of China
cyliou@csie.ntu.edu.tw

Abstract. This paper presents two manifold training techniques to reconstruct high dynamic range images from a set of low dynamic range images which have different exposure times. It provides the performance on noisy images.

Keywords: SIR algorithm, SOM, HDR image, High dynamic range.

1 Introduction

The ordinary digital camera is a low dynamic range device. The intensity of the environment scene may have a very wide dynamic range. It may exceed the camera's range limit, 255. Those intensity values which exceed the limit will be set to 0 or 255. Many efforts have been done to recover the high dynamic range (HDR) images with varying degrees of success. Many camera systems transform the sensor exposure value of the CCD (Charge-Coupled Device) through a non-linear function which is called the camera response function (CRF) and record the transformed value as the restored scene intensity. Since this function may not be available from the manufacturer, the key to obtain the HDR image is to recover the CRF. With this CRF one can produce the 'real' time-invariant irradiance of the scene. The method in [1] shows how to reconstruct the CRF from a series of images which are taken from the same scene with different exposures. It develops a parametric model for the CRF. The method in [2] uses a series of digital pictures and solves a set of linear equations to estimate the inverse of the CRF. Those pictures are taken with a fixed aperture and different known shutter speeds. Debevec's method [2] is not a parametric model. It assumes the inverse of the CRF is smooth. Mitsunaga [3] proposed an iterative method to adjust the coefficients of a high-order polynomial to fit the CRF. In this work, we devise two manifold training techniques to obtain the HDR without any irradiance information. One is based on SIR method [4,5], the second is a relaxation method similar to SOM [6]. The technique based on SIR will not use the continuous polynomial [3] and not use the smooth assumption [2]. Without matrix decomposition, the SIR and SOM are relatively easy in implementation.

* Corresponding author.

F. Sun et al. (Eds.): ISNN 2008, Part II, LNCS 5264, pp. 402–409, 2008.

2 Camera Model

Suppose there are N pictures with different exposures taken from the same scene. We assume these images are aligned. Therefore the same pixel location of all images should correspond to the same point of the scene. Each image has P pixels. For an 800×600 image, P is equal to 480000. The difference among images is the shutter speed setting. All images are taken with the same aperture setting. The different exposure time, Δt, can be obtained by varying the shutter speed. Let Δt_j denote the exposure time of the jth image. The sensor exposure value, X_{ij}, of the ith pixel in the jth image can be modeled as

$$X_{ij} = E_i \Delta t_j,\ i \in \{1, \ldots, P\}\,,\ j \in \{1, \ldots, N\}\,. \tag{1}$$

The E_i is the sensor irradiance of the ith pixel. The X_{ij} is the output of the ith CCD unit during the jth photo image. The unit of X_{ij} is Jm^{-2}. After cutting out all large and small intensity values of the outputs of the CCD that exceed the range limits, the rest values are passed through the CRF and are digitalized (quantized). A function f is used for representing the whole quantization process,

$$Z_{ij} = f\left(X_{ij}\right) = f\left(E_i \Delta t_j\right),\ Z_{ij} \in \{0, \ldots, 255\}\,. \tag{2}$$

The Z_{ij} is the intensity value which is finally stored in the storage device. We can rewrite (2) with an inverse function and take log of both sides,

$$\ln f^{-1}\left(Z_{ij}\right) = \ln X_{ij} = \ln E_i + \ln \Delta t_j\,. \tag{3}$$

Defining a $g = \ln f^{-1}$, (3) can be written as

$$g\left(Z_{ij}\right) = \ln E_i + \ln \Delta t_j\,. \tag{4}$$

3 Manifold Training

3.1 SIR Method

We use the SIR method [4,5] to solve the function g=$\ln f^{-1}$. For the ith pixel, the SIR energy function is

$$O_i = \frac{1}{4}\sum_{k=1}^{N}\sum_{r=1}^{N}\left(\left(g\left(Z_{ik}\right) - g\left(Z_{ir}\right)\right)^2 - \left(\ln \Delta t_k - \ln \Delta t_r\right)^2\right)^2, \tag{5}$$

where $Z_{ik}, Z_{ir} \in \{0, \ldots, 255\}$. In this energy, O_i, we assume that the image pixels at the same location have the same or similar irradiance value. Then we have the idea form

$$\begin{aligned} & g\left(Z_{ik}\right) - g\left(Z_{ir}\right) \\ &= \ln E_i + \ln \Delta t_k - \ln E_i - \ln \Delta t_r \\ &= \ln \Delta t_k - \ln \Delta t_r\,, \end{aligned} \tag{6}$$

where $k, r \in \{1, \ldots, N\}$. We plan to utilize the energy O_i to seek a solution $g(Z_{ij})$ that satisfies the idea form. One way to minimize (5) is to adjust $g(Z_{ik})$ and $g(Z_{ir})$ toward the gradient descent direction. Differentiate O_i with respect to $g(Z_{ik})$,

$$\frac{\partial O_i}{\partial g(Z_{ik})} = \sum_{r=1}^{N} \left((g(Z_{ik}) - g(Z_{ir}))^2 - (\ln \Delta t_k - \ln \Delta t_r)^2 \right) (g(Z_{ik}) - g(Z_{ir})) . \tag{7}$$

Differentiate O_i with respect to $g(Z_{ir})$,

$$\frac{\partial O_i}{\partial g(Z_{ir})} = -\sum_{k=1}^{N} \left((g(Z_{ik}) - g(Z_{ir}))^2 - (\ln \Delta t_k - \ln \Delta t_r)^2 \right) (g(Z_{ik}) - g(Z_{ir})) . \tag{8}$$

The SIR method is briefly described as follows:

1. Randomly initialize the function g with a discrete form.
2. Randomly select a pixel i from $\{1, \ldots, P\}$.
3. Update $g^{T+1}(Z_{ik}) = g^T(Z_{ik}) - \eta \frac{\partial O_i}{\partial g(Z_{ik})}$ and $g^{T+1}(Z_{ir}) = g^T(Z_{ir}) - \eta \frac{\partial O_i}{\partial g(Z_{ir})}$, for a pair (k, r) selected from the N images, $k, r \in \{1, \ldots, N\}$. η is the training rate and T is the number of training epoch.
4. Gradually decrease η and repeat Step 2-4.

Note that the discrete form g is much more flexible to operate with by the SIR method than the continuous high-order polynomial and smooth functions used in other methods. We expect that g will approximate the 'real' $\ln f^{-1}$ when the training time is long enough. The concept is illustrated in Fig. 1(a). Once the whole discrete form function $g(Z_{ij})$ is determined, we can calculate the irradiance, E_i, for every pixel by using the formula in [2],

$$\ln E_i = \frac{\sum_{j=1}^{N} w(Z_{ij}) (g(Z_{ij}) - \ln \Delta t_j)}{\sum_{j=1}^{N} w(Z_{ij})}, \; i \in \{1, \ldots, P\} . \tag{9}$$

The w is a weighting function. We set $w(x) = e^{\frac{-1}{80}\|x-127\|}$ in this paper. The recovered HDR image will include the irradiance maps, $\{E_i, \; i = 1, \ldots, P\}$. We will use the tone mapping [7] to display the HDR images in all experiments.

3.2 Relaxation Using the Self-organization Method

The CRF can also be obtained by a relaxation method similar to the self-organizing map (SOM) [6]. Suppose there are 256 cells regularly aligned on a straight line, the marked horizontal axis in Fig. 1(b). Each cell has a single weight. The mth weight value is $g(m)$. The neighborhood function h in SOM is set as

$$h(u, v) = e^{-\left(\frac{u-v}{\sigma}\right)^2}, \tag{10}$$

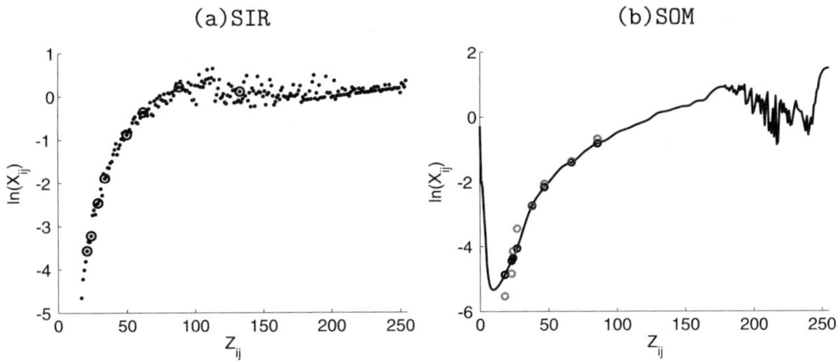

Fig. 1. (a) The concept of SIR training. The 256 points are obtained and updated during each training epoch. The eight black circle dots are those points for the ith pixel, $\{(Z_{ij}, g^{50}(Z_{ij})), j \in \{1, 2, \ldots, 8\}\}$. (b) The concept of SOM training. The grey circles are the values $\{(\ln E_i)^{50} + \ln \Delta t_j, j \in \{1, \ldots, N\}\}$.

where $u, v \in \{0, \ldots, 255\}$ and $\sigma \in R$. The σ is a parameter which controls the size of neighborhood. We suppose each pixel i has it own irradiance, E_i. In each epoch T, the current estimate of $\ln E_i$ is

$$(\ln E_i)^T = \frac{1}{N} \sum_j \left(g^T(Z_{ij}) - \ln \Delta t_j\right). \tag{11}$$

Based on this estimate, the g can be updated by

$$\Delta g^T(m) = h(m, Z_{ij}) \left[\left((\ln E_i)^T + \ln \Delta t_j\right) - g^T(m)\right], \tag{12}$$

and

$$g^{T+1}(m) = g^T(m) + \eta \Delta g^T(m), \; m \in \{0, \ldots, 255\}, \tag{13}$$

where η is the training rate. Fig. 1(b) shows an example of the self-organizing CRF curve during the 50th training epoch. The eight black circles denote the pairs, $\{(Z_{ij}, g^{50}(Z_{ij})), j = 1, \ldots, 8\}$ for a specific pixel i. The eight gray circles are the values $\{(\ln E_i)^{50} + \ln \Delta t_j, j \in \{1, \ldots, N\}\}$. The SOM method randomly selects a pixel i from the jth image, then uses the Z_{ij} to update the function g by (12) and (13). The training epoches are repeated until the curve g is converged. The irradiance maps, $\{E_i, i = 1, \ldots, P\}$, are then calculated using the (9).

4 Experiments

We have two sets of images. One is a scene of buildings and the other is of natural scenery. Fig. 2 and Fig. 3 plot the inverse CRFs using the building images and the natural scenery images. The red, green and blue lines (points) represent the three inverse CRFs of the RGB channel, respectively. The vertical axis, $\ln X_{ij}$, is defined in (3). Fig. 2 and Fig. 3 also show the HDR images obtained by the

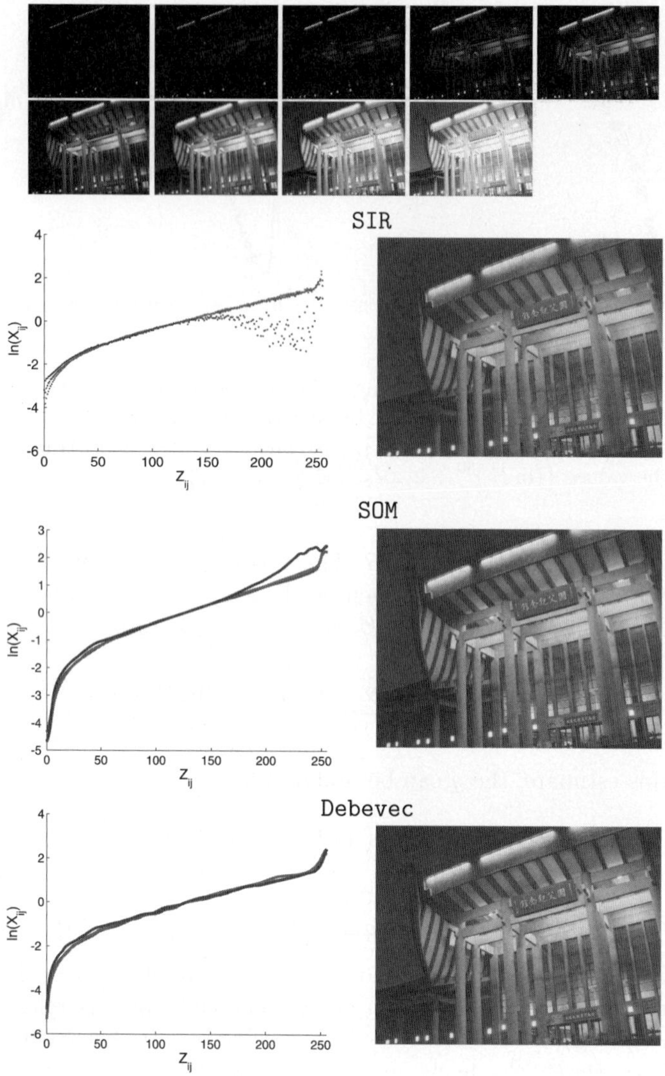

Fig. 2. These CRFs are obtained by using the same image set and the same camera settings. Three inverse CRFs obtained by SIR, SOM and Debevec's method. The nine small size images on top are taken with different exposure times in the night. The three HDR images on right are obtained by the three methods.

two methods. The result of Debevec's algorithm is presented for comparison. We randomly sample 300 pixels to solve the linear equations and the parameter λ is set to 15, $\lambda = 15$, in Debevec's algorithm [2]. Note that the SIR and SOM use all pixels to solve the CRF.

We show that the SIR method can recover the inverse CRF when the images are corrupted. The noisy images contain normal distributed noises whose

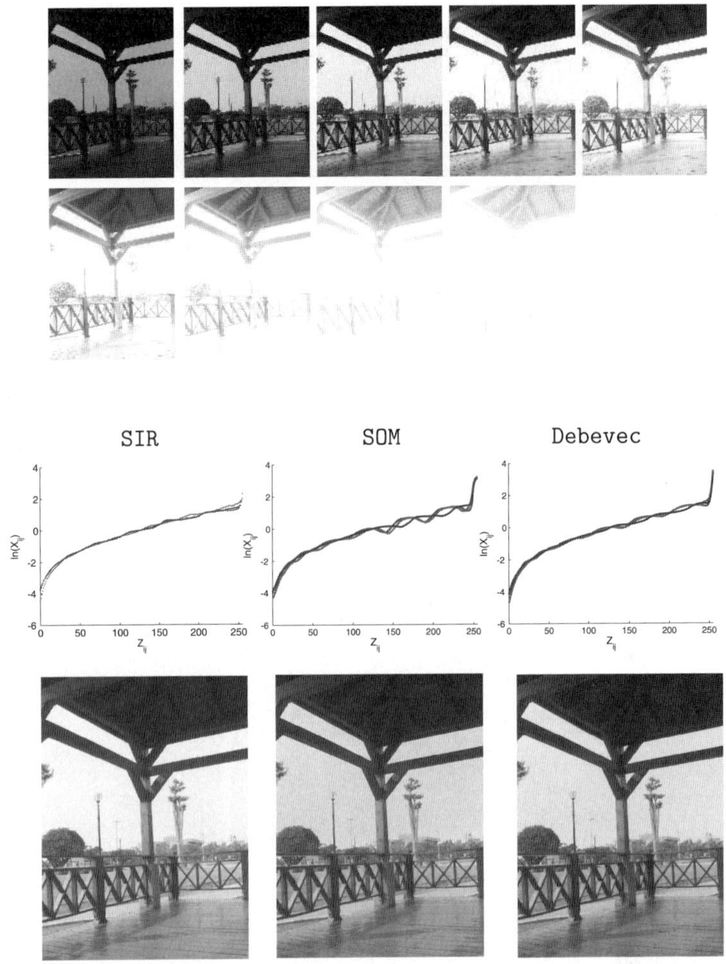

Fig. 3. The images are taken from the nature scenery. The nine small size images on top are the scenery images during sunlight. Three HDR images are reconstructed. The color of the sky in Debevec's image and in SOM image tends to be bluer than the SIR image.

variance is $\sigma = 0.089$, see Fig. 4(a). Fig. 4(b) is the CRF trained by the SIR method using the images in Fig. 4(a). Fig. 4(c) is obtained by using the method in [2]. The parameter λ in [2] is set to 15, $\lambda = 15$. Three hundred selected pixels are used in solving the linear equations. Figs. 4(d,e) show the HDR images constructed by using the noisy images in Fig. 4(a) and the CRFs in Fig. 4(b,c). Figs. 4(f,g) show the HDR images constructed from the images without noise, $\sigma = 0$, and the CRF in Fig. 4(b,c). Fig. 4(f) shows better image on the right-top dark corner. We also used the software [8] to solve [3] the noise images, it can not recover the three CRFs from noisy images.

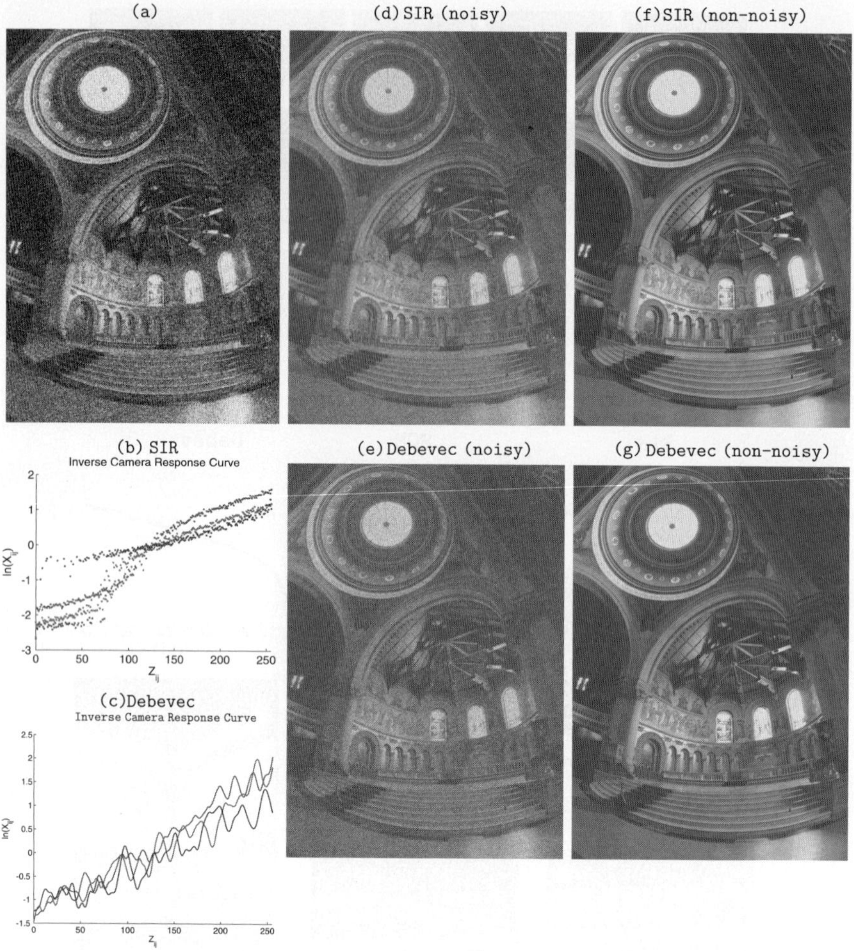

Fig. 4. (a) One noisy image in a series of photos. **(b)** The CRF reconstructed by SIR. **(c)** The CRF recovered by Debevec's method. **(d)** The HDR image by SIR. **(e)** The HDR image by Debevec's method. **(f)** The HDR image using the clean image, $\sigma = 0$, and the CRF in (b). **(g)** The HDR image using the clean image, $\sigma = 0$, and the CRF in (c).

In summary, this paper proposes two manifold techniques to reconstruct HDR images. The trained CRF can be used in estimating the irradiance value from a series of photos with different exposures. Furthermore, we test the performance of the SIR method using images with heavy noise. The experimental results show that the SIR method can recover the CRF from noisy images. The reconstructed HDR image has many potential applications, such as film, astronomy image, and medical imaging.

References

1. Mann, S., Picard, R.: On Being 'undigital' with Digital Cameras: Extending Dynamic Range by Combining Differently Exposed Pictures. In: IS&T's 46th Annual Conference, pp. 422–428 (1995)
2. Debevec, P.E., Malik, J.: Recovering High Dynamic Range Radiance Maps from Photographs. In: 24th Annual Conference on Computer Graphics and Interactive Techniques, pp. 369–378. ACM Press, New York (1997)
3. Mitsunaga, T., Nayar, S.K.: Radiometric Self Calibration. In: IEEE Conference on Computer Vision and Pattern Recognition, pp. 374–380 (1999)
4. Liou, C.-Y., Chen, H.-T., Huang, J.-C.: Separation of Internal Representations of the Hidden Layer. In: International Computer Symposium, Workshop on Artificial Intelligence, pp. 26–34 (2000)
5. Liou, C.-Y., Cheng, W.-C.: Manifold Construction by Local Neighborhood Preservation. In: Ishikawa, M., Doya, K., Miyamoto, H., Yamakawa, T. (eds.) ICONIP 2008, Part II. LNCS, vol. 4985, pp. 683–692. Springer, Heidelberg (2008)
6. Kohonen, T.: Self-organized Formation of Topologically Correct Feature Maps. Biological Cybernetics 43, 59–69 (1982)
7. Reinhard, E., Stark, M., Shirley, P., Ferwerda, J.: Photographic Tone Reproduction for Digital Images. In: 29th Annual Conference on Computer Graphics and Interactive Techniques, pp. 267–276 (2002)
8. `http://www1.cs.columbia.edu/CAVE/software/rascal/rrhome.php`

Face Hallucination Based on CSGT and PCA

Xiaoling Wang[1], Ju Liu[1], Jianping Qiao[1], Jinyu Chu[1], and Yujun Li[2]

[1] School of Information Science and Engineering, Shandong University,
Jinan 250100, P.R. China
[2] Hisense Group,
Qingdao 26607, P.R. China
juliu@sdu.edu.cn
http://202.194.26.100/liuju/index.htm

Abstract. In this paper, based on Circularly Symmetrical Gabor Transform (CSGT) and Principal Component Analysis (PCA), we propose a face hallucination approach. In this approach, all of the face images (both input face image and original training database) are transformed through CSGT at first and then local extremes criteria is utilized to extract the intrinsic features of the faces. Based on these features, we calculate Euclidean distances between the input face image and every face image in the original training database, and then Euclidean distances are used as criteria to choose the reasonable training database. Once the training database is chosen, PCA is applied to hallucinate the input face image as the linear combination of the chosen training images. Experimental results show that our approach can choose training database automatically according to the input face image and get high quality super-resolution image.

Keywords: Face Hallucination, CSGT, PCA, Training Database.

1 Introduction

Super-resolution is a kind of technique which can generate a high-resolution image from a set of low-resolution images. In several important applications like video surveillance and medical applications, images with high resolution can offer more information. Hence, super-resolution has become an active research area in recent years.

Typically, super-resolution algorithms can be classified into many categories based on different criteria such as single/multiple images reconstruction, frequency/spatial domain algorithms. Frequency domain algorithm was first proposed by Tsai and Huang [1], and this algorithm was based on the shift and aliasing properties of Fourier transforms. After that, many researchers have proposed spatial domain algorithms like reconstruction-based [2,3,4] and learning-based algorithms [5,6,7] etc.

Among these algorithms, learning-based algorithms seem to be the most promising ones. Recently, a great amount of work has been done in this field. Freeman [6] proposed an example-based learning algorithm to restore the high-resolution

F. Sun et al. (Eds.): ISNN 2008, Part II, LNCS 5264, pp. 410–418, 2008.

image up to a zoom of 8. In [8], manifold learning method particularly locally linear embedding (LLE) was applied to reconstruct the high-resolution image. And similar local geometry was utilized to build the connections between high- and low-resolution images. In [9], PCA was used to model the global face image (low-frequency parts) and patch-based Markov network was applied to estimate the missing details (high-frequency parts). Xiaogang Wang [10] proposed a face hallucination approach using PCA to represent the structural similarity of face images and the high-resolution face image was synthesized from the linear combination of the training samples. Nevertheless, the hallucinated face image restored by this approach has strong dependence on the training database. Given the same low-resolution face image, different training databases may give different hallucinated results. Little work has been done to solve this problem. [11] applied histogram matching to choose the training database for manifold learning and this approach can choose image which is the most relevant to the input image. However, for most learning-based algorithms in face hallucination, how to choose the reasonable training database from face images of different expressions is the prime problem encountered in practical applications.

In this paper, we propose an approach where the training database is chosen automatically from the original training database based on Euclidean distances calculated upon Circularly Symmetrical Gabor Transform. Once the training database is chosen, PCA is employed to realize the hallucination. Experimental results show that our approach can make the chosen training database more reasonable and get better hallucinated face images. What's more, since the training database is selected automatically, our approach is helpful for real-time process.

2 Circularly Symmetrical Gabor Transform

Circular Symmetrical Gabor Transform (CSGT) is defined as follows:

$$\psi(k, \mathbf{r}) = \frac{k^2}{\sigma^2} exp(-\frac{k^2 \mathbf{r}^2}{2\sigma^2}) e^{ik|\mathbf{r}|}. \tag{1}$$

where $k = \pi/(\sqrt{2})^i$, $i = 1, 2, \ldots$, $\sigma = \sqrt{2 \ln 2}(\frac{2^{\Delta w}+1}{2^{\Delta w}-1})$, $\mathbf{r} = (x, y)$ is the coordinate vector in spatial domain and $\Delta w = 1$ in our experiment. Given a face image $\mathbf{f}(x, y)$, its CSGT can be written as follows:

$$g_w(x, y) = r_w(x, y) + j i_w(x, y) = a_w(x, y) exp(j \varphi_w(x, y)). \tag{2}$$

where $w = 1, 2, \ldots, K$ is the scale of the transform and in our experiment, we choose the first scale ($w = 1$). Due to the good locality in frequency/spatial domain and perfect agreement with mammalian visual characteristics, CSGT has been widely used in texture segmentation, classification, object matching etc. In [12], CSGT was first applied to face recognition and good recognition results were obtained.

3 Face Hallucination Based on CSGT and PCA

3.1 The Necessity of Choosing Reasonable Training Database

In learning-based algorithms, high-resolution image is generated from a single low-resolution image with the help of a set of one or more training images from scenes of the same or similar types. PCA based face hallucination [10] is an effective learning-based algorithm to build the connections between high- and low-resolution faces using the structural similarity of face images. However, this approach is very sensitive to the training database, as shown in Fig. 1.

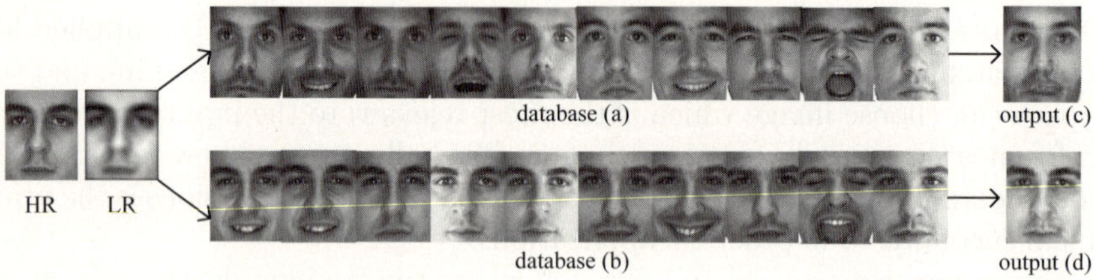

Fig. 1. This figure shows the influence of training database on hallucinated results. We give two training databases (a) and (b) to hallucinate the same input face image and the results are (c) and (d).

To solve this problem, in our face hallucination approach, Circularly Symmetrical Gabor Transform (CSGT) is employed to choose the reasonable training database automatically and then PCA is utilized for face hallucination.

3.2 Database Training through CSGT

For each training person x_i, $i = 1, 2, \ldots, M$, in the original training database, he/she has N_i images of different expressions named $x_{i_1}, x_{i_2}, \ldots, x_{i_{N_i}}$. Given an input low-resolution face image y_0, a more reasonable training database should contain more images which are similar to y_0. For training person x_i, steps of the selection process are listed as follows.

1. Down-sample the face images $x_{i_1}, x_{i_2}, \ldots, x_{i_{N_i}}$ into low-resolution images $y_{i_1}, y_{i_2}, \ldots, y_{i_{N_i}}$, and $y_{i_1}, y_{i_2}, \ldots, y_{i_{N_i}}$ have the same size as y_0
2. Transform $y_0, y_{i_1}, y_{i_2}, \ldots, y_{i_{N_i}}$ through CSGT and the transform results are $Gy_0, Gy_{i_1}, Gy_{i_2}, \ldots, Gy_{i_{N_i}}$. $Gy_0, Gy_{i_1}, Gy_{i_2}, \ldots, Gy_{i_{N_i}}$ refer to the amplitudes of CSGT, which is $a_w(x, y)$ in (2)
3. Divide $Gy_0, Gy_{i_1}, Gy_{i_2}, \ldots, Gy_{i_{N_i}}$ into small patches and local extremes criteria is utilized to extract the local features patch by patch [12]. For each face image, features extracted from their patches are put into one vector to represent the whole image
4. Based on these extracted features, Euclidean distances are calculated between Gy_0 and $Gy_{i_1}, Gy_{i_2}, \ldots, Gy_{i_{N_i}}$. Then these Euclidean distances are

ranged from the smallest to the largest, and images with larger Euclidean distances are excluded from the original training database while the ones with smaller Euclidean distances are saved to build the chosen training database. This is reasonable because Euclidean distances represent the similarity between different face images

For each training person in the original training database, follow Step1-Step4 and images which are not similar to the input face image will be excluded from the original training database. Therefore, a more reasonable training database is chosen through CSGT.

3.3 Face Hallucination

Given an input low-resolution face image, we apply the approach in Sect.3.2 to find its corresponding training database. Once the reasonable training database is chosen, the desired high-resolution face image is rendered through PCA based face hallucination [10]. The block diagram of the proposed approach is shown in Fig. 2.

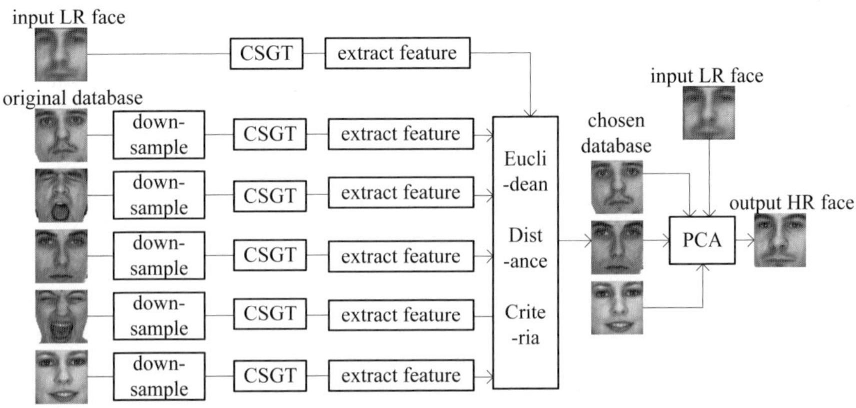

Fig. 2. This figure shows the block diagram of our proposed approach and we just give five images to represent the original training database for convenience

It can be seen that the whole process of the proposed approach is completely automatically finished by machine, which means that given any input low-resolution face image, we don't have to select its training database from large amount of training images by hand. What's more, the approach can also promise the selected training database to be the reasonable one and hence the better hallucinated results.

4 Experimental Results

In this section, two related experiments are conducted: one to verify that the approach proposed in Sect.3.2 can choose the reasonable training database, and

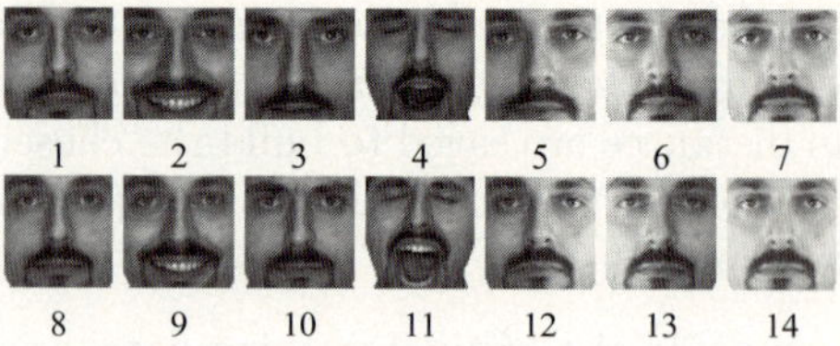

Fig. 3. Each person in the original database has 14 expressions: number 1 to number 14

the other one to show the hallucinated results obtained by the proposed approach in Sect.3.3. Our experiments are based on AR database [14]. For each person, he/she has 14 expressions and we use number 1 to number 14 to represent them, as shown in Fig. 3.

4.1 Training Database Selection through CSGT

In this experiment, 20 persons with expression 1 are chosen as the test images and the original training database has 56×14 images. Besides, we cut every face image to 64×64 pixels to be the low-resolution image and equalization proposed in [13] is utilized to decrease the CSGT's sensitivity to luminance. After that, the experiment totally follows the steps in Sect.3.2. In Step 3, the patch size is 8×8 and the first ten percent of amplitudes are saved to represent the local features when using local extremes criteria. In Step 4, the first ten expressions of each training person are saved to build the chosen training database according to the ranged Euclidean distances. For the test images, images of expression 4 and 11 are the most dissimilar ones and will definitely ruin the hallucinated results. And in the original training database, there are 112 (56×2) images of expression 4 and 11. Our work is to exclude these images as many as possible and therefore to form a more reasonable training database. Table 1 shows the number of images of expression 4 and 11 which have been successfully excluded from the original training database for each test person. And number 1 to number 20 is used to represent the 20 test persons. Euclidean distances between one test person and training person1-person5 are listed in Table 2, where the Euclidean distances are ranged from the smallest to the largest.

Table 1. Number indicates how many images of expression 4 and 11 have been successfully excluded from the original training database for each test person and rate is calculated through number/112

person	1	2	3	4	5	6	7	8	9	10
number	112	80	77	103	100	93	106	88	13	106
rate	100%	71.43%	68.75%	91.96%	89.29%	83.04%	94.64%	78.57%	11.61%	94.64%
person	11	12	13	14	15	16	17	18	19	20
number	109	97	98	109	31	89	79	56	96	94
rate	97.32%	86.61%	87.50%	97.32%	27.68%	79.46%	70.54%	50.00%	85.71%	83.93%

Table 2. This table shows the Euclidean distances between one test person and training person1-person5 with 14 expressions for each training person. The first ten expressions of each training person are saved to build the chosen training database according to the ranged Euclidean distances. num represents that each training person has 14 expressions, number 1 to number 14. ED1-ED5 represents the corresponding Euclidean distances between test person and training person1-person5.

order	num	ED1	num	ED2	num	ED3	num	ED4	num	ED5
1	5	801.97	1	819.89	1	813.59	8	817.93	1	861.36
2	13	896.65	3	835.07	3	863.19	14	899.29	8	867.29
3	3	940.85	10	840.08	10	868.54	9	933.10	12	906.69
4	1	952.10	8	893.14	8	889.55	5	995.35	5	920.64
5	8	953.78	5	899.33	13	899.93	1	1015.9	3	922.76
6	12	963.30	12	935.73	5	916.58	7	1025.6	10	945.47
7	10	966.43	13	947.21	12	958.14	12	1032.7	2	958.59
8	7	966.63	7	954.19	2	970.81	13	1037.3	6	1006.1
9	6	991.60	14	980.90	9	996.94	10	1038.3	7	1013.9
10	14	1046.0	9	988.31	14	1031.7	3	1067.9	13	1016.9
11	2	1067.4	2	1032.5	7	1064.1	2	1078.4	14	1026.0
12	9	1081.7	11	1058.2	6	1082.7	6	1106.3	9	1076.3
13	11	1193.3	6	1080.7	11	1129.6	4	1227.1	11	1120.5
14	4	1208.3	4	1225.8	4	1205.0	11	1238.6	4	1238.3

As shown in Table 1, for most of the test images, images of expression 4 and 11 are mostly excluded. And Table 2 shows that our approach can not only exclude images of expression 4 and 11, but also exclude some images of expression 2 and 9. These tables indicate that given an input low-resolution face image, a reasonable training database can be chosen effectively by our approach. However, there are still some exceptions like test image 9, 15 and 18 in the Table 1. This flaw is caused by CSGT's sensitivity to the postures and the influences of beard and hair. And these test images are shown in Fig. 4.

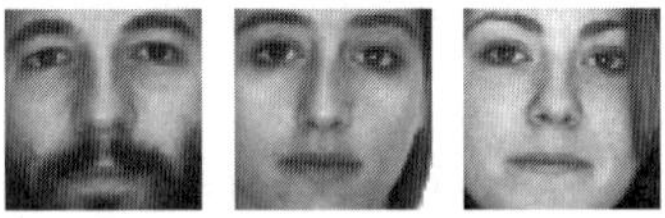

(a) (b) (c)

Fig. 4. (a) test image 9 (b) test image 15 (c) test image 18

4.2 Hallucinated Results

In this experiment, we align all the images into fixed positions like eyes and mouths and cut them to 144 × 120 pixels as high-resolution face images. We choose 20 persons with expression 1 to be test images and degrade images through down-sampling and Gaussian blur. The low-resolution images are of

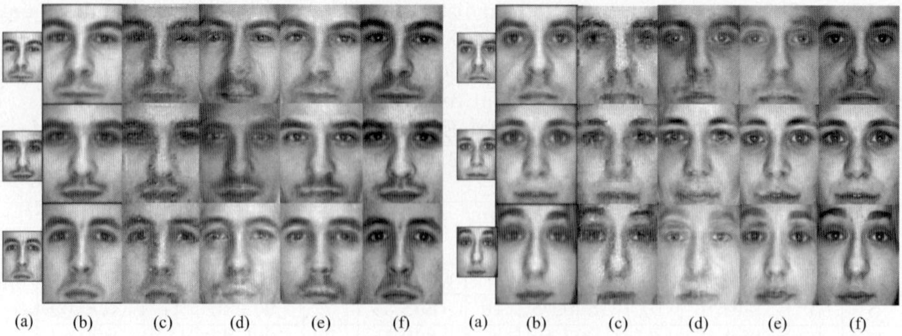

Fig. 5. This figure shows the hallucinated results. (a) the input low-resolution face image (b) bicubic interpolation (c) neighbor embedding (d) PCA based face hallucination (e) our proposed approach (f) the original high-resolution face image.

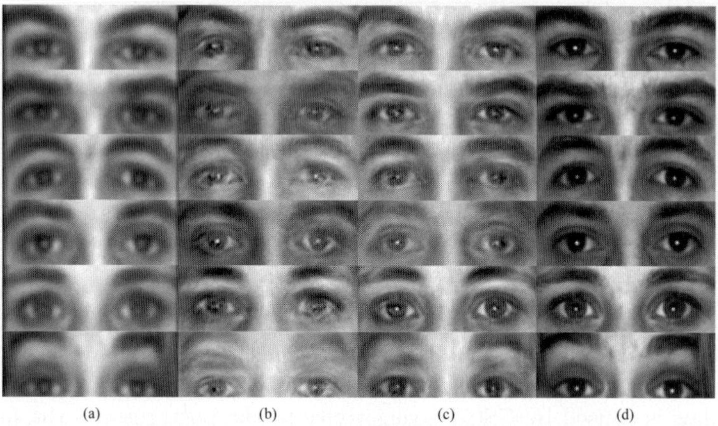

Fig. 6. This figure shows the eyes of persons in Fig. 5. (a) bicubic interpolation (b) PCA based face hallucinatio, (c) our proposed approach (d) the original image. Neighbor embedding is not shown for its obvious blocking effect.

size 72×60. For comparison, hallucinated results obtained by bicubic interpolation, approaches in [8] and [10] are given. For neighbor embedding in [8], only two images are chosen as the training images. For PCA based face hallucination in [10] and our approach, the original training database (14 expressions for each training person) is the same for each test person. For PCA based face hallucination in [10], the training database is chosen at random from the original training database (10 expressions for each training person) while in our approach, the training database is chosen based on the approach in Sect.3.2 (the first 10 expressions for each training person). The hallucinated results are shown in Fig. 5 and details of the hallucinated results such as eyes and mouths are shown in Fig. 6 and Fig. 7.

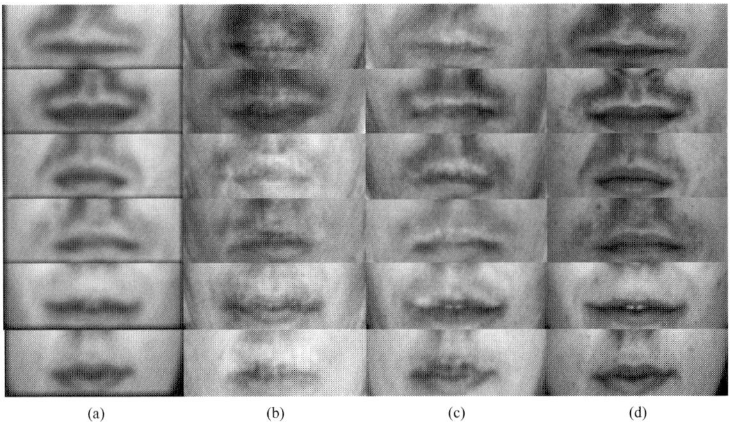

Fig. 7. This figure shows the mouths of persons in Fig. 5. (a) bicubic interpolation (b) PCA based face hallucination (c) our proposed approach (d) the original image. Neighbor embedding is not shown for its obvious blocking effect.

From Fig. 5, Fig. 6 and Fig. 7, we can see that the implementation of [8] is completed patch by patch, therefore the hallucinated results are affected by blocking effect. Due the disturbances of expressions 2, 4, 9 and 11 in the training database, hallucinated results based on [10] are not ideal for the blurring on the faces, and do not look like the original images especially on some details like eyes, mouths and noses. On the contrary, in our approach, most of these expressions (especially expression 4 and 11) are excluded from the training database, and therefore better hallucinated results are obtained. The hallucinated results in our approach are satisfying considering the validity and efficiency on the selection of training database.

5 Conclusion

Learning-based algorithms are restricted by their dependence on the training database. In this paper, CSGT and PCA are applied for face hallucination. CSGT is used as a tool to choose the training database and PCA is applied to hallucinate the high-resolution face image. In our approach, reasonable training database can be chosen automatically according to different input face image, which not only promise the high-quality hallucinated face images but also save the labor on choosing training database by hand in practical applications. For persons whose training database can not be chosen effectively through CSGT, the improvements of hallucinated results are limited, and this needs to be studied in our future work.

Acknowledgments. This work is supported by Program for New Century Excellent Talents in University, Education Ministry of China (No. NCET-05-0582),

Specialized Research Fund for the Doctoral Program of Higher Education (No. 20050422017), Natural Science Foundation of Shandong Province (No.Y2007G04), SRF for ROCS, SEM (No:[2005]55) and the Excellent Youth Scientist Award Foundation of Shandong Province (No.2007BS01023). The corresponding author is Ju Liu. (email: juliu@sdu.edu.cn)

References

1. Tsai, R.Y., Huang, T.S.: Multipleframe Image Restoration and Registration. In: Advances in Computer Vision and Image Processing, CT, pp. 317–339. JAI Press Inc., Greenwich (1984)
2. Irani, M., Peleg, S.: Improving Resolution by Image Registration. Computer Vision, Graphic, and Image Processing 53(5), 231–239 (1991)
3. Stark, H., Oskoui, P.: High Resolution Image Recovery from Image-plane Arrays, Using Convex Projections. Journal of Optical Society of America 6(11), 1715–1726 (1989)
4. Schulz, R.R., Stevenson, R.L.: Extraction of High-Resolution Frames from Video Sequences. IEEE Transactions on Image Processing 5(6), 996–1011 (1996)
5. Baker, S., Kanade, T.: Hallucinating Faces. In: 4th IEEE International Conference on Automatic Face and Gesture Recognition, pp. 83–88 (2000)
6. Freeman, W.T., Jones, T.R., Pasztor, E.C.: Example-based Super-resolution. IEEE Computer Graphics and Applications 22(2), 56–65 (2002)
7. Capel, D., Zisserman, A.: Super-resolution from Multiple Views using Learnt Image Models. In: 12th IEEE Computer Society Conference on Computer Vision and Pattern Recognition, vol. 2, pp. 627–634 (2001)
8. Chang, H., Yeung, D.Y., Xiong, Y.: Super-resolution through Neighbor Embedding. In: 14th IEEE Computer Society Conference on Computer Vision and Pattern Recognition, vol. 1, pp. 275–282 (2004)
9. Liu, C., Hum, H., Zhang, C.S.: A Two-step Approach to Hallucinating Faces: Global Parametric Model and Local Non-parametric Model. In: 12th IEEE Computer Society Conference on Computer Vision and Pattern Recognition, vol. 1, pp. 192–198 (2001)
10. Wang, X.G., Tang, X.O.: Hallucinating Face by Eigentransformation. IEEE Transactions on Systems, Man and Cybernetics 35(3), 425–434 (2005)
11. Ming, C.T., Zhang, J.P.: An Improved Super-Resolution with Manifold Learning and Histogram Matching. In: Zhang, D., Jain, A.K. (eds.) ICB 2005. LNCS, vol. 3832, pp. 756–762. Springer, Heidelberg (2005)
12. Wang, H.Y.: Face Recognition Approaches Based on Linear Subspace and Circularly Symmetrical Gabor Transforms. Ph.D. Thesis, Shandong University (2007)
13. Leng, Y., Wang, H.Y., Guo, K., Wang, Z.F.: Face Recognition Based on Bit Planes and Generalized PCA. Computer Engineering 33(10), 203–205 (2007)
14. Martinez, A.M., Benavente, R.: The AR Face Database. CVC Technical Report #24 (1998)

Complex Effects Simulation Based Large Particles System on GPU

Xingquan Cai, Jinhong Li, and Zhitong Su

College of Information Engineering, North China University of Technology,
Beijing, 100144, China
xingquancai@126.com

Abstract. In this paper, we present a new method to implement complex effects simulation based large particles system on GPU. Our method could be used in 3D games to simulate the photorealist effects. Our particles system is a state-preserving simulation system. We update the dynamic attributes and render the particles in batches on GPU. The most important is that we handle the collisions between particles and other models on GPU. We also compare with the CPU particles system method and implement complex effects on GPU. Finally, we give the implementation results.

Keywords: GPU (Graphics Processing Unit), complex effects simulation, particles system, state-preserving simulation, collision detection.

1 Introduction

The simulation of natural sceneries becomes a hot topic in the research field of computer graphics. Usually, the particles system could be used to simulate complex natural sceneries on CPU. However, if the number of particles is above 10K, the particles system on CPU is difficult to run on real-time. It is required that plenty of particles more than 10K should be need in simulate system of photorealist natural sceneries effects. Today, with the development of GPU, we could deal with complex computing and programming on GPUs.

In this paper, we present a new method to implement efficient complex effects simulation based large particles system on GPU. Our particles system is a state-preserving simulation system. We update the dynamic attributes and render the particles in batches on GPU. The most important is that we handle the collisions between particles and other models on GPU. Our method could be used in 3D games to simulate the photorealist effects.

In this paper, after exploring the related work of particles system, we present our particles system method on GPU. In Section 4, we show the results using our method before we draw the conclusion in Section 5.

2 Related Work

Particles system has a long history in video games and computer graphics. In 1983, Reeves [1] first described the basic motion operations and the basic data representing

F. Sun et al. (Eds.): ISNN 2008, Part II, LNCS 5264, pp. 419–428, 2008.

a particle, which both have not been altered much since being presented. The latest descriptions of CPU-based particle system for using in video games and photorealist natural sceneries have been done by Wang et al [2], Liu et al [3], Guan et al [4], Burg [5].

With the development of GPU, several forms of physical simulation have recently been developed for modern GPU. In 2003, Harris [6] has used GPU to perform fluid simulations and cellular automata with similar texture-based iterative computation. Recently, Schneider et al [7], Livny et al [8], and Eric et al [9] have used GPU to render large scale terrain scene. Christopher et al [10] also provide the method of real-time mesh simplification using GPU. As GPU could deal with complex computing so fast, we want to implement particles system on GPU.

Some particles systems have been implemented with vertex shaders (also called vertex programs) on programmable GPUs in NVIDIA SDK [11]. However, these particles systems are stateless. They do not store the current attributes of the particles, including current position, current velocity, etc. To determine a particle's position, the system needs to find a closed form function for computing the current position only from initial values and the current time [12,13]. Stateless particles are not meant to collide with the environment. They are only influenced by global gravity acceleration and could be simulated quite easily with a simple function. As a consequence, such particles system can hardly react to a dynamic environment.

So we provide a state-preserving particles system method in this paper. We also implement collision detection for state-preserving particle system on GPU.

3 Data Storage of Particles on GPU

Position is one of the most important attributes of a particle. In our system, positions of all active particles are stored in a floating point texture with three color components that will be treated as x, y and z coordinates. Each texture is conceptually treated as a one-dimensional array, texture coordinates representing the array index. However, the actual textures need to be two-dimensional for the size restrictions of current hardware. The texture itself is also a render target, so it can be updated with the computed positions. In the stream processing model [14], which is the programming model in graphics hardware, it represents either the input or the output data stream. As a texture cannot be used as input and output at the same time, we use a pair of these textures and a double buffering technique to compute new data from the previous values.

If other particle attributes, such as velocity, orientation, size, color, and opacity, were to be simulated with the iterative integration method, they would need texture double buffers as well. And other static attributes just need one texture buffer.

4 State-Preserving Particles System Method on GPU

The following subsections describe the algorithm of our state-preserving particles system on GPU in detail.

The algorithm consists of five basic steps:

1. Processing birth and death
2. Updating attributes of particles

3. Collision detection with surface of other models
4. Transferring texture data to vertex data
5. Rendering particles in batches.

4.1 Processing Birth and Death of Particles

The particles system must process the birth of a new particle, such as its allocation, the death of a particle and its deallocation. The birth of a particle requires associating new data with an available index in the attribute textures. Since allocation problems are serial by nature, this can not be done efficiently with a data-parallel algorithm on the GPU.

In our method, the death of particle is processed independently on the CPU and GPU. The CPU registers the death of a particle and adds the freed index to the allocator. The GPU does an extra pass over the particle data: The death of a particle is determined by the time of birth and the computed age. The dead particle's position is simply moved to invisible areas, e.g. infinity. As particles usually fade out or fall out of visible areas anyway at the end of their lifetime, the extra pass rarely really needs to be done. It is a basically clean-up step to increase rendering efficiency.

4.2 Updating Attributes of Particles

The most important attributes of a particle are its position and velocity. So we just deal with position and velocity of particles. The actual program code for the attributes simulation is a pixel shader which is used with the stream processing algorithm. The shader is executed for each pixel of the render target by rendering a screen-sized quad. The current render target is set to one of the double buffer attribute textures. The other texture of the double buffer is used as input data stream and contains the attributes from the previous time step. Other particle data, either from inside the attribute textures or as general constants, is set before the shader is executed.

Updating Velocities: There are several velocity operations that can be combined as desired: global forces (e.g. gravity, wind), local forces (attraction, repulsion), and velocity dampening. For our GPU-based particles system these operations need to be parameterized via pixel shader constants. Their dynamic combination is a typical problem of real-time graphics. Comparable to the problem of light sources and material combinations, it could be solved in similar ways. Typical operation combinations are to be prepared in several variations beforehand. Other operations could be applied in separate passes, as all operations are completely independent.

Global and local forces are accumulated into a single force vector. The acceleration can then be calculated with Newtonian physics as Equation (1). In Equation (1), a is the acceleration vector, F is the accumulated force and m is the mass of the particle.

The velocity is then updated from the acceleration with a simple Euler integration in the form of Equation (2). In Equation (2), v is the current velocity, $\overline{v}$ is the previous velocity and $\triangle t$ is the time step.

$$a = \frac{F}{m} \tag{1}$$

$$v = \overline{v} + a \cdot \Delta t \tag{2}$$

Updating Positions: Euler integration has already been used to update the velocity by using the acceleration. The computed velocity can be applied to all particles in just the same way. We use Equation (3) to update the position. In Equation (3), p is the current position and $\overline{p}$ is the previous position.

$$p = \overline{p} + v \cdot \Delta t \tag{3}$$

4.3 Collision Detection with Surface of Other Models

Collision Detection with surface of other models is the most important step in our method. In practical projects, particle may collide with regular surface models, such as plane, bounding sphere surface, ellipsoid surface, etc. Particle also may collide with irregular surface models, like terrain, Stanford Bunny, Dragon, and so on. Collision with regular models or irregular models, we can compute the Collision Detection using the normal vector and the tangential vector of the collision point.

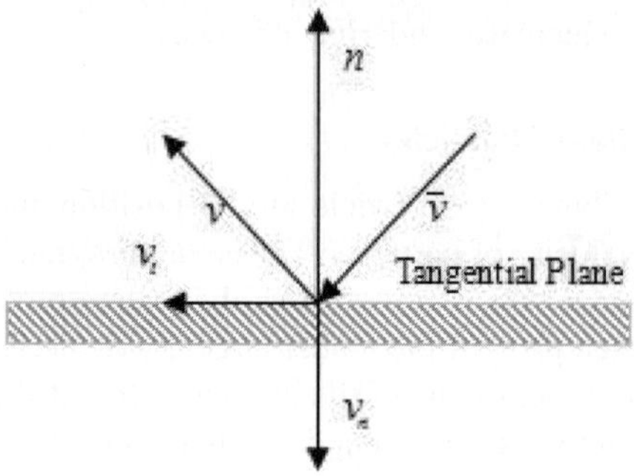

Fig. 1. Collision detection with surface of other models

In Fig. 1, the normal vector of Tangential Plane is n and the previous velocity is $\overline{v}$. If $\overline{v} \cdot n \geq 0$, the particle could not collide with the surface. If $\overline{v} \cdot n < 0$, the particle may collide with the surface. So the previous velocity $\overline{v}$ could be divided into the normal component of the velocity v_n and the tangential one v_t. v_n and v_t could be computed by Equation (4) and Equation (5). And Equation (6) show the current velocity v under the ideal condition. If we concern the friction μ and the resilience ε, we also can computed v using Equation (7). All the computing of collision detection is dealt with in fragment shader of GPU.

$$v_n = \frac{\overline{v} \cdot n}{|\overline{v}| \cdot |n|} \tag{4}$$

$$v_t = \overline{v} - v_n \tag{5}$$

$$v = v_t - v_n \tag{6}$$

$$v = (1 - \mu)v_t - \varepsilon v_n \tag{7}$$

4.4 Transfer Texture Data to Vertex Data

Before rendering particles, we should copy the particle data from the floating point texture to vertex data. Copying the particles data from a texture to vertex data is a hardware feature that is only just coming up in PC GPUs. OpenGL [15] offers vertex textures with ARB_vertex_shader extension. OpenGL also provides two functions, glReadBuffer and glReadPixels. These functions could copy the particle data from the floating point texture to vertex data.

4.5 Rendering Particles in Batches

The particles can be rendered as point sprites, triangles or quads. If the particle is rendered as triangles or quads, the particle may have three vertices or more vertices. So it is required that we should recompute the vertices position of particle before rendering. Because we need complex effects, we must select the quads method.

In order to use the ability of GPUs rendering triangles in batches and be convenient to manage the particle system, we import the theory of advanced particles system [5]. We divide the particles system into three layers, including Particles Manager, Particles Cluster and Particles. Particles Cluster is the batches of particles having the similar attributes, such as velocity, color, texture, etc. Particles Manager manages the Particles Clusters and is responsible for the birth of a new Particles Cluster, the death of a Particles Cluster and its deallocation. In this way, we could use the ability of GPUs rendering triangles in batches and implement plenty of particles in video games and photorealist natural sceneries.

5 Results

We have implemented our algorithm. Our implementations are running on a Intel PIV 2.8GHz computer with 1GB RAM, and NVIDIA GeForce7650 graphics card with 256M RAM, under Windows XP, Visual C++ 6.0, OpenGL and Cg 2.0 environment, while running smoothly in real time. The rendering system has a real viewport size of 1024×768 .

5.1 Comparison with CPU Particles System

We implement a particles system on CPU and a particles system on GPU to simulate flowing magma. There is only one Particles Cluster in each system. The particle just has the gravity and we do not concern other forces and the collision. At the same number of particles, we note the rendering frame rate. In order to ensure the objectivity of the experiment data, we sample continuous 2000 frames, note the FPS (Frames Per Second) and compute the average FPS.

Just as Fig. 2 shows, in our experiment, when the number of particles is 100,000, the FPS of particles system on GPU is above 60. But at the similar condition, the FPS of particles system on CPU is below 18. When the number of particles is 200,000, the FPS of particle system on GPU is 36, but the FPS of particles system on CPU is below 8. All these prove that particles system on GPU is higher performance than particles system on CPU.

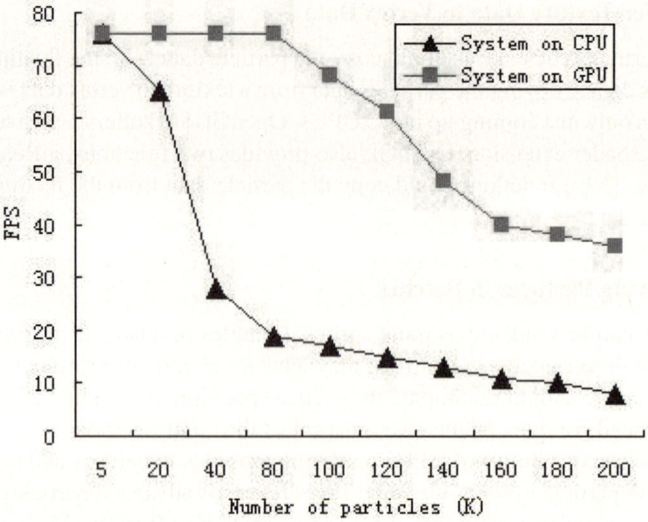

Fig. 2. Comparison between particles system on CPU and particles system on GPU

5.2 Collision Detection with Surface of Other Models on GPU

We implement a particles system on GPU to simulate flowing magma. There is also only one Particles Cluster in each system. The particle also just has the gravity and we do not concern other forces. But we concern the collisions between particles and other models.

As Fig. 3 shows, in our implementation, the particle collides with the blue sphere, the red sphere and the flat plane in turn. All the computing of Collision Detection is dealt with on GPU. In our implementation, there are 65,536 particles, but our system is running smoothly in 28 fps. We also implement the Collision Detection on CPU with the same conditions, but the system is running below 10fps and is hard to render smoothly.

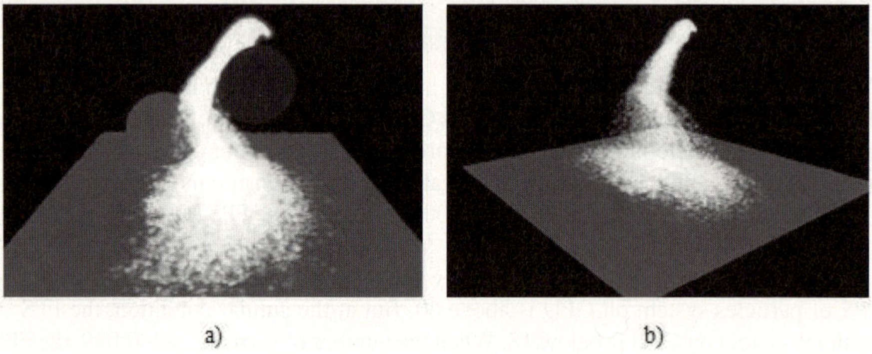

Fig. 3. Collision with blue sphere, red sphere and flat plane in turn. a) Rendering two spheres. b) Not rendering two spheres.

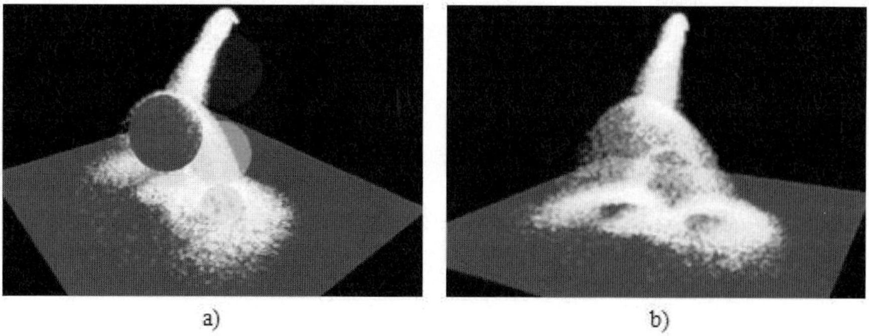

Fig. 4. Collision with five spheres and flat plane in turn. a) Rendering five spheres. b) Not rendering five spheres.

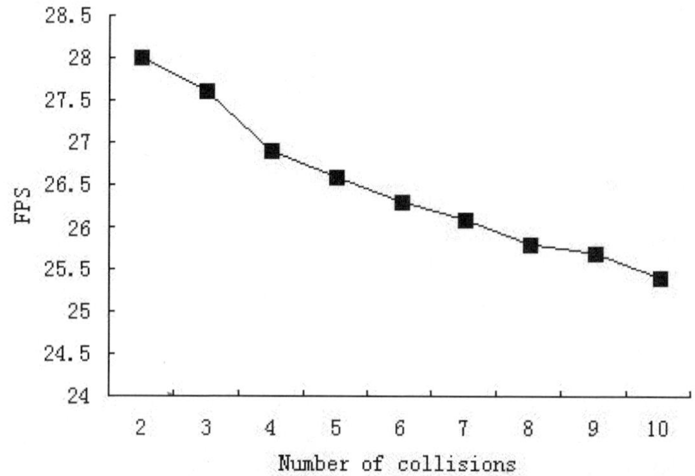

Fig. 5. FPS of particles system with collision detection on GPU

As Fig. 4 shows, in our next implementation, the particle collides with five spheres and the flat plane in turn. And our system is running in 26.6 fps. And the implementation on CPU with the same conditions is also hard to render smoothly. Fig. 5 is our FPS line of particles system with collision detection on GPU. The FPS line shows that the collision computing on GPU is very high performance.

However, in stateless particle systems, it is difficult to deal with Collision Detection so many times, because stateless particle systems need a close form function with initial values and current time. But this function always is hard to find.

5.3 Complex Effects with Particles in Patches

We have implemented our method to simulate the Mushroom Cloud of atomic bomb explosion effect. The mushroom cloud has five Particles Clusters. Five Particles Clusters

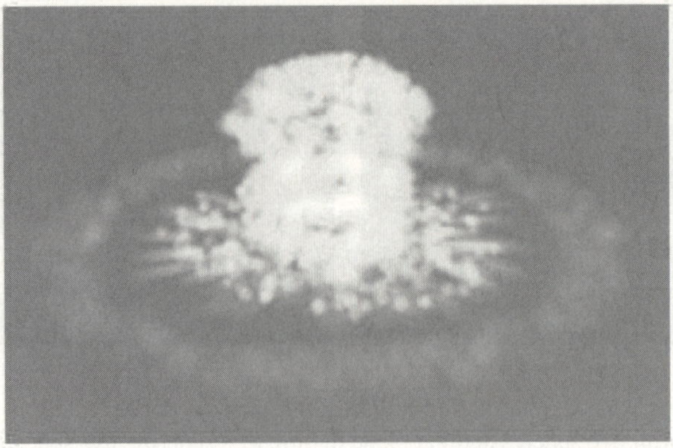

Fig. 6. Mushroom Cloud effect of atomic bomb explosion

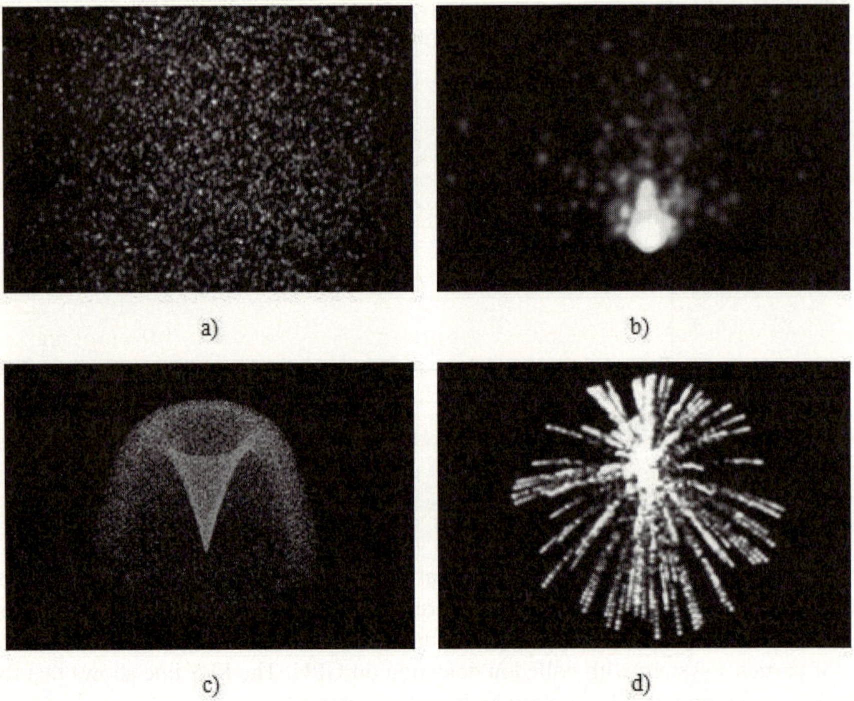

Fig. 7. Other complex effects with particles in patches. a)Snow scene. b)Flame scene. c)Fountain scene. d)Fireworks.

stand for Bottom Wave portion, Ground Shock Wave portion, Column portion, Ring portion and Core portion of the mushroom cloud. All the particles of five Particles Clusters are above 60,000, and the system is running smoothly and is above 25 fps. Fig. 6 shows the Mushroom Cloud effect.

We also have used our method to simulate other complex effects, such as snow scene, flame scene, fountain scene, and fireworks. And our system is running smoothly. Fig. 7 shows these complex effects in turn.

6 Conclusion and Future Work

In this paper, we present a new method to implement efficient complex effects simulation based large particles system on GPU. Our particles system is a state-preserving simulation system. We update the dynamic attributes and render the particles in batches on GPU. The most important is that we also handle the collision between particles and other models on GPU. We make a comparison with CPU particles system. And we also implement complex effects using our method, such as Mushroom Cloud, snow scene, flame scene, fountain scene, and fireworks. Our system is running smoothly and photorealist. The experiments prove that our method is feasible and high performance. Our method could be used in 3D games to simulate the photorealist effects. And our method has also been used in the practical projects.

As a future possibility, we are working on using our method to implement other complex natural phenomenon, and developing the method to deal with the collision between particles and other particles.

Acknowledgments. This work was supported by PHR(IHLB) Grant, by Funding Project of Beijing Municipal Education Committee (No. KM200710009006), and by Funding Project of North China University of Technology (No. 20080018). We would like to thank those who care of this paper and our projects. Also, we would like to thank everyone who spent time on reading early versions of this paper, including the anonymous reviewers. And thanks to those who devote themselves into studies on graphics and 3D games, they gave me inspirations as well as wonderful demos of their works.

References

1. Reeves, W.T.: Particle Systems-Technique for Modeling a Class of Fuzzy Objects. In: Proceedings of SIGGRAPH 1983 (1983)
2. Wang, C., Wang, Z., Peng, Q.: Real-time Snowing Simulation. The Visual Computer 22(5), 315–323 (2006)
3. Liu, X., Yu, Y., Chen, H., et al.: Real-time simulation of special effects in navigation scene. Journal of Engineering Graphics 3, 44–49 (2007)
4. Guan, Y., Zou, L., Chen, W., Peng, Q.: Real Time Waterfall Simulation Based Particle System. Journal of System Simulation 16(11), 2471–2474 (2004)
5. Burg, V.D.: Building an Advanced Particle System. Game Developer Magazine (2000), http://www.gamasutra.com/features/20000623/vanderburg_pfv.htm
6. Harris, M.: Real-Time Cloud Simulation and Rendering. PhD thesis, University of North Carolina at Chapel Hill (2003)
7. Schneider, J., Westermann, R.: GPU-Friendly High-Quality Terrain Rendering. Journal of WSCG 14(1), 49–56 (2006)
8. Livny, Y., Kogan, Z., El-Sana, J.: Seamless Patches for GPU-based Terrain Rendering. Journal of WSCG 15(1), 201–208 (2007)

9. Eric, B., Fabrice, N.: Real-time rendering and editing of vector-based terrains. In: Proceedings of Eurographics 2008, pp. 311–320 (2008)
10. Christopher, D., Natalya, T.: Real-time Mesh Simplification Using the GPU. In: Proceedings of Symposium on Interactive 3D Graphics 2007 (I3D 2007), p. 6 (2007)
11. NVIDIA Corporation: NVIDIA SDK (2004), `http://developer.nvidia.com`
12. Latta, L.: Building a Million Particle System. In: Proceedings of Game Developers Conference 2004 (GDC 2004) (2004)
13. Kolb, A., Latta, L., et al.: Hardware-based Simulation and Collision Detection for Large Particle Systems. In: Proceedings of Graphics Hardware 2004, pp. 123–132 (2004)
14. Ian, B.: Data Parallel Computing on Graphics Hardware. Stanford University (2003)
15. SGI. OpenGL ARB: OpenGL Extension ARB_vertex_shader (2003)
16. Cai, X., Li, F., et al.: Research of Dynamic Terrain in Complex Battlefield Environments. In: Pan, Z., Aylett, R.S., Diener, H., Jin, X., Göbel, S., Li, L. (eds.) Edutainment 2006. LNCS, vol. 3942, pp. 903–912. Springer, Heidelberg (2006)
17. Fernando, R.: GPU Gems: Programming Techniques, Tips, and Tricks for Real-Time Graphics. Addison Wesley Publishing, Reading (2004)
18. Matt, P.: GPU Gems 2: Programming Techniques for High-Performance Graphics and General-Purpose Computation. Addison Wesley Publishing, Reading (2005)

A Selective Attention Computational Model for Perceiving Textures

Woobeom Lee

School of Computer&Information Engineering, Sangji University
660 Woosan-dong, Wonju-si, Kangwon-do 220-702, Republic of Korea
beomlee@sangji.ac.kr

Abstract. This paper presents a biologically-inspired method of perceiving textures from various texture images. Our approach is motivated by a computational model of neuron cells found in the cerebral visual cortex. An unsupervised learning schemes of SOM(: Self-Organizing Map) is used for the block-based textures clustering, plus a selective attention computational model tuning to the response frequency properties of texture is used for perceiving any texture from the clustered texture. To evaluate the effectiveness of the proposed method, various texture images were built, and the quality of the perceived TROI(: Texture Region Of Interest) was measured according to the discrepancies. Our experimental results demonstrated a very successful performance.

Keywords: A selective attention, Cerebral visual cortex, Texture peception, Self-organizing net, Gabor scheme.

1 Introduction

Texture analysis using the 2D spatial filters is the most effective technique in the sate-of-the-arts. As a Gabor scheme among the filtering approaches is motivated by a computational model of neuron cells in biological vision, most of these schemes have focused on optimizing a Gabor filter.

With respect to this subject, two major approaches have been studied for using merely one filter in this literature. One is the supervised method that refers a bank of Gabor filters[1,2]. It is obtained by saving multi-channel filters which needs to be sufficient to analyze textures. Although these methods are effective for segmentation, previous works are restricted as regards computational complexity and supervised problems. The other is the unsupervised method that designs a single Gabor filter, which is distinctly responding to the specific texture component[3,4]. Although it is with in the unsupervised methods, an optimal filtering has focused on detecting the only pertinent texture component, and using the texture information is inherent to a particular image with pre-knowledge.

Consequently, there is currently no completely unsupervised method like the human behavior recognizing the texture in an image without pre-knowledge, and providing a useful information for object recognition and retrieval systems

F. Sun et al. (Eds.): ISNN 2008, Part II, LNCS 5264, pp. 429–438, 2008.

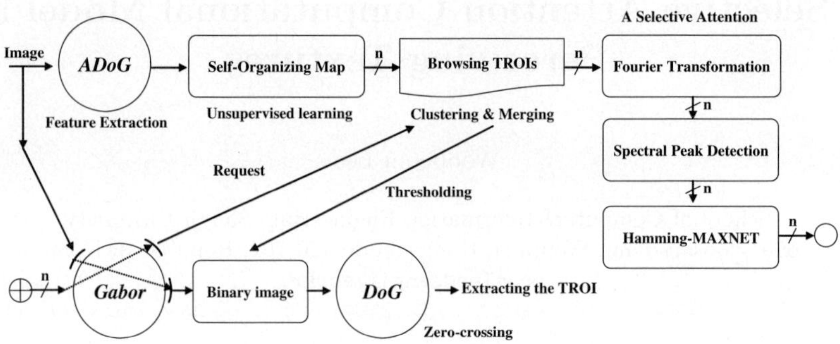

Fig. 1. A selective attention computational model used in our approach

that use a query image. Accordingly, this paper proposes a biologically-inspired method of perceiving textures.

This paper focuses on implementing a biological computational model corresponding to the receptive field of neuron cells such as ganglion cell and various simple cell found in the human visual pathway from the retina to the cerebral visual cortex, proposing an unsupervised learning scheme for clustering textures without pre-knowledge, and segmenting any TROI from the clustered results automatically. A Self-Organizing Map uses the preferred-orientation response properties of a simple cell for a texture clustering, plus a selective attention computational model is based on the Hamming-MAXNET and the spatial frequency response properties of another simple cell. Finally, the zero-crossing principle of a ganglion cell existed in the retina is applied to segment the TROI from image. Then the threshold value for segmenting the TROI is automatically determined based on the selective attended Gabor filtering response.

2 A Self-organized Textures Clustering

The proposed method uses a SOM for clustering textures from an image automatically. Then an orientation selective feature of a ADoG filter is used for the input vector of SOM, and an image is clustered into the block-based parts by the unsupervised learning scheme of SOM. A selective attention computational model used in our approach is outlined in Fig. 1.

2.1 Spatial Feature Extraction by ADoG Filters

It is very important that the extracted feature must be separative in the intra-cluster and constant in the inter-cluster, as the clustering performance of a SOM depends on the spatial features extracted from original image.

Thus, an orientation selective feature is used for the input vector of SOM in our approach. In order to yield the orientation selective features, ADoG

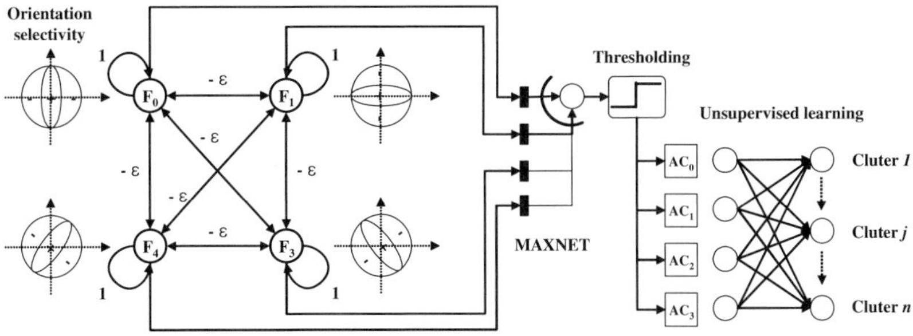

Fig. 2. A the 4-preferred orientation feature extraction by the ADoG filters

(: Asymmetrical Difference Of two Gaussian) function with a preferred orientation ϕ is defined by

$$ADoG(x', y', \phi) = \Big(\exp(-\frac{x'^2}{2\sigma_e^2}) - \frac{\sigma_e}{\sigma_i}\exp(-\frac{x'^2}{2\sigma_i^2})\Big) \cdot \exp(-\frac{y'^2}{2\sigma_{en}^2}) \quad (1)$$

where $(x', y') = (x\cos\phi + y\sin\phi, -x\sin\phi + y\cos\phi)$ are rotated coordinates, and σ_{en} determines the sensitivity of preferred orientation of the filter. This filter corresponds to a simple cell receptive field found in mammal's visual cortex domain[7]. Simple cells are excellent at detecting the presence of simple visual features, such as lines and edges of a particular orientation.

After the 4-preferred orientation feature ($\phi = 0, \pi/4, 2\pi/4, 3\pi/4$) for a sampling point in an image is calculated respectively by using Eq. (1), a competitive learning among each values is implemented by the MAXNET algorithm[9]. As a results, only the orientation node with the largest value is allowed to remain, which is submitted into each accumulator for the input vector of a SOM. Then if the value is over any threshold value, it becomes a valid data. Otherwise, it is ignored.

2.2 An Unsupervised Learning by SOM

After acquiring of the preferred orientation features is completed, the image is divided into the equal-sized block for the block-based image clustering. The input vector x_B of a SOM for any block B in an image is defined by as follows:

$$x_B = \frac{1}{N}\,[AC_0,\ AC_1,\ AC_2,\ AC_3]^T \quad (2)$$

where AC_i is a accumulative value of the preferred orientation $\phi = i \times \pi/4$ for any block B, and N is the total number of pixel in the block B.

After yielding the input vectors for all block in an image, the SOM algorithm proposed by Kohonen[8] is implemented. After a complete scan of the block-based clustering, similar blocks can be assumed to belong to the same cluster. Nonetheless, since unique labels are assigned to each block. The textures can be split into several parts, causing a fragmentation problem despite a homogeneous texture region in the image. Thus, to overcome the problem of a fragmented texture region, a block-based merging procedure is proposed. If one block is identified as similar to a neighboring block, the same label is assigned and the blocks are merged, yielding a number of distinct regions in the texture image.

As shown in the dash-line squares of Fig. 5(c), merged map provides a number of the block-based bounding boxes for browsing the TROIs. And, this bounding box corresponds to the maximum square including the TROI, and preserving the block-based connectivity with respect to the same label. Thus, the regions (Fig. 5(d)) of bounding square extracted from the original image are then presented to the user monitor for selecting the wanted TROI.

3 A Selective Attention of Texture Clusters

After the textures is clustered by using SOM, if one of the clustered textures is selected as the TROI, its spatial frequency is analyzed for optimizing the Gabor filter. The analyzed frequency of each textures is tuned to the optimal frequency for a selective attention behavior.

3.1 The Relatively Optimal Frequency Perception

Each TROI, corresponding to the results of clustering textures, are then transformed into the frequency domain using the Fourier transformation method. As a result, each transformed region has a number of optimal frequency candidates, corresponding to the sorting-searching frequencies of the spectral peaks detected in the Fourier spectrum images of each regions.

To perceive the TROI_t from the clustered TROIs, The Hamming-MAXNET neural network is proposed in Fig. 3. where X is the highest frequency set of the clustered textures except the TROI_t, and $e_i(j)$ is the j-th optimal frequency candidate in the TROI_i.

To determine the optimal frequency from the optimal frequency candidates in the TROI_t, Weight vector w for the Hamming net is defined as:

$$w = \frac{1}{2}\begin{bmatrix} e_t(1) \\ e_t(2) \\ \vdots \\ e_t(m-1) \\ e_t(m) \end{bmatrix} = \frac{1}{2}\begin{bmatrix} u_t^1 \; u_t^2 \ldots u_t^{m-1} \; u_t^m \\ v_t^1 \; v_t^2 \ldots v_t^{m-1} \; v_t^m \end{bmatrix} \tag{3}$$

And the biases are initiated:

$$b_j = \frac{l}{2}, \quad (j = 1, \ldots\ldots m). \tag{4}$$

where l is the the number of input nodes, and m is the number of the output nodes that means the number of the optimal freqeuncy candidates in the TROI_t.

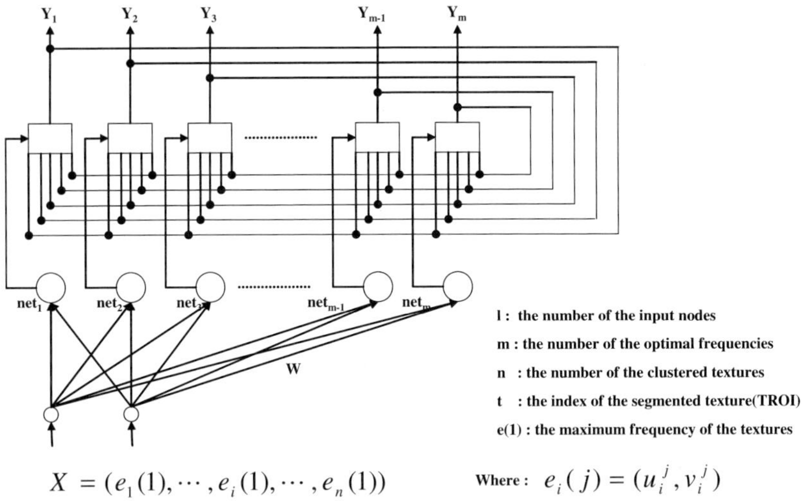

Fig. 3. The proposed Hamming-MAXNET neural network for perceiving the optimal frequency

After defining the input vector X, the weight vector w, and the bias vector b, Each node net_j of the Hamming net is define as follow Eq. (5):

$$net_j = \sum_{i=1 \neq t}^{n} \left(b_j + \sum_{k=1}^{l} e_t^k(1) \cdot w_{kj} \right) \tag{5}$$

where

$$e_i(j) = [e_i^1(j), e_i^2(j)] = [u_i^j, v_i^j].$$

After each node net_j is computed, the activation nodes $Y_j(0)$ for MAXNET is initialized by using Eq. (6).

$$Y_j(0) = f(net_j) = \text{ELV} - net_j, \quad (j = 1, \ldots\ldots m). \tag{6}$$

where ELV is the Enough Large Value for the MINNET transformation. Then the MAXNET is iterated to find the worst match mode, which is the most distinct spatial frequency for perceiving when compared to the highest frequencies of the TROIs in the image. As a result of the MAXNET competition, if the node Y_w is Winner-Take-ALl, $e_t(w)$ is determined as the optimal frequency.

Therefore, the parameter set of Gabor function for perceiving a TROI$_t$ is given by

$$Gp\{u_0,\ v_0,\ \lambda,\ \phi\} = \left\{ \frac{u_t^w}{N},\ \frac{v_t^w}{M},\ \{a, b\},\ \theta \right\} \tag{7}$$

where N, M, considering as $N = M$ generally, are the spatial resolution of the TROI, and 1/N is the frequency sample interval. The center frequency (u_0, v_0) is tuned according to the optimal frequency $e_t(w)$, and the orientation θ is used to consider as the rotated parameter ϕ in the rotated coordinate system.

3.2 A Selective Attention Using 2D Gabor Filtering

2D Gabor filters corresponds to another simple cell receptive field found in mammal's visual cortex domain[7]. They can be an appropriate computational model for a selective attention to the very specific frequency and orientation characteristic, as they have a tunable orientation, center frequency and radial frequency bandwidth. The 2D Gabor function as a spatial filter in image processing is defined in the form Eq. (8)[6].

$$\begin{aligned} Gabor(\ x,\ y;\sigma,u_0,v_0,\lambda,\phi) \\ &= g(x',y';\sigma)\cdot\exp\big(-2\pi i(u_0x+v_0y)\big) = g(x',y';\sigma)\cdot\exp(-2\pi i f_0 x') \\ &= g(x',y';\sigma)\cdot\big(\cos(2\pi f_0x') - i sin(2\pi f_0 x')\big), \end{aligned} \tag{8}$$

where

$$g(x,y;\sigma) = \Big(\frac{1}{2\pi\lambda\sigma^2}\Big)\cdot\exp\Big(-\frac{(x/\lambda)^2+y^2}{2\sigma^2}\Big),$$

$(x',y') = (x\cos\phi + y\sin\phi, -x\sin\phi + y\cos\phi)$ are rotated coordinates, $\lambda(= b/a)$ specifies the aspect ratio, σ is the standard deviation of the Gaussian envelope. Also, the radial center frequency f_0 can be calculated as $f_0 = \sqrt{u_0^2+v_0^2}$, and λ, ϕ, the center frequency (u_0, v_0) of the Gabor function is defined by the above Eq. (7) as follow:

$$u_0 = u_t^w/N,\quad v_0 = v_t^w/M,\quad \lambda = b/a,\quad \phi = \theta(=\tan^{-1}(v_0/u_0)). \tag{9}$$

Here the effective method is proposed for analyzing the filtering response. In a discrete spatial convolution, the Gabor function in Eq. (10) has real and imaginary components respectively given by

$$Gabor_R(x,y) = g(x',y')\cdot\cos(2\pi Fx'),\ \ Gabor_I(x,y) = g(x',y')\cdot\sin(2\pi Fx') \tag{10}$$

For simplicity, ϕ is not considered with an aspect ratio $\lambda = 1$. Since the function $Gabor_R(\cdot)$ and $Gabor_I(\cdot)$ are symmetrically even and odd , respectively, along the preferred orientation direction, the convolution results are approximately identical, other than a difference of $\pi/2$ in the phase spectra. Therefore, a more uniform response can be obtained by considering the real and imaginary parts simultaneously. The analog response of the optimized Gabor filter, $u_g(x,y)$ can be defined in the form of Eq. (11).

$$\begin{aligned} u_g(x,y) = \Bigg[\Big(\int\int_A Gabor_R(\xi,\eta)\cdot t(x+\xi,y+\eta)d\xi d\eta\Big)^2 \\ +\Big(\int\int_A Gabor_I(\xi,\eta)\cdot t(x+\xi,y+\eta)d\xi d\eta\Big)^2\Bigg]^{1/2} \end{aligned} \tag{11}$$

where A denotes the distance of receptive field satisfying $|\xi/a|^2 + |\eta/b|^2 \leqq |A|^2$, $t(x,y)$ is reagrded as the texture model TROI, and $Gabor_R(\cdot)$ and $Gabor_I(\cdot)$ represent the strength coefficients of the real and imaginary parts respectively.

The optimized Gabor filtering then results in essentially uniform responses in the similar texture regions, which means that the selective attended Gabor filter can be effective for segmenting distinct texture regions in an image. Accordingly, after applying the optimized Gabor filter to the original image, the segmentation is then completed by extracting uniform regions from the response image.

4 Texture Segmentation by DoG Filter

The unsupervised segmentation of a TROI requires a threshold value for creating a binary image $B(x, y)$ before the segmentation. Therefore, after applying the selective attended Gabor filter to the corresponding TROI of the clustering results, the H and L attributes for extracting the TROI are registered in the Look-up table. H and L are the highest and lowest response value, respectively, for $u_g(\cdot)$, in the TROI. Table 1 is an example of the Look-up table used in experimental of Fig. 5.

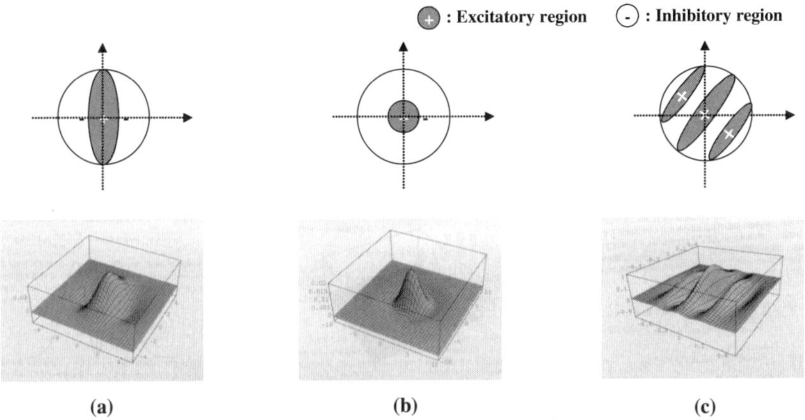

Fig. 4. 2D profile of biological-spatial filter used in our approach (a) ADoG(·) filter - Simple cell : the sensitivity of preferred orientation, (b) DoG(·) filter - Ganglion cell : contrast detection, (c) Gabor(·) filter - Another simple cell : selective attention of frequency

Table 1. The contents of Look-up table used in experimental of Fig. 5

No.	Parameters Set for Gabor function				Thresholding value	
	u_0	v_0	λ={ a, b }	Φ	H	L
1 : SAND	-0.0156250	0.0937500	{ 2, 2 }	-80.5785276°	0.5129187	0.5098017
2 : D112	0.0625000	-0.0312500	{ 2, 2 }	-26.5785252°	0.4703219	0.4613092
3 : D24	-0.0937500	-0.0156250	{ 2, 2 }	9.4671216°	0.4924787	0.4881189

Thus, without any pre-knowledge or heuristic decision, the upper and lower bound for the binary image transformation can be automatically determined by searching for the threshold value in the look-up table, and binary image $B(x, y)$ for segmentation can be created by Eq. (12).

$$B(x,y) = \varphi[u_g(x,y)] = \begin{cases} 1 \ \text{if} \ (\lceil \omega \times H \rceil / \omega) \leqq u_g(x,y) \leqq (\lfloor \omega \times L \rfloor / \omega) \\ 0 \ \text{otherwise.} \end{cases} \tag{12}$$

where ω is the precision coefficient, and $\lceil \cdot \rceil$ and $\lfloor \cdot \rfloor$ denote the *ceiling*() function and *floor*() function, respectively, for the truncation using the integer transformation.

The final segmentation is achieved by applying an efficient edge detection algorithm to the binary image $B(x, y)$. The approach by D. Marr is also applied for the edge detection algorithm, as follows[7]:

$$\nabla^2 DoG(x,y) = \frac{1}{2\pi\sigma_e^2} \exp\left(-\frac{x^2+y^2}{2\sigma_e^2}\right) - A\frac{1}{2\pi\sigma_i^2} \exp\left(-\frac{x^2+y^2}{2\sigma_i^2}\right) \tag{13}$$

where σ_e and σ_i represent the space constants of the excitatory and inhibitory regions, respectively, and the ratio of the space constants $\sigma_e/\sigma_i = 1.6$. The ratio yields a good approximation of the ideal Laplacian operator. This filter corresponds to an on-center and off-surrend receptive field of ganglion cells found in the retina of the visual pathways[7]. The segmentation is accomplished by finding the zero crossing points of the spatial filter Eq. (13) applied to the binary image $B(x, y)$.

5 Experimental Results

To demonstrate the performance of the proposed approach, experiments were tested using various texture images obtained from the Brodaz texture book [10]. The proposed method was implemented using C language under the X-Window environment in the SUN SPARC workstation.

It should be note that the experiments is limited to two or three-texture problems when using one filter. Nonetheless, the experimental results in Fig. 5 show that the performance of the proposed system was very successful. To evaluate the quality of the segmentation performance, given more than 100 texture images, the segmentation quality was measured according to the discrepancies based on the number of mis-segmented pixels[11], as defined below:

$$D_k = 100 \times \left[\left(\sum_{i=1}^{N} C_{ik}\right) - C_{kk}\right] / \left[\sum_{i=1}^{N} C_{ik}\right] \tag{14}$$

where C_{ij} represents the number of cluster j pixels classified as cluster i in the segmentation results. The results were measured as close to 5% for the two-texture problem, and 7% for three-texture problem. This means that the proposed method preserves the segmentation quality in spite of reducing the constraint problems.

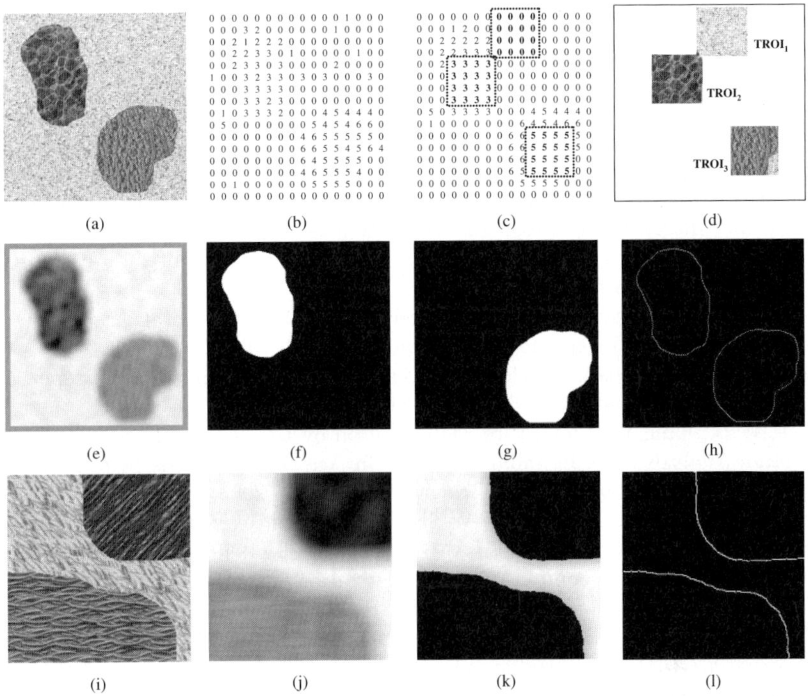

Fig. 5. Experimental Result I: (a) Collage of the Brodaz textures Background(Sand), D112(Plastic bubbles) and D24(Pressed calf leather). (b) Clustered map. (c) Merged map, where the size of original image is 512x512 pixel, and one of the block unit is scaled to a size of 32x32 pixel. (d) Results of browsing the TROIs. (e) Selective attended Gabor filtering image for the TROI(D112). (f) Image of extracted the TROI(D112). (g) Image of extracted the TROI(D24). (h) Image of segmented the TROIs(D112 and D24) by zero-crossing. Experimental Result II: (i) Collage of the textures slat, wheat, wire. (j) Selective attended Gabor filtering image for the TROI(wheat). (k) Image of extracted the TROI(wheat). (l) Image of segmented the TROI(wheat) by zero-crossing.

6 Conclusions

A biologically-inspired computational model was presented for automatic clustering and perceiving textures. This paper focuses on (1) implementing a biological filters corresponding to the receptive field of neuron cells such as ganglion cell and various simple cell found in the visual pathway from the retina to the cerebral visual cortex, (2) proposing an unsupervised learning scheme for clustering textures without pre-knowledge, and (3) composing a full neural scheme.

However, several problems remain for future work: the sensitivity of preferred orientation by simple cell. In particular, the selection method of the appropriate parameters, such as the orientation, phase, and aspect ratio, is an important task when using a Gabor filter for a selective attention. Consequently, when these problems are solved, the proposed method has a potential application for the neuro-vision system development.

References

1. Manthalkar, R., et al.: Rotation invarient texture classification using even symmetric Gabor filters. Pattern Recognition Letters 24, 2061–2068 (2003)
2. Idrissa, M., Acheroy, M.: Texture classification using Gabor filters. Pattern Recognition Letters 23, 1095–1102 (2002)
3. Tsai, D., et al.: Optimal Gabor filter design for texture segmentation using stochastic optimazation. Image and Vision Computing 19, 299–316 (2001)
4. Clausi, D.A., Jernigan, M.: Designing Gabor filters for optimal texture seperability. Pattern Recognition 33, 1835–1849 (2000)
5. Lee, W.B., Kim, W.H.: Texture Segmentation by Unsupervised Learning and Histogram Analysis using Boundary Tracing. In: Yue, H., et al. (eds.) CIS 2005. LNCS (LNAI), vol. 3801, pp. 25–32. Springer, Heidelberg (2005)
6. Bovik, A.C., Clark, M., Geisler, W.S.: Multichannel texuture analysis using localized spatial filter. IEEE Trans. PAMI 12 (1), 55–73 (1990)
7. Marr, D.: Vision: A computational investigation into the human representation and processing of visual information. W. H. Freedom & Company (1982)
8. Kohonen, T.: The self-organizing map. Proc. IEEE 78 (9), 1464–1480 (1990)
9. Lippmann, R.P.: An introduction to computing with neural nets. IEEE ASSP Magagine 4, 4–22 (1987)
10. Brodatz, P.: Texture: A photographic album for artists and designer. Dover Publication (1966)
11. Zhang, Y.J.: A survey on evaluation methods for image segmentation. Pattern Recognition 29 (8), 1335–1346 (1996)

Classifications of Liver Diseases from Medical Digital Images

Lequan Min[1], Yongan Ye[2], and Shubiao Gao[2]

[1] Applied Science School/Information Engineering School
University of Science and Technology Beijing
Beijing 100083, P.R. China
[2] Beijing University of Chinese Medicine
Beijing 100700, P.R. China
minlequan@sina.com, yeyongan2002@yahoo.com.cn

Abstract. Hepatitis B/C virus (HBV/HCV) infections are serious problems of world-wide, which cause over million die each year. Most of HBV/HCV patients need long term therapy. Side effects and virus mutations make difficult to determine the durations and endpoints of treatments. Medical images of livers provide evaluating tools for effectiveness of anti-virus treatments. This paper presents a liver hepatitis progression model. Each class C_i in the model consists of three characteristic qualities: gray-scale characteristic interval $I_{G,i}$, non-homogenous degree $N_{h,i}$ and entropy $Entro_i$. This model aims to describe both digitally and visually a patient's liver damage. Examples are given to explain how to use the liver hepatitis progress model to classify people with normal livers, healthy HBV carriers, light chronic HBV patients and chronic cirrhosis HBV patients. The results show that our analysis results are in agreement with the clinic diagnoses and provide quantitative and visual interpretations.

Keywords: Hepatitis, liver medical digital images, classifications.

1 Introduction

It is estimated that 2 billion/170million people worldwide have been infected with HBV/HCV. Over 400 millions have chronic (lifelong) HBV or HCV infection; 25%~40% of these chronic infection carriers will die from liver cirrhosis or primary hepatocellular carcinoma. One million die each year from complication of infection, including cirrhosis, hepatocellular carcinoma, or both [1]. In China, there are over 120/4.1 millions HBV/HCV carriers. $20 \sim 30$ millions have developed chronic liver disease. Each year about 270 thousand people die from liver cirrhosis or primary hepatocellular carcinoma [2].

Effective treatment of chronic HBV patients aims to prevent progression of chronic hepatitis B (CHB) to cirrhosis, hepatocellular carcinoma, and eventually death. It has been recognized that the effects of monotherapy with a single antiviral agent are limited in controlling HBV or HCV infection in majority of patients. For example, only about 20% or 14% HBeAg-positive patients profit from

F. Sun et al. (Eds.): ISNN 2008, Part II, LNCS 5264, pp. 439–448, 2008.

IFN α or Adefovir dipivoxil treatments, seroconvert to anti-HBe and lose serum HBV-DNA ([3], [4]). Long time treatment may be well tolerated and produced significant, increasing improvement in hepatic fibrosis, durable suppression of HBV replication [5]. However treatment side effects and virus mutations make it be difficult to determine the choices of durations and endpoints of therapy.

The liver damage of a patient with hepatitis infection is almost independent on the patient's serum virus level. Therefore the medical-image-based classifications of liver diseases are important not only for monitoring the development of the disease but also for evaluating the effectiveness of therapies.

The traditional method to judge whether a liver tissue is normal is decided by physicians. However some researches have shown that the accuracy of the decision of diffused liver disease is about 72% via simple visual interpretation [6]. Liver biopsies are also usually done to evaluate liver damages.However it is an invasive method and costly. Furthermore the abnormal cells in a liver with diffused disease have not been homogenous in the liver. Very thin areas detected by liver biopsies may not represent the damage level of whole liver tissues.

An obvious application of computer-aided detection (CAD) system is in the interpretation of screening digital medical images. Recently, several digital analysis techniques based on texture analysis and neural network have been developed in tissue classification ([7]-[9]).

Most researches are concentrate on the classifications of liver cyst, hepatoma, and cavernous. However, it is also important for doctors and patients to know the progression form "normal" hepatocyte to fibrosis and cirrhosis.

To give a minimal model describing digital images of liver with disease, we make a number of assumptions for different disease stages:

1) The gray levels of abnormal livers cells are different.
2) The non-homogenous degrees of hepatocyte are different. This means that a patient's liver cells have different quantities which are represented via gray levels describing the degrees of abnormal.
3) The random degrees of gray levels of digital images of patients's livers. are different.
4) In medical images, there are random noises.

The paper is organized as follows. Section 2 proposes three fundamental feature qualities which describe the features of the ALHPM. Practical examples are given in Section 3 to illustrate the effectiveness of our theory. Some concluding remarks are stated in Section 4.

2 Liver Hepatitis Progression Model

We can classify liver disease into l classes:

$$C_1, C_2, \cdots, C_l$$

where C_1 represents a normal liver and C_l stand for the most abnormal liver.

For simplicity, the gray levels of C_i mean the gray levels of the pixels in corresponding digital images of livers.

First, denote the mean and standard deviation of the gray levels of C_i as

$$G_{m,i} \text{ and } G_{\sigma,i},$$

respectively. Hence we can assume that for a liver belonging to i^{th} class, if the pixels with the gray levels of the liver which are in the interval

$$I_{G,i} = [G_{m,i} - 2G_{\sigma,i}, G_{m,i} + 2G_{\sigma,i}] \tag{1}$$

have high probabilities. The reason is that for many case (random perturbations) but not all cases, about 95% random variables are in the above interval (for example see [10]).

Let g *be the gray level of the pixel in an liver digital image.* We call the pixel belong to class C_i if

$$g \in [G_{m,i} - 2G_{\sigma,i}, G_{m,i} + 2G_{\sigma,i}].$$

Consequently an individual's different liver cells (represented by pixels in corresponding digital image) may belong to different classes.

Second, for the digital image of a liver with $M \times N$ pixels, we use the following formula to stand for the value of the non-homogenous degree of the class C_i:

$$N_{h,i} = \frac{1}{(M-1)\times(N-1)} \sum_{m=2}^{M-1}\sum_{n=2}^{N-1} |g_{m-1,n-1} + g_{m-1,n+1} + g_{m-1,n} + g_{m+1,n} + g_{m,n-1} + g_{m+1,n-1} + g_{m,n+1} + g_{m+1,n+1} - 8g_{i,j}| \tag{2}$$

where $g_{m,n}$ is the gray level of the $(m,n)th$ pixel.

Third, we use the entropy to represent the random degree of gray levels of the class C_i:

$$Entro_i = \sum_{l=0}^{G} P(l) \log_2 P(l) \tag{3}$$

where G is the maximum gray level of the digital image; $P(l)$ is the probability that the gray levels equal to l.

In summary, we can use the three feature quantities $I_{G,i}$, $N_{h,i}$ and $Entro_i$ to describe the feature of the class C_i.

Let the abstract liver hepatitis progression model (ALHPM) consist of l classes: $C_1, C_2, \cdots, C_l$. Each class C_i can be describe by the three quantities. Therefore we can denote

$$C_i = \{I_{G,i},\ N_{h,i},\ Entro_i\}. \tag{4}$$

For a liver with liver disease, different liver cells may belong to different classes. Therefore the above quantities can determine that the percentages of a patient's liver cells belong to different classes. Consequently our classification are different from traditional ones.

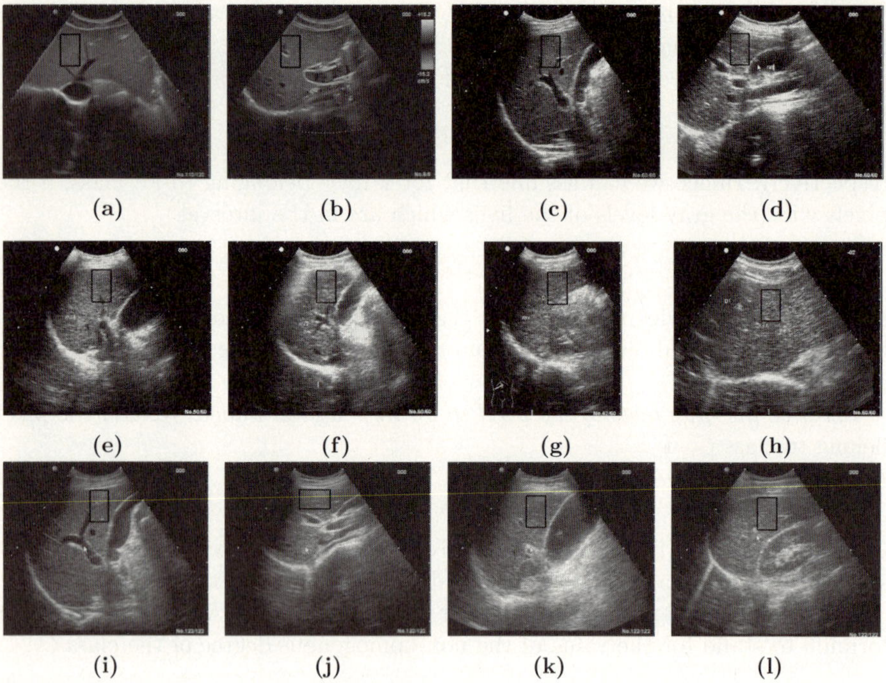

Fig. 1. Liver tissue B-scan digital images. Normal persons: (a) No.1 and (b) No.2. HBV infection patients: (c) No.3a, (d) No.4a, (e) No.5a , (f) No.6a, (g) No.7a, (h) No.7b, (i) No.3b, (j) No. 4b, (k) No. 5b, and (l) No. 6b.

In practical applications, we need to solve the following problems in advance.

1) We need to select large enough efficient region(s) of liver. Since the pathology of the liver disease caused by hepatitis virus infection is distributed all over the liver volume, whole liver is not, in fact, suitable for analysis because of the complexity of liver structure. We can expect that analysis for large enough efficient region(s) of a liver may be more efficient than the analysis for very thin area detected by liver biopsy.
2) We need enough patient samples (samples are larger than 30) for determine the values $C_i's$. On the other hand, for different instruments which obtain digital liver images, the $C_i's$ are different. For a specific instrument, we must make sure that the repeatability and the accuracy of the instrument. And then decide how many classes need to be classified.

In the next section, we are not going to discuss the second issue. We only provide some practical examples to illustrate how to implement our classifications. And hope readers to develop or amend our approaches for better classifications to liver diseases caused by Hepatitis virus infections.

3 Examples

Figures 1(a)-1(l) are seven individuals' liver tissue B-scan digital images (BSDIs), which are scanned by an EUB-8500 Ultrasound Scanner. These images are not taken in the same technique conditions and designed specially for this research. Even so we will see that our theory can success to interpret these images.

Figures 1(a) and 1(b) are two persons's BSDIs, who have healthy livers (numbered by 1 and 2). Figures 1(c) and 1(i) are a patient's BSDIs, who is clinically diagnosed as a healthy HBV carrier (Numbered by 3a and 3b). The others are the BSDIs of four chronic HBV patients' liver tissues (numbered by 4a-6a and 4b-6b). The clinic diagnoses for the four patients are listed as follows.

1) The No.4a liver tissue has some liver damages but not fibrosis (Figs. 1(d)) yet.
2) The No.5a liver tissue has mild fibrosis (Figure 1(e)).
3) The No.6a liver tissue has mild cirrhosis (Figure 1(f)).
4) Figures 1(g) and 1(h) are the same patient's BSDIs (numbered by No.7a and No.7b), scanned in a seven months' interval– in Jan 2007 and August 2007, respectively.
   a) The image shown in Figure 1(g) has been diagnosed to have liver damage without fibrosis.
   b) The image shown in Figure 1(h) has been diagnosed to develop to mild cirrhosis.
5) Figures 1(i) $\sim$ 1(l) are the same patient's (No. 3a $\sim$ No. 6a) BSDIs (numbered by No.3b $\sim$ No.6b), scanned in a seven months' interval– in Sept 2007 and April 2008, respectively. The clinic diagnoses for the BSDIS of Nos. 3b, 4b, 6b are the same as those stated in the above. However, the BSDIS of No. 5 is diagnosed as cirrhosis under the condition without referring Figure 1(e).

Now let us take a rectangle with 80×50 pixels in each BSDI shown in Figure 1 (see Figure 2) for analyzing.

Discussions

1) The mean of the gray-scales of No.1's BSDI is 16 less than the other patients. However it does not give us a general measurement to compare directly the patients' liver damages.
2) We assume that the I_G of No.1's BSDI represent the gray-scales corresponding undamaged liver tissue. The cells with gray scales larger than I_G are abnormal cells (having damages); the cells with gray scales less than I_G may be not very normal cells.
3) We take No.1's $[0 \;\; I_G]$ as a threshold interval, that is, we transform the gray-scale images Figs 2(a) - 2(l) to binary images by the following approaches. For a gray-scale $g(i,j)$ of a pixel at $(i,j)th$ position, we define a new gray-scale

$$g^*(i,j) = \begin{cases} 0 \;\; if \; g(i,j) \in [0, 102.41] \\ 255 \; otherwise \end{cases} \tag{5}$$

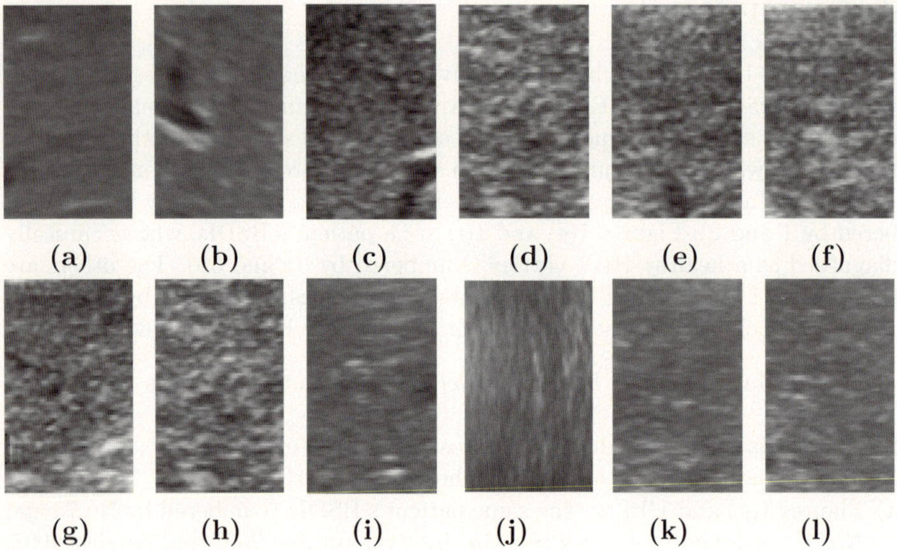

Fig. 2. Selected rectangle areas for liver tissue B-scan digital images. Normal persons: (a) No.1 and (b) No.2. HBV infection patients: (c) No.3a, (d) No. 4a, (e) No.5a , (f) No.6a, (g) No.7a, and (h) No.7b, (i) No.3b, (j) No. 4b (rotated 90° anticlockwise), (k) No. 5b, (l) No. 6b.

The generating binary images are shown in Figure3. Observe that the black pixels shown in Figure3 corresponding normal or not normal but without damage tissues, and the white pixels corresponding damaged tissues.

4) Using the formulas (1)-(3) calculates the $I_G, N_h, Entro$ for the twelve rectangle images. The results are shown in Table 1, in which the percentages represent the pixel numbers whose gray scale are less than 102.41. From Table 1, it follows that No.6 and No.7b have less 42% undamaged hepatocyte. Hence we can understand why No.6 and No.7b have been diagnosed clinically to have mild cirrhosis.
5) However that No.5b is diagnosed clinically to have mild cirrhosis seems to be questionable because the data given in Table 1 show that No. 4b and No.5b do not have essentially differences.
6) Table 1 shows also that the non-homogenous degrees of the healthy persons No.1 and No.2 are less than 22.5. However the the HBV infection patients' non-homogenous degrees are larger than 34.5 for $3b \sim 6b$ series, and larger than 40 for $3a \sim 6a$ series, which do not have significant differences. For the entropies, the chronic HBV patients' situations are the similar.
7) Observing Figure 1, we can conclude that the differences of the I_G, N_h between $3a \sim 6a$ and $3b \sim 6b$ may not show that the 4 patients's situation become better. It may be caused by different scan parameters. However the data listed in the 4th column in Table 1 and Figure 3 may imply that No. 3 $\sim$ No. 6 patients's illnesses do not get worse also.

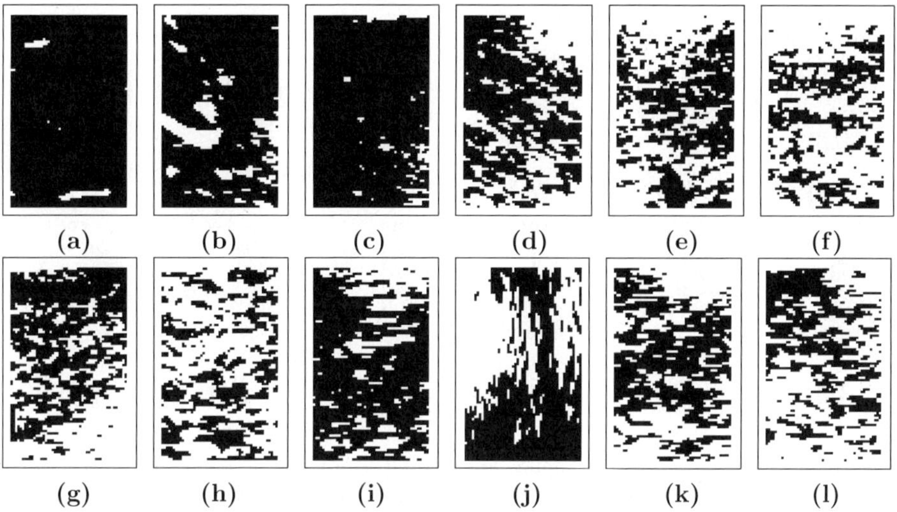

Fig. 3. Selected rectangle areas for individuals' liver tissue B-scan binary digital images: (a) No.1, (b) No.2, (c) No.3a, (d) No. 4a, (e) No.5a , (f) No.6a, (g) No.7a, (h) No.7b, (i) No.3b, (j) No. 4b (rotated 90° anticlockwise), (k) No.5b, (l) No.6b.

8) It is well-known that a host's immune responding to HBV is in charge of the damages of livers. The above facts imply that after immune respondence begins for some time, the non-homogenous degrees of liver cells may be vary in a small threshold interval [0, 15] (see the data of $3 \sim 7$ given in Table 1), which may not represent the developments of liver diseases accurately.
9) Figure 4 gives a visual description for the seven persons' liver states. Combining Figs. 3 and 4, we can interpret more visually the meanings of the characteristic quality $I_{G,i}$.

From Figure 1, we can assume that the BSDIs shown in Fig1. (a), (b), (i) $\sim$ (l) are taken under the similar technique parameter values of the EUB-8500 Ulyrasound Scanner. The other BSDIs are taken under another set technique parameter values. we can classify the above 12 BSDIs into four catalogs: C_1, C_2, C_3 and C_4 as follows, representing the healthy persons, healthy HBV carriers, chronic HBV patients without fibrosis, and chronic HBV patients with mild cirrhosis.

$$C_1 = \{[64.29 \quad 102.41], 19.379, 5.1402\} \tag{6}$$

$$C_2 = \{[63.10 \quad 126.57], 40.263, 5.839\} \; or \; \{[22.43 \quad 134.32], 65.418, 5.8018\} \tag{7}$$

$$C_3 = \{[63.05 \quad 136.57], 34.688, 6.431\} \; or \; \{[33.69 \quad 165.90], 77.049, 6.9256\} \tag{8}$$

$$C_4 = \{[77.70 \quad 135.05], 41.545, 5.739\} \; or \; \{[57.80 \quad 172.83], 70.81, 6.7915\} \tag{9}$$

Denote $N_G^j, N_h^j, Entro^j$ as No.j patient's the characteristic interval, non-homogenous degree and entropy, respectively.

$$V_j = \{N_G^j, N_h^j, Entro^j\} \tag{10}$$

is said to be No.i and No.j patient's characteristic vector.

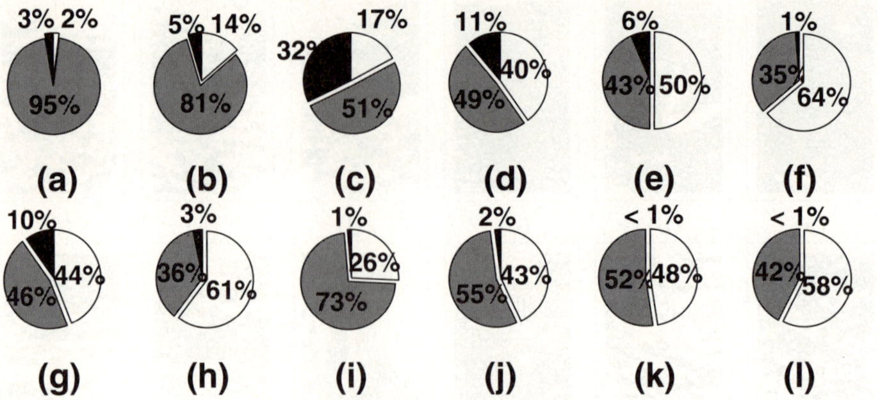

Fig. 4. Percentages of normal tissues are colored by brown. Percentages of abnormal tissues whose gray-scale are larger than 103 are colored by white. Percentages of abnormal tissues whose gray-scale are less than 64 are colored by black. (a) No.1, (b) No.2, (c) No.3a, (d) No.4a, (e) No. 5a, (f) No.6a, (g) No.7a, (h) No.7b, (i) No.3b, (j) No.4b (rotated 90° anticlockwise), (k) No.5b, (l) No.6b.

Table 1. Three feature quantities for the 12 rectangle images. The percentages listed in the fourth column represent how many percentages pixels whose gray-scales are lie in the interval [0 102.41] of the No.1. The means of the gray-scales of the 12 images are listed in the second column.

Nos.	mean	I_G	%	N_h	Entro
1	83.349	64.29, 102.41	98%	19.379	5.1402
2	90.533	56.29, 124.78	86%	22.481	5.7763
3a	78.392	22.43, 134.32	83%	65.418	5.8018
3b	90.533	63.1, 126.57	74%	40.263	5.839
4a	99.793	33.69, 165.90	60%	77.049	6.9256
4b	99.81	63.051, 136.57	57%	34.688	6.4306
5a	103.51	47.65, 159.38	49%	71.857	6.8043
5b	106.38	75.647, 130.91	52%	38.961	5.7177
6a	115.31	57.80, 172.83	36%	70.813	6.7915
6b	106.38	77.703 135.05	42%	41.545	5.739
7a	102.98	32.48, 173.47	56%	76.564	7.0327
7b	111.55	53.45, 169.66	39%	79.830	6.8545

For No. j patient, we can calculate the 2-norm:

$$\Delta_{i,j} = \|V_j - V_i\|_2 \tag{11}$$

to determine deviation between V_i and V_j. The calculation results are shown in Table 2.

Table 2. The 2-norms of patients' characteristic vectors V_i - V_j

$\Delta_{i,j}$	V_1	V_2	V_{3a}	V_{3b}	V_{4a}	V_{4b}	V_{5a}	V_{5b}	V_{6a}	V_{6b}	V_{7a}	V_{7b}
V_1	0	27.743	71.691	40.606	100.04	57.92	95.309	63.017	111.86	73.505	107.15	112.12
V_2	27.743	0	56.899	22.578	77.078	35.508	72.438	45.727	88.096	55.808	83.172	89.229
V_{3a}	71.691	56.899	0	50.747	47.396	61.108	55.61	72.718	79.593	78.038	55.602	73.853
V_{3b}	40.606	22.578	50.747	0	63.614	22.498	55.739	30.217	71.828	39.521	70.242	71.987
V_{4a}	100.04	77.078	47.396	63.614	0	59.378	19.979	67.4	38.535	67.207	9.2283	31.489
V_{4b}	57.92	35.508	61.108	22.498	59.378	0	47.082	16.647	57.694	23.069	63.725	60.719
V_{5a}	95.309	72.438	55.61	55.739	19.979	47.082	0	51.901	24.357	49.711	22.363	19.171
V_{5b}	63.017	45.727	72.718	30.217	67.4	16.647	51.901	0	58.533	11.316	71.528	62.131
V_{6a}	111.86	88.096	79.593	71.828	38.535	57.694	24.357	58.533	0	52.876	35.023	11.551
V_{6b}	73.505	55.808	78.038	39.521	67.207	23.069	49.711	11.316	52.876	0	70.393	57.337
V_{7a}	107.15	83.172	55.602	70.242	9.2283	63.725	22.363	71.528	35.023	70.393	0	28.764
V_{7b}	112.12	89.229	73.853	71.987	31.489	60.719	19.171	62.131	11.551	57.337	28.764	0

We assume that if

$$\Delta_{1,j} \leq 30$$

then No.j patient belongs to class C_1. If

$$\Delta_{i,j} \leq 20, i = 2, 3, 4$$

then No.j patient belongs to class C_i. Hence we can obtain the following conclusions.

(i) The two healthy persons belong to class C_1.
(ii) Only healthy HBV carrier (No. 3) himself belongs to class C_2.
(iii) No.4 patient and No. 5 patient belong to class C_3.
(iv) No. 5, No.6 and No.7b patients belong to class C_4.
(v) Observe that No.5 patient belongs to both C_3 and C_4. Hence we can understand why this patient is difficult for classification.

In summary, the above classifications are in agreement with the clinic diagnosis. Furthermore Figure 3 and Figure 4, Table 1 and Table 2 give us more clear graphical and mathematical descriptions both.

4 Concluding Remarks

HBV/HCV infections are worldwide problems. For most HBV/HCV patients need long term therapies. However the side effects of long term treatments and virus mutations caused by drugs, for example lamivudine, make difficult to determine the choice of durations and endpoints of therapies.

Monitoring developments of hepatic damages can provide evaluations on the effectiveness of anti-hepatitis infection therapies. Medical digital images of livers are important tools for recording the states of liver tissues.

Computer-aided analysis for medical digital images of livers may provide more object criteria than simple visual interpretations even invasive method—liver biopsies.

This paper firstly presents a liver hepatitis progress model described via three quantities: gray-scale characteristic interval $I_{G,i}$, non-homogenous degree $N_{h,i}$ and entropy $Entro_i$. This model aims to describe both numerically and visually patients' disease states.

The examples are given to explain how to use the liver hepatitis progress model to classify healthy HBV carriers, light chronic HBV patients and chronic cirrhosis HBV patients. The results show that our analysis results are able to be in agreement with the clinic diagnoses and provide quantitative and visual interpretations. The important issue is to provide uniform examine technology standards.

Acknowledgments. This project is jointly supported by the National Natural Science Foundations of China (Grant No. 60674095), and the Key Discipline Cooperation Establishing Program of Education Committee of Beijing (Grant No. XK100080537).

References

1. World Health Organization: Hepatitis B Fact Sheet no. 204. Geneva (October 2000)
2. Xu, D.: The Current Clinic Situations of Hepatitis B in China (in Chinese) (December 15, 2003), `http://www.ganbing.net/disparticle.asp?classid=3\&id=15`
3. Lok, A.S., McMahon, B.J.: Chronic Hepatitis B. Hepatology 45(2), 507–539 (2007)
4. Lau, G.K.K., Piratvisuth, T., Luo, K.X., et al.: Peginterferon Alfa-2a, Lamivudine, and the Combination for HBeAg-Positive Chronic Hepatitis B. New England Journal of Medicine 352(26), 2682–2695 (2005)
5. Hadziyannis, J.S., Tassopoulos, N.C., Heathcote, E.J., et al.: Long-term Therapy with Adefovir Dipivoxil for HBeAg-Negative Chronic Hepatitis B for Up to 5 Years. Gastroenterology 131(6), 1743–1751 (2006)
6. Pavlopoulos, S., Kyriacou, E., Koutsouris, D., et al.: Fuzzy Neural Network-Based Texture Analysis of Ultrasonic Images. IEEE Engineering in Medicine and Biology 19(1), 39–47 (2000)
7. Lee, W.L., Chen, Y.C., Hsieh, K.S.: Ultrasonic Liver Tissues Classification by Fractal Feature Vector Based on M-band Wavelet Transform. IEEE Trans. Med. Image. 22(3), 382–391 (2003)
8. Kadah, Y.M., Frag, A.A., Zurada, J.M., et al.: Classification Algorithms for Quantitative Tissue Characteration of Diffuse Liver Disease from Ultrasound Images. IEEE Trans. Med. Image. 15(4), 466–478 (1996)
9. Gletsos, G., Mougiakakou, S.G., Matsopoulos, G.K., et al.: A Computer-Aided Diagnosic System to Characterize CT Focal Liver Lesion: Design and Optimization of a Neural Network Classifier. IEEE Trans. Information Technolm Biomed. 7(3), 153–162 (2003)
10. Rosner, B.: Fundamentals of Biostatistic, 5th edn. Thomson Learning and Science Press (2000)

A Global Contour-Grouping Algorithm Based on Spectral Clustering

Hui Yin, Siwei Luo, and Yaping Huang

School of Computer and Information Technology, Beijing Jiaotong University, Beijing, 100044, P.R.China
hyin@bjtu.edu.cn

Abstract. Perceptual organization has two essential factors that affect the grouping result directly: how to extract grouping cues and how to grouping. In this paper, a global contour-grouping algorithm based on spectral clustering is presented. First, a new grouping cue called wavelet edge is obtained in multi-scale space, which not only has the property of intensity and direction, but also has the property of singularity measured by lipschitz exponent. Thus grouping cues carry the information of both areas and edges. Secondly, a global grouping approach is presented by use of spectral clustering that has no limitation of neighborhood. Furthermore, the Gestalt principles are used to optimize the grouping result by adding penalty item in iterative process. The experiments show that this algorithm will be effective on condition that the singularities of the edges that belong to one object are equal or close, especially for partially occluded object.

Keywords: Contour grouping, Spectral clustering, Singularity, Wavelet edge, Global feature.

1 Instruction

Perceptual organization is one of the most basic processes of biology vision [1], which is the problem of grouping local image features that project from a common structure (e.g., an object) in the scene [2]. It is not only concerned in the primary visual processing but also concerned in the high lever tasks, Indeed, perceptual organization is used in many levels and domains, starting from low-level processes such as smoothness-based figure-ground discrimination, through motion-based grouping (mid-level processes), to high-level vision processes such as object recognition [3]. Perceptual organization is a problem that can be solved in complete generality, in bottom-up fashion or top-down fashion [2]. The former assumed that visual context and history and the higher-level knowledge and goals of the perceiver have no role to play and the latter says that human visual system does make use of higher-level knowledge to speed and simplify the task of perceptual organization and to resolve ambiguities [4]. Our focus in this paper is to study the important ability of perceptual organization, which is clustering and organizing under conditions of being ignorant of the prior knowledge of input. The Gestalt psychologists noticed that humans use some basic properties to recognize

F. Sun et al. (Eds.): ISNN 2008, Part II, LNCS 5264, pp. 449–456, 2008.

the existence of certain perceptual structures in a scene and to extract the image elements associated with such structures, even before they are recognized as meaningful objects [3]. These properties have been called perceptual grouping cues.

Perceptual grouping cues, such as pixels, edge fragments or blocks and so on, are used in most perceptual organization method in bottom-up fashion. Generally, various perceptual grouping cues are combined to get an effective grouping cue by use of two methods, one is searching with heuristic function [5], and the other is clustering with statistical property. Searching approaches group the image features by detecting the distinct structures making use of the local relationship [6]. Generally, edge snippets are used as grouping cues and the grouping result is effective contour, which is called contour grouping. As one of the important method of perceptual organization, the goal of contour grouping is to obtain a sequence of local edge elements that belong to different objects with clear visual meanings. Contour-based representations are the abstraction of objects, which contain both shape features and topological structures of original objects. There are numerous existing algorithms which grouping contour elements into disjoint contour fragments [7]. However, it is very difficult to compute the complete bounding contour of an object of arbitrary shape in a complex natural image without any prior knowledge. Recently, approaches exploiting the property of closure and proximity have yielded limited success [8]. Whereas searching methods are local because most of them search the appropriate grouping in a definite neighborhood by use of a heuristic function no matter what physical cues are used. In this case, the global topological structures are ignored because of the limitation of neighborhood in actual algorithm although the global features are considered theoretically.

In this paper, a new grouping cue called wavelet edge is used for contour grouping, which has an excellent property of singularity. And we try to eliminate the limitation of neighborhood by use of global spectral clustering. Furthermore, the Gestalt principles are used to optimize the grouping result by adding penalty item in iterative process.

2 Wavelet Edge

Visual system extracts features from an input image by receptive fields with different shape. So the information obtained by cerebra is a multi channels and multi scales expression of actual scene. And the response of cerebra is fast because of the parallel process of the information with different expressions and the cooperation between different expressions [9].

Multi-scale analysis is one of the basic features of human vision. The theory of scale space is the method of multi-scale analysis that is developed in recent years [10]. Since the primary framework of signal could be reflected by its extreme points, the profile or the feature region of the signal could be obtained by locating the extreme points in multi-scale and tracking the characteristics of the extreme points in different scales. The lifecycles of the extreme points could be used for merging unstable regions. Therefore, the theory of scale space is appropriate for study on the human visual system.

Wavelet transform has good local spatial frequency characteristics that are effective especially for analyzing the singularity of the input. And it is feasible to locate the singular points and measure the degree of singularities.

The edges in an image are often the most important features for recognition because they are often the locations of object contour. Edge points are often located where the image intensity has sharp transitions. We define two wavelets that are, respectively, the partial derivatives along $\boldsymbol{x}$ and y of a two-dimensional smoothing function $\theta(x, y)$:

$$\begin{cases} \psi^1(x, y) = \dfrac{\partial \theta(x, y)}{\partial x} \\ \psi^2(x, y) = \dfrac{\partial \theta(x, y)}{\partial y} \end{cases} \tag{1}$$

Let

$$\begin{cases} \psi^1(s, x, y) = \left(\dfrac{1}{s^2}\right)\psi^1\left(\dfrac{x}{s}, \dfrac{y}{s}\right) \\ \psi^2(s, x, y) = \left(\dfrac{1}{s^2}\right)\psi^2\left(\dfrac{x}{s}, \dfrac{y}{s}\right) \end{cases} \tag{2}$$

Let $f(x, y) \in L^2(R^2)$, the wavelet transform defined with respect to $\psi^1(s,x,y)$ and $\psi^2(s,x,y)$ has two components:

$$\begin{cases} W^1 f(s, x, y) = f(x, y) * \psi^1(s, x, y) \\ W^2 f(s, x, y) = f(x, y) * \psi^2(s, x, y) \end{cases} \tag{3}$$

Let

$$|M_s f(s,x,y)| = \sqrt{|W^1 f(s,x,y)|^2 + |W^2 f(s,x,y)|^2} \tag{4}$$

The edges could be extracted by detecting the local modulus maxima at different scales in a similar way. Although the local modulus maxima of wavelet transform could reflect all the singularities of the image, it is more valuable to measure these singularities for perceptual grouping tasks.

Perceptual grouping ought to take more features into account besides the locations of edges. Edges are the singular points that reflect the irregularities of the input. The singular points of the signal could be detected by the discontinuous of the amplitude and the discontinuous of the first order differential simply. To get more precise characterization of the singularities, the Lipschitz exponent is more appropriate actually.

In mathematics, Lipschitz exponent is used to measure the local singularities. The local maxima of the wavelet transform modulus locate the acute change in the input that corresponds to the singularities of the input [10]. There are several possibilities for these singular points. Maybe they're real changes, or burrs, or noise. Wavelet transform is particularly well adapted to estimate the local regularity of the input. The singular points could be classified by their different characteristics of Lipschitz exponent.

The definition of Lipschitz exponent is: Let n be a positive integer and $n \le \alpha \le n+1$. A function f (x) is said to be Lipschitz α, at x_0, if and only if there exists two constants A and $h_0 > 0$, and a polynomial of order n, $P_n(x)$, such that for $h < h_0$:

$$\left|f(x_0+h)-P_n(h)\right| \le A\left|h\right|^{\alpha} \tag{5}$$

Let $x_0 \in R$ and let $f(x) \in L^2(R)$. We suppose that there exists a neighborhood $]a,b[$ of x_0 and a scale $s_0 > 0$, such that the wavelet transform Wf(s,x) has a constant sign for $s < s_0$ and $x \in]a,b[$.Let us also suppose that there exists a constant B and $\varepsilon > 0$, such that for all points $x \in]a,b[$ and any scale s:

$$\left|Wf\ (s,x)\right| \le Bs^{\varepsilon} \tag{6}$$

Let x=X(s) be a curve in the scale space(s,x) such that $\left|x_0 - X(s)\right| \le Cs$,with C<K. If there exists a constant A such that for any scale $s < s_0$, the wavelet transform satisfies:

$$\left|Wf\ (s, X(s))\right| \le As^{\gamma} \text{ With } 0 \le \gamma \le n \tag{7}$$

Then f (x) is Lipschitz α at x_0, for any $\alpha < \gamma$.

According to the above theorem, the Lipschitz regularity of the singularities could be computed from the evolution of the wavelet transform modulus maxima. α and A could be estimated by minimizing the following equation:

$$U = \sum_{j=1}^{J} (\log a_j - \log A - \alpha \log s) \tag{8}$$

Where $a_j = \max_{\Lambda} \left|Wf\,(s,x)\right|$, $s = 2^j$, and the largest scale is 2^J.

Thus, a new kind of perceptual grouping cue is obtained which is called wavelet-edge. It has the property of intensity, direction, location and singularity,

3 Grouping Algorithm

Here, a clustering approach operates on the obtained grouping cues, wavelet edge. Taking the global feature into account, a spectral method based on the graph partitioning is presented. Node of the graph is wavelet edge. Link weight between nodes is determined by Gestalt principles such as parallelism, perpendicularity, proximity, continuity and common region [5]. These Gestalt relations are expressed by distance between end points of edges, intensity and so on. One of the important problems of graph partitioning is how to calculate the affinity between the nodes. There are many researches on transforming the Gestalt principles into computational models [8]. Most of them construct the affinity function according to the models and group cues by the obtained affinity. However, some principles, such as closure, are difficult to measure pairwise. Here, an iterative algorithm that meets the Gestalt principles is used for grouping [11].

First, the affinity of the grouping cues is measured by the proximity and similarity, which are basic parts of the Gestalt principles.

We construct a weighted graph G= (V,E), by taking each wavelet edge as a node and connecting each pair of wavelet edge by an edge. That is $V = \{v_1, v_2, v_3, \ldots\ldots, v_n,\}$,where v_i is the grouping cues. In this algorithm, $V = \{v_1, v_2, v_3, \ldots\ldots, v_n,\}$ is the set of the obtained wavelet edge. Let $F_i\ ,i = 1,\ldots\ldots,n$ be the feature vector of $v_i\ ,i = 1,\ldots\ldots,\ n$. The weight w_{ij} on each edge $e_{ij} \subset E$ represent the affinity between nodes i and j, which could be calculated by the following function:

$$w_{ij} = e^{-\frac{\|F_i - F_j\|^2}{\sigma}} \qquad i, j = 1,..n \tag{9}$$

Here, $F_i\ ,i = 1,\ldots\ldots,n$ are vectors that consist of five elements, σ is a scale factor. $F_i = \{f_1, f_2, f_3, f_4, f_5,\}_i\ ,i = 1,\ldots\ldots,n$, where f_1 measures the average singularity of the edge that could be calculated by the lipschitz exponent, $\{f_2, f_3\}$ is the coordinate of the midpoint of the edge which represent the physical location of it, and . f_4, f_5 represent the direction and the average intensity of the edge.

Thus, the affinity function could be written as following:

$$\begin{aligned} w_{ij} &= e^{-\frac{\|F_i - F_j\|^2}{\sigma}} \\ &= e^{-\frac{\left(|f_{i1}-f_{j1}| + \sqrt{(f_{i2}-f_{j2})^2 + (f_{i3}-f_{j3})^2} + |f_{i4}-f_{j4}| + |f_{i5}-f_{j5}|\right)^2}{\sigma}} \\ & \qquad i, j = 1,..n \end{aligned} \tag{10}$$

Then spectral clustering approach such as Ncut [11] should be operated on the weighted graph G= (V, E). Solve for the eigenvectors with the smallest eigenvalues of the system:

$$(D - W)y = \lambda Dy$$

$$\text{where, } D = diag(d_1, d_2, \ldots, d_n)$$

$$d_i = \sum_{j=1}^{n} w_{ij} \tag{11}$$

Once the eigenvectors are computed, a partition of graph will be obtained by using the second smallest eigenvector as Ncut[11] is minimized.

Thus, a primary contour grouping is obtained because the partition result is a sequence of edges. As discussed above, some Gestalt principles are difficult to be quantified between the pairwise edges. But it is liable to do it after a primary grouping is obtained. We define a penalty item of the affinity by measure the closure of the obtained grouping. A saliency measure of closure [8] is used to obtain the penalty item. The approach could be described as following:

Step1, a partition result is obtained by clustering the wavelet edge.
Step 2, Detect a closed contour in the partition result.
Step 3, if some edges in a partition compose a closed contour, a positive penalty item will be added to the weighted value between the edges that belong to the closed contour.
Step 4, A negative penalty item will be added to the weighted value between the edges on the closed contour and those not on it.
Step 5, if latest two partitions are same, the iteration stops, otherwise, return to step 1.

4 Experiments

To explain the property of the wavelet edge, an experiment on an artificial image is shown in Fig1. As we can see, edges detected by canny operator have no information of the area on both side of the edge and the edges shown in Fig1 (b) have same property. Thus it is difficult for grouping process to distinguish them to three separate parts as human vision will do. However, the wavelet edges have the property of accurate singularity as shown in Table 1. So it is easy to clustering them into three groups.

An experiment for grouping on an artificial image that has a partially occluded object is shown in Fig 2. As shown in Fig2.(d), e_3 and e_{16} are very close. It is likely that they will be grouped into the same object by using conventional grouping algorithm that takes the contiguity into account. e_7 and e_1, e_{18},e_{19} and e_{11} , are in a similar way However ,the grouping algorithm based on spectral clustering acquires a satisfactory result by using the wavelet edge as cues and iterative process by detecting closure contour.

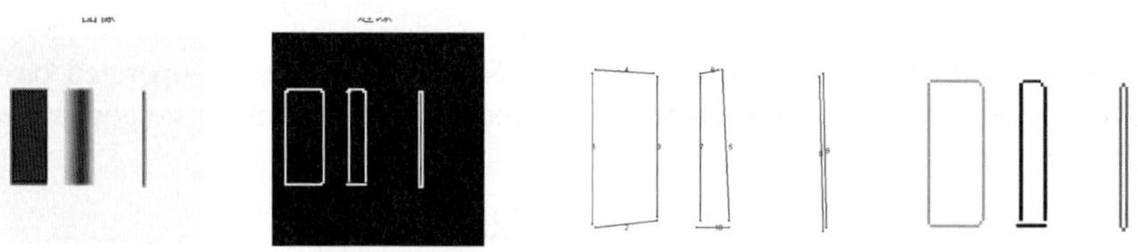

(a) (b) (c) (d)

Fig. 1. The property of wavelet edge. (a) original image. (b) Edges detected by canny operator. (c) The label of the edges (d) the image of lipschitz exponent

Table 1. The property of wavelet edge

label	Lipschitz exponent	x_{middle}	y_{middle}
1	0.82017	49	7
2	0.7959	72	16
3	0.80438	49	25
4	0.83799	27	16
5	0.94113	49	45
6	0.92758	27	40
7	1	49	37
8	0.5872	51	70
9	0.58231	50	72
10	0.95706	72	41

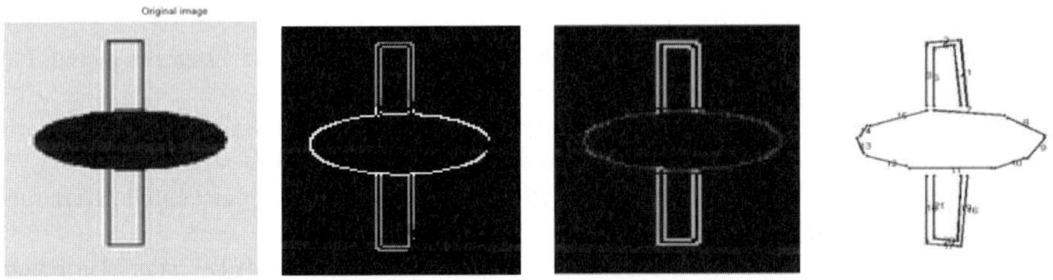

(a) (b) (c) (d)

Fig. 2. Grouping by wavelet edge (a) original image.(b) the image of lipschitz exponent.(c) grouping result (d) The label of wavelet edge.

5 Conclusions

In this paper, a global contour-grouping algorithm based on spectral clustering is presented. As a new perceptual grouping cue, wavelet edge has the excellent properties that carry the information of both areas and edges. And Gestalt principles are used to optimize the grouping result by adding penalty item in iterative process which is liable to realize by modeling the Gestalt principles. The advantages of this algorithm are groping cues with more rich information and global features by means of Gestalt principles. The experiments show that our algorithm is effective on the condition that the singularities of the edges that belong to one object are equal or close. Especially, it will be more excellent when the object is partially occluded and the grouping cues that belong to the same object are disconnected. It is worth to apply this algorithm to real image for grouping task.

Acknowledgments. This work is supported by National Nature Science Foundation of China (60773016) and the National High Technology Research and Development Program of China (2007AA01Z168).

References

1. Marr, D.: Vision. W.H. Freeman, San Francisco (1982)
2. Elder, J.H., Krupnik, A., Johnston, L.A.: Contour Grouping with Prior Models. IEEE Trans. on Pattern Analysis and Machine Intelligence 25(6), 661–674 (2003)
3. Amir, A., Lindenbaum, M.: A Generic Grouping Algorithm and Quantitative Analysis. IEEE Trans. on Pattern Analysis and Machine Intelligence 20(2), 168–185 (1998)
4. Cavanagh, P.: Top-down processing in vision. In: Wilson, R.A., Keil, F.C. (eds.) MIT Encyclopedia of Cognitive Science, pp. 844–845. MIT Press, Cambridge (1999)
5. Golubchyck, R., Lindenbaum, M.: The Analysis of Saliency Processes and Its Application to Grouping Cues Design. In: IEEE Trans. on CBMI, pp. 18–24 (2007)
6. Jitendra, M., Serge, B., Thomas, L., Jianbo, S.: Contour and Texture Analysis for Image Segmentation. Int. J. of Comp. Vision 43(1), 7–27 (2001)
7. Jacobs, D.W., Lindenbaum, M.: Guest Editors' Introduction to the Special Section on Perceptual Organization in Computer Vision. IEEE Trans. on Pattern Analysis and Machine Intelligence 25(6), 641 (2003)
8. Qi, Z.: Research on Computational Model of Contour Grouping and Attention model. PhD thesis, Beijing Jiaotong University (2006)
9. Hui, Y., Siwei, L., Yu, Z.: Visual Perceptual Grouping Algorithm Based on the Singularity of Edge. Journal of Computational Information Systems 2(4) (2006)
10. Witkin, A.P.: Scale Space Filtering. In: Proc. Int. Joint Conf., Artificial Intell., Karlsruhe, pp. 1019–1022 (1983)
11. Jianbo, S., Malik, J.: Normalized Cuts and Image Segmentation. IEEE Trans. on Pattern Analysis and Machine Intelligence 22(8) (2000)

Emotion Recognition in Chinese Natural Speech by Combining Prosody and Voice Quality Features

Shiqing Zhang

School of Physics and Electronic Engineering, Taizhou University
318000 Taizhou, China
tzczsq@163.com

Abstract. Nowadays, recognition of human emotion is a challenging yet important speech technology. In this paper, based on deriving prosody features from emotional speech, some voice quality features are proposed to be extracted as new emotional features to improve emotion recognition. Utilizing support vector machines classifier, four emotions from Chinese natural emotional speech corpus including anger, joy, sadness and neutral are discriminated by combining prosody and voice quality features. The experiment results show that combining prosody and voice quality features yields an overall accuracy of 76% for emotion recognition, which makes approximately 10% improvement compared with using the single prosody features. It also shows that voice quality features in speech are effective emotional features and can promote prosody features for improving emotion recognition results.

Keywords: Prosody features, Voice quality features, Emotion recognition.

1 Introduction

The recognition of emotions in human speech has recently received increasing attention for building more intuitive human-computer interaction [1]. Speech usually comes to mind first when thinking about possible methods to recognize human emotions. Such applications are useful in areas where human interact with automated systems like call centre [2], interactive movies [3], etc.

Most previous studies [4,5] aiming at the automatic recognition of emotions in speech have made much use of prosody features, such as pitch, intensity and duration, which are easier to handle but give mainly information related to the arousal dimension of emotion. But there is mounting evidence that shows it's difficult to distinguish these emotions like anger and joy with the almost same level in the arousal dimension of emotion when only using prosody features in speech [6].

Recently, the importance of voice quality features in the valence dimension of emotion has been highlighted in the literatures [7,8]. Voice quality features changed when different utterances occurred due to the different phonation types such as breathy, creaky, harsh, etc. Accordingly, in this paper we will extract not only basic prosody features but also several voice quality features and then combine them to identify four emotions including anger, joy, sadness as well as neutral in Chinese

F. Sun et al. (Eds.): ISNN 2008, Part II, LNCS 5264, pp. 457–464, 2008.

mandarin. These proposed voice quality features are first three formants, spectral energy distribution in four different frequency bands, harmonics to noise ratio, jitter and shimmer.

This paper is structured as follows. In Section 2, the distribution of emotions in arousal-valence space is described. The natural speech corpus and feature extraction are detailed in Section 3. Section 4 introduces the experiment study. Section 5 discusses a number of experiment results with different emotional features. Finally, in Section 6, the conclusions are presented.

2 Distribution of Emotions in Arousal-Valence Space

The description of emotions can be conceptualized as a three continuous dimensional space model called the arousal-valence-power space model [6]. The arousal dimension refers to how active or passive the emotion is. The valence dimension refers to how positive or negative the emotion is. The power dimension refers to the degree of power or sense of control over the emotion. In practice, the two dimensions of arousal and valence model is usually utilized to describe different emotions as shown in figure 1. From the figure 1, it can be seen that joy and anger show nearly the same highest rating on the arousal dimension, but very different high rating on the valence dimension. It's thus clear that only using the prosody features reflecting the arousal dimension could not classify joy and anger well and voice quality features reflecting the valence dimension are needed to extract.

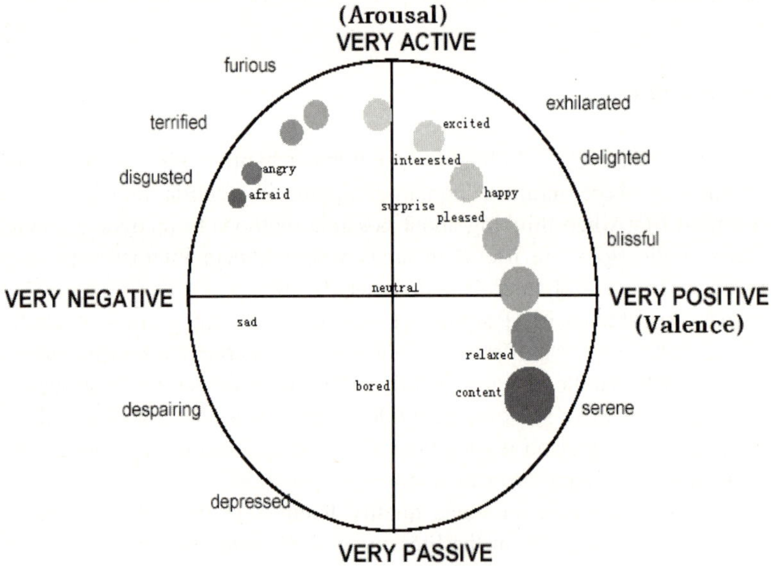

Fig. 1. Two dimensions of emotion space model

3 Speech Corpus and Feature Extraction

3.1 Natural Emotional Speech Corpus

To study the emotion recognition on authentic emotion in speech, the natural emotional speech corpus used in this study was collected from 20 different Chinese dialogue episodes from a talk-show on TV. In each talk-show, two or three persons discuss the problems such as social typical issues, family conflict, inspiring deeds, etc. Due to the spontaneous and unscripted manner of the episodes, the emotional expressions can be considered authentic. Because of the limited topics, the speech corpus consists of four kinds of common emotion including anger, joy, sadness and neutral. This corpus collected in total contains 800 emotional utterances from 53 different speakers (16 male / 37 female), speaker-independent, each of four emotion for about 200 utterances. All utterances recorded at a sample rate of 16 kHz and 16 bits resolution with mono-phonic Windows WAV format stored in computer. In addition, another 4 different speakers listened to test the effectiveness of the whole utterances in random order so that the utterances for human bad test were eliminated and recollected.

Feature Extraction

After extracting basic 25 prosody features, 23 voice quality features are also extracted from the natural emotional speech corpus. These extracted 48 features in total are statistical as follows:

♦ Prosody features:

(1-10)Pitch: maximum, minimum, range, mean, std (standard deviation), first quartile, median, third quartile, inter-quartile range, mean-absolute-slope.

(11-19)Intensity: maximum, minimum, range, mean, std, first quartile, median, third quartile, inter-quartile range.

(20-25)Duration: total-frames, voiced-frames, unvoiced-frames, ratio of voiced to unvoiced frames, ratio of voiced-frames to total-frames, ratio of unvoiced-frames to total-frames (20ms/ frame).

♦ Voice quality features:

(26-37)First three formants F1-F3: mean of F1, std of F1, median of F1, bandwidth of median of F1, mean of F2, std of F2, median of F2, bandwidth of median of F2, mean of F3, std of F3, median of F3, bandwidth of median of F3.

(38-41)Spectral energy distribution in 4 different frequency bands: band energy from 0 Hz to 500 Hz, band energy from 500 Hz to 1000 Hz, band energy from 2500 Hz to 4000 Hz, band energy from 4000 Hz to 5000 Hz.

(42-46)Harmonics-to-noise-ratio: maximum, minimum, range, mean, std.

(47)Jitter: pitch perturbation in vocal chords vibration

Jitter is calculated with the following equation (1), in which T_i is the i -th peak-to-peak interval and N is the number of intervals:

$$Jitter\ (\%) = \sum_{i=2}^{N-1} (2T_i - T_{i-1} - T_{i+1}) \Big/ \sum_{i=2}^{N-1} T_i \tag{1}$$

(48)Shimmer: perturbation cycle to cycle of the energy

Shimmer is calculated similarly to Jitter as shown in equation (2), in which E_i is the i -th peak-to-peak energy values and N is the number of intervals:

$$Shimmer(\%) = \sum_{i=2}^{N-1}(2E_i - E_{i-1} - E_{i+1}) \Big/ \sum_{i=2}^{N-1} E_i \,. \tag{2}$$

4 Experiment Study

4.1 Feature Parameter Importance Analysis

To illustrate which prosody and voice quality feature parameters are the most important to distinguish different emotion categories, a method of feature selection with information-gain [9] was implemented for evaluating the worth of an feature attribute by measuring the information gain with respect to the class. The feature parameter importance analysis by conducting information-gain showed the 24 most important feature parameters to distinguish the emotions from the corpus in Table 1. From the results in Table 1, we can see that many prosody feature parameters are taken into account, but also several voice quality feature parameters, especially shimmer, band energy from 4000 Hz to 5000 Hz, F2-F3 related feature, and so on, are used. Thus, the features ranking shows that voice quality features and prosody features are both important for emotion recognition.

4.2 Classifier Selection

Initially, a number of classifiers—including K-Nearest Neighborhood (KNN), Gaussian Mixture Models (GMM) and Support Vector Machines (SVM)—were tested in speech emotion classification tasks. SVM [10] yielded the best performance and thus were selected in the following experiments. SVM is based on the statistical learning theory of structural risk management which aims to limit the empirical risk on the training data and on the capacity of the decision function. SVM is built by mapping the training patterns into a higher dimensional feature space where the points can be separated using a hyperplane. In WEKA toolkit used in this study, SVM are implemented as the Sequential Minimal Optimization (SMO) algorithm [11]. The Radial Basis Function (RBF) kernel were used for its better performance compared with other kernels. The RBF kernel is defined as

$$K(x_i \cdot x_j) = \exp(-\gamma \|x_i - x_j\|^2), \gamma > 0 \tag{3}$$

For all classification experiments, we used SVM classifier to employed 10-fold stratified cross validations over the data sets so as to achieve more reliable experiment results. In other words, each classification model is trained on nine tenths of the total data and tested on the remaining tenth. This process is repeated ten times, each with a different partitioning seed, in order to account for variance between the partitions.

Table 1. Feature parameter importance analysis

Information gain	Feature parameters
1.990	shimmer
1.858	band energy from 4000 Hz to 5000 Hz
1.048	ratio of voiced to unvoiced frames
0.505	third quartile of pitch
0.482	first quartile of F3
0.447	mean of pitch
0.378	mean of intensity
0.357	band energy from 2500 Hz to 4000 Hz
0.349	maximum of intensity
0.319	third quartile of intensity
0.318	inter-quartile range of pitch
0.289	std of F3
0.277	maximum of pitch
0.269	mean-absolute-slope of pitch
0.259	band energy from 500 Hz to 1000 Hz
0.246	std of F2
0.232	std of pitch
0.223	median of intensity
0.204	median of F3
0.184	range of pitch
0.138	unvoiced-frames
0.136	minimum of pitch
0.135	mean of harmonics-to-noise-ratio
0.134	bandwidth of median of F1

5 Experiment Results

After scaling all extracted features data to the range [0, 1], firstly, we utilized SVM classifier to experiment for four emotions identification with the single 25 prosody features related to the arousal dimension. The emotion recognition results were showed in Table 2 with confusion matrix. As shown in Table 2, the classification results indicated that anger and neutral could be discriminated well with an accuracy of 80% and 86%, respectively. While other two emotions, joy and sadness, only could be classified with low recognition rate of 48%, 50%, respectively. The main reasons are that "joy" has a high confusion to "anger" and "sadness" for each other. The average recognition accuracy with prosody features only comes up to 66%.

Table 2. Confusion matrix with the single 25 prosody features

	Anger	Joy	Sadness	Neutral
Anger	0.80	0.13	0.04	0.03
Joy	0.21	0.48	0.26	0.05
Sadness	0.17	0.26	0.50	0.07
Neutral	0.00	0.03	0.11	0.86

Secondly, we experimented with the single 23 voice quality features, reflecting the valence dimension, for four emotions recognition and showed the recognition results with confusion matrix in Table 3. From the results in Table 3, we can see that different emotion classification accuracies achieved are, 58% for anger, 44% for joy, 56% for sadness, 80% for neutral. The average recognition rate with voice quality features has come up to about 60%, a little lower to the results with prosody features. In consequence, the results show that voice quality features can get classifying performance nearly as the same as prosody features.

Table 3. Confusion matrix with the single 23 voice quality features

	Anger	Joy	Sadness	Neutral
Anger	0.58	0.21	0.02	0.19
Joy	0.37	0.44	0.05	0.14
Sadness	0.09	0.15	0.56	0.20
Neutral	0.07	0.08	0.05	0.80

Finally, we combined the 25 prosody features and 23 voice quality features to experiment and showed the recognition results with confusion matrix in Table 4. As shown in Table 4, combining prosody and voice quality features, the average accuracy achieved is up to about 76%, in detail, anger for 84%, joy for 70%, sadness for 62% and neutral for 86%, respectively. It's clear emotion classification results using prosody and voice quality features outperformed the previous two experiment results showed in Table 2 and Table 3. The three experiment results indicate that, the single prosody features and the single voice quality features yield an overall accuracy of 66%, 60%, respectively. Whereas combining voice quality and prosody features yields an overall accuracy of 76%, making approximately 10%-16% improvement for emotion recognition, compared with the single prosody features and the single voice quality features.

Table 4. Confusion matrix with prosody and voice quality features

	Anger	Joy	Sadness	Neutral
Anger	0.84	0.11	0.05	0.00
Joy	0.17	0.70	0.10	0.03
Sadness	0.07	0.21	0.62	0.10
Neutral	0.00	0.10	0.03	0.87

6 Conclusions

This paper presents a comparative study of prosody and voice quality features used for emotion recognition from Chinese natural speech corpus. From the results obtained, we may conclude that voice quality features can also make high contribution to discriminate different emotion categories as well as prosody features. Thus combining prosody and voice quality features is found to be more effective than the single kind of features for speech emotion identification described in this study. In the future, we aim to extend our natural emotional speech corpus and identify more other different emotion categories such as disgust, afraid, surprise and so on.

References

1. Cowie, R., Douglas-Cowie, E., Tsapatsoulis, N., Votsis, G., Kollias, S., Fellenz, W., Taylor, J.G.: Emotion Recognition in Human-Computer Interaction. IEEE Signal Processing Magazine 18(01), 32–80 (2001)
2. Lee, C.M., Narayanan, S.S.: Toward Detecting Emotions in Spoken Dialogs. IEEE Transactions on Speech and Audio Processing 13(2), 293–303 (2005)
3. Nakatsu, R., Nicholson, J., Tosa, N.: Emotion Recognition and Its Application to Computer Agents with Spontaneous Interactive Capabilities. Knowledge-Based Systems 13(7-8), 497–504 (2000)
4. Ang, J., Dhillon, R., Krupski, A., Shriberg, E., Stolcke, A.: Prosody-Based Automatic Detection of Annoyance and Frustration in Human-Computer Dialog. In: Proceedings of the ICSLP, Denver, Colorado, pp. 2037–2039 (2002)
5. Schuller, B., Rigoll, G., Lang, M.: Hidden Markov Model-Based Speech Emotion Recognition. In: Proceedings of the ICASSP, Hong Kong, vol. 2, pp. 1–4 (2003)
6. Tato, R., Santos, R., Kompe, R., Pardo, J.M.: Emotional Space Improves Emotion Recognition. In: Proceedings of the ICSLP, Denver, Colorado, pp. 2029–2032 (2002)
7. Gobl, C., Ni-Chasaide, A.: The Role of Voice Quality in Communicating Emotion, Mood, and Attitude. Speech Communication 40, 189–212 (2003)

8. Johnstone, T., Scherer, K.R.: The Effects of Emotions on Voice Quality. In: Proceedings of the XIVth International Congress of Phonetic Science, San Francisco, pp. 2029–2032 (1999)
9. Yang, Y., Pedersen, J.O.: A Comparative Study on Feature Selection in Text Categorization. In: Proceedings of the 14th International Conference on Machine Learning ICML 1997, pp. 412–420 (1997)
10. Vapnik, V.: The Nature of Statistical Learning Theory. Springer, New York (1995)
11. Platt, J.: Fast training of support vector machines using sequential minimal optimization. In: Schoelkopf, B., Burges, C., Smola, A. (eds.) Advances in Kernel Methods-Support Vector Learning. MIT Press, Cambridge (1998)

On-Line Diagnosis of Faulty Insulators Based on Improved ART2 Neural Network

Hailong Zhang, Weimin Guan, and Genzhi Guan

Department of Electrical Engineering, Wuhan University, Wuhan 430072, China
Longyan04@163.com

Abstract. In this paper, an improved ART2 neural network is applied to on-line diagnosis of faulty insulators. Deterioration of insulator is a gradual phenomenon. Thus, the cluster centers tend to drift during the process of recognizing on-line monitor signals. The drifts of cluster centers cause wrong judgments. In order to solve this problem, an initial layer is settled in layer F_2. The improved ART2 neural network divides layer F_2 into upper-sublayer F_{22} and lower-sublayer F_{21}. When insulators start working for the first time, the upper-sublayer F_{22} store initial states and typical-malfunction patterns. Furthermore, the improved ART2 structure of the network map, and detailed algorithm flowchart are also illustrated in this paper.

1 Introduction

Adaptive Resonance Theory (ART) was developed by Grossberg and Carpenter in 1987[1-2]. The network ART1 was designed for clustering binary vectors, and ART2 was designed for clusters continuous valued vectors. Attentional subsystem and orienting subsystem of ART2 are more complicated than ART1. The ART2 network can cluster the inputted patterns according to any precision by modifying vigilance parameter—ρ.

Insulators are essential apparatus for electricity transmission line. Faulty insulators seriously threat the security of power supply. According to statistic data, power transmission outage time caused by insulator fault can occupy over 50 percent in the whole power failure time. Thus, the immediate detection of potential faults of insulators is very important for transmission lines' safe operation. At present, there are many methods for detecting faulty insulators, such as measuring insulation resistance, using spark gap, pulse count of corona current, detecting voltage distribution and IR imaging of insulators. Whereas, the pulse count of corona discharge current is a relatively effective method, that doesn't require operators to check the insulators by climbing the power towers. Further more, it can realize remote monitoring the insulators on transmission lines through GPRS and satellite communication network.

2 The Principle of Insulator On-Line Diagnosis

For composite insulators, the degradation of machinery and electric properties results from initial flaw and silicone rubber aging owing to long-term field operation. In

F. Sun et al. (Eds.): ISNN 2008, Part II, LNCS 5264, pp. 465–472, 2008.

transmission line operation, potential distribute on each insulator. However, with the appearing of faulty insulators, voltage distribution on the insulators deviates. Simultaneously, the insulation resistance of faulty insulators will drop obviously, and the voltage distribute on the faulty insulators are lower than others. Thus, normal insulators will bear higher voltage and the corona discharge aggravate. If the corona discharge intensity exceeds certain degree, the insulators will be damaged. We obtain pulse corona discharge signals from sensors, which are installed in ironware between the power tower and the insulators. The frequency spectrum of corona discharge pulse is less than 20MHz, AD9237 had been chosen as A/D convert of the equipments to collect the current pulse of corona discharge. AD9237 meets the requirement of data sampling very well in the experiment. A cycle of power frequency voltage is divided into 360 small zones, and every small zone can be regarded as a unit. Accounting the number of corona discharge pulse current that exceed a pre-set current value in each small unit, an N-φ diagram was drawn. Finally, through ART2, the insulators have been judged whether fault or not.

3 Traditional ART2 Neural Network

3.1 Introduction to Traditional ART2 Neural Network

The basic principle of transitional ART2 neural network is on competitive learning and self-stabilized learning. The ART2 is made up of one reset module and LTM (long time memory) that is between F_1 and F_2 and two STM (short time memory) layers that are settled in F_1 and F_2. The layer F_1 is divided into three sublayers: F_{11}, the lower-sublayer, inputs signals; F_{12}, the middle-sublayer, mixes the inputted signals and feedback; F_{13}, upper-sublayers, receives feedback signals from layer F_2. The structure of ART2 neural network has obtained a fine balance between stability and plasticity. The 'fast learning' of ART2 can fully fulfill the requirement of insulator on-line monitoring. Layer F_1 adopts positive feedback and nonlinear transformation to strengthen the essential character of inputted patterns, suppress noise and decrease the unmatched possibilities caused by the nonessential difference.

ART2 neural network can realize both supervised learning and unsupervised learning. If a new inputted pattern is similar enough to the existing patterns, the ART2 will classify the new one into the existing; if not, it will create a new cluster for the inputted pattern. This is the basic function of the ART2 network.

3.2 Deficiency of Traditional ART2 Network

According to the principle of traditional ART2 network, most signals collected by insulator on-line monitoring equipments are gradually changing. Traditional ART2 network takes the patterns which have been storied in the LTM as the only cluster standard. With the continuously inputting patterns into ART2 networks, the focus of a certain cluster would be so drifting that it would be far from the initial cluster focus. At last ART2 would give a wrong conclusion. Therefore, traditional ART2 networks can't be directly applied to the pattern recognition of faulty insulators. So, according to the feature of insulator on-line diagnosis, a scheme on improved ART2network has been given in this article.

4 Improved Structure of ART2 Neural Network

The improved ART2 neural network can greatly compensate for the lack of typical-malfunction pattern. It is designed for supervised learning and unsupervised learning. Improved ART2 network not only makes a judgment according to the typical-malfunction pattern, but also can continuously learn new malfunction patterns. It can realize 'primary form' and 'advanced form' of faulty insulator diagnosis. 'Primary form' can judge whether the insulator has broken down, but can't diagnose which kind of trouble it belongs to. After summing up various faulty patterns happened in the same batch, the improved ART2 network can realize 'advanced form' of fault insulator diagnosis. Then it can more accurately judge the type of the malfunction such as One Zero Value Insulator or Two Zero Value Insulator in the whole insulator.

4.1 The Construction of The Improvement ART2 Neural Network

The improved network divides layer F_2 into upper-sublayer (F_{22}) and lower-sublayer (F_{21}). F_{22} is the initial condition layer. F_{21} is the general condition layer whose function is equivalent to F_2 in traditional ART2 network. The new construction is shown in Fig.1.

4.2 Initialization of F22

During the new insulator's first working time, supposing the electricity performance and machinery performance of the insulator is good, a series of pulse signals caused by corona discharge should be recoded. Inputting the batch of signals into the ART2 network, we will get top-down and bottom-up weights and save them into the LTM

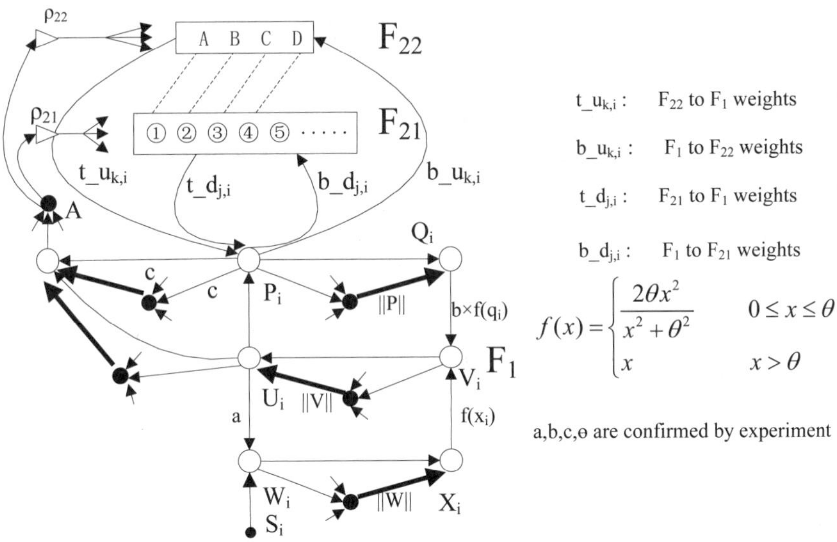

Fig. 1. Structure chart of improved ART2 neural network

which is between layer F_{22} and layer F_1. Then, a new cluster can be built, which represents layer F_{22} insulators in working order. Through the identical method, some representative fault signal templates have been saved into the LTM which is between layer F_{22} and layer F_1. At length, the initializing of layer F_{22} has finished. As shown in figure 1, there are four clusters in F_{22}, one is normal and three is faulty. The LTM between F_{22} and F_1 is not updated when learning a new inputted pattern, and only participates in the mode recognition. During the long-termed operation, no permission is given to increase or modify LTM until ART2 network find some new representative fault signal patterns. Obviously, it shows the man's absolute control of the neural network, and effectively avoids the misjudgment caused by the learning at will.

4.3 Initialization of F_{21} and Other Parameter

In improved ART2 network, supposing layer F_{22} has five kinds of clusters during initializing and the five kinds of F_{21} is one-to-one correspondence with the other five kinds of F_{22}. We copy weights (between F_{22} and F_1) to LTM (between F_{21} and F_1). The bellow is the modified formula

$$\begin{cases} t\_d_{j,i} = t\_u_{k,i} \\ b\_d_{j,i} = b\_u_{k,i} \end{cases} \quad \text{j=1 to g} \quad \text{k=1 to g} \quad (\text{g is the number of clusters in } F_{22}) \tag{1}$$

$$\begin{cases} b\_d_{j,i} \le \dfrac{1}{(1-d)\times\sqrt{n}} \\ t\_d_{j,i} = 0 \end{cases} \quad \text{j=g+1 to m (m is the number of clusters in } F_{21} \tag{2}$$

i=1 to n (n is the number of nodes in F_1

The choice of other parameter is the same with traditional ART2 networks.

4.4 Updated Weights

In improved ART2 networks, both the kinds of F_{22} and the kinds of F_{21} are only modifying the weights between F_{21} and F_1. For instance, when the inputted pattern is classified to be k_F_{22}_max cluster in layer F_{22}, the ART2 network will only update the corresponding weights in F_{21}. The formula of updated weights is:

$$t\_d_{k\_F_{22}\_\max,i}(l+1) = d\times u_i + t\_d_{k\_F_{22}\_\max,i}(l)\times\left[1+d\times(d-1)\right] \tag{3}$$

$$b\_d_{k\_F_{22}\_\max,i}(l+1) = d\times u_i + b\_d_{k\_F_{22}\_\max,i}(l)\times\left[1+d\times(d-1)\right] \tag{4}$$

k=1 to g; i=1 to n

When the inputted pattern is classified to be j_F_{21}_max cluster in layer F_{21}, the formula of updated weights is:

$$t\_d_{j\_F_{21}\_\max,i}(l+1) = d\times u_i + t\_d_{j\_F_{21}\_\max,i}(l)\times\left[1+d\times(d-1)\right] \tag{5}$$

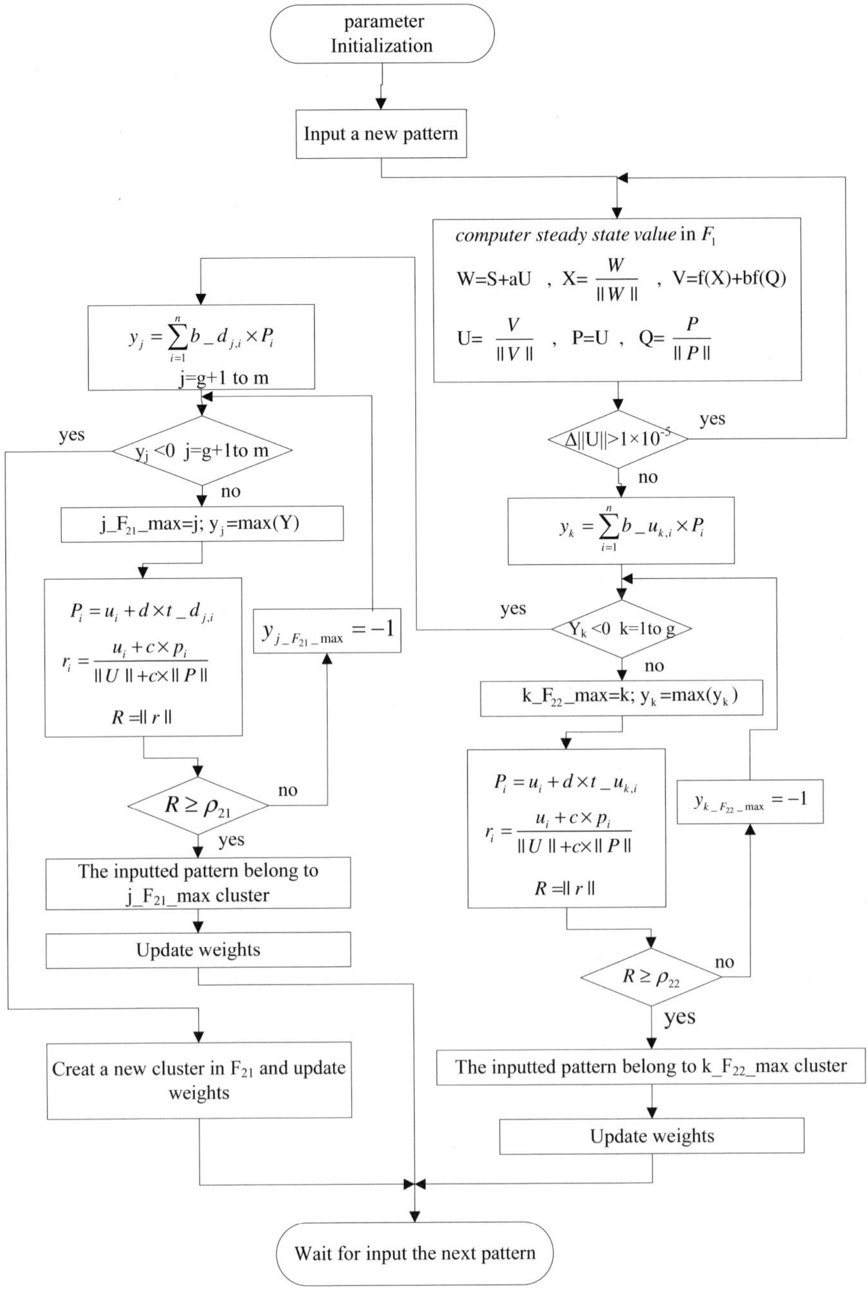

Fig. 2. Flow chart of algorithms of the improved ART2 neural network

$$b\_d_{j\_F_{21}\_\max,i}(l+1) = d \times u_i + b\_d_{j\_F_{21}\_\max,i}(l) \times \left[1 + d \times (d-1)\right] \tag{6}$$

j=g+1 to m; i=1 to n

This improved ART2 network can not only guarantee the initial cluster focus not so drifting as to misjudge, but also can learn new patterns and prevent from misjudging.

4.5 The Algorithms of Improved ART2 Neural Network

There are some difference between improved ART2 algorithms and traditional ART2. The detailed description is as followed. First, initialize parameters, input a new signal template and compute steady value of layer F_1. Second, judge the inputted pattern belongs to which kind of the existing clusters and update the corresponding weights. When the inputted pattern doesn't belong to any cluster of layer F_{22}, it is start to search layer F_{21}. If the similar cluster has been found in layer F_{21}, the corresponding weights will be updated; if not, a new cluster will be created in layer F_{21}. The flow-chart of improved ART2 algorithms is shown in Fig.2.

5 Simulation

Based on the above improved ART2 network, we have carried on simulation research. The corona discharge of insulators is faint under the good electricity performance and machinery performance. The n-φ map is shown in figure 3. Corona discharge begins to aggravate when the insulators have some bug. It accelerates the deterioration process of insulator. Ultimately, insulator flashover has happened. Figure 4-(a) and 4-(b) are two typical-malfunction patterns saved in F_{22}_A and F_{22}_B.

A set of N-φ maps, which are results of simulating the process of insulator deterioration, is shown in Fig.5. Traditional ART2 neural network classify them as a same cluster, while the classify result of improved ART2 is show in table 1.

From the results of pattern recognition, it is obviously that the improved ART2 network can recognise whether the insulator is fault or not, which the traditional ART2 network cannot.

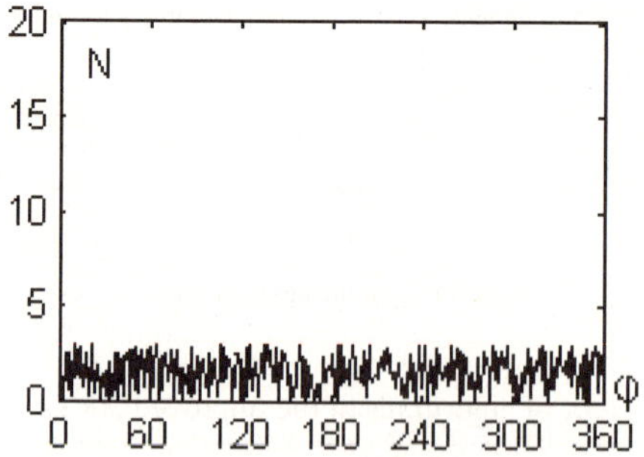

Fig. 3. N-φ map of normal insulator

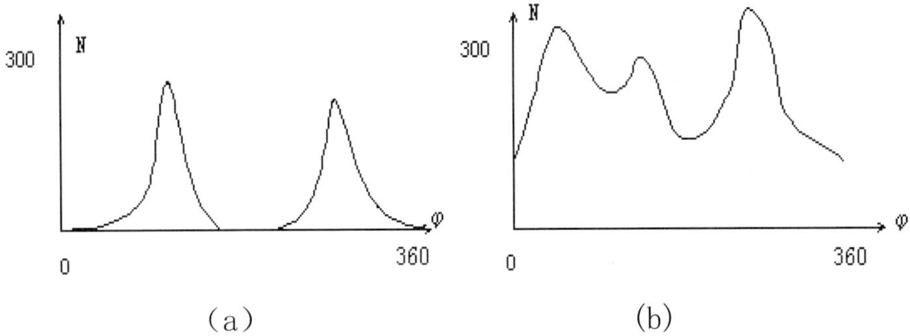

Fig. 4. Two typical-malfunction patterns of fault insulators

Table 1. The classify result of improved ART2

Cluster	F22_A	F22_B	F22_C	F22_D	F21_⑤
serial number of in-putted pattern	1,2,3	10-20			4-9

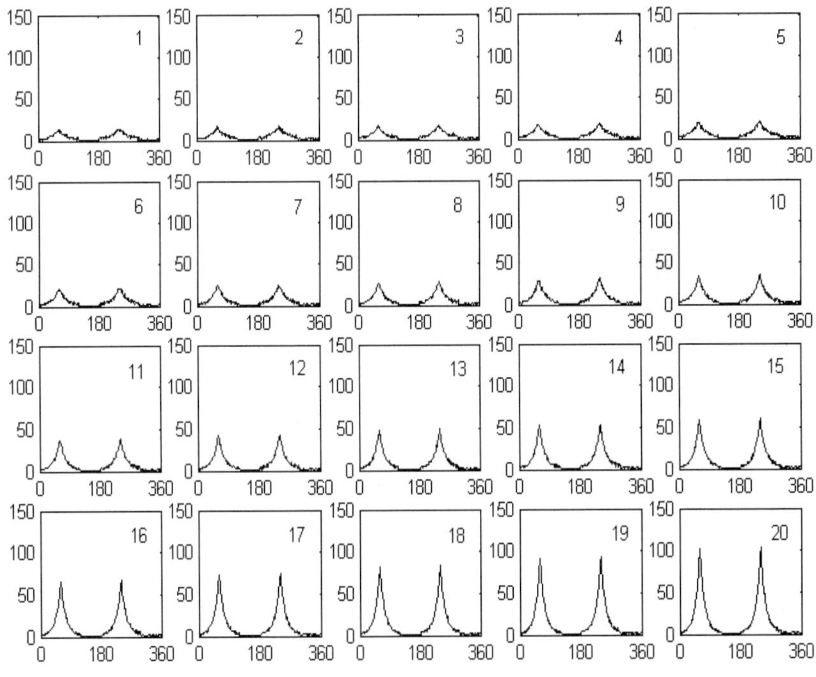

Fig. 5. The N-φ map of insulator during its insulation property gradually deteriorates

6 Conclusions

In some cases, the traditional ART2 network can result in a drift of cluster center, and it is not proper for accurate on-line diagnosis. The improved ART2 network described in this paper establish initial layer F_{22} and adopts the corresponding method of layer initializing, parameter setting, weight modifying and computing method, so it completely overcomes the cluster misjudgment problem. The results of simulation experiment show that the improved ART2 network is effective for on-line diagnosis of faulty insulators.

Prospectively, on-line diagnosis of faulty insulators should take weather factors and electromagnetic interference around insulators into consideration. In addition, with adopting a combined neural network, which is constituted by several of different neural networks, the network will make a better on-line diagnosis of faulty insulator.

References

1. Carpenter, G.A., Grossberg, S.: A Massively Parallel Architecture for A Self-organizing Neural Pattern Recognition Machine. Computer Vision, Graphics and Image Processing 37, 54–115 (1987)
2. Carpenter, G.A., Grossberg, S.: ART2: Self-organization of Stable Category Recognition Codes for Analog Input Patterns. Applied Optics 26, 4919–4930 (1987)
3. Yang, X.J., Zheng, J.L.: Artificial Neural Networks. Higher Education Press (1992)
4. Zhang, Q.G.: Introduction to Artificial Neural Networks. China Water Power Press (2004)
5. Wan, S.L., Cheng, Y.C., Li, C.G.: Diagnosis of Faulty Porcelain Insulators Based on AN Artificial Neural Network. High Voltage Engineering 28, 6–10 (2002)
6. Ma, J.S.: A New ART2 Neural Network Model. Pattern Reorganization and Artificial Intelligence 13, 231–234 (2000)
7. Kishore, L., Subrata, C., Chang, T.C.: Feature Recognition Using ART2: A Self-Organizing Neural Network. Journal of Intelligent Manufacturing 8, 203–214 (1997)
8. Cheng, Y.C., Li, C.G.: Detecting the Insulator String with Faulty Insulator on Ground Based on Corona Discharge Finger Prints. Proceedings of the Chinese Society of Electrical Engineering 20, 6–10 (2000)

Diagnosis Method for Gear Equipment by Sequential Fuzzy Neural Network

Xiong Zhou[1], Huaqing Wang[2,3], Peng Chen[2,*], and Jingwei Song[4]

[1] College of Mechanical Engineering, Chongqing University, Chongqing, China
[2] Graduate School of Bioresources, Mie University, Tsu, 514-8507 Mie, Japan
Tel. & Fax: +81 59-2319592
chen@bio.mie-u.ac.jp, wanghq_buct@hotmail.com
[3] School of Mech. & Elec. Eng., Beijing University of Chemical Technology, Beijing, China
[4] Huadong Jiaotong University, China

Abstract. This paper proposes a new method called "sequential fuzzy neural network" to diagnose fault of gear equipment automatically and precisely. Symptom parameters in time domain, by which fault of the gear equipment can be detected and distinguished, are selected according to its values calculated from the signals measured in each state of gear equipment. The probability density functions are translated to possibility distribution functions by possibility theory to express the relationship between the gear condition and the symptom parameters. The fuzzy neural networks proposed in this paper can sequentially distinguish fault types of gear equipment. Examples of practical diagnosis are shown to verify the efficiency of this method.

Keywords: Gear equipment, Fault diagnosis, Fuzzy inference, Possibility distribution function, Neural network.

1 Introduction

In the fault states of rotating machinery, the faults of gear equipment account for about 20% of total faults [1]. Therefore, in order to detect various faults of gear equipment as earlier as possible, and prevent production loss due to breakdown accidents, it is required to raise the level of "fault diagnosis technology of gear equipment". Diagnosis process for detection and identification of faults is largely depending on "intuition, experience and individual ability" of the inspector. Therefore, it is necessary to establish the method of automatic diagnosis to raise the efficiency and accuracy of the condition diagnosis. The most difficulty to construct a plant machinery management system is, establishing online condition monitoring system and intelligent diagnosis system of rotating machinery under operation [2]. Due to the complexity of plant machinery conditions, it is not always able to find out the SP that can identify all fault types. In addition, in most practical cases of plant, it is difficult to establish the exacting accurate mathematical models for machinery because the fault mechanisms of machinery and the features of fault types cannot be perfectly

* Corresponding author.

F. Sun et al. (Eds.): ISNN 2008, Part II, LNCS 5264, pp. 473–482, 2008.

clarified adequately perfectly explained by theory. Furthermore, the statistical objectivity of measured date data cannot be always satisfied because of the measuring techniques and manner of the inspectors.

For the above reasons, a sequential fuzzy inference method is proposed in the work. In order to explain relationship between SP and fault types for gear diagnosis, the method converting probability distribution function of SP calculated from measured signal into possibility function by possibility theory is also proposed for precise diagnosis. Furthermore, the membership functions of several SPs are established by the certainty factor (CF) [3] of MYCIN, and used for building a sequential fuzzy neutral network to efficiently carry out sequential fuzzy inference. Practical examples of diagnosis for gear equipment are shown to verify the efficiency of the method.

2 Sequential Fuzzy Neutral Network

2.1 Principle of Sequential Fuzzy Inference

When carrying out sequential fuzzy inference, if *M* of states to be identified, the condition diagnosis must be inferred in *M-1* times. That is to say, to diagnose a normal state or abnormal states firstly, if it is judged as a normal state, it need not carry out following diagnosis. Similarly, if symptom parameters for identifying fault *j* (*j=1-M*) or non fault *j* are all found out, it can diagnose all of state sequentially.

At this time, in the various diagnosis steps, it is just required to diagnose two states. A SP for identification of two states is quite easy to find out. In addition, in order to improve state identification ability, several SPs may be selected, which may be supplemental for each other through fuzzy inference. A basic principle of a sequential fuzzy inference is described as following.

Vector *C* of *M* kinds of states to be diagnosed should be:

$$C = \{C_1, C_2, \cdots, C_j, \cdots, C_M\} . \tag{1}$$

F_j, a membership function set of *N* of symptom parameters by which state C_j can be identified, is defined as follows

$$F_j = \{\mu_{1j}, \cdots, \mu_{ij}, \cdots, \mu_{Nj}\}^T . \tag{2}$$

In addition, when diagnosing a membership function set F' of symptom parameters calculated from a signal in time domain shall be:

$$F' = \{\mu'_i\}_{i=1\sim N}^{\mathrm{T}} . \tag{3}$$

Based on above sets, following descriptions are made to fuzzy inference formula for diagnosing C_j state:

Premise: if *F* is $F_j = \{\mu_{ij}\}_{i=1\sim N}^{T}$, then *C* is C_j (j=1-M)

Input: *F* is $F' = \{\mu'_i\}_{i=1\sim N}^{T}$

$$\text{Conclusion: } C \text{ is } C_j' . \tag{4}$$

When carrying out sequential fuzzy inference, *j* inference stage is the stage to identify C_j, $\overline{C}_j$ (non C_j). In this stage, the SPs by which can identify the C_j, $\overline{C}_j$, and a corresponding membership function should be obtained.

2.2 Selection of Symptom Parameter

In this paper, using an acceleration signal of gear equipment, the NSPs in time domain are defined for identifying the fault types [2].

$$SF = x_{rms} / \overline{x}_{abs} \quad \text{(Shape Factor)} . \tag{5}$$

where, $x_{rms} = \sqrt{\sum_{i=1}^{N} x_i^2 / N}$ (RMS value) 。 $\overline{x}_{abs}$ is the absolute average value.

$$CF = x_p / x_{rms} \quad \text{(Crest Factor)} . \tag{6}$$

where, x_p is the peak value.

$$I_p = x_p / \overline{x}_{abs} \quad \text{(Impact index)} . \tag{7}$$

$$\beta_1 = \left\{ \sum_{i=1}^{N} (x_i - \overline{x})^3 / (N-1) \right\} / s^3 \quad \text{(Skewness)} . \tag{8}$$

where, s is standard deviation.

$$\beta_2 = \left\{ \sum_{i=1}^{N} (x_i - \overline{x})^4 / (N-1) \right\} / s^4 - 3 \quad \text{(Kurtosis)} . \tag{9}$$

In addition, when a symptom parameter x is used to identify state 0 and state 1 (two states in total), its identification ability can be shown through the Distinction Index (DI) as follows according to Mahalanobis Generalized Distance [4],

$$DI = \left| \overline{x}_1 - \overline{x}_0 \right| / \sqrt{s_1^2 + s_0^2} \ . \tag{10}$$

where, $\overline{x}_i$ and s_i are the average value and the standard deviation of a SP in state *i*, respectively. It is obvious that the larger the value of the *DI*, the higher the sensitivity of a SP. When *DI* is more than 2, distinction probability is more than 95%.

2.3 Membership Function for State Identification

According to the possibility theory, whatever, SPs are distributed in any probability, a possibility distribution function can be calculated, and a membership function for fault diagnosis can be obtained by a possibility distribution function. In general, when experimental data are used to obtain a probability density function of SP, it is able to obtain possibility distribution function [6]. For example, when SP x conforms to the normal distribution, it can be changed to a possibility function $P_0(x_i)$ by the following formula.

$$p_o(x_i) = \sum_{k=1}^{n} \min\{\lambda_i, \lambda_k\} . \tag{11}$$

$$\lambda_i = \int_{x_{i-1}}^{x_i} \frac{1}{s\sqrt{2\pi}} \exp\left\{-\frac{(x-\bar{x})^2}{2s^2}\right\} dx \, , \lambda_k = \int_{x_{k-1}}^{x_k} \frac{1}{s\sqrt{2\pi}} \exp\left\{-\frac{(x-\bar{x})^2}{2s^2}\right\} dx \, . \tag{12}$$

where, $x=\bar{x}-3s\sim\bar{x}+3s$, $\bar{x}$ and s is the average value and the standard deviation of x, respectively. Figure 1 shows an example of the probability density function and the possibility distribution function.

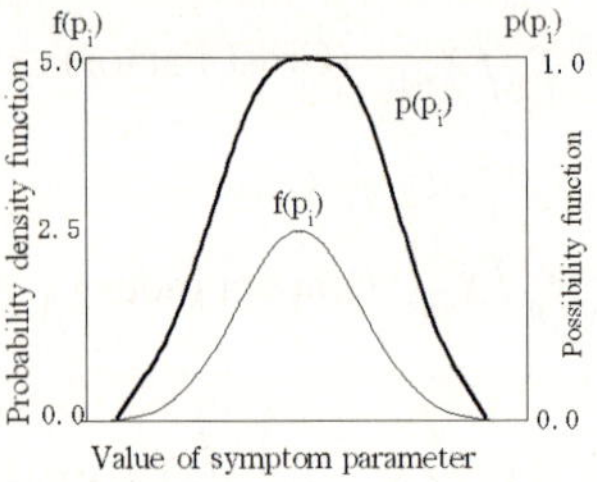

Fig. 1. An example of probability density function and possibility distribution function

2.4 Structure of Sequential Fuzzy Neutral Network

Membership Function of Learning Neural Network. Through selected SP, it is to obtain membership function under various states and acquire teaching data for state identification. According to the certainty factor (CF) of MYCIN, it is to infer a membership function of complex multiple SPs for diagnosing different faults, which is taken as learning data of neural network.

In case of N (N> 1) SPs, it is able to obtain various possible states as follows according to the CF. Namely, set W_{ij} as possibility of state i obtained from jth SP, probability $W_i(i=1, 2\cdots)$ of state i can be calculated by (13)-(16).

$$W'_1 = W_{11} + W_{12}(1 - W_{11}) . \tag{13}$$

$$W'_2 = W_{21} + W_{22}(1 - W_{21}) . \tag{14}$$

Then, following formulas are used to normalize W'_1 and W'_2.

$$W_1 = W'_1 / (W'_1 + W'_2) . \tag{15}$$

$$W_2 = W'_2 / (W'_1 + W'_2) . \tag{16}$$

where, W_1 and W_2 are possibility degree of state 1 and state 2, respectively.

2.4.1 Sequential Fuzzy Neural Network

Figure 2 shows a sub-network for inference at jth stage mentioned in formula (4). Input of sub-network shall be applied with symptom parameter *d1* and *d2*, and output applied with possibility degree of state C_j and $\overline{C}_j$. In addition, the sub-network should be created in first half part and second half part. The first half part is used to learn the membership functions, and the second half part is used to synthesis possibility degrees of *d1* and *d2*. Then, according to the number of faults to be diagnosed, the sequential fuzzy neutral network is constructed through integrating respective sub-network. Thus, after entire network is divided into several sub-networks, it is very easy for learning. When it is need to correct a membership function or an inference rule, we can only correct a relative sub-network.

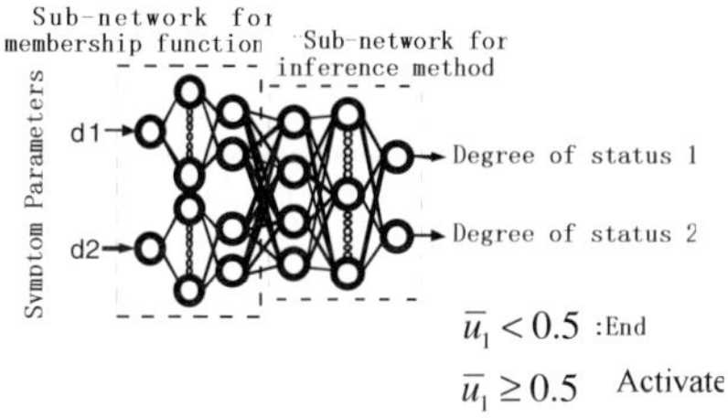

Fig. 2. Sub-network at jth inference stage

When diagnosing, symptom parameters by which C_j and $\overline{C}_j$ can be identified, should be calculated. Once being input in the network, final C_j and $\overline{C}_j$ should be output through sub-networks of the membership function and inference rule. If the possibility of C_j is larger than the possibility of $\overline{C}_j$, diagnosis should finish and it is unnecessary to diagnose other states. If the possibility $\overline{C}_j$ is larger than the possibility of C_j, it means probability of other faults is larger, and it is necessary to diagnose other states.

Whether it is to diagnose other states is depending on $\mu_{\overline{C}1}$. Here, $\mu_{\overline{C}1}$ with a threshold of 0.5, is named as activating signal of next inference. Namely, it will end when $\mu_{\overline{C}1} \geq 0.5$, and activate diagnosis of other states when $\mu_{\overline{C}1} < 0.5$.

As shown in Fig.5, a sequential fuzzy neutral network is created through an inference of above sub-networks.

3 Application for Fault Diagnosis of Gear Equipment

3.1 Selection of Symptom Parameter and Evaluation of Identification Sensitivity

In order to evaluate the identification sensitivity of a SP, the distinction index (DI) should be calculated using a signal measured under tow states to be identified, and a SP used for diagnosis can be determined by the *DI*.

In this research, Filters are used to perform noise cancellation for feature extraction of the vibration signal across an optimum frequency region. Identification sensitivity of symptom parameter should be evaluated for five states, such as normal, scar, spot, eccentricity and misalignment.

Filters used at this time are high-pass filtering (HPF, above 200 Hz), band-pass filtering (BPF, 100~500Hz) and low-pass filtering (LPF, below 200Hz). According to the calculated *DI*, if only one filtering and one symptom parameter is used, it is unable to identify all states; even if it is under the same state, identification sensitivity of the SP should also be different due to variation of rotating speed.

Based on above methods, a normal state and each fault state (spot, eccentricity, scar and misalignment), filters, symptom parameters and the *DIs* of faults are shown in Table 1, respectively.

Table 1. Symptom parameter and DI used at various diagnosis stage

Inference stage	Symptom parameter	*SF* of HPF	β_1 of LPF
	Normal:—Spot	2.71	1.80
Normal: Faults	Normal: Eccentricity	3.24	1.78
	Normal: Scar	1.55	2.06
	Normal: Misalignment	4.15	0.23
	Symptom parameter	*SF* of HPF	β_1 of HPF
Spot: Other	Spot: Eccentricity	3.31	2.39
	Spot: Scar	6.01	3.91
	Spot: Misalignment	5.81	3.47
	Symptom parameter	*SF* of HPF	*Ip* of HPF
Eccentricity: Other	Eccentricity: Scar	2.94	1.88
	Eccentricity: Misalignment	3.83	1.73
Scar: Misalignment	Symptom parameter	β_1 of BPF	Non-filtering β_1
	Scar: Misalignment	3.43	2.89

3.2 Acquiring of Training Data

An experimental system used for diagnosis is shown in Fig. 3. An accelerometer is mounted on the bearing housing at the output end of the motor, by which vibration signals of each state can be measured for obtaining teacher data. SPs and membership functions can be obtained using those signals of each state (48 in total, 4096 points for

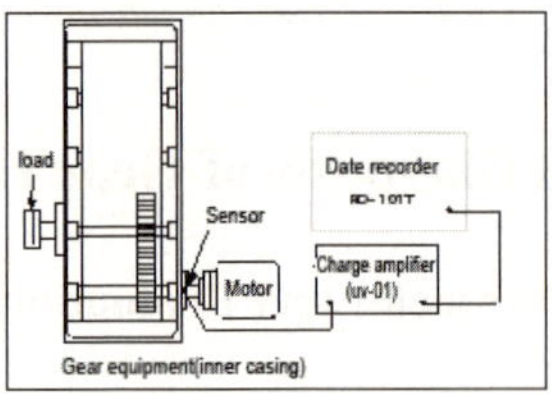

Fig. 3. Gear equipment for fault diagnosis

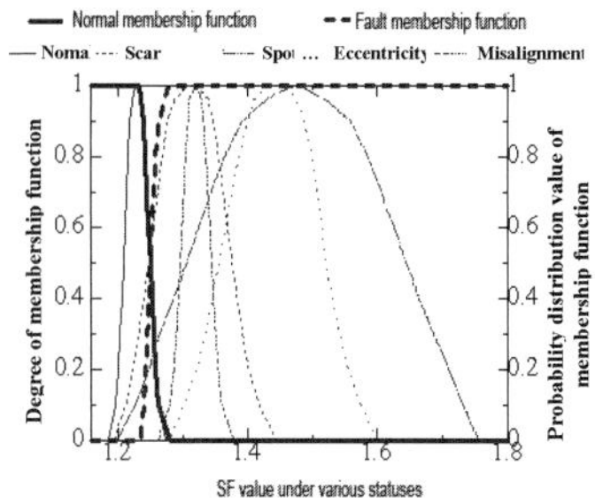

(a) Membership function of *SF*

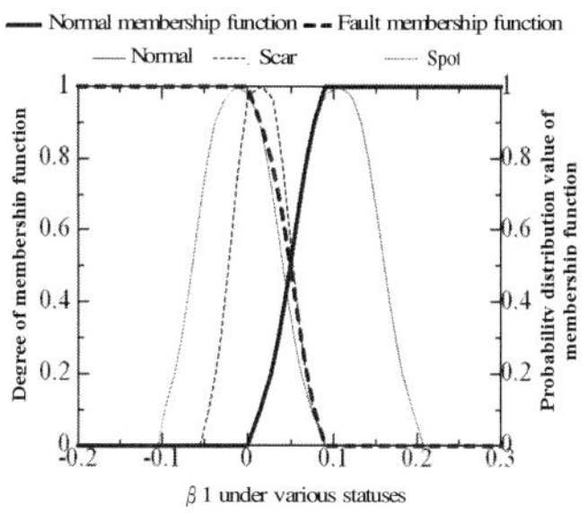

(b) Membership function of β_l

Fig. 4. Membership functions of normal and fault identification

each), and then training data for identification states should be acquired. As examples, membership functions of the *SF* and the β_l are used to distinguish normal state from fault states, as shown in Fig. 4.

It means that identification sensitivity of the *SF* for distinguishing normal state from two fault states is low, and the fault types cannot be distinguished perfectly. Therefore, we have to supplement another parameter β_l, as shown in Fig. 4(b). By two of those parameters, each state can be identified. As regard to sub-network structure used for membership functions, the network consists of the first layer, one hidden layer, and the last layer. The number of neurons in the first layer is one, the number of neurons in the hidden layer is fifteen, and the number of neurons in output layer is two. In the learned sub-network, symptom parameters obtained from diagnosis data should be input and membership functions of symptom parameters should be output.

Table 2. Example of sub-network teaching data used by membership functions

(a) Sub-network teaching data used for identification of normal state and fault state (SF)

Input d1	1.160	1.235	1.240	1.243	1.246	1.248	1.251	1.253	1.256	1.259	1.264	1.400
Output W_{11}	1.0	1.0	0.9	0.8	0.7	0.6	0.5	0.4	0.3	0.2	0.1	0.0
Output W_{21}	0.0	0.0	0.1	0.2	0.3	0.4	0.5	0.6	0.7	0.8	0.9	1.0

(b) Sub-network teaching data used for identification of normal state and fault state (β1)

Input d2	-0.01	0.000	0.015	0.027	0.036	0.043	0.050	0.056	0.062	0.069	0.077	0.105	0.200
Output W_{12}	0.0	0.0	0.1	0.2	0.3	0.4	0.5	0.6	0.7	0.8	0.9	1.0	1.0
Output W_{22}	1.0	1.0	0.9	0.8	0.7	0.6	0.5	0.4	0.3	0.2	0.1	0.0	0.0

Table 2 show teaching data of sub-network used for the membership functions when identifying normal state from fault states. As regard to structure of sub-network used by the inference rule, the number of neurons in input layer is four (except activating signal), the number of hidden layer is one, the number of neurons in hidden layer is fifteen, and the number of neurons in output layer is two. Table 3 show teaching data of inference rule obtained through the *CF*.

Table 3. Teaching data of inference method

Input W_{11}	0.0	0.0	0.0	0.0	0.0	0.0	0.2	0.2	0.2	0.2	0.2	0.4	0.4	0.4	0.4
Input W_{21}	1.0	1.0	1.0	1.0	1.0	1.0	0.8	0.8	0.8	0.8	0.8	0.6	0.6	0.6	0.6
Input W_{12}	0.0	0.2	0.4	0.6	0.8	1.0	0.2	0.4	0.6	0.8	1.0	0.0	0.6	0.8	1.0
Input W_{22}	1.0	0.8	0.6	0.4	0.2	0.0	0.8	0.6	0.4	0.2	0.0	1.0	0.4	0.2	0.0
Output W_1	0.00	0.17	0.29	0.37	0.44	0.50	0.27	0.36	0.44	0.50	0.56	0.29	0.50	0.56	0.63
Output W_2	1.00	0.83	0.71	0.63	0.56	0.50	0.73	0.64	0.56	0.50	0.44	0.71	0.50	0.44	0.37
Input W_{11}	0.6	0.6	0.6	0.6	0.6	0.6	0.8	0.8	0.8	0.8	0.8	1.0	1.0	1.0	1.0
Input W_{21}	0.4	0.4	0.4	0.4	0.4	0.4	0.2	0.2	0.2	0.2	0.2	0.0	0.0	0.0	0.0
Input W_{12}	0.0	0.2	0.4	0.6	0.8	1.0	0.2	0.4	0.6	0.8	1.0	0.0	0.6	0.8	1.0
Input W_{22}	1.0	0.8	0.6	0.4	0.2	0.0	0.8	0.6	0.4	0.2	0.0	1.0	0.4	0.2	0.0
Output W_1	0.37	0.44	0.50	0.57	0.64	0.71	0.50	0.56	0.64	0.73	0.83	0.50	0.71	0.83	1.00
Output W_2	0.63	0.56	0.50	0.43	0.36	0.29	0.50	0.44	0.36	0.27	0.17	0.50	0.29	0.17	0.00

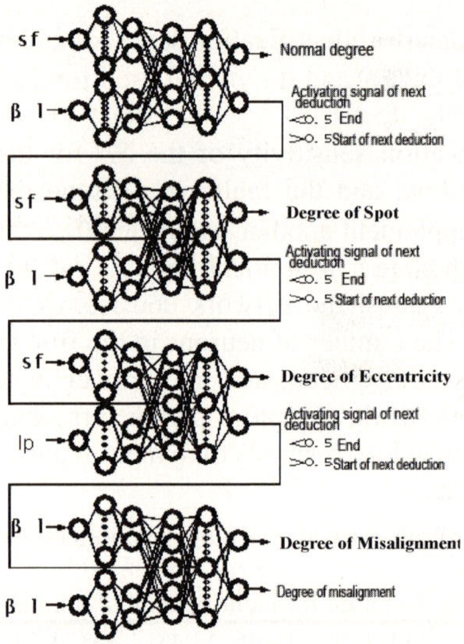

Fig. 5. Sequential fuzzy neutral networks

A sequential fuzzy neutral network is shown in Fig. 5. As regard to diagnosis flowchart, a normal state should be identified from fault states in the first stage, the spot state should be identified from other fault states in the second stage, the eccentricity state should be identified from other faults in the third stage, and the scar state should be identified from misalignment in the fourth stage.

We used the data measured in each state that had not been learned by the neural network to verify the diagnosis capability of the neural network. Table 4 shows examples of gear equipment for fault diagnosis. According to the verification results, the possibilities output by the network show correct judgments in each state.

Table 4. Example of data diagnosis result under various states

(a) Data diagnosis result under normal state

	Symptom parameter	Diagnosis result
Stage I inference	HPFSF=1.2095	μ_{c1} =0.9147
	LPF β_1=0.1444	$\bar{\mu}_{c1}$ =0.0863

(b) Data diagnosis result under spot state

	Symptom parameter	Diagnosis result
Stage I inference	HPFSF=1.7739	μ_{c1} =0.006
	LPF β_1=−0.0163	$\bar{\mu}_{c1}$ =0.9934
Stage II inference	HPFSF=1.7739	μ_{c2} =0.9953
	HPF β_1=−1.3112	$\bar{\mu}_{c2}$ =0.0051

(c) Data diagnosis result under centrifugal state

	Symptom parameter	Diagnosis result
Stage I inference	HPFSF=1.3797	μ_{c1} =0.0056
	LPF β_1=−0.0738	$\bar{\mu}_{c1}$ =0.9938
Stage II inference	HPFSF=1.3797	μ_{c2} =0.2375
	HPF β_1=0.4402	$\bar{\mu}_{c2}$ =0.7583
Stage III inference	HPFSF=1.3797	μ_{c3} =0.7762
	HPFI_p=9.8342	$\bar{\mu}_{c3}$ =0.2235

(d) Data diagnosis result under pitting corrosion state

	Symptom parameter	Diagnosis result
Stage I inference	HPFSF=1.2833	μ_{c1} =0.0212
	LPF β_1=0.0167	$\bar{\mu}_{c1}$ =0.9776
Stage II inference	HPFSF=1.2833	μ_{c2} =0.131
	HPF β_1=0.1009	$\bar{\mu}_{c2}$ =0.8676
Stage III inference	HPFSF=1.2833	μ_{c3} =0.0119
	HPFI_p=5.2373	$\bar{\mu}_{c3}$ =0.9872
Stage IV inference	BPF β_1=0.3425	μ_{c4} =0.9956
	β_1=−0.3293	$\bar{\mu}_{c4}$ =0.0048

4 Conclusion

In this paper, following diagnosis algorithms are presented to solve fuzzy diagnosis through a neutral network.

(1) When it is hard to identify multiple fault states through several symptom parameters in one time, it is recommended to identify fault through sequential fuzzy inference approach.

(2) Non-dimensional symptom parameters in time domain are used in sequential fuzzy inference method for fault diagnosis.

(3) The possibility function of the symptom parameter is used to show the ambiguous relationship between a symptom parameter and a fault, which can be obtained from its probability density function.

(4) The membership function of several symptom parameters can be obtained by the certainty factor (CF) to acquire the diagnosis knowledge for the training of the sequential fuzzy neural network.

In conclusion, this paper proposed a concept of sequential fuzzy diagnosis integrating sequential fuzzy inference and neutral network. A sequential fuzzy neutral network for gear fault diagnosis is also constructed, by which the fault states can be distinguished automatically and sequentially. Examples of practical diagnosis for gear equipment are shown to verify the efficiency of this method.

References

1. Kumar, S.: Vibration and Oil Analysis Detect a Gearbox Problem. Vibrations 30, 7–13 (2004)
2. Matuyama, H.: Diagnosis Algorithm. Journal of JSPE 75, 35–37 (1991)
3. Shortliffe, E.H.: Computer-Based Medical Consultation. MYCIN. Elsevier, Amsterdam (1976)
4. Chen, P., Toyota, T., He, Z.J.: Automated Function Generation of Symptom Parameters and Application to Fault Diagnosis of Machinery in Variable Operation-conditions. IEEE Transactions on System, Man, and Cybernetics (Part A) 31, 775–781 (2001)
5. Chen, P., Toyota, T.: Sequential Fuzzy Diagnosis for Plant Machinery. JSME International Journal, Series C 43, 1121–1129 (2003)
6. Dubois, D., Prade, H.: Possibility Theory: An Approach to Computerized Processing. Plenum Press, N.Y (1988)

Study of Punch Die Condition Discrimination Based on Wavelet Packet and Genetic Neural Network

Zhigao Luo, Xiang Wang, Ju Li, Binbin Fan, and Xiaodong Guo

College of Mechanical Engineering, Jiangsu University, Zhenjiang 212013, China
lzg3359@163.com, wxiang1983@126.com

Abstract. According to the characteristics of the acoustic emission signal which was induced by punch die when It fails, the characteristic parameters of failure signal is determined. The energy eigenvector of signal failure die is extracted by wavelet packet analysis technology, and the comparison between the energy in different frequency bands and total energy is taken as the characteristic parameters. Then a BP neural network is established in which the time factor is considered based on genetic algorithm. The characteristic parameters are used as input specimen, learning and training the network to complete the pattern recognition of model working state. Experiments show that the method can quickly and reliably discriminate the conditions of the punch die and has strong practicability.

Keywords: Punch die, acoustic emission, wavelet packet, genetic algorithm, neural network

1 Introduction

The extremely bad working environment of punch die include not only high contact pressure and intensity friction, but also stress, strain and temperature cyclical changes caused by cyclic loading which make die fatigue failure. Enterprises have to cease production for maintenance, and suffer enormous economic losses. One of the key ways to solve punch die's reliability and security is the condition identification [1].Due to complex background noise of die's acoustic emission signal and the very weak signal, it is hard to extract the signal parameters which reflect the actual physical state. The show between the state and sign is also a very complex nonlinear relationship, so it brings great difficulties to state identification on die. In this paper, the wavelet packet and genetic neural network method is used to extract the failure parameters of the online tested acoustic emission signal on punch die, and identify condition. Experiments show that it not only inherited the local feature of the wavelet packet's analysis and neural network's learning and promotion capability, but also the genetic algorithm's optimization characteristics of general superiority, rapidness, adaptability, robust, which can effectively distinguish the working state of punch die.

2 Extract the Characteristics of Acoustic Emission Signal

2.1 Energy Analysis of Crack Propagation on Punch Die

According to the law of conservation of energy, the load power dW should be equivalent to the sum of the variable of elastic strain d_μ, plastic energy $d\delta$ and surface crack

F. Sun et al. (Eds.): ISNN 2008, Part II, LNCS 5264, pp. 483–491, 2008.

energy d_σ. That is $d_\mu + d_\delta + d_\sigma = dW$. In the loading process, the system release energy $d_q = dW - d_\mu = d_\delta + d_\sigma$, when crack expands. The energy released from the unit area d_s of crack propagation could be calculated in the formula (1) below:

$$G = \frac{\partial q}{\partial s} = \frac{\partial W}{\partial s} - \frac{\partial \mu}{\partial s} = \frac{\partial \mu}{\partial s} \tag{1}$$

When the punch die works, the force on the punch and convex consists of the pressure perpendicular to the contacting surface p and shear force τ. The strain exits under the pressure and shear force could be described as below:

$$\mu = \int_V (\bar{\mu}_p + \bar{\mu}_\tau) dV = \int_V \frac{1}{2}\left(\frac{\xi^2}{E} + \frac{\tau^2}{G_\tau}\right) dV = V \cdot \frac{1}{2}\left(\frac{\xi^2}{E} + \frac{\tau^2}{G_\tau}\right) \tag{2}$$

When crack appears, the elasticity modulus E and G_τ will reduce near the crack area. Elastic strain changes as below:

$$d\mu = V\{(\frac{\xi}{E} d\xi + \frac{\tau}{G_\tau} d\tau - \frac{1}{2}(\frac{\xi^2}{E^2} dE + \frac{\tau^2}{G_\tau^2} dG_\tau)\} \tag{3}$$

Therefore,

$$G = \frac{\partial \mu}{\partial s} = V \cdot [\frac{\xi}{E}\frac{\partial \xi}{\partial s} + \frac{\tau}{G_\tau}\frac{\partial \tau}{\partial s} - (\frac{\xi^2}{E^2}\frac{\partial E}{\partial s} + \frac{\tau^2}{G_\tau^2}\frac{\partial G_\tau}{\partial s})] \tag{4}$$

The energy consumed from the unit area d_s of crack growth could be calculated in the formula (5) below:

$$G_c = \frac{\partial \delta}{\partial s} + \frac{\partial \sigma}{\partial s} \tag{5}$$

When $G > G_c$, crack begins to propagate, remainder energy is released in the mode elastic wave. Acoustic emission signal with consistent ratio is generated. The signal energy of relevant frequency released during unit time will be larger as the extending of the crack.

2.2 Determine the Parameters of Acoustic Emission Signal

The operational status of punch die can be determined by using the difference between the acoustic emission signal in the event of failure or damage and the normal state. When the die failures occur, it would have a greater influence on the energy of every frequency band signal. Therefore, a feature vector can be constructed with energy as an element. The energy value as the dimensionless parameter is easy to be impacted by the conditions and working conditions, so this paper uses the energy percentage of each frequent band as characteristic parameters. It can effectively avoid the above impact.

2.3 Extract the Characteristics of Acoustic Emission Signal

2.3.1 The Wavelet Packet Analysis Technique [2, 3]

Assume that the wavelet packet decomposition signal as j level, the frequency domain of $0 \sim f_{\max}$ in the j level is shared averagely. In a three-level decomposition, its wavelet

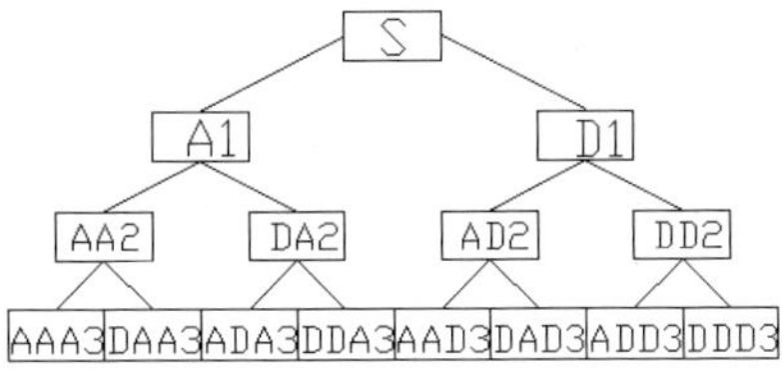

Fig. 1. The wavelet packet decomposition tree

packet decomposition tree is shown in Fig.1. A is low frequency, D is high frequency, the serial number at the end is the wavelet packet levels.

The relation of the decompositions:

$$S = AAA3 + DAA3 + ADA3 + DDA3 + AAD3 + DAD3 + ADD3 + DDD3$$

Suppose that $\{u_n(t)\}_{n\in z}$ is the wavelet packet group on h_k , $g_j^n(t)\in U_j^n$, then, $g_j^n(t)$ can be expressed as :

$$g_j^n(t) = \sum_l d_l^{j,n} u_n(2^i t - 1) \tag{6}$$

Here, $d_l^{j,n}$ is the projection coefficient of the function $g_j^n(t)$ in space U_j^n . Know from $U_{j+1}^n = U_j^{2n} \oplus U_j^{2n+1}$, the wavelet packet decomposition is to divide $g_{j+1}^n(t)$ into $g_j^{2n}(t)$ and $g_j^{2n+1}(t)$, thus the wavelet packet decomposition algorithm is:

$$\left\{ \begin{aligned} d_l^{j,2n} &= \sum_k^n a_{k-2l} d_k^{j+1,n} \\ d_l^{j,2n+1} &= \sum_k^n b_{k-2l} d_k^{j+1,n} \end{aligned} \right\} \tag{7}$$

Where, $a_k = \frac{1}{2}\overline{h_0}(k)$, $b_k = \frac{1}{2}\overline{h_1}(k)$, $\overline{h}$ is the dual operator of h .

2.3.2 Extract the Signal Eigenvector

Suppose that $E_{j,k}$ is the corresponding energy of $S_{j,k}$, then

$$E_{j,k} = \int \left|S_{j,k}(t)\right|^2 dt = \sum_{k=1}^n \left|x_{k,m}\right|^2 \tag{8}$$

Where $x_{k,m}(k = 0,1,2,\cdots,2^j - 1; m = 1,2,\cdots,n,$ are the signal sampling points) is the signal amplitude of the reconstruction signal $S_{j,k}$. Suppose that the total energy of the analyzed signal is E, then:

$$E = \sum_{k=0}^{2^j} E_{j,k} \tag{9}$$

The corresponding eigenvector is:

$$T = [E_{j0}, E_{j1}, ..., E_{jk}] \tag{10}$$

The proportion energy of the band account for the total energy of signal analysis is:

$$E_k = \frac{E_{j,k}}{E} \times 100\% \tag{11}$$

Where, $k = 0,1,2,\cdots,2^j - 1$.

3 Pattern Recognition Based on Genetic Neural Network

3.1 BP Neural Network Theory [4]

BP neural network includes input layer, hidden layer and output layer. When signal is input, firstly it transfers from input layer to hidden layer. Under the effect of the hidden layer node activation function, output signals of the hidden node transmit to the output layer nodes. After the process of the output layer, the output value of the network can be obtained, as Fig.2 shows [5, 6].

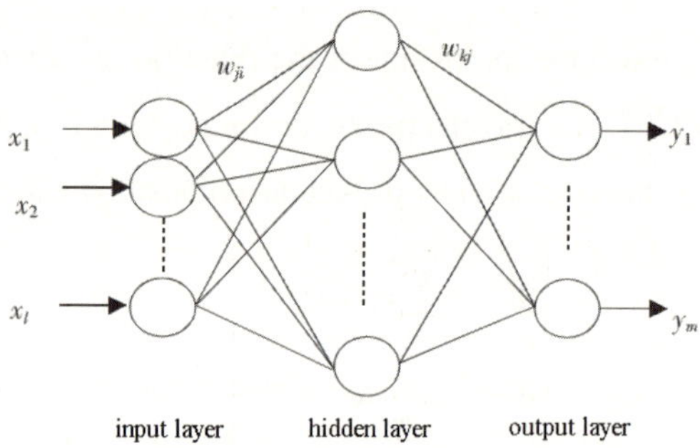

Fig. 2. Three-layer BP network topology

Output of the neuron j of the hidden layer or output layer of BP neural network can be determined by the following formula:

$$O_j = f_j = (\sum_j \omega_{ji} x_i + \theta_i) \tag{12}$$

In the formula, f_j represents the neuronal excitation function of the neuron j; θ_i represents the neurons threshold of the neuron j; x_i represents the input of the neuron j; ω_{ji} represents the connection weights from the neuron j to the neuron i. Nowadays excited function usually adopts continuous micro-nonlinear *sigmoid* function: $f(x) = \frac{1}{1+e^{-x}}$.

3.2 The Basic Principles of Genetic Algorithms

The main operation of genetic algorithm is: selection operator. Random leagues matches are used to select, and arbitrary choice of a certain number of individuals is made from the group. The individual of highest adaptation is preserved to the next generation. Repeating this process until the next generation of individuals meets the requirements; crossover operator: according to the crossover probability set p_c, by using two-point cross method, individual coding of the population undergoes respective cross-operation each time. Father generation of the cross-operation is produced by random method; mutation operator: according to the mutation rate set p_m, the probability of p_m of the group individuals' five parts is respectively mutated to generate the son generation group.

3.3 BP Neural Network Learning Algorithm Based on Genetic Algorithm

3.3.1 The Basic Theory of GA-BP Neural Network

Firstly, using the BP arithmetic to pre-train the initial value of weights and the GA to optimize the net structure, the right connections, threshold, and the learning rate, as well as momentum factors, the searching space can be better located. And then, reuses the BP arithmetic to optimize the net weights and threshold values in this space to search the optimal solution.

3.3.2 The Adaptive Function's Determination of GA-BP Neural Network [7]

The GA is only based on the adaptive function in searching evolutionary process and then utilizes the fitness value of individuals to search. Under normal circumstances, the GA-BP neural network takes the difference value between the actual output of the network and desired output as fitness function. In order to improve the learning speed of neural work, and considering the time factor, the fitness function may be as follows:

$$F = \frac{1}{E_{av} + kt} \tag{13}$$

Where: $$E_{av} = \frac{1}{2N}\sum_{n=1}^{N}\left\{\left[\sum_{i\in Y}\left(y_i(n) - d_i(n)\right)^2\right]\right\}$$

And, E_{av} is the average square error of the neural work's output

k is the time coefficient

t is the designated study time of the network training to the required precision.

3.3.3 The Design of GA-BP Neural Network

To avoid the restriction of father, the float encoding is replaced by binary code. The program flow chart is as Fig.3:

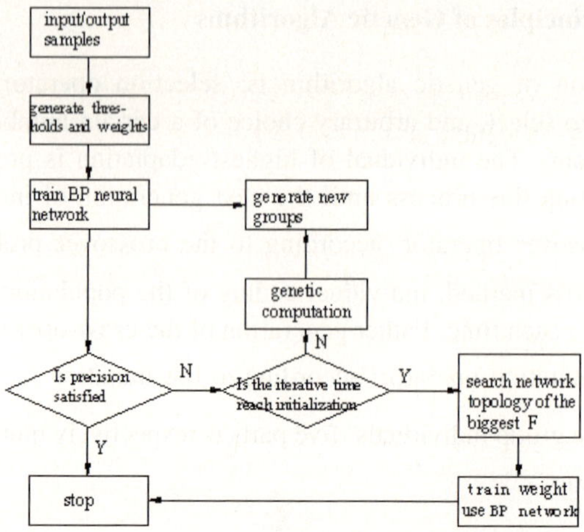

Fig. 3. The flow chart of GA's BP neural work

Specific steps of algorithm are as follows:

1) Given network input/output samples, take the randomly generated n groups of weights as the initial groups

2) Use the BP algorithm to pre-train and calculate the N groups of initial values respectively. If E_{av} meets the requirements, the calculation can stop, otherwise utilizing the fitness function to calculate every gene's fitness value F .

3) Arrange the calculated values F from large to small, and withhold the former m=n/2 values of the individuals.

4) Take the m individuals as father generation and m individuals generated by alone point crossing as offspring, and randomly change some bit of genes according to cross-rate $p_c = 0.3$ and mutation rate $p_m = 0.1$.

5) Compose the new groups by m father individuals and m offspring individuals and repeat the step 2) until E_{av} reaches required demand or the iterative time reaches initialization. If the iterative time reaches initialization while the E_{av} doesn't reach required demand, take the individuals corresponding to the biggest fitness values F as most optimal network topology of this stylebook's parallelism, and use this BP network to train network values.

4 Example

4.1 Experimental Device

The experiment is carried out in composite die which is used for the diamond saw blade matrix has been done. The matrix's material is 50 steel and the die's material is GCr15. The model of hydraulic permeability punch machine is YF-315A, and stamping frequency is 20 times per minute. The Fig.4 is the simple diagram of the experimental device.

The four-channel acoustic emission produced by Beijing Pengxiang Company is used for collecting signals. Its components are as follow: 1. The 1045S acoustic emission sensors are fixed on the side of die by coupling. 2. The model of 2/4/6 preamplifier is PXPA II; the magnification is set at 40 db. 3. The model of AE's data acquisition card is PXDAQ12204.

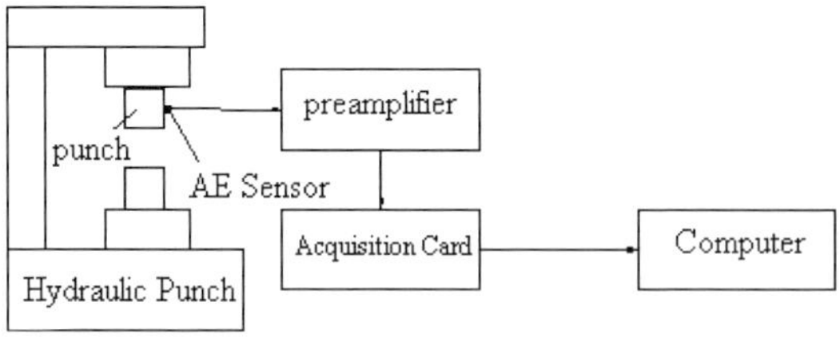

Fig. 4. The simple diagram of the experimental device

4.2 Signal Collection

In the metal, the wave frequencies of AE signals are up to 100 KHz and above. In order to reserve the signals farthest, sampling frequency is set at 2MHz, stamping frequency band is set at 20K~1MHz. Chart3 is the signal waves of cracks free and cracked die

4.3 Signal Processing and Result Analysis

According to the massive experiences, the paper introduces the mother wavelet function 'db3' to decompose the acoustic emission signals into 3 levels. Then the frequency bandwidth per layer node is 125 KHz. From the Equation (10) and (11), the each frequency band of energy and energy percentage can be obtained.

Three-layer neural network is chosen to study and train, and it is the 8×24×2 model. The input layer has eight nodes. The input vector is defined by

$$X = [E_{0,} E_{1,} E_{2,} E_{3,} E_{4,} E_{5,} E_{6,} E_{7}]^T$$

where, E_k ($k = 0,1,\cdots,7$) is the energy percentage of the frequency band. The output layer has two nodes, which indicates the normal and failure to punch die. The number of the hidden nodes is set at 24. The output vector defined by $Y = [y_1, y_2]$, when the punch die is normal, the expected output value is $Y = [1,0]$; when the punch die fails, the expected output value is $Y = [0,1]$. Set the Error: 0.001, the Learning Rate: 0.95, the momentum coefficient: 0.5, the crossover rate: 0.4, the mutation rate: 0.02, the iteration number: 1000. Give 24 groups of sample data to train, half of which are obtained by normal die. Six groups of sample data for training are shown in table 1. The unknown sample data are shown in table 2. Using the trained network to compute, and the result are shown in table 3. Fig.5 is the error curve of trained by GA-BP network and general BP network. As shown in table 3 and Fig5, the unknown sample data are recognized quickly and accurately. Experiment shows the validity, the accuracy can reach 95%.

Table 1. Training sample

Condition	The vectors of characteristic parameters	Expectation value
normal	[8.99,11.82,20.98,23.6,19.73,8.64,3.59,2.65]	[1,0]
normal	[12.56,13.65,18.54,26.38,13.68,10.95,2.57,1.67]	[1,0]
normal	[7.25,12.89,20.69,24.69,18.65,9.99,2.95,1.89]	[1,0]
failure	[2.95,2.36,12.8,58.96,13.62,4.53,2.93,1.85]	[0,1]
failure	[3.25,3.95,10.62,55.64,14.62,5.69,3.37,2.86]	[0,1]
failure	[3.59,2.95,13.65,56.94,12.75,6.98,2.2,0.94]	[0,1]

Table 2. The unknown sample

Number	The vectors of characteristic parameters
1	[10.65,10.99,19.62,26.75,19.35,9.16,2.36,1.12]
2	[6.65,9.31,17.91,36.38,16.86,8.51,3.09,1.29]
3	[4.03,3.63,15.62,51.31,15.62,6.54,2.89,0.36]

Table 3. Recognition results

Number	Result
1	[1.003, -0.003]
2	[0.995, 0.005]
3	[0.001, 0.999]

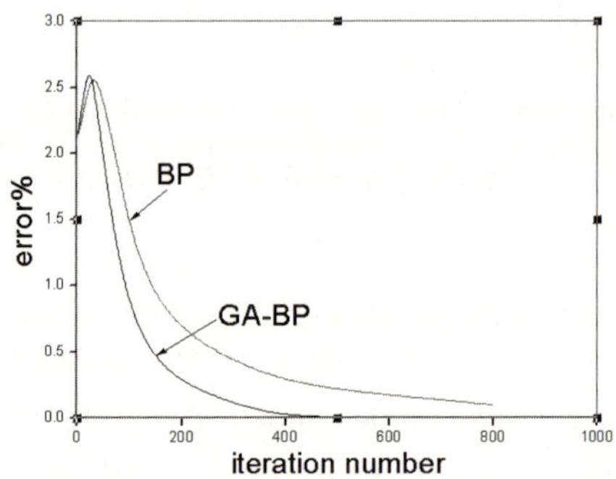

Fig. 5. The error curve of trained by GA-BP network and general BP network

5 Conclusion

This paper presents a novelty punch die condition determination method based on wavelet packet analysis, genetic algorithm and BP neural network technology. Conclusions are as follow:

1). Acoustic emission signal is always accompanied by some unsteady components because of its complex characteristics, so it is necessary to be extracted by an advanced analysis tool. Wavelet packets transform can extract the energy characteristics of original signal, which is superior to traditional flitting method;

2). the GA- BP neural network is an effective tool to recognize the condition of punch die. The genetic algorithm and the BP neural network are combined together to bring out the best in each other. Global searching is done by genetic algorithm first, then improved BP algorithm is used to carry through precision training, so convergence speed of the network is quickened and the problem of local minimum is avoided. In the fitness function to add the time factor to enhance the ability to optimize the overall network and real-time increased significantly. The training results indicate that both precision and speed of the temperature prediction model are satisfying.

References

1. Irving, S., Liu, Y.: An effective method for improving IC package die failure during assembly punch processing. In: Proceedings of the 6th International Conference on Thermal, Mechanical and Multi-Physics Simulation and Experiments in Micro-Electronics and Micro-Systems - EuroSimE 2005, pp. 227–233 (2005)
2. Kaewkongka, T., Au, Y.H.J.: Application of acoustic emission to condition monitoring of rolling element bearings. Measurement and Contorl 34(8), 245–247 (2001)
3. Velayudham, A., Krishnamurthy, R., Soundarapandian, T.: Polymeric composite using wavelet packet transform. Materials Science and Engineering A 412(1-2), 141–145 (2005)
4. Lin, S.T., Mcfadden, P.D.: Gear vibration analysis by b-spline wavelet-based linear transform. Mechanical Systems and Signal Processing 11(4), 603–609 (1997)
5. Tian, B., Azimi-Sadjadi, M.R., Vonder-Haar, T.H., Reinke, D.L.: Temporal updating scheme for probabilistic nNeural networks with application to satellite cloud classification. IEEE Transactions on Neural Network 11(7), 903–920 (2000)
6. Tao, Q., Fang, T., Qiao, H.: A novel continuous-time neural network for realizing associative memory. IEEE Transactions on Neural Networks 12(2), 418–423 (2001)
7. Jin, J.L., Yang, X.B., Ding, J.: Real coding based acceleration genetic algorithm. Journal of Sichuan university (engineering science edition) 32(3), 20–24 (2000)

Data Reconstruction Based on Factor Analysis

Zhong-Gai Zhao and Fei Liu

Institute of Automation, Southern Yangtze University
Wuxi, 214122, P.R. China
gaizihao@yahoo.com.cn

Abstract. In industrial process, the method based on principal component analysis (PCA) for data reconstruction is popular during dealing with missing data. In the method, the objective for data reconstruction is to make minimum square prediction error (SPE), an index measuring the relationship among process variables. However, PCA is a special case of factor analysis (FA), it has more limitations than FA, above all, SPE is a Euclidian distance, which is a suitable measurement for variables meeting uniform distribution rather than normal distribution, while process variables often satisfy the latter distribution. Due to the extensive sense of FA, the paper proposes a Mahalanobis distance as the index to measure the degree that the sample accords with the FA model, and then introduces FA into data reconstruction. The proposed index can more reflect the relationship among variables and the estimation of missing data with more precision can be achieved by making the new index minimum than by SPE.

1 Introduction

In modern chemical industry process, there are thousands upon thousands sensors to take various variables. The readings of sensors directly tell the process operation state, so sensors play a vital role in process control, evaluation, optimization and monitoring. However, due to long time running, sensors will always be bias and even work abnormally. In order to ensure sensors work normally, sensors are often taken off-line periodically for routine maintenance. Besides these, outliers will be moved away from normal data. Therefore, inaccurate sensors' readings or the cases of missing data are common cases and data reconstructions are often needed in real industrial process. In the monitoring method based on principal component analysis (PCA), the statistical index, square prediction error (SPE) is used to measure the correlation among process variables and appraise whether the current sample data accords with PCA model [1]. Recently, missing data reconstruction is usually achieved by making minimum the SPE value of the sample with missing variable [2-5], which means that the reconstructed data can make the sample data closest to PCA model. The method performs well and is widely used [6-9].

Unfortunately, SPE is a Euclidian distance from the prediction error to its normal center, it assumes all variables have uniform property in the PCA model

F. Sun et al. (Eds.): ISNN 2008, Part II, LNCS 5264, pp. 492–497, 2008.

and doesn't take the covariance of prediction error into account, which results in that data reconstruction can not fully reflect the property of PCA model and some information is lost. On the other hand, PCA algorithm is the special case of factor analysis (FA) and FA can transform into PCA if some limitations are put on FA algorithm, then FA is more extensive. The paper focuses on the improvement of data reconstruction based on FA, its contributions are as follows: 1) propose an index based on Mahalanobis distance in FA to describe the relationship among process variables; 2) propose a novelty reconstruction method based on FA to take place of PCA-based method, and the proposed method performs better than the latter one. At last, the paper applies the new reconstruction method into Tennessee -Eastman Process (TEP) and compares with the PCA-based method.

2 FA Model

Given a $M \times N$ measurement data set X, where M is the number of process variables and N is the number of the samples. After normalization of X, the generative model is defined as $X = PT + E$, where $P \in \Re^{M \times K}$ is loading matrix, $K < M$ is the number of factors, $T \sim N(0, I)$ is factors matrix and I is unit matrix, $E \sim N(0, \Psi)$ is a noise matrix, Ψ represents an arbitrary diagonal matrix, therefore $X \sim N(0, PP^T + \Psi)$. In PCA, $\Psi = \lambda I$ and $\lambda \to 0$, so PCA is a special case of FA. The united probability distribution of T is $p(T) = (2\pi)^{-K/2} \exp\{-XX^T/2\}$, then the probability distribution over X-space for a given T is $p(X|T) = (2\pi)^{-M/2}|\Psi|^{-1/2} \exp\{-(X - PT)^T \Psi^{-1}(X - PT)/2\}$, and the probability distribution of measurement data is as follows:

$$p(X) = \int p(X|T)p(T)dX = (2\pi)^{-M/2}|C|^{-1/2} \exp\{-X^T C^{-1} X/2\} \quad (1)$$

Where $C = \Psi + PP^T$. According to Bayes' rule, the posterior distribution of the latent variable is as follows:

$$p(T|X) = (2\pi)^{-K/2}|M|^{1/2} \exp\{-(T - \beta X)^T M(T - \beta X)/2\} \quad (2)$$

Where $M = (I - P^T(\Psi + PP^T)^{-1}P)^{-1}$, $\beta = P^T(\Psi + PP^T)^{-1}$. Expectation maximization (EM) algorithm is a valid method to calculate P and Ψ by treating the factors as missing data and making the likelihood function of complete data maximum, it is an iterative algorithm including expectation step (E_step) and maximization step (M_step). Through the two steps, the following iterative formula is educed:

$$\widetilde{P} = (\sum_{i=1}^{N} x_i \mathbf{E}[T|x_i]^T)(\sum_{i=1}^{N} \mathbf{E}[TT^T|x_i])^{-1} \quad (3)$$

$$\widetilde{\Psi} = (1/N) diag(\sum_{i=1}^{N} x_i x_i^T - \widetilde{P}\mathbf{E}[T|x_i]x_i) \quad (4)$$

Where x_i is the ith sample, i.e. the ith column of X, $\mathbf{E}[T|x_i] = \beta x_i$, $\mathbf{E}[TT^T|x_i] = I - \beta P + \beta x_i x_i^T \beta^T$. Repeatedly iterate equation (3) and equation (4) until convergence, P and Ψ are calculated and FA model is developed.

3 Data Reconstruction Based on FA

3.1 Data Reconstruction Method Based on PCA

Assume the jth variable of the ith sample data needs reconstructing, i.e. x_{ij} is missing. x_{ij} can be linearly reconstructed by the rest variables like $\widetilde{x_{ij}} = \sum_{k=1\neq j}^{m} \alpha_k x_{ik}$, where α_k is linear coefficient. Therefore, the complete sample data including the reconstruction term is $\widetilde{x}_i = (x_{i1}, x_{i2}, \cdots, x_{i,j-1}, \sum_{k=1\neq j}^{m} \alpha_k x_{ik}, x_{i,j+1}, \cdots, x_{im})$, the prediction error is equal to $\widetilde{e}_i = (\widetilde{x}_i - \widetilde{T}_i P^T) = (\widetilde{x}_i - \widetilde{x}_i PP^T) = \widetilde{x}_i(I - PP^T)$. The SPE is as follows:

$$\begin{aligned} SPE = \widetilde{e}_i \widetilde{e}_i^{\,T} &= \widetilde{x}_i(I - PP^T)(I - PP^T)^T \widetilde{x}_i^{\,T} = \widetilde{x}_i(I - PP^T)\widetilde{x}_i^{\,T} \\ &= [\sum_{k=1\neq j}^{m} \alpha_k x_{ik}]^2 (1 - V_{jj}) - 2 \sum_{k=1\neq j}^{m} \sum_{a=1\neq j}^{m} \alpha_k V_{aj} x_{ik} x_{ia} + u \end{aligned} \tag{5}$$

Where u is a remaining term which does not depend on α_k, V_{aj} is the jth element of the ath row of matrix $V = PP^T$. The main idea of this method is to make SPE minimum, i.e. to make the Euclidian distance between sample data and PCA model shortest, then α_k is calculated. The derivative of equation (5) with respect to α_k is determined as:

$$\frac{\partial SPE}{\partial \alpha_k} = 2(1 - V_{ii}) x_{ik} \sum_{k=1\neq j}^{m} \alpha_k x_{ik} - 2x_{ik} \sum_{k=1\neq j}^{m} V_{kj} x_{ik} = 0 \tag{6}$$

The above formula leads to:

$$\alpha_k = \frac{V_{jk}}{1 - V_{jj}}, k = 1, 2, \cdots, j-1, j+1, \cdots, m \tag{7}$$

So the reconstructed value is:

$$\widetilde{x_{ij}} = \sum_{k=1\neq j}^{m} \frac{V_{jk}}{1 - V_{jj}} x_{ik} \tag{8}$$

3.2 Data Reconstruction Based on FA

As an index to measure the information in error space, SPE can reflect the degree that the sample accords with the PCA model. The smaller SPE is, the tighter the sample accords with model. While SPE is a Euclidian distance according to equation (5), it can be transformed into:

$$SPE = \widetilde{e}_i \widetilde{e}_i^{\,T} = (\widetilde{e}_i - 0)(\widetilde{e}_i - 0)^T \tag{9}$$

Since measurements have been normalized, the mean of normal error is **0** vector. From equation (9), SPE represents the Euclidian distance from $\widetilde{e_i}$ to the mean. Euclidian distance treats the properties of all $\widetilde{e_i}$ as uniform ones, however, the different property of $\widetilde{e_i}$ is important to distinguish them from the others.

In statistical process monitoring, the main principle is the process variables meet normal distribution under working order, then SPE is not suitable for the evaluation of process operation. Unlike SPE, the paper proposes a generalized SPE (GSPE) based on Mahalanobis distance to measure the error space. Compare with Euclidian distance, Mahalanobis distance is more suitable to be used for the measurement of normal process variables. The squared Mahalanobis norm of the noise $\widetilde{e_i}$ is $\widetilde{e_i}^T \Psi^{-1} \widetilde{e_i}$, while $\widetilde{e_i}$ is undetectable, and its posterior distribution is

$$p(E|X) = (2\pi)^{-K/2} |\Sigma^{E|X}|^{-1/2} \exp\{-\frac{1}{2}[E - \mu^{E|X}]^T (\Sigma^{E|X})^{-1} [E - \mu^{E|X}]\} \quad (10)$$

Where $\mu^{E|X} = (I - PMP^T)X$, $\Sigma^{E|X} = \Psi PMP^T$. Substitute $\widetilde{e_i}$ with $\mu^{E|x_i}$, then the proposed GSPE is like:

$$\begin{aligned} GSPE_i &= \|\Psi^{-1/2} \mu^{E|x_i}\|^2 = (\mu^{E|x_i})^T \Psi^{-1} \mu^{E|x_i} \\ &= ((I - PMP^T)\widetilde{x_i})^T \Psi^{-1} (I - PMP^T)\widetilde{x_i} \\ &= \widetilde{x_i}^T (I - PMP^T)^T \Psi^{-1} (I - PMP^T)\widetilde{x_i} \\ &= \widetilde{x_i}^T R \widetilde{x_i} \end{aligned} \quad (11)$$

where $(I - PMP^T)^T \Psi^{-1} (I - PMP^T)$ is a constant matrix. In order to make GSPE minimum, set the derivative of equation (11) with respect to $\widetilde{x_{ij}}$ to 0, then leads to:

$$\begin{aligned} \frac{\partial GSPE_i}{\partial \widetilde{x_{ij}}} &= [0, 0, \cdots, 0, 1, 0, \cdots, 0]^T R \widetilde{x_{ij}} + \widetilde{x_{ij}}^T R [0, 0, \cdots, 0, 1, 0, \cdots, 0] \\ &= 2R_{jj}\widetilde{x_{ij}} + \sum_{k=1 \neq j}^{m} R_{jk} x_{ik} + \sum_{k=1 \neq j}^{m} R_{kj} x_{ik} = 0 \end{aligned} \quad (12)$$

The missing data is estimated as:

$$\widetilde{x_{ij}} = -(\sum_{k=1 \neq j}^{m} R_{jk} x_{ik} + \sum_{k=1 \neq j}^{m} R_{kj} x_{ik}) / 2R_{jj} \quad (13)$$

SPE does not take into account the information among $\widetilde{e_i}$ and the other normal errors produced by model data, however, the tightness that the prediction error $\widetilde{e_i}$ associates with normal error can help to decide if $\widetilde{e_i}$ accords with the model. If E satisfies uniform distribution, SPE is a valid tool to detect the changes in the relations between variables. While in normal distribution, GSPE is better selection. Fortunately, E meets normal distribution.

4 Application Case

In this section, Tennessee-Eastman process (TEP) is used to study the data reconstruction, the performances of PCA-based method and FA-based method are compared.

TEP is created by the Eastman Chemical Company to provide a realistic industrial process for evaluating process control and monitoring method, it has been widely used as a benchmark simulation plane. The process consists of five major units: a reactor, condense, compressor, separator, stripper. TEP has 52 process variables and is set 21 fault modes. Its detailed introduction and the process flow sheet consults [10]. The whole operation time is 48h and sample interval is 3min. Generate 960 samples from normal process, use the former 700 samples to build FA model and the rest 260 normal samples to study on data reconstruction.

Select 15 factors and build the FA model according to the section 1, then save P and Ψ. Assume the 16th variable (stripper pressure) is missing. The performances of popular method and the new approach to construct the 16th variable are shown in Fig.1.

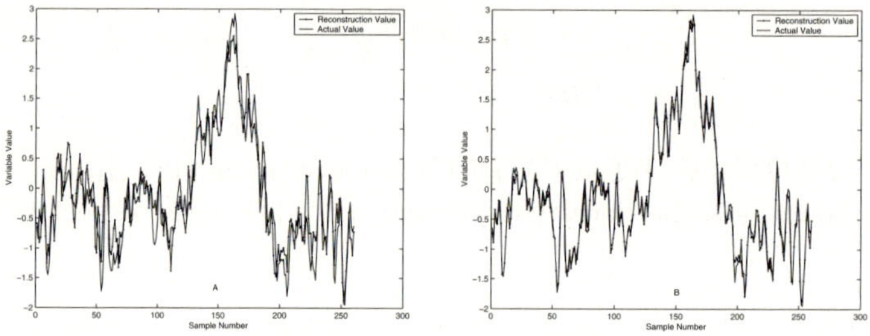

Fig. 1. Reconstruction data charts of variable 16. (A) PCA-based method; (B) FA-based method.

Compare Fig.1 (A) with Fig.1 (B), it is obvious that the latter is better and almost identical with the actual value. Assume the data from variable 13 to variable 20 is missing in turn, the mean square error (MSE) $MSE = \|x_{:,j} - \widetilde{x_{:,j}}\|^2/260$ of reconstruction value is proposed to appraise reconstruction performance, where $x_{:,j} \in \Re^{260\times 1}$ is the actual value, $\widetilde{x_{:,j}}$ is the estimation value corresponding to $x_{:,j}$. MSE by PCA and FA is shown in table 1. For simplicity, the performances of reconstruction for the other variables are not listed.

From Tab.1, exception variable 14, both of the two methods perform well, the MSE by the FA-based method is smaller than that by the PCA-based method, which proves that the former method performs better than the latter method. To the 14th variable, the two methods perform badly and the estimation values completely deviate from the actual value, then the estimation values are not worth depending on. Because the methods for data reconstruction are all based on the linear relationships among the missing variable and the rest variables,

Table 1. The MSE value by PCA and FA in TEP

Variable NO	13	14	15	16	17	18	19	20
MSE by PCA	0.070	3.161	0.043	0.152	0.049	0.091	0.129	0.186
MSE by FA	0.0035	28.713	0.000	0.016	0.000	0.041	0.025	0.102

the almost independent variable can not be estimated by others, which is why variable 14 is estimated poorly.

5 Conclusion

The paper proposes a novelty data reconstruction method based on FA. The results of the application of the proposed method in TEP indicate that the proposed method can more represent the intrinsic property among variables and is superior to the popular one based on PCA. On the other hand, the estimation of variable 14 shows that the precondition to achieve better performance of data reconstruction is the remaining variables closely correlate with the missing variable, otherwise the performance is bad.

References

1. Goulding, P.R., et al.: Fault Detection in Continuous Processes Using Multivariate Statistical Methods. International Journal of Systems Science 31(11), 1459–1471 (2000)
2. Ricardo, D., et al.: Use of Principal Component Analysis for Sensor Fault Identification. Computers and Chemical Engineering 20(suppl.), S713–S718 (1996)
3. Ricardo, D., et al.: Identification of Faulty Sensors Using Principal Component Analysis. AIChE Journal 10(10), 2797–2812 (1996)
4. Fuat, D., et al.: A Strategy for Detection and Isolation of Sensor Failures and Process upset. Chemometrics and Intelligent Laboratory Systems 20, 109–123 (2001)
5. Ricardo, D., Qin, S.J.: A Unified Geometric Approach to Process and Sensor Fault Identification and Reconstruction: the Unidimensional Fault Case. Computers and Chemical Engineering 22(7-8), 927–943 (1998)
6. Walczak, B., Massart, D.L.: Dealing with Missing Data: Part I. Chemometrics and Intelligent Laboratory Systems 58, 15–27 (2001)
7. Walczak, B., Massart, D.L.: Dealing with Missing Data: Part II. Chemometrics and Intelligent Laboratory Systems 58, 29–42 (2001)
8. Philip, R.C.N., et al.: Missing Data Methods in PCA and PLS: Score Calculations with Incomplete Observation. Chemometrics and Intelligent Laboratory Systems 35, 45–65 (2001)
9. Dirk, N., et al.: Improved Diagnosis of Sensor Faults Using Multivariate Statistics. In: Proceeding of the 2004 American control conference, Boston, Massachusetts, June 30-July 2, pp. 4403–4407 (2004)
10. Chiang, L.H., Russell, E.L., Braatz, R.D.: Fault Detection and Diagnosis in Industrial Systems. Springer, London (2001)

Synthetic Fault Diagnosis Method of Power Transformer Based on Rough Set Theory and Bayesian Network

Yongqiang Wang, Fangcheng Lu, and Heming Li

Key Laboratory of Power System Protection and Dynamic Security Monitoring and Control under Ministry of Education, North China Electric Power University, 071003 Baoding, China
wyq_hd_bd@163.com

Abstract. Power transformer is very important in power system. In this paper, according to complementary strategy, a new transformer fault diagnosis method based on rough sets theory (RST) and Bayesian network (BN) is present. Through reduction approach of RST information table to simplify expert knowledge and to reduce fault symptoms, the minimal diagnostic rules can be mined. According to the minimal rules, complexity of BN structure and difficulties of fault symptom acquisition are largely decreased. At the same time, probability reasoning can be realized by BN, which can be used to describe changes of fault symptoms and analyze fault reasons of transformer. Finally, the correctness and effectiveness of this method are validated by the result of practical fault diagnosis examples.

Keywords: Transformer; Fault diagnosis; Rough set theory; Bayesian network.

1 Introduction

Power Transformer is very important in power system. But it's very difficult to diagnosis the fault exactly because of power transformer's complexity of configuration. At present, based on difference symptoms, three-ratio method, fuzzy technique, support vector machine (SVM), neural network (NN) and degree of area incidence analysis has already been used in power transformer fault diagnosis[1-6]. But because of the limited precision and amount of practical sampling and the complexity of transformer fault, and because of the limited of single method, these methods still have many problems. So, according to complementary strategy, in this paper, a new transformer fault diagnosis method based on rough sets theory (RST) and Bayesian network (BN) is present. Through reduction approach of RST information table to simplify expert knowledge and to reduce fault symptoms, the minimal diagnostic rules can be mined. According to the minimal rules, complexity of BN structure and difficulties of fault symptom acquisition are largely decreased. At the same time, probability reasoning can be realized by BN, which can be used to describe changes of fault symptoms and analyze fault reasons of transformer. Finally, the application examples in the fault diagnosis of transformer are given which shows that this method is effective.

F. Sun et al. (Eds.): ISNN 2008, Part II, LNCS 5264, pp. 498–505, 2008.

2 Rough Set Theory Application

2.1 Introduction of Rough Set Theory

Rough set theory is advanced by Professor Z. Pawlak of Poland in 1980's. It can be used to deal with fuzzy and uncertainty question and to reduce the complexity of question.

RST is main study the information systems made of object set and property set.

Definition 1. Let $S = <U, C, D, V, f>$ be knowledge system. U is the object set and $A = C \cup D$ is property set. Subset C is causal set and D is effect set. $V = \bigcup_{q \in A} V_q$ is property value set and V_q is the value of q. f is the function which appoint the property value of every object of U, i.e., $f: U \times A \rightarrow V_q$. A two dimension table named information table (also named decision table) can be gotten from this formula.

In the RST, knowledge is described by the property and property value of object. If we use its causal property to show the causal part of rule and use its effect property to show the effect part of rule, then, a produce rule can be described by every object.

2.2 Reduction Approach of Decision Table

It isn't necessary to the every information in the decision table created by RST. Many information can be deleted from decision table and cannot influence the effect of expression. So, we can get a minimal decision table through reduction of RST decision table. The reducing steps is shown as following. (1) Delete the redundant causal property, i.e., delete a row from the decision table.(2) Delete the repeated row.(3) Delete the redundancies of every decision rule.

The minimal decision table would be gotten through upwards reduction and it isn't a full information table as before. It only includes the necessary information.

Table 1, 2, 3 and 4 show us the process of the decision table reduction.

Table 1. Original decision table

U	a	b	c	d
1	1	0	0	d_1
2	0	0	0	d_1
3	0	0	1	d_1
4	0	1	0	d_2
5	1	2	0	d_2
6	0	2	1	d_1
7	0	2	0	d_2
8	1	2	1	d_1

Table 2. Get rid of property ***a***

U	b	c	d
1	0	0	d_1
2	0	0	d_1
3	0	1	d_1
4	1	0	d_2
5	2	0	d_2
6	2	1	d_1
7	2	0	d_2
8	0	1	d_1

Table 3. Core table

U	b	c	d
2	0	0	d_1
4	1	0	d_2
5	2	0	d_2
6	2	1	d_1

Table 4. Minimal table

U	b	c	d
1	0	*	d_1
2	0	*	d_1
3	*	*	d_1
4	1	*	d_2
5	2	0	d_2
6	*	1	d_1
7	2	0	d_2
8	*	1	d_1

Through reduction approach of RST information table to simplify expert knowledge and to reduce fault symptoms, the minimal diagnostic rules can be mined. It can simplify the fault diagnosis rules and make the reasoning process fast.

3 Bayesian Network and Its Construction

Bayesian network is also called probabilities causation network, knowledge map. It is a directed acyclic graph (DAG) [5]. It shows the causations of variables by directed graph, and these causations are to be shown by probabilities value. It is made of nodes and directed lines. The node is the independence randomicity variable. The line is the probability relationship between the nodes.

Assume $U=\{X_1, X_2,..., X_n\}$. In the formula, every variable X_i has limited values.

Now, a Bayesian network can be described as $B=\langle G,\Theta\rangle$. In the formula, G is a directed acyclic graph, and the nodes in the graph are corresponding to the $X_1, X_2,...X_n$, the directed lines are the probability relationship between the nodes. Θ is the conditional probabilities of BN, $\Theta=\{\theta_{xi|pai}\}$. And $\theta_{xi|pai}=P(x_i \mid pa_i)$ showing the conditional probability when the value of Pa_i is equal to pa_i, and Pa_i is the father nodes of every value of node X_i. The probability distributing that scored by Bayesian network B and U can be described by the following formula.

$$P(X_1, X_2,..., X_n)=\prod_i P(X_i \mid Pa_i)$$

Fig. 1 shows the frame of a simple Bayesian Network. (The conditional probabilities have been ignored in this figure.)

The probabilities of Bayesian Network are Bayesian probabilities or physical probabilities. It is Bayesian probability when it only gets from knowledge, and it is physical probability when it only learns from data [6,7].

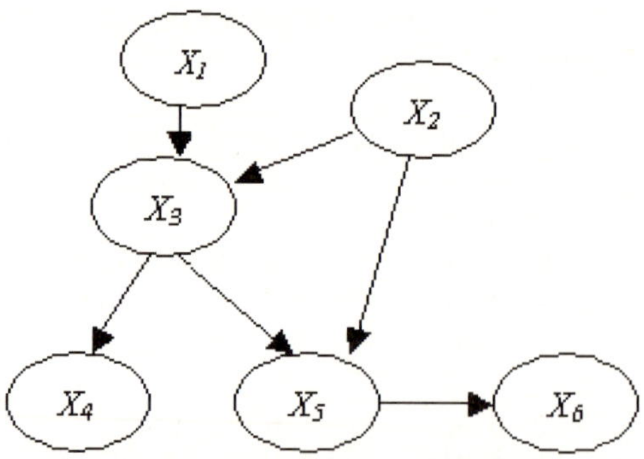

Fig. 1. Bayesian network

We can construct a BN as the following steps.

(1) The variables that can be used to construct a BN model and its meaning must be given first. This step needs to confirm the goal and its meaning of the model, to confirm the possible value of goal and subclass of model, to confirm the independence and mutually exclusive variables and these variables end all states of the problem.

(2) Construct a directed acyclic graph to show the conditional connection of the variables. As probability multiplication formula, we can get the following formula,

$$P(X) = \prod_{i=1}^{n} P(X_i \mid x_1, x_2, \dots, x_{i-1}) \tag{1}$$

Pa_i is the father nodes of variables.

Then,

$$P(X_1, X_2, \dots, X_n) = \prod_i P(X_i \mid Pa_i) \tag{2}$$

In order to confirm the structure of BN, we must arrange the $X_1, X_2, \dots X_n$ with given sequence and ensure the Pa_i which is decided by formula (2).

(3) Confirm the partial probabilities $P(X_i \mid Pa_i)$. Every father node of X_i must be given a probability.

The steps described above can be done alternately.

4 Fault Diagnosis Model of Power Transformer Based on Rough Set Theory and Bayesian Network

Through reduction approach of RST information table to simplify expert knowledge and to reduce fault symptoms, the minimal diagnostic rules can be mined. According to the minimal rules, we can construct the most compact Bayesian network. And use the causal reasoning of Bayesian network, we can diagnose the transformer fault quikly. This shows, it is complementary between RST and BN.

Based on the literature search, we get the swatch sets of transformer fault[1-3]. Symptom set (***M***) is showing in Table 5. Fault set (***D***) is showing in Table 6. The intensity table of cause (***R***) is showing in Table 7. and we use $\boldsymbol{R}_{ij}$ describe the probability $\boldsymbol{P}(m_j|d_i)$.

From RST theory, Table 7 is a decision table. In order to disperse the Table 7, we make following definition.

(1) If $\boldsymbol{R}_{ij}>0.5$, then its discrete value is 2.
(2) If $\boldsymbol{R}_{ij}<0.5$, then its discrete value is 1.
(3) If $\boldsymbol{R}_{ij}=0$, then its discrete value is 0.

The discrete value 2 means the fault cause the symptom strongly, value 1 means not very strong, and value 0 means no relation. Through the conversion based on these definition, we get the decision table of fault diagnosis, and it is shown as Table 8.

Table 5. Table of Symptom set M

Code	**Symptom**
m_1	Grounding Current of Core
m_2	Overheat
m_3	Imbalance of three phase Winding Resisters
m_4	Damped
m_5	Discharge
m_6	Ratio Changed
m_7	Partial Discharge
m_8	CO/CO_2
m_9	Absorptance

Table 6. Table of fault set D and its prior probability

Faults	Types of fault	Probability (P%)
d_1	Multi-grounding of Core	22.71
d_2	Aging	5.27
d_3	Overhet	6.29
d_4	Short Circuit Between Circles	6.12
d_5	Damped	5.06
d_6	Fault of Switcher	13.17
d_7	Discharge of Mote	7.98
d_8	Flashover	14.21
d_9	Distortion of Winding	12.02
d_{10}	Discharge in the oil	7.17

Table 7. Table of intensity of cause set R

	m_1	m_2	m_3	m_4	m_5	m_6	m_7	m_8	m_9
d_1	0.90	0.818			0.189		0.30		
d_2		0.219		0.267				0.816	
d_3		0.713			0.289		0.35		
d_4					0.515	0.80	0.90	0.681	
d_5				0.718					0.75
d_6		0.674	0.87		0.231				
d_7					0.863		0.90		
d_8				0.416	0.879		0.90	0.759	
d_9		0.149			0.681	0.80	0.75	0.721	
d_{10}		0.20		0.60	0.70		0.90		

Table 8. The decision table of fault diagnosis

	m_1	m_2	m_3	m_4	m_5	m_6	m_7	m_8	m_9
d_1	2	2	0	0	1	0	1	0	0
d_2	0	1	0	1	0	0	0	2	0
d_3	0	2	0	0	1	0	1	0	0
d_4	0	0	0	0	2	2	2	2	0
d_5	0	0	0	2	0	0	0	0	2
d_6	0	2	2	0	1	0	0	0	0
d_7	0	0	0	0	2	0	2	0	0
d_8	0	0	0	1	2	0	2	2	0
d_9	0	1	0	0	2	2	2	2	0
d_{10}	0	1	0	2	2	0	2	0	0

Table 9. Minimal decision table of fault diagnosis

	m_1	m_2	m_4	m_7	m_8
d1	2	2	0	1	0
d2	0	1	1	0	2
d3	0	2	0	1	0
d4	0	0	0	2	2
d5	0	0	2	0	0
d6	0	2	0	0	0
d7	0	0	0	2	0
d8	0	0	1	2	2
d9	0	1	0	2	2
d10	0	1	2	2	0

Through reduction approach of decision table, the minimal property sets are found. They are $\{m_1,m_2,m_3,m_4,m_6\}$, $\{m_1,m_2,m_3,m_4,m_8\}$, $\{m_1,m_2,m_4,m_6,m_7\}$ and $\{m_1,m_2,m_4,m_7,m_8\}$. Obviously, $\{m_1,m_2,m_3,m_4,m_6\}$ and $\{m_1,m_2,m_3,m_4,m_8\}$ have a row which value are all zero. So both of them can not make a minimal decision table. Consider the difficulty of symptoms, we select $\{m_1,m_2,m_4,m_7,m_8\}$ as the minimal decision table as Table 9. Based on Table 9, we construct the most simple Bayesian network as Fig. 2.

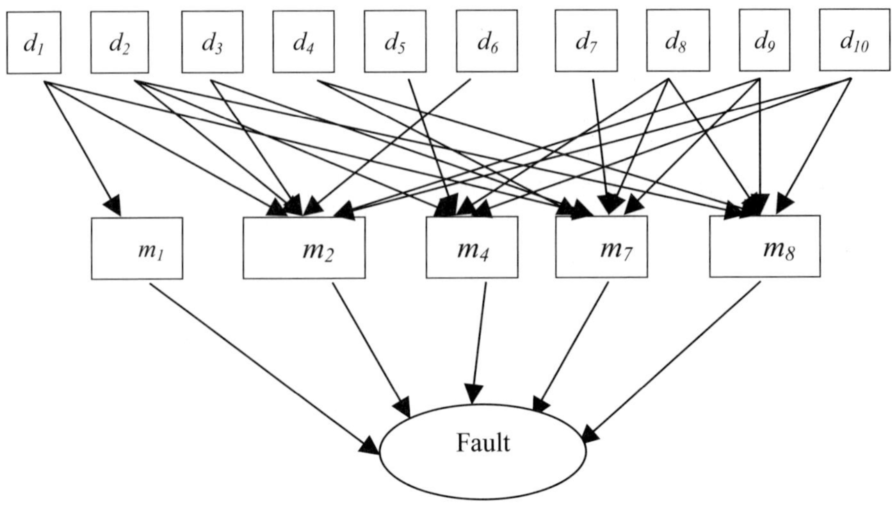

Fig. 2. The transformer fault diagnosis model based on Bayesian network

5 Fault Diagnosis Method Based on BN

The reasoning process of BN is fast calculating process of probability based on already existent information.

Before the calculating of the probability, we must define the fault symptom set as following.

1) The existent fault symptom set M^+, i.e., every fault symptom in M^+ is proved to existent.

2) The inexistent fault symptom set M^-, i.e., every fault symptom in M^- is proved to inexistent. In the model of this paper, $M=M^++M^-$.

The calculating process of equipment fault diagnosis based on BN model is shown the following.

1) Collect information, and apply it to Bayesian network.

2) Calculate the probabilities of father nodes based on Bayesian network.

3) Select the father node which has most probability as the main fault, if it has father nodes then analyses its father nodes as step 2) and step 3), until find the finally main fault.

4) If the probability of the finally main fault reason is more than the twice of the average probability, then stop diagnosis and output the result, at the same time, strengthen Bayesian network. Otherwise, note this fault and eliminate this fault, then analyses the other fault reason as step 2) to step 4), finally, output alternately the all results which probability is more than average.

6 Fault Diagnosis Example of Transformer

A transformer is fault. The dissolved gas analysis (DGA) result is shown in Table 10. Its iron core grounding current is 0.1A. The moisture value in the oil and partial discharge value are all normal. $CO/CO_2=0.21$, $CH_4/H_2=0.99$, $C_2H_2/C_2H_4=0.043$, $C_2H_4/C_2H_6=8.35$, so the three-ratio code is "002". Then we can find it is overheat fault. The iron core grounding current is normal, and CO/CO_2 is normal too.

Table 10. Components of dissolved gas in a transformer 10^{-6}

Gas	H_2	CH_4	C_2H_2	C_2H_4	C_2H_6	CO_2	CO
Value	70.4	69.5	10.4	241	28.9	3350	704

The result of Bayesian reasoning is shown as Table 11. From Table 11, we can draw the conclusion that the diagnosis result is tap changer or down-lead fault (d_6). The practice examine result the third tap nod of A phase was overheat, it became black and had thawy phenomenon in the 35kV side. The examine result is same as the diagnosis result of this paper method. This practical fault diagnosis example validate the correctness and effectiveness of this paper method. This method can use prior probability to set uo model, so it needn't many samples to study the parameter like NN. The result of 20 calculation examples in Table 12 prove this method is more quick and exact than the normal BN method like the paper [7].

Table 11. The probabilities of fault causes ($P(d_i \mid M^+ \cap M^-)$)

Reason of Fault	d_1	d_2	d_3	d_4	d_5	d_6	d_7	d_8	d_9	d_{10}
Probability $P(d_i \mid M^+ \cap M^-)$	0.0130	0.0016	0.0336	0	0	0.0879	0	0	0.0012	0.0006

Table 12. The calculation result of this paper method and paper [7] method

Diagnosis result	Average calculation speed	Correct	wrong	Cannot diagnosis
This paper method	0.00018 S	20	0	0
Paper [7] method	0.00243 S	17	2	1

7 Conclusion

Power transformer is very important in power system. In this paper, according to complementary strategy, a new transformer fault diagnosis method based on rough sets theory (RST) and Bayesian network (BN) is present. Through reduction approach of RST information table to simplify expert knowledge and to reduce fault symptoms, the minimal diagnostic rules can be mined. According to the minimal rules, complexity of BN structure and difficulties of fault symptom acquisition are largely decreased. At the same time, probability reasoning can be realized by BN, which can be used to describe changes of fault symptoms and analyze fault reasons of transformer. Finally, the correctness and effectiveness of this method are validated by the result of practical fault diagnosis examples.

Referrences

1. Dong, M., Meng, Y.Y., Xu, C.X., Yan, Z.: Fault Diagnosis Model for Power Transformer Based on Support Vector Machine and Dissolved Gas Analysis. In: Proceedings of the CSEE, vol. 23, pp. 88–92 (2003)
2. Sun, C.X., Li, J., Zheng, H.P., et al.: A New Method of Faulty Insulation Diagnosis in Power Transformer Based on Degree of Area Incidence Analysis. Power System Technology 26, 24–29 (2002)
3. Yang, L., Shang, Y., Zhou, Y.F., Yan, Z.: Probability Reasoning and Fuzzy Technique Applied for Identifying Power Transformer Malfunction. In: Proceedings of the CSEE, vol. 20, pp. 19–23 (2000)
4. Zhang, W.X.: Rough Set Theory. Publishing Company of Science, Beijing (2001)
5. Lin, S.M., Tian, F.Z., Lu, Y.C.: Construction and Applications in Data Mining of Bayesian Networks. Journal of Tsinghua University (Sci. & Tech.) 41, 49–52 (2001)
6. Liu, Z.Q.: Causation, Bayesian Network, and Cognitive Maps. Acta Automatica Sinica 27, 552–566 (2001)
7. Wang, Y.Q., Lu, F.C., Li, H.M.: Fault Diagnosis for Power Transformer Based on BN and DGA. Transactions of China Electrotechnical Society 19, 74–77 (2004)

Fuzzy Information Fusion Algorithm of Fault Diagnosis Based on Similarity Measure of Evidence

Chenglin Wen[1], Yingchang Wang[1,⋆], and Xiaobin Xu[1,2]

[1] Institute of Information and Control, Hangzhou Dianzi University, 310018 Hangzhou, China
[2] Department of Electrical and Automation, Shanghai Maritime University, 200135 Shanghai, China
wangyingwang2006@163.com

Abstract. In this paper, a fuzzy information fusion method of fault diagnosis based on evidence similarity measure is presented. First, because of the fuzziness of information received by sensors, membership functions are introduced to describe the fault template mode in model database and features extracted from sensor observations; then the degrees of matching between them are obtained using random set model of fuzzy information, which can be transformed into BPAs. Second, cosine similarity measure of evidence is introduced to compute confidence degree of evidence. Finally, original evidences are modified according to the confidence degree. The diagnosis results of rotor system show that the proposed method can improve the accuracy of decision-making.

Keywords: Fault diagnosis, Similarity measure, Confidence degree.

1 Introduction

Multi-source information fusion is critical to the delivery of effective decision system.As an important decision-making fusion method, Dempster-Shafer evidence theory has been widely used in fault diagnosis because of its simple operations and ability of dealing with uncertainty [1-4]. It is general extension of probability theory and can robustly deal with incomplete data. Also, it allows the representation of both imprecision and uncertainty.

In fault diagnosis based on evidence reasoning, elements in discernment framework denote fault modes. BPAs of them can be gotten by matching the fault templates in model database with the features extracted from sensor observations. Because of differences of observation time and place, as well as variety of environments, sensor observations are often incomplete and uncertain, especially fuzzy, so fault templates and features usually are characterized by fuzziness at the same time. However, so far there are hardly suitable methods for dealing

⋆ This work has been supported by the National Natural Science Foundation of China (No. 60434020, 60772006)

F. Sun et al. (Eds.): ISNN 2008, Part II, LNCS 5264, pp. 506–515, 2008.

with the acquirement of reasonable BPA from fuzzy fault templates and fuzzy features.

On the other hand, it has been found that Dempster's combination rule gives counterintuitive results when evidences conflict each other. In order to solve this problem, many scholars have proposed alternative combination rules and modification methods. Murphy has thought that evidence sources should be modified before combination, and suggested incorporating average belief into combining [5]. Fan has used membership function and weights of features to modify the BPAs of evidence and proposed modification of Dempster's combination rule and decision rule [6]. Deng thought that the importance of each body of evidence could be different and it should have different effect on the final combination results . He proposed a modified combination rule based on the distance of evidence [7]. In the actual fault diagnosis, because of unreasonable measure positions, mistakes in feature extraction, sudden accidents of sensors or other factors, evidences often conflict each other. If they are combined directly, decision is often not consistent with practical case, and even false. Therefore, before making decisions, it is necessary to check the consensus of evidences, give confidence degree of evidence so as to modify conflicting evidences.

In order to solve above two common problems in fault diagnosis based on evidence reasoning, this paper presents a fuzzy information fusion method based on similarity measure of evidence. First, membership functions are used to describe the fault templates in model database and features extracted from sensor observations. Secondly, the degrees of matching between them can be computed by random set model of fuzzy information, which is equivalent to BPA in value. Thirdly, we propose the cosine similarity measure of evidence in defined measure space, by which, confidence degree of evidence is given to modify original evidences. Fourthly, fault decisions can be obtained from evidence combination. Finally, the diagnosis results of machine rotor show that the proposed method can enhance diagnostic accuracy and reliability.

2 Evidence Theory

The Dempster-Shafer theory of evidence is a mechanism formalized by Shafer information. It is based on Dempster's original work on the modeling of uncertainty in terms of upper and lower probabilities that are induced by a multivalued mapping rather than as a single probability value.

2.1 Dempster-Shafer Theory [8]

A set is called a frame of discernment if it contains mutually exclusive and exhaustive possible answers to a question. It is usually denoted as Θ . The set is required that at any time one and only one element in the set is true. The power set 2^{Θ} composed with all of subset of Θ. A function m: $2^{\Theta} \rightarrow [0,1]$ is called a mass function on frame Θ if it satisfies the following conditions: $m(\emptyset) = 0$; and $\sum_{A\subseteq\Theta} = 1$, where $\emptyset$ is an empty set and A is a subset of Θ. A mass function is also called a Basic Probability Assignment, denoted as BPA. A subset A with $m(A) > 0$ is called a focal element of frame Θ.

When more than one mass function is given on the same frame of discernment, the combined result of these pieces of evidence is obtained using a mathematical formula called Dempster's combination rule. If m_1 and m_2 are two mass functions on frame Θ, then m is the mass function after combining m_1 and m_2:

$$m(C) = \begin{cases} \frac{\sum\limits_{A \cap B = C} m_1(A) \cdot m_2(B)}{1 - \sum\limits_{A \cap B = \emptyset} m_1(A) \cdot m_2(B)} & C \neq \emptyset, \\ 0 & C = \emptyset. \end{cases} \tag{1}$$

where $k = \sum\limits_{A \cap B = \emptyset} m_1(A) \cdot m_2(B)$ reflects the conflict between two mass functions.

2.2 The Improved Method

In fault diagnosis based on multi-source information fusion, once there is an unrealistic determination, conflict will appear. If we treat all evidences equally, fusion decision will lead to error, and even misjudgments. Therefore, it is important to assign an appropriate weight to evidence according to its confidence degree before combining. This section proposes cosine similarity measure of evidence that computes confidence degree of evidence so as to modify original evidences.

Definition 1(*similarity measure function*). Let A, B are two vectors of R^n, a function S(·): $A \times B \rightarrow$ [0,1] is called similarity measure function, if it satisfies the following three conditions:

1. $0 \leq S(A, B) \leq 1$
2. $S(A, B) = S(B, A)$
3. $S(A, B) = 1$ if and only if $A{=}B$; $S(A, B) = 0$ if and only if $A \perp B$.

Definition 2 (*similarity measure space*). Suppose $S(\cdot, \cdot)$ is a binary real function on set Γ, for any x, $y \in \Gamma$, if it satisfies the three basic conditions in definition 1, then Γ is called similarity measure space.

In fault diagnosis, information obtained from sensor p is considered as evidence m_p for p=1, 2,. . .,n in evidence theory, it is a decision about equipment's running condition, where n is quantities of sensors in diagnosis system. In order to verify evidences' consistency, we regard m_p as a vector of evidence, whose component represents a BPA corresponding to certain hypothesis in 2^Θ, all of evidences compose a set which contains n vectors.

Definition3 (*cosine similarity measure of evidence*). In evidence theory, let m_1 and m_2 are two mass functions, then the similarity between them is

$$Sim(m_1, m_2) = \cos\theta = \frac{m_1 \cdot m_2^T}{\|m_1\| \cdot \|m_2\|}. \tag{2}$$

where $m_1 \cdot m_2 = \sum\limits_{j=1}^{k} m_{1,j} \cdot m_{2,j}$ is inner product of m_1 and m_2 , k is the dimension of vector; $\|\cdot\|$ represents the norm of vector. In fact, (2) measures the cosine of

m_1 and m_2. If angle θ between m_1 and m_2 is $0°$, then their similarity is 1, it notes that m_1 and m_2 have the same support to all hypothesis in frame . On the contrary, if angle θ between them is $90°$, then similarity is 0, that is to say, these two evidences conflict completely. The geometric explanation of cosine similarity measure is shown in Fig. 1.

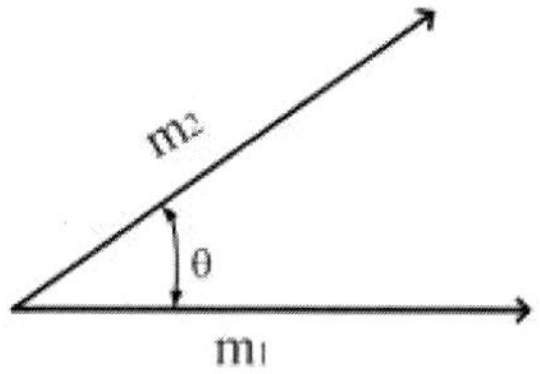

Fig. 1. Geometric explanation of cosine similarity measure

Obviously, $Sim(m_1, m_2)$ satisfies the condition in definition 2, so in fault diagnosis, all information obtained from sensors can compose an evidence vector space, which is a similarity measure space.

For n-sources fusion system, similarity measure matrix is denoted as

$$SIM = \begin{pmatrix} S_{11} & S_{12} & \cdots & S_{1n} \\ S_{21} & S_{22} & \cdots & S_{2n} \\ \vdots & & & \\ S_{n1} & S_{n2} & \cdots & S_{nn} \end{pmatrix}. \tag{3}$$

where, $S_{pq} = Sim(m_p, m_q)$ is similarity of m_p and m_q, $S_{pp} = 1$, $S_{pq} = S_{qp}$ for p, q=1,2, ...,n. The support degree of evidence is defined as:

$$\mathbf{Support}(m_p) = \sum_{p \neq q} S_{pq}, p, q = 1, 2, \ldots, n. \tag{4}$$

The confidence degree of evidence is defined as:

$$\omega_p = \frac{Support(m_p)}{\mathbf{max} Support_p(m_p)}. \tag{5}$$

The new modification of BPAs considering the confidence degree is shown:

$$m'_p(A) = \begin{cases} \omega_p \cdot m_p(A) & A \subset \Theta, \\ 1 - \sum\limits_{B \subset \Theta} \omega_p \cdot m_p(B) & A = \Theta. \end{cases} \tag{6}$$

By using confidence degree,. the hypothesis's BPAs are reduced and more uncertain is assigned to the Θ's BPA, finally the conflicts are weaken and the reasonability of decision is improved. Also, it has found that confidence degree $0 \leq \omega_p \leq 1$, when the similarity is bigger, the ω_p becomes bigger, and the ω_p will be more effect in reducing the conflict.

3 Compute BPA Based on Fuzzy Information

Because of differences of observation time and place, as well as variety of environments, sensor observations are often incomplete and uncertain, especially fuzzy. In this case, fault features extracted from observations are described by fuzzy membership function [3-4]. In order to deal with the fuzziness of fault templates and features extracted from sensor observations, in this section, membership functions of them are obtained by method of experiment statistics, and then BPAs are given based on random set model of fuzzy information.

3.1 Determinate the Membership Function of Fault Templates

Let U be a characteristic parameter space about certain equipment, the membership function of fault template is defined as:

$$\mu_F(x) : U \to [0,1], x \in U, \tag{7}$$

where, F is fault mode in template base. The process of determination of $\mu_F(x)$ is shown as following:

(1)Simulate several classic fault modes, including normal condition, and under each mode observe equipment running conditions at the same time interval, l groups data are collected, and denoted as $x_i = (x_{i1}, x_{i2}, , x_{im})$ for i=1,2,,l. Generally, $m \in [30, 50]$.

(2) For x_i, compute mean M_i and standard deviation σ_i.

$$M_i = \frac{1}{n}\sum_j x_{ij},$$
$$\sigma_i = \sqrt{\frac{1}{n}\sum_j (x_{ij} - M_j)^2}, i, 1, 2, \dots, l; j = 1, 2, \dots, m. \tag{8}$$

(3) Construct membership function based on x_i using M_i, σ_i.

$$\mu_{F,i}(x) = exp(-\frac{(x_i - M_i)^2}{2\sigma_i^2}). \tag{9}$$

(4) Determinate the membership functions of fault templates according to $\mu_{F,i}(x_i)$

$$\mu_F(x) = \begin{cases} exp(-\frac{(x-M_a)^2}{2\sigma_a^2}) & x < M_a, \\ 1 & M_a \le x \le M_b, \\ exp(-\frac{(x-M_b)^2}{2\sigma_b^2}) & x > M_b. \end{cases} \tag{10}$$

where, x denotes sensor's measured value, M_a=min(M_i), M_b=max(M_i) for i=1, 2,...,l are standard deviations corresponding to M_a, M_b.

It notes that the membership functions of fault templates also can be obtained from the historical maintenance data or expert experiences in condition monitoring and diagnosis to equipment.

3.2 Determinate the Membership Function of Fuzzy Features

Now suppose equipment runs steadily at time interval $\triangle t$, write down equipment running status value, and the membership function of features extracted from sensor observations can be obtained according Subsection 3.1, denoted as $\mu_o(x)$: $U \to [0,1]$, subscript 'o' represents observations.

$$\mu_o(x) = exp(-\frac{(x-M)^2}{2\sigma^2}). \tag{11}$$

where, M, σ are mean and standard deviation corresponding to the observation, respectively.

3.3 Determinate BPAs

In this section, the degree of matching between the fault templates and features extracted from sensor observations are given by random set model of fuzzy information, which is BPA. Let δ be a random number in interval [0, 1], and define

$$\sum_F \triangleq \sum_\delta(\mu_F) = \{x \in U \mid \delta \leq \mu_F(x)\}. \tag{12}$$

$\sum_F$ is a set that all of its elements' value are more larger than δ in characteristic parameter space, which is a kind of **random set** [9-10], the number of its elements changes with δ, and $\sum_F$ is a determined set when δ was fixed, namely, a **cut set** in fuzzy set theory. In the same way, the random set model of fuzzy features extracted from sensors is defined as:

$$\theta \triangleq \sum_\delta(\mu_o) = \{x \in U \mid \delta \leq \mu_o(x)\}. \tag{13}$$

If fuzzy observation θ and $\sum_F$ match better, it is thought that θ is corresponding to the fault mode F. θ and $\sum_F$ don't conflict with each other when they match, in this case, $\theta \cap \sum_F \neq \emptyset$. Matching based on this definition is a probability phenomenon. θ and $\sum_F$ change randomly, when they are disjointed, measured features and fault templates mismatch; on the contrary, when they don't conflict, it notes that measurement and fault template mode have similar aspects. Intuitively, if θ and $\sum_F$ match frequently, it is thought that θ is caused by F; otherwise, θ and F are unrelated. So a plausibility measure is introduced to quantity the degree of matching between θ and F [10].

$$\rho(\theta \mid F) = Pr(\theta \cap \sum_F \neq \emptyset). \tag{14}$$

Form above, we can obtain

$$\begin{aligned}\rho(\theta \mid F) &= Pr(\theta \cap \sum_F \neq \emptyset) \\ &= p(\delta \leq (\mu_F \wedge \mu_o)(x)) = sup_x min\{\mu_F(x), \mu_o(x)\}.\end{aligned} \tag{15}$$

For each variable x, the smaller of $\mu_F(x)$ and $\mu_o(x)$ are computed at first, then maximize the above result, finally, denote the maximization as plausibility value between θ and F.

For instance, suppose there are 4 classic running conditions for rotor system, $F_0 = \{normal\}$, $F_1 = \{unbalance\}$, $F_2 = \{eccentricity\}$ and $F_3 = \{baseloosening\}$, their common characteristic are the peak value of vibration acceleration. The membership function in normal condition is shown in (16):

$$\mu_{F_0}(x) = \begin{cases} 1 & x \leq M_0, \\ exp(-\frac{(x-M_0)^2}{2\sigma_0^2}) & x \geq M_0. \end{cases} \tag{16}$$

M_0, σ_0 are mean and standard deviation respectively. The membership functions of fault template and fuzzy feature extracted from sensors following (10) and (11). Let practical parameter values

$$\begin{aligned} M_0 &= 1.5, \sigma_0 = 0.05; \\ M_{a,1} &= 1.55, \sigma_{a,1} = 0.04, M_{b,1} = 1.6, \sigma_{b,1} = 0.05; \\ M_{a,2} &= 1.65, \sigma_{a,2} = 0.04, M_{b,2} = 1.7, \sigma_{b,2} = 0.05; \\ M_{a,3} &= 1.55, \sigma_{a,3} = 0.04, M_{b,3} = 1.6, \sigma_{b,3} = 0.05; \\ M_o &= 1.52, \sigma_o = 0.05. \end{aligned} \tag{17}$$

Then $\mu_o(x)$ and $\mu_F(x)$ are shown in following Fig. 2.

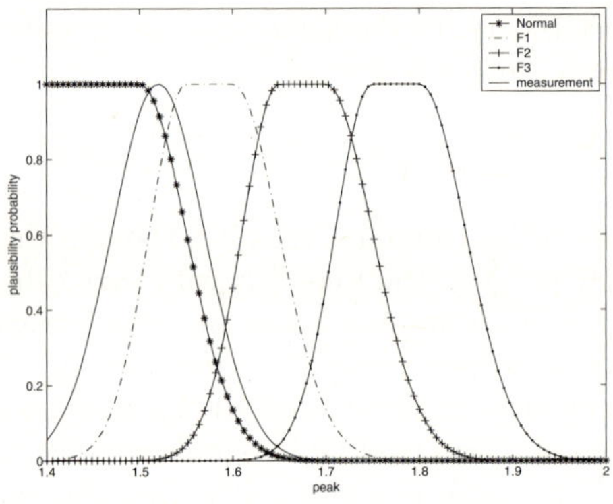

Fig. 2. Matching between fault templates and measurement

Based on (16), the values of $\rho(\theta \mid F_j)$, j=0,1,2,3 are ordinate of intersection points of observations and fault templates, that is, "×" shown in Fig. 2, and $\rho(\theta \mid F_j)$ for j=0,1,2,3 are 0.98, 0.9458, 0.3522, 0.0381, respectively.

Intuitively, $\rho(\theta \mid F_j)$, for j=0,1,2,3 has measured the intersecting degree of $\mu_o(x)$ and $\mu_{F,i}(x)$. The bigger is their intersecting degree, the greater is the degree of matching. In evidence theory, the degrees of matching are evidences provided by sensors. In the view of value: if $\mu_{F,i}(x)$ and $\mu_o(x)$ are membership functions of fault templates and features extracted from sensors respectively, then the probability of fault F_j characterized by sensors is $\rho(\theta \mid F_j)$, it has probability properties, responds the support degree to certain hypothesis, and can be denoted as BPA.

In order to reduce computational complexity, this paper only involves single element and universal set Θ. Each BPA is assigned by the previous presented method, and $\rho(\Theta)$denoted the degree of ignorance. It notes that plausibility probability don't meet the restriction weighted sum is 1. So, it is necessary to normalize when transforming $\rho(\cdot)$ to BPA. From the above example, we can get:

$$\begin{aligned} m(F_0) &= 0.98, m(F_1) = 0.9458, m(F_2) = 0.3522, m(F_3) = 0.0381, \\ m(\Theta) &= 1 - max(m(F_0), m(F_1), m(F_2), m(F_3)) = 0.02. \end{aligned} \tag{18}$$

After normalization

$$\begin{aligned} m(F_0) &= 0.4195, m(F_1) = 0.4049, m(F_2) = 0.1507, \\ m(F_3) &= 0.0163, m(\Theta) = 0.0085. \end{aligned} \tag{19}$$

For the multi-sensor system, each BPA can be assigned by the method presented in this section, and then Dempster's combination rule is used to combine them.

4 Decision Criteria

In fault diagnosis based on evidence reasoning, it is necessary to have certain criteria to diagnosis equipment running conditions. Fault decision principle is presented as following.

(1) The determinated fault type corresponding to the maximal BPA value, and the value is greater than the threshold 0.6.
(2) $m(\Theta)$ must be smaller than the threshold 0.3.
(3) Difference between the determinated fault type and other types must be larger than the threshold 0.15.

5 Experiments

The proposed method has been verified by experiments that was done on machine rotor. First, 4 equipment running conditions are set, such as $F_0 = \{normal\}$, $F_1 = \{unbalance\}$, $F_2 = \{eccentricity\}$ and $F_3 = \{baseloosening\}$, the vibration signals about rotor system are collected by acceleration sensors, velocity sensors, and displacement sensors. The specific procedure is as follows:

(1) ***compute membership functions of fault templates***
For each mode, we write down the vibration signals 30 times at time interval 40s, and denote it as a group, and collect 5 groups in all. Then the membership functions are determinated using the method presented in Subsection 3.1.

(2) ***compute membership function of fuzzy features***
Secondly, after returning to normal, reset the unbalance fault on rotor system, and write down the vibration signals at time interval 20s. Then membership functions $\mu_{o,i}(x), i = 1, 2, 3$ are computed by the method presented in Subsection 3.2
(3) ***match and assign BPA***
According to Subsection 3.3, compute BPA based on three sensors respectively, which compose 3 evidence vectors.

$$\begin{aligned} m_1 &= (0.1226, 0.6923, 0.0327, 0.0908, 0.0616); \\ m_2 &= (0.1148, 0.7230, 0.0424, 0.0598, 0.060); \\ m_3 &= (0.1429, 0.5626, 0.1045, 0.0925, 0.0975); \end{aligned} \tag{20}$$

(4) ***Weighted fusion***
According to Subsection 2.2, give the confidence degree of evidence so as to modify original BPAs, and combine evidences using (2).
Fusion result

$$\begin{aligned} &m(F_0) = 0.0188, m(F_1) = 0.9682, m(F_2) = 0.0041, \\ &m(F_3) = 0.008, m(\Theta) = 0.001. \end{aligned} \tag{21}$$

(5) ***Decision-making***
According to fusion result obtained from step 4 and criteria in Section 4, it has been found that unbalance fault appears truly. In this case, suitable measures should be taken to control the situation in time.

In order to verify the validity of proposed method in dealing with conflicting evidences, some interferences are added to velocity sensor in experiments. In this case, the velocity sensor can't reflect correctly equipment's running conditions. The combining results are shown in Table 1.

Table 1. BPAs and fusion results

	Basic Probability Assignment					result
	normal	*unbalance*	*eccentricity*	*loosening*	ignorance	
acceleration sensor	0.012	0.6954	0.047	0.0435	0.2021	*unbalance*
velocity sensor	0.0245	0.1027	0.6803	0.0503	0.1422	*eccentricity*
displacement sensor	0.0161	0.5251	0.2029	0.0404	0.2155	*uncertain*
Dempster's rule	0.0080	0.5708	0.3743	0.0229	0.0240	*uncertain*
proposed method	0.0096	0.6680	0.228	0.0276	0.0668	*unbalance*
Deng's method	0.0091	0.6611	0.2487	0.0264	0.0547	*unbalance*

Seen from Table 1, different single sensor gets different conclusion, so we can't judge the specific fault from them. And Dempster's combination rule can't decide which fault has occurred too. According to the proposed method we get ω_1=0.8740, ω_2=0.5762, ω_3=1. After combining the modified the BPAs, m(*unbalance*)=0.6680>0.6, m(*unbalance*)-m(*eccentricity*)=0.44>0.15, m(Θ)= 0.0668<0.3. So the fault is F_1={*unbalance*}. What's more, the belief of F_1 is much bigger than Deng's. Therefore, the proposed method treat validly conflict among evidences and ensure the reasonable decision-making.

6 Conclusions and Summary

Reasonable BPAs and conflict problems limit the utilization of Dempster's combination rule, so many methods have been proposed to increase the accuracy of decision-making. Fan has used membership function and weights of features to modify the BPAs of evidence and proposed modification of Dempster's combination rule and decision rule. Deng has introduced the consensus index of evidence to evidence theory by using distance of evidences.

In this paper, a fuzzy information fusion method based on similarity measure of evidence has been proposed. Random set model of fuzzy information and cosine similarity measure of evidence are introduced into the improvement of D-S evidence theory. A practical example shows that the method is valid and easy to use. From the result obtained in this paper, it has been concluded that the improved method can resolve two common problems in fault diagnosis and improve the accuracy of decision-making using multi-source information.

References

1. Zhu, D.Q., Liu, R.A.: Information Fusion Method for Fault Diagnosis. Control and Decision 22, 1321–1328 (2007)
2. Zhu, D.Q.: The Principle and Practice of Electronic Equipment Diagnosis. Electronic Industry Press, Beijing (2004)
3. Zhu, D.Q.: Data Fusion Algorithm Based on D-S Evidence Theory and Its Application for Circuit Fault Diagnosis. Chinese Journal of Electroics 30, 221–223 (2002)
4. Han, J.: Multi-Sensor Data Fusion Algorithm Based on D-S Evidence and Fuzzy Mathematics. Chinese Journal of Scientific Instrument 21, 644–647 (2000)
5. Murphy, C.K.: Combining Belief Functions Based on Distance of Evidence. Decision Support Systems 29, 1–9 (2000)
6. Fan, X.F., Zuo, M.J.: Fault Diagnosis of Machines Based on D-S Evidence Theory, Part 1: D-S Evidence Theory and Its Improvement. Pattern Recognition Letter 27, 366–376 (2006)
7. Deng, Y., Shi, W., Zhu, Z., Liu, Q.: Combining Belief Function Based on Distance of Evidence. Decision Support Systems 38, 489–493 (2004)
8. Liu, W.R., Hong, J.: Reinvestigating Dempster's Idea on Evidence Combination. Knowledge and Information Systems 2, 223–241 (2000)
9. Wen, C.L., Xu, X.B., Li, Z.L.: Research on Unified Description and Extension of Combination Rules of Evidence Based on Random Set Theory. The Chinese Journal of Electronics 17, 279–284 (2008)
10. Deng, Y., Zhu, Z.F., Zhong, S.: Fuzzy Information Fusion Based on Evidence Theory and Its Application in Target Recognition. Acta aeronautica et Astronautica Sinica 26, 754–758 (2005)

NN-Based Near Real Time Load Prediction for Optimal Generation Control

Dingguo Chen

Siemens Power Transmission and Distribution Inc.,
10900 Wayzata Blvd., Minnetonka, Minnesota 55305, USA

Abstract. In the environment of ongoing deregulated power industry, traditional automatic generation control (AGC) has become a set of ancillary services traded in separate markets which are different than the energy market. The performance of AGC is mandated to meet the NERC control performance standards (CPS). The new CPS criteria allow the over-compliant power utilities to loosen control of their generating units. The competition introduced by the deregulation process provides the opportunities for the over-compliant power utilities to sell their excess regulating capabilities. In addition, load following service is often priced lower than regulation service. All these lead generation companies to optimizing the portfolio of their generating assets to achieve better economy. The optimization process involves economic allocation of generation over a consecutive set of time intervals, which requires the load profile to be predicted for the dispatch period of minute level. This paper addresses the importance of very short term load prediction in this context, and proposes a new approach to make load predictions. Procedures involved in this approach are presented. Case studies are presented to demonstrate the effectiveness of the proposed approach.

Keywords: Very Short Term Load Prediction (VSTLP), Hierarchical Neural Network, Load Dynamics, Automatic Generation Control (AGC), Dynamic Economic Dispatch, Control Performance Standard (CPS).

1 Introduction

In the recent years, the electricity industry has witnessed significant deregulation from vertically integrated power utilities to independent market entities such as Generation Companies (GenCo), Transmission Companies (TransCo) and Distribution Companies (DisCo). Deregulation consists of business privatization, institutional restructuring, and profit redistribution. As a consequence, competitive market is established which allows competition in both generation and demand sides, which in turn leads to the ultimate goal of deregulation - improvement of the overall economy of the overall electricity industry such that all the market participants benefit from this deregulation process. Driven by the profit, each market participant has the flexibility to adjust its investment approach and operation strategies to maximize its profit while meeting the operation rules of electric power systems to together ensure stable and reliable system operations.

F. Sun et al. (Eds.): ISNN 2008, Part II, LNCS 5264, pp. 516–525, 2008.

In particular, the objectives of Automatic Generation Control (AGC) remain the same as before the deregulation although it is considered as a set of ancillary services that can be traded in the competitive power markets. These ancillary services include regulation, load following, spinning reserve, operating reserve, ramping and emergency support and black start capability. Regulation service is primarily for regulating interconnection frequency and meeting the scheduled interchange to minimize mismatch between generation and load demand. Load following service is mainly for meeting the energy mismatch. Please note that regulation and load following services are traded in separate competitive markets, and regulation service is generally priced much higher than load following service. Careful coordination of regulation and load following services may lead to significantly improved economy over a period of time. On the other hand, given the requirements of regulating reserve and load following reserve, dynamic economic dispatch of generation results in even better economy. With the introduction of the newly adopted NERC mandated control performance standards (CPS) CPS1 and CPS2 [14,15], control areas no longer have the obligation to return the area control error (ACE) to zero every 10 minutes, which in turn results in a much less tight unit control and a smaller number of unit reversals. These, however, depend on an accurate load profile for a time frame of at least a few minutes. The importance of an accurate load profile and how the overall system performance and economy benefits from its use can be seen more clearly in the diagram, as shown in Fig. 1. The very short term load prediction module calculates the minute-by-minute load values for the next 15 minutes and present these projections to the dynamic economic dispatch module. With the interchange schedule information available, and for the set of committed generating units, the dynamic economic dispatch module identifies the dispatchable generating units for the next 15 minutes, determines the profile of total generation to be dispatched based on the predicted load profile, and calculates the optimal power output trajectories for each dispatchable generating units to minimize the production cost of all the dispatchable generating units while respecting the ramp rate limits of all the dispatchable generating units for a 15 minute time period at 1 minute intervals. It is worth emphasizing that the key feature of this AGC scheme is the incorporation of the very short term load prediction module and the dynamic economic dispatch module. Dynamic economic dispatch distinguishes from conventional economic dispatch which is only executed in a snapshot manner. On the other hand, the optimal power output trajectories for all the generating units are presented to the Load Frequency Control (LFC - the control core of AGC). The LFC adjusts the setpoints of the generating units through time according to the optimal power output trajectories and use respective regulating ranges to regulate the power outputs of regulating units as per the control decision that is based on CPS criteria (CPS1 and CPS2).

Traditional short term load forecast (STLF) can not meet the minute-by-minute load projection requirement although it is very important for planning system operations for a longer time frame from an hour to a week [1,2,3,4,5,12,13]. Very short term load forecast has been studied by a few researchers [6,7]. In particular,

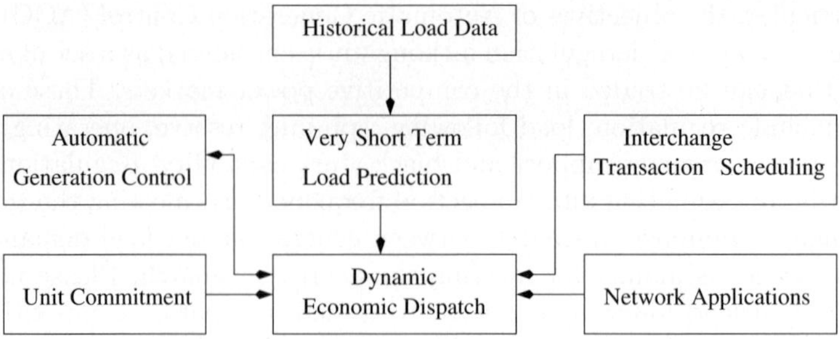

Fig. 1. VSTLP in AGC

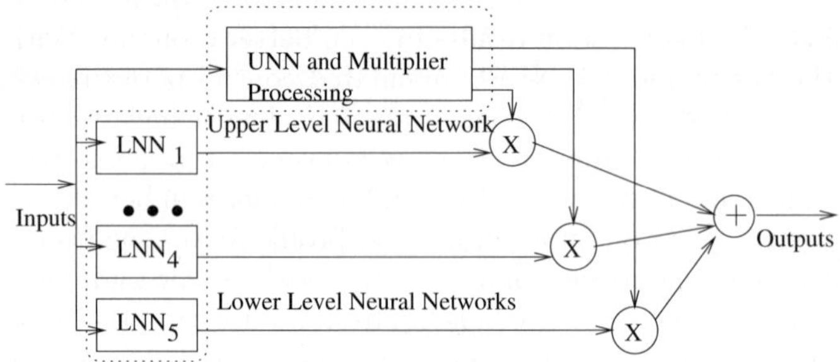

Fig. 2. Hierarchical Neural Network

neural-network-based very short term load prediction schemes were implemented to provide for minute-by-minute load projection (VSTLP) in a previous attempt with a dependency on STLF. The objective of this paper is to provide a sound approach while eliminating the dependency on STLF. An innovative approach is proposed for the purpose of very short term load prediction, which consists of providing load increment estimation based on the previous available load increments, an appropriate normalization scheme, an overall prediction architecture to ensure consistent prediction precision for both individual time segments and around boundaries of time segments. This approach is different than the one proposed in [7] in that this approach estimates individual minute-by-minute load increments and uses them additively to provide the ultimate load predictions while the approach in [7] uses individual load increments in a multiplicative manner; and in addition, this approach utilizes a hierarchical neural network structure [16,17] to enhance the ability to deal with the uncertainty of load dynamics, as shown in Fig. 2, where a hierarchical neural network consist of lower-level neural networks corresponding to several nominal groups of load dynamics and upper-level neural networks that coordinates the lower level neural networks to make a more robust and more accurate prediction scheme. As a result, the approach proposed in this paper results in a better precision. An analysis will be provided in a later section. The rest of the paper is organized as follows: Section 2 provides the mathematical formulation of the load prediction model on which the proposed approach is based. The overall architecture of the very short term load prediction is presented in Section 3. An comparison on the prediction accuracy provided by the proposed approach and the approach in [7] is made in Section 4. To demonstrate the effectiveness of the proposed approach, Section 5 presents the major procedures involved in VSTLP with remarks, and the experimental results. An analysis is provided as to how AGC benefits from VSTLP with a good prediction accuracy. This is presented in Section 6 followed by the concluding remarks.

2 Model Formulation for Very Short Term Load Prediction

Dynamic load modeling has been investigated in [11] where high-order load dynamics are approximated by properly trained neural networks. Similarly, for load prediction purpose, assume that a dynamic load model exists in the context of automatic generation control, and may be expressed in the following form:

$$g(P^{(N)}, \cdots, \dot{P}, P, t) = 0 \tag{1}$$

where N denotes the order of load dynamics, P the load, and t the time.

For the discrete case,

$$h(P_{n-N}, \cdots, P_{n-1}, P_n, n) = 0 \tag{2}$$

With some algebra and conventional AGC considerations, it can be shown that the load increments satisfy the following:

$$\begin{bmatrix} \Delta P_n \\ \Delta P_{n+1} \\ \cdots \\ \Delta P_{n+M-1} \end{bmatrix} = \begin{bmatrix} \psi_1(\Delta P_{n-N+1}, \cdots, \Delta P_{n-1}) \\ \psi_2(\Delta P_{n-N+1}, \cdots, \Delta P_{n-1}) \\ \cdots \\ \psi_M(\Delta P_{n-N+1}, \cdots, \Delta P_{n-1}) \end{bmatrix} \tag{3}$$

Since the exact forms of functions ψ_i's are unknown, with historical load data available, a feedforward neural network with proper layers may be trained to approximate such functions. As is well known, neural networks with certain structure have the capability to approximate any continuous function defined on a compact support with any arbitrary small error $\epsilon > 0$ in some sense [8,10,9]. Though the actual bound for ΔP_k is not known explicitly, it is always practical to assume that ΔP_{n-1}, $\cdots$, ΔP_{n-N+1} are all bounded. In other words, it is always reasonable to assume that the support of functions ψ_i's is compact.

Thus, there exist a neural network such that

$$\begin{bmatrix} \Delta P_n \\ \Delta P_{n+1} \\ \cdots \\ \Delta P_{n+M-1} \end{bmatrix} = NN(\Delta P_{n-N+1}, \cdots, \Delta P_{n-1}; \theta) \tag{4}$$

Where θ is a parameter vector that contains weights between neighboring layers, and biases for all the hidden neurons, and is to be tuned in a way that the discrepancy between the calculated values for the future times and actual values is minimized in terms of a performance index. Neural networks are trained off-line using historical load data. After the completion of neural network training and validation, they are ready to be used on-line.

3 Architecture of Very Short Term Load Predictor

Load shape changes with seasons, from weekdays to weekends to holidays, from off-peak times to on-peak times. However, load shapes for both weekdays and weekends are roughly recurrent. The load shapes of holidays are quite different from those of normal weekdays/weekends. Special care must be given to the treatment of load data of holidays. Consequently, a number of neural networks will be trained for each to capture the load patterns that occur in a specific time period of a specific day.

For each individual day, 6 neural networks (NN1, NN2, NN3, NN4, NN5 and NN6) will be used to cover the 24-hour period. Each neural network is responsible for a 4-hour period. For instance, NN1, NN2, NN3, NN4, NN5 and NN6 cover 12:00am to 4:00am, 4:00am to 8:00am, 8:00am to 12:00pm, 12:00pm to 16:00pm, 16:00pm to 20:00pm, and 20:00pm to 12:00am, respectively. To ensure the smooth transition from one 4-hour period to the next 4-hour period, one additional quarter-hour is added to the both ends of each 4-hour period. For instance, NN1 covers 11:45pm to 4:15am, and NN2 covers 3:45am to 8:15am. The split of the whole day into 6 different 4-hour periods reflects the fact that load displays different dynamics during the day, and use of multiple neural networks for different periods of the day allows for predictions with better accuracy. Such a split may be changed to comply with the patterns of the real situations. As such, each NN-based VSTLP module shown in the NN-based VSTLP Architecture diagram will have several NNs to predict load corresponding to a specific time period.

To obtain a full hour minutely load forecasted values for monitoring purpose, estimates generated from NN-based VSTLP at times when actual load values are not available yet will be used as inputs to VSTLP instead, and further future load values can be forecasted. This can be done iteratively till all the minutely forecasted load values for a whole hour are computed.

4 Performance Analysis

In this section, we first compare the proposed approach with the one proposed in [7]. It will be observed from the following analysis that our proposed approach will outperform the one in [7].

First let us perform an perturbation analysis on the approach in [7] to reveal the accuracy of that approach. As defined in that paper, the increment or decrement ratio is $r_t = \frac{P_{t+1}-P_t}{P_t}$. As pointed out in the model formulation, we are predicting the future minute-by-minute load values for 15 minutes based on the previous 15 1-minute load values. Then $M = N = 15$.

At any point of time n, we have to predict $r_{n+1}, \cdots, r_{n+M}$ based on $r_{n-N+1}, \cdots, r_n$. Denote the estimates of $\hat{r}_{n+1}, \cdots, \hat{r}_{n+M}$. These estimates are obtained as follows:

$$\hat{r}_{n+i} = NN(r_{n-N+1}, \cdots, r_n) \tag{5}$$

where $NN(.)$ is the neural network model.

To obtain the future load values based on the predictions of $r_{n+1}, \cdots, r_{n+M}$, and already available P_n, we have the following:

$$\begin{aligned} \hat{P}_{n+1} &= P_n(1 + r_n) \\ \hat{P}_{n+2} &= \hat{P}_{n+1}(1 + r_{n+1}) \\ &\cdots \\ \hat{P}_{n+M} &= \hat{P}_{n+M-1}(1 + r_{n+M-1}) \end{aligned} \tag{6}$$

On the other hand, the estimates of $P_{n+1}, \cdots, P_{n+M}$ at time n, by our proposed approach, can be obtained as follows:

$$\hat{P}_{n+i} = \hat{P}_{n+i-1} + \Delta P_{n+i-1} \tag{7}$$

where $i = 1, 2, \cdots, M$.

For the purpose of clarity, denote the actual load value at time j by P_j^0, the actual load increment by $\Delta P_j^0 = P_j^0 - P_{j-1}^0$, and the actual increment ratio by $r_j^0 = \frac{P_j^0 - P_{j-1}^0}{P_{j-1}^0}$. To make distinction for $\hat{P}_{n+i}$ between equations (6) and (7), equation (6) uses $\hat{P}_{n+i,r}$ for $\hat{P}_{n+i}$ while equation (7) uses $\hat{P}_{n+i,d}$.

Assume an estimate error rate μ for P_{n+i-1} so that $\hat{P}_{n+i-1} = P_{n+i-1}^0(1+\mu)$, an estimate error rate δ for ΔP_{n+i-1} so that $\Delta P_{n+i-1} = \Delta P_{n+i-1}^0(1+\delta)$, and an estimate rate δ for r_{n+i-1} so that $r_{n+i-1} = r_{n+i-1}^0(1+\delta)$. Please note that same estimate rate for outputs resulting from two different approaches is assumed to make the comparison fair.

Then with some substitutions, the following equation can be derived from equation (6):

$$\hat{P}_{n+i,r} = P_{n+i-1}^0(1+\mu)(1 + r_{n+i-1}^0(1+\delta)) \tag{8}$$

Similarly, equation (7) can be rewritten as

$$\hat{P}_{n+i,d} = P_{n+i-1}^0(1+\mu) + \Delta P_{n+i-1}^0(1+\delta) \tag{9}$$

Notice that $\Delta P_{n+i-1}^0 = P_{n+i-1}^0 r_{n+i-1}^0$. It follows from equation (8) that

$$\begin{aligned} \hat{P}_{n+i,r} = P_{n+i-1}^0(1+\mu) + \Delta P_{n+i-1}^0(1+\delta) + \\ \Delta P_{n+i-1}^0(\mu + \mu\delta) \end{aligned} \tag{10}$$

Since $P_{n+i}^0 = P_{n+i-1}^0 + \Delta P_{n+i-1}^0$, then $\hat{P}_{n+i,d} = P_{n+i}^0 + \mu P_{n+i-1}^0 + \delta \Delta P_{n+i-1}^0$. This implies, $|\hat{P}_{n+i,d} - P_{n+i}^0| \le |\mu| P_{n+i-1}^0 + |\delta||\Delta P_{n+i-1}^0|$. On the other hand, $|\hat{P}_{n+i,r} - P_{n+i}^0| \le |\mu| P_{n+i-1}^0 + |\delta||\Delta P_{n+i-1}^0| + |\Delta P_{n+i-1}^0||\mu|(1+|\delta|)$.

It is observed that at time n, $\hat{P}_{n+i,d}$ is closer to $\hat{P}_{n+i,r}$ for all $i = 2, \cdots, M$. Therefore, we conclude that the proposed approach in this paper statistically performs better than the existing one [7].

5 Simulation Results

Minutely load values collected from a major power utility for a whole summer month - June, 2001, are used to evaluate the performance of the load increment based VSTLP scheme.

Table 1 lists the NN training performance statistics that correspond to a whole month worth of load data. These performance statistics include 15-minute RMS error percentage, 15-minute MAPE, 15-minute worst MAPE, 1-minute RMS error percentage, 1-minute MAPE, 1-minute worst MAPE for each time segment of a day and a whole day on a monthly average. The original load data are filtered before they are used for NN training. The original load data is used with the outputs of the trained NNs for calculating these performance statistics.

Table 1. NN Training Performance Over Filtered Summer Load Data and Against the Actual Load Data

Time	15-min RMS(%)	15-min MAPE(%)	15-min Worst MAPE(%)	1-min RMS(%)	1-min MAPE(%)	1-min Worst MAPE(%)
(0:00-4:00)	0.6075	0.4281	4.5025	0.2947	0.1867	2.6930
(4:00-8:00)	0.6394	0.4667	3.6105	0.2974	0.2063	1.4488
(8:00-12:00)	0.4714	0.3253	2.7702	0.2190	0.1440	1.6773
(12:00-16:00)	0.3991	0.2869	2.3951	0.1878	0.1225	1.2241
(16:00-20:00)	0.4579	0.3241	2.8601	0.2117	0.1362	1.6409
(20:00-0:00)	0.5518	0.4073	3.2160	0.2534	0.1794	2.1108
All Day	0.5081	0.3721	3.2257	0.2377	0.1621	1.7992

Table 2 lists the performance statistics of the trained NNs that correspond to a whole month worth of load data. These performance statistics include 15-minute RMS error percentage, 15-minute MAPE, 15-minute worst MAPE, 1-minute RMS error percentage, 1-minute MAPE, 1-minute worst MAPE for each time segment of a day and a whole day on a monthly average. The original load data is used with the outputs of the trained NNs for calculating these performance statistics.

The experimental experience suggests that smoothing the original load data reduces the temporal fluctuation, helps neural networks during training to capture the major load shapes instead of more details and improves the prediction accuracy. Using the filtered load data when calculating the performance statistics can achieve better overall prediction accuracy than directly using the original load data. This seems to suggest that it might be better to use the filtered version of the original load data instead of the original load data themselves in calculating the total generation to be dispatched among the dispatchable generating units. This makes more sense when CPS criteria are considered and CPS-based AGC is implemented in which the control objective is no longer to drive the area control error to zero, but to keep the area control error within a specified neighborhood of zero.

Table 2. Performance of Trained NNs Against the Actual Data

Time	15-min RMS(%)	15-min MAPE(%)	15-min Worst MAPE(%)	1-min RMS(%)	1-min MAPE(%)	1-min Worst MAPE(%)
(0:00-4:00)	0.8908	0.6295	4.9278	0.4120	0.2543	2.2218
(4:00-8:00)	0.9457	0.6919	5.1505	0.4205	0.2667	1.8942
(8:00-12:00)	0.6943	0.4823	3.7951	0.3082	0.1911	1.4960
(12:00-16:00)	0.5492	0.4062	2.9555	0.2600	0.1617	1.3146
(16:00-20:00)	0.6558	0.4608	3.5054	0.2851	0.1778	1.3504
(20:00-0:00)	0.7710	0.5683	3.6870	0.3267	0.2143	1.6144
All Day	0.7379	0.5386	4.0035	0.3305	0.2105	1.6486

6 Benefit of Very Short Term Load Prediction

The benefit of using very short term load prediction in the CPS-based AGC scheme can be demonstrated in the sequel. With a predicted load profile with a good accuracy, economically dispatching this load profile is used by the dynamic economic module to dispatch the power output of each generating unit.

For the conventional economic dispatch with one step look-ahead, it does not recognize the difference among the ramping capabilities of generating units. It allocates generation economically for each time interval, but it is unable to consider the limits of ramp rates and may not be able to recognize the need to ramp up/down the generation units that have slow ramping capabilities well ahead of time. Consequently, when the generating units that have fast ramping capabilities reach their generation limits, and the system demand is well above the total generation that fast moving generating units can produce, the system is significantly short of certain amount of generation to balance its load demand, and can only wait until the slow moving generating units ramp up their generation sufficiently long. On the other hand, dynamic economic dispatch looks ahead with the very short term load predictions, and recognizes the ramping up/down requirements in the dispatch time period comprised by a consecutive set of intervals. It then adjusts the generation outputs of fast moving generating units and the slow moving generation units, allows the slow moving generating units to ramp up/down and the fast moving generating units to ramp down/up when it foresees a possible load peak/valley in the specified look-ahead time period. Consequently, dynamic dispatch enables the system generation to balance the system load demand in a timely manner. Dynamic economic dispatch may not reach the optimization for each individual time interval like the static economic dispatch, and but for the whole look-ahead time period, it arrives at the optimization while the static economic dispatch does not.

If the predicted load profile is not accurate enough, dynamic economic dispatch sets the base points of the dispatchable generating units with a marginal accuracy. Significant system imbalance is produced, and CPS-based AGC is required to actively respond, which incurs additional control efforts of more costly regulation.

7 Conclusions and Outlook

Load prediction for a future time period at minute level plays an important role CPS-based AGC, in which predicted minute-by-minute load values can be used to economically allocate generation among dispatchable generating units over this time period while ramp rate limits of generating units are respected. It is true that without very short term load prediction, the conventional economic dispatch executed even in an anticipatory mode may have the system operate economically at one moment, and leave the system in significant power imbalance at the very next moment if load changes significantly and no load prediction is made. This is because that momentary economic operation does not guarantee that the system will have sufficient ramping capability to meet the significant load change in the next time interval due to the system's ramping rate limits.

The proposed load prediction approach, following the mathematical formulation of very short term load forecast, makes use of the previous load increments and a hierarchical neural network structure, and computes the future load increments, which in turn are used to derive the predicted minute-by-minute load values for the next 15 minutes. It does not depend on the availability of the conventional STLF for some statistical parameters. On the other hand, accuracy analysis is performed to show that the proposed approach supplies prediction results with better accuracy than the one in [7] which has the tendency to incur bigger prediction errors in all of its predictions.

Simulations are performed and results indicate that under normal situations, forecasted minutely load values by the proposed VSTLP for the future 15 minutes are provided with good accuracy. On the other hand, to improve the accuracy, careful selection of the days whose load data are collected for offline training of a particular neural network is of critical importance. The benefit of using very short term load prediction in AGC is discussed.

References

1. Khotanzad, A.: Hourly Load Forecasting Using Artificial Neural Networks. EPRI TR-105278 (1995)
2. Lowther, R.A., Shoults, R.R., Maratukulam, D.: Advanced Concepts in Automatic Generation Control. Canadian Electrical Association, Engineering and Operating Division, Montreal (1996)
3. Papalexopoulos, A.D., Hao, S., Peng, T.: An Implementation of a Neural Network Based Load Forecasting Model for the EMS. IEEE Trans. Power Systems 9, 1956–1962 (1994)
4. Mohammed, O., Park, D., Merchant, R., Dinh, T., Tong, C., Azeem, A., Farah, J., Drake, C.: Practical Experiences with An Adaptive Neural Network Short Term Load Forecasting System. IEEE Trans. Power Systems 10, 254–260 (1995)
5. Kim, K., Park, J., Hwang, K., Kim, S.: Implementation of Hybrid Short-Term Load Forecasting System Using Artificial Neural Networks and Fuzzy Expert Systems. IEEE Trans. Power Systems 10, 1534–1539 (1995)
6. Liu, K., Subbarayan, S., Shoults, R., Manry, K., Kwan, C., Lewis, F., Naccarino, J.: Comparison of Very Short-Term Load Forecasting Techniques. IEEE Trans. Power Systems 11, 877–882 (1996)

7. Charytoniuk, W., Chen, M.: Very Short-Term Load Forecasting Using Artificial Neural Networks. IEEE Trans. Power Systems 15, 263–268 (2000)
8. Funahashi, K.: On the Approximation Realization of Continuous Mappings by Neural Networks. Neural Networks 2, 183–192 (1989)
9. Hornik, K.: Approximation Capabilities of Multilayer Feedforward Networks. Neural Networks 4, 251–257 (1991)
10. Hornik, K., Stinchcombe, M., White, H.: Multilayer Feedforward Networks Are Universal Approximators. Neural Networks 2, 359–366 (1989)
11. Chen, D., Mohler, R.: Load Modelling and Voltage Stability Analysis by Neural Networks. In: American Control Conf., pp. 1086–1090. Omni Press, Albuquerque (1997)
12. Amjady, N.: Short-Term Hourly Load Forecasting Using Time-Series Modeling with Peak Load Estimation Capability. IEEE Trans. Power Systems 16, 498–505 (2001)
13. Mori, H., Yuihara, A.: Deterministic Annealing Clustering for ANN-Based Short-Term Load Forecasting. IEEE Trans. Power Systems 16, 545–551 (2001)
14. Jaleeli, N., VanSlyck, L.: Control Performance Standards and Procedures for Interconnected Operation. Technical Report, EPRI TR-107 803 (1997)
15. Jaleeli, N., VanSlyck, L.: NERC's New Control Performance Standards. IEEE Trans. Power Systems 14, 1092–1099 (1999)
16. Chen, D.: Nonlinear Neural Control with Power Systems Applications. Ph.D. Dissertation, Oregon State University (1998)
17. Chen, D., Mohler, R., Chen, L.: Synthesis of Neural Controller Applied to Power Systems. IEEE Trans. Circuits and Systems I 47, 376–388 (2000)

A Fuzzy Neural-Network-Driven Weighting System for Electric Shovel

Yingkui Gu, Luheng Wu, and Shuyun Tang

School of Mechanical and Electronical Engineering,
Jiangxi University of Science and Technology,
341000 Ganzhou, China
guyingkui@163.com

Abstract. In order to improve the weighting precision and optimize the loading of trucks and the production efficiency of electric shovel, an online weighting model is developed by using fuzzy logic and improved T-S neural network in this paper. The weighting model is established based on the mechanics analysis of the electric shovel firstly. Then, a T-S fuzzy neural network model is established to obtain the influence coefficient through training large numbers of samples. Applications show that by using the presented weighting model, it not only can decrease the fuzzy and uncertain factors in the weighting process, but also can improve the production and management efficiency.

Keywords: Fuzzy neural network, T-S model, Electric shovel, Online weighting.

1 Introduction

Electric shovel loading, truck transit is the major production way of open-cut mine. Nowadays, the calculation and statistic of the transport quantity still adopts the manual method and automobile weighting which have many disadvantages as following.

(1) The manual method results in low data reliability and slow information transmission. Moreover, some artificial factors cause the phenomenon "lose ton" in production acceptance frequently, which brings great negative effect to mining production operation.

(2) In order to make the metal recovery in ore as far as possible ample, the average grade of ore should be controlled as far as possible consistent. Traditionally, it is to rely on automobile weighting system to calculate the ore collocation proportion. However, because of the greater error of the automobile weighting system, it is hard to control the collocation proportion.

(3) The match between electric shovel and the trucks is very important to the mine production, which could reflect the match between the minimum number of electric shovel buckets and the best loading weight of each truck. However, because of lacking electric shovel online weighting system, it can't make the process of "shovel – load–transit" optimum and develop the biggest production usefulness of the equipments.

Therefore, in order to ensure electric shovel and trucks under optimum operating mode fully, realize the optimum match of the equipments, avoid the phenomenon

F. Sun et al. (Eds.): ISNN 2008, Part II, LNCS 5264, pp. 526–532, 2008.

"lose ton" occur and make mining production realize automation, it is very necessary and urgent to develop the electric shovel online weighting system

It is well known that accurately dynamic weighting is a difficult problem in industry measure field. Because electric shovel weighting system is concerned with high sensitivity sensor technology, complex sampling technology, wireless data transmission technology as well as the other technologies such as interface with information management network, the difficulty of research and development is very great. But because the economic benefits and wide market prospect of this project are very large, many companies and research institutes have carried out long-term research with plenty of manpower and material resources.

In order to improve the weighting precision and optimize the loading of haul trucks and the production efficiency of electric shovel and trucks, computing intelligence, such as fuzzy set theory, neural network, genetic algorithms, rough set theory, and so on, is used to establish the weighting model. Computing intelligence provides a stronger tool for weighting modeling. With the fast development of computing intelligence, many new weighting models have been presented in recent years, such as fuzzy-logic-based model, neural network model, and rough-set-based model, and so on[1-3].

The fuzzy modeling has been the research hot spot of the nonlinear modeling in recent years. Compared with the other non-linear modeling methods, the advantages of fuzzy modeling are that the structure of the model and the physical meaning of the parameters are easily to understand.

The reason for incorporating neural network into fuzzy logic system is that neural network has the characteristics of self-learning capacity, fault tolerance and model-free. The neural network and fuzzy logic have been put to use in the following ways [4]:

(1) Using neural network to replace fuzzy rule evaluation in fuzzy logic system.
(2) Neural network as correcting mechanisms for fuzzy system.
(3) Using neural network to simulate membership functions in fuzzy logic system.
(4) Combination of neural network and fuzzy system.

In this paper, an online weighting model is developed by using fuzzy logic and improved T-S neural network. The structures of this paper are as follows: the mechanics model of the weighting system is developed in Section 2, a fuzzy-neural-network-based model is developed to solve the influence coefficient of the weighting model is established in Section 3. Conclusions are given in Section 4.

2 The Mechanics Model of the Weighting System

There exist three sub-processes during the working process of the electric shovel. That is the promotion process of the steel wire, the push and press process of the gear-rack mechanism and the turning process of the shovel respectively. The movement of the bucket is the integration of the three movements above. At one moment of the working process, if the influence of libration, impact and noise and the centrifugal effect of additive motorial load are not taken into account, the mechanism composed by the arm of electric shovel, the lifting pulley at the top of the arm, the steel wire and the gear-rack mechanism can be simplified as a quadrilateral mechanism as shown in Fig. 1.

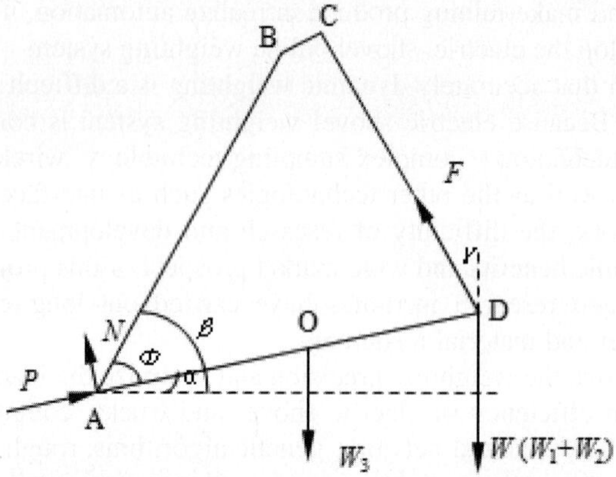

Fig. 1. The mechanics analysis of the electric shovel

In Fig. 1, W_1 is the weight of the shovel bucket, and W_2 is the weight of the ore. Before the ore is unloaded, $W = W_1 + W_2$. After the ore is unloaded, $W = W_1$. W_3 is the weight of gear-rack and bucket handle. F is the tensile force of the steel wire. P is the push force of the gear-rack mechanism. N is the force of the push and press center. α is the obliquity of the arm. β is the obliquity of the bucket handle. Φ is the angle between the arm and the bucket handle. γ is the angle between the steel wire and the plumb line. Point A is the push and press center of the gear-rack mechanism. Point B is the axial center of the pulley. Point C is the cutting point between the lifting steel wire and the pulley. Point D is the crunode between the lifting steel wire and the bucket. Point O is the quality center of the gear-rack and the bucket handle.

The weight model of the ore in the bucket is established as follows.

$$W_2 = f \cdot (F_1 - F_2). \tag{1}$$

where F_1 and F_2 is the tensile force of the steel wire before and after the ore is unloaded respectively. f is the influence coefficient. Many factors, such as structure parameters, position parameters, dynamics parameters, environment parameters, noise, libration, impact, etc., have different influence on the influence coefficient. Some of the parameter can be measured by the intelligent instruments or the sensors. But some of the parameters are random, uncertain and fuzzy, which can't be measured. In order to solve this problem, the T-S fuzzy neural network model is established to obtain the influence coefficient through training large number of samples.

The tensile force of the steel wire is as follows.

$$F = 2\left[\left(T_n \cdot I_f \cdot I_a\right) / \left(I_{fn} \cdot I_{an}\right) - J\frac{\mathrm{d}\omega}{\mathrm{d}t} \right] / (\beta \cdot D). \tag{2}$$

where T_n is the rated electromagnetic torque of the lifting electromotor. I_{an} and I_{fn} is the rated armature current and rated exciting current of the lifting electromotor respectively. D is the diameter of the rolling drum. β is the rotate speed ration between the rolling drum and the electromotor spindle. J is the dynamics parameter of the electric shovel. I_f is the exciting current. I_a is the armature current. ω is the rotate speed of electromotor spindle. T_n, I_{an} and I_{fn} are the rated parameters of the electromotor. I_f and I_a can be measured by current sensors. ω can be measured by the photoelectric coder.

3 Solving Influence Coefficient by Using Fuzzy T-S Neural Network

Fuzzy rule-based models and especially Takagi-Sugeno (TS) fuzzy models have gained significant impetus due to their flexibility and computational efficiency. They have a quasi-linear nature and use the idea of approximation of a nonlinear system by a collection of fuzzily mixed local linear models. The TS fuzzy model is attractive because of its ability to approximate nonlinear dynamics, multiple operating modes and significant parameter and structure variations.

The T-S fuzzy model approximates a nonlinear system with a combination of several linear systems. The overall T-S fuzzy model is formed by fuzzy partitioning of the input space. The premise of a fuzzy implication indicates a fuzzy subspace of the input space and each consequent expresses a local input-output relation in the subspace corresponding to the premise part. The set of fuzzy implications shown in the Takagi-Sugeno (T-S) model can express a highly nonlinear functional relation in spite of a small number of fuzzy implication rules [5-7].

Many factors, such as structure parameters, position parameters, dynamics parameters, environment parameters, noise, libration, impact, etc., have different influence on the influence coefficient. Some of the parameter can be measured by the intelligent instruments or the sensors. But some of the parameters are random, uncertain and fuzzy, which can't be measured. Because neural network has the characteristics of self-learning capacity and fault tolerance and model-free, it is often used to model complicated system. Therefore, an improved fuzzy T-S neural network model as shown in Figure.2 is established to obtain the influence coefficient through training large numbers of samples. The improved fuzzy T-S network have some advantages, such as simple structure, quick convergence, high precision, good study ability and fault-tolerant ability, and so on. By using this fuzzy neural network, the influence coefficient could be obtained based on the state monitoring data.

The fuzzy neural network has four layers. The first layer is the input layer, the input is the structure parameters, position parameters, dynamics parameters, environment parameters, etc. Here the number of input nodes is 4, and input nodes are $\alpha\,(x_1)$, $\beta\,(x_2)$, $s\,(x_3)$ and $\omega\,(x_4)$.

The second layer is the membership function production layer, the input is the output of the first layer, and the number of input nodes is 4×4, where 4 indicates that there are 4 technical state, i.e. good, better, general, bad. The membership function adopts the Gauss function form.

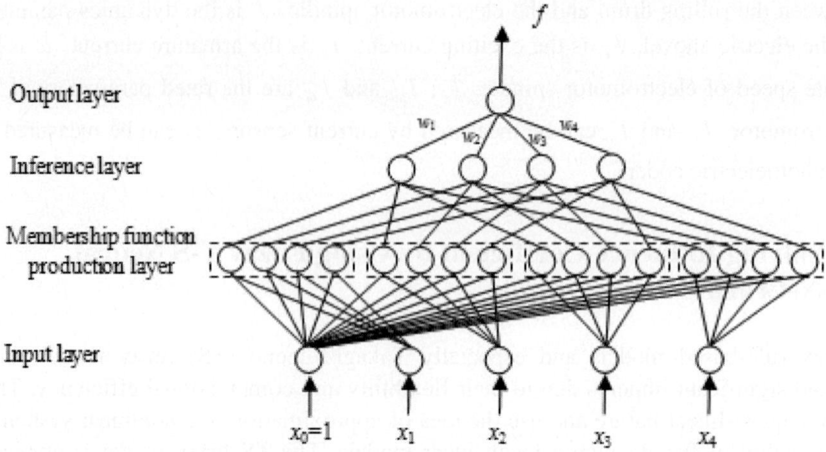

Fig. 2. Fuzzy-neural-network-based weighting model for electric shovel

Table 1. The input-output of the fuzzy-neural-network-based weighting model

Layer	The number of the nodes	Input	Output
The First layer(input layer)	4	$I_i^{(1)} = x_i$, $i = 1,2,\cdots,n$	$O_i^{(1)} = x_i$, $i = 1,2,\cdots,n$
The second layer(membership function production laye)	16	$I_{ij}^{(2)} = O_i^{(1)}$ $i = 1,2,\cdots,n$, $j = 1,2,3,4$	$O_{ij}^{(2)} = \exp\left(-\left(\frac{x_i - m_{ij}}{\delta_{ij}}\right)^2\right)$ $i = 1,2,\cdots,n$, $j = 1,2,3,4$
The third layer(reasoning layer)	4	$I_{ij}^{(3)} = O_{ij}^{(2)}$ $i = 1,2,\cdots,n$, $j = 1,2,3,4$	$O_j^{(3)} = \prod_{i=1}^{n} \mu_{ij}(x_i)$ $i = 1,2,\cdots,n$, $j = 1,2,3,4$
The fourth layer(output layer))	1	$I_{ij}^{(4)} = O_j^{(3)}$	$O^{(4)} = \sum_{j=1}^{4} w_j I_j^{(4)}$, $j = 1,2,3,4$

$$\mu_{ij} = \exp\left(-\left(\frac{x_i - m_{ij}}{\delta_{ij}}\right)^2\right), \; 1 \le i \le n, \; 1 \le j \le n. \tag{3}$$

The third layer is the reasoning layer. The fourth layer is the output layer. The output of this layer is the influence coefficient f. The input and output of the fuzzy neural sub-network are as shown in Table 1. By training the fuzzy neural network, the influence coefficient f can be obtained, and the weight of the ore can be obtained also.

The study rule of fuzzy neural network adopts the error function adjustment method. The error objective function is

$$E = 0.5(y - \bar{y})^2 . \tag{4}$$

where y is network actual output, $\bar{y}$ is expectation output.

Take 20 weighting results of the electric shovel P&H6100 as example to illustrate the established fuzzy TS-neural-network model-driven weighting model. The network precision of training parameter is 10-3 and the study rate is 0.001. We can take one sample as examination sample, and the other 19 samples can be as training samples. The network errors are as shown in Figure 3. From Figure 3 we can see, the expectation output and the actual output of each sample are very close. The absolute error is smaller than 0.1. Moreover, the system average error is in 10-3 magnitude.

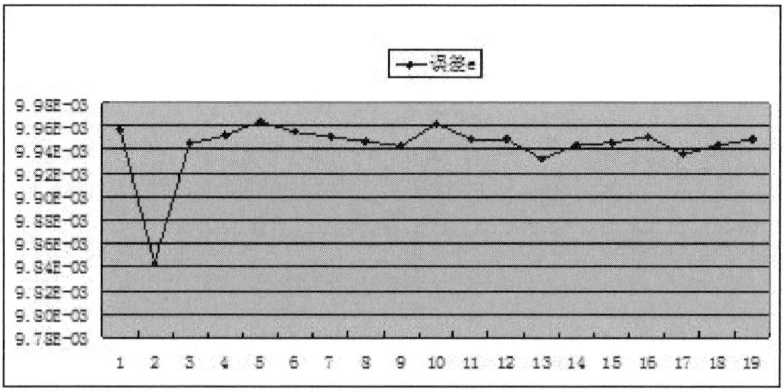

Fig. 3. The drawing of the network training

4 Conclusions

In order to optimize the production process of mine and improve the efficiency of the mine equipments, the online weighting system is developed by using T-S fuzzy neural network in this paper. The weighting model is established firstly based on the mechanics analysis of the electric shovel. Applications show that applying the weighting system can improve the efficiency and level of management and production of mine, facilitate and support the leader to make decision, and establish important foundation for the mine to realize production and management automation.

Acknowledgments

This research was partially supported by the Jiangxi Provincial Natural Science Foundation under the contract number 2007GQC0654 and the Science Foundation of Education Commission of Jiangxi Province under the contract number 2007[201].

References

1. Yue, Y.F., Mao, J.Q.: An Adaptive Modeling Method Based on T-S Fuzzy Models. Control and Decision 17, 155–158 (2002)
2. Sun, Z.Q., Xu, H.B.: Neuro-Fuzzy Network Based on T-S Fuzzy Models. Journal of Tsinghua University 37, 76–80 (1997)
3. Xu, C.M., Er, L.J., Hu, H.J.: TSK-FNN and Its Constrained Optimization Algorithm. Journal of Beijing University of Aeronautics and Astronautics 31, 595–600 (2005)
4. Du, T.C., Wolfe, P.M.: Implementation of Fuzzy Logic Systems and Neural Network in Industry. Computer in Industry 32, 261–272 (1997)
5. Gu, Y.K., Yang, Z.Y.: TS-neural-network-based Maintenance Decision Model for Diesel Engine. In: Liu, D., Fei, S., Hou, Z.-G., Zhang, H., Sun, C. (eds.) ISNN 2007. LNCS, vol. 4491, pp. 557–565. Springer, Heidelberg (2007)
6. Jang, C.H.: A TSK-Type Recurrent Fuzzy Network for Dynamic Systems Processing by Neural Network and Genetic Algorithms. IEEE Transaction on Fuzzy Systems 10, 155–169 (2002)
7. Tor, A.J.: On the Interpretation and Identification of Dynamic Takagi–Sugeno Fuzzy Models. IEEE Transactions on Fuzzy Systems 8, 213–297 (2000)

Neural-Network-Based Maintenance Decision Model for Diesel Engine

Yingkui Gu, Juanjuan Liu, and Shuyun Tang

School of Mechanical and Electronical Engineering,
Jiangxi University of Science and Technology,
341000 Ganzhou, China
guyingkui@163.com

Abstract. To decrease the fuzzy and uncertain factors in the maintenance decision models of diesel engine, a combination BP-neural-network-based maintenance decision model for diesel engine is presented in this paper. It can make the maintenance of diesel engine follow the prevention policy and take the technology and economy into account at the same time. In the presented model, the maintenance decision theory of diesel engine was analyzed in detail firstly. Then, the combination neural network model for maintenance decision was established, including an entire network and two modules sub-networks. Finally, an example was given to verify the effective feasibility of the proposed method. By training the network, the deterioration degree of the diesel engine and its parts can be obtained to make the right maintenance decision.

Keywords: Neural network, Maintenance decision, Diesel engine, Deterioration degree.

1 Introduction

Because the diesel engine has many advantages, such as high thermal efficiency, energy conservation and good energy use efficiency, it already becomes the main power of automobile, agricultural machinery, construction machinery, ship, internal combustion locomotion, drilling machinery, power generation, and so on. However, the technology state of the diesel engine would become bad and its performance would decrease in the using process because of the weary, fatigue, deformation or the damage. Maintenance is one of the most important steps in the equipment management. It is also an important guarantee to prolong the life of the equipment and to prevent the accident [1]. Therefore, modern enterprises pay more attention to the high efficiency and low consumption of the equipment than any time before, and regard the equipment management as the important part of the business management. Moreover, the equipment management mainly is the equipment maintenance management, and modern enterprises need more advanced maintenance management patterns to realize the scientific management.

In order to make the diesel engine maintenance work under the prevention policy, it is necessary to make each kind of maintenance decision correctly at the right moment. During the maintenance decision process, one important task is to identify the

F. Sun et al. (Eds.): ISNN 2008, Part II, LNCS 5264, pp. 533–541, 2008.

specific time for maintenance and repair. Nowadays, point to the diesel engine maintenance decision, many maintenance management modes appeared, such as TNPM (Total normalized productive maintenance), SMS (Selective maintenance system), PMS (Preventive maintenance system), and so on [2]. However, maintenance decision is a time-consuming process, and there exist many fuzzy and uncertain factors which have important influence on the decision-making.

In reference [3], we present a fuzzy neural network decision model for diesel engine using fuzzy logic and improved T-S neural network. But there exist some disadvantages in the presented model, such as modeling difficult, time consuming, and so on. In this paper, we present a multi-layer neural network decision model for diesel engine. The structures of this paper are as follows: the maintenance decision theory for the diesel engine are analyzed in detail in section 2, a multi-layer neural network maintenance decision model is developed in section 3, an example is given in section 4 to verify the feasibly of the model. Conclusions are given in section 5.

2 Maintenance Decision Theory of Diesel Engine

2.1 The Maintenance Decision Process of Diesel Engine [2,3]

The maintenance decision theory of diesel engine is as follows. According to the oil monitor results and performance parameter monitor data, we can decide whether the main engine needs to be serviced or not, or whether the examination cycle needs to be prolonged or not. When approach legal examination cycle, if parameter monitoring and oil analysis results indicate that the technical state of the main engine is normal, the examination cycle should be prolonged. Otherwise, the engine and its parts should be serviced. We can decide whether the engine or its parts needs to maintenance based on the monitor results.

The maintenance decision process generally include: (1) determination the state monitor performance parameters and standards of maintenance object. (2). Calculating the deterioration degree of performance parameters of maintenance object based on the technical parameter monitor value. (3) Carrying on the maintenance decision. Because the diesel engine technical state is mainly decided by the technical state of the piston-cylinder liner module and the crankshaft-bearing module, the diesel engine state monitor parameters main includes the piston-cylinder liner module and the crankshaft-bearing module. In these state monitor parameters, the technical standards of performance parameters of oil monitor is mainly decided by the experience and statistical method. Other technical standards of performance parameters generally are decided by the inspection standard, the conventional statistical value of the diesel engine, the instruction and maintenance instruction of the diesel engine.

2.2 Diesel Engine Technical State Factors Analysis

Diesel engine technical state is decided by the technical state of piston-cylinder liner module and crankshaft-bearing module. The technical state of the piston-cylinder liner module and the crankshaft-bearing module is decided by the oil analysis results and some performance parameters. The factors of diesel engine technical state can be analyzed and cataloged as shown in Figure 1.

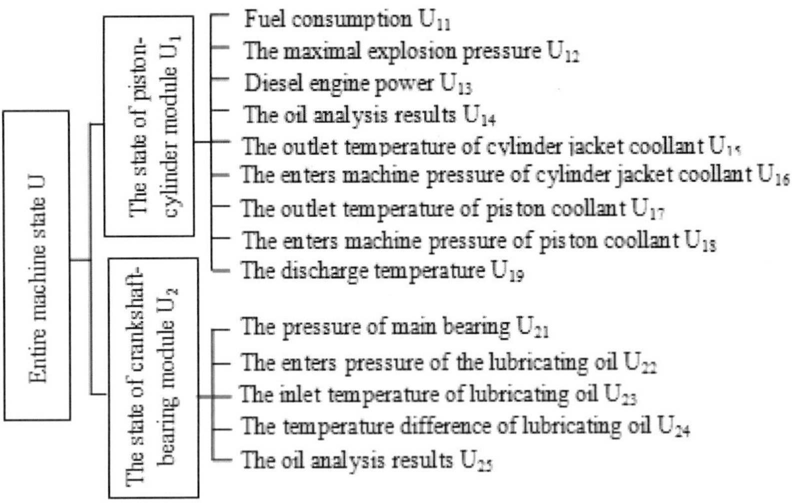

Fig. 1. The condition factors of diesel engine technology

In Figure 1, U is the deterioration degree of the diesel engine, U_1 is the deterioration degree of the piston-cylinder liner module, U_2 is the deterioration degree of the crankshaft-bearing module; U_{ii} is the deterioration degree of the performance parameter in each module. From Figure 1 we can see, oil analysis results can reflect the weary of piston-cylinder liner and crankshaft-bearing, so they are the important influence factors and should exist in the two modules.

2.3 Deterioration Degree Computation

The equipment deterioration degree indicates the deviation degree from good state to the fault state. The value of deterioration degree varies from 0 to 1. If the value of deterioration degree is 0, it shows that the equipment state is good. If the value of deterioration degree is 1, it shows that the fault appears, and the equipment state is worse. It is well known, each kind of equipment has a series of state variable x_1, x_2,..., x_n, and can be written as $x_i(t)$. The function of the equipment can be regarded as the set of the normal work state; on the contrary, the failure can be regarded as the set of the state which surpasses the normal. We call the state as "the middle state" when the equipment state deviated the good state but not surpass the limit.

The parameter deterioration degree can be expressed by the following methods.

(1) Calculating the deterioration degree according to the state parameters which can be detected. The formula is as follows [2]

$$l_i = \begin{cases} 1 & (Q \geq N) \\ [(Q-M)/(N-M)]^K & (M \leq Q \leq N) \\ 0 & (Q \leq M) \end{cases} . \tag{1}$$

where l_i is the deterioration degree of the ith technical parameter. M is the permissible value (good value) of the technical parameter. N is the limit value of the technical parameter. Q is the actual value of technical parameter inspection. K is an index. The value of K varies from 0.5 to 1.

(2) Calculating the deterioration degree according to the actual using period. The deterioration degree can be expressed as follows

$$l_i = \left(t/T \right)^K . \tag{2}$$

where t is the using period of the part. T is the mean time between failures (MTBF). K is an index, its value is 2.

(3) Evaluating the deterioration degree by technician, examinator and user of the equipment. The deterioration degree of this maintenance object can be expressed as follows

$$L = \left(A \times P_1 + B \times P_2 + C \times P_3 \right) / \left(P_1 + P_2 + P_3 \right) . \tag{3}$$

where A, B and C is the evaluation value of technician, examinator and user respectively, and its value is between 0 and 1. P_1, P_2 and P_3 is the weight of technician, examinator and user respectively.

2.4 Maintenance Decision for Diesel Engine

When the deterioration degree of the diesel engine and its modules is obtained, the maintenance decision can be carried on, the decision steps are as follows:

(1) Let b be the deterioration degree of entire machine and b_i ($i > 1$) be the deterioration degree of its each module.

(2) Set threshold value. The threshold value can be regarded as the criteria to evaluate the technical state of maintenance object.

(3) Let the threshold value be 0.6. If b <0.6, it indicates that the technical state of the diesel engine is good, and does not need to maintenance. If b >0.8, it indicates that the technical state of the diesel engine is bad, and needs to maintenance. If 0.8> b >0.6, it indicates that whether the diesel engine needs to maintenance or not depends on the state of the module. If b_{imax}>0.8 and 0.6< b_{iover}<0.8 (where b_{imax} is the maximal value of the deterioration degree of each module and b_{iover} is the average value of the deterioration degree), it indicates that the state of most parts of the engine is worse, and needs to maintenance. Otherwise, it only needs to strengthen the monitoring of the engine. When the main engine needs to maintenance, which parts need to maintenance is decided by the technical state of the engine, the parts which need to maintenance are that their deterioration degree are bigger than 0.8, i.e. b_i>0.8.

3 The Neural Network Model of the Maintenance Decision for Diesel Engine

A back-propagation neural network is a multi-layer network with an input layer, an output layer, and some hidden layers between the input and output layers. Each layer has a number of proceeding unit, called neurons. A neuron simply computes the sum of their weighted inputs, subtracts its threshold from the sum, and passes the results through its transfer function [4-6]. One of the most important characteristics of BP neural networks is their ability to learn by examples. With proper training, the network can memorize the knowledge in the problem solving of a particular domain.

Because the technical state monitor parameters of the diesel engine mainly are the performance parameters of the crankshaft-bearing module and piston-cylinder module, and these performance parameters exist some overlapping, an important neural network is proposed as shown in Figure 2. By using this BP neural network, the maintenance decision of diesel engine could be realized based on the state monitoring data.

The maintenance decision model of the diesel engine includes three sub-networks and one combination network. The inputs of sub-network are the deterioration degree of the correlation performance parameters, the output are the technical state deterioration degree of the modules. The input of the combination network are each sub-network, the output of the combination network is the technical state deterioration degree of the main engine. The structure of each neural network is similar. The differences of the networks lie in the number of the input nodes. Take the sub-network of cylinder-piston module as an example to illustrate the neural network which has three layers.

The first layer is the input layer, the input is the deterioration degree of each performance parameter, and the number of input nodes is 9. The second layer is the

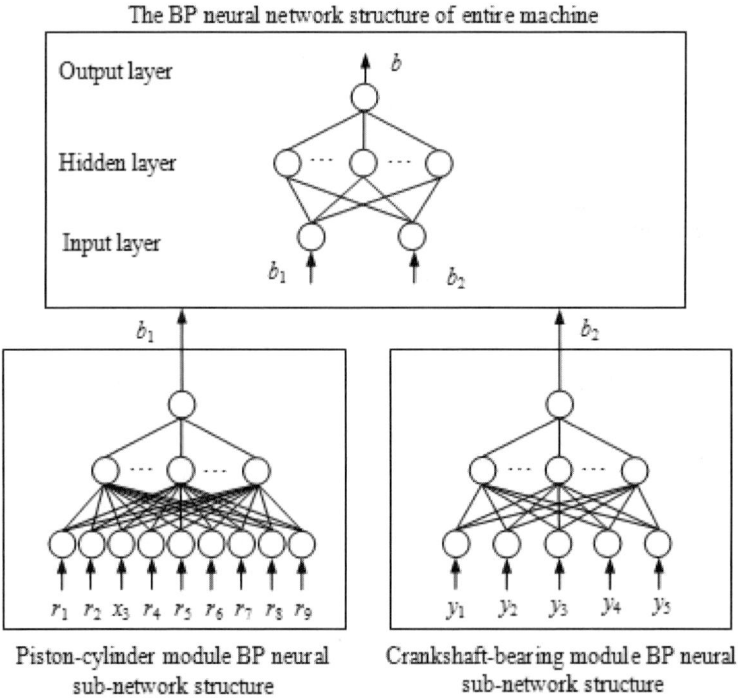

Fig. 2. Neural-network-based maintenance decision model for diesel engine

Table 1. The input and output of piston-cylinder module BP neural network

Nodes	Input	Output	Joint weight
The i th node of input layer	r_i	$u_i = r_i$	
The k th node of hidden layer	$I_k = \sum_{i=1}^{m} w_{ik} r_i$	$u_k = \dfrac{1}{1+\left[\left(\sum_{i=1}^{m} w_{ik} r_i\right)^{-1} - 1\right]^2}$	$\sum_{i=1}^{m} w_{ik} = 1$, $w_{ik} \geq 0$
The node p of output layer	$I_p = \sum_{k=1}^{l} w_{kp} u_k$	$u_p = \dfrac{1}{1+\left[\left(\sum_{k=1}^{l} w_{kp} u_k\right)^{-1} - 1\right]^2}$	$\sum_{k=1}^{l} w_{kp} = 1$, $w_{kp} \geq 0$

hidden layer. The third layer is the output layer. The output of this layer is the technical state deterioration degree of the maintenance object.

The input and output of piston-cylinder module neural sub-network are as shown in Table 1. r_i and u_i is the input and output of the i th node in input layer respectively, where $i = 1,2,\cdots,m$. I_k and u_k is the input and output of the k th node in hidden layer respectively. w_{ik} is the joint weight between the i th node and the k th node. There is only one node p in the output layer, and the input and out is I_p and u_p respectively. w_{kp} is the joint weight between hidden layer and output layer. By training the neural network, the technical state deterioration degree of maintenance object can be obtained, and the maintenance decision can be carried on based on these deterioration degrees.

4 Case Study

Take the maintenance decision of YC6100 as an example to illustrate the established BP-neural-network model. According to the maintenance record, there are 9 performance parameters. i.e. the input of the cylinder-piston module subnet are the deterioration degree of the piston coollant enters machine pressure x_1, the deterioration degree of the cylinder liner coollant enters machine pressure x_2, the deterioration degree of the air cylinder coollant outlet temperature x_3, the deterioration degree of the piston coollant outlet temperature x_4, the deterioration degree of the discharge temperature x_5, the deterioration degree of the main engine power x_6, the output of this subnetwork is the deterioration degree of the cylinder-piston module b_1; The input of the crankshaft bearing module subnet are the deterioration degree of the main bearing the lubricating oil enters pressure y_1, the deterioration degree of the crosshead bearing the lubricating oil enters pressure y_2, the deterioration degree of the lubricating oil cooler inlet temperature y_3, the deterioration degree of the lubricating oil import and export temperature difference y_4, the outputs is the deterioration degree of crankshaft bearing

Table 2. The samples of the piston- cylinder module

	The piston- cylinder module performance parameter deterioration degree						The module deterioration degree
	x_1	x_2	x_3	x_4	x_5	x_6	b_1
1	0.3921	0.4115	0.3723	0.5652	0.6245	0.4587	0.5125
2	0.3372	0.4123	0.3745	0.5647	0.6268	0.4557	0.5351
3	0.4011	0.4271	0.3787	0.5635	0.6287	0.4547	0.4012
4	0.4121	0.4223	0.3758	0.564	0.6284	0.4568	0.4546
5	0.3984	0.4121	0.3765	0.5668	0.6278	0.4575	0.5231
6	0.3887	0.4126	0.3732	0.5735	0.6214	0.4565	0.6411
7	0.3889	0.4127	0.3737	0.5642	0.6224	0.4522	0.4352
8	0.3967	0.4111	0.3768	0.5667	0.6284	0.4536	0.6524
9	0.3954	0.4225	0.3721	0.5654	0.6221	0.4547	0.6352
10	0.3955	0.4119	0.3723	0.5631	0.6234	0.4578	0.5215

module b_2; The input of the combination network are the output of the two sub-network, i.e. b_1, b_2. The output is the deterioration degree of the main engine, i. e. b. The sample sets are shown in Table 2 and Table 3.

The network precision of training parameter is 10^{-6} and the study rate is 0.001. We can take one sample as examination sample, such as sample 10, and the other nine samples can be as training samples. The training results including expectation output, network error and system error are listed in Table 4. The network errors are as shown in Figure 3. From Table 4 we can see, the expectation output and the actual output of each sample are very close. The absolute error is smaller than 0.01. Moreover, the system average error is in 10^{-6} magnitude.

Table 3. The samples of the crankshaft-bearing module and the entire engine

	The crankshaft-bearing module					Deterioration degree of the entire engine
	The technical parameter deterioration degree of the crankshaft-bearing module				Deterioration degree of the module	
	y_1	y_2	y_3	y_4	b_2	b
1	0.4858	0.2114	0.1565	0.2321	0.4839	0.5157
2	0.4645	0.2135	0.1554	0.2324	0.5172	0.5442
3	0.4758	0.2127	0.1524	0.2325	0.5288	0.5339
4	0.4614	0.2128	0.1512	0.2374	0.5553	0.5412
5	0.4678	0.2136	0.1532	0.2357	0.5362	0.5556
6	0.4665	0.2124	0.1521	0.2312	0.5225	0.6324
7	0.4725	0.2127	0.1515	0.2335	0.5375	0.6425
8	0.4710	0.2135	0.1521	0.2347	0.5121	0.5100
9	0.4721	0.2118	0.1545	0.2336	0.5459	0.6232
10	0.4735	0.2123	0.1515	0.2335	0.5612	0.6545

Table 4. The training results of the entire engine fuzzy neural network

	The training results of the piston- cylinder module		The training results of the crankshaft-bearing module		The training results of the entire engine	
	b_1	error_1	b_2	error_2	b	error
1	0.5125	9.962E-07	0.4839	9.957E-07	0.5157	9.936E-07
2	0.5351	9.975E-07	0.5172	9.976E-07	0.5442	9.955E-07
3	0.4012	9.974E-07	0.5288	9.957E-07	0.5339	9.978E-07
4	0.4546	9.973E-07	0.5553	9.969E-07	0.5412	9.936E-07
5	0.5231	9.979E-07	0.5362	9.968E-07	0.5556	9.949E-07
6	0.6411	9.976E-07	0.5225	9.954E-07	0.6324	9.955E-07
7	0.4352	9.968E-07	0.5375	9.947E-07	0.6425	9.938E-07
8	0.6524	9.941E-07	0.5121	9.938E-07	0.5100	9.975E-07
9	0.6352	9.943E-07	0.5459	9.948E-07	0.6232	9.944E-07

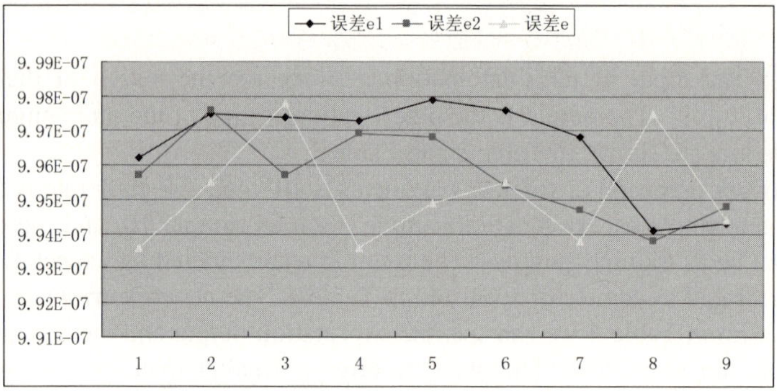

Fig. 3. The drawing of the network training

5 Conclusions

From the simulation result we can see, the combination BP neural network model not only could reflect the logic behavior of the main engine system structure, but also could assist the decision-maker to carry on the maintenance decision. The training results are very accurate. Therefore the combination BP neural network model could offer theory base for the maintenance decision for the diesel engine.

Acknowledgments

This research was partially supported by the Jiangxi Provincial Natural Science Foundation under the contract number 2007GQC0654 and the Science Foundation of Education Commission of Jiangxi Province under the contract number 2007[201].

References

1. Yan, L., Zhu, X.H., Yan, Z.J., Xu, J.J.: The Status and Prospect of Transportation Machine Equipment Repairing in 21th Century. Journal of Traffic and Transportation Engineering 1, 47–51 (2001)
2. Hu, Y.P., Yan, L., Zhu, X.H.: Application of Neuro-net in Maintenace Decision for Ship Diesel Engine. Journal of Traffic and Transportation Engineering 1, 69–73 (2001)
3. Gu, Y.K., Yang, Z.Y.: TS-neural-network-based Maintenance Decision Model for Diesel Engine. In: Liu, D., Fei, S., Hou, Z.-G., Zhang, H., Sun, C. (eds.) ISNN 2007. LNCS, vol. 4491, pp. 557–565. Springer, Heidelberg (2007)
4. Gu, Y.K., Huang, H.Z.: Neural-network-driven Fuzzy Reasoning of Dependency Relationships Among Product Development. In: Wang, J., Liao, X.-F., Yi, Z. (eds.) ISNN 2005. LNCS, vol. 3498, pp. 927–932. Springer, Heidelberg (2005)
5. Gu, Y.K., He, X.W.: Neural-network-driven Fuzzy Optimum Selection for Mechanism Schemes. In: Liu, D., Fei, S., Hou, Z., Zhang, H., Sun, C. (eds.) ISNN 2007. LNCS, vol. 4492, pp. 275–283. Springer, Heidelberg (2007)
6. Sun, J., Kalenchuk, D.K., Xue, D., Gu, P.: Design Candidate Identification Using Neural Network-Based Fuzzy Reasoning. Robotics and Computer Integrated Manufacturing 16, 383–396 (2000)

Design of Intelligent PID Controller Based on Adaptive Genetic Algorithm and Implementation of FPGA*,**

Liguo Qu, Yourui Huang, and Liuyi Ling

School of Electronic and Information EngineeringAnhui University of Science and Technology, Huainan 232001, Anhui, China

qlgcool@sina.com.cn

Abstract. Now, many systems are time-varying, nonlinear and real-time in industrial processes. According to PID parameters adjustment problems of these systems, firstly, the paper adopts adaptive genetic algorithm (AGA) to optimize the parameters of PID controller and introduces altera FPGA 1P1C6F256C8 to implement PID controller. Secondly, the closed-loop test system is constructed by DSP builder. At last, TCL script file which is generated by signal compiler is run in modelsim to verify VHDL code which is compiled in Quartus II. The results show that AGA improves the precision of PID parameters optimization and the adaptability of control system, simultaneously demonstrate the feasibility and practicability of intelligent PID controller based on FPGA.

Keywords: PID parameters optimization, Adaptive genetic algorithm, FPGA, DSP builder.

1 Introduction

PID control is the most ancient, the most common and the strongest control method in automatic control. With the development of the industry, control objects become more and more complex, particularly time-varying, nonlinear and real-time. Conventional PID control has become powerless for these systems [1]. According to these flaws of traditional PID controller, some people make a great improvement on approach of PID controller parameters optimization. In recent years, intelligent PID controller, such as fuzzy PID controller, neural network PID controller, expert PID, has become a popular control method. The intelligent PID controller simplifies the modeling procedure and has self-adaptive, self-organization and self-leaning abilities. Simultaneously it has the advantages of traditional PID controller, such as simple structure, strong robustness. Besides, the integration and technology level of VLSI have a great improvement. It is possible to finish system design in one chip. FPGA can quickly complete massive complex calculations because of its ability of parallel calculation and processing. Moreover, the design of controller based on FPGA has high development efficiency because of its flexibility and universality.

* The work has been supported by the youth fund of Anhui University of Science and Technology (DG726).

** All authors have used the western naming convention, with given names preceding surnames.

F. Sun et al. (Eds.): ISNN 2008, Part II, LNCS 5264, pp. 542–551, 2008.

A concrete implementation scheme of intelligent PID controller based on AGA is introduced in detail in this article. Hardware implementation is based on Altera FPGA 1P1C6F256C8. When implementing the design in hardware, the VHDL language is used to describe the whole algorithm.

2 The Principle of Intelligent PID Controller

Parameters tuning of PID controller is looking for a group of the optimal or near-optimal combination of k_p, k_i and k_d to make system obtain better control quality [2]. The principle schematics of intelligent PID controller is as shown in Figure 1. Here AGA is used to find optimal result among all possible combinations of the unknown parameters k_p,k_i and k_d to obtain the smallest value of fitness function .

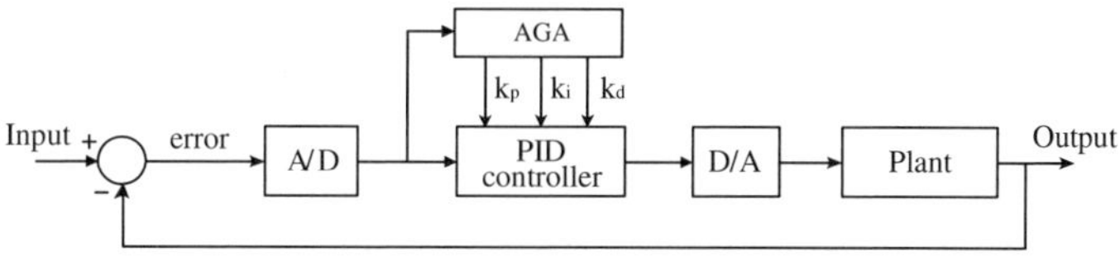

Fig. 1. System of PID controller based on AGA

3 Design of Algorithm

3.1 The Algorithm of Incremental PID Controller

For saving hardware source of FPGA, the PID controller uses incremental PID algorithm , such as formula (1). From formula (1), we can calculate u(k) if we know u(k-1), e(k), e(k-1) and e(k-2) [3].

$$u(k)=u(k-1)+(k_p+k_i+k_d)e(k)-(k_p+2k_d)e(k-1)+k_d e(k-2). \quad (1)$$

Where, k_p —proportional coefficient, k_i—integral coefficient, k_d—differential coefficient, u(k) —the output of controller at k time; e(k) —the error of sampling input at k time; e (k-1) —the input error of control system at k-1 time.

3.2 Adaptive Genetic Algorithm (AGA)

AGA adopts variational crossover probability (p_c) and mutation probability (p_m) which depend on the fitness value and the iteration number to avoid premature convergence[4]. The main aspects of AGA are as follows:

(1) Real coding is applied in AGA. it can enhance the precision of solution and the convergence speed.

(2) In order to facilitate FPGA Implementation, AGA selects randomly two different individuals from group at a time and then retains an optimal individual in accordance with their fitness values.

(3) In AGA, adaptive cross operator is adopted. Crossover probability is p_c.

$$p_c = \begin{cases} p_{c\_max} - ((p_{c\_max} - p_{c\_min})/itmax) * iter & f' < f_{avg} \\ p_{c\_max} & f' \geq f_{avg} \end{cases} \tag{2}$$

(4) In AGA, adaptive mutation operator is adopted. Mutation probability is p_m.

$$p_m = \begin{cases} p_{m\_min} + ((p_{m\_max} - p_{m\_min})/itmax) * iter & f < f_{avg} \\ p_{m\_min} & f \geq f_{avg} \end{cases} \tag{3}$$

Where, p_{c_max} —the biggest crossover probability, p_{c_min}—the smallest crossover probability, p_{m_max}—the biggest mutation probability, p_{m_min}—the smallest mutation probability, itmax—the biggest generation, *iter*—current generation, f_{avg} —the average fitness value in current generation. f'— the bigger fitness value among two crossover individual , f—the fitness value of mutation individual.

(5) Computation of fitness function

In order to obtain the satisfied dynamic process characteristic, we use ITAE performance index as the object function, ITAE can be express as follows:

$$ITAE = \int_0^\infty t|e(t)|dt, \tag{4}$$

Discrete ITAE can be express：

$$ITAE = \Delta t^2 \sum_{K=1}^{N} K|e(k)|. \tag{5}$$

Where, Δt is sampling period, N is sampling number.

4 The Hardware Design of Intelligent PID Controller

After having analyzed the PID control method and the characteristic of genetic algorithm, firstly, we divide the function modules of the system into the sequential circuits and the combinational logic circuits [5]. Secondly, we carry on the timing design of the system. Timing is just like a commander of system and commands all modules work in an orderly manner. In the process of hardware implementation, we consider comprehensively the speed of operation and the occupation of resource, at last, make them reach a good balance [6].

4.1 Hardware Structure

Hardware structure of intelligent PID controller is as shown in Figure 2. It mainly includes initialization module, selection module, crossover and mutation module, storage module, multiplexer (MUX) module, random number module, dual-port RAM(RAM1 and RAM2), fitness calculation module, control module and incremental PID controller. Here, RAM1 is used to store individual and RAM2 is used to store fitness value of individual.

Figure 3 is the hardware structure of incremental PID controller. It mainly includes register (reg1~reg10), adder, multiplier and control module. Here, incremental PID controller adopts parallel format for improving speed and the data type of adder and multiplier adopts sign binary fraction (SBF) for saving source. The error(k) is stored in reg1.The error(k-1) is stored in reg2. The error(k-2) is stored in reg3.The middle results are stored in reg4 to reg9.

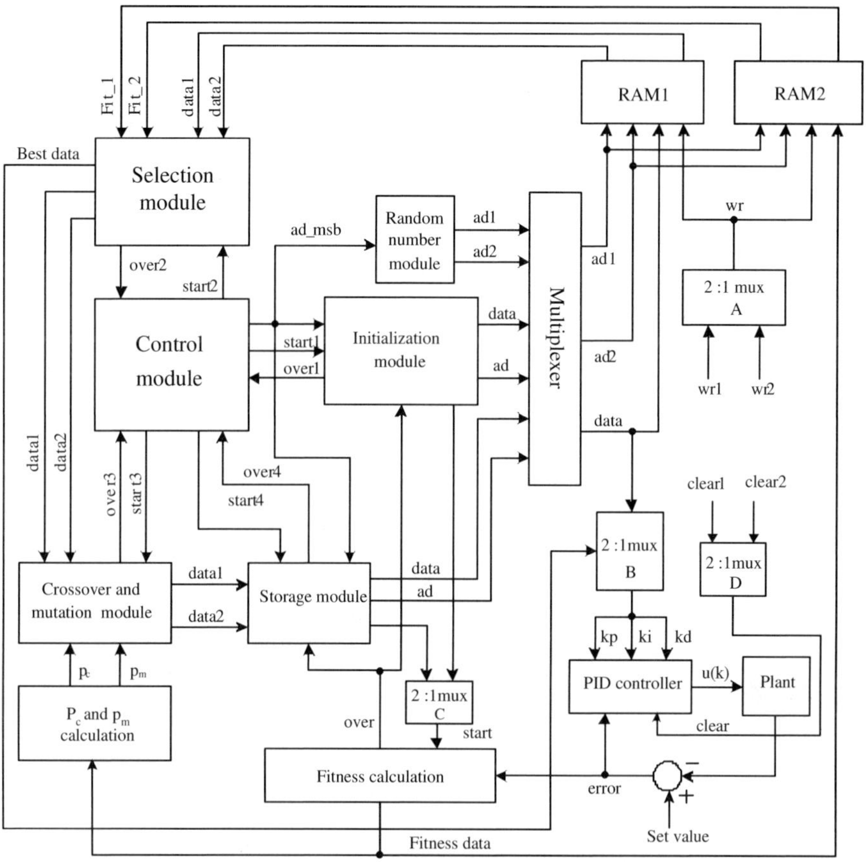

Fig. 2. Structure of intelligent PID controller based on AGA

4.2 State Machine Design

If we want to all modules work orderly, timing control is very important. The state machine is a good tool for timing control design. Because the sequential circuits and the combinational logic circuits are in same process in single process moore state machine, it may avoid the glitch phenomenon as far as possible. Here, we use a single process moore state machine to realize state machine design of control modules [7].

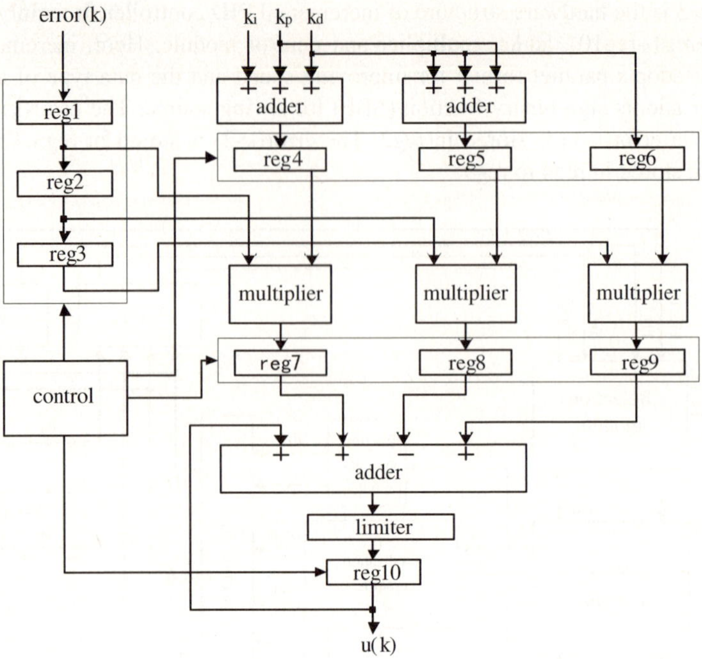

Fig. 3. The circuit structure of PID controller

4.2.1 Design of Main Control Module State Machine

Figure 4 is state machine of main control module in figure 2. It consists of six states (idle,st1,st2,st3,st4,stop). State “idle” is a reset state and state “stop” is an over state. The others are four work states. The work process of each states is as follows:

Idle: When asynchronous reset signal (reset) changes from “1” to “0”, state machine jumps to state “idle” immediately. At this state, all modules begin to initialize some signals. We call this process system reset. When reset is over, state machine will jump to state “st1”.

St1: System finishes mainly group initialization at this state. After entering this state, control module sets signal “start1” to “1” and then starts up the operation of initialization. Under the clock of FPGA, initialization module generates a new individual “data” and a new address signal “ad” at a time. The individual “data” and address signal “ad1” that are output of multiplexer come form initialization module under control of the main control module. At the same time new individual acts on PID controller by 2:1 MUX B and then initialization module starts up computation of fitness by 2:1 MUX C. When computation of fitness is over, fitness module will generate an end signal “over”. After receiving end signal “over”, initialization module will generate a writing enable signal “wr1” which acts on RAM1 and RAM2 by 2:1 MUX A. Then the new individual and the value of fitness will be stored in RAM1 and RAM2 respectively at one time. After an individual storage end, individual counter plus one. Then initialization module judges whether individual number is up to group size. If not, initialization module will generate new individual, otherwise initialization module will generate an initialization end signal "over1" and change signal "start1" to "0" for closing the initialization module, and the state machine will jump to state "st2".

St2: System finishes mainly selection operation at this state. After entering this state, the control module sets signal "start2" to "1" and then starts up the operation of selection. Random number module generates randomly two different address signal "ad1" and "ad2" which act on RAM1 and RAM2 by multiplexer. Then the selection module selects two different individuals from group at a time and then retains an optimal individual in accordance with their fitness values. Repeating the selection process again, two optimal individuals are generated. After selection finish, the selection module will generate a selection end signal "over2" and change signal "start2" to "0" for closing the selection module, and the state machine will jump to state "st3".

St3: System finishes mainly crossover and mutation operation at this state. After entering this state, the control module sets signal "start3" to "1" and then starts up the operation of crossover and mutation. Two individuals which come from selection module will be executed firstly crossover operation according to p_c and then mutation operation according to p_m. After we obtain two new individuals, the crossover and mutation module will generate a selection end signal "over3" and change signal "start3" to "0" for closing the crossover and mutation module, and the state machine will jump to state "st4".

St4: System finishes mainly storage operation at this state. After entering this state, the control module sets signal "start4" to "1" and then starts up the operation of storage. The process of storage is similar to the storage process of initialization. After an individual storage end, individual counter plus one. If new individual number is up to group size, iteration counter plus one. Then control module judges whether it meets the iterative requirements. If not, the state machine will jump to state "st2" to continue genetic operation, otherwise the state machine will jump to state "stop".

Stop: When coming into state "stop", genetic algorithm is over, then the optimal individual "Best data" will be outputted and act on PID controller by 2:1 MUX B. State "stop" doesn't exist next state. State machine is in the state of deadlock at this time. When only asynchronous reset signal "reset" become available, state machine can jump to state "idle". We don't judge "reset" at state "stop" because it is asynchronous reset signal.

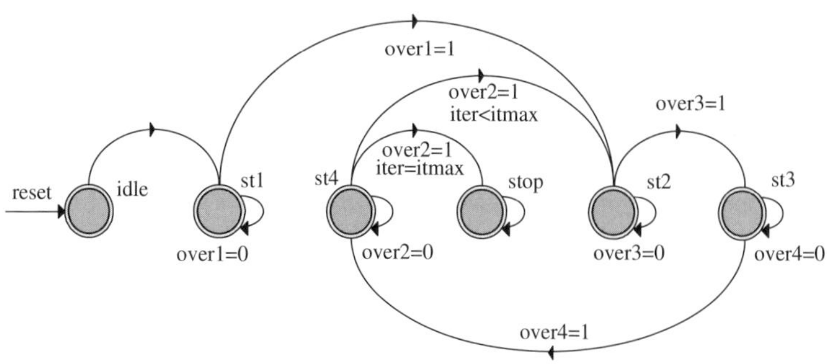

Fig. 4. State machine of main control module

4.2.2 State Machine Design of Control Module in Incremental PID Controller

Figure 5 is state machine of control module in figure 3. State "idle" is a reset state. State "st1", "st2", "st3"and "st4"are four work states. At rising edge of each clock, state machine will change from current state to next state. The work process of these states is as follows.

Idle: After control signal "clear" changes to"1", system comes into the state "idle". Some interior register will be initialized at this state.

Stl: Regl, reg2, reg3 can be shifted at this state. The value of reg2 is shifted in reg3. The value of regl is shifted in reg2. The error(k) is shifted in regl.The sum of k_p, k_i, k_d is saved in reg4. The sum of k_p and $2k_d$ is saved in reg5. The k_d is saved in reg6.

St2: The control module enables reg7, reg8, reg9. The value of $[(k_p+k_i+k_d)* e(k)]$ is saved in reg7. The value of $[(k_p+2k_d)* e(k-1)]$ is saved in reg8. The value of $[k_d*e(k-2)]$ is saved in reg9.

St3: The control module enables reg10 which outputs u(k).

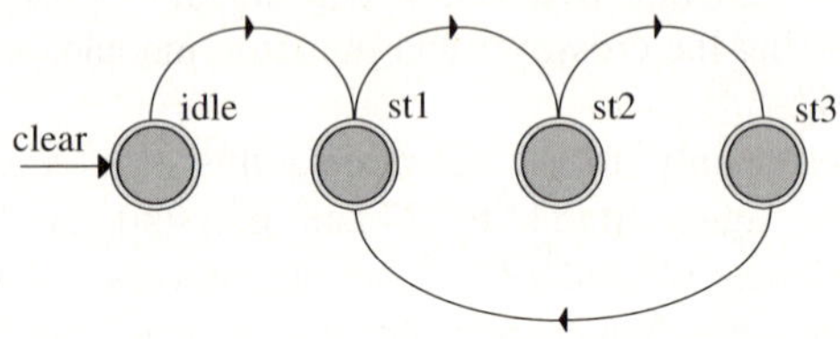

Fig. 5. The state machine of control module in PID controller

5 The Simulation and Implementation

Because the open-loop simulation of the intelligent PID controller can't simulate input signal accurately, this paper uses closed-loop simulation based on QuartusII6.0, DSP builder6.0, modelsim SE 6.1f and simulink6.0. The test steps are as follows [8]:

Step 1: Using HDL import tool of DSP builder to import the VHDL code to simulink and designing test module by DSP builder and simulink.

Step 2: After running test module correctly, conversing file(.mdl) to VHDL file and TCL script file which can be recognized by modelsim by signal compiler.

Step 3: Setting VHDL file and TCL script file.

Step 4: Running test bench file in modelsim and examining the test results.

Figure 6 is test plan of intelligent PID controller in DSP builder. The transfer function of the plant is $Y(s)=0.15/(s^2+2.5s+1.5)$. The parameters of AGA are as follows. The group size is 32. The p_{c_max} is 0.875 and the p_{c_min} is 0.125. The p_{m_max} is 0.5 and the p_{m_min} is 0.0625. The itmax is 128. As shown in Figure 7, we could obtain that kp, ki, kd are 3.25, 5.5, 0.625 respectively after parameters tuning. Figure 8(a) is simulation curve of Scopel in figure 6. From figure 8(a), we can see the output of system has a little overshoot slightly, but ,at last the system is stable at 1. Figure 8(b) is simulation curve of Scope3 in figure 6. Figure 8(c) is simulation curve of Scope2 in figure 6. It is output curve of PID controller and is stable at 10 eventually. Figure 9 is

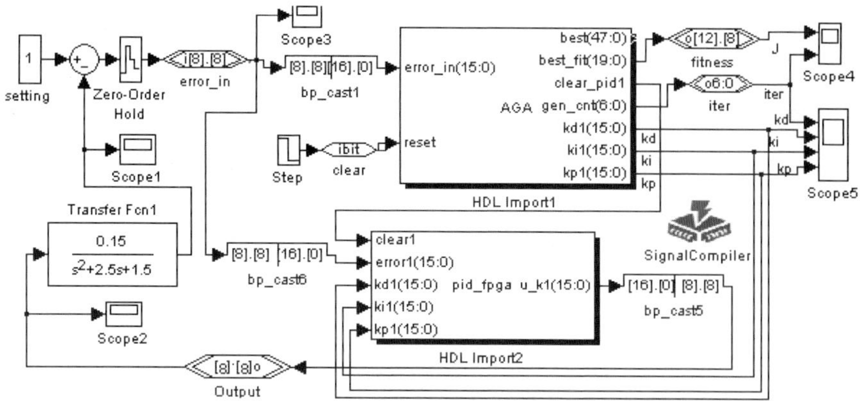

Fig. 6. Intelligent PID controller test

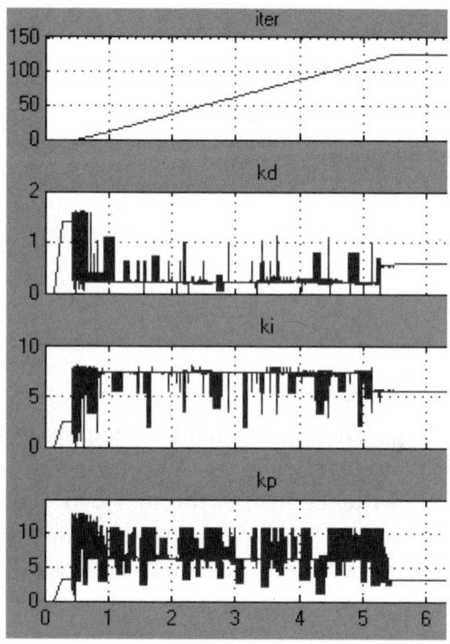

Fig. 7. Curve of PID parameters tuning

the simulation curve of RTL level circuit of intelligent PID controller in modelsim. The output of the controller and input error are showed as analog signal. From the figure 9, we can see that the output value ascends from 0 and stabilizes at 9F9 (HEX), at same time the error signal declines from 1 to 0. Because the output results are expressed with SBF (the high 8-bit are the integral parts and the low 8-bit are the fractional parts), the decimal format of 9F9 is 9.97265625. The test results are basically same as results that simulated by MATLAB. After completing the above work correctly, we will download the design to FPGA 1P1C6F256C8 by the quartus II programer environment.

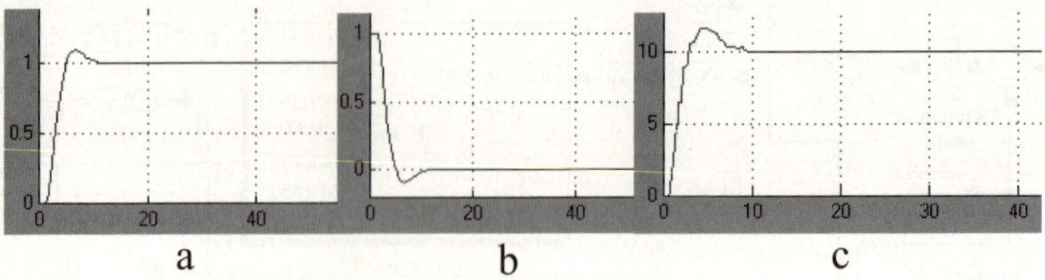

Fig. 8. Simulation curve of system

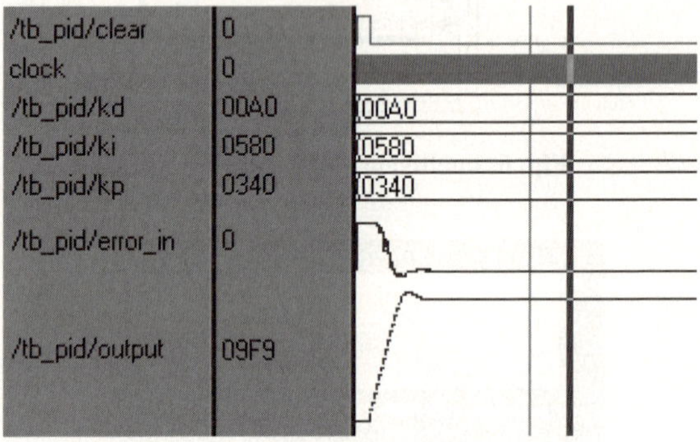

Fig. 9. Controller output curve in modelSim

6 Conclusion

(1) AGA is used to optimize parameters of PID controller. The simulation results show that AGA improves global search capabilities and convergence speed of GA, that this method is feasible and practical.

(2) The closed loop test of PID controller through DSP builder, simulink and modelsim solves the sample source of test input as well as the input sample extraction of the controller. It can simulate the input behavior of controller effectively, enhanced the flexibility of design and test. Simultaneously, the test results are reliable and persuasive.

(3) Intelligence controller based on FPGA has some advantages such as flexible design, self-tuning on line, high reliability, low development cycle and high speed.

References

1. Liu, J.K.: Control of Advanced PID and Matlab Simulation. Electronic Industry Press (2004)
2. Hu, B.G., George, K.I.M., Raymond, G.G.: A Systematic Study of Fuzzy PID Controllers-Function-Based Evalution Approach. IEEE Trans. on Fuzzy Systems 9, 699–708 (2001)
3. Cirstea.: FPGA Fuzzy Logic Controller for Variable Speed Generators. In: Proceedings of the IEEE International Conference on Control Application, pp. 301–304 (2001)

4. Hou, Z.X., Shen, Q.T., Li, H.Q.: Adjustment of PID Parameters Based on Improved Genetic Algorithms and Its Application on Heating Furnace. Computer of Engineering 30, 165–167 (2004)
5. John, L.H., David, A.P.: Computer Architecture, A Quantitative Approach. Mechanical Industry Press, Beijing (2002)
6. Sharma, C.A., DeMara, R.F.: A Combinatorial Group Testing Method for FPGA Fault Location. In: Proc. International Conference on Advances in Computer Science and Technology, Puerto Vallarta, Mexico, pp. 23–25 (2006)
7. Lund, T.: The Architecture of An FPGA- Style Programmable Fuzzy Logic Controller Chip. In: 5th Australasian Computer Architecture Conference (2000)
8. Chen, J.W., Zhou, Y.J.: Implementation of FPGA and Hardware and Software Co-simulation of Digital PID Controller. Information Technology 9, 38–41 (2005)

Fragile Watermarking Schemes for Tamperproof Web Pages

Xiangyang Liu[1,2] and Hongtao Lu[1]

[1] Department of Computer Science and Engineering, Shanghai Jiao Tong University
Shanghai 200240, China
lu-ht@cs.sjtu.edu.cn
[2] College of Science, Hohai University
Nanjing 210098, China
liuxy@sjtu.edu.cn

Abstract. In this paper, to improve the security, sensitivity and efficiency, we propose two novel fragile watermarking schemes for tamperproof web pages, in which a message authentication method and a sparse coding method are used to generate the watermarks for a web page, respectively. The former ensures high security and good sensitivity of the watermarking scheme. The latter attaches more importance to satisfy all the requirements of security, sensitivity and efficiency well. The generated watermarks are then embedded into the web page through the upper and lower cases of letters in HTML tags. The properties of the schemes are illustrated with several experiments, and the results show that the proposed schemes can be good practical tools for the tamperproof of web pages.

Keywords: Fragile watermark, Hash, Keyed-Hashing for Message Authentication(HMAC), Non-negative Sparse Coding(NNSC).

1 Introduction

The tamperproof of web pages becomes very important as much information are provided through web pages, and the ever-increasing complexity of Web-Service environments and platforms needs the self-protection of web pages. Nowadays, despite under various forms of protections, web pages often suffer from tampering (the contents of web pages are altered, deleted or added), which brings forward the need of a secure authentication system to verify the integrity of the web pages. At the same time, we expect the system to automatically detect the tampering. Fragile digital watermarking provides a convenient tool for authentication of digital information. A fragile watermark is very sensitive to a modification of a digital document, and a fragile watermarking scheme should be able to detect any possible change [1]. Thus, fragile digital watermarking [2] provides a convenient tool for authentication and tamper detection of digital information.

Recently, many fragile watermarking schemes for tamperproof of images are presented. A popular idea involved is to combine fragile watermark with cryptography [3], other researchers focused on content-based methods [4]. But for

F. Sun et al. (Eds.): ISNN 2008, Part II, LNCS 5264, pp. 552–559, 2008.

tamperproof of web pages, only the PCA-watermarking scheme (PCAW) is proposed [5][6], where watermarks are generated from web pages through the principle component analysis (PCA) technique. The PCAW is a good method for tamperproof of web pages, but can not satisfy all the requirements well such as sensitivity and time-efficiency. It is not sensitive because much information will be lost after the features are extracted through the PCA method, and the low-efficiency of PCAW because the PCA method cost much time [7]. For tamperproof of web pages, any tampering with a web page should be detectable by watermarks embedded in it, and the time that the scheme consumes in watermark generating, embedding and validating should be as short as possible.

In this paper, to improve the security, sensitivity and efficiency, we developed two new fragile watermark methods for the tamper detection of web pages based on the above work. Keyed-Hashing for Message Authentication (HMAC) [8] and Nonnegative Sparse Coding (NNSC) [9] are used to generate a fragile watermark from a web page, respectively, and the watermark is then embedded into the web page. The proposed method to store the watermark in the web pages does not add additional data to the file itself. Whenever the document is tampered, the watermark in it will be destroyed or become inconsistent with the content. Thus the tampering can be simply detected in this scheme.

The remainder of this paper is organized as follows. The message authentication method and the sparse coding method of fragile watermarking scheme are described in section 2 and 3, respectively. In section 4 we show the experimental results and discussions. Finally, we conclude this paper in section 5, where further research directions are given.

2 The Message Authentication Based Fragile Watermark (MAFW)

2.1 Hash Functions and HMAC

A hash function has an arbitrary length binary vector x as its input, and produces a fixed length binary output h [10]. Keyed-Hashing for Message Authentication (HMAC) [8] is a mechanism for message authentication using cryptographic hash functions. HMAC algorithms take two functionally distinct inputs, a message x and a secret key, and produce a fixed-size output, with the design intent that it be infeasible in practice to produce the same output without knowledge of the key.

$$HMAC(x) = h((k \oplus opad)||h((k \oplus ipad)||x)) \tag{1}$$

where h is the hash function, $\oplus$ denotes the XOR operation, $||$ is the link operation, k is a secret key, *opad* and *ipad* are distinct strings of sufficient length to pad k out to a full block for the compression function. The overall construction is quite efficient despite two calls to h, since the outer execution processes only a two-block input, independent of the length of x [8].

2.2 Watermark Generating

In order to generate a watermark from a web page, each character in the source code of the web page is first mapped to integer according to its index in the code chart adopted by the document, ASCII or UNICODE, for example. Because of the case-insensitive feature of HTML [11], we first change all the English letters in the HTML tags to upper cases, and then we map all the source code of the web page to a data vector. For example, the text "<title> Watermarking </title>" is mapped to the data vector <60 84 73 84 76 69 62 87 97 116 101 114 109 97 114 107 105 110 103 60 47 84 73 84 76 69 62 10> by ASCII.

In doing so, an integer vector D is obtained from the source code of the web page. We can see that any modifications to the web page will result in a different vector D. We then use the keyed-hash function $HMAC$ to generate the watermark W from the vector D as follows,

$$W = HMAC(D). \tag{2}$$

If the hash function is SHA-256 in $HMAC$, W is a 256 bits binary sequence. W is to be embedded as a fragile watermark in the following. Since W has close relation with the content of the web page and the diffusion property of the hash function, the tampering of the web page will result in a completely different W, i.e., W has good sensitivity. Moreover, the hash function can also ensure the high security of this scheme.

2.3 Watermark Embedding and Validating

Katzenbeisser et al. [12] proposed to insert spaces and tabs into the source code of a web page for embedding information. Because of the case-insensitiveness of HTML [11], we embed watermarks through altering the case of letters in HTML tags [5][6]. The advantage of the method is that it will not increase the file size of web pages. Specifically, the 0s and 1s in the watermark W determine the case of letters in the HTML tags: the ith letter is set to be the lower case if the ith element of W is '0', otherwise the upper case. For example, taking W =<0100101010> and the web page "<title>Web Page Watermarking </title>", after being watermarked, it becomes "<tItlE>Web Page Watermarking </tItLe>".

To check whether a web page has been illegally tampered, we first use the same method to generate a watermark as the one introduced in section 2.2. Afterward, another watermark is extracted from the letters in the HTML tags of the web page: '1' is retrieved from the upper case letter and '0' from the lower case. Because we change all the English letters in HTML tags as upper case letters in watermark generating, the data vector D extracted from the source code of the original web page is identical with the one extracted from the watermarked web page. Therefore, if the generated watermark and the extracted one are identical, the web page is legal; otherwise tampered.

3 The Sparse Coding Based Fragile Watermark (SCFW)

3.1 NNSC

Nonnegative Matrix Factorization (NMF) [13] is a part-based representation and dimensionality reduction method which factorizes a nonnegative matrix V into the product of two nonnegative matrices as $V \approx WH$. In order to increase the sparseness of the factorization, sparse coding and NMF can be combined to generate Nonnegative matrix factorization with Sparseness Constraints (NNSC) [9][14]. The NNSC has good properties of nonnegativity, sparseness and dimensionality reduction.

3.2 Watermark Generating, Embedding and Validating

In the watermark generating of this scheme, NNSC is used to generate watermark from source code of a web page. A vector D is obtained from the source code of the web page in the similar way as in the MAFW scheme, and then is reshaped to a matrix $T \in R^{n \times n}$ ($R^{n \times n}$ denotes the set of all $n \times n$ real matrices) whose elements are taken columnwise from the vector D. In this scheme, a key K is also needed in the watermark generating. It can be either an image or a random sequence. Assume it is an $m \times m$ matrix, we will use the operation of convolution to diffuse the data of web page and combine the key by the following equation.

$$V = T \otimes K \tag{3}$$

where $\otimes$ denotes the convolution operation, and $V \in R^{(n+m)\times(n+m)}$.

One of the most useful properties of NNSC is that it produces a very sparse representation of the data. Such a representation encodes much of the data using few 'active' components. In order to sparsely represent the matrix V, we use NNSC to factorize V as the product of a most smooth basis matrix W and a most sparse coefficient matrix H. In this study, we use the sparseness measure (4) to control the sparsenesses of basis matrix W and coefficient matrix H.

$$sparseness(x) = \frac{\sqrt{n} - (\sum |x_i|)/\sqrt{\sum x_i^2}}{\sqrt{n} - 1}, \tag{4}$$

where x is a n dimensional vector. See [9] for more details. The desired sparsenesses of the basis W and the coefficient H are set to 0 and 0.95, respectively.

If we choose a r-dimensional ($r << m+n$) projection, then every $r \times (m+n)$ elements in H are converted to binary form, i.e. a sequence of '0' and '1'. All of them are then linked together to generate a binary sequence W, which is taken as the watermark for the whole web page. Because of the high level of sparseness of the matrix H, r may be set higher than it in the PCAW scheme. Thus the efficiency and sensitivity of this scheme are better than the PCAW method.

The watermark embedding and validating of this scheme are the same as those in MAFW scheme.

```
<html><head>
<meta http-equiv="Content-Language" content="zh-cn">
<title>Web pages watermarking</title></head>
<body><p><a href="http://www.watermarkingworld.org/">
Watermarkingworld 2007</a></p>
<p>Fragile digital watermarking: <br>
is a well-known method for authentication of digital informat
```

(a)

```
Watermarkingworld 2007

Fragile digital watermarking:
is a well-known method for authentication of digital information.
security, sensitivity and efficiency, we present two novel fragil
watermarking schemes for this, in which a message authentication
method and a sparse coding method are used to generate the
watermarks for a web page, respectively. The former ensures high
security and good sensitivity of the watermarking scheme through
keyed-hashing for message authentication. The latter attaches mor
importance to sensitivity and efficiency by the non-negative matr
```

(b)

Fig. 1. Effectiveness experiment. (a) Source code of an original web page. (b) The web page compiled by IE from (a).

4 Experiment Results

We conduct a series of experiments on the effectiveness, sensitivity and time-efficiency of our proposed schemes, all of the experiments are based on the same machine configuration.

4.1 Effectiveness

Firstly, we testify the effectiveness of the proposed schemes. As mentioned in the previous section, we use the HMAC function and NNSC algorithm to generate a fragile watermark from a web page, respectively, and the watermark is then embedded into the web page. Whenever the document is tampered, the watermark in it will be destroyed or become inconsistent with the content. Thus the tampering can be detected.

```
<HTMl><hEAd>
<META http-equiv="Content-Language" content="zh-cn">
<TiTle>Web pages watermarking</TItle></HeAD>
<bOdY><P><a href="http://www.watermarkingworld.org/">
Watermarkingworld 2007</a></P>
<P>Fragile digital watermarking: <Br>
is a well-known method for authentication of digital informat
```

(a)

```
<HTML><hEaD>
<Meta http-equiv="Content-Language" content="zh-cn">
<titlE>Web pages watermarking</tItLE></hEad>
<BODY><p><A href="http://www.watermarkingworld.org/">
Watermarkingworld 2007</A></P>
<p>Fragile digital watermarking: <bR>
is a well-known method for authentication of digital informat
```

(b)

Fig. 2. Effectiveness experiment. (a) Source code of watermarked web page of Fig. 1 (a) based on MAFW. (b) Source code of watermarked web page of Fig. 1 (a) based on SCFW.

To demonstrate that our proposed schemes are effective, we carried out a simple experiment with an artificial web page. Fig. 1-3 presents the experimental results. Fig. 1 (a) is the source code of the original web page, and it can be compiled to the web page Fig. 1 (b). The watermarked web pages based on MAFW and SCFW are shown in Fig. 2 (a) and (b), respectively. We can see that watermarks are embedded in the upper and lower cases of the letters in HTML tags and the embedding algorithm does not increase the file size of web

```
<HTMl><hEAd>
<META http-equiv="Content-Language" content="zh-cn">
<TiTle>Web pages watermarking</TItle></HeAD>
<bOdY><P><a href="http://www.watermarkingworld.org/">
Watermarkingworld 2006</a></P>
<P>Fragile digital watermarking: <Br>
is a well-known method for authentication of digital informat
```

(a)

```
<HTML><hEaD>
<Meta http-equiv="Content-Language" content="zh-cn">
<titlE>Web pages watermarking</tItLE></hEad>
<BODY><p><A href="http://www.matermarkingworld.org/">
Watermarkingworld 2007</A></P>
<p>Fragile digital watermarking: <bR>
is a well-known method for authentication of digital informat
```

(b)

Fig. 3. Effectiveness experiment. (a) Tampered source code of watermarked web page from Fig. 2 (a). (b) Tampered source code of watermarked web page from Fig. 2 (b).

pages from Fig. 2 (a)(b). Finally they all can be compiled to the same web page Fig. 1 (b) by IE. Fig. 3 (a) and (b) are the arbitrary tampered source code of watermarked web page Fig. 2 (a) and (b), respectively. They all can be validated by our two schemes. Even the tampering only makes a little change of number or address of the web pages, the validating process of the proposed schemes can find out both of these attacks.

4.2 Sensitivity

In order to test the sensitivity of the schemes, we firstly construct two artificial data sets of web pages. The Dataset1 is composed of one hundred web pages which are tampered as follows: randomly select a line L in the watermarked source code of a web page, then select a random character of the line L, and finally alter it into another random character. For example, the watermarked source code of the web page in Fig. 2 (a) is tampered to the Fig. 3 (a). The Dataset2 also consists of one hundred web pages, in which two places of the watermarked source codes are tampered.

Table 1. Sensitivity of the three schemes on the Dataset1 and Dataset2

	Probability of successful detection	
	Dataset1	Dataset2
PCAW	85%	98%
SCFW	99%	100%
MAFW	100%	100%

We use the validating algorithms of the three schemes to detect the tampering. Because of the diffusion property of hash function, the MAFW scheme can detect all the tampering in the two data sets. In the SCFW scheme, the operation of convolution is used to diffuse the data of the web page, and the features can be selected more than those of the PCAW scheme because of the sparseness of coefficient matrix, so we also have a high probability of successful detection. Table 1 shows the results of the successful detection probability, and the results show that the MAFW and SCFW schemes are very sensitive and they are more sensitive than PCAW scheme as analyzed in the previous section.

4.3 Efficiency

In order to testify the efficiency of the proposed schemes, we have done a series of experiments on BankSearch Dataset [15], which is a dataset containing 11,000 web pages. Although this dataset was designed for unsupervised clustering experiments, it can be used in our research for its randomly sampling from the Internet. For more information see the web page [15].

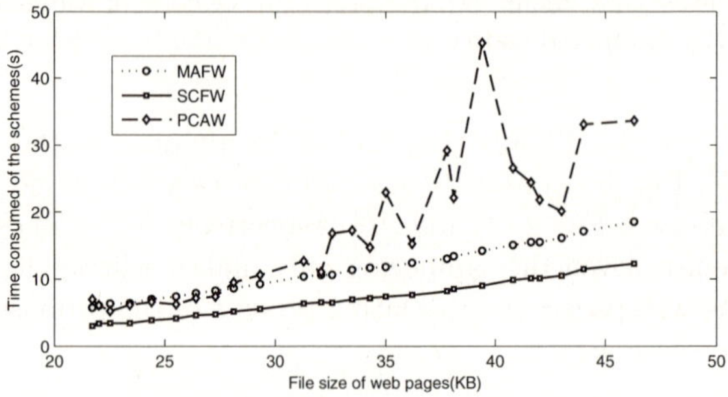

Fig. 4. Time consumed of the three schemes

Fig. 4 shows the time spent on watermark generating and embedding with the different file size of web pages which are randomly sampled from the BankSearch Dataset. Because the time of the three schemes is almost same for the small web pages and the time of our two schemes is much lower than that of the PCAW schemes for the prodigious web pages, we only show the result of web pages from around 20 KB to around 50 KB in Fig. 4. We can see that the time spent on our two schemes is almost linear to the file size of web pages, but for the PCAW scheme, the time-consumed is very fluctuant because of the watermark generating algorithm. We also can see that the two schemes have a better time-efficiency than PCAW for the big web pages from Fig. 4.

5 Conclusions

We have presented two fragile watermarking schemes for tamperproof web pages. The MAFW scheme has the best security, and successfully detects all alterations made to web pages. The second approach, SCFW, has good security, better sensitivity and efficiency. They all have non-destructivity to web pages.

The SCFW scheme can be a good practical tool for tamperproof web pages. But for applications requiring high security and sensitivity, the MAFW scheme may be preferred to SCFW and PCAW. The two schemes can be easily modified to use in text document for authentication and other Markup Language for tamperproof, such as XML(Extensible Markup Language). In the future work we will give the analysis of security and sensitivity of the schemes.

Acknowledgment

This work was supported by NSFC under project No.60573033, Program for New Century Excellent Talents in University (NCET-05-0397) and Research Fund for the Doctoral Program of Higher Education of China (No.20050248048). The authors would like to thank the reviewers for their constructive advice, and thank StatLib and Banksearchdataset.info for the BankSearch dataset.

References

1. Podilchuk, C., Delp, E.: Digital watermarking: algorithms and applications. Signal Processing Magazine, IEEE 18(4), 33–46 (2001)
2. Barreto, P., Kim, H., Rijmen, V.: Toward secure public-key blockwise fragile authentication watermarking. Vision, Image and Signal Processing, IEEE 149(2), 57–62 (2002)
3. Li, C., Lou, D., Chen, T.: Image authentication and integrity verification via content-basedwatermarks and a public key cryptosystem. In: IEEE Int. Conf. Image Processing, vol. 3, pp. 694–697 (September 2000)
4. Yeh, F.H., Lee, G.C.: Content-based watermarking in image authentication allowing remedying of tampered images. Optical Engineering 45(7) (2006)
5. Zhao, Q., Lu, H.: A PCA-based watermarking scheme for tamper-proof of web pages. Pattern recognition 38(8), 1321–1323 (2005)
6. Zhao, Q., Lu, H.: PCA-based web page watermarking. Pattern Recognition 40(4), 1334–1341 (2007)
7. Jolliffe, I.: Principal Component Analysis (2002)
8. Menezes, A.: Handbook of Applied Cryptography. CRC Press, New York (1997)
9. Hoyer, P.O.: Non-negative matrix factorization with sparseness constraints. Journal of Machine Learning Research 5, 1457–1469 (2004)
10. Stinson, D.: Cryptography: Theory and Practice. CRC Press, Boca Raton (2002)
11. Powell, T.: References of HTML. McGraw-Hill Education Co., New York (2002)
12. Katzenbeisser, S., Petitcolas, F.: Information Hiding Techniques for Steganography and Digital Watermarking. Artech House, Inc., Norwood (2000)
13. Lee, D.D., Seung, H.S.: Learning the parts of objects by non-negative matrix factorization. Nature 401(6755), 788–791 (1999)
14. Liu, W., Zheng, N., Lu, X.: Non-negative matrix factorization for visual coding. In: IEEE Int. Conf. Acoustics, Speech, and Signal Processing (ICASSP 2003), vol. 3, pp. 293–296 (2003)
15. Sinka, M.P.: Banksearchdataset.info, `http://www.banksearchdataset.info/`

Real-Time Short-Term Traffic Flow Forecasting Based on Process Neural Network

Shan He[1,⋆], Cheng Hu[1], Guo-jie Song[1,⋆⋆], Kun-qing Xie[1], and Yi-zhou Sun[2]

[1] Key Laboratory of Machine Perception (Peking University), Ministry of Education, Beijing, 100871
hhh@pku.edu.cn, {hucheng,gjsong,kunqing}@cis.pku.edu.cn
http://www.cis.pku.edu.cn

[2] Department of Computer Science, University of Illinois at Urbana-Champaign, Urbana, IL, USA
sun22@uiuc.edu
http://www.cs.uiuc.edu

Abstract. Existing short-term Traffic flow forecasting models have not fully considered the characteristics of spatio-temporal process and online analysis, we imported the process neural network which can model spatio-temporal process well into short-term traffic forecasting. The model use wavelet radix as weighted function expanding radix of process neurons to deal with the inputs on multi-scale. By using principal component analysis to consider the space affect of traffic flow, the model was optimized. In addition, online learning algorithm of the model was proposed. The experimental results show that the forecasting accuracy of the model is better than ordinary neural networks, and the model can meet the demand of real-time forecasting of short-term traffic flow.

Keywords: Process neural network, Short-time traffic flow forecasting, Traffic engineering.

1 Introduction

The traffic congestion has become a worldwide problem of the development of economy and society. The Intelligent Transportation System (ITS) may effectively resolve the problem. Being the center part of ITS, Dynamic Route Guidance System can provide "Induced Travel", which may achieve the purpose to reduce the travel time, to relax the traffic congestion and to save energy. Comparing with apperceiving and gathering the real-time traffic data, the real-time, correct and reliable traffic flow forecasting is more important.

The traffic system is an uncertain, time-varied and nonlinear complex system. Along with the time for forecasting, the probability of events taking place will

---

⋆ The research is sponsored by the National Natural Science Foundation of China under Grant No. 60703066 and supported by the National High-Tech Research and Development Plan of China (863) under Grant No.2006AA12Z217 and supported by the Wiser Foundation of IDC-Peking University (No. W08SI07).

⋆⋆ Correspondent author.

F. Sun et al. (Eds.): ISNN 2008, Part II, LNCS 5264, pp. 560–569, 2008.

mount up. This makes the forecasting more difficult. The short-term traffic flow forecasting with a time of 5 to 30 minutes has drawn more and more attention for its higher veracity. Since the change of short-term traffic flow is ill-defined and it is greatly influenced by uncertain factors, short-term traffic flow forecasting is hardest. The real-time and reliable short-term traffic flow forecasting has been one of the hotspots and difficulties of ITS research.

The problem of short-term traffic flow forecasting is to determine the traffic flow data in the next time interval, usually in the range of 5 min to half an hour [1]. The research of short-term traffic flow forecasting can be classified into three categories: (1) models based on statistic, such as switching autoregressive integrated moving average (ARIMA) model [2], time series methods [3], Kalman filter theory [4], etc. These methods use simple models and easily to be comprehended, but all of them are built on a linear basis. Because of the linear basis all these models can not image the uncertainty and nonlinearity of the traffic, and can not deal with stochastic noises. All of these drop the forecasting effect of the methods. (2) methods of traffic simulation based on dynamic network assignment. The methods can get the forecasted traffic flow information and reflect the relations of traffic variables, but they over emphasize exact optimum assignment result of the system or the user and this makes the models more complex to be solved. Because the optimizing process is very time-consuming the methods can not do real-time forecasting and use in a large road network. (3) Netural networks [6][7][8] as the representative of the artificial intelligence methods. Because of their characteristic of recognizing nonlinear complex system and the black-box learning method netural networks have become the very widely used models of traffic flow forecasting. In the traditional netural network models all the inputs to systems are constants that are independent of time procedure, but the inputs of real traffic system are functions (or processes) dependent on time. Existing models that aim at dynamic signal processing, e.g. time delay netural network [9], dynamic recurrent netural network [10], etc. all use the external time delay to achieve the delay between inputs and outputs and can not well depict the process character. The precision of the models are greatly limited.

The data that are detected by all kinds of the traffic sensors have obvious characteristic of real-time, dynamic stream and continuous process. According to analysis, existing researches can not well meet the demand of the real-time, reliable and adaptive traffic flow forecasting system. It is unbefitting to all appearances that using the traditional modeling methods and forecasting technologies that based on static data environment to deal with the real-time, dynamic and variational data stream scene. In this paper, we imported a short-term traffic forecasting model which based on the data stream and the process neural network (PNN) [11][12]. In the forecasting model, we use PNN capturing the process character of the traffic flow, use wavelet transform and principal component analysis to considered the characteristics of spatio-temporal process of traffic flow series. The model was optimized to ensure the veracity. By using model online learning based on data stream technology we ensure the forecasting to be real-time and adaptive. The experimental results show that the forecasting accuracy of the model is excelled than neural networks.

In the second section of this paper, the rationale of the forecasting model is described. In the third section the analysis of the model and the experiment example is considered. Last, the fourth section concludes the paper.

2 Rationale of the Forecasting Model

2.1 Process Neural Network

In 2000, He Xingui proposed a novel neural network model - Process Neutral Network (PNN). In recent years, PNN has been greatly developed and well applied in many fields [13][14]. PNN is generalized model of traditional neural networks. The inputs to system relax synchronization instantaneous conditions, which limit inputs in the traditional neural network models. Therefore, PNN makes neural network models more fit to the practical situations and many applied problems can come down to this kind of model.

Process neural consists of four parts: time-varied process inputs, space aggregation operation, time aggregation operation, energizing threshold and energizing outputs. A single process neuron's structure is shown in Fig.1.

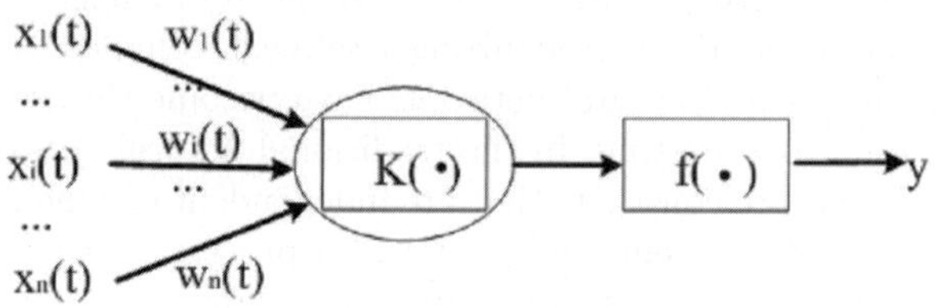

Fig. 1. Process neural network model

In Fig.1 where $x_i(t)$ (i=1,2,...,n) are time-varied functions inputted to process neuron. While $w_i(t)$ are corresponding time-varied weight functions. $K(\cdot)$ is the aggregating kernel function that do some transform process to the input signals. $f(\cdot)$ is the transfer function. The process neuron has a converge operation includes the multi-inputs converge not only to space but also to time.

The inputs, outputs of PNN and the connection weighting can all be the functions of time. Because the form of network connection weighting and the parameter contained are random, the connection weighting function is very hard to be confirmed by the training. The form of connection weighting functions must be limited. The connection weighting functions can be expanded on a group of certain base functions. Using a group of base functions $B(t) = b_1(t), b_2(t), \ldots, b_L(t)$, L is the number of base functions. So, $w_i(t) = \sum_{i=1}^{L} w_{ij} b_l(t)$, where w_{ij} is the expansion coefficient of connection weighting function $w_i(t)$ to the base function $b_l(t)$.

For simplicity, suppose space-aggregation operation is sum, time-aggregation operation is integral, $K(\cdot) = 1$, PPN model can be expanded as following.

$$y = f(\sum_{i=1}^{n}\sum_{l=1}^{L} w_{ij} \int_{0}^{T} b_l(t)x_i(t)dt - \theta). \quad (1)$$

2.2 Process Neural Network Based on Wavelet Transform

The feature extraction of traffic flow may sometimes drop the performance of PPN model. Meanwhile, considering the traffic flow is a dynamic and non-stationary time series and has the period of one week or one day. This paper chose wavelet radix that can extract feature of time series data on multi-scale as weighted function expanding radix. The traffic flow series were expanded on the wavelet radix that ensure not only the veracity but also the high performance of the model. In formula (1), $\int_0^T b_l(t)x_i(t)dt$is the expansion coefficient of input function $x_i(t)$ to the base function $b_i(t)$, and we marked it as c_{il}. w_{il} can be trained using learning sample set by ordinary methods such as the gradient descent algorithm.

In the road networks, there is space influence between different nodes. Usuallythere are many nodes that are correlative with traffic flow of the current node and may offer information. Building single forecasting model for every node can not well use the space-characteristic of traffic data. Therefore, this paper built a model for the whole road networkwhich we called network PNN (NPNN). We import space operation into the model. This model well imaged the space and procedure characteristic of traffic flow data. The model is shown in Fig.2.

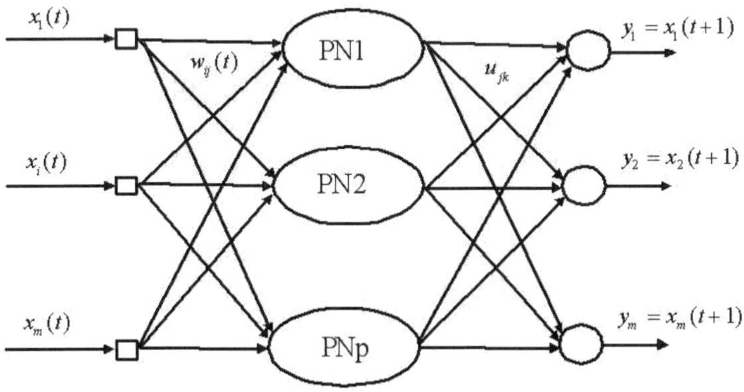

Fig. 2. Topological structure of NPNN model

In the model, the output of the *k-th* loop detector is:

$$y_k = g(\sum_{j=1}^{p} u_{jk} f(\int_{t=t_0-T-1}^{t_0} (\sum_{i=1}^{m} w_{ij}(t)x_i(t))dt - \theta_j) - \theta'). \quad (2)$$

In formula (2), here $x_i(t)$ is the input time series of the *k-th* loop detector, and T is the time window of wavelet transform.

2.3 Model Optimization

The NPNN model for the whole road network is too complex to meet the demand of real-time forecasting because its cost of training and forecasting is too high. Using principal component analysis to consider the space of the loop detectors, this paper optimized the NPNN model and reduced the scale of network.

Let m loops compose m-dimension random variable X, the traffic parameters of every time is an observation value of X. According to n observation value, we use samples to solve principal component. Supposing the *i-th* eigenvalue of X is λ_i (sort ascending), the information with the largest p principal components can be measured by degree of cumulate contribution:

$$\sum_{k=1}^{p} \lambda_k / \sum_{i=1}^{n} \lambda_i.$$

Choosing the first p principal components order by degree of cumulate contributionthe original m-dimension data $X = (x_1, x_2, \ldots, x_m)$ transform to p-dimension vector $Y = (y_1, y_2, \ldots, y_p)$. Every components of Y is linear assembly of different loops.

We use the space radix getting with principal component analysis to optimize the model. According to the number of principal componentwe decide the number of process neuron in the hidden layer. Every process neuron deal with the corresponding principal component.

The input of the j-th process neuron is the j-th principal component, which is $PCA_j(t) = \sum_{i=1}^{m} a_{ij} x_i(t)$, here a_{ij} is the conversion coefficient of the i-th component and the j-th principal component. So, the local induced domain of the j-th process neuron is expanded on the wavelet radix

$$O_j(t) = \sum_{l=1}^{L} \int w_j^{(l)} \sum_{i=1}^{m} a_{ij} x_i(t) b_l(t))dt - \theta. \tag{3}$$

Comparing with formula (1), we can know $w_{il} = a_{ij} \cdot w_j^{(l)}$. This is a process that the connection weighting is decomposed by space radix. The number of parameters is reduced. The outputs from the hidden layer to the output layer can be considered as that the coordinate on the space radix is converse transform to the original coordinate. So, in formula (2), $\mu_{jk} = a_{kj}$. The parameter that need to be trained in the whole model is $w_j^{(l)}$, j= *1, 2,. . . ,p;* l = *1, 2,. . . ,* L, and its number is $p \times L$.

2.4 Online Learning

In the strict sense of the word, the traditional forecasting models such as neural network have less capacity of online learning. These models were trained by the static sampling traffic data set, then began to apply. The real traffic flow is dynamic, so the model must keep adaptively adjusting and learning online to meet the demand of real-time and adaptive forecasting.

The form of learning sample is $\{< X_{t0}(t), Y_{t0}(t) >\}, t_0 \in N$, here $X_{t0}(t)$ is the time series of the time window corresponding the t_0 time. Y_{t0} is the values of traffic flow parameters in t_0+1 time. Every step of time window moving forward product a new sample, and the model learns by the new sample. Because there only need to store the time series of the current time window and only need to retrieve just once, the model suit for the environment of data stream.

Every time, model forecasting the parameters of next time according to the input functions of the current time window. From formula (2) we can know that the input functions must do wavelet transform in time for forecasting. Every new time, the time window steps forward and there need the wavelet coefficients of last time input functions and the wavelet coefficients of the current traffic data to online refresh the weighting function coefficients parameter w_{il}. Therefore, the incremental computation of wavelet coefficients is very important to real-time computation and online learning of the model.

This paper chose the most popular Harr Wavelet [15] for the wavelet transform and provide a method to incremental compute the wavelet coefficients. Considering *à* trous discret wavelet transform [16], to the time series $\{X_t\}$,

$$x_t = c_{J,t} + \sum_{j=1}^{J} w_{j,t}. \tag{4}$$

Here $c_{j,t}$ is the low frequency part after discret of series. $w_{j,t}$ is the high frequency part of the scale. So there are the recursion formulas of the wavelet coefficients

$$c_{j+1,t} = \frac{1}{2}(c_{j,t} + c_{j,t-2^j}). \tag{5}$$

$$w_{j+1,t} = c_{j,t} - c_{j+1,t}. \tag{6}$$

According to formula (5) and (6), when the time window refreshed, only all the current scale coefficients need to be refresh. This problem have a complex rate of $O(J)$, here J is the largest considered scale. The incremental computation of wavelet coefficients can be done by just storing all the wavelet coefficient values of the J layer in the last 2^J time. If the values of last $2^J + 1$ time were stored, the time window with the length of 2^{J+1} could be described.

3 Model Application and Experiment Analysis

3.1 Experimental Data

The PeMS system [17] of University of California is a performance measurement system of freeway, which purpose is to collect and to deal with the loops' detecting data of freeway. The system provide two series of freeway traffic data in different time distances. One time distance is 30 seconds, the other is 5 minutes.

The data contained by the system such as traffic volume, occupancy and speed can be real-time downloaded. This paper chose six weeks traffic volume data in 5 minutes time distance of the District 7 from 2007/2/8 to 2007/3/21.

3.2 Length of Time Series

The traffic flow length of the model may greatly influence the veracity and efficiency of forecasting. On one hand, since the traffic data have a periodicity of one week long, the wrong choosing of time window will make the long-period information lost, then may influence the veracity of forecasting. On the other hand, the longer time window we chose, the larger influence historical data had. To the high-time-varied traffic network, because the influence of short-term uncertainty may be weakened, the veracity of forecasting may be influenced, too. In addition, with the limit of wavelet transform, the time window must be a power of 2. Considering synthetically, this paper set the time window longer than a week and containing the number of time points like power of 2.

3.3 Parameters

This paper chose the feed-forward PNN with single hidden layer. After data preprocessing, there are 2500 high quality loop detectors, so the inputs are 2500 time series. Using the data of a week from 2007/2/22 to 2007/2/28 as the observational samples, computing the principal components, the cumulate contribution of first 17 principal components have come to 90%, and the cumulate contribution of first 146 principal components have come to 95%. Choosing the first 100 principal components, the number of hidden layer is 100. In traffic flow forecasting, the nearer scale detail information may have bigger help to current forecasting. This paper chose only the nearest 2 wavelet coefficients of every scale layer, so there were only $2 \times 11 = 22$ weighting function radixes. The number of the parameters that need to be trained in the model was only $100 \times 22 = 2200$. When online learning, we use Widrow-Hoff learning rule to reduce the training error, and use the difference between the forecasting values and the real values of the new time to refresh the model.

Using the online learning algorithm, we did online forecasting with two weeks traffic volume data from 2007/2/22 to 2007/3/8, containing 4032 time distances. The transit-time is 1983 time distances.

3.4 Model Comparison

For validating the excellence of the NPNN model, we compared the forecasting performance of NPNN model with single node artifical netural network (ANN) model and single node process netural network (SPNN) model.

Fig.3 shows the traffic volume forecasting result of a loop detector in one day. We can see that the NPNN model and the SPNN model both can image the trend of traffic volume varying with time better than traditional ANN method.

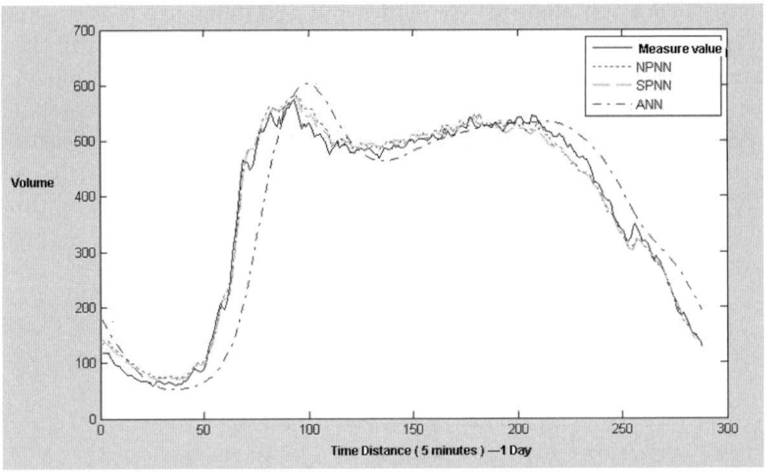

Fig. 3. Accuracy comparison within one day

For result comparison, let us consider the following two targets: mean absolute percent error (MAPE) and root of mean square error (RMSE).

$$MPAE = \frac{1}{n} \sum_{t=1}^{n} \mid \frac{y_t - \hat{y}_t}{y_t} \mid . \qquad (7)$$

$$RMSE = \sqrt{\frac{1}{n} \sum_{t=1}^{n} (y_t - \hat{y}_t)^2}. \qquad (8)$$

In formula (7) and (8), y_t is the real observational value of traffic volume at t time, $\hat{y}_t$ is the forecasting value of traffic volume at t time, n is the number of time distances containing in the forecasting period. Selecting 50 loop detectors randomly, computing the mean of MAPEs and the mean of RMSEs of the forecasting result of the 50 loop detectors in the two weeks, we can get Table 1. Because the real traffic volume data may be 0 or very small in some time, MAPE can not well weigh the performance of model. But it can use to compare the models. From Table 1, we can see that forecasting performance of NPNN model is obviously better than the ANN model. MAPE and RMSE of NPNN

Table 1. Comparison of forecasting performance

	ANN	SPNN	NPNN
MAPE	0.2268	0.1403	0.1645
RMSE	38.622	26.185	28.04

model are both worse than SPNN model, but the difference is very small. Although the principal component analysis may reduce the complexity of the model and may consider the space affect, the precision of the forecasting model would fall with the length of forecasting period because the principal component of time-varied traffic data may change with time. For real-time forecasting, the forecasting time is an important performance indicator. Using NPNN as unit time, the comparison of forecasting time of three models is shown as Table 2.

Table 2. Comparison of forecasting time

	ANN	SPNN	NPNN
Time	76.7	24.9	1

Comparing with the other two models, the NPNN model can save a great deal of time. Although the optimized NPNN model abnegated a part of forecasting precision, the model can achieve better forecasting result for the non-stationary and non-linear traffic flow series. The model did not substantially influence the real application. The suitable forecasting precision and the high forecasting speed meet the demand of the short-term traffic flow forecasting.

4 Conclusions

The generation of traffic flow is a spatio-temporal process, this paper used a forecasting model based on data stream and PNN to forecast the short-term traffic flow of the next 5 minutes. The model made sufficient use of the characteristic of time series of traffic flow. Meanwhile, by using the space affect the model was optimized. The experimental results show that the whole forecasting performance of the model is better than ordinary neural networks, and the model can save the forecasting time to meet the demand of real-time forecasting of short-term traffic flow. Because of the time-varied traffic data, along with the time process, the principal components of traffic data may change continually. The next stress of the research is how to find an incremental computation method of the principal component by using the characteristic of traffic flow to achieve the adaptive update of the model.

References

1. Sun, S.L., Zhang, C.S.: The Selective Random Subspace Predictor for Traffic Flow Forecasting. IEEE Transanctions on Intelligent Transportation System 8(2), 367–373 (2006)
2. Yu, G.Q., Zhang, C.S.: Switching ARIMA Model Based Forecasting for Traffic Flow. In: IEEE International Conference on Acoustics, Speech, and Signal Processing, pp. 429–432. ETATS-UNIS, Monographie (2004)

3. Moorthy, C.K., Ratcliffe, B.G.: Short Term Traffic Forecasting Using Time Series Methods. Transp. Plan. Technol. 12(1), 45–56 (1988)
4. Okutani, I., Stephanedes, Y.J.: Dynamic Prediction of Traffic Volume through Kalman Filter Theory. Transp. Res., Part B. Methodol. 18(1), 1–11 (1984)
5. Moshe, B., Michel, B.: DynaMIT: A Simulation-based System for Traffic Prediction. Massachusetts Institute of Technology ITS Program (1998)
6. Sun, S.L., Zhang, C.S., Yu, G.Q.: A Bayesian Network Approach to Traffic Flow Forecasting. IEEE Trans. Intell. Transp. Syst. 7(1), 124–132 (2006)
7. Park, B., Messer, C.J., Urbanik, T.: Short-term Freeway Traffic Volume Forecasting Using Radial Basis Function Neural Network. Transportation Research Record: Journal of the Transportation Research Board 1651, 39–46 (1998)
8. Abdulhai, B., Porwal, H., Recker, W.: Short-term Freeway Traffic Flow Prediction Using Genetically Optimized Time-Delay-based Neural Networks. California Partners for Advanced Transit and Highways (PATH). Working Papers: Paper UCB-ITS-PWP-99-1 (1999), `http://repositories.cdlib.org/its/path/papers/UCB-ITS-PWP-99-1`
9. Waibel, A., Hanazawa, T., Hinton, G., Shikano, K., Lang, K.J.: Phoneme Recognition Using Time Delay Neural Networks. IEEE Trans. ASSP 37(2), 328–339 (1989)
10. Draye, J.S., Pavisic, D.A., Cheron, G.A., Libert, G.A.: Dynamic Recurrent Neural Networks: A Dynamical Analysis. IEEE Trans. SMC(B) 26, 692–706 (1996)
11. He, X.G., Liang, J.Z.: Process Neural Network. In: The 16th World Computer Congress 2000, Proceedings of the Conference on Intelligent Information Processing, pp. 143–146. Publishing House of Electronics Industry, Beijing (2000)
12. He, X.G., Liang, J.Z.: Some Theoretical Issues on Process Neural Networks. Engineering Science 2(12), 40–44 (2000)
13. He, X.G., Liang, J.Z.: Process Neural Network with Time-Varied Input and Output Functions and Its Applications. Journal of software 14(4), 764–769 (2003)
14. Song, G.J., Yang, D.Q., Wu, L., Wang, T.J., Tang, S.W.: A Mixed Process Neural Network and its Application to Churn Prediction in Mobile Communications. In: ICDM Workshops 2006, pp. 798–802. IEEE Computer Society Press, Washington (2006)
15. Renaud, O., Starck, J.L., Murtagh, F.: Prediction Based on a Multiscale Decomposition. Internation Journal of Wavelets. Multiresolusion and Information Processing 1(2), 217–232 (2003)
16. Shensa, M.J.: Discret Wavelet Transform: Wedding the Atrous and Mallat Algorithms. IEEE Transanctions on Signal Processing 40(10), 2464–2482 (1992)
17. Berkeley, U.C.: Freeway Performance Measurement System(pems), `http://pems.eecs.berkeley.edu`

Fuzzy Expert System to Estimate Ignition Timing for Hydrogen Car

Tien Ho and Vishy Karri

School of Engineering,
University of Tasmania, GPO Box 252-65, Hobart, Tasmania, 7001, Australia
{Vishy.Karri,ntho}@utas.edu.au

Abstract. This paper presents the application of fuzzy expert system technique as a basis to estimate ignition timing for subsequent tuning of a Toyota Corolla 4 cylinder, 1.8l hydrogen powered car. Ignition timing prediction is a typical problem to which decision support fuzzy system can be used. Based on extensive experiments, the basic fuzzy rules on ignition timing have been constructed, in which the engine speed, throttle position, manifold air pressure, fuel pulse width, engine power, lambda value were chosen as fuzzy sets of the linguistic input variables, and ignition advance is selected as performance output of the fuzzy system. The constructed fuzzy system initially mapped 136 basic rules based on physical theories and extensive experimentation. For all the input parameters various triangular, trapezoidal and generalized bell-shaped membership functions were successfully applied to best represent the ignition timing output from the expert system. The results have shown that the minimum ignition advance for maximum torque without detonation was achieved. The estimation of ignition advance achieved from fuzzy expert system was ± 5% root mean square error.

Keywords: Hydrogen powered car, Ignition timing, Ignition advance, Fuzzy expert system, Hydrogen engine tuning.

1 Introduction

The use of fossil fuels leads to harmful emissions which have detrimental effect on climate changes and contributes towards green house effect. Most of the fossil fuels when burnt produce carbon monoxide, carbon dioxide, nitrous oxides, sulphur oxides and other harmful emissions. The pollutions from automobiles also have numerous health and environmental impacts, including urban smog and the green house effect. In order to minimize the environmental damage, it is necessary to ensure reduced harmful emissions. With limited fossil fuel resources and their depletion in the near future, the selection and use of alternative fuels is becoming even more imperative [1].

Hydrogen is an abundant element and unlike most other energy carriers is carbon free. Hydrogen is investigated as alternative fuel for internal combustion engines and fuel cell systems. It is also shown that Hydrogen has the potential for automotive applications and stationary applications such as powering generators and motors [1].

While the combustion 'know how' of a gasoline engine is a very established science, little or no information, in the public domain, is available that comprehensively

F. Sun et al. (Eds.): ISNN 2008, Part II, LNCS 5264, pp. 570–579, 2008.

explains the H2 powered internal combustion (IC) engines. Any matured knowledge related to H2 IC engines is in the domain of major automotive companies such as BMW, Mazda, Toyota and more recently Ford [1].

The economic justification for building a full-fledged H2 IC engine plant for automotive applications is questionable with current prices associated with H2 production and unit production cost. Nevertheless, a good understanding and parallel progress of the H2 IC engine technology is essential when the unit cost of H2 production is affordable. One of the major aspects of IC engine running on hydrogen is to build appropriate engine tuning maps for ignition timing and injection timing for smoother knock free combustion. This aspect is particularly exasperated when tailor made engine management systems are to be integrated with the H2-IC engines without prior matrix of data to refer for online tuning. While the process of exhaustive experimentation and subsequent tabulation are time-consuming, an intelligent system to estimate ignition timing will be of great use in fine tuning of these engines. In this work, a fuzzy expert system is proposed that uses the on-line engine parameters such as engine speed, throttle position, manifold air pressure, fuel pulse width, engine power and lambda values were chosen as inputs to estimate ignition timing. The following section details a brief description of H2 powered car as the test rig.

2 Brief Description of Hydrogen Powered Car as Test Rig

A Toyota Corolla was successfully converted to use hydrogen as a fuel in its internal combustion engine. Certain characteristics of hydrogen make it unique for application as an automotive fuel. The wide flammability limits of hydrogen allow for a larger range of air to fuel mixtures to be used at different engine operating conditions. This means that very lean mixtures may be used for lower emissions while enriched mixtures could be used when additional power is required [2]. Hydrogen also has a very high flame propagation rate even with lean mixtures providing a very sharp rise in pressure immediately after spark ignition [3]. The combination of the ability to run at very lean mixtures and fast flame propagation allow hydrogen engines to run very efficiently. However, there is an overall decrease in power output when using lean mixtures [2]. In addition, the fast burn characteristics of hydrogen also enables it to be able to operate well at higher engine speed while its gaseous form and ease of combustion can help when performing engine cold starts [2]. In general, hydrogen has the following dedicated advantages over gasoline [4]: reduce engine oil dilation, reduce engine wear, reduce the emissions as well as increase the fuel economy.

The modified hydrogen powered car was installed with new engine control unit (ECU) M400 from Motec. Initially, the basic tuning of ignition timing was figured out based on the initial "detune" data from supplier. The final tuning of ignition timing data from the hydrogen-powered car that probably would have correlated well with the "detune" data and saved considerable time. Therefore, the basic rules of fuzzy expert system were based on extensive experimental database, which was recorded from supplier basic "detune" file.

With the complexity of the configuration and tuning techniques of an aftermarket ECU for a new fuel such as hydrogen, the work to make the engine to start properly is a very difficult task. Meanwhile in the basic terms, the providing of the correct amount of hydrogen at the right time with appropriate ignition timing needs all of ECU functions to work correctly. Idle condition was seemed to be the most difficult task for tuning for hydrogen engine which requires load to be placed on. Moreover, the mixture of air/fuel ratio or fuel quality needs to be measured to keep the mixture is neither too rich nor too lean which are the main reason of exhaust gas excessive heat or severe engine damage and increasing NOx emission gases [5]. A best tuning engine will show the best performance for any given conditions including rich mixture at high power, lean or stoichiometric when idle or slow speed. Beside that, hydrogen injection systems should be tuned correctly to precisely control the mixture air/fuel ratio to suit different operating conditions. Therefore, this has online instrumentation to measure the input parameters were all installed for subsequent data acquisition and modelling.

3 Brief Description of Record Data Method and Developed Fuzzy Expert System

Much effort was made to reduce the number of required linguistic input variables parameters as well as simplify the data acquisition process for developed ignition advance fuzzy expert system. Although fuzzy expert system can handle several linguistic inputs, it was considered to reduce redundant inputs so that the results could give a clearer effect of individual linguistic membership function on ignition timing. In order to acquire data for database to construct basic rules of fuzzy inference system, a list of engine parameters as well as the sensors required measuring the desired parameters that were recorded as listed in Table 1 below.

Table 1. Engine parameters and appropriate source of measurement sensor

Parameter	Designation	Sensor/Source
Engine speed	RPM	Stock ECU
Throttle Position	TP	Stock ECU
Manifold Air Pressure	MAP	MAP Sensor via Stock ECU
Fuel Actual Pulse Width	FAPW	Motec ECU lookup table
Lambda	La1	Lambda Sensor via PLM
Output Power	PWR	Dynamometer
Ignition Advance	IA	Motec ECU lookup table

The problem faced with ignition timing tuning data is the ability for the user to accurately and quickly classify a particular value of ignition advance without having to undertake a detailed analysis, which requires a thoroughly experience knowledge as well as time consuming. A solution to this dilemma is to develop a system where the large majority of commonly found situation on hydrogen engine operating conditions are defined. The ability to provide this fuzzy system to professional tuners and amateurs alike would provide them with a valuable tool for ignition timing tuning. By filling the right number of engine operating parameters to gain specific information, the ignition timing can be identified quickly and easily. Using a fuzzy expert system allows the user to assemble all their "rules of thumb" into one easy database of knowledge, as well as having the option of refining, changing and adding rules to enlarge, or specify the program.

The overall domain of the fact tuning a hydrogen engine problem may be considered to be quite complex process, especially considering how many related parameters such as injection timing, valve control, fuel quality (lambda value), engine temperature and air temperature associated with fuel trim functions, etc.

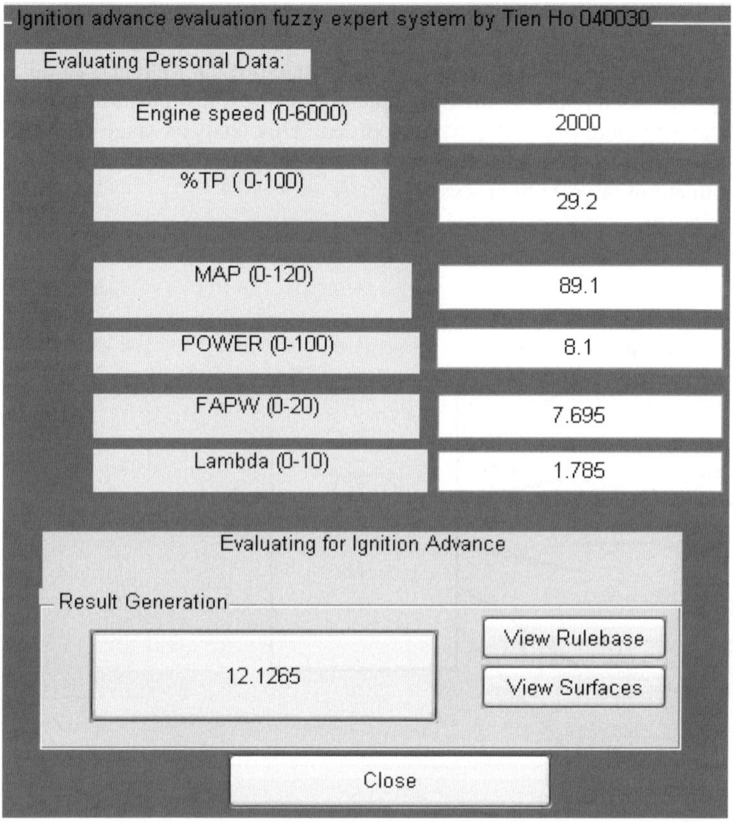

Fig. 1. Ignition timing fuzzy expert system user interface

The development of a ignition timing fuzzy expert system have the friendly user interface as shown in Fig. 1 which is simple enough such that any tuner could use the system to try and identify an ignition advance and then cross reference this result with a proven analysis.

It is quite obvious that the created fuzzy expert system that has been presented here could be expanded to encompass a broader range of data tuning, including injection timing and lambda tuning. The programs could be run independently of each other, or they could be run concurrently so that the information stored in the various databases could be cross-checked, depending on the filled data response to a particular parameter. A system like this would also be invaluable to an interactive hydrogen engine-tuning exhibition where visiting patrons would have the opportunity to become involved in 'being an expert tuner' for a period of time and helping to identify various ignition timing. Therefore, we can say that the particular domain problem selected was very appropriate for fuzzy expert system implementation. There was more than enough information in regard to the ignition advance tuning and also adequate 'expert' information available so that there a basic series of parameters could be filled in order to obtain a specific solution.

Once a suitable problem was identified, research was undertaken to develop knowledge which was thought should to be included in the system. Initially the system was designed on paper by drawing up and drafting a few basic simple rules based on practical knowledge and database recorded. This analysis stage blended into the design stage easily by means of a 'key to develop fuzzy expert system' that was drawn up to aid in the design process.

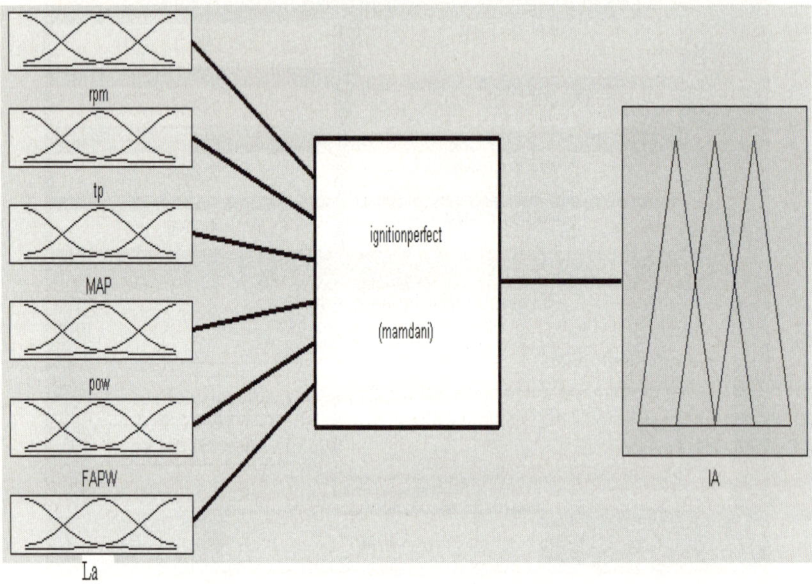

Fig. 2. Developed fuzzy inference system

It was necessary to develop 6 linguistic input variables as shown in Fig. 2 including the engine speed, throttle position, manifold air pressure, fuel pulse width, engine power, lambda value and ignition advance as target output variable for the fuzzy system within the Matlab environment, as some experience was required to understand the operating system. These prototypes also allowed a time to improve the initial design solution that was devised. The constructed fuzzy system was represented by 136 basic rules and concepts base on physical theories and extensive experiments. Triangular, trapezoidal and generalized bell-shaped membership functions were successfully applied to represent the knowledge of the ignition timing tuning expert system. The rule base of developed system is as shown in Fig. 3 below.

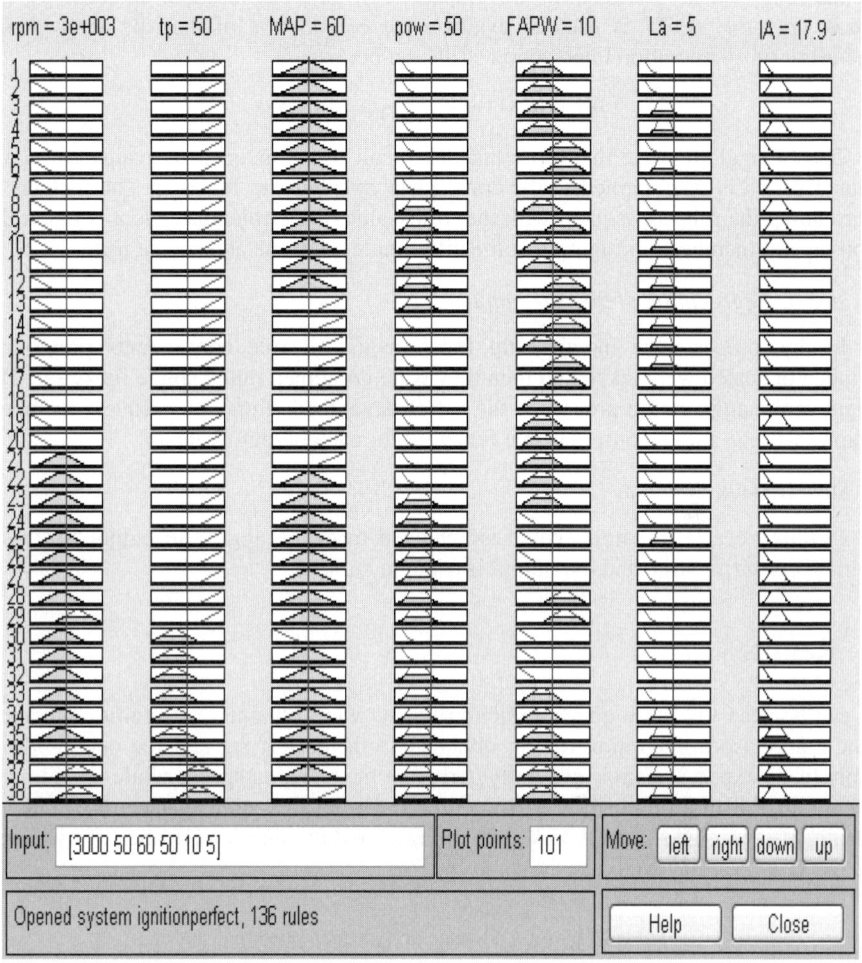

Fig. 3. Rule base of developed fuzzy expert system

The used fuzzy inference technique is the so-called Mamdani method in which the performance process includes four steps: fuzzification of the input variables, rule evaluation, aggregation of the rule outputs, and finally, defuzzification. This method is built on four steps as discussed below [6]:

Step 1: Fuzzification

The first step is to take the 6 crisp inputs including engine speed, throttle position, manifold air pressure, fuel pulse width, engine power, lambda value, then determine the degree to which these inputs belong to each of the appropriate fuzzy sets.

Step 2: Rule Evaluation

The second step is to take the fuzzified inputs, and apply them to the antecedents of the fuzzy rules. Because the developed fuzzy rule has multiple antecedents, the fuzzy operator 'AND' is used to evaluate the conjunction of the rule antecedents, therefore fuzzy operation intersection as shown below:

$$m_{A \cap B}(x) = \min\,[m_A(x), m_B(x)]$$

Then a single number that represents the result of the antecedent evaluation is obtained which is then applied to the consequent membership function. The method of correlating the rule consequent with the truth-value of the rule antecedent is to cut the consequent membership function at the different level of the antecedent truth.

Step 3: Aggregation of the rule outputs

In this process, the membership functions of all rule consequents previously clipped or scaled were taken and then they were combined into a single fuzzy set. The input of the aggregation process is the list of developed clipped consequent membership functions, and the output is one fuzzy set for each output variable.

Step 4: Defuzzification

In this process, the input which is achieved from the aggregate output fuzzy set will be defuzzification and the output is a single number.

4 Results

It can be seen that very good prediction results were obtained for ignition advance. The comparison of ignition timing obtained from the test rig and the one achieved from fuzzy expert system is generally correct at most engine speeds while operating at wide range of throttle position as shown in Table 2. The curve surface of calculated results of ignition advance is as shown in Fig. 4 and the ignition advance of hydrogen car at 75% throttle position is shown in Fig. 5.

Table 2. Ignition advance of hydrogen powered car

		Engine Speed						
Throttle position		1040	1470	2022	3054	4038	5052	5840
99.80%	Fuzzy caculated	7.2	15.4	15.2	18.2	21.2	24.5	26.5
	Experimental	6.9	15.2	14.5	17.8	22.3	23.1	25.4
74.80%	Fuzzy caculated	10.5	19.5	18	20.6	25	27.6	29.8
	Experimental	10.3	20.4	18.4	21.2	24.8	26.7	28.9
50.20%	Fuzzy caculated	13.2	22.1	19.5	23	27.4	30.2	32.5
	Experimental	12.6	21.7	20.3	22	25.9	31.3	31.2
25.80%	Fuzzy caculated	15.7	24.2	22.3	24.5	32.2	33.1	36.2
	Experimental	16.5	25.1	22.8	23.4	33.6	34.1	36.7

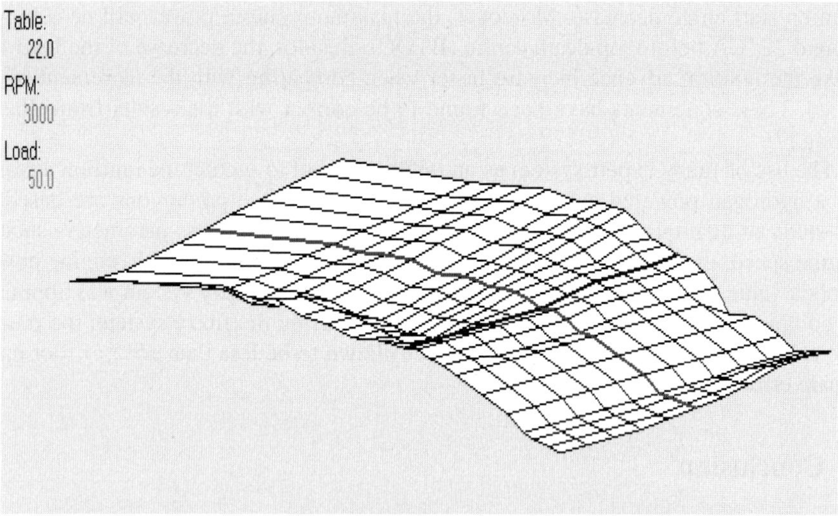

Fig. 4. Curved surface of hydrogen car ignition advance

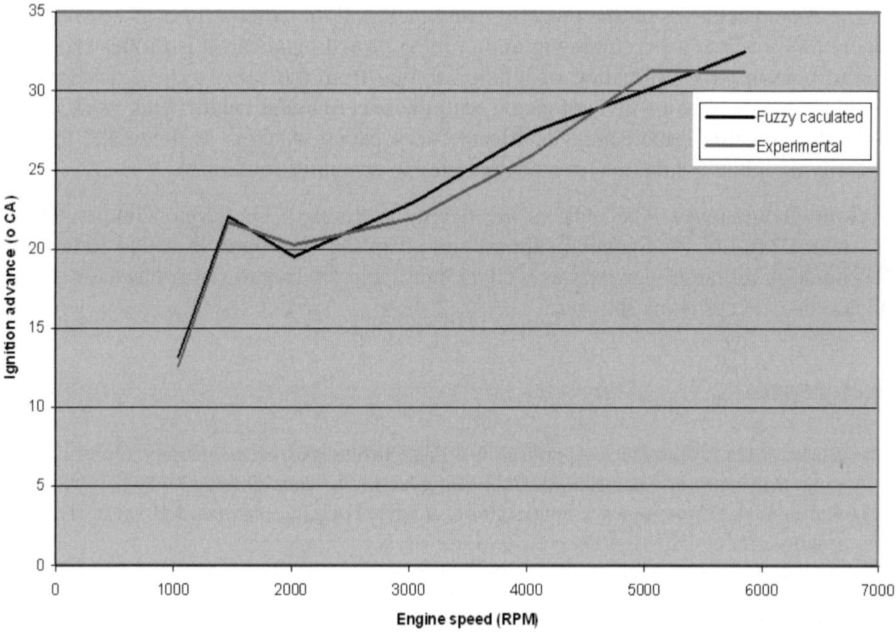

Fig. 5. Ignition advance at 75% throttle position

The created fuzzy expert systems which are mapped a comprehensive range of data variation through many different testing engine conditions. In each of the studied prediction hydrogen powered car ignition advance parameters, the intelligent technique was achieved. The results show that when the load (throttle position) increases, ignition start angle decrease. Moreover, the maximum output power will be achieved around 22 °CA before top death centre (BTDC). Besides, the decrease of the load will make the ignition advance increase faster when comparing with the increment of the speed. These statements have been found to be correct with the results from Zhengzhong [7].

The use of fuzzy expert system as an intelligent tool to predict the ignition advance for a hydrogen powered car has been demonstrated. These predictions are based on the study of qualitative and quantitative effects of engine process parameters such as engine speed, throttle position, manifold air pressure, fuel pulse width, engine power, lambda value on the ignition timing of hydrogen car. The fuzzy system was appraised by comparing with test rig. It has been shown that using the fuzzy system, the predictive capability for ignition advance has been shown to be less than ±5% of root mean square error.

5 Conclusion

It has been shown that using the input parameters from the engine, the fuzzy model to estimate the ignition timing is developed. Each ignition advance within the database was tested and the system was found to be correct and reliable. In conclusion, the developed fuzzy system achieves all the objectives and meets all the criteria for a successful intelligent tool for later use in tuning ignition timing of hydrogen car. It is further encouraging to note that the estimation of ignition timing from the fuzzy system developed is within ± 5% accuracy when compared with the experimental values. This work is seen as a step towards establishing intelligent fuzzy expert systems as predictive tools in identifying initial conditions for smoother engine operating conditions.

Acknowledgements. The authors are deeply grateful to Dr Sergio Giudici, Hydro Tasmania Pty Ltd for financial support and all of the Hydrogen & Allied Renewable Technology research members as well as Intelligent Hydrogen Car project for sharing ideas and concept along the way.

References

1. Burke, P.H.: Performance Appraisal of a Four Stroke Hydrogen Internal Combustion Engine. University of Tasmania, Master of Engineering Science Thesis (2005)
2. Karim, G.A.: Hydrogen as a Spark Ignition Engine Fuel. International Journal of Hydrogen Energy 28, 569–577 (2003)
3. Alousi, A.: Examination of the Combustion Processes and Performance of a Spark Ignition Engine Using a Data Acquisition System. University of Calgary, Calgary (1982)
4. Jehad, A., Yamin, A., Gupta, H.N., Bansal, B.B., Srivastava, O.N.: Effect of Combustion Duration on the Performance and Emission Characteristics of a Spark Ignition Engine Using Hydrogen as a Fuel. International Journal of Hydrogen Energy 25, 581–589 (2000)
5. Butler, D.A.: Enhancing Automotive Stability Control with Artificial Neural Networks. University of Tasmania, PhD of Engineering Science Thesis (2006)
6. Negnevitsky, M.: Artificial Intelligence: a Guide to Intelligent Systems, 2nd edn. Addison-Wesley, Reading (2005)

7. Yang, Z.Z., Wang, L.J., Xiong, S.S., Li, J.D.: Research on Optimizing Control Technology Based on Fuzzy-Neural Network for Hydrogen Fuelled Engines. International Journal of Hydrogen Energy 31, 2370–2377 (2006)
8. Yang, Z.Z., Wei, J.Q., Li, J.D.: An Investigation of Optimum Control of Ignition Timing and Injection System in an In-Cylinder Injection Type Hydrogen Fueled Engine. International Journal of Hydrogen Energy 27, 213–217 (2002)
9. Australia Standard 2875: Alloy Steel Cylinders for Compressed Gases-Seamless-0.1 kg to 500 kg. Standards Australia (1995)
10. Australia Standard 4838: Gas Cylinders-High Pressure Cylinders for the On Board Storage of Natural Gas as a Fuel for Automotive Vehicles. Standards Australia (2002)
11. MATLAB: Fuzzy Toolbox Help. MATLAB Inc. (2001)
12. Lim, J.S.D.: Development of a Hydrogen Car and Emissions Modeling Using Artificial Intelligence Tools. University of Tasmania, Master of Engineering Science Thesis (2007)
13. Barrett, D.: Study of the Performance of a Four Cylinder Hydrogen-Fuelled Internal Combustion Engine. University of Tasmania, Master of Engineering Science Thesis (2007)

Circuitry Analog and Synchronization of Hyperchaotic Neuron Model

Shukai Duan[1,2] and Lidan Wang[1]

[1] School of Electronics & Information Engineering, Southwest University, 400715, Chongqing, P.R. China
[2] College of Automation, Chongqing University, 400044, Chongqing, P.R. China
duansk@swu.edu.cn, ldwang@swu.edu.cn

Abstract. Recently, a general model of nonlinear systems with distributed delays is studied and a four-dimensional continuous autonomous chaotic neuron model is found and analyzed. In this paper, circuitry analog of the chaotic neuron model is introduced. Adaptive chaos synchronization scheme is presented and analyzed. A series of computer simulations demonstrated the effectiveness of the proposed synchronization schemes.

Keywords: Chaotic neuron model; circuitry analog; chaos synchronization; adaptive chaos synchronization.

1 Introduction

Over last four decades, chaos has been intensively studied and chaotic system provided a rich mechanism for signal design, generation and processing [1]. Recently, Liao and Chen proposed a generalized Lorenz system and derived a new four-dimensional continuous autonomous chaotic neuron model [2]. In our previous work, Duan et al [3] implemented the chaotic neuron with electronic circuit. This circuit can be used as a high-dimensional chaos generator.

On the other hand, chaos synchronization plays an important role in secure communication which based on chaos systems. Chaos synchronization means that one system will converge to the same values as the other and they will remain in step with each other [4]. But in practical situations, some or all of the system's parameters are unknown. Hence, the derivation of adaptive chaos synchronization in the presence of unknown system parameters is an interesting and attractive issue [5].

In this paper, hyperchaotic behavior of the chaotic neuron and circuitry analog are introduced. Adaptive chaos synchronization schemes are presented and analyzed and demonstrated by a series of computer simulations. The organization of this paper is as follows. In section II, the mathematical model of four-dimensional hyperchaotic system and its circuitry analog are introduced. In section III, adaptive chaos synchronization scheme is presented proved. In section IV, a series of computer simulations demonstrated the effectiveness of the proposed synchronization schemes. Finally, some concluding remarks are given in the last section.

F. Sun et al. (Eds.): ISNN 2008, Part II, LNCS 5264, pp. 580–587, 2008.

2 The Chaotic Neuron Description and Circuitry Analog

In this section, the mathematical model of the four-dimensional hyperchaotic system and its circuitry analog are introduced.

2.1 The Mathematical Description of the Chaotic Neuron Model

A general model of nonlinear system with distributed delays is given as follows [2]:

$$\begin{aligned} \dot{x}_1(t) &= (c-a)\int_{-\infty}^{0} x_1(\tau)k(t-\tau)d\tau - x_2(t)\int_{-\infty}^{0} x_1(\tau)k(t-\tau)d\tau + cx_1(t) \\ \dot{x}_2(t) &= x_1(t)\int_{-\infty}^{0} x_1(\tau)k(t-\tau)d\tau - bx_2(t) \end{aligned} \tag{1}$$

in which a, b and c are positive constants and k is the delay kernel. If the weak kernel $k(s) = ae^{-as}$ is applied, then model (1) becomes the original Chen system. However, if the strong kernel $k(s) = a^2 se^{-as}$ is considered, the model (1) becomes the following four-dimensional chaos neuron model.

$$\begin{cases} \dot{x}_1 = dx_4 - x_2x_4 + cx_1 \\ \dot{x}_2 = x_1x_4 - bx_2 \\ \dot{x}_3 = a(x_1 - x_3) \\ \dot{x}_4 = a(x_3 - x_4) \end{cases} \tag{2}$$

where a, b, c and d are parameters. This four-dimensional autonomous system have exhibited much rich chaotic behaviors while a=3.1, b=4, c=2.45, d=-0.65 [2]. By calculating Lyaponov exponents (see Fig. 1), the results are (0.310756, 0.036703, -3.124290,-4.973169). It has two positive Lyaponov exponents and this means it is a hyperchaotic neuron model.

2.2 The Circuitry Analog of the Chaotic Neuron Model

Operational amplifiers, analog multipliers and linear resistors and capacitors are employed to analog the four-dimensional chaos neuron model [3]. The circuit consists of four channels. The first channel performs the first equation in system (2) (see Fig. 2). It has four inputs x_1, x_2, x_4 and $-x_4$. By applying standard node analysis techniques, a state variable equation of the circuit may be obtained as following

$$\dot{x}_1 = \frac{(R_1+R_5)R_2R_3R_4R_7}{R_1R_6R_8C_1(R_3R_4+R_2R_4+R_2R_3)}\left(\frac{1}{R_3}x_1 - \frac{R_{25}}{R_2R_{24}}x_4\right) - \frac{R_5R_7}{R_1R_6R_8C_1}x_2x_4 \tag{3}$$

In same way, the rest equations in system (2) may be realized by electronic circuits (see Fig. 3, **F**ig.4 and Fig.5).

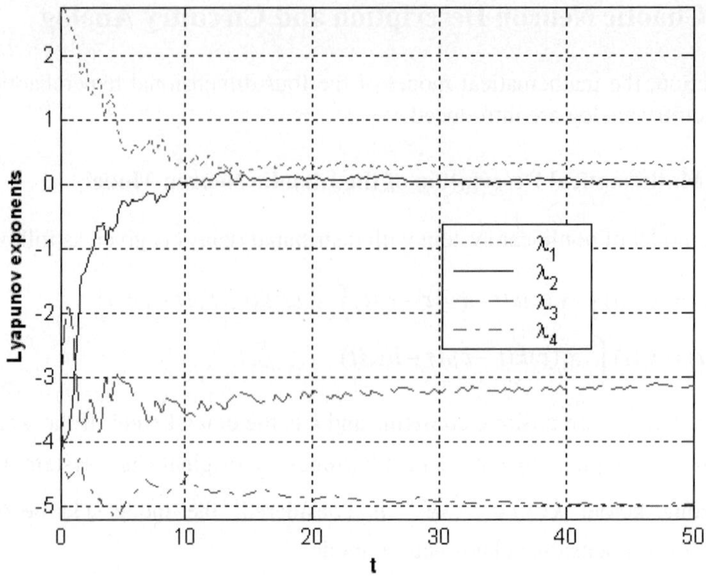

Fig. 1. Lyaponov exponent spectrum of system (2)

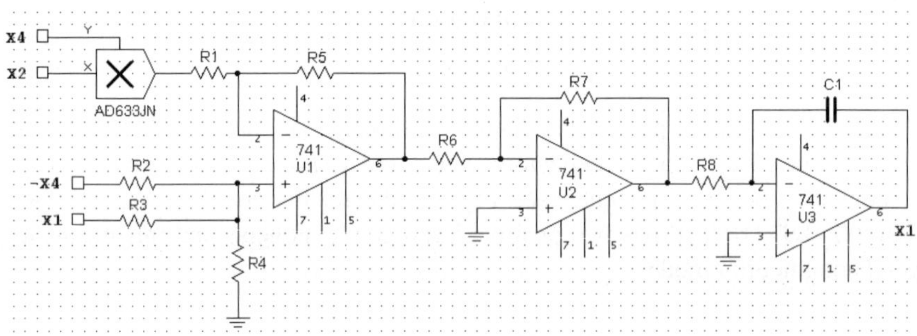

Fig. 2. Circuitry analog of the first equation in system (2)

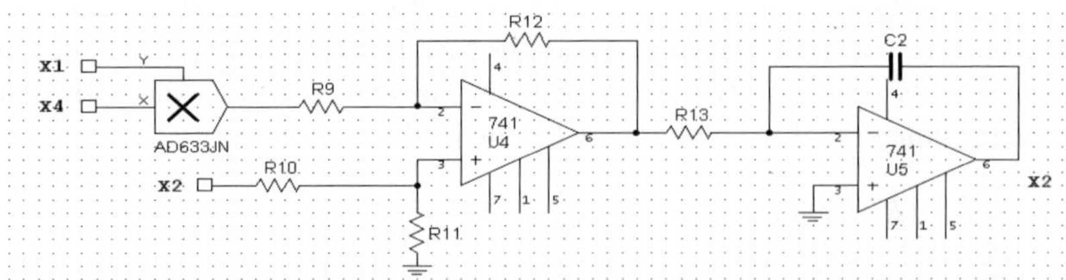

Fig. 3. Circuitry analog of the second equation in system (2)

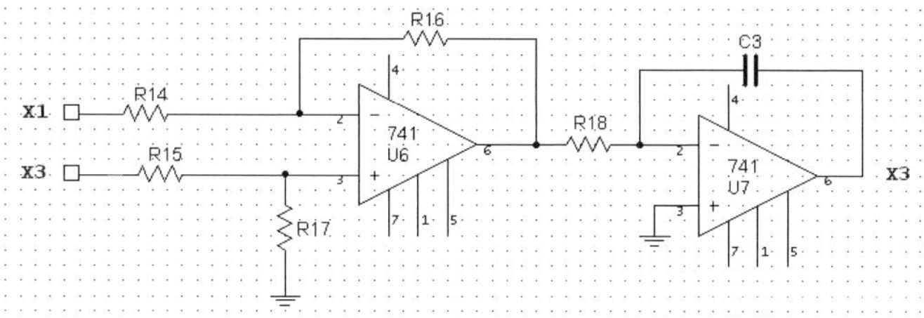

Fig. 4. Circuitry analog of the third equation in system (2)

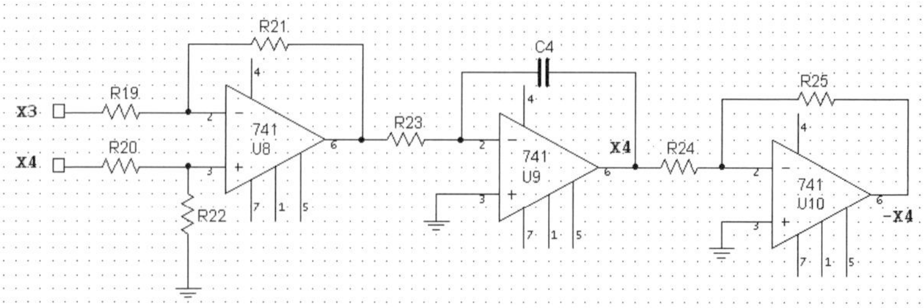

Fig. 5. Circuitry analog of the forth equation in system (2)

By applying standard node analysis techniques, state variable equations of the above circuits may be obtained as following.

The second state variable equation is

$$\dot{x}_2 = -\frac{(R_9 + R_{12})R_{11}}{R_9 R_{13} C_2 (R_{10} + R_{11})} x_2 + \frac{m_2 R_{12}}{R_9 R_{13} C_2} x_1 x_4 \tag{4}$$

The third state variable equation is

$$\dot{x}_3 = \frac{1}{R_{14} R_{18} C_3} (R_{16} x_1 - \frac{(R_{14} + R_{16})R_{17}}{R_{15} + R_{17}} x_3) \tag{5}$$

and the fourth state variable equation is

$$\dot{x}_4 = \frac{1}{R_{19} R_{23} C_4} (R_{21} x_3 - \frac{(R_{19} + R_{21})R_{22}}{R_{20} + R_{22}} x_4) \tag{6}$$

Circuit parameters values are set as: R_1=3 $K\Omega$, R_2=20 $K\Omega$, R_3=5.31 $K\Omega$, R_4=32.5 Ω, R_5=R_{16}=R_{17}=R_{21}=R_{22}=1 $K\Omega$, R_6=R_7=R_9= R_{10}=R_{12}=R_{14}=R_{15}=R_{19}=R_{20}=R_{24}= R_{25}=10 $K\Omega$, R_8=3.3 $K\Omega$, R_{11}=204.1 Ω, R_{13}=10 $K\Omega$, R_{18}=R_{23}= 32.3 $K\Omega$, C_1= C_2=C_3=C_4=1nF. *AD633JN* and *uA741* are employed as operational amplifiers and analog multipliers, respectively.

3 Adaptive Chaos Synchronization Scheme

We assume that we have two chaotic neurons where the drive system with the sub-script 1 drives the response system having identical equations denoted by the subscript 2. The initial conditions and partial parameters on the drive system are different from that of the response system. The drive and response systems are defined as following equations

$$\begin{cases} \dot{x}_1 = dx_4 - x_2x_4 + cx_1 , \\ \dot{x}_2 = x_1x_4 - bx_2 , \\ \dot{x}_3 = a(x_1 - x_3) , \\ \dot{x}_4 = a(x_3 - x_4) . \end{cases} \tag{7}$$

and

$$\begin{cases} \dot{y}_1 = dy_4 - y_2y_4 + cy_1 + u_1 , \\ \dot{y}_2 = y_1y_4 - by_2 + u_2 , \\ \dot{y}_3 = a(y_1 - y_3) + u_3 , \\ \dot{y}_4 = a(y_3 - y_4) + u_4 . \end{cases} \tag{8}$$

where controllers u_1, u_2, u_3 and u_4 are to be determined for synchronizing the two systems with the unknown parameters .n spite of the differences in initial conditions. In order to estimate the control functions, we may subtract (7) from (8) and obtain error system

$$\begin{cases} e_1 = y_1 - x_1 , \\ e_2 = y_2 - x_2 , \\ e_3 = y_3 - x_3 , \\ e_4 = y_4 - x_4 . \end{cases} \tag{9}$$

Obviously, error system (9) is governed by the following differential system

$$\begin{cases} \dot{e}_1 = -e_2e_4 - ce_1 + de_4 - e_2x_4 - e_4x_2 + u_1 , \\ \dot{e}_2 = e_1e_4 - be_2 + e_1x_4 + e_4x_1 + u_2 , \\ \dot{e}_3 = ae_1 - ae_3 + u_3 , \\ \dot{e}_4 = ae_3 - ae_4 + u_4 . \end{cases} \tag{10}$$

The goal of control is to find controllers and a parameter estimation update law such that the states of response system and the drive system are globally synchronized asymptotically, i.e.,

$$\lim_{t\to\infty} \|e(t)\| = 0, \quad \forall a,b,c,d \in R, \tag{11}$$

where $e(t) = [e_1, e_2, e_3, e_4]^T$. A Lyapunov function for (10) is considered as following

$$V(e, \tilde{a}, \tilde{b}, \tilde{c}, \tilde{d}) = \frac{1}{2} e^T e + \frac{1}{2} \tilde{a}^2 + \frac{1}{2} \tilde{b}^2 + \frac{1}{2} \tilde{c}^2 + \frac{1}{2} \tilde{d}^2 \tag{12}$$

where $\tilde{a}=a-\hat{a}, \tilde{b}=b-\hat{b}, \tilde{c}=c-\hat{c}, \tilde{d}=d-\hat{d}$. $\hat{a}, \hat{b}, \hat{c}$ and $\hat{d}$ are estimate values of the unknown parameter . .. . and . , respectively.

In order to the time-derivative of Lyapunov function (12) along the solutions of (10) satisfying the following inequality

$$\dot{V}(e, \tilde{a}, \tilde{b}, \tilde{c}, \tilde{d}) \leq -W(e), \tag{13}$$

where $W(e)=e_1^2+e_2^2+e_3^2+e_4^2$ .We therefore need to find a controller u and a parameter estimation update law $\hat{a}, \hat{b}, \hat{c}$ and $\hat{d}$ to guarantee that for all $e \in R^4$ the inequality (13) holds. The following type of controller and updating law are adopted.

$$\begin{aligned}
&u_1=(\hat{c}-1)e_1+e_2e_4+e_2x_4+e_4(x_2-\hat{d}),\ u_2=-(1-\hat{b})e_2-e_1e_4-e_1x_4-e_4x_1 \\
&u_3=(\hat{a}-1)e_3-\hat{a}e_1,\ u_4=(\hat{a}-1)e_4-\hat{a}e_3 \\
&\dot{\hat{a}}=e_1e_3+e_3e_4-e_3^2-e_4^2,\ \dot{\hat{b}}=-e_2^2,\ \dot{\hat{c}}=-e_1^2,\ \dot{\hat{d}}=e_1e_4
\end{aligned} \tag{14}$$

With those choices, we have the time derivative of $V(e, \tilde{a}, \tilde{b}, \tilde{c}, \tilde{d})$. along the solutions of (10)

$$\begin{aligned}
\dot{V}(e, \tilde{a}, \tilde{b}, \tilde{c}, \tilde{d}) &= e^T\dot{e}+\tilde{a}\dot{\tilde{a}}+\tilde{b}\dot{\tilde{b}}+\tilde{c}\dot{\tilde{c}}+\tilde{d}\dot{\tilde{d}} \\
&= e_1(-e_2e_4-ce_1+de_4-e_2x_4-e_4x_2+u_1)+e_2(e_1e_4-be_2+e_1x_4+e_4x_1+u_2) \\
&\quad +e_3(ae_1-ae_3+u_3)+e_4(ae_3-ae_4+u_4)+\tilde{a}(-\dot{\hat{a}})+\tilde{b}(-\dot{\hat{b}})+\tilde{c}(-\dot{\hat{c}})+\tilde{d}(-\dot{\hat{d}}) \\
&= -e_1^2-e_2^2-e_3^2-e_4^2
\end{aligned} \tag{15}$$

This leads to

$$\lim_{t\to\infty}\|e(t)\|=0, \quad \forall a,b,c,d \in R \tag{16}$$

Hence, adaptive synchronization of two chaos systems with unknown parameters is achieved.

4 Computer Simulations

To demonstrate the effectiveness of the proposed synchronization approach, numerical simulations are given when unknown uncertain parameters *a*, *b* and *c* and *d* exist. The true values of the unknown parameters are selected as a=3.1, b=4, c=2.45, d=-0.65 in simulations. The initial values of the drive system states, initial values of the response system states and initial value of estimate for "unknown" parameters are taken as $x_1(0)$ =0.5, $y_1(0)$=10, $x_2(0)$=0.5, $y_2(0)$=10, $x_3(0)$=0.5, $y_3(0)$=10, $x_4(0)$=0.5, $y_4(0)$=10 and $\hat{a}(0)=0.5$, $\hat{b}(0)=0.5$, $\hat{c}(0)=0.5$ and $\hat{d}(0)=0.5$, respectively.

Under the adaptive controller and a parameter estimation update law, the synchronization of two chaos systems with unknown parameters is achieved immediately, as is shown in Fig. 6.

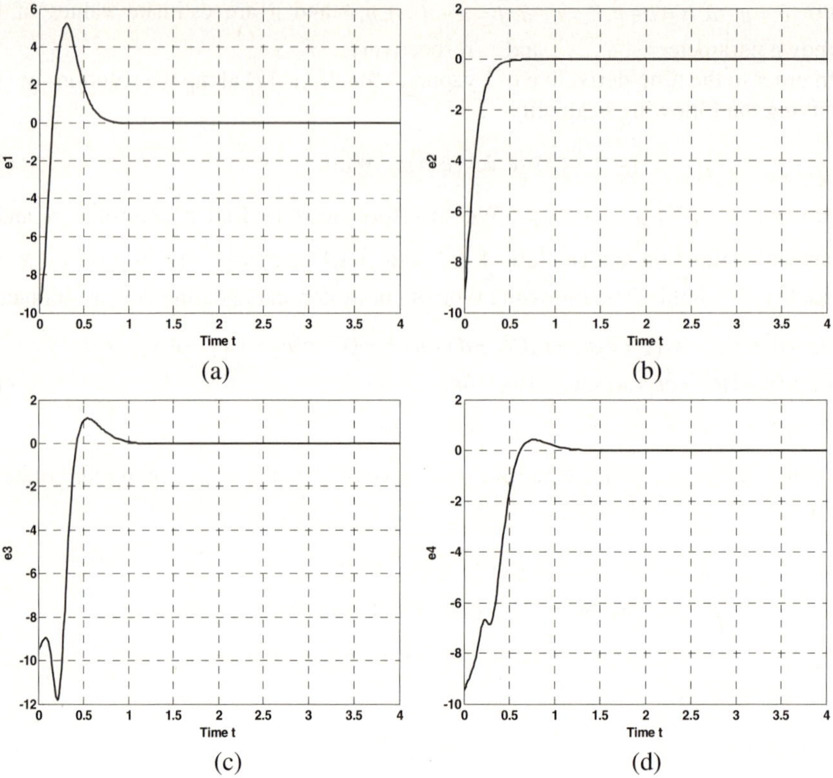

Fig. 4. The trajectories of error system: (a)e_1, (b) e_2, (c) e_3 and (d) e_4

5 Conclusions

In this paper, hyperchaotic behavior of the chaotic neuron and circuitry analog are introduced. Adaptive chaos synchronization scheme is presented and analyzed. Numerical simulations are given to demonstrate the effectiveness of the proposed synchronization approach when unknown uncertain parameters.

Acknowledgments. This research was supported by Chongqing Natural Science Foundation of CQ CSTC under Grant 2007BB2331, the doctoral Foundation of the Southwest University under Grant SWUB2007008.

References

1. Cuomo, K.M., Oppenheim, A.V., Strogatz, S.H.: Synchronization of Lorenz-based chaotic circuits with applicationsto communications. IEEE Transactions on Circuits and Systems II: Analog and Digital Signal Processing 40(10), 626–633 (1993)
2. Liao, X.F., Chen, G.R.: Hopf bifurcation and chaos analysis of Chen's system with distributed delays. Chaos, Solitons and Fractals 25(1), 197–220 (2005)

3. Duan, S.K., Wang, L.D., Liao, X.F.: Circuitry implementation of a new four-dimensional chaotic system. In: Dynamics of Continuous Discrete and Impulsive Systems-Series B, vol. 14 (s2), pp. 762–765. Watam Press (2007)
4. Pecora, L.M., Carroll, T.L.: Synchronization in chaotic systems. Physical Review Letters 64(8), 821–824 (1990)
5. Yassen, M.T.: Adaptive chaos control and synchronization for uncertain new chaotic dynamical system. Physics Letters A 350, 36–43 (2006)

A Genetic-Neural Method of Optimizing Cut-Off Grade and Grade of Crude Ore

Yong He, Sixin Xu, Kejun Zhu, Ting Liu, and Yue Li

School of Economics and Management, China University of Geosciences,
430074 Wuhan, China
jiabeisheng@sina.com

Abstract. Cut-off grade and grade of crude ore is crucial to the economic benefit of enterprise and sustainable utilization of resource in mining system. Generally, they are determined by experiment data or worker's experience, and they can't be widely used. In this work, we use genetic algorithm and neural networks nesting method to simulate the highly complexity and non-linear relationship of mine system, to optimize the cut-off grade and grade of crude ore. The inner layer of nesting is neural networks, which is used to compute loss rate, amount of tailing ore and total cost; the outer layer is genetic algorithm, with cut-off grade and grade of crude ore as chromosome, which is used to get the revenue. These two layers carry out the optimization of cut-off grade and grade of crude ore jointly. Take Daye Iron Mine as a case, the result shows that: During the period of August to November in the year 2007, the optimal cut-off grade is 15.8%, and optimal grade of crude ore is 43.7762-44.1387%. Comparing with the present scheme (cut-off grade is 18%, grade of crude ore is 41-43%), the optimized scheme can improve the present value by 9.01-9.44 million Yuan.

Keywords: Genetic-neural optimization; Cut-off grade; Grade of crude ore.

1 Introduction

Cut-off grade is the grade of ore in the last time of ore drawing during sublevel caving with no sill pillar. The Loss rate and dilution rate in mining are two significant indexes in mine system, and they have much influence on the benefit of mine. Actually the loss rate and dilution rate of ore directly depend on the cut-off grade. If the cut-off grade is low, the dilution rate of ore will be high. This will increase the cost of ore processing, and will decrease the amount of the concentrate in the condition of a certain mineral processing ability. Contrarily, high cut-off grade will not only lead to waste of mine resource, but also increase the cost of fundamental construction. Grade of crude ore plays a role as a link between mining and milling. It is determined by the average grade of ore bodies, cut-off grade and dilution rate. And it affects the concentrate grade and milling recovery. The optimum grade of crude ore can improve the benefit of mine and make use of mine resource. Many scholars proposed the concepts and calculate methods of the cut-off grade and grade of crude ore from different aspects, including Xie,Y L [1], Dong, C H[2], M. Osanloo and M. Ataei [3], Mishra.B [4], Bascetin, A. [5], Asad, M.W.A. [6] etc. They studied on computation method

F. Sun et al. (Eds.): ISNN 2008, Part II, LNCS 5264, pp. 588–597, 2008.

from the aspects of mining and milling, these computations include the multi-factor analysis, the linear and non-linear regression, the dynamic optimization analysis, etc. They can guide mine production and management. Regrettably, because of the particularity of production environment, multiplicity of ore drawing and extension of mine management, until now, cut-off grade and grade of crude ore are determined by experiment data or worker's experience. Moreover, because these methods aim at the specific ore deposit and production technique, they lack robustness. Along with the production development, increase of resources' scarcity, enhancement of technical and management level, experts and managers gradually realize that, the determining method based on workers' experience, easy and feasible though, greatly increases the mining and milling cost, and wastes resource seriously. Obviously, cut-off grade and grade of crude ore are two concepts related with income, cost, geological grade, loss rate and dilution rate. The mapping function between them are highly-complex and highly-nonlinear, and it is difficult to directly or indirectly find out the mathematics expression. In this paper, we use genetic algorithms and neural networks nesting method to simulate the highly complexity and non-linear relationship of variables in mine system, in order to optimize the cut-off grade and grade of crude ore. This paper gave a case study on Daye Iron mine.

2 Descriptions of Algorithms

2.1 Genetic Algorithm

Genetic Algorithm (GA) is an adaptive artificial intelligence technology developed by Holland [7]. A standard Genetic Algorithm consists of

(1). Coding: Data from the solution space must be coded into genetic data in the inheritance space. For example, variable $x = 13,8,2$ can be coded into binary data 01101, 01000, and 00010, respectively, with the length l=5;

(2). Formation of species: Genetic operation is conducted on a colony consisting of certain chromosomes, which is called a species;

(3). Fitness function f: This function is used to measure the goodness of individuals (chromosomes) and serves as the target function in the GA;

(4). Selection/copying: The selection and copying are aimed at selecting optimal individuals from current species. The goodness of individuals is determined by the fitness value: the higher the value is, the greater opportunity the individuals to be selected. The standard GA selects the individuals based on relevant probabilities;

(5). Crossover operation: A simple crossover operation consists of two steps of random pairing of individuals followed by random setup of crossover positions with exchange of partial information between paired individuals. The crossover results in generation of two new individuals;

(6). Mutation: This is another way of genetic operation to produce new individuals, and can be done by random complementary operation of a certain character of a given individual from 1 to 0 or vice versa. The probability of mutation p_m is generally very low (0.001~0.01).

Genetic operation is repeated iteratively. It progressively approaches but will never probably arrive at the best solution. Thus conditions of termination are needed. The most common way of termination is to set up a maximum iterative time (e.g., 100-1000). Once the iterative time reaches the maximum, termination can be done. The second method of termination is to monitor the variation of fitness of the best individuals. Once it shows negligible variation in the later stage of genetic operation as compared to the previous stage, termination can be done.

2.2 Neural Networks

Neural networks (NN) are numerical algorithms inspired in the functioning of biological neurons. This concept was introduced by McCulloch and Pitts [8], who proposed a mathematical model to simulate neuron behavior. The neuron model is shown in Fig.1, representing the neuron m that receives an input vector $x = [x_1, x_2, \cdots x_l]^T$. It then computes the weighted sum of the components of x multiplying each component x_k by a coefficient w_{mk} . The neuron m activation, a_m, is given by the expression

$$a_m = \sum_{k=1}^{l} w_{mk} x_k + b_m \tag{1}$$

Here b_m called bias. The output signal of neuron m , s_m is the numerical value that results from the computation of an activation function $f(a_m)$.

$$s_m = f(a_m) = \frac{1 - e^{-2a_m}}{1 + e^{-2a_m}} \tag{2}$$

Several neurons can be assembled to form a network. In this work, it was decided to adopt a well-known network arrangement, the multilayer perceptron. It is possible to prove [9] that this type of network with sigmoid activation functions in the hidden and output layers can approximate any continuous function with satisfactory precision, provided that it has enough neurons in the hidden layer. The process of obtaining the unknown coefficients w_{mk} and b_m required to approximate the prescribed function is

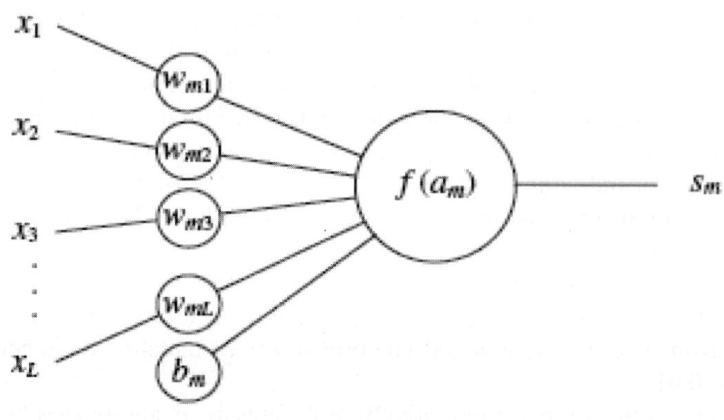

Fig. 1. Artificial neuron

called training and it is a somewhat challenging task. The most common training process, called supervised training [10], consists of proposing some initial values to the coefficients and then adjusting those values in order to minimize the error between the predicted output produced by the NN and the exact value of the function. In this work, back propagation (BP) [11] is used to train the coefficients, this type of multilayer perceptron is called BP neural network (BPNN).

3 Problem Description and Modeling

Optimal cut-off grade and grade of crude ore is based on maximizing the NPV. Considering the relationship among profit, cost, geological reserves, tailing ore and all kinds of grades, we establish a model as follows

$$\begin{cases} MaxNPV = \sum_{t=1}^{n} \dfrac{R_t - C(a_t, q_t, a_j, a_r)}{(1+i)^t} \\ s.t \quad R_t = \dfrac{a_t q_t (1-\phi(a_j)) - a_w Q_w (a_t, q_t, a_j, a_r)}{\beta} J \\ 0 \le a_j \le a_r \le 1 \end{cases} \tag{3}$$

Where, NPV denotes net present value, R_t denotes revenue of the ore body in the t th month, a_t denotes geological grade in the t th month, a_j denotes cut-off grade, a_r denotes grade of crude ore, q_t denotes geological reserves in the t th month, $C(a_t, q_t, a_j, a_r)$ denotes the function of cost including mining and milling, $\phi(a_j)$ denotes the function of loss rate related to cut-off grade, a_w denotes tailing ore grade, $Q_w(a_t, q_t, a_j, a_r)$ denotes the function of tailing ore amount, β denotes the concentrate grade, J denotes the current price of concentrate, i denotes rate of discount.

The process of solving formula (3) is searching for a combination among several grades (geological grade, cut-off grade and crude ore grade) to maximize the NPV. First, we should construct three functions including $\phi(a_j)$, $Q_w(a_t, q_t, a_j, a_r)$, $C(a_t, q_t, a_j, a_r)$.

4 Genetic-Neural Optimization

We make a nesting of genetic algorithms and neural networks to be a genetic-neural model, and use it to optimize the cut-off grade a_j and crude ore grade a_r. The inner layer of nesting is neural networks, which is used to compute loss rate, amount of tailing ore and total cost; the outer layer is genetic algorithm, with cut-off grade and grade of crude ore as chromosome, which is used to get the NPV. The details are as follows: Join cut-off grade and grade of crude ore together as chromosome of population for evolution computation. Then make self-adaptive neural network to get the local connection between the income value (fitness function) and chromosome. Finally, use genetic algorithms globally to search the optimal cut-off grade and grade of crude ore to maximize fitness function. The steps of algorithm are shown as follows. (Flowchart is shown in Fig.2).

① Define coding type and the length of individual strings;
② Define GA operation parameters;
③ Generate initial population with population size N;
④ Decoding individual strings to get different groups of grade values;
⑤ Input the individual to REG regression, BPNN1 and BPNN2 models to calculate loss rate, the amount of tailing ore and the cost respectively;
⑥ Calculate the NPV and convert it into fitness;
⑦ Carry out selection operation according to the fitness;
⑧ Carry out crossover and mutation operation to obtain a new generation;
⑨ Return to step ④ until the group of grades meets the termination conditions;
⑩ Calculate and output the NPV with the optimized grades.

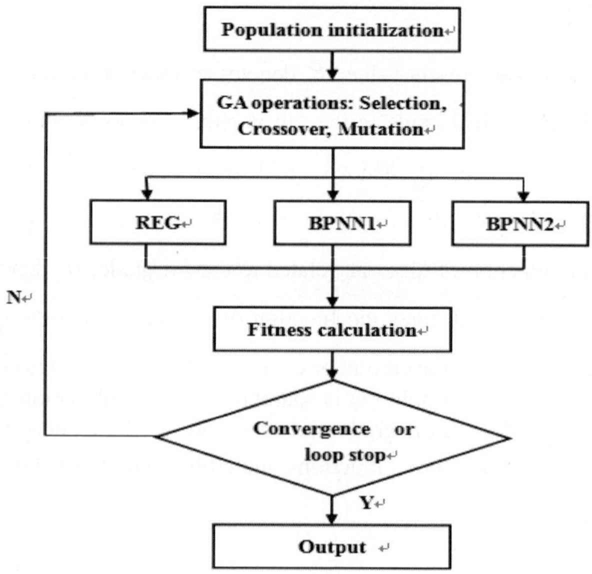

Fig. 2. Algorithm flowchart

5 Case Study

Daye iron mine is the main supply base of the Wuhan steel and iron (Group) Corp., which is a very famous large enterprise in China. Daye iron mine has six big ore bodies from west to east, which are divided in two mining workshops. Now it faces the following problems. First, the cut-off grade scheme was established according to the mining technique, milling technology and price of the concentrate in 1990s, but now whether the scheme is reasonable need to be studied. Second, the geological condition, mining and milling technology have changed greatly. Along with reconstruction of milling process, it is necessary to find out the optimum grade of crude ore. Therefore, it is urgent to optimize cut-off grade and grade of crude ore, in order to guide mining and milling production. According to geology, production and cost report forms from Jan.2005 to Nov.2007, we acquire the relative data which are omitted in this paper.

5.1 Computation Models of Loss Rate, Tailing Ore and Cost

According to the data of cut-off grade and loss rate, use the command "postreg" in the matlab language[12], and obtain the regression fit chart in Fig.3, the correlation coefficient is 0.97419, the function relationship of loss rate and cut-off grade can be expressed as $\phi = 1.7a_j - 12$.

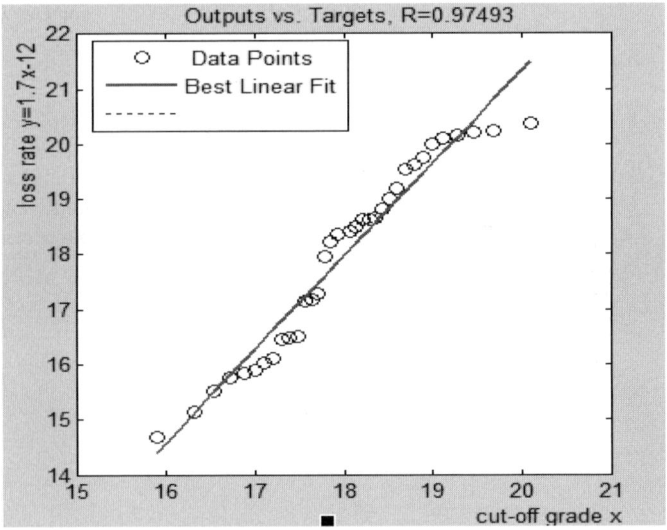

Fig. 3. Regression Fit chart

We establish the BP network mapping relationship from cut-off grade a_j, grade of crude ore a_r, geological reserves q_t, geological grade a_t to the amount of tailing ore Q_w and the total cost C. The sample simulation diagram is shown in Fig.4 and Fig.5. It shows that the constructed BP networks have prefect simulation performance.

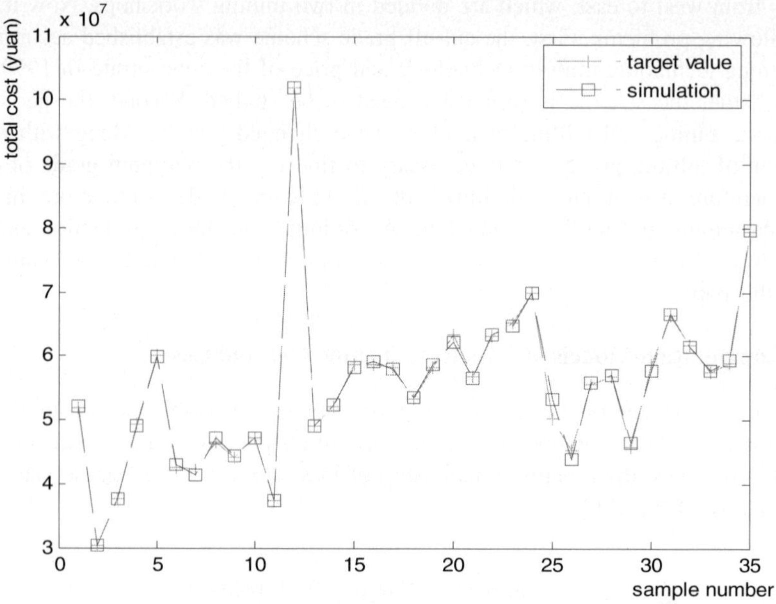

Fig. 4. Tailing ore simulation

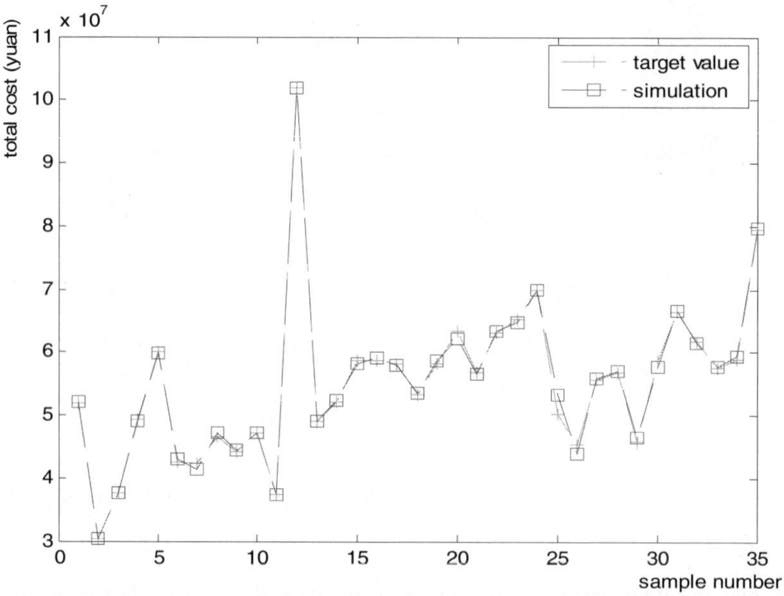

Fig. 5. Cost simulation

5.2 Genetic-Neural Optimization Integration

Fig.6 is the technical diagram, which is adopted to optimize cut-off grade (COG) and grade of crude ore (GCO) to maximize the net present value (NPV). The system is constituted with a main module and three sub-modules. The main module realizes the optimization of two grades, which is used for ore drawing and ore mixing, and then transmits two grades information and the relative geological data to three sub-modules. Once the sub-modules receive the input information, they calculate the loss rate, the amount of tailing ore and the cost respectively. These are used for ore drawing management (ODM), tailing ore management (TOM) and cost control (CC), and compute income (CI) simultaneously, together with market parameters (MP) and cost, finally get the NPV.

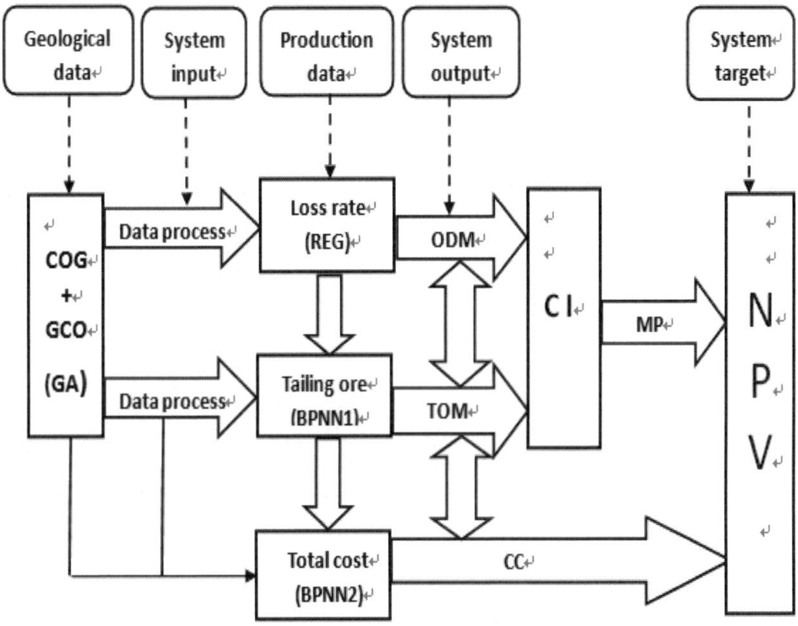

Fig. 6. Grade optimization system

The population contains 80 chromosomes. We use the binary encoding method, with the value range of cut-off grade being 15-20.1%, the value range of crude ore grade being 40.7-46.4%, the crossover probability being 0.7, and the mutation probability being 0.008. The fitness function is NPV, as shown in formula (3). Here, a_w=9.77%, β=64%, J=517.12, i=0.05. In the following we optimize cut-off grade and grade of crude ore grade from Aug. to Nov.2007. Table1 shows the result and comparison with present schemes (No1-5 are the optimum schemes, n1 and n2 are the present schemes).

Table1. Optimization result and comparison

Scheme	Cut-off grade (%)	Grade of crude ore (%)	NPV (10^8Yuan)
1*	15.8	43.9571	1.8811
2*	15.8	43.8667	1.8811
3*	15.8	44.0476	1.8811
4*	15.8	43.7762	1.8811
5*	15.8	44.1387	1.8811
n1	18	41	1.7867
n2	18	43	1.7910

From Table1, we can see that, if Daye mine adopts the optimized scheme (cut-off grade is 15.8%, grade of crude ore is 43.7762-44.1387%), the NPV from Aug. to Nov.2007 is 188.11million Yuan, and the NPV of present scheme (cut-off grade is 18%, grade of crude ore is 41-43%) is 178.67-179.10 million Yuan. Therefore, the optimized scheme can improve the NPV by 9.01-9.44 million Yuan.

6 Conclusions

This paper introduces the intelligence technology into the field of mine system, which uses a nesting of genetic algorithm and neural networks to optimize the cut-off grade and grade of crude ore. Neural network is used to construct mapping relationship from cut-off grade, grade of crude ore, geological reserves and geological grade to the amount of tailing ore and total cost. Genetic algorithm is used to reach dynamic optimization of cut-off grade and grade of crude ore, with NPV as fitness function to evaluate the two grades. Take Daye Iron Mine as an example, we obtain an approving result. The proposed method of optimizing grades provides a brand-new idea to the metal mine system, and could be used widely. In the further studies, we will combine the revenue with the utilization of resource to judge the cut-off grades and grade of crude ore, and fuse more intelligence algorithms, such as FNN, PSO, ACO, SA, to optimize the two grades.

Acknowledgments

This research was supported by National Natural Science Foundation grant NO: 70573101 of the People's Republic of China and Research fund grant NO: 070429 of Wuhan Steel and Iron (Group) Corp.

References

1. Xie, Y.L.: Optimization of Cut-off Grade in Open-pit Based on Control Theory. Tran. Nonferrous Metals Society of China 8, 353–356 (1998)
2. Dong, C.H.: Application of Ore Grade Optimization Method on Erfengshan Iron Mine. Metal Mine 4, 14–17 (2002)

3. Osanloo, M., Ataei, M.: Using Equivalent Grade Factors to Find the Optimum Cut-off Grades of Multiple Metal Deposits. Minerals Engineering 16, 771–776 (2003)
4. Mishra, B.: Development of a Computer Model for Determination of Cut off Grade for Metalliferous Deposits. J. Mines, Metals and Fuels 54, 147–152 (2006)
5. Bascetin., A.: Determination of Optimal Cut-off Grade Policy to Optimize NPV Using a New Approach with Optimization Factor. J. the South African Institute of Mining and Metallurgy. 107, 87–94 (2007)
6. Asad, M.W.A.: Optimum Cut-off Grade Policy for Open Pit Mining Operations through Net Present Value Algorithm Considering Metal Price and Cost Escalation. Engineering Computations 24, 723–736 (2007)
7. Holland, J.H.: Adaptation in Nature and Artificial Systems. The University of Michigan Press/MIT Press, Ann Arbor/Cambridge (1975)
8. McCulloch, W.S., Pitts, W.: A Logical Calculus of Ideas Immanent in Nervous Activity. Bull. Math. Biophys. 5, 15–33 (1943)
9. Cybenko, G.: Approximation by Superposition of Sigmoidal Functions. Math. Control. Signal Syst. 2, 303–314 (1989)
10. Rumelhart, D.E., Hinton, G., Williams, R.: Learning Internal Representations by Error Propagation. PDP Research Group. MIT Press, Cambridge (1986)
11. Martin, T.H., Howard, B.D., Mark, B.: Neural Network Design. PWS Publishing Company, New York (1996)
12. Xu., D.: System Analysis and Design Based on MATLAB——Neural Networks. Xidian Press, Xi'an (2002)

A SPN-Based Delay Analysis of LEO Satellite Networks*

Zhiguo Hong[1], Yongbin Wang[2], and Minyong Shi[3]

[1] Postdoctoral Station, Communication University of China, Beijing 100024, China
[2] School of Computer and Software, Communication University of China, Beijing 100024, China
[3] School of Animation, Communication University of China, Beijing 100024, China
hongzhiguo1977@yahoo.com.cn, {ybwang,myshi}@cuc.edu.cn

Abstract. In this paper, we focus on the average time delay analysis of Low Earth Orbit (LEO) satellite networks. Firstly, a geometrical analysis of LEO satellite networks is given according to the characteristics of communication process. By taking the communication process of LEO satellite networks into account, average time delay is formulated by four parameters, i.e. average altitude of LEO satellite networks, uplink bandwidth, Inter-satellite links (ISLs) bandwidth and downlink bandwidth. Then, a Stochastic Petri Net (SPN) model is constructed. Accordingly, in order to represent the characteristics of average time delay, the arrival rate in the SPN model is mapped from the four above-mentioned parameters. Furthermore, the effects of these parameters on average time delay are also analyzed with Stochastic Petri Net Package (SPNP) 6.0. Because of its easiness and accuracy, the proposed approach has great benefit to the design and performance analysis of satellite networks.

Keywords: LEO satellite networks, Geometrical analysis, SPN model, SPNP.

1 Introduction

Recent years have seen the development that satellite networks hold the promise of providing effective and inexpensive global coverage, providing connectivity in the areas where existing terrestrial networks are either infeasible or impractical to deploy [1],[2].

The feasibility of physical experiment of satellite networks is slim because of the expensiveness of satellites and so on, mathematical model-based performance analysis would be a good choice. The queue theory was used to analyze the propagation delay in the satellite networks with inter-satellite links (ISLs)[3]. However, the theory has its limitation in modeling relatively complicated structures like

* The paper is supported by the Project of China Next Generation Internet under Grant No. CNGI-04-12-2A and the science and technology Project for colleges and universities of State Administration of Radio Film and Television (SARFT) under Grant No. BG0206.

F. Sun et al. (Eds.): ISNN 2008, Part II, LNCS 5264, pp. 598–606, 2008.

blocking, locking, simultaneous resource possession and synchronization in the satellite networks.

Petri nets, which was first developed in 1962 by C.A. Petri in his PhD. dissertation, is powerful in modeling concurrent, distributed, asynchronous behaviors of a system [4]. With algebra theory and the net theory as its mathematical basis, the Petri nets theory has been successfully employed to describe various relations and behaviors of the discrete event system and communication networks [5],[6],[7].

However, different from terrestrial networks, there exist long propagation delay, dynamic topology variation and rare on-board resources in satellite networks. Wang C. J. analyzed a class of LEO satellite networks using Stochastic Petri Nets (SPN) models and calculated the performance of LEO satellite networks, assuming packet loss, processing delay and propagation delay to be negligible [8]. However, for actual satellite networks, the effect of propagation delay should not be omitted for it strongly affects the delay analysis in LEO satellite networks. Hong Z. G. et al. constructed a Generalized Stochastic Petri Net (GSPN) model to analyze the performance of LEO satellite networks in the case of packet loss. Further, the impact of buffer size's effect on time delay and packet loss probability in LEO satellite networks is also analyzed with SPNP6.0 software package [9]. Wu F. G. et al. contructed GSPN models to carry out the delay analysis of a double-layered satellite network. Through two sets of experiments the delay analysis of satellite networks were conducted, and it showed that the double-layered satellite network outperforms single-layered ones on the heavy traffic load. Furthermore, the feasibility and effectiveness of the proposed approach is verified by simulation experiments [10]. Nevertheless, these papers neglected the analysis of geometrical characteristics during communication process in LEO satellite networks. How to model and analyze satellite networks performance accurately is becoming an urgent and important issue in the field of satellite networks.

In this paper, a Stochastic Petri Nets (SPN) model is constructed to analyze the impact of some parameters on the delay performance of LEO satellite networks by taking the geometrical characteristics into account. Furthermore, the effects of average altitude of LEO satellite networks, uplink bandwidth, ISLs bandwidth and downlink bandwidth on average time delay are also analyzed with Stochastic Petri Net Package (SPNP) 6.0 [11].

The paper is organized as follows. Section 2 shows the architecture of LEO satellite networks and analyzes the composition of delay in satellite networks. Section 3 presents the constructed SPN model of LEO satellite networks by assuming communication mode to be full duplex. Section 4 calculates the networks average time delay through four experiments with SPNP 6.0 for different parameters. Finally, Section 5 concludes the paper.

2 The Architecture of LEO Satellite Networks

2.1 The Structure of LEO Satellite Networks

Communication satellites can be classified into three groups according to their altitudes of orbit: high earth orbit (HEO), medium earth orbit (MEO) and low earth orbit (LEO). Because of having the shortest distance from the earth, LEO satellite network has some obvious advantages, e.g. in time delay, over HEO and MEO satellite network.

The height of LEO varies from 500 to 2000Km [12] and the LEO network is composed of satellites, which are located at different orbit planes but connected with each other through interplane and intraplane satellite link. ISLs are the bidirectional communication links among satellite networks. For satellite networks' constellation of multiple orbits, ISLs can be divided into intra-PlaneISLs and inter-PlaneISLs.

In our model, we assume that the LEO satellites from a Walker Polar constellation offer terrestrial access services, and all the orbits of this Walker pattern satellite constellation have the same inclination. Consequently, the satellites can keep each other's position relatively static so as to simplify the routing strategy [13]. The Walker Parameter of the LEO satellite networks in this paper is 64/8/3. Figure 1 shows the LEO satellite networks with T/P/F=64/8/3.

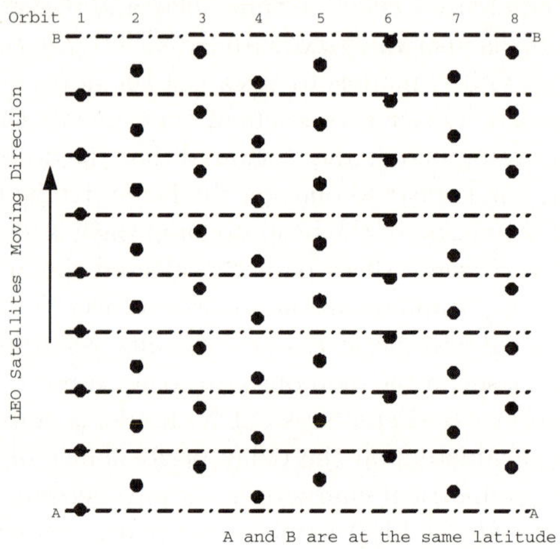

Fig. 1. LEO satellite networks with T/P/F=64/8/3

2.2 Geomety-Based Analysis of Delay in LEO Satellite Networks

In order to improve the accuracy of modeling and simulating LEO satellite networks, it is necessary to analyze the composition of delay in satellite networks.

The end-to-end delay experienced by a data packet traversing the satellite network is the sum of the transmission delay, the uplink and downlink ground segment to satellite propagation delay and the buffering delay etc. Here, we consider the buffering delay to be omitted. There exist three phases in the whole communication process, i.e. uplink period, period of link among satellite networks and downlink period.

Propagation Delay. Propagation delay is the linear function of propagation distance. According to the data packet's propagation direction, the propagation delay (T_{prop}) is the sum of uplink User Data Links (UDLs) propagation delay (T_{prop_up}), links propagation delay among LEO satellite networks (T_{prop_sat}) and downlink UDLs propagation delay (T_{prop_down}).

(1) Calculation of uplink and downlink's propagation delay
By considering the symmetric characteristic of LEO satellite networks, we simplify the calculation of T_{prop_up} and T_{prop_down} as the linear function of LEO satellite networks' average altitude:

$T_{prop\_up} = T_{prop\_down} = h/c$,where c is the propagation speed of electromagnetic wave as a constant value being 3×10^{8}m/s and h represents the average altitude of LEO satellite networks.

(2) Calculation of link propagation delay among LEO satellite networks
From geometrical analysis, the distance of intra-PlaneISLs is formulated as:

$d_{intra} = \sqrt{2}(R_e + h)\sqrt{1 - \cos(\frac{360^\circ}{M})}$,where R_e is the radius of the earth as a constant value being 6378.14Km and M represents the number of satellites per plane.

Similarly, the distance of inter-PlaneISLs is formulated as:

$d_{inter} = \sqrt{2}(R_e + h)\cos\theta\sqrt{1 - \cos(\frac{360^\circ}{2\times N})}$,where θ denotes the latitude of satellite node and Nrepresents the number of orbit planes.

In the case of multiple satellite orbit planes, data packets would probability traverse several hops of intra-PlaneISLs and inter-PlaneISLs in the satellite networks. Let hop_{intra} and hop_{inter} denotes the hops number of intra-PlaneISLs and inter-PlaneISLs respectively, then data packets' total propagation delay via ISLs is given by:

$$T_{prop\_sat} = (hop_{intra} \times d_{intra} + hop_{inter} \times d_{inter})/c \tag{1}$$

Transmission Delay. Transmission delay (T_{trans}) is the time generated by transmitting data packets at the speed of network rate and can be denoted by: $T_{trans} = ps/data\ rate$, where ps is the packet size and *data rate* represents data packet's transmitting rate. Generally, equals network's bandwidth numerically.

Accordingly, T_{trans} is the sum of uplink User Data Links (UDLs) transmission delay (T_{trans_up}), links transmission delay among LEO satellite networks (T_{trans_sat}) and downlink UDLs transmission delay (T_{trans_down}).

Here, we use B_{up}, B_{sat}, B_{down} to denote the bandwidth of uplink, links among LEO satellite networks, downlink respectively. So, the counterpart delays are formulated by:

$$T_{trans\_up} = ps/B_{up}, \quad T_{trans\_sat} = ps/B_{sat}, \quad T_{trans\_down} = ps/B_{down}$$

Let T_{up}, T_{sat}, T_{down} symbolize the uplink bandwidth, average bandwidth of ISLs, downlink bandwidth respectively. We can have:

$$T_{up} = T_{prop\_up} + T_{trans\_up} = h/c + ps/B_{up} \tag{2}$$

$$T_{down} = T_{prop\_down} + T_{trans\_down} = h/c + ps/B_{down} \quad (3)$$

$$T_{sat} = T_{prop\_sat} + T_{trans\_sat} \\ = \sqrt{2} \times (R_e + \overline{h}) \times c^{-1} \\ \times \left(hop_{intra} \times \sqrt{1 - \cos(\frac{360°}{M})} + hop_{inter} \times \cos\theta \times \sqrt{1 - \cos(\frac{360°}{2 \times N})} \right) + ps/B_{sat} \quad (4)$$

3 SPN Model for LEO Satellite Networks

According to the analysis of satellite node's geometrical characteristics, composition of delay and communication mode, we make the following assumptions:

1. The performance of the LEO satellites is alike.
2. Messages received from ground stations follow the Poisson process.
3. Messages are transmitted independently in the satellite networks.
4. The satellite networks link is of full duplex type.

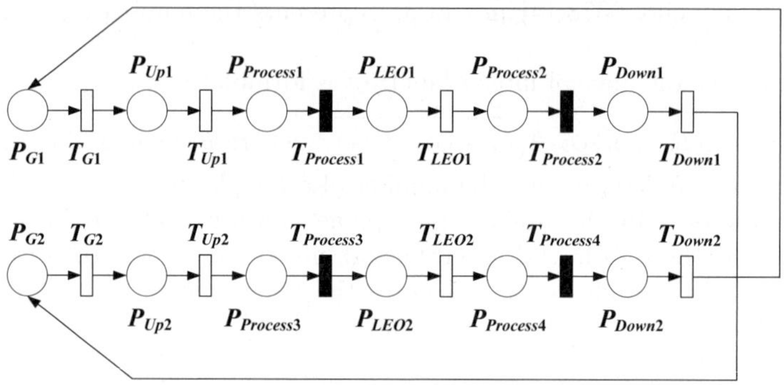

Fig. 2. SPN model for LEO satellite networks

Figure 2 shows our constructed SPN model. The objects of the SPN model are listed in Table 1.

In the SPN model, we associate the Poisson process with the arrival of data packets. For Poisson arrival process, the interval of average arrival is the reciprocal of arrival rate numerically. Here, we use it as a basis for us to set parameters of the SPN model properly. By taking the full duplex mode of delivering data packets and the symmetry in the SPN model, we set the related parameters as follows: $\lambda_1 = 1/T_{up}$, $\lambda_2 = 1/T_{sat}$, $\lambda_3 = 1/T_{down}$.

4 SPN-Based Delay Analysis of LEO Satellite Networks

Average time delay is an important performance index of the satellite networks. We evaluate networks performance using SPNP 6.0 software to concern average

Table 1. List of SPN Objects in Figure 2

Name	Marking/rate	Meaning
P_{G1},P_{G2}	P	Data packets arriving from the ground station
T_{G1},T_{G2}	β	Delivering data packets
P_{Up1},P_{Up2}	0	Waiting to access transmitting links
T_{Up1},T_{Up2}	λ_1	Transmitting data packets via uplink from the ground station to LEO satellite networks
$P_{processi}$ (i=1,2,3,4)	0	Acquiring transmitting links
$T_{processj}$ (j=1,2,3,4)	0	Delivering data packets
P_{LEO1},P_{LEO2}	0	Waiting to access transmitting links
T_{LEO1},T_{LEO2}	λ_2	Transmitting data packets via ISLs among LEO satellite networks
P_{Down1},P_{Down2}	0	Waiting to access transmitting links
T_{Down1},T_{Down2}	λ_3	Transmitting data packets via downlink from LEO satellite networks to the ground station

time delay in the SPN model. For the delay analysis of LEO satellite networks, four experiments are designed to study different parameters' effects on average time delay.

***Case* 1:** In the case of θ=$\pi/3$, B_{up}=100Kbps, B_{sat}=5Mbps, B_{down}=10Mbps, hop_{intra}=2, hop_{inter}=2, N=6, M=11, and β=100, the effect of LEO satellite networks' average altitude h on average time delay is shown in Figure 3. For the different parameters of h=500Km, 1200Km and 2000Km, three different series are depicted. Due to the increase of h, the distance between terrestrial node and satellite networks becomes longer which results in the addition of propagation delay. Consequently, it leads to the increase of average time delay. Figure 3 also shows that for a fixed average altitude, with the argument of data packet number average time delay seems to increase rapidly.

***Case* 2:** In the case of θ=$\pi/3$, h=1200Km, B_{sat}=5Mbps, B_{down}=10Mbps, hop_{intra}=2, hop_{inter}=2, N=6, M=11, and β=100, the effect of uplink bandwidth on average time delay is shown in Figure 4. For the different parameters of B_{up}=10Kbps, 100Kbps and 1Mbps, three different series are depicted. Due to the upgrade of B_{up}, the transmission delay of data packets via uplink tends to become smaller and the propagation delay is constant. As a result, it leads to the decline of average time delay. What's more, for larger available uplink bandwidth the transmission delay is rather smaller. Then the propagation delay via uplink during the communication process would account for a larger proportion and become the main part of average time delay. So we can also observe from Figure 4 that the curve of Bup=100kbps is slightly above that of B_{up}=1Mbps, i.e. there is a little improvement of average time delay due to the increase of Bup when the uplink bandwidth reaches Mbps level.

***Case* 3:** In the case of θ=$\pi/3$, h=1200Km,B_{up}=100Kbps, B_{down}=10Mbps, hop_{intra}=2, hop_{inter}=2, N=6, M=11, and β=100, the effect average ISLs bandwidth on average time delay is shown in Figure 5. For the different parameters of Bsat=50Kbps, 5Mbps and 500Mbps, three different series are depicted. Due

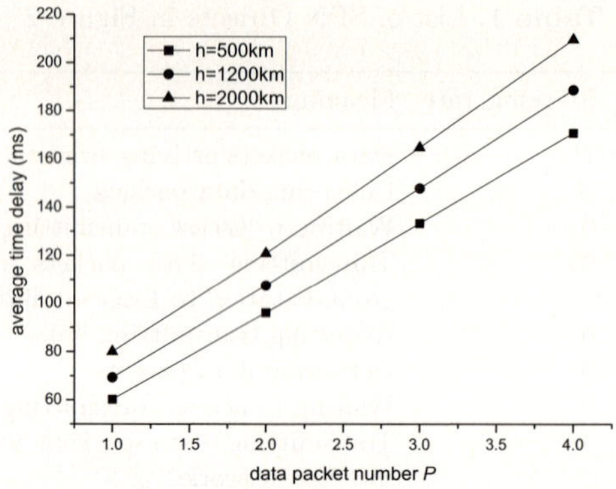

Fig. 3. Effect of LEO satellite networks average altitude on average time delay

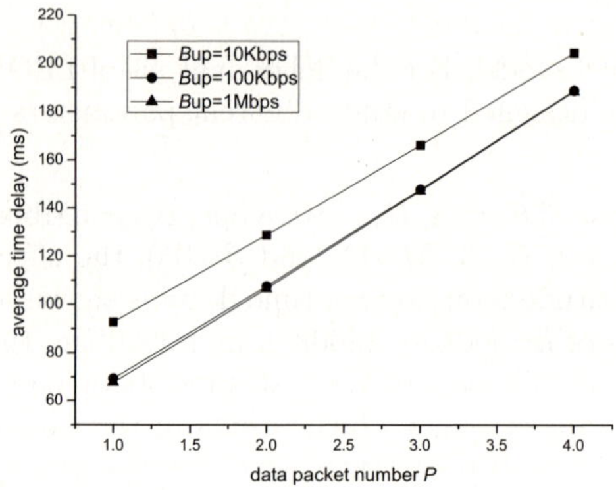

Fig. 4. Effect of uplink bandwidth on average time delay

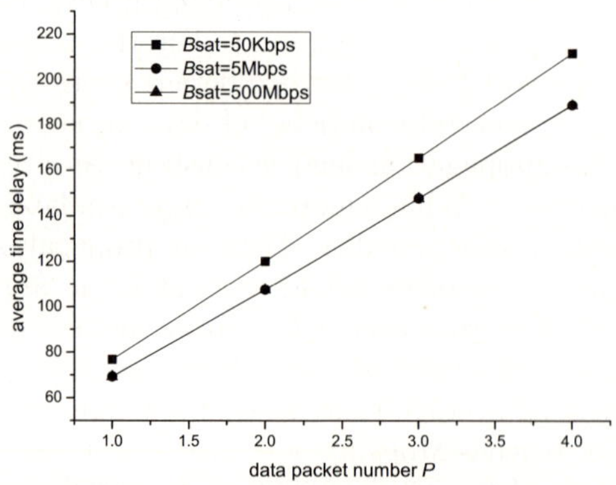

Fig. 5. Effect of average ISLs bandwidth on average time delay

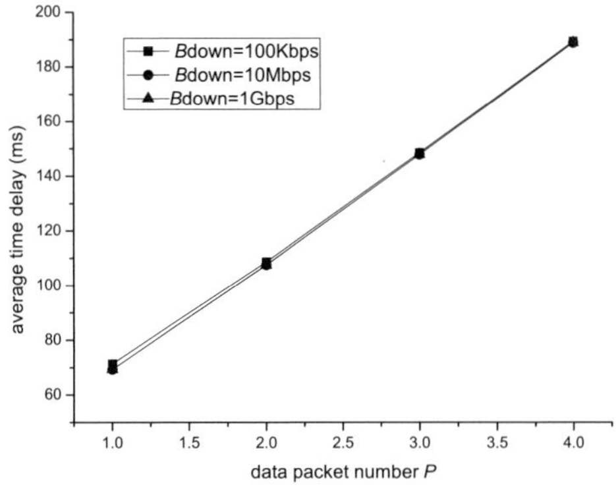

Fig. 6. Effect of downlink bandwidth on average time delay

to the upgrade of B_{sat}, the transmission delay of data packets via ISLs tends to become smaller and the propagation delay is constant. As a result, it leads to the reduction of average time delay. Additionally, for larger available ISLs bandwidth the transmission delay is considerable smaller. Then the propagation delay via ISLs during the communication process would account for a larger proportion and become the major part of average time delay. So it can also be seen from Figure 5 that the curve of B_{sat}=5Mbps is slightly above that of B_{sat}=500Mbps, i.e. there is a little improvement of average time delay due to the increase of $B_{sat}$when the ISLs bandwidth is in Mbps level.

***Case* 4:** In the case of $\theta=\pi/3$, h=1200Km, B_{up}=100Kbps, B_{sat}=5Mbps, =10Mbps, hop_{intra}=2, hop_{inter}=2, N=6, M=11, and β=100, the effect of downlink bandwidth on average time delay is shown in Figure 6. For the different parameters of B_{down}=100Kbps, 10Mbps and 1Gbps, three different series are depicted. Due to the upgrade of B_{down}, the transmission delay of data packets via downlink seems to become smaller and the propagation delay is constant. As a result, it leads to the reduction of average time delay. Since in our experiment the size of each packet is set to 32 bytes, when B_{down}exceeds 100Kbps the transmission delay would become ms level or 10^{-3}ms or even small. It can result in appreciable improvement with the argument of B_{down}. That's the reason why we can observe from Figure 6 that the three series of B_{down}=100Kbps, B_{down}=10Mbps and B_{down}=1Gbps tend to overlap each other.

5 Conclusion

In this paper, we construct a SPN-based model to analyze the delay performance of LEO satellite networks. The effects of average altitude, uplink bandwidth, ISLs bandwidth and downlink bandwidth on average time delay are given with four examples respectively. It can be concluded that increasing average altitude of LEO satellite networks would argument average time delay, whereas enhancing uplink bandwidth or ISLs bandwidth or downlink bandwidth respectively would decrease average time delay. Further, it can be derived for a certain uplink bandwidth or ISLs bandwidth or downlink bandwidth, average time delay would become bigger with the addition of data packet number. The results we get can be used for the design and performance optimization of satellite networks.

On the basis of current work, further work on the simulation and performance analysis of multi-layered satellite networks can be carried on.

References

1. Iera, A., Molinaro, A., Marano, S., Petrone, M.: QoS for multimedia applications in satellite systems. IEEE Multimedia, 46–53 (October-December1999)
2. Hu, Y.F., Maral, G., Ferro, E.: Service efficient network interconnection via satellite. John Wiley & Sons, Chichester (2002)
3. Wang, C.J.: Delivery time analysis of low earth orbit satellite work for seamless PCS, IEEE. Journal on Selected Areas in Communications 13, 389–396 (1995)
4. Petri, C.A.: Communication with automata, Tech. Rep. RADC-TR-65-377, Rome Air Dev. Center, New York (1966)
5. Chen, D.Y., Soong, B.H., Trivedi, K.S.: Optimal call admission control policy for wireless communication networks. In: Proc. Int'l Conf. on Information, Communication and Signal Processing (ICICS), Singapore (October 2001)
6. Xiong, C., Murata, T., Tsai, J.: Modeling and simulation of routing protocol for mobile ad hoc networks using Colored Petri Nets, Research and Practice in Information Technol. Australian Computer Society 12, 145–153 (2002)
7. Wise, J., Xia, J., Chang, C.K., Huang, J.C.: Performance analysis based on requirements traceability. Tech. Rep. 05-04, Dept. Computer Science, Iowa State University (2005)
8. Wang, C.J.: Performance modeling of a class of low earth orbit satellite networks. In: IEEE GLOBECOM 1993, Houston, pp. 569–573 (1993)
9. Hong, Z.G., Fan, Z.H., Li, L., Xu, F.J., et al.: A GSPN-based performance analysis of LEO satellite networks. Journal of Information and Computational Science 1(1), 47–51 (2004)
10. Wu, F.G., Sun, F.C., Yu, K., Zheng, C.W.: Performance evaluation on a double-layered satellite network. International Journal of Satellite Communications and Networking 23(6), 359–371 (2005)
11. Hirel, C., Tuffin, B., Trivedi, K.S.: SPNP: Stochastic Petri Nets. In: Haverkort, B.R., et al. (eds.) TOOLS 2000. LNCS, vol. 1786, pp. 354–357. Springer, Heidelberg (2000)
12. Vatalaro, F., Corazza, G.E., Caini, C., Ferrarelli, C.: Analysis of LEO, MEO, and GEO global mobile satellite systems in the presence of interference and fading. IEEE Journal on Select Areas in Communication 13, 291–300 (1995)
13. Walker, J.G.: Satellite constellations. Journal of the British Interplanetary Society 37, 559–571 (1984)

Research on the Factors of the Urban System Influenced Post-development of the Olympics' Venues

Changzheng Liu[1], Qian Ding[1], and Yao Sun[2]

[1] College of Computer Science and Technology, Harbin University of Science and Technology, Harbin, Hei Longjiang, 150080, P.R. China
[2] Automation College, Harbin Engineering University, Harbin, Hei Longjiang, 150001, P.R. China
liuchangzhengy@gmail.com

Abstract. Post-development of the Olympics' venues is a big challenge which the Games host cities should face with the continuing development of the Olympic Games. Based on the interactive mechanism between the urban system and the Olympics' venue projects as well as the demand theory, with the relevant statistic materials of the recent Games host cities in the Games year, this paper explores the factors of the urban system which influence post-development of the Olympics' venues, meanwhile, applies two methods of ward cluster analysis and k-mean cluster analysis to verify the validity of the factors, per capita GDP, population, ratio of the service industries of the urban system which influence post-development of the Olympics' venues. And the conclusion is given that these 3 factors are the important determinants of the urban system which support post-development of the Olympics' venues, and also the base of making the strategic and tactical plans to improve the venues after the Games.

Keywords: Olympics, Venues, Urban system, Post-development.

1 Introduction

With the rapid development of the Olympics, the quantity of its venues is increasing, and the various standards of the venues are also rising higher and higher. Therefore, with the gorgeous firework cleared off in the closing ceremony of the Olympic Games, the future of the huge and ingenious venues would be on focus more and more. In the light of the relevant regulations of IOC, the construction cost of the permanent venues would not be shouldered by the OCOG, and should be funded or guaranteed by the city or the region, even the central or federal government. The huge construction of the venues brings big economic burden to the host cities, the memory of the Montreal trap is not faded, Sydney of Australia and Athens of Greece, none of them escaped the predicament [1], especially post-development of those venues is more difficult, which inhabitants of the host cities are not familiar with the events that are conducted [2]. Some scholars believe that the main reasons for this situation are that the Olympics' venues are expanding and very exquisite, luxurious, the feasibility study and master planning are not enough in-depth, and post-development and

F. Sun et al. (Eds.): ISNN 2008, Part II, LNCS 5264, pp. 607–614, 2008.

demand are not analyzed carefully [3]. However, the relative studies don't explain the reasons of building the "luxurious venues" and the basic theory of the feasibility study and master planning, post-development and demand analysis. Some researchers provided the strategies for post-development of the Olympics' venue projects, within the context of life circle, for instant, the master planning being made rationally, public and private investment cooperating, and with mixed management of the venues after the Games [4]. It is also an important trend for the venues to stage more and more international mega sport events to improve the tourism, and establish the venue legacy foundations [5]. But there is not enough analysis about the interactive mechanism between the urban system and the venue projects. Therefore, the relevant strategies provided are lack of integrality and pertinence. Under the urban system conditions, this paper studies the factors of the urban system which influence post-development of the Olympics' venues, and establishes the base for making the strategic and tactical plans to improve the venues after the Games, and it is of great significance for sustainable development of the Olympics' venues, host cities, and even the Games itself.

2 Methods

2.1 Establishing the Factors of the Urban System

The venues of the Olympic Games are very exquisite and luxurious, because the customers are very fastidious. The functions of the venues should meet the requirements of the IFs, the appearance of the venues would reflect the high-tech and the culture of the host cities to show their beauty, improve their awareness before different countries' athletes, coaches, officials, media, audiences, etc. But both the construction of the Olympics' venue projects before the Games and post-development of the venue projects after the Games will need to be nourished by the urban system. According to the interactive mechanism between the urban system and the venue projects, the urban system interacts with the venue projects depending on the relevant industries previous, during, and after the Games, within the context of life circle, and impacts post-development of the Olympics' venue projects. Before the Games, the factors of the urban system that influence post-development of the venue projects include macro economy, the development level of the relevant industries (mainly service industries), which will determine the industries layout in the urban system, and influence the venues location, planning, designing, as well as the functions after the Games. This phase is the planning period of the venue projects. After the Games, the factors of the urban system that influence post-development of the venue projects still include macro economy, the development level of the relevant industries (mainly service industries), which will influence the demand for the venues directly after the Games, reflect the support of the relevant industries to post-development of the venue projects. This phase is the realization period of functions of the venue projects, and influencing post-development of the venue projects directly.

Under market economy condition, post-development of the Olympics' venue projects will relay on the market environments that the urban systems determine. On the base of the demand theory, demand is the quantity of a good or service that customers are willing and able to purchase during a specified period under a given set of economic conditions. Conditions to be considered are price of the goods, price and availability of relevant goods, consumer incomes, consumer tastes and preferences, advertising expenditure, expectations of price changes, population, etc. Among these factors, advertising expenditure is leaned upon decisions of an individual firm, price of the good, price and availability of relevant goods are formed by the market that the urban system shapes, consumer tastes and preferences are the part of the society psychology of the urban system, expectations of price changes mainly affect the durable goods and luxury goods, consumer incomes and population are decided by the macro economy and the natural situation of the urban system. So, the factors of the urban system which influence demand for the venue projects after the Games are mainly price of the goods, price and availability of relevant goods, consumer incomes, and population.

The macro economy and consumer income are positive related, and both of them can be reflected by per capita GDP. Higher or lower of price of the goods, price and availability of relevant goods are relative under various market environments of different urban systems, and determined by the development level of the relevant industries. With the development level of the relevant industries rising, price of the goods, price and availability of relevant goods will decrease, and leading the demand curve to an upward or rightward shift. Meanwhile, ratio of the service industries can replace the development level of the relevant industries to reflect price of the goods, price and availability of relevant goods. Therefore, per capita GDP, population, ratio of the service industries, these 3 factors of the urban system are the important determinants which influence post-development of the Olympics' venues.

2.2 Verifying the Factors of the Urban System

Table 1 and Table 2 are the statistic materials of per capita GDP, population, ratio of the service industries of the recent Olympic Summer & Winter Games host cities in the Games year respectively, and analysis of these statistic materials can verify the validity of the 3 factors of the urban system which influence post-development of the Olympics' venue projects. Based on availability of the statistic materials, the some data of population are the statistic materials in the year which closed the Games year, because census may not be done for the host cities in the Games year, the years of the statistic materials are within brackets. Meanwhile, in order to compare the different urban systems on the same stage, per capita GDP is calculated the present value in 2000 US dollars.

With the 3 factors as the variables, cluster analysis is applied to cluster the different urban systems to verify the validity of the 3 factors of the urban system which influence post-development of the Olympics' venue projects. In other words, with the 3 factors as the variables and the urban systems as cases, if some urban systems can be cluster in one group which post-development of the Olympics' venues is well, and some urban systems can be cluster in another group which post-development of the Olympics' venues is bad, the validity of the 3 factors of the urban system influenced post-development of the Olympics' venues is proved.

The dollars is the present value in 2000. Source: X.D. Cheng, "The Competitiveness of Cities Bidding the International Mega Sports Events" in press.

The dollars is the present value in 2000. Source: X.D. Cheng, "The Competitiveness of Cities Bidding the International Mega Sports Events" in press.

Table 1. The relative statistic materials of recent Olympic Summer Games host cities

Year	Host Country & City	Per capita GDP (dollars)*	Population [Year] (persons)	Ratio of the service industries (%)
1988	South Korea Seoul	10,240	10,800,000 [1989]	0.563
1992	Spain Barcelona	25,201	1,643,542 [1991]	0.679
1996	United States Atlanta	42,000	416,474 [2000]	0.783
2000	Australia Sydney	32,000	4069,093 [2000]	0.696
2004	Greece Athens	20,100	772,072 [1991]	0.610

Table 2. The relative statistic materials of recent Olympic Winter Games host cities

Year	Host Country & City	Per capita GDP (dollars)*	Population (persons)	Ratio of the service industries (%)
1994	Norway Lillehammer	42609	24,170 [1994]	0.606
1998	Japan Nagano	30700	360,000 [1998]	0.735
2002	United States Salt Lake City	41975	178,097 [2002]	0.783

The detailed procedure of the methods is just as following steps. Firstly, the standardization of the statistic materials in Table 1 and Table 2 is performed, because the measurement scales are not same across the 3 dimensions. Then, the ward cluster analysis is used to cluster the recent Olympics host cities, meanwhile, both the agglomeration schedule and the dendrogram obtained with the ward procedure are inspected to identify the distance between the different urban systems. After that, k-mean cluster analysis is undertaken by specifying the cluster number to examine the cluster members and the distance between the different clusters. The aim of using two different methods is to minimize the dependence of the solution on the method chosen. Lastly, depending on the results from the 2 different methods with the discriminate strategies discussed above, the validity of the 3 factors of the urban systems which influence post-development of the Olympics' venue projects can be verified. SPSS, version 13.0, is used to perform the analysis.

3 Results

3.1 Ward Cluster Analyses

Table 3 is the results of the ward cluster analysis of the recent summer & winter Olympic Games host cities with the 3 factors as the variables. And the Euclidean distance is used.

Table 3. The results of the ward cluster analysis of the recent Olympics host cities

Case		Number of Clusters						
6	Norway Lillehammer	X	X	X	X	X	X	X
		X	X	X	X	X	X	
8	United States Salt Lake City	X	X	X	X	X	X	X
		X	X	X	X	X	X	X
3	United States Atlanta	X	X	X	X	X	X	X
		X	X					
4	Australia Sydney	X	X	X	X	X	X	X
		X	X	X				
7	Japan Nagano	X	X	X	X	X	X	X
		X	X	X	X			
5	Greece Athens	X	X	X	X	X	X	X
		X	X	X	X	X		
2	Spain Barcelona	X	X	X	X	X	X	X
		X						
1	South Korea Seoul	X	X	X	X	X	X	X

From the results of the ward cluster analysis, Salt Lake City and Atlanta of US are clustered together first, then Lillehammer of Norway joins them. And the relative research materials uncovered that these urban systems could provide better conditions for post-development of the Olympics' venues, and the Olympics' venues developed better after the Games indeed. However, Seoul of South Korea is last one which is clustered with the other cities, and it is believed that the urban system of Seoul supported post-development of the Olympics' venues little, and post-development of the venues in Seoul was depressed.

Because the scales of the winter Games are smaller than the summer Games, the winter Games host cities and the summer Games host cities are clustered separately to verify the validity of the 3 factors of the urban system which influence post-development of the Olympics' venues under the condition of the different scales of the Games. The cluster results are in Tables 4 and 6.

From the results in the Table 4, Salt Lake City of US and Lillehammer of Norway are clustered together first, then Nagano of Japan joins them. Table 5 is the proximity matrix of the ward cluster analysis. From the proximity matrix, the distances between Nagano and the two other cities are large, 1.091 (Salt Lake City) and 1.420 (Lillehammer) separately, and the distance between Salt Lake City and Lillehammer is only 0.492. In other words, the urban systems of Salt Lake City and Lillehammer are more similar. In fact, the Olympics' venues perform better in these urban systems, such as Salt Lake City and Lillehammer, post-development of the venues in Nagano urban system could only follow them, which had been proved.

From the cluster results in Table 6, it is clarified that Barcelona of Spain and Athens of Greece are clustered together first, then Sydney of Australia joins them. There are Atlanta of United States, Seoul of South Korea which follow the 3 cities. From Table 7, the distance between Barcelona and Athens is 0.194, and the distance between Sydney and Athens is 0.499, the distance between Sydney and Barcelona is 0.317. This means that the abilities of these urban systems are alike, which influence

Table 4. The results of the ward cluster analysis of the recent Winter Olympics host cities

Case		Number of Clusters	
1	Norway Lillehammer	X	X
		X	X
3	United States Salt Lake City	X	X
		X	
2	Japan Nagano	X	X

Table 5. Proximity matrix of the ward cluster analysis

Case	Euclidean Distance		
	1	2	3
1	0.000	1.420	0.492
2	1.420	0.000	1.091
3	0.492	1.091	0.000

Table 6. The results of the ward cluster analysis of the recent Summer Olympics host cities

Case		Number of Clusters			
3	United States Atlanta	X	X	X	X
		X	X		
4	Australia Sydney	X	X	X	X
		X	X	X	
5	Greece Athens	X	X	X	X
		X	X	X	X
2	Spain Barcelona	X	X	X	X
		X			
1	South Korea Seoul	X	X	X	X

post-development of the Olympics' venues. The min distance between Seoul and other cities is 0.953 (Sydney), and the max distance is 1.431 (Atlanta), which indicate that Seoul is most far away from other cities. And Atlanta is also a city which is far away form the other cities, the min distance between Atlanta and other cities is 0.480 (Sydney), and the max distance is 1.431 (Seoul), but its urban system supported post-development of the Olympics' venues strongly, and the support from Seoul urban system to post-development of the Olympics' venues is little, performance of support from the other 3 urban systems are between Atlanta and Seoul, just as the relevant researches had uncovered.

Although the summer Games host cities and the winter Games host cities are discussed separately, the results are similar with the results which all host cities are discussed together. It shows that no matter large or small are the scales of the Games, the factors of the urban system which influence post-development of the Olympics' venues are same.

Table 7. Proximity matrix of the ward cluster analysis

Case	Euclidean Distance				
	1	2	3	4	5
1	0.000	1.006	1.431	0.953	1.016
2	1.006	0.000	0.552	0.317	0.194
3	1.431	0.552	0.000	0.480	0.712
4	0.953	0.317	0.480	0.000	0.499
5	1.016	0.194	0.712	0.499	0.000

3.2 K-Mean Cluster Analysis

The results are given in the Table 8, which the recent Games host cities are clustered by k-mean cluster analysis in the light of number of clusters from 2 to 5 separately, with the 3 factors as variables and cities as cases. And the Euclidean distance is used.

Table 8. The results of the K-Means cluster analysis of the recent Olympics host cities

Case		2 clusters	3 clusters	4 clusters	5 clusters
1	South Korea Seoul	1	1	1	1
2	Spain Barcelona	2	2	2	5
3	United States Atlanta	2	3	3	3
4	Australia Sydney	2	3	4	4
5	Greece Athens	2	2	2	5
6	Norway Lillehammer	2	3	3	3
7	Japan Nagano	2	3	4	2
8	United States Salt Lake City	2	3	3	3

From the results of k-mean cluster analysis, when 2 clusters are determined, Seoul is far away from the other cities, and appeared that the support from the urban system is weak. When 5 clusters are determined, Seoul is still in a single group, Nagano is in a single group, Sydney is also in a single group, Salt Lake City, Atlanta and Lillehammer are in one group, Athens and Barcelona are in one group. From the distances between the final cluster centers in Table 9, it can be identified that Salt Lake City, Atlanta and Lillehammer have the similar abilities to support post-development of the Olympics' venues. Although Nagano and Sydney are in a single group separately, both of them are close the 5th cluster, which Barcelona and Athens are in the group. And the distance between Nagano and the 5th cluster is 0.278, and the distance between Sydney and the 5th cluster is 0.396. So, all of the 4 cities have the similar abilities to support post-development of the Olympics' venues. Although they are better than Seoul, there is still a long way to go for them to chase the cities in the 3rd cluster on the abilities to support post-development of the Olympics' venues, which was proved by the relevant researches.

From discussion above, the cluster results by two different methods are similar, and both of them verify the validity of the 3 factors of the urban system which influence post-development of the Olympics' venues.

Table 9. Distances between final cluster centres of the K-Means cluster analysis

Cluster	1	2	3	4	5
1		1.170	1.403	0.927	0.973
2	1.170		0.356	0.349	0.278
3	1.403	0.356		0.478	0.616
4	0.927	0.349	0.478		0.396
5	0.973	0.278	0.616	0.396	

4 Conclusions

According to the cluster results discussed above, the conclusion can be given that the urban system influences post-development of the Olympics' venues by the 3 factors, per capita GDP, population, ratio of the service industries. Therefore, it is important to study these 3 factors of the urban system when making the strategic and tactical plans for post-development of the Olympics' venues.

Acknowledgments. This paper is supported by the Natural Sciences Foundation of Heilongjiang Province under Grant QC06C028, and Scientific Research Fund of Heilongjiang Provincial Education Department (NO□11531326) , Doctor Foundation 20050217021.

References

1. Gratton, C., Shibli, S., Coleman, R.: Sport and Economic Regeneration in Cities. Urban Studies 42, 985–999 (2005)
2. Foster, I., Kesselman, C.: Old Locals and New Spaces. Leisure Studies 24, 415–434 (2005)
3. Lin, X.: Research on the Construction and Post-Games Utilization of Modern Olympic Venues. Journal of Beijing Sport University 28, 1441–1444 (2005)
4. Luo, P.: Research of the Management Profit of Large Sports Entertainment Yards in China. Journal of Xian University of Physical Education 24, 17–20 (2007)
5. Steinke, T.: Olympic Games Host City Marketing: An Exploration of Expectations and Outcomes. Sport Marketing Quarterly 12, 37–47 (2003)

A Stock Portfolio Selection Method through Fuzzy Delphi

Mehdi Fasanghari[1] and Gholam Ali Montazer[2]

[1] MSc Student of IT Eng., Tarbiat Modares University, Tehran, Iran, P.O. Box: 14115-179
fasanghari@gmail.com
[2] Assist. Professor of IT Eng., Tarbiat Modares University, Tehran, Iran, P.O. Box: 14115-179
montazer@modares.ac.ir

Abstract. The evaluation and selection of stocks is one of the most important decision issues for stock market managers. Owing to vague concept frequently represented in decision data, an agent-based decision-making approach is proposed to solve the stock portfolio selection problem. In the proposed method, the experts' opinions are described by trapezoidal fuzzy numbers, and the Fuzzy Delphi method is adopted to adjust each expert's opinion to achieve the consensus condition. Finally, a practical example of stock portfolio selection in Tehran Stock Exchange (TSE) is demonstrated its engineering. The results show that the method is flexible and credible in stock market decision-making.

Keywords: Fuzzy system, Intelligent agent, Tehran Stock Exchange (TSE), Fuzzy Delphi, Fuzzy number.

1 Introduction

Evaluating the stocks is a complex system of interacting elements and a multiple criteria decision-making problem [1]. During the last decade, the agent-based systems have become one of the most widely used methods for evaluating social management problems and stock portfolio selection [2-5].

In addition, conflict always occurs in group decision making since members in a group generally cannot reach the same decision. How to resolve conflicts becomes an important issue in group decision making. For group decision making, the main approach is collective individual decision-making [6]. This approach focuses on group benefit consideration.

Frequently, real world decision-making problems are ill defined, i.e., their objectives and parameters are not precisely known. These obstacles have been dealt with using the probabilistic approach, but, due to the fact that the requirements on the data and on the environment are very high and that many real world problems are fuzzy by nature and not random, the probability applications have not been very satisfactory in a lot of cases. On the other hand, the application of fuzzy set theory in real world decision-making problems has given very good results. Its main feature is that it provides a more flexible framework, where it is possible to solve satisfactorily many of the obstacles of lack of precision.

Usually, experts express their opinions by means of numerical values (numerical setting). When experts are not able to give exact numerical value to express their

F. Sun et al. (Eds.): ISNN 2008, Part II, LNCS 5264, pp. 615–623, 2008.

opinions, then a more realistic alternative option is using linguistic assessments instead of numerical values [7, 8]. In such a situation, for each variable of problem domain an appropriate linguistic label set is chosen and used by individuals participating in the decision-making process to express their opinions. This setting is known as the linguistic setting.

In this paper, we will propose a fuzzy agent-based decision-making method for evaluating stocks. The experts' opinions are described by linguistic terms, which can be expressed in trapezoidal fuzzy numbers, to make the consensus of the experts are consistent. We use Fuzzy Delphi method to adjust the fuzzy ranking of every expert to achieve the consensus condition. Therefore, we can get the best selection for evaluating system by our proposed method.

The rest of this paper is organized as follows. In next section, intelligent agent is defined. Then, the preliminaries of trapezoidal fuzzy number are briefly introduced. The Fuzzy Delphi is adopted to obtain the consensus of the experts' opinions in next section. Afterward, the architecture of fuzzy-agent-based stock portfolio selection for TSE was presented. Finally, case study in TSE is put forward to illustrate our method, and conclusions are drawn in the last section.

2 Intelligent Agent

Intelligent software agents are probably one of the fastest growing areas of information technology. They are being used and touted for applications as diverse as personalized information management, electronic commerce, computer games, etc. An agent can be thought of as a computer program that simulates a human relationship by doing something that another person could do for you [9] and performs a specific task on behalf of a user, independently or with a little guidance [10]. An agent is a self-contained program capable of controlling its own decision making and acting based on its perception of environment in pursuit of one or more objectives [11].

More than one type of agent is possible. In its simplest form it is a software object that sifts through large amounts of data and presents a subset of these data as useful information to another agent, user, or system. These types of agents are called static agents. Mobile agents have the ability to migrate across nodes of a network in order to perform their tasks and report back their findings. These agents typically gather and analyze data from a multitude of nodes on the network, and present a subset of this data as information to a user, agent, or system. Mobile agents can also act as brokers for users; for example, a single sign-on agent can sign on to many different systems, relieving the user from typing in his/her password for every system [12]. An agent that roams the stock exchanges of the world and trades shares on its user's behalf can, using fixed rules, build up a valuable portfolio of shares for a user, but in this paper we use static agent to prepare recommendation to users.

3 Trapezoidal Fuzzy Number

The fuzzy set theory was introduced by Zadeh [13] to deal with problems in which a source of vagueness is involved. A trapezoidal fuzzy number can be defined by a quadruplet $\tilde{A} = (a_1, a_2, a_3, a_4)$, where $a_1 \le a_2 \le a_3 \le a_4$, its member function represented as follows

$$\mu_{\tilde{A}} = \begin{cases} 0 & x < a_1 \\ (x-a_1)\big/(a_2-a_1) & a_1 \le x \le a_2 \\ 1 & a_2 \le x \le a_3 \\ (x-a_4)\big/(a_3-a_4) & a_3 \le x \le a_4 \\ 0 & x > a_4 \end{cases} . \tag{1}$$

Trapezoidal fuzzy numbers are appropriate for quantifying the vague information about most decision problems [14], and the primary reason for using trapezoidal fuzzy numbers can be stated as their intuitive and computational-efficient representation.

In this paper, the trapezoidal fuzzy number is used for ranking the stock with the purpose of stock portfolio selection. More details about arithmetic operations laws of trapezoidal fuzzy number can be seen in [4].

4 Fuzzy Delphi

Because the decisions made by the experts rely on their individual competence, their decisions are subjective, and it is more appropriate to present the data by trapezoidal fuzzy numbers instead of crisp numbers, we utilize Fuzzy Delphi method [4] to adjust the fuzzy evaluation of each expert for achieving the consensus condition of the all experts consistent. The Fuzzy Delphi method consists of the following steps.

Step1. Considering experts E_i provide the possible realization ranking of a certain event. The evaluation value given by each expert E_i are presented in the form of a trapezoidal fuzzy

$$\tilde{A}^{(i)} = (a_1^{(i)}, a_2^{(i)}, a_3^{(i)}, a_4^{(i)}), \qquad i = 1, 2, \ldots, n . \tag{2}$$

Step2. Firstly, the average $\tilde{A}_m$ of all $\tilde{A}^{(i)}$ is computed using average

$$\tilde{A}_m = (a_{m1}, a_{m2}, a_{m3}, a_{m4}) = (\frac{1}{n}\sum_{i=1}^{n} a_1^{(i)}, \frac{1}{n}\sum_{i=1}^{n} a_2^{(i)}, \frac{1}{n}\sum_{i=1}^{n} a_3^{(i)}, \frac{1}{n}\sum_{i=1}^{n} a_4^{(i)}). \tag{3}$$

Then, for each expert E_i , the differences

$$\Delta\tilde{A}^{(i)} = \tilde{A}_m - \tilde{A}^{(i)} = (a_{m1} - a_1^{(i)}, a_{m2} - a_2^{(i)}, a_{m3} - a_3^{(i)}, a_{m4} - a_4^{(i)}) \tag{4}$$

are calculated and sent back to the expert E_i for reexamination.

Step3. Each expert E_i presents a revised evaluation value $\tilde{B}^{(i)} = (b_1^{(i)}, b_2^{(i)}, b_3^{(i)}, b_4^{(i)})$. This process starting with Step2 is repeated. The average $\tilde{B}_m$ is calculated and the differences that now $\Delta\tilde{A}^{(i)}$ are substituted corresponding by $\Delta B^{(i)}$. If it still necessary new evaluation value $\tilde{C}^{(i)} = (c_1^{(i)}, c_2^{(i)}, c_3^{(i)}, c_4^{(i)})$ are presented, and their average

$\tilde{C}_m$ is calculated. The process could be repeated again and again until successive means $\tilde{A}_m, \tilde{B}_m, \tilde{C}_m, \ldots$ become reasonably close (We can define the distance of two fuzzy numbers, $d_i \leq 0.2$).

Step4. At a later time, the same process may reexamine the ranking if there is important information available due to new discoveries.

5 The Architecture of Fuzzy-Agent-Based System

Our approach for stock portfolio selection in the fuzzy agent-based decision making method is a three phase process, which has been shown in Fig. 1. Accordingly the stock portfolio selection application was developed using Matlab 7.0.4. The functional architecture of the application can be divided into six modules, as illustrated in Fig. 2: (1) user interface module, (2) fuzzification module, (3) defuzzification module, (4) fuzzy inference engine module, (5) fuzzy rule-base module, and (6) optimal project selection module.

User interface module collects the off-line information for the subsequent modules. The information includes the alternatives for evaluation. The major roles of the stock portfolio selection module include the preparation and setup of the selection. Hence, the stock portfolio selection module can assist in loading the off-line information through the interaction with the user.

The user interface module collects scores of all of the stock's parameters and turns them to the corresponding trapezoidal fuzzy numbers. Fuzzification refers to the process of taking a crisp input value and transforming it into the degree required by the terms [15]. Since our inputs, as signals, don't have variance fuzzy singleton has been chosen as the fuzzification method.

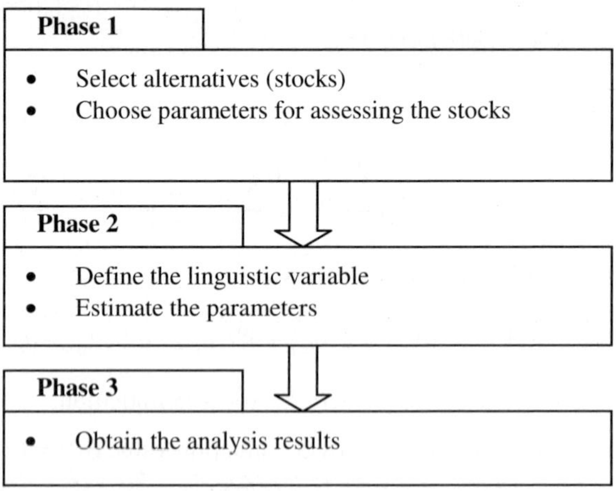

Fig. 1. Process of project selection

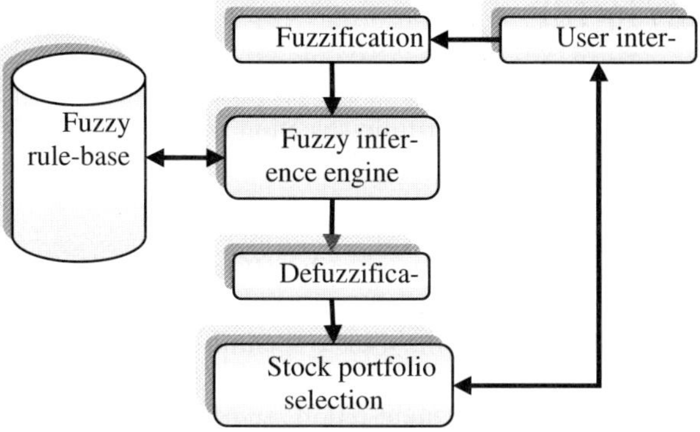

Fig. 2. The architecture of fuzzy-agent-based for project selection

In the max–min composition fuzzy inference method, the min operator is used for the AND conjunction (set intersection) to make conservative recommendation and the max operator is used for the OR disjunction (set union) in order to evaluate the grade of membership of the antecedent clause in each rule. Cause of local rules and conservative method, the Mamdani inference is used. As a result, the standard max–min inference algorithm by

$$\mu_{B'}(y) = \max_{l=1}^{M} \sup_{x \in U} \min[(\mu_{A'}, \ldots, \mu_{A'_n}(x_n), \mu_{B'_y}(y))] \tag{5}$$

is used in the fuzzy inference process, as it is a commonly used fuzzy inference strategy [15].

The process of computing a single number that best represents the outcome of the fuzzy set portfolio matrix parameter is called defuzzification [15]. There are several existing methods that can be used for defuzzification. We chose the centroid method which is

$$y^* = \frac{\int_v y \cdot \mu_{B'}(y)\, dy}{\int_v \mu_{B'}(y)\, dy} \tag{6}$$

since it is more accurate than other methods.

Fuzzy system makes decisions and generates output values for stock portfolio selection based on knowledge provided by the experts in the form of IF _ THEN rules. The rule base consists of 219 rules like: IF market of selected stock is good, its sale's rules is medium, its EPS is good, its project is good, its stockholders are good, its legal audit report is bad, and its float shares is bad then this stock is suitable for selection, that is collected through literature review and consult with 42 stock company.

After every decision-maker completes each round of data acquisition, the stock portfolio selection module, runs the system to obtain the results.

With the scores representing the complete information, the stocks which have the higher score are selected to make the stock portfolio.

6 Case Study in TSE

In order to evaluate the applicability of the proposed method, we implemented it in TSE. The three phases for stock portfolio selection are described as follows.

6.1 Phase 1

In order to reduce the difficulty of case study, 10stock was chosen. We collected the important parameters from literature review and consulted with different experts for their opinions. This step can help the user to identify the relevant parameters of positioning the stocks that consist of 7 parameters: market of stock, sale's rules, EPS and DPS, projects, stockholders, legal audit report, and float shares. We then conducted a group that included 3 experts in the field of stock market to determine the parameters and assess the stocks.

Two basic assumptions are declared:

1. All stocks are independent of one another.
2. Each stock can be selected without any constraint in number of stocks.

6.2 Phase 2

The main activity in this phase is collecting the definition of the linguistic variables that was used in fuzzy rule-base, and the corresponding trapezoidal fuzzy numbers with Fuzzy Delphi. Here, the three managers defined the linguistic variables and trapezoidal fuzzy numbers for every one of the parameters (result of last round in Fuzzy Delphi is shown in Table 1 to Table 7.

Table 1. Membership functions for linguistic value of market of stock

Linguistic values	Fuzzy numbers
Low	(0,0,0.5,7.25)
Medium	(5.25,6.25,6.5,7.75)
High	(6.25,9.75,10,10)

Table 2. Membership functions for linguistic value of sale's rules

Linguistic values	Fuzzy numbers
Low	(0,0,2,4.5)
Medium	(3,6.25,7.25,8)
High	(7.25,9.25,10,10)

Table 3. Membership functions for linguistic value of EPS

Linguistic values	Fuzzy numbers
Low	(0,0,0.25,6.25)
Medium	(3.75,5.75,6.5,8.5)
High	(5.25,9,10,10)

Table 4. Membership functions for linguistic value of projects

Linguistic values	Fuzzy numbers
Low	(0,0,1,7.75)
Medium	(5.5,6,6.25,8)
High	(5,9.25,10,10)

Table 5. Membership functions for linguistic value of stockholders.

Linguistic values	Fuzzy numbers
Low	(0,0,1.75,7.25)
Medium	(4.75,5.25,6,7.75)
High	(6,8.75,10,10)

Table 6. Membership functions for linguistic value of legal audit report

Linguistic values	Fuzzy numbers
Low	(0,0,1.25,5.5)
Medium	(3.25,5.75,6.25,8.25)
High	(7.25,8.5)

Table 7. Membership functions for linguistic value of float shares

Linguistic values	Fuzzy numbers
Low	(0,0,1.5,5.75)
Medium	(4,5.75,6,7.25)
High	(7,8.25,10,10)

Furthermore, experts inputted the parameters value for all of 10 selected stocks as inputs of one of our experts is illustrated in Table 8.

Table 8. Fuzzy scores of selected stocks by one of expert of TSE

Stock	Fuzzy Scores
Stock 1	(3,5,7)
Stock 2	(3,5,8)
Stock 3	(1,2,3)
Stock 4	(2,4,5)
Stock 5	(5,6,7)
Stock 6	(4,6,8)
Stock 7	(2,5,7)
Stock 8	(2,3,4)
Stock 9	(5,6,8)
Stock 10	(4,5,7)

6.3 Phase 3

In this phase, the fuzzy agent-based has been run, and solution is shown in Table 9. This shows that in sequence the stocks in higher level are better for selection in stock portfolio than lower level stocks. For example, the stock number 5 is better than stock number 4.

The result implies that stock portfolio selection with this sequence, can maximize the outcome of selected stock portfolio.

Table 9. Ranking of selected stocks of TSE

Stocks	Stock defuzzification mean	Stock defuzzification variance	Stock ranking
Stock 5	6.6	0.25	1
Stock 4	6.45	0.01	2
Stock 8	6.44	0.3	3
Stock 10	5.91	0.21	4
Stock 1	5.74	0.12	5
Stock 6	4.2	0.03	6
Stock 7	5.74	0.42	7
Stock 3	3.81	0.09	8
Stock 9	5.48	0.00	9
Stock 2	3.57	0.20	10

7 Conclusion

In this study, the proposed fuzzy-agent-based system is a flexible system that: (a) simultaneously considers all the different criteria in determining the most suitable stocks, (b) takes advantage of the best characteristics of the existing methods, (c) involves the full participation of the users in deciding the alternative stocks, and evaluation parameters, and (d) provides that users can investigate the impact of changes in certain parameters on the solution and quickly receive feedback on the consequences of such changes. The implementation of the proposed fuzzy-agent-based system for a stock portfolio selection undertaken by TSE confirmed the above considerations. The experts of the TSE were pleased and agreed with our recommendations since the fuzzy-agent-based system can reduce the decision-making time, and are a practical tool for dealing with the uncertainty problem of linguistic terms.

References

1. Yang, C.W., Huang, K.S., Yu, G., Jan, D.Y.: Using queuing theory to estimate the storage space of stocker in automated material handling systems. In: Semiconductor Manufacturing Technology Workshop, pp. 102–104 (2002)
2. Tseng, C.C.: Portfolio management using hybrid recommendation system. In: IEEE International Conference on e-Technology, e-Commerce and e-Service, pp. 202–206 (2004)
3. Kendall, G., Su, Y.: A multi-agent based simulated stock market - testing on different types of stocks. Evolutionary Computation 4, 2298–2305 (2003)

4. Lee, J.W., Hong, E., Park, J.: A Q-learning based approach to design of intelligent stock trading agents. In: IEEE International Engineering Management Conference, vol. 3, pp. 1289–1292 (2004)
5. Pandey, V., Ng, W.K., Lim, E.P.: Financial advisor agent in a multi-agent financial trading system. In: 11th International Workshop on Database and Expert Systems Applications, pp. 482–486 (2000)
6. French, S.: Decision Theory, England (1986)
7. Herrera, F., Viedma, E.H., Verdegay, I.L.: A linguistic decision process in group decision making. Group Decision Negotiation 5, 165–176 (1996)
8. Herrera, F., Viedma, E.H., Verdegay, J.L.: Choice processes for non-homogeneous goup declsion making in linguistic setting. Fuzzy Sets and System 94, 287–308 (1998)
9. Selker, T.: A teaching agent that learns. Communications of the ACM 37, 92–99 (1994)
10. Bui, T., Lee, J.: An agent-based framework for building decision support systems. Decision Support Systems 25, 225–237 (1999)
11. Jennings, N., Wooldridge, M.: Software Agents. IEEE Review, 17–20 (1996)
12. Merwe, J.V.D., Solms, S.H.V.: Electronic commerce with secure intelligent trade agents. Computers & security 17, 435–446 (1998)
13. Zadeh, L.A.: Fuzzy Sets. Information and Control 8, 338–353 (1965)
14. Cheng, C.H., Lin, Y.: Evaluating the best main battle tank using fuzzy decision theory with linguistic criteria evaluation. European journal of operation research 142, 174–176 (2002)
15. Ngai, E.W.T., Wat, F.K.T.: Design and development of a fuzzy expert system for hotel selection. Omega (The International Journal of Management Science) 31, 275–286 (2003)

A Prediction Algorithm Based on Time Series Analysis

JianPing Qiu, Lichao Chen, and Yingjun Zhang

Intelligent Software Technology Laboratory, School of Computer Science,
Taiyuan University of Science and Technology,
030024, Taiyuan, Shanxi, China
choujianp@yahoo.com.cn

Abstract. In the context of the Semantic Web, it may be beneficial for a user to receive a forecast regarding the reliability of an information source. We offer an algorithm for building more effective social networks of trust by using CLRM (classic linear regression models). For managing uncertainty, we introduce some random variables which neither the consumer nor the provider can control its value. Such random variables that can be successively accumulated from each stage of multi-stage forecasts are reduced through the use of analytical tools that combine statistical methods with advances in time series analysis. Time series analysis can relate 'current' values of a critical variable to its past values and to the values of current. Moreover, to model real world scenario, VAR-GARCH (Vector Auto Regression Generalized Autoregressive Conditional Heteroskedasticity) model is used to represent forecasting results which are generally influenced by interactions between decision makers.

Keywords: Semantic Web of Trust, CLRM, Time Series Analysis, VAR-GARCH.

1 Introduction

The vision of the Semantic Web is to construct a common semantic interpretation for World Wide Web pages, to one day reliably run software to interpret the information conveyed in any of its documents. In building the Semantic Web, however, information may be supplied by a wide selection of sources, with the result that a user seeking information will need to judge whether the content of any given source is in fact trustworthy. It is, therefore, important to develop models for trust in the context of the Semantic Web. Various approaches to date have been formulated about how best to form a Web of Trust, to share information and selectively choose trustworthy partners from whom information may be obtained. In this paper, an algorithm is presented which is used to predict the reliability on the Semantic Web.

2 Background and Related Work

We motivate the need to acquire information about the reliability of sources and then briefly outline some current research on modeling the trustworthiness of sources. This includes some discussion of approaches to communicate with other users to obtain

F. Sun et al. (Eds.): ISNN 2008, Part II, LNCS 5264, pp. 624–631, 2008.

advice about sources, sometimes referred to as a Web of Trust, as well as an approach for addressing the problem that some users may provide untruthful advice. The challenge of trusting information providers in a Web-based environment is discussed in [1]. Paolucci provided valuable insights into the need for trust on the Web, in the context of Web services, where Web sites dynamically exchange information using XML descriptions, but where it is difficult to ensure that the meaning of the messages being sent is well understood, without human intervention. The Semantic Web contributes by providing ontologies for Web services to interpret meanings in exchanged messages. According to [1], with the Semantic Web, the interaction between users and providers needs a process of capability matching to link users with providers of Web services. Specifically, providers advertise their capabilities, a user sends a request for the type of service he requires, a registry matches the capabilities of providers and the capabilities expected by the user, and finally the user selects the most suitable provider. However, in their advertisements, providers may lie about their capabilities in order to be selected by the user. To avoid selection of an untruthful provider, there is a need to properly model the trustworthiness of providers. In [2] this problem is reinforced for the Semantic Web: whether to trust the content of a Web resource, depending on the source. Richardson [3] explain further that due to the great diversity of the Web, it is difficult to expect the content to be consistent and of high quality. It then becomes important to decide how trustworthy each information source is. Maximilien and Singh [3][4][5] adopt an agent-based approach for modeling trust on the Semantic Web. Their work focuses on representing multiple qualities of services (QoS) for automatic runtime Web service selection. This trust model is based on a shared conceptualization of QoS and takes into account providers' quality advertisement, consumers' quality preferences, quality relationships, and consumers' quality tradeoffs. To select a Web service implementation, a consumer dynamically associates a trust value with each service Implementation and selects the service implementation with the highest assigned level of trust. The trust value of each service implementation partially depends on its reputation value, which is determined by the set of quality values from other users who previously selected that provider. Kagal [6] use a DAML + OIL trust ontology in a multi-agent system, which is based on a distributed trust and delegation mechanism verifying that a user's credentials are acceptable. The trust ontology is built for specifying credentials and checking if the credentials conform to policies. A policy maps credentials to a certain ability or right. The mechanism allows propagation of trust beliefs exchanged between users and avoids repeated checking of users' credentials. The research of Gil and Ratnakar [2] provides a framework for users to express their trust about a source and the statements the source contains, by annotating each part of the source to indicate their views. The focus of the work is on how to provide an effective interface for users to record their annotations. This TRELLIS system ultimately averages the ratings provided over many users and many analyses, to present a reflection of the trustworthiness of the source. A credibility-reliability pair emerges for each source-statement pair, to derive an overall rating of a single source, based on each of the associated statements provided by the source. Modeling trust on the Semantic Web, as discussed so far in this section, includes a reliance on the beliefs or ratings provided by third parties to be truthful. In fact, it is important to address the problem of possibly unfair or unreliable ratings. One approach that explores this possibility is that of Richardson [3]. In this work, each user first explicitly specifies a small set of users whom he trusts, leading to a Web of Trust. This arrangement allows any user to compute the trustworthiness of a possible provider, based on the ratings supplied by others in his social network. The trust value of a provider is computed locally by combining the trust ratings provided by other users. One feature of this approach is to recursively propagate trust through the user's social network. In effect, trust in a provider is derived using some aggregating functions along each possible chain of trust from the user to the provider. One concern with this approach, however, is that this method of propagating trust may be computationally intractable, as there may be many different paths, of various lengths, which need to be aggregated. In our own research, we are developing a model for representing the trustworthiness of a provider. This framework is sufficiently general to operate in a variety of environments including electronic commerce.

3 Modeling Trustworthiness [7]

In simplest form:

$C = \mu + \xi + \varepsilon$

Where,

C = sum of three forecasts
μ = Third-party forecast
ξ = user information
ε = uncertainty
μ is predicted by analysts.

The user has sources and/or experience to derive private forecast information ξ that is not known to the provider in a decentralized system (information asymmetry).

However, the provider can categorize the user into certain "types" based on prior actions or credibility of the user. Thus, the provider updates its "belief" of the user and may select a value of ξ represented by a normal distribution.

This introduces a random (stochastic) variable. Uncertainty is given by epsilon (ε) and neither the user nor the provider can control its value.

To determine μ, planners and analysts may use one or more statistical tools that may include:

[1] Classic linear regression models (CLRM)
[2] Auto regression (AR)
[3] Moving averages (MA)
[4] ARMA (AR+MA)
[5] Vector auto regression (VAR)

Classic linear regression models (CLRM) have been around for a century and widely used for a variety of purposes. CLRM may be expressed as an equation for a straight line:

$$y_t = \beta_0 + \beta_1 x_t + \varepsilon_t \quad (1)$$

Where,

y = dependent variable of interest to be modeled for forecast (for example, credibility of a provider)

t = time period (frequency of observation, for example, t-1 may indicate prior week 1)

β = coefficients to be estimated (based on values of y and x)

x = explanatory variable that is used to 'explain' variations in the dependent variable y (for example, low credibility of a provider may be explained by low credit rating {x} of a provider)

ε = random (stochastic) error term

This simple technique can model multiple explanatory variables, that is, multiple x's, since the variation in y, say, credibility of a provider, is dependent on multiple parameters, such as credit rating (x_1), occupation (x_2), age (x_3). The choice of x's (number of explanatory variables) will drive the validity and accuracy of the model. Therefore, x's may be based on underlying theoretical principles and/or practical logic. However, no matter how many x's are included, there may be an inherent randomness that cannot be explained by the model. Thus, the random error term (ε) is included in the equation (admission of the fact that the dependent variable (y) cannot be modeled perfectly). To solve for y, a bold assumption is made that ε is characterized by a normal distribution with a mean = 0 and variance = σ^2 for all time periods (t):

$$\varepsilon_t \sim N(0, \sigma^2)$$

The objective of CLRM is to estimate the parameters (β_0,β_1) of the model (from data on y and x), depending on the sample of observations on y and x. Therefore, there can be multiple sets of (β_0,β_1) that can, when plotted, produce straight lines with varying slopes (gradient). This statistical procedure introduces two sources of error. First, taking sample data from a large number of observations inherits sampling errors. One reason for use of sample data (as practiced by the Census) may stem from lack of granular data acquisition tools. Another reason may be a lack of computing power. With low cost yet powerful microprocessors and the emergence of Grid computing, we may be increasingly better prepared to process exabytes of raw data. Second, given the multiple sets of (β_0,β_1) that may be estimated, the objective of CLRM is to choose that pair of (β_0,β_1) which minimizes the sum of squared residuals ($e_1^2, e_2^2, ..., e_n^2$):

$$\sum_{t=1}^{n} e_t^2$$

Where, $e_t$is the random error term for the sample and ε_t represents the random error term of the 'population' data. This technique is known as the principle of ordinary least squares (OLS). The sole objective of OLS is to minimize forecast errors by selecting the most suitable (β_0,β_1), thus ignoring the volatility of the sample.

The attractiveness of CLRM based forecasting stems from the fact that we can model cross variable linkages. The regression model is an explicit multi-variate model. Hence, forecasts are made not only on the basis of the variable's own historical data (for example, credibility of a provider, y, the dependent variable) but also takes into account the historical data of other related and relevant explanatory variables, x_1 through x_K, that is, any number of x's(credit rating (x_1), occupation (x_2), age (x_3). In our example, credibility of a provider, may be modeled by the analysts of a provider not only based on the history of its own credit rating (x_1), occupation (x_2) and age (x_3) but also taking into account the historical data with respect to credit rating (x_4), occupation (x_5) and age (x_6) of its competitor provider $x_{4t}, x_{4t-1}, x_{4t-2}, ..., x_{4t-n}$.

With time series analysis, let us develop the concept by starting with a basic CLRM equation:

$$y_t = \beta_0 + \beta_1 x_{1t} + \beta_2 x_{2t} + ... + \beta_K x_{Kt} + \varepsilon_t \tag{2}$$

The effect of change in only one explanatory variable ($x_1, ..., x_K$) may be analyzed at a time (all else remains constant). Therefore, in building this model, the choice of x is a process decision based on the model builder's knowledge about the operation.

We start by forecasting the values of x's to obtain an unconditional forecast for y. Instead of inserting arbitrary values for future x's, we use forecasted values based on historical data. To forecast x, we fit an univariate model to x where we use past (lagged) values of x to forecast x, as given in equation 3 (for $x_{1t}, ..., x_{Kt}$):

$$\begin{aligned} x_{1t} &= \alpha_{01} + \alpha_{11} x_{1t-1} + \alpha_{12} x_{1t-2} + ... + \alpha_{1N_{x_{1t}}} x_{1t-N_{x_{1t}}} + \mu_{x_{1t}} \\ &\vdots \\ x_{Kt} &= \alpha_{01} + \alpha_{11} x_{kt-1} + \alpha_{12} x_{kt-2} + ... + \alpha_{1N_{x_{kt}}} x_{1t-N_{x_{kt}}} + \mu_{x_{kt}} \end{aligned} \tag{3}$$

$$y_t = \beta_0 + \sum_{i=1}^{N_{x_1}} \alpha_{1i} x_{1t-i} + ... + \sum_{i=1}^{N_{x_{kt}}} \alpha_{Ki} x_{Kt-i} + \varepsilon_t \tag{4}$$

$$y_t = \beta_0 + \sum_{k=1}^{K} \sum_{i=1}^{N_{x_{kt}}} \alpha_{ki} x_{kt-i} + \varepsilon_t \tag{5}$$

Where,

x_{it} = variable x_1 at time t (for example, we used x_1 for credit rating thus x_{1t} is credit rating at time t)

x_{Kt} = variable x_K at time t (up to K number of x's)

x_{1t-1} = value of x_1 at time t-1 (referred to as the lagged value by one period)

N = period up to which the lagged values of x_{1t}will be used in the equation

μ = random error term

In equation 3, $\alpha_{11}, \alpha_{22}$are coefficients of $x_{1t-1}, x_{1t-2}$and are referred to as lagged weights. An important distinction is that instead of arbitrarily assigning weights, these coefficients are estimated using OLS technique. The error term in equation 3 represented by $\mu$is analogous to εin equation 2. Depending on the number of x's ($x_1, ...x_K$) that adequately represents the process being modeled in equation 1, there will be K number of equations (of the type equation 3) that must be estimated to forecast the x's ($x_1, ...x_K$) which will then be used to obtain an unconditional forecast of y. Thus, to simplify the task, we can estimate all the parameters (α, β) simultaneously by re-writing equation 2, the basic CLRM equation, as equation 4 or its shortened version, as in equation 5 (above).

Equation 5 is another step toward forecasting the dependent variable (y) with greater accuracy using forecasts of x's based on historical data of x's (lagged values). But no sooner, we have moved a step ahead, it is clear that equation 5 ignores the impact on y of the past values of y itself (lagged values). Consequently, a preferable model will include not only lagged values of x but also lagged values of y, as shown in equation 6 (below).

$$y_t = \beta_0 + \sum_{j=1}^{N_y} \varphi_j y_{t-j} + \sum_{k=1}^{K} \sum_{i=1}^{N_{x_{kt}}} \alpha_{ki} x_{kt-i} + \varepsilon_t \tag{6}$$

Present observation of a variable to its past history, for example:

y_t to $y_{t-1}, y_{t-1}, ..., y_{t-p}$

Where, p indicates the order of the autoregressive process AR (p) or the period up to which the historical data will be used (a determination made by using other statistical tools). Thus, AR is a technique by which a variable can be regressed on its own lagged values. For example, today's credibility (y_t) may depend on credibility from yesterday (y_{t-1}) and the day before (y_{t-2}). AR (p) is appealing to forecasters because a real-world model must link the present to the past (yet remain dynamic). MA expresses current observations of a variable in terms of current and lagged values of the random error $\varepsilon_t, \varepsilon_{t-1}, \varepsilon_{t-2}, ..., \varepsilon_{t-q}$where q is the order of the moving average process MA (q). Combining AR (p) and MA (q) we get ARMA (p, q) where p and q represents the lagging order of AR and MA. Engle used this ARMA technique to model the time varying volatility and proposed the Autoregressive Conditional Heteroskedasticity model or ARCH. The 'conditional' nature of non-constant variance (heteroskedasticity) refers to the forecasting of variance conditional upon the information set available up to a time period (t). Modeling variance in this fashion allows us to forecast the volatility of the random error term (ε). Using ARCH technique, the variance of the random error term (ε_t) in equation 6 can be expanded in terms of current and lagged values ($\varepsilon_{t-1}, \varepsilon_{t-2}, ..., \varepsilon_{t-q}$), as follows:

$$\sigma^2 = \theta_0 + \theta_1 \varepsilon_{t-1}^2 + \theta_2 \varepsilon_{t-2}^2 + ... + \theta_q \varepsilon_{t-q}^2$$

Where,

σ^2 is the variant of ε_t[var (ε_t)].

This MA (q) representation of σ_t^2 was later generalized to an ARMA representation of σ_t^2 and is referred to as GARCH. GARCH evolved when Bollerslev extended Engle's MA (q) representation of σ_t^2 (the ARCH model) by combining the existing MA (q) with an AR (p) process, that is, regressing a variable (σ_t^2) on its own (past) lagged values ($\sigma_{t-1}^2, \sigma_{t-2}^2, ..., \sigma_{t-p}^2$). Thus, variance of the random error term (ε) in a certain period (ε_t) depends not only on previous errors ($\varepsilon_{t-1}, \varepsilon_{t-2}, ..., \varepsilon_{t-p}$) but also on the lagged value of the variance ($\sigma_{t-1}, \sigma_{t-2}, ..., \sigma_{t-p}$). Thus, GARCH may be represented by equation 7:

$$\begin{gathered} y_t = \beta_0 + \textstyle\sum_{j=1}^{N_y} \varphi_j y_{t-j} + \sum_{k=1}^{K} \sum_{i=1}^{N_{x_{kt}}} \alpha_{ki} x_{kt-i} + \varepsilon_t \\ \sigma^2 = \theta_0 + \theta_1 \varepsilon_{t-1}^2 + \theta_2 \varepsilon_{t-2}^2 + ... + \theta_q \varepsilon_{t-q}^2 \end{gathered} \tag{7}$$

Variance of the random error term depends not only lagged values of ε (t-1, t-2… t-q) but also on lagged values of the variance δ^2 (t-1, t-2… t-q)

$$\begin{aligned} y_t = \beta_0 + \sum_{j=1}^{N_y} \varphi_j y_{t-j} + \sum_{k=1}^{K} \sum_{i=1}^{N_{x_{kt}}} \alpha_{ki} x_{kt-i} + \varepsilon_t \\ \sigma_t^2 = \theta_0 + \sum_{i=1}^{q} \theta_i \varepsilon_{t-i}^2 + \sum_{j=1}^{p} \tau_j \sigma_{t-j}^2 \end{aligned} \tag{8}$$

Given the data up to time period t, we can use equation 8 to predict, h periods ahead (1, 2, 3… h) where h may be hours or days or weeks. The process model is a key to precision forecasting, hence, tools to create and validate such models may be a pre-requisite. In developing the GARCH model, equation 8 takes into account the lagged values of the dependent variable (Credibility), the impact of multiple explanatory variables (K number of x's that influence Credibility, such as credit rating, occupation), the heteroskedasticity of the error term and the lagged values of the variance of the error term. But, we have not considered the fact that to predict Credibility h periods ahead, it is also crucial to model the interaction between the entities (user, provider, Third-party) in the value network. Interaction between partners can impact any outcome. Collaborative strategies such as CPFR (collaborative planning forecasting and replenishment) may be still plagued by lack of trust between entities and efforts to share data or information, may be, sluggish, at best. It may be essential to model the dynamics between entities. The VAR-GARCH technique captures this dynamics through incorporation of VAR or vector auto regression in addition to GARCH. Previously we discussed AR (p), which is a univariate model. In contrast, VAR (p) is a n-variate (multi-variate) model where we can estimate n different equations and in each equation we regress a variable on p lags of itself as well as p lags of every other variable. The real-world cross-variable dynamics captured by VAR models enables each variable to be related not only to its own past but also to the past values of all other variables in the model.

There are at least two parties in this example. To model this multi-variate dynamics of n=2 using VAR (p), let us assume that p=1 (lagged by 1 period). Equation 8 can be extended to the VAR-GARCH type to model two entities and consider only one lag period (n=2, p=1) as shown in equation 9.

$$\begin{aligned} y_{1t} = \beta_0 + \sum_{k=1}^{K} \sum_{i=1}^{N_{x_{kt}}} \alpha_{ki} x_{kt-i} + \varphi(11) y_{1t-1} + \varphi_{12} y_{2t-1} + \varepsilon_{1t} \\ y_{2t} = \beta_0 + \sum_{k=1}^{K} \sum_{i=1}^{N_{x_{kt}}} \alpha_{ki} x_{kt-i} + \varphi(21) y_{1t-1} + \varphi_{22} y_{2t-1} + \varepsilon_{2t} \\ \sigma_{1t}^2 = \theta_0 + \sum_{i=1}^{q} \theta_i \varepsilon_{1t-i}^2 + \sum_{j=1}^{p} \tau_j \sigma_{1t-j}^2 \\ \sigma_{2t}^2 = \theta_0 + \sum_{i=1}^{q} \theta_i \varepsilon_{2t-i}^2 + \sum_{j=1}^{p} \tau_j \sigma_{2t-j}^2 \end{aligned} \tag{9}$$

Real world results or outcomes are generally influenced by events or interactions between decision makers. In the VAR-GARCH model represented by equation 9 (above), this dynamics is captured by estimating the coefficient φ_{ij} that refers to changes iny_i with respect toy_j. For example, if y_1 represents Credibility in China store and y_2 represents Credibility in USA, then the parameter φ_{12}refers to changes in China (y_1) with respect to in USA (y_2). If any one of the two random error terms (ε_{1t} and ε_{2t} changes, it will impact both the dependent variables (y_1 and y_2). For example, changes in China may impact in USA. Thus, the VAR component, in VAR-GARCH, brings us closer to the real world scenario by making it possible to quantify cross-variable dynamics. For example, if ε_{1t} changes, it will change y_{1t} and since y_{1t}also appears as one of the regressors (explanatory variable) for y_{2t}in the equation, the change in any error term impacts both dependent variables in this VAR representation. These changes have been thus far ignored by current practices.

4 Conclusions

In this paper, we first introduce the Semantic Web setting for sharing information about sources. Due to the fact that any user on the Web can become an information source, there is a need to form a Web of Trust. Some current research on modeling the trustworthiness of information sources on the Semantic Web relies on the unrealistic assumption that an information source is truthful. We offer an algorithm for predicting the trustworthiness of a provider.

References

1. Paolucci, M., Sycara, K., Nishimura, T., Srinivasan, N.: Toward a Semantic Web Ecommerce. In: 6th Conference on Business Information Systems, pp. 153–161. Springs Press, Colorado (2003)
2. Gil, Y., Ratnakar, V.: Trusting Information Sources One Citizen at a Time. In: 1st International Semantic Web Conference, pp. 162–176. Springs Press, Italy (2002)
3. Richardson, M., Agrawal, R., Domingos, P.: Trust Management for the Semantic web. In: 2nd International Semantic Web Conference, pp. 351–368. Springs Press, Island (2003)
4. Maximilten, E.M., Singh, M.P.: Toward Autonomic Web Services Trust and Selection. In: 2nd International Conference on Service Oriented Computing, pp. 212–221. Springs Press, New York (2004)
5. Maximilten, E.M., Singh, M.P.: Agent-based Trust Model Using Multiple Qualities. In: 4th International Autonomous Agents and Multi Agent Systems, pp. 519–526. Springs Press, Netherlands (2005)
6. Kagal, L., Finin, T., Joshi, A.: Developing Secure Agent Systems Using Delegation Based Trust Management. In: 1st International Autonomous Agents and Multi Agent Systems, pp. 27–34. Springs Press, Italy (2002)
7. Qiu, J.P.: Supply Chain Management System Based on Radio Frequency Identification. Journal of Computer Applications 175, 734–735 (2005)

An Estimating Traffic Scheme Based on Adaline

Fengjun Shang

College of Computer Science and Technology
Chongqing University of Posts and Telecommunications
Chongqing 400065, China
shangfj@cqupt.edu.cn

Abstract. In order to manage the whole network, a novel estimating traffic scheme is proposed, called the Adaline model. In this model, there are two parts of works. Firstly, an Adaline network is proposed and implemented for estimating traffic. By using the Adaline network, we are able to overcome the challenge of ill-posed nature of the problem. Secondly, in order to optimize solutions, a prior knowledge is introduced. By using prior, the error is reduced markedly. Numeral result shows our model achieves better performance than the existing representative methods.

Keywords: Traffic matrix, Adaline, Topology.

1 Introduction

OD(Origin-Destination) pairs traffic is important for many network design, engineering, and management functions. In this paper, the traffic matrix is introduced to describe *OD* traffic distribution. However they are often difficult to measure directly. Because networks are dynamic, analysis tools must be adaptive and computationally light weight.It is very important to measure traffic matrix in the network. Firstly, traffic matrix is measured to justify whether to block and evaluate route compotation. In using traffic engineer, the load distribution may be known by the whole network traffic matrix and control traffic and evaluate result. We may also optimize route computation by traffic matrix knowing load distribution. And then traffic matrix may balance load. Furthermore, according to long time traffic measurement, we may know traffic demand trend to be convenient for network planning.

This paper is organized as follows. We discuss the related work in Section 2. In Section 3, the traffic matrix model is introduced. We present performance evaluation and analysis in Section 4. In Section 5, conclusions are presented.

2 Related Traffic Model

The traffic model mainly depicts the traffic distribution in the network, in this paper, we use traffic matrix to depict the traffic distribution. Traffic matrix includes two parts: *OD* pairs traffic and *OD* pair. We introduce interrelated traffic model as follows.

F. Sun et al. (Eds.): ISNN 2008, Part II, LNCS 5264, pp. 632–641, 2008.

In [1], the author uses active method to measure LSP and update by delay and the dropping rate of packet. In [2], it synthesizes user connect information, SLA, traffic forecasting, historical data, network topology etc. to estimate traffic matrix. In [3], Path flow estimator(PFE) is a one-stage network observer proposed in the transportation literature to estimate path flows and path travel times from traffic counts in a transportation network. The estimated path flows can further be aggregated to obtain the origin–destination flows, which are usually required in many transportation applications. Vanderebei and Iannone (1994) assumed independent Poisson traffic counts for the entries of the *OD* matrix, developed three equivalent formulations of the *EM* (Expectation Maximization) algorithm and studied the fixed points of the *EM* operator. Cao et al. (2000) assumed independent Gaussian *OD* pairs traffic flows, and used *EM* algorithm [4] to derive estimates for the parameters, that depended on t. In the traffic mapping, the most measurement systems based on flow, [5][6] works on network boundary. The traffic matrix can be made up according to the measuring result of boundary and synthesizing user information, network topology, forecasting information of traffic demand. In [7], traffic matrix is gotten by sampling method, which adopt hash function and acquire measurement packet. No information about network topology and route is needed in this method. Shortcoming of model need acquire right hash sampling function, but this can't be applied in the reality. Traffic measurement, especially traffic matrix measurement is the most basic demand [8]. In [9], a least-square modeling approach is considered here for solving the *OD* estimation and prediction problem in ITS(Intelligent Transport Systems), which seems to offer convenient and exible algorithms. The dynamic nature of the problem is represented by an auto-regressive process, capturing the serial correlations of the state variables.

In the paper, we research on acquiring *OD* pairs traffic by traffic matrix. We have two contributions listed as follows. Firstly, an Adaline method is proposed to estimate traffic matrix. Secondly, a prior constraint condition is introduced so that it is profitable to solve the equation.

3 Traffic Matrix Model

Traffic matrix includes *OD* flows and traffic distribution for each *OD* flow. Let c be the number of *OD* pairs. If the network has n nodes, then $c = n \times (n-1)$. Although conceptually traffic demands are represented in a matrix X, with the amount of data transmitted from node i to node j as element X_{ij}, it is more convenient to use a vector representation. Thus, we order the *OD* pairs and let X_j be the amount of data transmitted by *OD* pair j. Let $Y = (y_1, \ldots, y_r)$ be the vector of link counts where y_l gives the link counts for link l, and r denotes the number of links in the network. The vectors X and Y are related through an r by c routing matrix A. A is a $\{0, 1\}$ matrix with rows representing the links of the network and columns representing the *OD* pairs. Element a_{ij} =1 if link i belongs to the path associated to *OD* pair j, and a_{ij}=0 otherwise. The *OD* flows are thus related to the link counts according to the following linear relation:

$$AX = Y \tag{1}$$

The routing matrix in *IP* networks can be obtained by gathering the *OSPF* or *IS-IS* links weights and computing the shortest-paths between all *OD* pairs. The link counts

Y are available from the *SNMP* data. The problem is thus to compute X, that is, to find a set of *OD* flows that would reproduce the link counts as closely as possible. The problem described by Equation (1) is highly underdetermined because in almost any network, the number of *OD* pairs is much higher than the number of links in the network, $r << c$. This means that there are an infinite number of feasible solutions for X, so an optimized solution must be found.

3.1 Solving the Equation

According to above analysis, estimating traffic matrix is equal to solving the equation (1). Yet the equation (1) is an ill-posed linear inverse problem, so there are the large numbers of roots for the equation (1) and then we can only estimate the optimal root. In order to acquire the optimal roots, we must make enough use of the prior, that is, we need find out the more constraints to make the equation have roots or make the equation be solved. This paper tries to resort from the current condition and direct estimate the solutions of the equation. The detailed description is as follows.

3.2 Introduction of Adaline

In this study, the neural network model was tested and optimized to obtain the best model configuration for the prediction of the traffic matrix of network. An Adaline network typically comprises three types of neuron layers: an input layer, one or more hidden layers and an output layer each including one or several neurons. As shown in Figure 1, nodes from one layer are connected to all nodes in the following layer, but no lateral connections within any layer, nor feed-back connections are possible. Several input neuron are used, each representing an environmental variable. The output layer comprises one neuron. With the exception of the input neurons, which only connect one input value with its associated weight values, the net input for each neuron is the sum of all input values x_n, each multiplied by its weight w_{jn}, and a bias term z_j which may be considered as the weight from a supplementary input equalling one:

$$a_j = \sum w_{ji} x_i + z_j \tag{2}$$

The output value, y_j, can be calculated by feeding the net input into the transfer function of the neuron:

$$y_j = f(a_j) \tag{3}$$

Many transfer functions can be used. In this study, two types of sigmoid functions have been compared: the tangential and logarithmic sigmoid transfer function[10].

In this paper, we use one model of neural networks i.e., Adaline network, which is selected among the main neural network architectures used in engineering. The basis of the model is neuron structure as shown in Figure 1, where r is the number of elements in input vector. Each input is weighted with an appropriate X. The sum of the weighted inputs and the bias, forms the input to the transfer function f. Neurons may

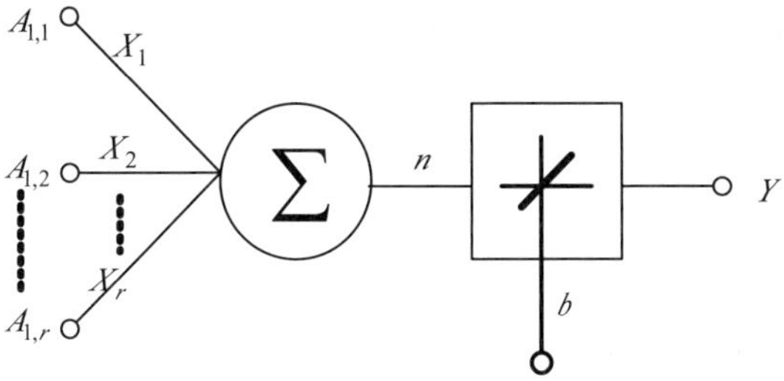

Fig. 1. An elemenrary neuron with r inputs

use any differentiable transfer function f to generate their output. The transfer function of Adaline is shown in Figure 2.

Artificial Neural Networks (ANNs) consist of a great number of processing elements (neurons), connected to each other. The strengths of the connections are called weights. A feedforward multilayer network is commonly used to model physical systems. It consists of a layer of input neurons, a layer of output neurons and one or more hidden layers.

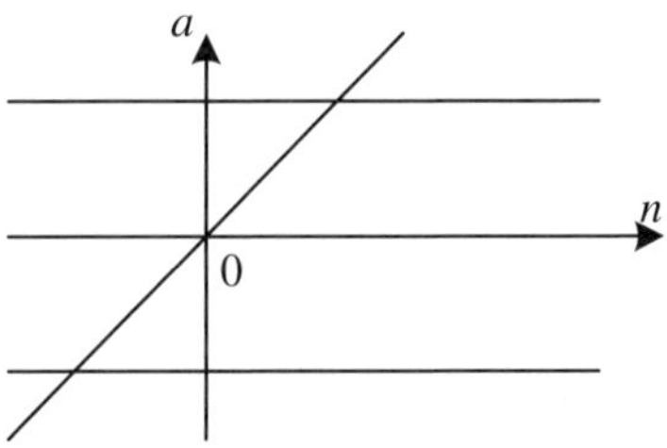

$a = purelin(n)$

Fig. 2. Linear transfer function

Widrow and Hoff's Adaline is one of the most effective, most well understood, and most widely used connectionist learning components (Widrow and Hoff, 1960). This paper introduces the Adaline model to estimate the traffic matrix, because it is linear between *OD* pair traffic and output link traffic.

The error function is defined as follows.

$$E(W,B) = \frac{1}{2}(T - A)^2 = \frac{1}{2}(T - WP)^2 \tag{4}$$

In order to minimize the error function (4), we use the W-H rules. At the same time, in order to optimize the network, we must exercise it, the process is as follows.

1) Calculating output vector $A = W \times P + B$ and error $E = T - A$ between expectation and average value;
2) Comparing between output error and expectation and minimizing the output error.
3) Introducing $W - H$ rule to calculate new weight and error and return (1).

3.3 Mining Subsistent Prior

In order to acquire the optimal solutions, the subsistent prior must be mined. For example, in [11], it is known about the *OD* pairs follows normal distribution, where $\mu = u_0$, $\sigma = 40$ etc., that is $X \sim N(\mu,\sigma)$ etc., the prior must be enough made use of.

4 Results and Algorithm Analysis

In order to explain our idea, two kinds of topology are selected: symmetrical topology and unsymmetrical topology.

4.1 Unsymmetrical Topology

We use unsymmetrical 4-node topology shown in Figure 3 to calculate the solution of equation (1) and compare between our method and the methods in [11]. In this topology, there are 7 direct links and 12 *OD* pairs.

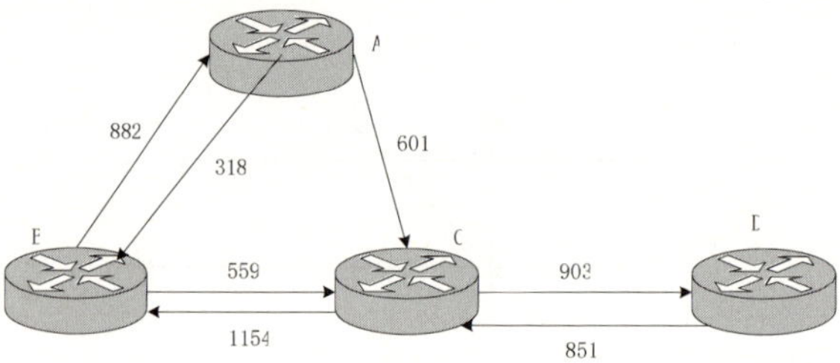

Fig. 3. 4-node network topology

We make use of this network, because the all *OD* pairs and link are known. It is beneficial for us to analysis traffic behavior and network performance. We introduce the data in [11] to calculate and compare among other methods.

Given the link measured value be real value without measured error. The route matrix *A* is invariable and observed link value is listed as follows.

$$Y=(318, 882, 601, 559, 1154, 903, 851)$$

LP, Bayesian and EM denote this method in [11]. Our method denotes Adaline and the results are shown on Table 1, where EST donates estimated value respectively.

Comparing results among four methods are shown on Table 1. The value in Table 1 includes estimated and absolute error. In **Poisson** distribution model, from Table 1, we may calculate average errors, where the errors are 98% for the LP method, 13% for the Bayesian method, 4.9% for the Adaline method respective. It is obvious that result of LP method is the worst in three kinds of methods. The reasons maybe are the network topology or no constraint. In next step, we will study how to improve the LP method make errors become little. For this small topology the Adaline method performs better than the Bayesian method in terms of both the average error and the worst case error. It is interesting that the worst case errors for the two statistical methods do not correspond to the same *OD* pairs. The Bayesian method makes its two biggest errors for *OD* pairs CB and CD, while the Adaline method makes its two biggest errors for *OD* pairs BD and AC. The link BD carries the largest number of *OD* pair flows, namely four, among all the links in the small network. We see that the Adaline method makes its worst error for the *OD* pair BD. This hints that estimated errors may be correlated to heavily shared links.

Table 1. Comparing results among LP, Bayesian, EM and Adaline

OD pairs Observation (**Poisson**)	LP		Bayesian		Adaline		*OD* pairs Observation (**Gauss**)	Adaline		EM	
	EST	Error %	EST	Error %	EST	Error %		EST	Error %	EST	Error %
AB:318	318	0	318	0	336.2	5.7	AB:318.65	297.94	6.5	318.65	0
AC:289	601	107	342	18	306.4	6	AC:329.48	310.13	5.9	286.98	13
AD:312	0	100	259	17	312.9	0.3	AD:277.18	274.58	0.9	318.36	15
BA:294	579	96	334	14	312.1	6.1	BA:298.14	250.74	15.9	298.14	0
BC:292	559	91	310	6	285.4	2.3	BC:354.81	361.87	2	360.97	1.6
BD:267	0	100	249	7	291.9	9.3	BD:355.39	326.32	8.2	347.94	2
CA:305	303	0.6	291	5	300.8	1.4	CA:327.20	287.74	12.1	317.34	3
CB:289	0	100	361	25	298.8	3.4	CB:330.04	337	2.1	373.65	13
CD:324	903	178	395	22	316.5	2.3	CD:253.01	264.45	4.5	217.32	14
DA:283	0	100	257	9	287.4	1.5	DA:320.50	286.74	10.5	329.07	3
DB:277	851	207	245	12	285.3	3	DB:291.52	336	15.3	246.6	15
DC:291	0	100	349	20	296.6	1.9	DC:310.40	299	3.7	344.82	11

In **Gauss** distribution model, from Table 1, we may calculate two kinds of average, 6.4% for Adaline method,7.6% for EM method respective. The errors are not equal in two kinds of method. The average errors of Adaline are better than EM method. From Table 1, it is known that the whole trend of results using Adaline is better than using EM method. The reasons maybe are that fluctuation of original traffic matrix produced Poisson model is bigger than Gauss model. In Table 1, the errors in BA traffic of *OD* pairs arrive at 16% and then it is the maximal error. The probable reasons are less constraint. Furthermore, affinity between BA and other two *OD* pairs is a key problem, because other two *OD* pairs pass by 3-4 hops, that is, the route information will maybe affect traffic distribution. This hints that estimated errors may be correlated to heavily shared links.

In order to justify the validity of Adaline algorithm, we design two kinds of real network traffic(**Poisson Distribution** and **Gauss Distribution**) and topology shown in Figure 3 using MATLAB and test 100 minutes respectively.

In case 1, network traffic follows **Poisson Distribution**. Figure 4 shows the average error over time. Figure 4 shows that the average error is below 8% for Adaline and 18% for Bayesian, so the Adaline approach is robust. From the average error, the Adaline is a little bit better than Bayesian method. The reasons maybe are that fluctuation of original traffic matrix produced Poisson model is less than other method. Figure 5 shows the maximum error over time. And then Figure 5 shows that maximum error is below 23% for Adaline and 33% for Bayesian, so the Adaline approach is robust. From Figure 5, it is known that maximum error trend of Bayesian is a little bit bigger than Adaline. From the maxmum error, Adaline is better than Bayesian approach.

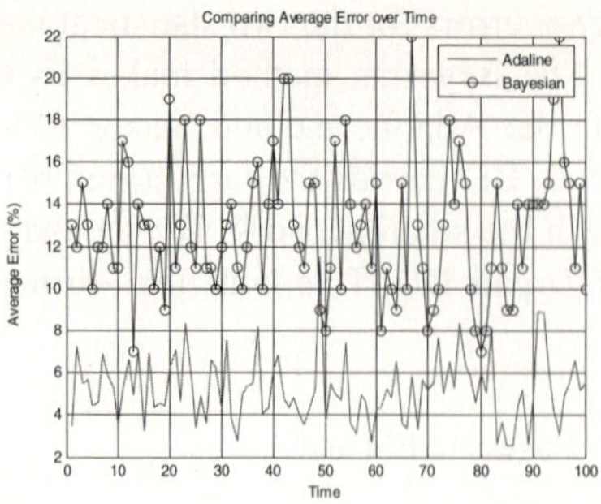

Fig. 4. Comparing the average error over time

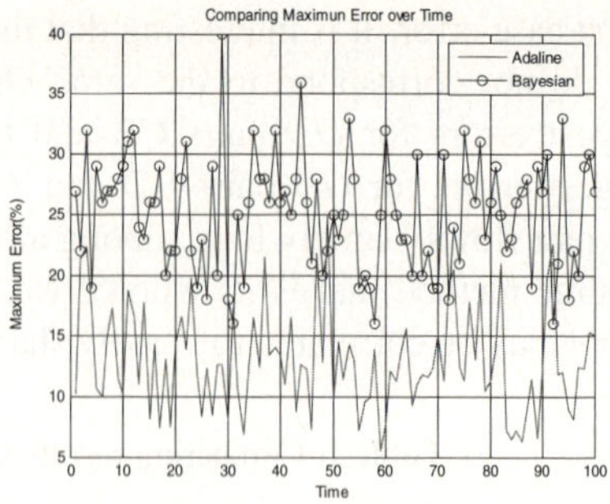

Fig. 5. Comparing the maximum error over time

In case 2, network traffic follows **Gauss Distribution**. Figure 6 shows the average error over time. Furthermore, Figure 6 shows that average error is below 8% for Adaline and 20% for *EM*, so the Adaline approach is robust. From the average error, the Adaline is better than *EM* method.

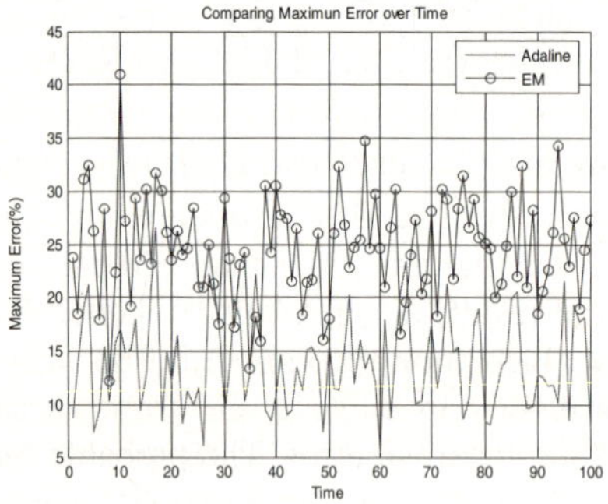

Fig. 6. Comparing the average error over time

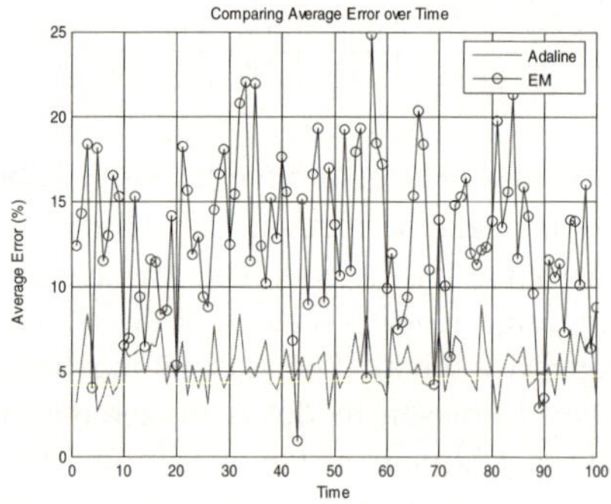

Fig. 7. Comparing the maximum error over time

Figure 7 shows the maximum error over time. At the same time, Figure 7 shows that maximum error is below 25% for Adaline and 35% for *EM*, so the Adaline approach is robust. From Figure 7, it is known that maximum error trend of *EM* is a little bit bigger than Adaline. From the maxmum error, Adaline is a little bit better than *EM* approach.

4.2 Symmetrical Topology

Since real traffic matrices are not available, we conduct our experiment with artificially constructed traffic matrices. We compare tomogravity and entropy maximization on a 4-node network. We did not consider the delay in our experiments as the estimation of traffic matrix is not performed online.

The experiment we carried out was on a very simple network shown in Figure 8. There are four vertices and ten directed edges in the graph. Each vertex is associated with two numbers which are respectively its incoming and outgoing traffic (e.g. at vertex 1 incoming and outgoing traffic are 14 and 20 respectively). Each directed edge or link is associated with one number which is the amount of traffic passing through the link (e.g. 6 unit of traffic flows through the link 0-1). For the four vertices there are twelve *OD* pairs in the network.

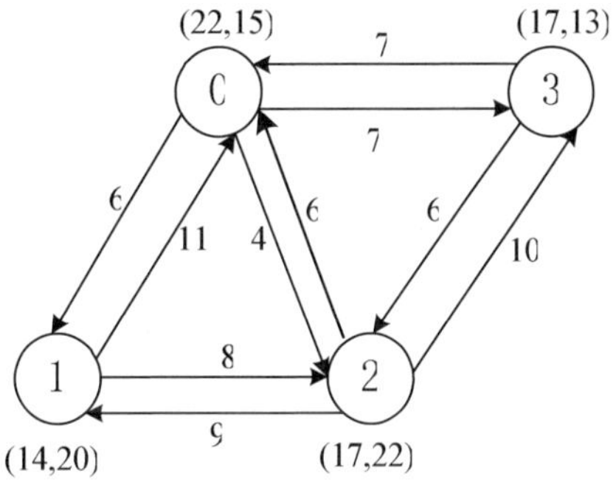

Fig. 8. 4-node unsymmetrical network topology

Given the link measured value be real value without measured error. The route matrix A is invariable and observed link value is listed as follows.

$$Y=(6, 11, 6, 4, 9, 8, 6, 10, 7,7)$$

In Table 2, we show the original *OD* and estimated *OD* traffic(**EST**) using WLSE, MAXENT and PDSCO [12]. From our experiment with 4-node network we find that Adaline method gives very good results. From Table 2, it is known that result of Adaline is robust. In spite of maximal error or average error, Adaline is best among WLSE, MAXENT, PDSCO method.

Table 2. Comparing traffic between Adaline and WLSE, MAXENT, PDSCO

OD pair	Original *OD* Traffic	WLSE		MAXENT		PDSCO		Adaline	
		EST	error%	EST	error%	EST	error%	EST	error%
0-1	5	4.493	10.1	2.108	57.8	3.959	20.8	6.199	24
0-2	4	4.000	0	4.000	0	4.000	0	5.5029	37.6
0-3	6	3.781	37	1.615	73.1	3.732	37.8	5.9489	0.9
1-0	10	7.781	22.2	5.615	43.9	7.732	22.7	9.9266	0.7
1-2	8	5.781	27.7	3.615	54.8	5.732	28.4	7.9444	0.7
1-3	2	6.438	221.9	10.771	438.6	6.537	226.9	5.1277	156.4
2-0	6	6.000	0	6.000	0	6.000	0	7.485	24.8
2-1	7	6.493	7.2	4.108	41.3	5.959	14.9	8.1811	16.9
2-3	9	6.781	24.7	4.615	48.7	6.732	25.2	8.9355	0.7
3-0	6	5.493	8.5	3.108	48.2	4.959	17.4	7.1766	19.6
3-1	2	3.014	50.7	7.785	289.3	4.081	104.1	2.6278	31.4
3-2	5	4.493	10.1	2.108	57.8	3.959	20.8	6.199	24

In Table 2, we give the result according to two above method. In Table 2, there are estimated and absolute error of traffic matrix. From Table 2, we may calculate four kinds of average, 35% for WLSE method, 96% for MAXENT, 43% for PDSCO method , 28% for Adaline method respective. The errors are not equal in four kinds of method. The average error of Adaline is best among above algorithm method.

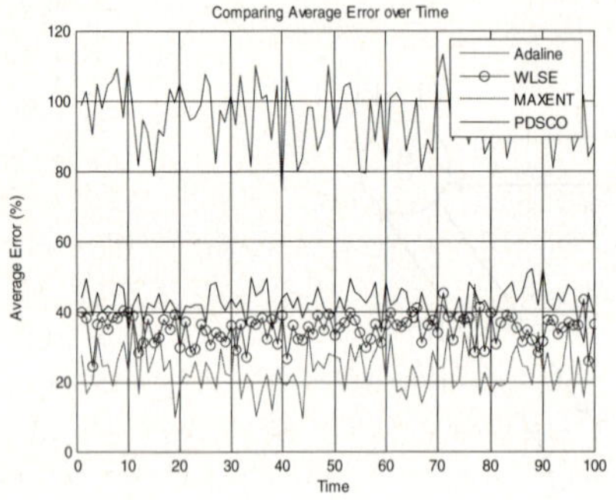

Fig. 9. Comparing the average error over time

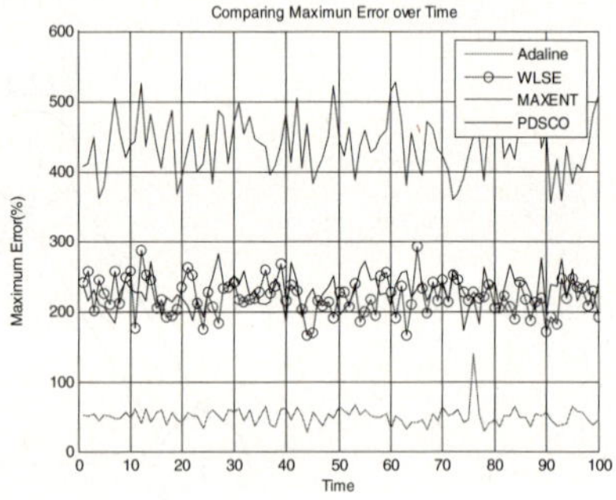

Fig. 10. Comparing the maximum error over time

Figure 9 shows the average error over time. At the same time, Figure 9 shows that the average error is below 30% for Adaline, 40% for WLSE, 50% for PDSCO and 110% for MAXENT, so the Adaline approach is robust. From the average error, the Adaline is the best among Adaline, WLSE, PDSCO and MAXENT. Figure 10 shows that the maxmum error is below 75% for Adaline, 250% for WLSE, 255% for PDSCO and 500% for MAXENT, so the Adaline approach is stable. From the maximum error, the Adaline is the best among Adaline, WLSE, PDSCO and MAXENT.

5 Conclusions

This paper has proposed an Adaline approach to estimate the traffic matrix. We apply the optimization theory to deduce the traffic matrix, with the aid of Adaline network. Through both theoretical analysis and numerical results, it is shown that the proposed algorithm achieves better performance than the existing representative methods.

Acknowledgment

The author would like to thank the Science and Technology Research Project of Chongqing Municipal Education Commission of China under Grant No. 080526 and Doctoral Research Fund of Chongqing University of Posts and Telecommunications (A2006-08). The author would also like to thank to MATLAB software.

Reference

1. Paxson, V., Almes, G., Mahdavi, J., Mathis, M.: Framework for IP Performance Metrics. IETF RFC 2330 (1998)
2. Cozzani, I., Giordano, S.: A Passive Test and Measurement System: Traffic Sampling for QoS Evaluation. In: IEEE Communications Society (ed.) Proceedings of the Global Telecommunications Conference, pp. 1236–1241. IEEE Press, Sydney (1998)
3. Chen, A., Chootinan, P., Recker, W.: Examining the Quality of Synthetic O-D Trip Table Estimated by Path Flow Estimator. Journal of Transportation Engineering 131, 506–513 (2005)
4. Elwalid, A., Jin, C., Low, S., Widjaja, I.: MATE: MPLS Adaptive Traffic Engineering. Proceedings of INFOCOM. 3, 1300–1309 (2001)
5. Trimintzios, P., Andrikopoulos, I.: A Management and Control Architecture for Providing IP Differentiated Services in MPLS-Based Networks. IEEE Communications Magazine 39, 80–88 (2001)
6. Vardi, Y.: Network Tomography: Estimation Tource-destination Traffic Intensities from Link Data. Journal of the American Statistical Association 91, 365–377 (1996)
7. Duffield, N.G., Grossglauser, M.: Trajectory Sampling for Direct Traffic Observation. IEEE/ACM Transactions on Networking 9, 280–292 (2001)
8. Feldmann, A., Greenberg, A., Lund, C., Reinqold, N., Rexford, J.: NetScope: Traffic Engineering for IP Networks. IEEE Network 14, 11–19 (2000)
9. Bierlaire, M., Crittin, F.: An Efficient Algorithm for Real-time Estimation and Prediction of Dynamic OD tables. Oper. Res. 52, 116–127 (2004)
10. Cong, S.: Object MATLAB Box Nerve Network Theory and Application, pp. 31–40. Science and Technology of University of China Press, Hefei (1998)
11. Medina, A., Taft, N., Salamatian, K., Bhattacharyya, S., Diot, C.: Traffic Matrix Estimation: Existing Techniques and New Directions. In: Paxson, V., Balakrishnan, H. (eds.) Proc. of the ACM SIGCOMM 2002 on Applications, Technologies, Architectures, and Protocols for Computer Communications, vol. 32, pp. 161–174. ACM Press, Pittsburgh (2002)
12. Rahman, M. M., Saha, S., Chengan, U., Alfa, A.S.: IP Traffic Matrix Estimation Methods: Comparisons and Improvements. Proceedings of IEEE ICC Istanbul 1, 90–96 (2006)

SVM Model Based on Particle Swarm Optimization for Short-Term Load Forecasting

Yongli Wang[2], Dongxiao Niu[1], and Weijun Wang[2]

[1] Professor, Institute of Business Management, North China Electric Power University, 102206, Beijing, China
wyl_2001_ren@163.com

[2] Doctor, Institute of Business Management, North China Electric Power University, 102206, Beijing, China
wyl_2001_ren@126.com

Abstract. A model integrating Particle Swarm Optimization (PSO) and support vector machines (SVM) is presented to forecast short-term load of electric power systems in this paper. PSO is a method for finding a solution of stochastic global optimizer based on swarm intelligence. Using the interaction of particles, PSO searches the solution space intelligently and finds out the best one. The PSO-SVM method proposed in this paper is based on the global optimization of PSO and local accurate searching of SVM. Practical example results indicate that the application of the PSO-SVM method to short term load forecasting of power systems is feasible and effective. And to prove the effectiveness of the model, other existing methods are used to compare with the result of SVM. The results show that the model is effective and highly accurate in the forecasting of short-term power load.

Keywords: SVM, Particle swarm optimization, Load forecasting.

1 Introduction

The short-term power load forecasting is very significant to the electric network's reliable and economic running. It can plan the open or stop of generators and keep the electric network running safely and stably with the exact load prediction. With the development of electric market, people have been paying more and more attention to load prediction. How to make the prediction on the short-term power load exactly becomes a hot point.

Basic operating functions such as unit commitment, economic dispatch, security assessment, fuel scheduling and unit maintenance etc. can be performed efficiently with an accurate load forecasting. Because of the importance of load forecasting, a wide variety of models have been proposed in the last two decades, such as the exponential smoothing model, state estimation model, multiple linear regression model and stochastic time series model .

Generally speaking, these techniques are based on statistic methods and extrapolated past load behavior while allowing for the effect of other influence factors such as the weather and day of week. However, the techniques employed for those models use a large number of complex and non-linear relationships between the load and these factors. A

F. Sun et al. (Eds.): ISNN 2008, Part II, LNCS 5264, pp. 642–649, 2008.

great amount of computational time is required and may result in numerical instabilities. Some deficiencies in the presence of an abrupt change in environment or weather variables are also believed to affect the load patterns. Therefore, some new forecasting models have been recently introduced. New methods, such as artificial intelligence (AI), artificial neural network (ANN), and support vector machines (SVM).

In this paper, a new thought which can improve the accuracy of load forecasting be presented, it believes the key which can improve the accuracy are the preprocessing of the history data and the improved forecasting model, so it present a new method which is Support Vector Machines based on Particle Swarm Optimization technology in power load forecasting model. The PSO is a powerful searching algorithm to handle optimization problems. It is particularly useful for complex optimization problems with a large number of tuned parameters.

2 The PSO Algorithm

PSO is a population based stochastic optimization technique developed by Dr. Eberhart and Dr. Kennedy in 1995, inspired by social behaviour of bird flocking or fish schooling. In each iteration, each particle is updated by following two "best" values. The first one is the best solution (fitness) achieved in the iteration processes so far. This value is called p_{best}. Another "best" value, which is tracked by the particle swarm optimizer, is also the best value, obtained so far by any particle in the population. This best value is a global best, which is called g_{best}. When a particle takes part of the population as its topological neighbors, the best value is a local best, which is called l_{best}.

After finding the two best values, the particle updates its velocity and positions using (1) and (2) .

$$v_i = wv_i + c_1 rand() \bullet (p_{best} - x_k) + c_2 rand() \bullet (g_{best} - x_k), \tag{1}$$

$$x_i = x_i + v_i, \tag{2}$$

where is the particle velocity, x_i is the current particle position, $rand()$ is a random number between (0,1), w is the inertia weight, usually $w \in (0,1)$, c_1, c_2 is the acceleration constant, generally $c_1 = c_2 = 2$

Particles' velocities on each dimension are clamped to a maximum velocity $v_{\max} (v_{\max} \succ 0)$. If the velocity that it is updated on that dimension exceeds $v_{\max}$, then the velocity on that dimension is limited to $v_{\max}$.

3 SVM Regression Theory

Suppose a set of data (x_i, y_i), $i = 1, 2 \cdots n$, $x_i \in \boldsymbol{R}^n$ are given as input, $y_i \in \boldsymbol{R}$ are the corresponding output. SVM regression theory is to find a nonlinear map from input space to output space and map the data to a higher dimensional feature space through the map, then the following estimate function is used to make linear regression[6]-[8].

$$f(x)=[\omega\cdot\phi(x)]+b \qquad \phi:\boldsymbol{R}^m\rightarrow\boldsymbol{F} \quad \omega\in\boldsymbol{F}\ , \tag{5}$$

b is the threshold value. The problem of the function approximate is equivalent with the minimizing the following problem:

$$\boldsymbol{R}_{reg}[f]=\boldsymbol{R}_{emp}[f]+\lambda\|\omega\|^2=\sum_{i=1}^{s}C(e_i)+\lambda\|\omega\|^2\ . \tag{6}$$

$\boldsymbol{R}_{reg}[f]$ is objective function and s is the number of the sample. $e(\bullet)$ is loss function and λ is adjusting constant meter. The following loss function can be gained concerning with the rarefaction character of the linear insensitive loss function ε .

$$|y-f(x)|_\varepsilon=\max\{0,|y-f(x)-\varepsilon|\}\ . \tag{7}$$

Empirical risk function is

$$\boldsymbol{R}_{emp}^{\varepsilon}[f]=\frac{1}{n}\sum_{i=1}^{n}|y-f(x)|_{\varepsilon}\ . \tag{8}$$

According to statistic theory, the regression function is determined by minimizing the following functions :

$$\min\left\{\frac{1}{2}\|\omega\|^2+\boldsymbol{C}\sum_{i=1}^{n}\left(\xi_i^*+\xi_i\right)\right\} \tag{9}$$

$$y_i-(\omega\cdot\phi(x)-b\le\varepsilon+\xi_i^* \tag{10}$$

$$(\omega,\phi(x))+b-y_i\le\varepsilon+\xi_i \tag{11}$$

$$\xi_i\ ,\xi_i^*\ge 0\ . \tag{12}$$

$\boldsymbol{C}$ is used to equalize the complicated item of the model and the parameters of the item of training error. ξ_i^* and ξ_i are relaxation factors and ε is insensitive loss function. The problem can be converted into the dual problem

$$\begin{aligned}&\max[-\frac{1}{2}\sum_{i,j=1}^{n}\left(a_i^*-a_i\right)\left(a_j^*-a_j\right)K(X_i,X_j)+\\&\sum_{i}^{l}a_i^*\left(y_i-\varepsilon\right)-\sum_{i=1}^{n}a_i\left(y_i-\varepsilon\right)]\end{aligned} \tag{13}$$

$$\sum_{i=1}^{n}a_i=\sum_{i=1}^{n}a_i^* \tag{14}$$

$$0\le a_i^*\le\boldsymbol{C} \tag{15}$$

$$0\le a_i\le\boldsymbol{C}\ . \tag{16}$$

Solve the problem, then the regression equation is

$$f(x)=\sum_{i=1}^{n}\left(a_i-a_i^*\right)k(X_i,X)+b\ . \tag{17}$$

4 SVM Based on Particle Swarm Optimization Method

Because of powerful ability of PSO algorithm in searching the global optimizer and the good adaptability of the back propagation algorithm in local accurate searching, the advantage of the two methods is integrated to used for the new method, named PSO-SVM method and proposed in this paper. The training steps of the SVM model is written as follows:

Step 1: Randomly initialize each connection weight of SVM as swarm particles. The initializing parameters include the stochastic position x_i and the velocity v_i. It is supposed that the scale of the particles is m, $x_i \in [v_{up}, v_{down}]$ is the change scope on each dimension of each particle, $v_i \in [-v_{\max}, v_{\max}]$, $v_{\max}$ is a maximum velocity of particles and it is a constant. Generally change scope on each dimension is 10%-20%.

Step 2: Given the inertia weight w, the acceleration constant c_1 and c_2, and initialized p_{best} and g_{best}. Input the forecasting elements to the input layer of SVM and begin the computation of front network. The fitness function is defined as:

$$f = M\left|u_j - u_j^0\right| . \tag{18}$$

where u_j^0 is the actual output value of the network, u_j is the expected output value of the network, M is a constant. The fitness function in (18) is used to evaluate the fitness of each particle.

Step 3: If the fitness value is better than the best fitness value (p_{best}), the current value is set as the new p_{best} and the particle with the best fitness value of all the particles is set as the g_{best}. For each particle, the particle velocity is updated according to (1) and the particle position is updated according to (2) while maximum iteration or minimum error criteria are not satisfied.

Step 4: PSO searches the solution space. If the fitness value is smaller than the threshold value, which established in advance, or if the iteration number is smaller than the maximum iteration $I_{\max}$, which is given in advance, it goes to the step 3, otherwise it goes to the step 5.

Step 5: In the searching the global optimisation, we can use small length back propagation algorithm to search the local best solution. If the anticipant convergent precision is obtained, then the network training is over.

5 Application and Analysis

The results of load forecasting of power systems is influenced by such factors as day sort, week sort, day weather sort, day temperature, etc. In this paper, these factors are included in one eigenvector at the day's value.

Power loading data in Inner Mongolia region is used to prove the effectiveness of the model. Because the variation regulation of power loading in that area is influenced explicitly by the weather and the temperature of seasonal variation is very obvious, year temperature difference is very visible, rainfall is relative centralized and agriculture loading occupies a big proportion, so it is very appropriate for this method. Comparing with the single SVM model and Neural Net Work which doesn't consider the weather factors, this new model has higher accuracy.

It is defined $X_i = [x_{i1}, x_{i2}, \cdots, x_{im}]$ (m is eigenvector number) as all eigenvector value of ith day, and $L_i = [l_{i1}, l_{i2}, \cdots, l_{iT}]$ as the load curve of ith day (T is data point number, such as T=24,T=96 , etc.). As a result, generalization values of the ith day are determined as $D_i = (X_i, L_i)$.

Table 1. Forecast results of electric power load using the PSO-FNN method and comparison with the FNN method MW

Time point	Actual load	SVM		PSO-SVM	
		Forecast load	Error	Forecast load	Error
T00_00	551.87	560.31	1.53%	556.78	0.89%
T01_00	540.42	537.99	-0.45%	546.20	1.07%
T02_00	523.73	531.01	1.39%	515.56	-1.56%
T03_00	518.40	528.09	1.87%	513.32	-0.98%
T04_00	518.49	510.25	-1.59%	520.82	0.45%
T05_00	558.00	552.53	-0.98%	570.95	2.32%
T06_00	635.72	650.79	2.37%	648.18	1.96%
T07_00	648.51	660.77	1.89%	655.38	1.06%
T08_00	682.36	668.64	-2.01%	692.94	1.55%
T09_00	762.83	775.26	1.63%	770.46	1.00%
T10_00	737.48	759.75	3.02%	752.30	2.01%
T11_00	759.07	739.26	-2.61%	782.15	3.04%
T12_00	686.83	664.78	-3.21%	676.32	-1.53%
T13_00	628.71	639.27	1.68%	623.30	-0.86%
T14_00	646.22	653.26	1.09%	630.97	-2.36%
T15_00	704.81	722.99	2.58%	712.77	1.13%
T16_00	748.12	733.53	-1.95%	755.45	0.98%
T17_00	764.69	788.47	3.11%	773.94	1.21%
T18_00	763.03	753.42	-1.26%	752.58	-1.37%
T19_00	772.51	758.14	-1.86%	766.02	-0.84%
T20_00	845.07	865.10	2.37%	864.59	2.31%
T21_00	882.00	868.33	-1.55%	894.35	1.40%
T22_00	773.68	794.03	2.63%	783.20	1.23%
T23_00	634.36	654.15	3.12%	655.29	3.30%
T24_00	575.67	588.97	2.31%	584.42	1.52%
RMSRE			2.17%		1.70%

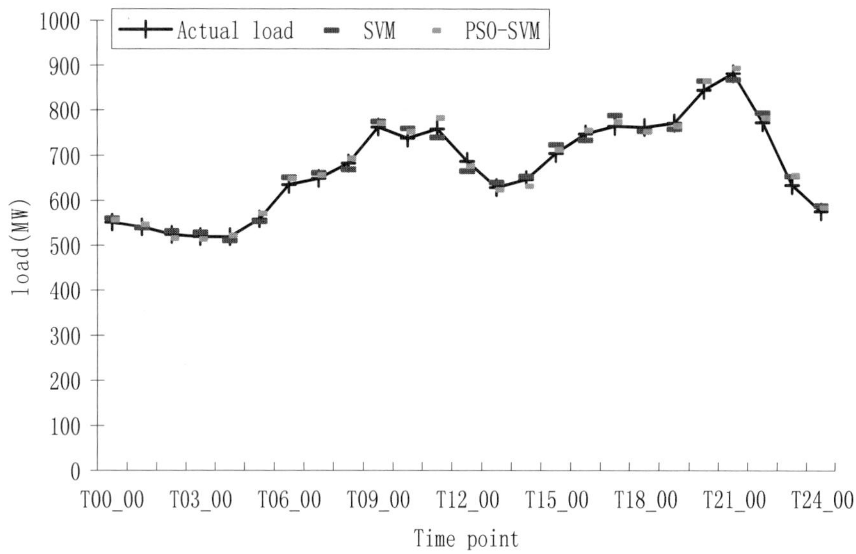

Fig. 1. The forecasting results using the proposed method

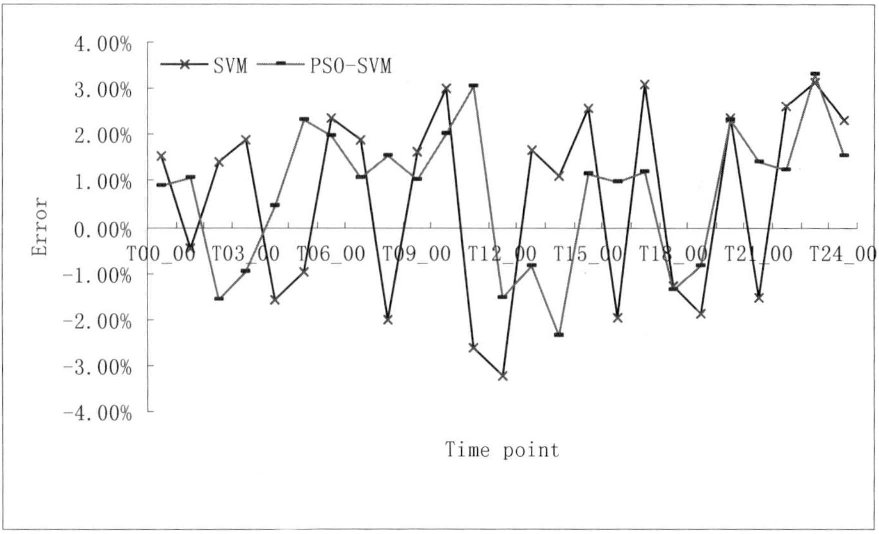

Fig. 2. Error analysis of the proposed method

A model integrating PSO and FNN is established. The parameters of PSO are given by: $a = 0.7, w = 0.5, c_1 = 2$ and $c_2 = 2$, $[v_{up}, v_{down}] \in [-100,100]$, $v_{\max} = 0.15 \times \left| v_{up} - v_{down} \right|$, $\eta = 0.04$. $\eta = 0.03$. The momentum operator is 0.48. We make use of MATLAB7.0 to simulate the PSO-SVM method. Training of the PSO-SVM is carried out using the data set obtained from Inner Mongolia Power Company, China.

Some data are chosen from the data bank of Inner Mongolia region. The power load data from 0:00 at 5/1/2005 to 12:00 at 4/30/2006 are as training sample and used to establish the single-variable time series. And the power load data from 13:00 at 4/30/2006 to 24:00 at 5/17/2006 as testing sample.

Relative error and root-mean-square relative error are used as the final evaluating indicators:

$$e = \frac{1}{n}\sum_{i=1}^{n}\left|\frac{A(i) - F(i)}{A(i)}\right| \times 100\% , \tag{19}$$

$$RMSRE = \sqrt{\frac{1}{N - n_{tr}}\sum_{t=n_{tr}}^{n}\left(\frac{x_t - y_t}{x_t}\right)^2} . \tag{20}$$

The 24-hours-ahead load prediction and the actual load at 5/26/2006 are shown in Table 1, the 24-hours-ahead load prediction curves and the actual load curve can be shown in Fig. 1.

Fig. 2 shows the error analysis of short-term load forecast of power systems using the propose method. In Table 1, a Relative Error comparison is made between the general SVM and PSO-SVM.

We can see from the above example: comparing with the SVM method, the PSO-SVM model has higher forecast precision. Evidently, the forecasting errors of the proposed model are lower than the other models.

6 Conclusion

The results show that SVM based on Particle Swarm Optimization has great effectiveness for short-term power load forecasting. And the conclusions are shown as the following:

1. It is necessary to preprocess the data which is influence very much by uncertain factors for short-term power load forecasting. It used Particle Swarm Optimization method to pretreatment the data, The load of power systems is influenced by such factors as day sort, week sort, day weather sort, day temperature, etc. In this paper, these factors are included in one eigenvector at the day's value. The real load data prediction shows that the model is effective in short-term power load forecasting.

2. The main influential factors are considered adequately to forecast in this method. Through data preprocessing it reduces training samples, picks up training speed and considers the weather factors.

3. Comparing with the single SVM, it can be proved that this method not only improves accuracy of short-term load forecasting, practicability of system, but also can be accomplished by software.

Acknowledgement

Project Supported by Doctor spot special Foundation of NCEPU(20040079008), Natural Science Foundation of China(50077007).

References

1. Kennedy, J., Eberhart, R.C.: Particle Swarm Optimization. In: Proceedings of the 1995 IEEE International Conference on Neural Networks, vol. 4, pp. 1942–1948 (1995)
2. Qiu, L., Chen, S.Y.: A Forecast Model of Fuzzy Recognition Neural Networks and Its Application. Advances in Water Science 9, 258–264 (1998)
3. Vapnik, V.N.: The Nature of Statistical Learning Theory. Springer, New York (1995)
4. Zhao, D.F., Wang, M., Zhang, J.S., et al.: A Support Vector Machine Approach for Short Term Load Forecasting. Proceedings of the CSEE 22, 26–30 (2002)
5. Yang, J.F., Cheng, H.Z.: Application of SVM to Power System Short-term Load Forecast. Electric Power Automation Equipment 24, 30–32 (2004)
6. Eberhart, R.C., Kennedy, J.: A New Optimizer Using Particle Swarm Theory. In: Proceedings of Sixth International Symposium Micro Machine and Human Science, Nagoya, Japan, pp. 39–43 (1995)
7. Shi, Y., Eberhart, R.C.: A Modified Particle Swarm Optimizer. In: Proceedings of the 1998 IEEE Conference on Evolutionary Computation, AK, Anchorage, pp. 69–73 (1998)
8. Chen, S.Y.: Power Sensitive Analysis of Multi-Objective Fuzzy Optimal Selection Model and Its Application. Journal of Dalian University of Technology 42, 477–479 (2002)
9. Chen, S.Y.: The Fuzzy Optimization Selection Neural Network and Multiobjective Decision Theory. Journal of Dalian University of Technology 37, 693–697 (1997)
10. Zhang, H.R., Han, Z.Z.: An Improved Sequential Minimal Optimization Learning Algorithm for Regression Support Vector Machine. Journal of Software 14, 2006–2013 (2003)
11. Shi, Y.H., Eberhart, R.C.: Parameter Selection in Particle Swarm Optimization. In: Proceedings of the Seventh Annual Conf. on Evolutionary Programming, pp. 591–601 (1998)
12. Kennedy, J.: The Particle Swarm: Social Adaptation of Knowledge. In: Proceedings of IEEE International Conference on Evolutionary Computation, Indianapolis, Indiana, pp. 303–308 (1997)

A New BSS Method of Single-Channel Mixture Signal Based on ISBF and Wavelet*

Xiefeng Cheng[1,2], Yewei Tao[1,2], Yufeng Guo[1], and Xuejun Zhang[1]

[1] Nanjing University of Posts & Telecommunications, Nanjing 210003, P.R. China
{chengxf,taoyw,guoyf,xjzhang}@njupt.edu.cn
[2] Jinan University, Jinan 250100, P.R. China

Abstract. A new BSS method based on independent sub-band function components (ISBF) and wavelet to separate single-channel mixture signal in noise was studied. Through obtaining sub-band functions with independent component characteristic in the time domain, 6-20 sub-band function by ICA were employed as the preparation knowledge for the blind source separation (BSS). By combining the independent sub-band function components (ISBF) into the single-channel mixture signal, a separation modeling of single-channel mixture signal was built based on ISBF. And the separation mathematics model of the single-channel signal in noise is investigated. The wavelet transform was used to eliminate the noise as well. Two simulation samples performed to verify the availability of the proposed methods. The results show that the methods played good role in one sensor source BSS, and had a capability to extract the sound signal feature.

Keywords: ISBF; Noise; Wavelet; ICA.

1 Introduction

Blind signal processing technique is a hot topic in information processing field which has been widely used in communication, biology and medical applications, image processing, speech processing etc. It aims at restoring the source signals as accurate as possible by reasonable assumption and approximation while the priori information of the observed signal is rare. Blind source separation given only single-channel mixture signal by a sensor observed in noisy is difficult to accomplish because there are many unknown factors [1].

The paper presents a new technique for accomplishing blind source separation (BSS) when given only single-channel mixture signal. The rest of this paper is organized as follows. In section one, we analyze the principle of makeup and independent sub-band function. One signal source can be generated by a set of weighted liner

* This research was supported by the Provincial Natural Science Foundation and Science and Technology Tackle Key Problem of Shandong (No.Y2006G03, No.Y2007G14, No.2006Gg3204005, No.2007Jy17), and Science Foundation of Nanjing University of Posts & Telecommunications (NY207139).

F. Sun et al. (Eds.): ISNN 2008, Part II, LNCS 5264, pp. 650–657, 2008.

superposition of the time domain sub-band functions with independent component characteristic. In section two, a novel technique based on sub-band functions is proposed. By combining the sub-band function components into the single-channel mixture signal, the single-channel mixture signal can be transformed into a multi-dimensional vector from one-dimensional. Thus we can apply these BSS method, for example ICA, to separate the extended mixture signal. In section three, simulation results and the analysis are shown. The simulation result verified the effectiveness and adaptability of the proposed method.

2 The Independent Sub-band Function Model for Signal Source

Let $s(t)=[s_1(t),s_2(t),...,s_k(t)]^T$, $x(t)=[x_1(t),x_2(t),...,x_m(t)]^T$, the mixing model can be expressed as follows.

$$x(t)=As(t) \tag{1}$$

Where A is the un-singular mixing matrix. Let $E[s_i(t_1),s_j(t_2)]=0$ $i\neq j(i=1,2,...,n),\forall t_1,t_2$, $E[s_i(t)]=0$ and $y(t)=[y_1(t),y_2(t),...,y_k(t)]^T$, W is the separation matrix, and the separation model is,

$$y(t)=Wx(t) \tag{2}$$

If we find out the separation matrix W , the source signals can be separated.

But our goal is to recover all $s_i(t)$ given only single-channel observed mixture signal source. The problem is too ill-conditioned for mathematically processed, since the number of unknowns variables is 2k given only one observation.

The method presented in this paper is that one signal source can be generated by a set of weighted liner superposition of the time domain sub-band functions with the independent component. The sub-band functions of the signal source are obtained by

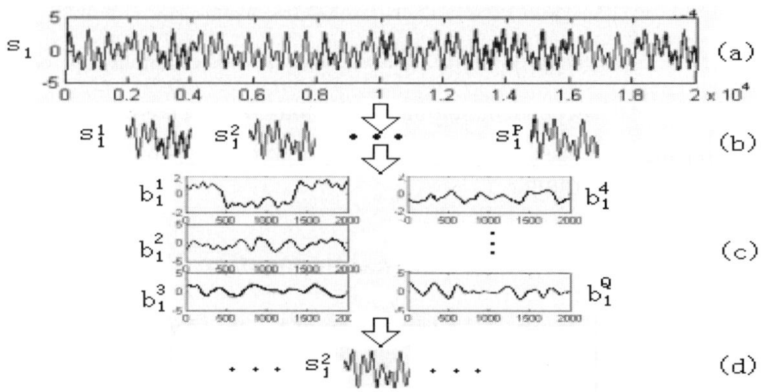

Fig. 1. The sub-band functions obtaining way and principle of makeup

learning from a training data set, and these sub-band functions are used to separate the unknown signal. For convenience to learn the sub-band functions of the sources, we assume k is 2. One signal source $s_1(t)$ is shown in Fig.1 (a).

For the training data $s_1(t)$, we adopt a decomposition approach to divide $s_1(t)$ into segments s_1^p $(p=1,\cdots,P)$ with the fixed length, which is shown in Fig.1 (b). The length of s_1^p is much shorter than that of $s_1(t)$. The s_1^p $(p=1,\cdots,P)$ can form a number of elementary patterns with time-varying independent component which is so-called independent sub-band functions. And s_1^p can be generated by a set of weighted liner superposition of the time domain independent sub-band functions b_1^q $(q=1,\cdots,Q)$, For example s_1^2, which are shown in Fig. 1(c) and (d)..

$$\begin{bmatrix} s_1^1 \\ s_1^2 \\ \vdots \\ s_1^P \end{bmatrix} = \begin{bmatrix} c_{11} & c_{12} & \cdots & c_{1Q} \\ c_{21} & c_{22} & \cdots & c_{2Q} \\ \vdots & \vdots & \vdots & \vdots \\ c_{P1} & c_{P2} & \cdots & c_{PQ} \end{bmatrix} \begin{bmatrix} b_1^1 \\ b_1^2 \\ \vdots \\ b_1^Q \end{bmatrix} \tag{3}$$

Where Q is the number of the independent sub-band functions, c_{pq} is coefficient of the q th sub-band functions of the p th signals segment. Let $C=\sum_{p=1}^{P}\sum_{q=1}^{Q}c_{pq}$, $P=Q$, then C is full rank, so that the transformation between s_1^p and b_1^q is reversible. The inverse matrix of matrix C is $W_c=C^{-1}$, which the BSS learning can be accomplished, just as depicted in the form (2). On the other word, maximizing the marginal probabilities of the transformed co-ordinates for the given training data s_1^p, we have

$$W_c^* = \arg\max_{W_c} \prod_p P_r(s_1^p, W_c) = \arg\max_{Wc} \prod_p \prod_q P_r(b_p^q) \tag{4}$$

Where $P_r(a)$ denotes the probability of a variable a. For example, by using JADE algorithm in [2], the independent sub-band functions b_1^q and c_{pq} can be obtained in advance. And they are used as prior information for the following single-channel mixture signal BSS algorithm.

3 A Single-Channel Mixture Signal Separation Technique

3.1 The Method of Expanding the Dimension of Single-Channel Signal Source

When m=1 and k=2, Eq. (1) reduce

$$x = a_1 s_1 + a_2 s_2 \tag{5}$$

If the length of x is T, it is chopped N segments, $x = \sum_{n=1}^{N} x^n$, length T of x^n is the same with length L of s_1^p, and $L \ll T$, we have

$$\sum_{n=1}^{N} x^n = a_2 \sum_{p=1}^{P} s_2^p + a_1 \sum_{p=1}^{P} s_1^p \tag{6}$$

Let $N = P$. By adding the sub-band function components $s_{1q}^p = a_1 c_{qp} b_1^q$ into the signal segments x^n, the one-dimensional observed mixture signal segments is transformed into a new multi-dimensional vector, and it can be expressed as

$$\begin{bmatrix} x^n \\ s_{11}^p \\ s_{12}^p \\ \vdots \\ s_{1Q}^p \end{bmatrix} = \begin{bmatrix} a_2 s_2^n + a_1 s_1^n \\ s_{11}^p \\ s_{12}^p \\ \vdots \\ s_{1Q}^p \end{bmatrix} = \begin{bmatrix} a_2 & a_1 c_{p1} & \cdots & a_1 c_{pQ} \\ 0 & c_{p1} & 0 & 0 \\ 0 & 0 & c_{p2} & 0 \\ 0 & 0 & \vdots & \vdots \\ 0 & 0 & \cdots & c_{pQ} \end{bmatrix} \begin{bmatrix} s_2^n \\ b_1^1 \\ b_1^2 \\ \vdots \\ b_1^Q \end{bmatrix} \tag{7}$$

The Eq. (7) satisfied all requirements in ICA. So we can obtain separate result $\hat{s}_2^n$ by using ICA learning algorithm. Then $\hat{s}_2^n$ and x^n combined to make a new two-dimensional vector, once more separate them by using the ICA technique. A good result would be carried out. The proposed algorithm can be implemented with the following steps.

[Step 1] According to (1), compute sub-band functions s_1^p.

[Step 2] Transform x^n into a multi-dimensional vector by (7).

[Step 3] Separate this vector, $(\hat{s}_2^n)_l$ can be obtained.

[Step 4] Transform $(\hat{s}_2^n)_l$ and x^n into a two-dimensional vector, and once more separate it.

[Step 5] Comparing the results in this and last iteration. If $(\hat{s}_2^n)_l$ and $(\hat{s}_2^n)_{l-1}$ are not similar, $l = l + 1$, repeat the step 4, otherwise $L^n = l$, go to the step 6.

[Step 6] n=n+1. If $n < N$, go to step2, otherwise output $\hat{s}_2 = \sum_{n=1}^{N} (\hat{s}_2^n)_{L^n}$.

3.2 The Methods to Determine the Number of the Independent Sub-wave Function

The basis of ICA manifest, variable K which describe the number of statistical independent state, impact the accuracy of system description of (3) and the property of output independent composition, directly. This paper suggest that by plus a known signal, we can use experimental analysis to determine the number of independent variable.

Let N as noise collection of every output signal, $N \in R^{M}$, so

$$\hat{x} = x + N \tag{8}$$

To the Eq. (8), we plus a known signal e_0, that is

$$\tilde{x} = (x + N) + e_0 \tag{9}$$

If e_0 and $(x+N)$ are statistical independence, and signal strength of e_0 is less than that of the $(x+N)$ .According to ICA theory, when estimating the number of independent variables, it is the same as the real, the best effectiveness can be reached, and mean-square-error of separation result $\hat{e}_0$ and input signal e_0 is smallest.

The method is as follows:

[step 1] According to the formula (9), estimate the noise strength of the sample , Q_N ;

[step 2] Add white noise e_0, and signal strength of e_0 is slightly larger than Q_N ;

[step 3] Set up $i = 1, \mathrm{K} = i + 1$, use ICA to separate the Eq. (11);

[step 4] Compare $\hat{e}_0^i$ and e_0, if $\left|e_0 - \hat{e}_0^i\right| > \varepsilon_N$, then $i+1$, repeat the step 3, otherwise go to the step 5. ε_N is given error threshold.

[step 5] Calculate the result. And the result of K is the independent variable number we estimate.

Let $\mathrm{P{=}K}$, this P is the number of the independent sub-wave function.

3.3 The Analysis of Wavelet De-noising

Through ICA ,the noise in the system can be eliminated by mutual information redundancy way, N_2 in Eq. (10) can be deleted, and the statistical independent variables of the system can be separated at the same time. But the noise N_1 contained in the statistical independent variables cannot be eliminated . In the other words, the self-related-information noise cannot be eliminated using ICA. Therefore we consider using wavelet transform to eliminate the noise N_1.

According to wavelet theory, Eq. (8) can be expressed as

$$\tilde{x}(t) = A_J(t) + \sum_{j=1}^{J} [D_j(t) + N_1^j] \tag{10}$$

In this equation, A_J is the low frequency components, D_j is the high frequency components, J is the decomposition series. In physical signals, the low-frequency signal characterizes the self-characteristic of the signal, while high-frequency characterize nuance of the signal. In voice signal, the noise N_1^j exists in high-frequency band, thus we could use Eq. (14) to de-noise.

According to the above analysis, we have found a blind separation method based on sub-band function and wavelet do-noising for one mixtures signal. First we do noise of one signal $\hat{x}$ with band-pass filter ,Eliminate background noise N_2 and then a signal x which is pure as far as possible can be got .Next we transform x into even segments , add sub-band function b_1^q (q=1、2... θ.) ,then the one-dimensional signal segments can be transformed into a new multi-dimensional vector. Processing it with Fast ICA to get dividing result s^1, s^2, ... s^n , and then we can connect each segment of signal to gain $\hat{s}$. Because residual noise will exist in the dividing segments, for this reason, we present to use wavelet transform to do-noising for ICA dividing signal $\hat{s}$, and finally gain much purer signal s.

4 Simulations

Experiment 1: Separation of Duffing's Mixture Signal.
Based on the Eq. (5), a sentence randomly chosen from TIMIT speech database is used for training. A male speech is shown in (a) of Fig. 2. The training data are segmented into ten segments, and it can form ten sub-band functions, shown in Fig. 2. (b).

$$\frac{d^2y}{dt^2}+\delta\frac{dx}{dt}-x+x^3 = fcon\omega t \tag{11}$$

According to the Duffing' equation, when δ =0.26, $f=2$, $\omega=2$, its one-dimension output can be used as the signal source $s_2(t)$, which is depicted in Fig. 2(c).

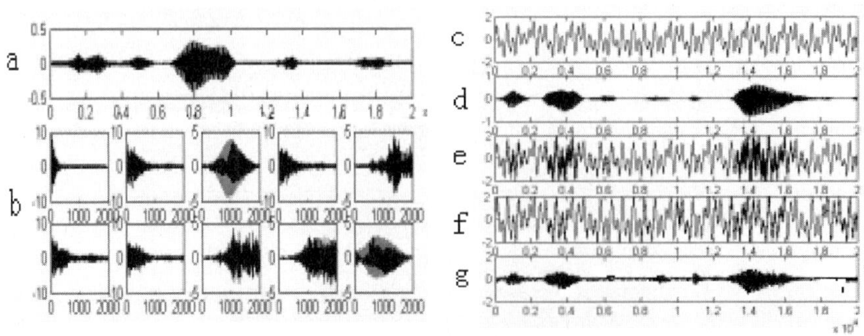

Fig. 2. Separation of Duffing's mixture signal

	$(\widehat{s}_2^1)_2$	$(\widehat{s}_2^2)_3$	$(\widehat{s}_2^3)_2$	$(\widehat{s}_2^4)_2$	$(\widehat{s}_2^5)_3$	$(\widehat{s}_2^6)_3$	$(\widehat{s}_2^7)_2$		$\hat{s}_2'$	$\hat{s}_2$
SC	0.9979	0.9917	0.9991	0.9990	0.9950	0.9898	0.9967		0.3849	0.9934
SPD								···		

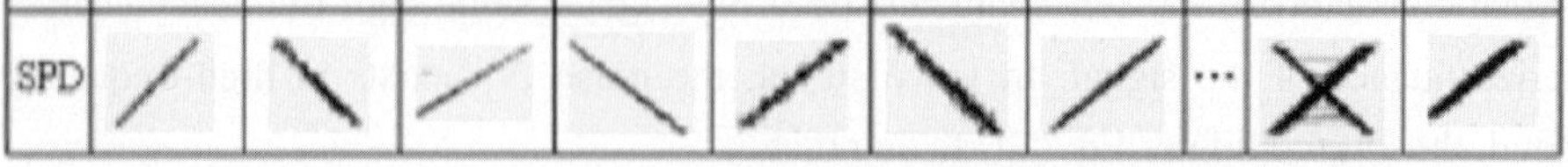

Fig. 3. Similitude coefficient (SC) and Similitude phase diagram (SPD) of about $\hat{s}_2$

A male speech signal by one microphone observed as $s_1(t)$, which is depicted in Fig. 2(d). $s_1(t)$ and $s_2(t)$ generate the mixture single x, shown in Fig. 2(e). The waveforms of separation result $\hat{s}_2(t)$ and $\hat{s}_1(t)$ which are shown in Fig. 2(f) (g), where $\hat{s}_1(t)$ is the difference between x and $\hat{s}_2(t)$. The similitude coefficient of $s_2(t)$ and $\hat{s}_2(t)$ is 0.9934. The similitude coefficient of $s_1(t)$ and $\hat{s}_1(t)$ is 0.8672.

Similitude coefficient (SC) and similitude phase diagram (SPD) of $\hat{s}_2(t)$ and every segment $\widehat{s}_2^p$ $(p=1,\cdots,10)$, and separation results are shown in Fig. 3. STP of $\widehat{s}_2^2, \widehat{s}_2^4, \widehat{s}_2^6$ show that that the differences can be found in magnitude and phase φ, where $\Delta\varphi = 180^\circ$. If obtaining $\hat{s}_2'$ by directly using the results to add, its SC is 0.3849, SPD is a X shape. After phase of $\widehat{s}_2^2, \widehat{s}_2^4, \widehat{s}_2^6$ are corrected, the SC of $\hat{s}_2$ is 0.9934, s_2 and $\hat{s}_2$ is same in phase.

Experiment 2: The Separation of the Transient Evoked Otoacoustic Emissions Signal The transient evoked otoacoustic emissions (TEOAEs) is a means all-pervading for hearing assessment in the clinic [4]. How to eliminate the artifact signal is a key in TEOAEs. Not less than two TEOAEs signals are needed in other techniques to eliminate artifact signal, for instance, DNLR method (Derived Nonlinear Response) [4]. Using our method, it needs only one TEOAEs signal. The un-pleasure of the patients

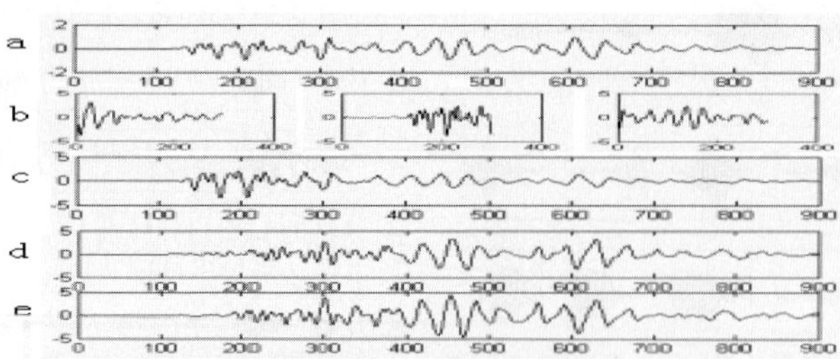

Fig. 4. The separation of TEOAEs

can be reduced. Separation result of TEOAEs signal is shown in Fig. 4. The artifact signal from the noise database is segmented into 3 segments, and it is regarded as the training data, which can form 3 independent sub-band functions, shown in Fig. 2(a) and (b). The Fig. 4 (c) shows one TEOAEs signal by one microphone observed. The results using DNLR method is shown in Fig. 4 (d). Using our method the result is shown in Fig. 4 (e). We can't see have much differences in them, but our method can be processed in real time.

5 Conclusion

We can obtain the conclusions stated as follows. First of all, the experiment results above are compared with that of the traditional DNLR arithmetic, our method need only one signal. So the method can be implemented efficient to detect the neonatal hearing based on a fixed stimulating signal. And the similar coefficient is near 1, this means that the separating method is successful. It proved that our method is reasonable and effective. In addition, the number of independent sub-band functions is near to the number of training data inhere ICA basis functions, so the separating results are better.

References

1. Comon, P.: Independent Component Analysis, A New Concept? Signal Processing 6(36), 287–314 (1994)
2. Cardoso, J.F.: Blind Beam Forming for Non-Gaussian Signals. IEEE Proceedings 8(12), 362–370 (1993)
3. Qin, H., Xie, S.: Blind Separation Algorithm Based on Covariance Matrix. Computer Engineeing 29, 36–38 (2003)
4. Ravazzani, P.: Evoked Otoacoustic Emissions: Nonlinearities and Response Interpretation. IEEE Trans Biomedical Engineering 2(40), 500–504 (1993)
5. Hyvarnena, O.: Independent Component Analysis: Arithmetic and Applications. Neural Networks 2(13), 411–430 (2000)
6. Qin, S.J., Dunia, R.: Determining the Number of Principal Components for Best Reconstruction. Journal of Process Control 10, 245–250 (2000)
7. Kundu, D.: Estimating the Number of Signals in the Presence of White Noise. Journal of Statistical Planning and Inference 5(90), 57–61 (2000)
8. Antoniadis, A., Pham, D.T.: Wavelet Regression for Random or Irregular Design. Comp. Stat. and Data Analysis 28, 353–359 (1998)

A Novel Pixel-Level and Feature-Level Combined Multisensor Image Fusion Scheme

Min Li[1], Gang Li[2], Wei Cai[1], and Xiao-yan Li[3]

[1] Xi'an Research Inst. of Hi-Tech Hongqing Town, 710025 Shaanxi Province, P.R.C.
[2] The Second Artillery Military Office in Li-Shan Microelectronics Company, 710075, Xi'an, P.R.C.
[3] Academy of Armored Force Engineering Department of Information Engineering, 100858, Beijing, P.R.C.
Prof.Limin@gmail.com

Abstract. This paper proposes a novel image fusion scheme, which combines the merits of pixel-level and feature-level fusion algorithms. It avoids some of the well-known problems in pixel-level fusion such as blurring effects and high sensitivity to noise and misregistration. The algorithm first segments images into several regions, then extract features from each segmented region to get the fused image. Two typical image segmentation methods, region growing and edge detection based are both presented in this paper. Experimental results have demonstrated that the proposed method has extensive application scope and it outperforms the multiscale decompositions (MSD) based fusion approaches, both in visual effect and objective evaluation criteria.

Keywords: Image fusion, Pixel level, Feature level.

1 Introduction

Image fusion refers to image processing techniques that produce a new, enhanced image by combining images from two or more sensors. This fused image should increase the performance of the subsequent processing tasks such as segmentation, feature extraction and object recognition. It is widely recognized as a valuable tool for improving overall system performance in image based application areas such as defence surveillance, remote sensing, medical imaging and computer vision [1].

The actual fusion process can take place at different levels of information representation. A common categorization is to distinguish between pixel, feature and symbol level. Currently, it seems that most image fusion applications employ pixel-based methods. The advantage of pixel fusion is that the images used contain the original information. Furthermore, the algorithms are rather easy to implement and time efficient. However, fusing data at pixel-level requires co-registered images at subpixel accuracy because the existing fusion methods are very sensitive to misregistration.

In this paper, we propose a pixel-level and feature-level combined image fusion scheme. This method is computationally simple and can be used in real-time

F. Sun et al. (Eds.): ISNN 2008, Part II, LNCS 5264, pp. 658–665, 2008.

applications. Moreover, it overcomes most image fusion algorithms split relationship among pixels and treat them more or less independently. Extensive experiments with multi-focus image fusion and different sensor image fusion were performed, all results show that the proposed method has extensive application scope and it avoids some of the well known problems in pixel-level fusion such as blurring effects and high sensitivity to noise and misregistration.

The rest of this paper is organized as follows. A brief introduction to the proposed scheme will be given in Section 2. The presented segmentation methods will be described in Section 3. Section 4 contains the introduction of extracted features. Experimental results will be presented in Section 5, and the last section gives some concluding remarks.

2 The Basic Algorithm

Fig. 1 shows a schematic diagram of the proposed image fusion scheme which combined the merits of pixel-level and feature-level image fusion algorithms.

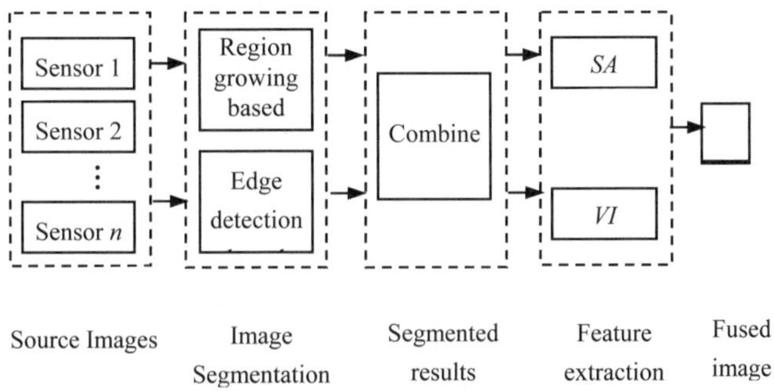

Fig. 1. Schematic diagram of the proposed fusion method

In detail, the algorithm consists of the following steps:

Step1. Segment the registered source images into different regions. (details in Section 3).

Step2. Combine the segmentation results with source images to determine the region each pixel belongs to. Suppose the multisensor source images are $S_1, S_2, \ldots, S_n$, and denote the ith region of image S_m (m=1, 2, …, n) by $DB_i(S_m)$.

Step3. From each image region $DB_i(S_m)$, extract two features, salience(*SA*) and visibility(*VI*), which reflect its clarity. Denote the feature vector for $DB_i(S_m)$ by ($SA_{DBi(Sm)}$, $VI_{DBi(Sm)}$) (details in Section 4).

Step4. Determine the fusion weight of $DB_i(S_m)$ according to ($SA_{DBi(Sm)}$, $VI_{DBi(Sm)}$). Denote the fusion weight for $DB_i(S_m)$ by $W_{DBi(Sm)}$ (details in Section 4).

Step5. Get the final fused image F by $DB_i(S_m)$ and $W_{DBi(Sm)}$.

3 Region Segmentation

Image segmentation is one of the most important and difficult tasks of digital image processing and analysis systems. Two related processes are involved: region segmentation (the splitting of an image into regions of a homogeneous image characteristic) and the extraction of prominent edges from the image. Methods of segmentation typically concentrate on one or other of these two approaches [2]. Here, we present edge detection based and region growing based region segmentation methods used in the proposed scheme.

3.1 Canny Edge-Based Image Segmentation

Edges are important features in an image since they represent significant local intensity changes. They provide important clues to separate regions within an object or to identify changes in illumination.

Edge detection is a method as significant as thresholding. A survey of the differences between particular edge detectors is presented by Michael [3].

The Canny edge detection [4] outperforms other edge detections in three aspects: firstly, the amplitude signal-to-noise ratio of the gradient is maximized to obtain a low probability of the failure to mark real edge points and a low probability of falsely marking non-edge points. Secondly, the edge points are identified as close as possible to the center of the edge (The closeness of the Canny edges to the real edges are surprisingly better than the edges detected by other edge detectors), and thirdly, the detected edges are of one pixel width.

Furthermore, the Canny edge detector can also detect small details of an object with much less noise if the threshold value is small. So the Canny edge detector is a good candidate for image segmentation method.

In the first step of our edge-based image segmentation procedure, the Canny edge detector is applied. Assume S_i is the ith multisensor source image and p is any pixel of S_i, whether p is on the edge is judged by:

$$BW_p = Canny\_Edge(S_i) \tag{1}$$

if pixel p is on the edge, the value of BW_p is 1, otherwise it is 0.

In the second step of our image segmentation procedure, we should determine lonely points and break points in the results after Canny edge detection (implemented by fomular. 2).

$$Sum(BW_{N_8(p)\cup p}) = \begin{cases} 1, & \text{p is a lonely point} \\ 2, & \text{p is a break point} \end{cases} \tag{2}$$

where $N_8\,(p)$ is the eight neighboring pixels of p.

Lonely points can be omitted directly for their little contributions to an image.

In the third step of our image segmentation procedure, break point should be linked with other point to form a close region.

Assume P is a set of edge points and boundary points. For every break point q, the linked point p will be found in P. Based on lots of experiments, we use the least amount of calculation and best performance method to determine the linked point:

$$D_E(p_k,q_k)=\min(D_E(p,q_k)) \quad p\in P \tag{3}$$

where q_k is break point and p_k is the linked point with it, $D_E(s,t)$ is Euclidean distance between point s and t. Considering q_k is edge point itself, the value of 0 is omitted in formula.(3).

After processed by these, every source images is segmented into several close regions. To satisfy the successive processing, we combine the segmented results together by logic "OR" to get more detailed segmented results.

3.2 Pulse Coupled Neural Network Based Image Segmentation

The PCNN model is a system composed of closely interacting nodes, with spiking neural behaviour. The theory is based on a neurophysiologic model, which evolved from studies of the cat's eye and the guinea pig [5, 6]. It finds many applications in image processing, including segmentation, edge extraction and so on [7].

Each PCNN neuron is divided into three compartments with characteristics of the receptive field, the modulation field, and the pulse generator. The receptive field is comprised of feeding field and linking field. Let N_{ij} denote the ijth neuron. The feeding field of N_{ij} receives the external stimulus S_{ij} and the pulses Y_{kl} from the neighboring neurons. A signal denoted by F_{ij} is outputted. The linking field receives the pulses from the neighboring neurons and outputs the signal denoted by L_{ij}. In the modulation field, F_{ij} and L_{ij} are inputted and modulated. The modulation result U_{ij} called internal activity signal is sent to the spike generator, in which U_{ij} is compared with the dynamic threshold θ_{ij} to form the pulse output Y_{ij}.

In the feeding field and linking field, there are six parameters, i.e., three time decay constants (α_F, α_L, α_θ) and three amplification factors (V_F, V_L, V_θ).

The following five equations are satisfied.

$$F_{ij}(n)=\exp(-\alpha_F)\cdot F_{ij}(n-1)+S_{ij}+V_F\cdot\sum M_{ijkl}Y_{kl}(n-1) \tag{4}$$

$$L_{ij}(n)=\exp(-\alpha_L)\cdot L_{ij}(n-1)+V_L\sum W_{ijkl}Y_{kl}(n-1)\ . \tag{5}$$

$$U_{ij}(n)=F_{ij}(n)(1+\beta\cdot L_{ij}(n))\ . \tag{6}$$

$$\theta_{ij}(n)=\exp(-\alpha_\theta)\theta_{ij}(n-1)+V_\theta Y_{ij}(n-1) \tag{7}$$

$$Y_{ij}(n)=step(U_{ij}(n)-\theta_{ij}(n))\ . \tag{8}$$

Where, M and W are the linking matrix, and normally W=M, β is the linking coefficient, *step(•)* is the unit step function.

In the application of image segmentation, each pixel corresponds to a single PCNN neuron. That is, a two dimensional intensity image (M×N) can be thought as a PCNN neuromime with M×N neurons, and the gray level of pixels can be thought as S_{ij}, the input of the neuron. The neurons are organized in a single layer network to perform the segmentation task. Considering M and W are the interior linking matrixes. When there are pixels whose gray levels are approximate in the neighborhood of M and W, one pixel's pulsating output can activate other corresponding pixels having the approximate gray level in the neighborhood and let them generate pulsating output sequence *Y(n)*. Obviously *Y* contains some information about this image such as regional information, edge, and texture features. Then the binary image constructed by *Y(n)*, the output of PCNN, is the segmented image.

To satisfy the demand of the following fusion process, all segmentation information in different source images is combined together. A simple way is to draw the contours for all segmented images, and then overlap all contour images together; thus each image can be divided into different regions.

Considering the lonely points have little contributions to image quality and probably belong to noise, we should get rid of them after PCNN segmentation.

4 Feature Extraction

We extract two features, the salience and visibility, from each image region to represent its clarity.

4.1 Salience (*SA*)

The salience proposed in this paper is mainly with a view to the difference between target and its neighboring regions. From experimental analysis, we can see that the clarity of a region edge in different source images directly determines the same area fusion weight in corresponding images. The clearer of the region edge, the bigger the fusion weight is.

Assume $DB_i(S_m)$ is *i*th region of the image S_m, p is any point of DB_i's edge and $N_8(p)$ is its eight neighboring pixels, we expand the edge by $N_8(p)$ to get an edge band about three pixels width. Take the edge band into account, we can work out the mean gray value of those pixels belong to DB_i and those not. The absolute difference is defined as the salience of DB_i.

$$SA = \left| \frac{1}{m}\sum_{i=1}^{m} f_i - \frac{1}{n}\sum_{j=1}^{n} g_j \right| \tag{9}$$

where f_i is the gray value of pixel which belongs to DB_i edge band, g_j is the gray value of pixel which not belongs to.

4.2 Visibility (*VI*)

This feature is inspired from the human visual system, and is defined as [8]. We rectify the formula as following.

$$VI(DB_i) = \frac{1}{\sum DB_i} \sum_{(x,y)\in DB_i} \left(\frac{1}{m_k}\right)^{\alpha} \cdot \frac{|f(x,y) - m_k|}{m_k} \tag{10}$$

where $\sum DB_i$ is the total number of pixels in DB_i, m_k is the mean gray value of the image region, and α is a visual constant ranging from 0.6 to 0.7.

Considering the different contributions to the fusion result between various source images, we use region fusion weight to denote these. Assume $W_{DBi}(S_m)$ is the fusion weight of region $DB_i(S_m)$, we use visibility and salience of the region as the two main factors to determine it. Based on extensive experiments, we define it as

$$W(DB_i) = e^{VI(DB_i)W_{VI}} + e^{SA(DB_i)} \tag{11}$$

where W_{VI} is a visibility constant.

5 Experimental Results

We do a lot of experiments to prove the validity and robustness of the proposed methods. Here, we give a typical one.

5.1 Setup for Quantitative Evaluation

To evaluate the performance of the proposed fusion method, extensive experiments with multi-focus image fusion and different sensor image fusion have been performed.

The objective evaluative criteria spatial frequency (*SF*) [9] is used.

$$SF = \sqrt{\frac{1}{MN}\left[\sum_{i=0}^{M-1}\sum_{j=1}^{N-1}[F(i,j)-F(i,j-1)]^2 + \sum_{j=0}^{N-1}\sum_{i=1}^{M-1}[F(i,j)-F(i-1,j)]^2\right]}$$

Here, *M*, *N* are the dimensions of the image, Notice that *SF* indicates the activity level of the whole image, and so a larger *SF* is preferred.

5.2 Subjective Evaluation

The experiment is performed on computed tomography (CT) image and magnetic resonance image (MRI) (showed as Fig.2). In this case, the source images come from different sensors. Quantitative comparisons of their performance are shown in Table1. For many multisensor image fusion applications, which source images come from different sensors, the ideal reference image can not be acquired, so here we only give *SF* evaluation criteria.

For comparison purpose, besides the fusion scheme proposed in this paper, another MSD-based fusion algorithm, discrete wavelet transform based (DWT), is also applied to fuse the same images. The wavelet basis "db8", together with a decomposition level of 3, is used. Similar to [10], we employ a region-based activity measurement for the active level of the decomposed wavelet coefficients, a maximum selection rule for coefficient combination, together with a window-based (5×5) consistency verification scheme, the value of α is 0.7.

Parameters in PCNN as: CT: β=0.4,α_F=0.28, α_L=0.28, α_θ=0.65, V_F=2, V_L=10, V_θ=10,r=1,N=2; MRI: β=0.2,α_F=69, α_L=0.1, α_θ=0.1, V_F=0.2, V_L=8, V_θ=10,r=1,N=14

Experimental results show that the proposed scheme outperforms the discrete wavelet transform approach with 64% improvement in space frequency. The objective evaluation results coincide with the visual effect very well. From visual effect and object

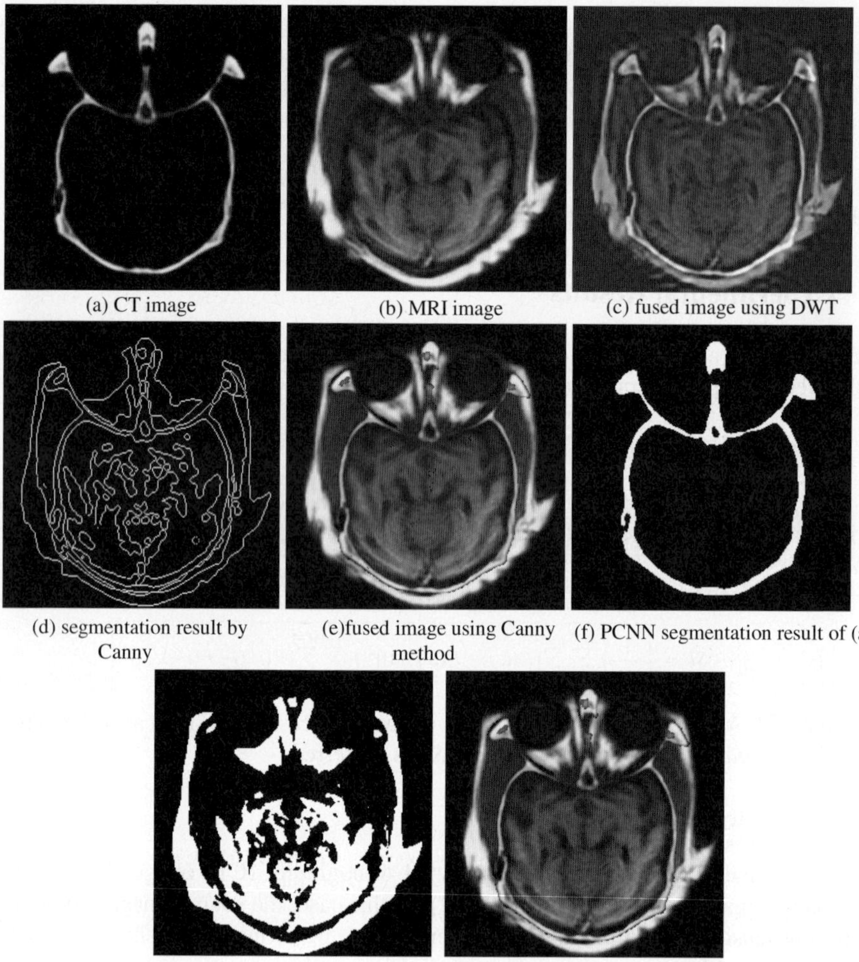

(a) CT image (b) MRI image (c) fused image using DWT

(d) segmentation result by Canny (e)fused image using Canny method (f) PCNN segmentation result of (a)

(g) PCNN segmentation result of (b) (h) fused image using PCNN

Fig. 2. The medical source imagesand fusion results

Table 1. Performance of the different fusion methods on processing Fig.2

Objective Criteria	DWT	Fusion method by Canny	Fusion method by PCNN
SF	16.9592	23.3304	23.3630

evaluation criteria, we can see the proposed fusion scheme also show significant improvement over the MSD-based method in the applications of source images come from different sensors.

6 Conclusion

In this paper, we propose a new image fusion scheme, which combines the merits of pixel-level and feature-level fusion algorithms. This approach has such advantage that the fusion process becomes more robust and avoids some of the well-known problems in pixel-level fusion such as blurring effects and high sensitivity to noise and misregistration. Lots of experiments on studying the fusion performance have been made and the results show that the proposed method can be used in image fusion applications, which source images come from the same type of sensor or different types of sensors, and it outperforms the MSD-based method, both in visual effect and objective evaluation criteria.

References

1. Piella, G.: A General Framework for Multiresolution Image Fusion: from Pixels to Regions. Information Fusion 4, 259–280 (2003)
2. Pal, N.R., Pal, S.K.: A Review on Image Segmentation Techniques. Pattern Recognition 26, 1277–1294 (1993)
3. Michael, H., Wilkinson, F.: Optimizing Edge Detectors for Robust Automatic Threshold Selection: Coping with Edge Curvature and Noise. Graphical Models and Image Processing 60, 385–401 (1998)
4. Canny, J.: A Computational Approach to Edge Detection. IEEE Trans. Pattern Recognition and Machine Analysis 8, 679–698 (1986)
5. Johnson, J.L., Padgett, M.L.: PCNN Model and Applications. IEEE Trans. on Neural Networks 10, 480–498 (1999)
6. Thhomas, L.: Inherent Features of Wavelets and Pulse Coupled Neural Networks. IEEE Trans. on Neural Networks 10, 9204–1092 (1999)
7. Kuntimad, G., Ranganath, H.S.: Perfect Image Segmentation Using Pulse Coupled Neural Networks. IEEE Trans. Neural Networks 10, 591–598 (1999)
8. Huang, J.W., Shi, Y.Q., Dai, X.H.: A Segmentation Based Image Coding Algorithm Using the Features of Human Vision System. IEEE Trans. on Neural Networks 4, 400–404 (1999)
9. Xydaes, C., Petrovi, V.: Objective Image Fusion Performance Measure. Electronic Letters 36, 308–309 (2000)
10. Zhang, Z., Blum, R.S.: A Categorization of Multiscale Decomposition based Image Fusion Schemes with a Performance Study for a Digital Camera Application. Proceedings of IEEE 87, 1315–1326 (1999)

Combining Multi Wavelet and Multi NN for Power Systems Load Forecasting

Zhigang Liu*, Qi Wang, and Yajun Zhang

Institute of Electrification & Automation, Southwest Jiaotong University, Chengdu, Sichuan, 610031, China
liuzg_cd@126.com

Abstract. In the paper, two pre-processing methods for load forecast sampling data including multiwavelet transformation and chaotic time series are introduced. In addition, multi neural network for load forecast including BP artificial neural network, RBF neural network and wavelet neural network are introduced, too. Then, a combination load forecasting model for power load based on chaotic time series, multiwavelet transformation and multi-neural networks is proposed and discussed in the paper. Firstly, the training sample is extracted from power load data through chaotic time series and multiwavelet decomposition. Then the obtained data is trained through BP network, RBF network and wavelet neural network. Lastly, the trained data from three neural networks are input a three-layer feedforward neural network based the variable weight combination load forecasting model. Simulation results show that accuracy of the combination load forecasting model proposed in the paper is higher than any one sole network model and the combination forecast model of three neural networks.

Keywords: Power system; Chaotic time series; Multiwavelet transformation; Combination load forecasting; Multi-neural networks.

1 Introduction

At present, there are a lot of models for power load forecasting, which is one of most important basis in planning and operation of electric power system. Only one model cannot totally reflect the changing rules and information of power load. Combination model of load forecasting has been a new direction of developing research [1-2]. Using the neural network for load forecasting is usual method. If main facts for the load are not considered, the input of neural network is a time serial sampling data generally. For the short term load forecasting of power system, if the sampling data is directly trained into the network, there is an obvious fault during the course of load forecasting. The fault is that only the time relativity of load sampling data at each time is considered during the course of network learning, but the space relativity of load sampling data is not considered. It will result into the poor precision for load forecasting. The chaotic time series is adopted as the pre-processing method for the input of load sampling data.

* This work is supported by Fok Ying Tung Education Foundation (No.101060) and Sichuan Province Distinguished Scholars Fund (No. 07ZQ026-012).

F. Sun et al. (Eds.): ISNN 2008, Part II, LNCS 5264, pp. 666–673, 2008.

In power load forecasting, the sample data is very important, as well as the pre-processing methods for load forecast. But in most conditions, the data is lack because of the inherent characteristic of load forecasting data. Based on the consideration above, multiwavelet transformation is adopted in the paper. Compared with the traditional wavelet transformation, more characteristic values with multiwavelet decomposition can be obtained than those with wavelet decomposition, which means more sample data can be acquired. Hence, multiwavelet transformation will become a pre-processing method before power load is forecasted. Multiwavelets have been applied in power system recently [3-5]. In order to obtain better accuracy of load forecasting, the multi-neural networks will be adopted in the paper. Because of the similarity of BP neural network, RBF neural network and wavelet neural network, we can put the three neural networks together to forecast the power load respectively. Then, the forecasting results of three neural networks are combined with nonlinear method, where a three-layer feedforward neural network is adopted.

The main idea in the paper is to get more characteristic values of power load data through the pre-processing of chaotic time series and multiwavelet decomposition, to train the characteristic values through multi-neural networks and to obtain the forecasting data in the end. The multiwavelet transformation and multi-neural networks will be introduced below.

2 Pre-processing Method

2.1 Chaotic Time Series

During the course of construction training data, with the technology of space reconstruction, load series $\{x_1, x_2, \cdots, x_n\}$ can be embedded into m dimension space, and the m dimension space track can be constructed.

$$\begin{aligned} y_1 &= (x_1, x_{1+\tau}, \cdots, x_{1+(m-1)\tau}) \\ y_2 &= (x_2, x_{2+\tau}, \cdots, x_{2+(m-1)\tau}) \\ &\vdots \\ y_{n-(m-1)\tau} &= (x_{n-(m-1)\tau}, x_{n-(m-1)\tau+1}, \cdots x_n) \end{aligned} \quad . \tag{1}$$

Where τ is delay parameter. If $\{x_1, x_2, \cdots, x_n\}$ can be obtained for the forecasting of load series, based on Takens Theorem here is a mapping $f: R^m \to R$ satisfies

$$f(x_t, x_{t-\tau}, \cdots, x_{t-m\tau}) = x_{t+\tau}, \quad t = m\tau + 1, m(\tau+1)+1, \cdots, n \ . \tag{2}$$

After the pre-processing of load sampling data with chaotic time series, the data can be input into the neural network for training.

2.2 Multiwavelet Transformation

The multi-resolution analysis is very important in traditional wavelets transformation. Likewise, there is multi-resolution in multiwavelets analysis. Different from wavelets, the multi-resolution analysis is produced through several scaling functions and the base of space $L^2(R)$ is constructed by the translation and dilation of multi-wavelet functions. These multi-wavelet functions are called as multiwavelet. Let $\Phi(t)=[\phi_1(t),\phi_2(t),...,\phi_r(t)]^T$, $\phi_l \in L^2(R)$, $l=1,2,...,r, r\in N$, Where $V_j=\overline{span\{2^{-j/2}\phi_l(2^{-j}t-k), n\in Z\}}$. Consider ϕ_l produces the multiplicity r multi-resolution analysis. The translation and dilation of orthogonal multiwavelets functions $\Psi(t)=[\psi_1(t),\psi_2(t),...,\psi_r(t)]^T (r\in N)$, namely $\Psi_{j,k}=\{\psi_1(2^{-j}x-k),...,\psi_r(2^{-j}x-k)\}^T$, $(j,k\in Z)$, we can construct the orthogonal base in orthogonal and supplemental subspace, which is the supplemental subspace of V_j in V_{j+1}. Multi-scaling function $\Phi(t)$ satisfies the two-scaling equation below.

$$\Phi(t)=\sum_{k=0}^{M} H_k \Phi(2t-k) \cdot \tag{3}$$

Where H_k, $k=0,1,..,M$ is $r\times r$ impulse response constant matrix. The following equation is given based on the multi-resolution analysis [7].

$$f(t)=\sum_{l=1}^{r}\sum_{k\in Z} c_{l,J,k}\phi_{l,J,k}(t)= \sum_{l=1}^{r}\sum_{k\in Z} c_{l,J_0,k}\phi_{l,J_0,k}(t)+\sum_{l=1}^{r}\sum_{J_0\le j<J}\sum_{k\in Z} d_{l,j,k}\psi_{l,j,k}(t) \tag{4}$$

Where $J_0<J$, $c_{l,j,k}=\int f(t)\phi_{l,j,k}(t)dt, d_{l,j,k}=\int f(t)\psi_{l,j,k}(t)dt$.

Let $c_{j,k}=(c_{1,j,k},...,c_{r,j,k})^T, d_{j,k}=(d_{1,j,k},...,d_{r,j,k})^T$. The decomposition equations are listed as following [7]:

$$c_{j-1,k}=\sqrt{2}\sum_n H_n c_{j,2k+n}, d_{j-1,k}=\sqrt{2}\sum_n G_n c_{j,2k+n} \cdot \tag{5}$$

3 Several Neural Networks

3.1 BP Neural Network

BP neural network belongs to feedforward neural networks. We suppose the layer number is m in BP neural network, and there are some neurons in each layer. The output relationship of the j th neuron in k th layer is list as following.

$$x_j^{(k)}=f_j^{(k)}[\sum_{i=1}^{n_{k-1}} W_{ij}^{(k)} x_i^{(k-1)} - H_j^{(k)}], \qquad (j=1,2,\cdots\cdots,n_k \quad k=1,2,\cdots\cdots,m) \cdot \tag{6}$$

Where $W_{ij}^{(k)}$ is the connection weight between i th neuron in $k-1$ th layer and j th neuron in k th layer. $H_j^{(k)}$ is the threshold of j th neuron in k th layer. $f_j^{(k)}$ is the transfer function. Function sigmoid is adopted in general in BP neural network. n_k is the number of neurons in k th layer.

3.2 RBF Neural Network

RBF neural network is similar to BP neural network except the function in hidden layer. The activation function in RBF neural network is a non-minus and nonlinear local distribution function with radial symmetry attenuation for the center point. The neural network can map the input space into a new space, where the input layer can be linearly combined. The adjustable parameters include the weight value and the parameter that controls the shape of activation function. In RBF neural network, Gauss function that is a nonlinear function is adopted in hidden layer neuron. The expression formula is shown below.

$$G(\|X(t)-T_i\|)=\exp\left[-\frac{1}{2}\left(\frac{\|X(t)-T_i\|}{R_i}\right)^2\right] . \tag{7}$$

Where $i=1,2,\cdots,m$, m is the number of neurons. $X(t)=(x_{t1},x_{t2},\cdots,x_{tn})$ is the t th input sample in training data. $T_i=(t_1,t_2,\cdots,t_n)$ is the center of i th hidden layer neuron. R_i is the unitary parameter of i th hidden layer neuron. The output of RBF neural network is shown below.

$$F(X(t),W_i,T_i)=\sum_{i=1}^{m}W_iG(\|X(t)-T_i\|) . \tag{8}$$

Where $t=1,2,...,N, N$ is the sum of training samples.

3.3 Wavelet Neural Network

In this paper, wavelet neural network (WNN) is adopted as a three-layer structure with an input layer, wavelet layer (hidden layer) and output layer. In the WNN structure, the hidden neurons have wavelet activation functions of different resolutions, and the output neurons have sigmoid activation functions. The activation functions of the wavelet nodes in the wavelet layer are derived from a mother wavelet $\psi(x)$. The output of the wavelet neural network Y is represented by the following equation [8]:

$$y_i(t)=\sigma(x_n)=\sigma\left(\sum_{j=0}^{M}v_{ij}\psi_{ab}\left(\sum_{k=0}^{L}w_{jk}x_k(t)\right)\right), \ (i=1,2,\cdots,N) . \tag{9}$$

Where, $\sigma(x_n)=1/(1+e^{-x_n})$; y_j denotes the jth component of the output vector; x_k denotes the kth component of the input vector; v_{ij} the connection weight between the output unit i and the hidden unit j; w_{jk} the weight between the hidden unit j and input unit k; a_j dilation coefficient of wavelons in hidden layer; b_j translation coefficient of wavelons in hidden layer; L, M, N the sum of input, hidden and output nodes respectively.

4 Multi-neural Networks Model

The structure of three neural networks above is generally similar. But the activation function of each neural network is quite different, which will produce different forecasting results. The improved BP algorithm is adopted in BP neural network. The forecasting results' accuracy is high, but training efficient is low and the algorithm easily gets into local optimization. RBF neural network can avoid local optimization and has quick training speed, but the forecasting accuracy is lower than one of BP neural network. Wavelet neural network has high forecasting accuracy. Because of random initialization method and training algorithm based on grads are adopted in wavelet neural network, the convergence of algorithms is poor. If the three neural networks are combined for power load forecasting, the localization of single model can be avoided. The forecasting accuracy can be better assured. Suppose f_{1i}, f_{2i}, f_{3i} be respectively i th forecasting result of BP neural network, RBF neural network and wavelet neural network, and f_i is i th actual load. We think the relationship of them be complex nonlinear, the expression is listed below.

$$\sum_{i=1}^{m}[f_i - g(f_{1i}, f_{2i}, f_{3i})]^2 . \tag{10}$$

Where m is the number of sample data. In order to get the minimum value of formula (8), the nonlinear function $g(\cdot)$ should be requested, but the correct root cannot be obtained with the traditional grads and genic algorithms. In the paper, a three-layer feedforward neural network is adopted to get the approximation of nonlinear mapping. The forecasting results from BP neural network, RBF neural network and wavelet neural network will be considered as the input of the neural network, and actual load acts as the output of the neural network for training samples. The structure of 3×7×1 BP neural networks is adopted. The variable weight combination of multi-neural networks is illustrated in Fig. 1.

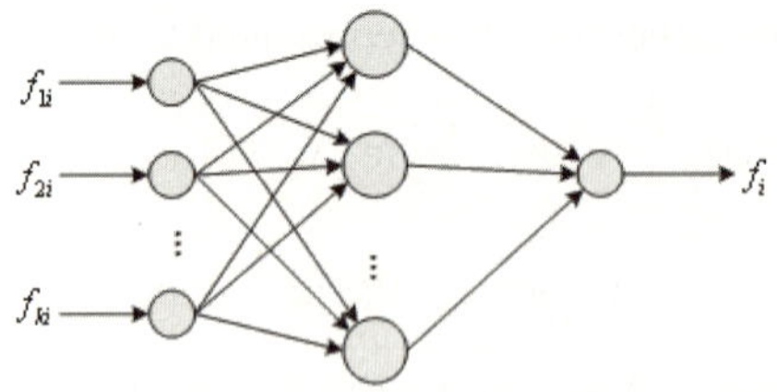

Fig. 1. The variable weight combination of multi-neural networks

5 Algorithm

The algorithm of combination forecasting model proposed in the paper mainly includes two sections. The first section is the course of pre-processing including sample extraction and data pre-processing, namely with chaotic time series and mutltiwavelet transformation. The second section is the sample training. In the course of sample training of power load forecasting, the input samples will be trained in multi-neural networks. The algorithm of combination forecasting model in the paper mainly includes several steps as following.

(1) The history data with chaotic characteristic can be pre-processed. The input data will be smoothed and normalized. With the chaotic time series, m and τ can be obtained and form the sampling data. The sampling data will be processed with mutltiwavelet transformation.

(2) Before the sample data is extracted, it is necessary that the pre-processing method should be discussed. For multiwavelet, the choice of pre-processing method is one of most important problems that result in the application effect. There are many pre-processing methods for different multiwavelets. Even if the same multiwavelet, the pre-processing methods may be different. Such as for GHM multiwavelet, the pre-processing methods include GHM init. method, Haar method, Deriv. method and so on [9]. Of course, their processing effect is not same for GHM multiwavelet. For the decomposition of multiwavelets, the pre-processing methods can be considered as prefilter matrix. GHM multiwavelet and GHM init. pre-processing method are adopted in the paper. The prefilter' matrix $Q(\omega)$ of GHM init. pre-processing method is listed.

$$Q(\omega)=\begin{bmatrix} -\dfrac{\phi_2(1/2)}{\phi_2(1)\phi_1(1/2)} & \dfrac{1}{\phi_1(1/2)}-\dfrac{\phi_2(1/2)}{\phi_2(1)\phi_1(1/2)}e^{-\frac{1}{2}i\omega} \\ \dfrac{1}{\phi_2(1)} & 0 \end{bmatrix}. \tag{11}$$

(3) After the pre-processing, the original data will be processed through multiwavelet transformation. Since there is more information of low and high frequency with multiwavelet decomposition than traditional wavelets, the more useful original data can be produced with multiwavelet transformation, which is very beneficial to obtain more training sample data in power load forecasting. The double coefficients after multiwavelet decomposition will be considered as the input samples of multi-neural networks.

(4) In the course of sample training of power load forecasting, the input samples will be trained in multi-neural networks. It means that the input samples will be input into BP neural network, RBF neural network and wavelet neural network at the same time. The three training results as sample input will be input into a three-layer feedforward neural network for the last load forecasting. In fact, it is a variable weight combination model for power load forecasting.

(5) The input samples are input into BP neural network, RBF neural network and wavelet neural network at the same time. The network structure (16×16×1) of three

neural networks is same. For BP neural network, activation function is Sigmoid and training algorithm is grads algorithm. For RBF neural network, activation function is Gauss function and training algorithm is standard RBF algorithm. For wavelet neural network, activation function is Morlet wavelet and training algorithm is minimizing energy algorithm.

(6) The three training results of BP neural network, RBF neural network and wavelet neural network are considered as the input of a three-layer feedforward neural network. After the training in the feedforward neural network, the power load forecasting data will be obtained in the end.

6 Example

Power load forecasting data of some district power network on 17th, June. 1997 is given in Table 1. ANN1 means the forecasting results with BP neural network solely. For the appointed forecasting hours, the input of 1-6 neurons means the nearest 5 hours' load values, the input of 7-12 neurons means the nearest 5 hours' load values of the two days before forecasting day, and the input of 13-16 neurons means the lowest and highest temperature values of the one and two days before forecasting day. ANN2 and ANN3 mean the forecasting results with RBF neural network and wavelet neural network respectively. COM1 means the forecasting results with combination forecasting model of BP neural network, RBF neural network and wavelet neural network. COM2 means the forecasting results with combination forecasting model proposed in the paper. It is obvious that the forecasting results of COM2 are better than COM1, ANN1, ANN2 and ANN3.

Table 1. The daily load forecast results on Oct 24, 1997 in a power network

Hours	Actual Load (MW)	Forecasting Load (MW)				
		ANN1	ANN2	ANN3	COM1	COM2
1	479.00	492.67	490.91	486.60	486.40	484.73
2	457.00	446.33	450.33	447.19	449.11	451.78
3	444.00	424.54	429.24	451.88	433.64	435.03
4	442.00	453.32	432.03	457.68	450.64	450.21
5	442.00	455.96	430.22	452.46	445.92	443.43
6	479.00	486.94	469.16	496.00	489.13	487.79
7	558.00	544.81	544.23	549.89	548.88	549.13
8	554.00	533.68	539.10	543.57	542.49	542.12
9	575.00	558.59	566.05	591.87	584.56	583.47
10	573.00	566.30	589.54	563.30	580.92	579.22
11	568.00	543.84	577.42	552.40	557.44	559.89
12	540.00	554.40	524.95	550.07	549.67	547.21
13	579.00	601.32	585.74	592.35	585.26	584.25
14	527.00	531.46	539.32	533.34	535.21	535.10
15	543.00	519.45	528.52	524.22	529.45	530.36

7 Conclusion

A combination load forecasting based on chaotic time series, multiwavelet transformation and multi-neural networks is proposed in the paper. Considering the advantages of chaotic time series and multiwavelet transformation, chaotic time series and multiwavelet decomposition are respectively used for the pre-processing and training sample extraction. The multi-neural networks including BP neural network, RBF neural network and wavelet neural network are adopted for the power load forecasting with variable weight combination forecasting model. The simulation results show that that accuracy of the method is higher than any one sole network model and the combination forecast model of three neural networks.

References

1. Kang, C.Q., Xia, Q., Zhang, B.M.: Review of Power System Load Forecasting and its Development. Automation of Electric Power Systems 28, 1–11 (2004)
2. Niu, D.X., Cao, S.H., Zhao, L.: Power Load Forecasting Technology and its Application. China Electric Power Press, Beijing (1998)
3. Liu, Z.G., Zeng, Y.D., Qian, Q.Q.: De-noising of Electric Power System Signals Based on Different Multiwavelets. In: Proceedings of the CSEE, vol. 24, pp. 30–34 (2004)
4. Liu, Z.G., Qian, Q.Q.: Compression of Fault Transient Data in Electric Power System Based on Multiwavelet. In: Proceedings of the CSEE, vol. 23, pp. 22–26 (2003)
5. Liu, Z.G., He, Z.Y., Qian, Q.Q.: A Fault Signal Data Compression Plan Based on Optimal Pre-processing Method of Multiwavelet. Power System Technology 29, 40–43 (2005)
6. Chui, C.K., Lian, J.A.: A Study of Orthonormal Multi-wavelets. Appl. Numer. Math. 20, 273–298 (1996)
7. Xia, X.G.: A New Prefilter Design for Discrete Multiwavelet Tansforms. IEEE Trans. on Signal Processing 46, 1558–1570 (1998)
8. Leopoldo, A., Pasquale, D.: Wavelet Network-Based Detection and Classification of Transients. IEEE Trans. Instruction and Measurement 50, 1425–1435 (2001)
9. Cotronei, M., Montefusco, L.B., Puccio, L.: Multiwavelet Analysis and Signal Processing. IEEE Trans. on Circuits and Systems-II: Analog and Digital Signal Processing 45, 970–987 (1998)

An Adaptive Algorithm Finding Multiple Roots of Polynomials

Wei Zhu[1,2,*], Zhe-zhao Zeng[1,*], and Dong-mei Lin[1]

[1] College of Electrical & Information Engineering, Changsha University of Science &Technology, Changsha, Hunan 410076, China
[2] College of Electrical & Information Engineering, Hunan University, Changsha, Hunan 410082, China
hncs6699@yahoo.com.cn

Abstract. An adaptive algorithm is proposed to find multiple roots of polynomials which were not well solved by the other methods. Its convergence was presented and proved. The computation is carried out by simple steepest descent rule with adaptive variable learning rate. The specific examples showed that the proposed method can find the multiple roots of polynomials at a very rapid convergence and very high accuracy with less computation.

Keywords: Adaptive Algorithm, Multiple Real or Complex Roots, Variable Learning Rate.

1 Introduction

Finding rapidly and accurately the roots of polynomials is an important problem in various areas of control and communication systems engineering, signal processing and in many other areas of science and technology. The problem of finding the zeros of a polynomial has fascinated mathematicians for centuries, and the literature is full of ingenious methods, analyses of these methods, and discussions of their merits [1-3]. Over the last decades, there exist a large number of different methods for finding all polynomial roots either iteratively or simultaneously. Most of them yield accurate results only for small degree or can treat only special polynomials, e.g., polynomials with simple real or complex roots [4].

So far, some better modified methods of finding roots of polynomials cover mainly the Jenkins/Traub method [5], the Markus/Frenzel method [4], the Laguerre method [6], the Routh method [7], the Truong, Jeng and Reed method [8], the Fedorenko method [9], the Halley method [10], and some modified Newton's methods [11-13], etc. Although the Laguerre method is faster convergent than all other methods mentioned above, it has more computation. Among other methods, some have low accuracy, and some have more computation, especially to say is the modified Newton's methods must have a good initial value near solution. Furthermore, it is very difficult for the all methods mentioned above to find multiple real or complex roots of polynomials.

[*] Corresponding authors.

F. Sun et al. (Eds.): ISNN 2008, Part II, LNCS 5264, pp. 674–681, 2008.

In order to solve the problems above, we propose an algorithm finding multiple real or complex roots of polynomials with adaptive variable learning rate. The approach can find multiple roots of polynomials with less computation, high accuracy and rapid convergence.

2 The Algorithm Finding Multiple Zeros of Polynomials

2.1 The Algorithm Description

We start by defining our typical polynomial of degree n as

$$f(x) = a_n x^n + a_{n-1} x^{n-1} + \cdots + a_1 x + a_0 \tag{1a}$$

$$= a_n (x - p_1)^{m_1} (x - p_2)^{m_2} \cdots (x - p_l)^{m_l} \tag{1b}$$

Where, $\sum_{j=1}^{l} m_j = n$, and $1 < m_i < n \, (i = 1, 2, \cdots, l)$.

Here we are given the coefficients, a_i ($a_n \neq 0$), and wish to find the multiple real or complex zeros: p_i . Usually, in science and engineering applications, the coefficients will all be real, and then the zeros will either be real or else occur in conjugate-complex pairs.

Let us then assume for the moment that all the p_i is real or complex and distinct, and numbered so that

$$\mathrm{Re}(p_1) < \mathrm{Re}(p_2) < \cdots \mathrm{Re}(p_l) \tag{2}$$

Also we will assume that we have made some real-valued or complex-valued guess, p_k , possibly quite crude, for one of the zeros, and that

$$\mathrm{Re}(p_m) < \mathrm{Re}(p_k) < \mathrm{Re}(p_{m+1}) \tag{3}$$

It is well known that the multiple root p_i is also the root of polynomial $f^{(m_i - 1)}(x)$ while p_i is the multiple m_i root of function $f(x)$. The principal feature of the algorithm proposed is to make $f^{(m_i - 1)}(x)$ satisfy

$$f^{(m_i - 1)}(x) = 0$$

by training the weighting variable x . The algorithm is as follows:

Given an arbitrary initial approximation weighting coefficient x_k , real or complex, an error function can be obtained:

$$e(k) = 0 - f^{(m_i - 1)}(x_k) = -f^{(m_i - 1)}(x_k) \tag{4}$$

Define an objective function J as

$$J(k)=\frac{1}{2}e^2(k) \tag{5}$$

To minimize the J , the weight x_k is recursively computed via using a simple gradient descent rule with variable learning rate:

$$x_{k+1}=x_k-\eta(k)\frac{dJ(k)}{dx_k} \tag{6}$$

Where $\eta(k)$ is learning rate and usually $0<\eta(k)<1$, and that

$$\Delta x_k=-\eta(k)\frac{dJ(k)}{dx_k} \tag{7}$$

On differentiating eqn.5 with respect to x_k , the gradient of $J(k)$ with respect to x_k is given by

$$\frac{dJ(k)}{dx_k}=\frac{dJ(k)}{de(k)}\frac{de(k)}{df^{(m_i-1)}(x_k)}\frac{df^{(m_i-1)}(x_k)}{dx_k}=-e(k)f^{(m_i)}(x_k) \tag{8}$$

Substituting (8) into (6), we have

$$x_{k+1}=x_k+\eta(k)e(k)f^{(m_i)}(x_k) \tag{9}$$

here, $\Delta x_k=e(k)\eta(k)f^{(m_i)}(x_k)$.

2.2 Research of the Algorithm Convergence

In order to ensure the absolute convergence of the algorithm above, it is important to select a proper learning rate: $\eta(k)$. In the section, we present and prove the theorem about convergence of the algorithm proposed as follows:

Theorem 1: the function $f\in C^{m_i}[a,b]$ has a zero of multiplicity m_i at p_i in (a, b) if and only if

$0=f(p_i)=f'(p_i)=f''(p_i)=\cdots=f^{(m_i-1)}(p_i)$, but $f^{(m_i)}(p_i)\neq 0$, then only when

$$0<\eta(k)<2/[f^{(m_i)}(x_k)]^2 \tag{10}$$

the algorithm proposed is convergent, where $\eta(k)$ is adaptive learning rate.

Proof: Define a Lyapunov function:

$$V(k)=\frac{1}{2}e^2(k) \tag{11}$$

Then

$$\Delta V(k) = \frac{1}{2}e^2(k+1) - \frac{1}{2}e^2(k) \tag{12}$$

Since

$$e(k+1) = e(k) + \Delta e(k) = e(k) + \frac{de(k)}{dx_k}\Delta x_k \tag{13}$$

and

$$\Delta x_k = -\eta(k)\frac{dJ(k)}{dx_k} = -\eta(k)e(k)\frac{de(k)}{dx_k} \tag{14}$$

According to (12), (13) and (14), we have

$$\Delta V(k) = \Delta e(k)[e(k) + \frac{1}{2}\Delta e(k)] = \left[\frac{de(k)}{dx_k}\right]^2 e^2(k)\left\{-\eta(k) + \frac{1}{2}\eta^2(k)\left[\frac{de(k)}{dx_k}\right]^2\right\} \tag{15}$$

Known from the (4) that

$$\frac{de(k)}{dx_k} = -f^{(m_i)}(x_k) \tag{16}$$

Substituting it into (15) gives

$$\Delta V(k) = \left[f^{(m_i)}(x_k)\right]^2 e^2(k)\left\{-\eta(k) + \frac{1}{2}\eta^2(k)\left[f^{(m_i)}(x_k)\right]^2\right\} \tag{17}$$

Also since $\left[f^{(m_i)}(x_k)\right]^2 e^2(k) \geq 0$, if the algorithm proposed is convergent, i.e. $\Delta V_k < 0$, then it is easy to see from (17) that

$$-\eta(k) + \frac{1}{2}\eta^2(k)\left[f^{(m_i)}(x_k)\right]^2 < 0 \tag{18}$$

Since $\eta(k) > 0$, thus we have

$$0 < \eta(k) < 2/\left[f^{(m_i)}(x_k)\right]^2 \tag{19}$$

2.3 Evaluation of the Optimal Learning Rate η_{opt}

It is important to determine the magnitude of the variable learning rate $\eta(k)$ during the training of the algorithm proposed. ***Theorem 1*** indicates the theory criterion determining the magnitude of the variable learning rate $\eta(k)$. If the $\eta(k)$ is too large, the algorithm may produce oscillation and is not convergent at all. If it is too small, the algorithm may be slowly convergent with more computation. Since $\eta(k)$ depends on the $f^{(m_i)}(x_k)$, hence it varies with the derivative evaluation of m_i th-degree: $f^{(m_i)}(x_k)$ at x_k . In order to make the algorithm be rapidly convergent, according to experience, the optimal learning rate should usually be

$$\eta_{opt}(k) = \begin{cases} 0.5, & \left|f^{(m_i)}(x_k)\right| \le 2 \\ (1.0-1.6)/\left[f^{(m_i)}(x_k)\right]^2, & other \end{cases} \quad (20)$$

2.4 Algorithm Steps

To find a zero of multiplicity m_i to $f(x)=0$ given one approximation x_0:

INPUT: x_0 (real number or complex number); tolerance Tol ; maximum number of iterations N ; let $k=0$;

OUTPUT: approximate solution x_{k+1} or message of failure.

Step 1: While $k \le N$ do Steps 2-5

Step 2: set $e(k) = -f^{(m_i-1)}(x_k)$; $J = \frac{1}{2}|e(k)|^2$

If $\left|f^{(m_i)}(x_k)\right| \le 2$ then $\eta_{opt}(k) = 0.5$

Else $\eta_{opt}(k) = 1/[f^{(m_i)}(x_k)]^2$

Step 3: $x_{k+1} = x_k + \eta_{opt}(k)e(k)f^{(m_i)}(x_k)$

Step 4: If $J \le Tol$ then

OUTPUT (x_{k+1}); (The procedure was successful.)

STOP

Step 5: Set $x_k = x_{k+1}$; $k = k+1$

Go back to step 2

Step 6: OUTPUT ('the method failed after N iterations, $n =$' k);

(The procedure was unsuccessful.)

STOP

3 Results and Discussion

In order to confirm the validity of the algorithm proposed, we will give three examples to evaluate the polynomial at the initial values.

Example 1: Consider $f(x) = e^x - x - 1$ [1].

Since $f(0) = e^0 - 0 - 1 = 0$ and $f'(0) = e^0 - 1 = 0$, but $f''(0) = e^0 = 1$, f has a zero of multiplicity two at $p = 0$.

The ***table 1*** shows the results of the method proposed and the modified Newton's method. The results in table1 illustrate that the algorithm proposed are much more accurate than the modified Newton's method. It is obvious that the method proposed can find zeros of nonlinear equation.

Example 2: In order to verify the validity finding double zeros using the method proposed, we give a polynomial as follows:

$$f(x) = (x+1)^2(x+2+j)^2(x+2-j)^2$$
$$= x^6 + 10x^5 + 43x^4 + 100x^3 + 131x^2 + 90x + 25$$

Table 1. The results of the example 1

Algorithm proposed		The modified Newton's method [1]	
k	x_k	k	x_k
0	1.0000000	0	1.0000000
1	3.6787944×10^{-1}	1	-2.3421061×10^{-1}
2	6.0080069×10^{-2}	2	-8.4582788×10^{-3}
3	1.7691994×10^{-3}	3	-1.1889524×10^{-5}
4	1.5641108×10^{-6}	4	-6.8638230×10^{-6}
5	1.2232654×10^{-12}	5	-2.8085217×10^{-7}
6	2.1676421×10^{-17}	-	-

Table 2. The results of the example 2

x_0	k	x_k	$\lvert x_k - p_i\rvert /10^{-12}$
-1000	42	-1.0000000000	0.00000000
1000	38	-1.0000000000	0.00000000
0.0	6	-0.99964102887403	358971126
	7	-0.99999961388124	386118.759
	8	-0.99999999999955	0.44730886
	9	-1.0000000000	0.00000000
-3+3i	12	-2.000000000 + 1.000000000i	0.00000000
-3-3i	12	-2.000000000 - 1.000000000i	0.00000000
-100+100i	30	-2.000000000 - 1.000000000i	0.00000000

Which has one double real zero, at -1, and two double conjugate-complex zeros, at $-2 \pm j$. Using the algorithm proposed produces the results in table 2.

The results in table 2 show that the algorithm proposed have very high accuracy in the field of finding double zeros of polynomials. It can find not only double real zeros, but also double conjugate-complex zeros.

Example 3: The eighth-degree polynomial

$$f(x) = (x+2)^4(x+1+2j)^2(x+1-2j)^2$$
$$= (x+2)^4[(x+1)^4 + 8(x+1)^2 + 16]$$

has one multiplicity four real zero at -2, and two double conjugate-complex zeros, at $-1 \pm 2j$. Using the method proposed produces the results in table3.

Table 3. The results of the example 3

Initial values x_0	k	x_k	$\lvert x_k - p_i \rvert \times 10^{-9}$
$\pm$ 1000	37	-2.0000000000	0.00000000
-3	8	-2.000000000	0.000000000
0+3i	10	-1.000000000 + 2.000000000i	0.000001510
	11	-1.000000000 + 2.000000000i	0.000000000
0-3i	11	-1.000000000 - 2.000000000i	0.000000000
1+10i	18	-1.000000000 +2.000000000i	0.000000000
1-10i	18	-1.000000000 - 2.000000000i	0.000000000

4 Concluding Remarks

We can know from the table 1 to table 3 that the algorithm proposed can rapidly and precisely calculate the multiple real and complex roots of polynomials or nonlinear equation which were not solved by other traditional methods at all. All the results in three examples have very high precise value with less computation. Especially, the results both in table 2 and table 3 can produce exact values. Furthermore, the algorithm proposed can select an initial approximation in a large scope. Hence, the algorithm proposed will play a very important role in the many fields of science and engineering practice.

References

1. Burden, R.L., Faires, J.D.: Numerical Analysis, 7th edn., pp. 47–103. Thomson Learning, Inc. (August 2001)
2. Zeng, Z.Z., Wen, H.: Numerical Computation, 1st edn., pp. 88–108. Qinghua University Press, Beijing (2005)
3. Xu, C.-F., Wang, M.M., Wang, N.H.: An Accelerated Iiteration Solution to Nonlinear Equation in Large Scope. J. Huazhong Univ. of Sci. & Tech(Nature Science Edition) 4, 122–124 (2006)
4. Markus, L., Frenzel, B.-C.: Polynomial Root Finding. IEEE Signal Processing Letters 10, 141–143 (1994)
5. Jenkins, M.A., Traub, J.F.: A Three-Stage Algorithm for Real Polynomials Using Quadratic Iiteration. SIAM Journal On Numerical Analysis 4, 545–566 (1970)
6. Orchard, H.J.: The Laguerre Method for Finding the Zeros of Polynomials. IEEE Trans. On Circuits and Systems 11, 1377–1381 (1989)
7. Lucas, T.N.: Finding Roots of Polynomials by Using the Routh Array. IEEE Electronics Letters 16, 1519–1521 (1996)
8. Truong, T.K., Jeng, J.H., Reed, I.S.: Fast Algorithm for Computing the Roots of Error Locator Polynomials up to Degree 11 in Reed-Solomon Decoders. IEEE Trans. Commun. 49, 779–783 (2001)
9. Sergei, V.F., Peter, V.T.: Finding Roots of Polynomials over Finite Fields. IEEE Trans. Commun. 50, 1709–1711 (2002)
10. Cui, X.-Z., Yang, D.-D., Long, Y.: The Fast Halley Algorithm for Finding All Zeros of a Polynomial. Chinese Journal of Engineering Mathematics 23, 511–517 (2006)
11. Ehrlich, L.W.: A Modified Newton Method for Polynomials. Comm. ACM 10, 107–108 (1967)
12. Huang, Q.-L.: An Improvement on a Modified Newton Method. Numerical Mathematics: A Journal of Chinese Universities 11, 313–319 (2002)
13. Huang, Q.-L., Wu, J.C.: On a Modified Newton Method for Simultaneous Finding Polynomial Zeros. Journal On Numerical Methods And Computer Applications (Beijing, China) 28, 292–298 (2006)

Robust Designs for Directed Edge Overstriking CNNs with Applications

Yongmei Su[1], Lequan Min[1], and Xinjian Zhuo[2]

[1] Applied Science School
University of Science and Technology Beijing
Beijing 100083, PR China
[2] School of Information Engineering
University of Post and Telecommunications
Beijing 100083, PR China
suym71@163.com, minlequan@sina.com, Zhuoxj@163.com

Abstract. A kind of templates of coupled **C**ellular **N**eural **N**etwork (CNN) are introduced, which are able to generate gray edges to a binary image and overstrike them "directionally". The robustness analysis gives the template parameter inequalities which guarantee the corresponding CNNs to work well for performing prescribed tasks. As applications, the CNNs may be used to generate art letters.

Keywords: Cellular neural network, Robust designs, Gray-scale image, Image processing.

1 Introduction

The CNN, first introduced by Chua & Yang [1] as an implementable alternative to fully-connected neural networks, has been widely studied for theoretic foundations and practical applications in image and video signal processing(see [2]-[6]), robotic [7] and biological visions (see [8], [9]), and higher brain visions (see[2] and cited references). Practically, although many methods used in image processing and pattern recognition can be easily implemented by CNN circuits ("programs") [2], the analysis of the behaviors of CNNs is not always an easy issue, in particular for determining the performances of coupled CNNs.

On the other hand, an engineer always hopes to design such a CNN that has both universality and robustness. This means that the CNN is able not only to perform its prescribed task for the "nominal (idea) model" but also to work well for a large set of perturbed models. In [2] and [10], the robustness analysis of a large kind of CNN-uncoupled Boolean CNNs has been addressed, which provides optimal design schemes for CNNs with prescribed tasks. In [11], an uncoupled CNN with a nonlinear B-template is introduced based on robust parameter design, which can detect convex corners in some gray-scale images. In two recent papers [12] and [13], two robust design schemes for a coupled $CNNs$ with symmetric templates are proposed, which have performances of global connectivity detection for binary and gray-scale images, respectively.

F. Sun et al. (Eds.): ISNN 2008, Part II, LNCS 5264, pp. 682–691, 2008.

In this paper, a kind of non-symmetric templates of coupled CNNs are introduced, which are able to generate gray edges to binary images and overstrike them “directionally”. The robust design of this kind of CNNs is stated by a theorem, which gives the template parameter inequalities to guarantee the corresponding CNNs to work well for performing prescribed tasks. A mathematical iterative formula calculates the gray-scale values along overstriking directions for the CNNs. As applications, the CNNs may be used to generate art letters.

2 Robust Design for CNN New Templates

The standard $M \times N$ CNN architecture is composed of cells $C(i,j)'s$. where $1 \le i \le M$ and $1 \le j \le N$. $M \times N$ can be understood as the dimension of the digital image $\mathbf{P}$ to be processed. The dynamics of each cell is given via the equation :

$$\begin{aligned}\dot{x}_{i,j} &= -x_{i,j} + \sum_{k,l \in S_{i,j}(r)} a_{k,l} y_{i+k,j+l} + \sum_{k,l \in S_{i,j}(r)} b_{k,l} u_{i+k,j+l} + z_{i,j} \\ &= -x_{i,j} + \sum_{k=-r}^{r} \sum_{l=-r}^{r} a_{k,l} y_{i+k,j+l} + \sum_{k=-r}^{r} \sum_{l=-r}^{r} b_{k,l} u_{i+k,j+l} + z_{i,j} \\ & \quad i = 1,2,\cdots,M; j = 1,2,\cdots,N\end{aligned}$$

where $x_{i,j}$, $y_{i,j}$, $u_{i,j}$ and $z_{i,j}$ represent state, output, input, and threshold respectively; $S_{i,j}(r)$ is the sphere of influence with radius r ; $a_{i,j}{}'s$ and $b_{k,l}{}'s$ are the elements of the $\mathbf{A}$-template and the $\mathbf{B}$-template respectively. The output $y_{i,j}$ is the piece-wise linear function given by

$$y_{i+k,j+l} = \frac{1}{2}(|x_{i+k,j+l} + 1| - |x_{i+k,j+l} - 1|) \text{ if } (i+k, j+l) \in [1, M] \times [1, N].$$

Generally speaking, there are no universal approaches to determine template parameters of $CNNs$ satisfying prescribed Local Rules. However, dynamic routs of $CNNs$ may be helpful to figure out template parameters, in particular for robustness designs of templates of $CNNs$.

Our templates of the edge directional overstriking (EDO) CNN have the following forms, which are discovered occasionally and can add white edge to black image and overstrike the white edge along 8 compass directions, respectively,

$$\mathbf{A}^{(\alpha,\beta)} = \begin{array}{|c|c|c|}\hline a_{-1,-1}^{(\alpha,\beta)} & a_{-1,0}^{(\alpha,\beta)} & a_{-1,1}^{(\alpha,\beta)} \\ \hline a_{0,-1}^{(\alpha,\beta)} & a & a_{0,1}^{(\alpha,\beta)} \\ \hline a_{1,-1}^{(\alpha,\beta)} & a_{1,0}^{(\alpha,\beta)} & a_{1,1}^{(\alpha,\beta)} \\ \hline\end{array}, \tag{1}$$

$$\mathbf{B} = \begin{array}{|c|c|c|}\hline \text{-b} & \text{-b} & \text{-b} \\ \hline \text{-b} & 8b & \text{-b} \\ \hline \text{-b} & \text{-b} & \text{-b} \\ \hline\end{array}, \quad \mathbf{Z} = z \tag{2}$$

where $\alpha, \beta \in \{-1, 0, 1\}$, and at least one of α and β does not equal to zero, and

$$a_{k,l}^{(\alpha,\beta)} = \begin{cases} c \text{ if } \ k = \alpha, \, l = \beta \\ 0 \ \text{otherwise.} \end{cases} \tag{3}$$

I. Global Task

Given : *static binary image* $\mathbf{P}$.
Input : $\mathbf{U} = (u_{i,j}) = \mathbf{P}$.
Initial state : $\mathbf{X}(0) = \mathbf{P}$.
Output : $\mathbf{Y}(\infty) = gray - scale$ *image with unchanged black pixels. A white pixel with at least one adjacentblack pixel keep unchanged, its adjacent or "nearby" white pixelsalong the direction* $(-\alpha, -\beta)$ *becomes gray one with prescribed gray scale value, Otherwise black.*

Boundary Conditions: Fixed type, $u_{i,j} = y_{i,j} = 1$ for all virtual cells.

II. Local Rules

$u_{i,j}(0) \longrightarrow y_{i,j}(\infty)$

1. *black pixel* *black, independent of neighbors.*
2. *white pixel* *white, if at least one of itsadjacent neighbor is black.*
3. *white pixel* *a prescribed value* $g \in [g_1, g_2] \subset (-1, 1)$, *if its adjacent white pixel along the direction* (α, β) *keep unchanged.*
4. *white pixel* *gray, if its adjacent white pixelalong the direction* (α, β) *has changed into a gray one with gray scale* g^* *and* $1 - a > cg^* + z > a - 1$. *Otherwise black.*

Our purpose is to determine general template parameter intervals such that the $CNNs$ satisfy the above global task and local rules. It can be described by the following robustness design theorem of the EDO $CNNs$.

Theorem 1. *Let the positions of* CNN *template parameters be described by (1)-(3). Assume that* $a < 1$, $b > 0$ *and* $c > 0$. *Then the* CNN *can perform the Local Rules, if the parameters satisfy the following parameter inequalities.*

1. $1 \leq a + 2b - c + z$.
2. $1 \leq a + c + z$.
3. $1 \leq a + 2b - c - z$.
4. $g_1 \leq \frac{z-c}{(1-a)} \leq g_2$.

Proof. The CNN EQ. has the form

$$\dot{x}_{i,j} = \underbrace{-x_{i,j} + ay_{i,j}}_{g(x_{i,j})} + \underbrace{cy_{i+\alpha,j+\beta} + bu_{i,j} + z}_{w_{i,j}(t)} \tag{4}$$
$$i = 1, 2, \cdots, M; \ \ j = 1, 2, \cdots, N.$$

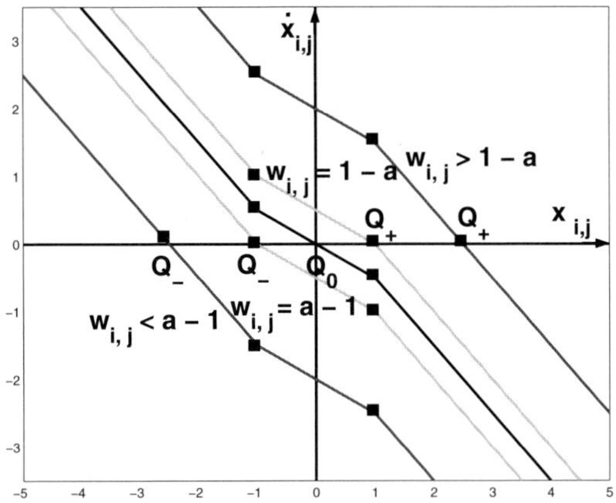

Fig. 1. The dynamic routs of the CNN with fixed $w_{i,j}(t)$

The dynamic routs of the CNN with fixed $w_{i,j}$ are shown in Fig. 1. From EQ. (4) and Fig. 1, it can be concluded that if $w_{i,j} \geq 1 - a$ then for any initial condition, $x_{i,j}$ will converge to an equilibrium point $x^*_{i,j} = Q_+ \geq 1$; if $w_{i,j} \leq a - 1$ then for any initial condition, $x_{i,j}$ will converge to an equilibrium point $x^*_{i,j} = Q_- \leq -1$; if $|w_{i,j}| < 1$ then for any initial condition, $x_{i,j}$ will converge to an equilibrium point $x^*_{i,j} = w_{i,j}/(1-a)$. Mathematically, we obtain the following conclusions that for $x_{i,j}(0) \in [-1, 1]$,

$$y_{i,j}(\infty) = \begin{cases} 1 & \text{if } w_{i,j}(t) \geq 1 - a \\ -1 & \text{if } w_{i,j}(t) \leq a - 1 \\ \frac{w_{i,j}}{(1-a)} & \text{if } 1 - a > |w_{i,j}| \end{cases} \tag{5}$$

$$\begin{cases} g_1 \leq y_{i,j}(\infty) \leq g_2 \\ \text{if } g_1(1-a) \leq w_{i,j}(t) \leq g_2(1-a). \end{cases} \tag{6}$$

Case 1. We show that condition 1,2 in the Theorem guarantees Local Rule 1 to hold. If $u_{i,j} = 1$, then

(1) If all its adjacent neighbors are black, Because $\mathbf{X}(0) = \mathbf{P} = (u_{i,j})$ implies

$$x_{i+\alpha,j+\beta}(0) = u_{i+\alpha,j+\beta} = 1,$$

Consequently,

$$\begin{aligned} w_{i,j}(0) &= cy_{i+\alpha,j+\beta}(0) + 8bu_{i,j} - b \sum_{k,l \neq (0,0)} u_{i+k,j+l} + z \\ &= cy_{i+\alpha,j+\beta}(t) + z \\ &= c + z \end{aligned}$$

Hence EQ. (5) and Fig. 1 show that if we choose parameters a, b, c, and z such that the following inequality holds

$$c + z \geq 1 - a,$$

that is,

$$a + c + z \geq 1,$$

then $y_{i,j}(t) \equiv 1$. Hence $y_{i,j}(t)(\infty) = 1$.

(2) If at least one of its adjacent neighbors is white,let $p_w$be the number of white pixels in the adjacent neighbor, then$p_w \geq 1$

$$\begin{aligned} w_{i,j}(t) &= cy_{i+\alpha,j+\beta}(t) + 8bu_{i,j} - b \sum_{k,l \neq (0,0)} u_{i+k,j+l} + z \\ &= cy_{i+\alpha,j+\beta}(t) + 2p_w b + z \\ &\geq 2b + z - c. \end{aligned}$$

Hence EQ. (5) and Fig. 1 show that if we choose parameters a, b, c, and z such that the following inequality holds

$$2b + z - c \geq 1 - a,$$

that is,

$$a + 2b - c + z \geq 1,$$

then $w_{i,j} \geq 1 - a$, $y_{i,j}(\infty) = 1$.

Case 2. Condition 3 in the Theorem guarantees Local Rule 2 holds. Indeed, $u_{i,j} = -1$, if at least one of its adjacent neighbors is black, let p_b be the number of black pixels in the adjacent neighbor, then $p_b \geq 1$

$$\begin{aligned} w_{i,j}(t) &= cy_{i+\alpha,j+\beta}(t) + 8bu_{i,j} - b \sum_{k,l \neq (0,0)} u_{i+k,j+l} + z \\ &= cy_{i+\alpha,j+\beta}(t) - 2p_b b + z \\ &\leq c + z - 2b. \end{aligned}$$

Hence EQ. (5) and Fig. 1 show that if we choose parameters a, b, c, and z such that the following inequality holds

$$c + z - 2b \leq a - 1.$$

We know $y_{i,j}(\infty) = -1$.

Case 3. If $u_{i,j} = -1$, and all adjacent neighbors are white, but $u_{i+\alpha,j+\beta}$ satisfies the conditions given in Case 2. Since $\mathbf{X}(0) = \mathbf{P} = (u_{i,j})$ implies

$$x_{i+\alpha,j+\beta}(0) = u_{i+\alpha,j+\beta} = -1,$$

we know, by Case 2 (also see Fig.1), $x_{i+\alpha,j+\beta}(t)$ decrease. Hence

$$y_{i+\alpha,j+\beta}(t) \equiv -1.$$

Consequently,

$$\begin{aligned} w_{i,j}(t) &= cy_{i+\alpha,j+\beta}(t) + 8bu_{i,j} - b \sum_{k,l \neq (0,0)} u_{i+k,j+l} + z \\ &= cy_{i+\alpha,j+\beta}(t) + z \\ &= -c + z \end{aligned}$$

It follows, from Fig.1 and EQ.(5), EQ.(6), that condition 4 in the Theorem guarantees

$$y_{i,j}(\infty) = g = \frac{z-c}{(1-a)} \in [g_1, g_2]$$

Case 4. If $u_{i,j} = -1$, and all its adjacent neighbors are white. But there exists sometime t^* such that when $t \geq t^*$, $y_{i+\alpha,j+\beta}(t) = y_{i+\alpha,j+\beta}(t^*) = g^*$. Therefore for $t \geq t^*$,

$$\begin{aligned} 1 - a > w_{i,j}(t) &\equiv w_{i,j}(t^*) \\ &= cy_{i+\alpha,j+\beta}(t^*) + z \\ &= cg^* + z \\ &> a - 1. \end{aligned} \tag{7}$$

Then EQ.(5), EQ.(6), and Fig.1 show that

$$1 > y_{i,j}(\infty) = \frac{cg^* + z}{1-a} > g > -1. \tag{8}$$

Hence $y_{i,j}(\infty)$ must be gray. Such process can be followed until at some step, for any t, $y_{i+\alpha,j+\beta}(t) \geq g^*$ and inequality (7) can not be satisfied, then EQ. (5) and Fig.1 implies that $y_{i,j}(\infty) = 1$. In summary, we complete the proof. ∎

Remark 1. The local rule 4 guarantees that along the (α, β) direction, white pixels will generate "domino-like" effect until inequality (7) is not able to be satisfied.

Remark 2. From inequalities (7) and (8), we can obtain an iterate formula to determine the overstruck gray-scale value g_n of the white pixel whose distance to the black pixels along the direction (α, β) equals n pixels as follows.

$$g_n = \frac{cg_{n-1} + z}{1-a}, n = 1, 2, 3, \ldots \tag{9}$$

where $g_0 = -1$.

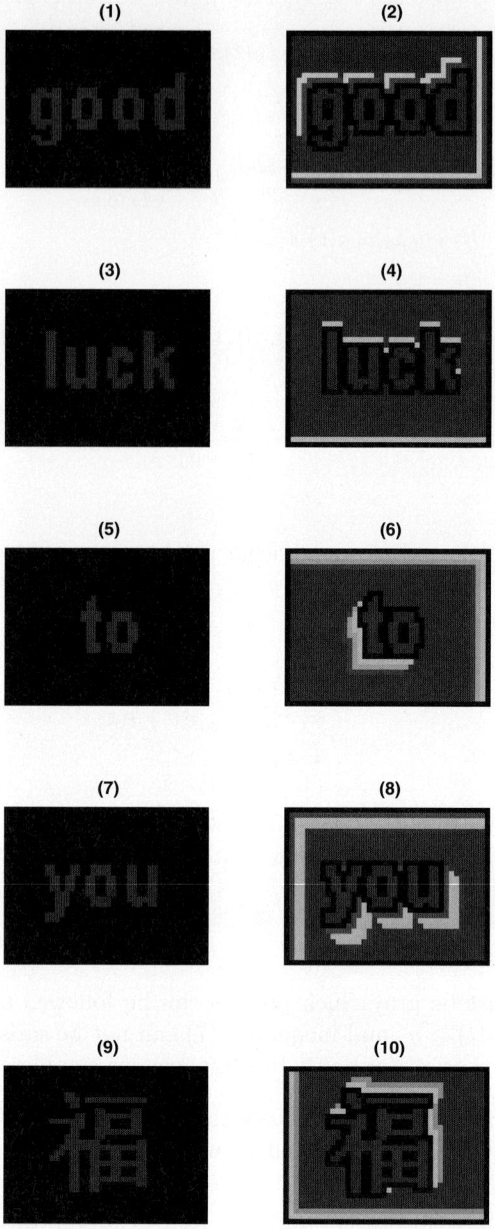

Fig. 2. Input binary images: (1), (3), (5), (7) and (9). The corresponding processed images via the *CNNs* listed in Table 1: (2) No. 1, (4) No. 2, (6) No. 3, (8) No. 4, and (10) No. 5.

Table 1. Five CNNs satisfying the conditions in Theorem 1

No.	α	β	a	b	c	z	g
1	1	1	0.60	1.00	0.60	0.40	-0.50
2	1	0	0.50	1.10	0.69	0.70	0.20
3	-1	1	0.30	1.20	0.40	0.40	0.00
4	-1	-1	-0.40	1.80	1.40	0.70	-0.20
5	1	-1	0.20	1.50	0.50	0.50	0

Fig. 3. Pseudo-color code used in Fig. 2.

3 Numerical Simulation

Now let us consider five CNNs with the parameters given in Table 1, where $g = (z - c)/(1 - a)$.

Using the five CNNs processes the English and Chinese words shown in Figs. 2 (1), (3), (5), (7) and (9), respectively. The corresponding processed graphs are given in Figs. 2 (2), (4), (6), (8) and (10), in which a pseudo-color code is used, as shown in Fig. 3.

As our analysis has expected, the pixels of the characters in the five graphs in the right column shown in Fig. 2 are "expanded" along 5 different directions. The numbers and the gray-scale values of the expanded pixels depend on the prescribed gray-scale value $g's$ given in Table 1 and equality (9). In fact, the theoretical formula and numerical simulation are agree well. For instance, formula (9) gives the overstruck values of the processed image shown in Fig.2(5):

$$g_1 = 0, \quad g_2 = 0.5714, \quad g_3 = 0.8980.$$

The corresponding CNN simulation results are

$$g_1 = -2.5675e - 5, g_2 = 0.5712, g_3 = 0.8972.$$

The numerical results show that the new CNNs can be used to "overstrike" some characters or generate art letters with some special requirements.

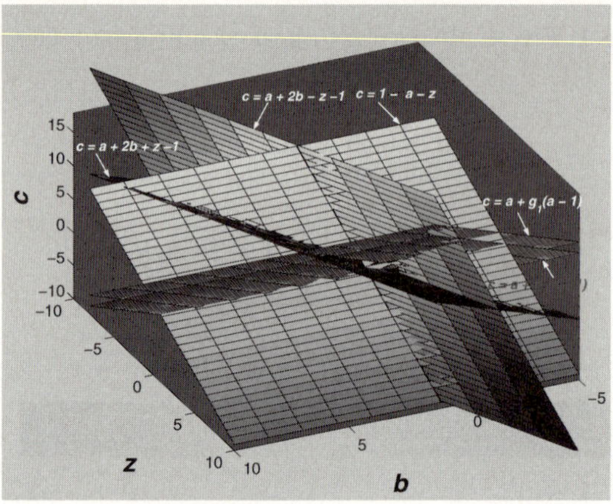

Fig. 4. Parameter domain determined by Theorem1, in which $g_1 = -0.9, g_2 = 0.9$, $a = 0.1$

4 Conclusions

Theorems 1 gives general constrain conditions of template parameters for the CNN, which guarantee the corresponding global task of the CNN overstriking, in 8 compass directions, the gray-scales of pixels in images, respectively.

The four inequalities given in Theorem 1 also imply algorithms that figure out the most robustness parameter group satisfying the Local rules of the CNN. These inequalities may provide selectable parameter domains of CNN templates to engineers for various purposes. Fig.4 gives such a parameter domain in which $g_1 = -0.9, g_2 = 0.9, a = 0.1$. The meaning of the parameters g_1, g_2 in condition 4 in the theorem can be seen virtually from Fig.4.

In summary, this paper introduces a kind of EDO $CNNs$ The numerical simulation examples confirmed that our theoretical analysis is efficient in practical applications for computer image processing.

Acknowledgments. This project is jointly supported by the National Natural Science Foundations of China (Grant No. 60674095) and the Science Foundations of University of Science and Technology.

References

1. Chua, L.O., Yang, L.: Cellular Neural Networks: Theory and Appilcations. IEEE Trans. Circuits Syst. 35, 1257–1290 (1988)
2. Chua, L.O.: CNN: A Vision of Complex. Int. J. Bifurcation and Chaos 7(10), 2219–2425 (1997)
3. Chua, L.O., Roska, T.: Cellular Neural Networks and Visual Computing. Cambridge University Press, Cambridge (2002)

4. kananen, A., Paasio, A., Laiho, M.: CNN Applications from the Hardware Point of View: Video Sequence Segmentation. Int. J. of Circuits Syst. I 30, 117–137 (2002)
5. Grssi, G., Grieco, L.A.: Object-orriected Image Analysis Using the CNN Univeral Machine: New Analogic CNN Algorithms for Motion Compensation, Image Synthesis and Considtency Observation. IEEE Transactions on Circuits and Systems I 50(4), 488–499 (2003)
6. Chang, C., Su, S.: Apply the Counter Propagation Neural Network to Digital Image Copyright Authentication. In: 9th IEEE Int. Workshop on Cellular Neural Networks and Their Applications, pp. 110–113 (2005)
7. Arena, P., Basile, A., Fortuna, L., et al.: CNN Wave Based Compution for Robot Navigation Planning. In: Proc. of the Int. Symposium on Circuit and Systems, vol. 5, pp. 500–503 (2004)
8. Werblin, F.S., Roska, T., Chua, L.O.: The Analogic Cellular Neural Network as a Bionic Eye. Int. J. of Circuit Theory and Applications 23, 541–569 (1995)
9. Bálya, D., Roska, B., Roska, T., Werblin, F.S.: A CNN Framework for Modeling Parallel Processing in a Mammalian Retina. Int. J. on Circuit Theory and Applications 30, 363–393 (2002)
10. Dogaru, R., Chua, L.O.: Universal CNN cells. Int. J. Bifurcation and Chaos 9(9), 1–48 (1999)
11. Min, L., Lei, M., Dong, X.: New Templates of CNN for Extracting Corners of Objects in Gray-scale Images. J. of Univ. of Sci. Technol. Beijing 10(3), 73–75 (2003)
12. Liu., J., Min, L.: Design for CNN Templates with Performance of Global Connectivity Detection. Commun Theor. Phys. 41(1), 151–156 (2004)
13. Liu, J., Min, L.: Robust Designs for Gray-scale Global Connectivity Detection CNN Template. INT. J. Bifurcationand Chaos 17(8), 2827–2838 (2007)

Application of Local Activity Theory of Cellular Neural Network to the Chen's System

Danling Wang, Lequan Min, and Yu Ji

Applied Science School
University of Science and Technology Beijing
Beijing 100083, PR China
wang_dan_ling@sina.com, minlequan@sina.com, jyslt@163.com

Abstract. The local activity theory introduced by Chua has provided a new tool for studying the complexity of high dimensional coupled nonlinear differential systems, in particular for reaction- diffusion cellular neural networks(R-D CNNs). In this paper some criteria for the local activity theory range from one-port to three-port cellular neural network cells with three local state variables are applied to Chen's system. Numerical simulations show that the dynamic behaviors of the Chen's CNN with one,two or three ports have the similar characteristics.

Keywords: Local activity principle, Edge of chaos, Cellular neural network.

1 Introduction

Nature abounds with complex patterns and structures emerging homogeneous media, many of these phenomena can be modelled and studied via the CNN paradigm [1]. The CNN, first introduced by Chua and Yang [2,3] as an implementable alternative to fully-connected Hopfield neural network, have been widely studied for image processing, robotic and biological versions and higher brain functions [4,5,6].

The local activity theory proposed by Chua [1,7] offering a constructive analytical tool, asserts that a wide spectrum of complex behaviors may exist if the corresponding cell parameters of the CNN's are chose in or nearby the edge of chaos [8,9,10]. In particular, some analytical criteria have been established and applied to the study of the dynamics of the CNN's related to the FitzHugh-Nagumo equation [8], the Brusselator equation [9], the Gierer-Meinhart equation [10], the Oregonator equation [11], the Hodgkin-Huxley equation [12], the biochemical model CNN [13], coupled excitable cell model [14], tumor growth and immune model [15], Lorenz-cell model [16], respectively.

Recently, Chua, has provided a self-contained mathematical proof of the local activity theorem [17] for the system of discrete reaction-diffusion equations. The local activity theorem provides the quantitative characterization of Prigogine's

F. Sun et al. (Eds.): ISNN 2008, Part II, LNCS 5264, pp. 692–701, 2008.

" instability of the homogeneous " and Smale's quest for an axiomatic principle on Turing instability [17]. Explicit mathematical criteria are given to identify a relatively small subset of the locally-active parameter region, called the edge of chaos, where most complex phenomena emerge.

2 Local Activity Principle and Analytical Criteria

Generally speaking, in a reaction-diffusion CNN, every $C_{i,j,k}$ has n state variables but only $m(\leq n)$ state variables coupled directly to their nearest neighbors via "reaction-diffusion".

In a component form, it has the form

$$\begin{aligned} \dot{\mathbf{V}}_{\mathbf{a}} &= \mathbf{f}_{\mathbf{a}}(\mathbf{V}_{\mathbf{a}}, \mathbf{V}_{\mathbf{b}}) + \mathbf{I}_{\mathbf{a}} \\ \dot{\mathbf{V}}_{\mathbf{b}} &= \mathbf{f}_{\mathbf{b}}(\mathbf{V}_{\mathbf{a}}, \mathbf{V}_{\mathbf{b}}) \end{aligned} \tag{1}$$

where

$$\begin{aligned} \mathbf{V}_{\mathbf{a}} &= [V_1, V_2, \ldots, V_m]^T \\ \mathbf{V}_{\mathbf{b}} &= [V_{m+1}, V_{m+2}, \ldots, V_n]^T \\ \mathbf{f}_{\mathbf{a}} &= [f_1(\cdot), f_2(\cdot), \ldots, f_m(\cdot)]^T \\ \mathbf{f}_{\mathbf{b}} &= [f_{m+1}(\cdot), f_{m+2}(\cdot), \ldots, f_n(\cdot)]^T \\ \mathbf{I}_{\mathbf{a}} &= \mathbf{D}_{\mathbf{a}} \nabla^2 \mathbf{V}_{\mathbf{a}} \\ &= [D_1 \nabla^2 V_1, D_2 \nabla^2 V_2, \ldots, D_m \nabla^2 V_m]^T \\ \mathbf{D}_{\mathbf{a}} &= diag[D_1, D_2, \ldots, D_m]. \end{aligned}$$

The cell equilibrium point $Q_i = (\overline{\mathbf{V}}_{\mathbf{a}}; \overline{\mathbf{V}}_{\mathbf{b}})(\in \mathbb{R}^{\mathbf{n}})$ of Eq.(1) for the restricted local activity domain can be determined numerically or analytically, via

$$\begin{aligned} \mathbf{f}_{\mathbf{a}}(\mathbf{V}_{\mathbf{a}}, \mathbf{V}_{\mathbf{b}}) &= \mathbf{0} \\ \mathbf{f}_{\mathbf{b}}(\mathbf{V}_{\mathbf{a}}, \mathbf{V}_{\mathbf{b}}) &= \mathbf{0} \end{aligned} \tag{2}$$

The Jacobian matrix at equilibrium point Q_i , for the restricted local activity domain, has the following form:

$$\mathbf{J}(Q_i) \triangleq [a_{lk}(Q_i)] \triangleq \begin{bmatrix} \mathbf{A}_{\mathbf{aa}}(Q_i) & \mathbf{A}_{\mathbf{ab}}(Q_i) \\ \mathbf{A}_{\mathbf{ba}}(Q_i) & \mathbf{A}_{\mathbf{bb}}(Q_i) \end{bmatrix}, \tag{3}$$

where

$$\mathbf{A}_{\mathbf{aa}}(Q_i) = \begin{pmatrix} \dfrac{\partial f_1}{\partial V_1} & \cdots & \dfrac{\partial f_1}{\partial V_m} \\ \vdots & \cdots & \vdots \\ \dfrac{\partial f_m}{\partial V_1} & \cdots & \dfrac{\partial f_m}{\partial V_m} \end{pmatrix}, \quad \mathbf{A}_{\mathbf{ab}}(Q_i) = \begin{pmatrix} \dfrac{\partial f_1}{\partial V_{m+1}} & \cdots & \dfrac{\partial f_1}{\partial V_n} \\ \vdots & \cdots & \vdots \\ \dfrac{\partial f_m}{\partial V_{m+1}} & \cdots & \dfrac{\partial f_m}{\partial V_n} \end{pmatrix},$$

$$\mathbf{A_{ba}}(Q_i) = \begin{pmatrix} \frac{\partial f_{m+1}}{\partial V_1} \cdots \frac{\partial f_{m+1}}{\partial V_m} \\ \vdots \quad \cdots \quad \vdots \\ \frac{\partial f_n}{\partial V_1} \cdots \frac{\partial f_n}{\partial V_m} \end{pmatrix}, \quad \mathbf{A_{bb}}(Q_i) = \begin{pmatrix} \frac{\partial f_{m+1}}{\partial V_{m+1}} \cdots \frac{\partial f_{m+1}}{\partial V_n} \\ \vdots \quad \cdots \quad \vdots \\ \frac{\partial f_n}{\partial V_{m+1}} \cdots \frac{\partial f_n}{\partial V_n} \end{pmatrix},$$

$a_{lk}(Q_i)'s$ are called cell parameters.

The local linearized state equations at the cell equilibrium point are defined [1]

$$\dot{\mathbf{V}}_{\mathbf{a}} = \mathbf{A_{aa}}\mathbf{V_a} + \mathbf{A_{ab}}\mathbf{V_b} + \mathbf{I_a} \tag{4}$$

$$\dot{\mathbf{V}}_{\mathbf{b}} = \mathbf{A_{ba}}\mathbf{V_a} + \mathbf{A_{bb}}\mathbf{V_b}. \tag{5}$$

$$Y_Q(s) \triangleq (s\boldsymbol{I} - \boldsymbol{A}_{\mathbf{aa}}) - \boldsymbol{A}_{\mathbf{ab}}(s\boldsymbol{I} - \boldsymbol{A}_{\mathbf{bb}})^{-1}\boldsymbol{A}_{\mathbf{ba}} \tag{6}$$

is called the admittance matrix at Q_i.

Based on the classic positive criteria for the passive linear network, Chua propose the following local activity principle.

Main Theorem on the Local Activity of CNN [1]. *A Reaction Diffusion CNN cell is locally active at a cell equilibrium point* $Q \triangleq (\bar{V}_a, \bar{V}_b, \bar{I}_a)$, *if and only if, its cell admittance matrix* $Y_Q(s)$ *satisfies at least one of the following four conditions:*

1. $Y_Q(s)$ *has a pole in* $Re[s] > 0$.
2. $Y_Q^H(i\omega) = Y_Q^*(i\omega) + Y_Q(i\omega)$ *is not a positive semi-definite matrix at some* $\omega = \omega_0$, *where* ω_0 *is an arbitrary real number,* $Y_Q^*(s)$ *is constructed by first taking the transpose of* $Y_Q(s)$, *and then by taking the complex conjugate operation .*
3. $Y_Q(s)$ *has a simple pole* $s = i\omega_\rho$ *on the imaginary axis, where its associate residue matrix*

$$k_1 \triangleq \lim_{s \to i\omega_\rho} (s - i\omega_\rho) Y_Q(s)$$

is either not a Hermitian matrix, or else not a positive semi-definite Hemitian matrix. number.
4. $Y_Q(s)$ *has a multiple pole on the imaginary axis.*

In the case of three state variables with one ports, we have

$$A_{11} = [a_{11}] \qquad A_{12} = [a_{12}a_{13}]$$

$$A_{21} = \begin{bmatrix} a_{21} \\ a_{31} \end{bmatrix} \qquad A_{22} = \begin{bmatrix} a_{22} \; a_{23} \\ a_{32} \; a_{33} \end{bmatrix}$$

the corresponding *complexity* matrix is

$$Y_Q(s) = s - a_{11} - \frac{T_1 s + \triangle_1}{s^2 + Ts + \triangle} \tag{7}$$

where

$$\begin{aligned} T_1 &= (a_{12}a_{21} + a_{13}a_{31}), \\ \triangle_1 &= (-a_{12}a_{21}a_{33} - a_{13}a_{31}a_{22} + a_{13}a_{32}a_{21} + a_{12}a_{23}a_{31}), \\ T &= -(a_{22} + a_{33}) \\ \triangle &= (a_{22}a_{33} - a_{23}a_{32}) \end{aligned}$$

We correct the theorems in the bibliography [18],and thus we have the following new theorems.

Theorem 1. *$Y_Q(s)$ has a pole in $Re[s] > 0$ if, and only if, at least one of the following three conditions holds*

1. $\triangle > 0, T < 0.$

2. $\triangle < 0, T_1 = 0,$ *and* $\triangle \neq 0.$

3. $\triangle < 0, T_1 \neq 0,$ *and* $s_1 = \frac{-T+\sqrt{T^2-4\triangle}}{2} \neq -\frac{\triangle_1}{T_1}.$
4. $\triangle = 0, T < 0, T_1 = 0.$

5. $\triangle = 0, T < 0, T_1 \neq 0, -TT_1 + \triangle_1 \neq 0$

Theorem 2. *[18] $Y_Q(s)$ has a multiple pole on the imaginary axis if, and only if, $T = 0$, $\triangle = 0, \triangle_1 \neq 0.$*

Theorem 3. *$Y_Q(s)$ satisfies condition 3 in Lemma 1 if, and only if, at least one of the following condition holds:*

1. $\triangle_1 = 0, T_1 > 0, \triangle > 0.$

2. $\triangle_1 \neq 0, \triangle > 0.$

3. $\triangle = 0, T \neq 0, T\triangle_1 > 0$

Theorem 4. *$Re[Y_Q(i\omega)] < 0$ for some $\omega \in (-\infty, \infty)$ if, and only if, at least one of the following condition holds:*

1. $a_{11} > 0.$

2. $a_{11} = 0, (-TT_1 + \triangle_1) > 0$ *and* $\triangle\triangle_1 > 0.$

3. $a_{11} = 0, (-TT_1 + \triangle_1) < 0$ *and* $\triangle\triangle_1 \leq 0.$

4. $a_{11} < 0, (-a_{11}T^2 + 2a_{11}\triangle - TT_1 + \triangle_1 > 0$ *and* $-a_{11}\triangle^2 - \triangle\triangle_1 < 0.$

5. $a_{11} < 0, (-a_{11}T^2 + 2a_{11}\triangle - TT_1 + \triangle_1) < 0,$ *and* $(-a_{11}T^2 + 2a_{11}\triangle - TT_1 + \triangle_1)^2 > -4a_{11}(-a_{11}\triangle^2 - \triangle\triangle_1).$

In the case of three state variables with two ports, we have

$$\boldsymbol{A_{11}} = \begin{bmatrix} a_{11} \, a_{12} \\ a_{21} \, a_{22} \end{bmatrix} \qquad \boldsymbol{A_{12}} = \begin{bmatrix} a_{13} \\ a_{23} \end{bmatrix}$$
$$\boldsymbol{A_{21}} = [a_{31} \, a_{32}] \qquad \boldsymbol{A_{22}} = [a_{33}]$$

the corresponding admittance matrix is

$$\begin{aligned} Y_Q(s) &= (s\boldsymbol{I} - \boldsymbol{A_{aa}}) - \boldsymbol{A_{ab}}(s\boldsymbol{I} - \boldsymbol{A_{bb}})^{-1}\boldsymbol{A_{ba}} \\ &= \begin{bmatrix} s - a_{11} - \dfrac{a_{13}a_{31}}{s - a_{33}} & -a_{12} - \dfrac{a_{13}a_{32}}{s - a_{33}} \\ -a_{21} - \dfrac{a_{23}a_{31}}{s - a_{33}} & s - a_{22} - \dfrac{a_{23}a_{32}}{s - a_{33}} \end{bmatrix}. \end{aligned}$$

Form the above formulas and lemma 1, the analytical criteria for testing the local activity of the CNN's with three state variables and two ports are stated as follows.

Theorem 5. *[13]* $Y_Q(s)$ has a pole in $\mathrm{Re}[s] > 0$. if, and only if, $a_{33} > 0$ and $\max\{|a_{13}a_{31}|, |a_{13}a_{32}|, |a_{23}a_{31}|, |a_{23}a_{32}|\} \neq 0$.

Theorem 6. *[13]* Let $b = a_{13}a_{31}a_{33}, c = a_{13}a_{32}a_{33}, d = a_{13}a_{31} - a_{23}a_{31}, e = a_{23}a_{32}a_{33}$, then $Y_Q(s)$ satisfies condition 2 in the main Theorem if, and only if, at least one of the following conditions holds.

1. $(a_{11} + a_{22}) > 0$.
2. $(a_{11} + a_{22}) \leq 0, a_{33} \neq 0$, and $(a_{11} + a_{22}) - (a_{13}a_{31} + a_{23}a_{32})/a_{33} > 0$.
3. $4a_{11}a_{22} - (a_{12} + a_{21})^2 < 0$.
4. $4(ba_{22} + ea_{11}) - 2c(a_{12} + a_{21}) + d^2 \neq 0$.

$$\omega^2 = \frac{2a_{33}^2 d^2 + 8be - 2c^2}{4(ba_{22} + ea_{11}) - 2c(a_{12} + a_{21}) + d^2} - a_{33}^2 > 0,$$
$$4\left(a_{11}a_{22} - \frac{ba_{22} + ea_{11}}{a_{33}^2 + \omega^2} + \frac{be}{(a_{33}^2 + \omega^2)^2}\right)$$
$$-\left(a_{12} + a_{21} - \frac{c}{a_{33}^2 + \omega^2}\right)^2 - \frac{\omega^2 d^2}{(a_{33}^2 + \omega^2)^2} < 0.$$

5. $a_{33} \neq 0$, and $4(a_{11}a_{33}^2 - b)(a_{22}a_{33}^2 - e) - [(a_{12} + a_{21})a_{33}^2 - c]^2 < 0$.
6. $a_{33} = 0$ and $4be - c^2 < 0$.

Theorem 7. *[13]* $Y_Q(s)$ satisfies condition 3 in the main Theorem if, and only if, at least one of the following conditions holds.

1. If $a_{33} = 0$, $\max\{|a_{13}a_{31}|, |a_{13}a_{32}|, |a_{23}a_{31}|, |a_{23}a_{32}|\} \neq 0$, and $a_{13}a_{32} \neq a_{23}a_{31}$.
2. If $a_{33} = 0$, $\max\{|a_{13}a_{31}|, |a_{13}a_{32}|, |a_{23}a_{31}|, |a_{23}a_{32}|\} \neq 0$, $a_{13}a_{32} = a_{23}a_{31}$ and

(a) $a_{13}a_{31} + a_{23}a_{32} > 0$ or
(b) $a_{13}a_{31}a_{23}a_{32} - a_{13}^2 a_{32}^2 < 0$.

Theorem 8. *[13]* $Y_Q(s)$ does not have a multiple pole on the imaginary axis.

Theorem 9. *A Reaction-Diffusion with* $m = n$ *(i.e. when the number of nonzero diffusion coefficients is equal to the number of state variables in the "kinetic" part)is locally active at* Q *if, and only if, the systematic part*

$$J(Q_I) = A_{aa}^T + A_{aa} \tag{8}$$

of the Jacobian matrix is not positive semi-definite.

Definition 4 ([8], [17]). Edge of Chaos. An uncoupled cell(with $I_a = 0$) of a reaction diffusion equation is said to be on the edge of chaos iff all of its cell equilibrium points are locally active but asymptotically stable. The set ε of all locally active parameters $\mu \in \mathbb{R}^\rho$ endowed with this property is called the edge of chaos parameter set.

3 Application of Local Activity Theory of CNN to the Chen's System

3.1 Chen's System

I Chen's System. Guanrong Chen introduced the three-variable system as a deterministic chaos model as following equations:

$$\begin{cases} \dot{x} = a(y - x) \\ \dot{y} = (c - a)x - xz + cy \\ \dot{z} = xy - bz \end{cases} \tag{9}$$

where a, b, c are three system parameters. We can obtain the equilibrium points of (9) as follows:

$$Q_0 = (0, 0, 0), Q_1 = (a, a, 2c - a), Q_2 = (-a, -a, 2c - a).$$

Therefore the Jacobian matrix of equilibrium points Q_1 and Q_2 are

$$A_J(Q_1) = A_J(Q_2) = \begin{bmatrix} -a & a & 0 \\ c - a - z & c & -x \\ y & x & -b \end{bmatrix}. \tag{10}$$

The Chen's system Eq. (9) can exhibit chaos behaviors when we choose $a = 35, b = 3, c = 28$.

3.2 Chen's CNN

Now the prototype Chen's equation (9) can be mapped to a Chen's CNN model.

$$\begin{cases} \dot{x}_{i,j} = a(y_{i,j} - x_{i,j}) + D_1[x_{i+1,j} + x_{i-1,j} \\ \qquad +x_{i,j+1} + x_{i,j-1} - 4x_{i,j}] \\ \dot{y}_{i,j} = (c-a)x_{i,j} - x_{i,j}z_{i,j} + cy_{i,j} + D_2[y_{i+1,j} + y_{i-1,j} \\ \qquad +y_{i,j+1} + y_{i,j-1} - 4y_{i,j}] \\ \dot{z}_{i,j} = x_{i,j}y_{i,j} - bz_{i,j} + D_3[z_{i+1,j} + z_{i-1,j} \\ \qquad +z_{i,j+1} + z_{i,j-1} - 4z_{i,j}] \end{cases} \quad i,j = 1,2,\cdots,21. \tag{11}$$

It can be proved easily that the bifurcation diagrams of the Chen's CNN with respect to Q_1 and Q_2 are the same. Numerical simulations show that the bifurcation diagrams of the Chen's CNN with respect to Q_0 are always locally active and unstable. Hence we need only to study the bifurcation diagrams of the Chen's CNN with respect to Q_1.

First let $D_2 = D_3 = 0$. Using Theorems 1 $\sim$ 4, the bifurcation diagram of the one-port Chen's CNN with respect to the equilibrium point Q_1 is shown in Fig.1. Observe that locally active unstable domain, edge of chaos and locally passive domain co-exist in the bifurcation diagram.

Second let $D_3 = 0$. Using Theorems 5 $\sim$ 8, the bifurcation diagram of the two-port Chen's CNN with respect to the equilibrium point Q_1 is shown in Fig.2. Observe that only locally active unstable domain, edge of chaos domain co-exist in the bifurcation diagram.

Third let D_1, D_2 and D_3 not equal to zero. Using Theorem 3, the bifurcation diagram of the three-port Chen's CNN with respect to the equilibrium point Q_1 is shown in Fig.3. Observe the bifurcation diagram is the same as that of the two-port Chen's CNN.

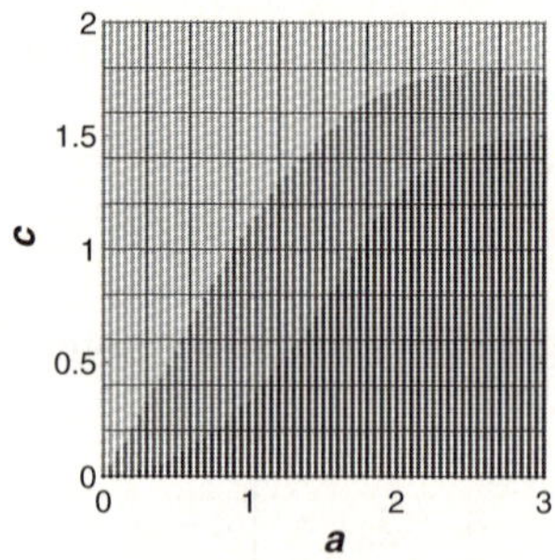

Fig. 1. Bifurcation diagram of the Chen's-CNN at cross section $a \in [0,3], c \in [0,2]$ with respect to the equilibrium point Q_0, The domains are coded as follows: edge of chaos (red), locally active unstable domain (green), passive domain(blue)

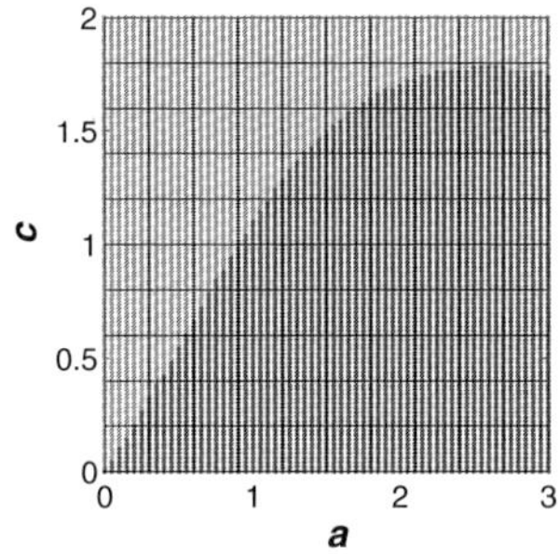

Fig. 2. Bifurcation diagram of the Chen's-CNN at cross section $a \in [0,3], c \in [0,2]$ with respect to the equilibrium point Q_1, The domains are coded as follows: edge of chaos (red), locally active unstable domain (green)

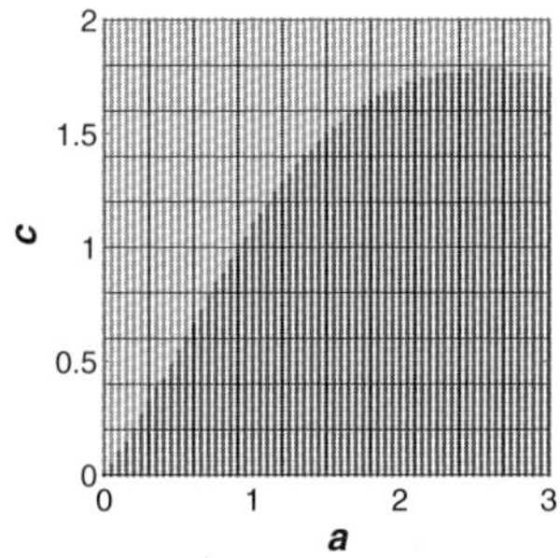

Fig. 3. Bifurcation diagram of the Chen's-CNN at cross section $a \in [0,3], c \in [0,2]$ with respect to the equilibrium point Q_2, The domains are coded as follows: edge of chaos (red), locally active unstable domain (green)

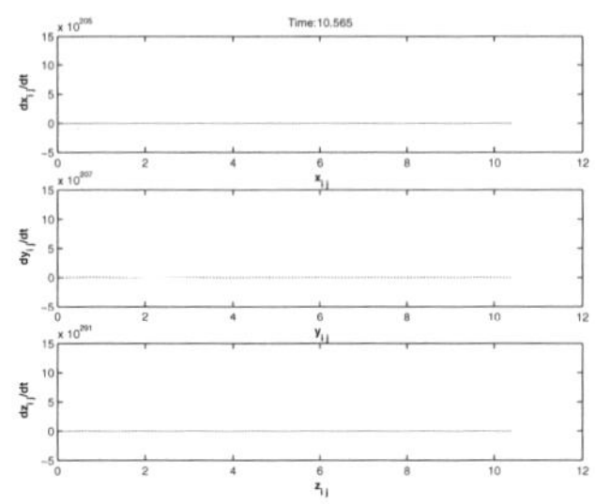

Fig. 4. The graphs of the time evolution of the state variables $x'_{i,j}$, $y'_{i,j}$, $z'_{i,j}$ with one port

3.3 Numerical Simulation

We simulate the Chen's CNN equation based on the brifucation graph given in Fig.1, Fig.2 and Fig.3, and choose the parameters which are located in such domain that is passive in Fig.1 but turns into the edge of chaos in Fig.2 and Fig.3. Numerical simulations show that the dynamic behaviors of the Chen's CNN with one,two or three ports have the similar characteristics, which still need further study.

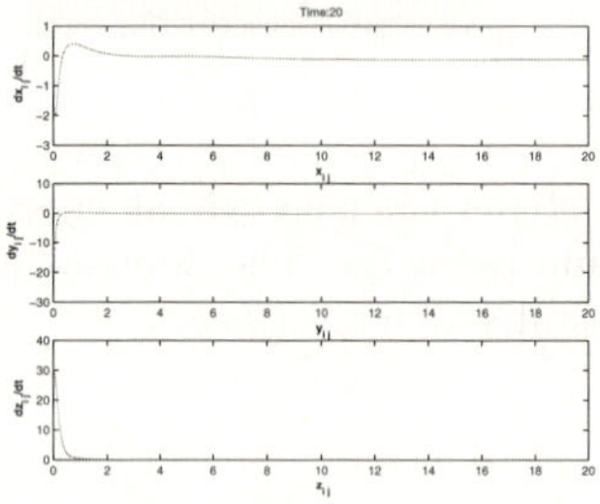

Fig. 5. The graphs of the time evolution of the state variables $x'_{i,j}$, $y'_{i,j}$, $z'_{i,j}$ with three ports

4 Concluding Remarks

Coupled nonlinear dynamical systems (CND's) have been widely studied in recent years. However, the dynamical properties of the CND's are difficult to be dealt with. The local activity criteria of CNN's provide a new tool to the research on the CND's cell nodels. This paper uses the criteria of three states with different numbers of port which ranging from one to three to study Chen's system. It has been found that the bifurcation diagrams of the Chen's-CNN at cross section$a \in [0, 3], c \in [0, 2]$ with respect to the equilibrium point both Q_1 and Q_2 turn out to be different when we add different numbers of ports to the original system. More specifically, locally active unstable domain, edge of chaos as well as locally passive domain co-exist in the bifurcation diagram when we let $D_2 = D_3 = 0$; while, there just exit locally active unstable domain and edge of chaos domain when we let $D_3 = 0$ and $D_i = 0$ $i = 1, 2, 3$ respectively.

Acknowledgments. This project is jointly supported by the National Natural Science Foundations of China (Grant No. 60674059), and the Science Foundation of USTB.

References

1. Chua, L.O.: CNN: Visions of Complexity. Int. J. Bifur. and Chaos 7, 2219–2425 (1997)
2. Chua, L.O., Yang, L.: Cellular Neural Networks: Theory. IEEE Trans. Circuits Syst. 35, 1257–1272 (1988)

3. Chua, L.O., Yang, L.: Cellular Neural Networks: Applications. IEEE Trans. Circuits Syst. 35, 1273–1290 (1988)
4. Weblin, F., Roska, T., Chua, L.O.: The Analigic Cellular Neural Network As Bionic eye. Int. J. Circuits Theor. Appl. 23, 541–569 (1994)
5. Chua, L.O., Roska, T.: Cellular Neural Networks and Visual Computing. Cambridge University Press, Cambridge (2002)
6. Chua, L.O., Min, L.: Design for Templates with Performance of Global Connectivity Detection. Commun. Theor. Phys. 41, 151–156 (2004)
7. Chua, L.O.: Passivity and Complexity. IEEE Trans. Circuits Syst. I. 46, 71–82 (1999)
8. Dogaru, R., Chua, L.O.: Edge of Chaos and Local Activity Domain of Fitzhugh-Nagumo Equation. Int. J. Bifur. and Chaos 8, 211–257 (1998)
9. Dogaru, R., Chua, L.O.: Edge of Chaos and Local Activity Domain of the Brusselator CNN. Int. J. Bifur. and Chaos 8, 1107–1130 (1998)
10. Dogaru, R., Chua, L.O.: Edge of Chaos and Local Activity Domain of Gierer-Meinhardt CNN. Int. J. Bifur. and Chaos 8, 2321–2340 (1998)
11. Min, L., Crounse, K.R., Chua, L.O.: Analytical Criteria for Local Activity and Applications to the Oregonator CNN. Int. J. Bifur. and Chaos 10, 25–71 (2000)
12. Min, L., Crounse, K.R., Chua, L.O.: Analytical Criteria for Local Activity of Reaction-diffusion CNN with Four State Variables and Applications to the Hodgkin-Huxley Equation. Int. J. Bifur. and Chaos 8, 1295–1343 (2000)
13. Min, L., Yu, N.: Analytical Criteria for Local Activity of CNN with Two-port and Applications to Biochemical Model. J. Univ. Sci. Technol. Beijing 7, 305–314 (2000)
14. Min, L., Yu, N.: Some Analytical Criteria for Local Activity of Two-port CNN with Three or Four State Variables: Analysis and Applications. Int. J. Bifur. and Chaos 12, 931–963 (2002)
15. Min, L., Wang, J., Dong, X., Chen, G.: Some Analytical Criteria for Local Activity of Three-port CNN with Four State Variables: Analysis and Applications. Int. J. Bifurc.and Chaos 13, 2189–2239 (2003)
16. Min, L., Yu, N.: Application of Local Activity Theory of the CNN With Two Ports to the Coupled Lorenz-cell Model. Communications in Theoretical Physics 37, 759–767 (2002)
17. Chua, L.O.: Local Activity is the Origin of Complexity. Int. J. Bifur. and Choas 15, 3435–3456 (2005)
18. Huang, H., Min, L., Su, Y.: Application of Local Activity Theory to Chaotic Chemical Reaction Model. Journal of Computational and Theoretical Nnoscience 4, 1269–1273 (2007)

Application of PID Controller Based on BP Neural Network Using Automatic Differentiation Method

Weiwei Yang, Yong Zhao, Li Yan, and Xiaoqian Chen

Multidisciplinary Aerospace Design Optimization Research Center, College of Aerospace and Materials Engineering, National University of Defense Technology, Changsha, China. 410073
nudtyww@163.com

Abstract. A simulation analysis of PID controller based on Back-Propagation Neural Network (BPNN) using Automatic Differentiation Method (ADM) is presented. As accurate partial differentiation can be acquired using ADM, the original meaning of learning rate is regained. By comparing with conventional PID controller, the simulation results of a simple tracking problem show that the new controller has a good adaptability for the nonlinear system, which benefits from on-line self-learning. Furthermore, experimental results are presented for an autonomous docking of the chaser simulator to the target, which validates the effectiveness and good robustness of the proposed controller.

1 Introduction

Conventional PID controller is still widely used in industrial control systems for its simple structure and understandability at present. But PID parameters have to be adjusted again and again to get a better control effect. It is not easy to evaluate them because the relationships between these control parameters are likely to be nonlinear and very complex especially when the model for the controlled plant is nonlinear. Neural Network (NN), as an intelligent control manner, can be applied to such complex (nonlinear) control problems which are very difficult to be solved by conventional methods. Explicit programming is not required when using NN. And NN is able to adjust itself appropriately when confronted with a new pattern because it has the adaptability to learn the relationships between a set of patterns. Consequently, when an external disturbance is introduced into a system, it will adjust the control logic accordingly [1]. The controller combined NN with conventional PID is well regarded and there have been many researches on it [2-8].

In section 2, the procedure for designing PID controller based on BPNN using ADM is presented, and three main problems in its implementation are also discussed. Section 3 summarizes the simulation results for tracking problems. The results are compared with those obtained from conventional PID controllers. Finally, the paper is concluded with section 4.

2 PID Controller Based on BPNN Using ADM

BP is a standard method used for NN training, which uses gradient-descent algorithm to obtain a solution for NN weights and bias constants. The PID control parameters

F. Sun et al. (Eds.): ISNN 2008, Part II, LNCS 5264, pp. 702–711, 2008.

K_p, K_i and K_d can be adjusted on-line through BPNN self-learning. And the best values of them can be obtained corresponding to the outputs of BPNN with a certain optimal control law.

The traditional increment PID algorithm is given by

$$\begin{aligned} &u(k) = u(k-1) + \Delta u(k) \\ &\Delta u(k) = K_p[e(k) - e(k-1)] + K_i e(k) + K_d[e(k) - 2e(k-1) + e(k-2)] \\ &e(k) = r(k) - y(k) \end{aligned} \tag{1}$$

Where u is the control variable, r is the expected output value and y is the actual output value obtained during evaluation.

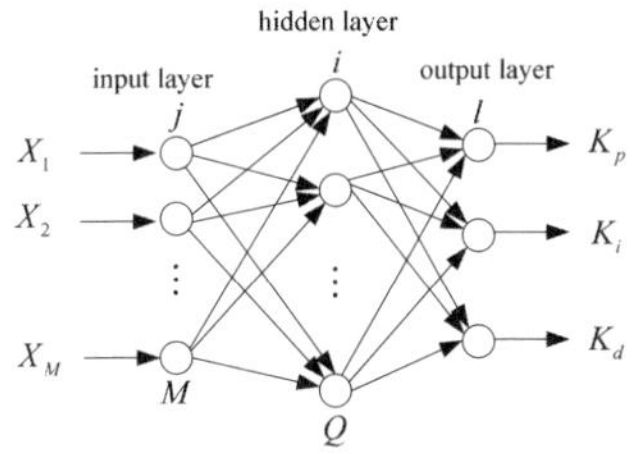

Fig. 1. Three-layer BPNN

A typical three-layer BPNN (one hidden layer) is shown in Fig.1. M and Q are the neuron numbers of the input layer and hidden layer respectively, while *j, i and l* are the indexes of neurons in each layer. M is selected according to the complexity of the controlled plant. But it is difficult to select Q which depends on the experiences of the designer. Three reference formulas are introduced to select Q in Literature [9]. One of them is given by

$$Q = \sqrt{M+3} + c \quad c \in [1,10] \tag{2}$$

Where c is a constant and is set to be 3 in this paper.

The inputs and outputs of the hidden layer and the output layer are given by

$$\begin{aligned} &net_i^{(2)}(k) = \sum_{j=0}^{M} \omega_{ij}^{(2)} O_j^{(1)}(k) \\ &O_i^{(2)}(k) = f(net_i^{(2)}(k)) \quad (i = 1,2,\cdots,Q) \\ &net_l^{(3)}(k) = \sum_{l=0}^{Q} \omega_{li}^{(3)} O_i^{(2)}(k) \\ &O_l^{(3)}(k) = g(net_l^{(3)}(k)) \quad (l = 1,2,3) \end{aligned} \tag{3}$$

Where ω is the weight, *net* is the input and O is the output. Superscript (1), (2) and (3) denote the input, the hidden and the output layer respectively. f and g are the activation functions of the hidden layer and the output layer. Obviously, $O^{(3)}$ are K_p, K_i and K_d.

A performance function is given by

$$E(k)=\frac{1}{2}[r(k)-y(k)]^2 \tag{4}$$

The weights of NN are updated according to gradient-descent algorithm.

$$\Delta\omega(k)=-\eta[\partial E(k)/\partial\omega(k)]+\alpha\Delta\omega(k-1) \tag{5}$$

Where η and α are defined as learning rate and inertia factor respectively. The inertia item $\alpha\Delta\omega(k-1)$ is added to accelerate convergence.

Most researchers use Manual Differentiation Method (MDM) to acquire the accurate value of $\partial E(k)/\partial\omega(k)$ in equation (5) [5-7]. The algorithm of weight update for the output layer is given by

$$\begin{aligned}&\Delta\omega_{li}^{(3)}(k)=\alpha\Delta\omega_{li}^{(3)}(k-1)+\eta\delta_l^{(3)}O_i^{(2)}(k)\\&\delta_l^{(3)}=e(k)\operatorname{sgn}\left(\frac{\partial y(k)}{\partial\Delta u(k)}\right)\frac{\partial\Delta u(k)}{\partial O_l^{3}(k)}\dot{g}(net_l^{(3)}(k))\quad(l=1,2,3)\end{aligned} \tag{6}$$

The process to acquire $\partial E(k)/\partial\omega(k)$ above is complex and fallible especially when the activation functions are complicated or Q is greater than 1. Besides that, according to equation (6), the value of the item $\partial y(k)/\partial\Delta u(k)$ is substituted for by its sign function. That is to say, it is set to be 1 or -1. The impreciseness produced here can be compensated by η [5-7]. But it is difficult to confirm the value of η. And owing to this compensation, η loses its original meaning. For example, if you want to increase the step-size along the direction of gradient-descent on-line, it is uncertain whether η should be increased or decreased because the alternation of impreciseness is unknown.

A method named Automatic Differentiation Method (ADM), by which the derivatives of the function can be evaluated exactly and economically, fits to solve the problem very well. The main idea of ADM is [10-11]: while computer programming is running, the function can be decomposed into a series of primary functions (such as trigonometric function etc.) and primary operations (such as plus, minus etc.) no matter how complex it is. Arbitrary order sensitivity can be gained by iterating these primary functions using chain rules given by

$$\frac{df(g(x),h(x))}{dx}=\frac{\partial f(s,r)}{\partial s}\times\frac{dg(x)}{dx}+\frac{\partial f(s,r)}{\partial r}\times\frac{dh(x)}{dx} \tag{7}$$

Accurate partial differentiations can be acquired using ADM, and the precision only depends on the computer [12]. It is remarkable that the computation cost will be much lower than that of usual methods when the number of variables increases.

The structure of the PID controller based on BPNN using ADM is shown in Fig.2 and the procedure for the controller is summarized as follows:

1. Initialize parameters for BPNN, e.g. α,η,M,Q,ω. Set k=1.
2. Select the activation functions f and g.

3. Obtain the training data $r(k)$ and $y(k)$ and calculate $e(k)$.
4. Calculate input and output values for each layer according to equation (3), the output value of output layer are K_p, K_i and K_d for PID controller.
5. Calculate $u(k)$ according to equation (1).
6. Update weights $\omega(k)$ on-line using ADM according to equation (5).
7. Stop if terminal condition is satisfied. Or else, set k=k+1 and go back to step 3.

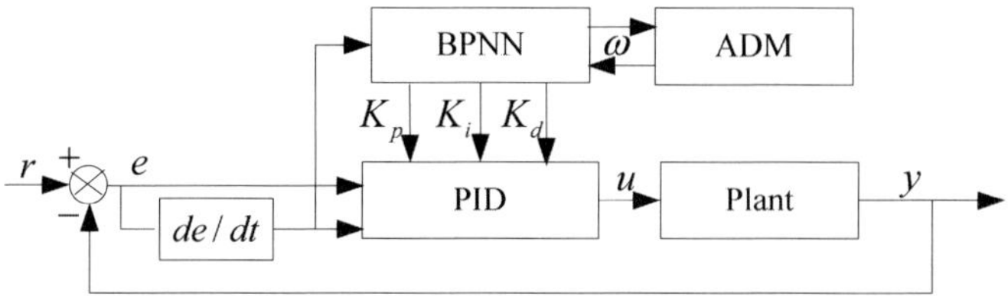

Fig. 2. Structure of PID controller based on BPNN using ADM

There are three main problems in its implementation which will be discussed as follows.

2.1 Weights Initialization

The learning speed can be improved by selecting appropriate initial values of adaptive weights [13]. A heuristic algorithm is often used to set the initial random values between -0.5 and 0.5. If there is only one hidden layer, another method is suggested by Nguyen and Widrow [14], in which the initial values of weights for the hidden layer are updated according to a scaling factor γ.

$$\omega_{ij} = \gamma \frac{\omega_{ij}}{\sqrt{\sum_{i=1}^{Q} \omega_{ij}^2}} = 0.7\sqrt[M]{Q} \frac{\omega_{ij}}{\sqrt{\sum_{i=1}^{Q} \omega_{ij}^2}} \tag{8}$$

Obviously, this method brings a correlation between weights and the neuron numbers of the input and hidden layer. It is more reasonable since the neuron number is one of the key parameters which affect the performance of NN.

2.2 Activation Function

It has been proved that the performance of NN does not depend on the type of activation functions very much, provided that the functions are nonlinear. Nonnegative Sigmoid function is usually selected for output layer since the PID control parameters are positive.

However, the value of Sigmoid function is no more than 1 for any variable x, which limits the range of control parameters. It can not meet the requirements of real systems. In order to improve the efficiency of the controller, an improved activation function for the output layer is presented here.

$$g(x) = \frac{e^x}{e^x + e^{-x}}(U - L) + L \tag{9}$$

Where U and L are the upper and lower bounds of the control parameters respectively. It can be evaluated before network training according to actual situations.

2.3 Update Learning Rate and Inertia Factor

As a gradient-based method, BPNN has marked disadvantages, such as local minima and low learning speed especially for high order systems. η is the key parameter to modify the speed of convergence. Learning speed would be low if η is too small while fluctuation may happen if it is too large. Many improved methods have been proposed to solve this problem. Usually, η and α are decreased gradually during the training. They can also be updated according to the change of the performance function value in BPNN [5]. If $E(k)$ is bigger than $E(k\text{-}1)$, both η and α should be reduced to avoid divergence, or else, they should be increased and the increase should be controlled to avoid vibration caused by overshoot. On the whole, there is not a certain rule to provide exact η and α owing to different characteristics of different systems.

3 Simulation and Results Discussion

PID controller based on BPNN using ADM presented above is applied to two examples. And the simulation results are analyzed by comparing with those obtained from conventional PID controllers.

3.1 Simple Tracking Problem

A model of a controlled plant is approximately given by

$$y(k+1) = \frac{0.8y(k) + 2u(k)}{1 + 1.5y(k-1)u(k-1)} \tag{10}$$

Input signal is given by $r = |\sin 5t\pi|$, which becomes the constant 1.0 when the time exceeds 0.8 second. The structure of BPNN is set to be 3-6-3, that is, M=3, Q=6. And set U=0.1, L=0 for three PID parameters. The tracking results and response curves are shown in fig.3 and fig.4. The output of controller tracks the input signal very well after short-time training even when the input signal shifts sharply. Overshoot didn't occur and the error is not very large in 0.8 second. It can be seen that at the moment when error changed suddenly, PID control parameters are altered accordingly. A conventional PID controller is also used with the parameters K_p=0.1, K_i=0.01, K_d=0.1 and the results are shown in Fig.5. It can be seen that the latter controller can not track so well as the former does with the nonlinear model of the controlled plant.

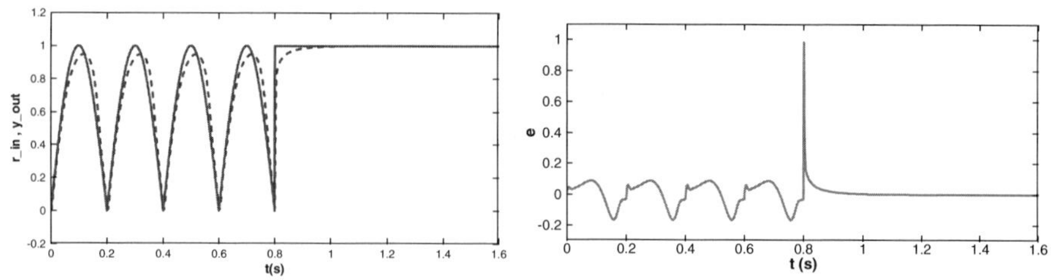

Fig. 3. Response curve and Error curve using PID controller based on BPNN using ADM

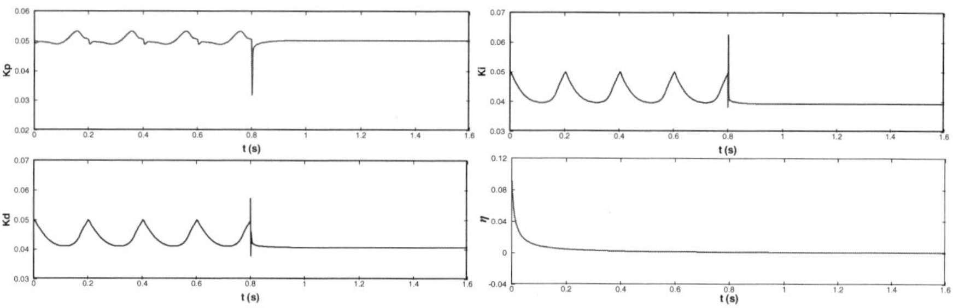

Fig. 4. Training curves of PID parameters and learning rate

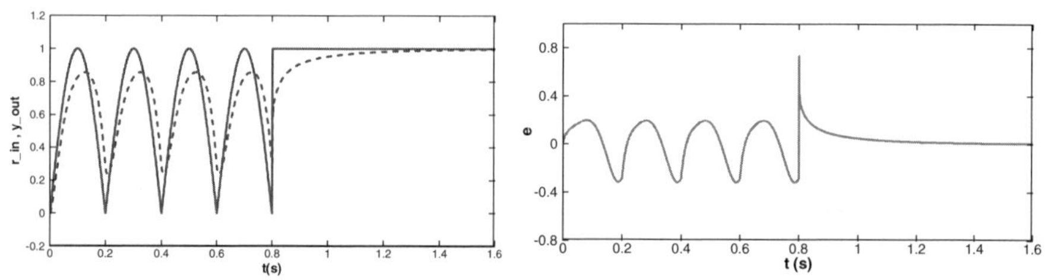

Fig. 5. Response curve and Error curve using conventional PID controller

3.2 Docking Problem

In order to examine its effectiveness and feasibility, PID controller based on BPNN using ADM is applied to Autonomous Docking test-bed which consists of three Degrees-of-Freedom (DoF) spacecraft simulators floating via air pads on a flat floor.

As regards the navigation of the chaser vehicle simulator, the data from the vision sensor are fused with data from other sensors by Kalman Filtering. For position and attitude control, three independent controllers are used. The flow of autonomous docking maneuver is shown in Fig.6. The chaser spacecraft simulator does not require any external reference for its navigation besides the light emitting diodes (LEDs) mounted on the target vehicle simulator.

The chaser weighs about 20*kg* and consists of six thrusters distributed as shown in Fig.7. Thruster 3 and Thruster 6 can provide thrust up to 0.4*N* while others 0.2*N*.

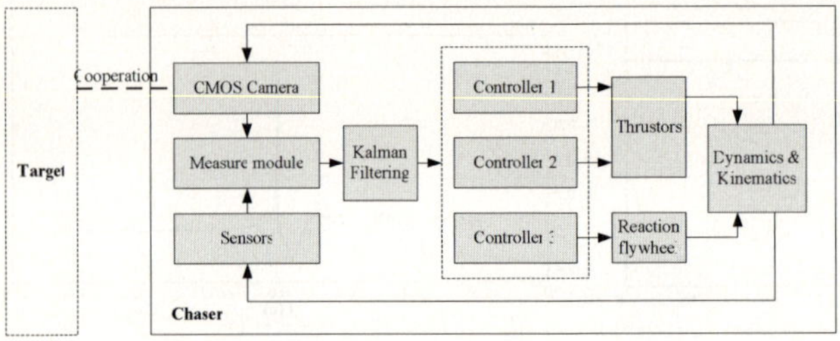

Fig. 6. Flow of autonomous docking maneuver

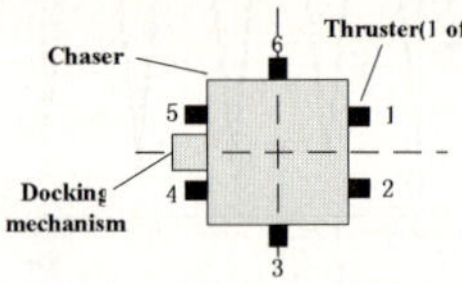

Fig. 7. Thrusters distribution

Since controller 1 and controller 2 don't work until the relative attitude of chaser and target satisfies the requirements of docking. Only plane translational motion is considered on the assumption that relative attitude fits for docking and can be maintained. Taking the relative position and relative velocity as the state variables, the continuous model of the system with one dimension translational motion is

$$\begin{aligned} \dot{\boldsymbol{X}} &= \boldsymbol{AX} + \boldsymbol{BU} \\ \boldsymbol{Y} &= \boldsymbol{CX} \end{aligned} \tag{11}$$

where

$$\boldsymbol{X} = \begin{bmatrix} x \\ \dot{x} \end{bmatrix}, \boldsymbol{A} = \begin{bmatrix} 0 & 1 \\ 0 & 0 \end{bmatrix}, \boldsymbol{B} = \begin{bmatrix} 0 \\ 1 \end{bmatrix}, \boldsymbol{C} = \begin{bmatrix} 1 & 0 \end{bmatrix}$$

In addition, the longitudinal direction x is defined as the main axis of the target while the lateral direction y is the vertical of x. The longitudinal controller begins to work while the relative lateral distance is lower than $0.1m$ to avoid missing docking mechanism in this problem.

Numerical Simulations. Numerical simulations have been carried out using both conventional PID controller and the new controller introduced above. The initial state in the present analysis is:

$$x_0=3m,\ y_0=0.8m,\ x^*=0.1m,\ y^*=0m$$

Where x_0 and y_0 are the initial relative position parameters, x^* and y^* are the expected relative position parameters for docking. It is relatively still to the target at the beginning.

The PID controller based on BPNN is a three-layered network with 4-6-3 structure. Set U=50, L=0 for K_p and K_d, U=0.1, L=0 for K_i. The curves of control process are shown in Fig.8. The parameters of a conventional PID controller are K_p=10, K_i=0.1, K_d=5 and the results are shown in Fig. 9. Both controllers are effective. However, oscillation and overshoot happen in the latter controller for both relative parameters. The convergence speed is much faster when using the former controller. It is obvious that the PID controller based on BPNN is more effective than the conventional PID controller in approaching the target.

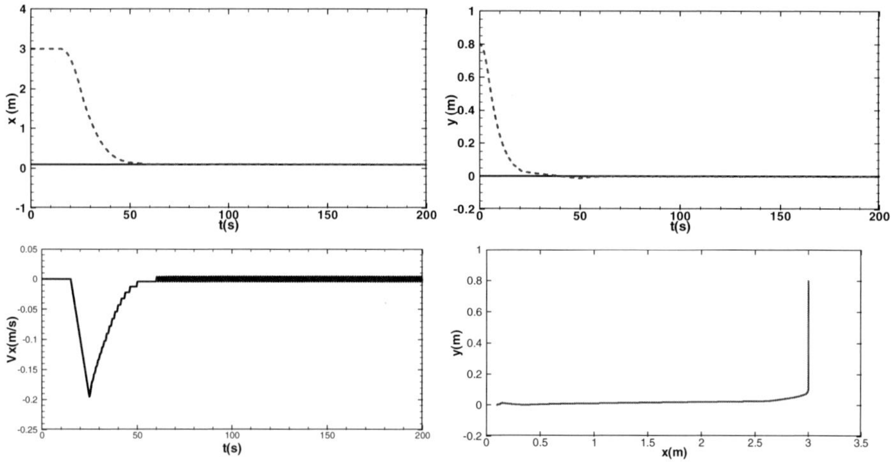

Fig. 8. Training curves of relative parameters using PID controller based on BPNN using ADM

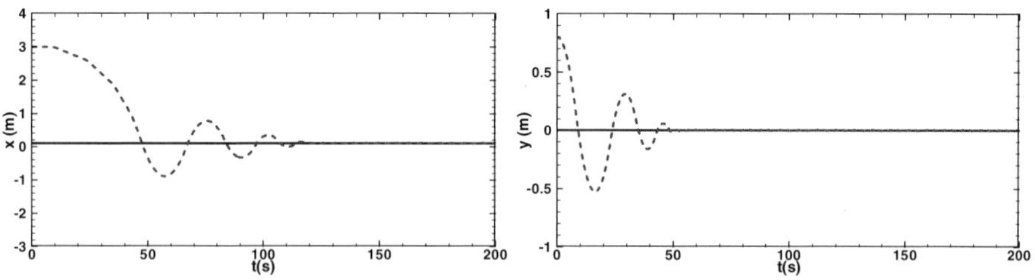

Fig. 9. Curves of relative parameters using conventional PID controller

Experimental Test. The PID controller based on BPNN has been coded in C++ and run in real time on the onboard computer of the chaser spacecraft simulator. During the experiment, the target vehicle is kept fixed. The maneuver consists of autonomously approaching the target and then docking to it. The results are shown in Fig.10. In this test, the chaser, which starts from an offset position and attitude, first reduces the angular error by attitude maneuver and then approaches the target. The process is shown in Fig.11. The entire maneuver lasts about 71s. The initial and the expected relative position parameters of the chaser are

$$x_0=1.7m,\ y_0=0.4m,\ x^*=0.1m,\ y^*=0m$$

During the first several seconds, the chaser is not floating until the Kalman filters converged. Then the attitude is controlled to point toward the target in about 20 seconds and maintains subsequently. Thus the process is recorded from time A.

It can be seen from the motion of the chaser shown in Fig.11, the PID controller based on BPNN is proved to be effective during docking. Lateral error is decreased while the longitudinal relative distance slightly decreased at the beginning owing to a little angle error which couples the two dimensions. The track is smooth except at

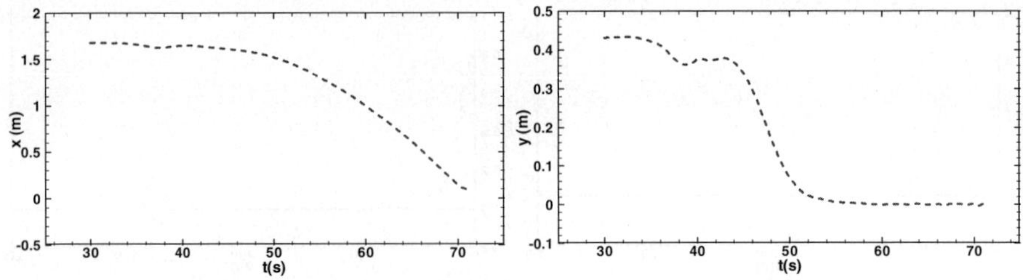

Fig. 10. Curves of relative parameters in experimental test

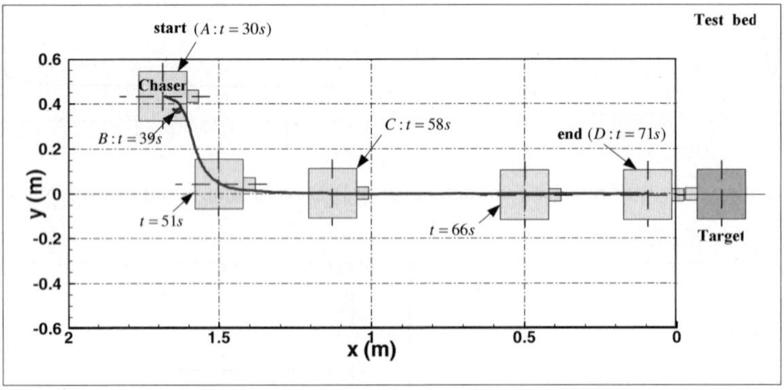

Fig. 11. The motion of the chaser viewed from the top

time B. It is possibly caused by disturbance of test bed roughness. As the velocity is not very high, effect of roughness on plane motion can be evident. After time B, the lateral error is controlled to zero gradually between time A and C and then it maintains within 1mm from time C to D. The longitudinal relative distance is decreased steadily from time B to D. The longitudinal relative velocity is reduced to about 1cm/s when the mechanism starts to work.

4 Conclusion

An active PID controller based on Back-Propagation Neural Network using automatic differentiation method is introduced and successfully implemented in simulations. The results of a simple tracking problem show its effectiveness on nonlinear system due to its good adaptability which benefits from on-line self-learning. Furthermore, by comparing with conventional PID controller in numerical simulation, PID controller based on BPNN using ADM is applied to an autonomous docking experimental test. The results show the validity and good robustness of the proposed control approach.

Acknowledgments. The authors wish to acknowledge the help of Ms. Hui Qin in preparing the final version of this paper.

References

1. Susan, J.R., Susan, R.B.: Neurocontrollers for complex systems. AIAA. 93-0005
2. Yue, Y., Li, Q., Yu, S.: A Survey of Intelligent PID Control (in Chinese). J. Programmable controller & factory automation 12, 9–13 (2006)
3. Cheng, H., Chen, R.: Application of Artificial Neural Network in PID Control Algorithm (in Chinese). Agriculture equipment and vehicle engineering 184, 42–45 (2006)
4. Li, Y., Wang, M.: Bleaching Temperature Control Algorithm Researching Based on BP Neural Network PID (in Chinese). Micro-computer information 22, 41–42 (2006)
5. Qin, P., Li, H., Zhang, D.: Research on the Controlling Methods of BLDCM Based on Improved BP Neural Network PID (in Chinese). Micro-electric machinery 39, 40–42 (2006)
6. Shi, C., Zhang, G.: Study of PID Control Based on Improved BP Neural Network (in Chinese). Computer Emulation 23, 156–159 (2006)
7. Hu, Z., Wang, J., Wang, H.: The Study and Simulation of PID Control Based on Optimized BP Neural Network (in Chinese). Micro-electronics and computer 23, 138–140 (2006)
8. Zhou, Y.: The Design of the Neural Network of PID Based on PLC (in Chinese). Micro-computer information 23, 97–98 (2007)
9. Fecit. Neural Network Theory and MATLAB Application (in Chinese). Publishing house of electronics industry (2005)
10. Zhang, H., Xue, Y.: Basic Principle and Implement of Automatic Differentiation (in Chinese). Journal of university of Beijing Industry 31, 332–336 (2005)
11. Yan, L.: Research on the theory and application of some key technologies in the multidisciplinary design optimization of flight vehicles (in Chinese). National University of Defense Technology (2006)
12. Louis, B.R.: Automatic Differentiation: Techniques and Applications. Springer, Berlin (1981)
13. Fredric, M., Ham, I.K.: Principles of Neurocomputing for science & engineering. China machine press (2007)
14. Nguyen, D., Widrow, B.: Improving the learning speed of the 2-layer Neural Networks by Choosing Initial Values of Adaptive Weights. In: Proceedings of the International Joint Conference on Neural Networks, vol. 3, pp. 21–26. IEEE Press, San Diego (1990)

Neuro-Identifier-Based Tracking Control of Uncertain Chaotic System

Wen Tan[1], Fuchun Sun[2], Yaonan Wang[3], and Shaowu Zhou[1]

[1] School of electrical and information engineering,Hunan University of Science and technology, 411201 Xiangtan, P.R. China
[2] Dept.of Computer Science and Technology, Tsinghua University, 100084 Beijing, P.R. China
[3] College of electrical and information engineering,Hunan University, 410082 Changsha, P.R. China

Abstract. A novel neuro-identifier-based tracking control of uncertain nonlinear chaotic system is presented. The algorithm is divided into two contributions. First, a dynamic neural networks is used to identify the unknown chaos, then a dynamic adaptive state feedback controller based on neuro-identifier is derived to direct the unknown chaotic system into desired reference model trajectories. Moreover, the identification error and trajectory error is theoretically verified to be bounded and converge to zero Computer simulations are shown to demonstrate the effectiveness of this proposed methodology.

Keywords: Chaos, Identification, Adaptive control, Neural networks.

1 Introduction

Recently, many different aspects of chaotic dynamics have attracted extensive interest from various disciplines [1-4]. Even though many chaotic systems are of simple model, they exhibit complicated dynamics. In particular, if the chaos system is partly known, for instance, the differential equation of it is knowable but some or all of the parameters are unknown, hence exact model-based control method may be infeasible. Adaptive strategy based on nonlinear model has been applied to solve the types of problems in recent years [5,6].

In this paper, a dynamical neural networks is first used to model the uncertain chaotic system, then adaptive state feedback controller is designed to guiding chaotic states to a given bounded input signal. Meanwhile, Lyapunov synthesis technique is applied to analyse stability and robustness of the networks identifier, and convergence of the error and boundedness for the total closed loop system are scheme, the experiment on forced Van der pol oscillator is performed, and some simulation results are presented.

2 Neuro-Identification Model

Let us consider continuous time nonlinear chaotic system which will be identified in the following form:

$$\dot{x} = f(x,t) + G(x)u \tag{1}$$

F. Sun et al. (Eds.): ISNN 2008, Part II, LNCS 5264, pp. 712–719, 2008.

where the state $x \in R^n$,is assumed to be measured, the control input $u \in \mu \subset R^n$,where μ is admissible inputs, f is an unknown smooth vector-fields, $G = [G_1, G_2 ... G_n]$ is a matrix with columns the unknown vectorfields $G_i, i = 1,2,..., n$. For analysis simplicity, we impose assumptions on system (1) as follows:

A1: Given a class μ ,then for any $u \in \mu$ and finite initial value $x(0)$,the trajectories are uniformly bounded for any finite $T > 0$,namely $|x(T)| < \infty$.

A2: f, G are continuous smooth function, and satisfy a local Lipschitz condition so that the solution $x(t)$ to the equation (1) is unique.

Obviously, in order to resolve our problem, it is necessary to more accurately model the unknown system. For this purpose, we apply dynamical neural networks to approximate (1)[7]:

$$\dot{x}_{nn} = Ax_{nn} + BW\,\sigma(x) + BW_{n+1}\vartheta(x)u \tag{2}$$

where $x_{nn} \in R^n$ is states of the neural networks, A, B are $n \times n$ diagonal matrices with elements $a_i, b_i, i = 1,2,...,n.$ $W \in R^{n \times n}$ with adjustable synaptic weights, W_{n+1} is $n \times n$ diagonal matrix of adjustable synaptic weights of the form $W_{n+1} = diag[w_{1,n+1}, w_{2,n+1}, ..., w_{n,n+1}]$. $\sigma(x) \in R^n$ and $\vartheta(x) \in R^{n \times n}$ are monotone increasing functions which are usually represented by sigmoids.

Now our task first is to derive learning algorithm of the neural networks model so that the unknown system (1) can be well approximated. In view of the approximation capability of the dynamical neural networks, we can assume there exists weight values W^*, W^*_{n+1} so that the system (1) can be represented as:

$$\dot{x} = Ax + BW^*\sigma(x) + W^*_{n+1}\vartheta(x)u \tag{3}$$

Define the error between network model and actual system as

$$e = x_{nn} - x$$

Then according to (2) and (3) we get error dynamics:

$$\dot{e} = Ae + BW_A\sigma(x) + BW_B\vartheta(x)u \tag{4}$$

where $W_A = W - W^*, W_B = W_{n+1} - W^*_{n+1}$.

Now we consider the Lyapunov function candidate

$$v(e, W_A, W_B) = \tfrac{1}{2}e^T Pe + \tfrac{1}{2}tr\{W_A^T W_A\} + \tfrac{1}{2}tr\{W_B^T W_B\} \tag{5}$$

where $P > 0$ satisfy Lyapunov equation

$$A^T P + PA = -I$$

Taking the time derivative of (5), we have

$\dot{v} = \tfrac{1}{2}(\dot{e}^T Pe + e^T P\dot{e}) + tr\{\dot{W}_A^T W_A\} + tr\{\dot{W}_B^T W_B\}$ or

$$\dot{v} = \tfrac{1}{2}(-e^T e + \sigma^T(x)W_A^T BPe + u^T \vartheta(x)W_B BPe + (\sigma^T(x)W_A^T BPe)^T$$
$$+ (u^T \vartheta(x)W_B BPe)^T) + tr\{\dot{W}_A^T W_A\} + tr\{\dot{W}_B^T W_B\}$$

Since $\sigma^T(x)W_A^T BPe, u^T \vartheta(x)W_B BPe$ are scalars, Thus

$$\sigma^T(x)W_A^T BPe = (\sigma^T(x)W_A^T BPe)^T, u^T \vartheta(x)W_B BPe = (u^T \vartheta(x)W_B BPe)^T$$

which leads to

$$\dot{v} = -\tfrac{1}{2}e^T e + \sigma^T(x)W_A^T BPe + u^T \vartheta(x)W_B BPe + tr\{\dot{W}_A^T W_A\} + tr\{\dot{W}_B^T W_B\} \quad (6)$$

therefore, we choose

$$tr\{\dot{W}_A^T W_A\} = -\sigma^T(x)W_A^T BPe \quad (7)$$

$$tr\{\dot{W}_B^T W_B\} = -u^T \vartheta(x)W_B BPe \quad (8)$$

Then (6) becomes

$$\dot{v} = -\tfrac{1}{2}e^T e \quad (9)$$

or

$$\dot{v} = -\tfrac{1}{2}\|e\|^2 \le 0 \quad (10)$$

From (7) and (8), after simple manipulation we easily obtain weights update laws of the network identifier

$$\begin{aligned} \dot{w}_{ij} &= -b_i p_i \sigma(x_j) e_i \\ \dot{w}_{i,n+1} &= -b_i \vartheta(x_i) p_i u_i e_i \end{aligned} \quad (11)$$

Accordingly we can prove the following theorem.

Theorem1: Consider both the error dynamical equation (4) and the update laws (11), the following properties can be held

- $e, x_{nn}, W_A, W_B \in L_\infty, e \in L_2$
- $\lim_{t\to\infty} e(t) = 0, \lim_{t\to\infty} \dot{W}_A(t) = 0, \lim_{t\to\infty} \dot{W}_B(t) = 0.$

proof: From (11) we have that $\dot{v} \le 0$, Hence $v \in L_\infty$ which implies $e, W_A, W_B \in L_\infty$, furthermore $x_{nn} = x + e$ is also bounded. Since v is a monotone decreasing function of time and bounded, namely $\lim_{t\to\infty} v = v_\infty$ exists. Therefore, we have

$$\int_0^\infty \|e\|^2 dt = 2[v(0) - v_\infty] < \infty.$$

which implies that $e \in L_\infty$. Due to the boundedness of both $\sigma(x_i)$ and all inputs of the neural networks, from (4) we infer $\dot{e} \in L_\infty$. Since $e \in L_2 \cap L_\infty$, this yields $\dot{e} \in L_\infty$, by the use of Barbalat Lemma[8], we conclude that $\lim_{t\to\infty} e(t) = 0$. Furthermore, using the boundedness of $u, \sigma(x), \vartheta(x)$ and the convergence of $e(t)$ to zero, we have that $\dot{W}_{n+1}$ also converges to zero. This completes proof.

3 Neuro-Identifier-Based Tracking Control

Based on identification model, we further consider the following problem that the real system states quickly follow the given reference model trajectories. we assume reference model can be written as

$$\dot{x}_m = \varphi(x_m, t), x_m \in R^n \tag{12}$$

Define the error between the states of neural networks identifier and that of reference model as

$$e_c = x_{nn} - x_m \tag{13}$$

Differentiating (13) , we obtain

$$\dot{e}_c = \dot{x}_{nn} - \dot{x}_m$$

or $$\dot{e}_c = Ax_{nn} + BW\sigma(x) + BW_{n+1}\vartheta(x)u - \varphi(x_m, t) \tag{14}$$

we choose control u as

$$u = -[BW_{n+1}\vartheta(x)]^{-1}[Ax_m + BW\sigma(x) - \varphi(x_m, t)] \tag{15}$$

Substituting (15) into (14), we have

$$\dot{e}_c = Ae_c \tag{16}$$

To continue, we utilize Lyapunov synthesis method to derive stable update laws. Weights learning laws (11) can be written in the format of matrix as :

$$\dot{W} = -EBPS_0, \dot{W}_{n+1} = -BP\vartheta UE$$

where all matrices are defined as follows

$$\begin{aligned} P &= diag[p_1, p_2, ..., p_n] \\ B &= diag[b_1, b_2, ..., b_n] \\ E &= diag[e_1, e_2, ..., e_n] \\ U &= diag[u_1, u_2, ..., u_n] \end{aligned} \qquad S_0 = \begin{bmatrix} \sigma(x_1) & \cdots & \sigma(x_n) \\ \vdots & & \vdots \\ \sigma(x_1) & \cdots & \sigma(x_n) \end{bmatrix}$$

It must be assured that $(BW_{n+1}\vartheta(x))^{-1}$ exists before applying control effort (15), namely $w_{i,n+1}(t) \neq 0, \forall i = 1,2,...,n$. Here, $W_{n+1}(t)$ is confined to the set $\Lambda = \{W_{n+1} : \|W_B\| \leq \omega_m\}$ by the usage of a projection algorithm[9,10],where $\omega_m > 0$ is constant, thus, learning laws (11),which are described in the form of matrix, can be modified as:

$$\dot{W} = -EBPS_0$$

$$\dot{W}_{n+1} = \begin{cases} -BP\vartheta UE \\ \quad if\, W_{n+1} \in \Lambda\, or \quad \{\|W_B\| = \omega_m \quad and \quad tr\{-BP\vartheta UEW_B\} \leq 0\} \\ -BP\vartheta UE + tr\{BP\vartheta\sigma UW_B\}(\frac{1+\|W_B\|}{\omega_m})^2 W_B \\ \quad if \quad \{\|W_B\| = \omega_m\, and \quad tr\{-BP\vartheta UEW_B\} > 0\} \end{cases} \tag{17}$$

Consequently, only the initial weights satisfy $\|W_B(0)\| \le \omega_m$, for $t \ge 0$, then we have that

$$\|W_{n+1}\| \le \omega_m \tag{18}$$

noting that whenever $\|W_B(t)\| = \omega_m$, then

$$\frac{d\|W_B\|^2}{dt} \le 0 \tag{19}$$

which implies that weights W_{n+1} are directed toward the inside or the ball $\{W_{n+1} : \|W_B\| \le \omega_m\}$. As to proof of (19), it can be referred to [7].

Theorem 2: Take the control scheme (4), (15) and (16) into consideration, the modified update laws (17) can guarantee the following properties:

- $e, e_c, x_{nn}, W_A, W_B \in L_\infty, e, e_c \in L_2$
- $\lim_{t\to\infty} e(t) = 0, \lim_{t\to\infty} e_c(t) = 0, \lim_{t\to\infty} \dot{W}_A(t) = 0, \lim_{t\to\infty} \dot{W}_B(t) = 0.$

Proof : consider Lyapunov function candidate:

$$v(e, e_c, W_A, W_B) = \tfrac{1}{2} e^T P e + \tfrac{1}{2} e_c^T P e_c + \tfrac{1}{2} tr\{W_A{}^T W_A\} + \tfrac{1}{2} tr\{W_B{}^T W_B\}$$

differentiating the above function, and using (17), we obtain

$$\begin{aligned}
\dot{v} &= -\tfrac{1}{2}\|e\|^2 - \tfrac{1}{2}\|e_c\|^2 + I_n tr\{tr\{BP\vartheta UEW_B\}\left(\frac{1+\|W_B\|}{\omega_m}\right)^2 W_B^T W_B\} \\
&\le -\tfrac{1}{2}\|e\|^2 - \tfrac{1}{2}\|e_c\|^2 + I_n tr\{BP\vartheta UEW_B\}\left(\frac{1+\|W_B\|}{\omega_m}\right)^2 tr\{W_B^T W_B\} \\
&\le -\tfrac{1}{2}\|e\|^2 - \tfrac{1}{2}\|e_c\|^2 + I_n tr\{BP\vartheta UEW_B\}\left(\frac{1+\omega_m}{\omega_m}\right)^2 \|W_B\|^2 \\
&\le -\tfrac{1}{2}\|e\|^2 - \tfrac{1}{2}\|e_c\|^2 + I_n tr\{BP\vartheta UEW_B\}(1+\omega_m)^2
\end{aligned}$$

where $I_n = \begin{cases} 1 & \|W_B\| = \omega_m \text{ and } tr\{-BP\vartheta UEW_B\} > 0 \\ 0 & otherwise \end{cases}$. Thus, $\dot{v} \le 0$. Therefore the additional terms introduced by the projection algorithm can only make $\dot{v}$ more negative, and negative semidefinite property of $\dot{v}$ leads to $v \in L_\infty$, which implies $e, e_c, W_A, W_B \in L_\infty$. Furthermore, $x_{nn} = e + x$ is also bounded. Since v is a non-increasing function of time and bounded, namely $\lim_{t\to 0} v = v_\infty$ exists, therefore, we obtain

$$\int_0^\infty \tfrac{1}{2}(\|e\|^2 + \|e_c\|^2)dt - I_n(1+\omega_m)^2 tr\{BP\vartheta UEW_B\}dt \le [v(0) - v_\infty] < \infty$$

which implies $e, e_c \in L_2$. Considering the boundedness of both $\sigma(x), \vartheta(x)$ and all inputs of reference model, hence from (15) we get bounded u, meanwhile, from (4) and (16), we have $\dot{e}, \dot{e}_c \in L_\infty$. Since $e, e_c \in L_2 \cap L_\infty$, through Barbalat Lemma[8], $\lim_{t\to\infty} e(t) = 0$, $\lim_{t\to\infty} e_c(t) = 0$ can be concluded. Then, employing the boundedness of $u, \sigma(x), \vartheta(x)$ and the convergence of $e(t)$ to zero, $\dot{W}$, $\dot{W}_{n+1}$ also converge to zero. The proof is completed.

4 Numerical Simulation

To verify the effectiveness of the proposed identification model-based control scheme, we consider well known forced chaotic oscillators, namely the Van der pol oscillator: $\ddot{x} + d(x^2 - 1)\dot{x} + x = a\cos(\omega t)$. For various values of $a, d, and\ w$, the oscillator exhibits a large number of nonlinear behaviors. Let us choose $x_1 = x, x_2 = \dot{x}$, then the controlled system can be rewritten as

$$\begin{cases} \dot{x}_1 = x_2 + u_1(t) \\ \dot{x}_2 = -d(x_1^2 - 1)x_2 - x_1 + a\cos(\omega t) + u_2(t) \end{cases} \tag{20}$$

When we select $u = 0, a = 2.5, d = 6, \omega = 3$, the corresponding system is chaotic. We first apply (2) to identify the system (20) before designing an adaptive controller. During the process of experiment, we choose

$$\sigma(x_i) = \frac{k}{1 + e^{-lx_i}},\ \vartheta(x_i) = \frac{k}{1 + e^{-lx_i}} + \lambda$$

where $k = 1.0, l = 0.8, \lambda = 0.3, a_i = -50, b_i = 12, i = 1,2$. The identification results are shown as Fig.1, where the solid lines represent the states of the uncontrolled system, while dashed lines is that of neuro-identifier. Then, drawing the support from the achievement of neural networks identification, we utilize adaptive (15) to regulate the system (20) so that its trajectory can track reference model. As a reference trajectory, we take $x_m = (\sin(5t), \sin(t))$. The resulting behavior is plotted in Fig.2, where the

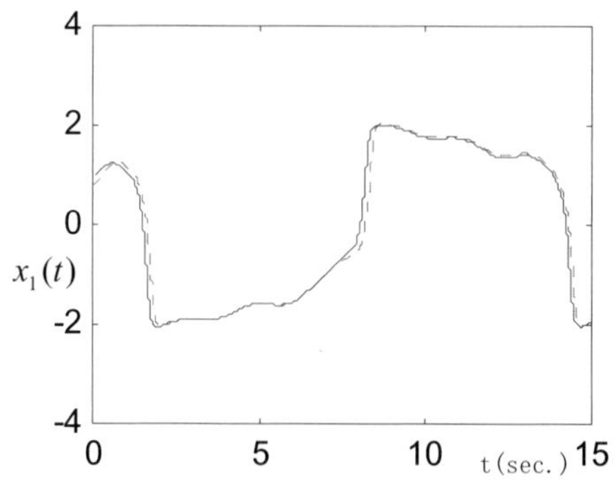

Fig. 1. (a) State trajectory of x_1(t) before control

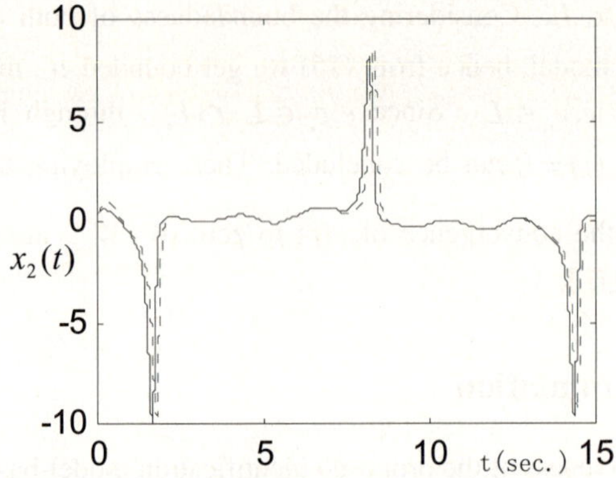

Fig. 1. (b) State trajectory of x_2(t) before control

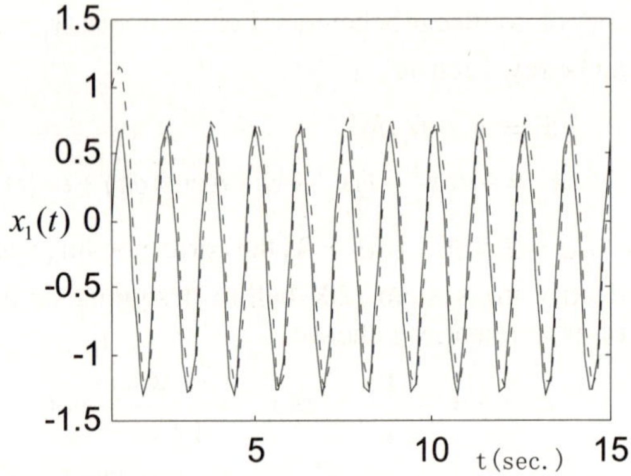

Fig. 2. (a) State trajectory of x_1(t) after control

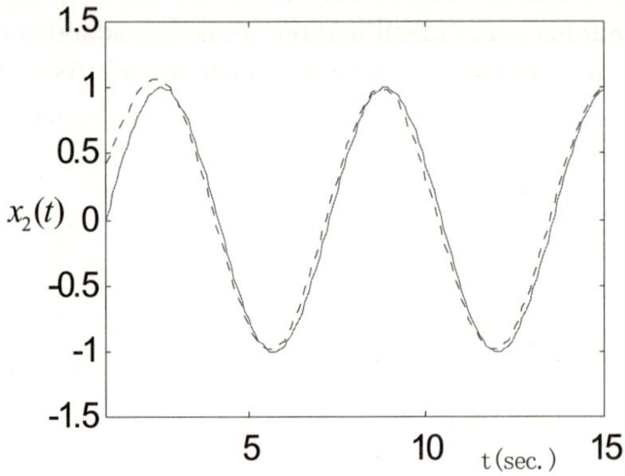

Fig. 2. (b) State trajectory of x_2(t) after control

solid lines denote reference model,trajectories,while dashed lines correspond to the tracking control trajectories. Through simulation results, it is easily found that the effectiveness is more satisfactory by employing dynamic neural networks to both identify unknown chaotic system and realize tracking problem.

5 Conclusions

Identification and control for uncertain nonlinear chaotic system is examined by employing dynamical neural networks. Neuro-identifier based adaptive state feedback controller is designed to achieve the tracking problem, i.e, to drive the solution of the chaotic system to follow a bounded reference input signal. Through the modification of update laws and control, both stability of the network identifier and the convergence of the error are assured. The numerical simulation results on forced Van der pol oscillator have confirmed the effectiveness of the suggested approach.

Acknowledgements. The authors gratefully acknowledge the support provided by the National Natural Science Foundation of China(Grant No. 60375001) and the Scienfitic Research Funds of Hunan Provincial Education Department, China(Grant No. 05B016).

References

1. Tanaka, T., Ikeda, T., Wang, H.O.: A Unified Approach to Controlling Chaos via an Lmi-based Fuzzy Control System Design. J. IEEE Transactions on Circuits and Systems 45, 1021–1040 (1998)
2. Wen, T., Yao, N.W.: Adaptive Regulation of Uncertain Chaos with Dynamical Neural Networks. J. Chinese Physics 13, 459–563 (2004)
3. Chen, G., Dong, X.: From Chaos to Order-Methodologies, Perspective and Applications. World Scientific, Singapore (1998)
4. Joo, Y.H., Shieh, L.S., Chen, G.: Hybrid State-space Fuzzy Model-based Controller with Dual-rate Sampling for Digital Control of Chaotic Systems. J. IEEE Transactions on Fuzzy Systems 7, 394–408 (1999)
5. Loria, A., Panteley, E.: Control of the Chaotic Duffing Equation with Uncertainty in All Parameters. J. IEEE Transactions on Circuits and Systems 45, 1252–1255 (1998)
6. Yu, X.: Tracking Inherent Periodic Orbits in Chaotic Dynamic Systems via Adaptive Variable Structure Time-delayed Self Control. J. IEEE Transactions on Circuits and Systems 46, 1408–1411 (1999)
7. George, A., Rovithakis Christodoulou, A.: Adaptive Control of Unknown Plants Using Dynamical Neural Networks. J. IEEE Transactions on Systems,Man and Cybernetics 24, 400–412 (1994)
8. Rouche, N., Habets, P., Laloy, M.: Stability Theory by Lyapunov's Direct Method. Springer, New York (1977)
9. Narendra, K.S., Annaswamy, A.M.: Stable Adaptive Systems. Prentice Hall, Englewood Cliffs (1989)
10. Goodwin, G.C., Mayne, D.Q.: A parameter Estimation Perspective of Continuous Time Model Reference Adaptive Control. J. Automatica 23, 57–70 (1987)

Robust Stability of Switched Recurrent Neural Networks with Discrete and Distributed Delays under Uncertainty

Shiping Wen, Zhigang Zeng, and Lingfa Zeng

School of Automation
Wuhan University of Technology, Wuhan, Hubei, 430070, China
zgzeng@gmail.com

Abstract. With the rapid development of intelligent control, switched systems have attracted great attention. In this letter, we try to introduce the idea of the switched systems into the field of recurrent neural networks (RNNs) with discrete and distributed delays under uncertainty which is considered to be norm bounded. At first, we establish the mathematical model of the switched RNNs in which a set of RNNs are used as the subsystems and an arbitrary switching rule is assumed. Secondly, for this kind of systems, robust analysis which is based on the Lyapunov-Krasovii approach is addressed, and for all admissible parametric uncertainties, some criteria which are derived in terms of a series of strict LMIs are presented to guarantee the switched RNNs to be globally exponentially stable. Finally, a specific example is shown to illustrate the applicability of the methodology.

Keywords: Recurrent neural networks, Switched systems, Stability.

1 Introduction

In recent years, recurrent neural networks have attracted huge attention and been studied widely [1]. For their numerous potentials of application prospective in different areas such as associative memory, knowledge acquisition and optimization problems [2, 3]. In optimization problems, the neural networks have to be designed to have only one globally stable equilibrium point and consider the discrete as well as distributed time. In addition, the uncertainty of the parameter which usually breaks the stability of systems can also cause the modeling inaccuracies and/or changes in the environment of the model. Many efforts have been made to deal with the difficulties caused by uncertainty. And robust stability analysis has achieved great progress [4].

From [5-7], we know that switching between two asymptotically stable systems can produce an unstable trajectory, on the other hand, the switching between two unstable systems can produce a stable trajectory. So it is obvious that the study of switched RNNs is very important and necessary. Some try has been made in [4], in which the switched Hopfield neural networks have been considered.

F. Sun et al. (Eds.): ISNN 2008, Part II, LNCS 5264, pp. 720–729, 2008.

In this paper, we are concerned with the issue of robust stability analysis of switched RNNs with discrete and distributed delays under uncertainty. All the subsystems are supposed to be norm bounded [8].

This paper is organized as follows, in Section 2, we will introduce the mathematical model of the switched RNNs. By using the linear matrix inequality (LMI), we get some sufficient conditions in Section 3. And the main results in Section 3 will be illustrated with a numerical example in Section 4. Finally, we make the conclusion in Section 5.

2 Preliminaries

Consider

$$\begin{aligned}\frac{du_k(t)}{dt} &= -(d_k+\Delta d_k)u_k(t)+\sum_{j=1}^{n}(a_{kj}+\Delta a_{kj})f_j'(u_j(t))\\ &\quad+\sum_{j=1}^{n}(b_{kj}+\Delta b_{kj})g_j'(u_j(t-\varsigma_1))\\ &\quad+\int_{t-\varsigma_2}^{t}\sum_{j=1}^{n}(c_{kj}+\Delta c_{kj})h_j'(u_j(s))ds+\mathcal{I}_k,\quad k=1,2,\ldots,n,\end{aligned} \tag{1}$$

where n denotes the number of the neurons in the neural network, $u_k(t)$ denotes the state of the k-th neural neuron at time t, $f_j'(u_j(t)), g_j'(u_j(t)), h_j'(u_j(t))$ are the activation functions of j-th neuron at time t. The constants a_{kj}, b_{kj}, c_{kj}, Δa_{kj}, Δb_{kj}, Δc_{kj} denote respectively, the connection weights, the discretely delayed connection weights, the distributed delay connection weights, the uncertainty connection weights, the uncertainty discrete delay connection weights, the uncertainty distributed delay connection weights, of the j-th neuron on the k-th neuron. $\mathcal{I}_k$ is the external bias on the k-th neuron and d_k denotes the rate with which the k-th neuron will reset its potential to the resting state in isolation, ς_1 is the constant discrete time delay, while ς_2 describes the distributed time delay. We can rewrite model (1) in the following matrix vector form:

$$\begin{aligned}\dot{u}(t) &= -(D+\Delta D)u(t)+(A+\Delta A)F'(u(t))+(B+\Delta B)G'(u(t-\zeta_1))\\ &\quad+(C+\Delta C)\int_{t-\varsigma_2}^{t}H'(u(s))ds+\mathcal{I},\end{aligned} \tag{2}$$

where $D=\mathrm{diag}(d_1,d_2,\ldots,d_n)$, $\Delta D=\mathrm{diag}(\Delta d_1,\Delta d_2,\ldots,\Delta d_n)$, $A=(a_{kj})_{n\times n}$, $\Delta A=(\Delta a_{kj})_{n\times n}$, $B=(b_{kj})_{n\times n}$, $\Delta B=(\Delta b_{kj})_{n\times n}$, $C=(c_{kj})_{n\times n}$, $\Delta C=(\Delta c_{kj})_{n\times n}$, $F'(u(t))=(f_1'(u_1),f_2'(u_2),\cdots,f_n'(u_n))^T$, $G'(u(t-\zeta_1))=(g_1'(u_1(t-\zeta_1)),g_2'(u_2(t-\zeta_1)),\cdots,g_n'(u_n(t-\zeta_1)))^T$, $H'(u(s))=(h_1'(u_1(s)),h_2'(u_2(s)),\cdots,h_n'(u_n(s)))^T$, $\mathcal{I}=(\mathcal{I}_1,\cdots,\mathcal{I}_n)^T$.

Assumption 1. The activation functions are assumed to be continuous differentiable and bounded monotonically increasing and satisfy the follow assumption: $\forall s_1,s_2\in\mathfrak{R}, s_1\neq s_2$, $k=1,2,\cdots,n$,

$$\begin{cases} l_k^- \leq \frac{f_k'(s_1)-f_k'(s_2)}{s_1-s_2} \leq l_k^+, \\ \alpha_k^- \leq \frac{g_k'(s_1)-g_k'(s_2)}{s_1-s_2} \leq \alpha_k^+, \\ \upsilon_k^- \leq \frac{h_k'(s_1)-h_k'(s_2)}{s_1-s_2} \leq \upsilon_k^+. \end{cases} \tag{3}$$

In addition, we assume that u^* is an equilibrium point of neural networks (1). Then by letting $x(t) = u(t) - u^*$, we get that

$$\begin{aligned} \frac{dx(t)}{dt} &= -(D + \Delta D)x(t) + (A + \Delta A)F(x(t)) + (B + \Delta B)G(x(t - \zeta_1)) \\ &\quad +(C + \Delta C) \int_{t-\zeta_2}^{t} H(x(s))ds, \end{aligned} \tag{4}$$

where $F(x(\cdot)) = (f_1(x_1(\cdot)), f_2(x_2(\cdot)), \cdots, f_n(x_n(\cdot)))^T$, $G(x(\cdot)) = (g_1(x_1(\cdot)), \cdots, g_n(x_n(\cdot)))^T$, $H(x(\cdot)) = (h_1(x_1(\cdot)), h_2(x_2(\cdot)), \cdots, h_n(x_n(\cdot)))^T$, and $f_k(x_k(\cdot)) = f_k'(x_k(\cdot)+u_k^*) - f_k'(u_k^*)$, $g_k(x_k(\cdot)) = g_k'(x_k(\cdot)+u_k^*) - g_k'(u_k^*)$, $h_k(x_k(\cdot)) = h_k'(x_k(\cdot)+u_k^*) - h_k'(u_k^*)$. Then from (3), $\forall s_1, s_2 \in \Re, s_1 \neq s_2$,

$$\begin{cases} l_k^- \leq \frac{f_k(s_1)-f_k(s_2)}{s_1-s_2} \leq l_k^+, \\ \alpha_k^- \leq \frac{g_k(s_1)-g_k(s_2)}{s_1-s_2} \leq \alpha_k^+, \\ \upsilon_k^- \leq \frac{h_k(s_1)-h_k(s_2)}{s_1-s_2} \leq \upsilon_k^+. \end{cases} \tag{5}$$

So the switched RNNs with discrete and distributed delays under uncertainty can be described as

$$\begin{aligned} \dot{x}(t) = &-(D_i + \Delta D_i)x(t) + (A_i + \Delta A_i)F(x(t)) \\ &+(B_i + \Delta B_i)G(x(t - \zeta_1)) + (C_i + \Delta C_i) \int_{t-\zeta_2}^{t} H(x(s))ds. \end{aligned} \tag{6}$$

where $i = 1, 2, \ldots, N$, N is the number of subsystem in the switched RNNs.

By assumption, the origin point $(0, 0, \cdots, 0)^T$ is an equilibrium point of system (6), define the indicator function $\xi(t) = (\xi_1(t), \xi_2(t), \cdots, \xi_N(t))^T$,

$$\xi_i(t) = \begin{cases} 1 & \text{when the switched system is described by the } i-\text{th mode}, \\ 0 & \text{otherwise} \end{cases} \tag{7}$$

and $\sum_{i=1}^{N} \xi_i(t) = 1$. So (6) can be rewritten as

$$\begin{aligned} \dot{x}(t) = \sum_{i=1}^{N} \xi_i(t)\big[&-(D_i + \Delta D_i)x(t) + (A_i + \Delta A_i)F(x(t)) \\ &+(B_i + \Delta B_i)G(x(t - \zeta_1)) + (C_i + \Delta C_i) \int_{t-\zeta_2}^{t} H(x(s))ds\big]. \end{aligned} \tag{8}$$

Denote

$$\mathfrak{L}_1 = diag\left(\iota_1^- \iota_1^+, \iota_2^- \iota_2^+, \cdots, \iota_n^- \iota_n^+\right),$$

$$\mathfrak{L}_2 = diag\left(\frac{\iota_1^- + \iota_1^+}{2}, \frac{\iota_2^- + \iota_2^+}{2}, \cdots, \frac{\iota_n^- + \iota_n^+}{2}\right),$$

$$\mathfrak{A}_1 = diag\left(\alpha_1^-\alpha_1^+, \alpha_2^-\alpha_2^+, \cdots, \alpha_n^-\alpha_n^+\right),$$

$$\mathfrak{A}_2 = diag\left(\frac{\alpha_1^- + \alpha_1^+}{2}, \frac{\alpha_2^- + \alpha_2^+}{2}, \cdots, \frac{\alpha_n^- + \alpha_n^+}{2}\right),$$

$$\mathfrak{V}_1 = diag\left(\upsilon_1^-\upsilon_1^+, \upsilon_2^-\upsilon_2^+, \cdots, \upsilon_n^-\upsilon_n^+\right),$$

$$\mathfrak{V}_2 = diag\left(\frac{\upsilon_1^- + \upsilon_1^+}{2}, \frac{\upsilon_2^- + \upsilon_2^+}{2}, \cdots, \frac{\upsilon_n^- + \upsilon_n^+}{2}\right).$$

Assumption 2. The parametric uncertainties $\Delta D_i(t), \Delta A_i(t), \Delta B_i(t), \Delta C_i(t)$ are time-variant and unknown, but norm bounded and they are in the following form $[\Delta D_i(t), \Delta A_i(t), \Delta B_i(t), \Delta C_i(t)] = MK(t)[E_i^D, E_i^A, E_i^B, E_i^C]$, and the uncertain matrix $K^T(t)K(t) \leq \mathrm{I}, \forall t \in \mathfrak{R}$, where I is a unit matrix.

Let the initial conditions associated with (6) be in the following form

$$x(s) = \phi(s), \forall s \in [-\zeta^*, 0], \zeta^* = \max[\zeta_1, \zeta_2], \tag{9}$$

where $\phi(s)$ is a continuous real-valued function.

If there exist $\alpha > 0, \beta > 0$, and each solution $x(t)$ of (6) satisfies

$$|x(t)| \leq \beta e^{-\alpha t} \sup_{s \in [-\zeta^*, 0]} |\phi(s)|, \forall t > 0, \tag{10}$$

then (6) can be said to be globally exponentially stable.

3 Main Results

The following lemmas are especially important in establishing our results.

Lemma 1 [9]. Given any matrices X, Y and Λ with appropriate dimensions such that $0 < \Lambda = \Lambda^T$ and we can obtain that

$$X^TY + Y^TX \leq X^T\Lambda X + Y^{\mathrm{T}}\Lambda^{-1}Y. \tag{11}$$

Lemma 2 [10]. For any symmetric positive definite matrix $Z > 0$, scalar $\gamma > 0$, vector function $\mu : [0, \gamma] \rightarrow \mathfrak{R}^n$, we have:

$$\left(\int_0^\gamma \mu(s)ds\right)^{\mathrm{T}} Z \left(\int_0^\gamma \mu(s)ds\right) \leq \gamma \left(\int_0^\gamma \mu^T(s)Z\mu(s)ds\right). \tag{12}$$

Lemma 3 [11]. The matrix

$$Q = \begin{bmatrix} Q_{11} & Q_{12} \\ Q_{12}^T & Q_{22} \end{bmatrix} < 0$$

is equivalent to one of the following conditions: 1) $Q_{11} < 0, Q_{22} - Q_{12}^T Q_{11}^{-1} Q_{12} < 0$, 2) $Q_{22} < 0, Q_{11} - Q_{12} Q_{22}^{-1} Q_{12}^T < 0$, where $Q_{11} = Q_{11}^T$, $Q_{12} = Q_{12}^T$, $Q_{22} = Q_{22}^T$.

Theorem 1. Assume the activation functions $F(x(t))$, $F(x(t))$, $H(x(t))$ and the parametric uncertainties satisfy Assumptions 1 and 2, ζ_1 and ζ_2 denote the discrete time delay and the distributed delay, $\xi_0 (0 < \xi_0 < 1)$ is a constant. The switched RNN (6) is globally exponentially stable if there exist $h > 0, m > 0, \eta > 0, \sigma > 0$, three symmetric positive definite matrices P_1, P_2, P_3, three diagonal matrices

$$\Lambda = diag\,(\lambda_1, \lambda_2, \cdots, \lambda_n) > 0,$$
$$\Gamma = diag\,(\gamma_1, \gamma_2, \cdots, \gamma_n) > 0,$$
$$\mathfrak{O} = diag\,(\delta_1, \delta_2, \cdots, \delta_n) > 0,$$

such that for $i = 1, 2, \ldots, N$, the following LMIs hold

$$\mathfrak{D}_i = \begin{pmatrix} \Theta_i & P_1 A_i + \Lambda \mathfrak{L}_2 & \Gamma \mathfrak{A}_2 & P_1 B_i & \mathfrak{O}\mathfrak{B}_2 & P_1 C_i \\ A_i^T P_1 + \Lambda \mathfrak{L}_2 & \mathfrak{D}_i^{(2)} & 0 & 0 & 0 & 0 \\ \Gamma \mathfrak{A}_2 & 0 & \mathfrak{D}_i^{(3)} & 0 & 0 & 0 \\ B_i^T P_1 & 0 & 0 & \mathfrak{D}_i^{(4)} & 0 & 0 \\ \mathfrak{O}\mathfrak{B}_2 & 0 & 0 & 0 & \mathfrak{D}_i^{(5)} & 0 \\ C_i^T P_1 & 0 & 0 & 0 & 0 & \mathfrak{D}_i^{(6)} \end{pmatrix} < 0,$$

where $\Theta_i = -P_1 D_i - D_i P_1 - \Lambda \mathfrak{L}_1 - \Gamma \mathfrak{A}_1 - \mathfrak{O}\mathfrak{B}_1 + (\sigma + \eta + h^{-1} + m^{-1}) P_1 (M_i)^T M_i P_1 + \sigma^{-1} (E_i^D)^T E_i^D$, $\mathfrak{D}_i^{(2)} = \eta^{-1} (E_i^A)^T E_i^A - \Lambda$, $\mathfrak{D}_i^{(3)} = (1 + \xi_0 \zeta_1) P_2 - \Gamma$, $\mathfrak{D}_i^{(4)} = h (E_i^B)^T E_i^B - P_2$, $\mathfrak{D}_i^{(5)} = \zeta_2 P_3 - \mathfrak{O}$, $\mathfrak{D}_i^{(6)} = m (E_i^c)^T E_i^c - \frac{1 - \xi_0}{\zeta_2} P_3$.

Proof. Define a Lyapunov-Krasovskii function candidate

$$\Pi(t) = e^{\beta t} V(t), \quad V(t) = \sum_{i=1}^{4} V_i(t), \tag{13}$$

$$V_1(t) = x^T(t) P_1 x(t), \qquad V_2(t) = \int_{t-\zeta_1}^{t} \left(G(x(s))\right)^T P_2 G(x(s)) ds,$$

$$V_3(t) = \xi_0 \int_0^{\zeta_1} \int_{t-s}^{t} \left(G(x(\eta))\right)^T P_2 G(x(\eta)) d\eta ds,$$

$$V_4(t) = \int_0^{\zeta_2} \int_{t-s}^{t} \left(H(x(\eta))\right)^T P_3 H(x(\eta)) d\eta ds.$$

Then we can get

$$\dot{V}_1(t) = 2(x(t))^T P_1 \sum_{i=1}^{N} \xi_i(t)\Big[-(D_i+\Delta D_i)x(t)+(A_i+\Delta A_i)F(x(t)) \\ +(B_i+\Delta B_i)G(x(t-\zeta_1))+(C_i+\Delta C_i)\int_{t-\zeta_2}^{t} H(x(s))ds\Big], \quad (14)$$

$$\dot{V}_2(t) = \big(G(x(t))\big)^T P_2 G(x(t)) - \big(G(x(t-\zeta_1))\big)^T P_2 G(x(t-\zeta_1)), \quad (15)$$

$$\dot{V}_3(t) = \xi_0\zeta_1\big(G(x(t))\big)^T P_2 G(x(t)) - \xi_0 \int_{t-\zeta_1}^{t} \big(G(x(s))\big)^T P_2 G(x(s))ds, \quad (16)$$

$$\dot{V}_4(t) \le \zeta_2\big(H(x(t))\big)^T P_3 H(x(t)) - \frac{1-\xi_0}{\zeta_2}\Big(\int_{t-\zeta_2}^{t} H(x(s))ds\Big)^T \\ \times P_3\Big(\int_{t-\zeta_2}^{t} H(x(s))ds\Big) - \xi_0\int_{t-\zeta_2}^{t}\big(H(x(s))\big)^T P_3 H(x(s))ds. \quad (17)$$

According to Lemma 1,

$$-P_1\Delta D_i - \Delta D_i P_1 \le \sigma P_1 M M^T P_1 + \sigma^{-1}\big(E_i^D\big)^T E_i^D, \quad (18)$$

$$\big(x(t)\big)^T (P_1\Delta A_i)\big(F(x(t))\big)^T + F(x(t))(\Delta {A_i}^T P_1)x(t) \\ \le \eta\big(x(t)\big)^T (P_1 M M^T P_1)x(t) + \eta^{-1}\big(F(x(t))\big)^T \big(E_i^A\big)^T E_i^A F(x(t)), \quad (19)$$

$$G(x(t-\zeta_1))^T(\Delta {B_i}^T P_1)x(t) + \big(x(t)\big)^T(P_1\Delta B_i)G(x(t-\zeta_1)) \\ \le h\big(G(x(t-\zeta_1))\big)^T(E_i^B)^T E_i^B G(x(t-\zeta_1)) + h^{-1}\big(x(t)\big)^T P_1 M M^T P_1 x(t) \quad (20)$$

$$\int_{t-\zeta_2}^{t}\big(H(x(s))\big)^T ds(\Delta {C_i}^T P_1)X(t) + \big(x(t)\big)^T(P_1\Delta C_i)\int_{t-\zeta_2}^{t} H(x(s))ds \\ \le m\int_{t-\zeta_2}^{t}\big(H(x(s))\big)^T ds(E_i^C)^T E_i^C \int_{t-\zeta_2}^{t} H(x(s))ds \\ + m^{-1}\big(x(t)\big)^T P_1 M M^T P_1 x(t). \quad (21)$$

From (7), (14), (15), (16), (17), (18), (19), (20), (21),

$$\dot{V}_1(t)+\dot{V}_2(t)+\dot{V}_3(t)+\dot{V}_4(t) \\ \le \sum_{i=1}^{N}\xi_i(t)\Big[\big(X(t)\big)^T \Xi_i X(t) - \xi_0\int_{t-\zeta_1}^{t} G(x(s))^T P_2 G(x(s))ds \\ - \xi_0\int_{t-\zeta_2}^{t}\big(H(x(s))\big)^T P_3 H(x(s))ds\Big], \quad (22)$$

where $X(t) = \Big((x(t))^T, (F(x(t)))^T, (G(x(t)))^T, (G(x(t-\zeta_1)))^T, (H(x(t)))^T, \int_{t-\zeta_2}^t (H(x(s)))^T ds\Big)^T$,

$$\Xi_i = \begin{pmatrix} \Omega_i & P_1A_i & 0 & P_1B_i & 0 & P_1C_i \\ A_i^TP_1 & \eta^{-1}(E_i^A)^TE_i^A & 0 & 0 & 0 & 0 \\ 0 & 0 & (1+\xi_0\zeta_1)P_2 & 0 & 0 & 0 \\ B_i^TP_1 & 0 & 0 & h(E_i^B)^TE_i^B - P_2 & 0 & 0 \\ 0 & 0 & 0 & 0 & \zeta_2P_3 & 0 \\ C_i^TP_1 & 0 & 0 & 0 & 0 & \Xi_i^{(6)} \end{pmatrix}, \quad (23)$$

$\Omega_i = -P_1D_i - D_iP_1 + (\sigma+\eta+h^{-1}+m^{-1})P_1M_i^TM_iP_1 + \sigma^{-1}(E_i^D)^TE_i^D$, $\Xi_i^{(6)} = m(E_i^c)^TE_i^c - \frac{1-\xi_0}{\zeta_2}P_3$. From (5), for $k = 1, 2, \cdots, n$,

$$\Big(\frac{f_k(x_k(t))}{x_k(t)} - \iota_k^-\Big)\Big(\frac{f_k(x_k(t))}{x_k(t)} - \iota_k^+\Big) \le 0,$$
$$\Big(\frac{g_k(x_k(t))}{x_k(t)} - \alpha_k^-\Big)\Big(\frac{f_k(x_k(t))}{x_k(t)} - \alpha_k^+\Big) \le 0,$$
$$\Big(\frac{h_k(x_k(t))}{x_k(t)} - \nu_k^-\Big)\Big(\frac{h_k(x_k(t))}{x_k(t)} - \nu_k^+\Big) \le 0,$$

which mean that for $k = 1, 2, \cdots, n$,

$$\begin{bmatrix} x(t) \\ F(x(t)) \end{bmatrix}^T \begin{bmatrix} l_k^-l_k^+e_ke_k^T & -\frac{l_k^-+l_k^+}{2}e_ke_k^T \\ -\frac{l_k^-+l_k^+}{2}e_ke_k^T & e_ke_k^T \end{bmatrix} \begin{bmatrix} x(t) \\ F(x(t)) \end{bmatrix} \le 0, \quad (24)$$

$$\begin{bmatrix} x(t) \\ H(x(t)) \end{bmatrix}^T \begin{bmatrix} \upsilon_k^-\upsilon_k^+e_ke_k^T & -\frac{\upsilon_k^-+\upsilon_k^+}{2}e_ke_k^T \\ -\frac{\upsilon_k^-+\upsilon_k^+}{2}e_ke_k^T & e_ke_k^T \end{bmatrix} \begin{bmatrix} x(t) \\ H(x(t)) \end{bmatrix} \le 0, \quad (25)$$

$$\begin{bmatrix} x(t) \\ G(x(t)) \end{bmatrix}^T \begin{bmatrix} \alpha_k^-\alpha_k^+e_ke_k^T & -\frac{\alpha_k^-+\alpha_k^+}{2}e_ke_k^T \\ -\frac{a_k^-+a_k^+}{2}e_ke_k^T & e_ke_k^T \end{bmatrix} \begin{bmatrix} x(t) \\ G(x(t)) \end{bmatrix} \le 0, \quad (26)$$

where $e_k = (e_k^{(1)}, e_k^{(2)}, \cdots, e_k^{(n)})^T$, $e_k^{(i)} = \begin{cases} 0, & i \ne k, \\ 1, & i = k. \end{cases}$ So

$$\sum_{k=1}^n \lambda_k \begin{bmatrix} x(t) \\ F(x(t)) \end{bmatrix}^T \begin{bmatrix} \iota_k^-\iota_k^+e_ke_k^T & -\frac{\iota_k^-+\iota_k^+}{2}e_ke_k^T \\ -\frac{\iota_k^-+\iota_k^+}{2}e_ke_k^T & e_ke_k^T \end{bmatrix} \begin{bmatrix} x(t) \\ F(x(t)) \end{bmatrix} \le 0, \quad (27)$$

$$\sum_{k=1}^n \gamma_k \begin{bmatrix} x(t) \\ G(x(t)) \end{bmatrix}^T \begin{bmatrix} \alpha_k^-\alpha_k^+e_ke_k^T & -\frac{\alpha_k^-+\alpha_k^+}{2}e_ke_k^T \\ -\frac{\alpha_k^-+\alpha_k^+}{2}e_ke_k^T & e_ke_k^T \end{bmatrix} \begin{bmatrix} x(t) \\ G(x(t)) \end{bmatrix} \le 0, \quad (28)$$

$$\sum_{k=1}^n \delta_k \begin{bmatrix} x(t) \\ H(x(t)) \end{bmatrix}^T \begin{bmatrix} \nu_k^-\nu_k^+e_ke_k^T & -\frac{\nu_k^-+\nu_k^+}{2}e_ke_k^T \\ -\frac{\nu_k^-+\nu_k^+}{2}e_ke_k^T & e_ke_k^T \end{bmatrix} \begin{bmatrix} x(t) \\ H(x(t)) \end{bmatrix} \le 0. \quad (29)$$

From (27), (28) and (29),

$$\left(X(t)\right)^T \Xi_i X(t) \leq \left(X(t)\right)^T \mathfrak{D}_i X(t) \leq \lambda_{\max}\{\mathfrak{D}_i\}|X(t)|^2 \leq 0. \tag{30}$$

Hence,

$$\begin{aligned}\dot{\Pi}(t) \leq& \sum_{i=1}^{N} \xi_i(t)e^{\beta t}\Big[\beta\sum_{i=1}^{4} V_i(t) + \left(X(t)\right)^T \mathfrak{D}_i X(t) - \\ &\xi_0 \int_{t-\zeta_1}^{t} G(x(s))^T P_2 G(x(s))ds - \xi_0 \int_{t-\zeta_2}^{t} \left(H(x(s))\right)^T P_3 H(x(s))ds\Big]\end{aligned} \tag{31}$$

In addition,

$$V_1(t) \leq \lambda_{\max}\{P_1\}|x(t)|^2, \tag{32}$$

$$V_3(t) \leq \xi_0\zeta_1 \int_{t-\zeta_1}^{t} G(x(s))^T P_2 G(x(s))ds, \tag{33}$$

$$V_4(t) \leq \zeta_2 \int_{t-\zeta_2}^{t} \left(H(x(s))\right)^T P_3 H(x(s))ds. \tag{34}$$

$$\begin{aligned}\sum_{i=1}^{4} V_i(t) \leq& \lambda_{\max}\{P_1\}|x(t)|^2 + (1+\xi_0\zeta_1)\int_{t-\zeta_1}^{t} G(x(s))^T P_2 G(x(s))ds \\ &+\zeta_2 \int_{t-\zeta_2}^{t} \left(H(x(s))\right)^T P_3 H(x(s))ds.\end{aligned} \tag{35}$$

So

$$\begin{aligned}\dot{\Pi}(t) \leq& \sum_{i=1}^{N} \xi_i(t)e^{\beta t}\Big[\beta\lambda_{\max}\{P_1\} + \lambda_{\max}\{\mathfrak{D}_i\}|x(t)|^2 \\ &+(\beta(1+\xi_0\zeta_1)-\xi_0)\int_{t-\zeta_1}^{t} G(x(s))^{\mathrm{T}} P_2 G(x(s))ds \\ &+(\beta\zeta_2-\xi_0)\int_{t-\zeta_2}^{t} \left(H(x(s))\right)^T P_3 H(x(s))ds\Big].\end{aligned} \tag{36}$$

Let $\beta_0 = \min\left\{-\frac{\lambda_{\max}\{\mathfrak{D}_i\}}{\lambda_{\max}\{P_1\}}, \frac{\xi_0}{1+\xi_0\zeta_1}, \frac{\xi_0}{\zeta_2}\right\}$ and $\beta \leq \beta_0$. Then $\dot{\Pi}(t) \leq 0$. We set that $\varpi_k = \max\{|\iota_k^-|, |\iota_k^+|\}$, and $\varpi = (\varpi_1, \varpi_2, \cdots, \varpi_n)^T$, $\partial_k = \max\{|\iota_k^-|, |\iota_k^+|\}$. And $\partial = (\partial_1, \partial_2, \cdots, \partial_n)^T$, $k = 1, 2, \ldots n$. We can get that

$$e^{\beta t}V(t) \leq V(0) \leq M_0 \sup_{-\zeta^* \leq s \leq 0} |x(s)|^2, \tag{37}$$

where $M_0 = \lambda_{\max}\{P_1\} + (1+\xi_0\zeta_1)\zeta_1\lambda_{\max}\{P_2\}\partial^T\partial + {\zeta_2}^2\lambda_{\max}\{P_3\}\varpi^T\varpi > 0$,

$$V(t) \leq M_0 e^{-\beta t} \sup_{-\zeta^* \leq s \leq 0} |x(s)|^2. \tag{38}$$

In addition,

$$V(t) \geq V_1(t) \geq \lambda_{\min}\{P_1\}|x(t)|^2 \tag{39}$$

then

$$|x(t)|^2 \leq \frac{V(t)}{\lambda_{\min}\{P_1\}} \leq \frac{M_0}{\lambda_{\min}\{P_1\}} e^{-\beta t} \sup_{-\zeta^* \leq s \leq 0} |x(s)|^2. \tag{40}$$

It means that the switched system of RNNs (6) is globally exponentially stable for all admissible uncertainties, the proof is completed.

4 An Illustrative Example

In this part, we will present a simple example to demonstrate the results derived above. Let $n = N = 2$, consider the following switched RNNs with discrete and ditributed delays under uncertainties

$$\begin{aligned} \dot{x}(t) = & \sum_{i=1}^{2} \xi_i(t)[-(D_i + \Delta D_i)x(t) + (A_i + \Delta A_i)F(x(t)) \\ & +(B_i + \Delta B_i)G(x(t-\zeta_1)) + (C_i + \Delta C_i)\int_{t-\zeta_2}^{t} H(x(s))ds]. \end{aligned} \tag{41}$$

where $\zeta_1 = 0.5, \zeta_2 = 0.8, \xi_0 = 0.5$, $f_1(x) = f_2(x) = \tan h(2x) = g_1(x) = g_2(x) = h_1(x) = h_2(x)$. So we can get that $\mathfrak{L}_1 = \mathfrak{A}_1 = \mathfrak{V}_1 = 0$, $\mathfrak{L}_2 = \mathfrak{A}_2 = \mathfrak{V}_2 = \begin{pmatrix} 1 & 0 \\ 0 & 1 \end{pmatrix}$. Let

$$M_1 = M_2 = \begin{pmatrix} 1 & 0 \\ 0 & 1 \end{pmatrix} \quad E_1^A = E_2^A = \begin{pmatrix} 1 & 0 \\ 0 & 1 \end{pmatrix}$$

$$E_1^C = E_2^C = \begin{pmatrix} -0.5 & 0 \\ 0 & 0.5 \end{pmatrix} \quad E_1^B = E_2^B = \begin{pmatrix} -0.5 & 0 \\ 0 & -0.5 \end{pmatrix} \quad E_1^A = E_2^A = \begin{pmatrix} 1 & 0 \\ 0 & -1 \end{pmatrix}$$

$$A_1 = \begin{pmatrix} -5 & 0 \\ 0 & -5 \end{pmatrix} \quad A_2 = \begin{pmatrix} -4 & 0 \\ 0 & -4 \end{pmatrix} \quad B_1 = \begin{pmatrix} 7 & 0 \\ 0 & 8 \end{pmatrix} \quad B_2 = \begin{pmatrix} 1 & 0 \\ 0 & 9 \end{pmatrix}$$

$$C_1 = \begin{pmatrix} -3 & 0 \\ 0 & 6 \end{pmatrix} \quad C_2 = \begin{pmatrix} 2 & 0 \\ 0 & 5 \end{pmatrix} \quad D_1 = \begin{pmatrix} 3 & 0 \\ 0 & 3 \end{pmatrix} \quad D_2 = \begin{pmatrix} 4 & 0 \\ 0 & 4 \end{pmatrix}.$$

We can obtain that $P_1 = P_2 = P_3 = I$, which make the matrix in Theorem 1 negative, then the switched RNNs (41) is globally exponentially stable.

5 Concluding Remarks

In this letter, a class of switched RNNs has been studied by integrating the theory of switched systems and neural networks. The mathematical model of the switched RNNs has been proposed and the globally exponential stability for the switched RNNs, by using the Lyapunov-Krasovskii approach, has been addressed under an arbitrary switching rule. Stability criteria have been derived in terms of a set of strict LMIs. The results about the switched RNNs we proposed can be used more widely than some delivered before it.

Acknowledgments. This work was supported by the Natural Science Foundation of China under Grant 60774051, Program for New Century Excellent Talents in Universities of China under Grant NCET-06-0658, the Fok Ying Tung Education Foundation under Grant 111068 and Major State Basic Research Development Program of China under Grant 2007CB311000.

References

1. Liu, Y., Wang, Z.D., Liu, X.H.: Global Exponential Stability of Generalized Recurrent Neural Networks with Discrete and Distributed Delays. Neural Networks 19, 667–675 (2006)
2. Mori, H.: Fuzzy Neural Network Applications to Power Systems. IEEE Powerengineering Society Winter Meeting 2, 1284–1288 (2000)
3. Riehle, D., Zullighoven, H.: Understanding and Using Patterns in Software Development. Theory and Practice of Object Systems 2, 3–13 (1996)
4. Huang, H., Qu, Y.Z., Li, H.X.: Robust Stability Analysis of Switched Hopfield Neural Networks with Time-varying Delay under Uncertainty. Physics Letters A 345, 345–354 (2005)
5. YfouLis, C.A., Shorten, R.: A Numerical Technique for the Stability Analysis of Linear Switched Systems. Int. J. Control 77, 1019–1039 (2004)
6. Liberzon, D., Morse, A.S.: Basic Problems in Stability and Design of Switched Systems. IEEE Cont. Sys. Mag. 19, 59–70 (1999)
7. Zhai, G., Kondo, H.: Hybrid Static Output Feedback Stabilization of Two-dimensional LTI Systems: Geometric Method. International Journal of Control 79, 982–990 (2006)
8. Ho, S.L., Xie, M., Tang, L.C., Xu, K., Goh, T.N.: Neural Network Modeling with Confidence Bounds: A Case Study on Thesolder Paste Deposition Process. IEEE Trans. Electronics Packaging Manufacturing 24, 323–332 (2001)
9. Sanchez, E.N., Perez, J.P.: Input-to-state Stability (ISS) Analysis for Dynamic Neural Networks. IEEE Trans. Circuits Systems 46, 1395–1398 (1999)
10. Gu, K.: An Integral Inequality in the Stability Problem of Time-delay Systems. In: Proceedings of 39th IEEE conference on decision and control, pp. 2805–2810. IEEE Press, New York (2000)
11. Boyd, S., Ghaoui, L.E., Feron, E., Balakrishnan, V.: Linear Matrix Inequalities in Systems and Control Theory. SLAM, Philadephia (1994)

WHFPMiner: Efficient Mining of Weighted Highly-Correlated Frequent Patterns Based on Weighted FP-Tree Approach

Runian Geng[1,2], Xiangjun Dong[2], Jing Zhao[1,2], and Wenbo Xu[1]

[1] School of Information Technology, Jiangnan University,
[2] School of Information Science and Technology, Shandong Institute of Light Industry
gengrnn@163.com, d-xj@163.com, lmzjmail@163.com,
xwb_sytu@hotmail.com

Abstract. Most algorithms for frequent pattern mining use a support-based pruning strategy to prune a combinatorial search space. However, they are not effective for finding correlated patterns with similar levels of support. In additional, traditional patterns mining algorithms rarely consider weighted pattern mining. In this paper, we present a new algorithm, *WHFPMiner* (Weighted Highly-correlated Frequent Patterns Miner) in which a new objective measure, called weighted h-confidence, is developed to mine weighted highly-correlated frequent patterns with similar levels of weighted support. Adopting an improved weighted FP-tree structure, this algorithm exploits both cross-weighted support and anti-monotone properties of the weighted h-confidence measure for the efficient discovery of weighted hyperclique patterns. A comprehensive performance study shows that *WHFPMiner* is efficient and fast for finding weighted highly-correlated frequent patterns. Moreover, it generates fewer but more valuable patterns with the high correlation.

Keywords: Data mining, Weighted h-confidence, Highly-correlated, Frequent pattern, FP-tree.

1 Introduction

Extensive growth of data gives the motivation to find meaningful patterns from the huge data. Frequent patterns mining has played an essential role in many applications mining association rules, correlations, sequences, graph patterns, closed frequent patterns and so on. Most previous algorithms use a support-based pruning strategy to prune the combinatorial search space. However, this strategy provides basic pruning methods but the resulting patterns usually have weak correlation after mining process whether the support threshold is low or high. In real life, we regularly mine patters from huge data with different importance (e.g., retail data, Web click-streams etc). To reflect the difference of data importance, we can assign a weight to each data. This mining method from dataset assigned weights is called weight constraint-based patterns mining. Constraint-based pattern mining can reduce the number of uninteresting patterns but it is not useful to detect patterns with the strong or weak affinity. For example, in real business, marketing managers would like to be interested in finding

F. Sun et al. (Eds.): ISNN 2008, Part II, LNCS 5264, pp. 730–739, 2008.

itemsets with similar levels of total profits (multiplying the profit of an item with the sale number of the item) or similar levels of total selling expenses. However, previous mining approaches could not detect this weak or strong affinity patterns.

In this paper, we propose an efficient mining algorithm called *WHFPMiner* (Weighted Highly-correlated Frequent Patterns Miner). The main goal for our work is to discover high affinity patterns with the similar levels of weighted support which cannot be detected by the conventional frequent pattern mining approaches, and remove spurious patterns with substantially different weighted support levels. In *WHFPMiner*, we define the concept of the weighted hyperclique pattern that uses a new objective measure, called weighted h-confidence (*whconf*), to detect correlated patterns with the weighted support affinity and prevent the generation of patterns with different weighted support levels. Moreover, *WHFPMiner* adopts an improved weighted FP-tree approach to more efficiently discover weighted support affinity patterns. A comprehensive performance analysis shows that *WHFPMiner* is efficient and fast for finding strong-correlation patterns.

The remainder of this paper is organized as follows. Section 2 reviews previous works. The related definitions are given in Section 3. Section 4 proposes the *WHFPMiner* algorithm. Experimental research and analysis of algorithm are reported in Section 5. Finally, Section 6 concludes the paper.

2 Related Works

There have been recent studies [1,2] to mine frequent patterns without support thresholds but the resulting patterns include weak affinity patterns. To mine correlated patterns, interesting measures [3,4] have been proposed, but they do not consider importance (weight) of the pattern. Most of previous weight constraint mining algorithms are related to the mining of association rules and frequent itemsets [5,6], and they are based on the Apriori algorithm [7] which has a poor performance. Although weight constraint-based patterns mining may greatly reduce the number of generated patterns and improve the computational performance, correlation patterns are not detected in the result patterns, i.e., they don not eliminate weakly-correlated patterns involving items with different support levels. Closed/maximal patterns mining algorithms are efficient for mining frequents patterns, especially for dense datasets [9,10]. However, the number of weakly-correlated patterns mined by them is still very large. Hyperclique Miner algorithm is an algorithm for mining hyperclique patterns involving items with a similar level of support [11]. However, it is based on the Apriori algorithm which is costly to generate and test all the candidates. In addition, Hyperclique Miner did not take weight into consideration.

3 Problem Statement

Let $I = \{i_1, i_2, ..., i_n\}$ be a unique set of items. A transaction database, TDB, is a set of transactions in which each transaction, denoted as a tuple <Tid, *X*>, contains a unique transaction identifier, Tid, and a set of items (i.e., *X*). A pattern *P* is called a *k*-pattern (denoted as P_k) if it contains *k* items, there *k* is the length of *P* (denoted as $|P|$). The support of itemset *Y* is the fraction of transactions in TDB containing itemset *P*, denoted as *supp*(*P*). Given a minimum support threshold, *minsup*, a pattern *P* is called frequent if *supp*(P) ≥ *minsup*. Support constraint is an *anti-monotone* constraint, i.e, if a pattern is infrequent pattern, all super patterns of the pattern must be infrequent patterns. Using the *anti-monotone* property, infrequent patterns can be pruned earlier.

3.1 Related Definitions of Weighted Pattern Mining

Definition 1 (*Weight of Pattern*). The weight of a pattern is an average value of weights of all items in it.

Given a weighted pattern $P=\{p_1,p_2,\dots,p_k\}$, $w(p_i)$ $(i=1,2...k)$ is the weight of each item in *P*, and the weight of *P* is represented as follows:

$$weight(P)=\sum_{i=1}^{k} w(p_i)\Big/|P| . \tag{1}$$

Definition 2 (*Weighted Support*). The weighted support of a pattern *P*, called *wsupp* (*P*), is defined as follows [6]:

$$wsupp(P) = weight(P)*(supp(P)). \tag{2}$$

Definition 3 (*Weighted Frequent*). A *P* is called a *WFP* (weighted frequent pattern) if *wsupp*(*P*) is no less than a given minimum weighted support threshold called *minwsup*, denoted as

$$wsupp(P) \geq minwsup . \tag{3}$$

Definition 4 (*h-confidence*). The h-confidence (denoted as *hconf*) of a pattern $P=\{p_1,p_2,\dots,p_k\}$ is defined as follows[13]:

$$hconf(P)=min\{conf\{p_1\rightarrow p_2,\cdots,p_k\},conf\{p_2\rightarrow p_1,p_3,\cdots,p_k\},\cdots,conf\{p_k\rightarrow p_1,\cdots,p_{k-1}\}\}, \tag{4}$$

There, $conf\{p_1\rightarrow p_2,\cdots,p_k\}=supp(P)/supp(\{p_1\})$, the other is similar to $conf(p_1\rightarrow p_2,\dots,p_k)$.

Inspired by the definition of *h-confidence* in [11], we devise the following definitions to be used to mine *WFP*s with similar levels of weighted support.

Definition 5 (*Weighted h-confidence*). The weighted h-confidence of a pattern $P=\{p_1,p_2,\dots,p_k\}$, denoted as *whconf*, is defined as follows:

$$\begin{aligned} whconf(P) &= min\{wsupp(P)/wsupp(\{p_1\}),wsupp(P)/wsupp(\{p_2\}),\cdots,wsupp(P)/wsupp(\{p_k\})\} \\ &= wsupp(P)/max\{wsupp_{1\le i\le k}(\{p_i\})\} \end{aligned} \tag{5}$$

Definition 6 (*Weighted Hyperclique Pattern*). Given a set of items $I=\{i_1, i_2,\dots, i_m\}$ and a minimum *whconf* threshold wh_c, a pattern $P\subseteq I$ is called a weighted hyperclique pattern if and only if $|P| > 0$ and $whconf(P)\geq wh_c$.

Definition 7 (*Cross-weighted Support Pattern*). Given a threshold *t*, a weighted pattern *P* is a cross-weighted support pattern with respect to *t* if *P* contains two items *x* and *y* such that $wsupp(\{x\})/wsupp(\{y\}) < t$, where $0 < t < 1$.

3.2 Revised Weighted Support

To let weighted support satisfy '*downward closure property*' so as to use this property to prune weighted infrequent patterns, we revise the representation of weighted support. We know that the weights of items in $I = \{i_1, i_2,\ldots, i_m\}$ containing m items , denoted as w_i $(i=1,2,\ldots,m)$, must satisfy: $\min(W) \le w_i \le \max(W)$, there, $W=\{w_1,w_2,\ldots,w_n\}$. To let weighted patterns satisfy the *anti-monotone* property (i.e., if *wsuppt* (*P*) < *minwsup* ⇒ *wsupport* (*P'*) < *minwsup*, there $P\subset P'$), we revise *weight* (p_i) as two following representations: $weight(p_i)=min(W)$ or $weight(p_i)=max(W)$. Accordingly, the weight of pattern is revised as two following representations.

$$weight(P)=\frac{\sum_{i=1}^{k} w(p_i)}{|P|}=\frac{\sum_{i=1}^{k} min(W)}{|P|}. \tag{6}$$

or

$$weight(P)=\frac{\sum_{i=1}^{k} w(p_i)}{|P|}=\frac{\sum_{i=1}^{k} max(W)}{|P|}. \tag{7}$$

However, if we adopt (6), we could prune some patterns which should have been weighted frequent to lead to incorrect mining results since we evaluate a lower the weight of pattern. To avoid this flaw, we adopt (7) to compute revised weight of the pattern However, the weighted support value computed by this is only an approximate value, in final step, we should check if the pattern is really a weighted frequent pattern with his real weight value, i.e., check if $\sum_{i=1}^{k} w(p_i)/k*(supp(P_k))\ge minwsup$.

After revising the weighted support, we can get the following property and lemma.

Lemma 1 (*Anti-monotone property of whconf*). Given a pattern *P* , its subset P'' , its super pattern P', and a minimum *whconf* threshold wh_c, if the *whconf*(*P*)≥wh_c, then *whconf*(P'')≥wh_c; if the *whconf*(*P*)<wh_c, then *whconf*(P')<wh_c,.

Proof: from (5), we know

$whconf(P)= wsupp(P)/max\{wsupp_{1\le i\le k}(\{p_i\})\}$, $whconf(P')= wsupp(P')/max\{wsupp_{1\le i\le j,j>k}(\{p_i\})\}$

$whconf(P'')= wsupp(P'')/max\{wsupp_{1\le i\le h,h<k}(\{p_i\})\}$. Because

$$max\{wsupp_{1\le i\le h,h<k}(\{p_i\})\}\le max\{wsupp_{1\le i\le k}(\{p_i\})\}\le max\{wsupp_{1\le i\le j,j>k}(\{p_i\})\} \tag{8}$$

By (8), $wsupp(P')\le wsupp(P)\le wsupp(P'')$ by (9) $\Rightarrow whconf(P')\le whconf(P)\le whconf(P'')$ $\Rightarrow if\ whconf(P)\le wh_c, whconf(P')\le wh_c; if\ whconf(P)\ge wh_c, whconf(P'')\ge wh_c$. □

Lemma 2 (*Cross-weighted support property of whconf*). Given a minimum *whconf* threshold *t*, for any cross-weighted support pattern *P* with respect to *t*, *whconf* (*P*) <*t*.

Proof: Since *P* is a cross-weighted support pattern with respect to *t*, by definition 8, we know *P* contains at least at least two items *x* and *y* such that *wsupp* ({*x*})/*wsupp* ({*y*}) < *t*. where 0<*t*<1. Without loss of generality, let *P*={…,*x*,…*y*,…}. By (6), we have the following:

$$whconf(P) = min\{\cdots, wsupp(P)/wsupp(x), \cdots, wsupp(P)/wsupp(y), \cdots\}$$
$$= wsupp(P)/max\{\cdots, wsupp(\{x\}), \cdots, wsupp(\{y\}), \cdots\} \le wsupp(\{x\})/wsupp(\{y\}) \le t. \quad \square$$

Corollary 1. Given an item *y*, all patterns that contain *y* and at least one item with weighted support less than $t*wsupp(\{y\})$ ($0<t<1$) are cross-weighted support patterns with respect to *t* and are guaranteed to have weighted h-confidence less than *t*.

Proof: This corollary follows from Definition 7 and Lemma 2. □

All such patterns can be automatically eliminated without computing their *whconf* if we are only interested in patterns with weighted h-confidence greater than *t*. Based on the above Lemmas and Corollary, we can push the weighted h-conference constraint into the mining algorithm and fast prune weak affinity patterns.

4 Mining Weighted Highly-Correlated Frequent Patterns

In this section, we define the weighted Highly-correlated Frequent Patterns and propose the *WHFPMiner* algorithm and its mining process.

Definition 8 (*Weighted Highly-correlated Frequent Patterns*). Given a set of items $I = \{i_1, i_2, \ldots, i_m\}$, a minimum *whconf* threshold wh_c, and a minimum weighted support threshold *minwsup*, a pattern $P \subseteq I$ is called a weighted highly-correlated frequent pattern if it satisfies: (1) $wsupp(P) \ge minsup$, and (2) $whconf(P) \ge wh_c$.

From above, we know the mining process contains two phases: one is the phase of mining *WFP*s; the other is that of mining weighted hyperclique patterns from the *WFP*s. As we described above, revised weighted setting has an *anti-monotone* property and *whconf* has the *anti-monotone* and *cross-weighted support property*. Base on the above property, we devise an efficient and scalable algorithm, called *WHFPMiner* which exploits a d*ivide-and-conquer* strategy to mine the weighted frequent patterns with highly-correlated affinity.

4.1 Weighted FP-Tree Construction

FP-tree is a compact representation of all relevant frequency information in a database [8]. To get a high performance, we adopt a weighted FP-tree as a compression structure based on pattern growth method. Each node in the weighted FP-tree has four fields: 'item-name', 'sup_count' 'weight' and 'node-link'. Additionally, for each weighted FP-tree, there is a header table which has there fields: 'item-id', 'support' and a 'headpointer'. Initially, a weighted FP-tree has only a root node. The weighted FP-trees in our algorithm are constructed as shown as Fig. 1.

Table 1 shows an example of a retail database, and Table 2 show the characteristics of each item in Table 1. Here, the attribute values such as prices (profits) of items can be used as weight factor and the prices of items are normalized within a specified weight range. For the TDB shown in table 1 with *minwsup*= 1.5, we can construct a global weighted FP-tree shown in Fig.2 by the method shown in Fig.1.

Table 1. Transaction database TDB

TID	Set of items
100	A,B,C,D
200	A,B,D
300	A,B,C,D, H
400	A,C, H
500	A,B,C,D
600	D,H
700	C,A
800	C,D
900	A,B
1000	H, C

Table 2. Example set of items with weight

Item	Price	Weight	Support	Wsupp
A	40$	4	0.7	2.8
B	60$	6	0.5	3.0
C	50$	5	0.7	3.5
D	30$	3	0.6	1.8
H	50$	5	0.4	2.0

1. First, scan the TDB once to identify the set of approximate weighted frequent item m that satisfies $max(W)*support(m) \geq minwsup$ and compute their corresponding support counts, and sort them by ***sup_count*** descending to form the set of ***f-list***, if two items' *sup_count* is equal, then sort them alphabetically.
2. Second, weighted infrequent items whose weighted supports are less than the *minwsup* are removed form ***f-list*** and each transaction.
3. Next, those remained weighted frequent items in each transaction are sorted by the ***sup_count*** descending order.
4. Lastly, scan the new TDB again and the sorted weighted frequent items in each transaction are sequentially inserted in the global weighted FP-tree along a path from the root to the corresponding node. If a new node is inserted in the path, the 'sup-count' field of this node is set to one. If an existing node in the FP-tree is met, the 'sup-count' of the node in the FP-tree is incremented by 1.
5. Stop when all the transactions are insert to the global weighted FP-tree.

Fig. 1. Process of constructing the weighted FP-tree

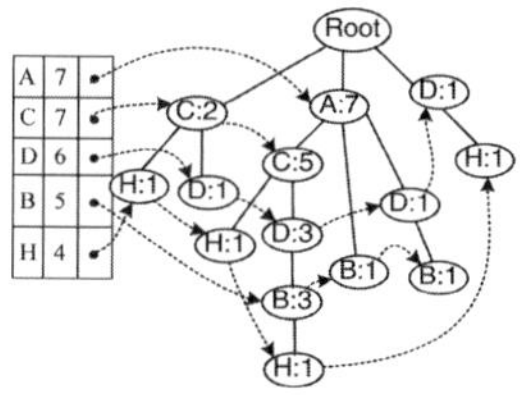

Fig. 2. The global weighted FP-tree

4.2 WHFPMiner Algorithm

In *WHFPMiner*, a *divide-and-conquer* traversal strategy is used to mine weighted FP-tree for mining weighted highly-correlated frequent patterns with a bottom-up manner. *WHFPMiner* algorithm is given in Fig. 3. Figure 4 gives the detail of procedure *WHFPM* (*Weighted FP-tree,*α*, WHFP*), in it, the set *WHFP* is used to store so far found real weighted highly-correlated frequent patterns.

WHFPMiner algorithm: mining ***W***eighted ***H***ighly-correlated ***F***requent ***P***atterns

Inputs: (1)A transaction database *TDB*, (2) minimum weighted support threshold *minwsup*, (3) minimum weighted h-confidence threshold wh_c, (4) weights of each item
Output: The complete set of weighted highly-correlated frequent patterns.
Method:
1. Scan *TDB* once to find the approximate weighted frequent items , and sort them by ***sup_count*** descending to form the set of ***f-list***. If two items' *sup_count* is equal, then sort them alphabetically.
2. Remove weighted infrequent items from ***f-list*** and each transaction. and sort the remaining weighted frequent items in each transaction by ***sup_count*** descending order.
3. Scan *TDB* again and build the global weighted FP-tree using order in ***f-list***.
4. Initialize the set of ***WHFP***. // ***WHFP*** is the set of weighted highly-correlated frequent patterns
5. Call ***WHFPM (Weighted FP-tree, null, WHFP)***.

Fig. 3. WHFPMiner algorithm

4.3 Bottom Up Traversal of Weighted FP-Tree with *Divide-and-Conquer* Scheme

WHFPMiner mines weighted hyperclique frequent patterns by adapting the *divide-and- conquer* approach from the global weighted FP-tree. For the TDB shown in Table 1 & 2 (*minwsup=1.5*, wh_c=0.8), it divides mining the FP-tree into mining smaller weighted FP-trees (i.e., each item's condition FP-tree) with *bottom up* traversal of the FP-tree, and mines first (1) the patterns suffixed by the item '*H*' and then (2) the patterns suffixed by the item '*B*',…and finally the patterns suffixed by the item '*A*'. According the algorithm *WHFPMiner*, the above mining process is shown in Table 3. Figure 5 is the conditional FP-tree of each node in head table *f_list*. After mining, we can get the final weighted hyperclique frequent patterns of each node in *f_list*. It is shown in Fig. 6. Cleary, from Table 1& 2, we can see the *wsupp* of 'A' and 'B' is more similar, the other items' *wsupp* have an obvious difference. This matches with the mined results shown in Fig.6.

Table 3. Process of mining weightes hyperclique frequent patterns from TDB in Tab.1

f_list_i	conditional DB	check #1	check #2	check #3	check #4	check #5
H:4	C:1; AC:1; ACDB:1; D:1	C:1; C:1; C:1 (prune A,D,B)	C:1; C:1; C:1	H:4; CH:3	H:4; CH:3 (*whconf*(CH)=1.5/3.5=0.43);	H:4
B:5	ACD:3; A:1; AD:1	ACD:3; A:1; AD:1	AD:3; A:1; AD:1 (prune "C")	B:5; DB:4; AB:5; ADB:4	B:5; DB:4(*whconf*(DB)=1.8/3=0.6); AB:5(*whconf*(AB)=2.5/3=0.83); ADB:4 (*whconf*(ADB)=1.73/3=0.58);	B:5; AB:5
D:6	C:1; AC:3; A:1	C:1; AC:3; A:1	C:1; AC:3; A:1	D:6; CD:4; ACD:4; AD:4	D:6; CD:4 (*whconf*(CD)=1.6/3.5=0.46); ACD:4 (*whconf*(ACD)=1.6/3.5=0.43); (prune {AD:4} wsupp(AD)=1.4);	D:6
C:7	A:5	A:5	A:5	C:7; AC:5	C:7; AC:5 (*whconf*(AC)=2.25/3.5=0.64)	C:7
A:7	∅	∅	∅	A:7	A:7	A:7

```
Procedure WHFPM (Weighted FP-tree,α, WHFP)
Inputs: (1) A weighted FP-tree constructed based on algorithm in Fig.1, (2) minwsup; (3)wh_c
Output: The complete set of weighted frequent patterns with highly-correlated affinity, WHFP
Method:
1: If FP-tree contains a single path B {
2: From the single path B, mine all approximate WFP set AWFP;// check if approximate WFPs (check #3)
3: for (∀P∈AWFP) {
4:    if (∑_{i=1,|P|}weight(p_i)/|P|*supp(P) ≥minwsup) { // check if real WFP (check #4)
5:      if (whconf(P)≥wh_c) { // check if weighted hyperclique frequent pattern (check #5)
6:       WHFP= WHFP ∪P;
7:      }// end of 'if' in line 5
8:    }// end of 'if' in line 4
9:  } // end of 'for' in line 3
10: }//end of 'if' in line 1
11: else { // weighted FP-tree has multiple paths
12:   for (∀f-list-i∈f-list in T.header) {  //from bottom-up, T is the weighted FP-tree, f-list-i is a item in f-list
13:     set β=α∪ {f-list-i};
14:     for (∀conof-i∈Con-DB_β ) { //Con-DB_β is β's conditional database,
15:        if (wsupp (conof-i)=supp(conof-i)*max(W) < minwsup)  delete conof-i from Con-DB_β ;//(check #1)
16:        if (wsupp (conof-i) <wsupp(f-list-i)*wh_c)   delete conof-i from Con-DB_β ; (check #2)
17:        X=conof-i β ;
18:         if (max(W)*supp(X)≥minwsup) { // check if approximate weighted frequent patterns(check #3)
19:              if (∑_{i=1,|X|}weight(x_i)/|X|*supp(X)≥minwsup) { // check if real weighted frequent pattern(check #4)
20:                 if (whconf(X)≥wh_c) { // check if weighted hyperclique frequent pattern(check #5)
21:                   WHFP= WHFP ∪X; } //end of 'if' in line 20
22:              } //end of 'if' in line 19
23:          } //end of 'if' in line 18
24:     } //end of 'for' in line 14
25:     WTree_β=Weighted FP-Tree-Construct (Con-DB_β); //construct the weighted FP-tree of  Con-DB_β
26:    If WTree_β≠ NULL
27:      Call WHFPM (WTree_β, β, WHFP);
28:  } // end of 'for' in line 12
29: } // end of 'else' in line 11
30: output WHFP;
```

Fig. 4. Procedure WHFPM

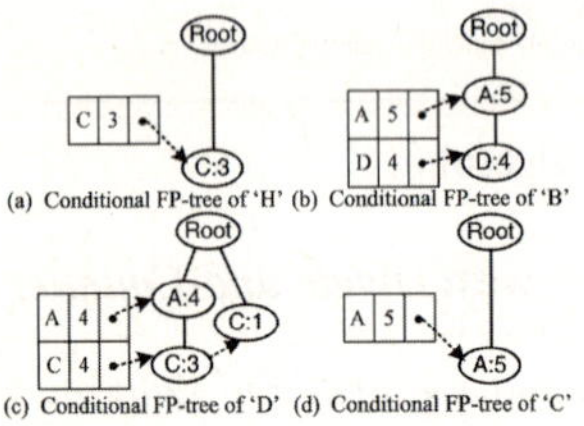

Fig. 5. Conditional weighted FP-trees

node in *f-list*	result patterns
H	{H}:4
B	{B}:5, {AB}:5
D	{D}:6
C	{C}:7
A	{A}:7

Fig. 6. The final mined results

5 Experimental Evaluations

5.1 Test Environment and Datasets

We used two real datasets to test the performance of *WHFPMiner* in comparison with Hyperclique Miner [11]. The two real datasets are Pumsb dataset, which is a dense dataset, and Mushroom dataset, which is a sparse dataset. These real datasets can be obtained from (http://fimi.cs.helsinki.fi/data/). Table 4 shows the characteristics of these datasets. We implemented our algorithm with C++ language, running under Microsoft VC++ 6.0. The experiments were performed on Pentium IV PC at 2.93 GHz and 768MB RAM with Windows XP Professional operating system. All the reported runtimes are in seconds.

Table 4. Data characteristics

Data sets	Size	#Trans	#Items	Maximum (average) transaction length
Pumsb	14.75M	49,046	2113	74 (74.00)
Mushroom	0.56M	8,124	120	23 (23.00)

Because the real datasets do not provide the weight values of items, so we randomly assign weigh to each item in datasets. The weight distributions of items in two real datasets is generated from Gauss distribution (μ=0.5, σ=0.125) shown as Fig.7.

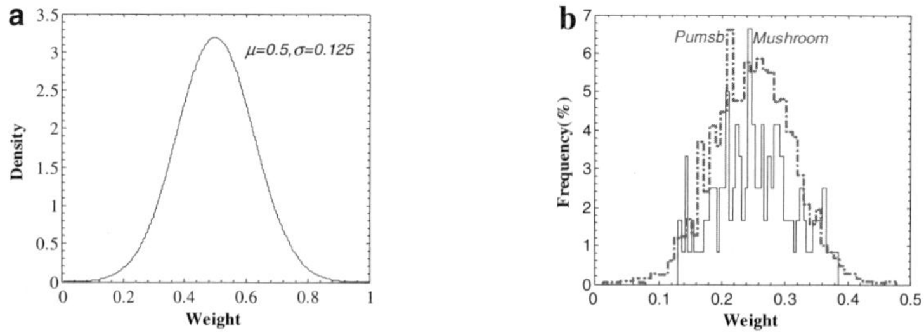

Fig. 7. (a) Gauss distribution density & (b) weight distributions of two real datasets

5.2 Experimental Results

In the test, we focused on testing the efficiency of *whconf.* To discover the effect of *whconf*, we changed the value of *whconf* with fixed *min-sup*. Figure 8 and 9 respectively give the evaluation result for Pumsb and Mushroom dataset. Fig. 8a shows: Given a specified minimum support threshold *min-sup*, with the *min-sup* increasing, fewer but correlated patterns are mined by *WHFPMiner* than Hyperclique Miner. The number of patterns at h_c=0.7 in Hyperclique Miner is unchanged, although the *minwsup* is changed. In Fig. 8b, *WHFPMiner* is fastest than *Hyperclique Miner* in all case of *min-sup*. The reason for that is that *WHFPMiner* mine not only weight-constraint frequent patterns but also *whconf*-constraint patterns. The more constraints, the fewer number of patterns would be mined. Meanwhile, from Fig. 8 and 9, we can see that the number of patterns mined by *WHFPMiner* is decreased with the wh_c becoming higher, and from Fig. 8b and 9b, we can see the *min-sup* and wh_c are higher, the more fast does *WHFPMiner* execute. This because wh_c is higher, the cross-weighted support patterns is larger, and *min-sup* is larger, the searching space of candidate patterns is smaller, so the *WHFP*s mined by *WHFPMiner* is smaller and the algorithm runs more fast.

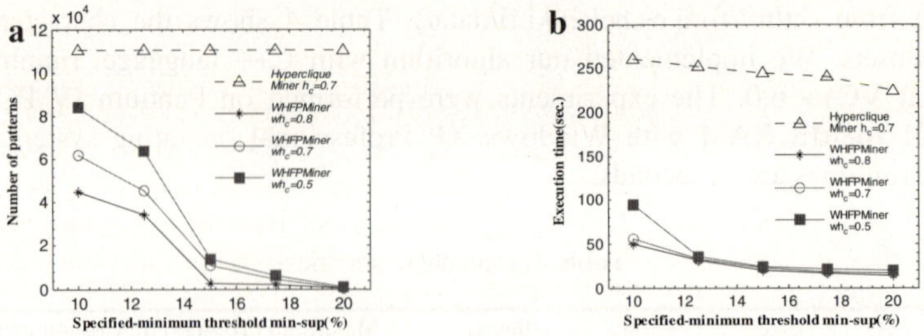

Fig. 8. On the Pumsb dataset (a) Number of patterns generated by WHFPMiner and Hyperclique Miner. (b) The runtime of WHFPMiner and Hyperclique Miner w.r.t different min-sup.

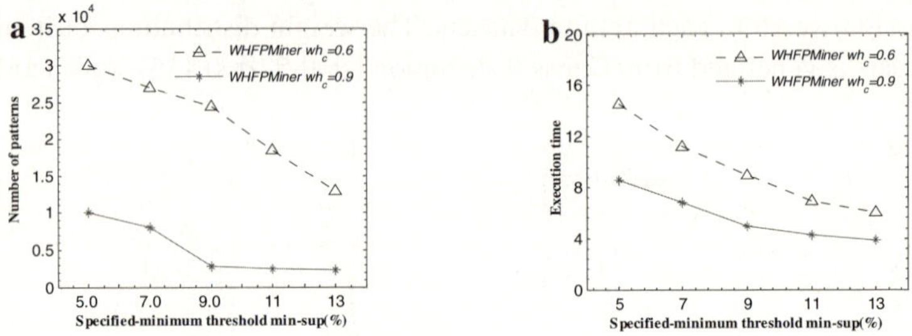

Fig. 9. On the Mushroom dataset (a) Number of patterns generated by WHFPMiner. (b) The execution time of WHFPMiner w.r.t different min-sup.

6 Conclusions

In this paper, we present a new algorithm *WHFPMiner* in which a new objective measure, called weighted h-confidence, is developed to mine weighted hyperclique patterns with similar levels of weighted support. By the revised weighted support, we proved that the weighted *h-confidence* satisfies both the *anti-monotone* property and *cross-weighted support* property, and showed how to prune the weakly-correlated patterns by these two properties. A comprehensive performance study shows that *WHFPMiner* is efficient and fast for finding highly-correlated weighted frequent patterns.

Acknowledgements

This work was supported in part by the Natural Science Fund of Shandong Province (No.Y2007G25), the Excellent Young Scientist Foundation of Shandong Province, China (No.2006BS01017) and the Scientific Research Development Project of Shandong Provincial Education Department, China (No. J06N06).

References

1. Cheung, Y.L., Fu, A.W.: Mining Frequent Itemsets without Support Threshold: with and without Item Constraints. IEEE Transactions on Knowledge and Data Engineering 16, 1052–1069 (2004)
2. Wang, K., He, Y., Cheung, D., Chin, Y.: Mining Confident Rules without Support Requirement. In: 2001 ACM CIKM International Conference on Information and Knowledge Management, pp. 89–96. ACM Press, New York (2001)
3. Omiecinski, E.R.: Alternative Interest Measures for Mining Associations in Databases. IEEE Transactions on Knowledge and Data Engineering 15, 57–69 (2003)
4. Tan, P.N., Kumar, V., Srivastava, J.: Selecting the Right Interestingness Measure for Association Patterns. In: 8th ACM SIGKDD International Conference on Knowledge Discovery and Data Mining, pp. 32–41. ACM, New York (2002)
5. Wang, W., Yang, J., Yu, P.S.: Efficient Mining of Weighted Association Rules (WAR). In: 6th ACM SIGKDD International Conference on Knowledge Discovery and Data Mining, pp. 270–274. ACM, New York (2000)
6. Cai, C.H., Fu, A.W.C., Cheng, C.H., Kwong, W.W.: Mining Association Rules with Weighted Items. In: 998 International Database Engineering and Applications Symposium, pp. 68–77. IEEE Computer Society, Washington (1998)
7. Agrawal, R., Srikant, R.: Fast Algorithms for Mining Association Rules in Large Databases. In: 20th International Conference on Very Large Data Bases, pp. 487–499. Morgan Kaufmann, San Francisco (1994)
8. Han, J., Pei, J., Yin, Y.: Mining Frequent Patterns without Candidate Generation. In: 2000 ACM SIGMOD International Conference on Management of Data, pp. 1–12. ACM, New York (2000)
9. Wang, J., Pei, J., Han, J.: Closet+: Searching for the Best Strategies for Mining Frequent Closed Itemsets. In: 8th ACM SIGKDD International Conference on Knowledge Discovery and Data Mining, pp. 236–245. ACM, New York (2003)
10. Grahne, G., Zhu, J.: High Performance Mining of Maximal Frequent Itemsets. In: 6th SIAM International Workshop on High Performance Data Mining, pp. 135–143. SIAM, Philadelphia (2003)
11. Xiong, H., Tan, P.N., Kumar, V.: Mining Strong Affinity Association Patterns in Data Sets with Skewed Support Distribution. In: 3rd IEEE International Conference on Data Mining, pp. 387–394. IEEE Computer Society, Washington (2003)

Towards a Categorical Matching Method to Process High-Dimensional Emergency Knowledge Structures

Qingquan Wang, Lili Rong, and Kai Yu

Institute of Systems Engineering, Dalian University of Technology, 116024 Dalian, China
dlwqq@hotmail.com, llrong@dlut.edu.cn, dlyukai@yahoo.cn

Abstract. To keep the original semantic information, Textual data in emergency knowledge acquisitions can be actually represented in categorical semantic structures based on typed category theory. These netted topological structures preserve high-dimensions to achieve higher reliability of knowledge processing that relates to the various scenarios of emergency responses. This paper presents a categorical matching method for effective processing on such high-dimensional structures of textual data. The quantification of the matching is achieved through the Greatest Common Subcategory between two categorical structures. Simulated experimental results show a reasonable matching rate for the semantic oriented high-dimensional knowledge processing.

Keywords: Matching Method; Categorical Knowledge Structure; Category Theory; High-dimensions; semantic information.

1 Introduction

There emerge more and more emergency documentation concern different types of disasters that include all kinds of terror attacks, epidemics, hurricanes, tsunamis, earthquakes, air crashes, collective food poisoning, snow damage and industrial accidents. There is an increasing need at the same time for emergency decision-making support on the acquisition of background knowledge. Dispersive textual data in these emergency documentation can be important guides for dealing with emergency problems, and Most of them need to be reorganized [1] according to the numerous and complicated emergency scenarios. Therefore we need suitable methods to deal with such semantic oriented textual data for higher precision and lower time consumed.

Emergency textual data can be represented with categorical knowledge structures, knowledge piece [2] based on typed category theory raised by Lu [3]. The knowledge pieces with categorical structures integrate the closed concepts into the mini-ontologies, the knowledge pieces, and can facilitate the construction of scalable knowledge bases. Actually a knowledge piece represents comparatively sufficient semantic features for higher reliability knowledge processing.

Knowledge pieces are actually high-dimensional discrete data. To deal with high-dimensional data, we generally lower their dimensions to simple structures for the convenience and feasibility in knowledge processing. However the reduction of dimensions can easily bring the losses of original semantic information. Despite knowledge pieces keep the high-dimensional characterize of emergency textual data, there

F. Sun et al. (Eds.): ISNN 2008, Part II, LNCS 5264, pp. 740–747, 2008.

still exists supervening problems that the processing of these high-dimensional knowledge pieces will be more difficult. More features should be selected from the categorical structures for knowledge processing. In such knowledge processing, knowledge matching is basic and can achieve more advanced knowledge processing on their categorical structures. In category theory, subcategories [4] usually mean some extracted objects together with the morphisms between them. However for two or more knowledge pieces, if there is a same subcategory of them, it is obviously that these knowledge pieces are similar or matched in a certain degree. We propose a matching method based on Greatest Common Subcategory (GCS) that is use to quantizing the knowledge matching on high-dimensional categorical structures.

Structures of this paper are defined over the type categorical knowledge structures. In section 2 we discuss the high-dimension categorical knowledge structures and the operations on them. In section 3, the matching method based on GCS is presented. Section 4 shows the simulation experiments to test the matching rate and computational efficiency. Finally, section 5 gives a summary and our further interests.

2 Structuralized High-Dimensional Emergency Textual Data

2.1 High-Dimension Structuralized Emergency Textual Data

Emergency textual data in all kinds of policy documentation actually cover the possible solutions for most of emergency problem solving. The high-dimensional characteristic of these textual data mainly reflected on that not only they have complex semantic relationships between domain concepts, but also these concepts and relationships generally are organized to generate corresponding solutions according to changing emergency scenarios. Such textual data can be represented as typed categorical knowledge structures, named knowledge pieces. This representation is based on category theory that is a relatively young branch of mathematics designed to describe various structural concepts from different mathematical fields in a uniform way [5]. Category theories have been employed to the field of computer science since early 1990s. This theory is also a mathematical underpinning of knowledge science and some of its constructions have been employed for ontology merging [6] and Syntactic Software Merging [7]. Knowledge piece is a typed category that consists of a class of objects, a class of morphisms, a class of types and a class of composition rules, and it has the topological of a network. Some basic notions and definitions about knowledge pieces can be found in [8]. Categorical knowledge structures can facilitate the reservation of comparatively plenty semantic information, and also can achieve some advanced knowledge processing, such as knowledge reasoning, through the computations on the structures. Therefore how to find meaningful information and extract correlations between these categorical structures is the fundamental issue of knowledge processing.

2.2 Content of Categorical Knowledge Matching

Among the possible semantic-based operations of knowledge science, knowledge matching is the elementary approach of other advanced knowledge processing, like intelligent information retrieval and automatic ontology construction. Just like pattern matching in other disciplines, knowledge matching also includes feature extraction for finding meaningful information, matching mode and method of measurement between two or more similar knowledge pieces; furthermore it is hard that knowledge matching is not only simply between textual data, but also between semantic information and even thinking contained. Since the special network topology of knowledge pieces, knowledge matching on categorical knowledge structures can be achieved through the matching between two networks that in the nature of type categories and their restrictions.

A matching between two knowledge pieces can not be achieved only by their common objects, but also by their common morphisms and other common structures that include types and compositions; i.e. matching between knowledge pieces can be converted into the matching between two netlike categorical structures with higher semantic complexity. On categorical knowledge structures, we attempt to accomplish the quantitative knowledge piece matching through the computation of their common subcategories that will be defined in section 3.

3 Matching Method Based on Greatest Common Subcategory

3.1 Definitions of GCS

Category theory creates relationships among categories using functors. However it is impossible to map every object or morphism between two knowledge pieces, because of the way of their semantic-based generation. The overlapped parts of knowledge pieces can be used for calculating the degree of their matching. We define such overlapped parts of knowledge pieces as common subcategories. If we got the maximization of common subcategories about the amount of objects and morphisms, we have found the Greatest Common Subcategory (GCS), similar with Largest Topological Subcategory [9], of these knowledge pieces, and so as to the value of knowledge piece matching. Therefore we have the definition of GCS as follows:

When two actual typed categories do have a possible common subcategory, then we can use their GCS to obtain their similarity value. It is obviously that a GCS from two knowledge pieces is unique under certain matching thresholds. In this paper we suppose that all the objects and morphisms can be computed their kindred similarities. There also exist a functor between GCS and the common commutative diagrams of the typed categories matched.

Definition 1. An object O_1 is a subobject O_2, if and only if O_1 is a concept that can be epitomized as O_2, denoted as $O_1 \prec O_2$. For examples, when *earth quake* can be epitomized as *natural disaster*, we correlate the concept *earth quake* to concept *natural disaster* as its subobject.

Definition 2. Given two morphisms f_1 and f_2, when ignoring the objects connected to them, f_1 still keeps a possible inclusive relationship to f_2, we define that f_1 is a submorphism of f_2, denoted as $f_1 \prec f_2$. For example, in the narrow sense of emergency response, morphism *evacuate* should be a sub-morphism of *rescue*, despite they connect different objects.

Definition 3. Given two types t_1 and t_2, if

$$\begin{cases} dom(t_1) \prec dom(t_2) \vee dom(t_1) = dom(t_2) \\ cod(t_1) \prec cod(t_2) \vee cod(t_1) = cod(t_2) \\ val(t_1) \prec val(t_2) \vee val(t_1) = val(t_2) \end{cases}, \tag{1}$$

Then we say t_1 is a subtype of t_2, denoted as $t_1 \prec t_2$, see figure 1. $dom(t)$ denotes the domain of type t, $cod(t)$ the codomain and $morp(t)$ the value of the type t.

For examples, $command\ \ center \xrightarrow{organize} rescue\ \ team$ can be a subset of $organization \xrightarrow{construct} organization$.

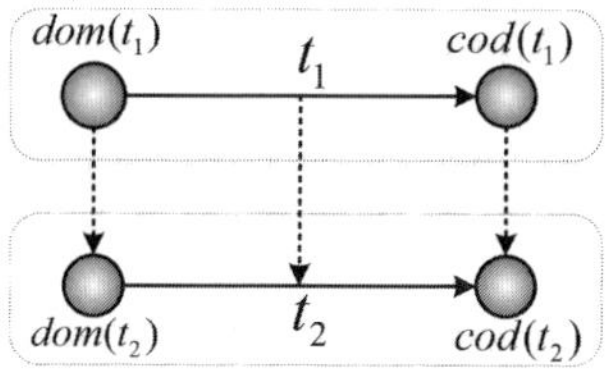

Fig. 1. Semantic-based Subtype

Definition 4. Given two typed categories, $\boldsymbol{K_1} = (O_1, M_1, T_1, R_1)$ and $\boldsymbol{K_2} = (O_2, M_2, T_2, R_2)$, $\boldsymbol{K_1}$ is a subcategory of $\boldsymbol{K_2}$, when

- For each object o_i in O_1, there is a unique object o_j in O_2 and have

$$o_i = o_j \vee o_i \prec o_j. \tag{2}$$

- For each object m_i in M_1, there is a unique object m_j in M_2 and have

$$m_i = m_j \vee m_i \prec m_j. \tag{3}$$

- For each object t_i in T_1, there is a unique object t_j in T_2 and have

$$t_i = t_j \vee t_i \prec t_j. \tag{4}$$

- And $R_1 \subset R_2$.

Definition 5. Given two typed categories, $\boldsymbol{K_1} = (O_1, M_1, T_1, R_1)$ and $\boldsymbol{K_2} = (O_2, M_2, T_2, R_2)$, the common subcategory $\boldsymbol{K_{12}} = (O_{12}, M_{12}, T_{12}, R_{12})$ is defined as follows:

- O_{12} is the intersection between O_1 and O_2, denoted as

$$O_{12} = O_1 \cap O_2, \tag{5}$$

where $o_i = o_j$ and $o_i \prec o_j (o_i \in O_{12}, o_j \in O_1, O_2)$ are both regard as equation. In the same way, we also can conclude that

$$M_{12} = M_1 \cap M_2, T_{12} = T_1 \cap T_2, \tag{6}$$

through the judgments of equation like objects. M_{12} includes the identity morphisms for each object in O_{12}. If all morphisms in $\boldsymbol{K_{12}}$ are identity morphisms, then the matching is only object oriented and can be degraded to keyword based matching.

Definition 6. Given two knowledge pieces, $\boldsymbol{K_1} = (O_1, M_1, T_1, R_1)$ and $\boldsymbol{K_2} = (O_2, M_2, T_2, R_2)$, $\boldsymbol{K_{12}} = (O_{12}, M_{12}, T_{12}, R_{12})$ is the greatest common subcategory of them, if and only if any additional objects and morphisms can result in the wrong results of common subcategory under matching threshold; i.e. Satisfying the largest sizes of objects and morphisms, a common subcategory is a greatest common subcategory (GCS).

Matching threshold is depends on the precision of equivalent judgment that takes the distance between typed categorical elements that include objects, morphisms and types. Generally more close distance between objects means higher precision and lower matching recall rate. We can build and adjust the parameters of the generation of GCS the precision also needs to adjust the degree of concept abstraction. GCS is also a category that is the full subcategories for both of matched typed categories.

3.2 A Sample of GCS Generation Steps

Given two knowledge pieces $\boldsymbol{K_1}$ and $\boldsymbol{K_2}$, their network topology are shown in figure 2. This matching can partly be a network matching [10], GCS generate is shown as follows.

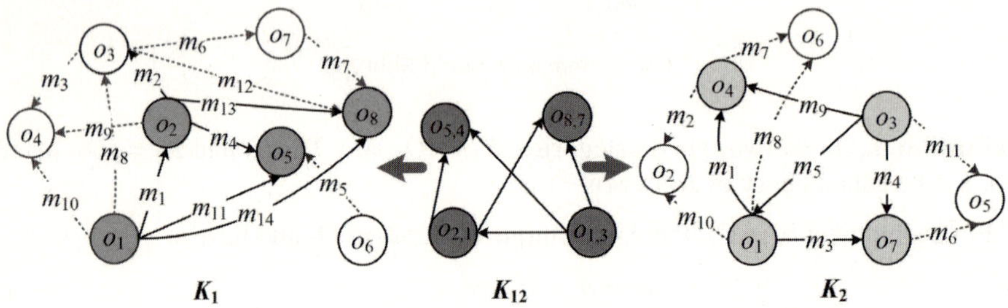

Fig. 2. GCS Generation

First we find correspondences objects (o_2, o_1), (o_1, o_3), (o_5, o_4) and (o_8, o_7) between these two knowledge pieces under there acceptable conceptual similarities; and then each type of these morphisms should be matching to judge whether the semantic features are close; at last connect these morphisms to the matched object to form the greatest common Subcategory. It is obviously that the result of generation GCS depends on the similarity computation of objects, morphisms and types. Suppose that there are no any morphism matched, the matching should be a simple matching that is keywords oriented.

3.3 Distance Threshold for Building GCS

The Distance threshold between two concepts has been used as an important method for the semantic-based knowledge matching. However only concept oriented semantic computation is not enough for high precision knowledge processing. Actually in categorical knowledge structures, their relations, types, or even more complex semantic structures, commutative diagrams, should be considered. As we have known, knowledge processing usually depends on the calculation of conceptual distances, and it can be measured by their distance in tree-like hierarchies that mainly concern the affiliations among concepts. Based on category theory and type category theory, we present an approach of measuring the distance between categorical objects, morphisms and types, so as to the similarity of two categorical knowledge structures. These elements of knowledge pieces can facilitate the higher precision knowledge matching for reliable knowledge supports for emergency decision-making.

There are still a lot of relations between them that do not belong to affiliations, such as the relation between *human* and *water*, however, we just conceder the similarity between knowledge pieces. Apparently they are close connected. Just like the organization of concepts in a hierarchy, relations are also able to be organized in a hierarchy mostly on their types.

Proposition 1. Given two objects o_1 and o_2, if there exists an inherited relationship under a semantic threshold ε between them, then o_1 and o_2 are similar, denoted as $(o_1 \approx o_2) < \varepsilon$. If there exists a close relationship except for affiliations under a semantic threshold φ between them, then o_1 and o_2 are close, denoted as $(a \cong b) < \varphi$.

Proposition 2. Given two objects o_1 and o_2, if $(o_1 \approx o_2) < \varepsilon$, then their similarity value $S_o\,(o_1, o_2)$ is defined as

$$S_o(o_1, o_2) = \begin{cases} 1 & o_1 = o_2 \\ \dfrac{L(o_1) + L(o_2) - 2L(o)}{\varepsilon} & 0 < L_c(o_1) + L_c(o_2) - 2L(o) \le \varepsilon \\ 0 & L_c(o_1) + L_c(o_2) - 2L(o) > \varepsilon \end{cases} \tag{7}$$

$L(o_i)$ is the layer of o_i in concept hierarchies. o is their nearest common parent concepts. ε is the threshold of the max possible distance between these two objects, generally given by human.

Similarly, we can also give the definition of the similarity between two $S_m\,(m_1, m_2)$.

Proposition 3. Given two types t_1 and t_2, similarity between them can be defined as the distance between them, denoted as

$$S_t(t_1, t_2) = \mu(\lambda S_o(d(t_1), d(t_2)) + (1-\lambda) S_o(c(t_1), c(t_2))) + (1-\mu) S_m(m(t_1), m(t_2)), \tag{8}$$

where λ is the significance of domain objects; μ is the significance of object compared with morphisms; $c(t)$, $i(t)$ and $m(t)$ are the abbreviations of $dom(t)$, $cod(t)$ and $morp(t)$ of the type t.

Proposition 4. Given two typed categories $\boldsymbol{K_1}$ and $\boldsymbol{K_2}$, then distance $d(K_1, K_2)$ between them can be defined as the Euclidean distance between the type set, denoted as

$$d(K_1, K_2) = \sqrt{\left(\sum_{i=1}^{M} (t_{1(i)} - t_{2(i)})^2\right)}, \tag{9}$$

where the M is the type amount of the GCS between $\boldsymbol{K_1}$ and $\boldsymbol{K_2}$, and $t_1(i)$ and $t_2(i)$ denote the ith types of $\boldsymbol{K_1}$ and $\boldsymbol{K_2}$ respectively. Then we have the distance between two categories through the middle GCS,

$$d(K_1, K_2) = d(K_1, K_{12}) + d(K_2, K_{12}), \tag{10}$$

where the $t_{12}(i)$ is the type amount of the GCS of $\boldsymbol{K_1}$ and $\boldsymbol{K_2}$,.

Therefore we have the similarity value S of matching two categories, denoted as

$$S = \begin{cases} 1 & d(K_1, K_2) = 0 \\ 1 - \dfrac{d(K_1, K_2)}{\delta} & d(K_1, K_2) < \delta \quad (\delta > 0) \\ 0 & d(K_1, K_2) > \delta \end{cases} \tag{11}$$

where the δ is the threshold of maximum distance between two categories assigned artificially.

4 Simulation Experiments

We make a simulated experiment to test the precision and recall rate of this categorical matching method. The knowledge pieces were generated stochastically from emergency corpus, and the concepts and relations in these materials are retrieved only by their relative positions for generating their hierarchies without the consideration of synonyms. The precision and recall rate of this matching are detected through changing the λ and μ with the step of 0.1 while the threshold of maximum distance δ is a constant 6. The results of these simulation experiments are shown in table 1 and it gives verification that this method can reach a relatively high precision and recall rate at one time.

Table 1. Matching Results of Simulation Experiment

λ	μ	δ	Precision	Recall
0.6	0.4	6	0.8253	0.8973
0.7	0.3	6	0.9225	0.8846
0.8	0.2	6	0.8604	0.8985

5 Conclusion

Knowledge matching is the basic processing of emergency knowledge that generally should be reorganized according to the various scenarios of emergency responses. The emergency knowledge is usually contained in emergency documentation with textual data. We represented the high-dimensional textual data to a categorical knowledge structures based on category theory and typed category theory. On these categorical knowledge structures, we proposed a categorical matching method to deal with high-dimensional textual data for higher reliable knowledge matching. The quantification of knowledge matching was achieved through the generation and computation of greatest common subcategory between two typed categorical structures, knowledge pieces. The computation of GCS depends on not only semantic concepts and their relationships, but also the computation on more complex knowledge structures, types and commutative diagrams. Experimental results show a reasonable matching rate on precision and recall. The computational efficiency is also acceptable for the knowledge processing with higher reliability.

Acknowledgement

This research is supported by the Natural Science Foundation of China (Grant No, 70571011, 70431001). We want give many Thanks to the teammates in Emergency Knowledge Management Group for their valuable discuss.

References

1. Rong, L.L.: A Method of Managing the Knowledge in Government Documents for Quick Response. International Journal of Knowledge and Systems Sciences 2, 67–73 (2005)
2. Wang, Q.Q., Rong, L.L.: Typed Category Theory-Based Micro-View Emergency Knowledge Representation. In: Zhang, Z., Siekmann, J.H. (eds.) KSEM 2007. LNCS (LNAI), vol. 4798, pp. 568–574. Springer, Heidelberg (2007)
3. Lu, R.Q.: Towards a mathematical theory of knowledge. Journal of Computer Science and Technology 20, 751–757 (2005)
4. Walters, R.F.C.: Categories and Computer Science. Cambridge Univ. Cambridge Univ. Press, London (1991)
5. ter Hofstede, A.H.M., Lippe, E., van der Weide, T.P.: Applications of a Categorical Framework for Conceptual Data Modeling. Acta Informatica 34, 927–963 (1997)
6. Menzel, C.: Basic Semantic Integration. Proceedings of the Semantic Integration 82 (2003)
7. Niu, N., Easterbrook, S., Sabetzadeh, M.: A Category-theoretic Approach to Syntactic Software Merging. In: 21st IEEE International Conference on Software Maintenance, pp. 197–206 (2005)
8. Wang, Q.Q., Rong, L.L.: A Structural Knowledge Representation Approach in Emergency Knowledge Reorganization. In: Proceedings of the 8TH International Symposium on Knowledge and Systems Sciences, Japan, pp. 179–186 (2007)
9. Menni, M., Simpson, A.K.: The Largest Topological Subcategory of Countably-based Equilogical Spaces. Electronic Notes in Theoretical Computer Science 20 (1999)
10. Xiong, D.M.: A three-stage computational approach to network matching. Transportation Research Part C 8, 71–89 (2000)

Identification and Extraction of Evoked Potentials Based on Borel Spectral Measure for Less Trial Mixtures*

Daifeng Zha

College of Electronic Engineering, Jiujiang University
332005 Jiujiang, China
zhadaifeng@jju.edu.cn

Abstract. A new method for identifying the independent components of an alpha-stable random vector for under-determined mixtures is proposed. The method is based on an estimate of the discrete Borel measure for the characteristic function of an alpha-stable random vector. Simulations demonstrate that the proposed method can identify the basis vectors of evoked potentials in the so-called under-determined case of more sources than mixtures.

Keywords: Alpha-stable distributions, Borel measure, Evoked potentials, Independent component analysis (ICA), Under-determined representation.

1 Introduction

In some applications, such as biomedical engineering, underwater acoustic signal processing, communications, radar system, etc., and for most conventional and linear-theory-based methods, it is reasonable to assume that the additive noise is usually assumed to be Gaussian distributed with finite second-order statistics. However in some scenarios, it is inappropriate to model the noise as Gaussian noise. Recent studies [1][2] show that the class of alpha stable distributions is better for modeling impulsive noise than Gaussian distribution in signal processing, which has some important characteristics and makes it very attractive. In general, Evoked potentials (EP) signals are always accompanied by ongoing electroencephalogram (EEG) signals which are considered noises in EP analysis. Often the EEG signals are assumed to be Gaussian white noise for mathematical convenience. However, the EEG signals are found to be non-Gaussian in other studies [3]. An analysis shows that the alpha stable model fits the EEG noises found in the impact acceleration experiment under study better than the Gaussian model [4].

Independent Component Analysis addresses the problem of reconstruction of sources from the observation of instantaneous linear combinations. The goal of ICA is to recover independent sources given only sensor observations and linear mixtures of the unobserved independent source signals are unknown. The standard formulation of ICA requires at least as many sensors as sources. Lewicki and Sejnowski [5][6] have

* This work is supported by National Science Foundation of China under Grant 60772037 and Science Foundation of Department of Health of Jiangxi province under Grant 20072048.

F. Sun et al. (Eds.): ISNN 2008, Part II, LNCS 5264, pp. 748–756, 2008.

proposed a generalized ICA method for learning under-determined representations of the data that allows for more basis vectors than dimensions in the input. The work of this paper considers estimation of the basis vectors of evoked potentials accompanied by ongoing electroencephalogram (EEG) signals in the so-called under-determined case of more sources than mixtures.

2 Vector with Discrete Borel Measure

Stable distribution is suitable for modeling random variables with tails of the probability density function that are *heavier* than the tails of the Gaussian density function. Stable distributions have found applications in signal processing, and also in processing of audio signals [2]. This is a kind of physical process with suddenly and short endurance high impulse in real world has no its second order or higher order statistics. It has no close form probability density function so that we can only describe it by its characteristic function:

$$\Phi(t) = \exp\{ j\mu t - \gamma|t|^{\alpha} [1 + j\beta \operatorname{sgn}(t)\omega(t,\alpha)]\} \tag{1}$$

where $\omega(t,\alpha) = \tan\frac{\alpha\pi}{2} (if\ \alpha \neq 1)$ *or* $\frac{2}{\pi}\log|t| (if\ \alpha = 1)$, $-\infty < \mu < \infty$, $\gamma > 0$, $0 < \alpha \leq 2$, $-1 \leq \beta \leq 1$, α is the characteristic exponent, it controls the thickness of the tail in the distribution. The Gaussian process is a special case of stable processes with $\alpha = 2$. The dispersion parameter γ is similar to the variance of Gaussian process and β is the symmetry parameter. If $\beta = 0$, the distribution is symmetric and the observation is referred to as the $S\alpha S$ (symmetry α -stable) distribution, i.e., it is symmetrical about μ , μ is the location parameter. When $\alpha = 2$ and $\beta = 0$, the stable distribution becomes the Gaussian distribution and $\gamma = \sigma^2/2$.

The characteristic function for a real random vector $\mathbf{x}$ is defined as $\mathrm{E}\exp(i\mathbf{t}^{\mathrm{T}}\mathbf{x})$. A random vector is an alpha stable vector if a finite measure Γ exists on the unit sphere S_M of $\mathbf{R}^M$ such that the characteristic function can be written on the form

$$\Phi(\mathbf{t}) = \begin{cases} \exp\{\mathbf{jt}^T\boldsymbol{\mu} - \mathbf{t}^T\mathbf{Dt}\}, & \alpha = 2 \\ \exp\{-\int_{S_M} \Psi_{\alpha}(\mathbf{t}^{\mathrm{T}}\mathbf{s})d\Gamma(\mathbf{s}) + i\mathbf{t}^{\mathrm{T}}\boldsymbol{\mu} + \mathbf{j}\beta_{\alpha}(\mathbf{t})\} & 0 < \alpha < 2 \end{cases} \tag{2}$$

where

$$\Psi_{\alpha}(u) = \begin{cases} |u|^{\alpha} (1 - i\tan(\alpha\pi/2)sign(u), & \alpha \neq 1 \\ |u|(1 + i\pi sign(u)\ln(u)/2), & \alpha = 1 \end{cases}$$

$$\beta_{\alpha}(\mathbf{t}) = \begin{cases} \tan\frac{\alpha\pi}{2} \int_{S_M} |\mathbf{t}^{\mathrm{T}}\mathbf{s}|^{\alpha}\ sign(\mathbf{t}^{\mathrm{T}}\mathbf{s})d\Gamma(\mathbf{s}), & \alpha \neq 1 \\ \int_{S_M} \mathbf{t}^{\mathrm{T}}\mathbf{s}\ln|\mathbf{t}^{\mathrm{T}}\mathbf{s}|\,d\Gamma(\mathbf{s}), & \alpha = 1 \end{cases}$$

where $\boldsymbol{\mu}, \mathbf{t} \in R^M$, $\mathbf{D}$ is positive semi-definite symmetric matrix. The measure Γ is called the Borel measure[1] of the αS random vector, and $\boldsymbol{\mu}$ is called the shift vector. The Borel measure is a real function, and the Borel measure representation $(\Gamma, \boldsymbol{\mu})$ is unique [1][7].

Consider a neural network ICA mixing model (Fig.1) and the real vector $\mathbf{v} = [v_1, v_2, ..., v_N]^{\mathrm{T}}$, where $v_1, v_2, ..., v_N$ are independent $\alpha S(\alpha, \beta_n, \gamma_n, \mu_n)$ random variables. Let $\mathbf{x} = [x_1, x_2, ..., x_M]^{\mathrm{T}}$ be a random vector, and $\mathbf{A} = [a_1, a_2, ..., a_n, ..., a_N]$ be $M \times N$ matrix.

The vector $\mathbf{x} = \mathbf{A}\mathbf{v}$ is then an αS random vector with characteristic function

$$E\exp(i\mathbf{t}^{\mathrm{T}}\mathbf{A}\mathbf{v}) = \prod_{n=1}^{N} E\exp(i\mathbf{t}^{\mathrm{T}}\mathbf{a}_n v_n) \tag{3}$$

where $\mathbf{a}_n$ is the n th column of $\mathbf{A}$. This is essentially the product of all the characteristic functions of v_n, and can be written as

$$E\exp(i\mathbf{t}^{\mathrm{T}}\mathbf{A}\mathbf{v}) = \exp(-\sum_{n=1}^{N} \gamma_n \Psi_\alpha(\mathbf{t}^{\mathrm{T}}\mathbf{a}_n) + i\sum_{n=1}^{N} \mu_n \mathbf{t}^{\mathrm{T}}\mathbf{a}_n) \tag{4}$$

Rewriting (4) to the form of (2) yields [7]

$$\Gamma(\mathbf{s}) = \sum_{n=1}^{N} \frac{1+\beta_n}{2} \gamma_n (\mathbf{a}_n^T \mathbf{a}_n)^{\alpha/2} \delta(\mathbf{s} - \mathbf{s}_n) + \sum_{n=1}^{N} \frac{1-\beta_n}{2} \gamma_n (\mathbf{a}_n^T \mathbf{a}_n)^{\alpha/2} \delta(\mathbf{s} + \mathbf{s}_n) \tag{5}$$

$$\boldsymbol{\mu} = \begin{cases} \sum_{n=1}^{N} \mathbf{a}_n \mu_n & \alpha \neq 1 \\ \sum_{n=1}^{N} \mathbf{a}_n (\mu_n - \beta_n \gamma_n \ln(\mathbf{a}_n^T \mathbf{a}_n)/\pi) & \alpha = 1 \end{cases} \tag{6}$$

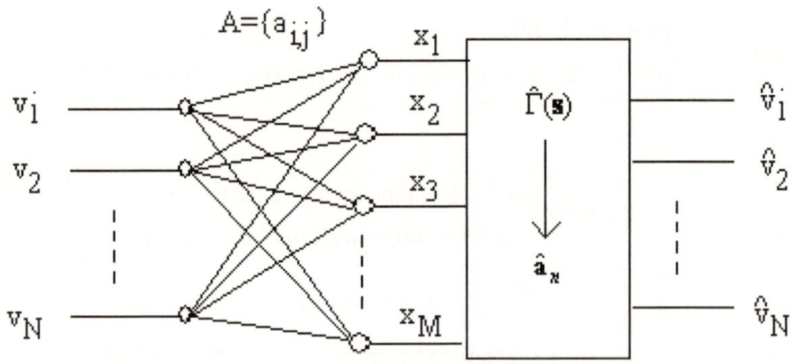

Fig. 1. ICA mixing model

where $\mathbf{s}_n = \mathbf{a}_n \big/ \sqrt{\mathbf{a}_n^T \mathbf{a}_n}$. Hence the Borel measure $\Gamma(\mathbf{s})$ of the random vector $\mathbf{x}$ is discrete and concentrated on N symmetric pairs of points $(\mathbf{s}_n, -\mathbf{s}_n), n = 1,2,..., N$. This result holds in general, thus the Borel measure $\Gamma(\mathbf{s})$ of an αS random vector $\mathbf{x}$ is discrete on the unit sphere S_N , if, and only if, $\mathbf{x}$ can be expressed as a linear transformation of independent αS random variables.

3 Estimation of Borel Measure $\Gamma(\mathbf{s})$

From [8] we have that for a d-dimensional αS variable with Borel measure $\Gamma(\mathbf{s})$ and density function $p(\mathbf{x})$, there is a discrete Borel measure $\Gamma_a(\mathbf{s})$ with corresponding density function $p_a(\mathbf{x})$ satisfying $\sup | p(\mathbf{x}) - p_a(\mathbf{x}) | \le \varepsilon, \quad \mathbf{x} \in R^d, \quad \varepsilon > 0$. Thus an arbitrary Borel measure $\Gamma(\mathbf{s})$ can be approximated by a discrete measure $\Gamma_a(\mathbf{s})$ such that the corresponding densities are arbitrarily close. The only requirement is that sampling of $\mathbf{s}$ is sufficiently dense. The characteristic function corresponding to the approximated discrete Borel measure $\Gamma_a(\mathbf{s})$ sampled in L points, can be written as

$$\hat{\Phi}_a(\mathbf{t}) = \exp(-\sum_{n=1}^{L} \Psi_\alpha(\mathbf{t}^{\mathrm{T}}\mathbf{s}_n)\Gamma_a(\mathbf{s}_n)) \tag{7}$$

Now define a vector $\mathbf{H} = [\Gamma_a(\mathbf{s}_1), \Gamma_a(\mathbf{s}_2),\ldots\ldots, \Gamma_a(\mathbf{s}_L)]^{\mathrm{T}}$ containing the L values of the approximated Borel measure. If we evaluate the approximated ch.f. for L values of $\mathbf{t}$ then we can formulate the set of linear equations

$$\begin{bmatrix} -\ln \Phi_a(\mathbf{t}_1) \\ -\ln \Phi_a(\mathbf{t}_2) \\ \vdots \\ -\ln \Phi_a(\mathbf{t}_L) \end{bmatrix} = \begin{bmatrix} \Psi_\alpha(\mathbf{t}_1^{\mathrm{T}}\mathbf{s}_1) & \Psi_\alpha(\mathbf{t}_1^{\mathrm{T}}\mathbf{s}_2) & .. & \Psi_\alpha(\mathbf{t}_1^{\mathrm{T}}\mathbf{s}_L) \\ \Psi_\alpha(\mathbf{t}_2^{\mathrm{T}}\mathbf{s}_1) & \Psi_\alpha(\mathbf{t}_2^{\mathrm{T}}\mathbf{s}_2) & .. & \Psi_\alpha(\mathbf{t}_2^{\mathrm{T}}\mathbf{s}_L) \\ \vdots & \vdots & \vdots & \vdots \\ \Psi_\alpha(\mathbf{t}_L^{\mathrm{T}}\mathbf{s}_1) & \Psi_\alpha(\mathbf{t}_L^{\mathrm{T}}\mathbf{s}_2) & .. & \Psi_\alpha(\mathbf{t}_L^{\mathrm{T}}\mathbf{s}_L) \end{bmatrix} \times \begin{bmatrix} \Gamma_a(\mathbf{s}_1) \\ \Gamma_a(\mathbf{s}_2) \\ \vdots \\ \Gamma_a(\mathbf{s}_L) \end{bmatrix} \tag{8}$$

Then the approximated Borel measure is given exact by the solution to (8).

In [3] a method for estimation of the stable Borel measure is proposed. The principle behind the estimation is based on (8). From this point, without loss of generality and to simplify the presentation, we will assume that $\mathbf{x}$ is $S\alpha S$. In the case of symmetric density function the Borel measure and the characteristic function is real valued and symmetric. From the definition of the characteristic function, an estimate based on samples of the random vector $\mathbf{x}$ can be obtained as

$$\hat{\Phi}(\mathbf{t}) = \frac{1}{K}\sum_{k=1}^{K} \exp(j\mathbf{t}^{\mathrm{T}}\mathbf{x}_k) \tag{9}$$

for K samples of $\mathbf{x}$. An estimate of the approximate discrete Borel measure is directly obtained from (8). The Borel measure is defined on the d-dimensional unit sphere. If no a priori knowledge about the Borel measure is available and if all directions are of equal importance, it is natural to sample uniformly on the unit sphere. The natural choice of the sampling grid of the characteristic function is to sample symmetrically on a d-dimensional sphere, again in the $S\alpha S$ case it suffices to sample on a half d-dimensional sphere. Thus the n th sample point in $\mathbf{t}$ is $\mathbf{t}_n = r\mathbf{s}_n, \mathbf{s}_n = \mathbf{a}_n \Big/ \sqrt{\mathbf{a}_n^T \mathbf{a}_n}$ where r is the sampling radius.

To avoid negative values in the estimate of the Borel measure, the solution in (8) is restated as

$$\min \left\| \begin{bmatrix} -\ln \Phi_a(\mathbf{t}_1) \\ -\ln \Phi_a(\mathbf{t}_2) \\ \vdots \\ -\ln \Phi_a(\mathbf{t}_L) \end{bmatrix} - 2Real \left\{ \begin{bmatrix} \Psi_\alpha(\mathbf{t}_1^T\mathbf{s}_1) & \Psi_\alpha(\mathbf{t}_1^T\mathbf{s}_2) & .. & \Psi_\alpha(\mathbf{t}_1^T\mathbf{s}_L) \\ \Psi_\alpha(\mathbf{t}_2^T\mathbf{s}_1) & \Psi_\alpha(\mathbf{t}_2^T\mathbf{s}_2) & .. & \Psi_\alpha(\mathbf{t}_2^T\mathbf{s}_L) \\ \vdots & \vdots & \vdots & \vdots \\ \Psi_\alpha(\mathbf{t}_L^T\mathbf{s}_1) & \Psi_\alpha(\mathbf{t}_L^T\mathbf{s}_2) & .. & \Psi_\alpha(\mathbf{t}_L^T\mathbf{s}_L) \end{bmatrix} \begin{bmatrix} \Gamma_a(\mathbf{s}_1) \\ \Gamma_a(\mathbf{s}_2) \\ \vdots \\ \Gamma_a(\mathbf{s}_L) \end{bmatrix} \right\} \right\|_2, \quad \begin{bmatrix} \Gamma_a(\mathbf{s}_1) \\ \Gamma_a(\mathbf{s}_2) \\ \vdots \\ \Gamma_a(\mathbf{s}_L) \end{bmatrix} \geq 0 \tag{10}$$

The estimation procedure for the Borel measure is now:

(i) determine the sampling radius r, a natural choice is $r = \gamma^{-1/\alpha}$; (ii) calculate $\Psi_\alpha(\mathbf{t}_i^T\mathbf{s}_j), \mathbf{t}_i = r\mathbf{s}_i, \mathbf{s}_j = \mathbf{a}_j \Big/ \sqrt{\mathbf{a}_j^T \mathbf{a}_j}$; (iii) estimate $\Phi_a(\mathbf{t}_n)$ according (9); (iv) solve the constrained least square problem in (10), we can get estimated Borel measure $\Gamma_a(\mathbf{s}_1)$, $\Gamma_a(\mathbf{s}_2)$,..., $\Gamma_a(\mathbf{s}_L)$.

4 Identification of Basis Vectors $\mathbf{a}_n$

Our objective is to estimate the basis vectors $\mathbf{a}_n$ of $\mathbf{A}$ based on observations of $\mathbf{x}$. Fundamentally blind identification has a scale ambiguity between γ_n and the norm of the columns $\| \mathbf{a}_n \|_2$, thus without loss of generality we can assume that $\| \mathbf{a}_n \|_2 = 1$. In the d=2 case the n th sample point is $\mathbf{s}_n = \mathbf{a}_n \Big/ \sqrt{\mathbf{a}_n^T \mathbf{a}_n} = \begin{bmatrix} \cos(\pi \frac{n-1}{L}) \\ \sin(\pi \frac{n-1}{L}) \end{bmatrix}$. Moreover, blind identification provides no order of the basis vectors, and for the case of symmetric densities there is a sign ambiguity of the basis vectors.

Now considering (5) leads to the conclusion that the identification of the basis vectors $\mathbf{a}_n$ is simply to determine the directions in which the Borel measure has maximum peaks (On the contrary, conventional ICA algorithms is based on minimization of contrast functions.). In general the Borel measure has maximum peaks in $2N$ directions (N directions if we only sample the half d-dimensional sphere).

Due to a finite number of samples, observation noise and possible deviations from the theoretical distributions, there will be some noise in the estimated Borel measure, i.e., $\mathbf{x} = \mathbf{A}\mathbf{v} + \mathbf{n}$, $\mathbf{n}$ is additive noise. In this case, the basis vectors should be determined as the directions in which the estimated Borel measure has dominating peaks (see section 5, Fig.2.(b), Fig.3.(b)). The number of sample points L determines the angular resolution of the basis vectors. In the d=2 case with sampling on the half 2-dimensional sphere the resolution is $\pm\pi / 2L$.

5 Experimental Results

Experiment 1: Let $\mathbf{v}$ be a random vector with four independent $S\alpha S(1.5,1)$ random variables, $\mathbf{x} = \mathbf{A}\mathbf{v}$ be the observable random vector:

$$\begin{bmatrix} x_1 \\ x_2 \end{bmatrix} = \begin{bmatrix} \cos(\theta_1)\cos(\theta_2) & \ldots & \cos(\theta_4) \\ \sin(\theta_1)\ \sin(\theta_2) & \ldots & \sin(\theta_4) \end{bmatrix} \begin{bmatrix} v_1 \\ v_2 \\ \vdots \\ v_4 \end{bmatrix} \tag{11}$$

where $\theta_1 = 0.2\pi$, $\theta_2 = 0.4\pi$, $\theta_3 = 0.6\pi$, $\theta_1 = 0.8\pi$. We can have scatter plot for the observations of $\mathbf{x}$ and the estimated typical Borel measure depicted in Fig.2. The basis vectors are identified as the directions in which the estimated Borel measure has significant peaks. Observe that the distribution of peaks for the Borel measure is very specific in the 4 directions corresponding to the directions $\theta_1 = 0.2\pi$, $\theta_2 = 0.4\pi$, $\theta_3 = 0.6\pi$, $\theta_1 = 0.8\pi$, and there are four directions of scatter plot of $\mathbf{x}$.

Experiment 2: In [3] it is demonstrated that $S\alpha S$ distributions are suitable for modeling a broad class of signals, including EEG noises found in the impact acceleration experiment. The proposed method for identification of basis vector in a mixture is applied to a mixture of EP and EEG noises. Let v be a random vector with EP and EEG noises. Let $\mathbf{x} = \mathbf{A}\mathbf{v}$ be the observable random vector:

$$\begin{bmatrix} x_1 \\ x_2 \end{bmatrix} = \begin{bmatrix} \cos(\theta_1) & \cos(\theta_2) \\ \sin(\theta_1) & \sin(\theta_2) \end{bmatrix} \begin{bmatrix} v_1 \\ v_2 \end{bmatrix} \tag{12}$$

$\theta_1 = 0.2\pi$, $\theta_2 = 0.8\pi$. We can have scatter plot for the observations of $\mathbf{x}$ and the estimated typical Borel measure depicted in Fig.3 In Fig.3, we observe that the distribution of peaks for the Borel measure is very specific in the direction $\theta_1 = 0.2\pi$, $\theta_2 = 0.8\pi$. And there are two directions of scatter plot of $\mathbf{x}$ and two peaks of Borel measure. According to (12), we can separate EP and EEG noises, see Fig.4.

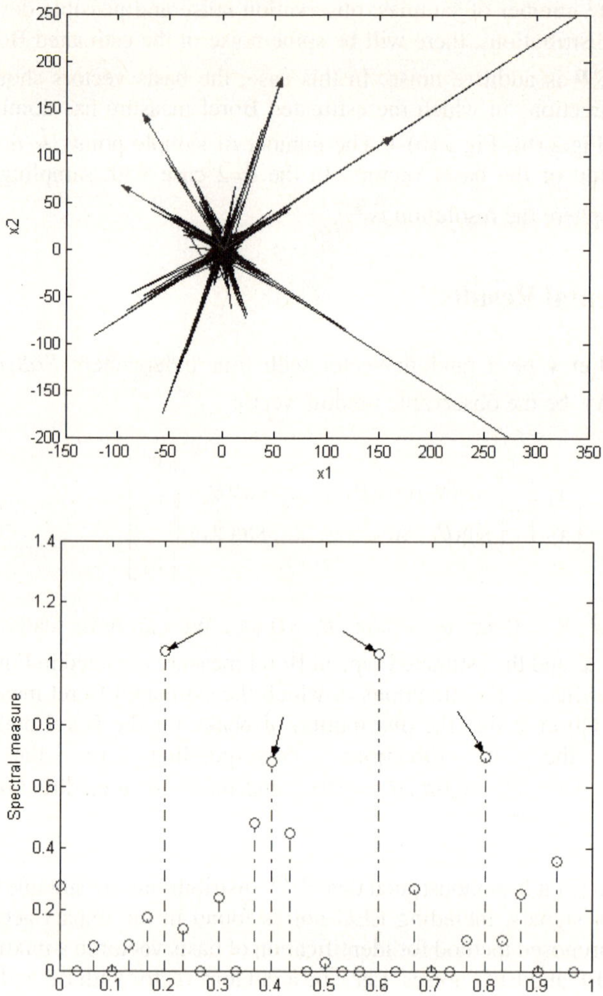

Fig. 2. Scatter plot of $\mathbf{x}$ and the estimated Borel measure

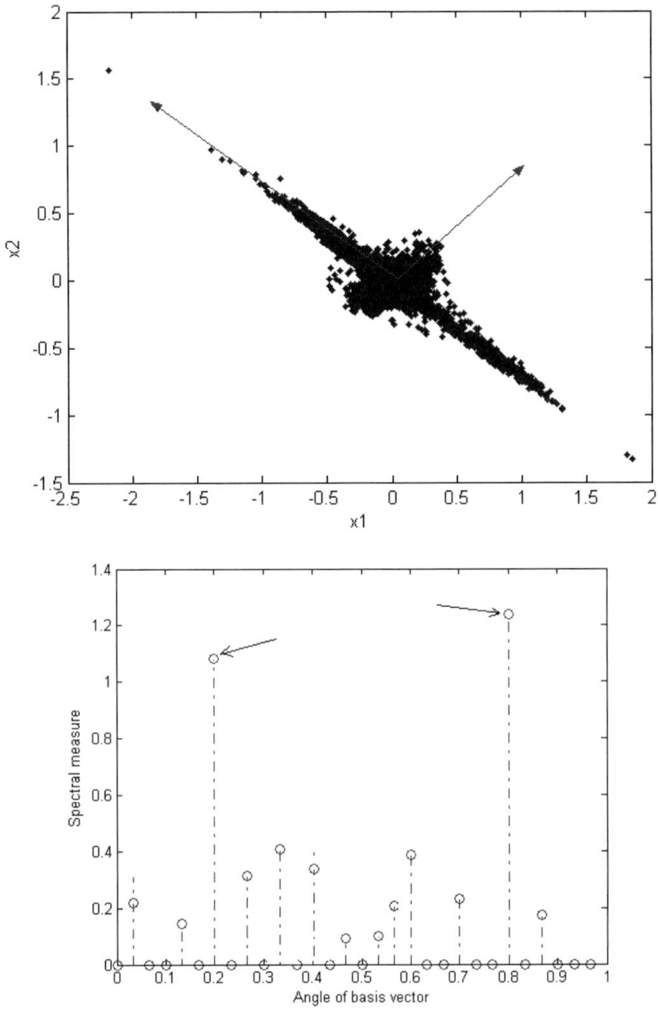

Fig. 3. Scatter plot of $\mathbf{x}$ and the estimated Borel measure

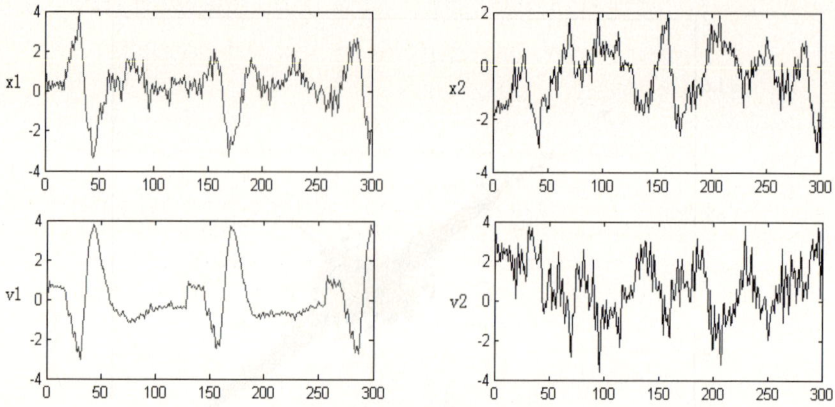

Fig. 4. Separated EP and EEG

6 Conclusion

In this paper, we propose an ICA method based on the observation that the Borel measure is discrete for stable random vectors with independent components. The method identifies the number of independent components and the non-orthogonal bases of the mixture. Simulations demonstrate that the method can identify the number of independent components and the bases of the under-determined mixtures.

References

1. Nikias, C.L., Shao, M.: Signal Processing with Alpha-Stable Distribution and Applications, 1st edn. Wiley, Chichester (1995)
2. Georgiou, P.G., Tsakalides, P., Kyriakakis, C.: Alpha-Stable Modeling of Noise and Robust Time-Delay Estimation in the Presence of Impulsive Noise. IEEE Trans. on Multimedia 1, 291–301 (1999)
3. Hazarika, N., Tsoi, A.C., Sergejew, A.A.: Nonlinear Considerations in EEG Signal Classification. IEEE Trans. on Signal Processing 45, 829–936 (1997)
4. Xuan, K., Tian, S.Q.: Adaptive Estimation of Latency Change in Evoked Potentials by Direct Least Mean p-Norm Time-Delay Estimation. IEEE Trans. on Biomedical Engineering 46 (1999)
5. Lee, T.W., Lewicki, M.S., Girolami, M., Sejnowski, T.J.: Blind Source Separation of More Sources than Mixtures Using Under-determined Representations. IEEE Signal Processing Letters 6, 87–90 (1999)
6. Lewicki, M., Sejnowski, T.J.: Learning Nonlinear Under-determined Representations for Efficient Coding. In: Advances in Neural Information Processing Systems, vol. 10, pp. 815–821. MIT Press, Cambridge (1998)
7. Samorodnitsky, G., Taqqu, M.S.: Stable Non-Gaussian Random Processes. Chapman & Hall, Boca Raton (1994)
8. Nolan, J.P., Panorska, A.K., McCulloch, J.H.: Estimation of Stable Borel Measures. Mathematical and Computer Modelling ((2001)

A Two-Step Blind Extraction Algorithm of Underdetermined Speech Mixtures

Ming Xiao[1,2], Fuquan Wang[1], and Jianping Xiong[1]

[1] Department of Electric and Information Engineering, Maoming University, Maoming, Guangdong, 525000, P. R. China
[2] School of Electric and Information Engineering, South China University of Technology, Guangzhou 510640, China
{xiaoming1968,wm2981138,jianping422}@163.com

Abstract. Underdetermined blind extraction problem of speech mixtures is discussed in this paper. A two-step blind extraction algorithm is proposed. It firstly estimates one basis vector of the mixing matrix using the samples in single source intervals (SSI), and then extracts or reconstructs the corresponding source. Thus, the algorithm can sequentially recovery partial sources using the SSI and decorrelative when the recoverability of the matrix is unknown. Compared with ICA algorithm, it is a non-iteration method. Several speech signals experiments demonstrate its performances.

Keywords: Independent component analysis, Under determined blind source separation (UBSS), Blind source exaction.

1 Introduction

Consider following noiseless instantaneous linear mixture model:

$$\mathbf{x}(t) = \mathbf{A}\mathbf{s}(t) \tag{1}$$

where the vector $\mathbf{x}(t)$is the observed signal from m sensors, the matrix $\mathbf{A} \in R^{m\times n}$ is mixing matrix, and the vector $\mathbf{s}(t)$is n sources. The mixing matrix can be represented as $\mathbf{A} = [\mathbf{a}^1, ..., \mathbf{a}^n]$, where the vector $\mathbf{a}^j$ is a basis vector of the j-th source and let $||\mathbf{a}^j|| = 1$, the symbol $||\cdot||$denotes the length of a vector.

In blind extraction, traditional algorithms often use independent component analysis [1-5]. Recently two-step method based on sparse representation has been a main approach to solve underdetermined blind separation problem [6-10]. Bofill's two-step method requests the mixing matrix can be recovery, but sometimes the matrix can't be estimated or it is largely different from the original matrix, so their sparse representation can't be an approximate source. Therefore, Can some columns of the matrix be recover precisely when the matrix can't be estimated? How to recover the sources if we estimate some columns of the mixing matrix?

F. Sun et al. (Eds.): ISNN 2008, Part II, LNCS 5264, pp. 757–763, 2008.

In view of the above problem, we propose a two-step blind exaction algorithm using sparsity and decorrelative. The first step is to estimate some basis vectors (i.e. some columns of the mixing matrix). The second step is to recover the sources correspond to the estimated columns in matrix. Several experimental results testify our algorithm's performance.

1.1 Single Source Intervals in Frequency Domain

After discrete Fourier transforms (DFT) of the mixture signals, we can obtain

$$\tilde{\mathbf{x}}(k) = \mathbf{A}\tilde{\mathbf{s}}(k) \tag{2}$$

where the vectors $\tilde{\mathbf{x}}(k)$ and $\tilde{\mathbf{s}}(k)$ are respectively the observation and the sources in frequency domain, $\tilde{s}_j^k$ denotes the DFT coefficients of *j-th* source in frequency bin k .

Definition 1: For a frequency interval $[k_1, k_2]$,only one of the sources is non-zero, the remaind sources is zero, the interval $[k_1, k_2]$ is called a single signal interval (SSI).
When the mixing matrix is expended into its columns, the expression (2) can be rewritten

$$\tilde{\mathbf{x}}(k) = \sum_{j=1}^{n} \mathbf{a}^j \tilde{s}_j^k \tag{3}$$

According to the definition of SSI, the frequency interval $[k_1, k_2]$ is a single source interval of the *jth* source, if $k \in [k_1, k_2]$, then

$$\tilde{\mathbf{x}}(k) = \mathbf{a}^j \tilde{s}_j^k \tag{4}$$

So $\mathrm{Re}[\tilde{\mathbf{x}}(k)] = \mathbf{a}^j \mathrm{Re}[\tilde{s}_j^k]$ and $\mathrm{Im}[\tilde{\mathbf{x}}(k)] = \mathbf{a}^j \mathrm{Im}[\tilde{s}_j^k]$, where $\mathrm{Re}[\cdot]$ and $\mathrm{Im}[\cdot]$ respectively denote the real and imaginary part of a complex.

Form $\mathrm{Re}[\tilde{\mathbf{x}}(k)] = \mathbf{a}^j \mathrm{Re}[\tilde{s}_j^k]$, the vector $\mathrm{Re}[\tilde{\mathbf{x}}(k)]$'s projection $\mathbf{u}_k$ on a unite hyper-sphere is equal the vector $\mathbf{a}^k$ or $-\mathbf{a}^k$, that is

$$\mathbf{u}^k = \frac{\mathrm{Re}[\tilde{\mathbf{x}}(k)]}{||\mathrm{Re}[\tilde{\mathbf{x}}(k)]||}. \tag{5}$$

where the symbol sign($\cdot$)denotes a signum function. Let $\mathbf{U} = (\mathbf{u}^0, \mathbf{u}^1, ..., \mathbf{u}^{N-1})$.

To detect all the samples in the SSI's, let $\mathbf{D} = (\mathbf{d}^0, ..., \mathbf{d}^{N-1})$, where

$$\mathbf{d}^k = \mathbf{u}^k \cdot \mathrm{sign}(u_l^k) - \mathbf{u}^{k+1} \mathrm{sign}(u_l^{k+1}) \quad (l = \arg\max_i |u_i^t|) \tag{6}$$

We give positive whole number $N_{\min}$, the interval $[k_1, k_2]$ can be seen as a single source interval (SSI) if $\mathbf{d}^k = 0$ for $k \in [k_1, k_2)$ and the length of the interval is more than $N_{\min}$. Therefore, the basis vector $\mathbf{a}^j$ can be estimated using the samples in SSI's.

Above the approach will search samples in the single source intervals and then average the columns in a cluster, we called it *searching-and-averaging method* (SAM)[12],[13].

2 Extraction or Reconstruction of the Sources

If the basis vector $\mathbf{a}^j$ is only estimated and other basis vectors can be unknown, how to extract or recover the *j-th* sources?

To recover the *jth* source, we will find a separation vector $\mathbf{w}_j = [w_{j1}, ..., w_{jm}]$ to make

$$\hat{s}_j(t) = \mathbf{w}_j \mathbf{x}(t) \tag{7}$$

where the signal $\hat{s}_j(t)$ is the estimated *j-th* source $s_j(t)$. Here, the signal $\hat{s}_j(t)$ can be the *j-th* source, it also can be approximate to the *j-th* source.

Let a submatrix be $\mathbf{A}_j = \left[\mathbf{a}^1 \cdots \mathbf{a}^{j-1}\ \mathbf{a}^{j+1} \cdots \mathbf{a}^n\right]$, which is matrix to delete the *jth* column in mixing matrix. It is clear the signal $\hat{s}_j(t)$ is equal to the *jth* source $s_j(t)$ if $\mathbf{w}_j \mathbf{A}_j = 0_{1\times(n-1)}$.

Now we assume $rank(\mathbf{A}) = m$, $rank(\mathbf{A}_j) = m - 1$.

The basis vector $\mathbf{a}^j$ is non-zero, so we assume its *k-th* element is non-zero, i.e. $a_{kj} \neq 0$. Now configure an m-dimensional hyperplane orthogonal to the basis vector $\mathbf{a}^j$. Let $\mathbf{B} = [\mathbf{b}_1, ..., \mathbf{b}_{m-1}]^T$, where

$$\mathbf{B} = \begin{bmatrix} -a_{kj} & 0 & \cdots & & a_{1j} & 0 & \cdots & 0 \\ \vdots & & \ddots & \vdots & \vdots & \vdots & \vdots & \vdots \\ 0 & & \cdots & -a_{kj} & a_{k-1,j} & 0 & \vdots & 0 \\ 0 & & \cdots & 0 & a_{k+1,j} & -a_{kj} & \cdots & 0 \\ \vdots & & \vdots & \vdots & \vdots & \vdots & \ddots & 0 \\ 0 & & \cdots & 0 & a_{mj} & 0 & \cdots & -a_{kj} \end{bmatrix}_{(m-1)\times m} . \tag{8}$$

Here, $rank(\mathbf{B}) = m - 1$. The row vectors $\mathbf{b}_i (i = 1, ..., m - 1)$ is orthogonal to the basis vector $\mathbf{a}^j$, and the row vectors $\mathbf{b}_i (i = 1, ..., m-1)$ are linear uncorraltive, so they produce a *m-1* dimensional hyperplane or subspace.

So we can obtain a signal vector $\mathbf{y}(t)$, that is

$$\mathbf{y}(t) = \mathbf{B}\mathbf{x}(t) \tag{9}$$

where $\mathbf{y}(t) = [y_1(t), ..., y_{m-1}(t)]^T$. The signal vector $\mathbf{y}(t)$ don't have the *j-th* source.

Set

$$x_0(t) = (\mathbf{a}^j)^T \mathbf{x}(t) \tag{10}$$

Let $\lambda = [\lambda_1, \lambda_2, ..., \lambda_{m-1}]$,

$$\hat{s}_j(t) = x_0(t) - \lambda \mathbf{y}(t) \tag{11}$$

To make $\hat{s}_j(t)$ be more approximate to the source $s_j(t)$, $E[\hat{s}_j(t), y_i(t)] = 0(i = 1, ..., m-1)$ in term of decorrelative. Therefore,

$$E[\hat{s}_j(t)\mathbf{y}(t)^T] = E[(x_0(t) - \lambda\mathbf{y}(t)]\mathbf{y}(t)^T) = 0 \tag{12}$$

and then get

$$\lambda = E[x_0(t)\mathbf{y}(t)^T][E(\mathbf{y}(t)\mathbf{y}(t)^T)]^{-1} \tag{13}$$

.

The recovered source is

$$\hat{s}_j(t) = (\mathbf{a}^j)^T\mathbf{x}(t) - (\mathbf{a}^j)^T E[\mathbf{x}(t)\mathbf{y}(t)^T][E(\mathbf{y}(t)\mathbf{y}(t)^T)]^{-1}\mathbf{y}(t) \tag{14}$$

and the separation vector is

$$\mathbf{w}_j = (\mathbf{a}^j)^T - (\mathbf{a}^j)^T E[\mathbf{x}(t)\mathbf{y}(t)^T][E(\mathbf{y}(t)\mathbf{y}(t)^T)]^{-1}\mathbf{B} \tag{15}$$

So we can obtain the following theorem.

Theorem 1. The estimated source $\hat{s}_j(t) = \mathbf{w}_j\mathbf{x}(t)$ is the true source when $\mathbf{w}_j = (\mathbf{a}^j)^T - (\mathbf{a}^j)^T E[\mathbf{x}(t)\mathbf{y}(t)^T][E(\mathbf{y}(t)\mathbf{y}(t)^T)]^{-1}\mathbf{B}$, if the mixing matrix satisfies $rank(\mathbf{A}) = m$ and $rank(\mathbf{A}_j) = m - 1$.

The theorem means we can extract some sources from the mixtures using a two-step blind extraction algorithm based on decorrelative and sparsity when the sources are extractable.

3 Experiments and Results

In this section simulation results of two examples are present to testify our algorithm. The error in estimated matrix was measure by the difference between the angles of the estimated and the actual basis vectors (columns of the mixing matrix). The vectors $\mathbf{a}^j$ and $\hat{\mathbf{a}}^j$ are unit length, the error can be computed as

$$e(j) = \cos^{-1}(\hat{\mathbf{a}}_j \cdot \mathbf{a}_j) \tag{16}$$

In sources recovery, the estimated signals were rescaled to minumate the interferences, so the estimated source $z_j(t)$ will multiply a coefficientα, its cost function to be

$$J(\alpha) = \min \sum_{t=0}^{N-1} |s_j(t) - \alpha z_j(t)|^2 \tag{17}$$

Then, $\frac{dJ(\alpha)}{d\alpha} = 0$, and get $\alpha = \frac{\sum_{t=0}^{N-1} z_j(t)s_j(t)}{\sum_{t=0}^{N-1} [z_j(t)]^2}$. A reconstruction index was defined as a signal-to-interference ratio (SIR),

$$\mathrm{SIR} = -10\log\frac{\sum_{t=0}^{N-1} |s_j(t) - \alpha z_j(t)|^2}{\sum_{t=0}^{N-1} |s_j(t)|^2}. \tag{18}$$

Experiment 1. Five speech signals (http://sassec.gforge.inria.fr/) and a highway car noise(from http://www.cmp.uea.ac.uk/Research/noise_db/series1/beach/beach.html) are mixed into three observations . The waveforms of the sources and observations see Fig.1. The mixing matrix is

$$\mathbf{A} = \begin{bmatrix} 0.2798 & 0.0766 & 0.9796 & 0.9496 & 0.2215 & 0.7391 \\ 0.9425 & 0.8958 & 0.1841 & 0.2853 & 0.8768 & 0.6080 \\ 0.1825 & 0.4377 & 0.0800 & 0.1299 & 0.4269 & 0.2900 \end{bmatrix}.$$

After detecting the samples in SSI's, we can obtain some estimated columns of the mixing matrix, i.e. $\hat{\mathbf{A}}_1 = \begin{bmatrix} 0.7392 & 0.9640 & 0.2205 & 0.3141 & 0.6220 \\ 0.6079 & 0.2423 & 0.8770 & 0.9309 & 0.7069 \\ 0.2899 & 0.1099 & 0.4268 & 0.1867 & 0.3367 \end{bmatrix}$.

From the above matrix, the second columns in the mixing matrix cannot be recovered, and the vector $\hat{\mathbf{a}}_4$ is the basis vector corresponding to the first source. the angle error between the vector $\hat{\mathbf{a}}_4$ and $\mathbf{a}_1$ is 0.45^0.

After the second step, get the vector $\mathbf{w}_1 = \begin{bmatrix} \text{-}0.001 & 0.4398 \text{ -}0.8981 \end{bmatrix}$ and the extracted signal $z(t)$(see Fig.1). The SIR between the extracted signal and the first source is 57.04 dB. As other sources don't satisfy the extractable conditions, their SIR respectively are 3.4874,8.3333,7.2037 and 6.339dB.

We compute $\mathbf{w}_1\mathbf{A}/(\mathbf{w}_1\hat{\mathbf{a}}_4) = \begin{bmatrix} 1 & 0.0032 & 0.0324 & 0.0313 & 0.0079 & 0.0247 \end{bmatrix}$, so the signal $z(t)$ is the extracted sources.

Experiment 2. Eight speech signal (from http://sassec.gforge.inria.fr/) are selected as the sources in Fig.2the mixing matrix is

$$\mathbf{A} = \begin{bmatrix} 0.8200 & 0.5555 & 0.1029 & 0.3642 & 0.2096 & 0.2849 & 0.4868 & 0.3378 \\ 0.3094 & 0.3972 & 0.9764 & 0.0476 & 0.9327 & 0.8794 & 0.5964 & 0.8285 \\ 0.2341 & 0.0582 & 0.1533 & 0.9189 & 0.1455 & 0.1365 & 0.0902 & 0.1281 \\ 0.4208 & 0.7282 & 0.1120 & 0.1442 & 0.2549 & 0.3563 & 0.6318 & 0.4279 \end{bmatrix}.$$

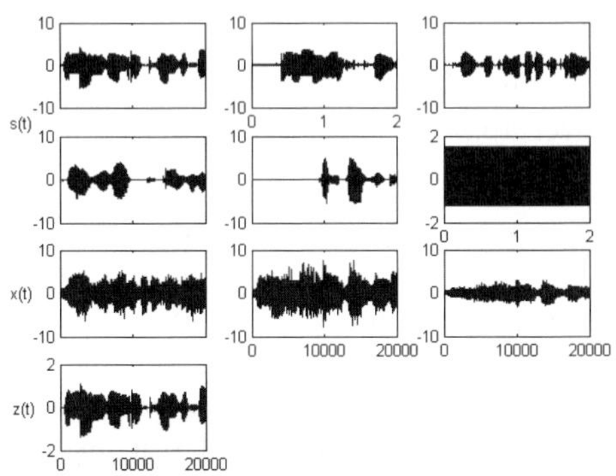

Fig. 1. The sources s(t), observations x(t) and the extracted signal z(t)

In the mixtures (see Fig.2), we know the first and fourth source are extractable. After detecting the samples in SSI's, the estimated matrix is

$$\hat{\mathbf{A}} = \begin{bmatrix} 0.8219\ 0.2952\ 0.4877\ 0.2083\ 0.4051\ 0.5570\ 0.3581\ 0.0971\ 0.3616 \\ 0.2998\ 0.8700\ 0.5945\ 0.9332\ 0.7447\ 0.4007\ 0.0302\ 0.9774\ 0.8046 \\ 0.2357\ 0.1348\ 0.0899\ 0.1440\ 0.1136\ 0.0603\ 0.9236\ 0.1572\ 0.1334 \\ 0.4231\ 0.3712\ 0.6329\ 0.2549\ 0.5181\ 0.7249\ 0.1338\ 0.1028\ 0.4517 \end{bmatrix}.$$

The vectors $\hat{\mathbf{a}}_1$ and $\hat{\mathbf{a}}_7$ respectively are the basis vectors corresponding to the first and fourth source, which respectively have 90 and 412 samples in SSI's, their angle error respectively are 0.58^0 and 0.32^0. Other basis vectors' angle errors are 1.03^0, 0, 8.08^0, 0.25^0 ,$1.14^0 0.69^0$and 2.36^0.

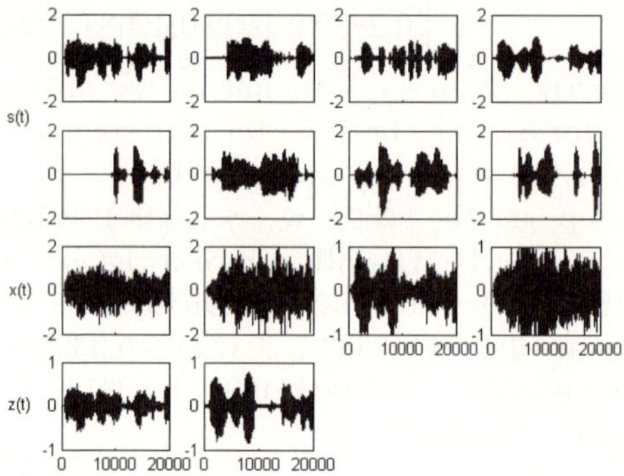

Fig. 2. The sources s(t), observations x(t) and the extracted signal z(t)

After the decorrelative,we obtain$\mathbf{w}_1 = [0.7996\ \text{-}\ 0.0042\ \text{-}\ 0.1390\ \text{-}\ 0.5842]$ and $\mathbf{w}_4 = [\ \text{-}\ 0.2779\ \text{-}\ 0.1424\ 0.9280\ 0.2033]$ and the reconstructed signals. Compute $\mathbf{w}_1\mathbf{A}/(\mathbf{w}_1\hat{\mathbf{a}}^1) = \left[1\ 0.024\ \text{-}0.0227\ 0.2103\ \text{-}0.0145\ \text{-}0.0079\ 0.0137\ \text{-}0.0030\right]$ and $\mathbf{w}_4\mathbf{A}/(\mathbf{w}_4\hat{\mathbf{a}}^7) = \left[0.0398\ \text{-}0.0115\ \text{-}0.0034\ 1\ \text{-}0.0055\ \text{-}0.0069\ \text{-}0.0105\ \text{-}0.0078\right]$. So we know the first and fourth source are extracted and their waveforms can be seen in Fig.2 . The SIR between the extracted signals and the sources respectively are 32.48 and 63.18dB. the SIR of the other the reconstructed signals are 10.03,7.13, 4.6,3.85,5.9 and 2.65dB.

Form above two experiments, our two-step extraction algorithm can extract some sources and recover approximately the sources.

4 Conclusion

The paper discuss the underdetermined blind extraction problem. Its main contribution is a two-step blind extraction algorithm. The algorithm can estimate only one basis vector and then compute one source sequentially. It needn't any iteration. It is very simple blind extraction algorithm; it also can resolve the sources recovery problem when the entire mixing matrix can't be estimated. several experimental results testify its good performance.

Acknowledgments. The work is supported by the National Natural Science Foundation of China for Excellent Youth (Grant 60325310), the National Natural Science Foundation of China (Grant 60674033 and 60505005),the Guangdong Province Science Foundation for Program (Grant 2006B20201037), the Natural Science Fund of Guangdong Province, China (Grant 0401152) and the Guangdong Province important project(Grant 2005B20101016).

References

1. Cardoso, J.F.: Blindsignal Separation, Statistical Principles. Proc. IEEE (Special Issue on Blind Identification and Estimation) 90, 2009–2026 (1998)
2. Cichocki, A., Amari, S.: Adaptive Blind Signal and Image Processing. Learning Algorithms and Applications. Wiley, New York (2002)
3. Li, Y., Zhang, X.: Sequential Blind Extraction Adopting Second-Order Statistics. IEEE Signal Processing Letters 14, 58–61 (2007)
4. Li, Y., Wang, J.: Sequential Blind Extraction of Instantaneously Mixed Sources. IEEE Trans. on Signal Processing 50, 997–1006 (2002)
5. Li, Y., Wang, J., Zurada, J.M.: Blind Extraction of Singularly Mixed Source Signals. IEEE Trans. Neural Netw. 11, 1413–1422 (2000)
6. Liu, D., Hu, S., Zhang, H.: Simultaneous Blind Separation of Instantaneous Mixtures with Arbitrary Rank. IEEE Trans. on Circuits and Systems I, Fundamental Theory and Applications 53, 2287–2298 (2006)
7. Lee, T.W., Lewicki, M., Girolami, M., et al.: Blind Source Separation of More Sources Than Mixtures Using Overcomplete Representations. IEEE Signal Processing Letters 6, 87–90 (1999)
8. Zibulevsky, M., Pearlmutter, B.A.: Blind Source Separation by Sparse Decomposition in Signal Dictionary. Neural Computation 13, 863–882 (2001)
9. Li, Y., Andrzej, C., Amari, S.: Analysis of Sparse Representation and Blind Source Separation. Neural Computation 16, 1193–1234 (2004)
10. Bofill, P., Zibulevsky, M.: Underdetermined Blind Source Separation Using Sparse Representations. Signal Processing 81, 2353–2362 (2001)
11. Theis, F.J., Lang, W.E., Puntonet, C.G.: A Geometric Algorithm for Overcomplete Linear ICA. Neurocomputing 56, 381–398 (2004)
12. Xiao, M., Xie, S.L., Fu, Y.L.: Searching-and-Averaging Method of Underdetermined Blind Speech Signal Separation in Time Domain. Sci. China Ser. F-Inf Sci. 50, 1–12 (2007)
13. Xiao, M., Xie, S.L., Fu, Y.L.: Underdetermined Blind Delayed Source Separation Based on Single Source Intervals in Frequency Domain. Acta Electronic China 35, 37–41 (2007)

A Semi-blind Complex ICA Algorithm for Extracting a Desired Signal Based on Kurtosis Maximization

Jun-Yu Chen and Qiu-Hua Lin

School of Electronic and Information Engineering,
Dalian University of Technology, Dalian 116023, China
goldfishwin@gmail.com, qhlin@dlut.edu.cn

Abstract. Semi-blind independent component analysis (ICA) incorporates some prior information into standard blind ICA, and thus solves some problems of blind ICA as well as provides improved performance. However, semi-blind algorithms thus far have been much focused on the separation of real-valued signals but little on separation of complex-valued signals. We propose in this paper a semi-blind complex ICA algorithm for extracting a complex-valued source of interest within the framework of constrained ICA. Specifically, magnitude information about the desired signal is utilized as inequality constraint to the cost function of kurtosis maximization algorithm, which is an efficient complex ICA algorithm for separating circular and noncircular sources. The simulation results demonstrate that the proposed algorithm can extract a desired complex signal with much improved performance and robustness.

Keywords: Independent component analysis (ICA); Constrained ICA ; Complex-valued signal ; Kurtosis maximization; Magnitude information.

1 Introduction

Independent component analysis (ICA) is a statistical and computational technique successfully used in blind source separation [1], and it has found fruitful applications in communications and biomedical signal processing. However, since utilizing no prior information about the sources or the mixing matrix, standard ICA has two major drawbacks: (1) order ambiguity, i.e., the order of the output estimations cannot be predefined, and (2) low efficiency, i.e., ICA usually recovers all of the independent components but only a few of them are of interest.

In order to overcome the above drawbacks, a number of semi-blind ICA algorithms have been developed to explicitly utilize available information about the sources or the mixing matrix. Among these algorithms, ICA with reference (ICA-R) gives a nice solution to the above-mentioned problems of standard ICA by utilizing the framework of constrained ICA [2, 3]. ICA-R not only can output signals of interest in a predefined order, but also can provide much improved performance compared to blind ICA by incorporating prior information about the desired source signals as references [2, 3].

Semi-blind ICA algorithms such as ICA-R thus far have been much focused on the separation of real-valued signals but little on separation of complex-valued signals. However, standard blind complex ICA for separating complex-valued signals suffers

F. Sun et al. (Eds.): ISNN 2008, Part II, LNCS 5264, pp. 764–771, 2008.

from the same drawbacks of order ambiguity and low efficiency. Thus it is essential to incorporate prior information to blind complex ICA to provide improved semi-blind algorithms. Since complex ICA is needed for convolutive source separation in the frequency domain [4], and for separation of complex-valued data in real world such as functional magnetic resonance imaging (fMRI) data [5], many blind algorithms have been developed. These algorithms can be roughly divided into two categories: (1) matrix diagonalization based algorithms such as joint approximate diagonalization of eigenmatrices (JADE) [6] and the strongly uncorrelating transform (SUT) [7], which have been extended to complex case straightforward, but cannot reflect specific characteristics of complex variable. (2) nonlinear functions based algorithms such as the complex fastICA algorithm (CfastICA) [8] and the kurtosis maximization (KM) algorithm [9, 10]. Compared to CfastICA suitable for separating circular sources, the KM algorithm can better separate both circular and noncircular sources by employing a complex nonlinear function. We thus propose in this paper a semi-blind KM algorithm to extract a complex-valued source of interest. Specifically, we utilized the magnitude information about the desired complex-valued signal within the framework of constrained ICA [3]. The results of simulations with circular and noncircular sources demonstrate that the proposed algorithm can extract a desired complex signal with much improved performance and robustness.

The rest of this paper is organized as follows. Section 2 introduces the basic complex-valued ICA model and the gradient-based KM algorithm (KM-G). In Section 3, we present our proposed algorithm in detail. Section 4 has the computer simulations and performance analyses. Finally, conclusions are given in Section 5.

2 Complex-Valued ICA Model and KM-G

2.1 Complex-Valued ICA Model

A general linear instantaneous complex-valued ICA model is described as follows [8]:

$$\mathbf{x} = \mathbf{A}\mathbf{s} \tag{1}$$

where $\mathbf{s} = [s_1, \ldots, s_n]^T$ are n unknown complex-valued independent source signals, $\mathbf{x} = [x_1, \ldots, x_n]^T$ are n observed mixtures. $\mathbf{A}$ is a complex $n \times n$ unknown mixing matrix. Complex-valued ICA is to find an unmixing matrix $\mathbf{W}$ such that:

$$\mathbf{y} = \mathbf{W}\mathbf{x} = \mathbf{W}\mathbf{A}\mathbf{s} = \mathbf{P}\mathbf{D}\mathbf{s} \tag{2}$$

where $\mathbf{P} \in \mathbf{R}^{n \times n}$ is a real permutation matrix, and $\mathbf{D} \in \mathbf{C}^{n \times n}$ is a complex diagonal scaling matrix. Consequently, the output has order, scaling and phase-shift ambiguities.

2.2 KM Algorithm with Gradient Optimization

Without loss of generality, the sources are typically assumed having unit variance ($E\{\mathbf{s}\mathbf{s}^H\} = \mathbf{I}$, H indicates conjugate transpose). The one-unit KM algorithm exploits the following cost function [9, 10]:

$$\begin{aligned} \max \ & J(\mathbf{w}) = |k(y)| \\ \text{s. t.} \ & E\{yy^H\} = 1 \end{aligned} \tag{3}$$

where $k(y)$ denotes the kurtosis of a zero mean complex random variable $y = \mathbf{w}^H \mathbf{x}$, $\mathbf{x}$ are the observed signals having been whitened as $E\{\mathbf{x}\mathbf{x}^H\} = \mathbf{I}$, thus the constraint $E\{yy^H\} = 1$ is equivalent to $\|\mathbf{w}^H\| = 1$, and:

$$k(y) = E\{(yy^*)^2\} - 2\left(E\{yy^*\}\right)^2 - E\{yy\}E\{y^*y^*\} \tag{4}$$

where y^* is the conjugate of y. By optimizing the cost function in Eq. (3) with the gradient optimization (named KM-G for simplicity), the weight vector is learnt by the following rules [9, 10]:

$$\begin{aligned} \mathbf{w}^+ &= sgn[k(\mathbf{y})]\lambda\left(E\{\mathbf{y}\mathbf{y}^*\mathbf{y}^*\mathbf{x}\} - 2E\{\mathbf{y}\mathbf{y}^*\}E\{\mathbf{y}^*\mathbf{x}\} - E\{\mathbf{y}^*\mathbf{y}^*\}E\{\mathbf{y}\mathbf{x}\}\right) \\ \mathbf{w} &= \frac{\mathbf{w}^+}{\|\mathbf{w}^+\|} \end{aligned} \tag{5}$$

where $sgn[k(\mathbf{y})]$ indicates the sign of $k(\mathbf{y})$, and λ is the learning rate.

Unlike the CfastICA algorithm assuming the sources are circular [8], KM-G based on the complex nonlinear function in Eq. (4) has no such assumption, i.e., it can separate both circular and non circular sources. However, the order of output signals is still unable to predict. Moreover, if some particular source signals are needed, we have to conduct post-processing using prior information.

3 The Proposed KM-G-R Algorithm

In this section, we present a semi-blind KM-G algorithm to extract a complex-valued source of interest by utilizing its magnitude information as a reference, which is thus called KM-G-R for simplicity. We first describe the block diagram of KM-G-R, and then present the algorithm in detail.

3.1 Block Diagram of KM-G-R

Fig. 1 shows the block diagram of the proposed algorithm, in which $x_1, x_2, \ldots, x_n$ are *n* complex-valued mixed signals, *y* is an estimated complex-valued output, *r* is a reference signal constructed from prior information about a desired source signal. Specifically, *r* is a real-valued vector correlated with $|y|$ when we utilize prior information about the magnitude of the desired signal. $\varepsilon(|y|, r)$ is a closeness measure between the estimated output *y* and the reference signal *r*, which is inserted into the cost function of the KM algorithm as a constraint to gain the desired estimation *y*.

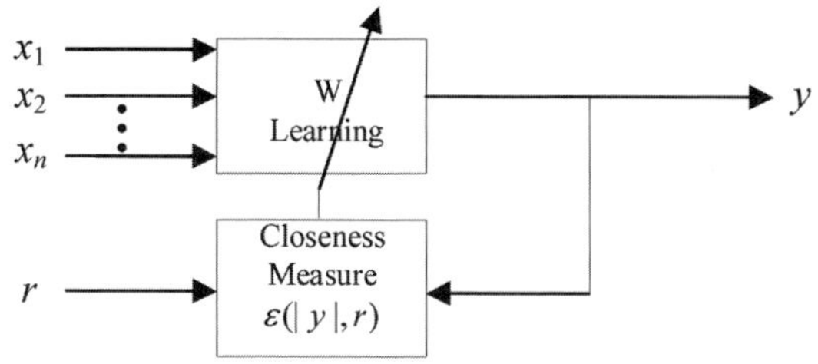

Fig. 1. Block diagram of the proposed algorithm

3.2 The KM-G-R Algorithm

The KM-G-R algorithm for extracting a desired complex-valued signal can be formulated within the framework of constrained ICA as follows:

$$\begin{aligned} &\text{maximize} \quad J(\mathbf{w}) = |k(y)| \\ &\text{s. t.} \qquad g(|\mathbf{w}^H\mathbf{x}|^2) \le 0, \ \left\|\mathbf{w}^H\right\| = 1 \end{aligned} \tag{6}$$

where $g(|\mathbf{w}^H\mathbf{x}|^2)$ is an inequality constraint to the contrast function in Eq. (3).

$$g(|\mathbf{w}^H\mathbf{x}|^2) = \varepsilon(|\mathbf{w}_i^H\mathbf{x}|, r) - \xi = -E\{|\mathbf{w}_i^H\mathbf{x}|^2 \cdot r\} - \xi \le 0 \tag{7}$$

and we define $\varepsilon(|y|, r)$ as:

$$\varepsilon(|y|, r) = -E\{|y|^2 \cdot r\} \tag{8}$$

ξ is a threshold. Assuming that the output y be extracted is the closest one to the reference r, thus $\varepsilon(|y|, r)$ achieves its minimum to guide the iteration, we have:

$$\varepsilon(|\mathbf{w}_i^H\mathbf{x}|, r) < \varepsilon(|\mathbf{w}_j^H\mathbf{x}|, r), j = 1, 2, i-1, \ldots, i+1, \ldots, n \tag{9}$$

Hence, a threshold ξ can be used to distinguish the desired source from the others.

To evaluate the inequality constraint, we transform it into an equality constraint $g(|\mathbf{w}^H\mathbf{x}|^2) + z = 0$ by a slack variable z. By explicitly manipulating the optimal z, an augmented Lagrangian function for Eq. (6) is then given by [2]:

$$L(\mathbf{w}, \gamma, \mu) = J(\mathbf{w}) - \frac{1}{2\gamma}\{\max{}^2\{\gamma g(|\mathbf{w}^H\mathbf{x}|^2) + \mu, 0\} - \mu^2\} \tag{10}$$

where γ is a penalty parameter, μ is a Lagrange multiplier. Correspondingly, a stochastic gradient descent algorithm for updating the weight vector can be obtained as:

$$\mathbf{w} = \mathbf{w} - \beta\nabla_{w^*} L(\mathbf{w}, \gamma, \mu) \tag{11}$$

where β is the learning rate in updates, $\nabla_{w^*} L(\mathbf{w}, \gamma, \mu)$ is the first derivative of $L(\mathbf{w}, \gamma, \mu)$ with respect to $\mathbf{w}^*$ (conjugate of $\mathbf{w}$) according to Brandwood's analyticity condition [9, 10]:

$$\nabla_{w^*} L(\mathbf{w},\gamma,\mu) = \frac{\partial k(\mathbf{w}^H \boldsymbol{x})}{\partial \mathbf{w}^*} - 0.5\mu_k E\{\mathbf{x}(\mathbf{w}^H \boldsymbol{x})^* g'(|\mathbf{w}^H \boldsymbol{x}|^2)\} \tag{12}$$

where the Lagrange multiplier μ_k is learned by the following gradient method:

$$\mu_{k+1} = \max\{0, \mu_k + g(|\mathbf{w}_k^H \mathbf{x}|^2)\} \tag{13}$$

Considering (5) and (12), the KM-G-R algorithm is obtained:

$$\begin{aligned} \Delta\mathbf{w} &= sgn[k(\mathbf{y})]\left(E\{\mathbf{y}\mathbf{y}^*\mathbf{y}^*\mathbf{x}\} - 2E\{\mathbf{y}\mathbf{y}^*\}E\{\mathbf{y}^*\mathbf{x}\} - E\{\mathbf{y}^*\mathbf{y}^*\}E\{\mathbf{y}\mathbf{x}\}\right) \\ &\quad - 0.5\mu_k E\{\mathbf{x}(\mathbf{w}^H \boldsymbol{x})^* g'(|\mathbf{w}^H \boldsymbol{x}|^2)\} \\ \mathbf{w} &= \frac{\mathbf{w} + \beta\Delta\mathbf{w}}{\|\mathbf{w} + \beta\Delta\mathbf{w}\|} \end{aligned} \tag{14}$$

where the normalization is to map $\mathbf{w}$ on the unit sphere to keep the variance of $\mathbf{w}^H\mathbf{x}$ constant.

4 Computer Simulations and Performance Analyses

In this section, we present three simulation examples using the proposed algorithm compared to KM-G, JADE, SUT, and CfastICA which used the real nonlinearity $G(|y|) = \log(0.1 + |y|)$. For the proposed algorithm, the learning rate β was gradually decreased to ensure convergency at the global maximum. The source signals include circular and noncircular signals. The mixing matrix $\mathbf{A}$ was generated randomly with real and imaginary entries from uniform distribution between -1 and 1. The reference signal was constructed with sign operation to the magnitude of each complex-valued signal of interest, and thus has the rough magnitude information about the desired source signal.

To quantitatively compare the performance of all the algorithms, we computed the intersymbol interference (ISI) defined in (15) as:

$$\mathrm{ISI(dB)} = 10\lg\left\{\frac{1}{2n}\sum_{k=1}^{n}\left(\sum_{l=1}^{n}\frac{|p_{kl}|^2}{(\max|p_k|)^2} - 1\right) + \frac{1}{2n}\sum_{l=1}^{n}\left(\sum_{k=1}^{n}\frac{|p_{kl}|^2}{(\max|p_l|)^2} - 1\right)\right\} \tag{15}$$

where p_{kl} denotes the entries of the global matrix $\mathbf{P} = \mathbf{WA}$. This performance index is always negative, and the negative infinity means the best performance. The final ISI results were the average of 20 independent trials with different sample sizes.

4.1 Simulation 1-Separation of Circular Sources

Eight random circular signals were artificially generated as $s_k = r_k(\cos\phi_k + i\sin\phi_k)$ where the radius r_k was drawn from eight different distributions and the phase angle ϕ_k was uniformly distributed on $[-\pi, \pi]$, which implied that $E\{ss^T\} = 0$. The eight distributions used are binomial, gamma, Poisson, hypergeometric, exponential, uniform, beta, geometric. Then we compared performance of the KM-G-R algorithm to KM-G and CfastICA which has proved superior performance for circular signals [9, 10]. Fig. 2 shows the results, and we can see that the KM-G-R yields a better performance than KM-G with about 5dB lower ISI, while CfastICA achieves better performance than KM-G. This demonstrates that the proposed algorithm can improve performance by utilizing the magnitude information.

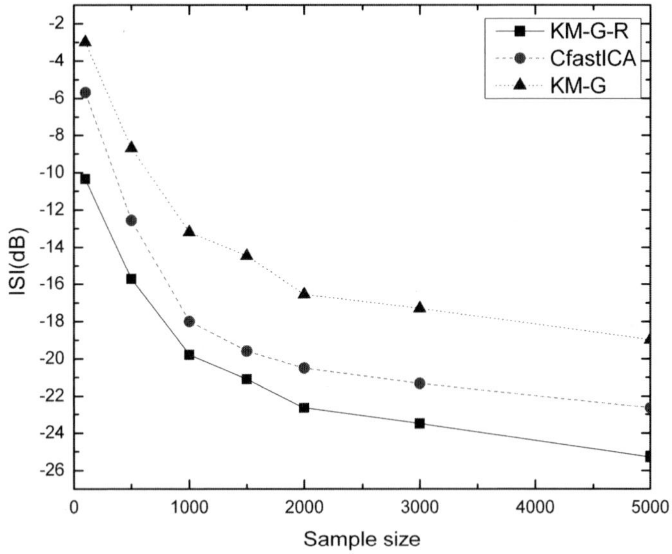

Fig. 2. Average ISI(dB) of various algorithms as a function of the sample size for the eight randomly generated circular sources. The negative infinity means the perfect performance.

4.2 Simulation 2-Separation of Noncircular Sources

In this simulation, we generated eight random noncircular signals as $s_k = s_{kr} + is_{ki}$ where the real part s_{kr} and imaginary part s_{ki} of each source have the same distribution such as one of the eight distributions given in subsection 4.1, but the variances of s_{kr} and s_{ki} are different.

Here we compare KM-G-R to KM-G and SUT which developed to recover noncircular sources with very good performance [7]. The separation results are given in Fig. 3. It can be seen that, the KM-G-R performs reliably with 4dB lower ISI than SUT and remarkably 10dB lower ISI than KM-G. This shows that KM-G-R is robust for departures from the circularity by utilizing magnitude information about the sources.

4.3 Simulation 3-Separation of Circular and Noncircular Sources

We conducted a more comprehensive test in this section. The two types of signals generated in subsection 4.1 and 4.2 were mixed together to further evaluate KM-G-R

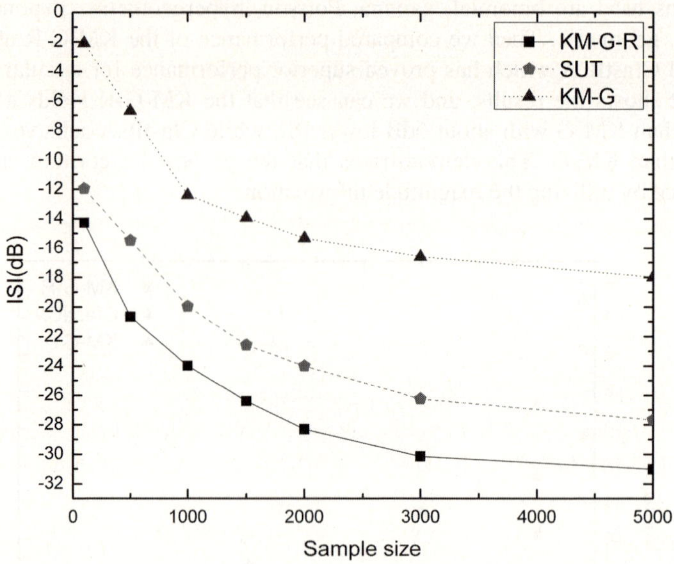

Fig. 3. Average ISI(dB) of various algorithms as a function of the sample size for the eight randomly generated noncircular sources

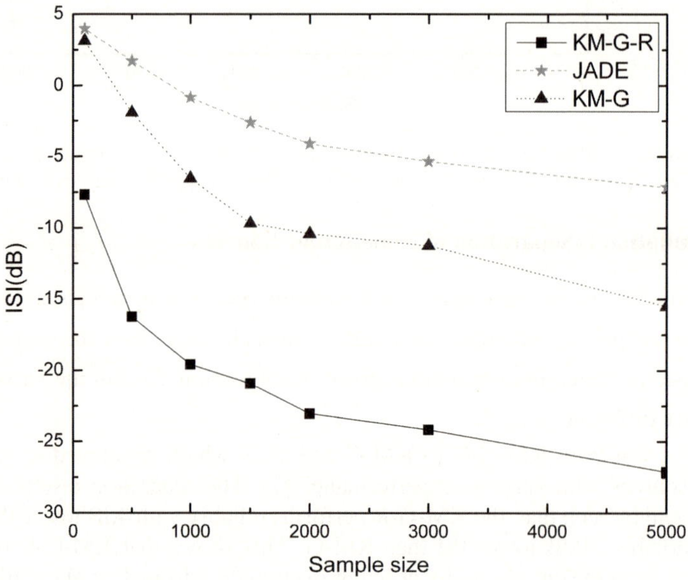

Fig. 4. Average ISI(dB) of various algorithms as a function of the sample size for the sixteen randomly generated sources, eight of which are circular, the others are noncircular

using prior information. In this case, circular and noncircular signals share a half of the sources, respectively, CfastICA and SUT thus cannot perform well for these signals. Therefore, we only compared KM-G-R to KM-G and JADE. Fig. 4 shows the results. It can be seen that KM-G-R improves ISI significantly compared to KM-G and JADE. Specifically, KM-G-R almost doubles its negative ISI compared to KM-G. Moreover, this algorithm reduces about 20dB than JADE when the sample size is increased to 5000.

5 Conclusion

By incorporating magnitude information into a blind complex ICA algorithm KM-G within the framework of constrained ICA, we propose a new semi-blind complex ICA approach for extracting a specific complex source signal using its prior information. Simulations results show that the proposed algorithm has much improved performance and robustness compared to the standard blind complex ICA algorithms due to use of prior information. Since the proposed algorithm can separate both circular and non-circular sources, it can be applied to extraction of desired signals from real complex data such as fMRI.

Acknowledgments. This work was supported by the National Natural Science Foundation of China under Grant No. 60402013, and the Liaoning Province Natural Science Foundation of China under Grant No. 20062174.

References

1. Comon, P.: Independent Component Analysis, a new concept? Signal Proc. 36, 287–314 (1994)
2. Lu, W., Rajapakse, J.C.: ICA with Reference. In: 3rd International Conference on Independent Component Analysis and Blind Source Separation (ICA2001), pp. 120–125 (2001)
3. Lu, W., Rajapakse, J.C.: Approach and Applications of Constrained ICA. IEEE Trans. Neural Nets 16, 203–212 (2005)
4. Sawada, H., Mukai, R., Araki, S., Makino, S.: Frequency-domain Blind Source Separation, Speech Enhancement. Springer, New York (2005)
5. Calhoun, V.D., Adali, T., Pearlson, G.D., Van Zijl, P.C., Pekar, J.J.: Independent Component Analysis of fMRI Data in the Complex Domain. Magn. Resonance Med. 48, 180–192 (2002)
6. Cardoso, J.F., Souloumiac, A.: Blind Beamforming for Non Gaussian signals. IEEE Proc. Radar Signal Proc. 140, 362–370 (1993)
7. Eriksson, J., Koivunen, V.: Complex-valued ICA using Second Order Statistics. In: 14th IEEE Signal Processing Society Workshop on Machine Learning for Signal Processing, Sao Luis, Brazil, pp. 183–191 (2004)
8. Bingham, E., Hyvärinen, A.: A fast fixed-point Algorithm for Independent Component Analysis of Complex Valued Signals. Int. J. Neural Systems 10, 1–8 (2000)
9. Li, H., Adali, T.: A Class of Complex ICA Algorithms Based on Kurtosis Maximization. IEEE Trans. Neural Nets 19, 408–420 (2008)
10. Li, H., Adali, T.: Gradient and Fixed-Point Complex ICA Algorithms Based on Kurtosis Maximization. In: 16th IEEE Signal Processing Society Workshop on Machine Learning for Signal Processing, Maynooth, Ireland, pp. 85–90 (2006)

Fast and Efficient Algorithms for Nonnegative Tucker Decomposition

Anh Huy Phan and Andrzej Cichocki*

RIKEN Brain Science Institute, Wako-shi, Saitama, Japan
{phan,cia}@brain.riken.jp

Abstract. In this paper, we propose new and efficient algorithms for nonnegative Tucker decomposition (NTD): Fast α-NTD algorithm which is much precise and faster than α-NTD [1]; and β-NTD algorithm based on the β divergence. These new algorithms include efficient normalization and initialization steps which help to reduce considerably the running time and increase dramatically the performance. Moreover, the multilevel NTD scheme is also presented, allowing further improvements (almost perfect reconstruction). The performance was also compared to other well-known algorithms (HONMF, HOOI, ALS algorithms) for synthetic and real-world data as well.

Keywords: Nonnegative Tucker decomposition (NTD), Nonnegative matrix factorization (NMF), Alpha divergence, Beta divergence, Hierarchical decomposition.

1 Introduction

Nonnegative Tucker decomposition (NTD), a type of Tucker decomposition [2] with nonnegative constraints, has many potential applications in neuroscience, bioinformatics, chemometrics etc. [1,3,4].

In this paper, we consider at first a simple nonnegative matrix factorization (NMF) model described by a decomposition of a known data matrix $\boldsymbol{Y} = [y_{ik}] \in \mathbb{R}_+^{I \times K}$ as follows: $\boldsymbol{Y} = \widehat{\boldsymbol{Y}} + \boldsymbol{R} = \boldsymbol{A}\boldsymbol{X} + \boldsymbol{R}$, where $\widehat{\boldsymbol{Y}}$ is an approximate matrix, $\boldsymbol{A} = [\boldsymbol{a}_1, \boldsymbol{a}_2, \ldots, \boldsymbol{a}_J] \in \mathbb{R}_+^{I \times J}$ is an unknown basis (mixing) matrix, $\boldsymbol{X} = [\boldsymbol{x}_1^\top, \boldsymbol{x}_2^\top, \ldots, \boldsymbol{x}_J^\top]^\top \in \mathbb{R}_+^{J \times K}$ is a matrix representing unknown nonnegative components $\boldsymbol{x}_j$ and $\boldsymbol{R} \in \mathbb{R}^{I \times K}$ represents errors or noise. The extended NTD model is described as a decomposition of a given N-th order tensor $\underline{\boldsymbol{Y}} \in \mathbb{R}_+^{I_1 \times I_2 \cdots \times I_N}$ into an unknown core tensor $\underline{\boldsymbol{G}} \in \mathbb{R}_+^{R_1 \times R_2 \cdots \times R_N}$ multiplied by a set of N unknown component matrices: $\boldsymbol{A}^{(n)} = [\boldsymbol{a}_1^{(n)}, \boldsymbol{a}_2^{(n)}, \ldots, \boldsymbol{a}_{R_n}^{(n)}] \in \mathbb{R}_+^{I_n \times R_n}$ $(n = 1, 2, \ldots, N)$ representing the common (or loading) factors [2,3,4]:

$$\underline{\boldsymbol{Y}} = \widehat{\underline{\boldsymbol{Y}}} + \underline{\boldsymbol{R}} = \sum_{r_1=1}^{R_1} \sum_{r_2=1}^{R_2} \cdots \sum_{r_N=1}^{R_N} g_{r_1 r_2 \cdots r_N}\, \boldsymbol{a}_{r_1}^{(1)} \circ \boldsymbol{a}_{r_2}^{(2)} \circ \cdots \circ \boldsymbol{a}_{r_N}^{(N)} + \underline{\boldsymbol{R}} \tag{1}$$

$$= \underline{\boldsymbol{G}} \times_1 \boldsymbol{A}^{(1)} \times_2 \boldsymbol{A}^{(2)} \cdots \times_N \boldsymbol{A}^{(N)} + \underline{\boldsymbol{R}} = \underline{\boldsymbol{G}} \times \{\boldsymbol{A}\} + \underline{\boldsymbol{R}}, \tag{2}$$

* Dr A. Cichocki is also from Systems Research Institute PAN Warsaw and University of Technology, Dept. of EE, Warsaw, POLAND.

F. Sun et al. (Eds.): ISNN 2008, Part II, LNCS 5264, pp. 772–782, 2008.

Table 1. Basic tensor operations and notations

$\circ$	outer product	$\boldsymbol{A}^{(n)}$	the $n-th$ factor
$\otimes$	Kronecker product	$\boldsymbol{a}_r^{(n)}$	r^{th} column vector of $\boldsymbol{A}^{(n)}$
$\circledast$	Hadamard product	$\boldsymbol{A}^{(n)\dagger}$	Moore pseudo inverse of $\boldsymbol{A}^{(n)}$
$\oslash$	element-wise division	$\underline{\boldsymbol{Y}}$	tensor
$\times_n$	$n-mode$ product of tensor and matrix	$\underline{\boldsymbol{Y}}_{(n)}$	$n-mode$ matricized version of $\underline{\boldsymbol{Y}}$
$\overline{\times}_n$	$n-mode$ product of tensor and vector	$\boldsymbol{A}^{\otimes}$	$\boldsymbol{A}^{(N)} \otimes \boldsymbol{A}^{(N-1)} \otimes \cdots \otimes \boldsymbol{A}^{(1)}$
$[\bullet]_r$	r^{th} column vector of matrix $[\bullet]$	$\boldsymbol{A}^{\otimes-n}$	$\boldsymbol{A}^{(N)} \otimes \cdots \otimes \boldsymbol{A}^{(n+1)} \otimes \boldsymbol{A}^{(n-1)} \otimes \cdots \otimes \boldsymbol{A}^{(1)}$
$\mathbf{1}$	column vector of ones	$\underline{\boldsymbol{G}} \times \{\boldsymbol{A}\}$	$\underline{\boldsymbol{G}} \times_1 \boldsymbol{A}^{(1)} \times_2 \boldsymbol{A}^{(2)} \cdots \times_N \boldsymbol{A}^{(N)}$
$\underline{\mathbf{1}}$	tensor of ones	$\underline{\boldsymbol{G}} \times_{-n} \{\boldsymbol{A}\}$	$\underline{\boldsymbol{G}} \times_1 \boldsymbol{A}^{(1)} \cdots \times_{n-1} \boldsymbol{A}^{(n-1)}$ $\times_{n+1} \boldsymbol{A}^{(n+1)} \cdots \times_N \boldsymbol{A}^{(N)}$
PSNR	$20\log_{10}$(Range of Signal/RMSE)		

where tensor $\widehat{\underline{\boldsymbol{Y}}}$ is an approximation of tensor $\underline{\boldsymbol{Y}}$, and tensor $\underline{\boldsymbol{R}}$ denotes the residue or error tensor. Throughout this paper, common standard notations are used as indicated in Table 1. When a core tensor $\underline{\boldsymbol{G}}$ is a cubic tensor with $R_1 = R_2 = \ldots = R_N$, and has nonzero elements only on its super-diagonal, the NTD model simplifies to nonnegative tensor factorization (NTF or Nonnegative PARAFAC) [5,6]. Note that matricized version and vectorized version of tensor $\underline{\boldsymbol{Y}}$ provide two NMF models for $\underline{\boldsymbol{Y}}_{(n)}$ and $\mathrm{vec}(\underline{\boldsymbol{Y}}_{(n)})$

$$\underline{\boldsymbol{Y}}_{(n)} = \boldsymbol{A}^{(n)}\, \underline{\boldsymbol{G}}_{(n)}\, \boldsymbol{A}^{\otimes-n\,\top}, \tag{3}$$

$$\mathrm{vec}(\underline{\boldsymbol{Y}}_{(n)}) = \mathrm{vec}(\boldsymbol{A}^{(n)}\, \underline{\boldsymbol{G}}_{(n)}\, \boldsymbol{A}^{\otimes-n\,\top}) = \left(\boldsymbol{A}^{\otimes-n} \otimes \boldsymbol{A}^{(n)}\right) \mathrm{vec}(\underline{\boldsymbol{G}}_{(n)}). \tag{4}$$

Although the α-NTD algorithm was already introduced [1], in this paper, we proposed an improved algorithm (referred as Fast α-NTD) which is much more precise and faster than α-NTD. In addition, we also propose a new flexible NTD algorithm coined as β-NTD based on the β divergence (as used in [7,8]). Moreover, the multilevel NTD scheme (see section (4)) allows further improvements (almost perfect reconstruction). The performance of new algorithms was also compared to well-known existing algorithms (HONMF [4], HOOI [9], ALS [10]).

2 Fast α-NTD Algorithm

We start from the so-called α-NTD algorithm derived recently in [1] with the update rules for factors $\boldsymbol{A}^{(n)}$ (5) and core tensor $\underline{\boldsymbol{G}}$ (6) given in the Table 2(left) and its pseudo code in Table 3(left). Based on the update rules (5) and (6) in the α-NTD algorithm, the approximated tensor $\widehat{\underline{\boldsymbol{Y}}}$ must be updated (step 6) after each iteration step. This causes high computational cost for this algorithm. Without this step, the algorithm cannot converge because some factor coefficients can achieve zero values while some others can be simultaneously very large.

In the next section, we propose a new algorithm (called the Fast α - NTD) which avoids updating the approximated tensor $\widehat{\underline{\boldsymbol{Y}}}$, and therefore reduces considerably running time and increases dramatically the performance, due to appropriate scaling and initialization.

Table 2. Learning rules of α-NTD and Fast α-NTD algorithms

α-NTD algorithm		Fast α-NTD algorithm	
$\boldsymbol{A}^{(n)} \leftarrow \boldsymbol{A}^{(n)} \circledast \left\{ \dfrac{\left[\left(\underline{\boldsymbol{Y}} \oslash \underline{\hat{\boldsymbol{Y}}}\right)^{.\alpha} \times_{-n} \{\boldsymbol{A}^{\top}\}\right]_{(n)} \underline{\boldsymbol{G}}_{(n)}^{\top}}{\mathbf{1}\left[\underline{\boldsymbol{G}} \times_{-n} \{\mathbf{1}^{\top}\boldsymbol{A}\}\right]_{(n)}^{\top}} \right\}^{.1/\alpha}$	(5)	$\boldsymbol{A}^{(n)} \leftarrow \boldsymbol{A}^{(n)} \circledast \left\{ \left[\underline{\boldsymbol{Y}} \oslash \underline{\hat{\boldsymbol{Y}}}\right]_{(n)}^{.\alpha} \hat{\underline{\boldsymbol{Y}}}_{(n)}^{\top} \boldsymbol{A}^{(n)\dagger\top} \right\}^{.1/\alpha}$	(13)
		$\boldsymbol{a}_{r_n}^{(n)} \leftarrow \boldsymbol{a}_{r_n}^{(n)} / \\|\boldsymbol{a}_{r_n}^{(n)}\\|_1$	(10)
$\underline{\boldsymbol{G}} \leftarrow \underline{\boldsymbol{G}} \circledast \left\{ \dfrac{\left(\underline{\boldsymbol{Y}} \oslash \underline{\hat{\boldsymbol{Y}}}\right)^{.\alpha} \times \{\boldsymbol{A}^{\top}\}}{\underline{\mathbf{1}} \times \{\boldsymbol{A}^{\top}\}} \right\}^{.1/\alpha}$	(6)	$\underline{\boldsymbol{G}} \leftarrow \underline{\boldsymbol{G}} \circledast \left\{ \left(\underline{\boldsymbol{Y}} \oslash \underline{\hat{\boldsymbol{Y}}}\right)^{.\alpha} \times \{\boldsymbol{A}^{\top}\} \right\}^{.1/\alpha}$	(18)

2.1 Learning Rule for Factors $\boldsymbol{A}^{(n)}$

The α-NMF algorithm for the NMF were proposed using the following learning rules [8]

$$\boldsymbol{A} \leftarrow \boldsymbol{A} \circledast \left\{ \left[\left(\boldsymbol{Y} \oslash \hat{\boldsymbol{Y}}\right)^{.\alpha} \boldsymbol{X}^{\top}\right] \oslash \left(\mathbf{1}\mathbf{1}^{\top}\boldsymbol{X}^{\top}\right)\right\}^{.1/\alpha}, \tag{7}$$

$$\boldsymbol{X} \leftarrow \boldsymbol{X} \circledast \left\{ \left[\boldsymbol{A}^{\top}\left(\boldsymbol{Y} \oslash \hat{\boldsymbol{Y}}\right)^{.\alpha}\right] \oslash \left(\boldsymbol{A}^{\top}\mathbf{1}\mathbf{1}^{\top}\right)\right\}^{.1/\alpha}. \tag{8}$$

From (3) and (7), the learning rules for factors $\boldsymbol{A}^{(n)}$ are derived as follows

$$\boldsymbol{A}^{(n)} \leftarrow \boldsymbol{A}^{(n)} \circledast \left\{ \left[\left(\underline{\boldsymbol{Y}}_{(n)} \oslash \hat{\underline{\boldsymbol{Y}}}_{(n)}\right)^{.\alpha} \boldsymbol{A}^{\otimes_{-n}} \underline{\boldsymbol{G}}_{(n)}^{\top}\right] \oslash \left(\mathbf{1}\mathbf{1}^{\top}\boldsymbol{A}^{\otimes_{-n}} \underline{\boldsymbol{G}}_{(\mathbf{n})}^{\top}\right)\right\}^{.1/\alpha}. \tag{9}$$

The term $\mathbf{1}^{\top}\boldsymbol{A}^{\otimes_{-n}}\underline{\boldsymbol{G}}_{(\mathbf{n})}^{\top}$ in the denominator in (9) returns a R_n-by-1 vector with the sum over each column of matrix $\boldsymbol{A}^{\otimes_{-n}}\underline{\boldsymbol{G}}_{(n)}^{\top}$. Hence, the denominator in (9) is represented by an I_n-by-R_n large matrix consisting of an I_n-by-1 tiling of copies of vector $\mathbf{1}^{\top}\boldsymbol{A}^{\otimes_{-n}}\underline{\boldsymbol{G}}_{(\mathbf{n})}^{\top}$. In other words, all elements in each column vector of this denominator matrix in (9) are identical. Therefore, this denominator can be omitted, if we normalize all vectors $\boldsymbol{a}_{r_n}^{(n)}$ to unit ℓ_1-norm. Here, we choose the $\ell_1 - norm$ normalization (see the next section for more justification)

$$\boldsymbol{a}_{r_n}^{(n)} \leftarrow \boldsymbol{a}_{r_n}^{(n)} / \|\boldsymbol{a}_{r_n}^{(n)}\|_1, \quad \forall n = 1, 2, \ldots, N, \quad \forall r_n = 1, 2, \ldots, R_N. \tag{10}$$

The $\ell_1 - norm$ normalization (10) enforces all factor coefficients into the range of [0,1], therefore eliminates the large differences between these coefficients, and especially allows us to avoid updating $\hat{\underline{\boldsymbol{Y}}}$ in each iteration step. Alternative technique which eliminates risk that the coefficients reach zeros is to apply a componentwise nonlinear operator to all factors defined as

$$[\boldsymbol{A}^{(n)}]_+ = \max\{\varepsilon, \boldsymbol{A}^{(n)}\}, \qquad \text{(typically in MATLAB } \varepsilon = 2^{-52}). \tag{11}$$

The term $\underline{\boldsymbol{G}}_{(n)}\boldsymbol{A}^{\otimes_{-n}\top}$ is exactly the $n - mode$ matricized version of tensor $\underline{\boldsymbol{G}} \times_{-n} \{\boldsymbol{A}\}$. Note that factors $\boldsymbol{A}^{(n)}$ are tall full-rank matrices, so $\boldsymbol{A}^{(n)\dagger}\boldsymbol{A}^{(n)} = \boldsymbol{I}$

Table 3. α-NTD and Fast α-NTD algorithms

Algorithm 1: α-NTD

Random non-negative initialization for all $\boldsymbol{A}^{(n)}$ and $\underline{\boldsymbol{G}}$
repeat
  Compute $\hat{\underline{\boldsymbol{Y}}} = \underline{\boldsymbol{G}} \times \{\boldsymbol{A}\}$
  for $n = 1$ to N **do**
    Compute $\boldsymbol{A}^{(n)}$ as in (5)
    Update $\hat{\underline{\boldsymbol{Y}}}$
  end for
  Compute $\underline{\boldsymbol{G}}$ as in (6)
until convergence criterion is reached

Algorithm 2: FAST α-NTD

Nonnegative ALS initialization for all $\boldsymbol{A}^{(n)}$ and $\underline{\boldsymbol{G}}$
repeat
  Compute $\hat{\underline{\boldsymbol{Y}}} = \underline{\boldsymbol{G}} \times \{\boldsymbol{A}\}$
  for $n = 1$ to N **do**
    Compute $\boldsymbol{A}^{(n)}$ as in (13)
    Normalize $\boldsymbol{A}^{(n)}$ to unit length
  end for
  Compute $\underline{\boldsymbol{G}}$ as in (18)
until convergence criterion is reached

is an identity matrix. Hence, we have:

$$\underline{\boldsymbol{G}}_{(n)} \boldsymbol{A}^{\otimes_{-n}\top} = \left[\underline{\boldsymbol{G}} \times_{-n} \{\boldsymbol{A}\} \times_n \boldsymbol{A}^{(n)} \times_n \boldsymbol{A}^{(n)\dagger}\right]_{(n)} = \left[\hat{\underline{\boldsymbol{Y}}} \times_n \boldsymbol{A}^{(n)\dagger}\right]_{(n)} = \boldsymbol{A}^{(n)\dagger} \hat{\underline{\boldsymbol{Y}}}_{(n)} \tag{12}$$

After some tedious mathematical transformations, we obtained the simplified learning rules for factors $\boldsymbol{A}^{(n)}$ as follows

$$\boldsymbol{A}^{(n)} \leftarrow \boldsymbol{A}^{(n)} \circledast \left\{ \left[\left(\underline{\boldsymbol{Y}} \oslash \hat{\underline{\boldsymbol{Y}}}\right)^{.\alpha}\right]_{(n)} \hat{\underline{\boldsymbol{Y}}}_{(n)}^{\top} \boldsymbol{A}^{(n)\dagger\top} \right\}^{.1/\alpha} . \tag{13}$$

2.2 Learning Rule for Core Tensor $\underline{\boldsymbol{G}}$

From (4) and (8), the core tensor $\underline{\boldsymbol{G}}$ can be estimated as follows

$$\mathrm{vec}(\underline{\boldsymbol{G}}_{(n)}) \leftarrow \mathrm{vec}(\underline{\boldsymbol{G}}_{(n)}) \circledast \left\{ \left[\left(\boldsymbol{A}^{\otimes_{-n}} \otimes \boldsymbol{A}^{(n)}\right)^{\top} \left(\mathrm{vec}(\underline{\boldsymbol{Y}}_{(n)}) \oslash \mathrm{vec}(\hat{\underline{\boldsymbol{Y}}}_{(n)})\right)^{.\alpha} \right] \oslash \left[\left(\boldsymbol{A}^{\otimes_{-n}} \otimes \boldsymbol{A}^{(n)}\right)^{\top} \mathbf{1} \right] \right\}^{.1/\alpha} \tag{14}$$

where $\left[\left(\boldsymbol{A}^{\otimes_{-n}} \otimes \boldsymbol{A}^{(n)}\right)^{\top} \mathbf{1}\right]$ was used instead of $\left(\left(\boldsymbol{A}^{\otimes_{-n}} \otimes \boldsymbol{A}^{(n)}\right)^{\top} \mathbf{1}\mathbf{1}^{\top}\right)$ because $\mathrm{vec}(\underline{\boldsymbol{Y}}_{(n)})$ is a column vector. Note that vector $\mathbf{1}$ in this expression has $I_1 I_2 \cdots I_N$ elements equal to 1, so it could be considered as a Kronecker product of N small vectors $\mathbf{1}^{(n)} \in \mathbb{R}^{I_n}$ $(n = 1, 2, \ldots, N)$ whose lengths are $I_1, I_2, \ldots, I_N$, respectively. With the assumption of unit-norm factors $\boldsymbol{A}^{(n)}$ $(\boldsymbol{A}^{(n)\top}\mathbf{1}^{(n)} = \mathbf{1})$, the denominator can be written in an equivalent form

$$\begin{aligned} \left(\boldsymbol{A}^{\otimes_{-n}} \otimes \boldsymbol{A}^{(n)}\right)^{\top} \mathbf{1} &= \left(\{\boldsymbol{A}^{\top}\}^{\otimes_{-n}} \otimes \boldsymbol{A}^{(n)\top}\right) \left(\mathbf{1}^{(N)} \otimes \cdots \otimes \mathbf{1}^{(n+1)} \otimes \mathbf{1}^{(n-1)} \otimes \cdots \otimes \mathbf{1}^{(1)} \otimes \mathbf{1}^{(n)}\right) \\ &= \left(\{\boldsymbol{A}^{\top}\mathbf{1}\}^{\otimes_{-n}}\right) \otimes \left(\boldsymbol{A}^{(n)\top}\mathbf{1}^{(n)}\right) = \left(\{\mathbf{1}\}^{\otimes_{-n}}\right) \otimes (\mathbf{1}) = \mathbf{1} \end{aligned} \tag{15}$$

It means that the denominator in (14) can be removed. This is the reason why $\ell_1 - norm$ normalization was selected. Note that the expression in (14) can be expressed as follows

$$\begin{aligned} \left(\boldsymbol{A}^{\otimes_{-n}} \otimes \boldsymbol{A}^{(n)}\right)^{\top} \left(\mathrm{vec}(\underline{\boldsymbol{Y}}_{(n)}) \oslash \mathrm{vec}(\hat{\underline{\boldsymbol{Y}}}_{(n)})\right)^{.\alpha} &= \left(\{\boldsymbol{A}^{\top}\}^{\otimes_{-n}} \otimes \boldsymbol{A}^{(n)\top}\right) \left(\mathrm{vec}(\underline{\boldsymbol{Y}}_{(n)} \oslash \hat{\underline{\boldsymbol{Y}}}_{(n)}^{.\alpha})\right)^{.\alpha} \\ = \mathrm{vec}\left(\boldsymbol{A}^{(n)\top} \left[\underline{\boldsymbol{Y}} \oslash \hat{\underline{\boldsymbol{Y}}}\right]_{(n)}^{.\alpha} \{\boldsymbol{A}^{\top}\}^{\otimes_{-n}\top}\right) &= \mathrm{vec}\left(\left[\left(\underline{\boldsymbol{Y}} \oslash \hat{\underline{\boldsymbol{Y}}}\right)^{.\alpha} \times \{\boldsymbol{A}^{\top}\}\right]_{(n)}\right) \end{aligned} \tag{16}$$

Hence, the learning rule (14) can be finally simplified as follows

$$\mathrm{vec}(\underline{\boldsymbol{G}}_{(n)}) \leftarrow \mathrm{vec}(\underline{\boldsymbol{G}}_{(n)}) \circledast \mathrm{vec}\left(\left[\left(\underline{\boldsymbol{Y}} \oslash \widehat{\underline{\boldsymbol{Y}}}\right)^{.\alpha} \times \{\boldsymbol{A}^{\top}\}\right]_{(n)}\right)^{.1/\alpha} \tag{17}$$

or in the compact tensor form as

$$\underline{\boldsymbol{G}} \leftarrow \underline{\boldsymbol{G}} \circledast \left\{\left(\underline{\boldsymbol{Y}} \oslash \widehat{\underline{\boldsymbol{Y}}}\right)^{.\alpha} \times \{\boldsymbol{A}^{\top}\}\right\}^{.1/\alpha} \tag{18}$$

2.3 Alternating Least Squares Algorithm as an Efficient Initialization Tool

ALS algorithm for Tucker decomposition is a very useful algorithm. However, it is not robust in respect to noise. This algorithm is also referred as Higher-order Orthogonal Iteration (HOOI) by De Lathauwer, De Moor and Vandewalle [9]. Recently, Kolda and Bader investigated this algorithm in details [11]. The idea of this algorithm is to apply SVD and to find R_n leading left singular vectors of the $n - mode$ matricized version of the product tensor $\underline{\boldsymbol{W}}^n = \underline{\boldsymbol{Y}} \times_{-n} \{\boldsymbol{A}^{\top}\}$. Adding the nonnegative constraints on all factors $\boldsymbol{A}^{(n)}$ and using only one or two iterations, ALS procedure becomes a very powerful and efficient initialization tool for our NTD algorithms.

procedure. $NonnegativeALSinitialization(\underline{\boldsymbol{Y}}, R_1, R_2, \ldots, R_N)$

Initialize randomly all non-negative factors $\boldsymbol{A}^{(n)} \in \mathbb{R}_+^{I_n \times R_n}$
for $n = 1$ to N **do**
  Compute $\underline{\boldsymbol{W}}^n = \underline{\boldsymbol{Y}} \times_{-n} \{\boldsymbol{A}^{\top}\}$ and perform SVD
  Form initial $\boldsymbol{A}^{(n)}$ by R_n leading left singular vectors of $\underline{\boldsymbol{W}}^n_{(n)}$
  Project[1] $\boldsymbol{A}^{(n)} = \left[\boldsymbol{A}^{(n)}\right]_+$
end for
$\underline{\boldsymbol{G}} \leftarrow \underline{\boldsymbol{Y}} \times \{\boldsymbol{A}^{\top}\}$

In summary, learning rules and detailed pseudo-code of the Fast α-NTD algorithm are given in Table (2)(right) and in Algorithm 2, Table (3)(right).

3 β-NTD Algorithm

In this section, we extend the β-NMF algorithm (proposed in [8] for the NMF using the β divergence)

$$\boldsymbol{A} \leftarrow \boldsymbol{A} \circledast \left[\left(\boldsymbol{Y} \circledast \widehat{\boldsymbol{Y}}^{.(\beta-1)}\right) \boldsymbol{X}^{\top}\right] \oslash \left(\widehat{\boldsymbol{Y}}^{.\beta} \boldsymbol{X}^{\top}\right), \tag{19}$$

$$\boldsymbol{X} \leftarrow \boldsymbol{X} \circledast \left[\boldsymbol{A}^{\top} \left(\boldsymbol{Y} \circledast \widehat{\boldsymbol{Y}}^{.(\beta-1)}\right)\right] \oslash \left(\boldsymbol{A}^{\top} \widehat{\boldsymbol{Y}}^{.\beta}\right), \tag{20}$$

[1] In practice, factors $\boldsymbol{A}^{(n)}$ are often fixed sign before rectifying.

to a new β-NTD algorithm. The derivation of this algorithm was done [2] by using the matricized expression (3) and learning rule (19) for factors $\boldsymbol{A}^{(n)}$

$$\boldsymbol{A}^{(n)} \leftarrow \boldsymbol{A}^{(n)} \circledast \left[\left(\underline{\boldsymbol{Y}}_{(n)} \circledast \underline{\widehat{\boldsymbol{Y}}}_{(n)}^{.(\beta-1)}\right) \boldsymbol{A}^{\otimes_{-n}} \underline{\boldsymbol{G}}_{(n)}^{\top}\right] \oslash \left(\underline{\widehat{\boldsymbol{Y}}}_{(n)}^{.\beta} \boldsymbol{A}^{\otimes_{-n}} \underline{\boldsymbol{G}}_{(n)}^{\top}\right) \quad (21)$$

and the vectorized expression (4) and learning rule (20) for a core tensor $\underline{\boldsymbol{G}}$

$$\mathrm{vec}(\underline{\boldsymbol{G}}_{(n)}) \leftarrow \mathrm{vec}(\underline{\boldsymbol{G}}_{(n)}) \circledast \mathrm{vec}\left(\left[\left(\underline{\boldsymbol{Y}} \circledast \underline{\widehat{\boldsymbol{Y}}}^{.(\beta-1)}\right) \times \{\boldsymbol{A}^{\top}\}\right]_{(n)}\right) \oslash \mathrm{vec}\left(\left[\underline{\widehat{\boldsymbol{Y}}}_{(n)}^{.\beta} \times \left\{\boldsymbol{A}^{\top}\right\}\right]_{(n)}\right) \quad (22)$$

or in the tensor form

$$\underline{\boldsymbol{G}} \leftarrow \underline{\boldsymbol{G}} \circledast \left[\left(\underline{\boldsymbol{Y}} \circledast \underline{\widehat{\boldsymbol{Y}}}^{.(\beta-1)}\right) \times \{\boldsymbol{A}^{\top}\}\right] \oslash \left[\underline{\widehat{\boldsymbol{Y}}}_{(n)}^{.\beta} \times \left\{\boldsymbol{A}^{\top}\right\}\right] \quad (23)$$

4 Multilevel Nonnegative Tensor Decomposition - High Accuracy Approximation

Performance of NTD can be further improved by multi-stage hierarchical tensor decomposition. Wu et al. proposed a hierarchical tensor approximation for multi-dimensional images using standard Tucker decomposition model in [12] based on approximations of all sub-blocks which are partitioned from the redundancies at each levels. In this section, we presents a different novel hierarchical multilevel scheme for NTD.

1. Approximate the raw given tensor $\underline{\boldsymbol{Y}}$ by the level-1 tensor $\underline{\widehat{\boldsymbol{Y}}}_1 = \underline{\widehat{\boldsymbol{G}}} \times \{\boldsymbol{A}\}$.
2. Compute the residue error tensor $\underline{\boldsymbol{R}}_1 = \underline{\boldsymbol{Y}}$- $\underline{\widehat{\boldsymbol{Y}}}_1$, and divide it into two parts by threshold values set up by its most frequent values (defined by the mode function): $\underline{\boldsymbol{R}}_{1up} = \max(\underline{\boldsymbol{R}}_1, \mathrm{mode}(\underline{\boldsymbol{R}}_1))$, $\underline{\boldsymbol{R}}_{1low} = \min(\underline{\boldsymbol{R}}_1, \mathrm{mode}(\underline{\boldsymbol{R}}_1))$. Then, we normalize these two tensors $\underline{\boldsymbol{R}}_{1up}$ and $\underline{\boldsymbol{R}}_{1low}$ to unit scale [0 1], and also invert $\underline{\boldsymbol{R}}_{1low} = \underline{\mathbf{1}} - \underline{\boldsymbol{R}}_{1low}$.
3. Decompose the two nonnegative residue tensors and get two new approximation tensors $\underline{\widehat{\boldsymbol{Y}}}_{1up}$ and $\underline{\widehat{\boldsymbol{Y}}}_{1low}$. Invert and scale two new tensors to the original ranges of their corresponding tensors $\underline{\boldsymbol{R}}_{1up}$ and $\underline{\boldsymbol{R}}_{1low}$.
4. Obtain the level-2 approximation tensor $\underline{\widehat{\boldsymbol{Y}}}_2$ and return to step 2 for the next level.

The residue tensor $\underline{\boldsymbol{R}}$ does not need to be splitted if we use standard or semi-nonnegative Tucker decomposition. Multilevel decomposition allows to achieve much smaller errors, and higher performance. Fig. 8 illustrates the approximated slices in multilevel scheme. The approximation accuracy increases proportionally to the number of decomposition level (Fig. 8(b)-8(h)). It should be noted that the approximated tensor obtained in the first layer is similar to the low−frequency data of its raw data. In the case of noisy data, to receive the high−frequency band details, we must trade-off between them and noise. If we combine some denoising techniques in the residue tensors, NTD will become

[2] We omit derivation of all algorithms due to space limit.

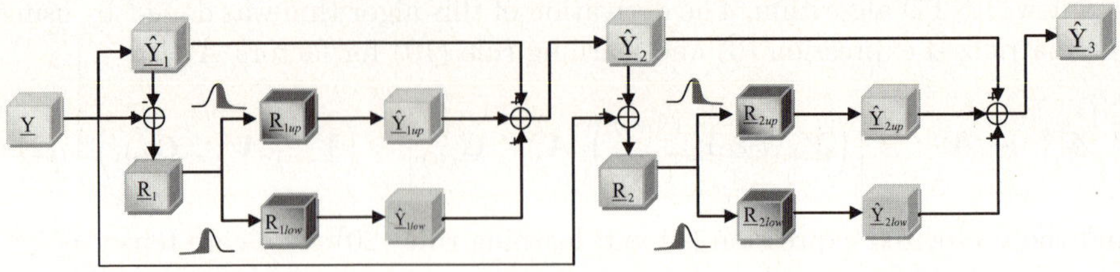

Fig. 1. Hierarchical multilevel nonnegative tensor decomposition

an efficient tool for multi−way restoration and compression. For reconstruction application, we take into account to an approximation tensor $\widehat{\underline{\boldsymbol{Y}}}$, but for compression application, factors $\boldsymbol{A}^{(n)}$ and core tensor $\underline{\boldsymbol{G}}$ are selected. Another advantage of this scheme is that we can avoid decomposition using a large core tensor since in hierarchical decomposition the dimension of core tensor can be much smaller.

5 Experiments

Extensive simulations were performed for synthetic and real−world data on a 2.66 GHz Quad−Core Windows 64−bit machine with 8GB memory. Results were compared with HONMF [4] algorithm and also with two standard Tucker algorithms: HOOI [9], ALS [10] algorithms under the same condition of difference of fit ratio (1e-5) (the explained variation) using Peak Signal to Noise Ratio (PSNR) for all frontal slices.

In Example 1, we consider a sample tensor $\underline{\boldsymbol{Y}} \in \mathbb{R}_+^{20\times19\times60}$ generated by benchmark `X_spectra_sparse` [13] and random core tensor $\underline{\boldsymbol{G}} \in \mathbb{R}_+^{5\times5\times5}$. Under the same desired fit ratio at 99%, α-NTD with $\alpha = 1$ took 397 iterations in 20.3877 seconds of running time and achieved PSNR = 45.9715dB; while fast α-NTD with the same parameter $\alpha = 1$ converged with only 6 iterations in time 0.2512 seconds, achieved PSNR = 60.8057dB. The slices of residual tensors shown in Fig. 2(a)-2(b) indicate the differences in performance between these two algorithms: [-0.017, 0.025] for α-NTD and [-4.5e-3, 6e-3] for fast α-NTD. Changing value of α parameter did not affect this difference. The performance of two other algorithms are also illustrated in this Fig. 2.

In Example 2, tensor $\underline{\boldsymbol{Y}} \in \mathbb{R}_+^{60\times60\times60}$ was generated by 3 benchmarks `ACPos-24sparse10`, `X_spectra_sparse.mat` and `X_spectra` [13] and random core tensor $\underline{\boldsymbol{G}} \in \mathbb{R}_+^{4\times5\times4}$. The performances of two examples are depicted in Fig. 5(a) with PSNR value in the left axis and running time in the right log scale axis. For the both examples, the proposed algorithms provided the highest performances with the fastest running times.

The next Example 3 illustrates reconstruction of tensor $\underline{\boldsymbol{Y}} \in \mathbb{R}_+^{20\times19\times60}$ degraded by additive Gaussian noise with SNR = 0dB. In this example, we used $\beta = 1.5$ and $\ell_2 - norm$ normalization for β-NTD algorithm, whereas $\alpha = 0.7$

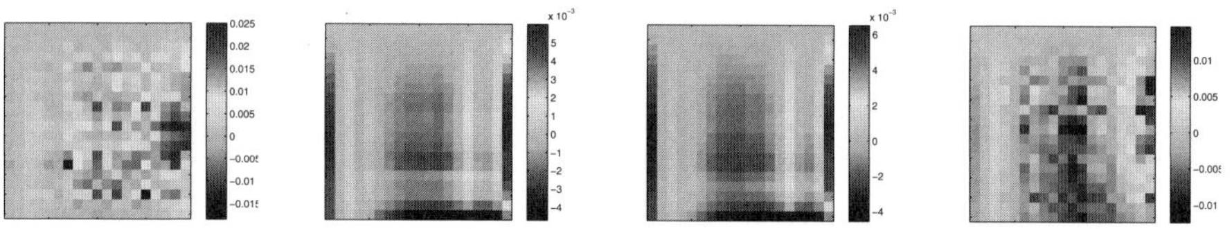

(a) α-NTD 45.97dB 20.39secs (b) Fast α-NTD 60.81dB 0.25secs (c) β-NTD 61.02dB 0.469secs (d) HONMF 47.11dB 3.79secs

Fig. 2. Illustration of the first residue slices for Example 1

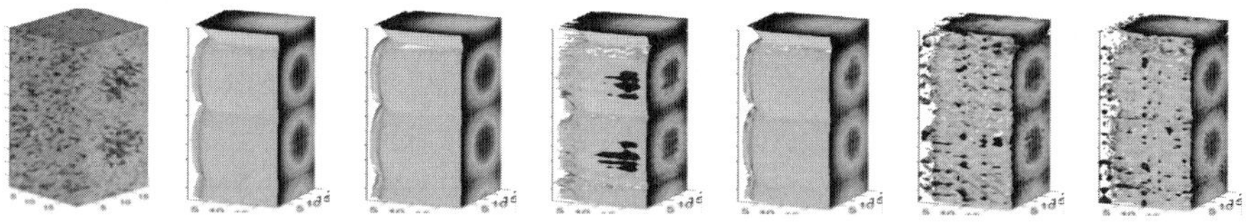

(a) Noisy data (b) Fast α-NTD 26.61dB 0.84secs (c) β-NTD 32.05dB 0.99secs (d) α-NTD 23.11dB 15.01secs (e) HONMF 25.56dB 4.22sec (f) HOOI 20.72dB 0.53secs (g) ALS 20.83dB 1.25sec

Fig. 3. Illustration of results for Example 1 with tensor $\underline{\boldsymbol{Y}} \in \mathbb{R}_+^{20\times 19\times 60}$ corrupted by Gaussian noise with SNR = 0dB in the iso-surface visualization

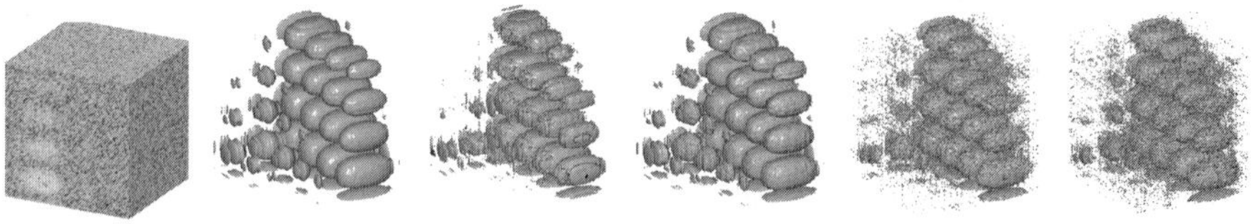

(a) Noisy raw data (b) β-NTD 40.82dB (c) Fast α-NTD 33.35dB (d) HONMF 36.75dB (e) HOOI 27.11dB (f) ALS 26.97dB

Fig. 4. Illustration of data reconstruction for tensor $\underline{\boldsymbol{Y}} \in \mathbb{R}_+^{100\times 100\times 100}$ corrupted by Gaussian noise with SNR = -10dB in iso-surface visualization

for Fast α-NTD algorithm with $\ell_1 - norm$ normalization. The iso-surface visualizations of the estimated tensors (Fig. 3(b)- 3(g)) and distributions of PSNR values depicted in the blue color in Fig. 5(b) show that our local algorithm provides the best performance with very consistent results. The comparison of performance is illustrated in the Fig. 5(b). The performance for Example 4 with tensor $\underline{\boldsymbol{Y}} \in \mathbb{R}_+^{100\times 100\times 100}$ corrupted by Gaussian noise with SNR = -10dB is illustrated in red color (see Fig. 4 for reconstructed tensors). HONMF, HOOI and ALS algorithms failed to reconstruct original data for such case.

In Example 5, a real-world data: Fluorescence excitation emission data from five samples containing tryptophan, phenylalanine, and tyrosine (claus.mat) $5 \times 201 \times 61$ [10] has been corrupted by Gaussian noise with SNR = 0dB (Fig. 6(a)) before reconstruction. The core tensor $\underline{\boldsymbol{G}}$ was chosen with tensor size of $5 \times 5 \times 5$. PSNR values and visualization in the Fig. 6(b)-6(f) show that our β-NTD algorithm (with β= 1.5) is robust to noise and provides high performance.

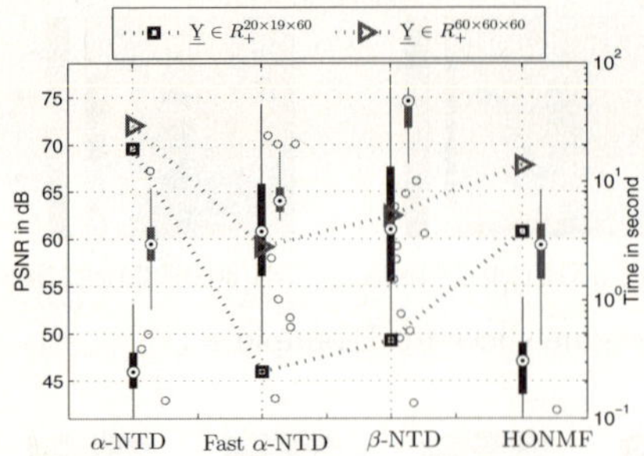

(a) $\underline{\boldsymbol{Y}} \in \mathbb{R}_+^{20\times19\times60}$ in blue color and $\underline{\boldsymbol{Y}} \in \mathbb{R}_+^{60\times60\times60}$ in red color

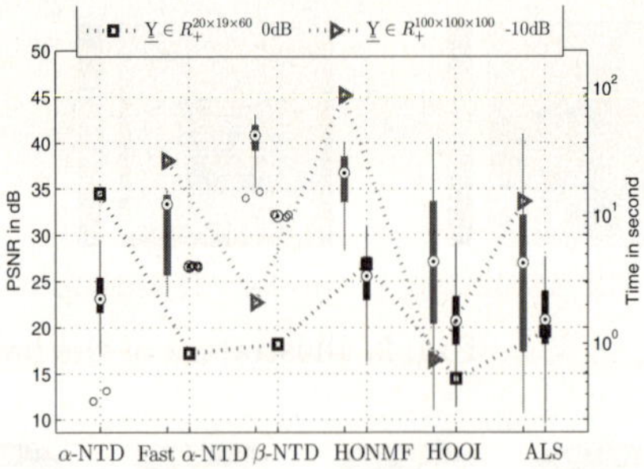

(b) $\underline{\boldsymbol{Y}} \in \mathbb{R}_+^{20\times19\times60}$ 0dB, SNR = 0dB Gaussian noise, blue color and $\underline{\boldsymbol{Y}} \in \mathbb{R}_+^{100\times100\times100}$, SNR = -10dB Gaussian noise, red color

Fig. 5. Comparison of PSNR values (in dB) and running time (in second, log scale) of Examples 1 and 2 (a), and Examples 3 and 4 (b)

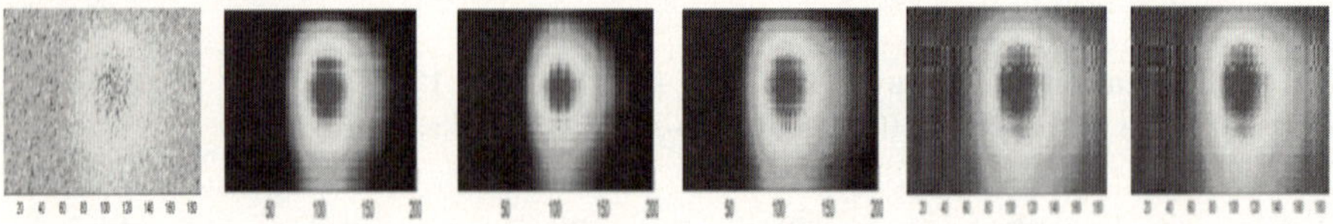

(a) Noisy data 0dB (b) β-NTD 31.74dB (c) Fast α-NTD 24.24dB (d) HONMF 30.67dB (e) HOOI 27.89dB (f) ALS 28.08dB

Fig. 6. Tensor reconstruction for real-world data `clauss.mat`, tensor size 5 × 201 × 61 corrupted by Gaussian noise with SNR= 0dB

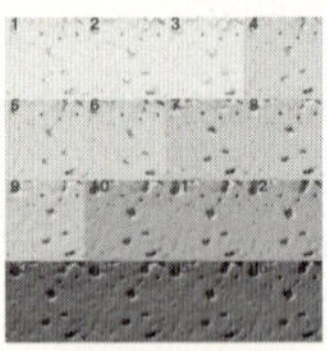

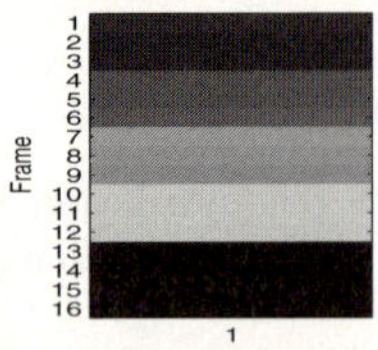

(a) 16 Ceiling textures (b) Clustering graph

Fig. 7. Illustration of texture clustering for 16 Ceiling textures (a) based on the factor $\boldsymbol{A}^{(3)}$, (b) vector $\boldsymbol{a}_1^{(3)}$ expresses the difference between textures, and forms 5 texture groups (1-3), (4-6), (7-9), (10-12) and (13-16)

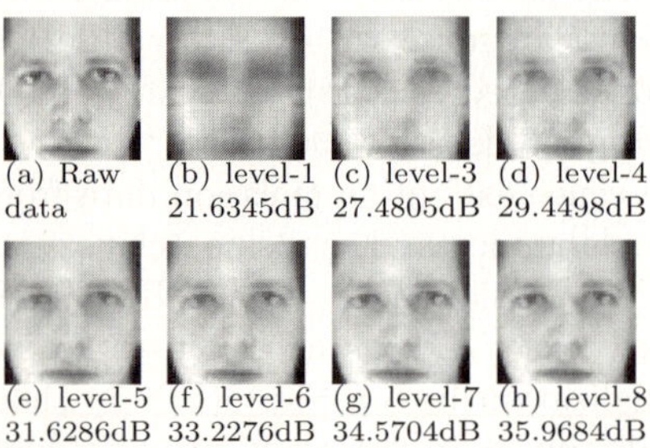

(a) Raw data (b) level-1 21.6345dB (c) level-3 27.4805dB (d) level-4 29.4498dB
(e) level-5 31.6286dB (f) level-6 33.2276dB (g) level-7 34.5704dB (h) level-8 35.9684dB

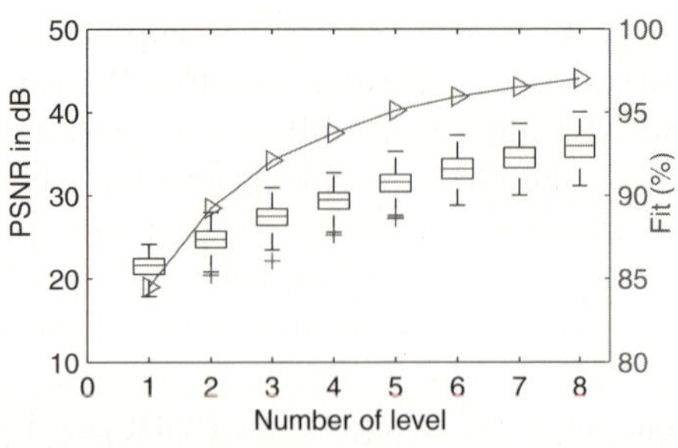

Fig. 8. Illustration of gradual improvement of reconstruction of face images by applying multilevel hierarchical decomposition with local NTD algorithm

For another real world data, 16 Ceiling textures [14] of size 128 × 128 were decomposed with a core tensor $\underline{\boldsymbol{G}} \in \mathbb{R}^{20\times 20\times 5}$ by β-NTD algorithm with $\beta = 1.5$. The 3rd factor $\boldsymbol{A}^{(3)}$ was used for texture clustering. The differences between 16 observed textures in Fig. 7(b) matches with the actual visualization in Fig. 7(a).

Another experiment was performed for the 3-way ORL face database (48x48 pixels and 400 images) using the hierarchical model in 8 levels, with core tensor $\underline{\boldsymbol{G}} \in \mathbb{R}^{24\times 24\times 50}$. The results of reconstruction are illustrated in the Fig. 8.

6 Conclusion

We presented a new fast α-NTD algorithm and β-NTD algorithm which are robust to noise for nonnegative Tucker decomposition. With the proper normalization and initialization, proposed NTD algorithms could achieve high and consistent performance. Moreover, the performance of NTD could be improved by our hierarchical scheme. Extensive experiments confirmed the validity and high performance of the developed algorithms.

References

1. Kim, Y.D., Cichocki, A., Choi, S.: Nonnegative Tucker Decomposition with Alpha Divergence. In: 2008 IEEE International Conference on Acoustics, Speech, and Signal Processing, ICASSP2008, Nevada (2008)
2. Tucker, L.R.: Some Mathematical Notes on Three–mode Factor Analysis. Psychometrika 31, 279–311 (1966)
3. Lathauwer, L.D., Moor, B.D., Vandewalle, J.: A Multilinear Singular Value Decomposition. SIAM J. Matrix Anal. Appl. 21, 1253–1278 (2000)
4. Mørup, M., Hansen, L.K., Arnfred, S.M.: Algorithms for Sparse Nonnegative Tucker Decompositions. Neural Computation (in print, 2008)
5. Carroll, J.D., Chang, J.J.: Analysis of Individual Differences in Multidimensional Scaling via an N-way Generalization of Eckart–Young Decomposition. Psychometrika 35, 283–319 (1970)
6. Phan, A.H., Cichocki, A.: Multi-way Nonnegative Tensor Factorization Using Fast Hierarchical Alternating Least Squares Algorithm (HALS). In: 2008 International Symposium on Nonlinear Theory and its Applications, Budapest (2008)
7. Cichocki, A., Amari, S., Zdunek, R., Kompass, R., Hori, G., He, Z.: Extended SMART Algorithms for Non-Negative Matrix Factorization. In: Rutkowski, L., Tadeusiewicz, R., Zadeh, L.A., Żurada, J.M. (eds.) ICAISC 2006. LNCS (LNAI), vol. 4029. Springer, Heidelberg (2006)
8. Cichocki, A., Zdunek, R., Choi, S., Plemmons, R., Amari, S.: Non-negative Tensor Factorization Using Alpha and Beta Divergences. In: International Conference on Acoustics, Speech, and Signal Processing (ICASSP 2007), pp. 1393–1396. IEEE Press, Honolulu (2007)
9. Lathauwer, L.D., Moor, B.D., Vandewalle, J.: On the Best Rank-1 and Rank-(R1,R2,. . . ,RN) Approximation of Higher-Order Tensors. SIAM J. Matrix Anal. Appl. 21, 1324–1342 (2000)

10. Andersson, C.A., Bro, R.: The N-way Toolbox for MATLAB. Chemometrics and Intelligent Laboratory Systems 52, 1–4 (2000)
11. Bader, B.W., Kolda, T.G.: MATLAB Tensor Toolbox Version 2.2 (2007), http://csmr.ca.sandia.gov/~tgkolda/TensorToolbox/
12. Wu, Q., Xia, T., Yu, Y.: Hierarchical Tensor Approximation of Multi–Dimensional Images. In: 14th IEEE International Conference on Image Processing, vol. 4, pp. 49–52 (2007)
13. Cichocki, A., Zdunek, R.: NMFLAB – NTFLAB for Signal and Image Processing. Technical Report, Laboratory for Advanced Brain Signal Processing, BSI, RIKEN (2006), http://www.bsp.brain.riken.jp
14. The BTF Database Bonn: CEILING Sample, http://btf.cs.uni-bonn.de/download.html

Neural Network Research Progress and Applications in Forecast

Shifei Ding[1,3], Weikuan Jia[2], Chunyang Su[1], Liwen Zhang[1], and Zhongzhi Shi[3]

[1] School of Computer Science and Technology, China University of Mining and Technoogy, Xuzhou 221008
[2] College of Plant Protection, Shandong Agricultural University, Taian 271018
[3] Key Laboratory of Intelligent Information Processing, Institute of Computing Technology, Chinese Academy of Science, Beijing 100080
dingsf@cumt.edu.cn, dingshifei@sina.com

Abstract. This paper roughly reviews the history of neural network, briefly introduces the principles, the features, and the applied fields of neural network, and puts emphasis on discussing the current situation of the latest research from parameters selection, algorithms improvement, network structure improvement and activation function; expounds the effect of neural network in modeling and forecast, from forward direction modeling and reverse direction modeling, describes the principles of modeling and forecast based on neural network in details, analyzes the basic steps of forecast using neural network, then discusses the latest research progress and the facing problems in this field, at last looks forward to the developing trend of this advancing front theory and its applied prospect in forecast.

1 Introduction

In 1943, W.S. McCulloch and W. Pitts established neural network (NN) and its mathematical model, which was called MP model[1]. Then they used MP model to put forward the neuron's formalization mathematical description and network construction method, and proved that each single neuron can perform logic functions, thereby started a new time of the NN research. During more than 60 years of development history, the research of NN can be roughly divided into the following three stages. The initial stage is from the MP model proposed to the 1960s, the main characteristic of this period was to produce the network model and confirm the learning algorithm; then in the developing process people met the essence difficulties which resulted from the electronic circuit overlapping limit, the development of the neural network entering a low ebb period; since the beginning of the 1980s, neural network theory development has entered a golden period. At present neural network has several hundreds of network model, as an active, marginal and cross subject, NN has been widely used in pattern recognition, economic management, control and decision-making, health and medical community, agriculture and many other fields, and has made tremendous achievements. NN theory and application has become the hotspot of intelligence science research.

F. Sun et al. (Eds.): ISNN 2008, Part II, LNCS 5264, pp. 783–793, 2008.

In forecasting work, the precision rate has been constraining the forecast reliability. Most of the previous forecast models are based on the classical statistical mathematical theory, the theory of the fuzzy principle etc., they played an important role in the former application, but they have their own limitations. For the actual need and in order to improve the forecast accuracy, the scholars were forced to look for and establish new theoretical models. Combining the actual problems needed to be processed and the current nonlinear mathematical theory, intelligent computing, powerful mathematical programming language, people proposed the forecast model based on wavelet transform, phase space reconstruction forecast model and the forecast model based on NN etc.. These new theoretical models combine the forefront of scientific theories, and open a new chapter of the study on forecast.

The natural problem of forecast is an input-output system, and the NN can approximate any nonlinear system, this exactly shows the superiority for solving these problems. Therefore, in these new models, the forecasting model which based on NN is the favorite of most people. As the advancing front of modern science, the research is full of challenges and attracts vast numbers of scholars to this field.

2 Neural Network

At present, the definition of NN is not unified. By integrating various views, the definition can be summarized as follows: artificial neural network is based on using computer network system to simulate the intelligent computing of biological NN, to process the information by means of simulating brain network processing, memory information processing through the nonlinear, self-adapted information processing system that is composed by a large number of processing units linked together. On the network each node can be seen as a neuron, it can memory (storage), process some information, and work with other nodes. Solving a problem is to input the information to some nodes, these nodes process and output the results to other nodes, and these nodes receive, process and output, until the whole NN works complete, then the final results are outputted.

2.1 NN Features

NN has the following basic characteristics, non-linear, non-restricted, non-invariance and non-convex. Although NN is different from the real biological neural network, but it has learnt some advantages of biological neural networks: its structure is essentially different from the current computer, it is composed by many small processing units, the function of each processing unit is very simple, but the network can get expected recognition by the congregate and parallel action of a large number of simple processing units, the computation has a high speed; and possess very strong fault-tolerance, that is, the damage of local neurons will not have a very big impact on the global actions; the memory information is stored in the connection weights between the neurons, the content of storage information is not clear from a single weight, it is a distributed storage; its learning function is very powerful, and its connection weights and connection structure can be gained through the study.

NN is a non-procedural, adaptability, cerebral style information procedure, its essence is to gain a parallel and distributed information process functions through the transformation of the network and the dynamics behavior, and simulate the information processing functions of human brain nervous system to some extent. It gets over the disadvantages of the traditional AI which based on logic symbol when dealing with intuition and in unstructured information, and has some characteristics of self-adapted, self-organization and real-time learning.

Based on the powerful fault-tolerance of NN, NN can easily achieve nonlinear mapping process, and has large-scale computing capacity, has been widely used in pattern recognition, economic management and optimization control, and other fields. In the health and medical, agriculture, automation, computer, artificial intelligence (AI) and other fields, NN has been widely applied solved a lot of problems that are hard to solve by traditional methods.

2.2 Current Research Statuses of NN

With the recovery of NN in the 1980s, the research upsurge is raised at home and abroad, and also received its due status in the international academic. Although currently there are several hundreds of NN models, these models have made a number of achievements in the practical application; however, the models themselves are not perfect. At present, the scholars mainly make improvement on the basis of the models, these improvements are embodied in four aspects which are parameter selection, algorithm, activation function, network structure[2]. The main index of evaluating the network performance are: training speed, astringency and generalization ability.

2.2.1 Parameter Selection

Parameter selection will directly affect the speed of the training network, as well as network convergence, convergence performance and etc. This study includes the selection of the initial weight, the identification of the number of hidden layer neurons, the selection of the learning rate.

If the initial selection is high, it may cause some or all of the neurons' net importation to be large; thereby the network runs in a saturation region where the slope of the transfer function is small, this causes that the adjustment range of the weight is small and network falls in a state of paralysis. There are no uniform standards for the selection of initial values, now we often use empirical methods. For instance, memory initial weight method, which sets the initial values of hidden nodes as the small random numbers that are equably distributed close to zero[3]. This can relieve the difficult problems caused by the high selected values.

The number of hidden layers' neurons is related to the network training speed and fault- tolerance directly, the number of the neurons should not be redundant, and the advantages of the network should be fully exerted. It was suggested that the number of hidden layers' neurons should equal half of the input and output neurons or the evolution, this is called "rules of thumb"; Also someone are in favor of the "pyramid rules", that is, the nodes is declining from the input layer to the output layer[4]. At present, people often use test method.

Learning rate also affects the convergence and stability; improper learning rate will lead to slow convergence, and even network falling into paralysis. At present, the common methods are fixed learning rate and mutative self-adapting learning rate, but they all follow the principles that to increase learning rate when the training speed is slow and to reduce the learning rate when the network is emanative.

These studies effectively improve the network convergence and the training speed to a certain extent, but these studies are carried through on the basis of gradient descent, can only use the first derivative of the performance function, and are ultimately "address symptoms but not root causes"

2.2.2 Algorithm Improvement

The improvements towards the algorithms mostly are based on the original algorithms, aiming at the inherent shortcomings of the traditional methods. In current studies, the methods which are more used to solve the problems include optimization algorithms, combining advanced algorithms etc.

Optimization algorithms Quasi-Newton algorithm, conjugate gradient algorithm, Levenberg-Marquardt algorithm, dynamic programming algorithm, Additional momentum algorithm etc., At present, mixed algorithm is used more, such as mixed-using gradient descent algorithm and conjugate gradient algorithm[5]; mixed-using gradient descent algorithm and Davidon-Fletcher-Powell Quasi-Newton algorithm[6]; "hypercube" sample training algorithm[7], and others. However, although these algorithms improve the convergence speed of the network, they waste the storage space.

A combination of advanced algorithms, each algorithm has its own advantages and disadvantages, for instance, back propagation (BP)[8] algorithm has a extensive mapping abilities, but it is easy to fall into local optimum; genetic algorithms[9] is used to deal with the non-smooth even discrete problems with relatively faster processing speed, but the coding is not easy; in addition it is easy to lead to premature phenomena; simulated annealing algorithm[10] doesn't need gradient, but its parameters are difficult to control, besides, there are chaotic algorithm[11], immune algorithm[12], and other methods don't depend on the performance index function very much, are widely applied and robust, and are suitable for parallel computing, but not suitable for global optimization. In view of each algorithm's advantages and disadvantages, combing different algorithms organically may enhance strong points and avoid weaknesses. Such as connecting GA algorithm and BP algorithm in series[13] or in parallel[14], organically combining simulated annealing algorithm and BP algorithm and so on, these combination algorithms have all been tested, and they perform well in the aspect of enhancing the training speed of network and the astringency.

In addition, supporting vector machine algorithm[15] and its model, are current research hot spots, many scholars have proposed new opinions in the aspect of performance function, for example, using "information entropy" to replace the mean square error in BP algorithm, defining the generalized performance index function and so on, this can make the algorithm consider the complexity of the network connection, then may have the possibility to delete the redundant connections and neurons.

At present, the above methods have solved the visible shortage, but it is only the improvement based on the algorithms, aiming at the problems that the algorithms meet. These methods don't change the algorithms essentially. In the research of network algorithms, in addition to the above improvement, there are also some new algorithms are proposed.

2.2.3 Network Structure Improvement

Neural network structure not only affects the generalization ability of the network, but also affects the network training speed. In the network structure, people more consider the information of the system's former state, while the system's dynamic performance is better, the complexity of the network increases. But in fact some of the NN model's computation (learning) time are not related with the number of the neuron very much, but obviously depends on the learning samples. Its substantive question still is the electronic circuit design and realization; the improvement of the network topologic structure[16] and so on. It is noteworthy that although it is the important feature of NN to use the massively parallel processor, but we should also seek for other efficacious methods, to process the complex computation, and to establish the computation theory with network fault-tolerance and toughness.

In addition, some new network models are also established, like Cellular neural network model[17], bidirectional associative memories model[18], Darwinism model, Optical neural network system[19] etc..

2.2.4 Activation Function

Neuron activation function reflects the input-output characteristics of the network unit, the weight revise uses the derivative of massive activation function, and therefore this affects the network's training speed even causes the network to be in the state of paralysis. People used other differentiable limited functions, the product Sigomodial function as activation function[20], have given the constructive description of using NN to approach nonlinear system, moreover, there are trigonometric function multinomial, Gaussian function, wavelet function[21], combination activation function[22] etc.

2.2.5 Other Mixed Theories

The neural network has its own superiority, but in specific applications it appears to be lack of the relevant theories, if organically combining its theory with other theories, then the superiority of the combination system will be exerted in the theory development and system application. This method can be more widely applied in practice. The theory in this aspect has made felicitous progress.

Unifying the fuzzy logic and the neural network, people establish fuzzy neural network[23]. In the network each node and all parameters have the obvious physics significance; the initial values of these parameters can be determined by the fuzzy or qualitative knowledge, and then the network may converge to the input-output relations very quickly by using learning algorithms. The fuzzy neural network provides more efficient and intelligent behavior, learning ability, self-adapting features, parallel mechanism and a high degree of flexibility; also enables it more successfully to deal with indefinite, complex, imprecise and approximate control issues.

Unifying the wavelet theory and the neural network[24], we can take the wavelet function as a primary function to form neural network, or process the status signal by wavelet transformation's multiresolution characteristic, to achieve signal-to-noise separation, and extract the state characteristic which affect the mismachining tolerance most, as the neural network's input.

The combination of the evolutionary algorithms and the neural network[25], evolutionary training of the network connection weight; evolutionary computation of the network structure; evolutionary design of training algorithms and so on, the design of the entire network can be devised through the evolutionary thought.

There're many examples in this aspect, like Chaos Neural Network[26], Rough Neural Network[27], Synergetic Neural Network[28] and so on. Organically combining NN, fuzzy math, wavelet transform, information theory and other intelligent technologies, with basic science like biology, physics, medicine, with economics, linguistics and other literae humaniores, and with all the experience and knowledge, their superiority has also been proved in applications.

The neural network theory is the current hot research in the field of intelligence, and its development both in theory and applications have gained great achievements. But in the research process we will meet new problems, the development of the neural network is also facing new challenges.

3 Forecast Based on Neural Networks

Lapedes et al. used the nonlinear neural network for the first time to carry on studies and forecasts[29] for the time series simulated data which are produced by the computer. Afterward using NN to forecast is gradually popularized in many fields. For example there were many scholars who did forecast studies on actual economic time series data[30], the annual average situation of sunspot activity[31], stock[32] and so on.

Recently, using the new scientific theory to explore new forecast technique is an important direction for the forecast topic research. And the artificial neural network progress quickly in the forecast domain research, it mainly includes the neural network forecast technique and the research of neural network as auxiliary means in the forecasting process. The neural network theory in the forecast technique and the applied area research mainly force on using the neural network to forecast and the regression forecast research for the time series, the neural network combination forecast method and its applied research, the economic early warning research using neural network, the research of nerve fuzzy network forecast model and the ARMA model structure recognition using the neural network etc..

3.1 The Principle of Neural Network Forecast

Artificial neural network's forecast principle[33] can be divided into two methods which are the forward modeling and the reversion modeling. The forward modeling is to training a neural network to express the system's forward dynamic process. In the forward model's structure, the connection of neural network and the under-identified system is parallel connection, the two's output error of the network can be used as the training signal, this is a supervised learning process. The actual system acts as the teacher, which provides the expected output the algorithm needs to the neural network. If the system is a traditional controller, using the multi-layered forward network form; when the system is a performance evaluation, selecting reinforcement learning algorithm, using the network that has global approach ability.

Reversion modeling is to take the inverse forecast's system output as the network's input, compare the network output and system's input, use the corresponding input error to train, and establish the system's inverse model through learning. When establishing the system inverse model, the system's reversibility should be ensured; selecting the sample set the network needs properly.

But it is very difficult to assign the input signal beforehand in practice, because the control objective is to enable the system output have expected movement, it is impossible to provide the expected input regarding the unknown controlled system, and in order to ensure the uniform convergence of parameters estimation algorithm in the system forecast, we must use certain continually active input signal.

3.2 The Forecast Steps of Neural Network

When the neural network is used to carry on the analysis and process to the predicted factors, the superiority will be shown in processing massive nonlinear system. It has the ability to approach any nonlinear systems through learning, using NN for the modeling and identification of nonlinear system, the NN won't be limited by the nonlinear model. The forecasting work can be essentially regard as an input-output system, the transformation relations include data fitness, fuzzy transformation and the logic inference, these may all be expressed by artificial neural network, the forecasting process can be separated to the following steps.

Step 1, regarding the actual problem, collect its related forecasting factors, pretreated these data. The pretreatment enables these data to suit the network's input, and enhance the network's training speed and the astringency.

Step 2, select the network structure according to the question that must be solved, and select some data to train the network and finally confirm the network model.

Step 3, take the data that need to be predicted as the network input, the network output is the forecasting results which we expect, and then carry on analysis and summary to these results.

3.3 The Progress of Neural Network Forecast

At present, the static multi-layer forward neural network is still the most popular system modeling and forecast method in applications; moreover the research of static multi-layer forward neural network is comparatively thorough. Using the static multi-layer forward neural network to establish an input/output model, and its essence is to know the position nonlinear function in the system difference equation by learning based on the approaching ability of the network. Multi-layer forward neural network can get good results in the forecast of static system modeling. But in practice the system that need to be modeled and forecasted mostly is nonlinear time-variable dynamic system, using multi-layer forward neural network should determine the category of the system model in advance, but it is difficult to achieve.

Using the feedback dynamic network to carry on the system modeling and forecast may solve the above difficult problems, it is taken seriously in the practical application, represents the trend of the neural network modeling and forecast. That is mainly because the dynamic network itself is a dynamic time-varying system, regarding the dynamic system modeling, it has the natural ability to reflect the system dynamic change, and it does not need to set the system model type and the order in advance. But studies based on the dynamic network modeling and forecast are far behind static multi-layer forward neural network research. Also the persuasive dynamic network models are lacked, rigorous theory achievement on the dynamic network approaches ability at present are rare, and the learning algorithms also need the further perfection.

Forecast based on neural network still has some blemish, this kind of theory is actually a "black box" method, it is unable to describe and analyze the relations between the forecasting system's input and output, and it is hard make any explanation to the obtained results or make a statistical test about the data; there is no standard method to choose an appropriate neural network structure when using the neural network to forecast, the only way we can take is to spend a lot of time to test, discover the "most appropriate" one from many experiments; so it takes more time and more machine-hours to forecast using neural network.

The main problems of the prediction theory based on NN which need further study from now on are: First, figure out the situation that is suitable to use the network to forecast; Second, how to select the appropriate structure of the neural network; Third, we must further study the basic theory[34] of the neural network forecast. The present neural network forecast is only a "convenience, but superficial" technology, and is not a basic method with solid theoretical principle.

Nevertheless, the study results of NN at the aspect of forecast is felicitous, like in control engineering[35], the power system load[36], earthquake prediction[37], transportation[38], economic prediction[39], health and medicine as well as agriculture, all successfully confirm its credibility.

4 Prospect

Neural network is an active, marginal, and cross subject, after the past half century's development, it has been widely used in pattern recognition, automatic control, machine learning, artificial intelligence and other numerous domains, but it is just a beginning to explore the relations among cerebrum-thought-computation, there is still a very long road, the study about brain's computation principle and complexity, the mechanism and simulation of learning, association and memory have attracted people's attention.

At the aspect of neuron extent of neural network model, we can construct the network with 1000 neurons at present, such extent is quite considerable, but it still has a very big disparity compared with the human body whose neuron number frequently is 10^{10}~10^{15} How to overcome the difficulty of connecting the network is still the key problem in the development process of NN, we are still facing the questions on how to enable the network function to approach the biological neural network as well as enhance the intelligibility of neural network, and how to connect the neurons.

We are waiting for the new breakthrough on the electronic circuit's physical realization and explanations to the human brain's complex behavior. The developments of related mathematics domain, the research of non-linear dynamics model and so on are affecting the neural network's further development. Also the learning rules that the present network uses mostly are from supervised learning and non-supervised learning, and their studies are quite perfect. Semi-supervised learning is the forefront question of the machine learning rule, has been applied in the domains like data mining, pattern recognition and so on, but it is a new attempt to neural network.

The research of neural network has a very wide development prospect, its future development will certainly be exciting, but is also full of challenges. Along with basic theory's progress and computer technology's maturity, we believe the neural network theory's forefront questions will seep in the challenging scientific questions of 21st century.

Along with the maturity of neural network research, the application is also getting more and more widespread. There will be a new upsurge in forecast field based on special superiority of neural network in processing non-linear problems. Forecast based on neural network has seeped in control engineering, economics and many other domains, and there're new breakthroughs of NN that are improved based on related knowledge in improving the forecast precision.

Acknowledgements

This work is supported by the National Natural Science Foundation of China under Grant no.40574001, the 863 National High-Tech Program under Grant no. 2006AA01Z128, and the Opening Foundation of Key Laboratory of Intelligent Information Processing of Chinese Academy of Sciences under Grant no.IIP2006-2.

References

[1] Mccllochw, S., Pitts, W.: A logical calculus of the ideas immanent in nervous activity. Bulletin of Mathematical Biophysics 5, 115–133 (1943)

[2] Lu, J.J., Chen, H.: Researching development on BP neural networks. Control Engineering of China 5, 449–451 (2006)

[3] Wang, W.C.: Neural network and application in automotive engineering. Beijing Institute Technology Publishing House, Beijing (1998)

[4] Gao, D.Q.: On Structures of Supervised Linear Basis Function Feed forward Three-layered Neural Networks. Chinese Journal of Computers 1, 80–86 (1998)

[5] Xu, X., Huang, D.: A new mixed algorithm based on feed-forward neural networks. Journal of East China University of Science and Technology 2, 175–178 (2004)

[6] Wang, Q.H.: Improvement on BP Algorithm in Artificial Neural Network. Journal of Qinghai University 3, 82–84 (2004)

[7] Jenkins, W.M.: Approximate analysis of structural grillages using a neural network. Proc. Instn. Civil Engrs. Structs. Buildings 122, 355–363 (1997)

[8] Mccleland, J.L., Rumlhard, D.E.: Exploration in parallel distributed processing. MIT Press, Cambridge (1986)

[9] Montana, D.J.: Training feed forward neural network using genetic algorithm. In: Proceeding of the Eleventh International Joint Conference on Artificial Intelligence, vol. 3, pp. 762–767 (1989)

[10] Kirpatrick, S., Graltt Jr., C.D.: Optimization by simulated annealing. Science 220, 671–680 (1983)

[11] Li, M.M., Ding, J., Qin, G.H.: BP neural networks model based on chaotic analysis and its application on power load forecasting. Journal of Sichuan University (Engineering Science Edition) 4, 15–18 (2004)

[12] Wang, L., Pan, J., Jiao, L.C.: The immune algorithm. Acta Electronica Sinica 7, 74–78 (2000)

[13] Pi, Y.M., Fu, Y.S., Huang, S.J.: Study on the learning algorithm of BP neural network based on evolutionary computation. Signal Processing 3, 261–264 (2002)
[14] Guo, L., Guo, B.L.: Neural inference based principle of motion decision. Chinese Journal of Computers 3, 225–230 (1995)
[15] Vapnik, V.N.: The nature of statistical learning theory. Springer, New York (1995)
[16] Hogan, M.T., Howard, B., Beale, D.M.: Neural network design. The Math Works Inc., Colorado (1996)
[17] Chua, L.O., Yang, L.: Cellular neural networks theory. IEEE Trans Circuits System 35, 1257–1272 (1988)
[18] Kosko, B.: Bidirectional associative memories. IEEE Transactions On Man System and Cybernitics 18, 49–59 (1988)
[19] Jenkins, B.K.: Optical architectures for neural network implementation, handbook of neural computing and neural networks. MIT Press, Boston (1995)
[20] Bulsari, A.: Some analytical solutions to the general approximation problem for feed-forward neural networks. Neural Networks 6, 991–996 (1993)
[21] Wu, Y.S.: How to choose an appropriate transfer function in designing a simplest ANN to solve specific problems. Science in China (Series E) 4, 105–109 (1996)
[22] Zhang, H.Y., Feng, T.J.: A study on BP networks with combined activation functions. Journal of Ocean University of Qingdao 4, 621–626 (2002)
[23] Satoru, I.: On neural approximation of fuzzy system. In: IEEE Proceedings of INCNN, New York, vol. 1, pp. 1263–1268 (1992)
[24] Zhang, Q., Benveniste, A.: Wavelet networks. IEEE Trans. on NN 4, 889–898 (1992)
[25] Goldberg, D.: Genetic algorithms in search, optimization and machine learning. Addison-Wesley Publish, Massacheusettes (1989)
[26] Ishii, S., Fukumizu, K., Watanabe, S.: A network of chaotic elements for information processing. Neural Networks 9, 25–40 (1996)
[27] Gu, X.P., Tao, S.K., Zhang, Q.: Combination of rough set theory and artificial neural networks for transient stability assessment. In: Proceeding of International Conference on Power system Technology, vol. 1, pp. 19–24 (2000)
[28] Kenneth, R.C., Chua, L.O.: A synergetics approach to image processing in cellular neural networks. In: IEEE International Symposium on Circuits and System, circuits & systems connecting the world proceedings, USA, vol. 3, pp. 134–137 (1996)
[29] Lapedes, A., Farber, R.: Nonlinear signal processing using neural networks: prediction and system modeling. Technical Report LA-UR-87-2662, Los Alamos National Laboratory. Los Alamos. NM (1987)
[30] Varfis, A., Versino, C.: Univariate economic time series forecasting by connectionist methods. In: IEEE ICNN 1990, pp. 342–345 (1990)
[31] Weigend, A.B.: Predicting the future: a connectionist approach. Intl, J. Neur. Sys. 1, 193–209 (1990)
[32] Han, W.L.: Nonlinear analysis and forecasting in China stock market. Northwestern Polytechnical University, Xian (2006)
[33] FECIT technical products R&D center: Neural network theory and application with MATLAB7. Publishing House of Electronic Industry, Beijing (2005)
[34] Liu, B., Hu, D.P.: Studies on applying artificial neural networks to some forecasting problems. Journal of Systems Engineering 4, 338–344 (1994)
[35] Andersen, K.: Artificial neural networks applied to arc welding process modeling and control. IEEE Transactions on Industry Applications 26, 824–830 (1990)

[36] Zhang, Y.X.: The study and application of neural network model optimization for short-term load forecasting. North China Electric Power University, Baoding (2007)
[37] Liu, Y.: Research on neural network ensemble and its application to earthquake prediction. Shanghai University, Shanghai (2005)
[38] Li, C.J.: The study on fusion prediction of traffic-flow volume in urban road based on integreted ann. Southwest Jiaotong University, Chengdu (2004)
[39] Wang, Q.B.: The China's macro-economy research based on artificial neural networks. Shandong University, Jinan (2006)

Adaptive Image Segmentation Using Modified Pulse Coupled Neural Network

Wei Cai[1], Gang Li[2], Min Li[1], and Xiaoyan Li[3]

[1] Xi'an Research Inst. of Hi-Tech, 710025, Xi'an, China
[2] The Second Artillery Military Office in Li-Shan Microelectronics Company, 710075, Xi'an, China
[3] Academy of Armored Force Engineering Department of Information Engineering, 100858, Beijing, China
clwn@163.com

Abstract. This paper proposes a novel image segmentation algorithm based on Pulse Coupled Neural Network (PCNN).Unlike the traditional PCNN image segmentation methods, the presented algorithm can achieve the optimum parameters automatically. Experimental results show its good performance and robustness. The research fruits have great importance both on the theory research and practical application of PCNN.

Keywords: Image segmentation, Pulse coupled neural network (PCNN).

1 Introduction

Image segmentation, as the pretreatment of the pattern recognition and image analysis, is regarded as the bottleneck of computer vision. Because the pulse-coupled neural network (PCNN) has great advantage in image segmentation, it has gained more and more attention in this research field.

PCNN is different from traditional artificial neural networks, models of which have biological background and are based on the experimental observations of synchronous pulse bursts in the cat visual cortex [1]. PCNN can be applied in many fields, such as image processing, image recognition, and optimization [2].

However, it is very difficult to determine the exact relationship between the parameters of PCNN model. Up to now, the parameters are most adjusted manually and it is a difficult task to determine PCNN parameters automatically. During recent years, some work on determining the optimal values of PCNN parameters has been done. Some of them are concentrated on optimizing single parameter while keeping others fixed [3, 4, 5]. Some train the parameters with desired images to achieve the optimal values [6].

Ma Y.D. *et al.* [3] have proposed a new PCNN algorithm of automatically determining the optimum iteration times N based on the entropy of segmented image. It is the criterion of maximal entropy of segmented binary image of PCNN output. According to this criterion, images can be segmented well when the pixel numbers of object and background are nearly the same.

F. Sun et al. (Eds.): ISNN 2008, Part II, LNCS 5264, pp. 794–800, 2008.

Liu, Q., *et al.* [4] have proposed an improved method, in which cross-entropy is put forward to replace maximal Shannon entropy as the criterion of cyclic iterations times *N*. However, the segmented results are lack of adaptability just as the approach in reference [3].

Karvonen, J.A. [6] has presented a method for segmentation and classification of Baltic Sea ice synthetic aperture radar images, based on PCNN. In that paper, Karvonen has adopted a modified PCNN model and given a detail discussion on how to determine parameters automatically. However, as the authors mentioned, a very large set of data representing different sea ice conditions should be required to optimize PCNN parameters, which is unfeasible in most applications.

Since image segmentation is an important step for image analysis and image interpretation, we focus on PCNN applications on image segmentation, establish a modified PCNN model, and propose a multi-threshold approach according to water valley area in histogram. Meanwhile, the adaptive determination method of PCNN parameters for image segmentation is presented.

The rest of this paper is organized as follows. The simple introduction of PCNN neuron model and its principle in image segmentation is described in Section 2. Section 3 proposes adaptive PCNN parameters determination scheme based on water valley area. Experiments will be presented in Section 4, and the last section gives some concluding remarks.

2 PCNN Neuron Model

As showed in Fig. 1, each PCNN neuron is divided into three compartments with characteristics of the receptive field, the modulation field, and the pulse generator.

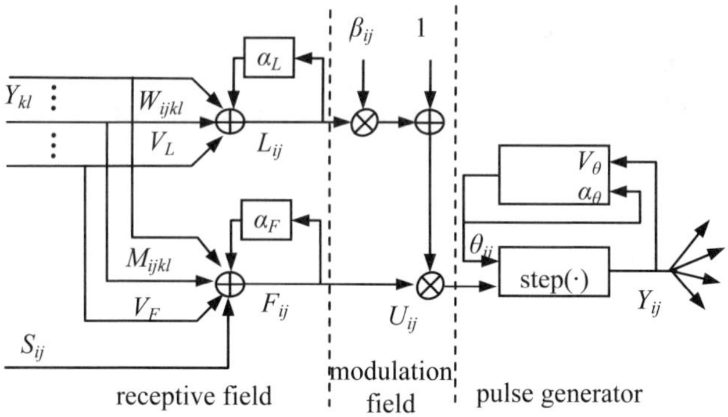

Fig. 1. Traditional PCNN neuron model

Each traditional PCNN neuron model has nine parameters to be determined, including three time decay constants (α_F, α_L, α_θ), three amplification factors (V_F, V_L, V_θ), linking coefficient β_{ij}, linking matrix *M* and *W*. The following five equations are satisfied.

$$F_{ij}(n) = \exp(-\alpha_F) \cdot F_{ij}(n-1) + S_{ij} + V_F \cdot \sum M_{ijkl} Y_{kl}(n-1) \ . \tag{1}$$

$$L_{ij}(n) = \exp(-\alpha_L) \cdot L_{ij}(n-1) + V_L \sum W_{ijkl} Y_{kl}(n-1) \ . \tag{2}$$

$$U_{ij}(n) = F_{ij}(n)(1 + \beta_{ij} \cdot L_{ij}(n)) \ . \tag{3}$$

$$\theta_{ij}(n) = \exp(-\alpha_\theta)\theta_{ij}(n-1) + V_\theta Y_{ij}(n-1) \ . \tag{4}$$

$$Y_{ij}(n) = step(U_{ij}(n) - \theta_{ij}(n)) \ . \tag{5}$$

Where *step*(•) is the unit step function. Moreover, to the whole neural network, the iteration times N should also be decided. The various parameters used in the PCNN model are of great significance when preparing the PCNN for a certain task.

In the application of image segmentation, each pixel corresponds to a single PCNN neuron. That is, a two dimensional intensity image (M×N) can be thought as a PCNN neuromime with M×N neurons, and the gray level of pixels can be thought as S_{ij}, the input of the neuron. The neurons are organized in a single layer network to perform the segmentation task. Considering M and W are the interior linking matrixes. When there are pixels whose gray levels are approximate in the neighborhood of M and W, one pixel's pulsating output can activate other corresponding pixels having the approximate gray level in the neighborhood and let them generate pulsating output sequence $Y(n)$. Obviously Y contains some information about this image such as regional information, edge, and texture features. Then the binary image constructed by $Y(n)$, the output of PCNN, is the segmented image. This is why the PCNN achieved the image segmentation.

The performance of segmentation results based on PCNN depends on the suitable PCNN parameters. It is necessary to determine the near optimal parameters of the network to achieve satisfactory segmentation results for different images. Up to now, the parameters are most adjusted manually and it is a difficult task to determine PCNN parameters automatically for different kinds of images.

3 Adaptive Parameters Determination Method of PCNN

3.1 Multi-threshold Approach Using Water Valley Area Method

In this paper, we propose the definition of 'water valley area' to determine multi-threshold in image segmentation. Assume $hist(f(x,y))$ is the histogram of image $f(x,y)$; S_i (i=1,2,…,K) is the maximum points on $hist(f(x,y))$; Q_j (j=1,2,…,N) is the minimum points on $hist(f(x,y))$; P_m (m=1,2,…,M+1) is the peak points, which satisfied with $P_1 < P_2 < \ldots < P_{M+1}$; T_n (n=1,2,…,M) is the multi-thresholds, which satisfied with $T_1 < T_2 < \ldots < T_M$. P_m and T_n are unknown and waiting for solution. Obviously, $P \subseteq S$ and $T \subseteq Q$.

Assume S_{i1} and S_{i2} is two maximum points of $hist(f(x,y))$, whose corresponding gray value is g_{Si1} and g_{Si2} respectively, and $g_{Si1} < g_{Si2}$. If there is no other maximum points in (g_{Si1}, g_{Si2}) or the value of existed maximum points is small than min{S_{i1}, S_{i2}}, we define a water valley between S_{i1} and S_{i2}. The bottom of water valley is the borderline of $hist(f(x,y))$, and the height of water valley is min{S_{i1}, S_{i2}}. Imagine we can use 'water' to abound the whole space, then the capacity can be defined as 'water valley area', $area(g_{Si1}, g_{Si2})$, the calculation formula is

$$area(g_{Si1}, g_{Si2}) = \frac{1}{2}\int_{g_{Si1}}^{g_{Si2}} \{[\min\{S_{i1}, S_{i2}\} - hist(x)] + |\min\{S_{i1}, S_{i2}\} - hist(x)|\} dx \quad (6)$$

Assume Q_j is the minimum point of (g_{Si1}, g_{Si2}), namely for $\forall g_x \in (g_{Si1}, g_{Si2})$, $hist(g_x) \le Q_j$ is satisfied, we use $valley(S_{i1}, Q_j, S_{i2})$ to denote water valley.

The detailed process to get peak points and multi-thresholds is given below.

*Step1.*Draw image histogram *hist(f(x,y))*and smooth it to decrease noise influence if necessary.

Step2. Seek all extremum points in the histogram, including maximum points S_i (i=1,2,…,K) and minimum points Q_j (j=1,2,…, N). For the need of building water valley, the extremum points on two sides of *hist(f(x,y))* must be maximum points, so $K = N + 1$.

Step3. From the left minimum point Q_1 and maximum points S_1、S_2 on its two sides($S_1<Q_1<S_2$), we built water valley, $valley(S_l, Q_c, S_r)$(l= 1, c=1, r=2), and calculate its area, $A=area(g_{Sl}, g_{Sr})$, by formula.(6).

Step4. Determine multi-thresholds and peak points, here, define Θ as a lower limitation ranging from 0.01 to 0.03. The smaller the value of Θ is, the more threshold points we will get.

(1) If $A \geqslant \Theta$, Q_c will be kept in threshold array T_n. Meanwhile, S_l will be kept in peak points array P_m. $S_l = S_r$, $Q_c = Q_r$, and $S_r = S_{r+1}$.

(2) If $A < \Theta$, the valley will be taken as invalid. At this situation, compare the value of S_l and S_r:

(i) if $S_l > S_r$, then S_l will be regarded as the new left maximum point, S_{r+1} is the new right maximum point. The smaller of Q_c and Q_r is minimum point in new water valley.

(ii) if $S_l \le S_r$, then S_r、Q_r、S_{r+1} will be the left maximum point, minimum point and right maximum point of new water valley.

Step5. Calculate water valley area, $A=area(g_{Sl}, g_{Sr})$, by formula.(6) and iteratively execute step 4 until all minimum points have been processed.

At last, we can get the threshold array T_n (n=1,…M and $T_1<\ldots<T_M$) and the corresponding peak array P_m(m=1,…M+1 and $P_1<\ldots<P_{M+1}$). Hence, a valid water valley $valley(P_m, T_m, P_{m+1})$ includes two neighboring peaks $\{P_m, P_{m+1}\}$.

3.2 Modified Pulse Coupled Neural Network

We have established a modified PCNN, which is implemented by applying iteratively the equations

$$L_{ij}[n] = \sum W_{ijkl} Y_{kl}[n-1] \quad (7)$$

$$U_{ij}[n] = S_{ij}(1 + \beta_{i,j}[n] L_{ij}[n]) \quad (8)$$

$$Y_{ij}[n] = \begin{cases} 1, & U_{ij}[n] > T_{ij}[n] \\ 0, & \text{otherwise.} \end{cases} \quad (9)$$

The indexes i and j refer to the pixel location in the image, indexes k and l refer to the dislocation in a symmetric neighborhood around a pixel, and n refers to the time (number of iteration). $L_{ij}[n]$ is linking from a neighborhood of the pixel at location (i,j), $U_{ij}[n]$ is internal activity at location (i,j) , which is dependent on the signal value S_{ij} at (i,j) and linking value. $\beta_{i,j}[n]$ is the PCNN linking parameter, and $Y_{ij}[n]$ is the output value of the PCNN element at (i,j). $T_{ij}[n]$ is a threshold value. We use a set of fixed threshold values, $T_n(n=1,\ldots M)$ determined by water valley area method mentioned above.

If $Y_{ij}[n]$ is 1 at location (i,j) at $n=t$, we say that the PCNN element at the location (i,j) fires at t. The firing due to the primary input S_{ij} is called the natural firing. The second type of firing, which occurs mainly due to the neighborhood firing at the previous iteration, we call the excitatory firing, or secondary firing.

Starting with the biggest threshold T_M, object whose mean gray value is larger than T_M will be picked out at the first iteration. We keep the threshold T_M fixed during the following iterations until no firing happens. At a certain threshold, the iteration times differ from image to image and a suitable amount of iterations in practice is 20-70. After the first iteration loop, both the natural firing pixels and excitatory firing pixels are collected, which is the first level PCNN segmented objects with the largest gray value. Then the second level objects can be got by the same algorithm using threshold T_{M-1}. Repeating this progress until all thresholds are processed, we will get M+1 levels of objects with different intensities at last.

In this PCNN algorithm, we are using the neighborhood with the radius r=1.5 (i.e., a usual 3×3 neighborhood, with the linking relative to the inverse of the squared distance from the midpixel and normalized to one, namely

$$W = \frac{1}{6.828}\begin{bmatrix} 0.707 & 1 & 0.707 \\ 1 & 0 & 1 \\ 0.707 & 1 & 0.707 \end{bmatrix}.$$

Considering those pixels whose intensities are smaller than peak point P_m ought not to be captured at T_n even if they have the largest linking value 1, so in the iteration loop at T_n, the value of β_m is chosen to be

$$\beta_m = \frac{T_n}{P_m} - 1 \tag{10}$$

Because P_1 may be 0, we choose the value of β_1 to be 0.1-0.3 at this situation.

4 Experiments

In order to comparing with the performance with traditional PCNN (showed as Fig.1), the experiments that PCNN used in image fusion applications are carried out.

As Fig. 2 shown, Fig. 2 (a) is the source image. Fig. 2 (b) is the segmented results by traditional PCNN. Fig. 2(c) shows the corresponding water valleys and multi-thresholds determined by water valley area. The segmented results by modified PCNN, which parameters are determined by water valley area, are showed as Fig.2(d).

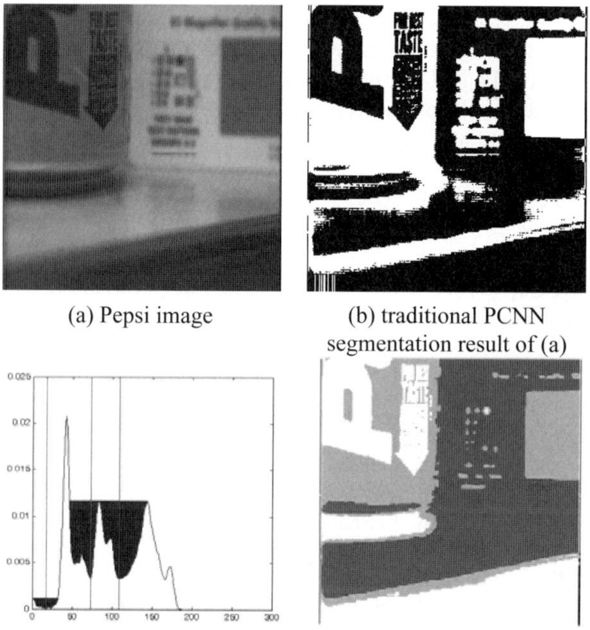

(a) Pepsi image

(b) traditional PCNN segmentation result of (a)

(c) water valleys and multi-hresholds of (a)

(d) modified PCNN segmentation result of (b)

Fig. 2. The "Pepsi" source images (256 level, size of 512×512) and segmented results

Table.1 shows the parameters of traditional PCNN which were tuned by trial in order to get perfect segmentation performance.

Table 1. Value of parameters in image segmentation using traditional PCNN

Parameters	β	α_F	α_L	α_θ	V_F	V_L	V_θ	r	N
Fig. 2(b)	0.3	1	4	2	10	10	100	1	2

From Fig. 2, we can see the images segmented by the proposed method provides more details and useful information. For example, the characters on the clock are all clearly in Fig. 2 (d) while '10' 、 '11'、 '12' is not clearly in Fig. 2 (b). The idea of multi-threshold makes segmented image more levels than traditional PCNN. We must mention that comparing with the parameters of traditional PCNN, which were tuned by trial, the parameters in our method can be determined automatically, this has great importance in expanding the application range of PCNN.

5 Conclusion

This paper brings forward an adaptive segmentation algorithm based on a modified PCNN with the multi-thresholds determined by water valley area method. The main contributions include establishing a modified PCNN, proposing adaptive PCNN parameters determination algorithm based on water valley area, and implementing the described methods on PCNN applications. Experimental results show its good performance and robustness. The research fruits have great importance both on the theory research and practical application of PCNN.

References

1. Eckhorn, R., ReitBoeck, H.J., et al.: Feature Linking via Synchronization among Distributed Assemblies: Simulation of Results Form Cat Visual Cortex. Neural Computation 2, 293–307 (1990)
2. Johnson, J.L., Padgett, M.L.: PCNN Model and Applications. IEEE Trans. on Neural Networks 10, 480–498 (1999)
3. Ma, Y.D., Shi, F., Li, L.: Gaussian Noise Filter Based on PCNN. In: IEEE Int. Conf. Neural Networks & Signal Processing, Nanjing, China, pp. 149–151 (2003)
4. Liu, Q., Ma, Y.D., Qian, Z.B.: Automated Image Segmentation Using Improved PCNN Model Based on Cross-Entropy. Journal of Image and Graphics 10, 579–584 (2005)
5. Gu, X.D.: Feature Extraction Using Unit-Linking Pulse Coupled Neural Network and Its Applications. Neural Processing Letters 27, 25–41 (2008)
6. Karvonen, J.A.: Baltic Sea Ice SAR Segmentation and Classification Using Modified Pulse-Coupled Neural Networks. IEEE Trans. Geoscience and Remote Sensing 42, 1566–1574 (2004)
7. Bi, Y.W., Qiu, T.S.: An Adaptive Image Segmentation Method Based on a Simplified PCNN. ACTA Electronica Sinica 33, 647–650 (2005)

Speech Emotion Recognition System Based on BP Neural Network in Matlab Environment

Guobao Zhang, Qinghua Song, and Shumin Fei

School of Automation, Southeast University, Nanjing 210096, P.R. China
guobaozh@seu.edu.cn, songqing1018@163.com

Abstract. An emotion recognition system based on BP neural network to recognize special human affective states existed in the speech signal is presented in this paper. About 600 short sentences with different contents in different emotional speeches from 4 speakers are collected for training and testing the feasibility of the system. The energy, pitch and speech rate characteristics are extracted from these speech signals. angry, calm, happy, sad, and surprise as the 5 typical emotions are classified with BP Neural network. In order to update automatically the emotion recognition system with time, an additional study step, just as the feed-back control, is adopted to train the finished network again according to the output. The experiments show that the system is of satisfactory emotion detection performance for some emotions.

Keywords: Emotional Speech, Recognition of Emotion, BP Neural Network, MATLAB.

1 Introduction

Emotion recognition by computer is a topic which has been researched recent years, and become more and more important with the development of the artificial intelligence. An effective human emotion recognition system will help to make the interaction between human and computer more natural and friendly. It has many potential applications in areas such as education, entertainment, custom service etc.

As one of major indicators of human affective state, speech plays an important roll in emotion recognition. With this and its potential uses in mind , we design a context independence system in speech-based emotion recognition in this paper. The way people covey their affective state will be changed with age, the state of health and the season, so the system must adapt itself to these aspects. So the system must have the ability to study, and update with times. The ultimate goal of this research is to make a Personal Robot Pet which can recognize its host's emotion.

The remainder of this paper is organized as follows. Section 2 describes the design and implementation of our system. Section 3 presents the experiments and the results. Conclusions and discussions are given in section 4.

F. Sun et al. (Eds.): ISNN 2008, Part II, LNCS 5264, pp. 801–808, 2008.

2 System Design

2.1 Processing Flow

Fig. 1 illustrates the processing flow of our system which is divided into two main parts: speech processing and emotion recognition. First, some pre-processing should be done to get the effective speech period by using the Cool Edit Pro 1.2a, Including filter and intercept the speech period(determine the beginning and end points in Fig. 3).Next, the features for each speech are extracted and compiled into a feature vector.

In the training stage, the feature vector is used to train the neural network. In the recognition stage the feature vector is applied to the already trained network and the output is a recognized emotion. According to the result, the system can study in different way. If the result is right, the "self-study" will work, if wrong, the "guide-study" will work.

These steps are explained further in the following sections.

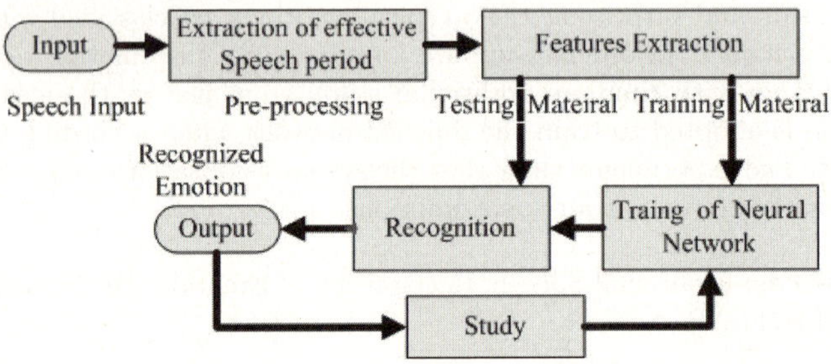

Fig. 1. The Processing Flow of Emotion Recognition System

2.2 Emotions and Features

Emotions: How to classify emotions is an interesting and difficult issue. A wide investigation on dimensions has been performed in the past, but researchers still haven't established a standard so far. Different researchers on emotion recognition diff on the num of categories and the kinds of categories to use [1-2].

Our motivation to do this study is to make a Personal Robot Pet which can recognize its host's emotion, so it is enough to select common basic emotions: angry, calm, happy, sad, surprise.

Features: Speech signal is short-term stable, so we estimate short-term acoustic features in this emotion recognition system. Some features have been proved to be useful for emotion recognition in many papers.

- speech power, pitch,12 LPC parameters, Delta LPC parameter [3].
- energy, median of F1,variance of duration of energy plateaus, minimum of F1, median of F0, mean F3, maximum/mean duration of energy plateaus, variance of F0 [4].

- signal energy, sub-band energies, spectral flux, zero- crossing rate, fundamental frequency, MFCC,FFBE [5-6].
- pitch average, pitch variance, intensity average, intensity variance, jitter pitch tremor, shimmer intensity tremor, speech rate [7-8].

After examining these examples, we have selected to use the following six features in this study:

speech rate, max-Energy, mean-Energy, mean-ZCR(zero- crossing rate), max-Pitch, mean-Pitch(fundamental frequency).

2.3 BP Neural Network Architecture

The network is composed of five sub-neural networks in this paper. with one network for each of the five emotions that are examined. As to each speech, the recognition processing flow is diagrammed in Fig. 2(a).The features vector is input into each of the five sub-networks, then each sub-network will give an output (v1,v2,v3..v5)that represents the likelihood to the sub-network' emotion. At last, the Logic Decision selects the "best" emotion based on these outputs.

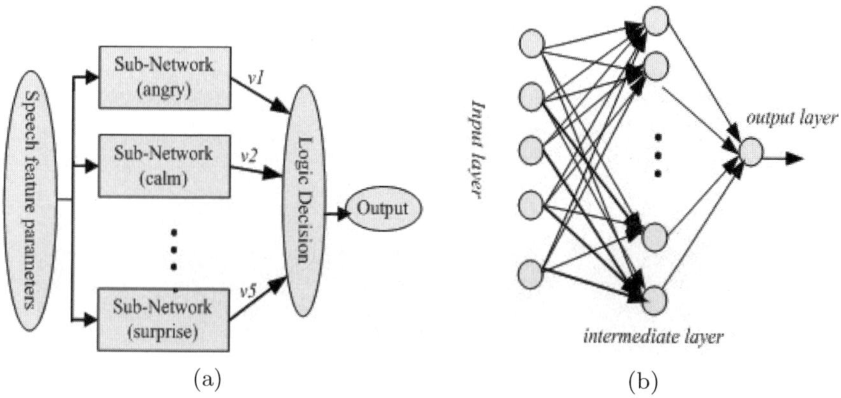

Fig. 2. (a)The Steps of Recogniton (b)Sub-Network Configuration

Backpropagation network can approximate a function, associate input vectors with specific output vectors, or classify input vectors in an appropriate way as defined by you. So we take the BP network as the sub-networks. Fig. 2(b) illustrates the configuration of each of the five sub-networks. Each network consists of three layers, one input layer six nodes, one intermediate layer with twelve nodes and one output layer with one nodes, which is an analog value 0.99 or 0.01 for training utterance. If the emotion of input utterance is the same as the sub-network, the value is 0.99, else 0.01.

Using separate sub-networks for the five emotions allows each network to be adjusted separately, and we are able to easily change each network separately without redesigning the entire system which is good for network's study and update for the feature.

Pre-processing should be done to get the effective speech period by using the Cool Edit Pro 1.2a, Including filter and intercept the speech period(determine the beginning and end points in Fig. 3).

3 Experiments and Results

3.1 Database

A record of emotional speech data collections is undoubtedly useful for researchers interested in emotional speech analysis. In this study, emotional speech recordings were made in a recording studio, using a professional-quality microphone and digital audio record software-Cool Edit Pro1.2a (Fig. 3), at a sampling rate of 11025, using a single channel 16-bit digitization. We have pleased four classmates-two male, two female, who are good at acting, to served as subjects. Each subject uttered a list of 30 Chinese sentences five times, one time for each of the five emotions. When finished, these speech database must be examined by a listening test, If disqualification, will be deleted and recorded again.

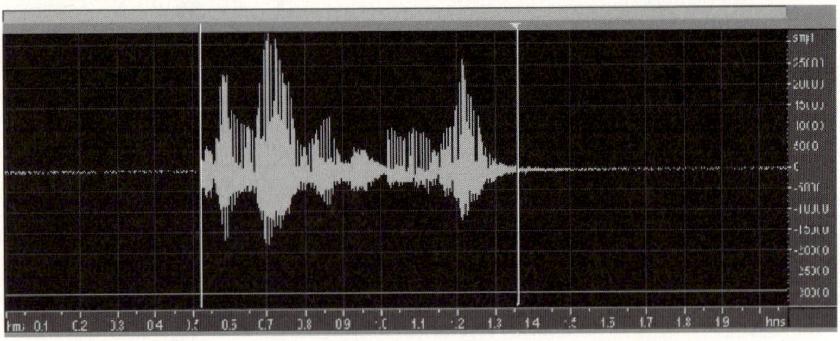

Fig. 3. Cool Edit Pro 1.2a

At last, some pre-processing should be done to get the effective speech period in the Cool Edit Pro, Including filter and determine the beginning and end points in Fig. 3.

3.2 Extract Features

We adopt the short-term analysis method to estimate acoustic features in this study. Short-term features are estimated on a frame basis:

$$f_s(n, m) = S(n)W(m - n) . \tag{1}$$

Where S(n) is the speech signal and is a window of length N ending at sample m[7]. Methods for estimating the fundamental phonation or pitch and speech energy are discussed as follows.

*Pitch:*The pitch signal, also known as the glottal wave-form, has information about emotion, because it depends on the tension of the vocal folds and the subglottal air pressure. The pitch signal is produced from the vibration of the vocal folds. Two features related to the pitch are widely used, namely the pitch frequency and the glottal air velocity. In this study, we compute the pitch frequency based on the cepstrum method with order 3 median post-filtering for every frame, and take elect the mean pitch and max pitch as features.

Energy: The short-time speech energy can be exploited for emotion recognition, because it is related to the arousal level of emotions. In this study, the short-time energy of the speech frame ending at m is:

$$E_s(m) = \sum_{n=N+1}^{m} |f_s(n,m)| \ . \qquad (2)$$

ZCR: zero-crossing rate , Compute the number of zero-crossings within a frame, based on the formula (3) and take the mean ZCR as the emotion feature.

$$Z_n = \frac{1}{2} \sum_{n=N+1}^{m} |sgn[s_n(m)] - sgn[s_n(m-1)]| \ . \qquad (3)$$

We can see the features clearly from the friend GUI in Fig. 4.Button down the button [**...**],you can select the speech file, then button down the [**Extract**],the extracting-feature program will be called back. In addition, the results will be return and show in the GUI in real time.

Click the [**Save**] button to save the features according to the emotion kind. and the label 'Num' reflects the number of the saved features. These features consist of a feature vector which is used as the input. The two upper Radio

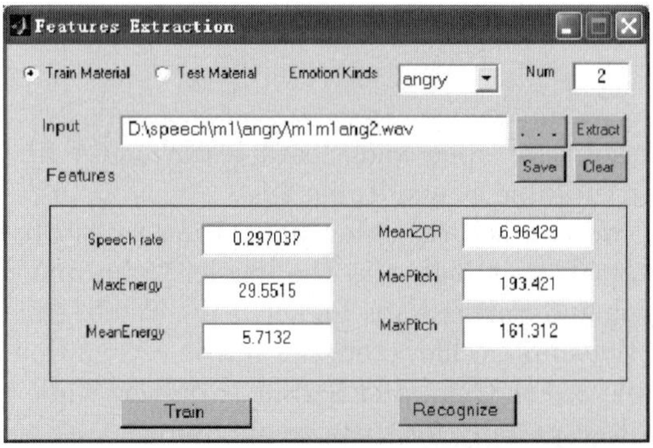

Fig. 4. The Extraction of Features

Button can decide that the feature vector is used for training or recognition. We choose 20 sentences as training materials for each emotion, and another 10 sentences as testing materials.

3.3 Train and Testing Network

The networks can be trained after getting the feature vectors, once the button [**Train**] is clicked, the five sub-networks will be trained automatically in order-angry, calm, happy, sad, surprise.

After training, the network is tested using both open and closed testing for the speakers separately. In the closed testing, the network is tested using the same set of data on which it was trained. In open testing, the net wok is tested using the remainder of data which was not used for training.

Fig. 5 presents the result as a active picture If the result is wrong, we can select the right emotion from the popup menu and click the [**guide-study**] to retrain the sub-network. If the result is right, we also can make the system memorize it through clicking the [**self-study**]. Therefore, this system isn't quiescent, it can update with the time.

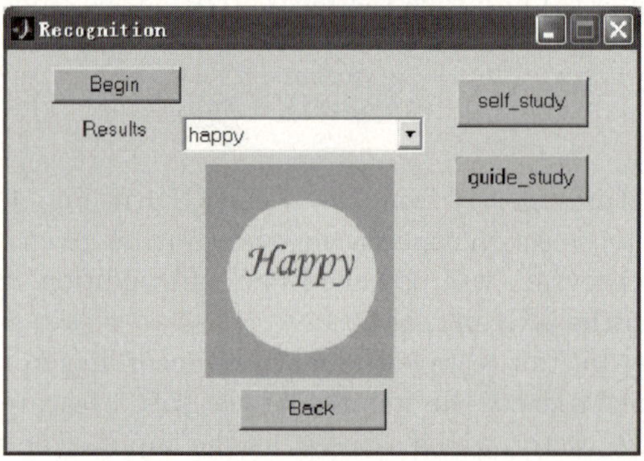

Fig. 5. The Result of Recognition

3.4 Results

The recognition results for open and closed testing after first training to one speaker are showed in Table 1. and Table 2.

From Table 1., we can see that the recognition rates are different for different emotions though It is higher in the whole. Calm and Sad 's rate is the highest,100%.Happy and Surprise's is higher, 95%. While Angry 90%.

Table 2. shows the similar characteristics in the recognition rate for different emotions with Table 2. For Calm and Sad ,it is higher, while it is lower for the other three emotions,especially for surprise only 60%.

Comparing the two tables, the closed recognition rates are very higher than the open recognition for a small number of Training subjects (20 every emotion).This

Table 1. Closed Results for One of Speakers

input/output	angry	calm	happy	sad	surprise	puzzled
angry	18	0	0	0	1	1
calm	0	20	0	0	0	0
happy	0	0	19	0	0	1
sad	0	0	0	20	0	0
surprise	1	0	0	0	19	0

Table 2. Open Results for One of Speakers

input/output	angry	calm	happy	sad	surprise	puzzled
angry	7	0	0	1	0	2
calm	0	10	0	0	0	0
happy	0	1	7	0	0	2
sad	0	0	0	9	0	1
surprise	0	1	2	0	6	1

is because the network is very finely tuned for that one speaker and cannot deal well with the general case. As the number of speech used for training increase, the network will becomes more general and the open recognition rates increase; however the closed recognition rates decrease.

4 Conclusions and Discussions

This paper proposes a personal context independent system for emotion recognition in speech using neural networks. We have designed and implemented it in Matlab environment. From the results, we can see that the recognition accuracy for some emotions is satisfying but others is low.

The results obtained in this study demonstrate that emotion recognition in speech is feasible and neural networks are well suited for this task. And the idea to make a robot pet with emotion which can communicate emotions with its host is possible to come true.

There is still more work to be done in the field of emotion recognition in speech. The speech features used in this study may allow the number of features to be reduced. And if we want to design a speaker independent system, we should increase other features, such as the first formant, the variance of the first formants. In addition, feature trials with different topologies of neural networks may also help improve performance. The combination of audio and visual data will convey more information about the human emotional state than speech alone. The complementary relationship of these two modalities on different emotion will help us to achieve higher recognition accuracy.

References

1. Bhatti, M.W., Wang, Y., Guan, L.: A Neural Network Approach for Human Emotion Recognition in Speech. In: Proceedings of the 2004 International Symposium on Circuits and Systems, pp. 181–184 (2004)
2. Murry, I.R., Arnott, J.L.: Applying an Analysis of Acted Vocal Emotions to Improve the Simulation of Synthetic Speech. Computer Speech and Lauguage, 107–129 (2008)
3. Nicholson, J., Takahashi, K., Nakatsu, R.: Emotion Recognition in Speech Using Neural Networks. Neural Computing & Applications, 290–296 (2000)
4. Zhongzhe, X., Dellandrea, E., Weibei, D., et al.: Features Extraction and Selection for Emotional Speech Classification. In: Conference on Advanced Video and Signal Based Surveillance, pp. 411–416 (2005)
5. Temko, A., Nadeu, C.: Classification of Acoustic Events Using SVM-based Clustering Schemes. Patttern Recognition, 682–694 (2006)
6. Fragopanagos, N., Taylor, J.G.: Emotion Recognition in Human-Computer Interaction. Neural Networks, 389–405 (2005)
7. Jiang, X.Q., Tain, L., Han, M.: Seperability and Recognition of Emotion States in Multilingual Speech. In: Proceedings on Communications, Ciruites and Systems, pp. 861–864 (2005)
8. Amir, N.: Classifying Emotions in Speech: a Comparison of Methods. In: EUROSPEECH 2001 Sandinavia 7th European Conference on Speech Communication and Technology 2th Interspeech Even, pp. 127–130 (2001)
9. Ververidis, D., Kotrropoulos, C.: Emotional Speech Recognition: Resource, Features, and Methods. Speech Communication, 1162–1181 (2006)

Broken Rotor Bars Fault Detection in Induction Motors Using Park's Vector Modulus and FWNN Approach

Qianjin Guo[1], Xiaoli Li[2], Haibin Yu[1], Wei Hu[1], and Jingtao Hu[1]

[1] Key Laboratory of Industrial Informatics, Shenyang Institute of Automation, Chinese Academy of Sciences, Shenyang 110016, China

[2] Department of Resource Engineering, University of Science and Technology Beijing, Beijing 100083, China

guoqianjin@sia.cn

Abstract. In this paper a new integrated diagnostic method based on the current Park's Vector modulus analysis and fuzzy wavelet neural network classifier is proposed for the diagnosis of rotor cage faults in operating three-phase induction motors. Detection of broken rotor bars has long been an important but difficult job in the detection area of induction motor faults. The characteristic frequency components of a faulted rotor in the stator current spectrum are very close to the power frequency component but by far less in amplitude, which brings about great difficulty for accurate detection. In order to overcome the shortage of broken rotor bars characteristic components being submerged by the fundamental one in the spectrum of the stator line current, Park's Vector modulus(PVM) analysis is used to detect the occurrence of broken rotor bar faults in our work. Simulation and experimental results are presented to show the merits of this novel approach for the detection of cage induction motor broken rotor bars.

Keywords: Fault Detection, Induction Motors, Park's Vector Modulus, FWNN.

1 Introduction

Online fault diagnostics of induction motors are crucial to ensure safe operation, timely maintenance and increased operation reliability. As new data processing techniques and new technologies for electrical drives emerge, monitoring, fault detection and diagnosis of electrical machines are becoming more and more important issues in the field of electrical machines.

In past decades, a number of different incipient fault detection methods and schemes have been presented and tested [1,2]. Special attention has been devoted to non-invasive methods, which are capable to detect faults using measured data without disassembling the machine or its structural parts. Motor Current Signature Analysis (MCSA) is a non-invasive and simple-to-implement detection method that can be used on-line to detect rotor bar failures by examining characteristic harmonic components in the frequency

[1] Foundation item: Project supported by the Shenyang Inst. of Automation, CAS Foundation for Youths((Grant No. O7A3110301).

F. Sun et al. (Eds.): ISNN 2008, Part II, LNCS 5264, pp. 809–821, 2008.

spectrum of the stator current of the motors using Fast Fourier Transformation (FFT) [1–9]. MCSA fully takes into account the operational conditions of the machines and can be implemented online non-invasively using existing current transducers. However, the validity of this technique relies on the assumption that the characteristic harmonics can be distinguished from the other harmonics. The slip of a usual induction motor under rated condition is small; it is smaller under light load or no load. It means that the characteristic frequency $(1\pm2s)f_l$ of broken rotor bars is very close to the supply frequency f_1, and the broken rotor bar characteristic components are always submerged by the fundamental component, and it is well known that it is sometimes difficult to filter out the fundamental component and highlight fault characteristics.

Recent efforts in motor current signature analysis (MCSA) are focused on sampling three phase currents simultaneously and converting them to other reference frames to highlight fault characteristics. In Park's Vector approach [5-7], three phase currents are converted to stationary reference frame (D, Q) to get components i_D and i_Q, and the track i_D and i_Q is related to the type and the extent of fault. The Extended Park's Vector Approach [7] is based on the spectrum analysis of the square of the current Park's Vector modulus. In this way, the spectrum of AC level of the spectrum of the square of the current Park's Vector Modulus is clear from any component at the fundamental supply frequency, making it more useful to detect the components directly related to the motor fault.

Conventional broken rotor bars fault detection schemes merely monitor the sideband sequence component and other higher harmonic components of line current and rely on the mathematical models for symmetrical induction machines, which neglect inherent asymmetries and lead to misdetection [8]. In recent several years, some methods such as parameters estimation and intelligent modeling have been proposed to diagnose fault for induction motors [9]. These methods work correctly under steady state but fail to detect the slowly developing fault under transient state due to the highly complex, non-linear nature of turn fault. Fuzzy wavelet neural network (FWNN) has the ability to achieve nonlinear dynamic mappings with simple structure, rapidly convergence and easily implementation [10], therefore it can be used to detect broken rotor bar fault in induction motors at various conditions. This paper focuses on fault diagnosis of three-phase squirrel cage induction motors with adjustable speed drive system applications by using a fuzzy wavelet network based system.

2 Current Park's Vector Modulus Analysis Approach

2.1 Current Park's Vector Modulus Analysis Approach

A two-dimensional representation can be used for describing three-phase induction motor phenomena, a suitable one being based on the stator current Park's vector [6]. As a function of mains phase variables (i_a, i_b, i_c), the current Park's vector components (i_d, i_q) are:

$$\begin{bmatrix} i_d \\ i_q \end{bmatrix} = \frac{N}{M} \begin{bmatrix} 1 & -0.5 & -0.5 \\ 0 & \frac{\sqrt{3}}{2} & -\frac{\sqrt{3}}{2} \end{bmatrix} \begin{bmatrix} i_a \\ i_b \\ i_c \end{bmatrix} \tag{1}$$

Under ideal conditions, i.e., when the motor supply currents constitute a positive-sequence system, the Park's Vector has the following components:

$$i_d = \left(\sqrt{6}/2\right) I_m \cos(\omega t) \tag{2}$$

$$i_q = \left(\sqrt{6}/2\right) I_m \sin(\omega t) \tag{3}$$

where "I_m," "ω," and "t" are, respectively, the peak value of the supply phase current (amperes), the angular supply frequency (radians per second) and, finally, the time (seconds). Given these conditions, the current's Park vector modulus is constant [7].

2.2 Broken Rotor Bars Spectral Analysis of the PVM of Motor Current

In the presence of rotor cage faults, such as broken rotor bars, the motor current spectrum will contain sideband components. These additional components at frequencies of $(1\text{-}2s)f_1$ and $(1\text{+}2s)f_1$ will also appear in both of the motor current's Park vector components (i_d, i_q). In fact, in the presence of a rotor fault, motor supply currents can be expressed as:

$$i_a = I_m \cos(\omega t - \alpha) + I_l \cos((1-2s)\omega t - \beta) + I_h \cos((1+2s) - \gamma) \tag{4}$$

$$i_b = I_m \cos(\omega t - \alpha - \frac{2\pi}{3}) + I_l \cos((1-2s)\omega t - \beta - \frac{2\pi}{3}) + I_h \cos((1+2s) - \gamma - \frac{2\pi}{3}) \tag{5}$$

$$i_c = I_m \cos(\omega t - \alpha + \frac{2\pi}{3}) + I_l \cos((1-2s)\omega t - \beta + \frac{2\pi}{3}) + I_h \cos((1+2s) - \gamma + \frac{2\pi}{3}) \tag{6}$$

where I_m, I_l and I_h are the peak values of, respectively, the fundamental supply phase current, the currents lower sideband component at frequency $(1\text{–}2s)f_1$, and the currents upper sideband component at frequency $(1\text{+}2s)f_1$, due to the torque oscillations; three angles α, β and r denote the initial phase angle, respectively, for fundamental supply current, the currents lower sideband component and, finally, the currents upper sideband component.

The fault signatures naturally will also appear in the i_d and i_q signals. The Parks vector transform is used to isolate the fault signature:

$$\begin{aligned} I_s(t) = \left| i_d + j i_q \right|^2 &= \frac{3}{2}(I_m^2 + I_l^2 + I_h^2) + 3 I_m I_l \cos(2s\omega t - \alpha + \beta) \\ &+ 3 I_m I_h \cos(2s\omega t + \alpha - \gamma) + 3 I_l I_h \cos(4s\omega t + \beta - \gamma) \end{aligned} \tag{7}$$

It is apparent that the spectrum of the current's Park vector modulus is the sum of a DC level, generated mainly from the fundamental component of the motor supply current, and two additional terms, at frequencies of $2sf_1$ and $4sf_1$, with no component at the fundamental supply frequency. This simple manipulation will help in identifying the level of the fault, just by rejecting the DC component of the signal and analyzing the AC component.

3 Broken Rotor Bars Fault Detection Using PVM and FWNN

A schematic diagram of the proposed fault detection strategy based on PVM and FWNN is shown in Fig.1. Key components of the fault detector are: 1)a data preprocessor and

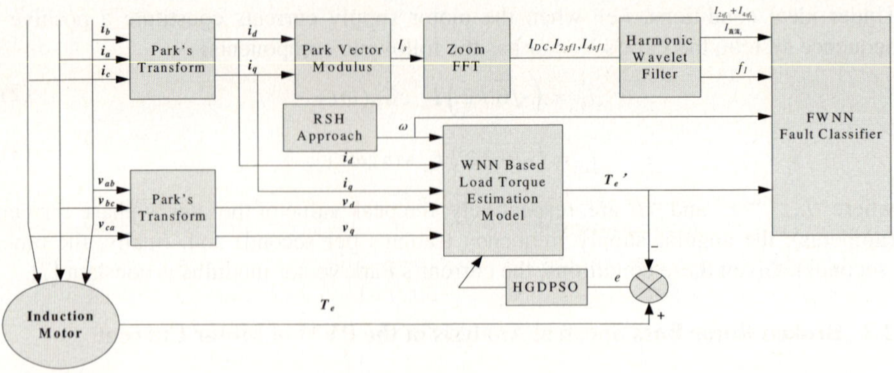

Fig. 1. The scheme of broken rotor bars fault detection using PVM and FWNN

filter by using zoom FFT and harmonic wavelet filter;2)a load torque estimation model using wavelet neural network (WNN), and 3) a fuzzy wavelet neural network model of the steady-state characteristics of a class of induction motors.

3.1 Speed Estimation—RSH Approach

When the processing of the motor diagnosis is included in the motor-drive systems, slip frequency is easily obtained from the outputs of speed sensors such as encoders. However, if diagnosis processing is independent of motor-drive systems, slip frequency is gathered only from the stator-current spectra in an MCSA-based diagnosis system. So, effective sensorless speed estimation is desirable for slip frequency detecting. One of the best ways to obtain the information of the slip frequency is to use rotor-slot harmonics (RSHs) as in [11].The mechanical speed information of an induction machine is embedded in the stator currents. The slots produce a continuous variation of the air-gap permeance in squirrel cage induction motors. During operation, the rotor slot MMF harmonics will interact with the fundamental component of the air-gap flux because of the rotor currents. Therefore, the air-gap flux will be modulated by the passing rotor slots, producing rotor slot harmonics (RSH).

A number of researches have been performed to extract rotor speed from RSH [12–14]. The induction motor speed ω, can be calculated using RSH at any slip condition from the following expression:

$$\omega = \frac{60}{R}(f_{sh} \pm v f_1) \tag{8}$$

where R is the number of rotor slots and is the number of poles, and v=1, 3, 5, …. Since the rotor slot harmonics are related to the rotor currents, their magnitudes reduce with decreasing load making it difficult to detect RSH. In the absence of detectable RSH, the use of eccentricity-related harmonics has been proposed [15].

3.2 Load Torque Estimation in Induction Motors

In this study, the wavelet neural network approach (WNNs) were applied to estimate torque demanded by the load coupled onto the induction motor shaft. The main objective here is in using wavelet neural networks to estimate load behavior on the motor shaft. The WNN employed in this study are designed as a three-layer structure with an input layer, wavelet layer (hidden layer) and output layer. The topological structure of the WNN is illustrated in Fig.2, where w_{jk} denotes the weights connecting the input layer and the hidden layer, u_{ij} denote the weights connecting the hidden layer and the output layer. In this WNN models, the hidden neurons have wavelet activation functions of different resolutions, the output neurons have sigmoid activation functions. A similar architecture can be used for general-purpose function approximation and system identification.

The activation functions of the wavelet nodes in the wavelet layer are derived from a mother wavelet $\psi(x)$, suppose $\psi(x) \in L^2(R)$, which represents the collection of all measurable functions in the real space, satisfies the admissibility condition [16]:

$$\int_{-\infty}^{+\infty} \frac{|\hat{\psi}(\omega)|^2}{\omega} d\omega < \infty \tag{9}$$

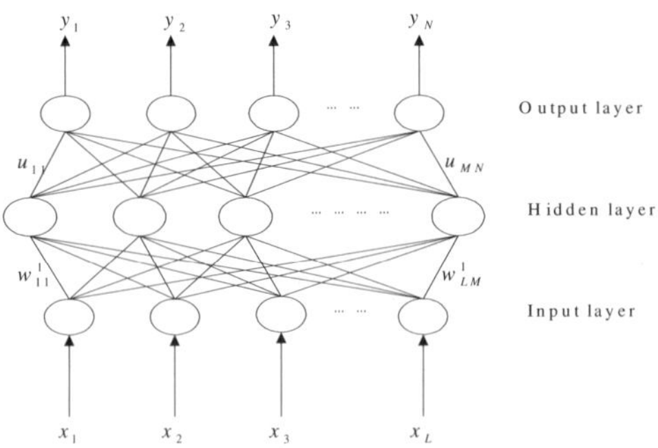

Fig. 2. The WNN topology structure

where $\hat{\psi}$ (x) indicates the Fourier transform of $\psi(x)$. The output of the wavelet neural network Y is represented by the following equation:

$$y_i(t) = \sigma(x_n) = \sigma\left(\sum_{j=0}^{M} u_{ij}\psi_{a,b}\left(\sum_{k=0}^{L} w_{jk} x_k(t)\right)\right)(i = 1,2,..., N) \tag{10}$$

where $\sigma(x_n) = 1/(1 + e^{-x_n})$, y_j denotes the j_{th} component of the output vector; x_k denotes the kth component of the input vector; u_{ij} denotes the connection weight between the output unit i and the hidden unit j; w_{jk} denotes the weight between the hidden unit j and input unit k; a_j , b_j denote dilation coefficient and translation coefficient of wavelons in hidden layer respectively; L,M, N denote the sum of input, hidden and output nodes respectively.

The network is trained with a hybrid algorithm integrating PSO with gradient descent algorithm in batch way [16]. As a result, the network model in this paper is constructed by using HGDPSO as training algorithm and Morlet mother wavelet basic function as node activation function [16].The input data of the wavelet neural network are motor speed, RMS current value and voltage. The output to be computed by the wavelet neural network is the load torque. The block diagram of this structure is presented in Fig.2, which illustrates the input/output relationship involved with the load torque estimation process. From this illustration, it is important to highlight that the type of load is not an explicit input to the wavelet neural network. However, in relation to each implemented simulation, a load type was assumed whose characteristics are implicitly represented by the input parameters of the network (voltage, current, speed), which are simulated from the transient to the steady state.

3.3 Broken Rotor Bars Fault Classifier Using FWNN

3.3.1 Symptoms Analysis and Model-Based Fault Detection

It is required that the model of broken bar faults should reflect the broken bar amount and various locations of broken bars, however, this process is very complicated. As a matter of fact, we often simplify the model. From Section 2, the ratio of additional components of $2sf_l$ to the DC component will increase when the rotor bar breakage increases. The ratio of additional components of $4sf_l$ to the DC component will increase also. Then, a severity effect factor can be defined as:

$$F_r = \frac{I_{2sf_1} + I_{4sf_1}}{I_{DC}} \tag{11}$$

where F_r is the effect severity rotor fault, I_{2sfl} and I_{4sfl} is the amplitude of $2sf_l$ and $4sf_l$sidebands respectively, and I_{DC} is the amplitude of the DC component of the current's Park vector modulus.

From Equation. (7) it is evident that the rotor bar frequencies are a function of the machine slip. The slip can be considered to be one of the symptoms for the diagnosis system of rotor bar broken faults. When the motor is healthy, the stator currents are not fully balanced. This unbalance depends on the load condition. As a result, the load condition should be considered as an indicator for the symptoms of the rotor bar breakage fault. So, when an induction motor is under analyzed, the slip and load condition must be considered, in order to improve the sensitivity and reliability of the diagnosis system. The practical task of induction motor fault diagnosis is to realize mapping between the symptoms and the faults, and this mapping can be implemented by function approximation. Function approximation can be described in mathematical terms as the following:

$$n_r = f(F_r, T_e, s) \tag{12}$$

$$s = \frac{\omega_1 - \omega}{\omega_1} \tag{13}$$

Where n_r is the severity rotor fault, $f(x)$ stands for a continuous real approximate function belonging to set X, F_r is the effect severity rotor fault, T_e is the load torque, s is the machine slip, ω is the rotor speed, ω_l is the synchronous speed.

There are two ways to solve problems concerning the approximate function. One is the parameter algorithm, that is, to get answers via analytical expression. The other is the nonparameter algorithm. The parameters estimation work correctly under steady state but fail to detect the slowly developing turn fault under transient state due to the highly complex, non-linear nature of turn fault. Fuzzy wavelet neural network (FWNN) has the ability to achieve nonlinear dynamic mappings with simple structure, rapidly convergence and easily implementation [17], therefore it can be used to detect rotor bar broken fault in induction motors at various conditions.

3.3.2 Broken Rotor Bars Fault Classifier Using FWNN

The basic configuration of the FWNN system includes a fuzzy rule base, which consists of a collection of fuzzy IF-THEN rules in the following form

$$R^l: \text{ IF } x_l \text{ is } F_l^l \text{ and ... and } x_m \text{ is } F_m^l \text{ THEN } y_l \text{ is } G_l^l \text{ and ... and } y_n \text{ is } G_n^l$$

where R^l is the lth rule ($(1 \le l \le S)$, $\{x_i\}_{i=1,\dots,m}$ are input variables, $\{y_j\}_{j=1,\dots,n}$ are the output variables of the FWNN system, respectively, F_i^l are the labels of the fuzzy sets characterized by the membership functions(MF) $\mu_{F_i^l}(x_i)$, G_1^l are the labels of the fuzzy sets in the output space.

A schematic diagram of the four-layered FWNN is shown in Fig.3[17]. Nodes in layer 1 are input nodes representing input linguistic variables. Nodes in layer 4 are output nodes representing output linguistic variables. Nodes in layer 2 are term nodes that act as membership functions. Each membership node is responsible for mapping an input linguistic variable into a possibility distribution for that variable. Thus, together all the layer 3 nodes formulate a fuzzy rule basis. Links between layers 3 and 4 function as a connectionist inference engine. The rule nodes reside in layer 3. The semantic meaning and function of the neurons in the proposed fuzzy neural network are as follows. A detailed description of neurofuzzy systems can be found in [17].

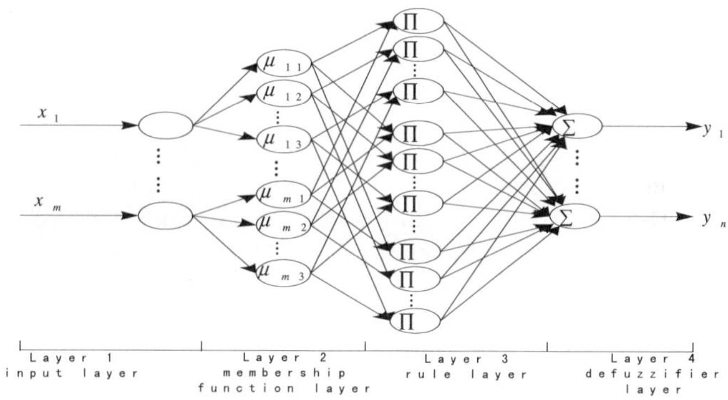

Fig. 3. FWNN architecture

In general, a good learning algorithm must have fast learning as well as good computational capacity and generalization capacity. Here, the gradient descent (GD) learning algorithm with adaptive learning rate is introduced [17]. The adaptive learning rate guarantees the convergence and speeds up the learning. The task of the learning algorithm for this architecture is to tune all the modifiable parameters, namely wavelet node parameters, *a, b* and FWNN weights, w_{ij}, to make the FWNN output match the training data, the cost function can be written as:

$$E = \frac{1}{2}(D - Y)^T(D - Y) \tag{14}$$

where *D* is the desired output acquired from specialists, and *Y* is the FWNN's current output. For the multi-output case, $Y=[y_1, y_{2,\ldots,} y_n]$.

4 Simulation and Experimental Study

4.1 Test-Bed Description

A special test bed has been designed in order to validate the proposed analytical expression of the stator current frequency components for the three-phase induction machine. The test bed has voltage (JSJW-6) and current (LZX-10) sensors connected to the same data acquisition board (PCI-1713) through voltage amplifiers to scale the magnitude and low-pass antialiasing filters to set the frequency bandwidth to a correct range (Fig.4).

The induction machine used in experimental tests was a three-phase, type Y100L2-4, 50 Hz, 2-pole,3kW, rated at 380 V, 6.82A, and 1420 rpm, driving a large DC motor via a flexible coupling. The DC motor acted as a generator and its power output was dissipated in a variable resistor bank. The characteristics of the 3-phase induction motor used in our experiment are listed in Table 1.

Fig. 4. Test bed used for experimental results

Table 1. Induction motor technical specification used in the experiment

Description	Value
Power	3kW
Input Voltage	380 V
Full Load Current	6.82A
Supply Frequency	50Hz
Number of Poles	2
Number of Rotor Slots	20
Number of Stator Slots	36
Full Load Torque	2.2kg.m
Full Load Speed	1420rpm

With the proposed experimental setup, the first objective is to validate the stator current harmonic analysis, at rated load in healthy conditions and under rotor faults. Several measurements were made, in which the stator current waveform was acquired for a given number of broken rotor bars. Current measurements were performed for a healthy motor and also for the same machine with different number of broken rotor bars.

4.2 Signal Processing for Diagnosis

Fig.5 shows elliptic pattern of an asynchronous motor used in the experimental investigation. Fig.5(b) shows the elliptic pattern of the current Park's Vector of the motor with healthy. Fig.5(c) shows the elliptic pattern of the motor with one broken rotor bar. We can learn from the two figures that the elliptic pattern of the motor with one broken rotor bar itself in the deformation of the current Park's vector pattern corresponding to a healthy condition. But the deformation is too slightly to judge the extent of the broken rotor bars.

Just as showed in Fig.6, Fig.6(a) shows the spectrum of the square of the Park's Vector modulus of the motor with healthy, at the same time, Fig.6(b) shows the spectrum of the motor with one broken rotor bars. Fig.6(c) shows the spectrum of the motor with two broken rotor bars. Fig.6 (d) shows the spectrum of the motor with nine broken rotor bars. We can learn from these figures that the spectrum of the motor with broken rotor bars has a stick up obviously in the $2ksf_l$ components, so we can judge the broken of the motor.

As comparison between Figures6 (a), (b) (c) and (d) shows that the components at $2sf_l$ and $4sf_l$ have increased respectively with the broken rotor bar increased. The results obtained show that there is an increase in the value of the components at $2sf_l$ and $4sf_l$ with the extension of the fault, making it a good indicator about the condition of the machine, these components can be used as the severity indicator of broken rotor bars.

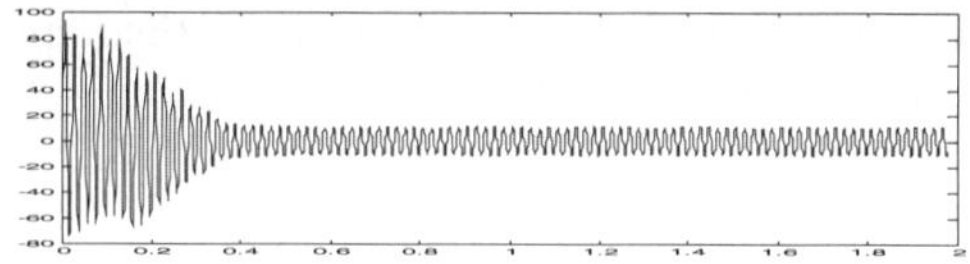

(a) Stator current in phase c at 50Hz with broken rotor bars under 11.9N.m load

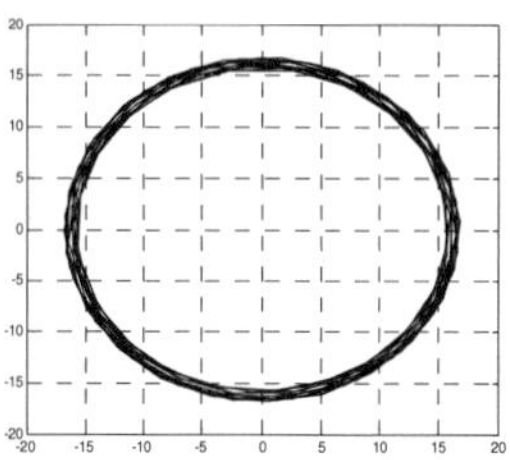

(b) Current Park's vector pattern for the healthy motor

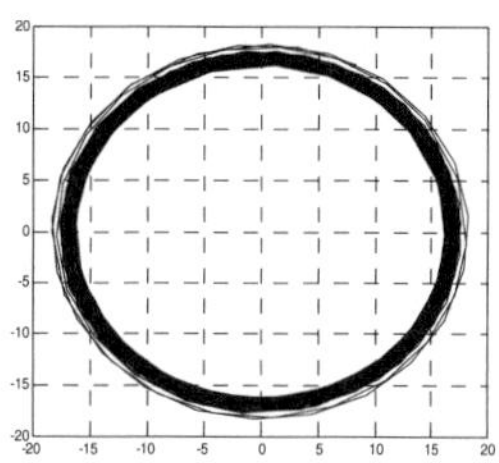

(c) Current Park's vector pattern with one broken rotor bar

Fig. 5. Current Park's Vector pattern for the case of rotor broken

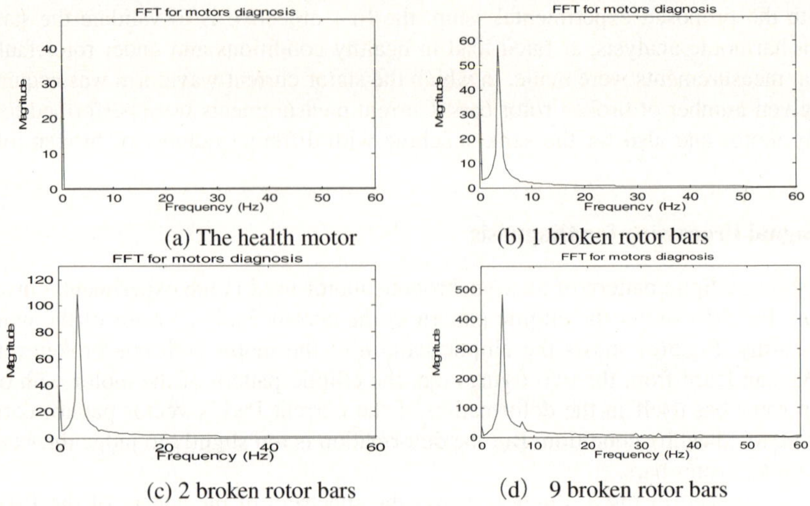

Fig. 6. The spectrum of the square of the Park's Vector modulus under 11.9N.m load

4.3 Feature Extraction

The feature extraction of the signal is a critical step in any monitoring and fault diagnosis system. Its accuracy directly affects the final monitoring results. Thus, the feature extraction should preserve the critical information for decision-making. In this paper, the statistical information of time-base data and frequency-base data were used for obtaining the feature information from the measured signals. As regards the amplitude of these current Park's Vector modulus sidebands, they depend on three factors: the motor's load inertia, the motor's load torque, and the severity of the fault (See Table.2).

When there are broken or even fissured bars, as it was theoretically and experimentally predicted, side bands reveal faults more clearly with high values of slip. Then it is recommended that the diagnosis were carried out with the motor running near its nominal load.

The final membership functions for the severity effect factor, slip ratio, and load torque are given in Fig.7 to Fig.9 for the FWNN system. The linguistic values of the slip ratio can be 'very small','small', 'medium', 'large', or 'very large' (labeled as

Table 2. Feature extraction by stator Park's Vector modulus spectrum analysis

Motor health status	Load Torque $T_e(N.m)$	Slip ratio $s(\%)$	Severity effect factor $F_r(\%)$
health	11.6	0.06	0.001
1 broken rotor bar	11.6	0.06	0.135
2 broken rotor bar	11.6	0.06	0.269
9 broken rotor bar	11.6	0.06	0.407

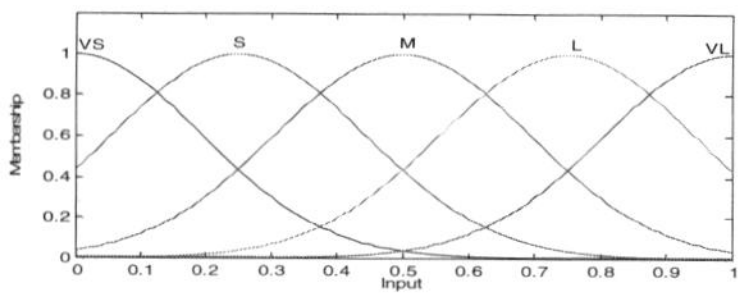

Fig. 7. The membership functions of the input linguistic value for severity effect factor F_r

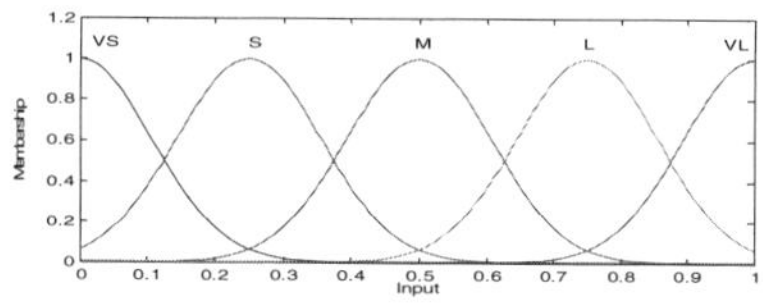

Fig. 8. The membership functions of the input linguistic value for slip ratio s

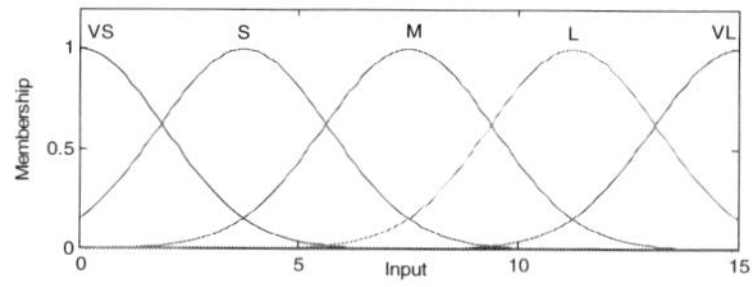

Fig. 9. The membership functions of the input linguistic value for load torque T_e

VS, S, M, L and VL respectively), denoting the heavy, medium and light slip ratio conditions, respectively. Morlet wavelet function is chosen as the membership functions for these five fuzzy sets (shown in Fig. 7). The linguistic values of the load toque can be 'very small','small', 'medium', 'large', or 'very large' (labeled as VS, S, M, L and VL respectively), denoting the heavy, medium and light load conditions, respectively. Morlet wavelet function is chosen as the membership functions for these five fuzzy sets (shown in Fig. 8). The linguistic values of the severity effect factor can be 'very small','small', 'medium', 'large', or 'very large' (labeled as VS, S, M, L and VL respectively), denoting the heavy, medium and light the severity effect factor conditions, respectively. Morlet wavelet function is chosen as the membership functions for these five fuzzy sets(shown in Fig. 9).

4.4 Fault Classification Using FWNN

Once the fuzzy wavelet neural network based motor fault detection system has been initialized, it is to be trained for motor faults by actual motor data. Through training, the system will modify the fuzzy membership functions and fuzzy rules based upon the input data and its initial fault detection heuristic knowledge through the fuzzy wavelet neural network based fault detector network weights. These modifications made by the system will provide valuable heuristic information about the actual form of the fuzzy membership functions and the validity of the fuzzy rule base, as well as providing the exact fault detection results.

At the end of the training period, when all of the operating conditions have been learned and classified, the network is switched to a fault sensing mode. When a spectral signature falls outside the trained clusters, it is tagged as a potential motor failure. Since a fault condition is not a spurious event but continues to degrade the machine, the postprocessor alarms the user only after multiple indications of a potential failure have occurred. In this way, the time history of the machine is incorporated into the monitoring system and protects the fuzzy wavelet neural network from alarming on random signals that have been incorrectly identified.

The results obtained with the Fuzzy systems can be seen in the second (T_e),third (F_r) and forth (f_l) column of Table 3 and corresponding values for the output variable nr is shown in the forth column. From these results it can be concluded that the Fuzzy systems can reliable detect whether the induction motor contains broken bars and reliable predict the number of broken bars.

Table 3. Detect broken rotor bars in induction motors by Park's Vector modulus

Motor health status	Torque T_e(N.m)	Severity effect factor F_r (%)	Supply frequency f_l(Hz)	The estimated number of broken rotor bar n_r
health	11.9	0.0004	50	0
1 broken rotor bar	11.9	0.0019	50	1
2 broken rotor bar	11.9	0.0035	50	2
9 broken rotor bar	11.9	0.0160	50	9

5 Conclusions

The paper described a system for broken bar detection of induction machine based on Park's Vector modulus and Fuzzy Wavelet Network. In order to overcome the shortage of broken rotor bars characteristic components being submerged by the fundamental one in the spectrum of the stator line current, Park's Vector modulus analysis is used to detect the occurrence of rotor faults in our work. The system was implemented and tested using a prototype having different number of broken bars in the rotor. Based on the results obtained with the systems it can be stated that the current Park's Vector modulus analysis and Fuzzy Wavelet Network based system developed for the broken bar detection proven to be are in good agreement with the practice, as it is capable to detect the correct number of rotor broken bars.

References

1. Thomson, W.T.: A Review of Online Condition Monitoring Techniques for Three-phase Squirrel Cage Induction Motors—Past Present and Future. In: IEEE SDEMPED 1999, pp. 3–18. IEEE Press, Spain (1999)
2. Schoen, R.B., Lin, T., et al.: An Unsupervised, On-line System for Induction Motor Fault Detection Using Stator Current Monitoring. J. IEEE Transactions on Industry Applications 31(6), 1280–1286 (1995)
3. Thomson, W.T.: On-line MCSA to Diagnose Shorted Turns in Low Voltage Stator Windings of 3-phase Induction Motors Prior to Failure. In: International Electric Machines and Drives Conference, Cambridge, MA, USA, pp. 891–898 (2001)
4. Kliman, G.B.R., Koegl, A.J., et al.: Noninvasive Detection of Broken Rotor Bars in Operating Induction Motors. J. IEEE Transactions on Energy Conversion (4), 873–879 (1988)

5. Cardoso, A.J.M., Cruz, S.M.A., Fonseca, D.S.B.: Interturn Stator Winding Fault Diagnosis in Three-phase Induction Motors by Park's Vector Approach. J.IEEE Trans. Energy Conversion 14, 595–598 (1999)
6. Cardoso, J.M.A., Mendes, M.S., Cruz, S.M.A.: The Park's Vector Approach: New Developments in On-line Fault Diagnosis of Electrical Machines, Power Electronics and Adjustable Speed Drives. In: Proc. IEEE Int. Symp. Diagnostics for Electrical Machines, Power Electronics and Drives, Gijón, Spain, pp. 89–97 (1999)
7. Cruz, S.M.A., Cardoso, A.J.M.: Stator Winding Fault Diagnosis in Three-phase Synchronous and Asynchronous Motors, by The Extended Park's Vector Approach. J. IEEE Trans. Industry Appl. 37(5), 1227–1233 (2001)
8. Kliman, G.B., Premelani, W.J., et al.: A New Approach to On-line Turn Fault Detection in AC Motors. In: IEEE-IAS Annual meeting, pp. 687–693. IEEE Press, New York (1996)
9. Nandi, S., Toliyat, H.A.: Condition Monitoring and Fault Diagnosis of Electrical Machines – A Review. In: IEEE-IAS Annual meeting, pp. 197–204. IEEE Press, Phoenix (1999)
10. Wang, W.Y., Lee, T.T., Liu, C.L.: Function Approximation Using Fuzzy-neural Networks with Robust Learning Algorithm. J. IEEE Trans. Syst. Man, Cyber. 27(4), 740–747 (1997)
11. Blasco-Gimenez, R., Asher, G.M., et al.: Performance of FFT Rotor Slot Harmonic Speed Detector for Sensorless Induction Motor Drives. Proc. Inst. Electr. Eng.—Electr. Power Appl. 143(3), 258–268 (1996)
12. Hurst, K.D., Habetler, T.G.: Sensorless Speed Measurement Using Current Harmonic Spectral Estimation in Induction Machine Drives. J. IEEE Trans. Power Electronics 11(1), 66–73 (1996)
13. Blasco-Gimenez, R., Hurst, K.D., Habetler, T.G.: Comments on 'Sensorless Speed Measurement Using Current Harmonic Spectral Estimation in Induction Machine Drives' [and reply]. J. IEEE Trans. Power Electronics 12(5), 938–940 (1997)
14. Ferrah, A., Bradley, K.J., et al.: A Speed Identifier for Induction Motor Drives Using Realtime Adaptive Digital Filtering. J. IEEE Trans. Industry Applications 34(1), 156–162 (1998)
15. Ferrah, P.J., Bradley, K.J., et al.: The Effect of Rotor Design on Sensorless Speed Estimation Using Rotor Slot Harmonics Identified by Adaptive Digital Filtering Using the Maximum Likelihood Approach. In: IEEE Thirty-Second IAS Annual Meeting, pp. 128–135. IEEE Press, Los Alamitos (1997)
16. Guo, Q.J., Yu, H.B., Xu, A.D.: A Hybrid PSO-GD Based Intelligent Method for Machine Diagnosis. J. Digital Signal Processing 16(4), 402–418 (2006)
17. Guo, Q.J., Yu, H.B., Xu, A.D.: Wavelet Fuzzy Neural Network for Fault Diagnosis. In: International Confer. of Communications, Circuits and Systems, pp. 993–998. IEEE Press, HongKong (2005)

Coal and Gas Outburst Prediction Combining a Neural Network with the Dempster-Shafter Evidence

Yanzi Miao[1,2], Jianwei Zhang[2], Houxiang Zhang[2], Xiaoping Ma[1], and Zhongxiang Zhao[1]

[1] China University of Mining and Technology, Xuzhou 221008, Jiangsu Prov. P.R.China
[2] TAMS Group, University of Hamburg, 22527 Hamburg, Germany
{miao,zhang}@informatik.uni-hamburg.de

Abstract. A novel prediction method combing a neural network with the D-S evidence theory for coal and gas outbursts is put forward in this paper. We take advantage of the fact that the non-linear input-output mapping function of the neural network can handle the non-linear parameters from coal and gas outburst monitor systems. And the output of the neural network is taken as the basic probability of the assignment function of the D-S evidence theory, which resolves the main problem of establishing the BPAF for the D-S evidence theory. The results from our experiments show that it is feasible and effective to combine the neural network with the D-S evidence theory for deciding on predictions. And using this method, we can make a more certain and credible prediction decision than witheach independent method.

Keywords: Neural Network, Dempster-Shafter Evidence Theory, Coal and Gas Outburst Prediction.

1 Introduction

Gas disasters, especially the outburst of gas, are very dangerous and complex phenomena in coal mines. Outbursts are hazardous through the mechanical effects of particle ejection and by asphyxiation from the gas produced. The violence of an outburst has frequently tossed miners back several meters from the face of a heading. In a number of cases the dislodged coal, which frequently consist of small particles, has engulfed the operators, preventing them from escaping while the released gas asphyxiated them [1]. So it is necessary and significant to develop a useful, accurate method to monitor the potential warning signs of outbursts, and to use an intelligent theory to predict the disaster in advance.

The coal bed conditions vary a great deal in China, and a coal and gas outburst is the synthetic result of stress, gas pressure, and the physical mechanics of coal and so on. Some researchers found that the mechanism of outbursts is very complicated, and there are many geological factors associated with it [1, 2]. In addition, the warning signs before an outburst are unexpected and differ from case to case. After analysing all

F. Sun et al. (Eds.): ISNN 2008, Part II, LNCS 5264, pp. 822–829, 2008.

existing methods of outburst prediction, we therefore conclude that the intelligent method of patterning the human brain is the only potent approach enabling us to consider the multiple associated factors and make a precise prediction.

The Artificial neural network (ANN) [3-5] method such as back propagation (BP) neural nets has been a highly feasible and creative technique to solve traditional problems of misreport or missing-report in the past. ANN can map highly nonlinear input-output relationships; the experiences it obtains include not only those of human knowledge, but also those which may be still unknown to human experts. Another important function of ANN is its ability to interpolate and extrapolate from its experiences. But the conventional BP neural nets converge slowly and have local minima and oscillation problems [6]. As a more effective uncertain reasoning method among the information fusion methods, the Dempster-Shafer evidence theory (D-S evidence theory) has been paid special attention due to the advantages of the uncertain denotation, measurement and combination, and its wider applications [7-10]. As it can be combined with other methods, the D-S Evidence Theory is more widely usable and can be extended very well in the future.

In this paper, we describe a novel prediction method of combing the advantages of the neural network with the D-S evidence theory for coal and gas outbursts. The non-linear input-output mapping function of a neural network can handle the non-linear parameters of coal and gas outburst signs. And the output of the neural network could be taken as the Basic Probability of the Assignment Function (BPAF) of the D-S evidence theory, which resolves the main problem of establishing the BPAF for the D-S evidence theory. After the introduction, in section 2, the model of our coal and gas outburst prediction based on an artificial network and the D-S evidence theory will be given, which describes the main structure of the prediction system. In Section 3, we analyze the experiments we carried out using our new method. In the end, the related experimental results and conclusion are given in Section 4.

2 Model of Coal and Gas Outburst Prediction Based on an Artificial Neural Network and the D-S Evidence Theory

As we know, neural networks are capable of nonlinear mapping and are widely used in pattern recognition [6]. As a neural network has a parallel structure and a parallel processor for nonlinear mapping, we chose some non-linear measurements from factors associated to the outburst as the input of the neural network, such as the type of roadway, the depth, the thickness and obliquity of the coal seam, geological structures, and mining methods. And the output of the neural network is taken as the basic probability of assignment (BPA), which resolves the difficulty of establishing the BPA function. Fig.1 illustrates the schematic diagram of the decision-making model for coal and gas outburst prediction based on a neural network and the D-S evidence theory. According to real-time monitoring data collected by security monitor systems in various situations, the pretreated records were selected as input features of the neural network. The error signals between patterns and dangerous levels are used as parameters of the D-S evidence theory to optimize the weights of neural networks and improve the training quality.

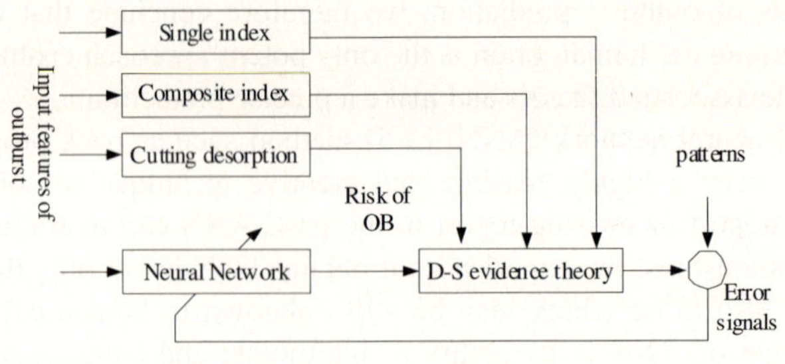

Fig. 1. Model of coal and gas prediction based on NN and the D-S evidence theory

2.1 Artificial Neural Network

For a quick overview and as a part model of the coal and gas outburst prediction system, Fig. 2 displays a model of an artificial neural network.

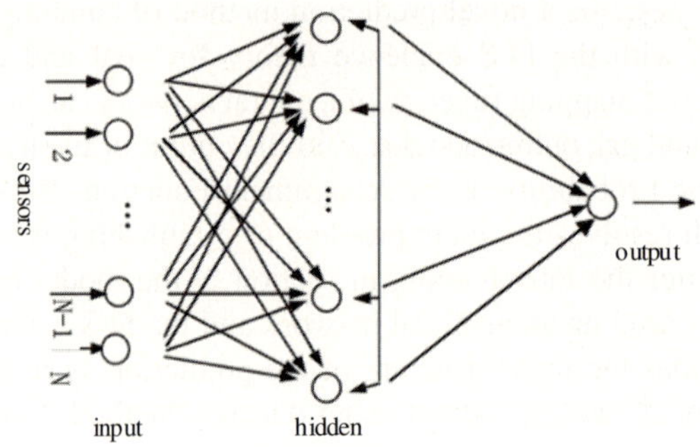

Fig. 2. Neural network model of coal and gas prediction

Commonly, the conventional back propagation algorithm is used to train multilayer perceptions as shown in [11]. The nodes in each layer receive input signals from the previous layer and pass the output to the subsequent layer. The nodes of the input layer receive a set of input signals from the outside system and directly deliver the input data to the hidden layer by the weighted links. In the following computation, a three -layered structure containing the input, hidden, and output layer will be adopted.

The input layer is the signal collection and processor, which is made up of n nodes. The input of NN is the data from sensors after pretreatment, which could be signal or image, even the feature level. In this prediction system the input of the NN is six selected non-linear parameters, such as type of roadway, depth, thickness, obliquity, geological structures, and mining methods.

The second layer is the hidden layer, which is made up of h hidden nodes. Besides the input signals from the previous layer, the nerve cells also receive the feedback output from other nerve cells in the same level, which results in a dynamic network. In the following computation, thirteen hidden nodes are chosen as the dynamic network.

The third output layer is made up of m output nodes, which is consistent with the whole information level fusion. Normally, it is only one output node, and the result of this output layer could supply a reference for decision-making. In our prediction system, two output nodes would be given, and the results from the neural network output layer would be used as the basic probability assignment function for the D-S evidence theory for further decision-making.

2.2 The Improved Combination Rule of the D-S Evidence Theory

The basic concept of the D-S evidence theory [12] is the frame of discrimination Θ, which is made up of a series of basic propositions which are endlessly mutually incompatible. The basic probability assignment function (BPAF) is defined as m: $2^{\Theta} \longrightarrow [0, 1]$, and it should satisfy two basic conditions:

$$m(\Phi)=0$$

$$\sum_{A\subseteq\Theta} m(A) = 1 \tag{1}$$

The definition of BPAF reflects what the evidence support the proposition A. evidence is made up of the item of evidence. The belief function Bel (A) quantifies the strength of the belief that event A occurs:

$$Bel(A) = \sum_{B\subseteq A} m(B) \quad (\forall A \subset \Theta) \tag{2}$$

And the plausibility function *Pl (A)* quantifies the strength to which we do not doubt the proposition A,

$$Pl(A) = \sum_{A\cap B\neq\varnothing} m(B) \tag{3}$$

So,

$$Pl(A) = 1 - Bel(A) \tag{4}$$

And the interzone of *[Bel (A), Pl (A)*] is the uncertainty zone, which quantifies the strength of the uncertainty degree of A. To abbreviate the length of the zone is one of the purposes of the D-S evidence theory.

The key of the evidence theory is that the single evidence BPAF may be combined to form an integrated BPAF, this is the evidence combination. As the kinds of evidence stem from different sensor sources, and as there is conflict and coherence between them, an improved combination rule of the D-S evidence theory has been described in [13] for resolving the evidence conflict problem. This effectiveness of this improved rule has been proved mathematically and by numerical experiments.

Thus, one of the problems of the D-S evidence theory, the evidence conflict, has been solved. However, another basic problem is the difficulty of establishing the BPAF, which is usually derived by statistics or an expert experimental formula. As this method is subjective, the BPAF is difficult to determine by the single D-S evidence theory. As a result, we took the output of the neural network as the BPAF for the D-S evidence in our experiments.

3 Experiments and Results

According to the monitoring data collected by security monitor systems and other data attained through other methods such as D-K composite indexes and the cutting desorption method, some non-linear parameters were selected as the input features of neural networks in our coal and gas outburst prediction system. In our experiments, we chose six main parameters as the input of the NN, such as the type of roadway, depth, thickness, obliquity, geological structures, and mining methods. The other main parameters, i.e damage classification, mining depth, initial desorption rate, and strength coefficient, are chosen as the indexes of the D-K composite index method. We got 23 groups of data from the typical coal mines, as shown in Table 1. The error signals between patterns and the risk of outburst from the D-S evidence theory decision-making are used for optimizing the weights of the neural network and improve the training quality.

Table 1. Learning and predicting samples of BP NN for coal and gas outburst prediction

No.	Type of roadway	Depth (m)	Thickness	Obliquity(°)	Geological structure	Mining methods	Actual results
1	0	544.5	6.5	40	1	0	1 0
2	0	842.2	4.0	26	1	1	1 0
3	0	733.6	5.5	50	3	1	1 0
4	1	875.7	2.5	15	2	1	1 0
5	1	987.5	2.9	35	2	1	0 1
6	1	710.4	1.8	27	2	1	1 0
7	0	807.8	3.0	35	1	1	1 0
8	0	716.5	8.0	55	3	1	1 0
9	0	1021.2	10.3	24	2	1	0 1
10	1	790.1	3.5	40	1	1	1 0
11	0	811.3	3.0	45	3	0	1 0
12	0	969.1	5.2	25	2	2	1 0
13	0	979.2	2.6	28	1	2	0 1
14	0	812.9	4.25	53	3	3	1 0
15	1	959.0	3.0	35	4	2	1 0
16	1	990.2	2.8	29	2	3	0 1
17	1	1052.3	5.2	25	4	2	1 0
18	0	919.0	3.0	25	4	1	1 0
19	0	788.4	5.45	55	3	3	1 0
20	1	787.0	3.4	20	2	1	1 0
21	1	992.8	2.4	30	2	3	0 1
22	1	959.4	12.2	23	4	2	1 0
23	0	1046.8	9.8	20	2	1	0 1

In the row for the type of roadway, 0 is oblique, 1 is flat; in the geological structure row, 1 is drape, 2 is faultage, 3 is escarpment. Four mining methods are chosen: 0 to 3 stand for unknown, firing a gun, drilling, and handpicking respectively.

A three-layer frame 6-13-2 was chosen as the neural network structure, the six parameters mentioned above were used as inputs, and the two prediction results, A [1, 0] and B [0, 1] as outputs. The first eighteen groups of learning sample data in Table 1 were used as training patterns for training the neural network, the epochs of the network training are up to 2000, and the trainrp function was chosen as the BP neural network fast training algorithm.

After learning training of the neural network, the error curve of network training was close to the ideal level, as shown in Fig.3. So we used the trained neural network to do the prediction with another five groups of sample data. The result of the neural network, i.e. of the first evidence source, are shown in Table 2.

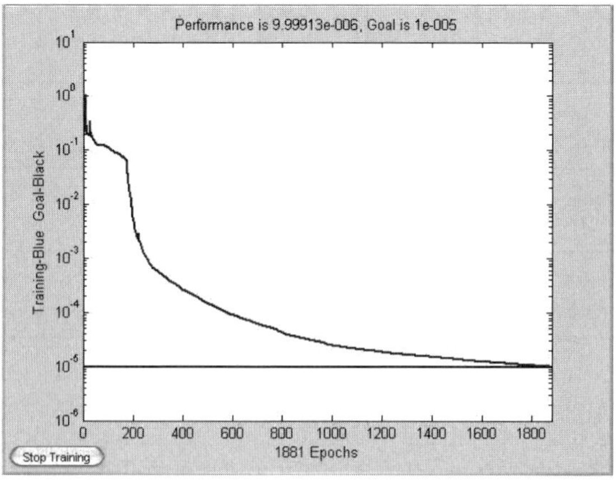

Fig. 3. The error curve of the BP neural network training

For future fusion, we need to make each prediction result unitary as the evidence stems from different sources. The unitary results from the BP neural network as evidence I are listed in Table 3.

Table 2. Prediction results of the BP neural network

No.	19	20	21	22	23
Actual Results	1 0	1 0	0 1	1 0	0 1
Prediction Results	1.1264 −0.1069	1.0007 0.0000	0.0137 1.0177	1.0007 0.0002	0.1152 0.8601

Table 3. Unitary results of the BP neural network

No.		19	20	21	22	23
BPAF	$m_1(A)$	0.9133	1	0.0133	0.9998	0.1181
	$m_1(B)$	0.0867	0	0.9867	0.0002	0.8819

Taking this result from the BP neural network as the basic probability assignment function (BPAF) of the first evidence for the D-S evidence theory, we also took the results from other methods like the single index, the composite index, and the cutting desorption method as the other kinds of evidence of the D-S evidence theory for decision-making. Then the finial prediction results were fused with the improved D-S evidence combination rules. For comparing the accuracy and credibility of the prediction, the results from different methods are given in Table 4 as follows:

Table 4. Results from different methods

No.		19	20	21	22	23
Neural Network	$m_1(A)$	0.9133	1	0.0133	0.9998	0.1181
	$m_1(B)$	0.0867	0	0.9867	0.0002	0.8819
Single Index	$m_2(A)$	1	1	0.0926	0.8606	0.1220
	$m_2(B)$	0	0	0.9074	0.1394	0.8780
Composite Index	$m_3(A)$	1	1	0.0620	0.9328	0.0588
	$m_3(B)$	0	0	0.9380	0.0672	0.9412
Cutting Desorption	$m_4(A)$	1	0.9524	0.3571	1	0.0816
	$m_4(B)$	0	0.0476	0.6429	0	0.9184
D-S Fusion	$m(A)$	0.9982	0.9994	0.0554	0.99	0.0316
	$m(B)$	0.0018	0.0006	0.9446	0.01	0.9684

From the table above, we can conclude that the result after fusion with the D-S evidence theory is more reasonable than the result from each single present prediction method, and it is feasible to use the result from the neural network as the basic probability of the assignment function of the D-S evidence theory. In addition, it is much easier to make a determinate decision from the fusion result. For example, according to the data from the area in 23, the result after the D-S fusion is clearer and more accurate than the result from the neural network, and we can give a highly credible prediction of the outburst.

4 Conclusion

In this paper, a novel method combing a neural network and the D-S evidence theory for coal and gas outburst prediction is put forward. This method is particularly suitable for establishing the basic probability of the assignment function of the D-S evidence theory using the result from a neural network. The results from our experiments show that our method is feasible and effective. Combing the non-linear mapping function of a neural network with the uncertain reasoning of the D-S evidence theory makes the prediction more reasonable and credible. Using this method, we can make a more certain decision than is possible when relying on each method independently.

References

1. Zhang, R.L.: Application of Advanced Information Technology on Coal and Gas Outburst Prediction. PhD thesis, Chongqing University (2004)
2. Li, C.W.: Research on Gay Classification and Prediction for Coal and Gas Outburst Quicksand. PhD thesis, China University of Mining and Technology (2005)
3. Hao, J.S.: Application of Improved BP Network in Prediction of Coal and Gas Outburst. Journal of Liaoning Technical University, 9–11 (2004)

4. Gao, L.F.: Prediction of Coal and Gas Outburst Disasters Based on Genetic and BP Algorithm. Journal of Liaoning Technical University, 408–410 (2002)
5. Zhang, T.G.: Prediction and Control of Coal and Gas Outburst in Pingdingshang Mining Area. Journal of China Coal Society, 173–177 (2001)
6. Haykin, S.: Neural Networks: A Comprehensive Foundation, 2nd edn. Prentice-Hall, New Jersey (1999)
7. Su, H., Zheng, G.: A Non-Intrusive Drowsiness Related Accident Prediction Model Based on D-S Evidence Theory. In: Proceedings of the 1st International Conference on Bioinformatics and Biomedical Engineering (ICBBE 2007), Wuhan, China, pp. 570–573 (2007)
8. Regis, S., Desachy, J., Doncescu, A.: Evaluation of Biochemical Sources Pertinence in Classification of Cell's Physiological States by Evidence Theory. In: Proceedings of 13th IEEE International Conference on Fuzzy Systems, Budapest, Hungary, pp. 867–872 (2004)
9. Tan, D., Yan, X., Gao, S., Liu, Z.: Fault Diagnosis for Spark Ignition Engine Based on Multi-sensor Data Fusion. In: Proceedings of IEEE International Conference on Vehicular Electronics and Safety, Xi'an, China, pp. 311–314 (2005)
10. Lu, W., Wu, Q., Shao, Q.: The D-S Evidence Theory and Application to Feasibility Evaluating of New Product Developing. Journal of Operational Research and Management, 111–115 (2004)
11. Cichocki, A., Unbehauen, R.: Neural Network for Optimization and Signal Processing. Wiley, Chichester (1993)
12. Duan, X.S.: Evidence Theory and Decision-making. Artificial Intelligent. China Rennin University Press, Beijing (1993)
13. Miao, Y.Z., Zhang, H.X., Zhang, J.W., Ma, X.P.: Improvement of the Combination Rules of the D-S Evidence Theory Based on Dealing with the Evidence Conflict. In: 4th IEEE International Conference on Information and Automation, Zhangjiajie, China (2008)

Using the Tandem Approach for AF Classification in an AVSR System

Tian Gan, Wolfgang Menzel, and Jianwei Zhang

Cinacs, Department of Informatics, University of Hamburg
Vogt-Koelln-Str. 30, 22527 Hamburg, Germany
{gan,menzel,zhang}@informatik.uni-hamburg.de
http://cinacs.informatik.uni-hamburg.de/

Abstract. This paper describes an audio visual speech recognition (AVSR) system based on articulatory features (AF). It implements a tandem approach where artificial neural networks (ANN), in particular multi-layer perceptrons (MLP), are used as posterior probability estimators for transforming raw input data into the more abstract articulatory features. Such an approach is particularly well suited if relatively few training data are available, a situation which is typical for AVSR. In addition, the MLP feature extraction results and some analysis in terms of recognition accuracy and confusions are presented. Our AF-based AVSR system has been trained on the audio-visual speech corpus VIDTIMIT, which contains conversational speech based on a medium size vocabulary including more than 1200 words.

Keywords: MLP, Articulatory Features, Audio Visual Speech Recognition.

1 Introduction

ASR has been well researched for almost 40 years. The technology has already been deployed in many areas of our daily life. Although some systems have achieved a relatively good performance under certain specific conditions, sensitivity to environmental and channel noise is still a major weakness. It seems very difficult to further improve the performance relying on acoustic evidence only.

Therefore, researchers try to combine the acoustic evidence with additional information channels to reach better recognition rates ([1,2,3,4]). In these approaches, the visual speech from lipreading is a natural candidate for compensating deficits of the noisy acoustic signal in ASR, because it is independent of the acoustic environment. Also, evidence from human speech perception convincingly shows that visual cues might considerably contribute to speech comprehension. Not surprisingly, since the first attempt by Petajan in 1984 [1], a range of AVSR systems have been developed ([2],[13–14]) which confirmed the initial assumption that lip reading information is particularly helpful for recognizing noisy speech. Although there are clear differences in how these systems process

F. Sun et al. (Eds.): ISNN 2008, Part II, LNCS 5264, pp. 830–839, 2008.

audio and visual information and combine it, they all share a quite similar system architecture based on a state-of-the-art approach to word recognition using phones as a subword modeling unit.

As an alternative, using articulatory information for ASR is an attempt to capture relevant characteristics of the speech signal about the procedure of speech production. It is also motivated by the expectation that certain aspects of articulation can be detected more robustly for ASR. Furthermore, with a small number of classes, each articulatory classifier should be better trained than a phone-based classifier if only sparse training data is available. The number of studies that make use of articulatory information for ASR has been steadily increasing over the years. These approaches can be categorized into three groups. 1) Modeling the true geometry of the articulators, by directly detecting and extracting features by means of physical or medical observations, such as X-Ray and EMA data [7,8,9]. 2) Using an abstract intermediate representation, namely articulatory features, which are transformed from low-level acoustic and visual speech features [5], [10–11]. 3) Articulatory decisions, which have been derived from decisions taken by a phone recognizer [12]. In this paper, we mainly focus on the second approach to deal with articulatory information.

Since there is an apparent correlation between some of the articulatory features and the visual shape of the lips while speaking, namely for the opening and rounding features which provide important cues to distinguish different kinds of vowels, classifying the same articulatory features from different data sources might provide complementary information. There are also some attempts to use visual information in articulatory feature-based ASR (e.g. [16–17]). This paper continues to investigate the use of AFs in an AVSR system. Using the MLP-based tandem approach to feature transformation, we achieved a classification accuracy for the acoustic features which is comparable with previous work. In addition, two visual articulatory feature groups were introduced.

The paper is organized as follows: Section 2 explains the MLP-based tandem approach. Section 3 decribes the articulatory feature-based audio-visual speech recognition framework and the design of the feature extractors. The experimental setup and an analysis of the results is presented in Section 4. Final conclusions are given in Section 5.

2 MLP-Based Feature Extraction

2.1 Informative and Discriminative Classification

The ASR problem is usually stated as the problem of finding the sequence of words W which maximizes $p(W|O)$, where O refers to a sequence of observations, i.e. feature vectors. According to Bayes' theorem:

$$p(W|O) = \frac{p(O|W)p(W)}{p(O)} \qquad (1)$$

given an acoustic observation sequence O, the effort of maximizing $p(W|O)$ can be converted to a search for the class W which maximizes $p(O|W)p(W)$.This approach is an instance of informative classification. $p(W)$ is known as the language model (LM) capturing high-level constraints and linguistic knowledge. $p(O|W)$ refers to the acoustic model, which describes the statistics of sequences of parameterized acoustic observations in the feature space given the corresponding uttered words (e.g. phone sequences). The most popular stochastic approach for acoustic modeling is HMM, where states of the hidden part represent phones (or other sub-phonetic units), whereas the observable part accounts for the statistical probabilities of the corresponding acoustic events. Due to their indirect modeling approach, HMM can introduce errors on both levels, class density and prior probabilities. This leads to poor discriminative power among different models.

In contrast, discriminative classification, as used in MLPs, is an approach to estimating the posterior probability directly. MLPs have been successfully applied to pattern classification in many areas. An MLP can discriminatively learn the unknown class boundaries and, when applied, can map the input within the class boundary to the target used in training. With large enough training data and sufficient network size, MLPs are effective at modeling unknown distributions, i.e. learning the posterior probability of a class given an observation, $p(W|O)$. In ASR, MLPs can only classify short-time acoustic-phonetic units, like phones. Due to their inability to capture long-term dependencies, ANNs are not suited as a general framework for ASR, especially with long sequences of acoustic observations, like words or sentences.

2.2 MLP Feature Extraction by Tandem Approach

The combination of discriminative and informative classification is an attempt to overcome the above-mentioned limitations. Hermansky et.al. proposed the most successful tandem approach [18]. The tandem approach has the advantage of discriminative acoustic model training, as well as being able to use the techniques for standard HMM systems. In the tandem approach, MLPs are trained as feature extractors for the HMM. MLPs attempt to transform input acoustic representations into compact but significant, low-dimensional representations which are more suitable to be modeled by the emission probabilities of the HMM than standard acoustic parameters.

As shown in Fig.1, after calculating low-level feature vectors, e.g. MFCC, MLPs can be regarded as nonlinear feature transformation components. The MLP output, which approximates the posterior probabilities for the given input features, can be further transformed by the post processing block into tandem features for better matching the gaussian mixtures of a continuous HMM. This can be implemented, for example, by taking the log of the MLP outputs. As an input for the HMM, either the tandem feature vectors alone or a combination with low-level feature vectors can be considered.

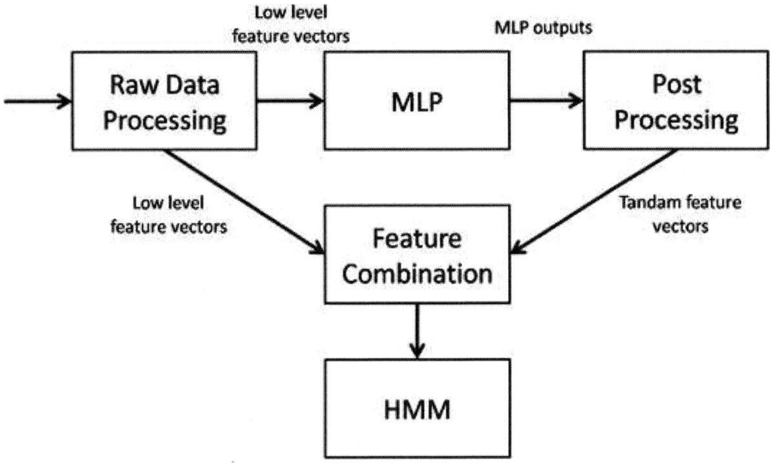

Fig. 1. Block diagram of a tandem approach for ASR

2.3 AF-Based MLP Feature Extraction

Most of the previous tandem approaches have been proposed to directly classify word sequences. In this paper, we investigate an indirect approach, which uses articulatory features as an intermediate representation. The articulatory features are extracted from the posterior estimates of a set of MLPs trained for articulatory feature classification, one for each feature.

The main motivation for using an AF-based tandem approach is trying to make more relevant characteristics of speech production available to the recognition system. In practice, the number of feature values from each individual AF feature group is much smaller than the number of phone classes. This means that it requires less effort to train an MLP network, and better classification results can be expected. Another reason to use AFs, while not considered in this paper, is to strive for a better design of an AVSR system. Using AFs as an intermediate abstract representation may contribute an attractive possibility for combining information from the audio and visual channel. Tabel 1 lists the articulatory features used for our experiments. *Voicing, manner, place, front-back,* and *rounding* are five articulatory feature groups for the acoustic channel, where the number of feature values per group varies between 3 and 10. *opening* and *rounding* are two additional feature groups for describing the movement of articulators based on visual information.

3 AF-Based AVSR Framework

Fig.2 shows a system framework consisting of three stages. Raw audio and video data are first processed in the feature extraction stage, where a series of parameterized feature vectors from both channels is generated. The feature classification

Table 1. AFs used in AVSR framework.

Features	Values
Voicing	+voice, -voice, sil
Rounding	+round, nil, -round, sil
Manner	vowel, nasal, lateral, stop, approximant, fricative, sil
Place	dental, labial, retroflex, coronal, velar, high, mid, low, glottal, sil
Front-Back	front, nil, back, sil
Visual opening	open, close, sil
Visual rounding	+round, nil, -round, sil

stage then transforms the feature vectors into the articulatory features. MLPs with three layers are applied to train a set of articulatory feature models. All the output activations from the different articulatory features classifiers can be simply concatenated into a large feature vector which contains all the articulatory information. A second-level classifier is applied in the third stage, which maps the articulatory features into a sequence of different phonetic units.

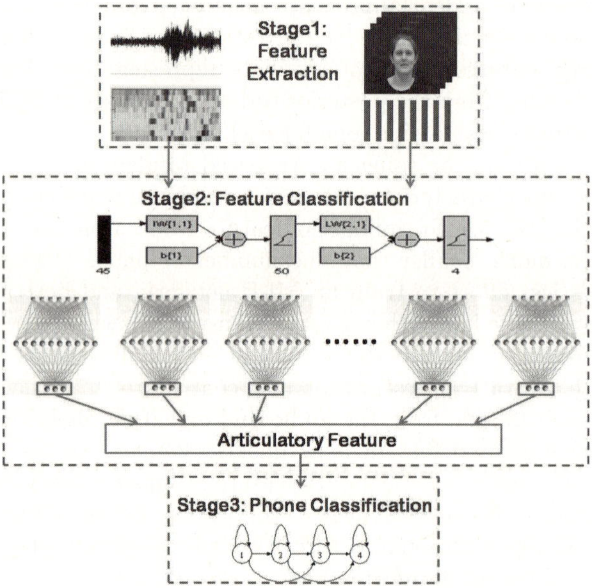

Fig. 2. AF-based AVSR system

Mapping low-level speech feature vectors into articulatory features requires non-linear classification methods. Similar to the work of [5], three-layered MLPs have been used in our experiment.The activation function of the hidden layer is the logistic function, which is one of the most common sigmoidal functions used for that purpose:

$$f(x) = \frac{1}{1 + exp(-ax)} \tag{2}$$

where a is a constant controlling the slope of the function. The output activation function is the softmax function:

$$f(x_i) = \frac{exp(x_i)}{\sum_{n=1}^{K} exp(x_n)} \tag{3}$$

where K is the number of units in the output layer. The softmax activation function is a logical extension of the logistic function, which is suitable for the case of categorical targets using 1-of-C coding. The output activation values are non-linearly mapped to the range [0,1] by equation . To Gaussianize the feature distribution, the logarithms of these output values are then concatenated and taken as the input feature vectors for the next level phone recognizer. The third stage can employ any classification method, such as HMM, for speech recognition.

Obviously, the final speech recognition results are highly dependent on the results of the MLP feature classification stage. Therefore, in this paper we focus on a comparison of different audio and video-based articulatory features to select the more efficient ones for combination. The method and results of the subsequent phone recognizer are not discussed here.

4 Experimental Results

The experimental data are based on the VIDTIMIT corpus [17], which is a continuous audio-visual speech corpus for an English medium-sized vocabulary task. It contains 10 sentences spoken by 34 speakers each. The sentences were chosen from the test section of the TIMIT corpus [20]. The recording was done in an office environment using a broadcast quality digital video camera. The corresponding audio signal is stored as a mono, 16 bit, 32 kHz WAV file.

The raw audio signal was converted into a sequence of vector parameters with a fixed 25ms frame and a frame rate of 10ms. A 9-dimensional RASTA-PLP [19] feature vector was calculated as raw audio data for each 10ms frame. More effort has been spent on robustly extracting lipreading information. Based on the Adaboost framework of Viola and Jones, we combine a relatively weak mouth detector with a more robust face detector to locate the mouth region-of-interest (ROI), using information about the general geometry of the face to guide the mouth region tracker. To determine a normalized size of the ROI, pre-extracted ROIs are modified according to the detected mouth corner. Following ROI extraction, the visual features can be selected based on pixel values, mouth shape information or a hybrid of both. In this experiment, the gray scaled pixel values are used as visual speech features. For dimensionality reduction, PCA is applied to taking the first 20 components representing the visual feature vectors.

Moreover, to complete the preparation of the visual articulatory features, an additional processing step is necessary. As the visual frame rate is only a quarter of the acoustic frame rate, visual vectors are carefully interpolated by averaging the values between two adjacent frames so that both signals are synchronously available.

Following the tandem approach, the 9-dimensional audio feature vectors and 20-dimensional visual feature vectors are then expanded to 45 dimensions and 100 dimensions by adding the two previous and two following frame vectors in order to introduce some context information. These feature vectors were used on the input layer of all MLPs. The number of units on the output layer of each MLP corresponds to the number of values in each articulatory group. The numbers of hidden units are shown in Table 2. They have been chosen according to the empirical selection made in [5].

Table 2. Number of hidden units used by each AF MLP.

Feature Group	Feature Values	Hidden Units
voicing	3	50
manner	7	100
place	10	100
front-back	4	100
acoustic rounding	4	100
visual rounding	4	100
visual opening	3	50

Table 3 shows a comparison of three AF-based feature extraction studies with respect to their frame level accuracy. Compared to the results of [6], our system obtained better results in all individual MLP classifiers. The accuracies reported by Kirchhoff [5] slightly outperformed our results. Considering that Kirchhoff used a small vocabulary corpus (OGI Numbers95), our medium vocabulary task resulted in a comparable performance.

Table 3. Frame level AF-based MLP classification accuracy.

Feature Group	Our Work	Work [6]	Work [5]
voicing	86.9%	61.7%	89.1%
manner	71.6%	66.6%	82.0%
place	67.5%	59.5%	77.2%
front-back	78.7%	-	82.9%
acoustic rounding	81.1%	70.1%	83.1%
visual rounding	63.2%	-	-
visual opening	66.5%	-	-

The detailed observations from our results are described as follows:

voicing. Out of the 3 classes *sil, +voice* and *-voice*, *+voice* has the highest recognition accuracy of 94%. With 81%, *-voice* has the lowest one. Many *-voice* values have been misclassified as *sil*. Voicing is the most reliable feature group of these parallel MLPs.

manner. The classification accuracy for the classes varies between 35% and 71% except for *lateral*, the lowest with 16%, and *vowel*, the highest with 78%. These results obviously are highly influenced by the prior probabilities of the values, since *vowel* and *lateral* correspond to the most and least frequent values in the training data.

place. The classification accuracy for the classes varies between 25% and 70% with *glott* as the lowest with 9% and *sil* as the highest with 76%.

front-back. *nil* and *sil* are better recognized compared to *front* and *back*. The latter two recognition accuracies were 71% and 65%.

acoustic rounding. *-round* has the lowest classification accuracy of 68%. The accuracy of the other classes varies between 72% and 86%.

visual rounding. The articulatory feature derived from visual information does not yield better recognition results than the corresponding AF from audio information. But the difference between the distribution of errors motivated us to fuse both channels for better results.

visual opening. Openness decisions from visual information are more reliable than the information from the visual rounding feature.

Poor results of some classes, e.g. *lateral* and *glott*, are probably due to insufficient training data, because they are relatively underrepresented in the corpus. The unbalanced distribution of training data led to better recognition results for the more frequent classes and to worse results for the rare ones. In particular, the original data contained approximately 30% of silence frames. To avoid an extreme bias this share has been reduced to 3% by removing non-speech segments in the beginning and end of the signal files. Otherwise the trained classifiers would perform as a kind of silence detector.

A major share of errors occurred at the feature boundaries. This is caused by the fact that the gold standard articulatory features for evaluation have been derived from phone level annotations. This procedure does not take the dynamic nature of articulatory features properly into account.

5 Conclusions

Our experiments focused on investigating the tandem approach for an articulatory feature-based audio visual speech recognition system. The design and training of a set of AF-based MLP classifiers has been presented. Lip movement information, namely *visual rounding* and *visual opening*, was also used as articulatory features. Compared to other acoustic-only AF feature classification tasks, our experiment showed a better performance considering its relatively large vocabulary. The visual AF feature classification results are still worse than the acoustic ones but seem to have enough potential for being successfully combined with the acoustic ones in a subsequent integrated phone recognizer.

Acknowledgments. This research was supported by the German Research Foundation (DFG) and the Ministry of Education of the People's Republic of China through the CINACS (Cross-Modal Interactions in Natural and Artificial Cognitive Systems) research school.

References

1. Petajan, E.D.: Automatic Lipreading to Enhance Speech Recognition. In: IEEE Conference on Computer Vision and Pattern Recognition, pp. 265–272. IEEE Press, San Francisco (1985)
2. Potamianos, G., Neti, L.J., Matthews, I.: Audio-Visual Automatic Speech Recognition: An Overview. In: Bailly, G., Vatikiotis-Bateson, E., Perrier, P. (eds.) Issues in Visual and Audio-Visual Speech Processing, pp. 356–396. MIT Press, Cambridge (2004)
3. Watson, R.: A Survey of Gesture Recognition Techniques. Technical report, Trinity College Dublin (1993)
4. Fasel, B., Luettin, J.: Automatic Facial Expression Analysis: A Survey. J. Pat. Rec. 36, 259–275 (2003)
5. Kirchhoff, K.: Robust Speech Recognition Using Articulatory Information. PhD Thesis, University of Bielefeld (1999)
6. Abu-Amer, T., Carson-Berndsen, J.: HARTFEX: A Multi-Dimensional System of HMM Based Recognizers for Articulatory Feature Extraction. In: 8th European Conference on Speech Communication and Technology, Geneva, Switzerland, pp. 2541–2544 (2003)
7. Papcun, J., Hochberg, J., Thomas, T.R., Laroche, F., Zacks, J., Levy, S.: Inferring Articulation and Recognizing Gestures from Acoustics with A Neural Network Trained on X-ray Microbeam Data. J. Acoust. Soc. Am. 92, 688–700 (1992)
8. Zacks, J., Thomas, T.R.: A new Neural Network for Articulatory Speech Recognition and Its Application to Vowel Identification. J. Com. Sp. Lan. 8, 189–209 (1994)
9. Frankel, J., King, S.: ASR - Articulatory Speech Recognition. In: 7th European Conference on Speech Communication and Technology, Scandinavia, Aalborg, pp. 599–602 (2003)
10. Eide, E., Rohlicek, J.R., Gish, H., Mitter, S.: A Linguistic Feature Representation of the Speech Waveform. In: 18th International Conference on Acoustics Speech and Signal Processing, pp. 483–486. IEEE Press, Minneapolis (1993)
11. Deng, L., Erler, K.: Microstructural Speech Units and Their HMM Representations for Discrete Utterance Speech Recognition. In: International Conference on Acoustics Speech and Signal Processing, pp. 193–196. IEEE Press, Washington (1991)
12. Gan, T., Menzel, W.: An Audio Visual Speech Recognition Framework Based on Articulatory Features. In: 7th International Coference on Auditory-Visual Speech Processing, pp. 137–141. Tilburg University, Tilburg (2007)

13. Hennecke, M.E., Stork, D.G., Prasad, K.V.: Visionary Speech: Looking ahead to Practical Speechreading Systems. J. Spe. Hum. Mach., 331–349 (1996)
14. Luettin, J.: Visual Speech and Speaker Recognition. PhD thesis, University of Sheffeld (1997)
15. Livescu, K., Cetin, O., Johnson, M.H., King, S., Bartels, C., Borges, N., Kantor, A., Lal, P., Yung, L., Bezman, A., Haggerty, S.D., Woods, B., Frankel, J., Doss, M.M., Saenko, K.: Articulatory Feature-based Methods for Acoustic and Audio-Visual Speech Recognition: JHU Summer Workshop Final Report. In: 32nd IEEE International Conference on Acoustics Speech and Signal Processing, pp. 621–624. IEEE Press, Honolulu (2007)
16. Saenko, K., Darrell, T., Glass, J.: Articulatory Features for Robust Visual Speech Recognition. In: 6th International Conference on Multimodal Interfaces, pp. 152–158. ACM, New York (2004)
17. Sanderson, C., Paliwal, K.K.: Identity Verification Using Speech and Face Information. J. Dig. Sig. Proc. 14, 449–480 (2004)
18. Hermansky, H., Ellis, D.I.W., Shamza, S.: Tandem Connectionist Feature Extraction for Conventional HMM Systems. In: 25th International Conference on Acoustics Speech and Signal Processing, pp. 1635–1638. IEEE Press, Istanbul (2000)
19. Hermansky, H., Morgan, N., Bayya, A., Kohn, P.: RASTA-PLP Speech Analysis Technique. In: Proceedings of the International Conference Acoustics Speech Signal Processing, San Francisco, California, pp. 1121–1124 (1991)
20. Fisher, W.M., Doddington, G.R., Goudie-Marshall, K.M.: The DARPA Speech Recognition Research Database: Specifications and Status. In: The DARPA Speech Recognition Workshop, Palo Alto, Canada, pp. 93–99 (1986)
21. Viola, P., Jones, M.: Robust Real-Time Face Detection. J. Com. Vis. 57, 137–154 (2004)

Author Index

Printing: Mercedes-Druck, Berlin
Binding: Stein+Lehmann, Berlin

Lecture Notes in Computer Science

Sublibrary 1: Theoretical Computer Science and General Issues

For information about Vols. 1–4971
please contact your bookseller or Springer

Vol. 5264: F. Sun, J. Zhang, Y. Tan, J. Cao, W. Yu (Eds.), Advances in Neural Networks, Part II. XXXI, 846 pages. 2008.

Vol. 5238: E. Börger, M. Butler, J. Bowen, P. Boca (Eds.), Abstract State Machines, B and Z. XII, 382 pages. 2008.

Vol. 5234: V. Adve, M.J. Garzarán, P. Petersen (Eds.), Languages and Compilers for Parallel Computing. XV, 354 pages. 2008.

Vol. 5228: A. Barbero (Ed.), Coding Theory and Applications. VIII, 197 pages. 2008.

Vol. 5226: D.-S. Huang, D.C. Wunsch II, D.S. Levine, K.-H. Jo (Eds.), Advanced Intelligent Computing Theories and Applications. XXVII, 1273 pages. 2008.

Vol. 5217: M. Dorigo, M. Birattari, C. Blum, M. Clerc, T. Stützle, A.F.T. Winfield (Eds.), Ant Colony Optimization and Swarm Intelligence. XV, 416 pages. 2008.

Vol. 5216: G.S. Hornby, L. Sekanina, P.C. Haddow (Eds.), Evolvable Systems: From Biology to Hardware. XV, 444 pages. 2008.

Vol. 5215: F. Cassez, C. Jard (Eds.), Formal Modeling and Analysis of Timed Systems. X, 295 pages. 2008.

Vol. 5214: H. Ehrig, R. Heckel, G. Rozenberg, G. Taentzer (Eds.), Graph Transformations. XIII, 523 pages. 2008.

Vol. 5213: M. Kaminski, S. Martini (Eds.), Computer Science Logic. XI, 513 pages. 2008.

Vol. 5204: C.S. Calude, J.F. Costa, R. Freund, M. Oswald, G. Rozenberg (Eds.), Unconventional Computing. X, 259 pages. 2008.

Vol. 5203: S.W. Golomb, M.G. Parker, A. Pott, A. Winterhof (Eds.), Sequences and Their Applications - SETA 2008. XII, 421 pages. 2008.

Vol. 5201: F. van Breugel, M. Chechik (Eds.), CONCUR 2008 - Concurrency Theory. XIII, 524 pages. 2008.

Vol. 5199: G. Rudolph, T. Jansen, S. Lucas, C. Poloni, N. Beume (Eds.), Parallel Problem Solving from Nature – PPSN X. XX, 1164 pages. 2008.

Vol. 5191: H. Umeo, S. Morishita, K. Nishinari, T. Komatsuzaki, S. Bandini (Eds.), Cellular Automata. XVI, 577 pages. 2008.

Vol. 5172: S. Dolev, T. Haist, M. Oltean (Eds.), Optical SuperComputing. IX, 129 pages. 2008.

Vol. 5171: A. Goel, K. Jansen, J.D.P. Rolim, R. Rubinfeld (Eds.), Approximation, Randomization and Combinatorial Optimization. XII, 604 pages. 2008.

Vol. 5170: O.A. Mohamed, C. Muñoz, S. Tahar (Eds.), Theorem Proving in Higher Order Logics. X, 321 pages. 2008.

Vol. 5169: J. Fong, R. Kwan, F.L. Wang (Eds.), Hybrid Learning and Education. XII, 474 pages. 2008.

Vol. 5168: E. Luque, T. Margalef, D. Benítez (Eds.), Euro-Par 2008 – Parallel Processing. XXVIII, 964 pages. 2008.

Vol. 5165: B. Yang, D.-Z. Du, C.A. Wang (Eds.), Combinatorial Optimization and Applications. XII, 480 pages. 2008.

Vol. 5164: J. Koutník, V. Kůrková, R. Neruda (Eds.), Artificial Neural Networks - ICANN 2008, Part II. XXIX, 986 pages. 2008.

Vol. 5163: J. Koutník, V. Kůrková, R. Neruda (Eds.), Artificial Neural Networks - ICANN 2008, Part I. XXIX, 1026 pages. 2008.

Vol. 5162: E. Ochmański, J. Tyszkiewicz (Eds.), Mathematical Foundations of Computer Science 2008. XIV, 626 pages. 2008.

Vol. 5160: J.S. Fitzgerald, A.E. Haxthausen, H. Yenigun (Eds.), Theoretical Aspects of Computing - ICTAC 2008. XI, 455 pages. 2008.

Vol. 5156: K. Havelund, R. Majumdar, J. Palsberg (Eds.), Model Checking Software. X, 343 pages. 2008.

Vol. 5148: O. Ibarra, B. Ravikumar (Eds.), Implementation and Applications of Automata. XII, 289 pages. 2008.

Vol. 5147: K. Horimoto, G. Regensburger, M. Rosenkranz, H. Yoshida (Eds.), Algebraic Biology. XII, 245 pages. 2008.

Vol. 5133: P. Audebaud, C. Paulin-Mohring (Eds.), Mathematics of Program Construction. X, 423 pages. 2008.

Vol. 5132: P.J. Bentley, D. Lee, S. Jung (Eds.), Artificial Immune Systems. XIV, 436 pages. 2008.

Vol. 5131: H.J. van den Herik, X. Xu, Z. Ma, M.H.M. Winands (Eds.), Computers and Games. XII, 275 pages. 2008.

Vol. 5130: J. von zur Gathen, J.L. Imaña, Ç.K. Koç (Eds.), Arithmetic of Finite Fields. X, 205 pages. 2008.

Vol. 5126: L. Aceto, I. Damgård, L.A. Goldberg, M.M. Halldórsson, A. Ingólfsdóttir, I. Walukiewicz (Eds.), Automata, Languages and Programming, Part II. XXII, 730 pages. 2008.

Vol. 5125: L. Aceto, I. Damgård, L.A. Goldberg, M.M. Halldórsson, A. Ingólfsdóttir, I. Walukiewicz (Eds.), Automata, Languages and Programming, Part I. XXIII, 896 pages. 2008.

Vol. 5124: J. Gudmundsson (Ed.), Algorithm Theory – SWAT 2008. XIII, 438 pages. 2008.

Vol. 5123: A. Gupta, S. Malik (Eds.), Computer Aided Verification. XVII, 558 pages. 2008.

Vol. 5117: A. Voronkov (Ed.), Rewriting Techniques and Applications. XIII, 457 pages. 2008.

Vol. 5114: M. Bereković, N. Dimopoulos, S. Wong (Eds.), Embedded Computer Systems: Architectures, Modeling, and Simulation. XVI, 300 pages. 2008.

Vol. 5104: F. Bello, E. Edwards (Eds.), Biomedical Simulation. XI, 228 pages. 2008.

Vol. 5103: M. Bubak, G.D. van Albada, J. Dongarra, P.M.A. Sloot (Eds.), Computational Science – ICCS 2008, Part III. XXVIII, 758 pages. 2008.

Vol. 5102: M. Bubak, G.D. van Albada, J. Dongarra, P.M.A. Sloot (Eds.), Computational Science – ICCS 2008, Part II. XXVIII, 752 pages. 2008.

Vol. 5101: M. Bubak, G.D. van Albada, J. Dongarra, P.M.A. Sloot (Eds.), Computational Science – ICCS 2008, Part I. XLVI, 1058 pages. 2008.

Vol. 5092: X. Hu, J. Wang (Eds.), Computing and Combinatorics. XIV, 680 pages. 2008.

Vol. 5090: R.T. Mittermeir, M.M. Sysło (Eds.), Informatics Education - Supporting Computational Thinking. XV, 357 pages. 2008.

Vol. 5084: J.F. Peters, A. Skowron (Eds.), Transactions on Rough Sets VIII. X, 521 pages. 2008.

Vol. 5083: O. Chitil, Z. Horváth, V. Zsók (Eds.), Implementation and Application of Functional Languages. XI, 272 pages. 2008.

Vol. 5073: O. Gervasi, B. Murgante, A. Laganà, D. Taniar, Y. Mun, M.L. Gavrilova (Eds.), Computational Science and Its Applications – ICCSA 2008, Part II. XXIX, 1280 pages. 2008.

Vol. 5072: O. Gervasi, B. Murgante, A. Laganà, D. Taniar, Y. Mun, M.L. Gavrilova (Eds.), Computational Science and Its Applications – ICCSA 2008, Part I. XXIX, 1266 pages. 2008.

Vol. 5065: P. Degano, R. De Nicola, J. Meseguer (Eds.), Concurrency, Graphs and Models. XV, 810 pages. 2008.

Vol. 5062: K.M. van Hee, R. Valk (Eds.), Applications and Theory of Petri Nets. XIII, 429 pages. 2008.

Vol. 5059: F.P. Preparata, X. Wu, J. Yin (Eds.), Frontiers in Algorithmics. XI, 350 pages. 2008.

Vol. 5058: A.A. Shvartsman, P. Felber (Eds.), Structural Information and Communication Complexity. X, 307 pages. 2008.

Vol. 5050: J.M. Zurada, G.G. Yen, J. Wang (Eds.), Computational Intelligence: Research Frontiers. XVI, 389 pages. 2008.

Vol. 5045: P. Hertling, C.M. Hoffmann, W. Luther, N. Revol (Eds.), Reliable Implementation of Real Number Algorithms: Theory and Practice. XI, 239 pages. 2008.

Vol. 5038: C.C. McGeoch (Ed.), Experimental Algorithms. X, 363 pages. 2008.

Vol. 5036: S. Wu, L.T. Yang, T.L. Xu (Eds.), Advances in Grid and Pervasive Computing. XV, 518 pages. 2008.

Vol. 5035: A. Lodi, A. Panconesi, G. Rinaldi (Eds.), Integer Programming and Combinatorial Optimization. XI, 477 pages. 2008.

Vol. 5029: P. Ferragina, G.M. Landau (Eds.), Combinatorial Pattern Matching. XIII, 317 pages. 2008.

Vol. 5028: A. Beckmann, C. Dimitracopoulos, B. Löwe (Eds.), Logic and Theory of Algorithms. XIX, 596 pages. 2008.

Vol. 5022: A.G. Bourgeois, S.Q. Zheng (Eds.), Algorithms and Architectures for Parallel Processing. XIII, 336 pages. 2008.

Vol. 5018: M. Grohe, R. Niedermeier (Eds.), Parameterized and Exact Computation. X, 227 pages. 2008.

Vol. 5015: L. Perron, M.A. Trick (Eds.), Integration of AI and OR Techniques in Constraint Programming for Combinatorial Optimization Problems. XII, 394 pages. 2008.

Vol. 5011: A.J. van der Poorten, A. Stein (Eds.), Algorithmic Number Theory. IX, 455 pages. 2008.

Vol. 5010: E.A. Hirsch, A.A. Razborov, A. Semenov, A. Slissenko (Eds.), Computer Science – Theory and Applications. XIII, 411 pages. 2008.

Vol. 5008: A. Gasteratos, M. Vincze, J.K. Tsotsos (Eds.), Computer Vision Systems. XV, 560 pages. 2008.

Vol. 5004: R. Eigenmann, B.R. de Supinski (Eds.), OpenMP in a New Era of Parallelism. X, 191 pages. 2008.

Vol. 5000: O. Grumberg, H. Veith (Eds.), 25 Years of Model Checking. VII, 231 pages. 2008.

Vol. 4996: H. Kleine Büning, X. Zhao (Eds.), Theory and Applications of Satisfiability Testing – SAT 2008. X, 305 pages. 2008.

Vol. 4988: R. Berghammer, B. Möller, G. Struth (Eds.), Relations and Kleene Algebra in Computer Science. X, 397 pages. 2008.

Vol. 4985: M. Ishikawa, K. Doya, H. Miyamoto, T. Yamakawa (Eds.), Neural Information Processing, Part II. XXX, 1091 pages. 2008.

Vol. 4984: M. Ishikawa, K. Doya, H. Miyamoto, T. Yamakawa (Eds.), Neural Information Processing, Part I. XXX, 1147 pages. 2008.

Vol. 4981: M. Egerstedt, B. Mishra (Eds.), Hybrid Systems: Computation and Control. XV, 680 pages. 2008.

Vol. 4978: M. Agrawal, D.-Z. Du, Z. Duan, A. Li (Eds.), Theory and Applications of Models of Computation. XII, 598 pages. 2008.

Vol. 4975: F. Chen, B. Jüttler (Eds.), Advances in Geometric Modeling and Processing. XV, 606 pages. 2008.

Vol. 4974: M. Giacobini, A. Brabazon, S. Cagnoni, G.A. Di Caro, R. Drechsler, A. Ekárt, A.I. Esparcia-Alcázar, M. Farooq, A. Fink, J. McCormack, M. O'Neill, J. Romero, F. Rothlauf, G. Squillero, A.Ş. Uyar, S. Yang (Eds.), Applications of Evolutionary Computing. XXV, 701 pages. 2008.

Vol. 4973: E. Marchiori, J.H. Moore (Eds.), Evolutionary Computation, Machine Learning and Data Mining in Bioinformatics. X, 213 pages. 2008.

Vol. 4972: J. van Hemert, C. Cotta (Eds.), Evolutionary Computation in Combinatorial Optimization. XII, 289 pages. 2008.